PHYSICAL CHEMISTRY

A MOLECULAR APPROACH

PHYSICAL CHEMISTRY

A MOLECULAR APPROACH

Donald A. McQuarrie
UNIVERSITY OF CALIFORNIA, DAVIS

John D. Simon
George B. Geller Professor of Chemistry
DUKE UNIVERSITY

University Science Books
www.uscibooks.com

University Science Books
www.uscibooks.com

Production manager: *Susanna Tadlock*
Manuscript editor: *Ann McGuire*
Designer: *Robert Ishi*
Illustrator: *John Choi*
Compositor: *Eigentype*
Printer & Binder: *Edwards Brothers, Inc.*

This book is printed on acid-free paper.

Library of Congress Cataloging-in-Publication Data

McQuarrie, Donald A. (Donald Allan)
    Physical chemistry : a molecular approach / Donald A.
McQuarrie, John D. Simon.
        p. cm.
    Includes bibliographical references and index.

ISBN 978-0-935702-99-6

    1. Chemistry, Physical and theoretical.   I. Simon, John
D. (John Douglas), 1957–   .   II. Title.
    QD453.2.M394   1997
    541—dc21   97-142
                                                       CIP

Printed in the United States of America
10

Contents

CHAPTER 15 / Lasers, Laser Spectroscopy, and Photochemistry   591

MATHCHAPTER G / Numerical Methods   627

CHAPTER 16 / The Properties of Gases   637

MATHCHAPTER H / Partial Differentiation   683

CHAPTER 17 / The Boltzmann Factor and Partition Functions   693

CHAPTER 27 / The Kinetic Theory of Gases   1101

CHAPTER 28 / Chemical Kinetics I: Rate Laws   1137

Preface

To the Student

You are about to begin your study of physical chemistry. You may have been told that physical chemistry is the most difficult chemistry course that you will take, or you may have even seen the bumper sticker that says "Honk if you passed P Chem". The anxiety that some students bring to their physical chemistry course has been eloquently addressed by the British professor E. Brian Smith in the preface of his introductory text, *Basic Chemical Thermodynamics*, Oxford University Press:

"The first time I heard about Chemical Thermodynamics was when a second-year undergraduate brought me the news in my freshman year. He told a spine-chilling story of endless lectures with almost three hundred numbered equations, all of which, it appeared, had to be committed to memory and reproduced in exactly the same form in subsequent examinations. Not only did these equations contain all the normal algebraic symbols but in addition they were liberally sprinkled with stars, daggers, and circles so as to stretch even the most powerful of minds. Few would wish to deny the mind-improving and indeed character-building qualities of such a subject! However, many young chemists have more urgent pressures on their time."

We certainly agree with this last sentence of Professor Smith's. The fact is, however, that every year thousands upon thousands of students take and pass physical chemistry, and many of them really enjoy it. You may be taking it only because it is required by your major, but you should be aware that many recent developments in physical chemistry are having a major impact in all the areas of science concerned with the behavior of molecules. For example, in biophysical chemistry, the application of both experimental and theoretical aspects of physical chemistry to biological problems has greatly advanced our understanding of the structure and reactivity of proteins and nucleic acids. The design of pharmaceutical drugs, which has seen great advances in recent years, is a direct product of physical chemical research.

Traditionally, there are three principal areas of physical chemistry: thermodynamics (which concerns the energetics of chemical reactions), quantum chemistry (which concerns the structures of molecules), and chemical kinetics (which concerns the rates of chemical reactions). Many physical chemistry courses begin with a study of thermodynamics, then discuss quantum chemistry, and treat chemical kinetics last. This order is a reflection of the historical development of the field, and both of us learned physical chemistry in this order. Today, however, physical chemistry is based on quantum mechanics, and so we begin our studies with this topic. We first discuss the underlying principles of quantum mechanics and then show how they can be applied to a number of model systems. Many of the rules you have learned in general chemistry and organic chemistry are natural results of the quantum theory. In organic chemistry, for example, you learned to assign molecular structures using infrared spectra and nuclear magnetic resonance spectra, and in Chapters 13 and 14 we explain how these spectra are governed by the quantum-mechanical properties of molecules.

Your education in chemistry has trained you to think in terms of molecules and their interactions, and we believe that a course in physical chemistry should reflect this viewpoint. The focus of modern physical chemistry is on the molecule. Current experimental research in physical chemistry uses equipment such as molecular beam machines to study the molecular details of gas-phase chemical reactions, high vacuum machines to study the structure and reactivity of molecules on solid interfaces, lasers to determine the structures of individual molecules and the dynamics of chemical reactions, and nuclear magnetic resonance spectrometers to learn about the structure and dynamics of molecules. Modern theoretical research in physical chemistry uses the tools of classical mechanics, quantum mechanics, and statistical mechanics along with computers to develop a detailed understanding of chemical phenomena in terms of the structure and dynamics of the molecules involved. For example, computer calculations of the electronic structure of molecules are providing fundamental insights into chemical bonding, and computer simulations of the dynamical interaction between molecules and proteins are being used to understand how proteins function.

In general chemistry, you learned about the three laws of thermodynamics and were introduced to the quantities enthalpy, entropy, and the Gibbs energy (formerly called the free energy). Thermodynamics is used to describe macroscopic chemical systems. Armed with the tools of quantum mechanics, you will learn that thermodynamics can be formulated in terms of the properties of the atoms and molecules that make up macroscopic chemical systems. Statistical thermodynamics provides a way to describe thermodynamics at a molecular level. You will see that the three laws of thermodynamics can be explained simply and beautifully in molecular terms. We believe that a modern introduction to physical chemistry should, from the outset, develop the field of thermodynamics from a molecular viewpoint. Our treatment of chemical kinetics, which constitutes the last five chapters, develops an understanding of chemical reactions from a molecular viewpoint. For example, we have devoted more than half of the chapter of gas-phase reactions (Chapter 30) to the reaction between a fluorine atom and a hydrogen molecule to form a hydrogen fluoride molecule and a hydrogen atom. Through our study of this seemingly simple reaction, many of the general molecular concepts of chemical reactivity are revealed. Again, quantum chemistry provides the necessary tools to develop a molecular understanding of the rates and dynamics of chemical reactions.

Perhaps the most intimidating aspect of physical chemistry is the liberal use of mathematical topics that you may have forgotten or never learned. As physicists say about physics, physical chemistry is difficult with mathematics; impossible without it.

You may not have taken a math course recently, and your understanding of topics such as determinants, vectors, series expansions, and probability may seem a bit fuzzy at this time. In our years of teaching physical chemistry, we have often found it helpful to review mathematical topics before using them to develop the physical chemical topics. Consequently, we have included a series of ten concise reviews of mathematical topics. We realize that not every one of these so-called reviews may actually be a review for you. Even if some of the topics are new to you (or seem that way), we discuss only the minimum amount that you need to know to understand the subsequent physical chemistry. We have positioned these reviews so that they immediately precede the chapter that uses them. By reading these reviews first (and doing the problems!), you will be able to spend less time worrying about the math, and more time learning the physical chemistry, which is, after all, your goal in this course.

To the Instructor

This text emphasizes a *molecular* approach to physical chemistry. Consequently, unlike most other physical chemistry books, this one discusses the principles of quantum mechanics first and then uses these ideas extensively in its subsequent development of thermodynamics and kinetics. For example, from the Contents, you will see that chapters titled The Boltzmann Factor and Partition Functions (Chapter 17) and Partition Functions and Ideal Gases (Chapter 18) come *before* The First Law of Thermodynamics (Chapter 19). This approach is pedagogically sound because we treat only energy, pressure, and heat capacity (all *mechanical* properties that the students have dealt with in their general chemistry and physics courses) in Chapters 17 and 18. This approach allows us to immediately give a molecular interpretation to the three laws of thermodynamics and to many thermodynamic relations. The molecular interpretation of entropy is an obvious example (an introduction to entropy without a molecular interpretation is strictly for purists and not for the faint of heart), but even the concepts of work and heat in the First Law of Thermodynamics have a nice, physical, molecular interpretation in terms of energy levels and their populations.

Research advances during the past few decades have changed the focus of physical chemistry and therefore should affect the topics covered in a modern physical chemistry course. To introduce the type of physical chemical research that is currently being done, we have included chapters such as Computational Quantum Chemistry (Chapter 11), Group Theory (Chapter 12), Nuclear Magnetic Resonance Spectroscopy (Chapter 14), Lasers , Laser Spectroscopy, and Photochemistry (Chapter 15), and Gas-Phase Reaction Dynamics (Chapter 30). The inclusion of new topics necessitated a rather large book, but one of the standard physical chemistry texts fifty years ago was Glasstone's *Textbook of Physical Chemistry*, which was considerably larger.

Keeping in mind that our purpose is to teach the next generation of chemists, the quantities, units, and symbols used in this text are those presented in the 1993 International Union of Pure and Applied Chemistry (IUPAC) publication *Quantities, Units, and Symbols in Physical Chemistry* by Ian Mills et al. (Blackwell Scientific Publications, Oxford). Our decision to follow the IUPAC recommendations means that some of the symbols, units, and standard states presented in this book may differ from those used in the literature and older textbooks and may be unfamiliar to some instructors. In some instances, we took some time ourselves to come to grips with the new notation and units, but it turned out that, indeed, there was an underlying logic to their use, and we found the effort to become facile with them worthwhile.

A unique feature of this text is the introduction of ten so-called MathChapters, which are short reviews of the mathematical topics used in subsequent chapters. Some of the topics covered that should be familiar to most students are complex numbers, vectors, spherical coordinates, determinants, partial derivatives, and Taylor and Maclaurin series. Some topics that may be new are probability, matrices (used only in the chapter on group theory), numerical methods, and binomial coefficients. In each case, however, the discussions are brief, elementary, and self-contained. After reading each MathChapter and doing the problems, a student should be able to focus on the following physical chemical material rather than having to cope with the physical chemistry and the mathematics simultaneously. We believe that this feature greatly enhances the pedagogy of our text.

Today's students are comfortable with computers. In the past few years, we have seen homework assignments turned in for which students used programs such as Math-Cad and Mathematica to solve problems, rather than pencil and paper. Data obtained in laboratory courses are now graphed and fit to functions using programs such as Excel, Lotus123, and Kaleidagraph. Almost all students have access to personal computers, and a modern course in the physical sciences should encourage students to take advantage of these tremendous resources. As a result, we have written a number of our problems with the use of computers in mind. For example, MathChapter G introduces the Newton-Raphson method for solving higher-order algebraic equations and transcedental equations numerically. We see no reason nowadays to limit calculations in a physical chemistry course to solving quadratic equations and other artificial examples. Students should graph data, explore expressions that fit experimental data, and plot functions that describe physical behavior. The understanding of physical concepts is greatly enhanced by exploring the properties of real data. Such exercises remove the abstractness of many theories and enable students to appreciate the mathematics of physical chemistry so that they can describe and predict the physical behavior of chemical systems.

Our Web Site

You can visit the Web site for our book by visiting the University Science Books website at http://www.uscibooks.com. We have posted various types of supplementary material on this site. For example, the figures (currently in .GIF format only) and the numerical tables in the book can be downloaded from the site. In addition, we are currently preparing a series of lecture slides to accompany the book. We will also be providing downloadable tables of spectroscopic, thermodynamic, and kinetic data. Instructors can use these data to prepare lecture presentations of the applications of the theoretical ideas; using programs such as Kaleidagraph and Mathematica, students can use these data to compare the predictions of equations derived in the text with real data for chemical systems. We encourage both students and instructors to send suggestions, comments, and (the inevitable) errors to us using the entry form posted on the site.

Acknowledgments

Many people have contributed to the writing and production of this book. We thank our colleagues Paul Barbara, James T. Hynes, Veronica Vaida, John Crowell, Andy Kummel, Robert Continetti, Amit Sinha, John Weare, Kim Baldridge, Jack Kyte, Bill Trogler, and Jim Ely for stimulating discussions on the topics that should be included in a modern physical chemistry course, and our students Bary Bolding, Peijun Cong, Robert Dunn, Scott Feller, Susan Forest, Jeff Greathouse, Kerry Hanson, Bulang Li, and Sunney Xie for reading portions of the manuscript and making many helpful suggestions. We are especially indebted to our superb reviewers Merv Hanson, John Frederick, Anne Meyers, George Shields, and Peter Rock, who read and commented on the entire manuscript; to Heather Cox, who also read the entire manuscript, made numerous insightful suggestions, and did every problem in the course of preparing the accompanying Solution Manual; to Carole McQuarrie, who spent many hours in the library and using the internet looking up experimental data and biographical data to write all the biographical sketches; and to Kenneth Pitzer and Karma Beal for supplying us with some critical biographical data. We also thank Susanna Tadlock for coordinating the entire project, Bob Ishi for designing what we think is a beautiful-looking book, Jane Ellis for competently dealing with many of the production details, John Choi for creatively handling all the artwork, Ann McGuire for a very helpful copyediting of the manuscript, and our publisher, Bruce Armbruster, for encouraging us to write our own book and for being an exemplary publisher and a good friend. Last, we thank our wives, Carole and Diane, both of whom are chemists, for being great colleagues as well as great wives.

PHYSICAL CHEMISTRY

A MOLECULAR APPROACH

Max Planck was born in Kiel, Germany (then Prussia) on April 23, 1858, and died in 1948.
He showed early talent in both music and science. He received his Ph.D. in theoretical physics
in 1879 at the University of Munich for his dissertation on the second law of thermodynamics.
He joined the faculty of the University of Kiel in 1885, and in 1888 he was appointed director
of the Institute of Theoretical Physics, which was formed for him at the University of Berlin,
where he remained until 1926. His application of thermodynamics to physical chemistry won
him an early international reputation. Planck was president of the Kaiser Wilhelm Society,
later renamed the Max Planck Society, from 1930 until 1937, when he was forced to retire
by the Nazi government. Planck is known as the father of the quantum theory because
of his theoretical work on blackbody radiation at the end of the 1890s, during which time
he introduced a quantum hypothesis to achieve agreement between his theoretical equations
and experimental data. He maintained his interest in thermodynamics throughout his long
career in physics. Planck was awarded the Nobel Prize in physics in 1918 "in recognition of
services he rendered to the advancement of physics by his discovery of energy quanta." Planck's
personal life was clouded by tragedy. His two daughters died in childbirth, one son died in
World War I, and another son was executed in World War II for his part in the assassination
attempt on Hitler in 1944.

CHAPTER 1

The Dawn of the Quantum Theory

Toward the end of the nineteenth century, many scientists believed that all the fundamental discoveries of science had been made and little remained but to clear up a few minor problems and to improve experimental methods to measure physical results to a greater number of decimal places. This attitude was somewhat justified by the great advances that had been made up to that time. Chemists had finally solved the seemingly insurmountable problem of assigning a self-consistent set of atomic masses to the elements. Stanislao Cannizzaro's concept of the molecule, while initially controversial, was then widely accepted. The great work of Dmitri Mendeleev had resulted in a periodic table of the elements, although the underlying reasons that such periodic behavior occurred in nature were not understood. Friedrich Kekulé had solved the great controversy concerning the structure of benzene. The fundamentals of chemical reactions had been elucidated by Svante Arrhenius, and the remaining work seemed to consist primarily of cataloging the various types of chemical reactions.

In the related field of physics, Newtonian mechanics had been extended by Comte Joseph Lagrange and Sir William Hamilton. The resulting theory was applied to planetary motion and could also explain other complicated natural phenomena such as elasticity and hydrodynamics. Count Rumford and James Joule had demonstrated the equivalence of heat and work, and investigations by Sadi Carnot resulted in the formulation of what is now entropy and the second law of thermodynamics. This work was followed by Josiah Gibbs' complete development of the field of thermodynamics. Shortly, scientists would discover that the laws of physics were also relevant to the understanding of chemical systems. The interface between these two seemingly unrelated disciplines formed the modern field of physical chemistry, the topic of this book. In fact, Gibbs's treatment of thermodynamics is so important to chemistry that it is taught in a form that is essentially unchanged from Gibbs's original formulation.

The related fields of optics and electromagnetic theory were undergoing similar maturation. The nineteenth century witnessed a continuing controversy as to whether light was wavelike or particlelike. Many diverse and important observations were unified by James Clerk Maxwell in a series of deceptively simple-looking equations

1

that bear his name. Not only did Maxwell's predictions of the electromagnetic behavior of light unify the fields of optics with electricity and magnetism, but their subsequent experimental demonstration by Heinrich Hertz in 1887 appeared to finally demonstrate that light was wavelike. The implications of these fields to chemistry would not be appreciated for several decades but are now important aspects of the discipline of physical chemistry, particularly in spectroscopy.

The body of these accomplishments in physics is considered the development of what we now call *classical physics*. Little did scientists realize in that justifiably heady era of success that the fundamental tenets of how the physical world works were to be shortly overturned. Fantastic discoveries were about to revolutionize not only physics, chemistry, biology, and engineering but would have significant effects on technology and politics as well. The early twentieth century saw the birth of the theory of relativity and quantum mechanics. The first, due to the work of Albert Einstein alone, completely altered scientist's ideas of space and time and was an extension of the classical ideas to include high velocities and astronomical distances. Quantum mechanics, the extension of classical ideas into the behavior of subatomic, atomic, and molecular species, on the other hand, resulted from the efforts of many creative scientists over several decades. To date, the effect of relativity on chemical systems has been limited. Although it is important in understanding electronic properties of heavy atoms, it does not play much of a role in molecular structure and reactivity and so is not generally taught in physical chemistry. Quantum mechanics, however, forms the foundation upon which all of chemistry is built. Our current understanding of atomic structure and molecular bonding is cast in terms of the fundamental principles of quantum mechanics and no understanding of chemical systems is possible without knowing the basics of this current theory of matter. For this reason, we begin this book with several chapters that focus on the fundamental principles of quantum mechanics. We then follow with a discussion of chemical bonding and spectroscopy, which clearly demonstrate the influence that quantum mechanics has had on the field of chemistry.

Great changes in science are spurred by observations and new creative ideas. Let us go back to the complacent final years of the nineteenth century to see just what were the events that so shook the world of science.

1–1. Blackbody Radiation Could Not Be Explained by Classical Physics

The series of experiments that revolutionized the concepts of physics had to do with the radiation given off by material bodies when they are heated. We all know, for instance, that when the burner of an electric stove is heated, it first turns a dull red and progressively becomes redder as the temperature increases. We also know that as a body is heated even further, the radiation becomes white and then blue as its temperature continues to increase. Thus, we see that there is a continual shift of the color of a heated body from red through white to blue as the body is heated to higher temperatures. In terms of frequency, the radiation emitted goes from a lower frequency to a higher frequency as the temperature increases, because red is in a lower frequency region of the spectrum than is blue. The exact frequency spectrum emitted by the body depends on the particular body itself, but an *ideal body*, which absorbs and emits all frequencies, is called a *blackbody* and serves as an idealization for any radiating material. The radiation emitted by a blackbody is called *blackbody radiation.*

A plot of the intensity of blackbody radiation versus frequency for several temperatures is given in Figure 1.1. Many theoretical physicists tried to derive expressions consistent with these experimental curves of intensity versus frequency, but they were all unsuccessful. In fact, the expression that is derived according to the laws of nineteenth century physics is

$$d\rho(\nu, T) = \rho_\nu(T)d\nu = \frac{8\pi k_B T}{c^3}\nu^2 d\nu \tag{1.1}$$

where $\rho_\nu(T)d\nu$ is the radiant energy density between the frequencies ν and $\nu + d\nu$ and has units of joules per cubic meter ($J\cdot m^{-3}$). In Equation 1.1, T is the absolute temperature, and c is the speed of light. The quantity k_B is called the *Boltzmann constant* and is equal to the molar gas constant R divided by the Avogadro constant (formerly called Avogadro's number). The units of k_B are $J\cdot K^{-1}\cdot particle^{-1}$, but $particle^{-1}$ is usually not expressed. (Another case is the Avogadro constant, 6.022×10^{23} particle$\cdot$mol^{-1}, which we will write as 6.022×10^{23} mol^{-1}; the unit "particle" is not expressed.) Equation 1.1 came from the work of Lord Rayleigh and J.H. Jeans and is called the *Rayleigh-Jeans law*. The dashed line in Figure 1.1 shows the prediction of the Rayleigh-Jeans law.

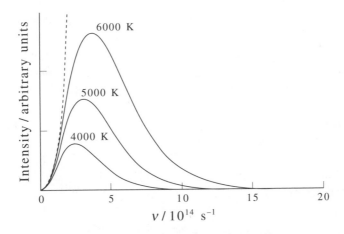

FIGURE 1.1
Spectral distribution of the intensity of blackbody radiation as a function of frequency for several temperatures. The intensity is given in arbitrary units. The dashed line is the prediction of classical physics. As the temperature increases, the maximum shifts to higher frequencies and the total radiated energy (the area under each curve) increases significantly. Note that the horizontal axis is labeled by $\nu/10^{14}$ s^{-1}. This notation means that the dimensionless numbers on that axis are frequencies divided by 10^{14} s^{-1}. We shall use this notation to label columns in tables and axes in figures because of its unambiguous nature and algebraic convenience.

Note that the Rayleigh-Jeans law reproduces the experimental data at low frequencies. At high frequencies, however, the Rayleigh-Jeans law predicts that the radiant energy density diverges as ν^2. Because the frequency increases as the radiation enters the ultraviolet region, this divergence was termed the *ultraviolet catastrophe*, a phenomenon that classical physics could not explain theoretically. This was the first such failure to explain an important naturally occuring phenomenon and therefore is of great historical interest. Rayleigh and Jeans did not simply make a mistake or misapply some of the ideas of physics; many other people reproduced the equation of Rayleigh and Jeans, showing that this equation was correct according to the physics of the time. This result was somewhat disconcerting and many people struggled to find a theoretical explanation of blackbody radiation.

1–2. Planck Used a Quantum Hypothesis to Derive the Blackbody Radiation Law

The first person to offer a successful explanation of blackbody radiation was the German physicist Max Planck in 1900. Like Rayleigh and Jeans before him, Planck assumed that the radiation emitted by the blackbody was caused by the oscillations of the electrons in the constituent particles of the material body. These electrons were pictured as oscillating in an atom much like electrons oscillate in an antenna to give off radio waves. In these "atomic antennae", however, the oscillations occur at a much higher frequency; hence, we find frequencies in the visible, infrared, and ultraviolet regions rather than in the radio-wave region of the spectrum. Implicit in the derivation of Rayleigh and Jeans is the assumption that the energies of the electronic oscillators responsible for the emission of the radiation could have any value whatsoever. This assumption is one of the basic assumptions of classical physics. In classical physics, the variables that represent observables (such as position, momentum, and energy) can take on a continuum of values. Planck had the great insight to realize that he had to break away from this mode of thinking to derive an expression that would reproduce experimental data such as those shown in Figure 1.1. He made the revolutionary assumption that the energies of the oscillators were discrete and had to be proportional to an integral multiple of the frequency or, in equation form, that $E = nh\nu$, where E is the energy of an oscillator, n is an integer, h is a proportionality constant, and ν is the frequency. Using this quantization of energy and statistical thermodynamic ideas that we will cover in Chapter 17, Planck derived the equation

$$d\rho(\nu, T) = \rho_\nu(T)d\nu = \frac{8\pi h}{c^3} \frac{\nu^3 d\nu}{e^{h\nu/k_B T} - 1} \tag{1.2}$$

All the symbols except h in Equation 1.2 have the same meaning as in Equation 1.1. The only undetermined constant in Equation 1.2 is h. Planck showed that this equation gives excellent agreement with the experimental data for all frequencies and temperatures if h has the value 6.626×10^{-34} joule·seconds (J·s). This constant is now one of the most famous and fundamental constants of physics and is called the *Planck constant*.

Equation 1.2 is known as the *Planck distribution law for blackbody radiation*. For small frequencies, Equations 1.1 and 1.2 become identical (Problem 1–4), but the Planck distribution does not diverge at large frequencies and, in fact, looks like the curves in Figure 1.1.

EXAMPLE 1–1

Show that $\rho_\nu(T)d\nu$ in both Equations 1.1 and 1.2 has units of energy per unit volume, $J\cdot m^{-3}$.

SOLUTION: The units of T are K, of k_B are $J\cdot K^{-1}$, of h are $J\cdot s$, of ν and $d\nu$ are s^{-1}, and of c are $m\cdot s^{-1}$. Therefore, for the Rayleigh-Jeans law (Equation 1.1),

$$d\rho(\nu, T) = \rho_\nu(T)d\nu = \frac{8\pi k_B T}{c^3}\nu^2 d\nu$$

$$\sim \frac{(J\cdot K^{-1})(K)}{(m\cdot s^{-1})^3}(s^{-1})^2(s^{-1}) = J\cdot m^{-3}$$

For the Planck distribution (Equation 1.2),

$$d\rho(\nu, T) = \rho_\nu(T)d\nu = \frac{8\pi h}{c^3}\frac{\nu^3 d\nu}{e^{h\nu/k_B T} - 1}$$

$$\sim \frac{(J\cdot s)(s^{-1})^3(s^{-1})}{(m\cdot s^{-1})^3} = J\cdot m^{-3}$$

Thus, we see that $\rho_\nu(T)d\nu$, the radiant energy density has units of energy per unit volume.

Equation 1.2 expresses the Planck distribution law in terms of frequency. Because wavelength (λ) and frequency (ν) are related by $\lambda\nu = c$, then $d\nu = -cd\lambda/\lambda^2$, and we can express the Planck distribution law in terms of wavelength rather than frequency (Problem 1–10):

$$d\rho(\lambda, T) = \rho_\lambda(T)d\lambda = \frac{8\pi hc}{\lambda^5}\frac{d\lambda}{e^{hc/\lambda k_B T} - 1} \tag{1.3}$$

The quantity $\rho_\lambda(T)d\lambda$ is the radiant energy density between λ and $\lambda + d\lambda$. The intensity corresponding to Equation 1.3 is plotted in Figure 1.2 for several values of T.

We can use Equation 1.3 to justify an empirical relationship known as the *Wien displacement law*. The Wien displacement law says that if λ_{max} is the wavelength at which $\rho_\lambda(T)$ is a maximum, then

$$\lambda_{max} T = 2.90 \times 10^{-3}\ m\cdot K \tag{1.4}$$

By differentiating $\rho_\lambda(T)$ with respect to λ, we can show (Problem 1–5) that

$$\lambda_{max} T = \frac{hc}{4.965 k_B} \tag{1.5}$$

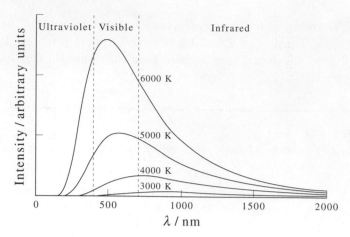

FIGURE 1.2
The distribution of the intensity of the radiation emitted by a blackbody versus wavelength for various temperatures. As the temperature increases, the total radiation emitted (the area under the curve) increases.

in accord with the Wein displacement law. Using the modern values of h, c, and k_B given inside the front cover, we obtain 2.899×10^{-3} m·K for the right side of Equation 1.5, in excellent agreement with the experimental value given in Equation 1.4.

The theory of blackbody radiation is used regularly in astronomy to estimate the surface temperatures of stars. Figure 1.3 shows the electromagnetic spectrum of the sun measured at the earth's upper atmosphere. A comparison of Figure 1.3 with Figure 1.2 suggests that the solar spectrum can be described by a blackbody at approximately 6000 K. If we estimate λ_{max} from Figure 1.3 to be 500 nm, then the Wein displacement law (Equation 1.4) gives the temperature of the surface of the sun to be

$$T = \frac{2.90 \times 10^{-3} \text{ m·K}}{500 \times 10^{-9} \text{ m}} = 5800 \text{ K}$$

The star Sirius, which appears blue, has a surface temperature of about 11 000 K (cf. Problem 1–7).

Certainly Planck's derivation of the blackbody distribution law was an impressive feat. Nevertheless, Planck's derivation and, in particular, his assumption that the energies of the oscillators have to be an integral multiple of $h\nu$ was not accepted by most scientists at the time and was considered simply an arbitrary derivation. Most believed that in time a satisfactory derivation would be found that obeyed the laws of classical physics. In a sense, Planck's derivation was little more than a curiosity. Just a few years later, however, in 1905, Einstein used the very same idea to explain the photoelectric effect.

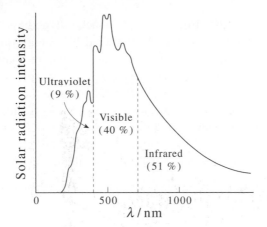

FIGURE 1.3
The electromagnetic spectrum of the sun as measured in the upper atmosphere of the earth. A comparison of this figure with Figure 1.2 shows that the sun's surface radiates as a blackbody at a temperature of about 6000 K.

1–3. Einstein Explained the Photoelectric Effect with a Quantum Hypothesis

In 1886 and 1887, while carrying out the experiments that supported Maxwell's theory of the electromagnetic nature of light, the German physicist Heinrich Hertz discovered that ultraviolet light causes electrons to be emitted from a metallic surface. The ejection of electrons from the surface of a metal by radiation is called the *photoelectric effect*. Two experimental observations of the photoelectric effect are in stark contrast with the classical wave theory of light. According to classical physics, electromagnetic radiation is an electric field oscillating perpendicular to its direction of propagation, and the intensity of the radiation is proportional to the square of the amplitude of the electric field. As the intensity increases, so does the amplitude of the oscillating electric field. The electrons at the surface of the metal should oscillate along with the field and so, as the intensity (amplitude) increases, the electrons oscillate more violently and eventually break away from the surface with a kinetic energy that depends on the amplitude (intensity) of the field. This nice classical picture is in complete disagreement with the experimental observations. Experimentally, the kinetic energy of the ejected electrons is independent of the intensity of the incident radiation. Furthermore, the classical picture predicts that the photoelectric effect should occur for any frequency of light as long as the intensity is sufficiently high. The experimental fact, however, is that there is a *threshold frequency*, v_0, characteristic of the metallic surface, below which no electrons are ejected, regardless of the intensity of the radiation. Above v_0, the kinetic energy of the ejected electrons varies linearly with the frequency v. These observations served as an embarrassing contradiction of classical theory.

To explain these results, Einstein used Planck's hypothesis but extended it in an important way. Recall that Planck had applied his energy quantization concept, $E = nh\nu$ or $\Delta E = h\nu$, to the emission and absorption mechanism of the atomic electronic oscillators. Planck believed that once the light energy was emitted, it behaved like a classical wave. Einstein proposed instead that the radiation itself existed as small packets of energy, $E = h\nu$, now known as *photons*. Using a simple conservation-of-energy argument, Einstein showed that the kinetic energy (KE) of an ejected electron is equal to the energy of the incident photon ($h\nu$) minus the minimum energy required to remove an electron from the surface of the particular metal (ϕ). In an equation,

$$\text{KE} = \frac{1}{2}mv^2 = h\nu - \phi \tag{1.6}$$

where ϕ, called the *work function* of the metal, is analogous to an ionization energy of an isolated atom. The left side of Equation 1.6 cannot be negative, so Equation 1.6 predicts that $h\nu \geq \phi$. The minimum frequency that will eject an electron is just the frequency required to overcome the work function of the metal, thus we see that there is a threshold frequency ν_0, given by

$$h\nu_0 = \phi \tag{1.7}$$

Using Equations 1.6 and 1.7, we can write

$$\text{KE} = h\nu - h\nu_0 \qquad \nu \geq \nu_0 \tag{1.8}$$

Equation 1.8 shows that a plot of KE versus ν should be linear and that the slope of the line should be h, in complete agreement with the data in Figure 1.4.

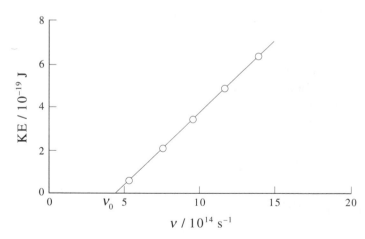

FIGURE 1.4
The kinetic energy of electrons ejected from the surface of sodium metal versus the frequency of the incident ultraviolet radiation. The threshold frequency here is 4.40×10^{14} Hz (1 Hz $= 1$ s^{-1}).

Before we can discuss Equation 1.8 numerically, we must consider the units involved. The work function ϕ is customarily expressed in units of electron volts (eV). One electron volt is the energy picked up by a particle with the same charge as an electron (or a proton) when it falls through a potential drop of one volt. If you recall that (1 coulomb) x (1 volt) = 1 joule and use the fact that the charge on a proton is 1.602×10^{-19} C, then

$$1 \text{ eV} = (1.602 \times 10^{-19} \text{ C})(1 \text{ V})$$
$$= 1.602 \times 10^{-19} \text{ J}$$

EXAMPLE 1–2
Given that the work function for sodium metal is 1.82 eV, calculate the threshold frequency ν_0 for sodium.

SOLUTION: We must first convert ϕ from electron volts to joules.

$$\phi = 1.82 \text{ eV} = (1.82 \text{ eV})(1.602 \times 10^{-19} \text{ J} \cdot \text{eV}^{-1})$$
$$= 2.92 \times 10^{-19} \text{ J}$$

Using Equation 1.7, we have

$$\nu_0 = \frac{2.92 \times 10^{-19} \text{ J}}{6.626 \times 10^{-34} \text{ J} \cdot \text{s}}$$
$$= 4.40 \times 10^{14} \text{ Hz}$$

In the last line here, we have introduced the unit hertz (Hz) for per second (s^{-1}).

EXAMPLE 1–3
When lithium is irradiated with light, the kinetic energy of the ejected electrons is 2.935×10^{-19} J for $\lambda = 300.0$ nm and 1.280×10^{-19} J for $\lambda = 400.0$ nm. Calculate (a) the Planck constant, (b) the threshold frequency, and (c) the work function of lithium from these data.

SOLUTION:
(a) From Equation 1.8, we write

$$(\text{KE})_1 - (\text{KE})_2 = h(\nu_1 - \nu_2) = hc\left(\frac{1}{\lambda_1} - \frac{1}{\lambda_2}\right)$$

or

$$1.655 \times 10^{-19} \text{ J} = h(2.998 \times 10^8 \text{ m} \cdot \text{s}^{-1})\left(\frac{1}{300.0 \times 10^{-9} \text{ m}} - \frac{1}{400.0 \times 10^{-9} \text{ m}}\right)$$

from which we obtain

$$h = \frac{1.655 \times 10^{-19} \text{ J}}{2.498 \times 10^{14} \text{ s}^{-1}} = 6.625 \times 10^{-34} \text{ J} \cdot \text{s}$$

(b) Using the $\lambda = 300.0$ nm data, we have

$$2.935 \times 10^{-19} \text{ J} = \frac{hc}{300.0 \times 10^{-9} \text{ m}} - h\nu_0$$

from which we find that $\nu_0 = 5.564 \times 10^{14}$ Hz

(c) Using Equation 1.7, we have

$$\phi = h\nu_0 = 3.687 \times 10^{-19} \text{ J} = 2.301 \text{ eV}$$

Einstein obtained a value of h in close agreement with Planck's value deduced from the blackbody radiation formula. This surely was a fantastic result because the whole business of energy quantization was quite mysterious and not well accepted by the scientific community of the day. Nevertheless, in two very different sets of experiments, blackbody radiation and the photoelectric effect, the very same quantization constant h, had arisen naturally. Scientists realized that perhaps there was something to all this after all.

1–4. The Hydrogen Atomic Spectrum Consists of Several Series of Lines

For some time scientists had known that every atom, when subjected to high temperatures or an electrical discharge, emits electromagnetic radiation of characteristic frequencies. In other words, each atom has a characteristic emission spectrum. Because the emission spectra of atoms consist of only certain discrete frequencies, they are called *line spectra*. Hydrogen, the lightest and simplest atom, has the simplest spectrum.

Figure 1.5 shows the part of the hydrogen atom emission spectrum that occurs in the visible and near ultraviolet region.

Because atomic spectra are characteristic of the atoms involved, it is reasonable to suspect that the spectrum depends on the electron distribution in the atom. A detailed analysis of the hydrogen atomic spectrum turned out to be a major step in the elucidation of the electronic structure of atoms. For many years, scientists had tried to find a pattern in the wavelengths or frequencies of the lines in the hydrogen atomic spectrum. Finally, in 1885, an amateur Swiss scientist, Johann Balmer, showed that a plot of the frequency of the lines versus $1/n^2 (n = 3, 4, 5, \ldots)$ is linear, as shown in Figure 1.6. In particular,

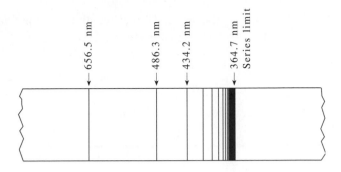

FIGURE 1.5
Emission spectrum of the hydrogen atom in the visible and the near ultraviolet region showing that the emission spectrum of atomic hydrogen is a line spectrum.

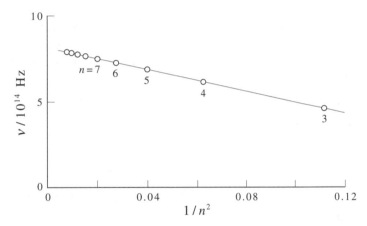

FIGURE 1.6
A plot of frequency versus $1/n^2$ ($n = 3, 4, 5, \ldots$) for the series of lines of the hydrogen atom spectrum that occurs in the visible and near ultraviolet regions. The actual spectrum is shown in Figure 1.5. The linear nature of this plot leads directly to Equation 1.9.

Balmer showed that the frequencies of the emission lines in the visible region of the spectrum could be described by the equation

$$\nu = 8.2202 \times 10^{14} \left(1 - \frac{4}{n^2}\right) \text{ Hz}$$

where $n = 3, 4, 5, \ldots$. This equation is customarily written in terms of the quantity $1/\lambda$ instead of ν. Reciprocal wavelength is called a *wavenumber*, whose SI units are m^{-1}. It turns out, however, that the use of the non-SI unit cm^{-1} is so prevalent in spectroscopy that we will use cm^{-1} in most of this book. Thus if we divide the previous equation by c and factor a 4 out of the two terms in parentheses, then we have

$$\tilde{\nu} = 109\,680 \left(\frac{1}{2^2} - \frac{1}{n^2} \right) cm^{-1} \qquad n = 3, \ 4, \ \ldots \qquad (1.9)$$

where $\tilde{\nu} = 1/\lambda = \nu/c$. This equation is called *Balmer's formula*.

EXAMPLE 1–4

Using Balmer's formula, calculate the wavelengths of the first few lines in the visible region of the hydrogen atomic spectrum and compare them to the experimental values given in Figure 1.5.

SOLUTION: The first line is obtained by setting $n = 3$, in which case we have

$$\tilde{\nu} = 109\,680 \left(\frac{1}{2^2} - \frac{1}{3^2} \right) cm^{-1}$$

$$= 1.523 \times 10^4 \ cm^{-1}$$

and

$$\lambda = 6.565 \times 10^{-5} \ cm = 656.5 \ nm$$

The next line is obtained by setting $n = 4$, and so

$$\tilde{\nu} = 109\,680 \left(\frac{1}{2^2} - \frac{1}{4^2} \right) cm^{-1}$$

$$= 2.056 \times 10^4 \ cm^{-1}$$

and

$$\lambda = 4.863 \times 10^{-5} \ cm = 486.3 \ nm$$

Thus, we see that the agreement with the experimental data (Figure 1.5) is excellent.

Note that Equation 1.9 predicts a series of lines as n takes on the values 3, 4, 5, ... This series of lines, the ones occurring in the visible and near ultraviolet regions of the hydrogen atomic spectrum and predicted by Balmer's formula, is called the *Balmer series*. The Balmer series is shown in Figure 1.5. Note also that Equation 1.9 predicts that the lines in the hydrogen atomic spectrum bunch up as n increases. As n increases, $1/n^2$ decreases and eventually we can ignore this term compared with the $\frac{1}{4}$ term and so in the limit $n \rightarrow \infty$ we have

$$\tilde{\nu} \longrightarrow 109\,680 \left(\frac{1}{4} \right) cm^{-1} = 2.742 \times 10^4 \ cm^{-1}$$

or $\lambda = 364.7$ nm, in excellent agreement with the data in Figure 1.5. This value is essentially that for the last line in the Balmer series and is called the *series limit*.

The Balmer series occurs in the visible and near ultraviolet regions. The hydrogen atomic spectrum has lines in other regions; in fact, series of lines similar to the Balmer series appear in the ultraviolet and in the infrared region (cf. Figure 1.7).

1–5. The Rydberg Formula Accounts for All the Lines in the Hydrogen Atomic Spectrum

The Swedish spectroscopist Johannes Rydberg accounted for all the lines in the hydrogen atomic spectrum by generalizing the Balmer formula to

$$\tilde{\nu} = \frac{1}{\lambda} = 109\,680 \left(\frac{1}{n_1^2} - \frac{1}{n_2^2} \right) \text{cm}^{-1} \quad (n_2 > n_1) \tag{1.10}$$

where both n_1 and n_2 are integers but n_2 is always greater than n_1. Equation 1.10 is called the *Rydberg formula*. Note that the Balmer series is recovered if we let $n_1 = 2$. The other series are obtained by letting n_1 be 1, 3, 4, ... The names associated with these various series are given in Figure 1.7 and Table 1.1. The constant in Equation 1.10 is called the *Rydberg constant* and Equation 1.10 is commonly written as

$$\tilde{\nu} = R_{\text{H}} \left(\frac{1}{n_1^2} - \frac{1}{n_2^2} \right) \tag{1.11}$$

where R_{H} is the Rydberg constant. The modern value of the Rydberg constant is $109\,677.581$ cm^{-1}; it is one of the most accurately known physical constants.

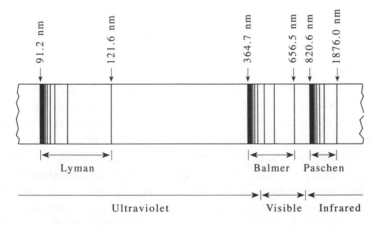

FIGURE 1.7
A schematic representation of the various series in the hydrogen atomic spectrum. The Lyman series lies in the ultraviolet region; the Balmer lies in the visible region; and the Paschen and Bracket series lie in the infrared region (see Table 1.1).

TABLE 1.1

The first four series of lines making up the hydrogen atomic spectrum. The term "near infrared" denotes the part of the infrared region of the spectrum that is near the visible region.

Series name	n_1	n_2	Region of spectrum
Lyman	1	2, 3, 4, ...	Ultraviolet
Balmer	2	3, 4, 5, ...	Visible
Paschen	3	4, 5, 6, ...	Near infrared
Bracket	4	5, 6, 7, ...	Infrared

EXAMPLE 1–5

Calculate the wavelength of the second line in the Paschen series, and show that this line lies in the near infrared, that is, in the infrared region near the visible.

SOLUTION: In the Paschen series, $n_1 = 3$ and $n_2 = 4,\ 5,\ 6,\ \ldots$ according to Table 1.1. Thus, the second line in the Paschen series is given by setting $n_1 = 3$ and $n_2 = 5$ in Equation 1.11:

$$\tilde{\nu} = 109\,677.57 \left(\frac{1}{3^2} - \frac{1}{5^2} \right) \text{cm}^{-1}$$

$$= 7.799 \times 10^3 \text{ cm}^{-1}$$

and

$$\lambda = 1.282 \times 10^{-4} \text{ cm} = 1282 \text{ nm}$$

which is in the near infrared region.

The fact that the formula describing the hydrogen spectrum is in a sense controlled by two integers is truly amazing. Why should a hydrogen atom care about our integers? We will see that integers play a special role in quantum theory.

The spectra of other atoms were also observed to consist of series of lines, and in the 1890s Rydberg found approximate empirical laws for many of them. The empirical laws for other atoms were generally more involved than Equation 1.11, but the really interesting feature is that all the observed lines could be expressed as the difference between terms such as those in Equation 1.11. This feature was known as the *Ritz combination rule*, and we will see that it follows immediately from our modern view of atomic structure. At the time, however, it was just an empirical rule waiting for a theoretical explanation.

1–6. Louis de Broglie Postulated That Matter Has Wavelike Properties

Although we have an intriguing partial insight into the electronic structure of atoms, something is missing. To explore this further, let us go back to a discussion of the nature of light.

Scientists have always had trouble describing the nature of light. In many experiments light shows a definite wavelike character, but in many others light seems to behave as a stream of photons. The dispersion of white light into its spectrum by a prism is an example of the first type of experiment, and the photoelectric effect is an example of the second. Because light appears wavelike in some instances and particle-like in others, this disparity is referred to as the *wave-particle duality of light*. In 1924, a young French scientist named Louis de Broglie reasoned that if light can display this wave-particle duality, then matter, which certainly appears particlelike, might also display wavelike properties under certain conditions. This proposal is rather strange at first, but it does suggest a nice symmetry in nature. Certainly if light can be particlelike at times, why should matter not be wavelike at times?

de Broglie was able to put his idea into a quantitative scheme. Einstein had shown from relativity theory that the wavelength, λ, and the momentum, p, of a photon are related by

$$\lambda = \frac{h}{p} \tag{1.12}$$

de Broglie argued that both light *and* matter obey this equation. Because the momentum of a particle is given by mv, this equation predicts that a particle of mass m moving with a velocity v will have a *de Broglie wavelength* given by $\lambda = h/mv$.

EXAMPLE 1–6
Calculate the de Broglie wavelength for a baseball (5.0 oz) traveling at 90 mph.

SOLUTION: Five ounces corresponds to

$$m = (5.0 \text{ oz}) \left(\frac{1 \text{ lb}}{16 \text{ oz}} \right) \left(\frac{0.454 \text{ kg}}{1 \text{ lb}} \right) = 0.14 \text{ kg}$$

and 90 mph corresponds to

$$v = \left(\frac{90 \text{ mi}}{1 \text{ hr}} \right) \left(\frac{1610 \text{ m}}{1 \text{ mi}} \right) \left(\frac{1 \text{ hr}}{3600 \text{ s}} \right) = 40 \text{ m} \cdot \text{s}^{-1}$$

The momentum of the baseball is

$$p = mv = (0.14 \text{ kg})(40 \text{ m} \cdot \text{s}^{-1}) = 5.6 \text{ kg} \cdot \text{m} \cdot \text{s}^{-1}$$

The de Broglie wavelength is

$$\lambda = \frac{h}{p} = \frac{6.626 \times 10^{-34} \text{ J·s}}{5.6 \text{ kg·m·s}^{-1}} = 1.2 \times 10^{-34} \text{ m}$$

a ridiculously small wavelength.

We see from Example 1.6 that the de Broglie wavelength of the baseball is so small as to be completely undetectable and of no practical consequence. The reason is the large value of m. What if we calculate the de Broglie wavelength of an electron instead of a baseball?

EXAMPLE 1–7
Calculate the de Broglie wavelength of an electron traveling at 1.00% of the speed of light.

SOLUTION: The mass of an electron is 9.109×10^{-31} kg. One percent of the speed of light is

$$v = (0.0100)(2.998 \times 10^8 \text{ m·s}^{-1}) = 2.998 \times 10^6 \text{ m·s}^{-1}$$

The momentum of the electron is given by

$$p = m_e v = (9.109 \times 10^{-31} \text{ kg})(2.998 \times 10^6 \text{ m·s}^{-1})$$
$$= 2.73 \times 10^{-24} \text{ kg·m·s}^{-1}$$

The de Broglie wavelength of this electron is

$$\lambda = \frac{h}{p} = \frac{6.626 \times 10^{-34} \text{ J·s}}{2.73 \times 10^{-24} \text{ kg·m·s}^{-1}} = 2.43 \times 10^{-10} \text{ m}$$
$$= 243 \text{ pm}$$

This wavelength is of atomic dimensions.

The wavelength of the electron calculated in Example 1–7 corresponds to the wavelength of X-rays. Thus, although Equation 1.12 is of no consequence for a macroscopic object such as a baseball, it predicts that electrons can be observed to act like X-rays. The wavelengths of some other moving objects are given in Table 1.2.

1–7. de Broglie Waves Are Observed Experimentally

When a beam of X rays is directed at a crystalline substance, the beam is scattered in a definite manner characteristic of the atomic structure of the crystalline substance. This phenomenon is called *X-ray diffraction* and occurs because the interatomic spacings in

TABLE 1.2
The de Broglie wavelengths of various moving objects.

Particle	Mass/kg	Speed/m·s^{-1}	Wavelength/pm
Electron accelerated through 100 V	9.11×10^{-31}	5.9×10^6	120
Electron accelerated through 10,000 V	9.29×10^{-31}	5.9×10^7	12
α particle ejected from radium	6.68×10^{-27}	1.5×10^7	6.6×10^{-3}
22-caliber rifle bullet	1.9×10^{-3}	3.2×10^2	1.1×10^{-21}
Golf ball	0.045	30	4.9×10^{-22}

the crystal are about the same as the wavelength of the X-rays. The X-ray diffraction pattern from aluminum foil is shown in Figure 1.8a. The X-rays scatter from the foil in rings of different diameters. The distances between the rings are determined by the interatomic spacing in the metal foil. Figure 1.8b shows an electron diffraction pattern from aluminum foil that results when a beam of electrons is similarly directed. The

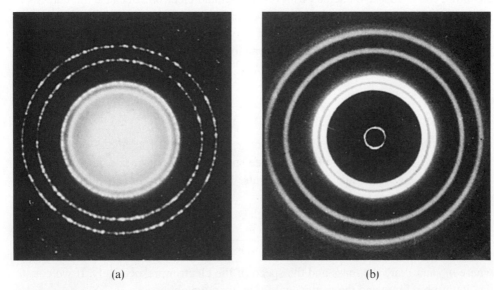

(a)   (b)

FIGURE 1.8
(a) The X-ray diffraction pattern of aluminum foil. (b) The electron diffraction pattern of aluminum foil. The similarity of these two patterns shows that electrons can behave like X-rays and display wavelike properties.

similarity of the two patterns shows that both X-rays and electrons do indeed behave analogously in these experiments.

The wavelike property of electrons is used in electron microscopes. The wavelengths of the electrons can be controlled through an applied voltage, and the small de Broglie wavelengths attainable offer a more precise probe than an ordinary light microscope. In addition, in contrast to electromagnetic radiation of similar wavelengths (X-rays and ultraviolet), the electron beam can be readily focused by using electric and magnetic fields, generating sharper images. Electron microscopes are used routinely in chemistry and biology to investigate atomic and molecular structures.

An interesting historical aside in the concept of the wave-particle duality of matter is that the first person to show that the electron was a subatomic particle was the English physicist Sir Joseph J. Thomson in 1895 and then his son Sir George P. Thomson was among the first to show experimentally in 1926 that the electron could act as a wave. The father won a Nobel Prize in 1906 for showing that the electron is a particle and the son won a Nobel Prize in 1937 for showing that it is a wave.

1–8. The Bohr Theory of the Hydrogen Atom Can Be Used to Derive the Rydberg Formula

In 1911, the Danish physicist Niels Bohr presented a theory of the hydrogen atom that gave a beautifully simple explanation of the hydrogen atomic spectrum. We present here a brief discussion of the Bohr theory.

According to the nuclear model of the atom, the hydrogen atom can be pictured as a central, rather massive nucleus with one associated electron. Because the nucleus is so much more massive than the electron, we can consider the nucleus to be fixed and the electron to be revolving about it. The force holding the electron in a circular orbit is supplied by the coulombic force of attraction between the proton and the electron (Coulomb's law):

$$f = \frac{e^2}{4\pi\varepsilon_0 r^2}$$

where r is the radius of the orbit, e is the charge on the electron, and $\varepsilon_0 = 8.85419 \times 10^{-12}$ $C^2 \cdot N^{-1} \cdot m^{-2}$ is the permittivity of free space. The occurrence of the factor $4\pi\varepsilon_0$ in Coulomb's law is a result of using SI units. The coulombic force is balanced by the centrifugal force (see Problem 1–41)

$$f = \frac{m_e v^2}{r} \tag{1.13}$$

where m_e and v are the mass and the speed of the electron, respectively. If we equate the coulombic force and the centrifugal force, then we obtain

$$\frac{e^2}{4\pi\varepsilon_0 r^2} = \frac{m_e v^2}{r} \tag{1.14}$$

We are tacitly assuming here that the electron is revolving around the fixed nucleus in a circular orbit of radius r. Classically, however, because the electron is constantly being accelerated according to Equation 1.13 (Problem 1–41), it should emit electromagnetic radiation and lose energy. Consequently, classical physics predicts that an electron revolving around a nucleus will lose energy and spiral into the nucleus, and so a stable orbit for the electron is classically forbidden. Bohr's great contribution was to make two nonclassical assumptions. The first was to assume the existence of stationary electron orbits, in defiance of classical physics. He then specified these orbits by the equivalent of assuming that the de Broglie waves of the orbiting electron must "match" or be in phase, as the electron makes one complete revolution. Without such matching, cancellation of some amplitude occurs during each revolution, and the wave will disappear (see Figure 1.9). For the wave pattern around an orbit to be stable, we are led to the condition that an integral number of complete wavelengths must fit around the circumference of the orbit. Because the circumference of a circle is $2\pi r$, we have the quantum condition

$$2\pi r = n\lambda \qquad n = 1,\ 2,\ 3,\ \ldots \tag{1.15}$$

If we substitute the de Broglie wavelength formula (Equation 1.12) into Equation 1.15, we obtain

$$m_e vr = \frac{nh}{2\pi}$$

or

$$m_e vr = n\hbar \qquad n = 1,\ 2,\ 3,\ \ldots \tag{1.16}$$

where we introduce the symbol $\hbar$ for $h/2\pi$. The short-hand notation is introduced because $\hbar$ appears in many of the equations of quantum chemistry. The quantity on the left side of Equation 1.16 is the angular momentum of the electron. Thus, another

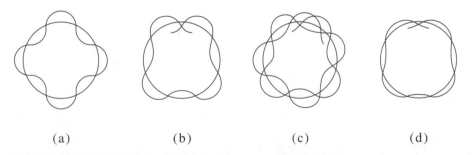

(a)	(b)	(c)	(d)

FIGURE 1.9
An illustration of matching and mismatching de Broglie waves travelling in Bohr orbits. If the wavelengths of the de Broglie waves are such that an integral number of them fit around the circle, then they match after a complete revolution (a). If a wave does not match after a complete revolution (b), cancellation will result and the wave will progressively disappear (c, d).

interpretation of Equation 1.15, and one more commonly attributed to Bohr, is that the angular momentum of the electron about the proton must be quantized; in other words, it can have only certain discrete values that satisfy Equation 1.16 for $n = 1, 2, 3, \ldots$.

If we solve Equation 1.16 for v and substitute it into Equation 1.14, we find that the radii of the orbits must satisfy

$$r = \frac{\varepsilon_0 h^2 n^2}{\pi m_e e^2} = \frac{4\pi \varepsilon_0 \hbar^2 n^2}{m_e e^2} \tag{1.17}$$

Thus, we see that the radii of the allowed orbits, or Bohr orbits, are quantized. According to this picture, the electron can move around the nucleus in circular orbits only with radii given by Equation 1.17. The orbit with the smallest radius is the one with $n = 1$, for which

$$r = \frac{4\pi(8.85419 \times 10^{-12} \text{ C}^2 \cdot \text{N}^{-1} \cdot \text{m}^{-2})(1.055 \times 10^{-34} \text{ J} \cdot \text{s})^2}{(9.109 \times 10^{-31} \text{ kg})(1.6022 \times 10^{-19} \text{ C})^2}$$

$$= 5.292 \times 10^{-11} \text{ m} = 52.92 \text{ pm} \tag{1.18}$$

The radius of the first Bohr orbit is often denoted by a_0.

The total energy of the electron in an atom is equal to the sum of its kinetic energy and potential energy. The potential energy of an electron and a proton separated by a distance r is given by Coulomb's law

$$V(r) = -\frac{e^2}{4\pi \varepsilon_0 r} \tag{1.19}$$

The negative sign here indicates that the proton and electron attract each other; their energy is less than it is when they are infinitely separated $[V(\infty) = 0]$. The total energy of the electron in a hydrogen atom is

$$E = \text{KE} + V(r) = \frac{1}{2} m_e v^2 - \frac{e^2}{4\pi \varepsilon_0 r} \tag{1.20}$$

Using Equation 1.14 to eliminate the $m_e v^2$ in the kinetic energy term, Equation 1.20 becomes

$$E = \frac{1}{2} \left(\frac{e^2}{4\pi \varepsilon_0 r} \right) - \frac{e^2}{4\pi \varepsilon_0 r} = -\frac{e^2}{8\pi \varepsilon_0 r} \tag{1.21}$$

The only allowed values of r are those given by Equation 1.17, and so if we substitute Equation 1.17 into Equation 1.21, we find that the only allowed energies are

$$E_n = -\frac{m_e e^4}{8\varepsilon_0^2 h^2} \frac{1}{n^2} \qquad n = 1, 2, \ldots \tag{1.22}$$

The negative sign in this equation indicates that the energy states are bound states; the energies given by Equation 1.22 are less than when the proton and electron are infinitely separated. Note that $n = 1$ in Equation 1.22 corresponds to the state of lowest energy.

This energy is called the *ground-state energy*. At ordinary temperatures, hydrogen atoms, as well as most other atoms and molecules, are found almost exclusively in their ground electronic states. The states of higher energy are called *excited states* and are generally unstable with respect to the ground state. An atom or a molecule in an excited state will usually relax back to the ground state and give off the energy as electromagnetic radiation.

We can display the energies given by Equation 1.22 in an energy-level diagram such as that in Figure 1.10. Note that the energy levels merge as $n \to \infty$. Bohr assumed that the observed spectrum of the hydrogen atom is due to transitions from one allowed energy state to another, and using Equation 1.22, he predicted that the allowed energy differences are given by

$$\Delta E = \frac{m_e e^4}{8\varepsilon_0^2 h^2} \left(\frac{1}{n_1^2} - \frac{1}{n_2^2} \right) = h\nu \tag{1.23}$$

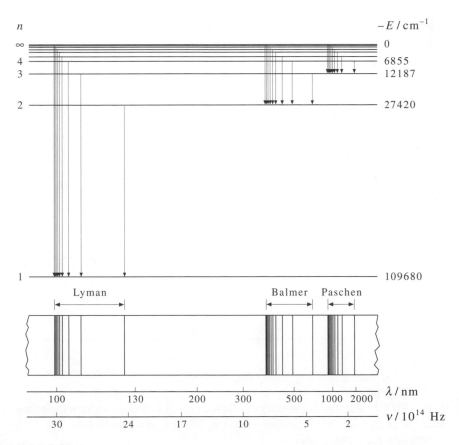

FIGURE 1.10
The energy-level diagram for the hydrogen atom, showing how transitions from higher states into some particular state lead to the observed spectral series for hydrogen.

The equation $\Delta E = h\nu$ is called the *Bohr frequency condition*. Bohr assumed that as the electron falls from one level to another, the energy evolved is given off as a photon of energy $E = h\nu$. Figure 1.10 groups the various transitions that occur according to the final state into which the electron falls. We can see, then, that the various observed spectral series arise in a natural way from the Bohr model. The Lyman series occurs when electrons that are excited to higher levels relax to the $n = 1$ state; the Balmer series occurs when excited electrons fall back into the $n = 2$ state, and so on.

We can write the theoretical formula (Equation 1.23) in the form of the empirical Rydberg formula by writing $h\nu = hc\tilde{\nu}$:

$$\tilde{\nu} = \frac{m_e e^4}{8\varepsilon_0^2 ch^3}\left(\frac{1}{n_1^2} - \frac{1}{n_2^2}\right) \tag{1.24}$$

If we compare Equations 1.11 and 1.24, we conclude that

$$R_\infty = \frac{m_e e^4}{8\varepsilon_0^2 ch^3} \tag{1.25}$$

should be equal to the Rydberg constant, Equation 1.11.

EXAMPLE 1–8

Using the values of the physical constants given inside the front cover of this book, calculate R_∞ and compare the result to its experimental value, $109\,677.6$ cm^{-1}.

SOLUTION:

$$R_\infty =$$
$$\frac{(9.10939 \times 10^{-31}\ \text{kg})(1.602177 \times 10^{-19}\ \text{C})^4}{(8)(8.85419 \times 10^{-12}\ \text{C}^2\cdot\text{N}^{-1}\cdot\text{m}^{-2})^2(2.99792 \times 10^8\ \text{m}\cdot\text{s}^{-1})(6.626076 \times 10^{-34}\ \text{J}\cdot\text{s})^3}$$
$$= 1.09737 \times 10^7\ \text{m}^{-1} = 109\,737\ \text{cm}^{-1}$$

which is within 0.05% of the experimental value of $109\,677.6$ cm^{-1}, surely a remarkable agreement.

EXAMPLE 1–9

Calculate the ionization energy of the hydrogen atom.

SOLUTION: The ionization energy IE is the energy required to take the electron from the ground state to the first unbound state, which is obtained by letting $n_2 = \infty$ in Equation 1.24. Thus, we write

$$IE = R_\infty\left(\frac{1}{1^2} - \frac{1}{\infty^2}\right)$$

or

$$IE = R_\infty = 109\,737 \text{ cm}^{-1}$$
$$= 2.1787 \times 10^{-18} \text{ J}$$
$$= 13.598 \text{ eV} = 1312.0 \text{ kJ} \cdot \text{mol}^{-1}$$

Note that we have expressed the energy in units of wave numbers (cm^{-1}). This unit is not strictly a unit of energy, but because of the simple relation between wave number and energy, $E = hc\tilde{\nu}$, energy is often expressed in this way (cf. Problem 1–1).

Despite a number of successes and the beautiful simplicity of the Bohr theory, the theory could not be extended successfully even to a two-electron system such as helium. Furthermore, even for a simple system such as hydrogen, it could not explain the spectra that arise when a magnetic field is applied to the system, nor could it predict the intensities of the spectral lines.

1–9. The Heisenberg Uncertainty Principle States That the Position and the Momentum of a Particle Cannot Be Specified Simultaneously with Unlimited Precision

We now know that we must consider light and matter as having the characteristics of both waves and particles. Let's consider a measurement of the position of an electron. If we wish to locate the electron within a distance Δx, then we must use a measuring device that has a spatial resolution less than Δx. One way to achieve this resolution is to use light with a wavelength on the order of $\lambda \approx \Delta x$. For the electron to be "seen", a photon must interact or collide in some way with the electron, for otherwise the photon will just pass right by and the electron will appear transparent. The photon has a momentum $p = h/\lambda$, and during the collision, some of this momentum will be transferred to the electron. The very act of locating the electron leads to a change in its momentum. If we wish to locate the electron more accurately, we must use light with a smaller wavelength. Consequently, the photons in the light beam will have greater momentum because of the relation $p = h/\lambda$. Because some of the photon's momentum must be transferred to the electron in the process of locating it, the momentum change of the electron becomes greater. A careful analysis of this process was carried out in the mid-1920s by the German physicist Werner Heisenberg, who showed that it is not possible to determine exactly how much momentum is transferred to the electron. This difficulty means that if we wish to locate an electron to within a region Δx, there will be an uncertainty in the momentum of the electron. Heisenberg was able to show that if Δp is the uncertainty in the momentum of the electron, then

$$\Delta x \, \Delta p \geq h \tag{1.26}$$

Equation 1.26 is called *Heisenberg's Uncertainty Principle* and is a fundamental principle of nature. The Uncertainty Principle states that if we wish to locate any particle to within a distance Δx, then we automatically introduce an uncertainty in the momentum of the particle and that the uncertainty is given by Equation 1.26. Note that this uncertainty does not stem from poor measurement or experimental technique but is a fundamental property of the act of measurement itself. The following two examples demonstrate the numerical consequences of the Uncertainty Principle.

EXAMPLE 1–10
Calculate the uncertainty in the position of a baseball thrown at 90 mph if we measure its momentum to a millionth of 1.0%.

SOLUTION: According to Example 1–6, a baseball traveling at 90 mph has a momentum of $5.6 \ \text{kg} \cdot \text{m} \cdot \text{s}^{-1}$. A millionth of 1.0% of this value is $5.6 \times 10^{-8} \ \text{kg} \cdot \text{m} \cdot \text{s}^{-1}$, so

$$\Delta p = 5.6 \times 10^{-8} \ \text{kg} \cdot \text{m} \cdot \text{s}^{-1}$$

The minimum uncertainty in the position of the baseball is

$$\Delta x = \frac{h}{\Delta p} = \frac{6.626 \times 10^{-34} \ \text{J} \cdot \text{s}}{5.6 \times 10^{-8} \text{kg} \cdot \text{m} \cdot \text{s}^{-1}}$$
$$= 1.2 \times 10^{-26} \ \text{m}$$

a completely inconsequential distance.

EXAMPLE 1–11
What is the uncertainty in momentum if we wish to locate an electron within an atom, say, so that Δx is approximately 50 pm?

SOLUTION:
$$\Delta p = \frac{h}{\Delta x} = \frac{6.626 \times 10^{-34} \ \text{J} \cdot \text{s}}{50 \times 10^{-12} \ \text{m}}$$
$$= 1.3 \times 10^{-23} \ \text{kg} \cdot \text{m} \cdot \text{s}^{-1}$$

Because $p = mv$ and the mass of an electron is 9.11×10^{-31} kg, this value of Δp corresponds to

$$\Delta v = \frac{\Delta p}{m_\text{e}} = \frac{1.3 \times 10^{-23} \ \text{kg} \cdot \text{m} \cdot \text{s}^{-1}}{9.11 \times 10^{-31} \ \text{kg}}$$
$$= 1.4 \times 10^7 \ \text{m} \cdot \text{s}^{-1}$$

which is a very large uncertainty in the speed.

These two examples show that although the Heisenberg Uncertainty Principle is of no consequence for everyday, macroscopic bodies, it has very important consequences in dealing with atomic and subatomic particles. This conclusion is similar to the one that we drew for the application of the de Broglie relation between wavelength and momentum. The Uncertainty Principle led to an awkward result. It turns out that the Bohr theory is inconsistent with the Uncertainty Principle. Fortunately, a new, more general quantum theory was soon presented that is consistent with the Uncertainty Principle. We will see that this theory is applicable to all atoms and molecules and forms the basis for our understanding of atomic and molecular structure. This theory was formulated by the Austrian physicist Erwin Schrödinger and will be discussed in Chapter 3. In preparation, in Chapter 2 we will discuss the classical wave equation, which serves as a useful and informative background to the Schrödinger equation.

Problems

1-1. Radiation in the ultraviolet region of the electromagnetic spectrum is usually described in terms of wavelength, λ, and is given in nanometers (10^{-9} m). Calculate the values of ν, $\tilde{\nu}$, and E for ultraviolet radiation with $\lambda = 200$ nm and compare your results with those in Figure 1.11.

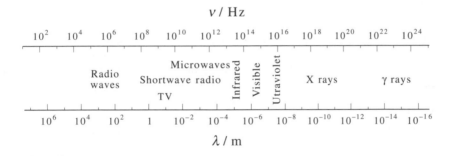

FIGURE 1.11
The regions of electromagnetic radiation.

1-2. Radiation in the infrared region is often expressed in terms of wave numbers, $\tilde{\nu} = 1/\lambda$. A typical value of $\tilde{\nu}$ in this region is 10^3 cm^{-1}. Calculate the values of ν, λ, and E for radiation with $\tilde{\nu} = 10^3$ cm^{-1} and compare your results with those in Figure 1.11.

1-3. Past the infrared region, in the direction of lower energies, is the microwave region. In this region, radiation is usually characterized by its frequency, ν, expressed in units of megahertz (MHz), where the unit, hertz (Hz), is a cycle per second. A typical microwave frequency is 2.0×10^4 MHz. Calculate the values of ν, λ, and E for this radiation and compare your results with those in Figure 1.11.

1-4. Planck's principal assumption was that the energies of the electronic oscillators can have only the values $E = nh\nu$ and that $\Delta E = h\nu$. As $\nu \to 0$, then $\Delta E \to 0$ and E is essentially continuous. Thus, we should expect the nonclassical Planck distribution to go over to the classical Rayleigh-Jeans distribution at low frequencies, where $\Delta E \to 0$. Show that Equation 1.2 reduces to Equation 1.1 as $\nu \to 0$. (Recall that $e^x = 1 + x + (x^2/2!) + \cdots$, or, in other words, that $e^x \approx 1 + x$ when x is small.)

1-5. Before Planck's theoretical work on blackbody radiation, Wien showed empirically that (Equation 1.4)

$$\lambda_{max} T = 2.90 \times 10^{-3} \text{ m·K}$$

where λ_{max} is the wavelength at which the blackbody spectrum has its maximum value at a temperature T. This expression is called the Wien displacement law; derive it from Planck's theoretical expression for the blackbody distribution by differentiating Equation 1.3 with respect to λ. *Hint*: Set $hc/\lambda_{max} k_B T = x$ and derive the intermediate result $e^{-x} + (x/5) = 1$. This problem cannot be solved for x analytically but must be solved numerically. Solve it by iteration on a hand calculator, and show that $x = 4.965$ is the solution.

1-6. At what wavelength does the maximum in the radiant energy density distribution function for a blackbody occur if (a) $T = 300$ K? (b) $T = 3000$ K? (c) $T = 10\,000$ K?

1-7. Sirius, one of the hottest known stars, has approximately a blackbody spectrum with $\lambda_{max} = 260$ nm. Estimate the surface temperature of Sirius.

1-8. The fireball in a thermonuclear explosion can reach temperatures of approximately 10^7 K. What value of λ_{max} does this correspond to? In what region of the spectrum is this wavelength found (cf. Figure 1.11)?

1-9. Calculate the energy of a photon for a wavelength of 100 pm (about one atomic diameter).

1-10. Express the Planck distribution law in terms of λ (and $d\lambda$) by using the relationship $\lambda\nu = c$.

1-11. Calculate the number of photons in a 2.00 mJ light pulse at (a) 1.06 μm, (b) 537 nm, and (c) 266 nm.

1-12. The mean temperature of the Earth's surface is 288 K. Calculate the wavelength at the maximum of the Earth's blackbody radiation. What part of the spectrum does this wavelength correspond to?

1-13. A helium-neon laser (used in supermarket scanners) emits light at 632.8 nm. Calculate the frequency of this light. What is the energy of a photon generated by this laser?

1-14. The power output of a laser is measured in units of watts (W), where one watt is equal to one joule per second. (1 W = 1 J·s^{-1}) What is the number of photons emitted per second by a 1.00 mW nitrogen laser? The wavelength emitted by a nitrogen laser is 337 nm.

1-15. A household lightbulb is a blackbody radiator. Many light bulbs use tungsten filaments that are heated by an electric current. What temperature is needed so that $\lambda_{max} = 550$ nm?

1-16. The threshold wavelength for potassium metal is 564 nm. What is its work function? What is the kinetic energy of electrons ejected if radiation of wavelength 410 nm is used?

1-17. Given that the work function of chromium is 4.40 eV, calculate the kinetic energy of electrons emitted from a chromium surface that is irradiated with ultraviolet radiation of wavelength 200 nm.

1-18. When a clean surface of silver is irradiated with light of wavelength 230 nm, the kinetic energy of the ejected electrons is found to be 0.805 eV. Calculate the work function and the threshold frequency of silver.

1-19. Some data for the kinetic energy of ejected electrons as a function of the wavelength of the incident radiation for the photoelectron effect for sodium metal are

λ/nm	100	200	300	400	500
KE/eV	10.1	3.94	1.88	0.842	0.222

Plot these data to obtain a straight line, and calculate h from the slope of the line and the work function ϕ from its intercept with the horizontal axis.

1-20. Use the Rydberg formula (Equation 1.10) to calculate the wavelengths of the first three lines of the Lyman series.

1-21. A line in the Lyman series of hydrogen has a wavelength of 1.03×10^{-7} m. Find the original energy level of the electron.

1-22. A ground-state hydrogen atom absorbs a photon of light that has a wavelength of 97.2 nm. It then gives off a photon that has a wavelength of 486 nm. What is the final state of the hydrogen atom?

1-23. Show that the Lyman series occurs between 91.2 nm and 121.6 nm, that the Balmer series occurs between 364.7 nm and 656.5 nm, and that the Paschen series occurs between 820.6 nm and 1876 nm. Identify the spectral regions to which these wavelengths correspond.

1-24. Calculate the wavelength and the energy of a photon associated with the series limit of the Lyman series.

1-25. Calculate the de Broglie wavelength for (a) an electron with a kinetic energy of 100 eV, (b) a proton with a kinetic energy of 100 eV, and (c) an electron in the first Bohr orbit of a hydrogen atom.

1-26. Calculate (a) the wavelength and kinetic energy of an electron in a beam of electrons accelerated by a voltage increment of 100 V and (b) the kinetic energy of an electron that has a de Broglie wavelength of 200 pm (1 picometer $= 10^{-12}$ m).

1-27. Through what potential must a proton initially at rest fall so that its de Broglie wavelength is 1.0×10^{-10} m?

1-28. Calculate the energy and wavelength associated with an α particle that has fallen through a potential difference of 4.0 V. Take the mass of an α particle to be 6.64×10^{-27} kg.

1-29. One of the most powerful modern techniques for studying structure is neutron diffraction. This technique involves generating a collimated beam of neutrons at a particular temperature from a high-energy neutron source and is accomplished at several accelerator facilities around the world. If the speed of a neutron is given by $v_n = (3k_B T/m)^{1/2}$, where m is the mass of a neutron, then what temperature is needed so that the neutrons have a de Broglie wavelength of 50 pm? Take the mass of a neutron to be 1.67×10^{-27} kg.

1-30. Show that a small change in the speed of a particle, Δv, causes a change in its de Broglie wavelength, $\Delta \lambda$, of

$$|\Delta \lambda| = \frac{|\Delta v| \lambda_0}{v_0}$$

where v_0 and λ_0 are its initial speed and de Broglie wavelength, respectively.

1-31. Derive the Bohr formula for $\tilde{\nu}$ for a nucleus of atomic number Z.

1-32. The series in the He^+ spectrum that corresponds to the set of transitions where the electron falls from a higher level into the $n = 4$ state is called the Pickering series, an important series in solar astronomy. Derive the formula for the wavelengths of the observed lines in this series. In what region of the spectrum does it occur? (See Problem 1–31.)

1-33. Using the Bohr theory, calculate the ionization energy (in electron volts and in $kJ \cdot mol^{-1}$) of singly ionized helium.

1-34. Show that the speed of an electron in the nth Bohr orbit is $v = e^2/2\varepsilon_0 nh$. Calculate the values of v for the first few Bohr orbits.

1-35. If we locate an electron to within 20 pm, then what is the uncertainty in its speed?

1-36. What is the uncertainty of the momentum of an electron if we know its position is somewhere in a 10 pm interval? How does the value compare to momentum of an electron in the first Bohr orbit?

1-37. There is also an uncertainty principle for energy and time:

$$\Delta E \Delta t \geq h$$

Show that both sides of this expression have the same units.

1-38. The relationship introduced in Problem 1–37 has been interpreted to mean that a particle of mass m $(E = mc^2)$ can materialize from nothing provided that it returns to nothing within a time $\Delta t \leq h/mc^2$. Particles that last for time Δt or more are called *real particles*; particles that last less than time Δt are called *virtual particles*. The mass of the charged pion, a subatomic particle, is 2.5×10^{-28} kg. What is the minimum lifetime if the pion is to be considered a real particle?

1-39. Another application of the relationship given in Problem 1–37 has to do with the excited state energies and lifetimes of atoms and molecules. If we know that the lifetime of an excited state is 10^{-9} s, then what is the uncertainty in the energy of this state?

1-40. When an excited nucleus decays, it emits a γ-ray. The lifetime of an excited state of a nucleus is of the order of 10^{-12} s. What is the uncertainty in the energy of the γ-ray produced? (See Problem 1–37.)

1-41. In this problem, we will prove that the inward force required to keep a mass revolving around a fixed center is $f = mv^2/r$. To prove this, let us look at the velocity and the acceleration of a revolving mass. Referring to Figure 1.12, we see that

$$|\Delta \mathbf{r}| \approx \Delta s = r\Delta\theta \tag{1.27}$$

if $\Delta\theta$ is small enough that the arc length Δs and the vector difference $|\Delta \mathbf{r}| = |\mathbf{r}_1 - \mathbf{r}_2|$ are essentially the same. In this case, then

$$v = \lim_{\Delta t \to 0} \frac{\Delta s}{\Delta t} = r \lim_{\Delta t \to 0} \frac{\Delta\theta}{\Delta t} = r\omega \tag{1.28}$$

where $\omega = d\theta/dt = v/r$.

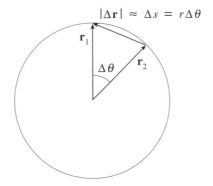

FIGURE 1.12
Diagram for defining angular speed.

If ω and r are constant, then $v = r\omega$ is constant, and because acceleration is $\lim_{t \to 0}(\Delta v/\Delta t)$, we might wonder if there is any acceleration. The answer is most definitely *yes* because velocity is a vector quantity and the direction of $\mathbf{v}$, which is the same as $\Delta \mathbf{r}$, is constantly changing even though its magnitude is not. To calculate this acceleration, draw a figure like Figure 1.12 but expressed in terms of v instead of r. From your figure, show that

$$\Delta v = |\Delta \mathbf{v}| = v\Delta\theta \tag{1.29}$$

is in direct analogy with Equation 1.27, and show that the particle experiences an acceleration given by

$$a = \lim_{\Delta t \to 0} \frac{\Delta v}{\Delta t} = v \lim_{\Delta t \to 0} \frac{\Delta\theta}{\Delta t} = v\omega \tag{1.30}$$

Thus we see that the particle experiences an acceleration and requires an inward force equal to $ma = mv\omega = mv^2/r$ to keep it moving in its circular orbit.

1-42. Planck's distribution (Equation 1.2) law gives the radiant energy density of electromagnetic radiation emitted between ν and $\nu + d\nu$. Integrate the Planck distribution over all frequencies to obtain the total energy emitted. What is its temperature dependence? Do you know whose law this is? You will need to use the integral

$$\int_0^\infty \frac{x^3 dx}{e^x - 1} = \frac{\pi^4}{15}$$

1-43. Can you derive the temperature dependence of the result in Problem 1–42 without evaluating the integral?

1-44. Ionizing a hydrogen atom in its electronic ground state requires 2.179×10^{-18} J of energy. The sun's surface has a temperature of ≈ 6000 K and is composed, in part, of atomic hydrogen. Is the hydrogen present as H(g) or H$^+$(g)? What is the temperature required so that the maximum wavelength of the emission of a blackbody ionizes atomic hydrogen? In what region of the electromagnetic spectrum is this wavelength found?

COMPLEX NUMBERS

Throughout physical chemistry, we frequently use complex numbers. In this mathchapter, we review some of the properties of complex numbers. Recall that complex numbers involve the imaginary unit, i, which is defined to be the square root of -1:

$$i = \sqrt{-1} \tag{A.1}$$

or

$$i^2 = -1 \tag{A.2}$$

Complex numbers arise naturally when solving certain quadratic equations. For example, the two solutions to

$$z^2 - 2z + 5 = 0$$

are given by

$$z = 1 \pm \sqrt{-4}$$

or

$$z = 1 \pm 2i$$

where 1 is said to be the real part and ± 2 the imaginary part of the complex number z. Generally, we write a complex number as

$$z = x + iy \tag{A.3}$$

with

$$x = \text{Re}(z) \qquad y = \text{Im}(z) \tag{A.4}$$

We add or subtract complex numbers by adding or subtracting their real and imaginary parts separately. For example, if $z_1 = 2 + 3i$ and $z_2 = 1 - 4i$, then

$$z_1 - z_2 = (2 - 1) + (3 + 4)i = 1 + 7i$$

Furthermore, we can write

$$2z_1 + 3z_2 = 2(2 + 3i) + 3(1 - 4i) = 4 + 6i + 3 - 12i = 7 - 6i$$

To multiply complex numbers together, we simply multiply the two quantities as binomials and use the fact that $i^2 = -1$. For example,

$$(2 - i)(-3 + 2i) = -6 + 3i + 4i - 2i^2$$
$$= -4 + 7i$$

To divide complex numbers, it is convenient to introduce the complex conjugate of z, which we denote by z^* and form by replacing i by $-i$. For example, if $z = x + iy$, then $z^* = x - iy$. Note that a complex number multiplied by its complex conjugate is a real quantity:

$$zz^* = (x + iy)(x - iy) = x^2 - i^2y^2 = x^2 + y^2 \qquad \text{(A.5)}$$

The square root of zz^* is called the magnitude or the absolute value of z, and is denoted by $|z|$.

Consider now the quotient of two complex numbers

$$z = \frac{2 + i}{1 + 2i}$$

This ratio can be written in the form $x + iy$ if we multiply both the numerator and the denominator by $1 - 2i$, the complex conjugate of the denominator:

$$z = \frac{2 + i}{1 + 2i}\left(\frac{1 - 2i}{1 - 2i}\right) = \frac{4 - 3i}{5} = \frac{4}{5} - \frac{3}{5}i$$

EXAMPLE A–1
Show that

$$z^{-1} = \frac{x}{x^2 + y^2} - \frac{iy}{x^2 + y^2}$$

SOLUTION: $z^{-1} = \dfrac{1}{z} = \dfrac{1}{x + iy} = \dfrac{1}{x + iy}\left(\dfrac{x - iy}{x - iy}\right) = \dfrac{x - iy}{x^2 + y^2}$

$$= \frac{x}{x^2 + y^2} - \frac{iy}{x^2 + y^2}$$

Because complex numbers consist of two parts, a real part and an imaginary part, we can represent a complex number by a point in a two-dimensional coordinate system where the real part is plotted along the horizontal (x) axes and the imaginary part is plotted along the vertical (y) axis, as in Figure A.1. The plane of such a figure is called the complex plane. If we draw a vector $\mathbf{r}$ from the origin of this figure to the point $z = (x, y)$, then the length of the vector, $r = (x^2 + y^2)^{1/2}$, is $|z|$, the magnitude or the absolute value of z. The angle θ that the vector $\mathbf{r}$ makes with the x-axis is the phase angle of z.

EXAMPLE A–2
Given $z = 1 + i$, determine the magnitude, $|z|$, and the phase angle, θ, of z.

SOLUTION: The magnitude of z is given by the square root of

$$zz^* = (1 + i)(1 - i) = 2$$

or $|z| = 2^{1/2}$. Figure A.1 shows that the tangent of the phase angle is given by

$$\tan \theta = \frac{y}{x} = 1$$

or $\theta = 45°$, or $\pi/4$ radians. (Recall that 1 radian $= 180°/\pi$, or $1° = \pi/180$ radian.)

We can always express $z = x + iy$ in terms of r and θ by using Euler's formula

$$e^{i\theta} = \cos \theta + i \sin \theta \tag{A.6}$$

which is derived in Problem A–10. Referring to Figure A.1, we see that

$$x = r \cos \theta \quad \text{and} \quad y = r \sin \theta$$

and so

$$z = x + iy = r \cos \theta + ir \sin \theta$$
$$= r(\cos \theta + i \sin \theta) = re^{i\theta} \tag{A.7}$$

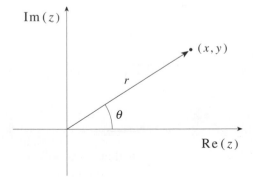

FIGURE A.1
Representation of a complex number $z = x + iy$ as a point in a two-dimensional coordinate system. The plane of this figure is called the complex plane.

where

$$r = \left(x^2 + y^2\right)^{1/2} \tag{A.8}$$

and

$$\tan \theta = \frac{y}{x} \tag{A.9}$$

Equation A.7, the polar representation of z, is often more convenient to use than Equation A.3, the Cartesian representation of z.

Note that

$$z^* = r e^{-i\theta} \tag{A.10}$$

and that

$$zz^* = \left(re^{i\theta}\right)\left(re^{-i\theta}\right) = r^2 \tag{A.11}$$

or $r = (zz^*)^{1/2}$. Also note that $z = e^{i\theta}$ is a unit vector in the complex plane because $r^2 = (e^{i\theta})(e^{-i\theta}) = 1$. The following example proves this result in another way.

EXAMPLE A–3

Show that $e^{-i\theta} = \cos\theta - i\sin\theta$ and use this result and the polar representation of z to show that $|e^{i\theta}| = 1$.

SOLUTION: To prove that $e^{-i\theta} = \cos\theta - i\sin\theta$, we use Equation A.6 and the fact that $\cos\theta$ is an even function of θ $[\cos(-\theta) = \cos\theta]$ and that $\sin\theta$ is an odd function of θ $[\sin(-\theta) = -\sin\theta]$. Therefore,

$$e^{-i\theta} = \cos\theta + i\sin(-\theta) = \cos\theta - i\sin\theta$$

Furthermore,

$$|e^{i\theta}| = [(\cos\theta + i\sin\theta)(\cos\theta - i\sin\theta)]^{1/2}$$
$$= (\cos^2\theta + \sin^2\theta)^{1/2} = 1$$

Problems

A-1. Find the real and imaginary parts of the following quantities:

    a. $(2 - i)^3$ **b.** $e^{\pi i/2}$ **c.** $e^{-2+i\pi/2}$ **d.** $(\sqrt{2}+2i)e^{-i\pi/2}$

A-2. If $z = x + 2iy$, then find

    a. $\text{Re}(z^*)$ **b.** $\text{Re}(z^2)$ **c.** $\text{Im}(z^2)$ **d.** $\text{Re}(zz^*)$

    e. $\text{Im}(zz^*)$

A-3. Express the following complex numbers in the form $re^{i\theta}$:

    a. $6i$ **b.** $4 - \sqrt{2}i$ **c.** $-1 - 2i$ **d.** $\pi + ei$

A-4. Express the following complex numbers in the form $x + iy$:

    a. $e^{i\pi/4}$ **b.** $6e^{2\pi i/3}$ **c.** $e^{-(\pi/4)i+\ln 2}$ **d.** $e^{-2\pi i}+e^{4\pi i}$

A-5. Prove that $e^{i\pi} = -1$. Comment on the nature of the numbers in this relation.

A-6. Show that

$$\cos \theta = \frac{e^{i\theta} + e^{-i\theta}}{2}$$

and that

$$\sin \theta = \frac{e^{i\theta} - e^{-i\theta}}{2i}$$

A-7. Use Equation A.7 to derive

$$z^n = r^n(\cos \theta + i \sin \theta)^n = r^n(\cos n\theta + i \sin n\theta)$$

and from this, the formula of De Moivre:

$$(\cos \theta + i \sin \theta)^n = \cos n\theta + i \sin n\theta$$

A-8. Use the formula of De Moivre, which is given in Problem A–7, to derive the trigonometric identities

$$\cos 2\theta = \cos^2 \theta - \sin^2 \theta$$
$$\sin 2\theta = 2 \sin \theta \cos \theta$$
$$\cos 3\theta = \cos^3 \theta - 3 \cos \theta \sin^2 \theta$$
$$= 4 \cos^3 \theta - 3 \cos \theta$$
$$\sin 3\theta = 3 \cos^2 \theta \sin \theta - \sin^3 \theta$$
$$= 3 \sin \theta - 4 \sin^3 \theta$$

A-9. Consider the set of functions

$$\Phi_m(\phi) = \frac{1}{\sqrt{2\pi}}e^{im\phi} \qquad m = 0, \pm1, \pm2, \ldots$$

$$0 \le \phi \le 2\pi$$

First show that

$$\int_0^{2\pi} d\phi\, \Phi_m(\phi) = 0 \qquad \text{for all value of } m \ne 0$$

$$= \sqrt{2\pi} \quad m = 0$$

Now show that

$$\int_0^{2\pi} d\phi\, \Phi_m^*(\phi)\Phi_n(\phi) = 0 \quad m \ne n$$

$$= 1 \quad m = n$$

A-10. This problem offers a derivation of Euler's formula. Start with

$$f(\theta) = \ln(\cos\theta + i\sin\theta) \tag{1}$$

Show that

$$\frac{df}{d\theta} = i \tag{2}$$

Now integrate both sides of Equation 2 to obtain

$$f(\theta) = \ln(\cos\theta + i\sin\theta) = i\theta + c \tag{3}$$

where c is a constant of integration. Show that $c = 0$ and then exponentiate Equation 3 to obtain Euler's formula.

A-11. We have seen that both the exponential and the natural logarithm functions (Problem A–10) can be extended to include complex arguments. This is generally true of most functions. Using Euler's formula and assuming that x represents a real number, show that $\cos ix$ and $-i\sin ix$ are equivalent to real functions of the real variable x. These functions are defined as the hyperbolic cosine and hyperbolic sine functions, $\cosh x$ and $\sinh x$, respectively. Sketch these functions. Do they oscillate like $\sin x$ and $\cos x$? Now show that $\sinh ix = i\sin x$ and that $\cosh ix = \cos x$.

A-12. Evaluate i^i.

A-13. The equation $x^2 = 1$ has two distinct roots, $x = \pm1$. The equation $x^N = 1$ has N distinct roots, called the N roots of unity. This problem shows how to find the N roots of unity. We shall see that some of the roots turn out to be complex, so let's write the equation as $z^N = 1$. Now let $z = re^{i\theta}$ and obtain $r^N e^{iN\theta} = 1$. Show that this must be equivalent to $e^{iN\theta} = 1$, or

$$\cos N\theta + i\sin N\theta = 1$$

Now argue that $N\theta = 2\pi n$, where n has the N distinct values $0, 1, 2, \ldots, N-1$ or that the N roots of units are given by

$$z = e^{2\pi in/N} \qquad n = 0, 1, 2, \ldots, N-1$$

Show that we obtain $z = 1$ and $z = \pm 1$, for $N = 1$ and $N = 2$, respectively. Now show that

$$z = 1, -\frac{1}{2} + i\frac{\sqrt{3}}{2}, \quad \text{and} \quad -\frac{1}{2} - i\frac{\sqrt{3}}{2}$$

for $N = 3$. Show that each of these roots is of unit magnitude. Plot these three roots in the complex plane. Now show that $z = 1, i, -1$, and $-i$ for $N = 4$ and that

$$z = 1, -1, \frac{1}{2} \pm i\frac{\sqrt{3}}{2}, \quad \text{and} \quad -\frac{1}{2} \pm i\frac{\sqrt{3}}{2}$$

for $N = 6$. Plot the four roots for $N = 4$ and the six roots for $N = 6$ in the complex plane. Compare the plots for $N = 3$, $N = 4$, and $N = 6$. Do you see a pattern?

A-14. Using the results of Problem A–13, find the three distinct roots of $x^3 = 8$.

Louis de Broglie was born on August 15, 1892 in Dieppe, France, into an aristocratic family and died in 1987. He studied history as an undergraduate in the early 1910s, but his interest turned to science as a result of his working with his older brother, Maurice, who had built his own private laboratory for X-ray research. de Broglie took up his formal studies in physics after World War I, receiving his Dr. Sc. from the University of Paris in 1924. His dissertation was on the wavelike properties of matter, a highly controversial and original proposal at that time. Using the special theory of relativity, de Broglie postulated that material particles should exhibit wavelike properties under certain conditions, just as radiation was known to exhibit particlelike properties. After receiving his Ph.D., he remained as a free lecturer at the Sorbonne and later was appointed professor of theoretical physics at the new Henri Poincaré Institute. He was professor of theoretical physics at the University of Paris from 1937 until his retirement in 1962. The wavelike properties he postulated were later demonstrated experimentally and are now exploited as a basis of the electron microscope. de Broglie spent the later part of his career trying to obtain a causal interpretation of the wave mechanics to replace the probabilistic theories. He was awarded the Nobel Prize for physics in 1929 "for his discovery of the wave nature of electrons."

CHAPTER 2

The Classical Wave Equation

In 1925, Erwin Schrödinger and Werner Heisenberg independently formulated a general quantum theory. At first sight, the two methods appeared different because Heisenberg's method is formulated in terms of matrices, whereas Schrödinger's method is formulated in terms of partial differential equations. Just a year later, however, Schrödinger showed that the two formulations are mathematically equivalent. Because most students of physical chemistry are not familiar with matrix algebra, quantum theory is customarily presented according to Schrödinger's formulation, the central feature of which is a partial differential equation now known as the *Schrödinger equation*. Partial differential equations may sound no more comforting than matrix algebra, but fortunately we require only elementary calculus to treat the problems in this book. The wave equation of classical physics describes various wave phenomena such as a vibrating string, a vibrating drum head, ocean waves, and acoustic waves. Not only does the classical wave equation provide a physical background to the Schrödinger equation, but, in addition, the mathematics involved in solving the classical wave equation are central to any discussion of quantum mechanics. Because most students of physical chemistry have little experience with classical wave equations, this chapter discusses this topic. In particular, we will solve the standard problem of a vibrating string because not only is the method of solving this problem similar to the method we will use to solve the Schrödinger equation, but it also gives us an excellent opportunity to relate the mathematical solution of a problem to the physical nature of the problem. Many of the problems at the end of the chapter illustrate the connection between physical problems and the mathematics developed in the chapter.

2–1. The One-Dimensional Wave Equation Describes the Motion of a Vibrating String

Consider a uniform string stretched between two fixed points as shown in Figure 2.1. The maximum displacement of the string from its equilibrium horizontal position is

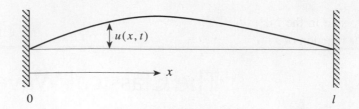

A vibrating string whose ends are fixed at 0 and l. The amplitude of the vibration at position x and time t is $u(x, t)$.

called its *amplitude*. If we let $u(x, t)$ be the displacement of the string, then $u(x, t)$ satisfies the equation

$$\frac{\partial^2 u(x, t)}{\partial x^2} = \frac{1}{v^2} \frac{\partial^2 u(x, t)}{\partial t^2} \tag{2.1}$$

where v is the speed with which a disturbance moves along the string. Equation 2.1 is the *classical wave equation*. Equation 2.1 is a *partial differential equation* because the unknown, $u(x, t)$ in this case, occurs in partial derivatives. The variables x and t are said to be the *independent variables* and $u(x, t)$, which depends upon x and t, is said to be the *dependent variable*. Equation 2.1 is a *linear partial differential equation* because $u(x, t)$ and its derivatives appear only to the first power and there are no cross terms.

In addition to having to satisfy Equation 2.1, the amplitude $u(x, t)$ must also satisfy certain physical conditions as well. Because the ends of the string are held fixed, the amplitude at these two points is always zero, and so we have the requirement that

$$u(0, t) = 0 \quad \text{and} \quad u(l, t) = 0 \qquad \text{(for all } t) \tag{2.2}$$

These two conditions are called *boundary conditions* because they specify the behavior of $u(x, t)$ at the boundaries. Generally, a partial differential equation must be solved subject to certain boundary conditions, the nature of which will be apparent on physical grounds.

2–2. The Wave Equation Can Be Solved by the Method of Separation of Variables

The classical wave equation, as well as the Schrödinger equation and many other partial differential equations that arise in physical chemistry, can often be solved by a method called *separation of variables*. We shall use the problem of a vibrating string to illustrate this method.

The key step in the method of separation of variables is to assume that $u(x, t)$ factors into a function of x, $X(x)$, times a function of t, $T(t)$, or that

$$u(x, t) = X(x)T(t) \tag{2.3}$$

If we substitute Equation 2.3 into Equation 2.1, we obtain

$$T(t)\frac{d^2 X(x)}{dx^2} = \frac{1}{v^2} X(x)\frac{d^2 T(t)}{dt^2} \tag{2.4}$$

Now we divide by $u(x, t) = X(x)T(t)$ and obtain

$$\frac{1}{X(x)}\frac{d^2 X(x)}{dx^2} = \frac{1}{v^2 T(t)}\frac{d^2 T(t)}{dt^2} \tag{2.5}$$

The left side of Equation 2.5 is a function of x only and the right side is a function of t only. Because x and t are independent variables, each side of Equation 2.5 can be varied independently. The only way for the equality of the two sides to be preserved under any variation of x and t is for each side to be equal to a constant. If we let this constant be K, we can write

$$\frac{1}{X(x)}\frac{d^2 X(x)}{dx^2} = K \tag{2.6}$$

and

$$\frac{1}{v^2 T(t)}\frac{d^2 T(t)}{dt^2} = K \tag{2.7}$$

where K is called the *separation constant* and will be determined later. Equations 2.6 and 2.7 can be written as

$$\frac{d^2 X(x)}{dx^2} - KX(x) = 0 \tag{2.8}$$

and

$$\frac{d^2 T(t)}{dt^2} - Kv^2 T(t) = 0 \tag{2.9}$$

Equations 2.8 and 2.9 are called *ordinary differential equations* (as opposed to partial differential equations) because the unknowns, $X(x)$ and $T(t)$ in this case, occur as ordinary derivatives. Both of these differential equations are linear because the unknowns and their derivatives appear only to the first power. Furthermore, the coefficients of every term involving the unknowns in these equations are constants; that is, 1 and $-K$ in Equation 2.8 and 1 and $-Kv^2$ in Equation 2.9. These equations are called *linear differential equations with constant coefficients* and are quite easy to solve, as we shall see.

The value of K in Equations 2.8 and 2.9 is yet to be determined. We do not know right now whether K is positive, negative, or even zero. Let us first assume that $K = 0$. In this case, Equations 2.8 and 2.9 can be integrated immediately to find

$$X(x) = a_1 x + b_1 \tag{2.10}$$

and

$$T(t) = a_2 t + b_2 \tag{2.11}$$

where the a's and b's are just integration constants, which can be determined by using the boundary conditions given in Equation 2.2. In terms of $X(x)$ and $T(t)$, the boundary conditions are

$$u(0, t) = X(0)T(t) = 0 \tag{2.12}$$

and

$$u(l, t) = X(l)T(t) = 0 \tag{2.13}$$

Because $T(t)$ certainly does not vanish for all t, we must have that

$$X(0) = 0 \qquad \text{and} \qquad X(l) = 0 \tag{2.14}$$

which is how the boundary conditions affect $X(x)$. Going back to Equation 2.10, we conclude that the only way to satisfy Equation 2.14 is for $a_1 = b_1 = 0$, which means that $X(x) = 0$ and that $u(x, t) = 0$ for all x. This is called a *trivial solution* to Equation 2.1 and is of no physical interest. (Throwing away solutions to mathematical equations should not disturb you. What we know from physics is that every physically acceptable solution $u(x, t)$ must satisfy Equation 2.1, *not* that every solution to the equation is physically acceptable.)

Let's look at Equations 2.8 and 2.9 for $K > 0$. Both equations are of the form

$$\frac{d^2 y}{dx^2} - k^2 y(x) = 0 \tag{2.15}$$

where k is a real constant. Experience shows that every solution to a linear differential equation with constant coefficients whose right side is equal to zero is of the form $y(x) = e^{\alpha x}$, where α is a constant to be determined. Therefore, we let $y(x) = e^{\alpha x}$ in Equation 2.15 and get

$$(\alpha^2 - k^2)y(x) = 0$$

Therefore, either $(\alpha^2 - k^2)$ or $y(x)$ must equal zero. The case $y(x) = 0$ is a trivial solution, and so $\alpha^2 - k^2$ must equal zero. Therefore,

$$\alpha = \pm k$$

Thus, there are two solutions: $y(x) = e^{kx}$ and e^{-kx}. We can easily prove that

$$y(x) = c_1 e^{kx} + c_2 e^{-kx}$$

(where c_1 and c_2 are constants) is also a solution. This is the general solution to all differential equations with the form of Equation 2.15. The fact that a sum of the two solutions, e^{kx} and e^{-kx}, is also a solution is a direct consequence of Equation 2.15 being a *linear* differential equation. Note that the highest derivative in Equation 2.15 is a second derivative, which implies that in some sense we are performing two integrations when we find its solution. When we do two integrations, we always obtain two constants of integration. The solution we have found has two constants, c_1 and c_2, which suggests that it is the most general solution.

The solution of other ordinary differential equations with constant coefficients is best illustrated by examples.

EXAMPLE 2–1
Solve the equation

$$\frac{d^2 y}{dx^2} - 3\frac{dy}{dx} + 2y = 0$$

SOLUTION: If we substitute $y(x) = e^{\alpha x}$ into this differential equation, we obtain

$$\alpha^2 y - 3\alpha y + 2y = 0$$
$$\alpha^2 - 3\alpha + 2 = 0$$
$$(\alpha - 2)(\alpha - 1) = 0$$

or that $\alpha = 1$ and 2. The two solutions are $y(x) = e^x$ and $y(x) = e^{2x}$ and the general solution is

$$y(x) = c_1 e^x + c_2 e^{2x}$$

Prove this by substituting this solution back into the original equation.

EXAMPLE 2–2
Solve the equation in Example 2–1 subject to the two boundary conditions $y(0) = 0$ and $dy/dx\,(\text{at } x = 0) = -1$.

SOLUTION: The general solution is

$$y(x) = c_1 e^x + c_2 e^{2x}$$

The two conditions given allow us to evaluate c_1 and c_2 and hence find a particular solution. Putting $x = 0$ into $y(x)$ and $x = 0$ into dy/dx gives

$$y(0) = c_1 + c_2 = 0$$

$$\frac{dy}{dx}(\text{at } x = 0) = c_1 + 2c_2 = -1$$

Solving these two equations simultaneously gives $c_1 = 1$ and $c_2 = -1$, and so

$$y(x) = e^x - e^{2x}$$

satisfies not only the differential equation, but also the two boundary conditions as well.

2–3. Some Differential Equations Have Oscillatory Solutions

Now let's consider the case where $K < 0$ in Equations 2.8 and 2.9. In this case, α will be imaginary. As a concrete example, consider the differential equation

$$\frac{d^2y}{dx^2} + y(x) = 0 \tag{2.16}$$

which is essentially Equation 2.8 with $K = -1$. If we let $y(x) = e^{\alpha x}$, we have

$$\left(\alpha^2 + 1\right) y(x) = 0$$

or that

$$\alpha = \pm i$$

(MathChapter A). The general solution to Equation 2.16 is

$$y(x) = c_1 e^{ix} + c_2 e^{-ix} \tag{2.17}$$

We can easily verify that this is a solution by substituting Equation 2.17 directly into Equation 2.16.

It is usually more convenient to rewrite expressions such as e^{ix} or e^{-ix} in Equation 2.17 using Euler's formula (Equation A.6):

$$e^{\pm i\theta} = \cos\theta \pm i\sin\theta$$

If we substitute Euler's formula into Equation 2.17, we find

$$y(x) = c_1(\cos x + i\sin x) + c_2(\cos x - i\sin x)$$
$$= (c_1 + c_2)\cos x + (ic_1 - ic_2)\sin x$$

But $c_1 + c_2$ and $ic_1 - ic_2$ are also just constants, and if we call them c_3 and c_4, respectively, we can write

$$y(x) = c_3 \cos x + c_4 \sin x$$

instead of

$$y(x) = c_1 e^{ix} + c_2 e^{-ix}$$

These two forms for $y(x)$ are equivalent.

EXAMPLE 2–3

Prove that

$$y(x) = A \cos x + B \sin x$$

(where A and B are constants), is a solution to the differential equation

$$\frac{d^2 y}{dx^2} + y(x) = 0$$

SOLUTION: The first derivative of $y(x)$ is

$$\frac{dy}{dx} = -A \sin x + B \cos x$$

and the second derivative is

$$\frac{d^2 y}{dx^2} = -A \cos x - B \sin x$$

Therefore, we see that

$$\frac{d^2 y}{dx^2} + y(x) = 0$$

or that $y(x) = A \cos x + B \sin x$ is a solution of the differential equation

$$\frac{d^2 y}{dx^2} + y(x) = 0$$

The next example is important and one whose general solution should be learned.

EXAMPLE 2–4

Solve the equation

$$\frac{d^2 x}{dt^2} + \omega^2 x(t) = 0$$

Subject to the initial conditions $x(0) = A$ and $dx/dt = 0$ at $t = 0$.

SOLUTION: In this case, we find $\alpha = \pm i\omega$ and

$$x(t) = c_1 e^{i\omega t} + c_2 e^{-i\omega t}$$

or

$$x(t) = c_3 \cos \omega t + c_4 \sin \omega t$$

Now

$$x(0) = c_3 = A$$

and

$$\left(\frac{dx}{dt} \right)_{t=0} = \omega c_4 = 0$$

implying that $c_4 = 0$ and that the particular solution we are seeking is

$$x(t) = A \cos \omega t$$

This solution is plotted in Figure 2.2. Note that it oscillates cosinusoidally in time, with an amplitude A. The wavelength of the oscillation is $2\pi/\omega$ and the frequency ν is given by (see Problem 2–3)

$$\nu = \frac{\omega}{2\pi}$$

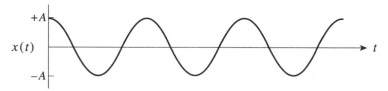

FIGURE 2.2
A plot of $x(t) = A \cos \omega t$, the solution to the problem in Example 2–4. The amplitude is A, the wavelength is $2\pi/\omega$, and the frequency is $\omega/2\pi$.

2–4. The General Solution to the Wave Equation Is a Superposition of Normal Modes

Let us assess where we are now. We have obtained Equations 2.8 and 2.9 by applying the method of separation of variables to the wave equation. We have already shown that if the separation constant K is zero, then only a trivial solution results. Now let's assume that K is positive. To this end, write K as β^2, where β is real. This assures that

K is positive because it is the square of a real number. In the case $K = \beta^2$, the general solution to Equation 2.8 is

$$X(x) = c_1 e^{\beta x} + c_2 e^{-\beta x}$$

We can easily show that the only way to satisfy the boundary conditions (Equation 2.14) is for $c_1 = c_2 = 0$, and so once again we find only a trivial solution.

Let's hope that assuming K to be negative gives us something interesting. If we set $K = -\beta^2$, then K is negative if β is real. In this case Equation 2.8 is

$$\frac{d^2 X(x)}{dx^2} + \beta^2 X(x) = 0$$

Referring to Example 2–4, we see that the general solution can be written as

$$X(x) = A \cos \beta x + B \sin \beta x$$

The boundary condition that $X(0) = 0$ implies that $A = 0$. The condition at the boundary $x = l$ says that

$$X(l) = B \sin \beta l = 0 \tag{2.18}$$

Equation 2.18 can be satisfied in two ways. One is that $B = 0$, but this along with the fact that $A = 0$ yields a trivial solution. The other way is to require that $\sin \beta l = 0$. Because $\sin \theta = 0$ when $\theta = 0,\ \pi,\ 2\pi,\ 3\pi,\ \ldots$, Equation 2.18 implies that

$$\beta l = n\pi \qquad n = 1,\ 2,\ 3,\ \ldots \tag{2.19}$$

where we have omitted the $n = 0$ case because it leads to $\beta = 0$, and a trivial solution. Equation 2.19 determines the parameter β and hence the separation constant $K = -\beta^2$. So far then, we have that

$$X(x) = B \sin \frac{n\pi x}{l} \tag{2.20}$$

Remember that we have Equation 2.9 to solve also. Equation 2.9 can be written as

$$\frac{d^2 T(t)}{dt^2} + \beta^2 v^2 T(t) = 0 \tag{2.21}$$

where Equation 2.19 says that $\beta = n\pi/l$. Referring to the result obtained in Example 2–4 again, the general solution to Equation 2.21 is

$$T(t) = D \cos \omega_n t + E \sin \omega_n t \tag{2.22}$$

where $\omega_n = \beta v = n\pi v/l$. We have no conditions to specify D and E, so the amplitude $u(x, t)$ is (cf. Equation 2.3)

$$
\begin{aligned}
u(x, t) &= X(x)T(t) \\
&= \left(B \sin \frac{n\pi x}{l} \right) \left(D \cos \omega_n t + E \sin \omega_n t \right) \\
&= \left(F \cos \omega_n t + G \sin \omega_n t \right) \sin \frac{n\pi x}{l} \qquad n = 1, 2, \ldots
\end{aligned}
$$

where we have let $F = DB$ and $G = EB$. Because there is a $u(x, t)$ for each integer n and because the values of F and G may depend on n, we should write $u(x, t)$ as

$$
u_n(x, t) = \left(F_n \cos \omega_n t + G_n \sin \omega_n t \right) \sin \frac{n\pi x}{l} \qquad n = 1, 2, \ldots \tag{2.23}
$$

Because each $u_n(x, t)$ is a solution to the linear differential equation, Equation 2.1, their sum is also a solution of Equation 2.1 and is, in fact, the general solution. Therefore, for the general solution we have

$$
u(x, t) = \sum_{n=1}^{\infty} \left(F_n \cos \omega_n t + G_n \sin \omega_n t \right) \sin \frac{n\pi x}{l} \qquad n = 1, 2, \ldots \tag{2.24}
$$

No matter how the string is plucked initially, its shape will evolve according to Equation 2.24. We can easily verify that Equation 2.24 is a solution to Equation 2.1 by direct substitution. Problem 2–5 shows that $F \cos \omega t + G \sin \omega t$ can be written in the equivalent form, $A \cos(\omega t + \phi)$, where A and ϕ are constants expressible in terms of F and G. The quantity A is the amplitude of the wave and ϕ is called the *phase angle*. Using this relation, we can write Equation 2.24 in the form

$$
u(x, t) = \sum_{n=1}^{\infty} A_n \cos \left(\omega_n t + \phi_n \right) \sin \frac{n\pi x}{l} = \sum_{n=1}^{\infty} u_n(x, t) \tag{2.25}
$$

Equation 2.25 has a nice physical interpretation. Each $u_n(x, t)$ is called a *normal mode*, and the time dependence of each normal mode represents harmonic motion of frequency

$$
\nu_n = \frac{\omega_n}{2\pi} = \frac{\upsilon n}{2l} \tag{2.26}
$$

where we have used the fact that $\omega_n = \beta v = n\pi v/l$ (cf. Equation 2.19). The spatial dependence of the first few terms in Equation 2.25 is shown in Figure 2.3. The first term, $u_1(x, t)$, called the *fundamental mode* or *first harmonic*, represents a sinusoidal (harmonic) time dependence of frequency $v/2l$ of the motion depicted in Figure 2.3a. The *second harmonic* or *first overtone*, $u_2(x, t)$, vibrates harmonically with frequency v/l and looks like the motion depicted in Figure 2.3b. Note that the midpoint of this harmonic is fixed at zero for all t. Such a point is called a *node*, a concept that arises in quantum mechanic as well. Notice that $u(0)$ and $u(l)$ are also equal to zero. These terms are not nodes because their values are fixed by the boundary conditions. Note that the

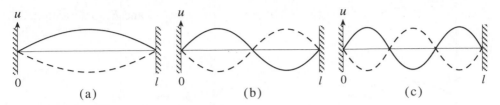

FIGURE 2.3
The first three normal modes of a vibrating string. Note that each normal mode is a standing wave and the the nth harmonic has $n - 1$ nodes.

second harmonic oscillates with twice the frequency of the first harmonic. Figure 2.3c shows that the *third harmonic* or *second overtone* has two nodes. It is easy to continue and show that the number of nodes is equal to $n - 1$ (Problem 2–10). The waves shown in Figure 2.3 are called *standing waves* because the positions of the nodes are fixed in time. Between the nodes, the string oscillates up and down.

Consider a simple case in which $u(x, t)$ consists of only the first two harmonics and is of the form (cf. Equation 2.25)

$$u(x, t) = \cos \omega_1 t \sin \frac{\pi x}{l} + \frac{1}{2} \cos \left(\omega_2 t + \frac{\pi}{2} \right) \sin \frac{2\pi x}{l} \tag{2.27}$$

Equation 2.27 is illustrated in Figure 2.4. The left side of Figure 2.4 shows the time dependence of each mode separately. Notice that $u_2(x, t)$ has gone through one complete oscillation in the time depicted while $u_1(x, t)$ has gone through only one-half cycle, nicely illustrating that $\omega_2 = 2\omega_1$. The right side of Figure 2.4 shows the sum of the two harmonics, or the actual motion of the string, as a function of time. You can see how a superposition of the standing waves in the the left side of the figure yields the *traveling wave* in the right side. The decomposition of any complicated, general wave motion into a sum or superposition of normal modes is a fundamental property of oscillatory behavior and follows from the fact that the wave equation is a linear equation.

Our path from the wave equation to its solution was fairly long because we had to learn to solve a certain class of ordinary differential equations on the way. The overall procedure is actually straightforward, and to illustrate this procedure, we will solve the problem of a vibrating rectangular membrane, a two-dimensional problem, in Section 2–5.

2–5. A Vibrating Membrane Is Described by a Two-Dimensional Wave Equation

The generalization of Equation 2.1 to two dimensions is

$$\frac{\partial^2 u}{\partial x^2} + \frac{\partial^2 u}{\partial y^2} = \frac{1}{v^2} \frac{\partial^2 u}{\partial t^2} \tag{2.28}$$

50

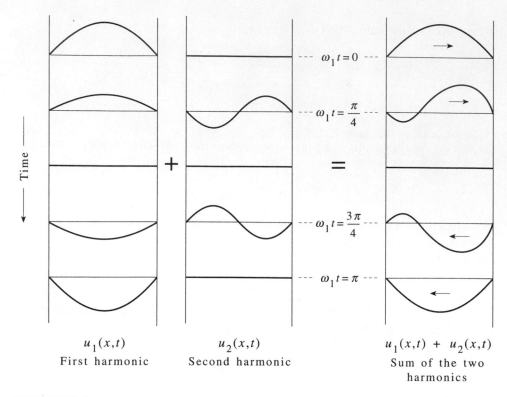

$u_1(x,t)$
First harmonic

$u_2(x,t)$
Second harmonic

$u_1(x,t) + u_2(x,t)$
Sum of the two
harmonics

FIGURE 2.4
An illustration of how two standing waves can combine to give a traveling wave. In both parts, time increases downward. The left portion shows the independent motion of the first two harmonics. Both harmonics are standing waves; the first harmonic goes through half a cycle and the second harmonic goes through one complete cycle in the time shown. The right side shows the sum of the two harmonics. The sum is not a standing wave. As shown the sum is a traveling wave that travels back and forth between the fixed ends. The traveling wave has gone through one-half a cycle in the time shown.

where $u = u(x, y, t)$ and x, y, and t are the independent variables. We will apply this equation to a rectangular membrane whose entire perimeter is clamped. By referring to the geometry in Figure 2.5, we see that the boundary conditions that $u(x, y, t)$ must

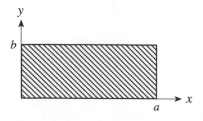

FIGURE 2.5
A rectangular membrane clamped along its perimeter.

satisfy because its four edges are clamped are

$$u(0, y) = u(a, y) = 0$$

$$\text{(for all } t)$$ (2.29)

$$u(x, 0) = u(x, b) = 0$$

By applying the method of separation of variables to Equation 2.28, we assume that $u(x, y, t)$ can be written as the product of a spatial part and a temporal part or that

$$u(x, y, t) = F(x, y)T(t)$$ (2.30)

We substitute Equation 2.30 into Equation 2.28 and divide by $F(x, y)T(t)$ to find

$$\frac{1}{v^2 T(t)} \frac{d^2 T}{dt^2} = \frac{1}{F(x, y)} \left(\frac{\partial^2 F}{\partial x^2} + \frac{\partial^2 F}{\partial y^2} \right)$$ (2.31)

The right side of Equation 2.31 is a function of x and y only and the left side is a function of t only. The equality can be true for all t, x, and y only if both sides are equal to a constant. Anticipating that the separation constant will be negative as it was in the previous sections, we write it as $-\beta^2$ and obtain the two separate equations

$$\frac{d^2 T}{dt^2} + v^2 \beta^2 T(t) = 0$$ (2.32)

and

$$\frac{\partial^2 F}{\partial x^2} + \frac{\partial^2 F}{\partial y^2} + \beta^2 F(x, y) = 0$$ (2.33)

Equation 2.33 is still a partial differential equation. To solve it, we once again use separation of variables. Substitute $F(x, y) = X(x)Y(y)$ into Equation 2.33 and divide by $X(x)Y(y)$ to obtain

$$\frac{1}{X(x)} \frac{d^2 X}{dx^2} + \frac{1}{Y(y)} \frac{d^2 Y}{dy^2} + \beta^2 = 0$$ (2.34)

Again we argue that because x and y are independent variables, the only way this equation can be valid is that

$$\frac{1}{X(x)} \frac{d^2 X}{dx^2} = -p^2$$ (2.35)

and

$$\frac{1}{Y(y)} \frac{d^2 Y}{dy^2} = -q^2$$ (2.36)

where p^2 and q^2 are separation constants, which according to Equation 2.34 must satisfy

$$p^2 + q^2 = \beta^2 \tag{2.37}$$

Equations 2.35 and 2.36 can be rewritten as

$$\frac{d^2 X}{dx^2} + p^2 X(x) = 0 \tag{2.38}$$

and

$$\frac{d^2 Y}{dy^2} + q^2 Y(y) = 0 \tag{2.39}$$

Equation 2.28, a partial differential equation in three variables, has been reduced to three ordinary differential equations (Equations 2.32, 2.38, and 2.39), each of which is exactly of the form discussed in Example 2–4. The solutions to Equations 2.38 and 2.39 are

$$X(x) = A \cos px + B \sin px \tag{2.40}$$

and

$$Y(y) = C \cos qy + D \sin qy \tag{2.41}$$

The boundary conditions, Equation 2.29, in terms of the functions $X(x)$ and $Y(y)$ are

$$X(0)Y(y) = X(a)Y(y) = 0$$

and

$$X(x)Y(0) = X(x)Y(b) = 0$$

which imply that

$$X(0) = X(a) = 0$$
$$Y(0) = Y(b) = 0 \tag{2.42}$$

Applying the first of Equation 2.42 to Equation 2.40 shows that $A = 0$ and $pa = n\pi$, so that

$$X(x) = B \sin \frac{n\pi x}{a} \qquad n = 1, \, 2, \, \ldots \tag{2.43}$$

In exactly the same manner, we find that $C = 0$ and $qb = m\pi$, where $m = 1, \, 2, \, \ldots$ and so

$$Y(y) = D \sin \frac{m\pi y}{b} \qquad m = 1, \, 2, \, \ldots \tag{2.44}$$

Recalling that $p^2 + q^2 = \beta^2$, we see that

$$\beta_{nm} = \pi \left(\frac{n^2}{a^2} + \frac{m^2}{b^2} \right)^{1/2} \qquad \begin{array}{l} n = 1, \, 2, \, \dots \\ m = 1, \, 2, \, \dots \end{array} \qquad (2.45)$$

where we have subscripted β to emphasize that it depends on the two integers n and m.

Finally, now we solve Equation 2.32 for the time dependence:

$$T_{nm}(t) = E_{nm} \cos \omega_{nm} t + F_{nm} \sin \omega_{nm} t \qquad (2.46)$$

where

$$\omega_{nm} = v\beta_{nm}$$

$$= v\pi \left(\frac{n^2}{a^2} + \frac{m^2}{b^2} \right)^{1/2} \qquad (2.47)$$

According to Problem 2–15, Equation 2.46 can be written as

$$T_{nm}(t) = G_{nm} \cos(\omega_{nm} t + \phi_{nm}) \qquad (2.48)$$

The complete solution to Equation 2.28 is

$$u(x, y, t) = \sum_{n=1}^{\infty} \sum_{m=1}^{\infty} u_{nm}(x, y, t)$$

$$= \sum_{n=1}^{\infty} \sum_{m=1}^{\infty} A_{nm} \cos(\omega_{nm} t + \phi_{nm}) \sin \frac{n\pi x}{a} \sin \frac{m\pi y}{b} \qquad (2.49)$$

As in the one-dimensional case of a vibrating string, we see that the general vibrational motion of a rectangular drum can be expressed as a superposition of normal modes, $u_{nm}(x, y, t)$. Some of these modes are shown in Figure 2.6. Note that in this two-dimensional problem we obtain *nodal lines*. In two-dimensional problems, the nodes are lines, as compared with points in one-dimensional problems. Figure 2.6 shows the normal modes for a case in which $a \neq b$. The case in which $a = b$ is an

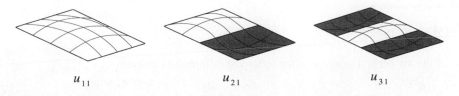

u_{11} $\qquad$ u_{21} $\qquad$ u_{31}

FIGURE 2.6
The first few normal modes of a rectangular membrane with shaded and clear sections having opposite sinusoidal displacements as indicated.

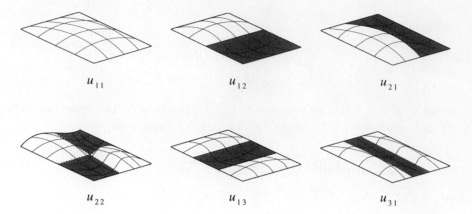

FIGURE 2.7
The normal modes of a square membrane, illustrating the occurrence of degeneracy in this system. The normal modes u_{12} and u_{21} have different orientations but the same frequency, given by Equation 2.50. The same is true for the normal modes u_{13} and u_{31}.

interesting one. The frequencies of the normal modes are given by Equation 2.47. When $a = b$ in Equation 2.47, we have

$$\omega_{nm} = \frac{v\pi}{a} \left(n^2 + m^2\right)^{1/2} \tag{2.50}$$

We see from Equation 2.50 that $\omega_{12} = \omega_{21} = 5^{1/2}/a$ in this case; yet the normal modes $u_{12}(x, y, t)$ and $u_{21}(x, y, t)$ are not the same, as seen from Figure 2.7. This is an example of a *degeneracy*, and we say that the frequency $\omega_{12} = \omega_{21}$ is *doubly degenerate* or *two-fold degenerate*. Note that the phenomenon of degeneracy arises because of the symmetry introduced when $a = b$. This phenomenon can be seen easily by comparing the modes u_{12} and u_{21} in Figure 2.7. Equation 2.50 shows that there will be at least a twofold degeneracy when $m \neq n$ because $m^2 + n^2 = n^2 + m^2$. We will see that the concept of degeneracy arises in quantum mechanics also.

This chapter has presented a discussion of the wave equation and its solutions. In Chapter 3, we will use the mathematical methods developed here, and so we recommend doing many of the problems at the end of this chapter before going on. Several problems involve physical systems and serve as refreshers or introductions to classical mechanics.

Problems

2-1. Find the general solutions to the following differential equations.

a. $\dfrac{d^2 y}{dx^2} - 4\dfrac{dy}{dx} + 3y = 0$ **b.** $\dfrac{d^2 y}{dx^2} + 6\dfrac{dy}{dx} = 0$ **c.** $\dfrac{dy}{dx} + 3y = 0$

d. $\dfrac{d^2 y}{dx^2} + 2\dfrac{dy}{dx} - y = 0$ **e.** $\dfrac{d^2 y}{dx^2} - 3\dfrac{dy}{dx} + 2y = 0$

2-2. Solve the following differential equations:

a. $\dfrac{d^2y}{dx^2} - 4y = 0$ $\qquad y(0) = 2;\ \dfrac{dy}{dx}(\text{at } x = 0) = 4$

b. $\dfrac{d^2y}{dx^2} - 5\dfrac{dy}{dx} + 6y = 0$ $\qquad y(0) = -1;\ \dfrac{dy}{dx}(\text{at } x = 0) = 0$

c. $\dfrac{dy}{dx} - 2y = 0$ $\qquad y(0) = 2$

2-3. Prove that $x(t) = \cos \omega t$ oscillates with a frequency $\nu = \omega/2\pi$. Prove that $x(t) = A \cos \omega t + B \sin \omega t$ oscillates with the same frequency, $\omega/2\pi$.

2-4. Solve the following differential equations:

a. $\dfrac{d^2x}{dt^2} + \omega^2 x(t) = 0$ $\qquad x(0) = 0;\ \dfrac{dx}{dt}(\text{at } t = 0) = v_0$

b. $\dfrac{d^2x}{dt^2} + \omega^2 x(t) = 0$ $\qquad x(0) = A;\ \dfrac{dx}{dt}(\text{at } t = 0) = v_0$

Prove in both cases that $x(t)$ oscillates with frequency $\omega/2\pi$.

2-5. The general solution to the differential equation

$$\frac{d^2x}{dt^2} + \omega^2 x(t) = 0$$

is

$$x(t) = c_1 \cos \omega t + c_2 \sin \omega t$$

For convenience, we often write this solution in the equivalent forms

$$x(t) = A \sin(\omega t + \phi)$$

or

$$x(t) = B \cos(\omega t + \psi)$$

Show that all three of these expressions for $x(t)$ are equivalent. Derive equations for A and ϕ in terms of c_1 and c_2, and for B and ψ in terms of c_1 and c_2. Show that all three forms of $x(t)$ oscillate with frequency $\omega/2\pi$. *Hint*: Use the trigonometric identities

$$\sin(\alpha + \beta) = \sin \alpha \cos \beta + \cos \alpha \sin \beta$$

and

$$\cos(\alpha + \beta) = \cos \alpha \cos \beta - \sin \alpha \sin \beta$$

2-6. In all the differential equations we have discussed so far, the values of the exponents α that we have found have been either real or purely imaginary. Let us consider a case in which α turns out to be complex. Consider the equation

$$\frac{d^2y}{dx^2} + 2\frac{dy}{dx} + 10y = 0$$

If we substitute $y(x) = e^{\alpha x}$ into this equation, we find that $\alpha^2 + 2\alpha + 10 = 0$ or that $\alpha = -1 \pm 3i$. The general solution is

$$y(x) = c_1 e^{(-1+3i)x} + c_2 e^{(-1-3i)x}$$
$$= c_1 e^{-x} e^{3ix} + c_2 e^{-x} e^{-3ix}$$

Show that $y(x)$ can be written in the equivalent form

$$y(x) = e^{-x}(c_3 \cos 3x + c_4 \sin 3x)$$

Thus we see that complex values of the α's lead to trigonometric solutions modulated by an exponential factor. Solve the following equations.

a. $\dfrac{d^2y}{dx^2} + 2\dfrac{dy}{dx} + 2y = 0$

b. $\dfrac{d^2y}{dx^2} - 6\dfrac{dy}{dx} + 25y = 0$

c. $\dfrac{d^2y}{dx^2} + 2\beta\dfrac{dy}{dx} + (\beta^2 + \omega^2)y = 0$

d. $\dfrac{d^2y}{dx^2} + 4\dfrac{dy}{dx} + 5y = 0 \qquad y(0) = 1; \quad \dfrac{dy}{dx}(\text{at } x = 0) = -3$

2-7. This problem develops the idea of a classical harmonic oscillator. Consider a mass m attached to a spring as shown in Figure 2.8. Suppose there is no gravitational force acting on m so that the only force is from the spring. Let the relaxed or undistorted length of the spring be x_0. Hooke's law says that the force acting on the mass m is $f = -k(x - x_0)$, where k is a constant characteristic of the spring and is called the force constant of the spring. Note that the minus sign indicates the direction of the force: to the left if $x > x_0$ (extended) and to the right if $x < x_0$ (compressed). The momentum of the mass is

$$p = m\frac{dx}{dt} = m\frac{d(x - x_0)}{dt}$$

Newton's second law says that the rate of change of momentum is equal to a force

$$\frac{dp}{dt} = f$$

Replacing $f(x)$ by Hooke's law, show that

$$m\frac{d^2(x - x_0)}{dt^2} = -k(x - x_0)$$

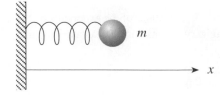

FIGURE 2.8
A body of mass m connected to a wall by a spring.

Upon letting $\xi = x - x_0$ be the displacement of the spring from its undistorted length, then

$$m\frac{d^2\xi}{dt^2} + k\xi = 0$$

Given that the mass starts at $\xi = 0$ with an intial velocity v_0, show that the displacement is given by

$$\xi(t) = v_0 \left(\frac{m}{k}\right)^{1/2} \sin\left[\left(\frac{k}{m}\right)^{1/2} t\right]$$

Interpret and discuss this solution. What does the motion look like? What is the frequency? What is the amplitude?

2-8. Consider the linear second-order differential equation

$$\frac{d^2y}{dx^2} + a_1(x)\frac{dy}{dx} + a_0(x)y(x) = 0$$

Note that this equation is linear because $y(x)$ and its derivatives appear only to the first power and there are no cross terms. It does not have constant coefficients, however, and there is no general, simple method for solving it like there is if the coefficients were constants. In fact, each equation of this type must be treated more or less individually. Nevertheless, because it is linear, we must have that if $y_1(x)$ and $y_2(x)$ are any two solutions, then a linear combination,

$$y(x) = c_1 y_1(x) + c_2 y_2(x)$$

where c_1 and c_2 are constants, is also a solution. Prove that $y(x)$ is a solution.

2-9. We will see in Chapter 3 that the Schrödinger equation for a particle of mass m that is constrained to move freely along a line between 0 and a is

$$\frac{d^2\psi}{dx^2} + \left(\frac{8\pi^2 mE}{h^2}\right)\psi(x) = 0$$

with the boundary condition

$$\psi(0) = \psi(a) = 0$$

In this equation, E is the energy of the particle and $\psi(x)$ is its wave function. Solve this differential equation for $\psi(x)$, apply the boundary conditions, and show that the energy can have only the values

$$E_n = \frac{n^2 h^2}{8ma^2} \qquad n = 1, 2, 3, \ldots$$

or that the energy is quantized.

2-10. Prove that the number of nodes for a vibrating string clamped at both ends is $n - 1$ for the nth harmonic.

2-11. Prove that

$$y(x, t) = A \sin\left[\frac{2\pi}{\lambda}(x - vt)\right]$$

is a wave of wavelength λ and frequency $\nu = v/\lambda$ traveling to the right with a velocity v.

2-12. Sketch the normal modes of a vibrating rectangular membrane and convince yourself that they look like those shown in Figure 2.6.

2-13. This problem is the extension of Problem 2–9 to two dimensions. In this case, the particle is constrained to move freely over the surface of a rectangle of sides a and b. The Schrödinger equation for this problem is

$$\frac{\partial^2 \psi}{\partial x^2} + \frac{\partial^2 \psi}{\partial y^2} + \left(\frac{8\pi^2 m E}{h^2}\right) \psi(x, y) = 0$$

with the boundary conditions

$$\psi(0, y) = \psi(a, y) = 0 \quad \text{for all } y, \quad 0 \leq y \leq b$$

$$\psi(x, 0) = \psi(x, b) = 0 \quad \text{for all } x, \quad 0 \leq x \leq a$$

Solve this equation for $\psi(x, y)$, apply the boundary conditions, and show that the energy is quantized according to

$$E_{n_x n_y} = \frac{n_x^2 h^2}{8ma^2} + \frac{n_y^2 h^2}{8mb^2} \qquad \begin{array}{l} n_x = 1, 2, 3, \dots \\ n_y = 1, 2, 3, \dots \end{array}$$

2-14. Extend Problems 2–9 and 2–13 to three dimensions, where a particle is constrained to move freely throughout a rectangular box of sides a, b, and c. The Schrödinger equation for this system is

$$\frac{\partial^2 \psi}{\partial x^2} + \frac{\partial^2 \psi}{\partial y^2} + \frac{\partial^2 \psi}{\partial z^2} + \left(\frac{8\pi^2 m E}{h^2}\right) \psi(x, y, z) = 0$$

and the boundary conditions are that $\psi(x, y, z)$ vanishes over all the surfaces of the box.

2-15. Show that Equations 2.46 and 2.48 are equivalent. How are G_{nm} and ϕ_{nm} in Equation 2.48 related to the quantities in Equation 2.46?

Problems 2–16 through 2–19 illustrate some other applications of differential equations to classical mechanics.

Many problems in classical mechanics can be reduced to the problem of solving a differential equation with constant coefficients (cf. Problem 2–7). The basic starting point is Newton's second law, which says that the rate of change of momentum is equal to the force acting on a body. Momentum p equals mv, and so if the mass is constant, then in one dimension we have

$$\frac{dp}{dt} = m\frac{dv}{dt} = m\frac{d^2x}{dt^2} = f$$

If we are given the force as a function of x, then this equation is a differential equation for $x(t)$, which is called the trajectory of the particle. Going back to the simple harmonic oscillator discussed in Problem 2–7, if we let x be the displacement of the mass from its equilibrium position, then Hooke's law says that $f(x) = -kx$, and the differential equation corresponding to Newton's second law is

$$\frac{d^2x}{dt^2} + kx(t) = 0$$

a differential equation that we have seen several times.

2-16. Consider a body falling freely from a height x_0 according to Figure 2.9a. If we neglect air resistance or viscous drag, the only force acting upon the body is the gravitational force mg. Using the coordinates in Figure 2.9a, mg acts in the same direction as x and so the differential equation corresponding to Newton's second law is

$$m\frac{d^2x}{dt^2} = mg$$

Show that

$$x(t) = \tfrac{1}{2}gt^2 + v_0t + x_0$$

where x_0 and v_0 are the initial values of x and v. According to Figure 2.9a, $x_0 = 0$ and so

$$x(t) = \tfrac{1}{2}gt^2 + v_0t$$

If the particle is just dropped, then $v_0 = 0$ and so

$$x(t) = \tfrac{1}{2}gt^2$$

Discuss this solution.

Now do the same problem using Figure 2.9b as the definition of the various quantities involved, and show that although the equations may look different from those above, they say exactly the same thing because the picture we draw to define the direction of x, v_0, and mg does not affect the falling body.

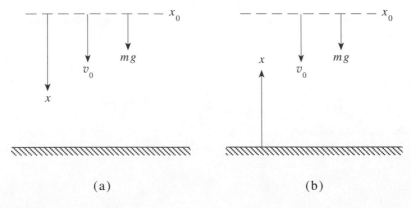

(a)   (b)

FIGURE 2.9
(a) A coordinate system for a body falling from a height x_0, and (b) a different coordinate system for a body falling from a height x_0.

2-17. Derive an equation for the maximum height a body will reach if it is shot straight upward with a velocity v_0. Refer to Figure 2.9b but realize that in this case v_0 points upward. How long will it take for the body to return to its initial position, $x = 0$?

2-18. Consider a simple pendulum as shown in Figure 2.10. We let the length of the pendulum be l and assume that all the mass of the pendulum is concentrated at its end as shown in Figure 2.10. A physical example of this case might be a mass suspended by a string. We assume that the motion of the pendulum is set up such that it oscillates within a plane so that we have a problem in plane polar coordinates. Let the distance along the arc in the figure describe the motion of the pendulum, so that its momentum is $mds/dt = mld\theta/dt$ and its rate of change of momentum is $mld^2\theta/dt^2$. Show that the component of force in the direction of motion is $-mg \sin\theta$, where the minus sign occurs because the direction of this force is opposite to that of the angle θ. Show that the equation of motion is

$$ml\frac{d^2\theta}{dt^2} = -mg \sin\theta$$

Now assume that the motion takes place only through very small angles and show that the motion becomes that of a simple harmonic oscillator. What is the natural frequency of this harmonic oscillator? *Hint*: Use the fact that $\sin\theta \approx \theta$ for small values of θ.

2-19. Consider the motion of a pendulum like that in Problem 2–18 but swinging in a viscous medium. Suppose that the viscous force is proportional to but oppositely directed to its velocity; that is,

$$f_{\text{viscous}} = -\lambda\frac{ds}{dt} = -\lambda l\frac{d\theta}{dt}$$

where λ is a viscous drag coefficient. Show that for small angles, Newton's equation is

$$ml\frac{d^2\theta}{dt^2} + \lambda l\frac{d\theta}{dt} + mg\theta = 0$$

Show that there is no harmonic motion if

$$\lambda^2 > \frac{4m^2g}{l}$$

Does it make physical sense that the medium can be so viscous that the pendulum undergoes no harmonic motion?

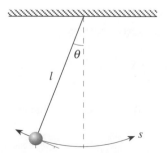

FIGURE 2.10
The coordinate system describing an oscillating pendulum.

2-20. Consider two pendulums of equal lengths and masses that are connected by a spring that obeys Hooke's law (Problem 2–7). This system is shown in Figure 2.11. Assuming that the motion takes place in a plane and that the angular displacement of each pendulum from the vertical is small, show that the equations of motion for this system are

$$m\frac{d^2x}{dt^2} = -m\omega_0^2 x - k(x - y)$$

$$m\frac{d^2y}{dt^2} = -m\omega_0^2 y - k(y - x)$$

where ω_0 is the natural vibrational frequency of each isolated pendulum, [i.e., $\omega_0 = (g/l)^{1/2}$] and k is the force constant of the connecting spring. In order to solve these two simultaneous differential equations, assume that the two pendulums swing harmonically and so try

$$x(t) = Ae^{i\omega t} \qquad y(t) = Be^{i\omega t}$$

Substitute these expressions into the two differential equations and obtain

$$\left(\omega^2 - \omega_0^2 - \frac{k}{m}\right) A = -\frac{k}{m} B$$

$$\left(\omega^2 - \omega_0^2 - \frac{k}{m}\right) B = -\frac{k}{m} A$$

Now we have two simultaneous linear homogeneous algebraic equations for the two amplitudes A and B. We shall learn in MathChapter E that the determinant of the coefficients must vanish in order for there to be a nontrivial solution. Show that this condition gives

$$\left(\omega^2 - \omega_0^2 - \frac{k}{m}\right)^2 = \left(\frac{k}{m}\right)^2$$

Now show that there are two natural frequencies for this system, namely,

$$\omega_1^2 = \omega_0^2 \quad \text{and} \quad \omega_2^2 = \omega_0^2 + \frac{2k}{m}$$

Interpret the motion associated with these frequencies by substituting ω_1^2 and ω_2^2 back into the two equations for A and B. The motion associated with these values of A and B are called *normal modes* and any complicated, general motion of this system can be written as a linear combination of these normal modes. Notice that there are two coordinates (x and y) in this problem and two normal modes. We shall see in Chapter 13 that the complicated

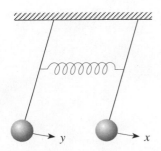

FIGURE 2.11
Two pendulums coupled by a spring that obeys Hooke's law.

vibrational motion of molecules can be resolved into a linear combination of natural, or normal, modes.

2-21. Problem 2–20 can be solved by introducing center-of-mass and relative coordinates (cf. Section 5–2). Add and subtract the differential equations for $x(t)$ and $y(t)$ and then introduce the new variables

$$\eta = x + y \quad \text{and} \quad \xi = x - y$$

Show that the differential equations for η and ξ are independent. Solve each one and compare your results to those of Problems 2–20.

PROBABILITY AND STATISTICS

In many of the following chapters, we will deal with probability distributions, average values, and standard deviations. Consequently, we take a few pages here to discuss some basic ideas of probability and show how to calculate average quantities in general.

Consider some experiment, such as the tossing of a coin or the rolling of a die, that has n possible outcomes, each with probability p_j, where $j = 1, 2, \ldots, n$. If the experiment is repeated indefinitely, we intuitively expect that

$$p_j = \lim_{N \to \infty} \frac{N_j}{N} \qquad j = 1, 2, \ldots, n \tag{B.1}$$

where N_j is the number of times that the outcome j occurs and N is the total number of repetitions of the experiment. Because $0 \le N_j \le N$, p_j must satisfy the condition

$$0 \le p_j \le 1 \tag{B.2}$$

When $p_j = 1$, we say the event j is a certainty and when $p_j = 0$, we say it is impossible. In addition, because

$$\sum_{j=1}^{n} N_j = N$$

we have the normalization condition,

$$\sum_{j=1}^{n} p_j = 1 \tag{B.3}$$

Equation B.3 means that the probability that some event occurs is a certainty. Suppose now that some number x_j is associated with the outcome j. Then we define the *average* of x or the *mean* of x to be

$$\langle x \rangle = \sum_{j=1}^{n} x_j p_j = \sum_{j=1}^{n} x_j p(x_j) \tag{B.4}$$

where in the last term we have used the expanded notation $p(x_j)$, meaning the probability of realizing the number x_j. We will denote an average of a quantity by enclosing the quantity in angular brackets.

EXAMPLE B–1
Suppose we are given the following data:

x	$p(x)$
1	0.20
3	0.25
4	0.55

Calculate the average value of x.

SOLUTION: Using Equation B.4, we have

$$\langle x \rangle = (1)(0.20) + (3)(0.25) + (4)(0.55) = 3.15$$

It is helpful to interpret a probability distribution like p_j as a distribution of a unit mass along the x axis in a discrete manner such that p_j is the fraction of mass located at the point x_j. This interpretation is shown in Figure B.1. According to this interpretation, the average value of x is the center of mass of this system.

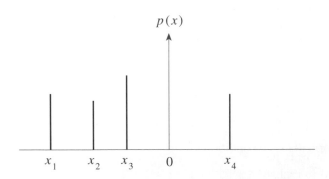

FIGURE B.1
The discrete probability frequency function or probability density, $p(x)$.

Another quantity of importance is

$$\langle x^2 \rangle = \sum_{j=1}^{n} x_j^2 p_j \tag{B.5}$$

The quantity $\langle x^2 \rangle$ is called the *second moment* of the distribution $\{p_j\}$ and is analogous to the moment of inertia.

EXAMPLE B–2

Calculate the second moment of the data given in Example B–1.

SOLUTION: Using Equation B.5, we have

$$\langle x^2 \rangle = (1)^2 (0.20) + (3)^2 (0.25) + (4)^2 (0.55) = 11.25$$

Note from Examples B–1 and B–2 that $\langle x^2 \rangle \neq \langle x \rangle^2$. This nonequality is a general result that we will prove below.

A physically more interesting quantity than $\langle x^2 \rangle$ is the *second central moment*, or the *variance*, defined by

$$\sigma_x^2 = \langle (x - \langle x \rangle)^2 \rangle = \sum_{j=1}^{n} (x_j - \langle x \rangle)^2 p_j \tag{B.6}$$

As the notation suggests, we denote the square root of the quantity in Equation B.6 by σ_x, which is called the *standard deviation*. From the summation in Equation B.6, we can see that σ_x^2 will be large if x_j is likely to differ from $\langle x \rangle$, because in that case $(x_j - \langle x \rangle)$ and so $(x_j - \langle x \rangle)^2$ will be large for the significant values of p_j. On the other hand, σ_x^2 will be small if x_j is not likely to differ from $\langle x \rangle$, or if the x_j cluster around $\langle x \rangle$, because then $(x_j - \langle x \rangle)^2$ will be small for the significant values of p_j. Thus, we see that either the variance or the standard deviation is a measure of the spread of the distribution about its mean.

Equation B.6 shows that σ_x^2 is a sum of positive terms, and so $\sigma_x^2 \geq 0$. Furthermore,

$$\sigma_x^2 = \sum_{j=1}^{n} (x_j - \langle x \rangle)^2 p_j = \sum_{j=1}^{n} (x_j^2 - 2\langle x \rangle x_j + \langle x \rangle^2) p_j$$

$$= \sum_{j=1}^{n} x_j^2 p_j - 2 \sum_{j=1}^{n} \langle x \rangle x_j p_j + \sum_{j=1}^{n} \langle x \rangle^2 p_j \tag{B.7}$$

The first term here is just $\langle x^2 \rangle$ (cf. Equation B.5). To evaluate the second and third terms, we need to realize that $\langle x \rangle$, the average of x_j, is just a number and so can be factored out of the summations, leaving a summation of the form $\sum x_j p_j$ in the second term and $\sum p_j$ in the third term. The summation $\sum x_j p_j$ is $\langle x \rangle$ by definition and the summation $\sum p_j$ is unity because of normalization (Equation B.3). Putting all this together, we find that

$$\sigma_x^2 = \langle x^2 \rangle - 2\langle x \rangle^2 + \langle x \rangle^2$$

$$= \langle x^2 \rangle - \langle x \rangle^2 \geq 0 \tag{B.8}$$

Because $\sigma_x^2 \geq 0$, we see that $\langle x^2 \rangle \geq \langle x \rangle^2$. A consideration of Equation B.6 shows that $\sigma_x^2 = 0$ or $\langle x \rangle^2 = \langle x^2 \rangle$ only when $x_j = \langle x \rangle$ with a probability of one, a case that is not really probabilistic because the event j occurs on every trial.

So far we have considered only discrete distributions, but continuous distributions are also important in physical chemistry. It is convenient to use the unit mass analogy. Consider a unit mass to be distributed continuously along the x axis, or along some interval on the x axis. We define the linear mass density $\rho(x)$ by

$$dm = \rho(x)dx$$

where dm is the fraction of the mass lying between x and $x + dx$. By analogy, then, we say that the probability that some quantity x, such as the position of a particle in a box, lies between x and $x + dx$ is

$$\text{Prob}(x, x + dx) = p(x)dx \tag{B.9}$$

and that

$$\text{Prob}(a \leq x \leq b) = \int_a^b p(x)dx \tag{B.10}$$

In the mass analogy, $\text{Prob}\{a \leq x \leq b\}$ is the fraction of mass that lies in the interval $a \leq x \leq b$. The normalization condition is

$$\int_{-\infty}^{\infty} p(x)dx = 1 \tag{B.11}$$

Following Equations B.4 through B.6, we have the definitions

$$\langle x \rangle = \int_{-\infty}^{\infty} xp(x)dx \tag{B.12}$$

$$\langle x^2 \rangle = \int_{-\infty}^{\infty} x^2 p(x)dx \tag{B.13}$$

and

$$\sigma_x^2 = \int_{-\infty}^{\infty} (x - \langle x \rangle)^2 p(x)dx \tag{B.14}$$

EXAMPLE B–3

Perhaps the simplest continuous distribution is the so-called uniform distribution, where

$$p(x) = \text{constant} = A \quad a \leq x \leq b$$
$$= 0 \quad \text{otherwise}$$

Show that A must equal $1/(b-a)$. Evaluate $\langle x \rangle$, $\langle x^2 \rangle$, σ_x^2, and σ_x for this distribution.

SOLUTION: Because $p(x)$ must be normalized,

$$\int_a^b p(x)dx = 1 = A \int_a^b dx = A(b-a)$$

Therefore, $A = 1/(b-a)$ and

$$p(x) = \frac{1}{b-a} \quad a \leq x \leq b$$
$$= 0 \quad \text{otherwise}$$

The mean of x is given by

$$\langle x \rangle = \int_a^b xp(x)dx = \frac{1}{b-a}\int_a^b xdx$$
$$= \frac{b^2 - a^2}{2(b-a)} = \frac{b+a}{2}$$

and the second moment of x by

$$\langle x^2 \rangle = \int_a^b x^2 p(x)dx = \frac{1}{b-a}\int_a^b x^2 dx$$
$$= \frac{b^3 - a^3}{3(b-a)} = \frac{b^2 + ab + a^2}{3}$$

Last, the variance is given by Equation B.6, and so

$$\sigma_x^2 = \langle x^2 \rangle - \langle x \rangle^2 = \frac{(b-a)^2}{12}$$

and the standard deviation is

$$\sigma_x = \frac{(b-a)}{\sqrt{12}}$$

EXAMPLE B–4

The most commonly occurring and most important continuous probability distribution is the *Gaussian distribution*, given by

$$p(x)dx = ce^{-x^2/2a^2}dx \quad -\infty < x < \infty$$

Find c, $\langle x \rangle$, σ_x^2 and σ_x.

SOLUTION: The constant c is determined by normalization:

$$\int_{-\infty}^{\infty} p(x)dx = 1 = c\int_{-\infty}^{\infty} e^{-x^2/2a^2} dx \tag{B.15}$$

If you look in a table of integrals (for example, *The CRC Standard Mathematical Tables* or *The CRC Handbook of Chemistry and Physics*, CRC Press), you won't find the above integral. However, you will find the integral

$$\int_{0}^{\infty} e^{-\alpha x^2} dx = \left(\frac{\pi}{4\alpha}\right)^{1/2} \tag{B.16}$$

The reason that you won't find the integral with the limts $(-\infty, \infty)$ is illustrated in Figure B.2(a), where $e^{-\alpha x^2}$ is plotted against x. Note that the graph is symmetric about the vertical axis, so that the corresponding areas on the two sides of the axis are equal. A function that has the mathematical property that $f(x) = f(-x)$ and is called an *even function*. For an even function

$$\int_{-A}^{A} f_{\text{even}}(x)dx = 2\int_{0}^{A} f_{\text{even}}(x)dx \tag{B.17}$$

If we recognize that $p(x) = ce^{-x^2/2a^2}$ is an even function and use Equation B.16, then we find that

$$c\int_{-\infty}^{\infty} e^{-x^2/2a^2} dx = 2c\int_{0}^{\infty} e^{-x^2/2a^2} dx$$

$$= 2c\left(\frac{\pi a^2}{2}\right)^{1/2} = 1$$

or $c = 1/(2\pi a^2)^{1/2}$.

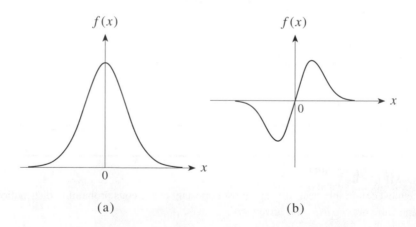

(a)   (b)

FIGURE B.2
(a) The function $f(x) = e^{-x^2}$ is an even function, $f(x) = f(-x)$. (b) The function $f(x) = xe^{-x^2}$ is an odd function, $f(x) = -f(-x)$.

The mean of x is given by

$$\langle x \rangle = \int_{-\infty}^{\infty} xp(x)dx = (2\pi a^2)^{-1/2} \int_{-\infty}^{\infty} xe^{-x^2/2a^2}dx \qquad \text{(B.18)}$$

The integrand in Equation B.18 is plotted in Figure B.2(b). Notice that this graph is antisymmetric about the vertical axis and that the area on one side of the vertical axis cancels the corresponding area on the other side. A function that has the mathematical property that $f(x) = -f(-x)$ is called an *odd function*. For an odd function,

$$\int_{-A}^{A} f_{\text{odd}}(x)dx = 0 \qquad \text{(B.19)}$$

The function $xe^{-x^2/2a^2}$ is an odd function, and so

$$\langle x \rangle = (2\pi a^2)^{-1/2} \int_{-\infty}^{\infty} xe^{-x^2/2a^2}dx = 0$$

The second moment of x is given by

$$\langle x^2 \rangle = (2\pi a^2)^{-1/2} \int_{-\infty}^{\infty} x^2 e^{-x^2/2a^2}dx$$

The integrand in this case is even because $y(x) = x^2 e^{-x^2/2a^2} = y(-x)$. Therefore,

$$\langle x^2 \rangle = 2(2\pi a^2)^{-1/2} \int_{0}^{\infty} x^2 e^{-x^2/2a^2}dx$$

The integral

$$\int_{0}^{\infty} x^2 e^{-\alpha x^2}dx = \frac{1}{4\alpha} \left(\frac{\pi}{\alpha} \right)^{1/2} \qquad \text{(B.20)}$$

can be found in integral tables, and so

$$\langle x^2 \rangle = \frac{2}{(2\pi a^2)^{1/2}} \frac{(2\pi a^2)^{1/2}a^2}{2} = a^2$$

Because $\langle x \rangle = 0$, $\sigma_x^2 = \langle x^2 \rangle$, and so σ_x is given by

$$\sigma_x = a$$

The standard deviation of a normal distribution is the parameter that appears in the exponential. The standard notation for a normalized Gaussian distribution function is

$$p(x)dx = (2\pi \sigma_x^2)^{-1/2} e^{-x^2/2\sigma_x^2}dx \qquad \text{(B.21)}$$

Figure B.3 shows Equation B.21 for various values of σ_x. Note that the curves become narrower and taller for smaller values of σ_x.

A more general version of a Gaussian distribution is

$$p(x)dx = (2\pi\sigma_x^2)^{-1/2}e^{-(x-\langle x\rangle)^2/2\sigma_x^2}dx \tag{B.22}$$

This expression looks like those in Figure B.3 except that the curves are centered at $x = \langle x \rangle$ rather than $x = 0$. A Gaussian distribution is one of the most important and commonly used probability distributions in all of science.

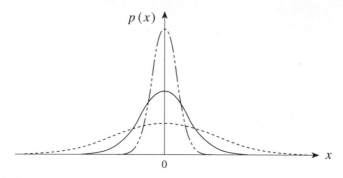

FIGURE B.3
A plot of a Gaussian distribution, $p(x)$, (Equation B.21) for three values of σ_x. The dotted curve corresponds to $\sigma_x = 2$, the solid curve to $\sigma_x = 1$, and the dash-dotted curve to $\sigma_x = 0.5$.

Problems

B-1. Consider a particle to be constrained to lie along a one-dimensional segment 0 to a. We will learn in the next chapter that the probability that the particle is found to lie between x and $x + dx$ is given by

$$p(x)dx = \frac{2}{a}\sin^2\frac{n\pi x}{a}dx$$

where $n = 1, 2, 3, \ldots$. First show that $p(x)$ is normalized. Now calculate the average position of the particle along the line segment. The integrals that you need are (*The CRC Handbook of Chemistry and Physics* or *The CRC Standard Mathematical Tables*, CRC Press)

$$\int \sin^2\alpha x dx = \frac{x}{2} - \frac{\sin 2\alpha x}{4\alpha}$$

and

$$\int x\sin^2\alpha x dx = \frac{x^2}{4} - \frac{x\sin 2\alpha x}{4\alpha} - \frac{\cos 2\alpha x}{8\alpha^2}$$

B-2. Calculate the variance associated with the probability distribution given in Problem B–1. The necessary integral is (*CRC tables*)

$$\int x^2 \sin^2 \alpha x \, dx = \frac{x^3}{6} - \left(\frac{x^2}{4\alpha} - \frac{1}{8\alpha^3}\right) \sin 2\alpha x - \frac{x \cos 2\alpha x}{4\alpha^2}$$

B-3. Using the probability distribution given in Problem B–1, calculate the probability that the particle will be found between 0 and $a/2$. The necessary integral is given in Problem B–1.

B-4. Prove explicitly that

$$\int_{-\infty}^{\infty} e^{-\alpha x^2} dx = 2 \int_{0}^{\infty} e^{-\alpha x^2} dx$$

by breaking the integral from $-\infty$ to ∞ into one from $-\infty$ to 0 and another from 0 to ∞. Now let $z = -x$ in the first integral and $z = x$ in the second to prove the above relation.

B-5. By using the procedure in Problem B–4, show explicitly that

$$\int_{-\infty}^{\infty} xe^{-\alpha x^2} dx = 0$$

B-6. We will learn in Chapter 27 that the molecules in a gas travel at various speeds, and that the probability that a molecule has a speed between v and $v + dv$ is given by

$$p(v)dv = 4\pi \left(\frac{m}{2\pi k_B T}\right)^{3/2} v^2 e^{-mv^2/2k_B T} dv \qquad 0 \leq v < \infty$$

where m is the mass of the particle, k_B is the Boltzmann constant (the molar gas constant R divided by the Avogadro constant), and T is the Kelvin temperature. The probability distribution of molecular speeds is called the Maxwell-Boltzmann distribution. First show that $p(v)$ is normalized, and then determine the average speed as a function of temperature. The necessary integrals are (*CRC tables*)

$$\int_{0}^{\infty} x^{2n} e^{-\alpha x^2} dx = \frac{1 \cdot 3 \cdot 5 \cdots (2n-1)}{2^{n+1} \alpha^n} \left(\frac{\pi}{\alpha}\right)^{1/2} \qquad n \geq 1$$

and

$$\int_{0}^{\infty} x^{2n+1} e^{-\alpha x^2} dx = \frac{n!}{2\alpha^{n+1}}$$

where $n!$ is n factorial, or $n! = n(n-1)(n-2)\cdots(1)$.

B-7. Use the Maxwell-Boltzmann distribution in Problem B–6 to determine the average kinetic energy of a gas-phase molecule as a function of temperature. The necessary integral is given in Problem B–6.

Erwin Schrödinger was born in Vienna, Austria, on August 12, 1887, and died there in 1961. He received his Ph.D. in theoretical physics in 1910 from the University of Vienna. He then held a number of positions in Germany and in 1927 succeeded Max Planck at the University of Berlin at Planck's request. Schrödinger left Berlin in 1933 because of his opposition to Hitler and Nazi policies and eventually moved to the University of Graz in Austria in 1936. After the invasion of Austria by Germany, he was forcibly removed from his professorship in 1936.
He then moved to the Institute of Advanced Studies, which was created for him, at the University College, Dublin, Ireland. He remained there for 17 years and then retired to his native Austria. Schrödinger shared the Nobel Prize for physics with Paul Dirac in 1933 for the "discovery of new productive forms of atomic theory." Schrödinger rejected the probabilistic interpretation of the wave equation, which led to serious disagreement with Max Born, but they remained warm friends in spite of their scientific disagreement. Schrödinger preferred to work alone, and so no school developed around him, as it did for several other developers of quantum mechanics. His influential book, *What is Life?*, caused a number of physicists to become interested in biology. His personal life, which was rather unconventional, has been engagingly related by Walter Moore in his book, *Schrödinger* (Cambridge University Press, 1989).

CHAPTER 3

The Schrödinger Equation and a Particle In a Box

The Schrödinger equation is our fundamental equation of quantum mechanics. The solutions to the Schrödinger equation are called *wave functions*. We will see that a wave function gives a complete description of any system. In this chapter, we present and discuss the version of the Schrödinger equation that does not contain time as a variable. Solutions to the time-independent Schrödinger equation are called *stationary-state wave functions* because they are independent of time. Many problems of interest to chemists can be treated by using only stationary-state wave functions. We do not consider any time dependence until Chapter 13, where we discuss molecular spectroscopy.

In this chapter, we present the time-independent Schrödinger equation and then apply it to a free particle of mass m that is restricted to lie along a one-dimensional interval of length a. This system is called a *particle in a box* and the calculation of its properties is a standard introductory problem in quantum mechanics. The particle-in-a-box problem is simple, yet very instructive. In the course of discussing this problem, we will introduce the probabilistic interpretation of wave functions. We use this interpretation to illustrate the application of the Uncertainty Principle to a particle in a box.

3–1. The Schrödinger Equation Is the Equation for Finding the Wave Function of a Particle

We cannot derive the Schrödinger equation any more than we can derive Newton's laws, and Newton's second law, $f = ma$, in particular. We shall regard the Schrödinger equation to be a fundamental postulate, or axiom, of quantum mechanics, just as Newton's laws are fundamental postulates of classical mechanics. Even though we cannot derive the Schrödinger equation, we can at least show that it is plausible and perhaps even trace Schrödinger's original line of thought. We finished Chapter 1 with a discussion of matter waves, arguing that matter has wavelike character in addition to its obvious particlelike character. As one story goes, at a meeting at which this new

73

idea of matter waves was being discussed, someone mentioned that if indeed matter does possess wavelike properties, then there must be some sort of wave equation that governs them.

Let us start with the classical one-dimensional wave equation for simplicity:

$$\frac{\partial^2 u}{\partial x^2} = \frac{1}{v^2} \frac{\partial^2 u}{\partial t^2} \tag{3.1}$$

We have seen in Chapter 2 that Equation 3.1 can be solved by the method of separation of variables and that $u(x, t)$ can be written as the product of a function of x and a harmonic or sinusoidal function of time. We will express the temporal part as $\cos \omega t$ (cf. Equation 2.25) and write $u(x, t)$ as

$$u(x, t) = \psi(x) \cos \omega t \tag{3.2}$$

Because $\psi(x)$ is the spatial factor of the amplitude $u(x, t)$, we will call $\psi(x)$ the *spatial amplitude* of the wave. If we substitute Equation 3.2 into Equation 3.1, we obtain an equation for the spatial amplitude $\psi(x)$,

$$\frac{d^2 \psi}{dx^2} + \frac{\omega^2}{v^2} \psi(x) = 0 \tag{3.3}$$

Using the fact that $\omega = 2\pi \nu$ and that $\nu\lambda = v$, Equation 3.3 becomes

$$\frac{d^2 \psi}{dx^2} + \frac{4\pi^2}{\lambda^2} \psi(x) = 0 \tag{3.4}$$

We now introduce the idea of de Broglie matter waves into Equation 3.4. The total energy of a particle is the sum of its kinetic energy and its potential energy,

$$E = \frac{p^2}{2m} + V(x) \tag{3.5}$$

where $p = mv$ is the momentum of the particle and $V(x)$ is its potential energy. If we solve Equation 3.5 for the momentum p, we find

$$p = \{2m[E - V(x)]\}^{1/2} \tag{3.6}$$

According to the de Broglie formula,

$$\lambda = \frac{h}{p} = \frac{h}{\{2m[E - V(x)]\}^{1/2}}$$

Substituting this into Equation 3.4, we find

$$\frac{d^2 \psi}{dx^2} + \frac{2m}{\hbar^2}[E - V(x)]\psi(x) = 0 \tag{3.7}$$

where $\hbar$ (called h bar) $= h/2\pi$.

Equation 3.7 is the famous *Schrödinger equation*, a differential equation whose solution, $\psi(x)$, describes a particle of mass m moving in a potential field described by $V(x)$. The exact nature of $\psi(x)$ is vague at this point, but in analogy to the classical wave equation, it is a measure of the amplitude of the matter wave and is called the wave function of the particle. Equation 3.7 does not contain time and is called the *time-independent Schrödinger equation*. The wave functions obtained from Equation 3.7 are called *stationary-state wave functions*. Although there is a more general Schrödinger equation that contains a time dependence (Section 4–4), we will see throughout this book that many problems of chemical interest can be described in terms of stationary-state wave functions.

Equation 3.7 can be rewritten in the form

$$-\frac{\hbar^2}{2m}\frac{d^2\psi}{dx^2} + V(x)\psi(x) = E\psi(x) \tag{3.8}$$

Equation 3.8 is a particularly nice way to write the Schrödinger equation when we introduce the idea of an operator in Section 3–2.

3–2. Classical-Mechanical Quantities Are Represented by Linear Operators in Quantum Mechanics

An *operator* is a symbol that tells you to do something to whatever follows the symbol. For example, we can consider dy/dx to be the d/dx operator operating on the function $y(x)$. Some other examples are SQR (square what follows), $\int_0^1$ (integrate from 0 to 1), 3 (multiply by 3), and $\partial/\partial y$ (partial derivative with respect to y). We usually denote an operator by a capital letter with a carat over it, e.g., $\hat{A}$. Thus, we write

$$\hat{A}f(x) = g(x)$$

to indicate that the operator $\hat{A}$ operates on $f(x)$ to give a new function $g(x)$.

EXAMPLE 3–1
Perform the following operations:

a. $\hat{A}(2x)$, $\hat{A} = \dfrac{d^2}{dx^2}$

b. $\hat{A}(x^2)$, $\hat{A} = \dfrac{d^2}{dx^2} + 2\dfrac{d}{dx} + 3$

c. $\hat{A}(xy^3)$, $\hat{A} = \dfrac{\partial}{\partial y}$

d. $\hat{A}(e^{ikx})$, $\hat{A} = -i\hbar\dfrac{d}{dx}$

SOLUTION:

a. $\hat{A}(2x) = \dfrac{d^2}{dx^2}(2x) = 0$

b. $\hat{A}(x^2) = \dfrac{d^2}{dx^2}x^2 + 2\dfrac{d}{dx}x^2 + 3x^2 = 2 + 4x + 3x^2$

c. $\hat{A}(xy^3) = \dfrac{\partial}{\partial y}xy^3 = 3xy^2$

d. $\hat{A}(e^{ikx}) = -i\hbar\dfrac{d}{dx}e^{ikx} = k\hbar e^{ikx}$

In quantum mechanics, we deal only with *linear operators*. An operator is said to be linear if

$$\hat{A}\left[c_1 f_1(x) + c_2 f_2(x)\right] = c_1\hat{A}f_1(x) + c_2\hat{A}f_2(x) \tag{3.9}$$

where c_1 and c_2 are (possibly complex) constants. Clearly the "differentiate" and "integrate" operators are linear because

$$\frac{d}{dx}\left[c_1 f_1(x) + c_2 f_2(x)\right] = c_1\frac{df_1}{dx} + c_2\frac{df_2}{dx}$$

and

$$\int \left[c_1 f_1(x) + c_2 f_2(x)\right]dx = c_1\int f_1(x)dx + c_2\int f_2(x)dx$$

The "square" operator, SQR, on the other hand, is nonlinear because

$$\mathrm{SQR}\left[c_1 f_1(x) + c_2 f_2(x)\right] = c_1^2 f_1^2(x) + c_2^2 f_2^2(x) + 2c_1 c_2 f_1(x)f_2(x)$$
$$\neq c_1 f_1^2(x) + c_2 f_2^2(x)$$

and therefore it does not satisfy the definition given by Equation 3.9.

EXAMPLE 3–2
Determine whether the following operators are linear or nonlinear:

a. $\hat{A}f(x) = \mathrm{SQRT}\ f(x)$ (take the square root)
b. $\hat{A}f(x) = x^2 f(x)$

SOLUTION:

a. $\hat{A}\left[c_1 f_1(x) + c_2 f_2(x)\right] = \text{SQRT}\left[c_1 f_1(x) + c_2 f_2(x)\right]$

$= \left[c_1 f_1(x) + c_2 f_2(x)\right]^{1/2} \neq c_1 f_1^{1/2}(x) + c_2 f_2^{1/2}(x)$

and so SQRT is a nonlinear operator.

b. $\hat{A}\left[c_1 f_1(x) + c_2 f_2(x)\right] = x^2\left[c_1 f_1(x) + c_2 f_2(x)\right]$

$= c_1 x^2 f_1(x) + c_2 x^2 f_2(x) = c_1 \hat{A} f_1(x) + c_2 \hat{A} f_2(x)$

and so x^2 (multiply by x^2) is a linear operator.

3–3. The Schrödinger Equation Can Be Formulated as an Eigenvalue Problem

A problem that we will frequently encounter in physical chemistry is the following: Given $\hat{A}$, find a function $\phi(x)$ and a constant a such that

$$\hat{A}\phi(x) = a\phi(x) \tag{3.10}$$

Note that the result of operating on the function $\phi(x)$ by $\hat{A}$ is simply to give $\phi(x)$ back again, only multipled by a constant factor. Clearly $\hat{A}$ and $\phi(x)$ have a very special relationship to each other. The function $\phi(x)$ is called an *eigenfunction* of the operator $\hat{A}$, and a is called an *eigenvalue*. The problem of determining $\phi(x)$ and a for a given $\hat{A}$ is called an *eigenvalue problem*.

EXAMPLE 3–3
Show that $e^{\alpha x}$ is an eigenfunction of the operator d^n/dx^n. What is the eigenvalue?

SOLUTION: We differentiate $e^{\alpha x}$ n times and obtain

$$\frac{d^n}{dx^n}e^{\alpha x} = \alpha^n e^{\alpha x}$$

and so the eigenvalue is α^n.

Operators can be imaginary or complex quantities. We will soon learn that the x component of the linear momentum can be represented in quantum mechanics by an operator of the form

$$\hat{P}_x = -i\hbar\frac{\partial}{\partial x} \tag{3.11}$$

EXAMPLE 3–4

Show that e^{ikx} is an eigenfunction of the operator, $\hat{P}_x = -i\hbar\dfrac{\partial}{\partial x}$. What is the eigenvalue?

SOLUTION: We apply $\hat{P}_x$ to e^{ikx} and find

$$\hat{P}_x e^{ikx} = -i\hbar\frac{\partial}{\partial x}e^{ikx} = \hbar k e^{ikx}$$

and so we see that e^{ikx} is an eigenfunction and $\hbar k$ is the eigenvalue of the operator $\hat{P}_x$.

Let's go back to Equation 3.8. We can write the left side of Equation 3.8 in the form

$$\left[-\frac{\hbar^2}{2m}\frac{d^2}{dx^2} + V(x)\right]\psi(x) = E\psi(x) \tag{3.12}$$

If we denote the operator in brackets by $\hat{H}$, then Equation 3.12 can be written as

$$\hat{H}\psi(x) = E\psi(x) \tag{3.13}$$

We have formulated the Schrödinger equation as an eigenvalue problem. The operator $\hat{H}$,

$$\hat{H} = -\frac{\hbar^2}{2m}\frac{d^2}{dx^2} + V(x) \tag{3.14}$$

is called the *Hamiltonian operator*. The wave function is an eigenfunction, and the energy is an eigenvalue of the Hamiltonian operator. This suggests a correspondence between the Hamiltonian operator and the energy. We will see that such correspondences of operators and classical-mechanical variables are fundamental to the formalism of quantum mechanics.

If $V(x) = 0$ in Equation 3.14, the energy is all kinetic energy and so we define a kinetic energy operator according to

$$\hat{K}_x = -\frac{\hbar^2}{2m}\frac{d^2}{dx^2} \tag{3.15}$$

(Strictly speaking, the derivative here should be a partial derivative, but we will consider only one-dimensional systems for the time being.) Furthermore, classically, $K = p^2/2m$, and so we conclude that the square of the momentum operator is given by $2m\hat{K}_x$, or

$$\hat{P}_x^2 = -\hbar^2\frac{d^2}{dx^2} \tag{3.16}$$

We can interpret the operator $\hat{P}_x^2$ by considering the case of two operators acting sequentially, as in $\hat{A}\hat{B}f(x)$. In cases such as this, we apply each operator in turn, working from right to left. Thus

$$\hat{A}\hat{B}f(x) = \hat{A}[\hat{B}f(x)] = \hat{A}h(x)$$

where $h(x) = \hat{B}f(x)$. Once again, we require that all the indicated operations be compatible. If $\hat{A} = \hat{B}$, we have $\hat{A}\hat{A}f(x)$ and denote this term as $\hat{A}^2 f(x)$. Note that $\hat{A}^2 f(x) \neq [\hat{A}f(x)]^2$ for arbitrary $f(x)$.

EXAMPLE 3–5
Given $\hat{A} = d/dx$ and $\hat{B} = x^2$ (multiply by x^2), show (a) that $\hat{A}^2 f(x) \neq [\hat{A}f(x)]^2$ and (b) that $\hat{A}\hat{B}f(x) \neq \hat{B}\hat{A}f(x)$ for arbitrary $f(x)$.

SOLUTION:

$$\hat{A}^2 f(x) = \frac{d}{dx}\left(\frac{df}{dx}\right) = \frac{d^2 f}{dx^2}$$

$$[\hat{A}f(x)]^2 = \left(\frac{df}{dx}\right)^2 \neq \frac{d^2 f}{dx^2}$$

for arbitrary $f(x)$.

$$\hat{A}\hat{B}f(x) = \frac{d}{dx}[x^2 f(x)] = 2xf(x) + x^2\frac{df}{dx}$$

$$\hat{B}\hat{A}f(x) = x^2\frac{df}{dx} \neq \hat{A}\hat{B}f(x)$$

for arbitrary $f(x)$. Thus, we see that the order of the application of operators must be specified. If $\hat{A}$ and $\hat{B}$ are such that

$$\hat{A}\hat{B}f(x) = \hat{B}\hat{A}f(x)$$

for any compatible $f(x)$, then the two operators are said to *commute*. The two operators in this example, however, do not commute.

Using the fact that $\hat{P}_x^2$ means two successive applications of $\hat{P}_x$, we see that the operator $\hat{P}_x^2$ in Equation 3.16 can be factored as

$$\hat{P}_x^2 = -\hbar^2\frac{d^2}{dx^2} = \left(-i\hbar\frac{d}{dx}\right)\left(-i\hbar\frac{d}{dx}\right)$$

so that we can say that $-i\hbar d/dx$ is equal to the momentum operator. Note that this definition is consistent with Equation 3.11.

80

3–4. Wave Functions Have a Probabilistic Interpretation

In this section, we will study the case of a free particle of mass m constrained to lie along the x-axis between $x = 0$ and $x = a$. This case is called the *problem of a particle in a one-dimensional box* (cf. Figure 3.1). It is mathematically a fairly simple problem, so we can study the solutions in great detail and extract and discuss their physical consequences, which carry over to more complicated problems. In addition, we will see that this simple model has at least a crude application to the π electrons in a linear conjugated hydrocarbon.

The terminology *free particle* means that the particle experiences no potential energy or that $V(x) = 0$. If we set $V(x) = 0$ in Equation 3.7, we see that the Schrödinger equation for a free particle in a one-dimensional box is

$$\frac{d^2\psi}{dx^2} + \frac{2mE}{\hbar^2}\psi(x) = 0 \qquad 0 \le x \le a \tag{3.17}$$

The particle is restricted to the region $0 \le x \le a$ and so cannot be found outside this region (see Figure 3.1). To implement the condition that the particle is restricted to the region $0 \le x \le a$, we must formulate an interpretation of the wave function $\psi(x)$. We have said that $\psi(x)$ represents the amplitude of the particle in some sense. Because the intensity of a wave is the square of the magnitude of the amplitude (cf. Problem 3–31), we can write that the "intensity of the particle" is proportional to $\psi^*(x)\psi(x)$, where the asterisk here denotes a complex conjugate [recall that $\psi^*(x)\psi(x)$ is a real quantity; see MathChapter A]. The problem lies in just what we mean by intensity. Schrödinger originally interpreted it in the following way. Suppose the particle to be an electron. Then Schrödinger considered $e\psi^*(x)\psi(x)$ to be the charge density and $e\psi^*(x)\psi(x)dx$ to be the amount of charge between x and $x + dx$. Thus, he presumably pictured the electron to be spread all over the region. A few years later, however, Max Born, a German physicist working in scattering theory, found that this interpretation led to logical difficulties and replaced Schrödinger's interpretation with $\psi^*(x)\psi(x)dx$ being the *probability that the particle is located between x and $x + dx$*. Born's view is now generally accepted.

Because the particle is restricted to the region $0 \le x \le a$, the probability that the particle is found outside this region is zero. Consequently, we shall require that

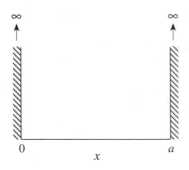

FIGURE 3.1
The geometry of the problem of a particle in a one-dimensional box.

$\psi(x) = 0$ outside the region $0 \leq x \leq a$, which is mathematically how we restrict the particle to this region. Furthermore, because $\psi(x)$ is a measure of the position of the particle, we shall require $\psi(x)$ to be a continuous function. If $\psi(x) = 0$ outside the interval $0 \leq x \leq a$ and is a continuous function, then

$$\psi(0) = \psi(a) = 0$$

These are boundary conditions that we impose on the problem.

3–5. The Energy of a Particle in a Box Is Quantized

The general solution of Equation 3.17 is (see Example 2–4)

$$\psi(x) = A\cos kx + B\sin kx$$

with

$$k = \frac{(2mE)^{1/2}}{\hbar} = \frac{2\pi(2mE)^{1/2}}{h} \tag{3.18}$$

The first boundary condition requires that $\psi(0) = 0$, which implies immediately that $A = 0$ because $\cos(0) = 1$ and $\sin(0) = 0$. The second boundary condition then gives us that

$$\psi(a) = B\sin ka = 0 \tag{3.19}$$

We reject the obvious choice that $B = 0$ because it yields a trivial or physically uninteresting solution, $\psi(x) = 0$, for all x. The other choice is that

$$ka = n\pi \qquad n = 1, 2, \ldots \tag{3.20}$$

(compare with Equations 2.18 through 2.20). By using Equation 3.18 for k, we find that

$$E_n = \frac{h^2 n^2}{8ma^2} \qquad n = 1, 2, \ldots \tag{3.21}$$

Thus, the energy turns out to have only the discrete values given by Equation 3.21 and no other values. The energy of the particle is said to be *quantized* and the integer n is called a *quantum number*. Note that the quantization arises naturally from the boundary conditions. We have gone beyond the stage of Planck and Bohr where quantum numbers are introduced in an *ad hoc* manner. The natural occurrence of quantum numbers was an exciting feature of the Schrödinger equation, and, in the introduction to the first of his now famous series of four papers published in 1926, Schrödinger says:

> In this communication I wish to show that the usual rules of quantization can be replaced by another postulate (the Schrödinger equation) in which there occurs

no mention of whole numbers. Instead, the introduction of integers arises in the same natural way as, for example, in a vibrating string, for which the number of nodes is integral. The new conception can be generalized, and I believe that it penetrates deeply into the true nature of the quantum rules.
[from *Annalen der Physik* **79**, 361 (1926)]

The wave function corresponding to E_n is

$$\psi_n(x) = B \sin kx$$
$$= B \sin \frac{n\pi x}{a} \qquad n = 1, 2, \ldots \qquad (3.22)$$

We will determine the constant B shortly. These wave functions are plotted in Figure 3.2. They look just like the standing waves set up in a vibrating string (cf. Figure 2.3). Note that the energy increases with the number of nodes.

The model of a particle in a one-dimensional box has been applied to the π electrons in linear conjugated hydrocarbons. Consider butadiene, $H_2C=CHCH=CH_2$, which has four π electrons. Although butadiene, like all polyenes, is not a linear molecule, we will assume for simplicity that the π electrons in butadiene move along a straight

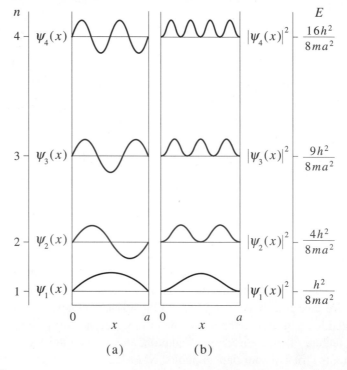

FIGURE 3.2
The energy levels, wave functions (a), and probability densities (b) for the particle in a box.

line whose length can be estimated as equal to two C=C bond lengths (2×135 pm) plus one C—C bond (154 pm) plus the distance of a carbon atom radius at each end (2×77.0 pm = 154 pm), giving a total distance of 578 pm. According to Equation 3.21, the allowed energies are given by

$$E_n = \frac{h^2 n^2}{8 m_e a^2} \qquad n = 1, \ 2, \ \ldots$$

But the Pauli Exclusion Principle (which we discuss later but is assumed here to be known from general chemistry) says that each of these states can hold only two electrons (with opposite spins) and so the four π electrons fill the first two levels as shown in Figure 3.3. The energy of the first excited state of this system of four π electrons is that which has one electron elevated to the $n = 3$ state (cf. Figure 3.3), and the energy to make a transition from the $n = 2$ state to the $n = 3$ state is

$$\Delta E = \frac{h^2}{8 m_e a^2}(3^2 - 2^2)$$

The mass m_e is that of an electron (9.109×10^{-31} kg), and the length of the box is given above to be 578 pm, or 578×10^{-12} m. Therefore,

$$\Delta E = \frac{(6.626 \times 10^{-34} \text{ J} \cdot \text{s})^2 5}{8(9.109 \times 10^{-31} \text{ kg})(578 \times 10^{-12} \text{ m})^2} = 9.02 \times 10^{-19} \text{ J}$$

and

$$\tilde{\nu} = \frac{\Delta E}{hc} = 4.54 \times 10^4 \text{ cm}^{-1}$$

Butadiene has an absorption band at 4.61×10^4 cm^{-1}, and so we see that this very simple model, called the *free-electron model*, can be somewhat successful at explaining the absorption spectrum of butadiene (cf. Problem 3–6).

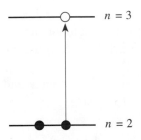

FIGURE 3.3
The free-electron model energy-level scheme for butadiene.

3–6. Wave Functions Must Be Normalized

According to the Born interpretation,

$$\psi_n^*(x)\psi_n(x)dx = B^*B \sin^2 \frac{n\pi x}{a} dx \tag{3.23}$$

is the probability that the particle is located between x and $x + dx$. Because the particle is restricted to the region $0 \le x \le a$, it is certain to be found there and so the probability that the particle lies between 0 and a is unity (Equation B.11), or

$$\int_0^a \psi_n^*(x)\psi_n(x)dx = 1 \tag{3.24}$$

If we substitute Equation 3.23 into Equation 3.24, we find that

$$|B|^2 \int_0^a \sin^2 \frac{n\pi x}{a} dx = 1 \tag{3.25}$$

We let $n\pi x/a$ be z in Equation 3.25 to obtain

$$\int_0^a \sin^2 \frac{n\pi x}{a} dx = \frac{a}{n\pi} \int_0^{n\pi} \sin^2 z\,dz = \frac{a}{n\pi}\left(\frac{n\pi}{2}\right) = \frac{a}{2} \tag{3.26}$$

Therefore, $B^2(a/2) = 1$, $B = (2/a)^{1/2}$, and

$$\psi_n(x) = \left(\frac{2}{a}\right)^{1/2} \sin \frac{n\pi x}{a} \qquad 0 \le x \le a \qquad n = 1, 2, \ldots \tag{3.27}$$

A wave function that satisfies Equation 3.24, and the one given by Equation 3.27 in particular, is said to be *normalized*. When the constant that multiplies a wave function is adjusted to assure that Equation 3.24 is satisfied, the resulting constant is called a *normalization constant*. Because the Hamiltonian operator is a linear operator, if ψ is a solution to $\hat{H}\psi = E\psi$, then any constant, say A, times ψ is also a solution, and A can always be chosen to produce a normalized solution to the Schrödinger equation, $\hat{H}\psi = E\psi$ (cf. Problem 3–7).

Because $\psi^*(x)\psi(x)dx$ is the probability of finding the particle between x and $x + dx$, the probability of finding the particle within the interval $x_1 \le x \le x_2$ is

$$\text{Prob}(x_1 \le x \le x_2) = \int_{x_1}^{x_2} \psi^*(x)\psi(x)dx \tag{3.28}$$

EXAMPLE 3–6
Calculate the probability that a particle in a one-dimensional box of length a is found to be between 0 and $a/2$.

SOLUTION: The probability that the particle will be found between 0 and $a/2$ is

$$\text{Prob}(0 \leq x \leq a/2) = \int_0^{a/2} \psi^*(x)\psi(x)dx = \frac{2}{a} \int_0^{a/2} \sin^2 \frac{n\pi x}{a} dx$$

If we let $n\pi x/a$ be z, then we find

$$\text{Prob}(0 \leq x \leq a/2) = \frac{2}{n\pi} \int_0^{n\pi/2} \sin^2 z\, dz = \frac{2}{n\pi} \left. \left| \frac{x}{2} - \frac{\sin 2x}{4} \right. \right|_0^{n\pi/2}$$

$$= \frac{2}{n\pi} \left(\frac{n\pi}{4} - \frac{\sin n\pi}{4} \right) = \frac{1}{2} \quad \text{(for all } n\text{)}$$

Thus, the probabiltiy that the particle lies in one-half of the interval $0 \leq x \leq a$ is $\frac{1}{2}$.

We can use Figure 3.2 and a slight variation of Example 3–6 to illustrate a fundamental principle of quantum mechanics. Figure 3.2 shows that the particle is more likely to be found near the center of the box for the $n = 1$ state but that the probability density becomes more uniformly distributed as n increases. Figure 3.4 shows that the probability density, $\psi_n^*(x)\psi_n(x) = (2/a) \sin^2 n\pi x/a$, for $n = 20$ is fairly uniformly distributed from 0 to a. In fact, a variation of Example 3–6 (Problem 3–8) gives

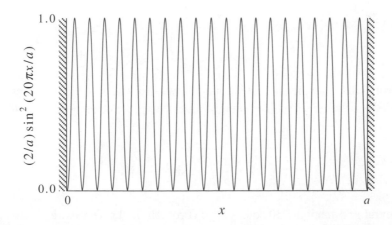

FIGURE 3.4
The probability density, $\psi_n^*(x)\psi_n(x) = (2/a) \sin^2 n\pi x/a$ for $n = 20$, illustrating the correspondence principle, which says that the particle tends to behave classically in the limit of large n.

$$\text{Prob}(0 \leq x \leq a/4) = \text{Prob}(3a/4 \leq x \leq a) = \begin{cases} \dfrac{1}{4} & n \text{ even} \\[2mm] \dfrac{1}{4} - \dfrac{(-1)^{\frac{n-1}{2}}}{2\pi n} & n \text{ odd} \end{cases}$$

and

$$\text{Prob}(a/4 \leq x \leq a/2) = \text{Prob}(a/2 \leq x \leq 3a/4) = \begin{cases} \dfrac{1}{4} & n \text{ even} \\[2mm] \dfrac{1}{4} + \dfrac{(-1)^{\frac{n-1}{2}}}{2\pi n} & n \text{ odd} \end{cases}$$

In both cases, the probabilities approach 1/4 for any large value of n. A similar result is found for any equi-sized intervals. In other words, the probability density becomes uniform as n increases, which is the expected behavior of a classical particle, which has no preferred position between 0 and a.

These results illustrate the *correspondence principle*, according to which quantum mechanical results and classical mechanical results tend to agree in the limit of large quantum numbers. The large quantum number limit is often called the classical limit.

3–7. The Average Momentum of a Particle in a Box Is Zero

We can use the probability distribution $\psi_n^*(x)\psi_n(x)$ to calculate averages and standard deviations (MathChapter B) of various physical quantities such as position and momentum. Using the example of a particle in a box, we see that

$$\psi_n^*(x)\psi_n(x)\,dx = \frac{2}{a}\sin^2\frac{n\pi x}{a}dx \qquad 0 \leq x \leq a \tag{3.29}$$
$$= 0 \qquad\qquad\qquad \text{otherwise}$$

is the probability that the particle is found between x and $x + dx$. These probabilities are plotted in Figure 3.2(b). The average value of x, or the mean position of the particle, is given by

$$\langle x \rangle = \frac{2}{a}\int_0^a x\sin^2\frac{n\pi x}{a}dx \tag{3.30}$$

The integral in Equation 3.30 equals $a^2/4$ (Problem B–1). Therefore,

$$\langle x \rangle = \frac{2}{a}\cdot\frac{a^2}{4} = \frac{a}{2} \qquad \text{(for all } n\text{)} \tag{3.31}$$

This is the physically expected result because the particle "sees" nothing except the walls at $x = 0$ and $x = a$, and so by symmetry $\langle x \rangle$ must be $a/2$.

We can calculate the spread about $\langle x \rangle$ by calculating the variance, σ_x^2. First we calculate $\langle x^2 \rangle$, which is (Problem B–2)

$$\langle x^2 \rangle = \frac{2}{a} \int_0^a x^2 \sin^2 \frac{n\pi x}{a} dx$$

$$= \left(\frac{a}{2\pi n} \right)^2 \left(\frac{4\pi^2 n^2}{3} - 2 \right) = \frac{a^2}{3} - \frac{a^2}{2n^2\pi^2} \tag{3.32}$$

The variance of x is given by

$$\sigma_x^2 = \langle x^2 \rangle - \langle x \rangle^2 = \frac{a^2}{12} - \frac{a^2}{2n^2\pi^2} = \left(\frac{a}{2\pi n} \right)^2 \left(\frac{\pi^2 n^2}{3} - 2 \right)$$

and so the standard deviation is

$$\sigma_x = \frac{a}{2\pi n} \left(\frac{\pi^2 n^2}{3} - 2 \right)^{1/2} \tag{3.33}$$

We shall see that σ_x is directly involved in the Heisenberg Uncertainty Principle.

A problem arises if we wish to calculate the average energy or momentum because they are represented by differential operators. Recall that the energy and momentum operators are

$$\hat{H} = -\frac{\hbar^2}{2m} \frac{d^2}{dx^2} + V(x)$$

and

$$\hat{P}_x = -i\hbar \frac{d}{dx}$$

The problem is that we must decide whether the operator works on $\psi^*(x)\psi(x)dx$ or on $\psi(x)$ or on $\psi^*(x)$ alone. To determine this, let's go back to the Schrödinger equation in operator notation:

$$\hat{H}\psi_n(x) = E_n \psi_n(x) \tag{3.34}$$

If we multiply this equation from the left (see Problem 3–19) by $\psi_n^*(x)$ and integrate over all values of x, we obtain

$$\int \psi_n^*(x) \hat{H} \psi_n(x) dx = \int \psi_n^*(x) E_n \psi_n(x) dx = E_n \int \psi_n^*(x)\psi_n(x) dx = E_n \tag{3.35}$$

where the second step follows because E_n is a number and the last step follows because $\psi_n(x)$ is normalized. Equation 3.35 suggests that we sandwich the operator between a wave function $\psi_n(x)$ and its complex conjugate $\psi_n^*(x)$ to calculate the average value of the physical quantity associated with that operator. We will set this up as a formal postulate in Chapter 4, but our assumption is that

$$\langle s \rangle = \int \psi_n^*(x) \hat{S} \psi_n(x) dx \qquad (3.36)$$

where $\hat{S}$ is the quantum-mechanical operator associated with the physical quantity s, and $\langle s \rangle$ is the average value of s in the state described by the wave function. For example, the average momentum of a particle in a box in the state described by $\psi_n(x)$ is

$$\langle p \rangle = \int_0^a \left[\left(\frac{2}{a} \right)^{1/2} \sin \frac{n\pi x}{a} \right] \left(-i\hbar \frac{d}{dx} \right) \left[\left(\frac{2}{a} \right)^{1/2} \sin \frac{n\pi x}{a} \right] dx \qquad (3.37)$$

In this particular case, $\psi_n(x)$ is real, but generally the operator is sandwiched in between $\psi_n^*(x)$ and $\psi_n(x)$ and so operates only on $\psi_n(x)$ because only $\psi_n(x)$ lies to the right of the operator. We did not have to worry about this when we calculated $\langle x \rangle$ above because the position operator $\hat{X}$ is simply the "multiply by x" operator and its placement in the integrand in Equation 3.36 makes no difference .

If we simplify Equation 3.37, then we find

$$\langle p \rangle = -i\hbar \frac{2\pi n}{a^2} \int_0^a \sin \frac{n\pi x}{a} \cos \frac{n\pi x}{a} dx$$

By consulting the table of integrals in the inside cover or Problem 3–14, we find that this integral is equal to zero, and so

$$\langle p \rangle = 0 \qquad (3.38)$$

Thus, a particle in a box is equally likely to be moving in either direction.

3–8. The Uncertainty Principle Says That $\sigma_p \sigma_x > \hbar/2$

Now let's calculate the variance of the momentum, $\sigma_p^2 = \langle p^2 \rangle - \langle p \rangle^2$, of a particle in a box. To calculate $\langle p^2 \rangle$, we use

$$\langle p^2 \rangle = \int \psi_n^*(x) \hat{P}_x^2 \psi_n(x) dx \qquad (3.39)$$

and remember that $\hat{P}_x^2$ means apply $\hat{P}_x$ twice in succession. Using Equation 3.36

$$\langle p^2 \rangle = \int_0^a \left[\left(\frac{2}{a} \right)^{1/2} \sin \frac{n\pi x}{a} \right] \left(-\hbar^2 \frac{d^2}{dx^2} \right) \left[\left(\frac{2}{a} \right)^{1/2} \sin \frac{n\pi x}{a} \right] dx$$

$$= \frac{2n^2 \pi^2 \hbar^2}{a^3} \int_0^a \sin \frac{n\pi x}{a} \sin \frac{n\pi x}{a} dx$$

$$= \frac{2n^2 \pi^2 \hbar^2}{a^3} \cdot \frac{a}{2} = \frac{n^2 \pi^2 \hbar^2}{a^2} \tag{3.40}$$

The square root of $\langle p^2 \rangle$ is called the *root-mean-square momentum*. Note how Equation 3.40 is consistent with the equation

$$\langle E \rangle = \left\langle \frac{p^2}{2m} \right\rangle = \frac{\langle p^2 \rangle}{2m} = \frac{n^2 h^2}{8ma^2} = \frac{n^2 \pi^2 \hbar^2}{2ma^2}$$

Using Equation 3.40 and 3.38, we see that

$$\sigma_p^2 = \frac{n^2 \pi^2 \hbar^2}{a^2}$$

and

$$\sigma_p = \frac{n\pi\hbar}{a} \tag{3.41}$$

Because the variance σ^2, and hence the standard deviation σ, is a measure of the spread of a distribution about its mean value, we can interpret σ as a measure of the uncertainty involved in any measurement. For the case of a particle in a box, we have been able to evaluate σ_x and σ_p explicitly in Equations 3.33 and 3.41. We interpret these quantities as the uncertainty involved when we measure the position or the momentum of the particle, respectively. We expect to obtain a distribution of measured values because the position of the particle is given by the probability distribution, Equation 3.29.

Equation 3.41 shows that the uncertainty in a measurement of p is inversely proportional to a. Thus, the more we try to localize the particle, the greater is the uncertainty in its momentum. The uncertainty in the position of the particle is directly proportional to a (Equation 3.33), which simply means that the larger the region over which the particle can be found, the greater is the uncertainty in its position. A particle that can range over the entire x-axis ($-\infty < x < \infty$) is called a *free particle*. In the case of a free particle, $a \to \infty$ in Equation 3.41, and there is no uncertainty in the momentum. The momentum of a free particle has a definite value (see Problem 3–32). The uncertainty in the position, however, is infinite. Thus, we see that there is a reciprocal relation between the uncertainty in momentum and position. If we take the product of σ_x and σ_p, then we have

$$\sigma_x \sigma_p = \frac{\hbar}{2} \left(\frac{\pi^2 n^2}{3} - 2 \right)^{1/2} \tag{3.42}$$

The value of the square-root term here is never less than 1, and so we write

$$\sigma_x \sigma_p > \frac{\hbar}{2} \tag{3.43}$$

Equation 3.43 is one version of the Heisenberg Uncertainty Principle. We have been able to derive Equation 3.43 explicitly here because the mathematical manipulations for a particle in a box are fairly simple.

Let's try to summarize what we have learned concerning the Uncertainty Principle. A free particle has a definite momentum, but its position is completely indefinite. When we localize a particle by restricting it to a region of length a, it no longer has a definite momentum, and the spread in its momentum is given by Equation 3.41. If we let the length a of the region go to zero, so that we have localized the particle precisely and there is no uncertainty in its position, then Equation 3.41 shows that there is an infinite uncertainty in the momentum. The Uncertainty Principle says that the minimum product of the two uncertainties is on the order of the Planck constant.

3–9. The Problem of a Particle in a Three-Dimensional Box Is a Simple Extension of the One-Dimensional Case

The simplest three-dimensional quantum-mechanical system is the three-dimensional version of a particle in a box. In this case, the particle is confined to lie within a rectangular parallelepiped with sides of lengths a, b, and c (Figure 3.5). The Schrödinger equation for this system is the three-dimensional extension of Equation 3.17.

$$-\frac{\hbar^2}{2m} \left(\frac{\partial^2 \psi}{\partial x^2} + \frac{\partial^2 \psi}{\partial y^2} + \frac{\partial^2 \psi}{\partial z^2} \right) = E\psi(x, y, z) \quad \begin{matrix} 0 \leq x \leq a \\ 0 \leq y \leq b \\ 0 \leq z \leq c \end{matrix} \tag{3.44}$$

Equation 3.44 is often written in the form

$$-\frac{\hbar^2}{2m} \nabla^2 \psi = E\psi$$

where the operator ("del squared"),

$$\nabla^2 = \frac{\partial^2}{\partial x^2} + \frac{\partial^2}{\partial y^2} + \frac{\partial^2}{\partial z^2} \tag{3.45}$$

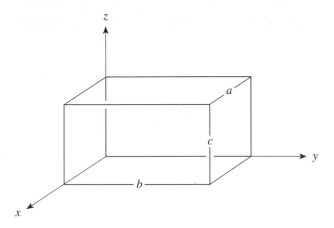

FIGURE 3.5
A rectangular parallelepiped of sides a, b, and c. In the problem of a particle in a three-dimensional box, the particle is restricted to lie within the region shown above.

is called the *Laplacian operator*. The Laplacian operator appears in many physical problems.

The wave function $\psi(x, y, z)$ satisfies the boundary conditions that it vanishes at all the walls of the box, and so

$$
\begin{aligned}
\psi(0, y, z) &= \psi(a, y, z) = 0 \quad &&\text{for all } y \text{ and } z \\
\psi(x, 0, z) &= \psi(x, b, z) = 0 \quad &&\text{for all } x \text{ and } z \\
\psi(x, y, 0) &= \psi(x, y, c) = 0 \quad &&\text{for all } x \text{ and } y
\end{aligned}
\tag{3.46}
$$

We will use the method of separation of variables to solve Equation 3.44. We write

$$
\psi(x, y, z) = X(x)Y(y)Z(z) \tag{3.47}
$$

Substitute Equation 3.47 into Equation 3.44, and then divide through by $\psi(x, y, z) = X(x)Y(y)Z(z)$ to obtain

$$
-\frac{\hbar^2}{2m}\frac{1}{X(x)}\frac{d^2X}{dx^2} - \frac{\hbar^2}{2m}\frac{1}{Y(y)}\frac{d^2Y}{dy^2} - \frac{\hbar^2}{2m}\frac{1}{Z(z)}\frac{d^2Z}{dz^2} = E \tag{3.48}
$$

Each of the three terms on the left side of Equation 3.48 is a function of only x, y, or z, respectively. Because x, y, and z are independent variables, the value of each term can be varied independently, and so each term must equal a constant for Equation 3.48 to be valid for all values of x, y, and z. Thus, we can write Equation 3.48 as

$$
E_x + E_y + E_z = E \tag{3.49}
$$

where E_x, E_y, and E_z are constants and where

$$
-\frac{\hbar^2}{2m} \frac{1}{X(x)} \frac{d^2 X}{dx^2} = E_x
$$

$$
-\frac{\hbar^2}{2m} \frac{1}{Y(y)} \frac{d^2 Y}{dy^2} = E_y \qquad (3.50)
$$

$$
-\frac{\hbar^2}{2m} \frac{1}{Z(z)} \frac{d^2 Z}{dz^2} = E_z
$$

From Equation 3.46, the boundary conditions associated with Equation 3.47 are that

$$
X(0) = X(a) = 0
$$
$$
Y(0) = Y(b) = 0 \qquad (3.51)
$$
$$
Z(0) = Z(c) = 0
$$

Thus, we see that Equations 3.50 and 3.51 are the same as for the one-dimensional case of a particle in a box. Following the same development as in Section 3–5, we obtain

$$
X(x) = A_x \sin \frac{n_x \pi x}{a} \qquad n_x = 1,\ 2,\ 3,\ \ldots
$$

$$
Y(y) = A_y \sin \frac{n_y \pi y}{b} \qquad n_y = 1,\ 2,\ 3,\ \ldots \qquad (3.52)
$$

$$
Z(z) = A_z \sin \frac{n_z \pi z}{c} \qquad n_z = 1,\ 2,\ 3,\ \ldots
$$

According to Equation 3.47, the solution to Equation 3.44 is

$$
\psi(x, y, z) = A_x A_y A_z \sin \frac{n_x \pi x}{a} \sin \frac{n_y \pi y}{b} \sin \frac{n_z \pi z}{c} \qquad (3.53)
$$

with n_x, n_y, and n_z independently assuming the values 1, 2, 3, ….. The normalization constant $A_x A_y A_z$ is found from the equation

$$
\int_0^a dx \int_0^b dy \int_0^c dz\, \psi^*(x, y, z) \psi(x, y, z) = 1 \qquad (3.54)
$$

Problem 3–24 shows that

$$
A_x A_y A_z = \left(\frac{8}{abc} \right)^{1/2} \qquad (3.55)
$$

Thus, the normalized wave functions of a particle in a three-dimensional box are

$$
\psi_{n_x n_y n_z} = \left(\frac{8}{abc} \right)^{1/2} \sin \frac{n_x \pi x}{a} \sin \frac{n_y \pi y}{b} \sin \frac{n_z \pi z}{c} \qquad \begin{array}{l} n_x = 1,\ 2,\ 3,\ \ldots \\ n_y = 1,\ 2,\ 3,\ \ldots \\ n_z = 1,\ 2,\ 3,\ \ldots \end{array} \qquad (3.56)
$$

If we substitute Equation 3.56 into Equation 3.44, then we obtain

$$E_{n_x n_y n_z} = \frac{h^2}{8m}\left(\frac{n_x^2}{a^2} + \frac{n_y^2}{b^2} + \frac{n_z^2}{c^2}\right) \quad \begin{array}{l} n_x = 1,\ 2,\ 3,\ \ldots \\ n_y = 1,\ 2,\ 3,\ \ldots \\ n_z = 1,\ 2,\ 3,\ \ldots \end{array} \quad (3.57)$$

Equation 3.57 is the three-dimensional extension of Equation 3.21.

We should expect by symmetry that the average position of a particle in a three-dimensional box is at the center of the box, but we can show this by direct calculation.

EXAMPLE 3–7

Show that the average position of a particle confined to the region shown in Figure 3.5 is the point $(a/2, b/2, c/2)$.

SOLUTION: The position operator in three dimensions is (see MathChapter C)

$$\hat{\mathbf{R}} = \hat{X}\mathbf{i} + \hat{Y}\mathbf{j} + \hat{Z}\mathbf{k}$$

where $\mathbf{i}$, $\mathbf{j}$, and $\mathbf{k}$ are unit vectors along the x-, y-, and z-axes, respectively. The average position is given by

$$\langle \mathbf{r} \rangle = \int_0^a dx \int_0^b dy \int_0^c dz\, \psi^*(x, y, z)\hat{\mathbf{R}}\psi(x, y, z)$$
$$= \mathbf{i}\langle x \rangle + \mathbf{j}\langle y \rangle + \mathbf{k}\langle z \rangle$$

Let's evaluate $\langle x \rangle$ first. Using Equation 3.55, we have

$$\langle x \rangle = \left[\frac{2}{a}\int_0^a x \sin^2\frac{n_x \pi x}{a}dx\right]\left[\frac{2}{b}\int_0^b \sin^2\frac{n_y \pi y}{b}dy\right]$$
$$\times \left[\frac{2}{c}\int_0^c \sin^2\frac{n_z \pi z}{c}dz\right]$$

The second and third terms in brackets are unity by the normalization condition of a particle in a one-dimensional box (Equation 3.27). The first integral is just $\langle x \rangle$ for a particle in a one-dimensional box. Referring to Equation 3.31, we see that $\langle x \rangle = a/2$. The calculation for $\langle y \rangle$ and $\langle z \rangle$ are similar, and so we see that

$$\langle \mathbf{r} \rangle = \frac{a}{2}\mathbf{i} + \frac{b}{2}\mathbf{j} + \frac{c}{2}\mathbf{k}$$

Thus, the average position of the particle is in the center of the box.

In a similar manner, based on the case of a particle in a one-dimensional box, we should expect that the average momentum of a particle in a three-dimensional box is zero. The momentum operator in three dimensions is

$$\hat{\mathbf{P}} = -i\hbar\left(\mathbf{i}\frac{\partial}{\partial x} + \mathbf{j}\frac{\partial}{\partial y} + \mathbf{k}\frac{\partial}{\partial z}\right) \quad (3.58)$$

and so

$$\langle \mathbf{p} \rangle = \int_0^a dx \int_0^b dy \int_0^c dz \psi^*(x, y, z) \hat{\mathbf{P}} \psi(x, y, z) \qquad (3.59)$$

It is a straightforward exercise to show that $\langle \mathbf{p} \rangle = 0$ (see Problem 3–25).

An interesting feature of a particle in a three-dimensional box occurs when the sides of the box are equal. In this case, $a = b = c$ in Equation 3.57, and so

$$E_{n_x n_y n_z} = \frac{h^2}{8ma^2}(n_x^2 + n_y^2 + n_z^2) \qquad (3.60)$$

Only one set of values n_x, n_y, and n_z corresponds to the lowest energy level. This level, E_{111}, is said to be nondegenerate. However, three sets of values of n_x, n_y, and n_z correspond to the second energy level, and we say that this level is threefold degenerate, or

$$E_{211} = E_{121} = E_{112} = \frac{6h^2}{8ma^2}$$

Figure 3.6 shows the distribution of the first few energy levels of a particle in a cube. Note that the degeneracy occurs because of the symmetry introduced when the general rectangular box becomes a cube and that the degeneracy is "lifted" when the symmetry is destroyed by making the sides of different lengths. A general principle of quantum

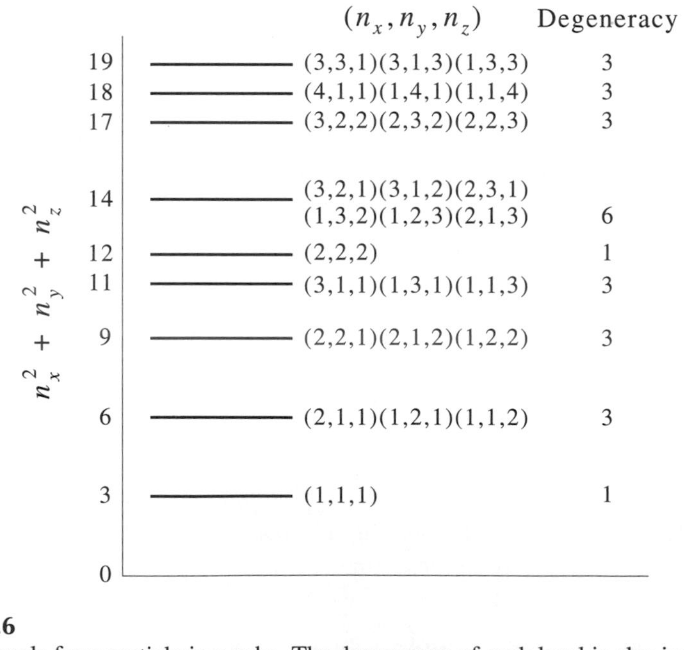

FIGURE 3.6
The energy levels for a particle in a cube. The degeneracy of each level is also indicated.

mechanics states that degeneracies are the result of underlying symmetry and are lifted when the symmetry is broken.

According to Equation 3.56, the wave functions for a particle in a three-dimensional box factor into products of wave functions for a particle in a one-dimensional box. In addition, Equation 3.57 shows that the energy eigenvalues are sums of terms corresponding to the x, y, and z directions. In other words, the problem of a particle in a three-dimensional box reduces to three one-dimensional problems. This is no accident. It is a direct result of the fact that the Hamiltonian operator for a particle in a three-dimensional box is a sum of three independent terms

$$\hat{H} = \hat{H}_x + \hat{H}_y + \hat{H}_z$$

where

$$\hat{H}_x = -\frac{\hbar^2}{2m}\frac{\partial^2}{\partial x^2} \qquad \hat{H}_y = -\frac{\hbar^2}{2m}\frac{\partial^2}{\partial y^2} \qquad \hat{H}_z = -\frac{\hbar^2}{2m}\frac{\partial^2}{\partial z^2}$$

In such a case, we say that the Hamiltonian operator is *separable*.

Thus, we see that if $\hat{H}$ is separable, that is, if $\hat{H}$ can be written as the sum of terms involving independent coordinates, say

$$\hat{H} = \hat{H}_1(s) + \hat{H}_2(w) \tag{3.61}$$

where s and w are the independent coordinates, then the eigenfunctions of $\hat{H}$ are given by the products of the eigenfunctions of $\hat{H}_1$ and $\hat{H}_2$,

$$\psi_{nm}(s, w) = \phi_n(s)\varphi_m(w) \tag{3.62}$$

where

$$\hat{H}_1(s)\phi_n(s) = E_n\phi_n(s)$$
$$\hat{H}_2(w)\varphi_m(w) = E_m\varphi_m(w) \tag{3.63}$$

and E_{nm}, the eigenvalues of $\hat{H}$, are the sums of the eigenvalues of $\hat{H}_1$ and $\hat{H}_2$,

$$E_{nm} = E_n + E_m \tag{3.64}$$

This important result provides a significant simplification because it reduces the original problem to several simpler problems.

We have used the simple case of a particle in a box to illustrate some of the general principles and results of quantum mechanics. In Chapter 4, we present and discuss a set of postulates that we use throughout the remainder of this book.

Problems

3-1. Evaluate $g = \hat{A}f$, where $\hat{A}$ and f are given below:

	$\hat{A}$	f
(a)	SQRT	x^4
(b)	$\dfrac{d^3}{dx^3} + x^3$	e^{-ax}
(c)	$\displaystyle\int_0^1 dx$	$x^3 - 2x + 3$
(d)	$\dfrac{\partial^2}{\partial x^2} + \dfrac{\partial^2}{\partial y^2} + \dfrac{\partial^2}{\partial z^2}$	$x^3 y^2 z^4$

3-2. Determine whether the following operators are linear or nonlinear:

a. $\hat{A}f(x) = \text{SQR}f(x)$ [square $f(x)$]
b. $\hat{A}f(x) = f^*(x)$ [form the complex conjugate of $f(x)$]
c. $\hat{A}f(x) = 0$ [multiply $f(x)$ by zero]
d. $\hat{A}f(x) = [f(x)]^{-1}$ [take the reciprocal of $f(x)$]
e. $\hat{A}f(x) = f(0)$ [evaluate $f(x)$ at $x = 0$]
f. $\hat{A}f(x) = \ln f(x)$ [take the logarithm of $f(x)$]

3-3. In each case, show that $f(x)$ is an eigenfunction of the operator given. Find the eigenvalue.

	$\hat{A}$	$f(x)$
(a)	$\dfrac{d^2}{dx^2}$	$\cos \omega x$
(b)	$\dfrac{d}{dt}$	$e^{i\omega t}$
(c)	$\dfrac{d^2}{dx^2} + 2\dfrac{d}{dx} + 3$	$e^{\alpha x}$
(d)	$\dfrac{\partial}{\partial y}$	$x^2 e^{6y}$

3-4. Show that $(\cos ax)(\cos by)(\cos cz)$ is an eigenfunction of the operator,

$$\nabla^2 = \frac{\partial^2}{\partial x^2} + \frac{\partial^2}{\partial y^2} + \frac{\partial^2}{\partial z^2}$$

which is called the Laplacian operator.

3-5. Write out the operator $\hat{A}^2$ for $\hat{A} =$

a. $\dfrac{d^2}{dx^2}$ **b.** $\dfrac{d}{dx} + x$ **c.** $\dfrac{d^2}{dx^2} - 2x\dfrac{d}{dx} + 1$

Hint: Be sure to include $f(x)$ before carrying out the operations.

3-6. In Section 3–5, we applied the equations for a particle in a box to the π electrons in butadiene. This simple model is called the free-electron model. Using the same argument, show that the length of hexatriene can be estimated to be 867 pm. Show that the first electronic transition is predicted to occur at 2.8×10^4 cm^{-1}. (Remember that hexatriene has six π electrons.)

3-7. Prove that if $\psi(x)$ is a solution to the Schrödinger equation, then any constant times $\psi(x)$ is also a solution.

3-8. Show that the probability associated with the state ψ_n for a particle in a one-dimensional box of length a obeys the following relationships:

$$\text{Prob}(0 \le x \le a/4) = \text{Prob}(3a/4 \le x \le a) = \begin{cases} \dfrac{1}{4} & n \text{ even} \\[2mm] \dfrac{1}{4} - \dfrac{(-1)^{\frac{n-1}{2}}}{2\pi n} & n \text{ odd} \end{cases}$$

and

$$\text{Prob}(a/4 \le x \le a/2) = \text{Prob}(a/2 \le x \le 3a/4) = \begin{cases} \dfrac{1}{4} & n \text{ even} \\[2mm] \dfrac{1}{4} + \dfrac{(-1)^{\frac{n-1}{2}}}{2\pi n} & n \text{ odd} \end{cases}$$

3-9. What are the units, if any, for the wave function of a particle in a one-dimensional box?

3-10. Using a table of integrals, show that

$$\int_0^a \sin^2 \frac{n\pi x}{a} dx = \frac{a}{2}$$

$$\int_0^a x \sin^2 \frac{n\pi x}{a} dx = \frac{a^2}{4}$$

and

$$\int_0^a x^2 \sin^2 \frac{n\pi x}{a} dx = \left(\frac{a}{2\pi n}\right)^3 \left(\frac{4\pi^3 n^3}{3} - 2n\pi\right)$$

All these integrals can be evaluated from

$$I(\beta) = \int_0^a e^{\beta x} \sin^2 \frac{n\pi x}{a} dx$$

Show that the above integrals are given by $I(0)$, $I'(0)$, and $I''(0)$, respectively, where the primes denote differentiation with respect to β. Using a table of integrals, evaluate $I(\beta)$ and then the above three integrals by differentiation.

3-11. Show that

$$\langle x \rangle = \frac{a}{2}$$

for all the states of a particle in a box. Is this result physically reasonable?

3-12. Show that $\langle p \rangle = 0$ for all states of a one-dimensional box of length a.

3-13. Show that

$$\sigma_x = (\langle x^2 \rangle - \langle x \rangle^2)^{1/2}$$

for a particle in a box is less than a, the width of the box, for any value of n. If σ_x is the uncertainty in the position of the particle, could σ_x ever be larger than a?

3-14. Using the trigonometric identity

$$\sin 2\theta = 2 \sin \theta \cos \theta$$

show that

$$\int_0^a \sin \frac{n\pi x}{a} \cos \frac{n\pi x}{a} dx = 0$$

3-15. Prove that

$$\int_0^a e^{\pm i 2\pi nx/a} dx = 0 \qquad n \neq 0$$

3-16. Using the trigonometric identity

$$\sin \alpha \sin \beta = \frac{1}{2} \cos(\alpha - \beta) - \frac{1}{2} \cos(\alpha + \beta)$$

show that the particle-in-a-box wave functions (Equations 3.27) satisfy the relation

$$\int_0^a \psi_n^*(x) \psi_m dx = 0 \qquad m \neq n$$

(The asterisk in this case is superfluous because the functions are real.) If a set of functions satisfies the above integral condition, we say that the set is *orthogonal* and, in particular, that $\psi_m(x)$ is orthogonal to $\psi_n(x)$. If, in addition, the functions are normalized, then we say that the set is *orthonormal*.

3-17. Prove that the set of functions

$$\psi_n(x) = (2a)^{-1/2} e^{i\pi nx/a} \qquad n = 0, \pm 1, \pm 2, \ldots$$

is orthonormal (cf. Problem 3–16) over the interval $-a \leq x \leq a$. A compact way to express orthonormality in the ψ_n is to write

$$\int_{-a}^a \psi_m^*(x) \psi_n dx = \delta_{mn}$$

The symbol δ_{mn} is called a Kroenecker delta and is defined by

$$\delta_{mn} = 1 \qquad \text{if } m = n$$
$$= 0 \qquad \text{if } m \neq n$$

3-18. Show that the set of functions

$$\phi_n(\theta) = (2\pi)^{-1/2} e^{in\theta} \qquad 0 \leq \theta \leq 2\pi$$

is orthonormal (Problem 3–16).

3-19. In going from Equation 3.34 to 3.35, we multiplied Equation 3.34 from the left by $\psi^*(x)$ and then integrated over all values of x to obtain Equation 3.35. Does it make any difference whether we multiplied from the left or the right?

3-20. Calculate $\langle x \rangle$ and $\langle x^2 \rangle$ for the $n = 2$ state of a particle in a one-dimensional box of length a. Show that

$$\sigma_x = \frac{a}{4\pi} \left(\frac{4\pi^2}{3} - 2 \right)^{1/2}$$

3-21. Calculate $\langle p \rangle$ and $\langle p^2 \rangle$ for the $n = 2$ state of a particle in a one-dimensional box of length a. Show that

$$\sigma_p = \frac{h}{a}$$

3-22. Consider a particle of mass m in a one-dimensional box of length a. Its average energy is given by

$$\langle E \rangle = \frac{1}{2m} \langle p^2 \rangle$$

Because $\langle p \rangle = 0$, $\langle p^2 \rangle = \sigma_p^2$, where σ_p can be called the uncertainty in p. Using the Uncertainty Principle, show that the energy must be at least as large as $\hbar^2/8ma^2$ because σ_x, the uncertainty in x, cannot be larger than a.

3-23. Discuss the degeneracies of the first few energy levels of a particle in a three-dimensional box when all three sides have a different length.

3-24. Show that the normalized wave function for a particle in a three-dimensional box with sides of length a, b, and c is

$$\psi(x, y, z) = \left(\frac{8}{abc} \right)^{1/2} \sin \frac{n_x \pi x}{a} \sin \frac{n_y \pi y}{b} \sin \frac{n_z \pi z}{c}$$

3-25. Show that $\langle \mathbf{p} \rangle = 0$ for the ground state of a particle in a three-dimensional box with sides of length a, b, and c.

3-26. What are the degeneracies of the first four energy levels for a particle in a three-dimensional box with $a = b = 1.5c$?

3-27. Many proteins contain metal porphyrin molecules. The general structure of the porphyrin molecule is

This molecule is planar and so we can approximate the π electrons as being confined inside a square. What are the energy levels and degeneracies of a particle in a square of side a? The porphyrin molecule has 18 π electrons. If we approximate the length of the molecule by 1000 pm, then what is the predicted lowest energy absorption of the porphyrin molecule? (The experimental value is $\approx 17\,000$ cm^{-1}.)

3-28. The Schrödinger equation for a particle of mass m constrained to move on a circle of radius a is

$$-\frac{\hbar^2}{2I}\frac{d^2\psi}{d\theta^2} = E\psi(\theta) \qquad 0 \leq \theta \leq 2\pi$$

where $I = ma^2$ is the moment of inertia and θ is the angle that describes the position of the particle around the ring. Show by direct substitution that the solutions to this equation are

$$\psi(\theta) = Ae^{in\theta}$$

where $n = \pm(2IE)^{1/2}/\hbar$. Argue that the appropriate boundary condition is $\psi(\theta) = \psi(\theta + 2\pi)$ and use this condition to show that

$$E = \frac{n^2\hbar^2}{2I} \qquad n = 0, \pm1, \pm2, \ldots$$

Show that the normalization constant A is $(2\pi)^{-1/2}$. Discuss how you might use these results for a free-electron model of benzene.

3-29. Set up the problem of a particle in a box with its walls located at $-a$ and $+a$. Show that the energies are equal to those of a box with walls located at 0 and $2a$. (These energies may be obtained from the results that we derived in the chapter simply by replacing a by $2a$.) Show, however, that the wave functions are not the same and in this case are given by

$$\psi_n(x) = \frac{1}{a^{1/2}}\sin\frac{n\pi x}{2a} \qquad n \text{ even}$$

$$= \frac{1}{a^{1/2}}\cos\frac{n\pi x}{2a} \qquad n \text{ odd}$$

Does it bother you that the wave functions seem to depend upon whether the walls are located at $\pm a$ or 0 and $2a$? Surely the particle "knows" only that it has a region of length $2a$ in which to move and cannot be affected by where you place the origin for the two sets of wave functions. What does this tell you? Do you think that any experimentally observable properties depend upon where you choose to place the origin of the x-axis? Show that $\sigma_x\sigma_p > \hbar/2$, exactly as we obtained in Section 3–8.

3-30. For a particle moving in a one-dimensional box, the mean value of x is $a/2$, and the mean square deviation is $\sigma_x^2 = (a^2/12)[1 - (6/\pi^2 n^2)]$. Show that as n becomes very large, this value agrees with the classical value. The classical probability distribution is uniform,

$$p(x)dx = \frac{1}{a}dx \qquad 0 \leq x \leq a$$

$$= 0 \qquad\qquad \text{otherwise}$$

3-31. This problem shows that the intensity of a wave is proportional to the square of its amplitude. Figure 3.7 illustrates the geometry of a vibrating string. Because the velocity at

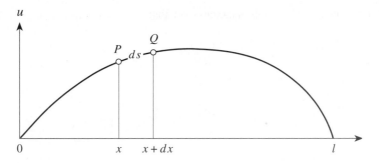

FIGURE 3.7
The geometry of a vibrating string.

any point of the string is $\partial u / \partial t$, the kinetic energy of the entire string is

$$K = \int_0^l \frac{1}{2} \rho \left(\frac{\partial u}{\partial t} \right)^2 dx$$

where ρ is the linear mass density of the string. The potential energy is found by considering the increase of length of the small arc PQ of length ds in Figure 3.7. The segment of the string along that arc has increased its length from dx to ds. Therefore, the potential energy associated with this increase is

$$V = \int_0^l T(ds - dx)$$

where T is the tension in the string. Using the fact that $(ds)^2 = (dx)^2 + (du)^2$, show that

$$V = \int_0^l T \left\{ \left[1 + \left(\frac{\partial u}{\partial x} \right)^2 \right]^{1/2} - 1 \right\} dx$$

Using the fact that $(1 + x)^{1/2} \approx 1 + (x/2)$ for small x, show that

$$V = \frac{1}{2} T \int_0^l \left(\frac{\partial u}{\partial x} \right)^2 dx$$

for small displacements.

The total energy of the vibrating string is the sum of K and V and so

$$E = \frac{\rho}{2} \int_0^l \left(\frac{\partial u}{\partial t} \right)^2 dx + \frac{T}{2} \int_0^l \left(\frac{\partial u}{\partial x} \right)^2 dx$$

We showed in Chapter 2 (Equations 2.23 through 2.25) that the nth normal mode can be written in the form

$$u_n(x, l) = D_n \cos(\omega_n t + \phi_n) \sin \frac{n \pi x}{l}$$

where $\omega_n = v n \pi / l$. Using this equation, show that

$$K_n = \frac{\pi^2 v^2 n^2 \rho}{4l} D_n^2 \sin^2(\omega_n t + \phi_n)$$

and

$$V_n = \frac{\pi^2 n^2 T}{4l} D_n^2 \cos^2(\omega_n t + \phi_n)$$

Using the fact that $v = (T/\rho)^{1/2}$, show that

$$E_n = \frac{\pi^2 v^2 n^2 \rho}{4l} D_n^2$$

Note that the total energy, or intensity, is proportional to the square of the amplitude. Although we have shown this proportionality only for the case of a vibrating string, it is a general result and shows that the intensity of a wave is proportional to the square of the amplitude. If we had carried everything through in complex notation instead of sines and cosines, then we would have found that E_n is proportional to $|D_n|^2$ instead of just D_n^2.

Generally, there are many normal modes present at the same time, and the complete solution is (Equation 2.25)

$$u(x, t) = \sum_{n=1}^{\infty} D_n \cos(\omega_n t + \phi_n) \sin \frac{n \pi x}{l}$$

Using the fact that (see Problem 3–16)

$$\int_0^l \sin \frac{n \pi x}{l} \sin \frac{m \pi x}{l} dx = 0 \qquad \text{if } m \neq n$$

show that

$$E_n = \frac{\pi^2 v^2 \rho}{4l} \sum_{n=1}^{\infty} n^2 D_n^2$$

3-32. The quantized energies of a particle in a box result from the boundary conditions, or from the fact that the particle is restricted to a finite region. In this problem, we investigate the quantum-mechanical problem of a free particle, one that is not restricted to a finite region. The potential energy $V(x)$ is equal to zero and the Schrödinger equation is

$$\frac{d^2 \psi}{dx^2} + \frac{2mE}{\hbar^2} \psi(x) = 0 \qquad -\infty < x < \infty$$

Note that the particle can lie anywhere along the x-axis in this problem. Show that the two solutions of this Schrödinger equation are

$$\psi_1(x) = A_1 e^{i(2mE)^{1/2} x / \hbar} = A_1 e^{ikx}$$

and

$$\psi_2(x) = A_2 e^{-i(2mE)^{1/2} x / \hbar} = A_2 e^{-ikx}$$

where

$$k = \frac{(2mE)^{1/2}}{\hbar}$$

Show that if E is allowed to take on negative values, then the wave functions become unbounded for large x. Therefore, we will require that the energy, E, be a positive quantity. We saw in our discussion of the Bohr atom that negative energies correspond to bound states and positive energies correspond to unbound states, and so our requirement that E be positive is consistent with the picture of a free particle.

To get a physical interpretation of the states that $\psi_1(x)$ and $\psi_2(x)$ describe, operate on $\psi_1(x)$ and $\psi_2(x)$ with the momentum operator $\hat{P}$ (Equation 3.11), and show that

$$\hat{P}\psi_1 = -i\hbar\frac{d\psi_1}{dx} = \hbar k \psi_1$$

and

$$\hat{P}\psi_2 = -i\hbar\frac{d\psi_2}{dx} = -\hbar k \psi_2$$

Notice that these are eigenvalue equations. Our interpretation of these two equations is that ψ_1 describes a free particle with fixed momentum $\hbar k$ and that ψ_2 describes a particle with fixed momentum $-\hbar k$. Thus, ψ_1 describes a particle moving to the right and ψ_2 describes a particle moving to the left, both with a fixed momentum. Notice also that there are no restrictions on k, and so the particle can have any value of momentum. Now show that

$$E = \frac{\hbar^2 k^2}{2m}$$

Notice that the energy is not quantized; the energy of the particle can have any positive value in this case because no boundaries are associated with this problem.

Last, show that $\psi_1^*(x)\psi_1(x) = A_1^* A_1 = |A_1|^2 = $ constant and that $\psi_2^*(x)\psi_2(x) = A_2^* A_2 = |A_2|^2 = $ constant. Discuss this result in terms of the probabilistic interpretation of $\psi^*\psi$. Also discuss the application of the Uncertainty Principle to this problem. What are σ_p and σ_x?

3-33. Derive the equation for the allowed energies of a particle in a one-dimensional box by assuming that the particle is described by standing de Broglie waves within the box.

3-34. We can use the Uncertainty Principle for a particle in a box to argue that free electrons cannot exist in a nucleus. Before the discovery of the neutron, one might have thought that a nucleus of atomic number Z and mass number A is made up of A protons and $A - Z$ electrons, that is, just enough electrons such that the net nuclear charge is $+Z$. Such a nucleus would have an atomic number Z and mass number A. In this problem, we will use Equation 3.41 to *estimate* the energy of an electron confined to a region of nuclear size. The diameter of a typical nucleus is approximately 10^{-14} m. Substitute $a = 10^{-14}$ m into Equation 3.41 and show that σ_p is

$$\sigma_p \gtrsim 3 \times 10^{-20} \text{ kg·m·s}^{-1}$$

Show that

$$E = \frac{\sigma_p^2}{2m} = 5 \times 10^{-10} \text{ J}$$
$$\approx 3000 \text{ MeV}$$

where millions of electron volts (MeV) is the common nuclear physics unit of energy. It is observed experimentally that electrons emitted from nuclei as β radiation have energies of only a few MeV, which is far less than the energy we have calculated above. Argue, then, that there can be no free electrons in nuclei because they should be ejected with much higher energies than are found experimentally.

3-35. We can use the wave functions of Problem 3–29 to illustrate some fundamental symmetry properties of wave functions. Show that the wave functions are alternately symmetric and antisymmetric or even and odd with respect to the operation $x \rightarrow -x$, which is a reflection through the $x = 0$ line. This symmetry property of the wave function is a consequence of the symmetry of the Hamiltonian operator, as we now show. The Schrödinger equation may be written as

$$\hat{H}(x)\psi_n(x) = E_n\psi_n(x)$$

Reflection through the $x = 0$ line gives $x \rightarrow -x$ and so

$$\hat{H}(-x)\psi_n(-x) = E_n\psi_n(-x)$$

Now show that $\hat{H}(x) = \hat{H}(-x)$ (i.e., that $\hat{H}$ is symmetric) for a particle in a box, and so show that

$$\hat{H}(x)\psi_n(-x) = E_n\psi_n(-x)$$

Thus, we see that $\psi_n(-x)$ is also an eigenfunction of $\hat{H}$ belonging to the same eigenvalue E_n. Now, if only one eigenfunction is associated with each eigenvalue (the state is nondegenerate), then argue that $\psi_n(x)$ and $\psi_n(-x)$ must differ only by a multiplicative constant [i.e., that $\psi_n(x) = c\psi_n(-x)$]. By applying the inversion operation again to this equation, show that $c = \pm 1$ and that all the wave functions must be either even or odd with respect to reflection through the $x = 0$ line because the Hamiltonian operator is symmetric. Thus, we see that the symmetry of the Hamiltonian operator influences the symmetry of the wave functions. A general study of symmetry uses group theory, and this example is actually an elementary application of group theory to quantum-mechanical problems. We will study group theory in Chapter 12.

VECTORS

A vector is a quantity that has both magnitude and direction. Examples of vectors are position, force, velocity, and momentum. We specify the position of something, for example, by giving not only its distance from a certain point but also its direction from that point. We often represent a vector by an arrow, where the length of the arrow is the magnitude of the vector and its direction is the same as the direction of the vector.

Two vectors can be added together to get a new vector. Consider the two vectors **A** and **B** in Figure C.1. (We denote vectors by boldface symbols.) To find $\mathbf{C} = \mathbf{A} + \mathbf{B}$, we place the tail of **B** at the tip of **A** and then draw **C** from the tail of **A** to the tip of **B** as shown in the figure. We could also have placed the tail of **A** at the tip of **B** and drawn **C** from the tail of **B** to the tip of **A**. As Figure C.1 indicates, we get the same result either way, so we see that

$$\mathbf{C} = \mathbf{A} + \mathbf{B} = \mathbf{B} + \mathbf{A} \qquad (\text{C.1})$$

Vector addition is commutative.

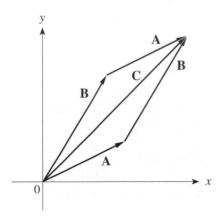

FIGURE C.1
An illustration of the addition of two vectors,
$\mathbf{A} + \mathbf{B} = \mathbf{B} + \mathbf{A} = \mathbf{C}$.

To subtract two vectors, we draw one of them in the opposite direction and then add it to the other. Writing a vector in its opposite direction is equivalent to forming the vector $-\mathbf{B}$. Thus, mathematically we have

$$\mathbf{D} = \mathbf{A} - \mathbf{B} = \mathbf{A} + (-\mathbf{B}) \tag{C.2}$$

A useful set of vectors are the vectors that are of unit length and point along the positive x-, y-, and z-axes of a Cartesian coordinate system. These *unit vectors* (unit length), which we designate by $\mathbf{i}$, $\mathbf{j}$, and $\mathbf{k}$, respectively, are shown in Figure C.2. We shall always draw a Cartesian coordinate system so that it is right-handed. A *right-handed coordinate system* is such that when you curl the four fingers of your right hand from $\mathbf{i}$ to $\mathbf{j}$, your thumb points along $\mathbf{k}$ (Figure C.3). Any three-dimensional vector $\mathbf{A}$ can be described in terms of these unit vectors

$$\mathbf{A} = A_x\mathbf{i} + A_y\mathbf{j} + A_z\mathbf{k} \tag{C.3}$$

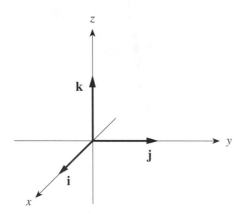

FIGURE C.2
The fundamental unit vectors $\mathbf{i}$, $\mathbf{j}$, and $\mathbf{k}$ of a Cartesian coordinate system.

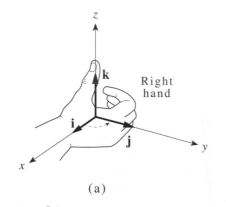

(a)   (b)

FIGURE C.3
(a) An illustration of a right-handed Cartesian coordinate system and (b) a left-handed Cartesian system. We use only a right-handed coordinate system in this book.

where, for example, $A_x\mathbf{i}$ is A_x units long and lies in the direction of $\mathbf{i}$. Generally, a number a times a vector $\mathbf{A}$ is a new vector that is parallel to $\mathbf{A}$ but whose length is a times the length of $\mathbf{A}$. If a is positive, then $a\mathbf{A}$ lies in the same direction as $\mathbf{A}$, but if a is negative, then $a\mathbf{A}$ lies in the opposite direction. The quantities A_x, A_y, and A_z in Equation C.3 are the *components* of $\mathbf{A}$. They are the projections of $\mathbf{A}$ along the respective Cartesian axes (Figure C.4). In terms of components, the sum or difference of two vectors is given by

$$\mathbf{A} \pm \mathbf{B} = (A_x \pm B_x)\mathbf{i} + (A_y \pm B_y)\mathbf{j} + (A_z \pm B_z)\mathbf{k} \tag{C.4}$$

Figure C.4 shows that the length of $\mathbf{A}$ is given by

$$A = |\mathbf{A}| = (A_x^2 + A_y^2 + A_z^2)^{1/2} \tag{C.5}$$

EXAMPLE C–1

If $\mathbf{A} = 2\mathbf{i} - \mathbf{j} + 3\mathbf{k}$ and $\mathbf{B} = -\mathbf{i} + 2\mathbf{j} - \mathbf{k}$, then what is the length of $\mathbf{A} + \mathbf{B}$?

SOLUTION: Using Equation C.4, we have that

$$\mathbf{A} + \mathbf{B} = (2 - 1)\mathbf{i} + (-1 + 2)\mathbf{j} + (3 - 1)\mathbf{k} = \mathbf{i} + \mathbf{j} + 2\mathbf{k}$$

and using Equation C.5 gives

$$|\mathbf{A} + \mathbf{B}| = (1^2 + 1^2 + 2^2)^{1/2} = \sqrt{6}$$

There are two ways to form the product of two vectors, and both have many applications in physical chemistry. One way yields a scalar quantity (in other words, just a number), and the other yields a vector. Not surprisingly, we call the result of the first method a *scalar product* and the result of the second method a *vector product*.

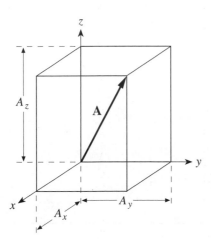

FIGURE C.4
The components of a vector $\mathbf{A}$ are its projections along the x-, y-, and z-axes, showing that the length of $\mathbf{A}$ is equal to $(A_x^2 + A_y^2 + A_z^2)^{1/2}$.

The scalar product of two vectors **A** and **B** is defined as

$$\mathbf{A} \cdot \mathbf{B} = |\mathbf{A}||\mathbf{B}| \cos \theta \tag{C.6}$$

where θ is the angle between **A** and **B**. Note from the definition that

$$\mathbf{A} \cdot \mathbf{B} = \mathbf{B} \cdot \mathbf{A} \tag{C.7}$$

Taking a scalar product is a *commutative operation*. The dot between **A** and **B** is such a standard notation that $\mathbf{A} \cdot \mathbf{B}$ is often called the *dot product* of **A** and **B**. The dot products of the unit vectors **i**, **j**, and **k** are

$$\mathbf{i} \cdot \mathbf{i} = \mathbf{j} \cdot \mathbf{j} = \mathbf{k} \cdot \mathbf{k} = |1||1| \cos 0° = 1$$

$$\mathbf{i} \cdot \mathbf{j} = \mathbf{j} \cdot \mathbf{i} = \mathbf{i} \cdot \mathbf{k} = \mathbf{k} \cdot \mathbf{i} = \mathbf{j} \cdot \mathbf{k} = \mathbf{k} \cdot \mathbf{j} = |1||1| \cos 90° = 0 \tag{C.8}$$

We can use Equations C.8 to evaluate the dot product of any two vectors:

$$\begin{aligned}
\mathbf{A} \cdot \mathbf{B} &= (A_x\mathbf{i} + A_y\mathbf{j} + A_z\mathbf{k}) \cdot (B_x\mathbf{i} + B_y\mathbf{j} + B_z\mathbf{k}) \\
&= A_x B_x \mathbf{i} \cdot \mathbf{i} + A_x B_y \mathbf{i} \cdot \mathbf{j} + A_x B_z \mathbf{i} \cdot \mathbf{k} \\
&\quad + A_y B_x \mathbf{j} \cdot \mathbf{i} + A_y B_y \mathbf{j} \cdot \mathbf{j} + A_y B_z \mathbf{j} \cdot \mathbf{k} \\
&\quad + A_z B_x \mathbf{k} \cdot \mathbf{i} + A_z B_y \mathbf{k} \cdot \mathbf{j} + A_z B_z \mathbf{k} \cdot \mathbf{k}
\end{aligned}$$

which simplifies to

$$\mathbf{A} \cdot \mathbf{B} = A_x B_x + A_y B_y + A_z B_z \tag{C.9}$$

EXAMPLE C–2
Find the length of $\mathbf{A} = 2\mathbf{i} - \mathbf{j} + 3\mathbf{k}$.

SOLUTION: Equation C.9 with $\mathbf{A} = \mathbf{B}$ gives

$$\mathbf{A} \cdot \mathbf{A} = A_x^2 + A_y^2 + A_z^2 = |\mathbf{A}|^2$$

Therefore,

$$|\mathbf{A}| = (\mathbf{A} \cdot \mathbf{A})^{1/2} = (4 + 1 + 9)^{1/2} = \sqrt{14}$$

EXAMPLE C–3

Find the angle between the two vectors $\mathbf{A} = \mathbf{i} + 3\mathbf{j} - \mathbf{k}$ and $\mathbf{B} = \mathbf{j} - \mathbf{k}$.

SOLUTION: We use Equation C.6, but first we must find

$$|\mathbf{A}| = (\mathbf{A} \cdot \mathbf{A})^{1/2} = (1 + 9 + 1)^{1/2} = \sqrt{11}$$

$$|\mathbf{B}| = (\mathbf{B} \cdot \mathbf{B})^{1/2} = (0 + 1 + 1)^{1/2} = \sqrt{2}$$

and

$$\mathbf{A} \cdot \mathbf{B} = 0 + 3 + 1 = 4$$

Therefore,

$$\cos\theta = \frac{\mathbf{A} \cdot \mathbf{B}}{|\mathbf{A}||\mathbf{B}|} = \frac{4}{\sqrt{22}} = 0.8528$$

or $\theta = 31.48°$.

One application of a dot product involves the definition of work. Recall that work is defined as force times distance, where "force" means the component of force that lies in the same direction as the displacement. If we let $\mathbf{F}$ be the force and $\mathbf{d}$ be the displacement, then work is defined as

$$\text{work} = \mathbf{F} \cdot \mathbf{d} \qquad (C.10)$$

We can write Equation C.10 as $(F\cos\theta)(d)$ to emphasize that $F\cos\theta$ is the component of $\mathbf{F}$ in the direction of $\mathbf{d}$ (Figure C.5).

Another important application of a dot product involves the interaction of a dipole moment with an electric field. You may have learned in organic chemistry that the separation of opposite charges in a molecule gives rise to a dipole moment, which is often indicated by an arrow crossed at its tail and pointing from the negative charge to the positive charge. For example, because a chlorine atom is more electronegative than a hydrogen atom, HCl has a dipole moment, which we indicate by writing $\overset{\longleftarrow}{\text{HCl}}$. Strictly speaking, a dipole moment is a vector quantity whose magnitude is equal to the product of the positive charge and the distance between the positive and negative

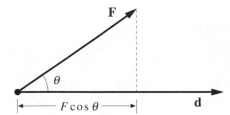

FIGURE C.5
Work is defined as $w = \mathbf{F} \cdot \mathbf{d}$, or $(F\cos\theta)d$, where $F\cos\theta$ is the component of $\mathbf{F}$ along $\mathbf{d}$.

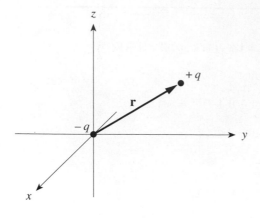

FIGURE C.6
A dipole moment is a vector that points from a negative charge, $-q$, to a positive charge, $+q$, and whose magnitude is $q\mathbf{r}$.

charges and whose direction is from the negative charge to the positive charge. Thus, for the two separated charges illustrated in Figure C.6, the dipole moment μ, is equal to

$$\mu = q\mathbf{r}$$

We will learn later that if we apply an electric field $\mathbf{E}$ to a dipole moment, then the potential energy of interaction will be

$$V = -\mu \cdot \mathbf{E} \tag{C.11}$$

The vector product of two vectors is a vector defined by

$$\mathbf{A} \times \mathbf{B} = |\mathbf{A}||\mathbf{B}|\mathbf{c}\sin\theta \tag{C.12}$$

where θ is the angle between $\mathbf{A}$ and $\mathbf{B}$ and $\mathbf{c}$ is a unit vector perpendicular to the plane formed by $\mathbf{A}$ and $\mathbf{B}$. The direction of $\mathbf{c}$ is given by the right-hand rule: If the four fingers of your right hand curl from $\mathbf{A}$ to $\mathbf{B}$, then $\mathbf{c}$ lies along the direction of your thumb. (See Figure C.3 for a similar construction.) The notation given in Equation C.12 is so commonly used that the vector product is usually called the *cross product*. Because the direction of $\mathbf{c}$ is given by the right-hand rule, the cross product operation is not commutative, and, in particular

$$\mathbf{A} \times \mathbf{B} = -\mathbf{B} \times \mathbf{A} \tag{C.13}$$

The cross products of the cartesian unit vectors are

$$\begin{aligned} \mathbf{i} \times \mathbf{i} = \mathbf{j} \times \mathbf{j} = \mathbf{k} \times \mathbf{k} &= |1||1|\mathbf{c}\sin 0° = 0 \\ \mathbf{i} \times \mathbf{j} = -\mathbf{j} \times \mathbf{i} &= |1||1|\mathbf{k}\sin 90° = \mathbf{k} \\ \mathbf{j} \times \mathbf{k} = -\mathbf{k} \times \mathbf{j} &= \mathbf{i} \\ \mathbf{k} \times \mathbf{i} = -\mathbf{i} \times \mathbf{k} &= \mathbf{j} \end{aligned} \tag{C.14}$$

In terms of components of **A** and **B**, we have (Problem C–9)

$$\mathbf{A} \times \mathbf{B} = (A_y B_z - A_z B_y)\mathbf{i} + (A_z B_x - A_x B_z)\mathbf{j} + (A_x B_y - A_y B_x)\mathbf{k} \qquad \text{(C.15)}$$

Equation C.15 can be conveniently expressed as a determinant (see MathChapter E)

$$\mathbf{A} \times \mathbf{B} = \begin{vmatrix} \mathbf{i} & \mathbf{j} & \mathbf{k} \\ A_x & A_y & A_z \\ B_x & B_y & B_z \end{vmatrix} \qquad \text{(C.16)}$$

Equations C.15 and C.16 are equivalent.

EXAMPLE C–4
Given $\mathbf{A} = -2\mathbf{i} + \mathbf{j} + \mathbf{k}$ and $\mathbf{B} = 3\mathbf{i} - \mathbf{j} + \mathbf{k}$, determine $\mathbf{C} = \mathbf{A} \times \mathbf{B}$.

SOLUTION: Using Equation C.15, we have

$$\mathbf{C} = [(1)(1) - (1)(-1)]\mathbf{i} + [(1)(3) - (-2)(1)]\mathbf{j} + [(-2)(-1) - (1)(3)]\mathbf{k}$$
$$= 2\mathbf{i} + 5\mathbf{j} - \mathbf{k}$$

One physically important application of a cross product involves the definition of angular momentum. If a particle has a momentum $\mathbf{p} = m\mathbf{v}$ at a position $\mathbf{r}$ from a fixed point (as in Figure C.7), then its *angular momentum* is defined by

$$\mathbf{L} = \mathbf{r} \times \mathbf{p} \qquad \text{(C.17)}$$

Note that the angular momentum is a vector perpendicular to the plane formed by $\mathbf{r}$ and $\mathbf{p}$ (Figure C.8). In terms of components, **L** is equal to (see Equation C.15)

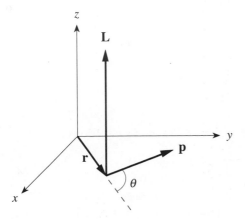

FIGURE C.7
The angular momentum of a particle of momentum **p** and position **r** from a fixed center is a vector perpendicular to the plane formed by **r** and **p** and in the direction of **r** × **p**.

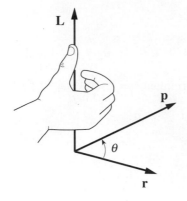

FIGURE C.8
Angular momentum is a vector quantity that lies perpendicular to the plane formed by **r** and **p** and is directed such that the vectors **r**, **p**, and **L** form a right-handed coordinate system.

$$L = (yp_z - zp_y)\mathbf{i} + (zp_x - xp_z)\mathbf{j} + (xp_y - yp_x)\mathbf{k} \tag{C.18}$$

We will learn that angular momentum plays an important role in quantum mechanics.

Another example that involves a cross product is the equation that gives the force **F** on a particle of charge q moving with velocity **v** through a magnetic field **B**:

$$\mathbf{F} = q(\mathbf{v} \times \mathbf{B})$$

Note that the force is perpendicular to **v**, and so the effect of **B** is to cause the motion of the particle to curve, not to speed up or slow down.

We can also take derivatives of vectors. Suppose that the components of momentum, **p**, depend upon time. Then

$$\frac{d\mathbf{p}(t)}{dt} = \frac{dp_x(t)}{dt}\mathbf{i} + \frac{dp_y(t)}{dt}\mathbf{j} + \frac{dp_z(t)}{dt}\mathbf{k} \tag{C.19}$$

(There are no derivatives of **i**, **j**, and **k** because they are fixed in space.) Newton's law of motion is

$$\frac{d\mathbf{p}}{dt} = \mathbf{F} \tag{C.20}$$

This law is actually three separate equations, one for each component. Because $\mathbf{p} = m\mathbf{v}$, if m is a constant, we can write Newton's equations as

$$m\frac{d\mathbf{v}}{dt} = \mathbf{F}$$

Furthermore, because $\mathbf{v} = d\mathbf{r}/dt$, we can also express Newton's equations as

$$m\frac{d^2\mathbf{r}}{dt^2} = \mathbf{F} \tag{C.21}$$

Once again, Equation C.21 represents a set of three equations, one for each component.

Problems

C-1. Find the length of the vector $\mathbf{v} = 2\mathbf{i} - \mathbf{j} + 3\mathbf{k}$.

C-2. Find the length of the vector $\mathbf{r} = x\mathbf{i} + y\mathbf{j}$ and of the vector $\mathbf{r} = x\mathbf{i} + y\mathbf{j} + z\mathbf{k}$.

C-3. Prove that $\mathbf{A} \cdot \mathbf{B} = 0$ if $\mathbf{A}$ and $\mathbf{B}$ are perpendicular to each other. Two vectors that are perpendicular to each other are said to be orthogonal.

C-4. Show that the vectors $\mathbf{A} = 2\mathbf{i} - 4\mathbf{j} - 2\mathbf{k}$ and $\mathbf{B} = 3\mathbf{i} + 4\mathbf{j} - 5\mathbf{k}$ are orthogonal.

C-5. Show that the vector $\mathbf{r} = 2\mathbf{i} - 3\mathbf{k}$ lies entirely in a plane perpendicular to the y axis.

C-6. Find the angle between the two vectors $\mathbf{A} = -\mathbf{i} + 2\mathbf{j} + \mathbf{k}$ and $\mathbf{B} = 3\mathbf{i} - \mathbf{j} + 2\mathbf{k}$.

C-7. Determine $\mathbf{C} = \mathbf{A} \times \mathbf{B}$ given that $\mathbf{A} = -\mathbf{i} + 2\mathbf{j} + \mathbf{k}$ and $\mathbf{B} = 3\mathbf{i} - \mathbf{j} + 2\mathbf{k}$. What is $\mathbf{B} \times \mathbf{A}$ equal to?

C-8. Show that $\mathbf{A} \times \mathbf{A} = 0$.

C-9. Using Equations C.14, prove that $\mathbf{A} \times \mathbf{B}$ is given by Equation C.15.

C-10. Show that $|\mathbf{L}| = mvr$ for circular motion.

C-11. Show that

$$\frac{d}{dt}(\mathbf{A} \cdot \mathbf{B}) = \frac{d\mathbf{A}}{dt} \cdot \mathbf{B} + \mathbf{A} \cdot \frac{d\mathbf{B}}{dt}$$

and

$$\frac{d}{dt}(\mathbf{A} \times \mathbf{B}) = \frac{d\mathbf{A}}{dt} \times \mathbf{B} + \mathbf{A} \times \frac{d\mathbf{B}}{dt}$$

C-12. Using the results of Problem C–11, prove that

$$\mathbf{A} \times \frac{d^2\mathbf{A}}{dt^2} = \frac{d}{dt}\left(\mathbf{A} \times \frac{d\mathbf{A}}{dt}\right)$$

C-13. In vector notation, Newton's equations for a single particle are

$$m\frac{d^2\mathbf{r}}{dt^2} = \mathbf{F}(x, y, z)$$

By operating on this equation from the left by $\mathbf{r} \times$ and using the result of Problem C–12, show that

$$\frac{d\mathbf{L}}{dt} = \mathbf{r} \times \mathbf{F}$$

where $\mathbf{L} = m\mathbf{r} \times d\mathbf{r}/dt = \mathbf{r} \times md\mathbf{r}/dt = \mathbf{r} \times m\mathbf{v} = \mathbf{r} \times \mathbf{p}$. This is the form of Newton's equations for a rotating system. Notice that $d\mathbf{L}/dt = 0$, or that angular momentum is conserved if $\mathbf{r} \times \mathbf{F} = 0$. Can you identify $\mathbf{r} \times \mathbf{F}$?

Werner Heisenberg was born on December 5, 1901 in Duisburg, Germany, grew up in Munich, and died in 1976. In 1923, Heisenberg received his Ph.D. in physics from the University of Munich. He then spent a year as an assistant to Max Born at the University of Göttingen and three years with Niels Bohr in Copenhagen. He was chair of theoretical physics at the University of Leipzig from 1927 to 1941, the youngest to have received such an appointment. Because of a deep loyalty to Germany, Heisenberg opted to stay in Germany when the Nazis came to power. During World War II, he was in charge of German research on the atomic bomb. After the war, he was named director of the Max Planck Institute for Physics, where he strove to rebuild German science. Heisenberg developed one of the first formulations of quantum mechanics, but it was based on matrix algebra, which was less easy to use than the wave equation of Schrödinger. The two formulations, however, were later shown to be equivalent. His Uncertainty Principle, which he published in 1927, illuminates a fundamental principle of nature involving the measurement and observation of physical quantities. Heisenberg was awarded the 1932 Nobel Prize for physics in 1933 "for the creation of quantum mechanics." His role in Nazi Germany is somewhat clouded, prompting one author (David Cassidy) to title his biography of Heisenberg *Uncertainty* (W.H. Freeman, 1993).

CHAPTER 4

Some Postulates and General Principles
of Quantum Mechanics

Up to now, we have made a number of conjectures concerning the formulation of quantum mechanics. For example, we have been led to view the variables of classical mechanics as represented in quantum mechanics by operators. These operate on wave functions to give the average or expected results of measurements. In this chapter, we formalize the various conjectures we made in Chapter 3 as a set of postulates and then discuss some general theorems that follow from these postulates. This formalization is similar to specifying a set of axioms in geometry and then logically deducing the consequences of these axioms. The ultimate test of whether the axioms or postulates are sensible is to compare the end results with experimental data. Here we present a fairly elementary set of postulates that will suffice for all the systems we discuss in this book and for almost all systems of interest in chemistry.

4–1. The State of a System Is Completely Specified by its Wave Function

Classical mechanics deals with quantities called *dynamical variables*, such as position, momentum, angular momentum, and energy. A measurable dynamical variable is called an *observable*. The classical-mechanical state of a particle at any particular time is specified completely by the three position coordinates (x, y, z) and the three momenta (p_x, p_y, p_z) or velocities (v_x, v_y, v_z) at that time. The time evolution of the system is governed by Newton's equations,

$$m\frac{d^2x}{dt^2} = F_x(x, y, z), \quad m\frac{d^2y}{dt^2} = F_y(x, y, z), \quad m\frac{d^2z}{dt^2} = F_z(x, y, z) \qquad (4.1)$$

where F_x, F_y, and F_z are the components of the force, $\mathbf{F}(x, y, z)$. Newton's equations, along with the initial position and momentum of a particle, give us $x(t)$, $y(t)$, and $z(t)$, which describe the position of the particle as a function of time. The three-dimensional path described by $x(t)$, $y(t)$, and $z(t)$ is called the *trajectory* of the particle. The trajectory of a particle offers a complete description of the state of the particle. Classical mechanics provides a method for calculating the trajectory of a particle in terms of the forces acting upon the particle through Newton's equations, Equations 4.1.

Newton's equations plus the forces involved enable us to deduce the entire history and predict the entire future behavior of the particle. We should suspect immediately that such predictions are not possible in quantum mechanics because the Uncertainty Principle tells us that we cannot specify or determine the position and momentum of a particle simultaneously to any desired precision. The Uncertainty Principle is of no practical importance for macroscopic bodies (see Example 1–10), and so classical mechanics is a perfectly adequate prescription for macroscopic bodies. For very small bodies, such as electrons, atoms, and molecules, however, the consequences of the Uncertainty Principle are far from negligible and the classical-mechanical picture is not valid. This leads us to our first postulate of quantum mechanics:

Postulate 1

The state of a quantum-mechanical system is completely specified by a function $\psi(x)$ that depends upon the coordinate of the particle. All possible information about the system can be derived from $\psi(x)$. This function, called the wave function or the state function, has the important property that $\psi^*(x)\psi(x)dx$ is the probability that the particle lies in the interval dx, located at the position x.

In Postulate 1 we have assumed, for simplicity, that only one coordinate is needed to specify the position of a particle, as in the case of a particle in a one-dimensional box. In three dimensions, we would have that $\psi^*(x, y, z)\psi(x, y, z)dxdydz$ is the probability that the particle described by $\psi(x, y, z)$ lies in the volume element $dxdydz$ located at the point (x, y, z). To keep the notation as simple as possible, we will express most of the equations to come in one dimension.

If there is more than one particle, say two, then $\psi^*(x_1, x_2)\psi(x_1, x_2)dx_1dx_2$ is the probability that particle 1 lies in the interval dx_1 located at x_1, and that particle 2 lies in the interval dx_2 located at x_2. Postulate 1 says that the state of a quantum-mechanical system such as two electrons is completely specified by this function and that nothing else is required.

Because the square of the wave function has a probabilistic interpretation, it must satisfy certain physical requirements. The total probabilty of finding a particle somewhere must be unity, thus

$$\int_{\text{all space}} \psi^*(x)\psi(x)dx = 1 \tag{4.2}$$

The notation "all space" here means that we integrate over all possible values of x. We have expressed Equation 4.2 for a one-dimensional system; for two- or three-dimensional systems, Equation 4.2 would be a double or a triple integral. Wave functions that satisfy Equation 4.2 are said to be *normalized*.

EXAMPLE 4–1

The wave functions for a particle restricted to lie in a rectangular region of lengths a and b (a particle in a two-dimensional box) are

$$\psi_{n_x n_y}(x, y) = \left(\frac{4}{ab}\right)^{1/2} \sin\frac{n_x \pi x}{a} \sin\frac{n_y \pi y}{b} \qquad \begin{matrix} n_x = 1, 2, \ldots \\ n_y = 1, 2, \ldots \end{matrix} \qquad \begin{matrix} 0 \le x \le a \\ 0 \le y \le b \end{matrix}$$

Show that these wave functions are normalized.

SOLUTION: We wish to show that

$$\int_0^a \int_0^b dx\, dy\, \psi^*(x, y)\psi(x, y) =$$

$$\frac{4}{ab} \int_0^a \int_0^b dx\, dy\, \sin^2\frac{n_x \pi x}{a} \sin^2\frac{n_y \pi y}{b} = 1$$

This double integral actually factors into a product of two single integrals:

$$\frac{4}{ab} \int_0^a dx\, \sin^2\frac{n_x \pi x}{a} \int_0^b dy\, \sin^2\frac{n_y \pi y}{b} \stackrel{?}{=} 1$$

Equation 3.26 shows that the first integral is equal to $a/2$ and that the second is equal to $b/2$, so that we have

$$\frac{4}{ab} \cdot \frac{a}{2} \cdot \frac{b}{2} = 1$$

and thus the above wave functions are normalized.

Even if the integral in Equation 4.2 equals some constant $A \ne 1$, we can divide $\psi(x)$ by $A^{1/2}$ to make it normalized. On the other hand, if the integral diverges (i.e. goes to infinity), normalizing $\psi(x)$ is not possible, and it is not acceptable as a state function (see Example 4–2b). Functions that can be normalized are said to be *normalizable*. Only normalizable functions are acceptable as state functions. Furthermore, for $\psi(x)$ to be a physically acceptable wave function, it and its first derivative must be single-valued, continuous, and finite (cf. Problem 4–4). We summarize these requirements by saying that $\psi(x)$ must be *well behaved*.

EXAMPLE 4–2

Determine whether each of the following functions is acceptable or not as a state function over the indicated intervals:

a. e^{-x} $(0, \infty)$

b. e^{-x} $(-\infty, \infty)$

c. $\sin^{-1} x$ $(-1, 1)$

d. $e^{-|x|}$ $(-\infty, \infty)$

SOLUTION:

a. acceptable; e^{-x} is single-valued, continuous, finite, and normalizable over the interval $(0, \infty)$.

b. Not acceptable; e^{-x} cannot be normalized over the interval $(-\infty, \infty)$ because e^{-x} diverges as $x \to -\infty$.

c. Not acceptable; $\sin^{-1} x$ is a multivalued function. For example,

$$\sin^{-1} 1 = \frac{\pi}{2}, \frac{\pi}{2} + 2\pi, \frac{\pi}{2} + 4\pi, \text{etc}$$

d. Not acceptable; the first derivative of $e^{-|x|}$ is not continuous at $x = 0$.

4–2. Quantum-Mechanical Operators Represent Classical-Mechanical Variables

In Chapter 3, we concluded that classical mechanical quantities are represented by linear operators in quantum mechanics. We now formalize this conclusion by our next postulate.

Postulate 2

To every observable in classical mechanics there corresponds a linear operator in quantum mechanics.

We have seen some examples of the correspondence between observables and operators in Chapter 3. These correspondences are listed in Table 4.1.

The only new entry in Table 4.1 is that for the angular momentum. Although we discussed angular momentum briefly in MathChapter C, we will discuss it more fully here. Linear momentum is given by $m\mathbf{v}$ and is usually denoted by the symbol $\mathbf{p}$. Now consider a particle rotating in a plane about a fixed center as in Figure 4.1. Let v_{rot}

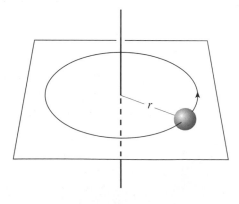

FIGURE 4.1
The rotation of a single particle about a fixed point.

TABLE 4.1

Classical-mechanical observables and their corresponding quantum-mechanical operators.

Observable		Operator	
Name	Symbol	Symbol	Operation
Position	x	$\hat{X}$	Multiply by x
	$\mathbf{r}$	$\hat{\mathbf{R}}$	Multiply by $\mathbf{r}$
Momentum	p_x	$\hat{P}_x$	$-i\hbar\dfrac{\partial}{\partial x}$
	$\mathbf{p}$	$\hat{\mathbf{P}}$	$-i\hbar\left(\mathbf{i}\dfrac{\partial}{\partial x}+\mathbf{j}\dfrac{\partial}{\partial y}+\mathbf{k}\dfrac{\partial}{\partial z}\right)$
Kinetic energy	K_x	$\hat{K}_x$	$-\dfrac{\hbar^2}{2m}\dfrac{\partial^2}{\partial x^2}$
	K	$\hat{K}$	$-\dfrac{\hbar^2}{2m}\left(\dfrac{\partial^2}{\partial x^2}+\dfrac{\partial^2}{\partial y^2}+\dfrac{\partial^2}{\partial z^2}\right)$
			$=-\dfrac{\hbar^2}{2m}\nabla^2$
Potential energy	$V(x)$	$\hat{V}(\hat{x})$	Multiply by $V(x)$
	$V(x,y,z)$	$\hat{V}(\hat{x},\hat{y},\hat{z})$	Multiply by $V(x,y,z)$
Total energy	E	$\hat{H}$	$-\dfrac{\hbar^2}{2m}\left(\dfrac{\partial^2}{\partial x^2}+\dfrac{\partial^2}{\partial y^2}+\dfrac{\partial^2}{\partial z^2}\right)$
			$+\,V(x,y,z)$
			$=-\dfrac{\hbar^2}{2m}\nabla^2+V(x,y,z)$
Angular momentum	$L_x=yp_z-zp_y$	$\hat{L}_x$	$-i\hbar\left(y\dfrac{\partial}{\partial z}-z\dfrac{\partial}{\partial y}\right)$
	$L_y=zp_x-xp_z$	$\hat{L}_y$	$-i\hbar\left(z\dfrac{\partial}{\partial x}-x\dfrac{\partial}{\partial z}\right)$
	$L_z=xp_y-yp_x$	$\hat{L}_z$	$-i\hbar\left(x\dfrac{\partial}{\partial y}-y\dfrac{\partial}{\partial x}\right)$

be the frequency of rotation (cycles per second). The speeed of the particle, then, is $v = 2\pi r v_{\text{rot}} = r\omega_{\text{rot}}$, where $\omega_{\text{rot}} = 2\pi v_{\text{rot}}$ has the units of radians per second and is called the *angular speed*. The kinetic energy of the revolving particle is

$$K = \tfrac{1}{2}mv^2 = \tfrac{1}{2}mr^2\omega^2 = \tfrac{1}{2}I\omega^2 \tag{4.3}$$

where the quantity $I = mr^2$ is the *moment of inertia*. By comparing the first and last expressions for the kinetic energy in Equation 4.3, we can make the correspondences $\omega \leftrightarrow v$ and $I \leftrightarrow m$, where ω and I are angular quantities and v and m are linear quantities. According to this correspondence, there should be a quantity $I\omega$ corresponding to the linear momentum mv, and in fact the quantity L, defined by

$$L = I\omega = (mr^2)\left(\frac{v}{r}\right) = mvr \tag{4.4}$$

is called the *angular momentum* and is a fundamental quantity associated with rotating systems, just as linear momentum is a fundamental quantity in linear systems.

Kinetic energy can be written in terms of momentum. For a linear system, we have

$$K = \frac{mv^2}{2} = \frac{(mv)^2}{2m} = \frac{p^2}{2m} \tag{4.5}$$

and for rotating systems,

$$K = \frac{I\omega^2}{2} = \frac{(I\omega)^2}{2I} = \frac{L^2}{2I} \tag{4.6}$$

The correspondences between linear systems and rotating systems are given in Table 4.2.

We learned in MathChapter C that the angular momentum of a particle is actually a vector quantity defined by $\mathbf{L} = \mathbf{r} \times \mathbf{p}$, where $\mathbf{r}$ is its position from a fixed point and $\mathbf{p} = m\mathbf{v}$ is its momentum (Figure C.8). Figure C.8 shows that the direction of $\mathbf{L}$ is perpendicular to the plane formed by $\mathbf{r}$ and $\mathbf{p}$. The components of $\mathbf{L}$ are (Equation C.18)

$$\begin{aligned}
L_x &= yp_z - zp_y \\
L_y &= zp_x - xp_z \\
L_z &= xp_y - yp_x
\end{aligned} \tag{4.7}$$

TABLE **4.2**
The correspondences between linear systems and rotating systems.

Linear motion	Angular motion
Mass (m)	Moment of inertia (I)
Speed (v)	Angular speed (ω)
Momentum ($\mathbf{p} = m\mathbf{v}$)	Angular momentum ($\mathbf{L} = I\boldsymbol{\omega}$)
Kinetic energy $\left(K = \dfrac{mv^2}{2} = \dfrac{p^2}{2m}\right)$	Rotational kinetic energy $\left(K = \dfrac{I\omega^2}{2} = \dfrac{L^2}{2I}\right)$

Note that the angular momentum operators given in Table 4.1 can be obtained from Equation 4.7 by letting the linear momenta, p_x, p_y, and p_z assume their operator equivalents.

According to Postulate 2, all quantum mechanical operators are linear. There is an important property of linear operators that we have not discussed yet. Consider an eigenvalue problem with a two-fold degeneracy; that is, consider the two equations

$$\hat{A}\phi_1 = a\phi_1 \quad \text{and} \quad \hat{A}\phi_2 = a\phi_2$$

Both ϕ_1 and ϕ_2 have the same eigenvalue a. If this is the case, then any linear combination of ϕ_1 and ϕ_2, say $c_1\phi_1 + c_2\phi_2$, is an eigenfunction of $\hat{A}$. The proof relies on the linear property of $\hat{A}$ (Section 3–2):

$$\hat{A}(c_1\phi_1 + c_2\phi_2) = c_1\hat{A}\phi_1 + c_2\hat{A}\phi_2$$
$$= c_1 a\phi_1 + c_2 a\phi_2 = a(c_1\phi_1 + c_2\phi_2)$$

EXAMPLE 4–3

Consider the eigenvalue problem

$$\frac{d^2\Phi(\phi)}{d\phi^2} = -m^2\Phi(\phi)$$

where m is a real (not imaginary nor complex) number. The two eigenfunctions of $\hat{A} = d^2/d\phi^2$ are

$$\Phi_m(\phi) = e^{im\phi} \quad \text{and} \quad \Phi_{-m}(\phi) = e^{-im\phi}$$

We can easily show that each of these eigenfunctions has the eigenvalue $-m^2$. Show that any linear combination of $\Phi_m(\phi)$ and $\Phi_{-m}(\phi)$ is also an eigenfunction of $\hat{A} = d^2/d\phi^2$.

SOLUTION:

$$\frac{d^2}{d\phi^2}(c_1 e^{im\phi} + c_2 e^{-im\phi}) = c_1\frac{d^2 e^{im\phi}}{d\phi^2} + c_2\frac{d^2 e^{-im\phi}}{d\phi^2}$$
$$= -c_1 m^2 e^{im\phi} - c_2 m^2 e^{-im\phi}$$
$$= -m^2(c_1 e^{im\phi} + c_2 e^{-im\phi})$$

Example 4–3 helps show that this result is directly due to the linear property of quantum-mechanical operators. Although we have considered only a two-fold degeneracy, the result is easily generalized. We will use this property of linear operators when we discuss the hydrogen atom in Chapter 6.

4–3. Observable Quantities Must Be Eigenvalues of Quantum Mechanical Operators

We now present our third postulate:

Postulate 3

In any measurement of the observable associated with the operator $\hat{A}$, the only values that will ever be observed are the eigenvalues a_n, which satisfy the eigenvalue equation

$$\hat{A}\psi_n = a_n\psi_n \tag{4.8}$$

Thus, in any experiment designed to measure the observable corresponding to $\hat{A}$, the only values we find are a_1, a_2, ... corresponding to the states ψ_1, ψ_2, ... No other values will ever be observed.

As a specific example, consider the measurement of the energy. The operator corresponding to the energy is the Hamiltonian operator, and its eigenvalue equation is

$$\hat{H}\psi_n = E_n\psi_n \tag{4.9}$$

This is just the Schrödinger equation. The solution of this equation gives the ψ_n and E_n. For the case of a particle in a box, $E_n = n^2h^2/8ma^2$ (Equation 3.21). Postulate 3 says that if we measure the energy of a particle in a box, we will find one of these energies and no others.

According to Postulate 1, wave functions have a probabilistic interpretation, and so we can use them to calculate average values of physical quantities. Recall from Section 3–7 that we argued that the average position of a particle in a box is given by

$$\langle x \rangle = \int_0^a \psi_n^*(x)x\psi_n(x)dx$$
$$= \frac{2}{a}\int_0^a x\sin^2\frac{n_x\pi x}{a}dx = \frac{a}{2} \quad \text{(for all } n) \tag{4.10}$$

This leads us to our fourth postulate.

Postulate 4

If a system is in a state described by a normalized wave function ψ, then the average value of the observable corresponding to $\hat{A}$ is given by

$$\langle a \rangle = \int_{\text{all space}} \psi^*\hat{A}\psi dx \tag{4.11}$$

EXAMPLE 4–4

We will learn in the next chapter that a good approximate wave function for the vibrational properties of a diatomic molecule in its lowest quantum state is

$$\psi_0(x) = \left(\frac{\alpha}{\pi}\right)^{1/4} e^{-\alpha x^2/2} \qquad -\infty < x < \infty$$

where x is the displacement of the nuclei from their equilibrium positions and α is a parameter characteristic of the molecule. Calculate the average value of the momentum associated with this wave function.

SOLUTION: From Postulate 4, we have

$$\langle p \rangle = \int_{-\infty}^{\infty} \psi_0^*(x) \hat{P}_x \psi_0(x) dx = \int_{-\infty}^{\infty} \psi_0^*(x) \left[-i\hbar \frac{d}{dx}\right] \psi_0(x) dx$$

$$= -i\hbar \left(\frac{\alpha}{\pi}\right)^{1/2} \int_{-\infty}^{\infty} e^{-\alpha x^2/2} \frac{d}{dx} e^{-\alpha x^2/2} dx$$

$$= i\hbar \left(\frac{\alpha}{\pi}\right)^{1/2} \alpha \int_{-\infty}^{\infty} x e^{-\alpha x^2} dx$$

The integrand here is an odd function and the limits are symmetric, and so we have (Equation B.19)

$$\langle p \rangle = 0$$

Suppose now that $\psi(x)$ in Postulate 4 just happens to be an eigenfunction of $\hat{A}$; that is, suppose that $\psi(x) = \psi_n(x)$ where

$$\hat{A}\psi_n(x) = a_n \psi_n(x)$$

Then

$$\langle a \rangle = \int_{-\infty}^{\infty} \psi_n^*(x) \hat{A} \psi_n(x) dx = \int_{-\infty}^{\infty} \psi_n^*(x) a_n \psi_n(x) dx = a_n \int_{-\infty}^{\infty} \psi_n^*(x) \psi_n(x) dx = a_n \tag{4.12}$$

Furthermore, if $\hat{A}\psi_n(x) = a_n \psi_n(x)$, then

$$\hat{A}^2 \psi_n(x) = \hat{A}[\hat{A}\psi_n(x)] = \hat{A}[a_n \psi_n(x)] = a_n[\hat{A}\psi_n(x)] = a_n^2 \psi_n(x)$$

and so

$$\langle a^2 \rangle = \int_{-\infty}^{\infty} \psi_n^*(x) \hat{A}^2 \psi_n(x) dx = a_n^2 \tag{4.13}$$

From Equations 4.12 and 4.13, we see that the variance of the measurements gives

$$\sigma_a^2 = \langle a^2 \rangle - \langle a \rangle^2 = a_n^2 - a_n^2 = 0 \tag{4.14}$$

Thus, as Postulate 3 says, the only value we measure is the value a_n.

EXAMPLE 4–5

Show that $\sigma_E^2 = \langle E^2 \rangle - \langle E \rangle^2 = 0$ for a particle in a box, for which

$$\psi_n(x) = \left(\frac{2}{a}\right)^{1/2} \sin\frac{n\pi x}{a} \qquad 0 \le x \le a$$

In other words, show that the only values of the energy that can be observed are the energy eigenvalues, $E_n = n^2 h^2 / 8ma^2$ (Equation 3.21).

SOLUTION: The operator that corresponds to the observable E is the Hamiltonian operator, which for a particle in a box is [Equation 3.14 with $V(x) = 0$]

$$\hat{H} = -\frac{\hbar^2}{2m}\frac{d^2}{dx^2}$$

The average energy is given by

$$\langle E \rangle = \int_0^a \psi_n^*(x)\hat{H}\psi_n(x)dx$$

$$= \frac{2}{a}\int_0^a \sin\frac{n\pi x}{a}\left[-\frac{\hbar^2}{2m}\frac{d^2}{dx^2}\right]\sin\frac{n\pi x}{a}dx$$

$$= \frac{\hbar^2}{2m}\cdot\frac{2}{a}\cdot\left(\frac{n\pi}{a}\right)^2\int_0^a \sin^2\frac{n\pi x}{a}dx = \frac{n^2 h^2}{8ma^2}$$

Similarly,

$$\langle E^2 \rangle = \int_0^a \psi_n^*(x)\hat{H}^2\psi_n(x)dx = \int_0^a \psi_n^*(x)\hat{H}[\hat{H}\psi_n(x)]dx$$

$$= \frac{2}{a}\int_0^a \sin\frac{n\pi x}{a}\left(-\frac{\hbar^2}{2m}\frac{d^2}{dx^2}\right)\left(-\frac{\hbar^2}{2m}\frac{d^2}{dx^2}\right)\sin\frac{n\pi x}{a}dx$$

$$= \frac{\hbar^4}{4m^2}\cdot\frac{2}{a}\int_0^a \sin\frac{n\pi x}{a}\left(\frac{d^4}{dx^4}\right)\sin\frac{n\pi x}{a}dx$$

$$= \frac{\hbar^4}{4m^2}\cdot\frac{2}{a}\cdot\left(\frac{n\pi}{a}\right)^4\int_0^a \sin^2\frac{n\pi x}{a}dx$$

$$= \frac{n^4 h^4}{64m^2 a^4} = \left(\frac{n^2 h^2}{8ma^2}\right)^2 = \langle E \rangle^2$$

Therefore, $\sigma_E^2 = \langle E^2 \rangle - \langle E \rangle^2 = 0$, and so we find that the energies of a particle in a box can be observed to have only the values E_1, E_2, ...

4–4. The Time Dependence of Wave Functions Is Governed by the Time-Dependent Schrödinger Equation

To this point, we have tacitly used all the given postulates in Chapter 3, and so our discussion so far should be fairly familiar. Now we must discuss the time dependence of wave functions. The time dependence of wave functions is governed by the time-dependent Schrödinger equation. We cannot derive the time-dependent Schrödinger equation any more than we can derive Newton's equation, so we will simply postulate its form and then show that it is consistent with the time-independent Schrödinger equation, $\hat{H}\psi_n = E_n\psi_n$.

Postulate 5

The wave function, or state function, of a system evolves in time according to the time-dependent Schrödinger equation

$$\hat{H}\Psi(x, t) = i\hbar\frac{\partial\Psi(x, t)}{\partial t} \tag{4.15}$$

Postulate 5 is the only one of the postulates that we did not use in Chapter 3 and thus is new. For most systems, $\hat{H}$ does not contain time explicitly, and in those cases we can apply the method of separation of variables to Equation 4.15 and write

$$\Psi(x, t) = \psi(x)f(t)$$

If we substitute this expression into Equation 4.15 and divide both sides by $\psi(x)f(t)$, we obtain

$$\frac{1}{\psi(x)}\hat{H}\psi(x) = \frac{i\hbar}{f(t)}\frac{df(t)}{dt} \tag{4.16}$$

If $\hat{H}$ does not contain time explicitly, then the left side in Equation 4.16 is a function of x only and the right side is a function of t only, and so both sides must equal a constant. If we denote the separation constant by E, then Equation 4.16 gives

$$\hat{H}\psi(x) = E\psi(x) \tag{4.17}$$

and

$$\frac{df(t)}{dt} = -\frac{i}{\hbar}Ef(t) \tag{4.18}$$

The first of these two equations is what we have been calling the Schrödinger equation. In view of Equation 4.15, Equation 4.17 is often called the *time-independent Schrödinger equation*.

Equation 4.18 can be integrated to give

$$f(t) = e^{-iEt/\hbar}$$

and so $\Psi(x, t)$ is of the form

$$\Psi(x, t) = \psi(x)e^{-iEt/\hbar} \tag{4.19}$$

If we use the relation $E = h\nu = \hbar\omega$, we can write Equation 4.19 as

$$\Psi(x, t) = \psi(x)e^{-i\omega t} \tag{4.20}$$

In almost all cases of interest to chemists, there is a set of solutions to Equation 4.17, so we write Equation 4.19 as

$$\Psi_n(x, t) = \psi_n(x)e^{-iE_n t/\hbar} \tag{4.21}$$

If the system happens to be in one of the eigenstates given by Equation 4.21, then

$$\Psi_n^*(x, t)\Psi_n(x, t)dx = \psi_n^*(x)e^{iE_n t/\hbar}\psi_n(x)e^{-iE_n t/\hbar}dx = \psi_n^*(x)\psi_n(x)\,dx \tag{4.22}$$

Thus, the probability density and the averages calculated from Equation 4.21 are independent of time, and the $\psi_n(x)$ are called *stationary-state* wave functions. Stationary states are of central importance in chemistry. For example, in later chapters we will deduce a set of stationary energy states for an atom or a molecule and express the spectroscopic properties of the system in terms of transitions from one stationary state to another. The Bohr model of the hydrogen atom is a simple illustration of this idea. The following example illustrates the stationary states of a model for a rotating diatomic molecule.

EXAMPLE 4–6

We will learn in Chapter 5 that a rotating diatomic molecule can be well approximated by a rigid rotator (essentially a dumbbell) and that the Schrödinger equation of a rigid rotator gives a set of stationary energy levels with energies

$$E_J = \frac{\hbar^2}{2I}J(J + 1) \qquad J = 0, 1, 2, \ldots$$

where I is the moment of inertia of the molecule. Given that transitions can occur only between adjacent levels, show that the rotational absorption spectrum of a diatomic molecule consists of a series of equally spaced lines.

SOLUTION: Absorption occurs for transitions from the level J to the level $J + 1$ (adjacent levels). The energy difference is

$$\Delta E = E_{J+1} - E_J = \frac{\hbar^2}{2I}[(J + 1)(J + 2) - J(J + 1)]$$

$$= \frac{\hbar^2}{I}(J + 1) \qquad J = 0, 1, 2, \ldots$$

Using the relation $\Delta E = h\nu$, we see that absorption occurs at the frequencies

$$\nu = \frac{\hbar}{2\pi I}(J+1) \qquad J = 0, \ 1, \ 2, \ \ldots$$

which corresponds to a series of lines separated by $\hbar/2\pi I$, from which one may obtain the moment of inertia and bond length of the molecule (Example 5–7).

4–5. The Eigenfunctions of Quantum Mechanical Operators Are Orthogonal

Table 4.1 contains a list of some commonly occurring quantum mechanical operators. We stated previously that these operators must have certain properties. We noticed they all are linear, and, in fact, linearity is a requirement we impose. A more subtle requirement arises if we consider Postulate 3, which says that, in any measurement of the observable associated with a quantum-mechanical operator, the only values that are ever observed are its eigenvalues. We have seen, however, that wave functions and quantum-mechanical operators can be complex quantities (see the expression for $\hat{P}_x$ in Table 4.1, for example), but certainly the eigenvalues must be real quantities if they are to correspond to the result of experimental measurement. In an equation, we have

$$\hat{A}\psi_n = a_n\psi_n \tag{4.23}$$

where $\hat{A}$ and ψ_n may be complex but a_n must be real. We will insist, then, that quantum-mechanical operators have only real eigenvalues. Clearly, this requirement places a certain restriction on the properties of quantum-mechanical operators. We will not elaborate on this restriction here (see Problems 4–28 and 4–29, however), but an important direct consequence of the fact that the eigenvalues of quantum-mechanical operators must be real is that their eigenfunctions satisfy the condition

$$\int_{-\infty}^{\infty} \psi_m^*(x)\psi_n(x)dx = 0 \qquad m \neq n \tag{4.24}$$

Let's see how this condition applies to the wave functions of a particle in a box. The wave functions for this system are (Equation 3.27)

$$\psi_n(x) = \left(\frac{2}{a}\right)^{1/2} \sin\frac{n\pi x}{a} \qquad n = 1, \ 2, \ \ldots \tag{4.25}$$

Proving that these functions satisfy Equation 4.24 is easy if you use the trigonometric identity (Problem 3–16)

$$\sin\alpha\sin\beta = \frac{1}{2}\cos(\alpha-\beta) - \frac{1}{2}\cos(\alpha+\beta)$$

Then

$$\frac{2}{a}\int_0^a \sin\frac{n\pi x}{a}\sin\frac{m\pi x}{a}dx = \frac{1}{a}\int_0^a \cos\frac{(n-m)\pi x}{a}dx - \frac{1}{a}\int_0^a \cos\frac{(n+m)\pi x}{a}dx$$

(4.26)

Because n and m are integers, both integrands on the right side of Equation 4.26 are of the form $\cos(N\pi x/a)$, where N is a nonzero integer, if $m \neq n$. Consequently, both integrals go over complete half cycles of the cosine and equal zero if $m \neq n$ (cf. Figure 4.2). Thus, we see that

$$\frac{2}{a}\int_0^a \sin\frac{n\pi x}{a}\sin\frac{m\pi x}{a}dx = 0 \qquad m \neq n$$

(4.27)

and that the particle-in-a-box wave functions satisfy Equation 4.24.

A set of wave functions that satisfy Equation 4.24 is said to be *orthogonal*, or we say that the wave functions are orthogonal to each other. The wave functions of a particle in a box are orthogonal to each other.

When $n = m$ in Equation 4.26, the integrand of the first integral on the right side of Equation 4.26 is unity because $\cos 0 = 1$. The second integral on the right side of Equation 4.26 vanishes, so we have that

$$\frac{2}{a}\int_0^a \sin^2\frac{n\pi x}{a}dx = 1$$

(4.28)

or that the particle-in-a-box wave functions are normalized. A set of functions that are both normalized and orthogonal to each other is called an *orthonormal* set. We can express the condition of orthonormality by writing

$$\int_{-\infty}^{\infty} \psi_i^* \psi_j dx = \delta_{ij}$$

(4.29)

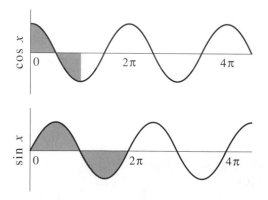

FIGURE 4.2
An illustratration of the fact that the integrals of $\cos x$ and $\sin x$ vanish if the limits of integration extend over the complete half cycles of $\cos x$ and complete cycles of $\sin x$.

where

$$\delta_{ij} = \begin{cases} 1 & i = j \\ 0 & i \neq j \end{cases} \qquad (4.30)$$

The symbol δ_{ij} is called the *Kroenecker delta* (cf. Problem 3–17).

EXAMPLE 4–7

According to Problem 3–28, the eigenfunctions of a particle constrained to move on a circular ring of radius a are

$$\psi_m(\theta) = (2\pi)^{-1/2} e^{im\theta} \qquad m = 0, \pm 1, \pm 2, \ldots$$

where θ describes the angular position of the particle about the ring. Clearly $0 \leq \theta \leq 2\pi$. Prove that these eigenfunctions form an orthonormal set.

SOLUTION: To prove that a set of functions forms an orthonormal set, we must show that they satisfy Equation 4.29. To see if they do, we have

$$\int_0^{2\pi} \psi_m^*(\theta) \psi_n(\theta) d\theta = \frac{1}{2\pi} \int_0^{2\pi} e^{-im\theta} e^{in\theta} d\theta$$

$$= \frac{1}{2\pi} \int_0^{2\pi} e^{i(n-m)\theta} d\theta$$

$$= \frac{1}{2\pi} \int_0^{2\pi} \cos(n-m)\theta d\theta + \frac{i}{2\pi} \int_0^{2\pi} \sin(n-m)\theta d\theta$$

For $n \neq m$, the final two integrals vanish because they are over complete cycles of the cosine and sine. For $n = m$, the first integral in the last expression gives 2π because $\cos 0 = 1$ and the second integral vanishes because $\sin 0 = 0$. Thus,

$$\int_0^{2\pi} \psi_m^*(\theta) \psi_n(\theta) d\theta = \delta_{mn}$$

and we have shown that the $\psi_m(\theta)$ form an orthonormal set.

Before we leave this section, we will discuss very briefly the property of quantum-mechanical operators that guarantees their eigenvalues will be real. In an equation, the property such an operator $\hat{A}$ must satisfy is that

$$\int_{\text{all space}} f^*(x) \hat{A} g(x) dx = \int_{\text{all space}} g(x) [\hat{A} f]^*(x) dx \qquad (4.31)$$

where $f(x)$ and $g(x)$ are any two state functions. Note that $\hat{A}$ operates on $g(x)$ on the left side of Equation 4.31 and that $\hat{A}^*$ operates on $f^*(x)$ on the right side. To see how this equation works, let $\hat{A}$ be the momentum operator $\hat{P}_x = -i\hbar d/dx$ and let

$$f(x) = \frac{1}{\pi^{1/4}} e^{-x^2/2} \qquad -\infty < x < \infty$$

and

$$g(x) = \frac{2^{1/2}}{\pi^{1/4}} x e^{-x^2/2} \qquad -\infty < x < \infty$$

[The constants in $f(x)$ and $g(x)$ are simply normalization constants. The functions $f(x)$ and $g(x)$ are solutions to the one-dimensional harmonic oscillator problem discussed in detail in the next chapter.] Therefore,

$$\hat{A}g(x) = -i\hbar \frac{d}{dx} \frac{2^{1/2}}{\pi^{1/4}} x e^{-x^2/2}$$

$$= -i\hbar \frac{2^{1/2}}{\pi^{1/4}} [e^{-x^2/2} - x^2 e^{-x^2/2}]$$

and

$$\int\limits_{\text{all space}} f^*(x)\hat{A}g(x)dx = -i\hbar \left(\frac{2}{\pi}\right)^{1/2} \int_{-\infty}^{\infty} (e^{-x^2} - x^2 e^{-x^2})dx$$

$$= -i\hbar \left(\frac{2}{\pi}\right)^{1/2} \left(\pi^{1/2} - \frac{\pi^{1/2}}{2}\right) = -\frac{i\hbar}{2^{1/2}}$$

Similarly,

$$\hat{A}^* f(x) = +i\hbar \frac{d}{dx} \frac{1}{\pi^{1/4}} e^{-x^2/2} = -\frac{i\hbar}{\pi^{1/4}} x e^{-x^2/2}$$

and

$$\int\limits_{\text{all space}} g(x)\hat{A}^* f^*(x)dx = -i\hbar \left(\frac{2}{\pi}\right)^{1/2} \int_{-\infty}^{\infty} x^2 e^{-x^2} dx$$

$$= -i\hbar \left(\frac{2}{\pi}\right)^{1/2} \cdot \frac{\pi^{1/2}}{2} = -\frac{i\hbar}{2^{1/2}}$$

Thus we see that $\hat{P}_x$ satisfies Equation 4.31. An operator that satisfies Equation 4.31 is said to be *Hermitian*. Thus, Postulate 2 should be amended to read

Postulate 2′

To every observable in classical mechanics there corresponds a linear, Hermitian operator in quantum mechanics.

Problems 4–28 and 4–29 take you through the proof that the eigenvalues of Hermitian operators are real and that their eigenfunctions are orthonormal.

4–6. The Physical Quantities Corresponding to Operators That Commute Can Be Measured Simultaneously to Any Precision

When two operators act sequentially on a function, $f(x)$, such as in $\hat{A}\hat{B}f(x)$, we apply each operator in turn, working from right to left (as in Example 3–5):

$$\hat{A}\hat{B}f(x) = \hat{A}[\hat{B}f(x)]$$

An important difference between operators and ordinary algebraic quantities is that operators do not necessarily *commute*. If

$$\hat{A}\hat{B}f(x) = \hat{B}\hat{A}f(x) \qquad \text{(commutative)} \qquad (4.32)$$

for arbitrary $f(x)$, then $\hat{A}$ and $\hat{B}$ are said to commute. If

$$\hat{A}\hat{B}f(x) \neq \hat{B}\hat{A}f(x) \qquad \text{(noncommutative)} \qquad (4.33)$$

for arbitrary $f(x)$, then $\hat{A}$ and $\hat{B}$ do not commute. For example, let $\hat{A}$ be the kinetic energy operator, $\hat{K}_x$, and $\hat{B}$ be the momentum operator, $\hat{P}_x$, for a one-dimensional system (Table 4.1). Then

$$\hat{K}_x\hat{P}_x\psi(x) = \left(-\frac{\hbar^2}{2m}\frac{d^2}{dx^2}\right)\left(-i\hbar\frac{d}{dx}\right)\psi(x)$$

$$= \frac{i\hbar^3}{2m}\frac{d^2}{dx^2}\left(\frac{d\psi}{dx}\right) = \frac{i\hbar^3}{2m}\frac{d^3\psi}{dx^3}$$

and

$$\hat{P}_x\hat{K}_x\psi(x) = \left(-i\hbar\frac{d}{dx}\right)\left(-\frac{\hbar^2}{2m}\frac{d^2}{dx^2}\right)\psi(x)$$

$$= \frac{i\hbar^3}{2m}\frac{d}{dx}\left(\frac{d^2\psi}{dx^2}\right) = \frac{i\hbar^3}{2m}\frac{d^3\psi}{dx^3}$$

Therefore,

$$\hat{K}_x\hat{P}_x\psi(x) = \hat{P}_x\hat{K}_x\psi(x) \qquad (4.34)$$

and we see that the kinetic energy operator and the momentum operator commute. We can write Equation 4.34 in the form

$$\hat{K}_x\hat{P}_x\psi(x) - \hat{P}_x\hat{K}_x\psi(x) = 0$$

or

$$(\hat{K}_x\hat{P}_x - \hat{P}_x\hat{K}_x)\psi(x) = \hat{O}\psi(x) \qquad (4.35)$$

where $\hat{O}$ is the "multiply by zero" operator. Because we have not used any special property of $\psi(x)$ to arrive at Equation 4.35, we can write it as an operator equation by suppressing $\psi(x)$ on both sides of the equation to give

$$\hat{K}_x \hat{P}_x - \hat{P}_x \hat{K}_x = \hat{O} \tag{4.36}$$

The left side of Equation 4.36 is called the *commutator* of $\hat{K}_x$ and $\hat{P}_x$ and is written as

$$[\hat{K}_x, \hat{P}_x] = \hat{K}_x \hat{P}_x - \hat{P}_x \hat{K}_x \tag{4.37}$$

and we can write Equation 4.36 as

$$[\hat{K}_x, \hat{P}_x] = \hat{O} \tag{4.38}$$

The commutator of commuting operators is the zero operator.

Now let $\hat{A}$ be the momentum operator $\hat{P}_x$ and $\hat{B}$ be the position operator $\hat{X} = x$ (multiply by x). In this case,

$$\hat{P}_x \hat{X} \psi(x) = \left(-i\hbar \frac{d}{dx} \right) x \psi(x)$$

$$= -i\hbar \psi(x) - i\hbar x \frac{d\psi}{dx}$$

and

$$\hat{X} \hat{P}_x \psi(x) = x \left(-i\hbar \frac{d}{dx} \right) \psi(x)$$

$$= -i\hbar x \frac{d\psi}{dx}$$

Note that

$$\hat{P}_x \hat{X} \psi(x) \neq \hat{X} \hat{P}_x \psi(x) \tag{4.39}$$

so $\hat{P}_x$ and $\hat{X}$ do not commute. In this particular case,

$$(\hat{P}_x \hat{X} - \hat{X} \hat{P}_x) \psi(x) = -i\hbar \psi(x)$$

or

$$(\hat{P}_x \hat{X} - \hat{X} \hat{P}_x) \psi(x) = -i\hbar \hat{I} \psi(x) \tag{4.40}$$

where we have introduced the identity operator $\hat{I}$, which is simply the "multiply by one" operator. Because we have not used any special property to arrive at Equation 4.40, we can write Equation 4.40 as an operator equation by suppressing $\psi(x)$ on both sides of the equation to give

$$\hat{P}_x \hat{X} - \hat{X} \hat{P}_x = -i\hbar \hat{I} \tag{4.41}$$

The left side here is the commutator of $\hat{P}_x$ and $\hat{X}$, so we can write Equation 4.41 as

$$[\hat{P}_x, \hat{X}] = -i\hbar \hat{I} \qquad (4.42)$$

We know from the Uncertainty Principle that both the momentum and the position of a particle cannot be measured simultaneously to any desired degree of accuracy. There is a direct relationship between the Uncertainty Principle and the commutator of two operators, which we give here without proof. Consider two operators, $\hat{A}$ and $\hat{B}$. The standard deviations, σ_a and σ_b, that correspond to these operators are quantitative statistical measures of the uncertainties in the observed values of these physical quantities. These standard deviations are given by (MathChapter B)

$$\sigma_a^2 = \langle A^2 \rangle - \langle A \rangle^2 = \int \psi^*(x)\hat{A}^2\psi(x)dx - \left[\int \psi^*(x)\hat{A}\psi(x)dx \right]^2 \qquad (4.43)$$

with a similar equation for σ_b^2. A rigorous expression of the Uncertainty Principle says that σ_a and σ_b (the uncertainties in the measurements of a and b) are related by

$$\sigma_a \sigma_b \geq \frac{1}{2} \left| \int \psi^*(x)[\hat{A}, \hat{B}]\psi(x)\, dx \right| \qquad (4.44)$$

where $[\hat{A}, \hat{B}] = \hat{A}\hat{B} - \hat{B}\hat{A}$ is the commutator of $\hat{A}$ and $\hat{B}$ and the vertical bars denote the absolute value of the integral.

If $\hat{A}$ and $\hat{B}$ commute, then the right side of Equation 4.44 is zero, so σ_a, σ_b, or both could equal zero simultaneously. There is no restriction on the uncertainties in the measurements of a and b. If, on the other hand, $\hat{A}$ and $\hat{B}$ do not commute, then the right side of Equation 4.44 will not equal zero. Thus, there is a reciprocal relation between σ_a and σ_b; one can approach zero only if the other approaches infinity. Therefore, both a and b cannot be measured simultaneously to arbitrary precision.

Let's consider as an example, the simultaneous measurement of the momentum and position of a particle, so that $\hat{A} = \hat{P}_x$ and $\hat{B} = \hat{X}$ in Equation 4.44. Equation 4.42 tells us that $[\hat{P}_x, \hat{X}] = -i\hbar \hat{I}$, and so Equation 4.44 gives

$$\sigma_p \sigma_x \geq \frac{1}{2} \left| \int \psi^*(x)(-i\hbar \hat{I})\psi(x)dx \right|$$

$$\geq \frac{1}{2}|-i\hbar| \geq \frac{\hbar}{2} \qquad (4.45)$$

Equation 4.45 is the usual expression given for the Uncertainty Principle for momentum and position. If σ_p is made to be small, then σ_x is necessarily large, and if σ_x is made to be small, then σ_p is necessarily large. Thus, the momentum and position cannot be measured simultaneously to arbitrary precision.

Thus, we see that there is an intimate connection between commuting operators and the Uncertainty Principle. If two operators $\hat{A}$ and $\hat{B}$ commute, then a and b can be measured simultaneously to any precision. If two operators $\hat{A}$ and $\hat{B}$ do not commute, then a and b cannot be measured simultaneously to arbitrary precision.

Problems

4-1. Which of the following candidates for wave functions are normalizable over the indicated intervals?

a. $e^{-x^2/2}$ $(-\infty, \infty)$ **b.** e^x $(0, \infty)$ **c.** $e^{i\theta}$ $(0, 2\pi)$ **d.** $\sinh x$ $(0, \infty)$

e. xe^{-x} $(0, \infty)$

Normalize those that can be normalized. Are the others suitable wave functions?

4-2. Which of the following wave functions are normalized over the indicated two-dimensional intervals?

a. $e^{-(x^2+y^2)/2}$ $\begin{matrix} 0 \le x < \infty \\ 0 \le y < \infty \end{matrix}$ **b.** $e^{-(x+y)/2}$ $\begin{matrix} 0 \le x < \infty \\ 0 \le y < \infty \end{matrix}$

c. $\left(\dfrac{4}{ab}\right)^{1/2} \sin\dfrac{\pi x}{a} \sin\dfrac{\pi y}{b}$ $\begin{matrix} 0 \le x \le a \\ 0 \le y \le b \end{matrix}$

Normalize those that aren't.

4-3. Why does $\psi^*\psi$ have to be everywhere real, nonnegative, finite, and of definite value?

4-4. In this problem, we will prove that the form of the Schrödinger equation imposes the condition that the first derivative of a wave function be continuous. The Schrödinger equation is

$$\frac{d^2\psi}{dx^2} + \frac{2m}{\hbar^2}[E - V(x)]\psi(x) = 0$$

If we integrate both sides from $a - \epsilon$ to $a + \epsilon$, where a is an arbitrary value of x and ϵ is infinitesimally small, then we have

$$\left.\frac{d\psi}{dx}\right|_{x=a+\epsilon} - \left.\frac{d\psi}{dx}\right|_{x=a-\epsilon} = \frac{2m}{\hbar^2}\int_{a-\epsilon}^{a+\epsilon}[V(x) - E]\psi(x)dx$$

Now show that $d\psi/dx$ is continuous if $V(x)$ is continuous.

Suppose now that $V(x)$ is *not* continuous at $x = a$, as in

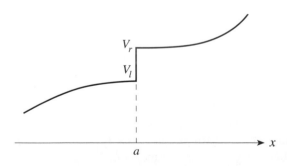

Show that

$$\left.\frac{d\psi}{dx}\right|_{x=a+\epsilon} - \left.\frac{d\psi}{dx}\right|_{x=a-\epsilon} = \frac{2m}{\hbar^2}[V_l + V_r - 2E]\psi(a)\epsilon$$

so that $d\psi/dx$ is continuous even if $V(x)$ has a *finite* discontinuity. What if $V(x)$ has an infinite discontinuity, as in the problem of a particle in a box? Are the first derivatives of the wave functions continuous at the boundaries of the box?

4-5. Determine whether the following functions are acceptable or not as state functions over the indicated intervals.

a. $\dfrac{1}{x}$ $(0, \infty)$
 b. $e^{-2x} \sinh x$ $(0, \infty)$

c. $e^{-x} \cos x$ $(0, \infty)$
 d. e^x $(-\infty, \infty)$

4-6. Calculate the values of $\sigma_E^2 = \langle E^2 \rangle - \langle E \rangle^2$ for a particle in a box in the state described by

$$\psi(x) = \left(\frac{630}{a^9}\right)^{1/2} x^2(a-x)^2 \qquad 0 \le x \le a$$

4-7. Consider a free particle constrained to move over the rectangular region $0 \le x \le a$, $0 \le y \le b$. The energy eigenfunctions of this system are

$$\psi_{n_x, n_y}(x, y) = \left(\frac{4}{ab}\right)^{1/2} \sin\frac{n_x \pi x}{a} \sin\frac{n_y \pi y}{b} \qquad \begin{array}{l} n_x = 1,\ 2,\ 3,\ \ldots \\ n_y = 1,\ 2,\ 3,\ \ldots \end{array}$$

The Hamiltonian operator for this system is

$$\hat{H} = -\frac{\hbar^2}{2m}\left(\frac{\partial^2}{\partial x^2} + \frac{\partial^2}{\partial y^2}\right)$$

Show that if the system is in one of its eigenstates, then

$$\sigma_E^2 = \langle E^2 \rangle - \langle E \rangle^2 = 0$$

4-8. The momentum operator in two dimensions is

$$\hat{P} = -i\hbar\left(\mathbf{i}\frac{\partial}{\partial x} + \mathbf{j}\frac{\partial}{\partial y}\right)$$

Using the wave function given in Problem 4–7, calculate the value of $\langle p \rangle$ and then

$$\sigma_p^2 = \langle p^2 \rangle - \langle p \rangle^2$$

Compare your result with σ_p^2 in the one-dimensional case.

4-9. Suppose that a particle in a two-dimensional box (cf. Problem 4–7) is in the state

$$\psi(x, y) = \frac{30}{(a^5 b^5)^{1/2}} x(a-x)y(b-y)$$

Show that $\psi(x, y)$ is normalized, and then calculate the value of $\langle E \rangle$ associated with the state described by $\psi(x, y)$.

4-10. Show that

$$\psi_0(x) = \pi^{-1/4}e^{-x^2/2}$$

$$\psi_1(x) = (4/\pi)^{1/4}xe^{-x^2/2}$$

$$\psi_2(x) = (4\pi)^{-1/4}(2x^2 - 1)e^{-x^2/2}$$

are orthonormal over the interval $-\infty < x < \infty$.

4-11. Show that the polynomials

$$P_0(x) = 1, \qquad P_1(x) = x, \qquad \text{and} \qquad P_2(x) = \tfrac{1}{2}(3x^2 - 1)$$

satisfy the orthogonality relation

$$\int_1^1 P_l(x)P_n(x)dx = \frac{2\delta_{ln}}{2l + 1}$$

where δ_{ln} is the Kroenecker delta (Equation 4.30).

4-12. Show that the set of functions $(2/a)^{1/2}\cos(n\pi x/a)$, $n = 0, 1, 2, \ldots$ is orthonormal over the interval $0 \leq x \leq a$.

4-13. Prove that if δ_{nm} is the Kroenecker delta

$$\delta_{nm} = \begin{cases} 1 & n = m \\ 0 & n \neq m \end{cases}$$

then

$$\sum_{n=1}^{\infty} c_n\delta_{nm} = c_m$$

and

$$\sum_n \sum_m a_n b_m \delta_{nm} = \sum_n a_n b_n$$

These results will be used later.

4-14. Determine whether or not the following pairs of operators commute.

	$\hat{A}$	$\hat{B}$
(a)	$\dfrac{d}{dx}$	$\dfrac{d^2}{dx^2} + 2\dfrac{d}{dx}$
(b)	x	$\dfrac{d}{dx}$
(c)	SQR	SQRT
(d)	$x^2\dfrac{d}{dx}$	$\dfrac{d^2}{dx^2}$

4-15. In ordinary algebra, $(P + Q)(P - Q) = P^2 - Q^2$. Expand $(\hat{P} + \hat{Q})(\hat{P} - \hat{Q})$. Under what conditions do we find the same result as in the case of ordinary algebra?

4-16. Evaluate the commutator $[\hat{A}, \hat{B}]$, where $\hat{A}$ and $\hat{B}$ are given below.

	$\hat{A}$	$\hat{B}$
(a)	$\dfrac{d^2}{dx^2}$	x
(b)	$\dfrac{d}{dx} - x$	$\dfrac{d}{dx} + x$
(c)	$\displaystyle\int_0^x dx$	$\dfrac{d}{dx}$
(d)	$\dfrac{d^2}{dx^2} - x$	$\dfrac{d}{dx} + x^2$

4-17. Referring to Table 4.1 for the operator expressions for angular momentum, show that

$$[\hat{L}_x, \hat{L}_y] = i\hbar \hat{L}_z$$

$$[\hat{L}_y, \hat{L}_z] = i\hbar \hat{L}_x$$

and

$$[\hat{L}_z, \hat{L}_x] = i\hbar \hat{L}_y$$

(Do you see a pattern here to help remember these commutation relations?) What do these expressions say about the ability to measure the components of angular momentum simultaneously?

4-18. Defining

$$\hat{L}^2 = \hat{L}_x^2 + \hat{L}_y^2 + \hat{L}_z^2$$

show that $\hat{L}^2$ commutes with each component separately. What does this result tell you about the ability to measure the square of the total angular momentum and its components simultaneously?

4-19. In Chapter 6 we will use the operators

$$\hat{L}_+ = \hat{L}_x + i\hat{L}_y$$

and

$$\hat{L}_- = \hat{L}_x - i\hat{L}_y$$

Show that

$$\hat{L}_+\hat{L}_- = \hat{L}^2 - \hat{L}_z^2 + \hbar\hat{L}_z$$

$$[\hat{L}_z, \hat{L}_+] = \hbar\hat{L}_+$$

and that

$$[\hat{L}_z, \hat{L}_-] = -\hbar\hat{L}_-$$

4-20. Consider a particle in a two-dimensional box. Determine $[\hat{X}, \hat{P}_y]$, $[\hat{X}, \hat{P}_x]$, $[\hat{Y}, \hat{P}_y]$, and $[\hat{Y}, \hat{P}_x]$.

4-21. Can the position and total angular momentum of any electron be measured simultaneously to arbitrary precision?

4-22. Can the angular momentum and kinetic energy of a particle be measured simultaneously to arbitrary precision?

4-23. Using the result of Problem 4–20, what are the "uncertainty relationships" $\Delta x \Delta p_y$ and $\Delta y \Delta p_x$ equal to?

4-24. We can define functions of operators through their Taylor series (MathChapter I). For example, we define the operator $\exp(\hat{S})$ by

$$e^{\hat{S}} = \sum_{n=0}^{\infty} \frac{(\hat{S})^n}{n!}$$

Under what conditions does the equality

$$e^{\hat{A}+\hat{B}} = e^{\hat{A}}e^{\hat{B}}$$

hold?

4-25. In this chapter, we learned that if ψ_n is an eigenfunction of the time-independent Schrödinger equation, then

$$\Psi_n(x, t) = \psi_n(x)e^{-iE_n t/\hbar}$$

Show that if ψ_m and ψ_n are both stationary states of $\hat{H}$, then the state

$$\Psi(x, t) = c_m \psi_m(x)e^{-iE_m t/\hbar} + c_n \psi_n(x)e^{-iE_n t/\hbar}$$

satisfies the time-dependent Schrödinger equation.

4-26. Starting with

$$\langle x \rangle = \int \Psi^*(x, t)x\Psi(x, t)dx$$

and the time-dependent Schrödinger equation, show that

$$\frac{d\langle x \rangle}{dt} = \int \Psi^* \frac{i}{\hbar}(\hat{H}x - x\hat{H})\Psi dx$$

Given that

$$\hat{H} = -\frac{\hbar^2}{2m}\frac{d^2}{dx^2} + V(x)$$

show that

$$\hat{H}x - x\hat{H} = -2\frac{\hbar^2}{2m}\frac{d}{dx} = -\frac{\hbar^2}{m}\frac{i}{\hbar}\hat{P}_x = -\frac{i\hbar}{m}\hat{P}_x$$

Finally, substitute this result into the equation for $d\langle x\rangle/dt$ to show that

$$m\frac{d\langle x\rangle}{dt} = \langle \hat{P}_x\rangle$$

Interpret this result.

4-27. Generalize the result of Problem 4–26 and show that if F is any dynamical quantity, then

$$\frac{d\langle F\rangle}{dt} = \int \Psi^* \frac{i}{\hbar}(\hat{H}\hat{F} - \hat{F}\hat{H})\Psi dx$$

Use this equation to show that

$$\frac{d\langle \hat{P}_x\rangle}{dt} = \left\langle -\frac{dV}{dx}\right\rangle$$

Interpret this result. This last equation is known as *Ehrenfest's theorem*.

4-28. The fact that eigenvalues, which correspond to physically observable quantities, must be real imposes a certain condition on quantum-mechanical operators. To see what this condition is, start with

$$\hat{A}\psi = a\psi \tag{1}$$

where $\hat{A}$ and ψ may be complex, but a must be real. Multiply Equation 1 from the left by ψ^* and then integrate to obtain

$$\int \psi^* \hat{A}\psi d\tau = a\int \psi^*\psi d\tau = a \tag{2}$$

Now take the complex conjugate of Equation 1, multiply from the left by ψ, and then integrate to obtain

$$\int \psi \hat{A}^* \psi^* d\tau = a^* = a \tag{3}$$

Equate the left sides of Equations 2 and 3 to give

$$\int \psi^* \hat{A}\psi d\tau = \int \psi \hat{A}^* \psi^* d\tau \tag{4}$$

This is the condition that an operator must satisfy if its eigenvalues are to be real. Such operators are called Hermitian operators.

4-29. In this problem, we will prove that not only are the eigenvalues of Hermitian operators real but that their eigenfunctions are orthogonal. Consider the two eigenvalue equations

$$\hat{A}\psi_n = a_n\psi_n \quad \text{and} \quad \hat{A}\psi_m = a_m\psi_m$$

Multiply the first equation by ψ_m^* and integrate; then take the complex conjugate of the second, multiply by ψ_n, and integrate. Subtract the two resulting equations from each other to get

$$\int_{-\infty}^{\infty} \psi_m^* \hat{A}\psi_n dx - \int_{-\infty}^{\infty} \psi_n \hat{A}^* \psi_m^* dx = (a_n - a_m^*)\int_{-\infty}^{\infty} \psi_m^*\psi_n dx$$

Because $\hat{A}$ is Hermitian, the left side is zero, and so

$$(a_n - a_m^*) \int_{-\infty}^{\infty} \psi_m^* \psi_n dx = 0$$

Discuss the two possibilities $n = m$ and $n \neq m$. Show that $a_n = a_n^*$, which is just another proof that the eigenvalues are real. When $n \neq m$, show that

$$\int_{-\infty}^{\infty} \psi_m^* \psi_n dx = 0 \qquad m \neq n$$

if the system is nondegenerate. Are ψ_m and ψ_n necessarily orthogonal if they are degenerate?

4-30. All the operators in Table 4.1 are Hermitian. In this problem, we show how to determine if an operator is Hermitian. Consider the operator $\hat{A} = d/dx$. If $\hat{A}$ is Hermitian, it will satisfy Equation 4 of Problem 4–28. Substitute $\hat{A} = d/dx$ into Equation 4 and integrate by parts to obtain

$$\int_{-\infty}^{\infty} \psi^* \frac{d\psi}{dx} dx = \left. \psi^* \psi \right|_{-\infty}^{\infty} - \int_{-\infty}^{\infty} \psi \frac{d\psi^*}{dx} dx$$

For a wave function to be normalizable, it must vanish at infinity, so the first term on the right side is zero. Therefore, we have

$$\int_{-\infty}^{\infty} \psi^* \frac{d}{dx} \psi dx = - \int_{-\infty}^{\infty} \psi \frac{d}{dx} \psi^* dx$$

For an arbitrary function $\psi(x)$, d/dx does *not* satisfy Equation 4 of Problem 4–28, so it is *not* Hermitian.

4-31. Following the procedure in Problem 4–30, show that the momentum operator is Hermitian.

4-32. Specify which of the following operators are Hermitian: id/dx, d^2/dx^2, and id^2/dx^2. Assume that $-\infty < x < \infty$ and that the functions on which these operators operate are appropriately well behaved at infinity.

Problems 4–33 through 4–38 examine systems with piece-wise constant potentials.

4-33. Consider a particle moving in the potential energy

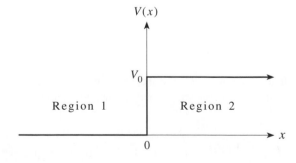

whose mathematical form is

$$V(x) = \begin{matrix} 0 & x < 0 \\ V_0 & x > 0 \end{matrix}$$

where V_0 is a constant. Show that if $E > V_0$, then the solutions to the Schrödinger equation in the two regions (1 and 2) are (see Problem 3–32)

$$\psi_1(x) = Ae^{ik_1x} + Be^{-ik_1x} \qquad x < 0 \tag{1}$$

and

$$\psi_2(x) = Ce^{ik_2x} + De^{-ik_2x} \qquad x > 0 \tag{2}$$

where

$$k_1 = \left(\frac{2mE}{\hbar^2}\right)^{1/2} \quad \text{and} \quad k_2 = \left(\frac{2m(E - V_0)}{\hbar^2}\right)^{1/2} \tag{3}$$

As we learned in Problem 3–32, e^{ikx} represents a particle traveling to the right and e^{-ikx} represents a particle traveling to the left. Let's consider a particle traveling to the right in region 1. If we wish to exclude the case of a particle traveling to the left in region 2, we set $D = 0$ in Equation 2. The physical problem we have set up is a particle of energy E incident on a potential barrier of height V_0. The squares of the coefficients in Equation 1 and 2 represent the probability that the particle is traveling in a certain direction in a given region. For example, $|A|^2$ is the probabillity that the particle is traveling with momentum $+\hbar k_1$ (Problem 3–32) in the region $x < 0$. If we consider many particles, N_0, instead of just one, then we can interpret $|A|^2 N_0$ to be the number of particles with momentum $\hbar k_1$ in the region $x < 0$. The number of these particles that pass a given point per unit time is given by $v|A|^2 N_0$, where the velocity v is given by $\hbar k_1/m$.

Now apply the conditions that $\psi(x)$ and $d\psi/dx$ must be continuous at $x = 0$ (see Problem 4–4) to obtain

$$A + B = C$$

and

$$k_1(A - B) = k_2 C$$

Now define a quantity

$$R = \frac{v_1|B|^2 N_0}{v_1|A|^2 N_0} = \frac{\hbar k_1|B|^2 N_0/m}{\hbar k_1|A|^2 N_0/m} = \frac{|B|^2}{|A|^2}$$

and show that

$$R = \left(\frac{k_1 - k_2}{k_1 + k_2}\right)^2$$

Similarly, define

$$T = \frac{v_2|C|^2 N_0}{v_1|A|^2 N_0} = \frac{\hbar k_2|C|^2 N_0/m}{\hbar k_1|A|^2 N_0/m} = \frac{k_2|C|^2}{k_1|A|^2}$$

and show that

$$T = \frac{4k_1 k_2}{(k_1 + k_2)^2}$$

The symbols R and T stand for reflection coefficient and transmission coefficient, respectively. Give a physical interpretation of these designations. Show that $R + T = 1$. Would you have expected the particle to have been reflected even though its energy, E, is greater than the barrier height, V_0? Show that $R \rightarrow 0$ and $T \rightarrow 1$ as $V_0 \rightarrow 0$.

4-34. Show that $R = 1$ for the system described in Problem 4–33 but with $E < V_0$. Discuss the physical interpretation of this result.

4-35. In this problem, we introduce the idea of *quantum-mechanical tunneling*, which plays a central role in such diverse processes as the α-decay of nuclei, electron-transfer reactions, and hydrogen bonding. Consider a particle in the potential energy regions as shown below.

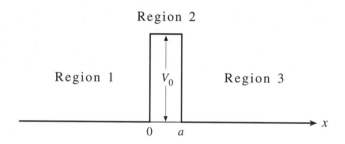

Mathematically, we have

$$V(x) = \begin{array}{ll} 0 & x < 0 \\ V_0 & 0 < x < a \\ 0 & x > a \end{array}$$

Show that if $E < V_0$, the solution to the Schrödinger equation in each region is given by

$$\psi_1(x) = Ae^{ik_1 x} + Be^{-ik_1 x} \qquad x < 0 \tag{1}$$

$$\psi_2(x) = Ce^{k_2 x} + De^{-k_2 x} \qquad 0 < x < a \tag{2}$$

and

$$\psi_3(x) = Ee^{ik_1 x} + Fe^{-ik_1 x} \qquad x > a \tag{3}$$

where

$$k_1 = \left(\frac{2mE}{\hbar^2}\right)^{1/2} \quad \text{and} \quad k_2 = \left(\frac{2m(V_0 - E)}{\hbar^2}\right)^{1/2} \tag{4}$$

If we exclude the situation of the particle coming from large positive values of x, then $F = 0$ in Equation 3. Following Problem 4–33, argue that the transmission coefficient, the probability the particle will get past the barrier, is given by

$$T = \frac{|E|^2}{|A|^2} \tag{5}$$

Now use the fact that $\psi(x)$ and $d\psi/dx$ must be continuous at $x = 0$ and $x = a$ to obtain

$$A + B = C + D \qquad ik_1(A - B) = k_2(C - D) \tag{6}$$

and

$$Ce^{k_2 a} + De^{-k_2 a} = Ee^{ik_1 a} \qquad k_2 Ce^{k_2 a} - k_2 De^{-k_2 a} = ik_1 Ee^{ik_1 a} \tag{7}$$

Eliminate B from Equations 6 to get A in terms of C and D. Then solve Equations 7 for C and D in terms of E. Substitute these results into the equation for A in terms of C and D to get the intermediate result

$$2ik_1 A = [(k_2^2 - k_1^2 + 2ik_1 k_2)e^{k_2 a} + (k_1^2 - k_2^2 + 2ik_1 k_2)e^{-k_2 a}]\frac{Ee^{ik_1 a}}{2k_2}$$

Now use the relations $\sinh x = (e^x - e^{-x})/2$ and $\cosh x = (e^x + e^{-x})/2$ (Problem A–11) to get

$$\frac{E}{A} = \frac{4ik_1 k_2 e^{-ik_1 a}}{2(k_2^2 - k_1^2)\sinh k_2 a + 4ik_1 k_2 \cosh k_2 a}$$

Now multiply the right side by its complex conjugate and use the relation $\cosh^2 x = 1 + \sinh^2 x$ to get

$$T = \left|\frac{E}{A}\right|^2 = \frac{4}{4 + \dfrac{(k_1^2 + k_2^2)^2}{k_1^2 k_2^2}\sinh^2 k_2 a}$$

Finally, use the definition of k_1 and k_2 to show that the probability the particle gets through the barrier (even though it does not have enough energy!) is

$$T = \frac{1}{1 + \dfrac{v_0^2}{4\varepsilon(v_0 - \varepsilon)}\sinh^2(v_0 - \varepsilon)^{1/2}} \tag{8}$$

or

$$T = \frac{1}{1 + \dfrac{\sinh^2[v_0^{1/2}(1 - r)^{1/2}]}{4r(1 - r)}} \tag{9}$$

where $v_0 = 2ma^2 V_0/\hbar^2$, $\varepsilon = 2ma^2 E/\hbar^2$, and $r = E/V_0 = \varepsilon/v_0$. Figure 4.3 shows a plot of T versus r. To plot T versus r for values of $r > 1$, you need to use the relation $\sinh ix = i\sin x$ (Problem A–11). What would the classical result look like?

144

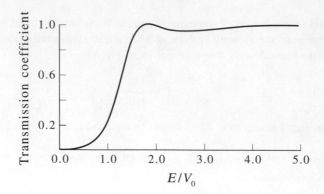

FIGURE 4.3
A plot of the probability that a particle of energy E will penetrate a barrier of height V_0 plotted against the ratio E/V_0 (Equation 9 of Problem 4–35).

4-36. Use the result of Problem 4–35 to determine the probability that an electron with a kinetic energy 8.0×10^{-21} J will tunnel through a 1.0 nm thick potential barrier with $V_0 = 12.0 \times 10^{-21}$ J.

4-37. Problem 4–35 gives that the probability of a particle of relative energy E/V_0 will penetrate a rectangular potential barrier of height V_0 and thickness a is

$$T = \frac{1}{1 + \dfrac{\sinh^2[v_0^{1/2}(1-r)^{1/2}]}{4r(1-r)}}$$

where $v_0 = 2mV_0a^2/\hbar^2$ and $r = E/V_0$. What is the limit of T as $r \to 1$? Plot T against r for $v_0 = 1/2$, 1, and 2. Interpret your results.

4-38. In this problem, we will consider a particle in a *finite* potential well

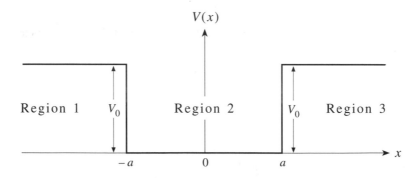

whose mathematical form is

$$V(x) = \begin{array}{ll} V_0 & x < -a \\ 0 & -a < x < a \\ V_0 & x > a \end{array} \qquad (1)$$

Note that this potential describes what we have called a "particle in a box" if $V_0 \to \infty$. Show that if $0 < E < V_0$, the solution to the Schrödinger equation in each region is

$$\begin{aligned}
\psi_1(x) &= Ae^{k_1 x} & x < -a \\
\psi_2(x) &= B\sin\alpha x + C\cos\alpha x & -a < x < a \\
\psi_3(x) &= De^{-k_1 x} & x > a
\end{aligned} \tag{2}$$

where

$$k_1 = \left(\frac{2m(V_0 - E)}{\hbar^2}\right)^{1/2} \quad \text{and} \quad \alpha = \left(\frac{2mE}{\hbar^2}\right)^{1/2} \tag{3}$$

Now apply the conditions that $\psi(x)$ and $d\psi/dx$ must be continuous at $x = -a$ and $x = a$ to obtain

$$Ae^{-k_1 a} = -B\sin\alpha a + C\cos\alpha a \tag{4}$$

$$De^{-k_1 a} = B\sin\alpha a + C\cos\alpha a \tag{5}$$

$$k_1 Ae^{-k_1 a} = \alpha B\cos\alpha a + \alpha C\sin\alpha a \tag{6}$$

and

$$-k_1 De^{-k_1 a} = \alpha B\cos\alpha a - \alpha C\sin\alpha a \tag{7}$$

Add and subtract Equations 4 and 5 and add and subtract Equations 6 and 7 to obtain

$$2C\cos\alpha a = (A + D)e^{-k_1 a} \tag{8}$$

$$2B\sin\alpha a = (D - A)e^{-k_1 a} \tag{9}$$

$$2\alpha C\sin\alpha a = k_1(A + D)e^{-k_1 a} \tag{10}$$

and

$$2\alpha B\cos\alpha a = -k_1(D - A)e^{-k_1 a} \tag{11}$$

Now divide Equation 10 by Equation 8 to get

$$\frac{\alpha\sin\alpha a}{\cos\alpha a} = \alpha\tan\alpha a = k_1 \qquad (D \neq -A \text{ and } C \neq 0) \tag{12}$$

and then divide Equation 11 by Equation 9 to get

$$\frac{\alpha\cos\alpha a}{\sin\alpha a} = \alpha\cot\alpha a = -k_1 \quad \text{and} \quad (D \neq A \text{ and } B \neq 0) \tag{13}$$

Referring back to Equation 3, note that Equations 12 and 13 give the allowed values of E in terms of V_0. It turns out that these two equations cannot be solved simultaneously, so we have two sets of equations

$$\alpha\tan\alpha a = k_1 \tag{14}$$

and

$$\alpha \cot \alpha a = -k_1 \tag{15}$$

Let's consider Equation 14 first. Multiply both sides by a and use the definitions of α and k_1 to get

$$\left(\frac{2ma^2 E}{\hbar^2}\right)^{1/2} \tan \left(\frac{2ma^2 E}{\hbar^2}\right)^{1/2} = \left[\frac{2ma^2}{\hbar^2}(V_0 - E)\right]^{1/2} \tag{16}$$

Show that this equation simplifies to

$$\varepsilon^{1/2} \tan \varepsilon^{1/2} = (v_0 - \varepsilon)^{1/2} \tag{17}$$

where $\varepsilon = 2ma^2 E/\hbar^2$ and $v_0 = 2ma^2 V_0/\hbar^2$. Thus, if we fix v_0 (actually $2ma^2 V_0/\hbar^2$), then we can use Equation 17 to solve for the allowed values of ε (actually $2ma^2 E/\hbar^2$). Equation 17 cannot be solved analytically, but if we plot both $\varepsilon^{1/2} \tan \varepsilon^{1/2}$ and $(v_0 - \varepsilon)^{1/2}$ versus ε on the same graph, then the solutions are given by the intersections of the two curves. Show that the intersections occur at $\varepsilon = 2ma^2 E/\hbar^2 = 1.47$ and 11.37 for $v_0 = 12$. The other value(s) of ε are given by the solutions to Equation 15, which are obtained by finding the intersection of $-\varepsilon^{1/2} \cot \varepsilon^{1/2}$ and $(v_0 - \varepsilon)^{1/2}$ plotted against ε. Show that $\varepsilon = 2ma^2 E/\hbar^2 = 5.68$ for $v_0 = 12$. Thus, we see there are only three bound states for a well of depth $V_0 = 12\hbar^2/2ma^2$. The important point here is not the numerical values of E, but the fact that there is only a finite number of bound states. Show that there are only two bound states for $v_0 = 2ma^2 V_0/\hbar^2 = 4$.

SPHERICAL COORDINATES

Although Cartesian coordinates (x, y, and z) are suitable for many problems, there are many other problems for which they prove to be cumbersome. A particularly important type of such a problem occurs when the system being described has some sort of a natural center, as in the case of an atom, where the (heavy) nucleus serves as one. In describing atomic systems, as well as many other systems, it is most convenient to use spherical coordinates (Figure D.1). Instead of locating a point in space by specifying the Cartesian coordinates x, y, and z, we can equally well locate the same point by specifying the spherical coordinates r, θ, and ϕ. From Figure D.1, we can see that the relations between the two sets of coordinates are given by

$$x = r \sin \theta \cos \phi$$
$$y = r \sin \theta \sin \phi \tag{D.1}$$
$$z = r \cos \theta$$

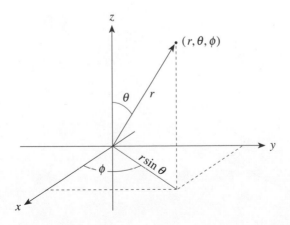

FIGURE D.1
A representation of a spherical coordinate system. A point is specified by the spherical coordinates r, θ, and ϕ.

147

This coordinate system is called a *spherical coordinate system* because the graph of the equation $r = c = $ constant is a sphere of radius c centered at the origin.

Occassionally we need to know r, θ, and ϕ in terms of x, y, and z. These relations are given by (Problem D–1)

$$r = \left(x^2 + y^2 + z^2\right)^{1/2}$$

$$\cos\theta = \frac{z}{(x^2 + y^2 + z^2)^{1/2}} \qquad (D.2)$$

$$\tan\phi = \frac{y}{x}$$

Any point on the surface of a sphere of unit radius can be specified by the values of θ and ϕ. The angle θ represents the declination from the north pole, and hence $0 \le \theta \le \pi$. The angle ϕ represents the angle about the equator, and so $0 \le \phi \le 2\pi$. Although there is a natural zero value for θ (along the north pole), there is none for ϕ. Conventionally, the angle ϕ is measured from the x-axis as illustrated in Figure D.1. Note that r, being the distance from the origin, is intrinsically a positive quantity. In mathematical terms, $0 \le r < \infty$.

In Chapter 6, we will encounter integrals involving spherical coordinates. The differential volume element in Cartesian coordinates is $dxdydz$, but it is not quite so

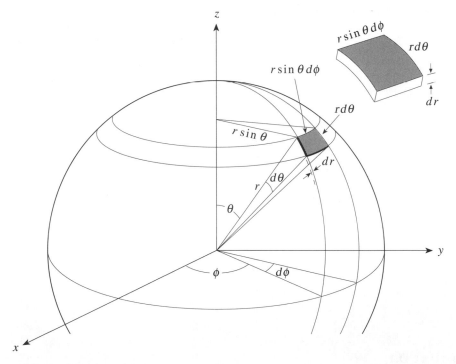

FIGURE D.2
A geometrical construction of the differential volume element in spherical coordinates.

simple in spherical coordinates. Figure D.2 shows a differential volume element in spherical coordinates, which can be seen to be

$$dV = (r\sin\theta d\phi)(rd\theta)dr = r^2\sin\theta drd\theta d\phi \tag{D.3}$$

Let's use Equation D.3 to evaluate the volume of a sphere of radius a. In this case, $0 \le r \le a$, $0 \le \theta \le \pi$, and $0 \le \phi \le 2\pi$. Therefore,

$$V = \int_0^a r^2 dr \int_0^\pi \sin\theta d\theta \int_0^{2\pi} d\phi = \left(\frac{a^3}{3}\right)(2)(2\pi) = \frac{4\pi a^3}{3}$$

Similarly, if we integrate only over θ and ϕ, then we obtain

$$dV = r^2 dr \int_0^\pi \sin\theta d\theta \int_0^{2\pi} d\phi = 4\pi r^2 dr \tag{D.4}$$

This quantity is the volume of a spherical shell of radius r and thickness dr (Figure D.3). The factor $4\pi r^2$ represents the surface area of the spherical shell and dr is its thickness.

The quantity

$$dA = r^2 \sin\theta d\theta d\phi \tag{D.5}$$

is the differential area on the surface of a sphere of radius r. (See Figure D.2.) If we integrate Equation D.5 over all values of θ and ϕ, then we obtain $A = 4\pi r^2$, the area of the surface of a sphere of radius r.

Often, the integral we need to evaluate will be of the form

$$I = \int_0^\infty \int_0^\pi \int_0^{2\pi} F(r,\theta,\phi)r^2 \sin\theta drd\theta d\phi \tag{D.6}$$

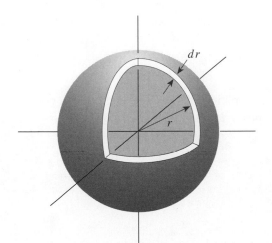

FIGURE D.3
A spherical shell of radius r and thickness dr. The volume of such a shell is $4\pi r^2 dr$, which is its area ($4\pi r^2$) times its thickness (dr).

When writing multiple integrals, for convenience we use a notation that treats an integral like an operator. To this end, we write the triple integral in Equation D.6 in the form

$$I = \int_0^\infty dr\, r^2 \int_0^\pi d\theta \sin\theta \int_0^{2\pi} d\phi\, F(r, \theta, \phi) \tag{D.7}$$

In Equation D.7, each integral "acts on" everything that lies to its right; in other words, we first integrate $F(r, \theta, \phi)$ over ϕ from 0 to 2π, then multiply the result by $\sin\theta$ and integrate over θ from 0 to π, and finally multiply that result by r^2 and integrate over r from 0 to ∞. The advantage of the notation in Equation D.7 is that the integration variable and its associated limits are always unambiguous. As an example of the application of this notation, let's evaluate Equation D.7 with

$$F(r, \theta, \phi) = \frac{1}{32\pi} r^2 e^{-r} \sin^2\theta \cos^2\phi$$

(We will learn in Chapter 6 that this function is the square of a $2p_x$ hydrogen atomic orbital.) If we substitute $F(r, \theta, \phi)$ into Equation D.7, we obtain

$$I = \frac{1}{32\pi} \int_0^\infty dr\, r^2 \int_0^\pi d\theta \sin\theta \int_0^{2\pi} d\phi\, r^2 e^{-r} \sin^2\theta \cos^2\phi$$

The integral over ϕ gives

$$\int_0^{2\pi} d\phi \cos^2\phi = \pi$$

so that

$$I = \frac{1}{32} \int_0^\infty dr\, r^2 \int_0^\pi d\theta \sin\theta\, r^2 e^{-r} \sin^2\theta \tag{D.8}$$

The integral over θ, I_θ, is

$$I_\theta = \int_0^\pi d\theta \sin^3\theta$$

It is often convenient to perform a transformation of variables and let $x = \cos\theta$ in integrals involving θ. Then $\sin\theta\, d\theta$ becomes $-dx$ and the limits become $+1$ to -1, so in this case we have

$$I_\theta = \int_0^\pi d\theta \sin^3\theta = -\int_1^{-1} dx(1 - x^2) = \int_{-1}^1 dx(1 - x^2) = 2 - \frac{2}{3} = \frac{4}{3}$$

Using this result in Equation D.8 gives

$$I = \frac{1}{24} \int_0^\infty dr\, r^4 e^{-r} = \frac{1}{24}(4!) = 1$$

where we have used the general integral

$$\int_0^\infty x^n e^{-x} dx = n!$$

This final result for I simply shows that our above expression for a $2p_x$ hydrogen atomic orbital is normalized.

Frequently the integrand in Equation D.7 will be a function only of r, in which case we say that the integrand is spherically symmetric. Let's look at Equation D.7 when $F(r, \theta, \phi) = f(r)$:

$$I = \int_0^\infty dr\, r^2 \int_0^\pi d\theta \sin\theta \int_0^{2\pi} d\phi\, f(r) \tag{D.9}$$

Because $f(r)$ is independent of θ and ϕ, we can integrate over ϕ to get 2π and then integrate over θ to get 2:

$$\int_0^\pi \sin\theta\, d\theta = \int_{-1}^1 dx = 2$$

Therefore, Equation D.9 becomes

$$I = \int_0^\infty f(r) 4\pi r^2 dr \tag{D.10}$$

The point here is that if $F(r, \theta, \phi) = f(r)$, then Equation D.7 becomes effectively a one-dimensional integral with a factor of $4\pi r^2 dr$ multiplying the integrand. The quantity $4\pi r^2 dr$ is the volume of a spherical shell of radius r and thickness dr.

EXAMPLE D–1

We will learn in Chapter 6 that a $1s$ hydrogen atomic orbital is given by

$$f(r) = \frac{1}{(\pi a_0^3)^{1/2}} e^{-r/a_0}$$

Show that the square of this function is normalized.

SOLUTION: Realize that $f(r)$ is a spherically symmetric function of x, y, and z, where $r = (x^2 + y^2 + z^2)^{1/2}$. Therefore, we use Equation D.10 and write

$$I = \int_0^\infty f^2(r) 4\pi r^2 dr = \frac{4\pi}{\pi a_0^3} \int_0^\infty r^2 e^{-2r/a_0} dr$$

$$= \frac{4}{a_0^3} \cdot \frac{2}{(2/a_0)^3} = 1$$

We need to discuss one final topic involving spherical coordinates. If we restrict ourselves to the surface of a sphere of unit radius, then the angular part of Equation D.5 gives us the differential surface area

$$dA = \sin\theta d\theta d\phi \tag{D.11}$$

If we integrate over the entire spherical surface ($0 \le \theta \le \pi, 0 \le \phi \le 2\pi$), then

$$A = \int_0^\pi \sin\theta d\theta \int_0^{2\pi} d\phi = 4\pi \tag{D.12}$$

which is the area of a sphere of unit radius.

We call the solid enclosed by the surface that connects the origin and the area dA a *solid angle*, as shown in Figure D.4. Because of Equation D.12, we say that a complete solid angle is 4π, just as we say that a complete angle of a circle is 2π. We often denote a solid angle by $d\Omega$, whereby

$$d\Omega = \sin\theta d\theta d\phi \tag{D.13}$$

and Equation D.12 becomes

$$\int_{\text{sphere}} d\Omega = 4\pi \tag{D.14}$$

In discussing the quantum theory of a hydrogen atom in Chapter 6, we will frequently encounter angular integrals of the form

$$I = \int_0^\pi d\theta \sin\theta \int_0^{2\pi} d\phi F(\theta, \phi) \tag{D.15}$$

Note that we are integrating $F(\theta, \phi)$ over the surface of a sphere. For example, we will encounter the integral

$$I = \frac{15}{8\pi} \int_0^\pi d\theta \sin\theta \int_0^{2\pi} d\phi (\sin^2\theta \cos^2\theta)$$

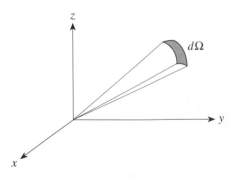

FIGURE D.4
The solid angle, $d\Omega$, subtended by the differential area element $dA = \sin\theta d\theta d\phi$.

The value of this integral is

$$I = \frac{15}{8\pi} \int_0^\pi d\theta \sin^2\theta \cos^2\theta \sin\theta \int_0^{2\pi} d\phi$$

$$= \frac{15}{4} \int_{-1}^1 (1-x^2)x^2 dx = \frac{15}{4}\left[\frac{2}{3} - \frac{2}{5}\right] = 1$$

EXAMPLE D–2
Show that

$$I = \int_0^\pi d\theta \sin\theta \int_0^{2\pi} d\phi Y_2(\theta,\phi)Y_1(\theta,\phi) = 0$$

where

$$Y_1(\theta,\phi) = \left(\frac{3}{4\pi}\right)^{1/2}\cos\theta$$

and

$$Y_2(\theta,\phi) = \left(\frac{5}{16\pi}\right)^{1/2}(3\cos^2\theta - 1)$$

SOLUTION: Because both Y_1 and Y_2 are independent of ϕ, the integration over ϕ gives 2π. The integral over θ is

$$I_\theta = \int_0^\pi \cos\theta(3\cos^2\theta - 1)\sin\theta d\theta$$

$$= \int_{-1}^1 x(3x^2 - 1)dx$$

But this is an odd function of x integrated between -1 and $+1$, so

$$I_\theta = 0$$

and therefore $I = 0$. We say that $Y_1(\theta,\phi)$ and $Y_2(\theta,\phi)$ are orthogonal over the surface of a unit sphere.

Problems

D-1. Derive Equations D.2 from D.1.

D-2. Express the following points given in Cartesian coordinates in terms of spherical coordinates.

$$(x, y, z): \quad (1, 0, 0); \quad (0, 1, 0); \quad (0, 0, 1); \quad (0, 0, -1)$$

D-3. Describe the graphs of the following equations:

   a. $r = 5$,   **b.** $\theta = \pi/4$,   **c.** $\phi = \pi/2$

D-4. Use Equation D.3 to determine the volume of a hemisphere.

D-5. Use Equation D.5 to determine the surface area of a hemisphere.

D-6. Evaluate the integral

$$I = \int_0^\pi \cos^2 \theta \sin^3 \theta d\theta$$

by letting $x = \cos \theta$.

D-7. We will learn in Chapter 6 that a $2p_y$ hydrogen atom orbital is given by

$$\psi_{2p_y} = \frac{1}{4\sqrt{2\pi}} re^{-r/2} \sin \theta \sin \phi$$

Show that ψ_{2p_y} is normalized. (Don't forget to square ψ_{2p_y} first.)

D-8. We will learn in Chapter 6 that a $2s$ hydrogen atomic orbital is given by

$$\psi_{2s} = \frac{1}{4\sqrt{2\pi}} (2 - r)e^{-r/2}$$

Show that ψ_{2s} is normalized. (Don't forget to square ψ_{2s} first.)

D-9. Show that

$$Y_1^0(\theta, \phi) = \left(\frac{3}{4\pi} \right)^{1/2} \cos \theta$$

$$Y_1^1(\theta, \phi) = \left(\frac{3}{8\pi} \right)^{1/2} e^{i\phi} \sin \theta$$

and

$$Y_1^{-1}(\theta, \phi) = \left(\frac{3}{8\pi} \right)^{1/2} e^{-i\phi} \sin \theta$$

are orthonormal over the surface of a sphere.

D-10. Evaluate the average of $\cos \theta$ and $\cos^2 \theta$ over the surface of a sphere.

D-11. We shall frequently use the notation $d\mathbf{r}$ to represent the volume element in spherical coordinates. Evaluate the integral

$$I = \int d\mathbf{r} e^{-r} \cos^2 \theta$$

where the integral is over all space (in other words, over all posible values of r, θ and ϕ).

D-12. Show that the two functions

$$f_1(r) = e^{-r} \cos \theta \qquad \text{and} \qquad f_2(r) = (2 - r)e^{-r/2} \cos \theta$$

are orthogonal over all space (in other words, over all possible values of r, θ and ϕ).

E. Bright Wilson, Jr. was born on December 18, 1908 in Gallatin, Tennessee, and died in 1992. Wilson received his Ph.D in 1933 from the California Institute of Technology, where he studied with Linus Pauling. In 1934, he went to Harvard University as a Junior Fellow and became a full professor just three years later. He was the Theodore Richards Professor of Chemistry from 1948 until his formal retirement in 1979. Wilson's experimental and theoretical work in microwave spectroscopy contributed to the understanding of the structure and dynamics of molecules. During World War II, he directed underwater explosives research at Woods Hole, Massachusetts. In the early 1950s, he spent a year at the Pentagon as a research director of the Weapons System Evaluation Group. In later years, he served on and chaired committees of the National Research Council seeking solutions to various environmental problems. Wilson wrote three books, all of which became classics. His book *Introduction to Quantum Mechanics*, written with Linus Pauling in 1935, was used by almost all physical chemistry graduate students for 20 years and *Molecular Vibrations: The Theory of Infrared and Raman Vibrational Spectra*, written with J.C. Decius and Paul Cross, was a standard reference for most of a generation of physical chemists. His *An Introduction to Scientific Research* is a model for both substance and clarity. One of his sons, Kenneth, was awarded the Nobel Prize in physics in 1982.

CHAPTER 5

The Harmonic Oscillator and the Rigid Rotator: Two Spectroscopic Models

The vibrational motion of a diatomic molecule can be approximated as a harmonic oscillator. In this chapter, we will first study a classical harmonic oscillator and then present and discuss the energies and the corresponding wave functions of a quantum-mechanical harmonic oscillator. We will use the quantum-mechanical energies to describe the infrared spectrum of a diatomic molecule and learn how to determine molecular force constants from vibrational spectra. Then we will model the rotational motion of a diatomic molecule by a rigid rotator. We will discuss the quantum-mechanical energies of a rigid rotator and show their relation to the rotational spectrum of a diatomic molecule. We will use the rotational spectrum of a diatomic molecule to determine the bond length of the molecule.

5–1. A Harmonic Oscillator Obeys Hooke's Law

Consider a mass m connected to a wall by a spring as shown in Figure 5.1. Suppose further that no gravitational force is acting on m so that the only force is due to

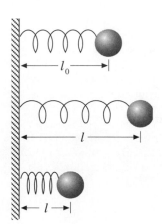

FIGURE 5.1
A mass connected to a wall by a spring. If the force acting upon the mass is directly proportional to the displacement of the spring from its undistorted length, then the force law is called Hooke's law.

the spring. If we let l_0 be the equilibrium, or undistorted, length of the spring, then the restoring force must be some function of the displacement of the spring from its equilibrium length. Let this displacement be denoted by $x = l - l_0$, where l is the length of the spring. The simplest assumption we can make about the force on m as a function of the displacement is that the force is directly proportional to the displacement and to write

$$f = -k(l - l_0) = -kx \tag{5.1}$$

The negative sign indicates that the force points to the right in Figure 5.1 if the spring is compressed ($l < l_0$) and points to the left if the spring is stretched ($l > l_0$). Equation 5.1 is called *Hooke's law* and the (positive) proportionality constant k is called the *force constant* of the spring. A small value of k implies a weak or loose spring, and a large value of k implies a stiff spring.

Newton's equation with a Hooke's law force is

$$m\frac{d^2l}{dt^2} = -k(l - l_0) \tag{5.2}$$

If we let $x = l - l_0$, then $d^2l/dt^2 = d^2x/dt^2$ (l_0 is a constant) and

$$m\frac{d^2x}{dt^2} + kx = 0 \tag{5.3}$$

According to Section 2–3, the general solution to this equation is (Problem 5–1)

$$x(t) = c_1 \sin \omega t + c_2 \cos \omega t \tag{5.4}$$

where

$$\omega = \left(\frac{k}{m}\right)^{1/2} \tag{5.5}$$

EXAMPLE 5–1
Show that Equation 5.4 can be written in the form

$$x(t) = A \sin(\omega t + \phi) \tag{5.6}$$

SOLUTION: The easiest way to prove this is to write

$$\sin(\omega t + \phi) = \sin \omega t \cos \phi + \cos \omega t \sin \phi$$

and substitute this into Equation 5.6 to obtain

$$x(t) = A \cos \phi \sin \omega t + A \sin \phi \cos \omega t$$
$$= c_1 \sin \omega t + c_2 \cos \omega t$$

where

$$c_1 = A \cos \phi \quad \text{and} \quad c_2 = A \sin \phi$$

Equation 5.6 shows that the displacement oscillates sinusoidally, or *harmonically*, with a natural frequency $\omega = (k/m)^{1/2}$. In Equation 5.6, A is the *amplitude* of the vibration and ϕ is the *phase angle*.

Suppose we initially stretch the spring so that its initial displacement is A and then let go. The initial velocity in this case is zero and so from Equation 5.4, we have

$$x(0) = c_2 = A$$

and

$$\left(\frac{dx}{dt}\right)_{t=0} = 0 = c_1 \omega$$

These two equations imply that $c_1 = 0$ and $c_2 = A$ in Equation 5.4, and so

$$x(t) = A \cos \omega t \tag{5.7}$$

The displacement versus time is plotted in Figure 5.2, which shows that the mass oscillates back and forth between A and $-A$ with a frequency of ω radians per second, or $\nu = \omega/2\pi$ cycles per second. The quantity A is called the *amplitude* of the vibration.

Let's look at the total energy of a harmonic oscillator. The force is given by Equation 5.1. Recall from physics that a force can be expressed as the negative derivative of a potential energy or that

$$f(x) = -\frac{dV}{dx} \tag{5.8}$$

so the potential energy is

$$V(x) = -\int f(x)dx + \text{constant} \tag{5.9}$$

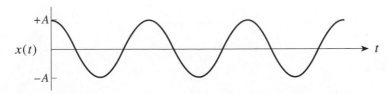

FIGURE 5.2
An illustration of the displacement of a harmonic oscillator versus time.

Using Equation 5.1 for $f(x)$, we see that

$$V(x) = \frac{k}{2}x^2 + \text{constant} \tag{5.10}$$

The constant term here is an arbitrary constant that can be used to fix the zero of energy. If we choose the potential energy of the system to be zero when the spring is undistorted ($x = 0$), then we have

$$V(x) = \frac{k}{2}x^2 \tag{5.11}$$

for the potential energy associated with a simple harmonic oscillator.

The kinetic energy is

$$K = \frac{1}{2}m\left(\frac{dl}{dt}\right)^2 = \frac{1}{2}m\left(\frac{dx}{dt}\right)^2 \tag{5.12}$$

Using Equation 5.7 for $x(t)$, we see that

$$K = \tfrac{1}{2}m\omega^2 A^2 \sin^2 \omega t \tag{5.13}$$

and

$$V = \tfrac{1}{2}kA^2 \cos^2 \omega t \tag{5.14}$$

Both K and V are plotted in Figure 5.3. The total energy is

$$E = K + V = \tfrac{1}{2}m\omega^2 A^2 \sin^2 \omega t + \tfrac{1}{2}kA^2 \cos^2 \omega t$$

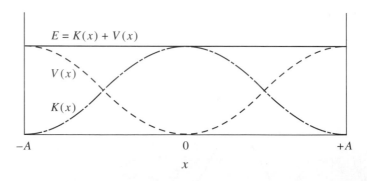

FIGURE 5.3
The kinetic energy [curve labelled $K(x)$] and the potential energy [curve labelled $V(x)$] of a harmonic oscillator during one oscillation. The spring is fully compressed at $-A$ and fully stretched at $+A$. The equilibrium length is $x = 0$. The total energy is the horizontal curve labelled E, which is the sum of $K(x)$ and $V(x)$.

If we recall that $\omega = (k/m)^{1/2}$, we see that the coefficient of the first term is $kA^2/2$, so that the total energy becomes

$$E = \frac{kA^2}{2}(\sin^2 \omega t + \cos^2 \omega t)$$

$$= \frac{kA^2}{2} \tag{5.15}$$

Thus, we see that the total energy is a constant and, in particular, is equal to the potential energy at its largest displacement, where the kinetic energy is zero. Figure 5.3 shows how the total energy is distributed between the kinetic energy and the potential energy. Each oscillates in time between zero and its maximum value but in such a way that their sum is always a constant. We say that the total energy is conserved and that the system is a *conservative system*.

5–2. The Equation for a Harmonic-Oscillator Model of a Diatomic Molecule Contains the Reduced Mass of the Molecule

The simple harmonic oscillator is a good model for a vibrating diatomic molecule. A diatomic molecule, however, does not look like the system pictured in Figure 5.1 but more like two masses connected by a spring as in Figure 5.4. In this case we have two equations of motion, one for each mass:

$$m_1 \frac{d^2 x_1}{dt^2} = k(x_2 - x_1 - l_0) \tag{5.16}$$

and

$$m_2 \frac{d^2 x_2}{dt^2} = -k(x_2 - x_1 - l_0) \tag{5.17}$$

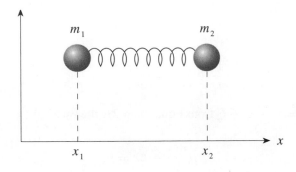

FIGURE 5.4
Two masses connected by a spring, which is a model used to describe the vibrational motion of a diatomic molecule.

where l_0 is the undistorted length of the spring. Note that if $x_2 - x_1 > l_0$, the spring is stretched and the force on mass m_1 is toward the right and that on mass m_2 is toward the left. This is why the force term in Equation 5.16 is positive and that in Equation 5.17 is negative. Note also that the force on m_1 is equal and opposite to the force on m_2, as it should be according to Newton's third law, action and reaction.

If we add Equations 5.16 and 5.17, we find that

$$\frac{d^2}{dt^2}(m_1 x_1 + m_2 x_2) = 0 \tag{5.18}$$

This form suggests that we introduce a *center-of-mass coordinate*

$$X = \frac{m_1 x_1 + m_2 x_2}{M} \tag{5.19}$$

where $M = m_1 + m_2$, so that we can write Equation 5.18 in the form

$$M\frac{d^2 X}{dt^2} = 0 \tag{5.20}$$

There is no force term here, so Equation 5.20 shows that the center of mass moves uniformly in time with a constant momentum.

The motion of the two-mass or two-body system in Figure 5.4 must depend upon only the *relative* separation of the two masses, or upon the *relative coordinate*

$$x = x_2 - x_1 - l_0 \tag{5.21}$$

If we divide Equation 5.17 by m_2 and then subtract Equation 5.16 divided by m_1 we find that

$$\frac{d^2 x_2}{dt^2} - \frac{d^2 x_1}{dt^2} = -\frac{k}{m_2}(x_2 - x_1 - l_0) - \frac{k}{m_1}(x_2 - x_1 - l_0)$$

or

$$\frac{d^2}{dt^2}(x_2 - x_1) = -k\left(\frac{1}{m_1} + \frac{1}{m_2}\right)(x_2 - x_1 - l_0)$$

If we let

$$\frac{1}{m_1} + \frac{1}{m_2} = \frac{m_1 + m_2}{m_1 m_2} = \frac{1}{\mu}$$

and introduce $x = x_2 - x_1 - l_0$ from Equation 5.21, then we have

$$\mu\frac{d^2 x}{dt^2} + kx = 0 \tag{5.22}$$

The quantity μ that we have defined is called the *reduced mass*.

Equation 5.22 is an important result with a nice physical interpretation. If we compare Equation 5.22 with Equation 5.3, we see that Equation 5.22 is the same except for the substitution of the reduced mass μ. Thus, the two-body system in Figure 5.4 can be treated as easily as the one-body problem in Figure 5.1 by using the reduced mass of the two-body system. In particular, the motion of the system is governed by Equation 5.6 but with $\omega = (k/\mu)^{1/2}$. Generally, if the potential energy depends upon only the *relative* distance between two bodies, then we can introduce relative coordinates such as $x_2 - x_1$ and reduce a two-body problem to a one-body problem. This important and useful theorem of classical mechanics is discussed in Problems 5–5 and 5–6.

5–3. The Harmonic-Oscillator Approximation Results from the Expansion of an Internuclear Potential Around its Minimum

Before we discuss the quantum-mechanical treatment of a harmonic oscillator, we should discuss how good an approximation it is for a vibrating diatomic molecule. The internuclear potential for a diatomic molecule is illustrated by the solid line in Figure 5.5. Notice that the curve rises steeply to the left of the minimum, indicating the difficulty of pushing the two nuclei closer together. The curve to the right side of the equilibrium position rises intially but eventually levels off. The potential energy at large separations is essentially the bond energy. The dashed line shows the potential $\frac{1}{2}k(l - l_0)^2$ associated with Hooke's law. Although the harmonic-oscillator potential may appear to be a terrible approximation to the experimental curve, note that it is, indeed, a good approximation in the region of the minimum. This region is the

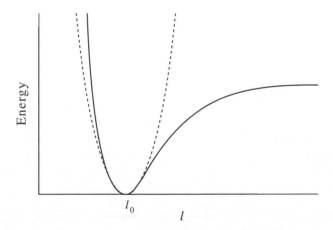

FIGURE 5.5
A comparison of the harmonic oscillator potential ($k(l - l_0)^2/2$; dashed line) with the complete internuclear potential (solid line) of a diatomic molecule. The harmonic oscillator potential is a satisfactory approximation at small displacements from the minimum.

physically important region for many molecules at room temperature. Although the harmonic oscillator unrealistically allows the displacement to vary from 0 to $+\infty$, these large displacements produce potential energies that are so large that they do not often occur in practice. The harmonic oscillator will be a good approximation for vibrations with small amplitudes.

We can put the previous discussion into mathematical terms by considering the Taylor expansion (see MathChapter I) of the potential energy $V(l)$ about the equilibrium bond length $l = l_0$. The first few terms in this expansion are

$$V(l) = V(l_0) + \left(\frac{dV}{dl}\right)_{l=l_0} (l - l_0) + \frac{1}{2!}\left(\frac{d^2V}{dl^2}\right)_{l=l_0} (l - l_0)^2$$
$$+ \frac{1}{3!}\left(\frac{d^3V}{dl^3}\right)_{l=l_0} (l - l_0)^3 + \cdots \tag{5.23}$$

The first term in Equation 5.23 is a constant and depends upon where we choose the zero of energy. It is convenient to choose the zero of energy such that $V(l_0)$ equals zero and relate $V(l)$ to this convention. The second term on the right side of Equation 5.23 involves the quantity $(dV/dl)_{l=l_0}$. Because the point $l = l_0$ is the minimum of the potential energy curve, dV/dl vanishes there, so there is no linear term in the displacement in Equation 5.23. Note that dV/dl is essentially the force acting between the two nuclei, and the fact that dV/dl vanishes at $l = l_0$ means that the force acting between the nuclei is zero at this point. This is why $l = l_0$ is called the *equilibrium bond length*.

If we denote $l - l_0$ by x, $(d^2V/dl^2)_{l=l_0}$ by k, and $(d^3V/dl^3)_{l=l_0}$ by γ, Equation 5.23 becomes

$$V(x) = \frac{1}{2}k(l - l_0)^2 + \frac{1}{6}\gamma(l - l_0)^3 + \cdots$$
$$= \frac{1}{2}kx^2 + \frac{1}{6}\gamma x^3 + \cdots \tag{5.24}$$

If we restrict ourselves to small displacements, then x will be small and we can neglect the terms beyond the quadratic term in Equation 5.24, showing that the general potential energy function $V(l)$ can be approximated by a harmonic-oscillator potential. Note that the force constant is equal to the curvature of $V(l)$ at the minimum. We can consider corrections or extensions of the harmonic-oscillator model by the higher-order terms in Equation 5.24. These are called *anharmonic terms* and will be considered in Chapter 13.

EXAMPLE 5–2
An analytic expression that is a good approximation to an intermolecular potential energy curve is a *Morse potential*

$$V(l) = D(1 - e^{-\beta(l-l_0)})^2$$

First let $x = l - l_0$ so that we can write

$$V(x) = D(1 - e^{-\beta x})^2$$

where D and β are parameters that depend upon the molecule. The parameter D is the dissociation energy of the molecule measured from the minimum of $V(l)$ and β is a measure of the curvature of $V(l)$ at its minimum. Figure 5.6 shows $V(l)$ plotted against l for H_2. Derive a relation between the force constant and the parameters D and β.

SOLUTION: We now expand $V(x)$ about $x = 0$ (Equation 5.23), using

$$V(0) = 0 \qquad \left(\frac{dV}{dx}\right)_{x=0} = \{2D\beta(e^{-\beta x} - e^{-2\beta x})\}_{x=0} = 0$$

and

$$\left(\frac{d^2V}{dx^2}\right)_{x=0} = \{-2D\beta(\beta e^{-\beta x} - 2\beta e^{-2\beta x})\}_{x=0} = 2D\beta^2$$

Therefore, we can write

$$V(x) = D\beta^2 x^2 + \cdots$$

Comparing this result with Equation 5.11 gives

$$k = 2D\beta^2$$

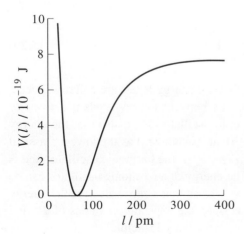

FIGURE 5.6
The Morse potential energy curve $V(l) = D(1 - e^{-\beta(l-l_0)})^2$ plotted against the internuclear displacement l for H_2. The values of the parameters for H_2 are $D = 7.61 \times 10^{-19}$ J, $\beta = 0.0193$ pm^{-1}, and $l_0 = 74.1$ pm.

5–4. The Energy Levels of a Quantum-Mechanical Harmonic Oscillator Are $E_v = \hbar\omega(v + \frac{1}{2})$ with $v = 0,\ 1,\ 2,\ \ldots$

The Schrödinger equation for a one-dimensional harmonic oscillator is

$$-\frac{\hbar^2}{2\mu}\frac{d^2\psi}{dx^2} + V(x)\psi(x) = E\psi(x) \tag{5.25}$$

with $V(x) = \frac{1}{2}kx^2$. Thus, we must solve the second-order differential equation

$$\frac{d^2\psi}{dx^2} + \frac{2\mu}{\hbar^2}\left(E - \frac{1}{2}kx^2\right)\psi(x) = 0 \tag{5.26}$$

This differential equation, however, does not have constant coefficients, so we cannot use the method we developed in Section 2–2. In fact, when a differential equation does not have constant coefficients, there is no simple, general technique for solving it, and each case must be considered individually.

When Equation 5.26 is solved, well-behaved, finite solutions can be obtained only if the energy is restricted to the quantized values

$$\begin{aligned}
E_v &= \hbar\left(\frac{k}{\mu}\right)^{1/2}\left(v + \frac{1}{2}\right) \\
&= \hbar\omega\left(v + \frac{1}{2}\right) = hv\left(v + \frac{1}{2}\right) \qquad v = 0,\ 1,\ 2,\ \ldots
\end{aligned} \tag{5.27}$$

where

$$\omega = \left(\frac{k}{\mu}\right)^{1/2} \tag{5.28}$$

and

$$v = \frac{1}{2\pi}\left(\frac{k}{\mu}\right)^{1/2} \tag{5.29}$$

The energies are plotted in Figure 5.7. Note that the energy levels are equally spaced, with a separation $\hbar\omega$ or hv. This uniform spacing between energy levels is a property peculiar to the quadratic potential of a harmonic oscillator. Note also that the energy of the ground state, the state with $v = 0$, is $\frac{1}{2}hv$ and is not zero as the lowest classical energy is. This energy is called the *zero-point energy* of the harmonic oscillator and is a direct result of the Uncertainty Principle. The energy of a harmonic oscillator can be written in the form $(p^2/2\mu) + (kx^2/2)$, and so we see that a zero value for the energy would require that both p and x or, more precisely, the expectation values of $\hat{P}^2$ and $\hat{X}^2$ be simultaneously zero, in violation of the Uncertainty Principle.

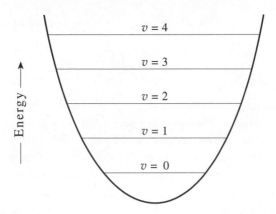

FIGURE 5.7
The energy levels of a quantum-mechanical harmonic oscillator.

5–5. The Harmonic Oscillator Accounts for the Infrared Spectrum of a Diatomic Molecule

We will discuss molecular spectroscopy in some detail in Chapter 13, but here we will discuss the spectroscopic predictions of a harmonic oscillator. If we model the potential energy function of a diatomic molecule as a harmonic oscillator, then according to Equation 5.27, the vibrational energy levels of the diatomic molecule are given by

$$E_v = \hbar \left(\frac{k}{\mu} \right)^{1/2} \left(v + \tfrac{1}{2} \right) \qquad v = 0, \ 1, \ 2, \ \ldots \tag{5.30}$$

A diatomic molecule can make a transition from one vibrational energy state to another by absorbing or emitting electromagnetic radiation whose observed frequency satisfies the Bohr frequency condition

$$\Delta E = h\nu_{\text{obs}} \tag{5.31}$$

We will prove in Chapter 13 that the harmonic-oscillator model allows transitions only between adjacent energy states, so that we have the condition that $\Delta v = \pm 1$. Such a condition is called a *selection rule*.

For absorption to occur, $\Delta v = +1$ and so

$$\Delta E = E_{v+1} - E_v = \hbar \left(\frac{k}{\mu} \right)^{1/2} \tag{5.32}$$

Thus, the observed frequency of the radiation absorbed is

$$\nu_{\text{obs}} = \frac{1}{2\pi} \left(\frac{k}{\mu} \right)^{1/2} \tag{5.33}$$

or

$$\tilde{\nu}_{obs} = \frac{1}{2\pi c} \left(\frac{k}{\mu}\right)^{1/2} \tag{5.34}$$

where the tilde indicates that the units are cm^{-1}. Furthermore, because successive energy states of a harmonic oscillator are separated by the same energy, ΔE is the same for all allowed transitions, so this model predicts that the spectrum consists of just one line whose frequency is given by Equation 5.34. This prediction is in good accord with experiment, and this line is called the *fundamental vibrational frequency*. For diatomic molecules, these lines occur at around 10^3 cm^{-1}, which is in the infrared region. Equation 5.34 enables us to determine force constants if the fundamental vibrational frequency is known. For example, for $H^{35}Cl$, $\tilde{\nu}_{obs}$ is 2.886×10^3 cm^{-1} and so, according to Equation 5.34, the force constant of $H^{35}Cl$ is

$$\begin{aligned}
k &= (2\pi c \tilde{\nu}_{obs})^2 \mu \\
&= [2\pi (2.998 \times 10^8 \text{ m} \cdot \text{s}^{-1})(2.886 \times 10^3 \text{ cm}^{-1})(100 \text{ cm} \cdot \text{m}^{-1})]^2 \\
&\quad \times \frac{(35.0 \text{ amu})(1.01 \text{ amu})}{(35.0 + 1.01) \text{ amu}}(1.661 \times 10^{-27} \text{ kg} \cdot \text{amu}^{-1}) \\
&= 4.78 \times 10^2 \text{ kg} \cdot \text{s}^{-2} = 4.78 \times 10^2 \text{ N} \cdot \text{m}^{-1}
\end{aligned}$$

EXAMPLE 5–3
The infrared spectrum of $^{75}Br^{19}F$ consists of an intense line at 380 cm^{-1}. Calculate the force constant of $^{75}Br^{19}F$.

SOLUTION: The force constant is given by

$$k = (2\pi c \tilde{\nu}_{obs})^2 \mu$$

The reduced mass is

$$\mu = \frac{(75.0 \text{ amu})(19.0 \text{ amu})}{(75.0 + 19.0) \text{ amu}}(1.661 \times 10^{-27} \text{ kg} \cdot \text{amu}^{-1}) = 2.52 \times 10^{-26} \text{ kg}$$

and so

$$\begin{aligned}
k &= [2\pi (2.998 \times 10^8 \text{ m} \cdot \text{s}^{-1})(380 \text{ cm}^{-1})(100 \text{ cm} \cdot \text{m}^{-1})]^2 (2.52 \times 10^{-26} \text{ kg}) \\
&= 129 \text{ kg} \cdot \text{s}^{-2} = 129 \text{ N} \cdot \text{m}^{-1}
\end{aligned}$$

Force constants for diatomic molecules are of the order of 10^2 $N \cdot m^{-1}$. Table 5.1 lists the fundamental vibrational frequencies, force constants, and bond lengths of some diatomic molecules. We will also see in Chapter 13 that not only must $\Delta v = \pm 1$ in the harmonic-oscillator model but the dipole moment of the molecule must change as the

TABLE 5.1
The fundamental vibrational frequencies, the force constants, and bond lengths of some diatomic molecules

Molecule	$\tilde{\nu}/\text{cm}^{-1}$	$k/\text{N·m}^{-1}$	Bond length/pm
H_2	4401	570	74.1
D_2	2990	527	74.1
$H^{35}Cl$	2886	478	127.5
$H^{79}Br$	2630	408	141.4
$H^{127}I$	2230	291	160.9
$^{35}Cl^{35}Cl$	554	319	198.8
$^{79}Br^{79}Br$	323	240	228.4
$^{127}I^{127}I$	213	170	266.7
$^{16}O^{16}O$	1556	1142	120.7
$^{14}N^{14}N$	2330	2243	109.4
$^{12}C^{16}O$	2143	1857	112.8
$^{14}N^{16}O$	1876	1550	115.1
$^{23}Na^{23}Na$	158	17	307.8
$^{23}Na^{35}Cl$	378	117	236.1
$^{39}K^{35}Cl$	278	84	266.7

molecule vibrates if the molecule is to absorb infrared radiation. Thus, the harmonic-oscillator model predicts that HCl absorbs in the infrared but N_2 does not. We will see that this prediction is in good agreement with experiment. There are, indeed, deviations from the harmonic-oscillator model, but we will see not only that they are fairly small but that we can systematically introduce corrections and extensions to account for them.

5–6. The Harmonic-Oscillator Wave Functions Involve Hermite Polynomials

The wave functions corresponding to the E_v for a harmonic oscillator are nondegenerate and are given by

$$\psi_v(x) = N_v H_v(\alpha^{1/2}x)e^{-\alpha x^2/2} \tag{5.35}$$

where

$$\alpha = \left(\frac{k\mu}{\hbar^2}\right)^{1/2} \tag{5.36}$$

TABLE 5.2
The first few Hermite polynomials.

$H_0(\xi) = 1$	$H_1(\xi) = 2\xi$
$H_2(\xi) = 4\xi^2 - 2$	$H_3(\xi) = 8\xi^3 - 12\xi$
$H_4(\xi) = 16\xi^4 - 48\xi^2 + 12$	$H_5(\xi) = 32\xi^5 - 160\xi^3 + 120\xi$

The normalization constant N_v is

$$N_v = \frac{1}{(2^v v!)^{1/2}} \left(\frac{\alpha}{\pi}\right)^{1/4} \tag{5.37}$$

and the $H_v(\alpha^{1/2}x)$ are polynomials called *Hermite polynomials*. The first few Hermite polynomials are listed in Table 5.2. Note that $H_v(\xi)$ is a vth-degree polynomial in ξ. The first few harmonic oscillator wave functions are listed in Table 5.3 and plotted in Figure 5.8.

Although we have not solved the Shrödinger equation for a harmonic oscillator (Equation 5.26), we can at least show that the functions given by Equation 5.35 are solutions. For example, let's consider $\psi_0(x)$, which according to Table 5.3 is

$$\psi_0(x) = \left(\frac{\alpha}{\pi}\right)^{1/4} e^{-\alpha x^2/2}$$

Substitution of this equation into Equation 5.26 with $E_0 = \frac{1}{2}\hbar\omega$ yields

$$\frac{d^2\psi_0}{dx^2} + \frac{2\mu}{\hbar^2}\left(E_0 - \frac{1}{2}kx^2\right)\psi_0(x) = 0$$

$$\left(\frac{\alpha}{\pi}\right)^{1/4}(\alpha^2 x^2 e^{-\alpha x^2/2} - \alpha e^{-\alpha x^2/2}) + \frac{2\mu}{\hbar^2}\left(\frac{\hbar\omega}{2} - \frac{kx^2}{2}\right)\left(\frac{\alpha}{\pi}\right)^{1/4} e^{-\alpha x^2/2} \overset{?}{=} 0$$

or

$$(\alpha^2 x^2 - \alpha) + \left(\frac{\mu\omega}{\hbar} - \frac{\mu k}{\hbar^2}x^2\right) \overset{?}{=} 0$$

TABLE 5.3
The first few harmonic-oscillator wave functions, Equation 5.35. The parameter $\alpha = (k\mu)^{1/2}/\hbar$.

$\psi_0(x) = \left(\dfrac{\alpha}{\pi}\right)^{1/4} e^{-\alpha x^2/2}$	$\psi_2(x) = \left(\dfrac{\alpha}{4\pi}\right)^{1/4}(2\alpha x^2 - 1)e^{-\alpha x^2/2}$
$\psi_1(x) = \left(\dfrac{4\alpha^3}{\pi}\right)^{1/4} xe^{-\alpha x^2/2}$	$\psi_3(x) = \left(\dfrac{\alpha^3}{9\pi}\right)^{1/4}(2\alpha x^3 - 3x)e^{-\alpha x^2/2}$

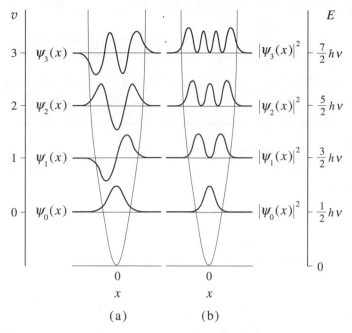

FIGURE 5.8
(a) The normalized harmonic-oscillator wave functions. (b) The probability densities for a harmonic oscillator.

Using the relations $\alpha = (k\mu/\hbar^2)^{1/2}$ and $\omega = (k/\mu)^{1/2}$, we see that everything cancels on the left side of the above expression. Thus, $\psi_0(x)$ is a solution to Equation 5.26. Problem 5–15 involves proving explicitly that $\psi_1(x)$ and $\psi_2(x)$ are solutions of Equation 5.26.

We can also show explicitly that the $\psi_v(x)$ are normalized, or that N_v given by Equation 5.37 is the normalization constant.

EXAMPLE 5–4
Show that $\psi_0(x)$ and $\psi_1(x)$ are normalized.

SOLUTION: According to Table 5.3,

$$\psi_0(x) = \left(\frac{\alpha}{\pi}\right)^{1/4} e^{-\alpha x^2/2} \quad \text{and} \quad \psi_1(x) = \left(\frac{4\alpha^3}{\pi}\right)^{1/4} x e^{-\alpha x^2/2}$$

Then

$$\int_{-\infty}^{\infty} \psi_0^2(x)dx = \left(\frac{\alpha}{\pi}\right)^{1/2} \int_{-\infty}^{\infty} e^{-\alpha x^2}dx = \left(\frac{\alpha}{\pi}\right)^{1/2} \left(\frac{\pi}{\alpha}\right)^{1/2} = 1$$

and

$$\int_{-\infty}^{\infty} \psi_1^2(x)dx = \left(\frac{4\alpha^3}{\pi}\right)^{1/2} \int_{-\infty}^{\infty} x^2 e^{-\alpha x^2}dx = \left(\frac{4\alpha^3}{\pi}\right)^{1/2} \left[\frac{1}{2\alpha}\left(\frac{\pi}{\alpha}\right)^{1/2}\right] = 1$$

The integrals here are given on the inside cover of this book and are evaluated in Problem 5–17.

We can appeal to the general results of Chapter 4 to argue that the harmonic-oscillator wave functions are orthogonal. The energy eigenvalues are nondegenerate, so we have that

$$\int_{-\infty}^{\infty} \psi_v(x)\psi_{v'}(x)dx = 0 \qquad v \neq v'$$

or, more explicitly, that

$$\int_{-\infty}^{\infty} H_v(\alpha^{1/2}x)H_{v'}(\alpha^{1/2}x)e^{-\alpha x^2}dx = 0 \qquad v \neq v'$$

EXAMPLE 5–5

Show explicitly that $\psi_0(x)$ and $\psi_1(x)$ for the harmonic oscillator are orthogonal.

SOLUTION:

$$\psi_0(x) = \left(\frac{\alpha}{\pi}\right)^{1/4} e^{-\alpha x^2/2} \quad \text{and} \quad \psi_1(x) = \left(\frac{4\alpha^3}{\pi}\right)^{1/4} x e^{-\alpha x^2/2}$$

so

$$\int_{-\infty}^{\infty} \psi_1(x)\psi_0(x)dx = \left(\frac{2\alpha^2}{\pi}\right)^{1/2} \int_{-\infty}^{\infty} x e^{-\alpha x^2}dx = 0$$

because the integrand is an odd function of x.

Problem 5–16 has you verify that the harmonic oscillator wave functions are orthogonal for a few other cases.

5–7. Hermite Polynomials Are Either Even or Odd Functions

Recall from MathChapter B that an even function is a function that satisfies

$$f(x) = f(-x) \qquad \text{(even)} \tag{5.38}$$

and an odd function is one that satisfies

$$f(x) = -f(-x) \qquad \text{(odd)} \tag{5.39}$$

EXAMPLE 5–6

Show that the Hermite polynomials $H_v(\xi)$ are even if v is even and odd if v is odd.

SOLUTION: Using Table 5.2,

$H_0(\xi) = 1$ is even.

$H_1(\xi) = 2\xi = -2(-\xi) = -H_1(-\xi)$ and so is odd.

$H_2(\xi) = 4\xi^2 - 2 = 4(-\xi)^2 - 2 = H_2(-\xi)$ is even.

$H_3(\xi) = 8\xi^3 - 12\xi = -[8(-\xi)^3 - 12(-\xi)] = -H_3(-\xi)$ and so is odd.

Recall that if $f(x)$ is an odd function, then

$$\int_{-A}^{A} f(x)dx = 0 \qquad f(x) \text{ odd} \tag{5.40}$$

because the areas from $-A$ to 0 and 0 to A cancel. According to Equation 5.35, the harmonic-oscillator wave functions are

$$\psi_v(x) = N_v H_v(\alpha^{1/2}x)e^{-\alpha x^2/2}$$

Because the $\psi_v(x)$ are even when v is an even integer and odd when v is an odd integer, $\psi_v^2(x)$ is an even function for any value of v. Therefore, $x\psi_v^2(x)$ is an odd function, and according to Equation 5.40, then,

$$\langle x \rangle = \int_{-\infty}^{\infty} \psi_v(x)x\psi_v(x)dx = 0 \tag{5.41}$$

Thus, the average displacement of a harmonic oscillator is zero for all the quantum states of a harmonic oscillator, or the average internuclear separation is the equilibrium bond length l_0.

The average momentum is given by

$$\langle p \rangle = \int_{-\infty}^{\infty} \psi_v(x) \left(-i\hbar \frac{d}{dx} \right) \psi_v(x)dx \tag{5.42}$$

The derivative of an odd (even) function is even (odd), so this integral vanishes because the integrand is the product of an odd and even function and hence is overall odd. Thus, we have that $\langle p \rangle = 0$ for a harmonic operator.

5–8. The Energy Levels of a Rigid Rotator Are $E = \hbar^2 J(J+1)/2I$

In this section we will discuss a simple model for a rotating diatomic molecule. The model consists of two point masses m_1 and m_2 at fixed distances r_1 and r_2 from their center of mass (cf. Figure 5.9). Because the distance between the two masses is fixed, this model is referred to as the *rigid-rotator model*. Even though a diatomic molecule

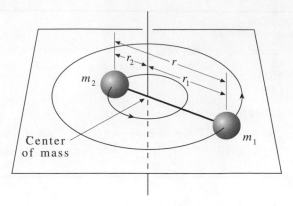

FIGURE 5.9
Two masses m_1 and m_2 shown rotating about their center of mass.

vibrates as it rotates, the vibrational amplitude is small compared with the bond length, so considering the bond length fixed is a good approximation (see Problem 5–22).

Let the molecule rotate about its center of mass at a frequency of ν_{rot} cycles per second. The velocities of the two masses are $v_1 = 2\pi r_1 \nu_{\text{rot}}$ and $v_2 = 2\pi r_2 \nu_{\text{rot}}$, which we write as $v_1 = r_1 \omega$ and $v_2 = r_2 \omega$, where ω (radians per second) $= 2\pi \nu_{\text{rot}}$ and is called the *angular speed* (Section 4–2). The kinetic energy of the rigid rotator is

$$K = \tfrac{1}{2} m_1 v_1^2 + \tfrac{1}{2} m_2 v_2^2 = \tfrac{1}{2}(m_1 r_1^2 + m_2 r_2^2)\omega^2$$
$$= \tfrac{1}{2} I \omega^2 \tag{5.43}$$

where I, the *moment of inertia*, is given by

$$I = m_1 r_1^2 + m_2 r_2^2 \tag{5.44}$$

Using the fact that the location of the center of mass is given by $m_1 r_1 = m_2 r_2$, the moment of inertia can be rewritten as (Problem 5–29)

$$I = \mu r^2 \tag{5.45}$$

where $r = r_1 + r_2$ (the fixed separation of the two masses) and μ is the *reduced mass* (Section 5–2). In Section 4–2, we discussed a single body of mass m rotating at a distance r from a fixed center. In that case, the moment of inertia, I, was equal to mr^2. By comparing Equation 5.45 with this result, we may consider Equation 5.45 to be an equation for the moment of inertia of a single body of mass μ rotating at a distance r from a fixed center. Thus, we have transformed a two-body problem into an equivalent one-body problem, just as we did for a harmonic oscillator in Section 5–2.

Following Equations 4.4 and 4.6, the angular momentum L is

$$L = I\omega \tag{5.46}$$

and the kinetic energy is

$$K = \frac{L^2}{2I} \tag{5.47}$$

There is no potential energy term because in the absence of any external forces (e.g., electric or magnetic), the energy of the molecule does not depend on its orientation in space. The Hamiltonian operator of a rigid rotator is therefore just the kinetic energy operator, and using the operator $\hat{K}$ given in Table 4.1 and the correspondences between linear and rotating systems given in Table 4.2, we can write the Hamiltonian operator of a rigid rotator as

$$\hat{H} = \hat{K} = -\frac{\hbar^2}{2\mu}\nabla^2 \qquad (r \text{ constant}) \tag{5.48}$$

where ∇^2 is the Laplacian operator. We encountered ∇^2 in Cartesian coordinates in Section 3–7, but if the system has a natural center of symmetry, such as one particle revolving around one fixed at the origin, then using spherical coordinates (Math-Chapter D) is much more convenient. Therefore, we must convert ∇^2 from Cartesian coordinates to spherical coordinates. This conversion involves a tedious exercise in the chain rule of partial differentiation, which is best left as problems (see Problems 5–30 through 5–32). The final result is

$$\nabla^2 = \frac{1}{r^2}\frac{\partial}{\partial r}\left(r^2\frac{\partial}{\partial r}\right)_{\theta,\phi} + \frac{1}{r^2\sin\theta}\frac{\partial}{\partial \theta}\left(\sin\theta\frac{\partial}{\partial\theta}\right)_{r,\phi} + \frac{1}{r^2\sin^2\theta}\left(\frac{\partial^2}{\partial\phi^2}\right)_{r,\theta} \tag{5.49}$$

The rigid rotator is a special case where r is a constant, so Equation 5.49 becomes

$$\nabla^2 = \frac{1}{r^2}\frac{1}{\sin\theta}\frac{\partial}{\partial\theta}\left(\sin\theta\frac{\partial}{\partial\theta}\right) + \frac{1}{r^2}\frac{1}{\sin^2\theta}\frac{\partial^2}{\partial\phi^2} \qquad (r \text{ constant}) \tag{5.50}$$

If we use this result in Equation 5.48, we obtain

$$\hat{H} = -\frac{\hbar^2}{2I}\left[\frac{1}{\sin\theta}\frac{\partial}{\partial\theta}\left(\sin\theta\frac{\partial}{\partial\theta}\right) + \frac{1}{\sin^2\theta}\left(\frac{\partial^2}{\partial\phi^2}\right)\right] \tag{5.51}$$

Because $\hat{H} = \hat{L}^2/2I$, we see we can make the correspondence

$$\hat{L}^2 = -\hbar^2\left[\frac{1}{\sin\theta}\frac{\partial}{\partial\theta}\left(\sin\theta\frac{\partial}{\partial\theta}\right) + \frac{1}{\sin^2\theta}\left(\frac{\partial^2}{\partial\phi^2}\right)\right] \tag{5.52}$$

Note that the square of the angular momentum is a naturally occurring operator in quantum mechanics. Both θ and ϕ are unitless, so Equation 5.52 shows that the natural units of angular momentum are $\hbar$ for atomic and molecular systems. We will make use of this fact later.

The orientation of a linear rigid rotator is completely specified by the two angles θ and ϕ, so rigid-rotator wave functions depend upon only these two variables. The rigid-rotator wave functions are customarily denoted by $Y(\theta, \phi)$, so the Schrödinger equation for a rigid rotator reads

$$\hat{H} Y(\theta, \phi) = E Y(\theta, \phi)$$

or

$$-\frac{\hbar^2}{2I} \left[\frac{1}{\sin \theta} \frac{\partial}{\partial \theta} \left(\sin \theta \frac{\partial}{\partial \theta} \right) + \frac{1}{\sin^2 \theta} \left(\frac{\partial^2}{\partial \phi^2} \right) \right] Y(\theta, \phi) = E Y(\theta, \phi) \tag{5.53}$$

If we multiply Equation 5.53 by $\sin^2 \theta$ and let

$$\beta = \frac{2IE}{\hbar^2} \tag{5.54}$$

we find the partial differential equation

$$\sin \theta \frac{\partial}{\partial \theta} \left(\sin \theta \frac{\partial Y}{\partial \theta} \right) + \frac{\partial^2 Y}{\partial \phi^2} + (\beta \sin^2 \theta) Y = 0 \tag{5.55}$$

The solutions to Equation 5.55 are the rigid-rotator wave functions, which we won't need in this chapter. We will encounter Equation 5.55 when we solve the Schrödinger equation for the hydrogen atom in Chapter 6. We therefore defer discussion of the rigid-rotator wave functions until we discuss the hydrogen atom in detail. Nevertheless, you might be interested to know that the solutions to Equation 5.55 are very closely related to the s, p, d, and f orbitals of a hydrogen atom.

When we solve Equation 5.55, it turns out naturally that β, given by Equation 5.54, must obey the condition

$$\beta = J(J+1) \qquad J = 0, 1, 2, \ldots \tag{5.56}$$

Using the definition of β (Equation 5.54), Equation 5.56 is equivalent to

$$E_J = \frac{\hbar^2}{2I} J(J+1) \qquad J = 0, 1, 2, \ldots \tag{5.57}$$

Once again, we obtain a set of discrete energy levels. In addition to the allowed energies given by Equation 5.57, we also find that each energy level has a degeneracy g_J given by $g_J = 2J + 1$.

5–9. The Rigid Rotator Is a Model for a Rotating Diatomic Molecule

The allowed energies of a rigid rotator are given by Equation 5.57. We will prove in Chapter 13 that electromagnetic radiation can cause a rigid rotator to undergo transitions from one state to another, and, in particular, we will prove that the selection rule for the rigid rotator says that transitions are allowed only between adjacent states or that

$$\Delta J = \pm 1 \tag{5.58}$$

In addition to the requirement that $\Delta J = \pm 1$, the molecule must also possess a permanent dipole moment to absorb electromagnetic radiation. Thus, HCl has a rotational spectrum, but N_2 does not. In the case of absorption of electromagnetic radiation, the molecule goes from a state with a quantum number J to one with $J + 1$. The energy difference, then, is

$$\Delta E = E_{J+1} - E_J = \frac{\hbar^2}{2I}[(J+1)(J+2) - J(J+1)]$$

$$= \frac{\hbar^2}{I}(J+1) = \frac{h^2}{4\pi^2 I}(J+1) \tag{5.59}$$

The energy levels and the absorption transitions are shown in Figure 5.10.

Using the Bohr frequency condition $\Delta E = h\nu$, the frequencies at which the absorption transitions occur are

$$\nu = \frac{h}{4\pi^2 I}(J+1) \qquad J = 0,\ 1,\ 2,\ \ldots \tag{5.60}$$

The reduced mass of a diatomic molecule is typically around 10^{-25} to 10^{-26} kg, and a typical bond distance is approximately 10^{-10} m (100 pm), so the moment of inertia of a diatomic molecule typically ranges from 10^{-45} to 10^{-46} kg·m². Substituting $I = 5 \times 10^{-46}$ kg·m² into Equation 5.60 gives that the absorption frequencies are about 2×10^{10} to 10^{11} Hz (cf. Problem 5–33). By referring to Figure 1.11 in Problem 1–1, we see that these frequencies lie in the microwave region. Consequently, rotational transitions of diatomic molecules occur in the microwave region, and the direct study of rotational transitions in molecules is called *microwave spectroscopy*.

It is common practice in microwave spectroscopy to write Equation 5.60 as

$$\nu = 2B(J+1) \qquad J = 0,\ 1,\ 2,\ \ldots \tag{5.61}$$

where

$$B = \frac{h}{8\pi^2 I} \tag{5.62}$$

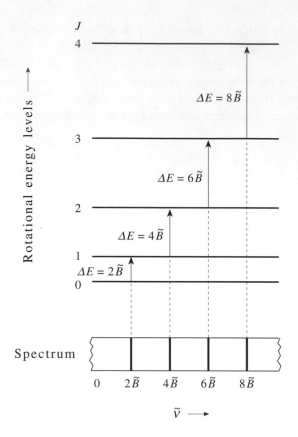

FIGURE 5.10
The energy levels and absorption transitions of a rigid rotator. The absorption transitions occur between adjacent levels, so the absorption spectrum shown below the energy levels consists of a series of equally spaced lines. The quantity $\tilde{B}$ is $h/8\pi^2 cI$ (Equation 5.64).

is called the *rotational constant* of the molecule. Also, the transition frequency is commonly expressed in terms of wave numbers (cm^{-1}) rather than hertz (Hz). If we use the relation $\tilde{\nu} = \nu/c$, then Equation 5.61 becomes

$$\tilde{\nu} = 2\tilde{B}(J+1) \qquad J = 0,\ 1,\ 2,\ \ldots \tag{5.63}$$

where $\tilde{B}$ is the rotational constant expressed in units of wave numbers

$$\tilde{B} = \frac{h}{8\pi^2 cI} \tag{5.64}$$

From either Equation 5.61 or 5.63, we see that the rigid-rotator model predicts that the microwave spectrum of a diatomic molecule consists of a series of equally spaced lines with a separation of $2B$ Hz or $2\tilde{B}$ cm^{-1} as shown in Figure 5.10. From the separation between the absorption frequencies, we can determine the rotational constant and hence the moment of inertia of the molecule. Furthermore, because $I = \mu r^2$, where r

is the internuclear distance or bond length, we can determine the bond length given the transition frequencies. This procedure is illustrated in Example 5–7.

EXAMPLE 5–7

To a good approximation, the microwave spectrum of $H^{35}Cl$ consists of a series of equally spaced lines, separated by 6.26×10^{11} Hz. Calculate the bond length of $H^{35}Cl$.

SOLUTION: According to Equation 5.61, the spacing of the lines in the microwave spectrum of $H^{35}Cl$ is given by

$$2B = \frac{h}{4\pi^2 I}$$

and so

$$\frac{h}{4\pi^2 I} = 6.26 \times 10^{11} \text{ Hz}$$

Solving this equation for I, we have

$$I = \frac{6.626 \times 10^{-34} \text{ J·s}}{4\pi^2 (6.26 \times 10^{11} \text{ s}^{-1})} = 2.68 \times 10^{-47} \text{ kg·m}^2$$

The reduced mass of $H^{35}Cl$ is

$$\mu = \frac{(1.01 \text{ amu})(35.0 \text{ amu})}{36.0 \text{ amu}}(1.661 \times 10^{-27} \text{ kg·amu}^{-1}) = 1.63 \times 10^{-27} \text{ kg}$$

Using the fact that $I = \mu r^2$, we obtain

$$r = \left(\frac{2.68 \times 10^{-47} \text{ kg·m}^2}{1.63 \times 10^{-27} \text{ kg}} \right)^{1/2} = 1.28 \times 10^{-10} \text{ m} = 128 \text{ pm}$$

Problems 5–34 and 5–35 give other examples of the determination of bond lengths from microwave data.

A diatomic molecule is not truly a rigid rotator, because it simultaneously vibrates, however small the amplitude. Consequently, we might expect that although the microwave spectrum of a diatomic molecule consists of a series of lines, their separation is not *exactly* constant. In Chapter 13, we will learn how to correct for the fact that the bond is not exactly rigid.

Problems

5-1. Verify that $x(t) = A \sin \omega t + B \cos \omega t$, where $\omega = (k/m)^{1/2}$ is a solution to Newton's equation for a harmonic oscillator.

5-2. Verify that $x(t) = C \sin(\omega t + \phi)$ is a solution to Newton's equation for a harmonic oscillator.

5-3. The general solution for the classical harmonic oscillator is $x(t) = C \sin(\omega t + \phi)$. Show that the displacement oscillates between $+C$ and $-C$ with a frequency ω radian·s^{-1} or $\nu = \omega/2\pi$ cycle·s^{-1}. What is the period of the oscillations; that is, how long does it take to undergo one cycle?

5-4. From Problem 5–3, we see that the period of a harmonic vibration is $\tau = 1/\nu$. The average of the kinetic energy over one cycle is given by

$$\langle K \rangle = \frac{1}{\tau} \int_0^\tau \frac{m\omega^2 C^2}{2} \cos^2(\omega t + \phi) dt$$

Show that $\langle K \rangle = E/2$ where E is the total energy. Show also that $\langle V \rangle = E/2$, where the instantaneous potential energy is given by

$$V = \frac{kC^2}{2} \sin^2(\omega t + \phi)$$

Interpret the result $\langle K \rangle = \langle V \rangle$.

5-5. Consider two masses m_1 and m_2 in one dimension, interacting through a potential that depends only upon their relative separation $(x_1 - x_2)$, so that $V(x_1, x_2) = V(x_1 - x_2)$. Given that the force acting upon the jth particle is $f_j = -(\partial V/\partial x_j)$, show that $f_1 = -f_2$. What law is this?

Newton's equations for m_1 and m_2 are

$$m_1 \frac{d^2 x_1}{dt^2} = -\frac{\partial V}{\partial x_1} \quad \text{and} \quad m_2 \frac{d^2 x_2}{dt^2} = -\frac{\partial V}{\partial x_2}$$

Now introduce center-of-mass and relative coordinates by

$$X = \frac{m_1 x_1 + m_2 x_2}{M} \qquad x = x_1 - x_2$$

where $M = m_1 + m_2$, and solve for x_1 and x_2 to obtain

$$x_1 = X + \frac{m_2}{M} x \quad \text{and} \quad x_2 = X - \frac{m_1}{M} x$$

Show that Newton's equations in these coordinates are

$$m_1 \frac{d^2 X}{dt^2} + \frac{m_1 m_2}{M} \frac{d^2 x}{dt^2} = -\frac{\partial V}{\partial x}$$

and

$$m_2 \frac{d^2 X}{dt^2} - \frac{m_1 m_2}{M} \frac{d^2 x}{dt^2} = +\frac{\partial V}{\partial x}$$

Now add these two equations to find

$$M \frac{d^2 X}{dt^2} = 0$$

Interpret this result. Now divide the first equation by m_1 and the second by m_2 and subtract to obtain

$$\frac{d^2x}{dt^2} = -\left(\frac{1}{m_1} + \frac{1}{m_2}\right)\frac{\partial V}{\partial x}$$

or

$$\mu\frac{d^2x}{dt^2} = -\frac{\partial V}{\partial x}$$

where $\mu = m_1 m_2/(m_1 + m_2)$ is the reduced mass. Interpret this result, and discuss how the original two-body problem has been reduced to two one-body problems.

5-6. Extend the results of Problem 5–5 to three dimensions. Realize that in three dimensions the relative separation is given by

$$r_{12} = [(x_1 - x_2)^2 + (y_1 - y_2)^2 + (z_1 - z_2)^2]^{1/2}$$

5-7. Calculate the value of the reduced mass of a hydrogen atom. Take the masses of the electron and proton to be $9.109\,390 \times 10^{-31}$ kg and $1.672\,623 \times 10^{-27}$ kg, respectively. What is the percent difference between this result and the rest mass of an electron?

5-8. Show that the reduced mass of two equal masses, m, is $m/2$.

5-9. Example 5–2 shows that a Maclaurin expansion of a Morse potential leads to

$$V(x) = D\beta^2 x^2 + \cdots$$

Given that $D = 7.31 \times 10^{-19}$ J·molecule^{-1} and $\beta = 1.81 \times 10^{10}$ m^{-1} for HCl, calculate the force constant of HCl. Plot the Morse potential for HCl, and plot the corresponding harmonic oscillator potential on the same graph (cf. Figure 5.5).

5-10. Use the result of Example 5–2 and Equation 5.34 to show that

$$\beta = 2\pi c\tilde{v}\left(\frac{\mu}{2D}\right)^{1/2}$$

Given that $\tilde{v} = 2886$ cm^{-1} and $D = 440.2$ kJ·mol^{-1} for H^{35}Cl, calculate β. Compare your result with that in Problem 5–9.

5-11. Carry out the Maclaurin expansion of the Morse potential in Example 5–2 through terms in x^4. Express γ in Equation 5.24 in terms of D and β.

5-12. It turns out that the solution of the Schrödinger equation for the Morse potential can be expressed as

$$\tilde{E}_v = \tilde{v}\left(v + \tfrac{1}{2}\right) - \tilde{v}\tilde{x}\left(v + \tfrac{1}{2}\right)^2$$

where

$$\tilde{x} = \frac{hc\tilde{v}}{4D}$$

Given that $\tilde{v} = 2886$ cm^{-1} and $D = 440.2$ kJ·mol^{-1} for H^{35}Cl, calculate $\tilde{x}$ and $\tilde{v}\tilde{x}$.

5-13. In the infrared spectrum of $H^{79}Br$, there is an intense line at 2630 cm^{-1}. Calculate the force constant of $H^{79}Br$ and the period of vibration of $H^{79}Br$.

5-14. The force constant of $^{79}Br^{79}Br$ is 240 N·m^{-1}. Calculate the fundamental vibrational frequency and the zero-point energy of $^{79}Br^{79}Br$.

5-15. Verify that $\psi_1(x)$ and $\psi_2(x)$ given in Table 5.3 satisfy the Schrödinger equation for a harmonic oscillator.

5-16. Show explicitly for a harmonic oscillator that $\psi_0(x)$ is orthogonal to $\psi_1(x)$, $\psi_2(x)$, and $\psi_3(x)$ and that $\psi_1(x)$ is orthogonal to $\psi_2(x)$ and $\psi_3(x)$ (see Table 5.3).

5-17. To normalize the harmonic-oscillator wave functions and calculate various expectation values, we must be able to evaluate integrals of the form

$$I_v(a) = \int_{-\infty}^{\infty} x^{2v} e^{-ax^2} dx \qquad v = 0, 1, 2, \ldots$$

We can simply either look them up in a table of integrals or continue this problem. First, show that

$$I_v(a) = 2 \int_0^{\infty} x^{2v} e^{-ax^2} dx$$

The case $v = 0$ can be handled by the following trick. Show that the square of $I_0(a)$ can be written in the form

$$I_0^2(a) = 4 \int_0^{\infty} \int_0^{\infty} dx dy e^{-a(x^2+y^2)}$$

Now convert to plane polar coordinates, letting

$$r^2 = x^2 + y^2 \quad \text{and} \quad dxdy = rdrd\theta$$

Show that the appropriate limits of integration are $0 \leq r < \infty$ and $0 \leq \theta \leq \pi/2$ and that

$$I_0^2(a) = 4 \int_0^{\pi/2} d\theta \int_0^{\infty} drre^{-ar^2}$$

which is elementary and gives

$$I_0^2(a) = 4 \cdot \frac{\pi}{2} \cdot \frac{1}{2a} = \frac{\pi}{a}$$

or that

$$I_0(a) = \left(\frac{\pi}{a}\right)^{1/2}$$

Now prove that the $I_v(a)$ may be obtained by repeated differentiation of $I_0(a)$ with respect to a and, in particular, that

$$\frac{d^v I_0(a)}{da^v} = (-1)^v I_v(a)$$

Use this result and the fact that $I_0(a) = (\pi/a)^{1/2}$ to generate $I_1(a)$, $I_2(a)$, and so forth.

5-18. Prove that the product of two even functions is even, that the product of two odd functions is even, and that the product of an even and an odd function is odd.

5-19. Prove that the derivative of an even (odd) function is odd (even).

5-20. Show that

$$\langle x^2 \rangle = \int_{-\infty}^{\infty} \psi_2(x) x^2 \psi_2(x) dx = \frac{5}{2} \frac{\hbar}{(\mu k)^{1/2}}$$

for a harmonic oscillator. Note that $\langle x^2 \rangle^{1/2}$ is the square root of the mean of the square of the displacement (the *root-mean-square displacement*) of the oscillator.

5-21. Show that

$$\langle p^2 \rangle = \int_{-\infty}^{\infty} \psi_2(x) \hat{P}^2 \psi_2(x) dx = \frac{5}{2} \hbar (\mu k)^{1/2}$$

for a harmonic oscillator.

5-22. Using the fundamental vibrational frequencies of some diatomic molecules given below, calculate the root-mean-square displacement (see Problem 5–20) in the $v = 0$ state and compare it with the equilibrium bond length (also given below).

Molecule	$\tilde{\nu}/cm^{-1}$	l_0/pm
H_2	4401	74.1
$^{35}Cl^{35}Cl$	554	198.8
$^{14}N^{14}N$	2330	109.4

5-23. Prove that

$$\langle K \rangle = \langle V(x) \rangle = \frac{E_v}{2}$$

for a one-dimensional harmonic oscillator for $v = 0$ and $v = 1$.

5-24. There are a number of general relations between the Hermite polynomials and their derivatives (which we will not derive). Some of these are

$$\frac{dH_v(\xi)}{d\xi} = 2\xi H_v(\xi) - H_{v+1}(\xi)$$

$$H_{v+1}(\xi) - 2\xi H_v(\xi) + 2v H_{v-1}(\xi) = 0$$

and

$$\frac{dH_v(\xi)}{d\xi} = 2v H_{v-1}(\xi)$$

Such connecting relations are called *recursion formulas*. Verify these formulas explicitly using the first few Hermite polynomials given in Table 5.2.

5-25. Use the recursion formulas for the Hermite polynomials given in Problem 5–24 to show that $\langle p \rangle = 0$ and $\langle p^2 \rangle = \hbar (\mu k)^{1/2} (v + \frac{1}{2})$. Remember that the momentum operator involves a differentiation with respect to x, not ξ.

5-26. It can be proved generally that

$$\langle x^2 \rangle = \frac{1}{\alpha}\left(v + \tfrac{1}{2}\right) = \frac{\hbar}{(\mu k)^{1/2}}\left(v + \tfrac{1}{2}\right)$$

and that

$$\langle x^4 \rangle = \frac{3}{4\alpha^2}(2v^2 + 2v + 1) = \frac{3\hbar^2}{4\mu k}(2v^2 + 2v + 1)$$

for a harmonic oscillator. Verify these formulas explicitly for the first two states of a harmonic oscillator.

5-27. This problem is similar to Problem 3–35. Show that the harmonic-oscillator wave functions are alternately even and odd functions of x because the Hamiltonian operator obeys the relation $\hat{H}(x) = \hat{H}(-x)$. Define a reflection operator $\hat{R}$ by

$$\hat{R}u(x) = u(-x)$$

Show that $\hat{R}$ is linear and that it commutes with $\hat{H}$. Show also that the eigenvalues of $\hat{R}$ are ± 1. What are its eigenfunctions? Show that the harmonic-oscillator wave functions are eigenfunctions of $\hat{R}$. Note that they are eigenfunctions of both $\hat{H}$ and $\hat{R}$. What does this observation say about $\hat{H}$ and $\hat{R}$?

5-28. Use Ehrenfest's theorem (Problem 4–27) to show that $\langle p_x \rangle$ does not depend upon time for a one-dimensional harmonic oscillator.

5-29. Show that the moment of inertia for a rigid rotator can be written as $I = \mu r^2$, where $r = r_1 + r_2$ (the fixed separation of the two masses) and μ is the reduced mass.

5-30. Consider the transformation from Cartesian coordinates to plane polar coordinates where

$$x = r\cos\theta \qquad r = (x^2 + y^2)^{1/2}$$
$$y = r\sin\theta \qquad \theta = \tan^{-1}\left(\frac{y}{x}\right) \tag{1}$$

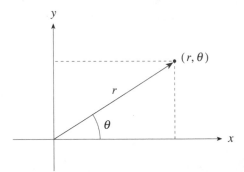

If a function $f(r, \theta)$ depends upon the polar coordinates r and θ, then the chain rule of partial differentiation says that

$$\left(\frac{\partial f}{\partial x}\right)_y = \left(\frac{\partial f}{\partial r}\right)_\theta \left(\frac{\partial r}{\partial x}\right)_y + \left(\frac{\partial f}{\partial \theta}\right)_r \left(\frac{\partial \theta}{\partial x}\right)_y \tag{2}$$

and that

$$\left(\frac{\partial f}{\partial y}\right)_x = \left(\frac{\partial f}{\partial r}\right)_\theta \left(\frac{\partial r}{\partial y}\right)_x + \left(\frac{\partial f}{\partial \theta}\right)_r \left(\frac{\partial \theta}{\partial y}\right)_x \tag{3}$$

For simplicity, we will assume r is constant so that we can ignore terms involving derivatives with respect to r. In other words, we will consider a particle that is constrained to move on the circumference of a circle. This system is sometimes called a *particle on a ring*. Using Equations 1 and 2, show that

$$\left(\frac{\partial f}{\partial x}\right)_y = -\frac{\sin\theta}{r}\left(\frac{\partial f}{\partial \theta}\right)_r \quad \text{and} \quad \left(\frac{\partial f}{\partial y}\right)_x = \frac{\cos\theta}{r}\left(\frac{\partial f}{\partial \theta}\right)_r \quad (r \text{ fixed}) \tag{4}$$

Now apply Equation 2 again to show that

$$\left(\frac{\partial^2 f}{\partial x^2}\right)_y = \left[\frac{\partial}{\partial x}\left(\frac{\partial f}{\partial x}\right)_y\right] = \left[\frac{\partial}{\partial \theta}\left(\frac{\partial f}{\partial x}\right)_y\right]_r \left(\frac{\partial \theta}{\partial x}\right)_y$$

$$= \left\{\frac{\partial}{\partial \theta}\left[-\frac{\sin\theta}{r}\left(\frac{\partial f}{\partial \theta}\right)_r\right]\right\}_r \left(-\frac{\sin\theta}{r}\right)$$

$$= \frac{\sin\theta\cos\theta}{r^2}\left(\frac{\partial f}{\partial \theta}\right)_r + \frac{\sin^2\theta}{r^2}\left(\frac{\partial^2 f}{\partial \theta^2}\right)_r \quad (r \text{ fixed})$$

Similarly, show that

$$\left(\frac{\partial^2 f}{\partial y^2}\right)_x = -\frac{\sin\theta\cos\theta}{r^2}\left(\frac{\partial f}{\partial \theta}\right)_r + \frac{\cos^2\theta}{r^2}\left(\frac{\partial^2 f}{\partial \theta^2}\right)_r \quad (r \text{ fixed})$$

and that

$$\nabla^2 f = \frac{\partial^2 f}{\partial x^2} + \frac{\partial^2 f}{\partial y^2} \longrightarrow \frac{1}{r^2}\left(\frac{\partial^2 f}{\partial \theta^2}\right)_r \quad (r \text{ fixed})$$

Now show that the Schrödinger equation for a particle of mass m constrained to move on a circle of radius r is (see Problem 3–28)

$$-\frac{\hbar^2}{2I}\frac{\partial^2 \psi(\theta)}{\partial \theta^2} = E\psi(\theta) \qquad 0 \le \theta \le 2\pi$$

where $I = mr^2$ is the moment of inertia.

5-31. Generalize Problem 5–30 to the case of a particle moving in a plane under the influence of a central force; in other words, convert

$$\nabla^2 = \frac{\partial^2}{\partial x^2} + \frac{\partial^2}{\partial y^2}$$

to plane polar coordinates, this time without assuming that r is a constant. Use the method of separation of variables to separate the equation for this problem. Solve the angular equation.

5-32. Using Problems 5–30 and 5–31 as a guide, convert ∇^2 from three-dimensional Cartesian coordinates to spherical coordinates.

5-33. Show that rotational transitions of a diatomic molecule occur in the microwave region or the far infrared region of the spectrum.

5-34. In the far infrared spectrum of $H^{79}Br$, there is a series of lines separated by 16.72 cm^{-1}. Calculate the values of the moment of inertia and the internuclear separation in $H^{79}Br$.

5-35. The $J = 0$ to $J = 1$ transition for carbon monoxide ($^{12}C^{16}O$) occurs at 1.153×10^5 MHz. Calculate the value of the bond length in carbon monoxide.

5-36. Figure 5.11 compares the probability distribution associated with the harmonic oscillator wave function $\psi_{10}(\xi)$ to the classical distribution. This problem illustrates what is meant by the classical distribution. Consider

$$x(t) = A \sin(\omega t + \phi)$$

which can be written as

$$\omega t = \sin^{-1}\left(\frac{x}{A}\right) - \phi$$

Now

$$dt = \frac{\omega^{-1} dx}{\sqrt{A^2 - x^2}} \tag{1}$$

This equation gives the time that the oscillator spends between x and $x + dx$. We can convert Equation 1 to a probability distribution in x by dividing by the time that it takes for the oscillator to go from $-A$ to A. Show that this time is π/ω and that the probability distribution in x is

$$p(x)dx = \frac{dx}{\pi\sqrt{A^2 - x^2}} \tag{2}$$

Show that $p(x)$ is normalized. Why does $p(x)$ achieve its maximum value at $x = \pm A$? Now use the fact that $\xi = \alpha^{1/2}x$, where $\alpha = (k\mu/\hbar^2)^{1/2}$, to show that

$$p(\xi)d\xi = \frac{d\xi}{\pi\sqrt{\alpha A^2 - \xi^2}} \tag{3}$$

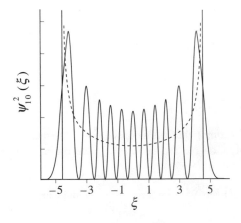

FIGURE 5.11
The probability distribution function of a harmonic oscillator in the $v = 10$ state. The dashed line is that for a classical harmonic oscillator with the same energy. The vertical lines at $\xi \approx \pm 4.6$ represents the extreme limits of the classical harmonic motion.

Show that the limits of ξ are $\pm(\alpha A^2)^{1/2} = \pm(21)^{1/2}$, and compare this result to the vertical lines shown in Figure 5.11. [*Hint*: You need to use the fact that $kA^2/2 = E_{10}$ ($v = 10$).] Finally, plot Equation 3 and compare your result with the curve in Figure 5.11.

5-37. Compute the value of $\hat{L}^2 Y(\theta, \phi)$ for the following functions:

a. $1/(4\pi)^{1/2}$ **b.** $(3/4\pi)^{1/2} \cos\theta$

c. $(3/8\pi)^{1/2} \sin\theta e^{i\phi}$ **d.** $(3/8\pi)^{1/2} \sin\theta e^{-i\phi}$

Do you find anything interesting about the results?

Problems 5–38 through 5–43 develop an alternative method for determining the eigenvalues and eigenfunctions of a one-dimensional harmonic oscillator.

5-38. The Schrödinger equation for a one-dimensional harmonic oscillator is

$$\hat{H}\psi(x) = E\psi(x)$$

where the Hamiltonian operator is given by

$$\hat{H} = -\frac{\hbar^2}{2\mu}\frac{d^2}{dx^2} + \frac{1}{2}kx^2$$

where $k = \mu\omega^2$ is the force constant. Let $\hat{P}$ and $\hat{X}$ be the operators for momentum and position, respectively. If we define $\hat{p} = (\mu\hbar\omega)^{-1/2}\hat{P}$ and $\hat{x} = (\mu\omega/\hbar)^{1/2}\hat{X}$, show that

$$\hat{H} = \frac{\hat{P}^2}{2\mu} + \frac{k}{2}\hat{X}^2 = \frac{\hbar\omega}{2}(\hat{p}^2 + \hat{x}^2)$$

Use the definitions of $\hat{p}$ and $\hat{x}$ to show that

$$\hat{p} = -i\frac{d}{dx}$$

and

$$\hat{p}\hat{x} - \hat{x}\hat{p} = [\hat{p}, \hat{x}] = -i$$

5-39. We will define the operators $\hat{a}_-$ and $\hat{a}_+$ to be

$$\hat{a}_- = \frac{1}{\sqrt{2}}(\hat{x} + i\hat{p}) \quad \text{and} \quad \hat{a}_+ = \frac{1}{\sqrt{2}}(\hat{x} - i\hat{p}) \tag{1}$$

where $\hat{x}$ and $\hat{p}$ are given in Problem 5–38. Show that

$$\hat{a}_-\hat{a}_+ = \tfrac{1}{2}(\hat{x}^2 + i[\hat{p}, \hat{x}] + \hat{p}^2) = \tfrac{1}{2}(\hat{p}^2 + \hat{x}^2 + 1) \tag{2}$$

and that

$$\hat{a}_+\hat{a}_- = \tfrac{1}{2}(\hat{p}^2 + \hat{x}^2 - 1) \tag{3}$$

Now show that the Hamiltonian operator for the one-dimensional harmonic oscillator can be written as

$$\hat{H} = \frac{\hbar\omega}{2}(\hat{a}_-\hat{a}_+ + \hat{a}_+\hat{a}_-)$$

Now show that $\hat{a}_-\hat{a}_+ + \hat{a}_+\hat{a}_-$ is equal to $2\hat{a}_+\hat{a}_- + 1$ so that the Hamiltonian operator can be written as

$$\hat{H} = \hbar\omega\left(\hat{a}_+\hat{a}_- + \tfrac{1}{2}\right)$$

The operator $\hat{a}_+\hat{a}_-$ is called the number operator, which we will denote by $\hat{v}$, and using this definition we obtain

$$\hat{H} = \hbar\omega\left(\hat{v} + \tfrac{1}{2}\right) \tag{4}$$

Comment on the functional form of this result. What do you expect are the eigenvalues of the number operator? Without doing any calculus, explain why $\hat{v}$ must be a Hermitian operator.

5-40. In this problem, we will explore some of the properties of the operators introduced in Problem 5–39. Let ψ_v and E_v be the wave functions and energies of the one-dimensional harmonic oscillator. Start with

$$\hat{H}\psi_v = \hbar\omega\left(\hat{a}_+\hat{a}_- + \tfrac{1}{2}\right)\psi_v = E_v\psi_v$$

Multiply from the left by $\hat{a}_-$ and use Equation 2 of Problem 5–39 to show that

$$\hat{H}(\hat{a}_-\psi_v) = (E_v - \hbar\omega)(\hat{a}_-\psi_v)$$

or that

$$\hat{a}_-\psi_v \propto \psi_{v-1}$$

Also show that

$$\hat{H}(\hat{a}_+\psi_v) = (E_v + \hbar\omega)(\hat{a}_+\psi_v)$$

or that

$$\hat{a}_+\psi_v \propto \psi_{v+1}$$

Thus, we see that $\hat{a}_+$ operating on ψ_v gives ψ_{v+1} (to within a constant) and that $\hat{a}_-\psi_v$ gives ψ_{v-1} to within a constant. The operators $\hat{a}_+$ and $\hat{a}_-$ are called *raising* or *lowering operators*, or simply *ladder operators*. If we think of each rung of a ladder as a quantum state, then the operators $\hat{a}_+$ and $\hat{a}_-$ enable us to move up and down the ladder once we know the wave function of a single rung.

5-41. Use the fact that $\hat{x}$ and $\hat{p}$ are Hermitian in the number operator defined in Problem 5–39 to show that

$$\int \psi_v^* \hat{v}\psi_v\, dx \geq 0$$

5-42. In Problem 5–41, we proved that $v \geq 0$. Because $\hat{a}_-\psi_v \propto \psi_{v-1}$ and $v \geq 0$, there must be some minimal value of v, v_{min}. Argue that $\hat{a}_-\psi_{v_{min}} = 0$. Now multiply $\hat{a}_-\psi_{v_{min}} = 0$ by $\hat{a}_+$ and use Equation 3 of Problem 5–39 to prove that $v_{min} = 0$, and that $v = 0, 1, 2, \ldots$.

5-43. Using the definition of $\hat{a}_-$ given in Problem 5–39 and the fact that $\hat{a}_-\psi_0 = 0$, determine the unnormalized wave function $\psi_0(x)$. Now determine the unnormalized wave function $\psi_1(x)$ using the operator $\hat{a}_+$.

Problems 5–44 through 5–47 apply the idea of reduced mass to the hydrogen atom.

5-44. Given the development of the concept of reduced mass in Section 5–2, how do you think the energy of a hydrogen atom (Equation 1.22) will change if we do not assume that the proton is fixed at the origin?

5-45. In Example 1–8, we calculated the value of the Rydberg constant to be $109\,737$ cm^{-1}. What is the calculated value if we replace m_e in Equation 1.25 by the reduced mass? Compare your answer with the experimental result, $109\,677.6$ cm^{-1}.

5-46. Calculate the reduced mass of a deuterium atom. Take the mass of a deuteron to be $3.343\,586 \times 10^{-27}$ kg. What is the value of the Rydberg constant for a deuterium atom?

5-47. Calculate the ratio of the frequencies of the lines in the spectra of atomic deuterium and atomic hydrogen.

Niels Bohr was born in Copenhagen, Denmark, on October 7, 1885 and died there in 1962.
In 1911, Bohr received his Ph.D. in physics from the University of Copenhagen. He then
spent a year with J.J. Thompson and Ernest Rutherford in England, where he formulated
his theory of the hydrogen atom and its atomic spectrum. In 1913, he returned to the University
of Copenhagen, where he remained for the rest of his life. In 1920, he was named director of
the Institute of Theoretical Physics, which was supported largely by the Carlsberg brewery.
The Institute was an international center for theoretical physics during the 1920s and 1930s,
when quantum mechanics was being developed. Almost every scientist who was active in the
development of quantum mechanics worked at Bohr's institute at one time or another. Bohr
later worked on the theory of the structure of the nucleus and its application to nuclear fission.
In 1943, because the Germans planned to arrest him to force him to work in their atomic bomb
project, Bohr and his family fled to England under great secrecy and spent the remaining war
years in the United States, where he participated in the Manhatten Project at Los Alamos.
After the war, Bohr worked energetically for peaceful uses of atomic energy. He organized the
first Atoms for Peace Conference in 1955 and received the first Atoms for Peace prize in 1957.
Bohr was awarded the Nobel Prize for physics in 1922 "for his investigation of the structure
of atoms and of the radiation emanating from them."

CHAPTER 6

The Hydrogen Atom

We are now ready to study the hydrogen atom, which is of particular interest to chemists because it serves as the prototype for more complex atoms and, therefore, molecules. In addition, probably every chemistry student has studied the results of a quantum-mechanical treatment of the hydrogen atom in general chemistry, and in this chapter we will see the familiar hydrogen atomic orbitals and their properties emerge naturally as solutions to the Schrödinger equation.

6–1. The Schrödinger Equation for the Hydrogen Atom Can Be Solved Exactly

For our model of a hydrogen atom, we will picture it as a proton fixed at the origin and an electron of mass m_e interacting with the proton through a Coulombic potential:

$$V(r) = -\frac{e^2}{4\pi\varepsilon_0 r} \tag{6.1}$$

where e is the charge on the proton, ε_0 is the permittivity of free space, and r is the distance between the electron and the proton. (We consider the case in which the nucleus is not fixed at the origin in Problem 6–35.) The factor $4\pi\varepsilon_0$ arises because we are using SI units. The spherical geometry of the model suggests that we use a spherical coordinate system with the proton at the origin. The Hamiltonian operator for a hydrogen atom is

$$\hat{H} = -\frac{\hbar^2}{2m_e}\nabla^2 - \frac{e^2}{4\pi\varepsilon_0 r} \tag{6.2}$$

where ∇^2 is the Laplacian operator in spherical coordinates (Equation 5.49):

$$\nabla^2 = \frac{1}{r^2} \frac{\partial}{\partial r} \left(r^2 \frac{\partial}{\partial r} \right) + \frac{1}{r^2 \sin \theta} \frac{\partial}{\partial \theta} \left(\sin \theta \frac{\partial}{\partial \theta} \right) + \frac{1}{r^2 \sin^2 \theta} \frac{\partial^2}{\partial \phi^2} \tag{6.3}$$

If we substitute Equation 6.3 into Equation 6.2, the Schrödinger equation for a hydrogen atom becomes

$$-\frac{\hbar^2}{2m_e} \left[\frac{1}{r^2} \frac{\partial}{\partial r} \left(r^2 \frac{\partial \psi}{\partial r} \right) + \frac{1}{r^2 \sin \theta} \frac{\partial}{\partial \theta} \left(\sin \theta \frac{\partial \psi}{\partial \theta} \right) + \frac{1}{r^2 \sin^2 \theta} \frac{\partial^2 \psi}{\partial \phi^2} \right]$$

$$-\frac{e^2}{4\pi \varepsilon_0 r} \psi(r, \theta, \phi) = E \psi(r, \theta, \phi) \tag{6.4}$$

At first sight, this partial differential equation looks exceedingly complicated. To bring Equation 6.4 into a more manageable form, first multiply through by $2m_e r^2$ to obtain

$$-\hbar^2 \left(\frac{\partial}{\partial r} r^2 \frac{\partial \psi}{\partial r} \right) - \hbar^2 \left[\frac{1}{\sin \theta} \left(\frac{\partial}{\partial \theta} \sin \theta \frac{\partial \psi}{\partial \theta} \right) + \frac{1}{\sin^2 \theta} \frac{\partial^2 \psi}{\partial \phi^2} \right]$$

$$- 2m_e r^2 \left[\frac{e^2}{4\pi \varepsilon_0 r} + E \right] \psi(r, \theta, \phi) = 0 \tag{6.5}$$

Notice that all the θ and ϕ dependence in Equation 6.5 occurs within the first large square brackets. The form of Equation 6.5 suggests that we use separation of variables and let

$$\psi(r, \theta, \phi) = R(r) Y(\theta, \phi) \tag{6.6}$$

If we substitute Equation 6.6 into Equation 6.5 and divide by $R(r) Y(\theta, \phi)$, we obtain

$$-\frac{\hbar^2}{R(r)} \left[\frac{d}{dr} \left(r^2 \frac{dR}{dr} \right) + \frac{2m_e r^2}{\hbar^2} \left(\frac{e^2}{4\pi \varepsilon_0 r} + E \right) R(r) \right]$$

$$-\frac{\hbar^2}{Y(\theta, \phi)} \left[\frac{1}{\sin \theta} \frac{\partial}{\partial \theta} \left(\sin \theta \frac{\partial Y}{\partial \theta} \right) + \frac{1}{\sin^2 \theta} \frac{\partial^2 Y}{\partial \phi^2} \right] = 0 \tag{6.7}$$

The terms in the first set of brackets are functions of only r, whereas the terms in the second set of brackets are functions of only θ and ϕ. Because r, θ, and ϕ are independent variables, we may write

$$-\frac{1}{R(r)} \left[\frac{d}{dr} \left(r^2 \frac{dR}{dr} \right) + \frac{2m_e r^2}{\hbar^2} \left(\frac{e^2}{4\pi \varepsilon_0 r} + E \right) R(r) \right] = -\beta \tag{6.8}$$

and

$$-\frac{1}{Y(\theta, \phi)} \left[\frac{1}{\sin \theta} \frac{\partial}{\partial \theta} \left(\sin \theta \frac{\partial Y}{\partial \theta} \right) + \frac{1}{\sin^2 \theta} \frac{\partial^2 Y}{\partial \phi^2} \right] = \beta \tag{6.9}$$

where β is a constant (we have incorporated $\hbar^2$ into β.). If we multiply Equation 6.9 through by the product of $\sin^2\theta$ and $Y(\theta, \phi)$ to get

$$\sin\theta \frac{\partial}{\partial\theta}\left(\sin\theta \frac{\partial Y}{\partial\theta}\right) + \frac{\partial^2 Y}{\partial\phi^2} + (\beta \sin^2\theta)Y = 0 \tag{6.10}$$

we see that Equation 6.10 is exactly the same as Equation 5.55, the equation for the rigid-rotator wave functions. Thus, the angular parts of hydrogen atomic orbitals are also rigid-rotator wave functions. Equation 6.8, which gives the radial dependence of hydrogen atomic orbitals, is called the *radial equation*. We will discuss the angular solutions first and then the radial solutions.

6–2. The Wave Functions of a Rigid Rotator Are Called Spherical Harmonics

To solve Equation 6.10, we again use the method of separation of variables and let

$$Y(\theta, \phi) = \Theta(\theta)\Phi(\phi) \tag{6.11}$$

If we substitute Equation 6.11 into Equation 6.10 and divide by $\Theta(\theta)\Phi(\phi)$, we find

$$\frac{\sin\theta}{\Theta(\theta)} \frac{d}{d\theta}\left(\sin\theta \frac{d\Theta}{d\theta}\right) + \beta \sin^2\theta + \frac{1}{\Phi(\phi)} \frac{d^2\Phi}{d\phi^2} = 0 \tag{6.12}$$

Because θ and ϕ are independent variables, we must have that

$$\frac{\sin\theta}{\Theta(\theta)} \frac{d}{d\theta}\left(\sin\theta \frac{d\Theta}{d\theta}\right) + \beta \sin^2\theta = m^2 \tag{6.13}$$

and

$$\frac{1}{\Phi(\phi)} \frac{d^2\Phi}{d\phi^2} = -m^2 \tag{6.14}$$

where m^2 is a constant. We use m^2 as a separation constant in anticipation of using the square root of the separation constant in later equations.

Because Equation 6.14 contains only constant coefficients, it is relatively easy to solve. Its solutions are

$$\Phi(\phi) = A_m e^{im\phi} \quad \text{and} \quad \Phi(\phi) = A_{-m}e^{-im\phi} \tag{6.15}$$

The requirement that $\Phi(\phi)$ be a single-valued function of ϕ is that

$$\Phi(\phi + 2\pi) = \Phi(\phi) \tag{6.16}$$

By substituting Equation 6.15 into Equation 6.16, we see that

$$A_m e^{im(\phi+2\pi)} = A_m e^{im\phi} \tag{6.17}$$

and that

$$A_{-m} e^{-im(\phi+2\pi)} = A_{-m} e^{-im\phi} \tag{6.18}$$

Equations 6.17 and 6.18 together imply that

$$e^{\pm i2\pi m} = 1 \tag{6.19}$$

In terms of sines and cosines, Equation 6.19 is (Equation A.6)

$$\cos(2\pi m) \pm i\sin(2\pi m) = 1$$

which implies that $m = 0, \pm 1, \pm 2, \ldots$, because $\cos 2\pi m = 1$ and $\sin 2\pi m = 0$ for $m = 0, \pm 1, \pm 2, \ldots$. Thus Equation 6.15 can be written as one equation

$$\Phi_m(\phi) = A_m e^{im\phi} \qquad m = 0, \pm 1, \pm 2, \ldots \tag{6.20}$$

We can find the value of A_m by requiring that the $\Phi_m(\phi)$ be normalized.

EXAMPLE 6–1
Determine the value of A_m so that the functions given by Equation 6.20 are normalized.

SOLUTION: The normalization condition is that

$$\int_0^{2\pi} \Phi_m^*(\phi)\Phi_m(\phi)d\phi = 1$$

Using Equation 6.20 for the $\Phi_m(\phi)$, we have

$$|A_m|^2 \int_0^{2\pi} d\phi = 1$$

or

$$|A_m|^2 2\pi = 1$$

or

$$A_m = (2\pi)^{-1/2}$$

Thus, the normalized functions of Equation 6.20 are

$$\Phi_m(\phi) = \frac{1}{(2\pi)^{1/2}} e^{im\phi} \qquad m = 0, \pm 1, \pm 2, \ldots \tag{6.21}$$

The differential equation for $\Theta(\theta)$, Equation 6.13, is not as easy to solve because it does not have constant coefficients. It is convenient to let $x = \cos\theta$ and $\Theta(\theta) = P(x)$ in Equation 6.13. (This x should not be confused with the Cartesian coordinate, x.) Because $0 \leq \theta \leq \pi$, the range of x is $-1 \leq x \leq +1$. Under the change of variable, $x = \cos\theta$, Equation 6.13 becomes (Problem 6–2)

$$(1 - x^2)\frac{d^2 P}{dx^2} - 2x\frac{dP}{dx} + \left[\beta - \frac{m^2}{1 - x^2}\right] P(x) = 0 \qquad (6.22)$$

with $m = 0, \pm 1, \pm 2, \ldots$. Equation 6.22 for $P(x)$ is called *Legendre's equation* and is a well-known equation in classical physics. It occurs in a variety of problems formulated in spherical coordinates. When Equation 6.22 is solved, it is found that β must equal $l(l + 1)$ with $l = 0, 1, 2, \ldots$ and that $|m| \leq l$, where $|m|$ denotes the magnitude of m, if the solutions are to remain finite. Thus, Equation 6.22 can be written as

$$(1 - x^2)\frac{d^2 P}{dx^2} - 2x\frac{dP}{dx} + \left[l(l + 1) - \frac{m^2}{1 - x^2}\right] P(x) = 0 \qquad (6.23)$$

with $l = 0, 1, 2, \ldots$ and $m = 0, \pm 1, \pm 2, \ldots, \pm l$.

The solutions to Equation 6.23 are most easily discussed by considering the $m = 0$ case first. When $m = 0$, the solutions to Equation 6.23 are called *Legendre polynomials* and are denoted by $P_l(x)$. Legendre polynomials arise in a number of physical problems. The first few Legendre polynomials are given in Table 6.1.

TABLE 6.1
The first few Legendre polynomials, which are the solutions to Equation 6.23 with $m = 0$. The subscript indexing the Legendre polynomials is the value of l in Equation 6.23.

$$P_0(x) = 1$$

$$P_1(x) = x$$

$$P_2(x) = \tfrac{1}{2}(3x^2 - 1)$$

$$P_3(x) = \tfrac{1}{2}(5x^3 - 3x)$$

$$P_4(x) = \tfrac{1}{8}(35x^4 - 30x^2 + 3)$$

EXAMPLE 6–2
Prove that the first three Legendre polynomials satisfy Equation 6.23 when $m = 0$.

SOLUTION: Equation 6.23 with $m = 0$ is

$$(1 - x^2)\frac{d^2 P}{dx^2} - 2x\frac{dP}{dx} + l(l + 1)P(x) = 0 \qquad (1)$$

The first Legendre polynomial $P_0(x) = 1$ is clearly a solution of Equation 1 with $l = 0$. When we substitute $P_1(x) = x$ into Equation 1 with $l = 1$, we obtain

$$-2x + 1(2)x = 0$$

and so $P_1(x)$ is a solution. For $P_2(x)$, Equation 1 is

$$(1 - x^2)(3) - 2x(3x) + 2(3)[\tfrac{1}{2}(3x^2 - 1)] = (3 - 3x^2) - 6x^2 + (9x^2 - 3) = 0$$

Notice from Table 6.1 that $P_l(x)$ is an even function if l is even and an odd function if l is odd. The factors in front of the $P_l(x)$ are chosen such that $P_l(1) = 1$. In addition, although we will not prove it, it can be shown generally that the $P_l(x)$ in Table 6.1 are orthogonal, or that

$$\int_{-1}^{1} P_l(x) P_n(x) dx = 0 \qquad l \neq n \tag{6.24}$$

Keep in mind here that the limits on x correspond to the natural, physical limits on θ (0 to π) in spherical coordinates because $x = \cos\theta$ (Problem 6–4). The Legendre polynomials are normalized by the general relation

$$\int_{-1}^{1} [P_l(x)]^2 dx = \frac{2}{2l + 1} \tag{6.25}$$

Equation 6.25 shows that the normalization constant of $P_l(x)$ is $[(2l + 1)/2]^{1/2}$.

Although the Legendre polynomials arise only in the case $m = 0$, they are customarily studied first because the solutions for the $m \neq 0$ case, called *associated Legendre functions*, are defined in terms of the Legendre polynomials. If we denote the associated Legendre functions by $P_l^{|m|}(x)$, then their defining relation is

$$P_l^{|m|}(x) = (1 - x^2)^{|m|/2} \frac{d^{|m|}}{dx^{|m|}} P_l(x) \tag{6.26}$$

Note that only the magnitude of m is relevant here because the defining differential equation, Equation 6.23, depends on only m^2. Because the leading term in $P_l(x)$ is x^l, Equation 6.26 shows that $P_l^{|m|}(x) = 0$ if $m > l$. The first few associated Legendre functions (Problem 6–6) are given in Table 6.2.

Before we discuss a few of the properties of the associated Legendre functions, let's be sure to realize that θ and not x is the variable of physical interest. Table 6.2 also lists the associated Legendre functions in terms of $\cos\theta$ and $\sin\theta$. Note that the factors $(1 - x^2)^{1/2}$ in Table 6.2 become $\sin\theta$ when the associated Legendre functions are expressed in the variable θ. Because $x = \cos\theta$, Equations 6.24 and 6.25 are

$$\int_{-1}^{1} P_l(x) P_n(x) dx = \int_{0}^{\pi} P_l(\cos\theta) P_n(\cos\theta) \sin\theta\, d\theta = \frac{2\delta_{ln}}{2l + 1} \tag{6.27}$$

TABLE 6.2
The first few associated Legendre functions $P_l^{|m|}(x)$

$P_0^0(x) = 1$

$P_1^0(x) = x = \cos\theta$

$P_1^1(x) = (1 - x^2)^{1/2} = \sin\theta$

$P_2^0(x) = \frac{1}{2}(3x^2 - 1) = \frac{1}{2}(3\cos^2\theta - 1)$

$P_2^1(x) = 3x(1 - x^2)^{1/2} = 3\cos\theta\sin\theta$

$P_2^2(x) = 3(1 - x^2) = 3\sin^2\theta$

$P_3^0(x) = \frac{1}{2}(5x^3 - 3x) = \frac{1}{2}(5\cos^3\theta - 3\cos\theta)$

$P_3^1(x) = \frac{3}{2}(5x^2 - 1)(1 - x^2)^{1/2} = \frac{3}{2}(5\cos^2\theta - 1)\sin\theta$

$P_3^2(x) = 15x(1 - x^2) = 15\cos\theta\sin^2\theta$

$P_3^3(x) = 15(1 - x^2)^{3/2} = 15\sin^3\theta$

Because the differential volume element in spherical coordinates is $d\tau = r^2\sin\theta\,dr\,d\theta\,d\phi$, we see that the factor $\sin\theta\,d\theta$ in Equation 6.27 is the "θ part" of $d\tau$ in spherical coordinates.

The associated Legendre functions satisfy the relation

$$\int_{-1}^{1} P_l^{|m|}(x) P_n^{|m|}(x)dx = \int_{0}^{\pi} P_l^{|m|}(\cos\theta) P_n^{|m|}(\cos\theta)\sin\theta\,d\theta$$

$$= \frac{2}{(2l + 1)}\frac{(l + |m|)!}{(l - |m|)!}\delta_{ln} \tag{6.28}$$

(Remember that $0! = 1$.) Equation 6.28 can be used to show that the normalization constant of the associated Legendre functions is

$$N_{lm} = \left[\frac{(2l + 1)}{2}\frac{(l - |m|)!}{(l + |m|)!}\right]^{1/2} \tag{6.29}$$

EXAMPLE 6–3
Use Equation 6.28 in both the x and θ variables and Table 6.2 to prove that $P_1^1(x)$ and $P_2^1(x)$ are orthogonal.

SOLUTION: According to Equation 6.28, we must prove that

$$\int_{-1}^{1} P_1^1(x) P_2^1(x)dx = 0$$

From Table 6.2, we have

$$\int_{-1}^{1} [(1 - x^2)^{1/2}][3x(1 - x^2)^{1/2}]dx = 3\int_{-1}^{1} x(1 - x^2)dx = 0$$

In terms of θ, we have from Equation 6.28 and Table 6.2

$$\int_{0}^{\pi} \sin\theta(3\cos\theta\sin\theta)\sin\theta d\theta = 3\int_{0}^{\pi} \sin^3\theta\cos\theta d\theta = 0$$

Returning to the original problem now, the solutions to Equation 6.10, which are not only the angular part of the hydrogen atomic orbitals but also the rigid-rotator wave functions, are $P_l^{|m|}(\cos\theta)\Phi_m(\phi)$. By referring to Equations 6.21 and 6.29, we see that the normalized functions

$$Y_l^m(\theta, \phi) = \left[\frac{(2l + 1)}{4\pi}\frac{(l - |m|)!}{(l + |m|)!}\right]^{1/2} P_l^{|m|}(\cos\theta)e^{im\phi} \qquad (6.30)$$

with $l = 0, 1, 2, \ldots$ and $m = 0, \pm 1, \pm 2, \ldots, \pm l$ satisfy Equation 6.10. The $Y_l^m(\theta, \phi)$ form an orthonormal set

$$\int_{0}^{\pi} d\theta\sin\theta \int_{0}^{2\pi} d\phi\, Y_l^m(\theta, \phi)^* Y_n^k(\theta, \phi) = \delta_{ln}\delta_{mk} \qquad (6.31)$$

Note that the $Y_l^m(\theta, \phi)$ are orthonormal with respect to $\sin\theta d\theta d\phi$, the angular part of the spherical coordinate volume element, and not just $d\theta d\phi$ (MathChapter D). According to Equation 6.31, the $Y_l^m(\theta, \phi)$ are orthonormal over the surface of a sphere and so are called *spherical harmonics*. The first few spherical harmonics are given in Table 6.3.

TABLE 6.3
The first few spherical harmonics.

$$Y_0^0 = \frac{1}{(4\pi)^{1/2}} \qquad\qquad Y_1^0 = \left(\frac{3}{4\pi}\right)^{1/2}\cos\theta$$

$$Y_1^1 = \left(\frac{3}{8\pi}\right)^{1/2}\sin\theta e^{i\phi} \qquad\qquad Y_1^{-1} = \left(\frac{3}{8\pi}\right)^{1/2}\sin\theta e^{-i\phi}$$

$$Y_2^0 = \left(\frac{5}{16\pi}\right)^{1/2}(3\cos^2\theta - 1) \qquad\qquad Y_2^1 = \left(\frac{15}{8\pi}\right)^{1/2}\sin\theta\cos\theta e^{i\phi}$$

$$Y_2^{-1} = \left(\frac{15}{8\pi}\right)^{1/2}\sin\theta\cos\theta e^{-i\phi} \qquad\qquad Y_2^2 = \left(\frac{15}{32\pi}\right)^{1/2}\sin^2\theta e^{2i\phi}$$

$$Y_2^{-2} = \left(\frac{15}{32\pi}\right)^{1/2}\sin^2\theta e^{-2i\phi}$$

EXAMPLE 6–4

Show that $Y_1^{-1}(\theta, \phi)$ is normalized and that it is orthogonal to $Y_2^1(\theta, \phi)$.

SOLUTION: Using $Y_1^{-1}(\theta, \phi)$ from Table 6.3, the normalization condition is

$$\int_0^\pi d\theta \sin\theta \int_0^{2\pi} d\phi\, Y_1^{-1}(\theta, \phi)^* Y_1^{-1}(\theta, \phi) = \frac{3}{8\pi} \int_0^\pi d\theta \sin\theta \sin^2\theta \int_0^{2\pi} d\phi \overset{?}{=} 1$$

Letting $x = \cos\theta$, we have

$$\frac{3}{8\pi} \cdot 2\pi \int_{-1}^1 (1 - x^2) dx = \frac{3}{4}\left(2 - \frac{2}{3}\right) = 1$$

The orthogonality condition is

$$\int_0^\pi d\theta \sin\theta \int_0^{2\pi} d\phi\, Y_2^1(\theta, \phi)^* Y_1^{-1}(\theta, \phi)$$

$$= \left(\frac{15}{8\pi}\right)^{1/2}\left(\frac{3}{8\pi}\right)^{1/2} \int_0^\pi d\theta \sin\theta \int_0^{2\pi} d\phi\, (e^{-i\phi}\sin\theta\cos\theta)(e^{-i\phi}\sin\theta)$$

$$= \left(\frac{45}{64\pi^2}\right)^{1/2} \int_0^\pi d\theta \sin^3\theta\cos\theta \int_0^{2\pi} d\phi\, e^{-2i\phi}$$

The integral over ϕ is zero because it is the integral of $\cos 2\phi$ and $\sin 2\phi$ over complete cycles. Thus, we see that $Y_1^{-1}(\theta, \phi)$ and $Y_2^1(\theta, \phi)$ are orthogonal.

According to Equation 5.52, the quantum-mechanical operator corresponding to the square of the angular momentum is

$$\hat{L}^2 = -\hbar^2 \left[\frac{1}{\sin\theta}\frac{\partial}{\partial\theta}\left(\sin\theta\frac{\partial}{\partial\theta}\right) + \frac{1}{\sin^2\theta}\frac{\partial^2}{\partial\phi^2}\right] \tag{6.32}$$

which is essentially the operator given in the square brackets in Equation 6.9. If we multiply both sides of Equation 6.9 with $\beta = l(l+1)$ by $\hbar^2 Y(\theta, \phi)$, we see that the spherical harmonics satisfy

$$\hat{L}^2 Y_l^m(\theta, \phi) = \hbar^2 l(l+1) Y_l^m(\theta, \phi) \tag{6.33}$$

Thus, we see that the spherical harmonics are also eigenfunctions of $\hat{L}^2$ and that the square of the angular momentum can have only the values given by

$$L^2 = \hbar^2 l(l+1) \qquad l = 0,\ 1,\ 2,\ \ldots \tag{6.34}$$

Because $\hat{H} = \hat{L}^2/2I$ for a rigid rotator (Equation 5.51), we also have

$$\hat{H} Y_l^m(\theta, \phi) = \frac{\hbar^2 l(l+1)}{2I} Y_l^m(\theta, \phi) \tag{6.35}$$

for a rigid rotator.

EXAMPLE 6–5

Show that $Y_1^1(\theta, \phi)$ is an eigenfunction of $\hat{L}^2$.

SOLUTION: From Table 6.3, we have

$$Y_1^1(\theta, \phi) = \left(\frac{3}{8\pi}\right)^{1/2} \sin\theta e^{i\phi}$$

The "θ" part of the differential operator in brackets in Equation 6.32 gives $e^{i\phi}(\cos^2\theta - \sin^2\theta)/\sin\theta$ and the "ϕ" part gives $-e^{i\phi}/\sin\theta$. If we add these two results, we get

$$\frac{(\cos^2\theta - \sin^2\theta - 1)e^{i\phi}}{\sin\theta} = -\frac{2e^{i\phi}\sin^2\theta}{\sin\theta} = -2e^{i\phi}\sin\theta$$

Therefore,

$$\hat{L}^2 Y_1^1(\theta, \phi) = 2\hbar^2 Y_1^1(\theta, \phi)$$

which is Equation 6.33 with $l = 1$.

6–3. Precise Values of the Three Components of Angular Momentum Cannot Be Measured Simultaneously

In this section, we will explore some of the quantum-mechanical properties of angular momentum. Recall that angular momentum is a vector quantity. The quantum-mechanical operators corresponding to the three components of angular momentum are given in Table 4.1. These operators are obtained from the classical expressions (Equation 4.7) by replacing the classical momenta by their quantum-mechanical equivalents to get

$$\hat{L}_x = y\hat{P}_z - z\hat{P}_y = -i\hbar\left(y\frac{\partial}{\partial z} - z\frac{\partial}{\partial y}\right)$$

$$\hat{L}_y = z\hat{P}_x - x\hat{P}_z = -i\hbar\left(z\frac{\partial}{\partial x} - x\frac{\partial}{\partial z}\right) \tag{6.36}$$

$$\hat{L}_z = x\hat{P}_y - y\hat{P}_x = -i\hbar\left(x\frac{\partial}{\partial y} - y\frac{\partial}{\partial x}\right)$$

Through a straightforward, but somewhat tedious, exercise in partial differentiation, we can convert Equations 6.36 into spherical coordinates (Problems 6–11 and 6–12)

to obtain

$$\hat{L}_x = -i\hbar \left(-\sin\phi \frac{\partial}{\partial\theta} - \cot\theta \cos\phi \frac{\partial}{\partial\phi} \right)$$

$$\hat{L}_y = -i\hbar \left(\cos\phi \frac{\partial}{\partial\theta} - \cot\theta \sin\phi \frac{\partial}{\partial\phi} \right) \qquad (6.37)$$

$$\hat{L}_z = -i\hbar \frac{\partial}{\partial\phi}$$

The equation for $\hat{L}_z$ turns out to be relatively simple. We can easily see that $e^{im\phi}$ is an eigenfunction of $\hat{L}_z$ or that

$$\hat{L}_z(e^{im\phi}) = -i\hbar \frac{\partial}{\partial\phi}(e^{im\phi}) = m\hbar(e^{im\phi})$$

All the ϕ dependence of the spherical harmonics occurs in the factor $e^{im\phi}$, and so the spherical harmonics are eigenfunctions of $\hat{L}_z$:

$$\begin{aligned} \hat{L}_z Y_l^m(\theta,\phi) &= N_{lm} \hat{L}_z P_l^{|m|}(\cos\theta)e^{im\phi} \\ &= N_{lm} P_l^{|m|}(\cos\theta)\hat{L}_z e^{im\phi} \\ &= \hbar m Y_l^m(\theta,\phi) \end{aligned} \qquad (6.38)$$

Equation 6.38 shows that measured values of L_z are integral multiples of $\hbar$. Notice that $\hbar$ is a fundamental measure of the angular momentum of a quantum-mechanical system.

The spherical harmonics are not eigenfunctions of $\hat{L}_x$ or $\hat{L}_y$, as the following example shows.

EXAMPLE 6–6
Use Equation 6.37 to show that $Y_1^{-1}(\theta,\phi)$ is not an eigenfunction of $\hat{L}_x$.

SOLUTION: From Table 6.3, $Y_1^{-1}(\theta,\phi) = (3/8\pi)^{1/2} \sin\theta e^{-i\phi}$. Using the first of Equations 6.37, we have

$$\hat{L}_x Y_1^{-1}(\theta,\phi) = -i\hbar \left(\frac{3}{8\pi} \right)^{1/2} [-\sin\phi \cos\theta e^{-i\phi} + i\cot\theta \cos\phi \sin\theta e^{-i\phi}]$$

$$= -i\hbar \left(\frac{3}{8\pi} \right)^{1/2} \cos\theta(-\sin\phi + i\cos\phi)e^{-i\phi}$$

But the term in parentheses is

$$-\sin\phi + i\cos\phi = -\frac{(e^{i\phi} - e^{-i\phi})}{2i} + i\frac{(e^{i\phi} + e^{-i\phi})}{2}$$

$$= +\frac{i}{2}(e^{i\phi} - e^{-i\phi}) + \frac{i}{2}(e^{i\phi} + e^{-i\phi}) = ie^{i\phi}$$

Therefore,

$$\hat{L}_x Y_1^{-1}(\theta, \phi) = \hbar \left(\frac{3}{8\pi}\right)^{1/2} \cos\theta = \frac{\hbar}{2^{1/2}} Y_1^0(\theta, \phi)$$

and $Y_1^{-1}(\theta, \phi)$ is not an eigenfunction of $\hat{L}_x$. Note that

$$\langle \hat{L}_x \rangle = \int_0^\pi d\theta \sin\theta \int_0^{2\pi} d\phi \, Y_1^{-1}(\theta, \phi)^* \hat{L}_x Y_1^{-1}(\theta, \phi)$$

$$= \frac{\hbar}{2^{1/2}} \int_0^{2\pi} d\phi \int_0^\pi d\theta \sin\theta \, Y_1^{-1}(\theta, \phi)^* Y_1^0(\theta, \phi) = 0$$

because of the orthogonality of $Y_1^{-1}(\theta, \phi)$ and $Y_1^0(\theta, \phi)$.

Equation 6.33 shows that the $Y_l^m(\theta, \phi)$ are eigenfunctions of $\hat{L}^2$. Because the spherical harmonics are eigenfunctions of both $\hat{L}^2$ and $\hat{L}_z$, we can determine precise values of L^2 and L_z simultaneously (Section 4–6), which implies that the operators $\hat{L}^2$ and $\hat{L}_z$ commute.

EXAMPLE 6–7
Prove that the operators $\hat{L}^2$ and $\hat{L}_z$ commute.

SOLUTION: Using $\hat{L}^2$ from Equation 6.32 and $\hat{L}_z$ from Equation 6.37, we have

$$\hat{L}^2 \hat{L}_z f = -\hbar^2 \left[\frac{1}{\sin\theta}\frac{\partial}{\partial\theta}\left(\sin\theta\frac{\partial}{\partial\theta}\right) + \frac{1}{\sin^2\theta}\frac{\partial^2}{\partial\phi^2}\right]\left(-i\hbar\frac{\partial f}{\partial\phi}\right)$$

$$= i\hbar^3 \left[\frac{1}{\sin\theta}\frac{\partial}{\partial\theta}\left(\sin\theta\frac{\partial^2 f}{\partial\theta\partial\phi}\right) + \frac{1}{\sin^2\theta}\frac{\partial^3 f}{\partial\phi^3}\right]$$

and

$$\hat{L}_z \hat{L}^2 f = \left(-i\hbar\frac{\partial}{\partial\phi}\right)\left\{-\hbar^2\left[\frac{1}{\sin\theta}\frac{\partial}{\partial\theta}\left(\sin\theta\frac{\partial}{\partial\theta}\right) + \frac{1}{\sin^2\theta}\frac{\partial^2}{\partial\phi^2}\right]\right\} f$$

$$= i\hbar^3 \left[\frac{1}{\sin\theta}\frac{\partial}{\partial\theta}\left(\sin\theta\frac{\partial^2 f}{\partial\phi\partial\theta}\right) + \frac{1}{\sin^2\theta}\frac{\partial^3 f}{\partial\phi^3}\right]$$

where in writing the last line here we have recognized that $(\partial/\partial\phi)$ does not affect terms involving θ. Because

$$\frac{\partial^2 f}{\partial\theta\partial\phi} = \frac{\partial^2 f}{\partial\phi\partial\theta}$$

for any function well enough behaved to be a wave function, we see that

$$\hat{L}^2 \hat{L}_z f = \hat{L}_z \hat{L}^2 f$$

or that

$$[\hat{L}^2, \hat{L}_z] = 0$$

because f is arbitrary.

We can use Equations 6.33 and 6.38 to prove that $|m| \le l$, or that $m = 0, \pm 1, \pm 2, \ldots, \pm l$. It follows from Equation 6.38 that

$$\hat{L}_z^2 Y_l^m(\theta, \phi) = m^2 \hbar^2 Y_l^m(\theta, \phi) \tag{6.39}$$

and because

$$\hat{L}^2 Y_l^m(\theta, \phi) = l(l + 1)\hbar^2 Y_l^m(\theta, \phi)$$

we obtain

$$(\hat{L}^2 - \hat{L}_z^2) Y_l^m(\theta, \phi) = [l(l + 1) - m^2] Y_l^m(\theta, \phi)$$

Furthermore, because

$$\hat{L}^2 = \hat{L}_x^2 + \hat{L}_y^2 + \hat{L}_z^2$$

then

$$(\hat{L}^2 - \hat{L}_z^2) Y_l^m(\theta, \phi) = (\hat{L}_x^2 + \hat{L}_y^2) Y_l^m(\theta, \phi) = [l(l + 1) - m^2]\hbar^2 Y_l^m(\theta, \phi) \tag{6.40}$$

Thus, the observed values of $L_x^2 + L_y^2$ are $[l(l + 1) - m^2]\hbar^2$. But because $L_x^2 + L_y^2$ is the sum of two squared terms, it cannot be negative, and so we have that

$$[l(l + 1) - m^2]\hbar^2 \ge 0$$

or that

$$l(l + 1) \ge m^2 \tag{6.41}$$

Because l and m are integers, Equation 6.41 says that

$$|m| \le l$$

or that the only possible values of the integer m are

$$m = 0, \pm 1, \pm 2, \ldots, \pm l \tag{6.42}$$

This result might be familiar as the condition of the magnetic quantum number associated with the hydrogen atom.

Equation 6.42 shows that there are $2l + 1$ values of m for each value of l. Let's look at the case of $l = 1$ for which $l(l + 1) = 2$. Because $l = 1$, m can have only the values 0 and ± 1. Using the equations

$$\hat{L}^2 Y_1^m(\theta, \phi) = 1(1 + 1)\hbar^2 Y_1^m(\theta, \phi) = 2\hbar^2 Y_1^m(\theta, \phi) \qquad m = 0, \pm 1$$

and

$$\hat{L}_z Y_1^m(\theta, \phi) = m\hbar Y_1^m(\theta, \phi) \qquad m = 0, \pm 1$$

we see that

$$|L| = (L^2)^{1/2} = \sqrt{2}\,\hbar$$

and

$$L_z = -\hbar, \ 0, \ +\hbar$$

where $|L|$ is the magnitude of the angular-momentum vector. Note that the maximum value of L_z is less than $|L|$, which implies that $\mathbf{L}$ and L_z cannot point in the same direction. This is illustrated in Figure 6.1, which shows L_z with a value $+\hbar$ and $|L|$ with its value $\sqrt{2}\,\hbar$. Now let's try to specify L_x and L_y. Problem 6–13 has you prove that $\hat{L}_x$, $\hat{L}_y$, and $\hat{L}_z$ commute with $\hat{L}^2$ but do not commute among themselves. This result implies that although it is possible to observe precise values of L^2 and one of the components of angular momentum simultaneously, it is not possible to observe precise values of the other two components at the same time. For example, we may observe

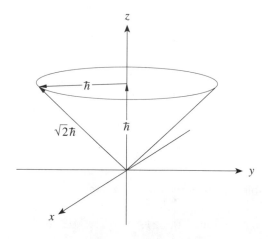

FIGURE 6.1

The $m = +1$ component of the angular-momentum state $l = 1$. The angular momentum describes a cone because the x and y components cannot be specified. The projection of the motion onto the $x–y$ plane is a circle of radius $\hbar$ centered at the origin (Example 6–8).

precise values of L^2 and L_z simultaneously (as shown above); in this case, it is not possible to observe precise values of either L_x or L_y.

Even though L_x and L_y do not have precise values, they do, of course, have average values, and Problem 6–14 shows that $\langle L_x \rangle = \langle L_y \rangle = 0$ (see also Example 6–6). These results are illustrated in Figure 6.1, which shows L_z with a value of $+\hbar$ and $|L|$ with a value of $\sqrt{2}\hbar$. A nice classical interpretation of these results is that L precesses about the z-axis, mapping out the surface of the cone shown in Figure 6.1. The projection of the motion onto the x–y plane is a circle of radius $\hbar$ centered at the origin.

EXAMPLE 6–8

Show that the projection onto the x–y plane of the motion of the angular momentum vector with $L^2 = 2\hbar^2$ and $L_z = \hbar$ is a circle of radius $\hbar$ in the x–y plane.

SOLUTION: From the cone in Figure 6.1, we see that the x–y projection will be a circle. To determine the radius, r, of the circle, consider the x, z cross section

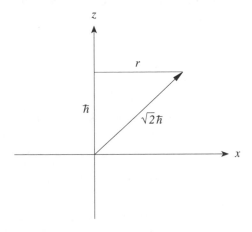

Because we have a right triangle, $r^2 + \hbar^2 = 2\hbar^2$ and so $r = \hbar$. Thus, we also see that while we know the magnitude of the angular momentum and its z-component, we do not know the direction in which the vector $L_x\mathbf{i} + L_y\mathbf{j}$ points.

According to Example 6–8, then, the average values of $\langle L_x \rangle$ and $\langle L_y \rangle$ are zero. This picture is in good accord with the Uncertainty Principle: By specifying L_z exactly, we have a complete uncertainty in the angle ϕ associated with L_z, so the angular momentum vector can lie anywhere on the rim of the cone.

Before leaving this section, we should address ourselves to the question: What is so special about the z direction? The answer is that nothing at all is special about the z direction. We could have chosen either the x or y direction as the unique direction and all the above results would be the same, except for exchanging x or y for z. For example, we can know both L^2 and L_x precisely simultaneously, in which case L_y and L_z do not have precise values. It is customary to choose the z direction because the expression for $\hat{L}_z$ in spherical coordinates is so much simpler than for $\hat{L}_x$ or $\hat{L}_y$ (cf. Equation 6.37). Clearly the rotating system does not know x from y from z and, in fact, this inability to distinguish between the three directions explains the $(2l + 1)$-fold degeneracy.

6–4. Hydrogen Atomic Orbitals Depend upon Three Quantum Numbers

Up to now we have solved Equation 6.9, giving the angular part of the hydrogen atomic orbitals. Now we will solve Equation 6.8, giving the radial part of the hydrogen atomic orbitals. Equation 6.8 with β set equal to $l(l + 1)$ can be written as

$$-\frac{\hbar^2}{2m_e r^2} \frac{d}{dr} \left(r^2 \frac{dR}{dr} \right) + \left[\frac{\hbar^2 l(l + 1)}{2m_e r^2} - \frac{e^2}{4\pi \varepsilon_0 r} - E \right] R(r) = 0 \qquad (6.43)$$

Equation 6.43 is an ordinary differential equation in r. It is somewhat tedious to solve, but once solved, we find that for solutions to be acceptable as the wave functions, the energy must be quantized according to

$$E_n = -\frac{m_e e^4}{8\varepsilon_0^2 h^2 n^2} = -\frac{m_e e^4}{32\pi^2 \varepsilon_0^2 \hbar^2 n^2} \qquad n = 1, 2, \dots \qquad (6.44)$$

If we introduce the Bohr radius from Section 1–8, $a_0 = \varepsilon_0 h^2 / \pi m_e e^2 = 4\pi \varepsilon_0 \hbar^2 / m_e e^2$, then Equation 6.44 becomes

$$E_n = -\frac{e^2}{8\pi \varepsilon_0 a_0 n^2} \qquad n = 1, 2, \dots \qquad (6.45)$$

It is surely remarkable that these are the same energies obtained from the Bohr model of the hydrogen atom. Of course, the electron now is not restricted to the sharply defined orbits of Bohr but is described by its wave function, $\psi(r, \theta, \phi)$.

In the course of solving Equation 6.43, we find not only that an integer n occurs naturally but that n must satisfy the condition that $n \geq l + 1$, which is usually written as

$$0 \leq l \leq n - 1 \qquad n = 1, 2, \dots \qquad (6.46)$$

because we have already seen that the smallest possible value of l is zero. (Equation 6.46 might be familiar from general chemistry.) The solutions to Equation 6.43 depend on two quantum numbers n and l and are given by

$$R_{nl}(r) = -\left\{ \frac{(n - l - 1)!}{2n[(n + l)!]^3} \right\}^{1/2} \left(\frac{2}{na_0} \right)^{l+3/2} r^l e^{-r/na_0} L_{n+l}^{2l+1} \left(\frac{2r}{na_0} \right) \qquad (6.47)$$

where the $L_{n+l}^{2l+1}(2r/na_0)$ are called *associated Laguerre polynomials*. The first few associated Laguerre polynomials are given in Table 6.4.

TABLE 6.4
The first few associated Laguerre polynomials.

$n = 1$,	$l = 0$	$L_1^1(x) = -1$
$n = 2$,	$l = 0$	$L_2^1(x) = -2!(2 - x)$
	$l = 1$	$L_3^3(x) = -3!$
$n = 3$,	$l = 0$	$L_3^1(x) = -3!(3 - 3x + \frac{1}{2}x^2)$
	$l = 1$	$L_4^3(x) = -4!(4 - x)$
	$l = 2$	$L_5^5(x) = -5!$
$n = 4$,	$l = 0$	$L_4^1(x) = -4!(4 - 6x + 2x^2 - \frac{1}{6}x^3)$
	$l = 1$	$L_5^3(x) = -5!(10 - 5x + \frac{1}{2}x^2)$
	$l = 2$	$L_6^5(x) = -6!(6 - x)$
	$l = 3$	$L_7^7(x) = -7!$

The functions given by Equation 6.47 may look complicated, but notice that each one is just a polynomial multiplied by an exponential. The combinatorial factor in front assures that the $R_{nl}(r)$ are normalized with respect to an integration over r, or that the $R_{nl}(r)$ satisfy

$$\int_0^\infty R_{nl}^*(r) R_{nl}(r) r^2 dr = 1 \tag{6.48}$$

Note that the volume element here is $r^2 dr$, which is the "r" part of the spherical coordinate volume element $r^2 \sin\theta dr d\theta d\phi$.

The complete hydrogen atomic wave functions are

$$\psi_{nlm}(r, \theta, \phi) = R_{nl}(r) Y_l^m(\theta, \phi) \tag{6.49}$$

The first few hydrogen atomic wave functions are given in Table 6.5. The normalization condition for hydrogen atomic wave functions is

$$\int_0^\infty dr r^2 \int_0^\pi d\theta \sin\theta \int_0^{2\pi} d\phi \, \psi_{nlm}^*(r, \theta, \phi) \psi_{nlm}(r, \theta, \phi) = 1 \tag{6.50}$$

Because $\hat{H}$ is Hermitian (Section 4–5), the functions ψ_{nlm} must also be orthogonal. This orthogonality relationship is given by

$$\int_0^\infty dr r^2 \int_0^\pi d\theta \sin\theta \int_0^{2\pi} d\phi \, \psi_{n'l'm'}^*(r, \theta, \phi) \psi_{nlm}(r, \theta, \phi) = \delta_{nn'} \delta_{ll'} \delta_{mm'} \tag{6.51}$$

where the δ's are Krönecker deltas (Equation 4.30).

TABLE 6.5

The complete hydrogenlike atomic wave functions for $n = 1$, 2, and 3. The quantity Z is the atomic number of the nucleus, and $\sigma = Zr/a_0$, where a_0 is the Bohr radius.

$n = 1$,	$l = 0$,	$m = 0$	$\psi_{100} = \dfrac{1}{\sqrt{\pi}}\left(\dfrac{Z}{a_0}\right)^{3/2} e^{-\sigma}$
$n = 2$,	$l = 0$,	$m = 0$	$\psi_{200} = \dfrac{1}{\sqrt{32\pi}}\left(\dfrac{Z}{a_0}\right)^{3/2}(2 - \sigma)e^{-\sigma/2}$
	$l = 1$,	$m = 0$	$\psi_{210} = \dfrac{1}{\sqrt{32\pi}}\left(\dfrac{Z}{a_0}\right)^{3/2}\sigma e^{-\sigma/2}\cos\theta$
	$l = 1$,	$m = \pm1$	$\psi_{21\pm1} = \dfrac{1}{\sqrt{64\pi}}\left(\dfrac{Z}{a_0}\right)^{3/2}\sigma e^{-\sigma/2}\sin\theta e^{\pm i\phi}$
$n = 3$,	$l = 0$,	$m = 0$	$\psi_{300} = \dfrac{1}{81\sqrt{3\pi}}\left(\dfrac{Z}{a_0}\right)^{3/2}(27 - 18\sigma + 2\sigma^2)e^{-\sigma/3}$
	$l = 1$,	$m = 0$	$\psi_{310} = \dfrac{1}{81}\left(\dfrac{2}{\pi}\right)^{1/2}\left(\dfrac{Z}{a_0}\right)^{3/2}(6\sigma - \sigma^2)e^{-\sigma/3}\cos\theta$
	$l = 1$,	$m = \pm1$	$\psi_{31\pm1} = \dfrac{1}{81\sqrt{\pi}}\left(\dfrac{Z}{a_0}\right)^{3/2}(6\sigma - \sigma^2)e^{-\sigma/3}\sin\theta e^{\pm i\phi}$
	$l = 2$,	$m = 0$	$\psi_{320} = \dfrac{1}{81\sqrt{6\pi}}\left(\dfrac{Z}{a_0}\right)^{3/2}\sigma^2 e^{-\sigma/3}(3\cos^2\theta - 1)$
	$l = 2$,	$m = \pm1$	$\psi_{32\pm1} = \dfrac{1}{81\sqrt{\pi}}\left(\dfrac{Z}{a_0}\right)^{3/2}\sigma^2 e^{-\sigma/3}\sin\theta\cos\theta e^{\pm i\phi}$
	$l = 2$,	$m = \pm2$	$\psi_{32\pm2} = \dfrac{1}{162\sqrt{\pi}}\left(\dfrac{Z}{a_0}\right)^{3/2}\sigma^2 e^{-\sigma/3}\sin^2\theta e^{\pm 2i\phi}$

EXAMPLE 6–9

Show that the hydrogenlike atomic wave function ψ_{210} in Table 6.5 is normalized and that it is orthogonal to ψ_{200}.

SOLUTION: The orthonormality condition is given by Equation 6.51. Using ψ_{210} from Table 6.5,

$$\int_0^\infty dr\, r^2 \int_0^\pi d\theta \sin\theta \int_0^{2\pi} d\phi \left[\frac{1}{\sqrt{32\pi}}\left(\frac{Z}{a_0}\right)^{3/2}\sigma e^{-\sigma/2}\cos\theta\right]^2$$

$$= \frac{1}{32\pi}\left(\frac{Z}{a_0}\right)^5 \int_0^\infty dr\, r^4 e^{-Zr/a_0} \int_0^\pi d\theta \sin\theta \cos^2\theta \int_0^{2\pi} d\phi$$

$$= \frac{1}{32\pi}\left(\frac{Z}{a_0}\right)^5 (2\pi)\left(\frac{2}{3}\right)\left[\left(\frac{a_0}{Z}\right)^5 24\right] = 1$$

and so ψ_{210} is normalized. To show that it is orthogonal to ψ_{200},

$$
\int_0^\infty dr\, r^2 \int_0^\pi d\theta \sin\theta \int_0^{2\pi} d\phi \left[\frac{1}{\sqrt{32\pi}} \left(\frac{Z}{a_0} \right)^{3/2} \left(\frac{Zr}{a_0} \right) e^{-Zr/2a_0} \cos\theta \right]
$$

$$
\times \left[\frac{1}{\sqrt{32\pi}} \left(\frac{Z}{a_0} \right)^{3/2} \left(2 - \frac{Zr}{a_0} \right) e^{-Zr/2a_0} \right]
$$

$$
= \frac{1}{32\pi} \left(\frac{Z}{a_0} \right)^4 \int_0^\infty dr\, r^3 \left(2 - \frac{Zr}{a_0} \right) e^{-Zr/a_0} \int_0^\pi d\theta \sin\theta \cos\theta \int_0^{2\pi} d\phi
$$

The integral over θ here vanishes, so we see that ψ_{210} and ψ_{200} are orthogonal.

6–5. *s* Orbitals Are Spherically Symmetric

Equation 6.49 tells us that the hydrogen atomic wave functions depend upon three quantum numbers, n, l, and m. The quantum number n is called the *principal quantum number* and has the values 1, 2, ... The energy of the hydrogen atom depends upon only the principal quantum number through the equation $E_n = -e^2/8\pi\varepsilon_0 a_0 n^2$. The quantum number l is called the *angular momentum quantum number* and has the values 0, 1, ..., $n - 1$. The angular momentum of the electron about the proton is determined completely by l through $|L| = \hbar\sqrt{l(l+1)}$. Note that the form of the radial wave functions depends upon both n and l. The value of l is customarily denoted by a letter, with $l = 0$ being denoted by s, $l = 1$ by p, $l = 2$ by d, $l = 3$ by f, with higher values of l denoted by the alphabetic sequence following f. The origin of the letters s, p, d, f is historic and has to do with the designation of the observed spectral lines of atomic sodium. (The letters s, p, d, and f stand for *sharp, principal, diffuse*, and *fundamental*.) A wave function with $n = 1$ and $l = 0$ is called a 1s wave function; one with $n = 2$ and $l = 0$ a 2s wave function, and so on.

The third quantum number m is called the *magnetic quantum number* and takes on the $2l + 1$ values $m = 0$, ± 1, ± 2, ... $\pm l$. The z component of the angular momentum is determined completely by m through $L_z = m\hbar$. The quantum number m is called the magnetic quantum number because the energy of a hydrogen atom in a magnetic field depends on m. In the absence of a magnetic field, each energy level has a degeneracy of $2l + 1$. In the presence of a magnetic field, these levels split, and the energy depends upon the particular value of m (Problems 6–43 through 6–47). This splitting is illustrated in Figure 6.2 and is called the *Zeeman effect* (see Problem 6–46). In this case, E is a function of both the quantum numbers n and m.

The complete hydrogen atomic wave functions depend on three variables (r, θ, ϕ), so plotting or displaying them is difficult. The radial and angular parts are commonly considered separately. The state of lowest energy of a hydrogen atom is the 1s state. The radial function associated with the 1s state is $(Z = 1)$

$$
R_{1s}(r) = \frac{2}{a_0^{3/2}} e^{-r/a_0}
$$

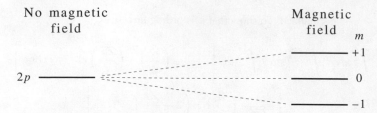

FIGURE 6.2
The splitting of the $2p$ state of a hydrogen atom in a magnetic field.

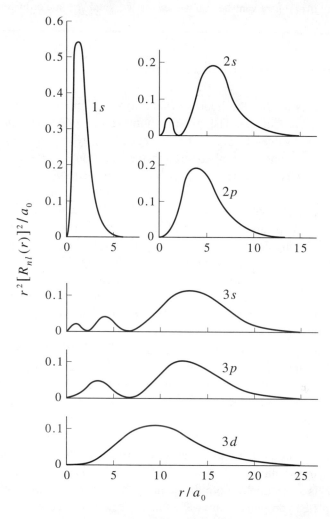

FIGURE 6.3
The probability densities $r^2[R_{nl}(r)]^2$ associated with the radial parts of the hydrogen atomic wave functions.

As mentioned above, the radial wave functions are normalized with respect to integration over r, so we have that

$$\int_0^\infty [R_{1s}(r)]^2 r^2 dr = \frac{4}{a_0^3} \int_0^\infty r^2 e^{-2r/a_0} dr = 1 \tag{6.52}$$

From Equation 6.52, we see that the probability that the electron lies between r and $r + dr$ is $[R_{nl}(r)]^2 r^2 dr$, and plots of $r^2 R_{nl}^2(r)$ are shown in Figure 6.3. An important observation from the plots in Figure 6.3 is that the number of nodes in the radial function is equal to $n - l - 1$. (Recall that the point $r = 0$ is not considered to be a node; see Section 2–4.)

For the $1s$ state, the probability that the electron lies between r and $r + dr$ is

$$\text{Prob} = \frac{4}{a_0^3} r^2 e^{-2r/a_0} dr \tag{6.53}$$

This result is contrary to the Bohr model in which the electron is incorrectly restricted to lie in fixed, well-defined orbits.

EXAMPLE 6–10
Calculate the probability that an electron described by a hydrogen atomic $1s$ wave function will be found within one Bohr radius of the nucleus.

SOLUTION: The probability that the electron will be found within one Bohr radius of the nucleus is obtained by integrating Equation 6.53 from 0 to a_0:

$$\text{Prob}(0 \le r \le a_0) = \frac{4}{a_0^3} \int_0^{a_0} r^2 e^{-2r/a_0} dr$$

$$= 4 \int_0^1 x^2 e^{-2x} dx$$

$$= 1 - 5e^{-2} = 0.323$$

We must keep in mind that we are dealing with only the radial parts of the total wave function here. The radial parts are easy to display because they depend on only the one coordinate r. The angular parts depend on both θ and ϕ and so are somewhat more difficult to display. The $l = 0$ case is easy, however, because when $l = 0$, m must equal zero and so we have $Y_0^0(\theta, \phi)$, which according to Table 6.3 is

$$Y_0^0(\theta, \phi) = \frac{1}{\sqrt{4\pi}}$$

$Y_0^0(\theta, \phi)$ is normalized with respect to integration over a spherical surface

$$\int_0^\pi d\theta \sin\theta \int_0^{2\pi} d\phi \, Y_0^0(\theta, \phi)^* Y_0^0(\theta, \phi) = \frac{1}{4\pi} \int_0^\pi d\theta \sin\theta \int_0^{2\pi} d\phi = 1$$

In this particular case, the angular dependence drops out and the wave function is spherically symmetric. The complete $1s$ wave function is

$$\psi_{1s}(r, \theta, \phi) = R_{10}(r) Y_0^0(\theta, \phi) = (\pi a_0^3)^{-1/2} e^{-r/a_0} \tag{6.54}$$

We have displayed the r, θ, and ϕ dependence on the left side of Equation 6.54 even though the θ and ϕ dependence drops out so we can emphasize that $\psi_{1s}(r, \theta, \phi)$ is the complete wave function. For example, the normalization condition is

$$\int_0^\infty dr r^2 \int_0^\pi d\theta \sin\theta \int_0^{2\pi} d\phi \, \psi_{1s}^*(r, \theta, \phi) \psi_{1s}(r, \theta, \phi) = 1$$

The hydrogen atomic wave functions are called *orbitals*, and, in particular, Equation 6.54 describes the $1s$ orbital; an electron in the $1s$ orbital is called a $1s$ electron.

The probability that a $1s$ electron lies between r and $r + dr$ from the nucleus is obtained by integrating $\psi_{1s}^*(r, \theta, \phi) \psi_{1s}(r, \theta, \phi)$ over all values of θ and ϕ according to

$$\text{Prob}(1s) = r^2 dr \int_0^\pi d\theta \sin\theta \int_0^{2\pi} d\phi \, \psi_{1s}^*(r, \theta, \phi) \psi_{1s}(r, \theta, \phi)$$

$$= \frac{4}{a_0^3} r^2 e^{-2r/a_0} dr \tag{6.55}$$

in agreement with Equation 6.53.

We can use Equation 6.55 to calculate average values of r. For example,

$$\langle r \rangle_{1s} = \frac{4}{a_0^3} \int_0^\infty r^3 e^{-2r/a_0} dr = \frac{3}{2} a_0 \tag{6.56}$$

Equation 6.55 can be used to determine the most probable distance of a $1s$ electron from the nucleus.

EXAMPLE 6–11

Show that the most probable value of r (r_{mp}) in a $1s$ state is a_0.

SOLUTION: To determine the most probable value of r, we find the value of r that maximizes the probability density of r or that maximizes

$$\text{Prob}(1s) = \frac{4}{a_0^3} r^2 e^{-2r/a_0}$$

If we differentiate Prob $(1s)$ and set the result equal to zero, we find that $r_{mp} = a_0$, the Bohr radius.

The next simplest orbital is the $2s$ orbital. A $2s$ orbital is given by

$$\psi_{2s}(r, \theta, \phi) = R_{20}(r) Y_0^0(\theta, \phi) \tag{6.57}$$

which is also spherically symmetric. In fact, because any s orbital will have the angular factor $Y_0^0(\theta, \phi)$, we see that all s orbitals are spherically symmetric. By referring to Table 6.5, we see that

$$\psi_{2s}(r, \theta, \phi) = \frac{1}{\sqrt{32\pi}} \left(\frac{1}{a_0}\right)^{3/2} \left(2 - \frac{r}{a_0}\right) e^{-r/2a_0} \tag{6.58}$$

Remember that ψ_{2s} is normalized with respect to an integration over r, θ, and ϕ. The average value of r in the $2s$ state is (cf. Problem 6–23)

$$\langle r \rangle_{2s} = \int_0^\infty dr\, r^3 \int_0^\pi d\theta \sin \theta \int_0^{2\pi} d\phi\, \psi_{2s}^*(r, \theta, \phi) \psi_{2s}(r, \theta, \phi) = 6a_0 \tag{6.59}$$

showing that a $2s$ electron is on the average a much greater distance from the nucleus than a $1s$ electron. In fact, using the general properties of the associated Laguerre polynomials, we can show that $\langle r \rangle = \frac{3}{2} a_0 n^2$ for an ns electron.

6–6. There Are Three p Orbitals for Each Value of the Principal Quantum Number, $n \geq 2$

When $l \neq 0$, the hydrogen atomic wave functions are not spherically symmetric; they depend on θ and ϕ. In this section, we will concentrate on the angular parts of the hydrogen wave functions. Let's first consider states with $l = 1$, or p orbitals. Because $m = 0$ or ± 1 when $l = 1$, there are three p orbitals for each value of n. The angular part of the p orbitals is given by the three spherical harmonics $Y_1^0(\theta, \phi)$ and $Y_1^{\pm 1}(\theta, \phi)$. The simplest of these spherical harmonics is

$$Y_1^0(\theta, \phi) = \left(\frac{3}{4\pi}\right)^{1/2} \cos \theta \tag{6.60}$$

which is readily shown to be normalized, because

$$\frac{3}{4\pi} \int_0^\pi d\theta \sin \theta \int_0^{2\pi} d\phi \cos^2 \theta = \frac{3}{2} \int_0^\pi \sin \theta \cos^2 \theta d\theta = \frac{3}{2} \int_{-1}^1 x^2 dx = 1$$

In the last step, we let $\cos \theta = x$, as in Example 6–4.

A common way to present the angular functions is as three-dimensional figures. Figure 6.4 is the familiar tangent sphere picture of a p orbital often presented in general chemistry texts. Although the tangent sphere picture represents the shape of the angular part of p orbitals, it is *not* a faithful representation of the shape of a p orbital because the radial functions are not included.

Because a complete wave function generally depends on three coordinates, wave functions are difficult to display clearly. One useful and instructive way, however, is the following. The quantity $\psi^* \psi d\tau$ is the probability that the electron is located within the volume element $d\tau$. Thus, we can divide space into little volume elements and compute the average or some representative value of $\psi^* \psi$ within each volume element

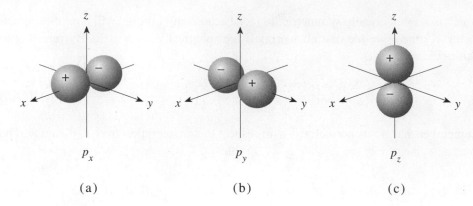

FIGURE 6.4
Three-dimensional polar plots of the angular part of the real representation of the hydrogen atomic wave functions for $l = 1$ (see Equations 6.62 for real representations of p_x and p_y.)

and then represent the value of $\psi^*\psi$ by the density of dots in a picture. Figure 6.5 shows such plots for several orbitals.

An alternate way to represent complete wave functions is as contour maps. Figure 6.6a shows a contour map for a $1s$ orbital. The nine contours shown in each case in Figure 6.6 enclose the 10%, 20%, ..., 90% probability of finding the electron within

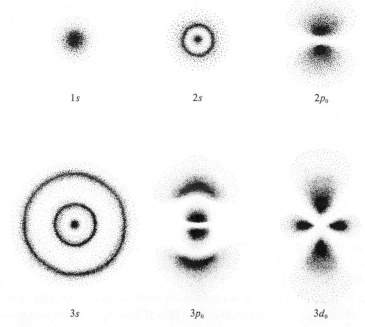

FIGURE 6.5
Probability density plots of some hydrogen atomic orbitals. The density of the dots is proportional to the probability of finding the electron in that region.

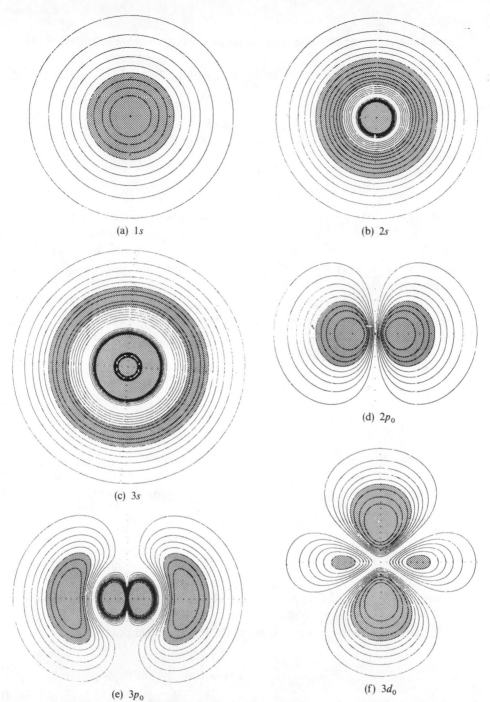

(a) 1s

(b) 2s

(c) 3s

(d) 2p₀

(e) 3p₀

(f) 3d₀

FIGURE 6.6

Probability contour maps for the hydrogen atomic orbitals. The nine contours shown in each case enclose the 10%, 20%, ..., 90% probability of finding the electron within each contour. The scale of the figure is indicated by hash marks: one mark corresponds to one Bohr radius a_0. Note that the different orbitals are presented on different scales. The shaded areas shown in each case indicates the highest 40% of the probability densities.

each contour. Note that the contour maps appear as cross sections of the plots in Figure 6.5.

It is interesting to compare the depictions of the $2p_0$ and $3p_0$ orbitals in Figures 6.5 and 6.6. The expressions for these orbitals are

$$\psi_{2p_0}(r, \theta, \phi) = R_{21}(r) Y_1^0(\theta, \phi)$$

and

$$\psi_{3p_0}(r, \theta, \phi) = R_{31}(r) Y_1^0(\theta, \phi)$$

Both orbitals have the same angular part, which is represented in Figure 6.4. The radial functions have $n - l - 1$ nodes, however, and so $R_{21}(r)$ has no nodes and $R_{31}(r)$ has one. The difference in the shapes of the $2p_0$ and $3p_0$ orbitals in Figures 6.5 and 6.6 is due to the node in $R_{31}(r)$. This example illustrates the inadequacy of the "tangent sphere" representation of p orbitals.

The angular functions with $m \neq 0$ are more difficult to represent pictorially because they not only depend on ϕ in addition to θ but are complex as well. In particular, the $l = 1$ states with $m \neq 0$ are

$$Y_1^1(\theta, \phi) = \left(\frac{3}{8\pi}\right)^{1/2} \sin\theta \, e^{i\phi}$$

$$Y_1^{-1}(\theta, \phi) = \left(\frac{3}{8\pi}\right)^{1/2} \sin\theta \, e^{-i\phi}$$

(6.61)

The probability densities associated with $Y_1^1(\theta, \phi)$ and $Y_1^{-1}(\theta, \phi)$ are the same because

$$|Y_1^1(\theta, \phi)|^2 = \frac{3}{8\pi} \sin^2\theta$$

and

$$|Y_1^{-1}(\theta, \phi)|^2 = \frac{3}{8\pi} \sin^2\theta$$

Because $Y_1^1(\theta, \phi)$ and $Y_1^{-1}(\theta, \phi)$ correspond to the same energy, we know from Section 4–2 that any linear combination of Y_1^1 and Y_1^{-1} is also an energy eigenfunction with the same energy. It is customary to use the combinations

$$p_x = \frac{1}{\sqrt{2}}(Y_1^1 + Y_1^{-1}) = \left(\frac{3}{4\pi}\right)^{1/2} \sin\theta \cos\phi$$

$$p_y = \frac{1}{\sqrt{2}i}(Y_1^1 - Y_1^{-1}) = \left(\frac{3}{4\pi}\right)^{1/2} \sin\theta \sin\phi$$

(6.62)

"Tangent sphere" plots of p_x and p_y are shown in Figure 6.4. They have the same shape as the p_z function except that they are directed along the x- and y-axes. The three functions p_x, p_y, and p_z are often used as the angular part of hydrogen atomic wave functions because they are real and have easily visualized directional properties.

For the $l = 2$ case, $m = 0$, ± 1, and ± 2, and so there are five d orbitals. For $m = \pm 1$ and ± 2, we take linear combinations as we did above for the p functions. The customary linear combinations are (Problem 6–42)

$$d_{z^2} = Y_2^0 = \left(\frac{5}{16\pi}\right)^{1/2} (3\cos^2\theta - 1)$$

$$d_{xz} = \frac{1}{\sqrt{2}}(Y_2^1 + Y_2^{-1}) = \left(\frac{15}{4\pi}\right)^{1/2} \sin\theta \cos\theta \cos\phi$$

$$d_{yz} = \frac{1}{\sqrt{2}i}(Y_2^1 - Y_2^{-1}) = \left(\frac{15}{4\pi}\right)^{1/2} \sin\theta \cos\theta \sin\phi \qquad (6.63)$$

$$d_{x^2-y^2} = \frac{1}{\sqrt{2}}(Y_2^2 + Y_2^{-2}) = \left(\frac{15}{16\pi}\right)^{1/2} \sin^2\theta \cos 2\phi$$

$$d_{xy} = \frac{1}{\sqrt{2}i}(Y_2^2 - Y_2^{-2}) = \left(\frac{15}{16\pi}\right)^{1/2} \sin^2\theta \sin 2\phi$$

The angular parts of the five d orbitals are shown in Figure 6.7. Note that the last four orbitals given in Equations 6.63 differ only in their orientation. Figure 6.7 suggests

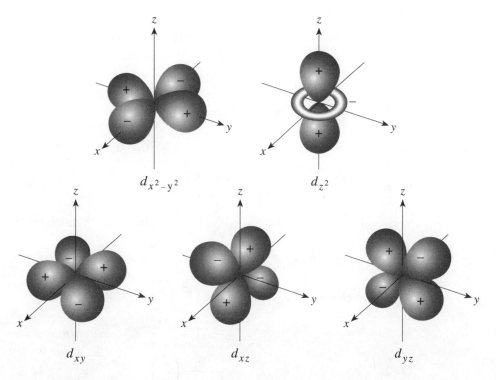

FIGURE 6.7
Three-dimensional plots of the angular part of the real representation of the hydrogen atomic wave functions for $l = 2$. Such plots show the directional character of these orbitals but are not good representations of the shape of these orbitals because the radial functions are not included.

the rationale of the notation of the d orbitals; d_{z^2} lies along the z-axis, $d_{x^2-y^2}$ lies along the x and y-axes; d_{xy} lies in the x–y plane; d_{xz} lies in the x–z plane, and d_{yz} lies in the y–z plane. There is no fundamental reason to choose these linear combinations of spherical harmonics over the spherical harmonics themselves, but most chemists use the five d orbitals given by Equations 6.63 because the functions in Equations 6.63 are real and have directional properties consistent with molecular structures. The real representations of the hydrogen atomic wave functions are given in Table 6.6. The functions in Table 6.6 are the linear combinations of the complex wave functions in

TABLE 6.6
The complete hydrogenlike atomic wave functions expressed as real functions for $n = 1, 2,$ and 3. The quantity Z is the atomic number of the nucleus and $\sigma = Zr/a_0$, where a_0 is the Bohr radius.

$n = 1,$	$l = 0,$	$m = 0$	$\psi_{1s} = \dfrac{1}{\sqrt{\pi}}\left(\dfrac{Z}{a_0}\right)^{3/2} e^{-\sigma}$
$n = 2,$	$l = 0,$	$m = 0$	$\psi_{2s} = \dfrac{1}{4\sqrt{2\pi}}\left(\dfrac{Z}{a_0}\right)^{3/2} (2-\sigma)e^{-\sigma/2}$
	$l = 1,$	$m = 0$	$\psi_{2p_z} = \dfrac{1}{4\sqrt{2\pi}}\left(\dfrac{Z}{a_0}\right)^{3/2} \sigma e^{-\sigma/2}\cos\theta$
	$l = 1,$	$m = \pm 1$	$\psi_{2p_x} = \dfrac{1}{4\sqrt{2\pi}}\left(\dfrac{Z}{a_0}\right)^{3/2} \sigma e^{-\sigma/2}\sin\theta\cos\phi$
			$\psi_{2p_y} = \dfrac{1}{4\sqrt{2\pi}}\left(\dfrac{Z}{a_0}\right)^{3/2} \sigma e^{-\sigma/2}\sin\theta\sin\phi$
$n = 3,$	$l = 0,$	$m = 0$	$\psi_{3s} = \dfrac{1}{81\sqrt{3\pi}}\left(\dfrac{Z}{a_0}\right)^{3/2} (27-18\sigma+2\sigma^2)e^{-\sigma/3}$
	$l = 1,$	$m = 0$	$\psi_{3p_z} = \dfrac{\sqrt{2}}{81\sqrt{\pi}}\left(\dfrac{Z}{a_0}\right)^{3/2} \sigma(6-\sigma)e^{-\sigma/3}\cos\theta$
	$l = 1,$	$m = \pm 1$	$\psi_{3p_x} = \dfrac{\sqrt{2}}{81\sqrt{\pi}}\left(\dfrac{Z}{a_0}\right)^{3/2} \sigma(6-\sigma)e^{-\sigma/3}\sin\theta\cos\phi$
			$\psi_{3p_y} = \dfrac{\sqrt{2}}{81\sqrt{\pi}}\left(\dfrac{Z}{a_0}\right)^{3/2} \sigma(6-\sigma)e^{-\sigma/3}\sin\theta\sin\phi$
	$l = 2,$	$m = 0$	$\psi_{3d_{z^2}} = \dfrac{1}{81\sqrt{6\pi}}\left(\dfrac{Z}{a_0}\right)^{3/2} \sigma^2 e^{-\sigma/3}(3\cos^2\theta-1)$
	$l = 2,$	$m = \pm 1$	$\psi_{3d_{xz}} = \dfrac{\sqrt{2}}{81\sqrt{\pi}}\left(\dfrac{Z}{a_0}\right)^{3/2} \sigma^2 e^{-\sigma/3}\sin\theta\cos\theta\cos\phi$
			$\psi_{3d_{yz}} = \dfrac{\sqrt{2}}{81\sqrt{\pi}}\left(\dfrac{Z}{a_0}\right)^{3/2} \sigma^2 e^{-\sigma/3}\sin\theta\cos\theta\sin\phi$
	$l = 2,$	$m = \pm 2$	$\psi_{3d_{x^2-y^2}} = \dfrac{1}{81\sqrt{2\pi}}\left(\dfrac{Z}{a_0}\right)^{3/2} \sigma^2 e^{-\sigma/3}\sin^2\theta\cos 2\phi$
			$\psi_{3d_{xy}} = \dfrac{1}{81\sqrt{2\pi}}\left(\dfrac{Z}{a_0}\right)^{3/2} \sigma^2 e^{-\sigma/3}\sin^2\theta\sin 2\phi$

Table 6.5. Both sets are equivalent, but chemists normally use the real functions in Table 6.6. We will see in later chapters that molecular wave functions can be built out of atomic orbitals and if the atomic orbitals have a definite directional character, we can use chemical intuition to decide which are the more important atomic orbitals to use to describe molecular orbitals.

6–7. The Schrödinger Equation for the Helium Atom Cannot Be Solved Exactly

The next system to study is clearly the helium atom, whose Schrödinger equation is

$$\left(-\frac{\hbar^2}{2M}\nabla^2 - \frac{\hbar^2}{2m_e}\nabla_1^2 - \frac{\hbar^2}{2m_e}\nabla_2^2\right)\psi(\mathbf{R}, \mathbf{r}_1, \mathbf{r}_2) +$$

$$\left(-\frac{2e^2}{4\pi\varepsilon_0|\mathbf{R} - \mathbf{r}_1|} - \frac{2e^2}{4\pi\varepsilon_0|\mathbf{R} - \mathbf{r}_2|} + \frac{e^2}{4\pi\varepsilon_0|\mathbf{r}_1 - \mathbf{r}_2|}\right)\psi(\mathbf{R}, \mathbf{r}_1, \mathbf{r}_2)$$

$$= E\psi(\mathbf{R}, \mathbf{r}_1, \mathbf{r}_2) \tag{6.64}$$

In this equation, $\mathbf{R}$ is the position of the helium nucleus and $\mathbf{r}_1$ and $\mathbf{r}_2$ are the positions of the two electrons; M is the mass of the nucleus and m_e is the electronic mass; ∇^2 is the Laplacian operator with respect to the position of the nucleus and ∇_1^2 and ∇_2^2 are the Laplacian operators with respect to the positions of the electronic coordinates. Realize that this is a three-body problem and *not* a two-body problem, and so the separation into center-of-mass and relative coordinates is much more complicated than it is for hydrogen. Because $M \gg m_e$, however, regarding the nucleus as fixed relative to the motion of the electrons is still an excellent approximation. Under this approximation, we can fix the nucleus at the origin of a spherical coordinate system and write the Schrödinger equation as

$$-\frac{\hbar^2}{2m_e}(\nabla_1^2 + \nabla_2^2)\psi(\mathbf{r}_1, \mathbf{r}_2) - \frac{2e^2}{4\pi\varepsilon_0}\left(\frac{1}{r_1} + \frac{1}{r_2}\right)\psi(\mathbf{r}_1, \mathbf{r}_2)$$

$$+ \frac{e^2}{4\pi\varepsilon_0|\mathbf{r}_1 - \mathbf{r}_2|}\psi(\mathbf{r}_1, \mathbf{r}_2) = E\psi(\mathbf{r}_1, \mathbf{r}_2) \tag{6.65}$$

Even this simplified equation cannot be solved exactly.

The $e^2/4\pi\varepsilon_0|\mathbf{r}_1 - \mathbf{r}_2|$ term is called the *interelectronic repulsion* term and is directly responsible for the difficulty associated with solving Equation 6.65. If this term were not there, the total Hamiltonian operator in Equation 6.65 would be the sum of the Hamiltonian operators of two hydrogenlike atoms. According to Equations 3.61 through 3.64, the total energy would be the sum of the energies of the two individual hydrogenlike atoms, and the wave function would be a product of two hydrogenlike atomic orbitals. To solve Equation 6.65, we must resort to some approximation method. Fortunately, two quite different approximation methods that can yield extremely good results have found wide use in quantum chemistry. These are called *perturbation theory* and the *variational method* and are presented in Chapter 7.

Problems

6-1. Show that both $\hbar^2 \nabla^2 / 2m_e$ and $e^2/4\pi\varepsilon_0 r$ have the units of energy (joules).

6-2. In terms of the variable θ, Legendre's equation is

$$\sin\theta \frac{d}{d\theta}\left(\sin\theta \frac{d\Theta(\theta)}{d\theta}\right) + (\beta^2 \sin^2\theta - m^2)\Theta(\theta) = 0$$

Let $x = \cos\theta$ and $P(x) = \Theta(\theta)$ and show that

$$(1 - x^2)\frac{d^2 P(x)}{dx^2} - 2x\frac{dP(x)}{dx} + \left[\beta - \frac{m^2}{1 - x^2}\right]P(x) = 0$$

6-3. Show that the Legendre polynomials given in Table 6.1 satisfy Equation 6.23 with $m = 0$.

6-4. Show that the orthogonality integral for the Legendre polynomials, Equation 6.24, is equivalent to

$$\int_0^\pi P_l(\cos\theta) P_n(\cos\theta) \sin\theta d\theta = 0 \qquad l \neq n$$

6-5. Show that the Legendre polynomials given in Table 6.1 satisfy the orthogonality and normalization conditions given by Equations 6.24 and 6.25.

6-6. Use Equation 6.26 to generate the associated Legendre functions in Table 6.2.

6-7. Show that the first few associated Legendre functions given in Table 6.2 are solutions to Equation 6.23 and that they satisfy the orthonormality condition, Equation 6.28.

6-8. There are a number of recursion formulas for the associated Legendre functions. One that we will have occasion to use in Section 13–12 is

$$(2l + 1)x P_l^{|m|}(x) = (l - |m| + 1)P_{l+1}^{|m|}(x) + (l + |m|)P_{l-1}^{|m|}(x)$$

Show that the first few associated Legendre functions in Table 6.2 satisfy this recursion formula.

6-9. Show that the first few spherical harmonics in Table 6.3 satisfy the orthonormality condition, Equation 6.31.

6-10. Using explicit expressions for $Y_l^m(\theta, \phi)$, show that

$$|Y_1^1(\theta, \phi)|^2 + |Y_1^0(\theta, \phi)|^2 + |Y_1^{-1}(\theta, \phi)|^2 = \text{constant}$$

This is a special case of the general theorem

$$\sum_{m=-l}^{+l} |Y_l^m(\theta, \phi)|^2 = \text{constant}$$

known as Unsöld's theorem. What is the physical significance of this result?

6-11. In Cartesian coordinates,

$$\hat{L}_z = -i\hbar\left(x\frac{\partial}{\partial y} - y\frac{\partial}{\partial x}\right)$$

Convert this equation to spherical coordinates, showing that

$$\hat{L}_z = -i\hbar\frac{\partial}{\partial\phi}$$

6-12. Convert $\hat{L}_x$ and $\hat{L}_y$ from Cartesian coordinates to spherical coordinates.

6-13. Prove that $\hat{L}^2$ commutes with $\hat{L}_x$, $\hat{L}_y$, and $\hat{L}_z$ but that

$$[\hat{L}_x, \hat{L}_y] = i\hbar\hat{L}_z \qquad [\hat{L}_y, \hat{L}_z] = i\hbar\hat{L}_x \qquad [\hat{L}_z, \hat{L}_x] = i\hbar\hat{L}_y$$

(*Hint:* Use Cartesian coordinates.) Do you see a pattern in these formulas?

6-14. It is a somewhat advanced exercise to prove generally that $\langle L_x \rangle = \langle L_y \rangle = 0$ (see, however, Problem 6–58), but prove that they are zero at least for the first few l, m states by using the spherical harmonics given in Table 6.3.

6-15. For an isolated hydrogen atom, why must the angular momentum vector **L** lie on a cone that is symmetric about the z-axis? Can the angular momentum operator ever point exactly along the z-axis?

6-16. Referring to Table 6.5, show that the first few hydrogen atomic wave functions are orthonormal.

6-17. Show explicitly that

$$\hat{H}\psi = -\frac{m_e e^4}{8\varepsilon_0^2 h^2}\psi$$

for the ground state of a hydrogen atom.

6-18. Show explicitly that

$$\hat{H}\psi = -\frac{m_e e^4}{32\varepsilon_0^2 h^2}\psi$$

for a $2p_0$ state of a hydrogen atom.

6-19. Given the first equality, show that the ground-state energy of a hydrogen atom can be written as

$$E_0 = -\frac{\hbar^2}{2m_e a_0^2} = -\frac{e^2}{8\pi\varepsilon_0 a_0} = -\frac{m_e e^4}{32\pi^2\varepsilon_0^2\hbar^2} = -\frac{m_e e^4}{8\varepsilon_0^2 h^2}$$

6-20. Calculate the probability that a hydrogen $1s$ electron will be found within a distance $2a_0$ from the nucleus.

6-21. Calculate the radius of the sphere that encloses a 50% probability of finding a hydrogen $1s$ electron. Repeat the calculation for a 90% probability.

6-22. Many problems involving the calculation of average values for the hydrogen atom require evaluating integrals of the form

$$I_n = \int_0^\infty r^n e^{-\beta r} dr$$

This integral can be evaluated readily by starting with the elementary integral

$$I_0(\beta) = \int_0^\infty e^{-\beta r} dr = \frac{1}{\beta}$$

Show that the derivatives of $I(\beta)$ are

$$\frac{dI_0}{d\beta} = -\int_0^\infty r e^{-\beta r} dr = -I_1$$

$$\frac{d^2 I_0}{d\beta^2} = \int_0^\infty r^2 e^{-\beta r} dr = I_2$$

and so on. Using the fact that $I_0(\beta) = 1/\beta$, show that the values of these two integrals are $-1/\beta^2$ and $2/\beta^3$, respectively. Show that, in general

$$\frac{d^n I_0}{d\beta^n} = (-1)^n \int_0^\infty r^n e^{-\beta r} dr = (-1)^n I_n$$

$$= (-1)^n \frac{n!}{\beta^{n+1}}$$

and that

$$I_n = \frac{n!}{\beta^{n+1}}$$

6-23. Prove that the average value of r in the $1s$ and $2s$ states for a hydrogenlike atom is $3a_0/2Z$ and $6a_0/Z$, respectively.

6-24. Prove that $\langle V \rangle = 2\langle E \rangle$ and, consequently, that $\langle \hat{K} \rangle = -\langle E \rangle$, for a $2s$ electron.

6-25. By evaluating the appropriate integrals, compute $\langle r \rangle$ in the $2s$, $2p$, and $3s$ states of the hydrogen atom; compare your results with the general formula

$$\langle r_{nl} \rangle = \frac{a_0}{2} [3n^2 - l(l+1)]$$

6-26. Show that the first few hydrogen atomic orbitals in Table 6.6 are orthonormal.

6-27. Show that the two maxima in the plot of $r^2 \psi_{2s}^2(r)$ against r occur at $(3 \pm \sqrt{5})a_0$. (See Figure 6.3.)

6-28. Calculate the value of $\langle r \rangle$ for the $n = 2$, $l = 1$ state and the $n = 2$, $l = 0$ state of the hydrogen atom. Are you surprised by the answers? Explain.

6-29. In Chapter 4, we learned that if ψ_1 and ψ_2 are solutions of the Schrödinger equation that have the same energy E_n, then $c_1\psi_1 + c_2\psi_2$ is also a solution. Let $\psi_1 = \psi_{210}$ and $\psi_2 = \psi_{211}$ (see Table 6.5). What is the energy corresponding to $\psi = c_1\psi_1 + c_2\psi_2$ where $c_1^2 + c_2^2 = 1$? What does this result tell you about the uniqueness of the three p orbitals, p_x, p_y, and p_z?

6-30. Show that the total probability density of the $2p$ orbitals is spherically symmetric by evaluating $\sum_{m=-1}^{1} \psi_{21m}^2$. (Use the wave functions in Table 6.6.)

6-31. Show that the total probability density of the $3d$ orbitals is spherically symmetric by evaluating $\sum_{m=-2}^{2} \psi_{32m}^2$. (Use the wave functions in Table 6.6.)

6-32. Show that the sum of the probability densities for the $n = 3$ states of the hydrogen atom is spherically symmetric. Do you expect this to be true for all values of n? Explain.

6-33. Determine the degeneracy of each of the hydrogen atomic energy levels.

6-34. Set up the Hamiltonian operator for the system of an electron interacting with a fixed nucleus of atomic number Z. The simplest such system is singly ionized helium, where $Z = 2$. We will call this a hydrogenlike system. Observe that the only difference between this Hamiltonian operator and the hydrogen Hamiltonian operator is the correspondance that e^2 for the hydrogen atom becomes Ze^2 for the hydrogenlike ion. Consequently, show that the energy becomes (cf. Equation 6.44)

$$E_n = -\frac{m_e Z^2 e^4}{8\varepsilon_0^2 h^2 n^2} \qquad n = 1, 2, \ldots$$

Furthermore, now show that the solutions to the radial equation, Equation 6.47, are

$$R_{nl}(r) = -\left\{ \frac{(n-l-1)!}{2n[(n+l)!]^3} \right\}^{1/2} \left(\frac{2Z}{na_0} \right)^{l+3/2} r^l e^{-Zr/na_0} L_{n+l}^{2l+1}\left(\frac{2Zr}{na_0} \right)$$

Show that the $1s$ orbital for this system is

$$\psi_{1s} = \frac{1}{\sqrt{\pi}} \left(\frac{Z}{a_0} \right)^{3/2} e^{-Zr/a_0}$$

and show that it is normalized. Show that

$$\langle r \rangle = \frac{3a_0}{2Z}$$

and that

$$r_{mp} = \frac{a_0}{Z}$$

Last, calculate the ionization energy of a hydrogen atom and a singly ionized helium atom. Express your answer in kilojoules per mole.

6-35. How does E_n for a hydrogen atom differ from Equation 6.44 if the nucleus is not considered to be fixed at the origin?

6-36. Determine the ratio of the ground-state energy of atomic hydrogen to that of atomic deuterium.

6-37. In this problem, we will prove the so-called *quantum-mechanical virial theorem*. Start with

$$\hat{H}\psi = E\psi$$

where

$$\hat{H} = -\frac{\hbar^2}{2m}\nabla^2 + V(x, y, z)$$

Using the fact that $\hat{H}$ is a Hermitian operator (Problem 4–28), show that

$$\int \psi^*[\hat{H}, \hat{A}]\psi \, d\tau = 0 \tag{1}$$

where $\hat{A}$ is any linear operator. Choose $\hat{A}$ to be

$$\hat{A} = -i\hbar \left(x\frac{\partial}{\partial x} + y\frac{\partial}{\partial y} + z\frac{\partial}{\partial z} \right) \tag{2}$$

and show that

$$[\hat{H}, \hat{A}] = i\hbar \left(x\frac{\partial V}{\partial x} + y\frac{\partial V}{\partial y} + z\frac{\partial V}{\partial z} \right) - \frac{i\hbar}{m}(\hat{P}_x^2 + \hat{P}_y^2 + \hat{P}_z^2)$$

$$= i\hbar \left(x\frac{\partial V}{\partial x} + y\frac{\partial V}{\partial y} + z\frac{\partial V}{\partial z} \right) - 2i\hbar\hat{K}$$

where $\hat{K}$ is the kinetic energy operator. Now use Equation 1 and show that

$$\left\langle x\frac{\partial V}{\partial x} + y\frac{\partial V}{\partial y} + z\frac{\partial V}{\partial z} \right\rangle = 2\langle \hat{K} \rangle \tag{3}$$

Equation 3 is the quantum-mechanical virial theorem.

Now show that if $V(x, y, z)$ is a Coulombic potential

$$V(x, y, z) = -\frac{Ze^2}{4\pi\varepsilon_0(x^2 + y^2 + z^2)^{1/2}}$$

then

$$\langle V \rangle = -2\langle \hat{K} \rangle = 2\langle E \rangle \tag{4}$$

where

$$\langle E \rangle = \langle \hat{K} \rangle + \langle V \rangle$$

In Problem 6–24 we proved that this result is valid for a $2s$ electron. Although we proved Equation 4 only for the case of one electron in the field of one nucleus, Equation 4 is valid for many-electron atoms and molecules. The proof is a straightforward extension of the proof developed in this problem.

6-38. Use the virial theorem (Problem 6–37) to prove that $\langle \hat{K} \rangle = \langle V \rangle = E/2$ for a harmonic oscillator (cf. Problem 5–23).

6-39. The average value of r for a hydrogenlike atom can be evaluated in general and is given by

$$\langle r \rangle_{nl} = \frac{n^2 a_0}{Z} \left\{ 1 + \frac{1}{2}\left[1 - \frac{l(l+1)}{n^2} \right] \right\}$$

Verify this formula explicitly for the ψ_{211} orbital.

6-40. The average value of r^2 for a hydrogenlike atom can be evaluated in general and is given by

$$\langle r^2 \rangle_{nl} = \frac{n^4 a_0^2}{Z^2} \left\{ 1 + \frac{3}{2} \left[1 - \frac{l(l+1) - \frac{1}{3}}{n^2} \right] \right\}$$

Verify this formula explicitly for the ψ_{210} orbital.

6-41. The average values of $1/r$, $1/r^2$, and $1/r^3$ for a hydrogenlike atom can be evaluated in general and are given by

$$\left\langle \frac{1}{r} \right\rangle_{nl} = \frac{Z}{a_0 n^2}$$

$$\left\langle \frac{1}{r^2} \right\rangle_{nl} = \frac{Z^2}{a_0^2 n^3 (l + \frac{1}{2})}$$

and

$$\left\langle \frac{1}{r^3} \right\rangle_{nl} = \frac{Z^3}{a_0^3 n^3 l(l + \frac{1}{2})(l + 1)}$$

Verify these formulas explicitly for the ψ_{210} orbital.

6-42. The designations of the d orbitals can be rationalized in the following way. Equation 6.63 shows that d_{xz} goes as $\sin\theta \cos\theta \cos\phi$. Using the relation between Cartesian and spherical coordinates, show that $\sin\theta \cos\theta \cos\phi$ is proportional to xz. Similarly, show that $\sin\theta \cos\theta \sin\phi$ (d_{yz}) is proportional to yz; that $\sin^2\theta \cos 2\phi$ ($d_{x^2-y^2}$) is proportional to $x^2 - y^2$; and that $\sin^2\theta \sin 2\phi$ (d_{xy}) is proportional to xy.

Problems 6–43 through 6–47 examine the energy levels for a hydrogen atom in an external magnetic field.

6-43. Recall from your course in physics that the motion of an electric charge around a closed loop produces a magnetic dipole, μ, whose direction is perpendicular to the loop and whose magnitude is given by

$$\mu = iA$$

where i is the current in amperes ($C \cdot s^{-1}$) and A is the area of the loop (m^2). Notice that the units of a magnetic dipole are coulombs·meters²·seconds⁻¹ ($C \cdot m^2 \cdot s^{-1}$), or amperes·meters² ($A \cdot m^2$). Show that

$$i = \frac{qv}{2\pi r}$$

for a circular loop, where v is the velocity of the charge q and r is the radius of the loop. Show that

$$\mu = \frac{qrv}{2}$$

for a circular loop. If the loop is not circular, then we must use vector calculus and the magnetic dipole is given by

$$\mu = \frac{q(\mathbf{r} \times \mathbf{v})}{2}$$

Show that this formula reduces to the preceding one for a circular loop. Last, using the relationship $\mathbf{L} = \mathbf{r} \times \mathbf{p}$, show that

$$\mu = \frac{q}{2m} \mathbf{L}$$

Thus, the orbital motion of an electron in an atom imparts a magnetic moment to the atom. For an electron, $q = -|e|$ and so

$$\mu = -\frac{|e|}{2m_e} \mathbf{L}$$

6-44. In Problem 6–43, we derived an expression for the magnetic moment of a hydrogen atom imparted by the orbital motion of its electron. Using the result that $L^2 = \hbar^2 l(l+1)$, show that the magnitude of the magnetic moment is

$$\mu = -\beta_B [l(l+1)]^{1/2}$$

where $\beta_B = \hbar|e|/2m_e$ is called the *Bohr magneton*. What are the units of β_B? What is its numerical value? A magnetic dipole in a magnetic field (**B**) has a potential energy

$$V = -\mu \cdot \mathbf{B}$$

(We will discuss magnetic fields when we study nuclear magnetic resonance, NMR, in Chapter 14.) Show that the units of the intensity of a magnetic field are $J \cdot A^{-1} \cdot m^{-2}$. This set of units is called a *tesla* (T), so that we have $1\ T = 1\ J \cdot A^{-1} \cdot m^{-2}$. In terms of teslas, the units of the Bohr magneton, β_B, are $J \cdot T^{-1}$.

6-45. Using the results of Problems 6–43 and 6–44, show that the Hamiltonian operator for a hydrogen atom in an external magnetic field where the field is in the z direction is given by

$$\hat{H} = \hat{H}_0 + \frac{\beta_B B_z}{\hbar} \hat{L}_z$$

where $\hat{H}_0$ is the Hamiltonian operator of a hydrogen atom in the absence of the magnetic field. Show that the wave functions of the Schrödinger equation for a hydrogen atom in a magnetic field are the same as those for the hydrogen atom in the absence of the field. Finally, show that the energy associated with the wave function ψ_{nlm} is

$$E = E_n^{(0)} + \beta_B B_z m \tag{1}$$

where $E_n^{(0)}$ is the energy in the absence of the magnetic field and m is the magnetic quantum number.

6-46. Equation 1 of Problem 6–45 shows that a state with given values of n and l is split into $2l+1$ levels by an external magnetic field. For example, Figure 6.8 shows the results for the $1s$ and $2p$ states of atomic hydrogen. The $1s$ state is not split ($2l+1 = 1$), but the $2p$ state is split into three levels ($2l+1 = 3$). Figure 6.8 also shows that the $2p$ to $1s$ transition in atomic hydrogen could (see Problem 6–47) be split into three distinct transitions instead

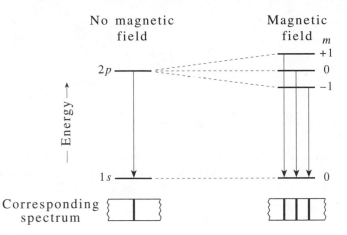

FIGURE 6.8
The splitting of the $2p$ state of the hydrogen atom in a magnetic field. The $2p$ state is split into three closely spaced levels. In a magnetic field, the $2p$ to $1s$ transition is split into three distinct transition frequencies.

of just one. Superconducting magnets have magnetic field strengths of the order of 15 T. Calculate the magnitude of the splitting shown in Figure 6.8 for a magnetic field of 15 T. Compare your result with the energy difference between the unperturbed $1s$ and $2p$ levels. Show that the three distinct transitions shown in Figure 6.8 lie very close together. We say that the $2p$ to $1s$ transition that occurs in the absence of a magnetic field becomes a *triplet* in the presence of the field. The occurrence of such multiplets when atoms are placed in magnetic fields is known as the *Zeeman effect*.

6-47. Consider a transition between the $l = 2$ and the $l = 3$ states of atomic hydrogen. What is the total number of conceivable transitions between these two states in an external magnetic field? For light whose electric field vector is parallel to the direction of the external magnetic field, the selection rule is $\Delta m = 0$. For light whose electric field vector is perpendicular to the direction of the external magnetic field, the selection rule is $\Delta m = \pm 1$. In each case, how many of the possible transitions are allowed?

Problems 6–48 through 6–57 develop the quantum-mechanical properties of angular momentum using operator notation, without solving the Schrödinger equation.

6-48. Define the two (not necessarily Hermitian) operators

$$\hat{L}_+ = \hat{L}_x + i\hat{L}_y \qquad \text{and} \qquad \hat{L}_- = \hat{L}_x - i\hat{L}_y$$

Using the results of Problem 6–13, show that

$$[\hat{L}_z, \hat{L}_+] = \hat{L}_z\hat{L}_+ - \hat{L}_+\hat{L}_z = \hbar\hat{L}_+$$
$$[\hat{L}_z, \hat{L}_-] = \hat{L}_z\hat{L}_- - \hat{L}_-\hat{L}_z = -\hbar\hat{L}_-$$

and

$$[\hat{L}^2, \hat{L}_+] = [\hat{L}^2, \hat{L}_-] = 0$$

6-49. Show that

$$\hat{L}_-\hat{L}_+ = \hat{L}_x^2 + \hat{L}_y^2 + i\hat{L}_x\hat{L}_y - i\hat{L}_y\hat{L}_x$$
$$= \hat{L}^2 - \hat{L}_z^2 - \hbar\hat{L}_z$$

and

$$\hat{L}_+\hat{L}_- = \hat{L}^2 - \hat{L}_z^2 + \hbar\hat{L}_z$$

6-50. Because $\hat{L}^2$ and $\hat{L}_z$ commute, they have mutual eigenfunctions. We know from the chapter that these mutual eigenfunctions are the spherical harmonics, $Y_l^m(\theta, \phi)$, but we really don't need that information here. To emphasize this point, let $\psi_{\alpha\beta}$ be the mutual eigenfunctions of $\hat{L}^2$ and $\hat{L}_z$ such that

$$\hat{L}^2\psi_{\alpha\beta} = \beta^2\psi_{\alpha\beta}$$

and

$$\hat{L}_z\psi_{\alpha\beta} = \alpha\psi_{\alpha\beta}$$

Now let

$$\psi_{\alpha\beta}^{+1} = \hat{L}_+\psi_{\alpha\beta}$$

Show that

$$\hat{L}_z\psi_{\alpha\beta}^{+1} = (\alpha + \hbar)\psi_{\alpha\beta}^{+1}$$

and

$$\hat{L}^2\psi_{\alpha\beta}^{+1} = \beta^2\psi_{\alpha\beta}^{+1}$$

Therefore, if α is an eigenvalue of $\hat{L}_z$, then $\alpha + \hbar$ is also an eigenvalue (unless $\psi_{\alpha\beta}^{+1}$ happens to be zero). In the notation for the spherical harmonics that we use in the chapter, $\hat{L}_+Y_l^m(\theta, \phi) \propto Y_l^{m+1}(\theta, \phi)$.

6-51. Using $\hat{L}_-$ instead of $\hat{L}_+$ in Problem 6–50, show that if α is an eigenvalue of $\hat{L}_z$, then $\alpha - \hbar$ is also an eigenvalue (unless $\psi_{\alpha\beta}^{-1} = \hat{L}_-\psi_{\alpha\beta}$ happens to be zero). In the notation for the spherical harmonics that we use in the chapter, $\hat{L}_-Y_l^m(\theta, \phi) \propto Y_l^{m-1}(\theta, \phi)$.

6-52. Show that each application of $\hat{L}_+$ to $\psi_{\alpha\beta}$ raises the eigenvalue by $\hbar$, so long as the result is nonzero.

6-53. Show that each application of $\hat{L}_-$ to $\psi_{\alpha\beta}$ lowers the eigenvalue by $\hbar$, so long as the result is nonzero.

6-54. According to Problem 6–48, $\hat{L}^2$ commutes with $\hat{L}_+$ and $\hat{L}_-$. Now prove that $\hat{L}^2$ commutes with $\hat{L}_+^2$ and $\hat{L}_-^2$. Now prove that

$$[\hat{L}^2, \hat{L}_\pm^m] = 0 \qquad m = 1, 2, 3, \ldots$$

6-55. In Problems 6–50 through 6–53, we proved that if $\psi_{\alpha\beta}^{\pm m} = \hat{L}_{\pm}^m \psi_{\alpha\beta}$, then

$$\hat{L}_z \psi_{\alpha\beta}^{\pm m} = (\alpha \pm m\hbar)\psi_{\alpha\beta}^{\pm m} \qquad m = 0,\ 1,\ 2,\ \ldots$$

so long as the result is non-zero. The operators $\hat{L}_{\pm}$ are called raising $(\hat{L}_{+})$ or lowering $(\hat{L}_{-})$ operators because they raise or lower the eigenvalues of $\hat{L}_z$. They are also called ladder operators because the set of eigenvalues $\alpha \pm m\hbar$ form a ladder of eigenvalues. Use the result of Problem 6–54 to show that

$$\hat{L}^2 \psi_{\alpha\beta}^{\pm m} = \beta^2 \psi_{\alpha\beta}^{\pm m}$$

6-56. Start with

$$\hat{L}_z \psi_{\alpha\beta}^{\pm m} = (\alpha \pm m\hbar)\psi_{\alpha\beta}^{\pm m}$$

Operate on both sides with $\hat{L}_z$ and subtract the result from (Problem 6–55)

$$\hat{L}^2 \psi_{\alpha\beta}^{\pm m} = \beta^2 \psi_{\alpha\beta}^{\pm m}$$

to get

$$(\hat{L}^2 - \hat{L}_z^2)\psi_{\alpha\beta}^{\pm m} = (\hat{L}_x^2 + \hat{L}_y^2)\psi_{\alpha\beta}^{\pm m} = [\beta^2 - (\alpha \pm m\hbar)^2]\psi_{\alpha\beta}^{\pm m}$$

Because the operator $\hat{L}_x^2 + \hat{L}_y^2$ corresponds to a nonnegative physical quantity, show that

$$\beta^2 - (\alpha \pm m\hbar)^2 \geq 0$$

or that

$$-\beta \leq \alpha \pm m\hbar \leq \beta \qquad m = 0,\ 1,\ 2,\ \ldots$$

Because β is fixed, the possible values of m must be finite in number.

6-57. Let α_{max} be the largest possible value of $\alpha \pm m\hbar$. By definition then, we have that

$$\hat{L}_z \psi_{\alpha_{\text{max}}\beta} = \alpha_{\text{max}} \psi_{\alpha_{\text{max}}\beta}$$

$$\hat{L}^2 \psi_{\alpha_{\text{max}}\beta} = \beta^2 \psi_{\alpha_{\text{max}}\beta}$$

and

$$\hat{L}_+ \psi_{\alpha_{\text{max}}\beta} = 0$$

Operate on the last equation with $\hat{L}_-$ to obtain

$$\hat{L}_- \hat{L}_+ \psi_{\alpha_{\text{max}}\beta} = 0$$
$$= (\hat{L}^2 - \hat{L}_z^2 - \hbar \hat{L}_z)\psi_{\alpha_{\text{max}}\beta}$$

and

$$\beta^2 = \alpha_{\text{max}}^2 + \hbar\alpha_{\text{max}}$$

Use a parallel procedure on $\psi_{\alpha_{\min}\beta}$ to obtain

$$\beta^2 = \alpha_{\min}^2 - \hbar\alpha_{\min}$$

Now show that $\alpha_{\max} = -\alpha_{\min}$, and then argue that the possible values of the eigenvalues α of $\hat{L}_z$ extend from $+\alpha_{\max}$ to $-\alpha_{\max}$ in steps of magnitude $\hbar$. This is possible only if $\alpha_{\max}$ is itself an integer (or perhaps a half-integer) times $\hbar$. Finally, show that this last result leads to

$$\beta^2 = l(l+1)\hbar^2 \qquad l = 0,\ 1,\ 2,\ \ldots$$

and

$$\alpha = m\hbar \qquad m = 0,\ \pm 1,\ \pm 2,\ \ldots,\ \pm l$$

6-58. According to Problems 6–50 and 6–51,

$$\hat{L}_+ Y_l^m(\theta, \phi) = \hbar c_{lm}^+ Y_l^{m+1}(\theta, \phi)$$

and

$$\hat{L}_- Y_l^m(\theta, \phi) = \hbar c_{lm}^- Y_l^{m-1}(\theta, \phi)$$

where we are using the notation $Y_l^m(\theta, \phi)$ instead of $\psi_{\alpha,\beta}$. Show that

$$\hat{L}_x Y_l^m(\theta, \phi) = \frac{\hbar c_{lm}^+}{2} Y_l^{m+1}(\theta, \phi) + \frac{\hbar c_{lm}^-}{2} Y_l^{m-1}(\theta, \phi)$$

and

$$\hat{L}_y Y_l^m(\theta, \phi) = \frac{\hbar c_{lm}^+}{2i} Y_l^{m+1}(\theta, \phi) - \frac{\hbar c_{lm}^-}{2i} Y_l^{m-1}(\theta, \phi)$$

Use this result to show that

$$\langle L_x \rangle = \langle L_y \rangle = 0$$

for any rotational state (see Problem 6–14).

6-59. Show that

$$\hat{L}_+ = \hbar e^{i\phi}\left[\frac{\partial}{\partial\theta} + i\cot\theta\,\frac{\partial}{\partial\phi}\right]$$

and

$$\hat{L}_- = \hbar e^{-i\phi}\left[-\frac{\partial}{\partial\theta} + i\cot\theta\,\frac{\partial}{\partial\phi}\right]$$

DETERMINANTS

In Chapter 7, we will encounter n linear algebraic equations in n unknowns. Such equations are best solved by means of determinants, which we discuss in this MathChapter. Consider the pair of linear algebraic equations

$$
\begin{aligned}
a_{11}x + a_{12}y &= d_1 \\
a_{21}x + a_{22}y &= d_2
\end{aligned}
\tag{E.1}
$$

If we multiply the first of these equations by a_{22} and the second by a_{12} and then subtract, we obtain

$$
(a_{11}a_{22} - a_{12}a_{21})x = d_1a_{22} - d_2a_{12}
$$

or

$$
x = \frac{a_{22}d_1 - a_{12}d_2}{a_{11}a_{22} - a_{12}a_{21}}
\tag{E.2}
$$

Similarly, if we multiply the first by a_{21} and the second by a_{11} and then subtract, we get

$$
y = \frac{a_{11}d_2 - a_{21}d_1}{a_{11}a_{22} - a_{12}a_{21}}
\tag{E.3}
$$

Notice that the denominators in both Equations E.2 and E.3 are the same. We represent $a_{11}a_{22} - a_{12}a_{21}$ by the quantity $\begin{vmatrix} a_{11} & a_{12} \\ a_{21} & a_{22} \end{vmatrix}$, which equals $a_{11}a_{22} - a_{12}a_{21}$ and is called a 2×2 *determinant*. The reason for introducing this notation is that it readily generalizes to the treatment of n linear algebraic equations in n unknowns. Generally, a $n \times n$

231

determinant is a square array of n^2 elements arranged in n rows and n columns. A 3×3 determinant is given by

$$\begin{vmatrix} a_{11} & a_{12} & a_{13} \\ a_{21} & a_{22} & a_{23} \\ a_{31} & a_{32} & a_{33} \end{vmatrix} = \begin{array}{l} a_{11}a_{22}a_{33} + a_{21}a_{32}a_{13} + a_{12}a_{23}a_{31} \\ -a_{31}a_{22}a_{13} - a_{21}a_{12}a_{33} - a_{11}a_{23}a_{32} \end{array} \qquad \text{(E.4)}$$

(We will prove this soon.) Notice that the element a_{ij} occurs at the intersection of the ith row and the jth column.

Equation E.4 and the corresponding equations for evaluating higher-order determinants can be obtained in a systematic manner. First we define a cofactor. The *cofactor*, A_{ij}, of an element a_{ij} is a $(n-1) \times (n-1)$ determinant obtained by deleting the ith row and the jth column, multiplied by $(-1)^{i+j}$. For example, A_{12}, the cofactor of element a_{12} of

$$D = \begin{vmatrix} a_{11} & a_{12} & a_{13} \\ a_{21} & a_{22} & a_{23} \\ a_{31} & a_{32} & a_{33} \end{vmatrix}$$

is

$$A_{12} = (-1)^{1+2} \begin{vmatrix} a_{21} & a_{23} \\ a_{31} & a_{33} \end{vmatrix}$$

EXAMPLE E–1
Evaluate the cofactor of each of the first-row elements in

$$D = \begin{vmatrix} 2 & -1 & 1 \\ 0 & 3 & -1 \\ 2 & -2 & 1 \end{vmatrix}$$

SOLUTION: The cofactor of a_{11} is

$$A_{11} = (-1)^{1+1} \begin{vmatrix} 3 & -1 \\ -2 & 1 \end{vmatrix} = 3 - 2 = 1$$

The cofactor of a_{12} is

$$A_{12} = (-1)^{1+2} \begin{vmatrix} 0 & -1 \\ 2 & 1 \end{vmatrix} = -2$$

and the cofactor of a_{13} is

$$A_{13} = (-1)^{1+3} \begin{vmatrix} 0 & 3 \\ 2 & -2 \end{vmatrix} = -6$$

We can use cofactors to evaluate determinants. The value of the 3×3 determinant in Equation E.4 can be obtained from the formula

$$
\begin{vmatrix} a_{11} & a_{12} & a_{13} \\ a_{21} & a_{22} & a_{23} \\ a_{31} & a_{32} & a_{33} \end{vmatrix} = a_{11}A_{11} + a_{12}A_{12} + a_{13}A_{13} \tag{E.5}
$$

Thus, the value of D in Example E–1 is

$$
D = (2)(1) + (-1)(-2) + (1)(-6) = -2
$$

EXAMPLE E–2

Evaluate D in Example E–1 by expanding in terms of the first column of elements instead of the first row.

SOLUTION: We will use the formula

$$
D = a_{11}A_{11} + a_{21}A_{21} + a_{31}A_{31}
$$

The various cofactors are

$$
A_{11} = (-1)^2 \begin{vmatrix} 3 & -1 \\ -2 & 1 \end{vmatrix} = 1
$$

$$
A_{21} = (-1)^3 \begin{vmatrix} -1 & 1 \\ -2 & 1 \end{vmatrix} = -1
$$

and

$$
A_{31} = (-1)^4 \begin{vmatrix} -1 & 1 \\ 3 & -1 \end{vmatrix} = -2
$$

and so

$$
D = (2)(1) + (0)(-1) + (2)(-2) = -2
$$

Notice that we obtained the same answer for D as we did for Example E–1. This result illustrates the general fact that a determinant may be evaluated by expanding in terms of the cofactors of the elements of any row or any column. If we choose the second row of D, then we obtain

$$
D = (0)(-1)^3 \begin{vmatrix} -1 & 1 \\ -2 & 1 \end{vmatrix} + (3)(-1)^4 \begin{vmatrix} 2 & 1 \\ 2 & 1 \end{vmatrix} + (-1)(-1)^5 \begin{vmatrix} 2 & -1 \\ 2 & -2 \end{vmatrix} = -2
$$

Although we have discussed only 3×3 determinants, the procedure is readily extended to determinants of any order.

EXAMPLE E–3

In Chapter 10 we will meet the *determinantal equation*

$$\begin{vmatrix} x & 1 & 0 & 0 \\ 1 & x & 1 & 0 \\ 0 & 1 & x & 1 \\ 0 & 0 & 1 & x \end{vmatrix} = 0$$

Expand this determinantal equation into a quartic equation for x.

SOLUTION: Expand about the first row of elements to obtain

$$x \begin{vmatrix} x & 1 & 0 \\ 1 & x & 1 \\ 0 & 1 & x \end{vmatrix} - \begin{vmatrix} 1 & 1 & 0 \\ 0 & x & 1 \\ 0 & 1 & x \end{vmatrix} = 0$$

Now expand about the first column of each of the 3×3 determinants to obtain

$$(x)(x) \begin{vmatrix} x & 1 \\ 1 & x \end{vmatrix} - (x)(1) \begin{vmatrix} 1 & 0 \\ 1 & x \end{vmatrix} - (1) \begin{vmatrix} x & 1 \\ 1 & x \end{vmatrix} = 0$$

or

$$x^2(x^2 - 1) - x(x) - (1)(x^2 - 1) = 0$$

or

$$x^4 - 3x^2 + 1 = 0$$

Note that although we can choose any row or column to expand the determinant, it is easiest to take the one with the most zeroes.

A number of properties of determinants are useful to know:

1. The value of a determinant is unchanged if the rows are made into columns in the same order; in other words, first row becomes first column, second row becomes second column, and so on. For example,

$$\begin{vmatrix} 1 & 2 & 5 \\ -1 & 0 & -1 \\ 3 & 1 & 2 \end{vmatrix} = \begin{vmatrix} 1 & -1 & 3 \\ 2 & 0 & 1 \\ 5 & -1 & 2 \end{vmatrix}$$

2. If any two rows or columns are the same, the value of the determinant is zero. For example,

$$\begin{vmatrix} 4 & 2 & 4 \\ -1 & 0 & -1 \\ 3 & 1 & 3 \end{vmatrix} = 0$$

3. If any two rows or columns are interchanged, the sign of the determinant is changed. For example,

$$
\begin{vmatrix} 3 & 1 & -1 \\ -6 & 4 & 5 \\ 1 & 2 & 2 \end{vmatrix} = - \begin{vmatrix} 1 & 3 & -1 \\ 4 & -6 & 5 \\ 2 & 1 & 2 \end{vmatrix}
$$

4. If every element in a row or column is multiplied by a factor k, the value of the determinant is multiplied by k. For example,

$$
\begin{vmatrix} 6 & 8 \\ -1 & 2 \end{vmatrix} = 2 \begin{vmatrix} 3 & 4 \\ -1 & 2 \end{vmatrix}
$$

5. If any row or column is written as the sum or difference of two or more terms, the determinant can be written as the sum or difference of two or more determinants according to

$$
\begin{vmatrix} a_{11} \pm a_{11}' & a_{12} & a_{13} \\ a_{21} \pm a_{21}' & a_{22} & a_{23} \\ a_{31} \pm a_{31}' & a_{32} & a_{33} \end{vmatrix} = \begin{vmatrix} a_{11} & a_{12} & a_{13} \\ a_{21} & a_{22} & a_{23} \\ a_{31} & a_{32} & a_{33} \end{vmatrix} \pm \begin{vmatrix} a_{11}' & a_{12} & a_{13} \\ a_{21}' & a_{22} & a_{23} \\ a_{31}' & a_{32} & a_{33} \end{vmatrix}
$$

For example,

$$
\begin{vmatrix} 3 & 3 \\ 2 & 6 \end{vmatrix} = \begin{vmatrix} 2+1 & 3 \\ -2+4 & 6 \end{vmatrix} = \begin{vmatrix} 2 & 3 \\ -2 & 6 \end{vmatrix} + \begin{vmatrix} 1 & 3 \\ 4 & 6 \end{vmatrix}
$$

6. The value of a determinant is unchanged if one row or column is added or subtracted to another, as in

$$
\begin{vmatrix} a_{11} & a_{12} & a_{13} \\ a_{21} & a_{22} & a_{23} \\ a_{31} & a_{32} & a_{33} \end{vmatrix} = \begin{vmatrix} a_{11}+a_{12} & a_{12} & a_{13} \\ a_{21}+a_{22} & a_{22} & a_{23} \\ a_{31}+a_{32} & a_{32} & a_{33} \end{vmatrix}
$$

For example

$$
\begin{vmatrix} 1 & -1 & 3 \\ 4 & 0 & 2 \\ 1 & 2 & 1 \end{vmatrix} = \begin{vmatrix} 0 & -1 & 3 \\ 4 & 0 & 2 \\ 3 & 2 & 1 \end{vmatrix} = \begin{vmatrix} 0 & -1 & 3 \\ 4 & 0 & 2 \\ 7 & 2 & 3 \end{vmatrix}
$$

In the first case we add column 2 to column 1, and in the second case we added row 2 to row 3. This procedure may be repeated n times to obtain

$$
\begin{vmatrix} a_{11} & a_{12} & a_{13} \\ a_{21} & a_{22} & a_{23} \\ a_{31} & a_{32} & a_{33} \end{vmatrix} = \begin{vmatrix} a_{11}+na_{12} & a_{12} & a_{13} \\ a_{21}+na_{22} & a_{22} & a_{23} \\ a_{31}+na_{32} & a_{32} & a_{33} \end{vmatrix} \tag{E.6}
$$

This result is easy to prove:

$$
\begin{vmatrix} a_{11}+na_{12} & a_{12} & a_{13} \\ a_{21}+na_{22} & a_{22} & a_{23} \\ a_{31}+na_{32} & a_{32} & a_{33} \end{vmatrix} = \begin{vmatrix} a_{11} & a_{12} & a_{13} \\ a_{21} & a_{22} & a_{23} \\ a_{31} & a_{32} & a_{33} \end{vmatrix} + n\begin{vmatrix} a_{12} & a_{12} & a_{13} \\ a_{22} & a_{22} & a_{23} \\ a_{32} & a_{32} & a_{33} \end{vmatrix}
$$

$$
= \begin{vmatrix} a_{11} & a_{12} & a_{13} \\ a_{21} & a_{22} & a_{23} \\ a_{31} & a_{32} & a_{33} \end{vmatrix} + 0
$$

where we used rule 5 to write the first line. The second determinant on the right side equals zero because two columns are the same (rule 2).

We provided these rules because simultaneous linear algebraic equations can be solved in terms of determinants. For simplicity, we will consider only a pair of equations but the final result is easy to generalize. Consider the two equations

$$
\begin{aligned}
a_{11}x + a_{12}y &= d_1 \\
a_{21}x + a_{22}y &= d_2
\end{aligned}
\tag{E.7}
$$

If $d_1 = d_2 = 0$, the equations are said to be *homogeneous*. Otherwise, they are called *inhomogeneous*. Let's assume at first that they are inhomogeneous. The determinant of the coefficients of x and y is

$$
D = \begin{vmatrix} a_{11} & a_{12} \\ a_{21} & a_{22} \end{vmatrix}
$$

According to Rule 4,

$$
\begin{vmatrix} a_{11}x & a_{12} \\ a_{21}x & a_{22} \end{vmatrix} = xD
$$

Furthermore, according to Rule 6,

$$
\begin{vmatrix} a_{11}x + a_{12}y & a_{12} \\ a_{21}x + a_{22}y & a_{22} \end{vmatrix} = xD
\tag{E.8}
$$

If we substitute Equation E.7 into Equation E.8, then we have

$$
\begin{vmatrix} d_1 & a_{12} \\ d_2 & a_{22} \end{vmatrix} = xD
$$

Solving for x gives

$$
x = \frac{\begin{vmatrix} d_1 & a_{12} \\ d_2 & a_{22} \end{vmatrix}}{\begin{vmatrix} a_{11} & a_{12} \\ a_{21} & a_{22} \end{vmatrix}}
\tag{E.9}
$$

Similarly, we get

$$y = \frac{\begin{vmatrix} a_{11} & d_1 \\ a_{21} & d_2 \end{vmatrix}}{\begin{vmatrix} a_{11} & a_{12} \\ a_{21} & a_{22} \end{vmatrix}} \qquad \text{(E.10)}$$

Notice that Equations E.9 and E.10 are identical to Equations E.2 and E.3. The solution for x and y in terms of determinants is called *Cramer's rule*. Note that the determinant in the numerator is obtained by replacing the column in D that is associated with the unknown quantity with the column associated with the right sides of Equation E.7. This result is readily extended to more than two simultaneous equations.

EXAMPLE E–4
Solve the equations

$$x + y + z = 2$$
$$2x - y - z = 1$$
$$x + 2y - z = -3$$

SOLUTION: The extension of Equations E.9 and E.10 is

$$x = \frac{\begin{vmatrix} 2 & 1 & 1 \\ 1 & -1 & -1 \\ -3 & 2 & -1 \end{vmatrix}}{\begin{vmatrix} 1 & 1 & 1 \\ 2 & -1 & -1 \\ 1 & 2 & -1 \end{vmatrix}} = \frac{9}{9} = 1$$

Similarly,

$$y = \frac{\begin{vmatrix} 1 & 2 & 1 \\ 2 & 1 & -1 \\ 1 & -3 & -1 \end{vmatrix}}{\begin{vmatrix} 1 & 1 & 1 \\ 2 & -1 & -1 \\ 1 & 2 & -1 \end{vmatrix}} = \frac{-9}{9} = -1$$

and

$$z = \frac{\begin{vmatrix} 1 & 1 & 2 \\ 2 & -1 & 1 \\ 1 & 2 & -3 \end{vmatrix}}{\begin{vmatrix} 1 & 1 & 1 \\ 2 & -1 & -1 \\ 1 & 2 & -1 \end{vmatrix}} = \frac{18}{9} = 2$$

What happens if $d_1 = d_2 = 0$ in Equation E.7? In that case, we find that $x = y = 0$, which is an obvious solution called a *trivial solution*. The only way that we could obtain a nontrivial solution for a set of homogeneous equations is for the denominator in Equations E.9 and E.10 to be zero, or for

$$D = \begin{vmatrix} a_{11} & a_{12} \\ a_{21} & a_{22} \end{vmatrix} = 0 \tag{E.11}$$

In Chapter 10, when we discuss ethene, we will meet the equations

$$c_1(\alpha - E) + c_2\beta = 0$$

and

$$c_1\beta + c_2(\alpha - E) = 0$$

where c_1 and c_2 are the unknowns (corresponding to x and y in Equation E.7), α and β are known quantities, and E is the energy of the π electrons. We can use Equation E.11 to derive an expression for the π-electron energies in ethene. Equation E.11 says that for a nontrivial solution (c_1, c_2) to exist, we must have that

$$\begin{vmatrix} \alpha - E & \beta \\ \beta & \alpha - E \end{vmatrix} = 0$$

or that $(\alpha - E)^2 - \beta^2 = 0$. Taking the square root of both sides and solving for E gives

$$E = \alpha \pm \beta$$

Although we considered only two simultaneous homogeneous algebraic equations, Equation E.11 is readily extended to any number. We will use this result in the next chapter.

Problems

E-1. Evaluate the determinant

$$D = \begin{vmatrix} 2 & 1 & 1 \\ -1 & 3 & 2 \\ 2 & 0 & 1 \end{vmatrix}$$

Add column 2 to column 1 to get

$$\begin{vmatrix} 3 & 1 & 1 \\ 2 & 3 & 2 \\ 2 & 0 & 1 \end{vmatrix}$$

and evaluate it. Compare your result with the value of D. Now add row 2 of D to row 1 of D to get

$$\begin{vmatrix} 1 & 4 & 3 \\ -1 & 3 & 2 \\ 2 & 0 & 1 \end{vmatrix}$$

and evaluate it. Compare your result with the value of D above.

E-2. Interchange columns 1 and 3 in D in Problem E–1 and evaluate the resulting determinant. Compare your result with the value of D. Interchange rows 1 and 2 of D and do the same.

E-3. Evaluate the determinant

$$D = \begin{vmatrix} 1 & 6 & 1 \\ -2 & 4 & -2 \\ 1 & -3 & 1 \end{vmatrix}$$

Can you determine its value by inspection? What about

$$D = \begin{vmatrix} 2 & 6 & 1 \\ -4 & 4 & -2 \\ 2 & -3 & 1 \end{vmatrix}$$

E-4. Find the values of x that satisfy the following determinantal equation

$$\begin{vmatrix} x & 1 & 1 & 1 \\ 1 & x & 0 & 0 \\ 1 & 0 & x & 0 \\ 1 & 0 & 0 & x \end{vmatrix} = 0$$

E-5. Find the values of x that satisfy the following determinantal equation

$$\begin{vmatrix} x & 1 & 0 & 1 \\ 1 & x & 1 & 0 \\ 0 & 1 & x & 1 \\ 1 & 0 & 1 & x \end{vmatrix} = 0$$

E-6. Show that

$$\begin{vmatrix} \cos\theta & -\sin\theta & 0 \\ \sin\theta & \cos\theta & 0 \\ 0 & 0 & 1 \end{vmatrix} = 1$$

E-7. Solve the following set of equations using Cramer's rule

$$x + y = 2$$
$$3x - 2y = 5$$

E-8. Solve the following set of equations using Cramer's rule

$$x + 2y + 3z = -5$$
$$-x - 3y + z = -14$$
$$2x + y + z = 1$$

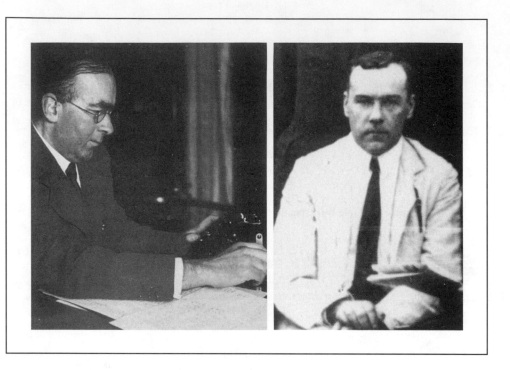

Douglas Hartree (left) and Vladimir Fock (right) formulated an approximate method for calculating atomic (and molecular) properties in the 1930s that is still used today. **Douglas Hartree** was born on March 27, 1897 in Cambridge, England, and died in 1958. After receiving his Ph.D. in applied mathematics from Cambridge University in 1926, he spent 1929 to 1937 as the chair of applied mathematics and 1937 to 1946 as professor of theoretical physics at the University of Manchester. From 1946 until his death, he was Plummer Professor of Mathematical Physics at Cambridge University. Hartree pioneered the use of computers in research in the United Kingdom. He developed powerful methods of numerical analysis, which he applied to problems in atomic structure, ballistics, atmospheric physics, and hydrodynamics. Hartree was also an accomplished pianist and drummer.

Vladimir Fock (also Fok) was born on December 22, 1898, in Petrograd (later Leningrad and now St. Petersburg), Russia, and died in 1974. After graduating from Petrograd University in 1922, he spent 1924 to 1936 at the Leningrad Institute of Physics and Technology. He spent 1936 to 1953 at the Institute of Physics, USSR Academy of Science, and then returned to Leningrad University, where he remained until his death. Fock generalized the equations of Hartree to include the fact that electronic wave functions must be antisymmetric under the interchange of any two electrons (the Pauli Exclusion Principle). Fock's research was in quantum electrodynamics, general relativity, and solid-state physics. He was almost completely deaf as an adult.

CHAPTER 7

Approximation Methods

We stated in Chapter 6 that the Schrödinger equation cannot be solved exactly for any atom or molecule more complicated than the hydrogen atom. At first thought, this statement would appear to certainly deprive quantum mechanics of any interest to chemists, but fortunately approximation methods can be used to solve the Schrödinger equation to almost any desired accuracy. In this chapter, we will present the two most widely used of these methods, the variational method and perturbation theory. We will present the basic equations of the variational method and perturbation theory and then apply them to a variety of problems.

7–1. The Variational Method Provides an Upper Bound to the Ground-State Energy of a System

We will first illustrate the *variational method*. Consider the ground state of some arbitrary system. The ground-state wave function ψ_0 and energy E_0 satisfy the Schrödinger equation

$$\hat{H}\psi_0 = E_0\psi_0 \tag{7.1}$$

Multiply Equation 7.1 from the left by ψ_0^* and integrate over all space to obtain

$$E_0 = \frac{\int \psi_0^* \hat{H}\psi_0 d\tau}{\int \psi_0^* \psi_0 d\tau} \tag{7.2}$$

where $d\tau$ represents the appropriate volume element. We have not set the denominator equal to unity in Equation 7.2 to allow for the possibility that ψ_0 is not normalized beforehand. A beautiful theorem says that if we substitute any other function ϕ for ψ_0 in Equation 7.2 and calculate the corresponding energy according to

$$E_\phi = \frac{\int \phi^* \hat{H} \phi d\tau}{\int \phi^* \phi d\tau} \tag{7.3}$$

then E_ϕ will be greater than the ground-state energy E_0. In an equation, we have the *variational principle*

$$E_\phi \geq E_0 \tag{7.4}$$

where the equality holds only if $\phi = \psi_0$, the exact wave function. We will not prove the variational principle here (although it is fairly easy), but Problem 7–1 takes you through the proof step by step.

The variational principle says that we can calculate an upper bound to E_0 by using any trial function we wish. The closer ϕ is to ψ_0 in some sense, the closer E_ϕ will be to E_0. We can choose a trial function ϕ such that it depends upon some arbitrary parameters, α, β, γ, ..., called *variational parameters*. The energy also will depend upon these variational parameters, and Equation 7.4 will read

$$E_\phi(\alpha, \beta, \gamma, \ldots) \geq E_0 \tag{7.5}$$

Now we can minimize E_ϕ with respect to each of the variational parameters and thereby determine the best possible ground-state energy that can be obtained from our trial wave function.

As a specific example, consider the ground state of the hydrogen atom. Although we know from Chapter 6 that we can solve this problem exactly, let's assume that we cannot and use the variational method. We will compare our variational result to the exact result. Because $l = 0$ in the ground state, the Hamiltonian operator is (cf. Equation 6.43)

$$\hat{H} = -\frac{\hbar^2}{2m_e r^2} \frac{d}{dr} \left(r^2 \frac{d}{dr} \right) - \frac{e^2}{4\pi \varepsilon_0 r} \tag{7.6}$$

Even if we did not know the exact solution, we would expect that the wave function decays to zero with increasing r. Consequently, as a *trial function*, we will try a Gaussian function of the form $\phi(r) = e^{-\alpha r^2}$, where α is a variational parameter. By a straightforward calculation, we can show that (cf. Problem 7–2)

$$4\pi \int_0^\infty \phi^*(r) \hat{H} \phi(r) r^2 dr = \frac{3\hbar^2 \pi^{3/2}}{4\sqrt{2} m_e \alpha^{1/2}} - \frac{e^2}{4\varepsilon_0 \alpha}$$

and that

$$4\pi \int_0^\infty \phi^*(r) \phi(r) r^2 dr = \left(\frac{\pi}{2\alpha} \right)^{3/2}$$

Therefore, from Equation 7.3,

$$E(\alpha) = \frac{3\hbar^2 \alpha}{2m_e} - \frac{e^2 \alpha^{1/2}}{2^{1/2} \varepsilon_0 \pi^{3/2}} \tag{7.7}$$

We now minimize $E(\alpha)$ with respect to α by differentiating $E(\alpha)$ with respect to α and setting the result equal to zero. We solve the equation

$$\frac{dE(\alpha)}{d\alpha} = \frac{3\hbar^2}{2m_e} - \frac{e^2}{(2\pi)^{3/2} \varepsilon_0 \alpha^{1/2}} = 0$$

for α to give

$$\alpha = \frac{m_e^2 e^4}{18\pi^3 \varepsilon_0^2 \hbar^4} \tag{7.8}$$

as the value of α that minimizes $E(\alpha)$. Substituting Equation 7.8 back into Equation 7.7, we find that

$$E_{\min} = -\frac{4}{3\pi} \left(\frac{m_e e^4}{16\pi^2 \varepsilon_0^2 \hbar^2} \right) = -0.424 \left(\frac{m_e e^4}{16\pi^2 \varepsilon_0^2 \hbar^2} \right) \tag{7.9}$$

compared with the exact value (Equation 6.44)

$$E_0 = -\frac{1}{2} \left(\frac{m_e e^4}{16\pi^2 \varepsilon_0^2 \hbar^2} \right) = -0.500 \left(\frac{m_e e^4}{16\pi^2 \varepsilon_0^2 \hbar^2} \right) \tag{7.10}$$

Note that $E_{\min} > E_0$, as the variational theorem assures us.

The normalized trial function is given by $\phi(r) = (2\alpha/\pi)^{3/4} e^{-\alpha r^2}$, where α is given by Equation 7.8, and the exact ground-state wave function (the hydrogen $1s$ orbital) is given by $(1/\pi a_0^3)^{1/2} e^{-r/a_0}$, where $a_0 = 4\pi \varepsilon_0 \hbar^2/m_e e^2$ is the Bohr radius. We can compare these two functions by first expressing α in terms of a_0, which comes out to be

$$\alpha = \frac{m_e^2 e^4}{18\pi^3 \varepsilon_0^2 \hbar^4} = \frac{16}{18\pi} \cdot \frac{m_e^2 e^4}{16\pi^2 \varepsilon_0^2 \hbar^4} = \frac{8}{9\pi} \cdot \frac{1}{a_0^2}$$

Thus, we can write the trial function as

$$\phi(r) = \frac{8}{3^{3/2}\pi} \left(\frac{1}{\pi a_0^3} \right)^{1/2} e^{-(8/9\pi)r^2/a_0^2}$$

This result is compared with ψ_{1s} in Figure 7.1.

Our variational calculation for the ground-state energy of a hydrogen atom is within 80% of the exact result. This result was obtained using a trial function with only one variational parameter. We can obtain progressively better results by using more flexible trial functions, containing more parameters. In fact, we will see such a progression (Table 7.1) that approaches the exact energy in Section 7–3.

244

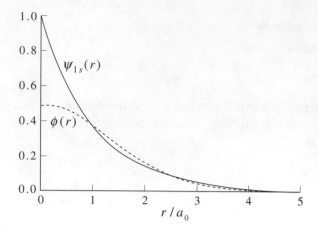

FIGURE 7.1
A comparison of the optimized Gaussian trial wave function $\phi(r) = (2\alpha/\pi)^{3/4}e^{-\alpha r^2}$, where α is given by Equation 7.8 (dashed line), and the exact ground-state hydrogen wave function, $\psi(r) = (1/\pi a_0^3)^{1/2}e^{-r/a_0}$, where $a_0 = 4\pi\varepsilon_0\hbar^2/m_e e^2$ is the Bohr radius (solid line). Both functions are plotted against the reduced distance, r/a_0, and the vertical axis is expressed in units of $1/(\pi a_0^3)^{1/2}$.

EXAMPLE 7–1
Use a trial function of the form $e^{-\alpha r}$ to calculate the ground-state energy of a hydrogen atom.

SOLUTION: The Hamiltonian operator for the ground state of hydrogen is given by Equation 7.6. Therefore,

$$\hat{H}e^{-\alpha r} = \frac{\alpha\hbar^2}{2m_e r^2}(2r - \alpha r^2)e^{-\alpha r} - \frac{e^2}{4\pi\varepsilon_0 r}e^{-\alpha r}$$

and the numerator of Equation 7.3 is

$$\text{numerator} = 4\pi \int_0^\infty e^{-\alpha r}\hat{H}e^{-\alpha r}r^2 dr$$

$$= \frac{2\pi\alpha\hbar^2}{m_e}\int_0^\infty (2r - \alpha r^2)e^{-2\alpha r}dr - \frac{e^2}{\varepsilon_0}\int_0^\infty e^{-2\alpha r}r dr$$

$$= \frac{2\pi\alpha\hbar^2}{m_e}\left[\frac{2}{(2\alpha)^2} - \frac{2\alpha}{(2\alpha)^3}\right] - \frac{e^2}{\varepsilon_0}\frac{1}{(2\alpha)^2}$$

$$= \frac{\pi\hbar^2}{2m_e\alpha} - \frac{e^2}{4\varepsilon_0\alpha^2}$$

We have used the fact that

$$\int_0^\infty x^n e^{-ax}dx = \frac{n!}{a^{n+1}}$$

to evaluate all the above integrals. Similarly, the denominator of Equation 7.3 is

$$\text{denominator} = 4\pi \int_0^\infty e^{-2\alpha r} r^2 dr = \frac{8\pi}{(2\alpha)^3} = \frac{\pi}{\alpha^3}$$

and so

$$E(\alpha) = \frac{\hbar^2 \alpha^2}{2m_e} - \frac{e^2 \alpha}{4\pi \varepsilon_0}$$

Setting $dE/d\alpha = 0$ gives

$$\alpha = \frac{m_e e^2}{4\pi \varepsilon_0 \hbar^2}$$

and substituting this result back into $E(\alpha)$ gives

$$E_{\text{min}} = -\frac{1}{2} \left(\frac{m_e e^4}{16\pi^2 \varepsilon_0^2 \hbar^2} \right)$$

This happens to be the exact ground-state energy of a hydrogen atom. We fortuitously chose the exact form of the ground-state wave function and so ended up with the exact energy.

EXAMPLE 7–2

Use the variational principle to estimate the ground-state energy of a harmonic oscillator using the trial function

$$\phi(x) = \frac{1}{1 + \beta x^2}$$

SOLUTION: The Hamiltonian operator of a harmonic oscillator is

$$\hat{H} = -\frac{\hbar^2}{2\mu} \frac{d^2}{dx^2} + \frac{k}{2} x^2$$

Therefore, we must first find $d^2\phi/dx^2$, which comes out to be

$$\frac{d^2\phi}{dx^2} = -\frac{2\beta}{(1 + \beta x^2)^2} + \frac{8\beta^2 x^2}{(1 + \beta x^2)^3}$$

The numerator of Equation 7.3 is

$$\text{numerator} = -\frac{\hbar^2}{2\mu} \int_{-\infty}^\infty \left[-\frac{2\beta}{(1 + \beta x^2)^3} + \frac{8\beta^2 x^2}{(1 + \beta x^2)^4} \right] dx$$

$$+ \frac{k}{2} \int_{-\infty}^\infty \frac{x^2 dx}{(1 + \beta x^2)^2}$$

The necessary integrals can be found in handbooks and are given by

$$\int_{-\infty}^{\infty} \frac{dx}{(1 + \beta x^2)^3} = \frac{3\pi}{8\beta^{1/2}}$$

$$\int_{-\infty}^{\infty} \frac{x^2 dx}{(1 + \beta x^2)^4} = \frac{\pi}{16\beta^{1/2}}$$

and

$$\int_{-\infty}^{\infty} \frac{x^2 dx}{(1 + \beta x^2)^2} = \frac{\pi}{2\beta^{3/2}}$$

Using these integrals, we get

$$\text{numerator} = \frac{\hbar^2 \beta}{\mu} \cdot \frac{3\pi}{8\beta^{1/2}} - \frac{4\hbar^2 \beta^2}{\mu} \cdot \frac{\pi}{16\beta^{1/2}} + \frac{k}{2} \cdot \frac{\pi}{2\beta^{3/2}}$$

$$= \frac{\hbar^2 \pi}{8\mu} \beta^{1/2} + \frac{k\pi}{4\beta^{3/2}}$$

The denominator of Equation 7.3 is

$$\text{denominator} = \int_{-\infty}^{\infty} \frac{dx}{(1 + \beta x^2)^2} = \frac{\pi}{2\beta^{1/2}}$$

and so $E(\beta)$ is given by

$$E(\beta) = \frac{\hbar^2}{4\mu} \beta + \frac{k}{2\beta}$$

To find the minimum value of $E(\beta)$, we use

$$\frac{dE}{d\beta} = \frac{\hbar^2}{4\mu} - \frac{k}{2\beta^2} = 0$$

and find that the optimum value of β is

$$\beta = \frac{(2\mu k)^{1/2}}{\hbar}$$

If we substitute this value back into the above equation for $E(\beta)$, we obtain

$$E_{\text{min}} = \frac{2^{1/2}}{4} \hbar \left(\frac{k}{\mu}\right)^{1/2} + \frac{1}{2^{3/2}} \hbar \left(\frac{k}{\mu}\right)^{1/2}$$

$$= \frac{\hbar}{2^{1/2}} \left(\frac{k}{\mu}\right)^{1/2} = 0.707 \, \hbar \left(\frac{k}{\mu}\right)^{1/2}$$

The exact value for the ground-state energy of a harmonic oscillator is (Equation 5.30)

$$E_{\text{exact}} = \frac{\hbar}{2} \left(\frac{k}{\mu}\right)^{1/2} = 0.500 \, \hbar \left(\frac{k}{\mu}\right)^{1/2}$$

so we see that our simple trial function gives a result that is about 40% too high.

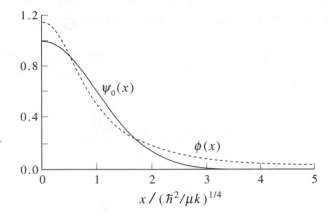

FIGURE 7.2
A comparison of the normalized, optimized trial function of Example 7–2 (dashed line) with the exact ground-state harmonic oscillator wave function (solid line). Both functions are plotted against $x/(\hbar^2/\mu k)^{1/4}$, and the vertical axis is expressed in units of $(\alpha/\pi)^{1/4}$, where $\alpha = (k\mu)^{1/2}/\hbar$ (see Example 7–3).

In Figure 7.2, the normalized, optimized trial function of Example 7–2 is compared with the exact ground-state harmonic oscillator wave function, $\psi_0(x) = (\alpha/\pi)^{1/4}e^{-\alpha x^2/2}$, where $\alpha = (k\mu)^{1/2}/\hbar$ (Section 5–6). Note that the trial function has greater amplitude at large displacements.

EXAMPLE 7–3
Determine the normalized, optimized trial function of Example 7–2.

SOLUTION: The (unnormalized) optimized trial function from Example 7–2 is

$$\phi(x) = \frac{1}{1 + \beta x^2}$$

with $\beta = (2\mu k)^{1/2}/\hbar$. To normalize $\phi(x)$, we need to evaluate

$$\int_{-\infty}^{\infty} \phi^2(x)dx = \int_{-\infty}^{\infty} \frac{dx}{(1 + \beta x^2)^2}$$

This integral is given in Example 7–2 as $\pi/2\beta^{1/2}$, so the normalized, optimized trial function is

$$\phi(x) = \frac{(4\beta/\pi^2)^{1/4}}{1 + \beta x^2}$$

In terms of the parameter $\alpha = (\mu k)^{1/2}/\hbar$ that occurs in the exact solution, $\beta = 2^{1/2}\alpha$, and so

$$\phi(x) = \left(\frac{\alpha}{\pi}\right)^{1/4} \frac{(2^{5/2}/\pi)^{1/4}}{1 + 2^{1/2}\alpha x^2} = \left(\frac{\alpha}{\pi}\right)^{1/4} \frac{1.158}{1 + 2^{1/2}\alpha x^2}$$

This function is plotted in Figure 7.2.

So far we have applied the variational method to two problems we actually know how to solve exactly. Now let's apply it to a problem for which we do not know the exact solution. We will use the variational method to estimate the ground-state energy of a helium atom. We saw at the end of Chapter 6 that the Hamiltonian operator for a helium atom is

$$\hat{H} = -\frac{\hbar^2}{2m_e}(\nabla_1^2 + \nabla_2^2) - \frac{2e^2}{4\pi\varepsilon_0}\left(\frac{1}{r_1} + \frac{1}{r_2}\right) + \frac{e^2}{4\pi\varepsilon_0}\frac{1}{r_{12}} \qquad (7.11)$$

The Schrödinger equation cannot be solved exactly for this system because of the term involving r_{12}. Equation 7.11 can be written in the form

$$\hat{H} = \hat{H}_{\mathrm{H}}(1) + \hat{H}_{\mathrm{H}}(2) + \frac{e^2}{4\pi\varepsilon_0}\frac{1}{r_{12}} \qquad (7.12)$$

where

$$\hat{H}_{\mathrm{H}}(j) = -\frac{\hbar^2}{2m_e}\nabla_j^2 - \frac{2e^2}{4\pi\varepsilon_0}\frac{1}{r_j} \qquad j = 1 \text{ and } 2 \qquad (7.13)$$

is the Hamiltonian operator for a single electron around a helium nucleus. Thus, $\hat{H}_{\mathrm{H}}(1)$ and $\hat{H}_{\mathrm{H}}(2)$ satisfy the equation

$$\hat{H}_{\mathrm{H}}(j)\psi_{\mathrm{H}}(r_j, \theta_j, \phi_j) = E_j\psi_{\mathrm{H}}(r_j, \theta_j, \phi_j) \qquad j = 1 \text{ or } 2 \qquad (7.14)$$

where $\psi_{\mathrm{H}}(r_j, \theta_j, \phi_j)$ is a hydrogenlike wave function with $Z = 2$ (Table 6.6) and where the E_j are given by (Problem 6–34)

$$E_j = -\frac{Z^2 m_e e^4}{32\pi^2\varepsilon_0^2\hbar^2 n_j^2} \qquad j = 1 \text{ or } 2 \qquad (7.15)$$

with $Z = 2$. If we ignore the interelectronic repulsion term $(e^2/4\pi\varepsilon_0 r_{12})$, then the Hamiltonian operator is separable and the ground-state wave function would be (Section 3–9)

$$\phi_0(\mathbf{r}_1, \mathbf{r}_2) = \psi_{1s}(\mathbf{r}_1)\psi_{1s}(\mathbf{r}_2) \qquad (7.16)$$

where (Table 6.5)

$$\psi_{1s}(\mathbf{r}_j) = \left(\frac{Z^3}{\pi a_0^3}\right)^{1/2} e^{-Zr_j/a_0} \qquad j = 1 \text{ or } 2 \qquad (7.17)$$

where $a_0 = 4\pi\varepsilon_0\hbar^2/m_e e^2$. We can use Equations 7.16 and 7.17 as a trial function using Z as a variational constant. Thus, we must evaluate

$$E(Z) = \int \phi_0(\mathbf{r}_1, \mathbf{r}_2) \hat{H} \phi_0(\mathbf{r}_1, \mathbf{r}_2) d\mathbf{r}_1 d\mathbf{r}_2 \tag{7.18}$$

with $\hat{H}$ given by Equation 7.11. The integral is a bit lengthy, albeit straightforward, to evaluate and is carried out step by step in Problem 7–32. The result is

$$E(Z) = \frac{m_e e^4}{16\pi^2 \varepsilon_0^2 \hbar^2} \left(Z^2 - \frac{27}{8} Z \right) \tag{7.19}$$

Equation 7.19 suggests that it is convenient to express E in units of $m_e e^4 / 16\pi^2 \varepsilon_0^2 \hbar^2$, and so we can write Equation 7.19 as

$$E(Z) = Z^2 - \frac{27}{8} Z \tag{7.20}$$

If we minimize $E(Z)$ with respect to Z, we find that $Z_{min} = 27/16$. We substitute this result back into Equation 7.20 to obtain

$$E_{min} = -\left(\frac{27}{16} \right)^2 = -2.8477 \tag{7.21}$$

compared with the most accurate calculated result of -2.9037 (in units of $m_e e^4 / 16\pi^2 \varepsilon_0^2 \hbar^2$), which is in excellent agreement with the experimental result (-2.9033). Thus, we achieve a fairly good result, considering the simplicity of the trial function.

The value of Z that minimizes E can be interpreted as an *effective nuclear charge*. The fact that Z comes out to be less than 2 reflects the fact that each electron partially screens the nucleus from the other, so that the net effective nuclear charge is reduced from 2 to 27/16.

7–2. A Trial Function That Depends Linearly on the Variational Parameters Leads to a Secular Determinant

As another example of the variational method, consider a particle in a one-dimensional box. Even without prior knowledge of the exact ground-state wave function, we should expect it to be symmetric about $x = a/2$ and to go to zero at the walls. One of the simplest functions with these properties is $x^n(a - x)^n$, where n is a positive integer. Consequently, let's estimate E_0 by using

$$\phi = c_1 x(a - x) + c_2 x^2(a - x)^2 \tag{7.22}$$

as a trial function, where c_1 and c_2 are to be determined variationally, that is, where c_1 and c_2 are the variational parameters. If ϕ in Equation 7.22 is used as a trial function, we find after quite a lengthy but straightforward calculation that

$$E_{\min} = 0.125002 \frac{h^2}{ma^2} \tag{7.23}$$

compared with

$$E_{\text{exact}} = \frac{h^2}{8ma^2} = 0.125000 \frac{h^2}{ma^2} \tag{7.24}$$

So we see that using a trial function with more than one parameter can produce impressive results. The price we pay is a correspondingly more lengthy calculation. Fortunately, there is a systematic way to handle a trial function such as Equation 7.22. Note that Equation 7.22 is a linear combination of functions. Such a trial function can be written generally as

$$\phi = \sum_{n=1}^{N} c_n f_n \tag{7.25}$$

where the c_n are variational parameters and the f_n are arbitrary known functions. We will use such a trial function often in later chapters. For simplicity, we will assume that $N = 2$ in Equation 7.25 and that the c_n and f_n are real. We relax these restrictions in Problem 7–17.

Consider

$$\phi = c_1 f_1 + c_2 f_2$$

Then,

$$\int \phi \hat{H} \phi d\tau = \int (c_1 f_1 + c_2 f_2) \hat{H} (c_1 f_1 + c_2 f_2) d\tau$$

$$= c_1^2 \int f_1 \hat{H} f_1 d\tau + c_1 c_2 \int f_1 \hat{H} f_2 d\tau$$

$$+ c_1 c_2 \int f_2 \hat{H} f_1 d\tau + c_2^2 \int f_2 \hat{H} f_2 d\tau$$

$$= c_1^2 H_{11} + c_1 c_2 H_{12} + c_1 c_2 H_{21} + c_2^2 H_{22} \tag{7.26}$$

where

$$H_{ij} = \int f_i \hat{H} f_j d\tau \tag{7.27}$$

We learned in Section 4–5 that quantum mechanical operators must be Hermitian to guarantee that their eigenvalues are real numbers. Equation 4.31 gives the relation that a Hermitian operator must satisfy, which in the notation we are using in Equation 7.27 becomes

$$\int f_i \hat{H} f_j d\tau = \int f_j \hat{H} f_i d\tau \tag{7.28}$$

In terms of the quantities H_{ij} given in Equation 7.27, Equation 7.28 says that $H_{ij} = H_{ji}$. Using this result, Equation 7.26 becomes

$$\int \phi \hat{H} \phi d\tau = c_1^2 H_{11} + 2c_1 c_2 H_{12} + c_2^2 H_{22} \tag{7.29}$$

Similarly, we have

$$\int \phi^2 d\tau = c_1^2 S_{11} + 2c_1 c_2 S_{12} + c_2^2 S_{22} \tag{7.30}$$

where

$$S_{ij} = S_{ji} = \int f_i f_j d\tau \tag{7.31}$$

The quantities H_{ij} and S_{ij} are called *matrix elements*. By substituting Equations 7.29 and 7.30 into Equation 7.3, we find

$$E(c_1, c_2) = \frac{c_1^2 H_{11} + 2c_1 c_2 H_{12} + c_2^2 H_{22}}{c_1^2 S_{11} + 2c_1 c_2 S_{12} + c_2^2 S_{22}} \tag{7.32}$$

where we emphasize here that E is a function of the variational parameters c_1 and c_2.

Before differentiating $E(c_1, c_2)$ in Equation 7.32 with respect to c_1 and c_2, it is convenient to write Equation 7.32 in the form

$$E(c_1, c_2)(c_1^2 S_{11} + 2c_1 c_2 S_{12} + c_2^2 S_{22}) = c_1^2 H_{11} + 2c_1 c_2 H_{12} + c_2^2 H_{22} \tag{7.33}$$

If we differentiate Equation 7.33 with respect to c_1, we find that

$$(2c_1 S_{11} + 2c_2 S_{12})E + \frac{\partial E}{\partial c_1}(c_1^2 S_{11} + 2c_1 c_2 S_{12} + c_2^2 S_{22}) = 2c_1 H_{11} + 2c_2 H_{12} \tag{7.34}$$

Because we are minimizing E with respect to c_1, $\partial E / \partial c_1 = 0$ and so Equation 7.34 becomes

$$c_1(H_{11} - ES_{11}) + c_2(H_{12} - ES_{12}) = 0 \tag{7.35}$$

Similarly, by differentiating $E(c_1, c_2)$ with respect to c_2 instead of c_1, we find

$$c_1(H_{12} - ES_{12}) + c_2(H_{22} - ES_{22}) = 0 \tag{7.36}$$

Equations 7.35 and 7.36 constitute a pair of linear algebraic equations for c_1 and c_2. There is a nontrivial solution, that is, a solution that is not simply $c_1 = c_2 = 0$, if and only if the determinant of the coefficients vanishes (MathChapter E), or if and only if

$$\begin{vmatrix} H_{11} - ES_{11} & H_{12} - ES_{12} \\ H_{12} - ES_{12} & H_{22} - ES_{22} \end{vmatrix} = 0 \tag{7.37}$$

This determinant is called a *secular determinant*. When this 2×2 determinant is expanded, we obtain a quadratic equation in E, called the *secular equation*. The quadratic secular equation gives two values for E, and we take the smaller of the two as our variational approximation for the ground-state energy.

To illustrate the use of Equation 7.37, let's go back to solving the problem of a particle in a one-dimensional box variationally using Equation 7.22 as a trial function. For convenience, we will set $a = 1$. In this case,

$$f_1 = x(1 - x) \qquad \text{and} \qquad f_2 = x^2(1 - x)^2 \tag{7.38}$$

and the matrix elements (see Equations 7.27 and 7.31) are (see Problem 7–26)

$$H_{11} = \frac{\hbar^2}{6m} \qquad\qquad S_{11} = \frac{1}{30}$$

$$H_{12} = H_{21} = \frac{\hbar^2}{30m} \qquad\qquad S_{12} = S_{21} = \frac{1}{140}$$

$$H_{22} = \frac{\hbar^2}{105m} \qquad\qquad S_{22} = \frac{1}{630}$$

EXAMPLE 7–4
Using Equation 7.38, show explicitly that $H_{12} = H_{21}$.

SOLUTION: Using the Hamiltonian operator of a particle in a box, we have

$$H_{12} = \int_0^1 f_1 \hat{H} f_2 dx$$

$$= \int_0^1 x(1 - x) \left[-\frac{\hbar^2}{2m} \frac{d^2}{dx^2} x^2(1 - x)^2 \right] dx$$

$$= -\frac{\hbar^2}{2m} \int_0^1 x(1 - x)[2 - 12x + 12x^2] dx$$

$$= -\frac{\hbar^2}{2m} \left(-\frac{1}{15} \right) = \frac{\hbar^2}{30m}$$

Similarly,

$$H_{21} = \int_0^1 f_2 \hat{H} f_1 dx$$

$$= \int_0^1 x^2 (1 - x)^2 \left[-\frac{\hbar^2}{2m} \frac{d^2}{dx^2} x(1 - x) \right] dx$$

$$= -\frac{\hbar^2}{2m} \int_0^1 x^2 (1 - x)^2 (-2) dx$$

$$= -\frac{\hbar^2}{2m} \left(-\frac{1}{15} \right) = \frac{\hbar^2}{30m}$$

Substituting the matrix elements H_{ij} and S_{ij} into the secular determinant (Equation 7.37) gives

$$\begin{vmatrix} \dfrac{1}{6} - \dfrac{E'}{30} & \dfrac{1}{30} - \dfrac{E'}{140} \\ \dfrac{1}{30} - \dfrac{E'}{140} & \dfrac{1}{105} - \dfrac{E'}{630} \end{vmatrix} = 0$$

where $E' = Em/\hbar^2$. The corresponding secular equation is

$$E'^2 - 56E' + 252 = 0$$

whose roots are

$$E' = \frac{56 \pm \sqrt{2128}}{2} = 51.065 \text{ and } 4.93487$$

We choose the smaller root and obtain

$$E_{min} = 4.93487 \frac{\hbar^2}{m} = 0.125002 \frac{\hbar^2}{m}$$

compared with (recall that $a = 1$)

$$E_{exact} = \frac{h^2}{8m} = 0.125000 \frac{h^2}{m}$$

The excellent agreement here is better than should be expected normally for such a simple trial function. Note that $E_{min} > E_{exact}$, as it must be.

EXAMPLE 7–5
Determine the normalized trial function for our variational treatment of a particle in a box.

SOLUTION: To determine the normalized trial function, we must determine c_1 and c_2 in Equation 7.22. These quantities are given by Equations 7.35 and 7.36, which are the two algebraic equations that lead to the secular determinant, Equation 7.37. These two equations are not independent of each other, so we will use the first to calculate the ratio c_2/c_1:

$$\frac{c_2}{c_1} = -\frac{H_{11} - ES_{11}}{H_{12} - ES_{12}} = -\frac{\dfrac{\hbar^2}{6m} - \left(4.93487\dfrac{\hbar^2}{m}\right)\dfrac{1}{30}}{\dfrac{\hbar^2}{30m} - \left(4.93487\dfrac{\hbar^2}{m}\right)\dfrac{1}{140}} = 1.133$$

or $c_2 = 1.133c_1$. So far, then we have

$$\phi(x) = c_1[x(1-x) + 1.133x^2(1-x)^2]$$

We now determine c_1 by requiring $\phi(x)$ to be normalized.

$$\int_0^1 \phi^2(x)dx = c_1^2 \int_0^1 \left[x^2(1-x)^2 + 2.266x^3(1-x)^3 + 1.284x^4(1-x)^4\right]dx = 1$$

Instead of expanding out each integral, it is more convenient to use (*CRC Handbook*)

$$\int_0^1 x^m(1-x)^n dx = \frac{m!n!}{(m+n+1)!}$$

in which case

$$\int_0^1 \phi^2(x)dx = c_1^2\left[\frac{2!2!}{5!} + 2.266\frac{3!3!}{7!} + 1.284\frac{4!4!}{9!}\right] = 0.05156c_1^2 = 1$$

giving us $c_1 = 4.404$. Thus, the normalized trial function is

$$\phi(x) = 4.404x(1-x) + 4.990x^2(1-x)^2$$

Figure 7.3 compares $\phi(x)$ with the exact ground-state particle-in-a-box wave function (with $a = 1$), $\psi_1(x) = 2^{1/2} \sin \pi x$.

You may wonder about the physical meaning of the other root to Equation 7.37. It turns out that it is an upper bound to the energy of the first excited state of a particle in a box. The value we calculated above is 1.2935 h^2/m, compared with the exact value of $4h^2/8ma^2$, or $0.5000h^2/m$. Thus, we see that although the second root is an upper bound to E_2, it is a fairly crude one. Although there are methods to give better upper bounds to excited-state energies, we will restrict ourselves to a determination of only ground-state energies.

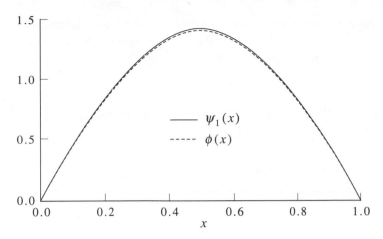

FIGURE 7.3
A comparison of the optimized and normalized trial function determined in Example 7–5 (dashed line) with the exact ground-state particle-in-a-box wave function, $\psi_1(x) = 2^{1/2} \sin \pi x$ (solid line). The width of the box is taken to be 1.

If we use a linear combination of N functions as in Equation 7.25 instead of using a linear combination of two functions as we have done so far, then we obtain N simultaneous linear algebraic equations for the c_js:

$$
\begin{aligned}
c_1(H_{11} - ES_{11}) + c_2(H_{12} - ES_{21}) &+ \cdots + c_N(H_{1N} - ES_{1N}) = 0 \\
c_1(H_{12} - ES_{12}) + c_2(H_{22} - ES_{22}) &+ \cdots + c_N(H_{2N} - ES_{2N}) = 0 \\
\vdots \qquad\qquad \vdots \qquad\qquad\qquad &\qquad\qquad \vdots
\end{aligned}
\tag{7.39}
$$

$$
c_1(H_{1N} - ES_{1N}) + c_2(H_{2N} - ES_{2N}) + \cdots + c_N(H_{NN} - ES_{NN}) = 0
$$

To have a nontrivial solution to this set of homogeneous equations, we must have that

$$
\begin{vmatrix}
H_{11} - ES_{11} & H_{12} - ES_{12} & \cdots & H_{1N} - ES_{1N} \\
H_{12} - ES_{12} & H_{22} - ES_{22} & \cdots & H_{2N} - ES_{2N} \\
\vdots & \vdots & & \vdots \\
H_{1N} - ES_{1N} & H_{2N} - ES_{2N} & \cdots & H_{NN} - ES_{NN}
\end{vmatrix} = 0
\tag{7.40}
$$

In writing Equation 7.40, we have used the fact that $\hat{H}$ is a Hermitian operator, so $H_{ij} = H_{ji}$. The secular equation associated with this secular determinant is an Nth-degree polynomial in E. We choose the smallest root of the Nth-order secular equation as an approximation to the ground-state energy. The determination of the smallest root must usually be done numerically for values of N larger than two. This is actually a standard numerical problem, and a number of packaged computer programs do this.

Once the smallest root of Equation 7.40 has been determined, we can substitute it back into Equation 7.39 to determine the c_js. As in Example 7–5, only $N - 1$ of these equations are independent, and so we can use them to determine only the ratios c_2/c_1, c_3/c_1, $\ldots$, c_N/c_1, for example. We can then determine c_1 by requiring that the trial function ϕ be normalized, as we did in Example 7–5.

7–3. Trial Functions Can Be Linear Combinations of Functions that Also Contain Variational Parameters

It is a fairly common practice to use a trial function of the form

$$\phi = \sum_{j=1}^{N} c_j f_j$$

where the f_j themselves contain variational parameters. An example of such a trial function for the hydrogen atom is

$$\phi = \sum_{j=1}^{N} c_j e^{-\alpha_j r^2}$$

where the c_js and the α_js are treated as variational parameters. We have seen in Section 7–1 that the use of one term gives an energy $-0.424(m_e e^4/16\pi^2\varepsilon_0^2\hbar^2)$ compared with the exact value of $-0.500(m_e e^4/16\pi^2\varepsilon_0^2\hbar^2)$. Table 7.1 shows the results for taking more terms. We can see that the exact value is approached as N increases. Realize in this case, however, that we do not obtain a simple secular determinant, because ϕ is linear only in the c_j but not in the α_j. The minimization of E with respect to the c_j and α_j is fairly complicated, involving $2N$ parameters, and must be done numerically. Fortunately, a number of readily available algorithms can be used to do this.

TABLE 7.1
The ground-state energy of a hydrogen atom using a trial function of the form $\phi = \sum_{j=1}^{N} c_j e^{-\alpha_j r^2}$, where the c_j and the α_j are treated as variational parameters. The exact value is $-0.500\,000$.

N	$E_{\min}/(m_e e^4/16\pi^2\varepsilon_0^2\hbar^2)$
1	-0.424413
2	-0.485813
3	-0.496967
4	-0.499276
5	-0.49976
6	-0.49988
8	-0.49992
16	-0.49998

7–4. Perturbation Theory Expresses the Solution to One Problem in Terms of Another Problem Solved Previously

The idea behind perturbation theory is the following. Suppose that we are unable to solve the Schrödinger equation

$$\hat{H}\psi = E\psi \tag{7.41}$$

for some system of interest but that we do know how to solve it for another system that is in some sense similar. We can write the Hamiltonian operator in Equation 7.41 in the form

$$\hat{H} = \hat{H}^{(0)} + \hat{H}^{(1)} \tag{7.42}$$

where

$$\hat{H}^{(0)}\psi^{(0)} = E^{(0)}\psi^{(0)} \tag{7.43}$$

is the Schrödinger equation we can solve exactly. We call the first term in Equation 7.42 the *unperturbed Hamiltonian operator* and the additional term the *perturbation*. You might expect intuitively that if the perturbation term is small in some sense, then the solution to Equation 7.41 should be close to the solution to Equation 7.43. For example, in the case of an anharmonic oscillator with

$$\hat{H} = -\frac{\hbar^2}{2\mu}\frac{d^2}{dx^2} + \frac{1}{2}kx^2 + \frac{1}{6}\gamma x^3 + \frac{b}{24}x^4$$

we treat the anharmonic terms, $\gamma x^3/6 + bx^4/24$, as a perturbation to a harmonic oscillator, and write (Chapter 5)

$$\hat{H}^{(0)} = -\frac{\hbar^2}{2\mu}\frac{d^2}{dx^2} + \frac{1}{2}kx^2$$

$$\psi_v^{(0)}(x) = \left[\left(\frac{\alpha}{\pi}\right)^{1/2}\frac{1}{2^v v!}\right]^{1/2} H_v(\alpha^{1/2}x)e^{-\alpha x^2/2} \tag{7.44}$$

$$E_v^{(0)} = \left(v + \frac{1}{2}\right)h\nu \qquad v = 0,\ 1,\ 2,\ \ldots$$

$$H^{(1)} = \frac{1}{6}\gamma x^3 + \frac{b}{24}x^4$$

where $\alpha = (k\mu/\hbar^2)^{1/2}$.

To apply perturbation theory to the solution of Equation 7.41 with $\hat{H}$ given by Equation 7.42, we write ψ and E in the form

$$\psi = \psi^{(0)} + \psi^{(1)} + \psi^{(2)} + \cdots \tag{7.45}$$

and

$$E = E^{(0)} + E^{(1)} + E^{(2)} + \cdots \qquad (7.46)$$

where $\psi^{(0)}$ and $E^{(0)}$ are given by the solution to the unperturbed problem (Equation 7.43) and $\psi^{(1)}$, $\psi^{(2)}$, ... are successive corrections to $\psi^{(0)}$ and $E^{(1)}$, $E^{(2)}$, ... are successive corrections to $E^{(0)}$. A basic assumption is that these successive corrections become increasingly less significant. Although we will not do so here, we can derive explicit expressions for these corrections. The only one we will use is the expression for $E^{(1)}$, which is

$$E^{(1)} = \int \psi^{(0)*} \hat{H}^{(1)} \psi^{(0)} d\tau \qquad (7.47)$$

(Problem 7–19 takes you step by step through the derivation of this result.) We say that $E^{(1)}$ is the first-order correction to $E^{(0)}$, and we write

$$E = E^{(0)} + E^{(1)} \qquad (7.48)$$

Equation 7.48 represents the energy through first-order perturbation theory. If we were to evaluate $\psi^{(1)}$ (which we will not), then

$$\psi = \psi^{(0)} + \psi^{(1)}$$

would represent ψ through first order. Similarly, if we were to evaluate $E^{(2)}$ (which we will not), then

$$E = E^{(0)} + E^{(1)} + E^{(2)}$$

would represent E through second-order perturbation theory. In this book, we evaluate E to first order only, using Equation 7.47.

Let's use Equation 7.47 to calculate the ground-state energy of the anharmonic oscillator described by Equation 7.44. In this case,

$$\psi^{(0)}(x) = \left(\frac{\alpha}{\pi}\right)^{1/4} e^{-\alpha x^2/2}$$

and so Equation 7.47 becomes

$$E^{(1)} = \int_{-\infty}^{\infty} \psi^{(0)}(x)^* \hat{H}^{(1)} \psi^{(0)}(x) dx$$

$$= \left(\frac{\alpha}{\pi}\right)^{1/2} \left[\frac{\gamma}{6} \int_{-\infty}^{\infty} x^3 e^{-\alpha x^2} dx + \frac{b}{24} \int_{-\infty}^{\infty} x^4 e^{-\alpha x^2} dx\right]$$

The first integral equals zero because the integrand is an odd function, and so

$$E^{(1)} = \frac{b}{12} \left(\frac{\alpha}{\pi}\right)^{1/2} \int_{0}^{\infty} x^4 e^{-\alpha x^2} dx$$

The integral here can be found in tables and is equal to $3\pi^{1/2}/8\alpha^{5/2}$, and so

$$E^{(1)} = \frac{b}{32\alpha^2} = \frac{\hbar^2 b}{32k\mu}$$

and the total ground-state energy through first order is

$$E = E^{(0)} + E^{(1)} = \frac{h\nu}{2} + \frac{\hbar^2 b}{32k\mu}$$

EXAMPLE 7–6
Use first-order perturbation theory to calculate the energy of a particle in a one-dimensional box from $x = 0$ to $x = a$ with a slanted bottom, such that

$$V(x) = \frac{V_0 x}{a} \qquad 0 \le x \le a$$

SOLUTION: In this case, the unperturbed problem is a particle in a box and so

$$\hat{H}^{(1)} = \frac{V_0}{a} x \qquad 0 \le x \le a$$

where V_0 is a constant. The wave functions and the energies for a particle in a box are

$$\psi^{(0)} = \left(\frac{2}{a}\right)^{1/2} \sin \frac{n\pi x}{a} \qquad 0 \le x \le a$$

and

$$E^{(0)} = \frac{n^2 h^2}{8ma^2}$$

According to Equation 7.47, the first-order correction to $E^{(0)}$ due to the perturbation is given by

$$E^{(1)} = \int_0^a \psi^{(0)*} \left(\frac{V_0}{a} x\right) \psi^{(0)} dx$$

$$= \frac{2V_0}{a^2} \int_0^a x \sin^2 \frac{n\pi x}{a} dx$$

This integral occurs in Equation 3.30 and is equal to $a^2/4$. Therefore, we find that

$$E^{(1)} = \frac{V_0}{2}$$

for all values of n. The energy levels are given through first order by

$$E = \frac{n^2 h^2}{8ma^2} + \frac{V_0}{2} + O(V_0^2) \qquad n = 1, 2, 3, \ldots$$

where the term $O(V_0^2)$ emphasizes that terms of order V_0^2 and higher have been dropped. Thus, we see in this case that each of the unperturbed energy levels is shifted by $V_0/2$.

We can apply perturbation theory to the helium atom whose Hamiltonian operator is given by Equation 7.11. For simplicity, we will consider only the ground-state energy. If we consider the interelectronic repulsion term, $e^2/4\pi\varepsilon_0 r_{12}$, to be the perturbation, then the unperturbed wave functions and energies are the hydrogenlike quantities given by

$$\hat{H}^{(0)} = \hat{H}_{\mathrm{H}}(1) + \hat{H}_{\mathrm{H}}(2)$$

$$\psi^{(0)} = \psi_{1s}(r_1, \theta_1, \phi_1)\psi_{1s}(r_2, \theta_2, \phi_2)$$

$$E^{(0)} = -\frac{Z^2 m_e e^4}{32\pi^2\varepsilon_0^2\hbar^2 n_1^2} - \frac{Z^2 m_e e^4}{32\pi^2\varepsilon_0^2\hbar^2 n_2^2}$$

(7.49)

and

$$\hat{H}^{(1)} = \frac{e^2}{4\pi\varepsilon_0 r_{12}}$$

with $Z = 2$. Using Equation 7.47, we have

$$E^{(1)} = \int\int d\mathbf{r}_1 d\mathbf{r}_2 \psi_{1s}(\mathbf{r}_1)\psi_{1s}(\mathbf{r}_2)\frac{e^2}{4\pi\varepsilon_0 r_{12}}\psi_{1s}(\mathbf{r}_1)\psi_{1s}(\mathbf{r}_2)$$

(7.50)

where $\psi_{1s}(\mathbf{r}_j)$ is given by Equation 7.17. The evaluation of the integral in Equation 7.50 is a little lengthy, but Problem 7–30 carries it out step by step. The final result is that

$$E^{(1)} = \frac{5Z}{8}\left(\frac{m_e e^4}{16\pi^2\varepsilon_0^2\hbar^2}\right)$$

(7.51)

or $E^{(1)} = 5Z/8$ in units of $m_e e^4/16\pi^2\varepsilon_0^2\hbar^2$. If we add this to $E^{(0)}$, with $n_1 = n_2 = 1$, then (in units of $m_e e^4/16\pi^2\varepsilon_0^2\hbar^2$)

$$E = E^{(0)} + E^{(1)} = -\frac{1}{2}Z^2 - \frac{1}{2}Z^2 + \frac{5}{8}Z$$

$$= -Z^2 + \frac{5}{8}Z$$

(7.52)

Letting $Z = 2$ gives -2.750 compared with our simple variational result (-2.8477) given by Equation 7.26 and the experimental result of -2.9033. So we see that first-order perturbation theory gives a result that is about 5% in error. It turns out that second-order perturbation theory gives -2.910 and that a higher-order calculation gives -2.9037. Thus, we see that both the variational method and the perturbation theory are able to achieve very good results.

Problems

7-1. This problem involves the proof of the variational principle, Equation 7.4. Let $\hat{H}\psi_n = E_n\psi_n$ be the problem of interest, and let ϕ be our approximation to ψ_0. Even though we do not know the ψ_n, we can express ϕ formally as

$$\phi = \sum_n c_n\psi_n \tag{1}$$

where the c_n are constants. Using the fact that the ψ_n are orthonormal, show that

$$c_n = \int \psi_n^*\phi\, d\tau$$

We do not know the ψ_n, however, so Equation 1 is what we call a formal expansion. Now substitute Equation 1 into

$$E_\phi = \frac{\int \phi^*\hat{H}\phi\, d\tau}{\int \phi^*\phi\, d\tau}$$

to obtain

$$E_\phi = \frac{\sum_n c_n^* c_n E_n}{\sum_n c_n^* c_n}$$

Subtract E_0 from the left side of the above equation and $E_0 \sum_n c_n^* c_n / \sum_n c_n^* c_n$ from the right side to obtain

$$E_\phi - E_0 = \frac{\sum_n c_n^* c_n (E_n - E_0)}{\sum_n c_n^* c_n}$$

Now explain why every term on the right side is positive, proving that $E_\phi \geq E_0$.

7-2. Using a Gaussian trial function $e^{-\alpha r^2}$ for the ground state of the hydrogen atom (see Equation 7.6 for $\hat{H}$), show that the ground-state energy is given by

$$E(\alpha) = \frac{3\hbar^2\alpha}{2m_e} - \frac{e^2\alpha^{1/2}}{2^{1/2}\varepsilon_0\pi^{3/2}}$$

and that

$$E_{min} = -\frac{4}{3\pi}\frac{m_e e^4}{16\pi^2\varepsilon_0^2\hbar^2}$$

7-3. Use a trial function $\phi(x) = 1/(1+\beta x^2)^2$ to calculate the ground-state energy of a harmonic oscillator variationally. The necessary integrals are

$$\int_{-\infty}^{\infty} \frac{dx}{(1+\beta x^2)^n} = \frac{(2n-3)(2n-5)(2n-7)\cdots(1)}{(2n-2)(2n-4)(2n-6)\cdots(2)}\frac{\pi}{\beta^{1/2}}$$

and

$$\int_{-\infty}^{\infty} \frac{x^2 dx}{(1+\beta x^2)^n} = \frac{(2n-5)(2n-7)\cdots(1)}{(2n-2)(2n-4)\cdots(2)}\frac{\pi}{\beta^{3/2}}$$

7-4. If you were to use a trial function of the form $\phi(x) = (1 + cax^2)e^{-ax^2/2}$, where $\alpha = (k\mu/\hbar^2)^{1/2}$ and c is a variational parameter to calculate the ground-state energy of a harmonic oscillator, what do you think the value of c will come out to be? Why?

7-5. Use a trial function of the form $\phi(r) = re^{-\alpha r}$ with α as a variational parameter to calculate the ground-state energy of a hydrogen atom.

7-6. Suppose we were to use a trial function of the form $\phi = c_1 e^{-\alpha r} + c_2 e^{-\beta r^2}$ to carry out a variational calculation for the ground-state energy of the hydrogen atom. Can you guess without doing any calculations what c_1, c_2, α, and E_{min} will be? What about a trial function of the form $\phi = \sum_{k=1}^{5} c_k e^{-\alpha_k r - \beta_k r^2}$?

7-7. Use a trial function of the form $e^{-\beta x^2}$ with β as a variational parameter to calculate the ground-state energy of a harmonic oscillator. Compare your result with the exact energy $h\nu/2$. Why is the agreement so good?

7-8. Consider a three-dimensional, spherically symmetric, isotropic harmonic oscillator with $V(r) = kr^2/2$. Using a trial function $e^{-\alpha r^2}$ with α as a variational parameter, estimate the ground-state energy. Do the same using $e^{-\alpha r}$. The Hamiltonian operator is

$$\hat{H} = -\frac{\hbar^2}{2\mu r^2}\frac{d}{dr}\left(r^2\frac{d}{dr}\right) + \frac{k}{2}r^2$$

Compare these results with the exact ground-state energy, $E = \frac{3}{2}h\nu$. Why is one of these so much better than the other?

7-9. Use a trial function of the form $e^{-\alpha x^2/2}$ to calculate the ground-state energy of a quartic oscillator, whose potential is $V(x) = cx^4$.

7-10. Use a trial function of the form $\phi = \cos \lambda x$ with $-\pi/2\lambda < x < \pi/2\lambda$ and with λ as a variational parameter to calculate the ground-state energy of a harmonic oscillator.

7-11. Use the variational method to calculate the ground-state energy of a particle constrained to move within the region $0 \leq x \leq a$ in a potential given by

$$V(x) = V_0 x \qquad\qquad 0 \leq x \leq \frac{a}{2}$$

$$= V_0(a - x) \qquad \frac{a}{2} \leq x \leq a$$

As a trial function, use a linear combination of the first two particle-in-a-box wave functions:

$$\phi(x) = c_1 \left(\frac{2}{a}\right)^{1/2} \sin\frac{\pi x}{a} + c_2 \left(\frac{2}{a}\right)^{1/2} \sin\frac{2\pi x}{a}$$

7-12. Consider a particle of mass m in the potential energy field described by

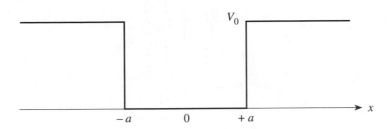

This problem describes a particle in a finite well. If $V_0 \to \infty$, then we have a particle in a box. Using $\phi(x) = l^2 - x^2$ for $-l < x < l$ and $\phi(x) = 0$ otherwise as a trial function with l as a variational parameter, calculate the ground-state energy of this system for $2m V_0 a^2/\hbar^2 = 4$ and 12. The exact ground-state energies are $0.530\,\hbar^2/ma^2$ and $0.736\,\hbar^2/ma^2$, respectively (see Problem 7–29).

7-13. Repeat the calculation in Problem 7–12 for a trial function $\phi(x) = \cos \lambda x$ for $-\pi/2\lambda < x < \pi/2\lambda$ and $\phi(x) = 0$ otherwise. Use λ as a variational parameter.

7-14. Consider a particle in a spherical box of radius a. The Hamiltonian operator for this system is (see Equation 6.43)

$$\hat{H} = -\frac{\hbar^2}{2mr^2}\frac{d}{dr}\left(r^2\frac{d}{dr}\right) + \frac{\hbar^2 l(l+1)}{2mr^2} \qquad 0 < r \le a$$

In the ground state, $l = 0$ and so

$$\hat{H} = -\frac{\hbar^2}{2mr^2}\frac{d}{dr}\left(r^2\frac{d}{dr}\right) \qquad 0 < r \le a$$

As in the case of a particle in a rectangular box, $\phi(a) = 0$. Use $\phi(r) = a - r$ to calculate an upper bound to the ground-state energy of this system. There is no variational parameter in this case, but the calculated energy is still an upper bound to the ground-state energy. The exact ground-state energy is $\pi^2\hbar^2/2ma^2$ (see Problem 7–28).

7-15. Repeat the calculation in Problem 7–14 using $\phi(r) = (a - r)^2$ as a trial function.

7-16. Consider a system subject to the potential

$$V(x) = \frac{k}{2}x^2 + \frac{\gamma}{6}x^3 + \frac{\delta}{24}x^4$$

Calculate the ground-state energy of this system using a trial function of the form

$$\phi = c_1\psi_0(x) + c_2\psi_2(x)$$

where $\psi_0(x)$ and $\psi_2(x)$ are the first two even wave functions of a harmonic oscillator. Why did we not include $\psi_1(x)$?

7-17. It is quite common to assume a trial function of the form

$$\phi = c_1\phi_1 + c_2\phi_2 + \cdots + c_n\phi_n$$

where the variational parameters and the ϕ_n may be complex. Using the simple, special case

$$\phi = c_1\phi_1 + c_2\phi_2$$

show that the variational method leads to

$$E_\phi = \frac{c_1^*c_1 H_{11} + c_1^*c_2 H_{12} + c_1 c_2^* H_{21} + c_2^*c_2 H_{22}}{c_1^*c_1 S_{11} + c_1^*c_2 S_{12} + c_1 c_2^* S_{21} + c_2^*c_2 S_{22}}$$

where

$$H_{ij} = \int \phi_i^* \hat{H}\phi_j\, d\tau = H_{ji}^*$$

and

$$S_{ij} = \int \phi_i^* \phi_j\, d\tau = S_{ji}^*$$

because $\hat{H}$ is a Hermitian operator. Now write the above equation for E_ϕ as

$$c_1^*c_1 H_{11} + c_1^*c_2 H_{12} + c_1 c_2^* H_{21} + c_2^*c_2 H_{22}$$
$$= E_\phi(c_1^*c_1 S_{11} + c_1^*c_2 S_{12} + c_1 c_2^* S_{21} + c_2^*c_2 S_{22})$$

and show that if we set

$$\frac{\partial E_\phi}{\partial c_1^*} = 0 \quad \text{and} \quad \frac{\partial E_\phi}{\partial c_2^*} = 0$$

we obtain

$$(H_{11} - E_\phi S_{11})c_1 + (H_{12} - E_\phi S_{12})c_2 = 0$$

and

$$(H_{21} - E_\phi S_{21})c_1 + (H_{22} - E_\phi S_{22})c_2 = 0$$

There is a nontrivial solution to this pair of equations if and only if the determinant

$$\begin{vmatrix} H_{11} - E_\phi S_{11} & H_{12} - E_\phi S_{12} \\ H_{21} - E_\phi S_{21} & H_{22} - E_\phi S_{22} \end{vmatrix} = 0$$

which gives a quadratic equation for E_ϕ. We choose the smaller solution as an approximation to the ground-state energy.

7-18. This problem shows that terms in a trial function that correspond to progressively higher energies contribute progressively less to the ground-state energy. For algebraic simplicity, assume that the Hamiltonian operator can be written in the form

$$\hat{H} = \hat{H}^{(0)} + \hat{H}^{(1)}$$

and choose a trial function

$$\phi = c_1\psi_1 + c_2\psi_2$$

where

$$\hat{H}^{(0)} \psi_j = E_j^{(0)} \psi_j \qquad j = 1, 2$$

Show that the secular equation associated with the trial function is

$$\begin{vmatrix} H_{11} - E & H_{12} \\ H_{12} & H_{22} - E \end{vmatrix} = \begin{vmatrix} E_1^{(0)} + E_1^{(1)} - E & H_{12} \\ H_{12} & E_2^{(0)} + E_2^{(1)} - E \end{vmatrix} = 0 \qquad (1)$$

where

$$E_j^{(1)} = \int \psi_j^* \hat{H}^{(1)} \psi_j \, d\tau \quad \text{and} \quad H_{12} = \int \psi_1^* \hat{H}^{(1)} \psi_2 \, d\tau$$

Solve Equation 1 for E to obtain

$$E = \frac{E_1^{(0)} + E_1^{(1)} + E_2^{(0)} + E_2^{(1)}}{2}$$
$$\pm \frac{1}{2} \left\{ [E_1^{(0)} + E_1^{(1)} - E_2^{(0)} - E_2^{(1)}]^2 + 4H_{12}^2 \right\}^{1/2} \qquad (2)$$

If we arbitrarily assume that $E_1^{(0)} + E_1^{(1)} < E_2^{(0)} + E_2^{(1)}$, then we take the positive sign in Equation 2 and write

$$E = \frac{E_1^{(0)} + E_1^{(1)} + E_2^{(0)} + E_2^{(1)}}{2} + \frac{E_1^{(0)} + E_1^{(1)} - E_2^{(0)} - E_2^{(1)}}{2}$$
$$\times \left\{ 1 + \frac{4H_{12}^2}{[E_1^{(0)} + E_1^{(1)} - E_2^{(0)} - E_2^{(1)}]^2} \right\}^{1/2}$$

Use the expansion $(1 + x)^{1/2} = 1 + x/2 + \cdots, \ x < 1$ to get

$$E = E_1^{(0)} + E_1^{(1)} + \frac{H_{12}^2}{E_1^{(0)} + E_1^{(1)} - E_2^{(0)} - E_2^{(1)}} + \cdots \qquad (3)$$

Note that if $E_1^{(0)} + E_1^{(1)}$ and $E_2^{(0)} + E_2^{(1)}$ are widely separated, the term involving H_{12}^2 in Equation 3 is small. Therefore, the energy is simply that calculated using ψ_1 alone; the ψ_2 part of the trial function contributes little to the overall energy. The general result is that terms in a trial function that correspond to higher and higher energies contribute less and less to the total ground-state energy.

7-19. We will derive the equations for first-order perturbation theory in this problem. The problem we want to solve is

$$\hat{H} \psi = E \psi \qquad (1)$$

where

$$\hat{H} = \hat{H}^{(0)} + \hat{H}^{(1)}$$

and where the problem

$$\hat{H}^{(0)} \psi^{(0)} = E^{(0)} \psi^{(0)} \qquad (2)$$

has been solved exactly previously, so that $\psi^{(0)}$ and $E^{(0)}$ are known. Assuming now that the effect of $\hat{H}^{(1)}$ is small, write

$$\psi = \psi^{(0)} + \Delta\psi$$

$$E = E^{(0)} + \Delta E \tag{3}$$

where we assume that $\Delta\psi$ and ΔE are small. Substitute Equations 3 into Equation 1 to obtain

$$\hat{H}^{(0)}\psi^{(0)} + \hat{H}^{(1)}\psi^{(0)} + \hat{H}^{(0)}\Delta\psi + \hat{H}^{(1)}\Delta\psi$$

$$= E^{(0)}\psi^{(0)} + \Delta E\psi^{(0)} + E^{(0)}\Delta\psi + \Delta E\Delta\psi \tag{4}$$

The first terms on each side of Equation 4 cancel because of Equation 2. In addition, we will neglect the last terms on each side because they represent the product of two small terms. Thus, Equation 4 becomes

$$\hat{H}^{(0)}\Delta\psi + \hat{H}^{(1)}\psi^{(0)} = E^{(0)}\Delta\psi + \Delta E\psi^{(0)} \tag{5}$$

Realize that $\Delta\psi$ and ΔE are the unknown quantities in this equation.

Note that all the terms in Equation (5) are of the same order, in the sense that each is the product of an unperturbed term and a small term. We say that this equation is first order in the perturbation and that we are using first-order perturbation theory. The two terms we have neglected in Equation 4 are second-order terms and lead to second-order (and higher) corrections.

Equation 5 can be simplified considerably. Multiply both sides from the left by $\psi^{(0)*}$ and integrate over all space to get

$$\int \psi^{(0)*}[\hat{H}^{(0)} - E^{(0)}]\Delta\psi\, d\tau + \int \psi^{(0)*}\hat{H}^{(1)}\psi^{(0)}\, d\tau = \Delta E \int \psi^{(0)*}\psi^{(0)}\, d\tau \tag{6}$$

The integral in the last term in Equation 6 is unity because $\psi^{(0)}$ is taken to be normalized. More important, however, is that the first term on the left side of Equation 6 is zero. Use the fact that $\hat{H}^{(0)} - E^{(0)}$ is Hermitian to show that

$$\int \psi^{(0)*}[\hat{H}^{(0)} - E^{(0)}]\Delta\psi\, d\tau = \int \{[\hat{H}^{(0)} - E^{(0)}]\psi^{(0)}\}^*\Delta\psi\, d\tau$$

But according to Equation 2, the integrand here vanishes. Thus, Equation 6 becomes

$$\Delta E = \int \psi^{(0)*}\hat{H}^{(1)}\psi^{(0)}\, d\tau \tag{7}$$

Equation 7 is called the *first-order correction* to $E^{(0)}$. To first order, the energy is

$$E = E^{(0)} + \int \psi^{(0)*}\hat{H}^{(1)}\psi^{(0)}\, d\tau + \text{higher order terms}$$

7-20. Identify $\hat{H}^{(0)}$, $\hat{H}^{(1)}$, $\psi^{(0)}$, and $E^{(0)}$ for the following problems:

  a. An oscillator governed by the potential

$$V(x) = \frac{k}{2}x^2 + \frac{\gamma}{6}x^3 + \frac{b}{24}x^4$$

b. A particle constrained to move in the region $0 \le x \le a$ with the potential

$$V(x) = 0 \qquad 0 < x < \frac{a}{2}$$

$$= b \qquad \frac{a}{2} < x < a$$

c. A helium atom

d. A hydrogen atom in an electric field of strength $\mathcal{E}$. The Hamiltonian operator for this system is

$$\hat{H} = -\frac{\hbar^2}{2m_e}\nabla^2 - \frac{e^2}{4\pi\varepsilon_0 r} + e\mathcal{E}r\cos\theta$$

e. A rigid rotator with a dipole moment μ in an electric field of strength $\mathcal{E}$. The Hamiltonian operator for this system is

$$\hat{H} = -\frac{\hbar^2}{2I}\nabla^2 + \mu\mathcal{E}\cos\theta$$

where ∇^2 is given by Equation 6.3.

7-21. Using a harmonic oscillator as the unperturbed problem, calculate the first-order correction to the energy of the $v = 0$ level for the system described in Problem 7–20(a).

7-22. Using a particle in a box as the unperturbed problem, calculate the first-order correction to the ground-state energy for the system described in Problem 7–20(b).

7-23. Using the result of Problem 7–20(d), calculate the first-order correction to the ground-state energy of a hydrogen atom in an external electric field of strength $\mathcal{E}$.

7-24. Calculate the first-order correction to the energy of a particle constrained to move within the region $0 \le x \le a$ in the potential

$$V(x) = V_0 x \qquad 0 \le x \le \frac{a}{2}$$

$$= V_0(a - x) \qquad \frac{a}{2} \le x \le a$$

where V_0 is a constant.

7-25. Use first-order perturbation theory to calculate the first-order correction to the ground-state energy of a quartic oscillator whose potential energy is

$$V(x) = cx^4$$

In this case, use a harmonic oscillator as the unperturbed system. What is the perturbing potential?

7-26. Use a trial function

$$\phi = c_1 x(a - x) + c_2 x^2(a - x)^2$$

for a particle in a one-dimensional box. For simplicity, let $a = 1$, which amounts to measuring all distances in units of a. Show that

$$H_{11} = \frac{\hbar^2}{6m} \qquad\qquad S_{11} = \frac{1}{30}$$

$$H_{12} = H_{21} = \frac{\hbar^2}{30m} \qquad\qquad S_{12} = S_{21} = \frac{1}{140}$$

$$H_{22} = \frac{\hbar^2}{105m} \qquad\qquad S_{22} = \frac{1}{630}$$

7-27. In Example 5–2, we introduced the Morse potential

$$V(x) = D(1 - e^{-\beta x})^2$$

as a description of the intramolecular potential energy of a diatomic molecule. The constants D and β are different for each molecule; for H_2, $D = 7.61 \times 10^{-19}$ J and $\beta = 0.0193$ pm^{-1}. First expand the Morse potential in a power series about x. (*Hint*: Use the expansion $e^x = 1 + x + \frac{x^2}{2} + \frac{x^3}{6} + \cdots$.) What is the Hamiltonian operator for the Morse potential? Show that the Hamiltonian operator can be written in the form

$$\hat{H} = -\frac{\hbar^2}{2\mu}\frac{d^2}{dx^2} + ax^2 + bx^3 + cx^4 + \cdots \tag{1}$$

How are the constants a, b, and c related to the constants D and β? What part of the Hamiltonian operator would you associate with $\hat{H}^{(0)}$, and what are the functions $\psi_n^{(0)}$ and energies $E_n^{(0)}$? Use perturbation theory to evaluate the first-order corrections to the energy of the first three states that arise from the cubic and quartic terms. Using these results, how different are the first two energy levels of H_2 if its intramolecular potential is described by a harmonic oscillator potential or the quartic expansion of the Morse potential (see Equation 1)?

7-28. In this problem, we will solve the Schrödinger equation for the ground-state wave function and energy of a particle in a spherical box of radius a. The Schrödinger equation is given by Equation 6.43 with $l = 0$ (ground state) and without the $e^2/4\pi\varepsilon_0 r$ term:

$$-\frac{\hbar^2}{2mr^2}\frac{d}{dr}\left(r^2\frac{d\psi}{dr}\right) = E\psi$$

Substitute $u = r\psi$ into this equation to get

$$\frac{d^2u}{dr^2} + \frac{2mE}{\hbar^2}u = 0$$

The general solution to this equation is

$$u(r) = A\cos\alpha r + B\sin\alpha r$$

or

$$\psi(r) = \frac{A\cos\alpha r}{r} + \frac{B\sin\alpha r}{r}$$

where $\alpha = (2mE/\hbar^2)^{1/2}$. Which of these terms is finite at $r = 0$? Now use the fact that $\psi(a) = 0$ to prove that

$$\alpha a = \pi$$

for the ground state, or that the ground-state energy is

$$E = \frac{\pi^2 \hbar^2}{2ma^2}$$

Show that the normalized ground-state wave function is

$$\psi(r) = (2\pi a)^{-1/2} \frac{\sin \pi r/a}{r}$$

7-29. In this problem, (see also Problem 4–38) we calculate the ground-state energy for the potential shown in Problem 7–12. The Schrödinger equation for this system is

$$-\frac{\hbar^2}{2m}\frac{d^2\psi}{dx^2} + V_0\psi = E\psi \qquad -\infty < x < -a$$

$$-\frac{\hbar^2}{2m}\frac{d^2\psi}{dx^2} = E\psi \qquad -a < x < a$$

$$-\frac{\hbar^2}{2m}\frac{d^2\psi}{dx^2} + V_0\psi = E\psi \qquad a < x < \infty$$

Label the three regions 1, 2, and 3. For the case $E < V_0$, show that

$$\psi_1(x) = Ae^{\beta x} + Be^{-\beta x}$$

$$\psi_2(x) = C \sin \alpha x + D \cos \alpha x$$

$$\psi_3(x) = Ee^{\beta x} + Fe^{-\beta x}$$

where $\beta = [2m(V_0 - E)/\hbar^2]^{1/2}$ is real and $\alpha = (2mE/\hbar^2)^{1/2}$. If $\psi_1(x)$ is to be finite as $x \to -\infty$ and $\psi_3(x)$ be finite as $x \to \infty$, we must have $B = 0$ and $E = 0$. Now there are four constants (A, C, D, and F) to be determined by the four boundary conditions

$$\psi_1(-a) = \psi_2(-a) \qquad \qquad \frac{d\psi_1}{dx}\bigg|_{x=-a} = \frac{d\psi_2}{dx}\bigg|_{x=-a}$$

$$\psi_2(a) = \psi_3(a) \qquad \qquad \frac{d\psi_2}{dx}\bigg|_{x=a} = \frac{d\psi_3}{dx}\bigg|_{x=a}$$

Before we go into all this algebra, let's remember that we are interested only in the ground-state energy. In this case, we expect $\psi_2(x)$ to be a cosine term because $\cos \alpha x$ has no nodes in region 2, whereas $\sin \alpha x$ does. Therefore, we will set $C = 0$. Show that the four boundary conditions give

$$Ae^{-\beta a} = D \cos \alpha a \qquad \qquad A\beta e^{-\beta a} = D\alpha \sin \alpha a$$

$$D \cos \alpha a = Fe^{-\beta a} \qquad \qquad -D\alpha \sin \alpha a = -F\beta e^{-\beta a}$$

These equations give $A = F$. Now divide $A\beta e^{-\beta a} = D\alpha \sin \alpha a$ by $Ae^{-\beta a} = D \cos \alpha a$ to get

$$\beta = \alpha \tan \alpha a$$

Now show that

$$\alpha^2 + \beta^2 = \frac{2m V_0}{\hbar^2}$$

and so

$$\alpha^2 + \alpha^2 \tan^2 \alpha a = \frac{2m V_0}{\hbar^2}$$

Multiply through by a^2 to get

$$\eta^2 (1 + \tan^2 \eta) = \frac{2m V_0 a^2}{\hbar^2} = \alpha$$

where $\eta = \alpha a$. Solve this equation numerically for η when $\alpha = 2m V_0 a^2 / \hbar^2 = 4$ and 12 to verify the exact energies given in Problem 7–12.

7-30. In applying first-order perturbation theory to the helium atom, we must evaluate the integral (Equation 7.50)

$$E^{(1)} = \frac{e^2}{4\pi\varepsilon_0} \int \int d\mathbf{r}_1 d\mathbf{r}_2 \psi_{1s}^*(\mathbf{r}_1)\psi_{1s}^*(\mathbf{r}_2)\frac{1}{r_{12}}\psi_{1s}(\mathbf{r}_1)\psi_{1s}(\mathbf{r}_2)$$

where

$$\psi_{1s}(\mathbf{r}_j) = \left(\frac{Z^3}{a_0^3\pi}\right)^{1/2} e^{-Zr_j/a_0}$$

and $Z = 2$ for the helium atom. This same integral occurs in a variational treatment of helium, where in that case the value of Z is left arbitrary. This problem proves that

$$E^{(1)} = \frac{5Z}{8}\left(\frac{m_e e^4}{16\pi^2\varepsilon_0^2\hbar^2}\right)$$

Let $\mathbf{r}_1$ and $\mathbf{r}_2$ be the radius vectors of electron 1 and 2, respectively, and let θ be the angle between these two vectors. Now this is generally *not* the θ of spherical coordinates, but if we choose one of the radius vectors, say $\mathbf{r}_1$, to be the z axis, then the two θ's are the same. Using the law of cosines,

$$r_{12} = (r_1^2 + r_2^2 - 2r_1 r_2 \cos \theta)^{1/2}$$

show that $E^{(1)}$ becomes

$$E^{(1)} = \frac{e^2}{4\pi\varepsilon_0}\frac{Z^6}{a_0^6\pi^2}\int_0^\infty dr_1 e^{-2Zr_1/a_0}4\pi r_1^2 \int_0^\infty dr_2 e^{-2Zr_2/a_0}r_2^2$$

$$\times \int_0^{2\pi} d\phi \int_0^\pi \frac{d\theta \sin \theta}{(r_1^2 + r_2^2 - 2r_1 r_2 \cos \theta)^{1/2}}$$

Letting $x = \cos\theta$, show that the integral over θ is

$$\int_0^\pi \frac{d\theta \sin\theta}{(r_1^2 + r_2^2 - 2r_1r_2\cos\theta)^{1/2}} = \int_{-1}^1 \frac{dx}{(r_1^2 + r_2^2 - 2r_1r_2 x)^{1/2}}$$

$$= \frac{2}{r_1} \qquad r_1 > r_2$$

$$= \frac{2}{r_2} \qquad r_1 < r_2$$

Substituting this result into $E^{(1)}$, show that

$$E^{(1)} = \frac{e^2}{4\pi\varepsilon_0} \frac{16Z^6}{a_0^6} \int_0^\infty dr_1 e^{-2Zr_1/a_0} r_1^2 \left(\frac{1}{r_1} \int_0^{r_1} dr_2 e^{-2Zr_2/a_0} r_2^2 \right.$$

$$\left. + \int_{r_1}^\infty dr_2 e^{-2Zr_2/a_0} r_2 \right)$$

$$= \frac{e^2}{4\pi\varepsilon_0} \frac{4Z^3}{a_0^3} \int_0^\infty dr_1 e^{-2Zr_1/a_0} r_1^2 \left[\frac{1}{r_1} - e^{-2Zr_1/a_0} \left(\frac{Z}{a_0} + \frac{1}{r_1} \right) \right]$$

$$= \frac{5}{8} Z \left(\frac{e^2}{4\pi\varepsilon_0 a_0} \right) = \frac{5}{8} Z \left(\frac{m_e e^4}{16\pi^2 \varepsilon_0^2 \hbar^2} \right)$$

Show that the energy through first order is

$$E^{(0)} + E^{(1)} = \left(-Z^2 + \frac{5}{8} Z \right) \left(\frac{m_e e^4}{16\pi^2 \varepsilon_0^2 \hbar^2} \right) = -\frac{11}{4} \left(\frac{m_e e^4}{16\pi^2 \varepsilon_0^2 \hbar^2} \right)$$

$$= -2.75 \left(\frac{m_e e^4}{16\pi^2 \varepsilon_0^2 \hbar^2} \right)$$

compared with the exact result, $E_{\text{exact}} = -2.9037(m_e e^4 / 16\pi^2 \varepsilon_0^2 \hbar^2)$.

7-31. In Problem 7–30 we evaluated the integral that occurs in the first-order perturbation theory treatment of helium (see Equation 7.50). In this problem we will evaluate the integral by another method, one that uses an expansion for $1/r_{12}$ that is useful in many applications. We can write $1/r_{12}$ as an expansion in terms of spherical harmonics

$$\frac{1}{r_{12}} = \frac{1}{|\mathbf{r}_1 - \mathbf{r}_2|} = \sum_{l=0}^\infty \sum_{m=-l}^{+l} \frac{4\pi}{2l+1} \frac{r_<^l}{r_>^{l+1}} Y_l^m(\theta_1, \phi_1) Y_l^{m*}(\theta_2, \phi_2)$$

where θ_i and ϕ_i are the angles that describe $\mathbf{r}_i$ in a spherical coordinate system and $r_<$ and $r_>$ are, respectively, the smaller and larger values of r_1 and r_2. In other words, if $r_1 < r_2$, then $r_< = r_1$ and $r_> = r_2$. Substitute $\psi_{1s}(r_i) = (Z^3/a_0^3\pi)^{1/2} e^{-Zr_i/a_0}$, and the above expansion for $1/r_{12}$ into Equation 7.50, integrate over the angles, and show that all the terms except for the $l = 0$, $m = 0$ term vanish. Show that

$$E^{(1)} = \frac{e^2}{4\pi\varepsilon_0} \frac{16Z^6}{a_0^6} \int_0^\infty dr_1 r_1^2 e^{-2Zr_1/a_0} \int_0^\infty dr_2 r_2^2 \frac{e^{-2Zr_2/a_0}}{r_>}$$

Now show that

$$E^{(1)} = \frac{e^2}{4\pi\varepsilon_0} \frac{16Z^6}{a_0^6} \int_0^\infty dr_1 r_1 e^{-2Zr_1/a_0} \int_0^{r_1} dr_2 r_2^2 e^{-2Zr_2/a_0}$$

$$+ \frac{e^2}{4\pi\varepsilon_0} \frac{16Z^6}{a_0^6} \int_0^\infty dr_1 r_1^2 e^{-2Zr_1/a_0} \int_{r_1}^\infty dr_2 r_2 e^{-2Zr_2/a_0}$$

$$= -\frac{e^2}{4\pi\varepsilon_0} \frac{4Z^3}{a_0^3} \int_0^\infty dr_1 r_1 e^{-2Zr_1/a_0} \left[e^{-2Zr_1/a_0} \left(\frac{2Z^2 r_1^2}{a_0^2} + \frac{2Zr_1}{a_0} + 1 \right) - 1 \right]$$

$$+ \frac{e^2}{4\pi\varepsilon_0} \frac{4Z^4}{a_0^4} \int_0^\infty dr_1 r_1^2 e^{-2Zr_1/a_0} \left[e^{-2Zr_1/a_0} \left(\frac{2Zr_1}{a_0} + 1 \right) \right]$$

$$= -\frac{e^2}{4\pi\varepsilon_0} \frac{4Z^6}{a_0^6} \int_0^\infty dr_1 e^{-4Zr_1/a_0} \left[\frac{r_1^2 a_0^2}{Z^2} + \frac{r_1 a_0^3}{Z^3} \right]$$

$$+ \frac{e^2}{4\pi\varepsilon_0} \frac{4Z^3}{a_0^3} \int_0^\infty dr_1 r_1 e^{-2Zr_1/a_0}$$

$$= \frac{5}{8} Z \left(\frac{e^2}{4\pi\varepsilon_0 a_0} \right)$$

as in Problem 7–30.

7-32. This problem fills in the steps of the variational treatment of helium. We use a trial function of the form

$$\phi(\mathbf{r}_1, \mathbf{r}_2) = \frac{Z^3}{a_0^3 \pi} e^{-Z(r_1 + r_2)/a_0}$$

with Z as an adjustable parameter. The Hamiltonian operator of the helium atom is

$$\hat{H} = -\frac{\hbar^2}{2m_e} \nabla_1^2 - \frac{\hbar^2}{2m_e} \nabla_2^2 - \frac{2e^2}{4\pi\varepsilon_0 r_1} - \frac{2e^2}{4\pi\varepsilon_0 r_2} + \frac{e^2}{4\pi\varepsilon_0 r_{12}}$$

We now evaluate

$$E(Z) = \int d\mathbf{r}_1 d\mathbf{r}_2 \phi^* \hat{H} \phi$$

The evaluation of this integral is greatly simplified if you recall that $\psi(r_j) = (Z^3/a_0^3\pi)^{1/2} e^{-Zr_j/a_0}$ is an eigenfunction of a hydrogenlike Hamiltonian operator, one for which the nucleus has a charge Z. Show that the helium atom Hamiltonian operator can be written as

$$\hat{H} = -\frac{\hbar^2}{2m_e} \nabla_1^2 - \frac{Ze^2}{4\pi\varepsilon_0 r_1} - \frac{\hbar^2}{2m_e} \nabla_2^2 - \frac{Ze^2}{4\pi\varepsilon_0 r_2} + \frac{(Z-2)e^2}{4\pi\varepsilon_0 r_1} + \frac{(Z-2)e^2}{4\pi\varepsilon_0 r_2} + \frac{e^2}{4\pi\varepsilon_0 r_{12}}$$

where

$$\left(-\frac{\hbar^2}{2m_e} \nabla^2 - \frac{Ze^2}{4\pi\varepsilon_0 r} \right) \left(\frac{Z^3}{a_0^3 \pi} \right)^{1/2} e^{-Zr/a_0} = -\frac{\hbar^2 Z^2}{2m_e a_0^2} \left(\frac{Z^3}{a_0^3 \pi} \right)^{1/2} e^{-Zr/a_0}$$

Show that

$$E(Z) = \frac{Z^6}{a_0^6 \pi^2} \int \int d\mathbf{r}_1 d\mathbf{r}_2 e^{-Z(r_1+r_2)/a_0} \left[-\frac{Z^2 e^2}{8\pi\varepsilon_0 a_0} - \frac{Z^2 e^2}{8\pi\varepsilon_0 a_0} + \frac{(Z-2)e^2}{4\pi\varepsilon_0 r_1} \right.$$

$$\left. + \frac{(Z-2)e^2}{4\pi\varepsilon_0 r_2} + \frac{e^2}{4\pi\varepsilon_0 r_{12}} \right] e^{-Z(r_1+r_2)/a_0}$$

The last integral is evaluated in Problem 7–30 or 7–31 and the others are elementary. Therefore, $E(Z)$, in units of $(m_e e^4/16\pi^2\varepsilon_0^2\hbar^2)$ is given by

$$E(Z) = -Z^2 + 2(Z-2)\frac{Z^3}{\pi} \int d\mathbf{r} \frac{e^{-2Zr}}{r} + \frac{5}{8}Z$$

$$= -Z^2 + 2(Z-2)Z + \frac{5}{8}Z$$

$$= Z^2 - \frac{27}{8}Z$$

Now minimize E with respect to Z and show that

$$E = -\left(\frac{27}{16}\right)^2 = -2.8477$$

in units of $m_e e^4/16\pi^2\varepsilon_0^2\hbar^2$. Interpret the value of Z that minimizes E.

Charlotte E. Moore was born in Ercildoun, Pennsylvania, on September 24, 1898 and died in 1990. After graduating from Swarthmore College in 1920, she worked at the Princeton University Observatory and the Mt. Wilson Observatory on stellar spectra and the determination of the Sun's chemical composition. She earned a Ph.D. in astronomy in 1931 from the University of California at Berkeley on a Lick Fellowship. In 1945, she moved from Princeton to Washington, D.C., to join the spectroscopy section at the National Bureau of Standards (now the National Institute of Standards and Technology) until her retirement in 1968. Moore was placed in charge of an Atomic Energy Level Program, a program whose mission was to produce a current and more complete compilation of spectral data and atomic energy levels. She not only compiled published data, but she critically analyzed the data for each spectrum. When the data were insufficient or of dubious quality, she persuaded competent spectroscopists to carry out new observations and analysis. The result of her effort, *Atomic Energy Levels* (1949–1958), is a classic work that provides data for 485 atomic species in a uniform, clear format with standardized notation. In 1949, Moore was elected as an Associate of the Royal Astronomical Society, the first woman to receive this honor, breaking a 129-year tradition. In 1937, she married a fellow astronomer, Bancroft Sitterly, but always published under her maiden name.

CHAPTER 8

Multielectron Atoms

We concluded Chapter 6 with an introduction to the helium atom. We showed there that if we considered the nucleus to be fixed at the origin, then the Schrödinger equation has the form

$$\left\{ \hat{H}_{\text{H}}(1) + \hat{H}_{\text{H}}(2) + \frac{e^2}{4\pi\varepsilon_0 r_{12}} \right\} \psi(\mathbf{r}_1, \mathbf{r}_2) = E\psi(\mathbf{r}_1, \mathbf{r}_2) \tag{8.1}$$

where $\hat{H}_{\text{H}}(j)$ is the hydrogenlike Hamiltonian operator of electron j (Equation 6.2). If it were not for the presence of the interelectronic repulsion term, Equation 8.1 would be immediately solvable. Its eigenfunctions would be products of hydrogenlike wave functions and its eigenvalues would be sums of the hydrogenlike energies of the two electrons (see Section 3–9). Helium is our first multielectron system, and although the helium atom may seem to be of minimal interest to chemists, we will discuss it in detail in this chapter because the solution of the helium atom illustrates the techniques used for more complex atoms. Then, after introducing the concept of electron spin and the Pauli Exclusion Principle, we will discuss the Hartree-Fock theory of complex atoms. Finally, we discuss the term symbols of atoms and ions and how they are used to label electronic states. This chapter illustrates the powerful utility of quantum mechanics in analyzing the electronic properties of atoms.

8–1. Atomic and Molecular Calculations Are Expressed in Atomic Units

We will apply both perturbation theory and the variational method to the helium atom, but before doing so, we will introduce a system of units, called atomic units, that is widely used in atomic and molecular calculations to simplify the equations. Natural units of mass and charge on an atomic or molecular scale are the mass of an electron and the magnitude of the charge on an electron (the charge on a proton). Recall in

Chapter 5 (Equation 5.52) that we saw that a natural unit of angular momentum on an atomic or molecular scale is $\hbar$. A natural unit of length on an atomic scale is the Bohr radius (see Sections 1–8 and 6–4)

$$a_0 = \frac{4\pi \varepsilon_0 \hbar^2}{m_e e^2} \tag{8.2}$$

and we saw repeatedly in Chapter 7 that a natural unit of energy is

$$E = \frac{m_e e^4}{16\pi^2 \varepsilon_0^2 \hbar^2} \tag{8.3}$$

It is convenient in atomic and molecular calculations to use units that are natural on that scale. The units that we will adopt for atomic and molecular calculations are given in Table 8.1. This set of units is called *atomic units*. The atomic unit of energy is called a *hartree* and is denoted by E_h. Note that in atomic units the ground-state energy of a hydrogen atom (in the fixed nucleus approximation) is $-E_h/2$ (cf. Equation 6.44).

TABLE 8.1
Atomic units and their SI equivalents.

Property	Atomic unit	SI Equivalent
mass	mass of an electron, m_e	9.1094×10^{-31} kg
charge	charge on a proton, e	1.6022×10^{-19} C
angular momentum	Planck constant divided by 2π, $\hbar$	1.0546×10^{-34} J·s
distance	Bohr radius, $a_0 = \dfrac{4\pi \varepsilon_0 \hbar^2}{m_e e^2}$	5.2918×10^{-11} m
energy	$\dfrac{m_e e^4}{16\pi^2 \varepsilon_0^2 \hbar^2} = \dfrac{e^2}{4\pi \varepsilon_0 a_0} = E_h$	4.3597×10^{-18} J
permittivity	$4\pi \varepsilon_0$	1.1127×10^{-10} C^2·J^{-1}·m^{-1}

EXAMPLE 8–1
The unit of energy in atomic units is given by

$$1E_h = \frac{m_e e^4}{16\pi^2 \varepsilon_0^2 \hbar^2}$$

Express $1E_h$ in units of joules (J), kilojoules per mole (kJ·mol^{-1}), wave numbers (cm^{-1}), and electron volts (eV).

SOLUTION: To find $1E_h$ expressed in joules, we substitute the SI values of m_e, e, $4\pi\varepsilon_0$, and $\hbar$ into the above equation. Using these values from Table 8.1, we find

$$1E_h = \frac{(9.1094 \times 10^{-31}\ \text{kg})(1.6022 \times 10^{-19}\ \text{C})^4}{(1.1127 \times 10^{-10}\ \text{C}^2\cdot\text{J}^{-1}\cdot\text{m}^{-1})^2(1.0546 \times 10^{-34}\ \text{J}\cdot\text{s})^2}$$

$$= 4.3597 \times 10^{-18}\ \text{J}$$

If we multiply this result by the Avogadro constant, we obtain

$$1E_h = 2625.5\ \text{kJ}\cdot\text{mol}^{-1}$$

To express $1E_h$ in wave numbers (cm^{-1}), we use the equation

$$\tilde{\nu} = \frac{1}{\lambda} = \frac{h\nu}{hc} = \frac{E}{ch} = \frac{4.3597 \times 10^{-18}\ \text{J}}{(2.9979 \times 10^8\ \text{m}\cdot\text{s}^{-1})(6.6261 \times 10^{-34}\ \text{J}\cdot\text{s})}$$

$$= 2.1947 \times 10^7\ \text{m}^{-1} = 2.1947 \times 10^5\ \text{cm}^{-1}$$

so that we can write

$$1E_h = 2.1947 \times 10^5\ \text{cm}^{-1}$$

Last, to express one E_h in terms of electron volts, we use the conversion factor

$$1\ \text{eV} = 1.6022 \times 10^{-19}\ \text{J}$$

Using the value of one E_h in joules obtained previously, we have

$$1E_h = (4.3597 \times 10^{-18}\ \text{J})\left(\frac{1\ \text{eV}}{1.6022 \times 10^{-19}\ \text{J}}\right)$$

$$= 27.211\ \text{eV}$$

The use of atomic units greatly simplifies most of the equations we will use in atomic and molecular calculations. For example, the Hamiltonian operator of a helium atom

$$\hat{H} = -\frac{\hbar^2}{2m_e}\nabla_1^2 - \frac{\hbar^2}{2m_e}\nabla_2^2 - \frac{2e^2}{4\pi\varepsilon_0 r_1} - \frac{2e^2}{4\pi\varepsilon_0 r_2} + \frac{e^2}{4\pi\varepsilon_0 r_{12}} \tag{8.4}$$

in atomic units becomes simply (Problem 8–7)

$$\hat{H} = -\frac{1}{2}\nabla_1^2 - \frac{1}{2}\nabla_2^2 - \frac{2}{r_1} - \frac{2}{r_2} + \frac{1}{r_{12}} \tag{8.5}$$

An important aspect of the use of atomic units in atomic and molecular calculations is that the calculated energies are independent of the values of physical constants such as the electron mass, the Planck constant, etc. As the values of physical constants are further refined by advances in experimental methodology, the energies calculated using atomic units will not be affected by these refinements. For example, we will see in the next section that the most accurate calculation of the ground-state energy of a helium atom gives $-2.903\,724\,375\ E_h$ (Table 8.2), which took months of computer time at the time the calculation was done. Because atomic units were used, this value will never have to be redetermined.

8–2. Both Perturbation Theory and the Variational Method Can Yield Excellent Results for Helium

The problem we want to solve is $\hat{H}\psi = E\psi$, where $\hat{H}$ is given by Equation 8.5. We applied perturbation theory to this problem at the end of Section 7–4 by considering the

TABLE 8.2
Ground-state energy of the helium atom[a].

Method	Energy/E_h	Ionization energy/E_h	Ionization energy/kJ·mol^{-1}
Perturbation calculations			
Complete neglect of the inter-electronic repulsion term	-4.0000	2.000	5250
First-order perturbation theory	-2.7500	0.7500	1969
Second-order perturbation theory	-2.9077	0.9077	2383
Thirteenth-order perturbation theory[b]	$-2.903\,724\,33$	$0.903\,724\,33$	2373
Variational calculations			
$(1s)^2$ with $\zeta = 1.6875$	-2.8477	0.8477	2226
$(ns)^2$ with $\zeta = 1.61162$ and $n = 0.995$	-2.8542	0.8542	2242
Hartree-Fock[c]	-2.8617	0.8617	2262
Hylleras[d], 10 parameters	$-2.903\,63$	$0.903\,63$	2372
Pekeris[e], 1078 parameters	$-2.903\,724\,375$	$0.903\,724\,375$	2373
Experimental value	-2.9033	0.9033	2373

[a] These are nonrelativistic, fixed nucleus-approximation energies. Corrections for nuclear motion and relativistic corrections are estimated to be $10^{-4}\ E_h$.

[b] C.W. Scheer and R.E. Knight, *Rev. Mod. Phys.* **35**, 426 (1963)

[c] E. Clementi and C. Roetti, *At. Data Nucl. Data Tables* **14**, 177 (1974)

[d] E.A. Hylleras, *Z. Physik* **54**, 347 (1929)

[e] C.L. Pekeris, *Phys. Rev.* **115**, 1216 (1959)

interelectronic repulsion term to be a perturbation and found that the energy through first order is given by

$$E = -Z^2 + \frac{5}{8}Z = -\frac{11}{4}E_h = -2.750\ E_h \tag{8.6}$$

or $-7220\ \text{kJ·mol}^{-1}$. The experimental value of the energy is $-2.9033\ E_h$, or $-7623\ \text{kJ·mol}^{-1}$, and so we see that first-order perturbation theory gives a result that is approximately 5% in error. Scheer and Knight (see Table 8.2) calculated the energy through many orders of perturbation theory and found that

$$E = -Z^2 + \frac{5}{8}Z - 0.15766 + \frac{0.00870}{Z} + \frac{0.000889}{Z^2} + \cdots \tag{8.7}$$

Equation 8.7 yields a value of $-2.9037\ E_h$, in good agreement with the experimental value of $-2.9033\ E_h$.

We can also use the variational method to calculate the ground-state energy of a helium atom. In Section 7–1, we used

$$\phi_0(\mathbf{r}_1, \mathbf{r}_2) = \psi_{1s}(\mathbf{r}_1)\psi_{1s}(\mathbf{r}_2) \tag{8.8}$$

where

$$\psi_{1s}(\mathbf{r}_j) = \left(\frac{Z^3}{\pi}\right)^{1/2} e^{-Zr_j} \tag{8.9}$$

in atomic units as a trial function with Z as a variational parameter and found that

$$E = -\left(\frac{27}{16}\right)^2 E_h = -2.8477\ E_h \tag{8.10}$$

compared with the first-order perturbation theory result of $-2.7500\ E_h$ and the higher-order result of $-2.9037\ E_h$.

The agreement we have found between first-order perturbation theory or our variational approximation and the experimental value of the energy may appear quite good, but let's examine this agreement more closely. The ionization energy (IE) of a helium atom is given by

$$\text{IE} = E_{\text{He}^+} - E_{\text{He}} \tag{8.11}$$

The energy of He^+ is $-2\ E_h$ (Problem 8–2), so we have

$$\text{IE} = \left(-2 + \frac{11}{4}\right)E_h = 0.7500\ E_h$$

$$= 1969\ \text{kJ·mol}^{-1} \quad \text{(first-order perturbation theory)}$$

or

$$\text{IE} = -2E_\text{h} + \left(\frac{27}{16}\right)^2 E_\text{h} = 0.8477\, E_\text{h}$$

$$= 2226\ \text{kJ}\cdot\text{mol}^{-1} \quad \text{(our variational result)}$$

whereas the experimental value of the ionization energy is $0.9033\, E_\text{h}$, or $2372\ \text{kJ}\cdot\text{mol}^{-1}$. Even our variational result, with its 6% discrepancy with the experimental total energy, is not too satisfactory if you realize that an error of $0.056\, E_\text{h}$ is equivalent to $150\ \text{kJ}\cdot\text{mol}^{-1}$, which is the same order of magnitude as the strength of a chemical bond. Clearly, we must be able to do better.

One way to improve our results is to use a more general trial function than Equation 8.8. Because a suitable trial function may be almost any function, we are not restricted to choosing a $1s$ hydrogenlike wave function. For example, in 1930 the American physicist John Slater introduced a set of orbitals, now called *Slater orbitals*, which are of the form

$$S_{nlm}(r, \theta, \phi) = N_{nl} r^{n-1} e^{-\zeta r} Y_l^m(\theta, \phi) \tag{8.12}$$

where $N_{nl} = (2\zeta)^{n+\frac{1}{2}}/[(2n)!]^{1/2}$ is a normalization constant and the $Y_l^m(\theta, \phi)$ are the spherical harmonics (Section 6–2 and Table 6.3). The parameter ζ (zeta) is taken to be arbitrary and is not necessarily equal to Z/n as in the hydrogenlike orbitals. Note that the radial parts of Slater orbitals do not have nodes like hydrogen atomic orbitals do.

EXAMPLE 8–2
Show that $S_{nlm}(r, \theta, \phi)$ is not orthogonal to $S_{n'lm}(r, \theta, \phi)$.

SOLUTION: We must show that $I \neq 0$ where

$$I = \int_0^\infty dr\, r^2 \int_0^\pi d\theta \sin\theta \int_0^{2\pi} d\phi\, S_{nlm}^*(r, \theta, \phi) S_{n'lm}(r, \theta, \phi)$$

$$= \int_0^\infty dr\, r^{n+n'} e^{-2\zeta r} \int_0^\pi d\theta \sin\theta \int_0^{2\pi} d\phi\, Y_l^m(\theta, \phi)^* Y_l^m(\theta, \phi)$$

The integral over θ and ϕ gives 1 by Equation 6.31, leaving

$$I = \int_0^\infty r^{n+n'} e^{-2\zeta r} dr \neq 0.$$

This integral cannot equal zero because the integrand is always positive.

If we use

$$\psi = S_{100}(r_1, \theta_1, \phi_1)S_{100}(r_2, \theta_2, \phi_2)$$

$$= \frac{\zeta^3}{\pi}e^{-2\zeta(r_1+r_2)} \tag{8.13}$$

as a trial function with ζ as the only variational parameter, then we have seen above that $\zeta = 1.6875 = (\frac{27}{16})$ and $E = -2.8477\ E_h$ (see also Table 8.2). This value of E gives an ionization energy of 2226 kJ·mol^{-1}, compared with the experimental value of 2373 kJ·mol^{-1}. If we let n also be a variational parameter so that the trial function is

$$\psi = S_{n00}(r_1, \theta_1, \phi_1)S_{n00}(r_2, \theta_2, \phi_2)$$

$$= \frac{(2\zeta)^{2n+1}}{4\pi(2n)!}r_1^{n-1}r_2^{n-1}e^{-\zeta(r_1+r_2)} \tag{8.14}$$

then we find that $n = 0.995, \zeta = 1.6116$, and $E = -2.8542\ E_h$, leading to an ionization energy of 2242 kJ·mol^{-1}.

If we use a more flexible trial function of the form in which $\psi(\mathbf{r}_1, \mathbf{r}_2)$ is a product of one-electron functions, or *orbitals*,

$$\psi(\mathbf{r}_1, \mathbf{r}_2) = \phi(\mathbf{r}_1)\phi(\mathbf{r}_2) \tag{8.15}$$

and allow $\phi(r)$ to be completely general, then we reach a limit that is both practical and theoretical. In this limit, $E = -2.8617E_h$ and the ionization energy is $0.8617E_h$, compared with the best variational values $-2.9037E_h$ and $0.9033E_h$, respectively. This limiting value is the best value of the energy that can be obtained using a trial function of the form of a product of one-electron wave functions (Equation 8.15). This limit is called the *Hartree-Fock limit*, and we will discuss it more fully in the next section. Note that the concept of electron orbitals is preserved in the *Hartree-Fock approximation*.

If we do not restrict the trial function to be a product of single-electron orbitals, then we can go on and obtain essentially the exact energy. It has been found to be advantageous to include terms containing the interelectronic distance r_{12} explicitly in the trial function. This was first done by Hylleras in 1930, who introduced a trial function of the (unnormalized) form

$$\psi(r_1, r_2, r_{12}) = e^{-Zr_1}e^{-Zr_2}[1 + cr_{12}] \tag{8.16}$$

Using Z and c as variational parameters, Hylleras obtained a value of $E = -2.8913\ E_h$, within less than 0.5% of the exact value. Using a computer, we could carry this procedure out to a larger number of terms to yield an energy that is essentially exact. The most extensive such calculation was carried out in 1959 by Pekeris, who obtained $E = -2.903\ 724\ 375\ E_h$ using 1078 parameters.

Although these calculations do show that we can obtain essentially exact energies by using the variational method with r_{12} in the trial function explicitly, these calculations are quite difficult computationally and do not readily lend themselves to large atoms and molecules. Furthermore, we have abandoned the orbital concept altogether. The orbital concept has been of great use to chemists, so the scheme nowadays is to find the Hartree-Fock orbitals mentioned above and correct them by some method such as perturbation theory. It is instructive to outline the Hartree-Fock procedure for helium because the equations are fairly simple for this two-electron case and provide a nice physical interpretation.

8–3. Hartree-Fock Equations Are Solved by the Self-Consistent Field Method

The starting point of the Hartree-Fock procedure for helium is to write the two-electron wave function as a product of orbitals, as in Equation 8.15:

$$\psi(\mathbf{r}_1, \mathbf{r}_2) = \phi(\mathbf{r}_1)\phi(\mathbf{r}_2) \tag{8.17}$$

The two functions on the right side of Equation 8.17 are the same because we are assuming that both electrons are in the same orbital, in accord with the Pauli Exclusion Principle. According to Equation 8.17, the probability distribution of electron 2 is $\phi^*(\mathbf{r}_2)\phi(\mathbf{r}_2)d\mathbf{r}_2$. We can also interpret this probability distribution classically as a charge density, and so we can say that the potential energy that electron 1 experiences at the point $\mathbf{r}_1$ due to electron 2 is (in atomic units)

$$V_1^{\text{eff}}(r_1) = \int \phi^*(\mathbf{r}_2)\frac{1}{r_{12}}\phi(\mathbf{r}_2)d\mathbf{r}_2 \tag{8.18}$$

where the superscript "eff" emphasizes that $V_1^{\text{eff}}(\mathbf{r}_1)$ is an effective, or average, potential. We now define an effective one-electron Hamiltonian operator by

$$\hat{H}_1^{\text{eff}}(\mathbf{r}_1) = -\frac{1}{2}\nabla_1^2 - \frac{2}{r_1} + V_1^{\text{eff}}(r_1) \tag{8.19}$$

The Schrödinger equation corresponding to this effective Hamiltonian operator is

$$\hat{H}_1^{\text{eff}}(\mathbf{r}_1)\phi(\mathbf{r}_1) = \epsilon_1\phi(\mathbf{r}_1) \tag{8.20}$$

There is a similar equation for $\phi(\mathbf{r}_2)$, but because $\phi(\mathbf{r}_1)$ and $\phi(\mathbf{r}_2)$ have the same functional form, we need to consider only one equation like Equation 8.20. Equation 8.20 is the *Hartree-Fock equation* for a helium atom, and its solution gives the best orbital wave function for helium. Note that $\hat{H}_1^{\text{eff}}(\mathbf{r}_1)$ depends upon $\phi(\mathbf{r}_2)$ through Equation 8.18. Thus, we must know the solution to Equation 8.20 before we even know the operator. The method of solving an equation like Equation 8.20 is by a scheme called the *self-consistent field method*, which can be implemented very easily on a computer.

First, we guess a form for $\phi(\mathbf{r}_2)$ and use it to evaluate $V_1^{\text{eff}}(r_1)$ by Equation 8.18. Then we solve Equation 8.20 for $\phi(\mathbf{r}_1)$. Usually, after one cycle the $\phi(\mathbf{r})$ that is used as input and the $\phi(\mathbf{r})$ obtained as output differ. [Remember that $\phi(\mathbf{r}_1)$ and $\phi(\mathbf{r}_2)$ have the same functional form.] Now we calculate $V_1^{\text{eff}}(r_1)$ with this new $\phi(\mathbf{r}_2)$ and then solve Equation 8.20 for a newer $\phi(\mathbf{r}_1)$. This cyclic process is continued until the $\phi(\mathbf{r}_2)$ used as input and the $\phi(\mathbf{r}_1)$ obtained from Equation 8.20 as output are sufficiently close, or are *self-consistent*. The orbitals obtained by this method are the *Hartree-Fock orbitals*.

In practice, we use linear combinations of Slater orbitals for $\phi(r)$, varying the parameters in each Slater orbital and the number of Slater orbitals used until convergence is obtained. For helium, the result obtained was

$$\phi_{1s}(r_1) = 0.75738e^{-1.4300r_1} + 0.43658e^{-2.4415r_1} + 0.17295e^{-4.0996r_1}$$
$$-0.02730e^{-6.4843r_1} + 0.06675e^{-7.978r_1}$$

with an identical equation for $\phi_{1s}(r_2)$. The Hartree-Fock limit gives $E_{\text{HF}} = -2.8617\ E_{\text{h}}$ compared with $E_{\text{exact}} = -2.9037\ E_{\text{h}}$. This procedure yields the best value of the energy under the orbital approximation, and the results seem to justify the use of the orbital concept for multielectron atoms (and molecules).

Investigating the discrepancy between the self-consistent field energy and the exact energy is interesting. Because $\psi(\mathbf{r}_1, \mathbf{r}_2) = \phi(\mathbf{r}_1)\phi(\mathbf{r}_2)$, the two electrons are taken to be independent of each other, or at least to interact only through some average, or effective, potential. We say, then, that the electrons are uncorrelated, and we define a *correlation energy* (CE) by the equation

$$\text{correlation energy} = \text{CE} = E_{\text{exact}} - E_{\text{HF}} \tag{8.21}$$

For helium, the correlation energy is (see Table 8.2)

$$\text{CE} = (-2.9037 + 2.8617)\ E_{\text{h}}$$
$$= -0.0420\ E_{\text{h}} = -110\ \text{kJ·mol}^{-1}$$

Although the Hartree-Fock energy is this case is almost 99% of the exact energy, the difference is still 110 kJ·mol^{-1}, which is unacceptably large because it is roughly of the same magnitude as the strength of a chemical bond. We will say more about this difference in Section 8–7 and Chapter 11.

So much for the ground state of the helium atom. Let's now consider the lithium atom. Following Equation 8.17, it would be "natural" to start a variational calculation with

$$\psi(\mathbf{r}_1, \mathbf{r}_2, \mathbf{r}_3) = \phi_{1s}(\mathbf{r}_1)\phi_{1s}(\mathbf{r}_2)\phi_{1s}(\mathbf{r}_3)$$

where ϕ_{1s} is a hydrogenlike or Slater 1s orbital, but we know from general chemistry that you cannot put three electrons into a 1s orbital. This fact leads us to a discussion of the Pauli Exclusion Principle and to the spin of the electron.

8–4. An Electron Has An Intrinsic Spin Angular Momentum

Although the Schödinger equation was amazingly successful in predicting or explaining the results of most experiments, it could not explain a few phenomena. One was the doublet yellow line in the atomic spectrum of sodium. The Schrödinger equation predicts that there should be one line around 590 nm, whereas two closely spaced lines (a doublet) are observed at 589.59 nm and 588.99 nm.

This observation and several others were explained in 1925 by two young Dutch physicists George Uhlenbeck and Samuel Goudsmit, who suggested that an electron behaves like a spinning top having z components of spin angular momentum of $\pm \hbar/2$. This suggestion amounts to introducing a fourth quantum number for an electron. This fourth quantum number, which represents the z component of the electron spin angular momentum, is now called the *spin quantum number*, m_s, and takes on the values $\pm 1/2$ in atomic units.

We are going to simply "graft" the concept of spin onto the quantum theory and onto the postulates we developed earlier. This may appear to be a somewhat unsatisfactory way to proceed, but it turns out to be quite satisfactory for our purposes. In the early 1930s, the English physicist Paul Dirac developed a relativistic extension of quantum mechanics, and one of its greatest successes is that spin arose in a perfectly natural way. We will introduce spin here, however, in an *ad hoc* manner.

To this point, we have restricted angular momentum to integer values of J in the case of the rigid rotator (Section 5–8) or l in the case of the orbital angular momentum in the hydrogen atom (Section 6–2). Following the suggestion of Uhlenbeck and Goudsmit, we will introduce half-integral angular momentum for electron spin. Just as we have the eigenvalue equations for $\hat{L}^2$ and $\hat{L}_z$ (Equations 6.33 and 6.38),

$$\hat{L}^2 Y_l^m(\theta, \phi) = \hbar^2 l(l+1) Y_l^m(\theta, \phi)$$

$$\hat{L}_z Y_l^m(\theta, \phi) = m\hbar Y_l^m(\theta, \phi)$$

(8.22)

we *define* the spin operators $\hat{S}^2$ and $\hat{S}_z$ and their eigenfunctions α and β by the equations

$$\hat{S}^2 \alpha = \tfrac{1}{2}\left(\tfrac{1}{2}+1\right)\hbar^2 \alpha \qquad \hat{S}^2 \beta = \tfrac{1}{2}\left(\tfrac{1}{2}+1\right)\hbar^2 \beta$$

(8.23)

and

$$\hat{S}_z \alpha = m_s \alpha = \tfrac{1}{2}\hbar\alpha \qquad \hat{S}_z \beta = m_s \beta = -\tfrac{1}{2}\hbar\beta$$

(8.24)

In analogy with Equations 8.22, $\alpha = Y_{1/2}^{1/2}$ and $\beta = Y_{1/2}^{-1/2}$, but this is a strictly *formal* association and α and β and even $\hat{S}^2$ and $\hat{S}_z$, for that matter, do not have to be specified any further.

Just as we can write that the value of the square of the orbital angular momentum of an electron in a hydrogen atom is given by

$$L^2 = \hbar^2 l(l+1) \tag{8.25}$$

we can say that the square of the spin angular momentum of an electron is

$$S^2 = \hbar^2 s(s+1) \tag{8.26}$$

Unlike l, which can range from 0 to ∞, s can have only the value $s = 1/2$. Note that because s is not allowed to assume large values, the spin angular momentum can never assume classical behavior. (See Section 3–6.) Spin is strictly a nonclassical concept. The functions α and β in Equations 8.23 and 8.24 are called *spin eigenfunctions*. Even though we do not know (or need to know) the forms of the operators $\hat{S}$ and $\hat{S}_z$, they must be Hermitian, and so α and β must be orthonormal, which we write *formally* as

$$\int \alpha^*(\sigma)\alpha(\sigma)d\sigma = \int \beta^*(\sigma)\beta(\sigma)d\sigma = 1$$

$$\int \alpha^*(\sigma)\beta(\sigma)d\sigma = \int \alpha(\sigma)\beta^*(\sigma)d\sigma = 0 \tag{8.27}$$

where σ is called the *spin variable*. The spin variable has no classical analog. We will use Equations 8.27 in a strictly formal sense.

8–5. Wave Functions Must Be Antisymmetric in the Interchange of Any Two Electrons

We must now include the spin function with the spatial wave function. We postulate that the spatial and spin parts of a wave function are independent and so write

$$\Psi(x, y, z, \sigma) = \psi(x, y, z)\alpha(\sigma) \quad \text{or} \quad \psi(x, y, z)\beta(\sigma) \tag{8.28}$$

The complete one-electron wave function Ψ is called a *spin orbital*. Using the hydrogenlike wave functions as specific examples, the first two spin orbitals of a hydrogenlike atom are

$$\Psi_{100\frac{1}{2}} = \left(\frac{Z^3}{\pi}\right)^{1/2} e^{-Zr}\alpha$$

$$\Psi_{100-\frac{1}{2}} = \left(\frac{Z^3}{\pi}\right)^{1/2} e^{-Zr}\beta \tag{8.29}$$

It follows that each of these spin orbitals is normalized because we can write

$$\int \Psi^*_{100\frac{1}{2}}(\mathbf{r}, \sigma)\Psi_{100\frac{1}{2}}(\mathbf{r}, \sigma)4\pi r^2 dr d\sigma = \int_0^\infty \frac{Z^3}{\pi}e^{-2Zr}4\pi r^2 dr \int \alpha^*\alpha d\sigma = 1 \tag{8.30}$$

where we have used Equation 8.27. The above two spin orbitals are orthogonal to each other because

$$\int \Psi_{100\frac{1}{2}}^*(\mathbf{r}, \sigma)\Psi_{100-\frac{1}{2}}(\mathbf{r}, \sigma)4\pi r^2 dr d\sigma = \int_0^\infty \frac{Z^3}{\pi}e^{-2Zr}4\pi r^2 dr \int \alpha^*\beta d\sigma = 0 \quad (8.31)$$

Note that even though the "100" part in Equation 8.31 is normalized, the two spin orbitals are orthogonal due to the spin parts.

You probably remember from general chemistry that no two electrons in an atom can have the same values of all four quantum numbers, n, l, m_l, and m_s. This restriction is called the *Pauli Exclusion Principle*. There is another, more fundamental statement of the Exclusion Principle that restricts the form of a multielectron wave function. We will present the Pauli Exclusion Principle as another postulate of quantum mechanics, but before doing so we must introduce the idea of an *antisymmetric wave function*. Let's go back to helium and write

$$\psi(1, 2) = 1s\alpha(1)1s\beta(2) \quad (8.32)$$

where $1s\alpha$ and $1s\beta$ are shorthand notation for $\Psi_{100\frac{1}{2}}$ and $\Psi_{100-\frac{1}{2}}$, respectively, and where the arguments 1 and 2 denote all four coordinates (x, y, z, and σ) of electrons 1 and 2, respectively. Note that Equation 8.32 corresponds to a product of the two wave functions given by Equation 8.29. Because no known experiment can distinguish one electron from another, we say that electrons are indistinguishable and, therefore, cannot be labelled. Thus, the wave function

$$\psi(2, 1) = 1s\alpha(2)1s\beta(1) \quad (8.33)$$

is equivalent to Equation 8.32. Mathematically, indistinguishability requires that we take linear combinations involving all possible labelings of the electrons. For a two-electron atom, we take the linear combinations of Equations 8.32 and 8.33:

$$\Psi_1 = \psi(1, 2) + \psi(2, 1) = 1s\alpha(1)1s\beta(2) + 1s\alpha(2)1s\beta(1) \quad (8.34)$$

and

$$\Psi_2 = \psi(1, 2) - \psi(2, 1) = 1s\alpha(1)1s\beta(2) - 1s\alpha(2)1s\beta(1) \quad (8.35)$$

Both Ψ_1 and Ψ_2 describe states in which there are two indistinguishable electrons; one electron is in the spin orbital $1s\alpha$ and the other is in $1s\beta$. Neither wave function specifies which electron is in each spin orbital, nor should they because the electrons are indistinguishable.

Both of the wave functions Ψ_1 and Ψ_2 appear to be acceptable wave functions for the ground state of a helium atom, but it turns out experimentally that we must use the wave function Ψ_2 to describe the ground state of a helium atom. Note that Ψ_2 has the property that it changes sign when the two electrons are interchanged because

$$\Psi_2(2, 1) = \psi(2, 1) - \psi(1, 2) = -\Psi_2(1, 2) \quad (8.36)$$

We say that $\Psi_2(1, 2)$ is *antisymmetric* under the interchange of the two electrons. The observation that the ground state of a helium atom is described by only Ψ_2 is but one example of the Pauli Exclusion Principle:

Postulate 6

All electronic wave functions must be antisymmetric under the interchange of any two electrons.

In Section 8–6, we will show that Postulate 6 implies the more familiar statement of the Pauli Exclusion Principle, that no two electrons in an atom can have the same values of the four quantum numbers, n, l, m_l, and m_s.

EXAMPLE 8–3
The wave function $\Psi_2(1, 2)$ given by Equation 8.35 is not normalized as it stands. Determine the normalization constant of $\Psi_2(1, 2)$ given that the "$1s$" parts are normalized.

SOLUTION: We want to find the constant c such that

$$I = c^2 \int \Psi_2^*(1, 2)\Psi_2(1, 2)dr_1 dr_2 d\sigma_1 d\sigma_2 = 1$$

First notice that $\Psi_2(1, 2)$ can be factored into the product of a spatial part and a spin part:

$$\Psi_2(1, 2) = 1s(1)1s(2)[\alpha(1)\beta(2) - \alpha(2)\beta(1)]$$
$$= 1s(\mathbf{r}_1)1s(\mathbf{r}_2)[\alpha(\sigma_1)\beta(\sigma_2) - \alpha(\sigma_2)\beta(\sigma_1)] \qquad (8.37)$$

The normalization integral becomes the product of three integrals

$$I = c^2 \int 1s^*(\mathbf{r}_1)1s(\mathbf{r}_1)dr_1 \int 1s^*(\mathbf{r}_2)1s(\mathbf{r}_2)dr_2 \times$$
$$\int \int [\alpha^*(\sigma_1)\beta^*(\sigma_2) - \alpha^*(\sigma_2)\beta^*(\sigma_1)][\alpha(\sigma_1)\beta(\sigma_2) - \alpha(\sigma_2)\beta(\sigma_1)]d\sigma_1 d\sigma_2$$

The spatial integrals are equal to 1 because we have taken the $1s$ orbitals to be normalized. Now let's look at the spin integrals. When the two terms in the integrand of the spin integral are multiplied we get four integrals. One of them is

$$\int \int \alpha^*(\sigma_1)\beta^*(\sigma_2)\alpha(\sigma_1)\beta(\sigma_2)d\sigma_1 d\sigma_2$$
$$= \int \alpha^*(\sigma_1)\alpha(\sigma_1)d\sigma_1 \int \beta^*(\sigma_2)\beta(\sigma_2)d\sigma_2 = 1$$

where we have used Equation 8.27. Another is

$$\int \int \alpha^*(\sigma_1)\beta^*(\sigma_2)\alpha(\sigma_2)\beta(\sigma_1)d\sigma_1 d\sigma_2$$
$$= \int \alpha^*(\sigma_1)\beta(\sigma_1)d\sigma_1 \int \beta^*(\sigma_2)\alpha(\sigma_2)d\sigma_2 = 0$$

The other two are equal to 1 and 0, and so

$$I = c^2 \int \Psi_2^*(1, 2)\Psi_2(1, 2)d\mathbf{r}_1 d\mathbf{r}_2 d\sigma_1 d\sigma_2$$

$$= 2c^2$$

In order that $I = 1$, $c = 1/\sqrt{2}$.

8–6. Antisymmetric Wave Functions Can Be Represented by Slater Determinants

Now that we have introduced spin and have seen that we must use antisymmetric wave functions, we must ask why we could ignore the spin part of the wave function when we treated the helium atom in Sections 7–1 and 8–2. The reason is that Ψ_2 can be factored into a spatial part and a spin part, as we saw in Equation 8.37 in Example 8–3. In Sections 7–1 and 8–2, we used only the spatial part of Ψ_2, and the spatial part is just a product of two $1s$ Slater orbitals. If we use Ψ_2 to calculate the ground-state energy of a helium atom, then we obtain

$$E = \frac{\int \Psi_2^*(1, 2)\hat{H}\Psi_2(1, 2)d\mathbf{r}_1 d\mathbf{r}_2 d\sigma_1 d\sigma_2}{\int \Psi_2^*(1, 2)\Psi_2(1, 2)d\mathbf{r}_1 d\mathbf{r}_2 d\sigma_1 d\sigma_2} \tag{8.38}$$

The numerator in Equation 8.38 is

$$\int 1s^*(\mathbf{r}_1)1s^*(\mathbf{r}_2)[\alpha^*(\sigma_1)\beta^*(\sigma_2) - \alpha^*(\sigma_2)\beta^*(\sigma_1)]$$
$$\times \hat{H}1s(\mathbf{r}_1)1s(\mathbf{r}_2)[\alpha(\sigma_1)\beta(\sigma_2) - \alpha(\sigma_2)\beta(\sigma_1)]d\mathbf{r}_1 d\mathbf{r}_2 d\sigma_1 d\sigma_2 \tag{8.39}$$

Because the Hamiltonian operator does not contain any spin operators, it does not affect the spin functions and so we can factor the integral in Equation 8.39 to give

$$\int 1s^*(\mathbf{r}_1)1s^*(\mathbf{r}_2)\hat{H}1s(\mathbf{r}_1)1s(\mathbf{r}_2)d\mathbf{r}_1 d\mathbf{r}_2$$
$$\times \int [\alpha^*(\sigma_1)\beta^*(\sigma_2) - \alpha^*(\sigma_2)\beta^*(\sigma_1)][\alpha(\sigma_1)\beta(\sigma_2) - \alpha(\sigma_2)\beta(\sigma_1)]d\sigma_1 d\sigma_2 \tag{8.40}$$

We showed in Example 8–3 that the total spin integral is equal to 2. It is a straightforward exercise (Problem 8–15) to show that the contribution of the spin integral to the denominator in Equation 8.38 is also equal to 2 and so Equation 8.38 becomes

$$E = \frac{\int \psi^*(\mathbf{r}_1, \mathbf{r}_2)\hat{H}\psi(\mathbf{r}_1, \mathbf{r}_2)d\mathbf{r}_1 d\mathbf{r}_2}{\int \psi^*(\mathbf{r}_1, \mathbf{r}_2)\psi(\mathbf{r}_1, \mathbf{r}_2)d\mathbf{r}_1 d\mathbf{r}_2} \tag{8.41}$$

where $\psi(\mathbf{r}_1, \mathbf{r}_2)$ is just the spatial part of $\Psi_2(1, 2)$. Equation 8.41 is equivalent to Equation 7.18 in Section 7–1. It is important to realize that a factorization into a spatial part and a spin part does *not* occur in general but that it does occur for two-electron systems.

It is fairly easy to write the antisymmetric two-electron wave function by in-spection, but what if we have a set of N spin orbitals and we need to construct an antisymmetric N-electron wave function? In the early 1930s, Slater introduced the use of determinants (MathChapter E) to construct antisymmetric wave functions. If we use Equation 8.35 as an example, then we see that we can write Ψ (we will drop the subscript 2) in the form

$$\Psi(1, 2) = \begin{vmatrix} 1s\alpha(1) & 1s\beta(1) \\ 1s\alpha(2) & 1s\beta(2) \end{vmatrix} \tag{8.42}$$

We obtain Equation 8.35 upon expanding this determinant. The wave function $\Psi(1, 2)$ given by Equation 8.42 is called a *determinantal wave function*.

Two properties of determinants are of particular importance to us. The first is that the value of a determinant changes sign when we interchange any two rows or any two columns of the determinant. The second is that a determinant is equal to zero if any two rows or any two columns are the same (MathChapter E).

Notice that when we interchange the two electrons in the determinantal wave function $\Psi(1, 2)$ (Equation 8.42), we interchange the two rows and so change the sign of $\Psi(1, 2)$. Furthermore, if we place both electrons in the same spin orbital, say the $1s\alpha$ spin orbital, then $\Psi(1, 2)$ becomes

$$\Psi(1, 2) = \begin{vmatrix} 1s\alpha(1) & 1s\alpha(1) \\ 1s\alpha(2) & 1s\alpha(2) \end{vmatrix} = 0$$

This determinant is equal to zero because the two columns are the same. Thus, we see that the determinantal representation of wave functions automatically satisfies the Pauli Exclusion Principle. Determinantal wave functions are always antisymmetric and vanish when any two electrons have the same four quantum numbers, that is, when both electrons occupy the same spin orbital.

We need to consider one more factor before our discussion of determinantal wave functions is complete. Recall from Example 8–3 that the normalization constant for $\Psi(1, 2)$ given by Equation 8.42 is $1/\sqrt{2}$. Therefore,

$$\Psi(1, 2) = \frac{1}{\sqrt{2}} \begin{vmatrix} 1s\alpha(1) & 1s\beta(1) \\ 1s\alpha(2) & 1s\beta(2) \end{vmatrix} \tag{8.43}$$

is a *normalized* two-electron determinantal wave function. The factor of $1/\sqrt{2}$ assures that $\Psi(1, 2)$ is normalized.

We have developed the determinantal representation of wave functions using a two-electron system as an example. To generalize this development for an N-electron system, we use an $N \times N$ determinant. Furthermore, one can show (Problem 8–21) that the normalization constant is $1/\sqrt{N!}$, and so we have the N-electron determinantal wave function

$$\Psi(1, 2, \ldots, N) = \frac{1}{\sqrt{N!}} \begin{vmatrix} u_1(1) & u_2(1) & \cdots & u_N(1) \\ u_1(2) & u_2(2) & \cdots & u_N(2) \\ \vdots & \vdots & \vdots & \vdots \\ u_1(N) & u_2(N) & \cdots & u_N(N) \end{vmatrix} \tag{8.44}$$

where the u's in Equation 8.44 are orthonormal spin orbitals. Notice that $\Psi(1, 2, \ldots, N)$ changes sign whenever two electrons (rows) are interchanged and vanishes if any two electrons occupy the same spin orbital (two identical columns).

We are now ready to go back to the problem that led us to discuss spin, that is, the lithium atom. Note that we cannot put all three electrons into $1s$ orbitals because two columns in the determinantal wave function would be the same. Thus, an appropriate wave function is

$$\Psi = \frac{1}{\sqrt{3!}} \begin{vmatrix} 1s\alpha(1) & 1s\beta(1) & 2s\alpha(1) \\ 1s\alpha(2) & 1s\beta(2) & 2s\alpha(2) \\ 1s\alpha(3) & 1s\beta(3) & 2s\alpha(3) \end{vmatrix} \tag{8.45}$$

The standard method for determining the optimal form of the spatial part of the spin orbitals in a determinantal wave function such as Equation 8.43 or 8.45 is the Hartree-Fock self-consistent field method, which we discuss in the next section.

8–7. Hartree-Fock Calculations Give Good Agreement with Experimental Data

In Section 8–3, we discussed the Hartree-Fock method for the helium atom. The Hartree-Fock equation for this system is given by Equation 8.20, where $\hat{H}_1^{\text{eff}}$ is given by Equation 8.19. The helium atom is a special case because the Slater determinant factors into a spatial part and a spin part, and so we were able to use Equation 8.17 as the helium atomic wave function. This factorization into a spatial part and a spin part does not occur for atoms with more than two electrons and so we must start with a complete Slater determinant such as Equation 8.45. This leads to an equation of the form

$$\hat{F}_i \phi_i = \epsilon_i \phi_i \tag{8.46}$$

where the effective Hamiltonian operator is called the *Fock operator* ($\hat{F}_i$). The use of a full Slater determinant instead of just a simple product of spatial orbitals like in Equation 8.17 makes $\hat{F}_i$ more complicated than $\hat{H}^{\text{eff}}$ given by Equation 8.19 for helium. We will not need an explicit expression for $\hat{F}_i$. It is sufficient to realize that Equation 8.46 must be solved in a self-consistent manner and that there are readily available computer programs to do this. The self-consistent orbitals obtained from

Equation 8.46 are called *Hartree-Fock orbitals*. The eigenvalues ϵ_i of Equation 8.46 are called *orbital energies*.

According to an approximation first introduced by Koopmans, ϵ_i in Equation 8.46 is the ionization energy of an electron from the ith orbital. Table 8.3 compares some ionization energies of neon and argon obtained by using Koopmans' approximation with those obtained by subtracting the Hartree-Fock energy of the neutral atom from that of the ion. You can see that Koopmans' approximation gives results that are almost as good as the direct calculation. Figure 8.1 shows the ionization energies of the elements hydrogen through xenon plotted against atomic number. Both ionization energies obtained by Koopmans' approximation and experimental data are shown in the figure. This plot clearly shows the shell and subshell structure that students first learn in general chemistry. Given that there are no adjustable parameters involved in the calculated values in Figure 8.1, the agreement with experimental data is remarkable.

Note that the order of the energies of the various subshells is in general agreement with observation for neutral atoms. In particular, the energies of the $2s$ and $2p$ orbitals are not the same as they are for the hydrogen atom. The degeneracy of the $2s$ and $2p$ orbitals or, more generally, the fact that the energy depends on only the principal quantum number is unique to the purely $1/r$ Coulombic potential in the hydrogen atom. In a Hartree-Fock calculation, the effective potential $V_j^{\text{eff}}(\mathbf{r}_j)$ is more complicated than $1/r$, and $V_j^{\text{eff}}(\mathbf{r}_j)$ breaks up the degeneracy found in the hydrogen atom, giving us the familiar ordering of the orbital energies we first learned in general chemistry.

TABLE 8.3

Ionization energies of neon and argon obtained from neutral atom orbital energies (Koopmans' approximation) and by subtracting the Hartree-Fock energy of the neutral atom from the Hartree-Fock energy of the appropriate state of the positive ion.

Electron removed	Resulting orbital occupancy	Ionization energies/MJ·mol^{-1}		
		Koopmans' approximation	Direct Hartree-Fock calculation	Experimental
Neon				
$1s$	$1s2s^22p^6$	86.0	83.80	83.96
$2s$	$1s^22s2p^6$	5.06	4.76	4.68
$2p$	$1s^22s^22p^5$	1.94	1.91	2.08
Argon				
$1s$	$1s2s^22p^63s^23p^6$	311.35	308.25	309.32
$2s$	$1s^22s2p^63s^23p^6$	32.35	31.33	
$2p$	$1s^22s^22p^53s^23p^6$	25.12	24.01	23.97
$3s$	$1s^22s^22p^63s3p^6$	3.36	3.20	2.82
$3p$	$1s^22s^22p^63s^23p^5$	1.65	1.43	1.52

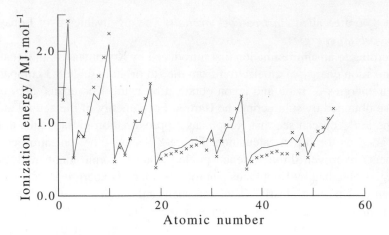

FIGURE 8.1
The ionization energies of neutral atoms of hydrogen through xenon plotted versus atomic number. The straight lines connect experimental data, and the crosses are calculated according to Koopmans' approximation.

Because the Hartree-Fock method uses determinantal wave functions, there is some correlation between electrons with the same spin, because two electrons with the same spin cannot occupy the same orbital. Nevertheless, the Hartree-Fock method is not exact, and so we define a correlation energy by (see Equation 8.21)

$$\text{correlation energy} = \text{CE} = E_{\text{exact}} - E_{\text{HF}}$$

Recall from Section 8–3 that the correlation energy of a helium atom is $0.042\ E_{\text{h}} = 110\ \text{kJ}\cdot\text{mol}^{-1}$. Although correlation energies appear to be small, they are significant when we realize that many quantities of chemical interest such as bond strengths and the energies associated with many chemical reactions are of the order of $100\ \text{kJ}\cdot\text{mol}^{-1}$. Consequently, much quantum-chemical research has been directed toward the calculation of correlation energies. For example, perturbation schemes have been developed that treat the Hartree-Fock orbitals as a zero-order wave function, so that the correlation energy can be calculated by perturbation theory.

8–8. A Term Symbol Gives a Detailed Description of an Electron Configuration

Electron configurations of atoms are ambiguous in the sense that a number of sets of m_l and m_s are consistent with a given electron configuration. For example, consider the ground-state electron configuration of a carbon atom, $1s^2 2s^2 2p^2$. The two $2p$ electrons could be in any of the three $2p$ orbitals ($2p_x$, $2p_y$, $2p_z$) and have any spins consistent with the Pauli Exclusion Principle. The energies of these different states may differ, and so we require a more detailed designation of the electronic states of atoms. The scheme we will present here is based upon the idea of determining the total orbital angular momentum **L** and the total spin angular momentum **S** and then adding **L** and **S** together vectorially to obtain the total angular momentum **J**. The result of such a calculation, called *Russell-Saunders coupling*, is presented as an *atomic term symbol*, which has the form

$$^{2S+1}L_J$$

In a term symbol, L is the total orbital angular momentum quantum number, S is the total spin quantum number, and J is the total angular momentum quantum number. We will see that L will necessarily have values such as 0, 1, 2, ... Similar to assigning the letters s, p, d, f to the values $l = 0, 1, 2, 3$ of the orbital angular momentum for the hydrogen atom, we will make the correspondence

$$L = 0 \quad 1 \quad 2 \quad 3 \quad 4 \quad 5 \quad \cdots$$
$$S \quad P \quad D \quad F \quad G \quad H \quad \cdots$$

We will also see that the total spin quantum number S will necessarily have values such as 0, $\frac{1}{2}$, 1, $\frac{3}{2}$, . . . and so the $2S + 1$ left superscript on a term symbol will have values such as 1 , 2, 3, ... The quantity $2S + 1$ is called the *spin multiplicity*. Thus, ignoring for now the subscript J, term symbols will be of the type

$$^3S \quad ^2D \quad ^1P$$

The total orbital angular momentum and the total spin angular momentum are given by the vector sums

$$\mathbf{L} = \sum_i \mathbf{l}_i \tag{8.47}$$

and

$$\mathbf{S} = \sum_i \mathbf{s}_i \tag{8.48}$$

where the summations are over the electrons in the atom. The z components of **L** and **S** are given by the scalar sums

$$L_z = \sum_i l_{zi} = \sum_i m_{li} = M_L \tag{8.49}$$

and

$$S_z = \sum_i s_{zi} = \sum_i m_{si} = M_S \tag{8.50}$$

Thus, although the angular momenta add vectorially as in Equations 8.47 and 8.48, the z components add as scalars (Figure 8.2). Just as the z component of **l** can assume the $2l + 1$ values $m_l = l$, $l - 1$, . . . , 0, . . . , $-l$, the z component of **L** can assume the $2L + 1$ values $M_L = L$, $L - 1$, . . . , 0, . . . , $-L$. Similarly, M_S can take on the

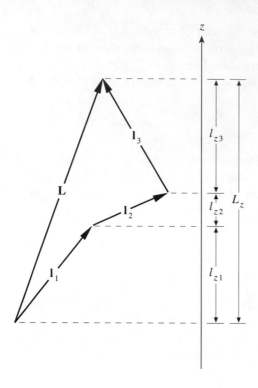

FIGURE 8.2
A schematic illustration of the addition of angular momentum vectors.

$2S + 1$ values S, $S - 1$, ..., $-S + 1$, $-S$. Thus, the spin multiplicity is simply the $2S + 1$ projections that the z component of $\mathbf{S}$ can assume.

Let's consider the electron configuration ns^2 (two electrons in a ns orbital). There is only one possible set of values of m_{l1}, m_{s1}, m_{l2}, and m_{s2}:

m_{l1}	m_{s1}	m_{l2}	m_{s2}	M_L	M_S
0	$+\frac{1}{2}$	0	$-\frac{1}{2}$	0	0

The fact that the only value of M_L is $M_L = 0$ implies that $L = 0$. Similarly, the fact that the only value of M_S is $M_S = 0$ implies that $S = 0$. The total angular momentum $\mathbf{J}$ is given by

$$\mathbf{J} = \mathbf{L} + \mathbf{S} \tag{8.51}$$

and its z component is

$$J_z = L_z + S_z = (M_L + M_S) = M_J = 0 \tag{8.52}$$

which implies that $J = 0$. Consequently, for an ns^2 electron configuration, $L = 0$, $S = 0$, and $J = 0$. The value $L = 0$ is written as S in the term symbol, and so we find that the term symbol corresponding to an ns^2 electron configuration is $^1S_0$ (singlet S zero). Because the two electrons have opposite spins, the total spin angular momentum is zero. Both electrons also occupy an orbital that has no angular momentum, thus the total angular momentum must be zero, which is what the $^1S_0$ term indicates.

An np^6 electron configuration also will have a $^1S_0$ term symbol. To understand this, realize that the six electrons in the three np orbitals have the quantum numbers $(n, 1, 1, \pm 1/2)$, $(n, 1, 0, \pm 1/2)$, and $(n, 1, -1, \pm 1/2)$. Therefore, when we add up all the m_{li} and m_{si}, we get $M_L = 0$ and $M_S = 0$, and we have $^1S_0$.

EXAMPLE 8–4
Show that the term symbol corresponding to an nd^{10} electron configuration is $^1S_0$.

SOLUTION: The ten d orbital electrons have the quantum numbers $(n, 2, 2, \pm 1/2)$, $(n, 2, 1, \pm 1/2)$, $(n, 2, 0, \pm 1/2)$, $(n, 2, -1, \pm 1/2)$, and $(n, 2, -2, \pm 1/2)$. Therefore, $M_L = 0$ and $M_S = 0$ as for ns^2 and np^6 electron configurations, and the term symbol is $^1S_0$.

Notice that M_L and M_S are necessarily equal to zero for completely filled subshells because for every electron with a negative value of m_{li}, there is another electron with a corresponding positive value to cancel it; the same holds true for the values of m_{si}. Thus, we can ignore the electrons in completely filled subshells when considering other electron configurations. For example, we can ignore the contributions of the $1s^2 2s^2$ orbitals to the $1s^2 2s^2 2p^2$ electron configurations of a carbon atom when we discuss a carbon atom later.

An electron configuration that has a term symbol other than $^1S_0$ is $ns^1 n's^1$, where $n \neq n'$. An example is a helium atom with the excited-state electron configuration $1s^1 2s^1$. To determine the possible values of m_{l1}, m_{s1}, m_{l2}, and m_{s2}, we set up a table in the following manner: Because m_{l1} and m_{l2} can both have a maximum value of 0, the maximum value of M_L is 0 (see Equation 8.49), and 0 is its only possible value. Similarly, because m_{s1} and m_{s2} can both have values of $\pm 1/2$, M_s can be -1, 0, or 1. We now set up a table with its columns headed by the possible values of M_S and its rows headed by the possible values of M_L, and we then fill in the sets of values of m_{l1}, m_{s1}, m_{l2}, and m_{s2} that are consistent with each value of M_L and M_S as shown

M_L	M_S		
	1	0	-1
0	$0^+, 0^+$	$0^+, 0^-$; $0^-, 0^+$	$0^-, 0^-$

The notation 0^+ means that $m_l = 0$ and $m_s = +1/2$ and 0^- means that $m_l = 0$ and $m_s = -1/2$. The possible sets of values of m_{l1}, m_{s1}, m_{l2}, and m_{s2} that are consistent with each value of M_L and M_S are called *microstates*.

There are four microstates in this table because there are two possible spins ($\pm 1/2$) for the electron in the ns orbital and two possible spins for the electron in the $n's$ orbital.

Note that we include both 0^+, 0^- and 0^-, 0^+ because the electrons are in *nonequivalent orbitals* (e.g., $1s$ and $2s$). Note that all the values of M_L in the above table are zero, so they all must correspond to $L = 0$. In addition, the largest value of M_S is 1. Consequently, S must equal 1 and the values $M_S = 1$, 0, and -1 correspond to $L = 0$, $S = 1$, corresponding to a ^3S state. This ^3S state accounts for one microstate from each column in the above table. The middle column contains two microstates, but it makes no difference which one we choose. After eliminating one microstate from each column $(0^+, 0^+;\ 0^-, 0^-;$ and either $0^+, 0^-$ or $0^-, 0^+)$, we are left with only the entry with $M_L = 0$, $M_S = 0$ (either $0^-, 0^+$ or $0^+, 0^-$), which implies that $L = 0$ and $S = 0$, corresponding to a ^1S state. These two pairs of $L = 0$, $S = 1$ and $L = 0$, $S = 0$ along with their possible values of M_J can be summarized as

$$
\begin{array}{ll}
L = 0,\ S = 1 & L = 0,\ S = 0 \\
M_L = 0,\ M_S = 1,\ 0,\ -1 & M_L = 0,\ M_S = 0 \\
M_J = M_L + M_S = 1,\ 0,\ -1 & M_J = M_L + M_S = 0
\end{array}
$$

The values of M_J here imply that $J = 1$ for the $L = 0$, $S = 1$ case and that $J = 0$ for the $L = 0$, $S = 0$ case. The two term symbols corresponding to the electron configuration $ns^1 n's^1$ are

$$^3S_1 \quad \text{and} \quad ^1S_0$$

The $^3S_1$ is called a *triplet S state*. These two term symbols correspond to two different electronic states with different energies. We will see below that the triplet state $(^3S_1)$ has a lower energy than the singlet state $(^1S_0)$.

8–9. The Allowed Values of J are $L + S,\ L + S - 1,\ \ldots,\ |L - S|$

As a final example of deducing atomic term symbols, we will consider a carbon atom, whose ground-state electron configuration is $1s^2 2s^2 2p^2$. We have shown previously that we do not need to consider completely filled subshells because M_L and M_S are necessarily zero for completely filled subshells. Consequently, we can focus on the electron configuration np^2. As for the case of $ns^1 n's^1$ above, we will make a table of possible values of m_{l1}, m_{s1}, m_{l2}, and m_{s2}. Before we do this, however, let's see how many entries there will be in the table for np^2. We are going to assign two electrons to two of six possible spin orbitals $(2p_x\alpha,\ 2p_x\beta,\ 2p_y\alpha,\ 2p_y\beta,\ 2p_z\alpha,\ 2p_z\beta)$. There are 6 choices for the first spin orbital and 5 choices for the second, giving a total of $6 \times 5 = 30$ choices. Because the electrons are indistinguishable, however, the order of the two spin orbitals chosen is irrelevant. Thus, we should divide the 30 choices by a factor of 2 to give 15 as the number of distinct ways of assigning the two electrons to the six spin orbitals. Generally, the number of distinct ways to assign N electrons to G spin orbitals belonging to the same subshell (*equivalent orbitals*) is given by

$$
\frac{G!}{N!(G-N)!} \quad \text{(equivalent orbitals)} \tag{8.53}
$$

Note that Equation 8.53 gives 15 if $G = 6$ and $N = 2$.

EXAMPLE 8–5

How many distinct ways are there of assigning two electrons to the nd orbitals? In other words, how many sets of m_{li} and m_{si} are there for an nd^2 electron configuration?

SOLUTION: There are five nd orbitals, or ten nd spin orbitals. Thus, the number of distinct ways of placing two electrons in nd orbitals is

$$\frac{10!}{2!8!} = 45$$

To determine the 15 possible sets of m_{l1}, m_{s1}, m_{l2}, and m_{s2} for an np^2 electron configuration, we first determine the possible values of M_L and M_S. Because m_{l1} and m_{l2} can both have a maximum value of 1, the maximum value of M_L is 2 (see Equation 8.49), and so its possible values are 2, 1, 0, −1, and −2. Similarly, because m_{s1} and m_{s2} can each have a maximum value of 1/2, the maximum value of M_S is 1 (see Equation 8.50), and so its possible values are 1, 0, and −1. Using this information, we set up a table with its columns headed by the possible values of M_S and its rows headed by the possible values of M_L, and then fill in the microstates consistent with each value of M_L and M_S, as shown:

M_L	M_S 1	0	−1
2	$\cancel{1^+, 1^+}$	$1^+, 1^-$	$\cancel{1^-, 1^-}$
1	$0^+, 1^+$	$1^+, 0^-$; $1^-, 0^+$	$0^-, 1^-$
0	$\cancel{0^+, 0^+}$; $1^+, -1^+$	$1^+, -1^-$; $-1^+, 1^-$; $0^+, 0^-$	$1^-, -1^-$; $\cancel{0^-, 0^-}$
−1	$0^+, -1^+$	$0^+, -1^-$; $0^-, -1^+$	$0^-, -1^-$
−2	$\cancel{-1^+, -1^+}$	$-1^+, -1^-$	$\cancel{-1^-, -1^-}$

where, for example, the notation $1^+, -1^-$ means that $m_{l1} = 1$, $m_{s1} = +1/2$, and $m_{l2} = -1$, $m_{s2} = -1/2$. Unlike the earlier example in which we treated nonequivalent orbitals, we do *not* include both $1^+, 0^-$ and $0^-, 1^+$ in the $M_S = 0$, $M_L = 1$ position because in this case, the orbitals are equivalent (two $2p$ orbitals). Consequently, the two microstates $1^+, 0^-$ and $0^-, 1^+$ are indistinguishable. The six microstates that are crossed out in the above table violate the Pauli Exclusion Principle. The remaining 15 microstates constitute all the possible microstates for an np^2 electron configuration.

We must now deduce the possible values of L and S from the tabulated values of M_L and M_S. The largest value of M_L is 2, which occurs only with $M_S = 0$. Therefore, there must be a state with $L = 2$ and $S = 0$ (^{1}D). Because $L = 2$, $M_L = 2$, 1, 0, −1, and −2, and so the ^{1}D state will account for one microstate in each row of the middle column of the above table. For those rows that contain more than one microstate (the second, third, and fourth rows), it makes no difference which microstate is chosen. We will arbitrarily choose the microstates $1^+, 0^-$; $1^+, -1^-$; and $0^+, -1^-$. If we eliminate these microstates from the table, we are left with the following table.

| | | M_S | |
M_L	1	0	−1
2			
1	$0^+, 1^+$	$1^-, 0^+$	$0^-, 1^-$
0	$1^+, -1^+$	$-1^+, 1^-$; $0^+, 0^-$	$1^-, -1^-$
−1	$0^+, -1^+$	$0^-, -1^+$	$0^-, -1^-$
−2			

The largest value of M_L remaining is $M_L = 1$, implying $L = 1$. There are microstates with $M_L = 1$, 0, −1 associated with $M_S = 1$ ($0^+, 1^+$; $1^+, -1^+$; $0^+, -1^+$), with $M_S = 0$, ($1^-, 0^+$; either $-1^+, 1^-$ or $0^+, 0^-$; $0^-, -1^+$) and with $M_S = -1$ ($0^-, 1^-$; $1^-, -1^-$; $0^-, -1^-$). Therefore, these nine microstates correspond to $L = 1$ and $S = 1$, or a ^{3}P (triplet P) state. If we eliminate these nine microstates from the table, then we are left with only one microstate with $M_L = 0$ and $M_S = 0$ at the center of the table, which implies $L = 0$ and $S = 0$ (^{1}S).

So far, we have found the partially specified term symbols, ^{1}D, ^{3}P, and ^{1}S. To complete the specification of these term symbols, we must determine the possible values of J in each case. Recall that $M_J = M_L + M_S$. For the five entries corresponding to the ^{1}D state, $M_S = 0$, and so the values of M_J are 2, 1, 0, −1, and −2, which implies that $J = 2$. Thus, the complete term symbol of the ^{1}D state is ^{1}D$_2$. Note that the degeneracy of this state is 5, or $2J + 1$. The values of M_J for the nine entries for the ^{3}P state are 2, 1, 1, 0, 0, −1, 0, −1, and −2. We clearly have one set of 2, 1, 0, −1, −2 corresponding to $J = 2$. If we eliminate these five values, then we are left with 1, 0, 0, −1, which corresponds to $J = 1$ and $J = 0$. Thus, the ^{3}P state has three possible values of J, so the term symbols are ^{3}P$_2$, ^{3}P$_1$, and ^{3}P$_0$. The ^{1}S state must be ^{1}S$_0$. In summary, then, the electronic states associated with an np^2 configuration are

$$^1\text{D}_2, \quad ^3\text{P}_0, \quad ^3\text{P}_1, \quad ^3\text{P}_2, \quad \text{and} \quad ^1\text{S}_0$$

The degeneracies of these states are $2J + 1 = 5$, 1, 3, 5, and 1, respectively. Table 8.4 lists the term symbols that arise from various electron configurations.

The values of J for the term symbols in Table 8.4 can be determined in terms of the values of L and S if we recall that

$$\mathbf{J} = \mathbf{L} + \mathbf{S}$$

TABLE 8.4
The possible term symbols (excluding the J subscript) for various electron configurations.

Electron configuration	Term symbol (excluding the J subscript)
s^1	^{2}S
p^1	^{2}P
p^2, p^4	^{1}S, ^{1}D, ^{3}P
p^3	^{2}P, ^{2}D, ^{4}S
p^1, p^5	^{2}P
d^1, d^9	^{2}D
d^2, d^8	^{1}S, ^{1}D, ^{1}G, ^{3}P, ^{3}F
d^3, d^7	^{2}P, ^{2}D (two), ^{2}F, ^{2}G, ^{2}H, ^{4}P, ^{4}F
d^4, d^6	^{1}S (two), ^{1}D (two), ^{1}F, ^{1}G (two), ^{1}I, ^{3}P (two), ^{3}D, ^{3}F (two), ^{3}G, ^{3}H, ^{5}D
d^5	^{2}S, ^{2}P, ^{2}D (three), ^{2}F (two), ^{2}G (two), ^{2}H, ^{2}I, ^{4}P, ^{4}D, ^{4}F, ^{4}G, ^{6}S

The largest value that J can have is in the case when both **L** and **S** are pointing in the same direction, so that $J = L + S$. The smallest value that J can have is when **L** and **S** are pointing in opposite directions, so that $J = |L - S|$. The values of J lying between $L + S$ and $|L - S|$ are obtained from

$$J = L + S, \ L + S - 1, \ L + S - 2, \ \ldots, \ |L - S| \tag{8.54}$$

Equation 8.54 has the following pictorial representation. The vectors **L** and **S** are added together in all ways such that their sum is a vector of length 0, 1, 2, ... if S is an integer, or 1/2, 3/2, 5/2,... if S is 1/2, 3/2, 5/2, and so on. For example, if $L = 2$ and $S = 1$, then **L** and **S** can be added vectorially as follows.

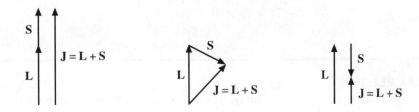

Note that the maximum value of J corresponds to $\mathbf{L}$ and $\mathbf{S}$ pointing in the same direction and that the minimum value of J corresponds to $\mathbf{L}$ and $\mathbf{S}$ pointing in opposite directions.

If we apply Equation 8.54 to the ^3P term symbol above, then we see that the values of J are given by

$$J = (1+1), \ (1+1)-1, \ 1-1$$

and so $J = 2, \ 1, \ 0$, as we deduced above.

EXAMPLE 8–6
Use Equation 8.54 to deduce the values of J associated with the term symbols ^2S, ^3D, and ^4F.

SOLUTION: For a ^2S state, $L = 0$ and $S = 1/2$. According to Equation 8.54, the only possible value of J is 1/2, and so the term symbol will be $^2S_{1/2}$. For a ^3D state, $L = 2$ and $S = 1$. Therefore, the values of J will be 3, 2, and 1, and so the term symbols will be

$$^3D_1, \quad ^3D_2, \quad \text{and} \quad ^3D_3$$

For a ^4F state, $L = 3$ and $S = 3/2$. Therefore, the values of J will be 9/2, 7/2, 5/2, and 3/2, and so the term symbols will be

$$^4F_{9/2}, \quad ^4F_{7/2}, \quad ^4F_{5/2}, \quad \text{and} \quad ^4F_{3/2}$$

Example 8–6 shows that the "L and S part" of a term symbol is sufficient to deduce the complete term symbol.

There is a useful consistency test between Equation 8.53 and the term symbols associated with a given electron configuration. A term symbol ^{2S+1}L will have $2S+1$ entries for each value of M_L in a table of possible values of the m_{li} and m_{si} (cf. the table of entries for np^2). Because there are $2L+1$ values of M_L for a given value of L, the total number of entries for each term symbol (excluding the J subscript) is $(2S+1)(2L+1)$. Applying this result to the np^2 case gives

$$\begin{array}{ccc} ^1S & ^3P & ^1D \\ (1 \times 1) + & (3 \times 3) + & (1 \times 5) = 15 \end{array}$$

EXAMPLE 8–7
Show that Equation 8.53 and the term symbols for nd^2 given in Table 8.4 are consistent.

SOLUTION: The total number of entries in a table of possible values of m_{li} and m_{si} for an nd^2 electron configuration is

$$\frac{G!}{N!(G-N)!} = \frac{10!}{2!8!} = 45$$

The term symbols given in Table 8.4 are

$$\begin{matrix} {}^1S & {}^1D & {}^1G & {}^3P & {}^3F \\ (1 \times 1) + (1 \times 5) + (1 \times 9) + (3 \times 3) + (3 \times 7) &=& 45 \end{matrix}$$

8–10. Hund's Rules Are Used to Determine the Term Symbol of the Ground Electronic State

Each of the states designated by a term symbol corresponds to a determinantal wave function that is an eigenfunction of $\hat{L}^2$ and $\hat{S}^2$, and each state corresponds to a certain energy. Although we could calculate the energy associated with each state, in practice, the various states are ordered according to three empirical rules formulated by the German spectroscopist Friederich Hund. Hund's rules are as follows:

1. The state with the largest value of S is the most stable (has the lowest energy), and stability decreases with decreasing S.
2. For states with the same value of S, the state with the largest value of L is the most stable.
3. If the states have the same value of L and S, then, for a subshell that is less than half filled, the state with the smallest value of J is the most stable; for a subshell that is more than half filled, the state with the largest value of J is the most stable.

EXAMPLE 8–8

Use Hund's rules to deduce the lowest energy state of an excited state of a beryllium atom whose electron configuration is $(1s)^2 2s^1 3s^1$ and of the ground state of a carbon atom.

SOLUTION: The term symbols for a $2s^1 3s^1$ configuration are (see Section 8–8)

$$ {}^3S_1 \quad \text{and} \quad {}^1S_0 $$

According to the first of Hund's rules, the more stable state is the ${}^3S_1$ state.

The ground-state electron configuration of a carbon atom is p^2. The term symbols for a p^2 configuration are (see Table 8.4)

$$ {}^1S_0, \quad {}^3P_0, \quad {}^3P_1, \quad {}^3P_2, \quad \text{and} \quad {}^1D_2 $$

According to the first of Hund's rules, the ground state is one of the ${}^3P$ states. According to Hund's third rule, the most stable state is the ${}^3P_0$ state.

8–11. Atomic Term Symbols Are Used to Describe Atomic Spectra

Atomic term symbols are sometimes called spectroscopic term symbols because atomic spectral lines can be assigned to transitions between states that are described by atomic term symbols. For example, the first few electronic states of atomic hydrogen are given in Table 8.5. The electron configuration $1s$ gives the term $^2S_{1/2}$, which is doubly degenerate, corresponding to $M_J = +1/2$ and $-1/2$. The $2s$ electron also gives rise to a doubly degenerate $^2S_{1/2}$ state. An electron in a $2p$ orbital gives rise to two states, $^2P_{1/2}$ and $^2P_{3/2}$. The first of these is two-fold degenerate, and the second is four-fold degenerate ($M_J = 3/2,\ 1/2,\ -1/2,\ -3/2$). The total degeneracy of the $n = 2$ level is eight-fold. When we solved the Schrödinger equation for the hydrogen atom in Chapter 6, we found that the electronic energy depended only upon the principal quantum number, n (Equation 6.44). The data in Table 8.5, however, show that the various n levels are split into sets of closely lying energy levels. The reason for this fine splitting is *spin-orbit coupling*, which we will discuss only briefly.

In addition to the usual kinetic energy and electrostatic terms in the Hamiltonian operator of a multielectron atom, there are a number of magnetic and spin terms. The most important of these terms is *the spin-orbit interaction* term, which represents the interaction of the magnetic moment associated with the spin of an electron with the magnetic field generated by the electric current produced by the electron's own orbital

TABLE 8.5
The first few electronic states of atomic hydrogen.[a]

Electron configuration	Term symbol	Energy/cm^{-1}
$1s$	$1s\ ^2S_{1/2}$	0.00
$2p$	$2p\ ^2P_{1/2}$	82 258.917
$2s$	$2s\ ^2S_{1/2}$	82 258.942
$2p$	$2p\ ^2P_{3/2}$	82 259.272
$3p$	$3p\ ^2P_{1/2}$	97 492.198
$3s$	$3s\ ^2S_{1/2}$	97 492.208
$3p, 3d$	$3p\ ^2P_{3/2}, 3d\ ^2D_{3/2}$	97 492.306
$3d$	$3d\ ^2D_{5/2}$	97 492.342
$4p$	$4p\ ^2P_{1/2}$	102 823.835
$4s$	$4s\ ^2S_{1/2}$	102 823.839
$4p, 4d$	$4p\ ^2P_{3/2}, 4d\ ^2D_{3/2}$	102 823.881
$4d, 4f$	$4d\ ^2D_{5/2}, 4f\ ^2F_{5/2}$	102 823.896
$4f$	$4f\ ^2F_{5/2}$	102 823.904

[a] From C.E. Moore, "Atomic Energy Levels", Natl. Bur. Std. Circ. No. 467 (U.S. Government Printing Office, Washington, D.C., 1949).

motion. Other terms include spin-spin interaction and orbit-orbit interaction, but these are numerically less important. The Hamiltonian operator for a multielectron atom can be written as

$$\hat{H} = -\frac{1}{2}\sum_j \nabla_j^2 - \sum_j \frac{Z}{r_j} + \sum_{i<j}\frac{1}{r_{ij}} + \sum_j \xi(r_j)\mathbf{l}_j \cdot \mathbf{s}_j \tag{8.55}$$

where $\mathbf{l}_j$ and $\mathbf{s}_j$ are the individual electronic orbital momenta and spin angular momenta, respectively, and where $\xi(r_j)$ is a scalar function of r, whose form is not necessary here (see Problem 8–46). We can abbreviate Equation 8.55 by writing

$$\hat{H} = \hat{H}_0 + \hat{H}_{so}^{(1)}$$

where $\hat{H}_0$ represents the first three terms, which we have treated in this chapter, and $\hat{H}_{so}^{(1)}$ represents the fourth term (spin-orbit coupling) in Equation 8.55. When $\hat{H}_{so}^{(1)}$ is small enough to be considered a small perturbation (particularly, for atoms whose atomic numbers are less than 30 or so), perturbation theory leads to the splitting we observe in Table 8.5 (Problem 8–46).

Let's use Table 8.5 to take a closer look at the atomic hydrogen spectrum. In particular, let's look at the Lyman series, which is the series of lines that arise from transitions from states with $n \geq 2$ to the $n = 1$ state. (See Figure 1.10.) As we did in Chapter 1, we can use the Rydberg formula to calculate the frequencies of the lines in the Lyman series. The frequencies of the lines in the Lyman series are given by

$$\tilde{\nu} = 109\,677.58\left(1 - \frac{1}{n^2}\right)\text{cm}^{-1} \qquad n = 2,\ 3,\ \ldots \tag{8.56}$$

If we express our results in terms of wave numbers, we obtain the following:

Transition	Frequency/cm^{-1}
$2 \to 1$	82 258.19
$3 \to 1$	97 491.18
$4 \to 1$	102 822.73
$5 \to 1$	105 290.48

If we use Table 8.5, we see that there are three states for $n = 2$. Not all these states can make a transition to the ground state because of selection rules. Recall from Chapter 5 that selection rules are restrictions that govern the possible, or *allowed*, transitions from one state to another. In the case of atomic spectra, the selection rules are

$$\begin{aligned}\Delta L &= 0,\ \pm 1 \\ \Delta S &= 0 \\ \Delta J &= 0,\ \pm 1\end{aligned} \tag{8.57}$$

except that a transition from a state with $J = 0$ to another state with $J = 0$ is not allowed (*forbidden*). The selection rules given by Equation 8.57 have been deduced experimentally and corroborated theoretically. We will derive some spectroscopic selection rules in Chapter 13, but here we will just accept them. (The rule $\Delta L = \pm 1$

follows from the principle of conservation of angular momentum because a photon has a spin angular momentum of $\hbar$.)

The selection rules given in Equations 8.57 tell us that $^2P \to {}^2S$ transitions are allowed, but that $^2S \to {}^2S$ transitions are not allowed because $\Delta L = 0$ and that $^2S \to {}^2D$ transitions and $^2F \to {}^2P$ transitions are not allowed because $\Delta L = \pm 2$, respectively, in these transitions. Thus, if we look closely at the Lyman series of atomic hydrogen, we see that the allowed transitions into the ground state are

$$np\,^2P_{1/2} \to 1s\,^2S_{1/2} \qquad \begin{pmatrix} \Delta L = 1 \\ \Delta S = 0 \\ \Delta J = 0 \end{pmatrix}$$

or

$$np\,^2P_{3/2} \to 1s\,^2S_{1/2} \qquad \begin{pmatrix} \Delta L = 1 \\ \Delta S = 0 \\ \Delta J = -1 \end{pmatrix}$$

No other transitions into the $1s\,^2S_{1/2}$ ground state are allowed.

The frequencies associated with the $2 \to 1$ transitions can be computed from Table 8.5; their values are

$$\tilde{\nu} = (82\,258.917 - 0.00)\ \text{cm}^{-1} = 82\,258.917\ \text{cm}^{-1}$$

and   (8.58)

$$\tilde{\nu} = (82\,259.272 - 0.00)\ \text{cm}^{-1} = 82\,259.272\ \text{cm}^{-1}$$

respectively. We see that the $n = 2$ to $n = 1$ transition occurs at a frequency $\tilde{\nu} = 82\,258.19\ \text{cm}^{-1}$ if we ignore spin-orbit coupling, but that it consists of two closely spaced lines whose frequencies are given by Equations 8.58 if spin-orbit coupling is included. This closely spaced pair of lines is called a *doublet*, and so we see that under high resolution, the first line of the Lyman series is a doublet. Table 8.5 shows that all the lines of the Lyman series are doublets and that the separation of the doublet lines decreases with increasing n. The increased spectral complexity caused by spin-orbit coupling is called *fine structure*.

EXAMPLE 8–9

Calculate the frequencies of the lines in the $3d\,^2D$ to $2p\,^2P$ transition for atomic hydrogen.

SOLUTION: There are two $2p\,^2P$ states in atomic hydrogen, $2p\,^2P_{1/2}$ and $2p\,^2P_{3/2}$. The allowed transition from the $3d\,^2D$ states into the $2p\,^2P_{1/2}$ state is

$$3d\,^2D_{3/2} \to 2p\,^2P_{1/2}$$
$$\tilde{\nu} = (97\,492.306 - 82\,258.917)\ \text{cm}^{-1} = 15\,233.389\ \text{cm}^{-1}$$

and the $3d\ ^2D \to 2p\ ^2P_{3/2}$ transitions are

$$3d\ ^2D_{3/2} \to 2p\ ^2P_{3/2}$$
$$\tilde{\nu} = (97\,492.306 - 82\,259.272)\ \text{cm}^{-1} = 15\,233.034\ \text{cm}^{-1}$$

and

$$3d\ ^2D_{5/2} \to 2p\ ^2P_{3/2}$$
$$\tilde{\nu} = (97\,492.342 - 82\,259.272)\ \text{cm}^{-1} = 15\,233.070\ \text{cm}^{-1}$$

These three transitions are illustrated in Figure 8.3. Note that the $3d\ ^2D_{5/2} \to 2p\ ^2P_{1/2}$ transition is not allowed because $\Delta J = 2$ is not allowed.

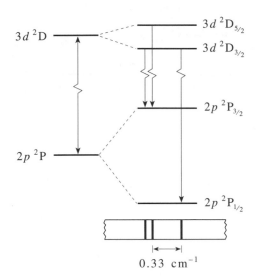

FIGURE 8.3
The fine structure of the spectral line associated with the $3d\ ^2D \to 2p\ ^2P$ transition in atomic hydrogen.

The type of data presented in Table 8.5 for atomic hydrogen have been tabulated for many atoms and ions in the publication *Atomic Energy Levels* by Charlotte E. Moore (see footnote to Table 8.5). Chemists usually refer to these tables as "Moore's tables". Table 8.6 is a direct copy of the energy-level data for the first few levels of atomic sodium, whose ground-state electron configuration is $1s^2 2s^2 2p^6 3s^1$. Figure 8.4 is an energy-level diagram of atomic sodium, showing the allowed electronic transitions. The $3s$, $3p$, etc. in front of the term symbols in Table 8.6 indicate that the electron configuration of those states are [Ne]$3s$, [Ne]$3p$, etc.

EXAMPLE 8–10
Use Table 8.6 to calculate the wavelengths of the two lines in the doublet associated with the $3p\ ^2P \to 3s\ ^2S$ transition in sodium, and compare your results with those of Figure 8.4.

TABLE 8.6

A reprint of a page from "Moore's tables", giving the energies (in cm^{-1}) of the first few states of atomic sodium.

Config.	Desig.	J	Level	Interval		Config.	Desig.	J	Level	Interval
			Na I						**Na I**	
$3s$	$3s\ ^2S$	$\frac{1}{2}$	0. 000			$5p$	$5p\ ^2P°$	$\frac{1}{2}$ $1\frac{1}{2}$	35040. 27 35042. 79	2. 52
$3p$	$3p\ ^2P°$	$\frac{1}{2}$ $1\frac{1}{2}$	16956. 183 16973. 379	17. 1963		$6s$	$6s\ ^2S$	$\frac{1}{2}$	36372. 647	
$4s$	$4s\ ^2S$	$\frac{1}{2}$	25739. 86			$5d$	$5d\ ^2D$	$2\frac{1}{2}$ $1\frac{1}{2}$	37036. 781 37036. 805	$-0. 0230$
$3d$	$3d\ ^2D$	$2\frac{1}{2}$ $1\frac{1}{2}$	29172. 855 29172. 904	$-0. 0494$		$5f$	$5f\ ^2F°$	$\left\{ \begin{matrix} 2\frac{1}{2} \\ 3\frac{1}{2} \end{matrix} \right\}$	37057. 6	
$4p$	$4p\ ^2P°$	$\frac{1}{2}$ $1\frac{1}{2}$	30266. 88 30272. 51	5. 63		$5g$	$5g\ ^2G$	$\left\{ \begin{matrix} 3\frac{1}{2} \\ 4\frac{1}{2} \end{matrix} \right\}$	37060. 2	
$5s$	$5s\ ^2S$	$\frac{1}{2}$	33200. 696			$6p$	$6p\ ^2P°$	$\frac{1}{2}$ $1\frac{1}{2}$	37296. 51 37297. 76	1. 25
$4d$	$4d\ ^2D$	$2\frac{1}{2}$ $1\frac{1}{2}$	34548. 754 34548. 789	$-0. 0346$		$7s$	$7s\ ^2S$	$\frac{1}{2}$	38012. 074	
$4f$	$4f\ ^2F°$	$\left\{ \begin{matrix} 2\frac{1}{2} \\ 3\frac{1}{2} \end{matrix} \right\}$	34588. 6			$6d$	$6d\ ^2D$	$2\frac{1}{2}$ $1\frac{1}{2}$	38387. 287 38387. 300	$=0. 0124$

*The last 14 members are not included because page proof had been prepared when the data were received.

From C. E. Moore, "Atomic Energy Levels," Natl. Bur. Std. U.S. Circ. No. 467 (U.S. Government Printing Office, Washington, D.C., 1949.

SOLUTION: The two transitions are

$$3p\ ^2P_{1/2} \rightarrow 3s\ ^2S_{1/2} \quad \tilde{\nu} = 16\ 956.183\ cm^{-1}$$

and

$$3p\ ^2P_{3/2} \rightarrow 3s\ ^2S_{1/2} \quad \tilde{\nu} = 16\ 973.379\ cm^{-1}$$

The wavelengths are given by $\lambda = 1/\tilde{\nu}$, or

$$\lambda = 5897.6\ \text{Å} \quad \text{and} \quad 5891.6\ \text{Å}$$

where Å represents the non SI but commonly used unit, Angstrom ($1\text{Å} = 10^{-10}$m).

If we compare these wavelengths with those in Figure 8.4, we see that there is a small discrepancy. This discrepancy is caused by the fact that wavelengths determined experimentally are measured in air, whereas the calculations using Table 8.6 provide wavelengths in a vacuum. We use the index of refraction in air (1.00029) to convert from one wavelength to another:

$$\lambda_{\text{vac}} = 1.00029\lambda_{\text{air}}$$

If we divide each of the wavelengths obtained above by 1.00029, then we obtain

$$\lambda_{\text{expt}} = 5895.9\ \text{Å} \quad \text{and} \quad 5889.9\ \text{Å}$$

in excellent agreement with Figure 8.4. These wavelengths occur in the yellow region of the spectrum and account for the intense yellow doublet, called the *sodium D line*, that is characteristic of the emission spectrum of sodium atoms.

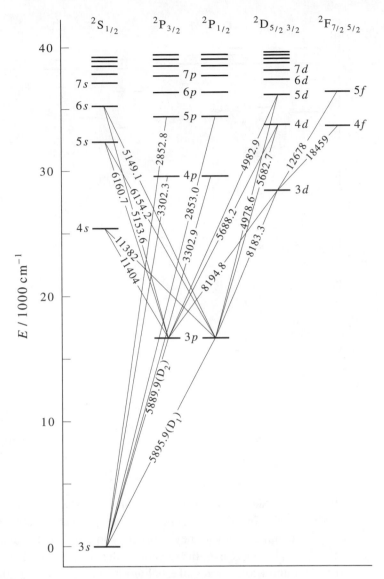

FIGURE 8.4

Energy-level diagram of atomic sodium, showing some of the allowed electronic transitions. The numbers on the lines are the wavelengths (measured in air) in angstroms ($1\ \text{Å} = 10^{-10}$ m) of the transitions.

Figure 8.5 shows the energy-level diagram of helium at a resolution at which the spin-orbit splittings are not significant. The principal feature of the helium energy-level diagram is that it indicates two separate sets of transitions. Notice from the figure that one set of transitions is among singlet states ($S = 0$) and the other is among triplet ($S = 1$) states. No transitions occur between the two sets of states because of the $\Delta S = 0$ selection rule. Thus, the only allowed transitions are between states with the

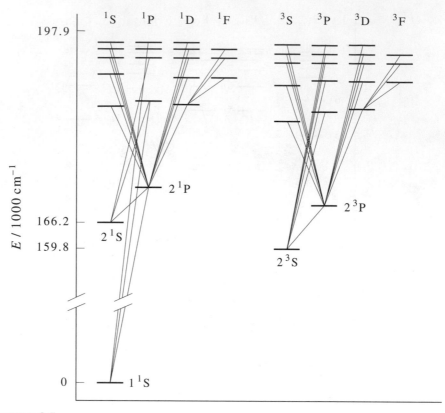

FIGURE 8.5
The energy-level diagram of helium, showing the two separate sets of singlet and triplet states.

same spin multiplicity. The observed spectrum of helium consists of two overlapping sets of lines. We should point out that the selection rules presented here are useful only for small spin-orbit coupling, and so they apply only to atoms with small atomic numbers. As the atomic number increases, the selection rules break down. For example, mercury has both singlet and triplet states like helium does, but many singlet-triplet state transitions are observed in the atomic spectrum of mercury.

We are now ready to discuss molecules. One of the great achievements of quantum mechanics is a detailed explanation of the stability of the chemical bond, such as in H_2. Because H_2 is the simplest molecule, we will discuss it in some quantitative detail like we did helium in this chapter and then discuss the results of similar calculations for more complicated molecules more qualitatively.

Problems

8-1. Show that the atomic unit of energy can be written as

$$E_h = \frac{\hbar^2}{m_e a_0^2} = \frac{e^2}{4\pi \varepsilon_0 a_0} = \frac{m_e e^4}{16\pi^2 \varepsilon_0^2 \hbar^2}$$

8-2. Show that the energy of a helium ion in atomic units is $-2E_{\text{h}}$.

8-3. The electric potential energy at a distance r from a charge q is

$$V = \frac{q}{4\pi\varepsilon_0 r}$$

Show that the atomic unit of potential energy is the potential energy at a distance of one Bohr radius from a proton (see Table 8.1).

8-4. Show that the speed of an electron in the first Bohr orbit is $e^2/4\pi\varepsilon_0\hbar = 2.188 \times 10^6$ m·s^{-1}. This speed is the unit of speed in atomic units.

8-5. Show that the speed of light is equal to 137 in atomic units.

8-6. Another way to introduce atomic units is to express mass as multiples of m_{e}, the mass of an electron (instead of kg); charge as multiples of e, the protonic charge (instead of C); angular momentum as multiples of $\hbar$ (instead of in J·s $=$ kg·m^2·s^{-1}); and permittivity as multiples of $4\pi\varepsilon_0$ (instead of in C^2·s^2·kg^{-1}·m^{-3}). This conversion can be achieved in all of our equations by letting $m_{\text{e}} = e = \hbar = 4\pi\varepsilon_0 = 1$. Show that this procedure is consistent with the definition of atomic units used in the chapter.

8-7. Derive Equation 8.5 from Equation 8.4. Be sure to remember that ∇^2 has units of (distance)$^{-2}$.

8-8. Show that the normalization constant for the radial part of Slater orbitals is $(2\zeta)^{n+\frac{1}{2}}/[(2n)!]^{1/2}$.

8-9. Use Equation 8.12 to write out the normalized $1s$, $2s$, and $2p$ Slater orbitals. How do they differ from the hydrogenlike orbitals?

8-10. Substitute Equation 8.5 for $\hat{H}$ into

$$E = \int\int d\mathbf{r}_1 d\mathbf{r}_2 \phi^*(\mathbf{r}_1)\phi^*(\mathbf{r}_2)\hat{H}\phi(\mathbf{r}_1)\phi(\mathbf{r}_2)$$

and show that

$$E = I_1 + I_2 + J_{12}$$

where

$$I_j = \int d\mathbf{r}_j \phi^*(\mathbf{r}_j)\left[-\frac{1}{2}\nabla_j^2 - \frac{Z}{r_j}\right]\phi(\mathbf{r}_j)$$

and

$$J_{12} = \int\int d\mathbf{r}_1 d\mathbf{r}_2 \phi^*(\mathbf{r}_1)\phi(\mathbf{r}_1)\frac{1}{r_{12}}\phi^*(\mathbf{r}_2)\phi(\mathbf{r}_2)$$

Why is J_{12} called a Coulomb integral?

In the next problem we use the above result to deduce a physical interpretation of the eigenvalue ϵ in the Hartree-Fock equation, Equation 8.20.

8-11. In this problem we will examine the physical significance of the eigenvalue ϵ in Equation 8.20. The quantity ϵ is called the orbital energy. Using Equation 8.19 for $\hat{H}_1^{\text{eff}}(\mathbf{r}_1)$, multiply Equation 8.20 from the left by $\phi^*(\mathbf{r}_1)$ and integrate to obtain

$$\epsilon_1 = I_1 + J_{12}$$

where I_1 and J_{12} are defined in the previous problem. Show that the total energy of a helium atom $E = I_1 + I_2 + J_{12}$ is *not* the sum of its orbital energies. In fact, show that

$$\epsilon_1 = E - I_2 \tag{1}$$

But according to the definition of I_2 in Problem 8–10, I_2 is the energy of a helium ion, calculated with the helium Hartree-Fock orbital $\phi(r)$. Thus, Equation 1 suggests that the orbital energy ϵ_1 is an approximation to the ionization energy of a helium atom or that

$$\text{IE} \approx -\epsilon_1 \qquad \text{(Koopmans' approximation)}$$

Even within the Hartree-Fock approximation, Koopmans' approximation is based upon the approximation that the same orbitals can be used to calculate the energy of the neutral atom and the energy of the ion. The value of $-\epsilon_1$ obtained by Clementi (see Table 8.2) is $0.91796E_{\text{h}}$, compared with the experimental value of $0.904E_{\text{h}}$.

8-12. Show that the two-term helium Hartree-Fock orbital

$$\phi(r) = 0.81839e^{-1.44608r} + 0.52072e^{-2.86222r}$$

is normalized.

8-13. The normalized variational helium orbital we determined in Chapter 7 is

$$\phi(r) = 1.2368e^{-27r/16}$$

A two-term Hartree-Fock orbital is given in Problem 8–12:

$$\phi(r) = 0.81839e^{-1.44608r} + 0.52072e^{-2.86222r}$$

and a five-term orbital given on page 283 is

$$\phi(r) = 0.75738e^{-1.4300r} + 0.43658e^{-2.4415r} + 0.17295e^{-4.0996r}$$
$$- 0.02730e^{-6.4843r} + 0.06675e^{-7.978r}$$

Plot these orbitals on the same graph and compare them.

8-14. Given that $\Psi(1, 2) = 1s\alpha(1)1s\beta(2) - 1s\alpha(2)1s\beta(1)$, prove that

$$\int d\tau_1 d\tau_2 \Psi^*(1, 2)\Psi(1, 2) = 2$$

if the spatial part is normalized.

8-15. Show that the spin integral in Equation 8.40 is equal to 2.

8-16. Why is it impossible to distinguish the two electrons in a helium atom, but not the two electrons in separated hydrogen atoms? Do you think the electrons are distinguishable in the diatomic H_2 molecule? Explain your reasoning.

8-17. Why is the angular dependence of multielectron atomic wave functions in the Hartree-Fock approximation the same as for hydrogen atomic wave functions?

8-18. Why is the radial dependence of multielectron atomic wave functions in the Hartree-Fock approximation different from the radial dependence of hydrogen atomic wave functions?

8-19. Show that the atomic determinantal wave function

$$\psi = \frac{1}{\sqrt{2}} \begin{vmatrix} 1s\alpha(1) & 1s\beta(1) \\ 1s\alpha(2) & 1s\beta(2) \end{vmatrix}$$

is normalized if the $1s$ orbitals are normalized.

8-20. Show that the two-electron determinantal wave function in Problem 8–19 factors into a spatial part and a spin part.

8-21. Argue that the normalization constant of an $N \times N$ Slater determinant of orthonormal spin orbitals is $1/\sqrt{N!}$.

8-22. The total z component of the spin angular momentum operator for an N-electron system is

$$\hat{S}_{z,\text{total}} = \sum_{j=1}^{N} \hat{S}_{zj}$$

Show that both

$$\psi = \frac{1}{\sqrt{2}} \begin{vmatrix} 1s\alpha(1) & 1s\beta(1) \\ 1s\alpha(2) & 1s\beta(2) \end{vmatrix}$$

and

$$\psi = \frac{1}{\sqrt{3!}} \begin{vmatrix} 1s\alpha(1) & 1s\beta(1) & 2s\alpha(1) \\ 1s\alpha(2) & 1s\beta(2) & 2s\alpha(2) \\ 1s\alpha(3) & 1s\beta(3) & 2s\alpha(3) \end{vmatrix}$$

are eigenfunctions of $\hat{S}_{z,\text{total}}$. What are the eigenvalues in each case?

8-23. Consider the determinantal atomic wave function

$$\Psi(1, 2) = \frac{1}{\sqrt{2}} \begin{vmatrix} \psi_{211}\alpha(1) & \psi_{21-1}\beta(1) \\ \psi_{211}\alpha(2) & \psi_{21-1}\beta(2) \end{vmatrix}$$

where $\psi_{21\pm1}$ is a hydrogenlike wave function. Show that $\Psi(1, 2)$ is an eigenfunction of

$$\hat{L}_{z,\text{total}} = \hat{L}_{z1} + \hat{L}_{z2}$$

and

$$S_{z,\text{total}} = \hat{S}_{z1} + \hat{S}_{z2}$$

What are the eigenvalues?

8-24. For a two-electron system, there are four possible spin functions:

1. $\alpha(1)\alpha(2)$ **2.** $\beta(1)\alpha(2)$

3. $\alpha(1)\beta(2)$ **4.** $\beta(1)\beta(2)$

The concept of indistinguishability forces us to consider only linear combinations of 2 and 3,

$$\psi_{\pm} = \frac{1}{\sqrt{2}}[\alpha(1)\beta(2) \pm \beta(1)\alpha(2)]$$

instead of 2 and 3 separately. Show that of the four acceptable spin functions, 1, 4, and $\psi_{\pm}$, three are symmetric and one is antisymmetric.

Now for a two-electron system, we can combine spatial wave functions with spin functions. Show that this combination leads to only four allowable combinations:

$$[\psi(1)\phi(2) + \psi(2)\phi(1)]\frac{1}{\sqrt{2}}[\alpha(1)\beta(2) - \alpha(2)\beta(1)]$$

$$[\psi(1)\phi(2) - \psi(2)\phi(1)][\alpha(1)\alpha(2)]$$

$$[\psi(1)\phi(2) - \psi(2)\phi(1)][\beta(1)\beta(2)]$$

and

$$[\psi(1)\phi(2) - \psi(2)\phi(1)]\frac{1}{\sqrt{2}}[\alpha(1)\beta(2) + \alpha(2)\beta(1)]$$

where ψ and ϕ are two spatial wave functions. Show that $M_S = m_{s1} + m_{s2} = 0$ for the first of these and that $M_S = 1$, -1, and 0 (in atomic units) for the next three, respectively.

Consider the first excited state of a helium atom, in which $\psi = 1s$ and $\phi = 2s$. The first of the four wave functions above, with the symmetric spatial part, will give a higher energy than the remaining three, which form a degenerate set of three. The first state is a singlet state and the degenerate set of three represents a triplet state. Because M_S equals zero and only zero for the singlet state, the singlet state corresponds to $S = 0$. The other three, with $M_S = \pm 1$, 0, corresponds to $S = 1$. Note that the degeneracy is $2S + 1$ in each case.

Putting all this information into a more mathematical form, given that $\hat{S}_{\text{total}} = \hat{S}_1 + \hat{S}_2$, we can show that (Problem 8–53)

$$\hat{S}^2_{\text{total}}[\alpha(1)\beta(2) - \alpha(2)\beta(1)] = 0$$

corresponding to $S = 0$, and that

$$\hat{S}^2_{\text{total}} \left[\begin{array}{c} \alpha(1)\alpha(2) \\ \frac{1}{\sqrt{2}}[\alpha(1)\beta(2) + \alpha(2)\beta(1)] \\ \beta(1)\beta(2) \end{array} \right] = 2\hbar^2 \left[\begin{array}{c} \alpha(1)\alpha(2) \\ \frac{1}{\sqrt{2}}[\alpha(1)\beta(2) + \alpha(2)\beta(1)] \\ \beta(1)\beta(2) \end{array} \right]$$

corresponding to $S = 1$.

8-25. Consider a helium atom in an excited state in which one of its $1s$ electrons is raised to the $2s$ level, so that its electron configuration is $1s2s$. Argue that because the two orbitals are different, there are four possible determinantal wave functions for this system:

$$\phi_1 = \frac{1}{\sqrt{2}} \begin{vmatrix} 1s\alpha(1) & 2s\alpha(1) \\ 1s\alpha(2) & 2s\alpha(2) \end{vmatrix}$$

$$\phi_2 = \frac{1}{\sqrt{2}} \begin{vmatrix} 1s\beta(1) & 2s\beta(1) \\ 1s\beta(2) & 2s\beta(2) \end{vmatrix}$$

$$\phi_3 = \frac{1}{\sqrt{2}} \begin{vmatrix} 1s\alpha(1) & 2s\beta(1) \\ 1s\alpha(2) & 2s\beta(2) \end{vmatrix}$$

$$\phi_4 = \frac{1}{\sqrt{2}} \begin{vmatrix} 1s\beta(1) & 2s\alpha(1) \\ 1s\beta(2) & 2s\alpha(2) \end{vmatrix}$$

To calculate the energy of the $1s2s$ configuration, assume the variational function

$$\psi = c_1\phi_1 + c_2\phi_2 + c_3\phi_3 + c_4\phi_4$$

Show that the secular equation associated with this linear combination trial function is (this is the only lengthy part of this problem and at least you have the answer in front of you; be sure to remember that the $1s$ and $2s$ orbitals here are eigenfunctions of the hydrogenlike Hamiltonian operator)

$$\begin{vmatrix} E_0 + J - K - E & 0 & 0 & 0 \\ 0 & E_0 + J - K - E & 0 & 0 \\ 0 & 0 & E_0 + J - E & -K \\ 0 & 0 & -K & E_0 + J - E \end{vmatrix} = 0$$

where

$$J = \int \int d\tau_1 d\tau_2\, 1s(1)1s(1)\left(\frac{1}{r_{12}}\right) 2s(2)2s(2)$$

$$K = \int \int d\tau_1 d\tau_2\, 1s(1)2s(1)\left(\frac{1}{r_{12}}\right) 1s(2)2s(2)$$

and E_0 is the energy without the $1/r_{12}$ term in the helium atom Hamiltonian operator. Show that

$$E_0 = -\frac{5}{2}E_h$$

Explain why J is called an atomic Coulombic integral and K is called an atomic exchange integral.

Even though the above secular determinant is 4×4 and appears to give a fourth-degree polynomial in E, note that it really consists of two 1×1 blocks and a 2×2 block. Show that this symmetry in the determinant reduces the determinantal equation to

$$(E_0 + J - K - E)^2 \begin{vmatrix} E_0 + J - E & -K \\ -K & E_0 + J - E \end{vmatrix} = 0$$

and that this equation gives the four roots

$$E = E_0 + J - K \qquad \text{(twice)}$$
$$= E_0 + J \pm K$$

Show that the wave function corresponding to the positive sign in E in the $E_0 + J \pm K$ is

$$\psi_3 = \frac{1}{\sqrt{2}} (\phi_3 - \phi_4)$$

and that corresponding to the negative sign in $E_0 + J \pm K$ is

$$\psi_4 = \frac{1}{\sqrt{2}} (\phi_3 + \phi_4)$$

Now show that both ψ_3 and ψ_4 can be factored into a spatial part and a spin part, even though ϕ_3 and ϕ_4 separately cannot. Furthermore, let

$$\psi_1 = \phi_1 \quad \text{and} \quad \psi_2 = \phi_2$$

and show that both of these can be factored also. Using the argument given in Problem 8–24, group these four wave functions (ψ_1, ψ_2, ψ_3, ψ_4) into a singlet state and a triplet state.

Now calculate the energy of the singlet and triplet states in terms of E_0, J, and K. Argue that $J > 0$. Given that $K > 0$ also, does the singlet state or the triplet state have the lower energy? The values of J and K when hydrogenlike wave functions with $Z = 2$ are used are $J = 34/81 E_h$ and $K = 32/(27)^2 E_h$. Using the ground-state wave function

$$\phi = \frac{1}{\sqrt{2}} \begin{vmatrix} 1s\alpha(1) & 1s\beta(1) \\ 1s\alpha(2) & 1s\beta(2) \end{vmatrix}$$

show that the first-order perturbation theory result is $E = -11/4 \, E_h$ if hydrogenlike wave functions with $Z = 2$ are used. Use this value of E to calculate the energy difference between the ground state and the first excited singlet state and the first triplet state of helium. The experimental values of these energy differences are $159\,700$ cm^{-1} and $166\,200$ cm^{-1}, respectively (cf. Figure 8.5).

8-26. Determine the term symbols associated with an np^1 electron configuration. Show that these term symbols are the same as for an np^5 electron configuration. Which term symbol represents the ground state?

8-27. Show that the term symbols for an np^4 electron configuration are the same as for an np^2 electron configuration.

8-28. Show that the number of sets of magnetic quantum numbers (m_l) and spin quantum numbers (m_s) associated with any term symbol is equal to $(2L + 1)(2S + 1)$. Apply this result to the np^2 case discussed in Section 8–9, and show that the term symbols ^1S, ^3P, and ^1D account for all the possible sets of magnetic quantum numbers and spin quantum numbers.

8-29. Calculate the number of sets of magnetic quantum numbers (m_l) and spin quantum numbers (m_s) for an nd^8 electron configuration. Prove that the term symbols ^1S, ^1D, ^3P, ^3F, and ^1G account for all possible term symbols.

8-30. Determine the term symbols for the electron configuration $nsnp$. Which term symbol corresponds to the lowest energy?

8-31. How many sets of magnetic quantum numbers (m_l) and spin quantum numbers (m_s) are there for an $nsnd$ electron configuration? What are the term symbols? Which term symbol corresponds to the lowest energy?

8-32. The term symbols for an nd^2 electron configuration are ^1S, ^1D, ^1G, ^3P, and ^3F. Calculate the values of J associated with each of these term symbols. Which term symbol represents the ground state?

8-33. The term symbols for an np^3 electron configuration are ^2P, ^2D, and ^4S. Calculate the values of J associated with each of these term symbols. Which term symbol represents the ground state?

8-34. Determine the electron configuration of a magnesium atom in its ground state, and its ground-state term symbol.

8-35. Given that the electron configuration of a zirconium atom is $[Kr](4d)^2(5s)^2$, determine the ground-state term symbol for Zr.

8-36. Given that the electronic configuration of a palladium atom is $[Kr](4d)^{10}$, determine the ground-state term symbol for Pd.

8-37. Consider the $1s2p$ electron configuration for helium. Determine the states (term symbols) that correspond to this electron configuration. Determine the degeneracies of each state. What will happen if you include the effect of spin orbit coupling?

8-38. Use Table 8.5 to calculate the separation of the doublets that occur in the Lyman series of atomic hydrogen.

8-39. Use Table 8.6 to calculate the wavelength of the $4f\,^2F \rightarrow 3d\,^2D$ transition in atomic sodium and compare your result with that given in Figure 8.4. Be sure to use the relation $\lambda_{vac} = 1.00029\lambda_{air}$ (see Example 8–10).

8-40. The orbital designations s, p, d, and f come from an analysis of the spectrum of atomic sodium. The series of lines due to $ns\,^2S \rightarrow 3p\,^2P$ transitions is called the *sharp* (*s*) series; the series due to $np\,^2P \rightarrow 3s\,^2S$ transitions is called the *principal* (*p*) series; the series due to $nd\,^2D \rightarrow 3p\,^2P$ transitions is called the *diffuse* (*d*) series; and the series due to $nf\,^2F \rightarrow 3d\,^2D$ transitions is called the *fundamental* (*f*) series. Identify each of these series in Figure 8.4, and tabulate the wavelengths of the first few lines in each series.

8-41. Problem 8–40 defines the sharp, principal, diffuse, and fundamental series in the spectrum of atomic sodium. Use Table 8.6 to calculate the wavelengths of the first few lines in each series and compare your results with those in Figure 8.4. Be sure to use the relation $\lambda_{vac} = 1.00029\lambda_{air}$ (see Example 8–10).

8-42. In this problem, we will derive an explicit expression for $V^{eff}(\mathbf{r}_1)$ given by Equation 8.18 using $\phi(\mathbf{r})$ of the form $(Z^3/\pi)^{1/2}e^{-Zr}$. (We have essentially done this problem in Problem 7–30.)

$$V^{eff}(\mathbf{r}_1) = \frac{Z^3}{\pi}\int d\mathbf{r}_2 \frac{e^{-2Zr_2}}{r_{12}}$$

As in Problem 7–30, we use the law of cosines to write

$$r_{12} = (r_1^2 + r_2^2 - 2r_1 r_2 \cos\theta)^{1/2}$$

and so V^{eff} becomes

$$V^{eff}(r_1) = \frac{Z^3}{\pi}\int_0^\infty dr_2 e^{-2Zr_2}r_2^2 \int_0^{2\pi} d\phi \int_0^\pi \frac{d\theta\sin\theta}{(r_1^2 + r_2^2 - 2r_1 r_2 \cos\theta)^{1/2}}$$

Problem 7–30 asks you to show that the integral over θ is equal to $2/r_1$ if $r_1 > r_2$ and equal to $2/r_2$ if $r_1 < r_2$. Thus, we have

$$V^{eff}(r_1) = 4Z^3\left[\frac{1}{r_1}\int_0^{r_1} e^{-Zr_2}r_2^2 dr_2 + \int_{r_1}^\infty e^{-2Zr_2}r_2 dr_2\right]$$

Now show that

$$V^{eff}(r_1) = \frac{1}{r_1} - e^{-2Zr_1}\left(Z + \frac{1}{r_1}\right)$$

8-43. Repeat Problem 8–42 using the expansion of $1/r_{12}$ given in Problem 7–31.

Problems 8–44 through 8–48 address the energy levels of one electron atoms that include the effect of spin-orbit coupling.

8-44. Show that $\hat{\mathbf{L}} \cdot \hat{\mathbf{S}} = \frac{1}{2}(\hat{J}^2 - \hat{L}^2 - \hat{S}^2)$.

8-45. Show that $[\hat{H}, \hat{L}^2] = [\hat{H}, \hat{S}^2] = [\hat{H}, \hat{J}^2] = 0$, where $\hat{H}$ is the Hamiltonian operator of a hydrogen atom. *Hint*: Use the result of Problem 8–44 and operate on a function that is a product of a spatial part and a spin part.

8-46. Because of the coupling of the spin and orbital angular momenta of the electron, the Hamiltonian operator for a hydrogenlike atom becomes

$$\hat{H} = \hat{H}^{(0)} + \hat{H}_{so}^{(1)}$$

where (in atomic units)

$$\hat{H}^{(0)} = -\frac{1}{2}\nabla^2 - \frac{Z}{r}$$

and

$$\hat{H}_{so}^{(1)} = \frac{Z}{2(137)^2}\frac{1}{r^3}\hat{\mathbf{l}} \cdot \hat{\mathbf{s}}$$

We will now use first-order perturbation theory to evaluate the first-order correction to the energy. Recall from Chapter 7 that

$$E_n^{(1)} = \int \psi_n^{(0)*} \hat{H}_{so}^{(1)} \psi_n^{(0)} d\tau$$

Using the result of Problem 8–44, show that

$$E_n^{(1)} = \frac{Z}{2(137)^2} \int \psi_n^{(0)*} \frac{1}{r^3} (\hat{\mathbf{l}} \cdot \hat{\mathbf{s}}) \psi_n^{(0)} d\tau$$

$$= \frac{1}{2} \{ j(j+1) - l(l+1) - s(s+1) \} \frac{Z}{2(137)^2} \left\langle \frac{1}{r^3} \right\rangle \tag{1}$$

where

$$\left\langle \frac{1}{r^3} \right\rangle = \int \psi_n^{(0)*} \left(\frac{1}{r^3} \right) \psi_n^{(0)} d\tau \tag{2}$$

Problem 6–41 shows that

$$\left\langle \frac{1}{r^3} \right\rangle = \frac{Z^3}{n^3 l(l+1)(l+\frac{1}{2})} \tag{3}$$

Now combine Equations 1 through 3 to obtain

$$E_n^{(1)}/E_h = \frac{Z^4}{2(137)^2 n^3} \frac{\{ j(j+1) - l(l+1) - s(s+1) \}}{2l(l+1)(l+\frac{1}{2})}$$

For $Z = 1$ and $l = 1$, what is the order of magnitude (in cm^{-1}) for the spin-orbit splitting between the two states as a function of n? (*Hint*: For a hydrogen atom $s = 1/2$ and j can be only $l \pm 1/2$.) Recall also that $1 E_h = 2.195 \times 10^5$ cm^{-1}. How does this energy compare with the energy separation between the energies for different values of n?

8-47. The two term symbols corresponding to the ns^2np^5 valence electron configuration of the halogens are $^2P_{1/2}$ and $^2P_{3/2}$. Which is the term symbol for the ground state? The energy difference between these two states for the different halogens is given below

Halogen	$[E(^2P_{1/2}) - E(^2P_{3/2})]/\text{cm}^{-1}$
F	404
Cl	880
Br	3685
I	7600

Suggest an explanation for this trend.

8-48. The photoionization spectra of the noble gases argon and krypton each show two closely spaced lines that correspond to the ionization of an electron from a $2p$ orbital. Explain why there are two closely spaced lines. (Assume the resulting ion is in its ground electronic state.)

The spin operators satisfy the same general equations that we developed for the angular momentum operators in Problems 6–48 through 6–57. Problems 8–49 through 8–53 review these results.

8-49. The spin operators, $\hat{S}_x$, $\hat{S}_y$, and $\hat{S}_z$, like all angular momentum operators, obey the commutation relations (Problem 6–13)

$$[\hat{S}_x, \hat{S}_y] = i\hbar \hat{S}_z \qquad [\hat{S}_y, \hat{S}_z] = i\hbar \hat{S}_x \qquad [\hat{S}_z, \hat{S}_x] = i\hbar \hat{S}_y$$

Define the (non-Hermitian) operators

$$\hat{S}_+ = \hat{S}_x + i\hat{S}_y \qquad \hat{S}_- = \hat{S}_x - i\hat{S}_y \tag{1}$$

and show that

$$[\hat{S}_z, \hat{S}_+] = \hbar \hat{S}_+ \tag{2}$$

and

$$[\hat{S}_z, \hat{S}_-] = -\hbar \hat{S}_- \tag{3}$$

Now show that

$$\hat{S}_+ \hat{S}_- = \hat{S}^2 - \hat{S}_z^2 + \hbar \hat{S}_z$$

and that

$$\hat{S}_- \hat{S}_+ = \hat{S}^2 - \hat{S}_z^2 - \hbar \hat{S}_z$$

where

$$\hat{S}^2 = \hat{S}_x^2 + \hat{S}_y^2 + \hat{S}_z^2$$

8-50. Use Equation 2 from Problem 8–49 and the fact that $\hat{S}_z \beta = -\frac{\hbar}{2}\beta$ to show that

$$\hat{S}_z \hat{S}_+ \beta = \hat{S}_+ \left(-\frac{\hbar}{2}\beta + \hbar\beta \right) = \frac{\hbar}{2} \hat{S}_+ \beta$$

Because $\hat{S}_z \alpha = \frac{\hbar}{2}\alpha$, this result shows that

$$\hat{S}_+ \beta \propto \alpha = c\alpha$$

where c is a proportionality constant. The following problem shows that $c = \hbar$, so that we have

$$\hat{S}_+ \beta = \hbar\alpha$$

Now use Equation 3 from Problem 8–49 and the fact that $\hat{S}_z \alpha = \frac{\hbar}{2}\alpha$ to show that

$$\hat{S}_- \alpha = c\beta \tag{1}$$

where c is a proportionality constant. The following problem shows that $c = \hbar$, so that we have

$$\hat{S}_+ \beta = \hbar\alpha \quad \text{and} \quad \hat{S}_- \alpha = \hbar\beta \tag{2}$$

Notice that $\hat{S}_+$ "raises" the spin function from β to α, whereas $\hat{S}_-$ "lowers" the spin function from α to β. The two operators $\hat{S}_+$ and $\hat{S}_-$ are called raising and lowering operators, respectively.

Now use Equation 2 to show that

$$\hat{S}_x\alpha = \frac{\hbar}{2}\beta \qquad \hat{S}_y\alpha = \frac{i\hbar}{2}\beta$$

$$\hat{S}_x\beta = \frac{\hbar}{2}\alpha \qquad \hat{S}_y\beta = -\frac{i\hbar}{2}\alpha$$

8-51. This problem shows that the proportionality constant c in

$$\hat{S}_+\beta = c\alpha \qquad \text{or} \qquad \hat{S}_-\alpha = c\beta$$

is equal to $\hbar$. Start with

$$\int \alpha^*\alpha d\tau = 1 = \frac{1}{|c|^2}\int (\hat{S}_+\beta)^*(\hat{S}_+\beta)d\tau$$

Let $\hat{S}_+ = \hat{S}_x + i\hat{S}_y$ in the second factor in the above integral and use the fact that $\hat{S}_x$ and $\hat{S}_y$ are Hermitian to get

$$\int (\hat{S}_x\hat{S}_+\beta)^*\beta d\tau + i\int (\hat{S}_y\hat{S}_+\beta)^*\beta d\tau = |c|^2$$

Now take the complex conjugate of both sides to get

$$\int \beta^*\hat{S}_x\hat{S}_+\beta d\tau - i\int \beta^*\hat{S}_y\hat{S}_+\beta d\tau = |c|^2$$

$$= \int \beta^*\hat{S}_-\hat{S}_+\beta d\tau$$

Now use the result in Problem 8–49 to show that

$$|c|^2 = \int \beta^*\hat{S}_-\hat{S}_+\beta d\tau = \int \beta^*(\hat{S}^2 - \hat{S}_z^2 - \hbar\hat{S}_z)\beta d\tau$$

$$= \int \beta^*\left(\frac{3}{4}\hbar^2 - \frac{1}{4}\hbar^2 + \frac{\hbar^2}{2}\right)\beta d\tau = \hbar^2$$

or that $c = \hbar$.

8-52. Use the result of Problem 8–50 along with the equations $\hat{S}_z\alpha = \frac{\hbar}{2}\alpha$ and $\hat{S}_z\beta = -\frac{\hbar}{2}\beta$ to show that

$$\hat{S}^2\alpha = \frac{3}{4}\hbar^2\alpha = \frac{1}{2}\left(\frac{1}{2} + 1\right)\hbar^2\alpha$$

and

$$\hat{S}^2\beta = \frac{3}{4}\hbar^2\beta = \frac{1}{2}\left(\frac{1}{2} + 1\right)\hbar^2\beta$$

8-53. In this problem, we will use the results of Problems 8–50 and 8–52 to verify the statements at the end of Problem 8–24. Because $\hat{S}_{\text{total}} = \hat{S}_1 + \hat{S}_2$, we have

$$\hat{S}_{\text{total}}^2 = (\hat{\mathbf{S}}_1 + \hat{\mathbf{S}}_2) \cdot (\hat{\mathbf{S}}_1 + \hat{\mathbf{S}}_2) = \hat{S}_1^2 + \hat{S}_2^2 + 2\hat{\mathbf{S}}_1 \cdot \hat{\mathbf{S}}_2$$

$$= \hat{S}_1^2 + \hat{S}_2^2 + 2(\hat{S}_{x1}\hat{S}_{x2} + \hat{S}_{y1}\hat{S}_{y2} + \hat{S}_{z1}\hat{S}_{z2})$$

Now show that

$$\hat{S}_{\text{total}}^2 \alpha(1)\alpha(2) = \alpha(2)\hat{S}_1^2\alpha(1) + \alpha(1)\hat{S}_2^2\alpha(2) + 2\hat{S}_{x1}\alpha(1)\hat{S}_{x2}\alpha(2)$$

$$+2\hat{S}_{y1}\alpha(1)\hat{S}_{y2}\alpha(2) + 2\hat{S}_{z1}\alpha(1)\hat{S}_{z2}\alpha(2)$$

$$= 2\hbar^2\alpha(1)\alpha(2)$$

Similarly, show that

$$\hat{S}_{\text{total}}^2 \beta(1)\beta(2) = 2\hbar^2\beta(1)\beta(2)$$

$$\hat{S}_{\text{total}}^2 [\alpha(1)\beta(2) + \beta(1)\alpha(2)] = 2\hbar^2[\alpha(1)\beta(2) + \beta(1)\alpha(2)]$$

and

$$\hat{S}_{\text{total}}^2 [\alpha(1)\beta(2) - \beta(1)\alpha(2)] = 0$$

8-54. We discussed the Hartree-Fock method for a helium atom in Section 8–3, but the application of the Hartree-Fock method to atoms that contain more than two electrons introduces new terms because of the determinantal nature of the wave functions. For simplicity, we shall consider only closed-shell systems, in which the wave functions are represented by N doubly occupied spatial orbitals. The Hamiltonian operator for a $2N$-electron atom is

$$\hat{H} = -\frac{1}{2}\sum_{j=1}^{2N}\nabla_j^2 - \sum_{j=1}^{2N}\frac{Z}{r_j} + \sum_{i=1}^{2N}\sum_{j>i}^{2N}\frac{1}{r_{ij}} \tag{1}$$

and the energy is given by

$$E = \int d\mathbf{r}_1 d\sigma_1 \cdots d\mathbf{r}_{2N} d\sigma_{2N} \Psi^*(1, 2, \ldots, 2N)\hat{H}\Psi(1, 2, \ldots, 2N) \tag{2}$$

Show that if Equation 1 and Equation 8.44 (with N replaced by $2N$ in this case) are substituted into Equation 2, then you obtain

$$E = 2\sum_{j=1}^{N} I_j + \sum_{i=1}^{N}\sum_{j=1}^{N}(2J_{ij} - K_{ij}) \tag{3}$$

where

$$I_j = \int d\mathbf{r}_j \phi_j^*(\mathbf{r}_j)\left[-\frac{1}{2}\nabla_j^2 - \frac{Z}{r_j}\right]\phi_j(\mathbf{r}_j) \tag{4}$$

$$J_{ij} = \int\int d\mathbf{r}_1 d\mathbf{r}_2 \phi_i^*(\mathbf{r}_1)\phi_i(\mathbf{r}_1)\frac{1}{r_{12}}\phi_j^*(\mathbf{r}_2)\phi_j(\mathbf{r}_2) \tag{5}$$

$$K_{ij} = \int\int d\mathbf{r}_1 d\mathbf{r}_2 \phi_i^*(\mathbf{r}_1)\phi_i(\mathbf{r}_2)\frac{1}{r_{12}}\phi_j^*(\mathbf{r}_2)\phi_j(\mathbf{r}_1) \tag{6}$$

Can you explain why the J_{ij} integrals are called *Coulomb integrals* and the K_{ij} integrals are called *exchange integrals* (if $i \neq j$)? Show that Equation 3 for a helium atom is the same as that given in Problem 8–10.

Robert S. Mulliken was born in Newburyport, Massachusetts, on June 25, 1896, and died in Chicago in 1986. He received his Ph.D. in physical chemistry from the University of Chicago in 1921, where his dissertation was on the separation of mercury isotopes by fractional distillation. He then went to Harvard University to continue his study of the behavior of isotopes. Realizing the importance of the new quantum theory to the understanding of atomic and molecular structure, he then spent a year in Europe studying quantum theory. After one year as an assistant professor at Washington Square College in New York City, he spent a year with Frederich Hund at the University of Göttingen, during which time they developed molecular orbital theory. Upon his return to the United States in 1928, Mulliken accepted a position at the University of Chicago, where he remained until his formal retirement in 1961. He continued working on molecular orbital theory at Chicago, and one of his early important contributions was the introduction of the LCAO approximation. After World War II, Mulliken and his collaborators pioneered the use of computers for calculating and elucidating molecular structure. He was known by all his associates as an unassuming and good-natured man. He was awarded the Nobel Prize in chemistry in 1966 "for his fundamental work concerning bonds and the electronic structure of molecules by the molecular orbital method."

CHAPTER 9

The Chemical Bond: Diatomic Molecules

One of the great achievements of quantum mechanics is its ability to describe the chemical bond. Before the development of quantum mechanics, chemists did not understand why two hydrogen atoms come together to form a stable chemical bond. We will see in this chapter that the existence of stable chemical bonds is described by quantum mechanics. Because the molecular ion H_2^+ involves the simplest chemical bond, we will discuss it in detail. Many of the ideas we will develop for H_2^+ are applicable to more complex molecules. After our discussion of H_2^+, we will learn how to construct molecular orbitals for diatomic molecules. We will place electrons into these orbitals in accord with the Pauli Exclusion Principle, just like we placed electrons into atomic orbitals for multielectron atoms. This chapter focuses on diatomic molecules and the next chapter examines polyatomic molecules.

9–1. The Born-Oppenheimer Approximation Simplifies the Schrödinger Equation for Molecules

For simplicity, let's consider the simplest neutral molecule, H_2. The Hamiltonian operator for a hydrogen molecule is given by

$$\hat{H} = -\frac{\hbar^2}{2M}(\nabla_A^2 + \nabla_B^2) - \frac{\hbar^2}{2m_e}(\nabla_1^2 + \nabla_2^2) - \frac{e^2}{4\pi\varepsilon_0 r_{1A}} - \frac{e^2}{4\pi\varepsilon_0 r_{1B}}$$

$$- \frac{e^2}{4\pi\varepsilon_0 r_{2A}} - \frac{e^2}{4\pi\varepsilon_0 r_{2B}} + \frac{e^2}{4\pi\varepsilon_0 r_{12}} + \frac{e^2}{4\pi\varepsilon_0 R} \tag{9.1}$$

In Equation 9.1, M is the mass of each hydrogen nucleus, m_e is the mass of the electron, the subscripts A and B refer to the nuclei of the individual atoms, the subscripts 1 and 2 refer to the individual electrons, and the various distances r_{1A}, r_{1B}, etc. are illustrated in

323

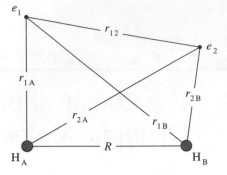

FIGURE 9.1
Definition of the distances between the nuclei
and the electrons involved in the Hamiltonian
operator for a hydrogen molecule (Equations 9.1
through 9.3)

Figure 9.1. The first two terms of the Hamiltonian operator in Equation 9.1 correspond
to the kinetic energy of the two nuclei; the next two terms represent the kinetic energy
of the two electrons; the four ensuing negative terms describe the contributions to the
potential energy that arise from the attraction between the nuclei and the electrons; and
the final two positive terms account for electron-electron and nuclear-nuclear repulsion,
respectively.

Because of the large difference between the masses of the nuclei and the electrons,
we can reasonably view the nuclei as fixed in position relative to the motion of the
electrons. Under such an approximation, the kinetic energy terms of the nuclei (the
first two terms in the Hamiltonian operator, Equation 9.1) can be treated separately.
This approximation of neglecting the nuclear motion is called the *Born-Oppenheimer*
approximation. Although the Born-Oppenheimer approximation will lead to approximate values of the energies and wave functions, it can be systematically corrected
using perturbation theory. For most practical purposes, however, these corrections are
on the order of the mass ratio ($\approx 10^{-3}$), and so the Born-Oppenheimer approximation
is a very good approximation. Neglecting the nuclear energy terms in Equation 9.1
gives the Hamiltonian operator for the motion of the electrons around the two nuclei
fixed at an internuclear separation:

$$\hat{H} = -\frac{\hbar^2}{2m_e}(\nabla_1^2 + \nabla_2^2) - \frac{e^2}{4\pi\varepsilon_0 r_{1A}} - \frac{e^2}{4\pi\varepsilon_0 r_{1B}} - \frac{e^2}{4\pi\varepsilon_0 r_{2A}} - \frac{e^2}{4\pi\varepsilon_0 r_{2B}}$$
$$+ \frac{e^2}{4\pi\varepsilon_0 r_{12}} + \frac{e^2}{4\pi\varepsilon_0 R} \tag{9.2}$$

Because the nuclei are considered to be fixed, the quantity R in Equation 9.2 is treated
as a parameter; the energy we will calculate using the above Hamiltonian operator will
depend upon R. As usual, we will express all our equations in atomic units (Section
8–1), and so Equation 9.2 becomes (Problem 9–1)

$$\hat{H} = -\frac{1}{2}(\nabla_1^2 + \nabla_2^2) - \frac{1}{r_{1A}} - \frac{1}{r_{1B}} - \frac{1}{r_{2A}} - \frac{1}{r_{2B}} + \frac{1}{r_{12}} + \frac{1}{R} \tag{9.3}$$

9–2. H_2^+ Is the Prototypical Species of Molecular-Orbital Theory

The method we will use to describe the bonding properties of molecules is called *molecular-orbital theory*. Molecular-orbital theory was developed in the early 1930s and is now the most commonly used method to calculate molecular properties. In molecular-orbital theory, we construct molecular wave functions in a manner similar to the way we constructed atomic wave functions in Chapter 8, where we expressed atomic wave functions in terms of determinants involving hydrogenlike, or single-electron, wave functions called atomic orbitals. Here we will express molecular wave functions in terms of determinants involving single-electron wave functions called molecular orbitals. The question that arises, then, is how to construct molecular orbitals. This problem is approached in an analogous way to what we did for multielectron atoms. There, the Schrödinger equation of the one-electron atom (hydrogen) was solved, and the orbitals obtained were used to construct wave functions for multielectron atoms. Here, for molecules, the Schrödinger equation for the one-electron molecular ion H_2^+ is solved, and the resulting orbitals are used to construct the wave functions for more complicated molecules. We should emphasize at this point that H_2^+ is a stable species that has been well studied spectroscopically. It has a bond length of 106 pm ($2.00\,a_0$) and a binding energy of $268\,kJ\cdot mol^{-1}$ ($0.103 E_h$).

Because H_2^+ is a one-electron species, the Schrödinger equation for H_2^+ can be solved exactly within the Born-Oppenheimer approximation. Nevertheless, the solutions are not easy to use, and their mathematical form does not give much physical insight as to how and why bonding occurs. Instead, it is more useful to solve H_2^+ approximately and use the resultant approximate molecular orbitals to build molecular wave functions. Although this approach may seem a crude way to proceed (after all, the problem can be solved exactly), it provides good physical insight into the nature of chemical bonds in molecules and yields results in good agreement with experimental observations. Furthermore, this approach can be systematically improved to give any desired degree of accuracy.

The Hamiltonian operator for H_2^+ in the Born-Oppenheimer approximation is

$$\hat{H} = -\frac{1}{2}\nabla^2 - \frac{1}{r_A} - \frac{1}{r_B} + \frac{1}{R} \tag{9.4}$$

where r_A and r_B are the distances of the electron from nucleus A and B, respectively, and R is the internuclear separation, which we treat as a variable parameter (Figure 9.2). The Schrödinger equation for H_2^+ is

$$\hat{H}\psi_j(r_A, r_B; R) = E_j\psi_j(r_A, r_B; R) \tag{9.5}$$

where the $\psi_j(r_A, r_B; R)$ are molecular orbitals, which extend over both nuclei. Recall that the variational principle (Chapter 7) says that we can get an excellent approximation

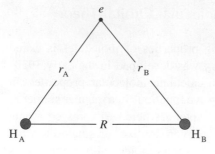

FIGURE 9.2
Definition of the distances involved in the Hamiltonian operator for H_2^+ (Equation 9.4).

to the energy if we use an appropriate trial function. As a trial function for $\psi_j(r_A, r_B; R)$, we take the linear combinations

$$\psi_{\pm} = c_1 1s_A \pm c_2 1s_B \tag{9.6}$$

where $1s_A$ and $1s_B$ are hydrogen atomic orbitals centered on nuclei A and B, respectively. The molecular orbital given by Equation 9.6 is a *Linear Combination of Atomic Orbitals*, and is called a *LCAO molecular orbital*.

Equation 9.6 for ψ_+ is sketched in Figure 9.3 for the case $c_1 = c_2$. Note that ψ_+ has the property you might expect of a molecular orbital in that it does indeed spread over both nuclei. Because the two nuclei in H_2^+ are identical, the weighting or the relative importance of $1s_A$ and $1s_B$ must be the same, and so c_1 must equal c_2. For simplicity, we will set $c_1 = c_2 = 1$, but note that before we can discuss a probability density associated with these molecular orbitals, $\psi_{\pm}$ must be normalized.

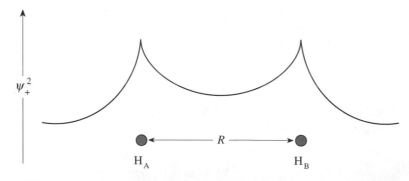

FIGURE 9.3
A sketch of the H_2^+ molecular orbital formed by a sum of hydrogen atomic $1s$ orbitals situated on each nucleus. Note that the molecular orbital spreads over both nuclei, or over the entire molecule.

9–3. The Overlap Integral Is a Quantitative Measure of the Overlap of Atomic Orbitals Situated on Different Atoms

Let's use ψ_+ given in Equation 9.6 to calculate the energy of H_2^+ as a function of the internuclear separation, R. (Problem 9–5 has you calculate the energy associated with ψ_-.) To determine E_+, the energy associated with ψ_+, we start with Equation 9.5:

$$\hat{H}\psi_+(\mathbf{r}; R) = E_+\psi_+(\mathbf{r}; R)$$

Multiplying on the left by $\psi_+^*(\mathbf{r}; R)$ and then integrating over the allowed values of $\mathbf{r}$ gives the full expression for E_+:

$$E_+ = \frac{\int d\mathbf{r}\psi_+^*\hat{H}\psi_+}{\int d\mathbf{r}\psi_+^*\psi_+} \tag{9.7}$$

Let's look at the denominator of Equation 9.7 first:

$$\int d\mathbf{r}\psi_+^*\psi_+ = \int d\mathbf{r}(1s_A^* + 1s_B^*)(1s_A + 1s_B)$$

$$= \int d\mathbf{r}1s_A^*1s_A + \int d\mathbf{r}1s_A^*1s_B + \int d\mathbf{r}1s_B^*1s_A + \int d\mathbf{r}1s_B^*1s_B \tag{9.8}$$

There are two types of integrals in Equation 9.8. The first and fourth integrals are simply the normalization expressions of the hydrogen atomic orbitals, so we have

$$\int d\mathbf{r}1s_A^*1s_A = \int d\mathbf{r}1s_B^*1s_B = 1 \tag{9.9}$$

The second and third integrals are another story, however. Because the hydrogen atomic $1s$ orbital is expressed by a real function, $1s^* = 1s$ and therefore, the two integrals are equal to each other (we denote them by S):

$$S = \int d\mathbf{r}1s_A^*1s_B = \int d\mathbf{r}1s_B^*1s_A = \int d\mathbf{r}1s_A1s_B \tag{9.10}$$

 Note that these integrals involve the product of a hydrogen atomic orbital situated on nucleus A and one situated on nucleus B. This product is significant only for regions where the two atomic orbitals have a large overlap (Figure 9.4). Consequently, S in Equation 9.10 is called an *overlap integral*. The extent of overlap and therefore the magnitude of the overlap integral depends upon the internuclear separation, R. For large

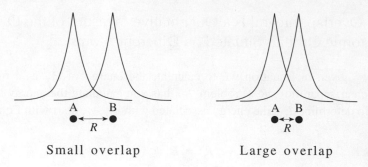

FIGURE 9.4
The overlap of the $1s$ orbitals centered on hydrogen nuclei located at A and B, a distance R apart.

internuclear separations, S is very close to zero. With decreasing internuclear distance, the value of S increases, approaching a value of one when $R = 0$ (Figure 9.5).

The integrals in Equation 9.10 are somewhat complicated to evaluate, but they can be evaluated analytically (Problems 9–3 and 9–42). The resulting function $S(R)$ for the overlap of two hydrogen $1s$ atomic orbitals is given by

$$S(R) = e^{-R}\left(1 + R + \frac{R^2}{3}\right) \tag{9.11}$$

and is plotted in Figure 9.5. Thus, we can write the denominator of Equation 9.7 as

$$\int d\mathbf{r}(1s_A^* + 1s_B^*)(1s_A + 1s_B) = 2 + 2S(R) \tag{9.12}$$

where we write $S(R)$ to emphasize that the function S depends upon the parameter R.

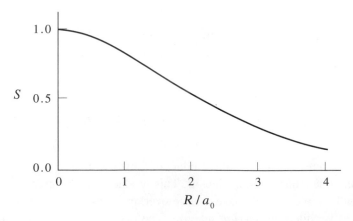

FIGURE 9.5
The overlap integral $S(R)$, Equation 9.11, for two hydrogen atom $1s$ orbitals plotted versus the internuclear separation in atomic units.

EXAMPLE 9–1

Determine the normalized wave function for ψ_+?

SOLUTION: The wave function for ψ_+ is given by $\psi_+ = c(1s_A + 1s_B)$. To normalize this function, we require that $\int d\mathbf{r}\psi_+^*\psi_+ = 1$. Thus

$$1 = c^2 \int d\mathbf{r}(1s_A^* + 1s_B^*)(1s_A + 1s_B)$$

$$= c^2 \left(\int d\mathbf{r}1s_A^*1s_A + \int d\mathbf{r}1s_A^*1s_B + \int d\mathbf{r}1s_B^*1s_A + \int d\mathbf{r}1s_B^*1s_B \right)$$

$$= c^2(1 + S + S + 1)$$

$$= c^2 2(1 + S)$$

whereby

$$c = \frac{1}{\sqrt{2(1+S)}}$$

and the normalized wave function is given by

$$\psi_+ = \frac{1}{\sqrt{2(1+S)}}(1s_A + 1s_B)$$

Using the same approach, you can show that the normalized wave function for ψ_- is given by

$$\psi_- = \frac{1}{\sqrt{2(1-S)}}(1s_A - 1s_B)$$

9–4. The Stability of a Chemical Bond Is a Quantum-Mechanical Effect

So far, we have evaluated the denominator of Equation 9.7. The evaluation of the numerator is more complicated. Using Equation 9.4 for the Hamiltonian operator, we obtain

$$\int d\mathbf{r}\psi_+^*\hat{H}\psi_+ = \int d\mathbf{r}(1s_A^* + 1s_B^*)\hat{H}(1s_A + 1s_B)$$

$$= \int d\mathbf{r}(1s_A^* + 1s_B^*)\left(-\frac{1}{2}\nabla^2 - \frac{1}{r_A} - \frac{1}{r_B} + \frac{1}{R}\right)(1s_A + 1s_B) \quad (9.13)$$

To see how this expression can be simplified, we expand the right side to give

$$\int d\mathbf{r}\psi_+^* \hat{H}\psi_+ = \int d\mathbf{r}(1s_A^* + 1s_B^*) \left(-\frac{1}{2}\nabla^2 - \frac{1}{r_A} - \frac{1}{r_B} + \frac{1}{R}\right) 1s_A$$

$$+ \int d\mathbf{r}(1s_A^* + 1s_B^*) \left(-\frac{1}{2}\nabla^2 - \frac{1}{r_A} - \frac{1}{r_B} + \frac{1}{R}\right) 1s_B \qquad (9.14)$$

We know that the $1s_A$ and $1s_B$ wave functions are solutions to the one-electron atomic Schrödinger equation for atom A and B, respectively, and so we have

$$\left(-\frac{1}{2}\nabla^2 - \frac{1}{r_A}\right) 1s_A = E_{1s} 1s_A \qquad (9.15)$$

and

$$\left(-\frac{1}{2}\nabla^2 - \frac{1}{r_B}\right) 1s_B = E_{1s} 1s_B \qquad (9.16)$$

where $E_{1s} = -\frac{1}{2}E_h$. Using Equation 9.15 in the first integral in Equation 9.14 and Equation 9.16 in the second gives

$$\int d\mathbf{r}\psi_+^* \hat{H}\psi_+ = \int d\mathbf{r}(1s_A^* + 1s_B^*) \left(E_{1s} - \frac{1}{r_B} + \frac{1}{R}\right) 1s_A$$

$$+ \int d\mathbf{r}(1s_A^* + 1s_B^*) \left(E_{1s} - \frac{1}{r_A} + \frac{1}{R}\right) 1s_B \qquad (9.17)$$

Now using Equations 9.9 and 9.10, we have

$$\int d\mathbf{r}\psi_+^* \hat{H}\psi_+ = 2E_{1s}(1 + S) + \int d\mathbf{r}1s_A^* \left(-\frac{1}{r_B} + \frac{1}{R}\right) 1s_A$$

$$+ \int d\mathbf{r}1s_B^* \left(-\frac{1}{r_B} + \frac{1}{R}\right) 1s_A + \int d\mathbf{r}1s_A^* \left(-\frac{1}{r_A} + \frac{1}{R}\right) 1s_B$$

$$+ \int d\mathbf{r}1s_B^* \left(-\frac{1}{r_A} + \frac{1}{R}\right) 1s_B \qquad (9.18)$$

Physically, the first integral of the right side of Equation 9.18 reflects (1) the charge density of the electron around nucleus A interacting with nucleus B via the Coulomb potential and (2) the internuclear repulsion. This integral is denoted by J and is called a *Coulomb integral*:

$$J = \int d\mathbf{r}1s_A^* \left(-\frac{1}{r_B} + \frac{1}{R}\right) 1s_A$$

$$= -\int \frac{d\mathbf{r}1s_A^*1s_A}{r_B} + \frac{1}{R} \qquad (9.19)$$

where we have used the fact that R is fixed during the integration over $\mathbf{r}$. The origin of the name follows by noting that the charge in the volume element $d\mathbf{r}$ is $d\mathbf{r}1s_A^*1s_A$, and the Coulomb energy of interaction with a proton separated a distance r_B is $d\mathbf{r}1s_A^*1s_A/r_B$.

If all the volume elements are added (integrated) and we add the Coulomb repulsion energy between the two protons ($1/R$), then we have J. The second integral in Equation 9.18 is denoted by K and is called an *exchange integral*:

$$K = \int d\mathbf{r} 1s_B^* \left(-\frac{1}{r_B} + \frac{1}{R} \right) 1s_A$$

$$= -\int \frac{d\mathbf{r} 1s_B^* 1s_A}{r_B} + \frac{S}{R} \tag{9.20}$$

where we have used the definition of S and the fact that R is constant during integration. We are lead naturally to seek a classical interpretation of Equation 9.20. It is, however, a purely quantum-mechanical effect and has no analogy in classical mechanics. Equation 9.20 is a direct result of the approximation that the molecular orbital is a linear combination of atomic orbitals centered on different atoms (Equation 9.6).

The final two integrals in Equation 9.18 are the same as the first two except that the indices of the two atoms, A and B, are swapped for each term. Because the atoms in this case are the same (both hydrogen), the mathematical expressions are identical, and so we have

$$\int d\mathbf{r} \psi_+^* \hat{H} \psi_+ = 2E_{1s}(1 + S) + 2J + 2K \tag{9.21}$$

Combining Equations 9.21 and 9.12, the energy (Equation 9.7) associated with the molecular-orbital wave function ψ_+ is $E_+ = E_{1s} + (J + K)/(1 + S)$ or

$$\Delta E_+ = E_+ - E_{1s} = \frac{J + K}{1 + S} \tag{9.22}$$

The quantity $\Delta E_+ = E_+ - E_{1s}$ represents the energy of H_2^+ relative to the completely dissociated species (i.e., H^+ and H). The integrals, J and K, can be evaluated analytically (Problem 9–6) and the results are

$$J = e^{-2R} \left(1 + \frac{1}{R} \right) \tag{9.23}$$

and

$$K = \frac{S}{R} - e^{-R}(1 + R) \tag{9.24}$$

Figure 9.6 shows a plot of the energy $\Delta E_+ = E_+ - E_{1s}$ of H_2^+ as a function of the internuclear separation, R. The plot of ΔE_+ against R describes a stable molecular species whose binding energy (the value of ΔE_+ at R_e) is $E_{binding} = 0.0648 E_h = 170$ kJ·mol^{-1} and whose bond length is $R_e = 2.50 a_0 = 132$ pm. The experimental values for these quantities are $E_{binding} = 0.102 E_h = 268$ kJ·mol^{-1} and $R_e = 2.00 a_0 = 106$ pm.

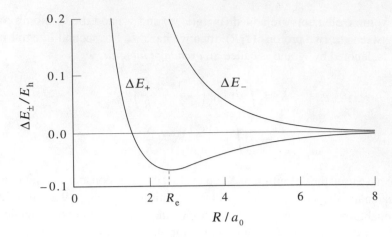

FIGURE 9.6
The energies $\Delta E_+ = E_+ - E_{1s}$ and $\Delta E_- = E_- - E_{1s}$ corresponding to the ψ_+ and ψ_- molecular orbital wave functions given in Equation 9.6 (with $c_1 = c_2$) plotted as a function of intermolecular separation R for H_2^+. The plot shows that ψ_+ leads to a bonding molecular orbital whereas ψ_- leads to an antibonding molecular orbital.

According to Equation 9.22, ΔE_+ is made up of two terms

$$\Delta E_+ = \frac{J}{1+S} + \frac{K}{1+S}$$

Figure 9.7 shows a plot of these two terms separately. Note that the Coulomb integral term is always positive (see also Equation 9.23), and therefore the exchange integral is entirely responsible for the existence of the chemical bond in H_2^+. Because the exchange

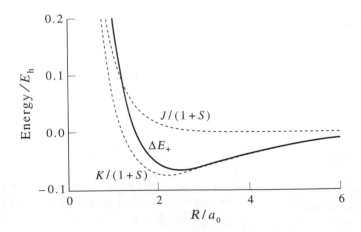

FIGURE 9.7
The separate contributions of the Coulomb integral, J, and the exchange integral, K, to the stability of H_2^+.

term has no classical analog, this result serves to highlight the quantum-mechanical nature of the chemical bond.

9–5. The Simplest Molecular Orbital Treatment of H_2^+ Yields a Bonding Orbital and an Antibonding Orbital

The two molecular orbitals ψ_+ and ψ_- describe quite different states. The orbital ψ_+ describes a state that exhibits a stable chemical bond and is called a *bonding orbital*. The other possible linear combination of the two $1s$ atomic orbitals is

$$\psi_- = c_1 1s_A - c_2 1s_B \tag{9.25}$$

and Problem 9–10 has you show that this molecular orbital results in an energy given by

$$\Delta E_- = E_- - E_{1s} = \frac{J - K}{1 - S} \tag{9.26}$$

Figure 9.6 also shows a plot of ΔE_- as a function of internuclear separation relative to that of the separated nuclei. The wave function ψ_- leads to a repulsive interaction between the two nuclei for all internuclear distances and is therefore called an *antibonding orbital*.

Figure 9.8 shows plots of the molecular orbitals ψ_+ and ψ_- and their squares. In the case of the bonding molecular orbital, ψ_+, electron density builds up in the region between the nuclei. For the antibonding molecular orbital, ψ_-, however, there is a node at the midpoint between the two nuclei and consequently a lack of electron

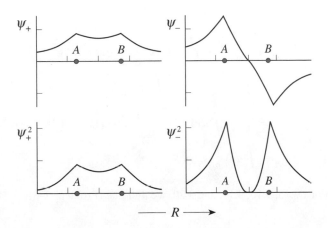

FIGURE 9.8
The molecular orbitals ψ_+ (bonding) and ψ_- (antibonding) and their squares are plotted along the internuclear axis.

density between them. We will indicate whether an orbital is a bonding orbital or an antibonding orbital by using a subscript "b" or an "a," respectively. Therefore, we write

$$\psi_b = \psi_+ = \frac{1}{\sqrt{2(1+S)}}(1s_A + 1s_B) \qquad (9.27)$$

and

$$\psi_a = \psi_- = \frac{1}{\sqrt{2(1-S)}}(1s_A - 1s_B) \qquad (9.28)$$

The bonding orbital describes the ground state of H_2^+ and the antibonding orbital describes an excited state. As the following example shows, the energies associated with the molecular orbital can also be determined through evaluation of the secular determinant.

EXAMPLE 9–2

Use the variational method to calculate the (two) energies associated with the trial function

$$\psi = c_1 1s_A + c_2 1s_B$$

SOLUTION: Equation 7.37 gives the secular determinant for a trial function that is a linear combination of two arbitrary known functions. For the trial function in this example, Equation 7.37 tells us that the secular determinant is given by

$$\begin{vmatrix} H_{AA} - ES_{AA} & H_{AB} - ES_{AB} \\ H_{AB} - ES_{AB} & H_{BB} - ES_{BB} \end{vmatrix} = 0$$

where

$$H_{AA} = H_{BB} = \int d\mathbf{r} 1s_A \hat{H} 1s_A = \int d\mathbf{r} 1s_B \hat{H} 1s_B$$

$$H_{AB} = \int d\mathbf{r} 1s_A \hat{H} 1s_B = \int d\mathbf{r} 1s_B \hat{H} 1s_A$$

$$S_{AA} = S_{BB} = \int d\mathbf{r} 1s_A 1s_A = \int d\mathbf{r} 1s_B 1s_B = 1$$

$$S_{AB} = \int d\mathbf{r} 1s_A 1s_B = S$$

Using Equations 9.15, 9.16, 9.19, and 9.20, we find that (Problem 9–8)

$$H_{AA} = H_{BB} = E_{1s} + J$$

$$H_{AB} = E_{1s} S + K$$

so the secular determinant becomes

$$\begin{vmatrix} E_{1s} + J - E & E_{1s}S + K - ES \\ E_{1s}S + K - ES & E_{1s} + J - E \end{vmatrix} = 0$$

or

$$(E_{1s} + J - E)^2 - (E_{1s}S + K - ES)^2 = 0$$

We can solve this equation to obtain the two values of E,

$$\Delta E_{\pm} = E_{\pm} - E_{1s} = \frac{J \pm K}{1 \pm S}$$

in agreement with our previous results. If we determine the values of c_1 and c_2 associated with each energy, we find they are equal in magnitude, $|c_1| = |c_2| = c$. In addition, we find that

$$\psi_+ = c(1s_A + 1s_B)$$

and

$$\psi_- = c(1s_A - 1s_B)$$

The contributions of the two atomic wave functions turn out to be identical ($c_1^2 = c_2^2$), reflecting that there is no physical reason to expect one of the two hydrogen atoms to dominate the wave function of this homonuclear diatomic molecule.

Note that the simple approach we have used here leads to only two molecular orbitals, one bonding, ψ_+, and one antibonding, ψ_-. We started this endeavor to find a set of molecular orbitals for H_2^+ that we can use to build molecular wave functions in much the same way that we built atomic wave functions for multielectron atoms from products of hydrogenlike atomic orbitals. You may wonder why in the atomic case there is an infinite set of atomic orbitals (e.g., the hydrogenlike atomic orbitals), whereas in the molecular case, the above treatments yielded only two molecular orbitals for H_2^+. We obtained only two molecular orbitals because we used a linear combination of only two atomic orbitals in Equation 9.6. This was done solely for simplicity. We could just as well have used a linear combination such as

$$\psi = c_1 1s_A + c_2 1s_B + c_3 2s_A + c_4 2s_B + c_5 2p_{zA} + c_6 2p_{zB}$$

Because such a trial wave function involves the linear combination of six atomic orbitals, it will lead to a set of six molecular orbitals and six energies. The energies will depend on the value chosen for R. Clearly, this procedure has no limit, and thus, a large set of molecular orbitals can be generated, with a corresponding improvement in our energy and wave-function estimates. For pedagogical reasons, however, we will develop the molecular-orbital treatment of H_2 using just the two normalized wave functions, ψ_+ and ψ_-, given in Example 9–2.

9–6. A Simple Molecular-Orbital Treatment of H_2 Places Both Electrons in a Bonding Orbital

Because ψ_b is the molecular orbital corresponding to the ground-state energy of H_2^+, we can describe the ground state of H_2 by placing two electrons with opposite spins in ψ_b, just as we place two electrons in a $1s$ atomic orbital to describe the helium atom. The Slater determinant corresponding to this assignment is

$$\psi = \frac{1}{\sqrt{2!}} \begin{vmatrix} \psi_b \alpha(1) & \psi_b \beta(1) \\ \psi_b \alpha(2) & \psi_b \beta(2) \end{vmatrix}$$

$$= \psi_b(1)\psi_b(2) \left\{ \frac{1}{\sqrt{2}} [\alpha(1)\beta(2) - \alpha(2)\beta(1)] \right\} \quad (9.29)$$

Once again, we see the spatial and spin parts of the wave function separate for a two-electron system (see Example 8–3). Note also that the two electrons have opposite spins, as expected. Because the Hamiltonian operator is taken to be independent of spin, we can calculate the energy using only the spatial part of Equation 9.29. Using Equation 9.27 for ψ_b, we have a molecular wave function, ψ_{MO}, of

$$\psi_{MO} = \frac{1}{2(1+S)} [1s_A(1) + 1s_B(1)][1s_A(2) + 1s_B(2)] \quad (9.30)$$

Note that ψ_{MO} is a product of *molecular orbitals*, which in turn are linear combinations of atomic orbitals. This method of constructing molecular wave functions is known as the *LCAO-MO (linear combination of atomic orbitals–molecular orbitals)* method and has been successfully extended and applied to a variety of molecules, as we will see in this and the following chapters.

To calculate the ground-state energy of H_2, we use

$$E_{MO} = \int d\mathbf{r}_1 d\mathbf{r}_2 \psi_{MO}^*(1,2) \hat{H} \psi_{MO}(1,2) \quad (9.31)$$

where $\hat{H}$ is given by Equation 9.3 and ψ_{MO} is given by Equation 9.30. The integrals in Equation 9.31 can be evaluated analytically, but they result in complicated functions of R, which we will not display. A plot of $\Delta E_+ = E_{MO} - 2E_{1s}$ versus R is shown in Figure 9.9. We find that $E_{binding} = \Delta E_+$ (at $R = R_e$) $= 0.0990 E_h = 260 \text{ kJ} \cdot \text{mol}^{-1}$, and $R_e = 1.61 a_0 = 85$ pm, compared with experimental values of $E_{binding} = 0.174 E_h = 457 \text{ kJ} \cdot \text{mol}^{-1}$, and $R_e = 1.40 a_0 = 74.1$ pm.

9–7. Molecular Orbitals Can Be Ordered According to Their Energies

In this section, we will construct a set of of molecular orbitals and assign electrons to them in accord with the Pauli Exclusion Principle. This procedure will generate electron configurations for molecules similar to those discussed for atoms in Chapter 8. We will

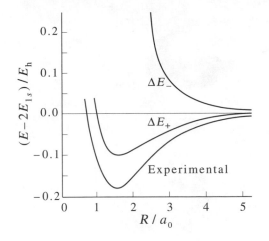

FIGURE 9.9
The ground-state energy of H_2 (relative to two separated ground-state hydrogen atoms) calculated according to molecular-orbital theory, Equation 9.31. The quantity ΔE_+ is the energy difference ($E_{MO} - 2E_{1s}$) corresponding to ψ_{MO} in Equation 9.30. The quantity ΔE_- is the energy difference corresponding to $\psi_{MO} = \psi_a(1)\psi_a(2)$.

illustrate this procedure in some detail for homonuclear diatomic molecules and then present some results for heteronuclear diatomic molecules.

We will use the LCAO-MO approximation and form molecular orbitals from linear combinations of atomic orbitals. In the simplest case, a molecular orbital consists of one atomic orbital centered on each atom. Starting with the $1s$ orbitals on each atom (as done in the treatment of H_2), the first two molecular orbitals we will discuss are

$$\psi_{\pm} = 1s_A \pm 1s_B \tag{9.32}$$

These two molecular orbitals are shown in Figure 9.10, which shows that the resulting molecular orbitals are cylindrically symmetric about the internuclear axis.

An orbital that is symmetric about the internuclear axis is called a σ *orbital*. Both ψ_+ and ψ_- are σ orbitals. Because many combinations of atomic orbitals lead to symmetric distributions about the internuclear axis, we must identify which atomic orbitals constitute a particular σ orbital. Molecular orbitals constructed from atomic $1s$ orbitals are denoted by $\sigma 1s$.

Remember that ψ_+ concentrates electron density in the region between the two nuclei, whereas ψ_- excludes electron density from that region and even has a nodal plane at the midpoint between the two nuclei (Figure 9.10). Consequently, as discussed in Section 9-5, ψ_+ is a bonding orbital and ψ_- is an antibonding orbital. Because a $\sigma 1s$ orbital may be a bonding orbital or an antibonding orbital, we need to distinguish between the two possibilities. There are two common ways to make this distinction. One is to use a superscript asterisk to denote an antibonding orbital, so that the two orbitals in Figure 9.10 are denoted by $\sigma 1s$ (bonding) and $\sigma^* 1s$ (antibonding). The other way is based upon the difference in the symmetry of the two molecular orbitals under an inversion of the orbital through the point midway between the two nuclei. If the orbital does not change its sign under this inversion, we label the wave function *gerade* after the German word for *even*, and we subscript the orbital with a g. Referring to Figure 9.10, we see that $\psi_+ = 1s_A + 1s_B$ does not change its sign under inversion, so we denote ψ_+ by $\sigma_g 1s$. We see from Figure 9.10 that $\psi_- = 1s_A - 1s_B$ changes sign

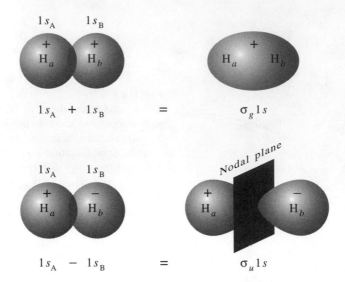

FIGURE 9.10
The linear combination of two 1s orbitals to give the bonding ($\sigma 1s$ or $\sigma_g 1s$) and antibonding molecular orbitals ($\sigma^* 1s$ or $\sigma_u 1s$).

under inversion, so we denote ψ_- by $\sigma_u 1s$, where the u stands for *ungerade*, the German word for *odd*. Thus, we have two designations of the $\sigma 1s$ orbitals: $\sigma 1s$ and $\sigma^* 1s$, or $\sigma_g 1s$ and $\sigma_u 1s$. Both designations are commonly used, but we will primarily use g, u notation for molecular orbitals. Note that in the case of $1s$ orbitals, the gerade symmetry leads to a bonding orbital and the ungerade symmetry leads to an antibonding orbital.

Molecular orbitals constructed from other kinds of atomic orbitals are generated in a similar way. In a first approximation, only atomic orbitals of similar energies are combined to give molecular orbitals (see Problem 9–40). Following the above approach, the next combinations to consider are $2s_A \pm 2s_B$. These two molecular orbitals look similar to those plotted in Figure 9.10 but are larger in extent because a $2s$ orbital is larger than a $1s$ orbital. In addition, there are spherical nodal planes about each nucleus reflecting the radial nodes of the individual $2s$ wave functions (see Figure 6.3). Following the notation introduced above, the two molecular orbitals $2s_A \pm 2s_B$ are designated $\sigma_g 2s$ and $\sigma_u 2s$. Because an atomic $2s$ orbital is associated with a higher energy than an atomic $1s$ orbital, the energy of the $\sigma_g 2s$ molecular orbital will be higher than that of the $\sigma_g 1s$ molecular orbital. This difference can be demonstrated rigorously by calculating the energies associated with these molecular orbitals, as done for the $\sigma_g 1s$ and $\sigma_u 1s$ molecular orbitals in Sections 9–4 and 9–5. In addition, bonding orbitals are lower in energy than corresponding antibonding orbitals. This then gives an energy ordering $\sigma_g 1s < \sigma_u 1s < \sigma_g 2s < \sigma_u 2s$ for the four molecular orbitals discussed so far.

Now consider linear combinations of the $2p$ orbitals. Although a $2p$ orbital has the same energy as a $2s$ orbital in the case of atomic hydrogen, this is not true for other atoms, in which case $E_{2p} > E_{2s}$. As a result, the molecular orbitals built from $2p$ orbitals will have a higher energy than the $\sigma_g 2s$ and $\sigma_u 2s$ orbitals. Defining the internuclear axis to be the z-axis, Figures 9.11 and 9.12 show that the atomic $2p_z$ orbitals combine

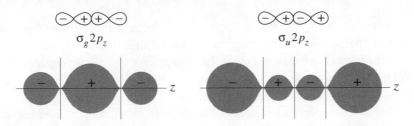

FIGURE 9.11

The $\sigma_g 2p_z$ and $\sigma_u 2p_z$ molecular orbitals formed from linear combinations of the $2p_z$ atomic orbitals. Note that the bonding orbital $(\sigma_g 2p_z)$ corresponds to the combination, $2p_{zA} - 2p_{zB}$, and that the antibonding orbital $(\sigma_u 2p_z)$ corresponds to the combination, $2p_{zA} + 2p_{zB}$, in contrast to the corresponding combinations of s orbitals.

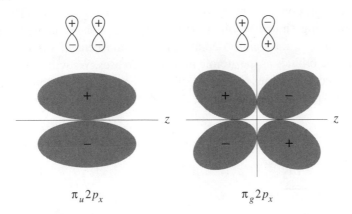

FIGURE 9.12

The bonding $\pi_u 2p_x$ and antibonding $\pi_g 2p_x$ molecular orbitals formed from linear combinations of the $2p_x$ atomic orbitals.

to give a differently shaped molecular orbital than that made by combining either the atomic $2p_x$ or $2p_y$ orbitals. The two molecular orbitals $2p_{zA} \pm 2p_{zB}$ are cylindrically symmetric about the internuclear axis and therefore are σ orbitals. Once again, both a bonding orbital and an antibonding molecular orbital are generated, and the two orbitals are designated by $\sigma_g 2p_z$ and $\sigma_u 2p_z$, respectively.

Unlike the $2p_z$ orbitals, the $2p_x$ and $2p_y$ orbitals combine to give molecular orbitals that are not cylindrically symmetric about the internuclear axis. Figure 9.12 shows that the y-z plane is a nodal plane in both the bonding and antibonding combinations of the $2p_x$ orbitals. Molecular orbitals with one nodal plane that contains the internuclear axis are called π orbitals. The bonding and antibonding molecular orbitals that arise from a combination of the $2p_x$ orbitals are denoted $\pi_u 2p_x$ and $\pi_g 2p_x$, respectively. Note that the antibonding orbital $\pi_g 2p_x$ also has a second nodal plane perpendicular to the internuclear axis that is not present in the $\pi_u 2p_x$ bonding orbital. The $2p_y$ orbitals combine in a similar manner, and the resulting molecular orbitals look like those in Figure 9.12 but are directed along the y-axis instead of the x-axis. The x-z plane is the nodal plane for the $\pi_u 2p_y$ and $\pi_g 2p_y$ orbitals. Because the $2p_x$ and $2p_y$ orbitals have identical energy and the resulting molecular orbitals differ only in their spatial orientation, the pairs of orbitals, $\pi_u 2p_x$, $\pi_u 2p_y$ and $\pi_g 2p_x$, $\pi_g 2p_y$, are degenerate. Note that unlike the bonding σ orbitals, the bonding π orbitals have ungerade symmetry and the antibonding π orbitals have gerade symmetry.

Now that we have developed a set of molecular orbitals by combining atomic $1s$, $2s$, and $2p$ orbitals, we need to know the order of these molecular orbitals with respect to energy. We can then determine the electronic configurations of molecules by placing electrons into these orbitals in accord with the Pauli Exclusion Principle and Hund's rules, just as we did for multielectron atoms in Chapter 8. The order of the various molecular orbitals depends upon the atomic number (nuclear charge) on the nuclei. As the atomic number increases from three for lithium to nine for fluorine, the energies of the $\sigma_g 2p_z$ and $\pi_u 2p_x$, $\pi_u 2p_y$ orbitals approach each other and actually interchange order in going from N_2 to O_2, as shown in Figure 9.13. The somewhat complicated ordering shown in Figure 9.13, which is consistent with calculations and experimental spectroscopic observations, is reminiscent of the ordering of the energies of atomic orbitals as the atomic number increases. We will use Figure 9.13 to deduce electron configurations of the second-row homonuclear diatomic molecules in Section 9–9, but first we will consider H_2 through He_2 in the next section.

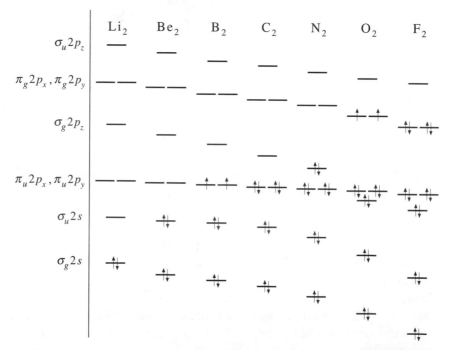

FIGURE 9.13
The relative energies (not to scale) of the molecular orbitals for the homonuclear diatomic molecules Li_2 through F_2. The $\pi_u 2p_x$ and $\pi_u 2p_y$ orbitals are degenerate, as are the $\pi_g 2p_x$ and $\pi_g 2p_y$ orbitals.

9–8. Molecular-Orbital Theory Predicts that a Stable Diatomic Helium Molecule Does Not Exist

For H_2 through He_2, we need to consider only the $\sigma_g 1s$ and $\sigma_u 1s$ orbitals, the two molecular orbitals of lowest energy. Consider the ground-state electron configuration of H_2. According to the Pauli Exclusion Principle, two electrons of opposite spin are placed in the $\sigma_g 1s$ orbital. The electron configuration of H_2 is written as $(\sigma_g 1s)^2$. The two electrons in the bonding orbital constitute a bonding pair of electrons and account for the single bond of H_2.

Now consider He_2. This molecule has four electrons, and its ground-state electron configuration is $(\sigma_g 1s)^2 (\sigma_u 1s)^2$. This assignment gives He_2 one pair of bonding electrons and one pair of antibonding electrons. Electrons in bonding orbitals tend to draw nuclei together, whereas those in antibonding orbitals tend to push them apart. The result of these opposing forces is that an electron in an antibonding orbital approximately cancels the effect of an electron in a bonding orbital. Thus, in the case of He_2, there is no net bonding. Simple molecular-orbital theory predicts that diatomic helium does not exist.

The above results are formalized by defining a quantity called *bond order* by

$$\begin{matrix} \text{bond} \\ \text{order} \end{matrix} = \frac{1}{2}\left[\left(\begin{matrix} \text{number of electrons} \\ \text{in bonding orbitals} \end{matrix}\right) - \left(\begin{matrix} \text{number of electrons} \\ \text{in antibonding orbitals} \end{matrix}\right)\right] \tag{9.33}$$

Single bonds have a bond order of one; double bonds have a bond order of two; and so on. The bond order for He_2 is zero. As the following example shows, the bond order does not have to be a whole number; it can be a half-integer.

EXAMPLE 9–3
Determine the bond order of He_2^+.

SOLUTION: The ground-state electron configuration of He_2^+ is $(\sigma_g 1s)^2 (\sigma_u 1s)^1$, and so the bond order is

$$\text{bond order} = \frac{1}{2}[(2) - (1)] = \frac{1}{2}$$

Table 9.1 gives the molecular-orbital theory results for H_2^+, H_2, He_2^+, and He_2.

TABLE 9.1
Molecular properties of H_2^+, H_2, He_2^+, and He_2.

Species	Number of electrons	Ground-state electron configuration	Bond order	Bond length/pm	Binding energy/ $kJ \cdot mol^{-1}$
H_2^+	1	$(\sigma_g 1s)^1$	1/2	106	268
H_2	2	$(\sigma_g 1s)^2$	1	74	457
He_2^+	3	$(\sigma_g 1s)^2(\sigma_u 1s)^1$	1/2	108	241
He_2	4	$(\sigma_g 1s)^2(\sigma_u 1s)^2$	0	≈ 6000	$\ll 1$

The molecular orbitals we are using here are very simple, being linear combinations of only two atomic orbitals in each case. This simple molecular orbital description predicts that the bond order of He_2 is zero and therefore should not exist. But in 1993, Gentry and his coworkers reported the spectroscopic observation of He_2 in a gas-phase sample of helium that had a temperature near 0.001 K. The He_2 bond, however, is by far the weakest chemical bond known, with $E_{binding} \approx 0.01$ $kJ \cdot mol^{-1}$. A more refined version of molecular-orbital theory predicts the weak bond in He_2.

9–9. Electrons Are Placed into Molecular Orbitals in Accord with the Pauli Exclusion Principle

Consider the homonuclear diatomic molecules Li_2 through Ne_2. Each lithium atom has three electrons, so the ground-state electron configuration for Li_2 is $(\sigma_g 1s)^2(\sigma_u 1s)^2 (\sigma_g 2s)^2$, and the bond order is one. We predict that the diatomic lithium molecule is stable relative to two separated lithium atoms. Lithium vapor is known to contain diatomic lithium molecules, which have a bond length of 267 pm and a bond energy of 105 $kJ \cdot mol^{-1}$.

Contour maps of the electron density in the individual molecular orbitals and the total electron density in Li_2 are shown in Figure 9.14. These contour maps were obtained by solving the Schrödinger equation for Li_2 to high accuracy using a computer. Each line in a contour map corresponds to a fixed value of electron density. Contours are generally plotted for fixed increments of electron density. Thus, the distance between contours provides information about how rapidly the electron density is changing. Figure 9.14 shows clearly that there is little difference between the electron densities of the $\sigma_g 1s$ and $\sigma_u 1s$ molecular orbitals of Li_2 and the electron densities of the two $1s$ atomic orbitals of the individual lithium atoms. This observation underlies the common assumption that only electrons in the valence shell need be included in discussions of chemical bonding. In the case of Li_2, the $1s$ electrons are held tightly about each nucleus and do not participate significantly in the bonding. The ground-state electron configuration of Li_2 can therefore be written as $KK(\sigma_g 2s)^2$, where K represents the filled $n = 1$ shell on a lithium atom.

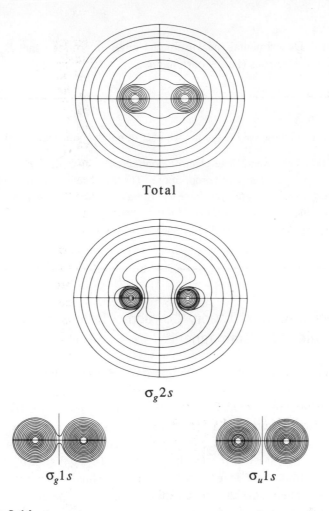

Total

$\sigma_g 2s$

$\sigma_g 1s$ $\qquad$ $\sigma_u 1s$

FIGURE 9.14
Electron density contours for the molecular orbitals of Li_2. Note that the electrons in the $\sigma_g 1s$ and $\sigma_u 1s$ are tightly held around the nucleus and do not participate to any large extent in the bonding. The electrons in the $\sigma_g 2s$ orbital are the ones responsible for the bonding in Li_2.

With increasing nuclear charge across the second row of the periodic table, the $1s$ electrons are held even more tightly than are the $1s$ electrons in lithium. Thus, to a first approximation, only the valence electrons need to be considered in writing electron configurations of diatomic molecules beyond He_2. The $\sigma_g 1s$ and $\sigma_u 1s$ molecular orbitals are equivalent to the filled K shell on each atom.

Diatomic boron is a particularly interesting case. This molecule has a total of six valence electrons (three from each boron atom). According to Figure 9.13, the ground-state electron configuration of B_2 is $KK(\sigma_g 2s)^2(\sigma_u 2s)^2(\pi_u 2p_x)^1(\pi_u 2p_y)^1$. As in the atomic case, Hund's rules apply, so we place one electron into each of the degenerate $\pi_u 2p$ orbitals such that their spins are parallel in order to achieve the greatest possible spin multiplicity. Experimental measurements have determined that B_2 does indeed have two unpaired electrons (i.e., is paramagnetic).

EXAMPLE 9–4

Use molecular-orbital theory to predict whether or not diatomic carbon exists.

SOLUTION: The ground-state electron configuration of C_2 is $KK(\sigma_g 2s)^2(\sigma_u 2s)^2$ $(\pi_u 2p_x)^2(\pi_u 2p_y)^2$, giving a bond order of two. Thus, we predict that diatomic carbon exists. Experimental measurements have determined that C_2 has no unpaired electrons (i.e., is diamagnetic). The correct prediction of the magnetic properties of B_2 and C_2 corroborates the ordering of the molecular orbital energies given in Figure 9.13 for $Z = 5$ and $Z = 6$.

9–10. Molecular-Orbital Theory Correctly Predicts that Oxygen Molecules Are Paramagnetic

The prediction of the correct electron configuration of an oxygen molecule is one of the most impressive successes of molecular-orbital theory. Oxygen molecules are paramagnetic; experimental measurements indicate that the net spin of the oxygen molecule corresponds to two unpaired electrons of the same spin. Let's see what molecular-orbital theory has to say about this. The predicted ground-state electron configuration of O_2 is $KK(\sigma_g 2s)^2(\sigma_u 2s)^2(\sigma_g 2p_z)^2(\pi_u 2p_x)^2(\pi_u 2p_y)^2(\pi_g 2p_x)^1(\pi_g 2p_y)^1$. Because the $\pi_g 2p_x$ and $\pi_g 2p_y$ orbitals are degenerate, according to Hund's rule, we place one electron in each orbital such that the spins of the electrons are parallel. The occupation of the other molecular orbitals, $KK(\sigma_g 2s)^2(\sigma_u 2s)^2(\sigma_g 2p_z)^2(\pi_u 2p_x)^2(\pi_u 2p_y)^2$, generates no net spin because all these occupied molecular orbitals contain two spin-paired electrons, so we predict that O_2 in its ground state has two unpaired electrons. Thus, the molecular-orbital configuration correctly accounts for the paramagnetic behavior of the O_2 molecule.

We can use molecular-orbital theory to predict relative bond lengths and bond energies, as shown in the following example.

EXAMPLE 9–5

Discuss the relative bond lengths and bond energies of O_2^+, O_2, O_2^-, and O_2^{2-}.

SOLUTION: O_2 has 12 valence electrons. According to Figure 9.13, the ground-state electron configurations and bond orders for these species are as follows:

	Ground-state electron configuration	Bond order
O_2^+	$KK(\sigma_g 2s)^2(\sigma_u 2s)^2(\sigma_g 2p_z)^2(\pi_u 2p_x)^2(\pi_u 2p_y)^2(\pi_g 2p_x)^1$	5/2
O_2	$KK(\sigma_g 2s)^2(\sigma_u 2s)^2(\sigma_g 2p_z)^2(\pi_u 2p_x)^2(\pi_u 2p_y)^2(\pi_g 2p_x)^1(\pi_g 2p_y)^1$	2
O_2^-	$KK(\sigma_g 2s)^2(\sigma_u 2s)^2(\sigma_g 2p_z)^2(\pi_u 2p_x)^2(\pi_u 2p_y)^2(\pi_g 2p_x)^2(\pi_g 2p_y)^1$	3/2
O_2^{2-}	$KK(\sigma_g 2s)^2(\sigma_u 2s)^2(\sigma_g 2p_z)^2(\pi_u 2p_x)^2(\pi_u 2p_y)^2(\pi_g 2p_x)^2(\pi_g 2p_y)^2$	1

We predict that the bond lengths decrease and that the bond energies increase with increasing bond order. This prediction is in nice agreement with the experimental values, which are

	Bond length/pm	Bond energy/kJ·mol^{-1}
O_2^+	112	643
O_2	121	494
O_2^-	135	395
O_2^{2-}	149	

Note that removing an electron from O_2 produces a stronger bond, in agreement with the prediction of molecular orbital theory.

Figure 9.15 illustrates the predicted bond orders and the experimentally measured bond lengths and bond energies for the homonuclear diatomic molecules B_2 through F_2. The results for the diatomic molecules of the elements in the second row of the periodic table are summarized in Table 9.2.

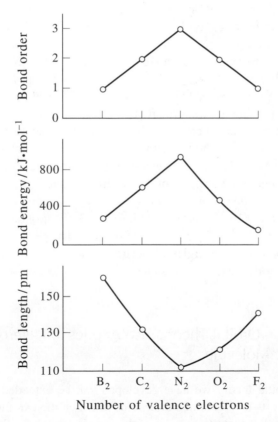

FIGURE 9.15
Plots of various bond properties for the homonuclear diatomic molecules B_2 through F_2.

TABLE 9.2

The ground-state electron configurations and various physical properties of homonuclear diatomic molecules of elements in the second row of the periodic table.

Species	Ground-state electron configuration	Bond order	Bond length/pm	Bond energy/ kJ·mol^{-1}
Li$_2$	$KK(\sigma_g 2s)^2$	1	267	105
Be$_2$	$KK(\sigma_g 2s)^2(\sigma_u 2s)^2$	0	245	≈ 9
B$_2$	$KK(\sigma_g 2s)^2(\sigma_u 2s)^2(\pi_u 2p_x)^1(\pi_u 2p_y)^1$	1	159	289
C$_2$	$KK(\sigma_g 2s)^2(\sigma_u 2s)^2(\pi_u 2p_x)^2(\pi_u 2p_y)^2$	2	124	599
N$_2$	$KK(\sigma_g 2s)^2(\sigma_u 2s)^2(\pi_u 2p_x)^2(\pi_u 2p_y)^2(\sigma_g 2p_z)^2$	3	110	942
O$_2$	$KK(\sigma_g 2s)^2(\sigma_u 2s)^2(\sigma_g 2p_z)^2(\pi_u 2p_x)^2(\pi_u 2p_y)^2(\pi_g 2p_x)^1(\pi_g 2p_y)^1$	2	121	494
F$_2$	$KK(\sigma_g 2s)^2(\sigma_u 2s)^2(\sigma_g 2p_z)^2(\pi_u 2p_x)^2(\pi_u 2p_y)^2(\pi_g 2p_x)^2(\pi_g 2p_y)^2$	1	141	154
Ne$_2$	$KK(\sigma_g 2s)^2(\sigma_u 2s)^2(\sigma_g 2p_z)^2(\pi_u 2p_x)^2(\pi_u 2p_y)^2(\pi_g 2p_x)^2(\pi_g 2p_y)^2(\sigma_u 2p_z)^2$	0	310	< 1

9–11. Photoelectron Spectra Support the Existence of Molecular Orbitals

The idea of atomic orbitals and molecular orbitals is rather abstract and sometimes appears far removed from reality. It so happens, however, that the electron configurations of molecules can be demonstrated experimentally. The approach used is very similar to the photoelectric effect discussed in Chapter 1. If high energy electromagnetic radiation is directed into a gas, electrons are ejected from the molecules in the gas. The energy required to eject an electron from a molecule, called the *ionization energy*, is a direct measure of how strongly bound the electron is within the molecule. The ionization energy of an electron within a molecule depends upon the molecular orbital the electron occupies; the lower the energy of the molecular orbital, the more energy needed to remove, or ionize, an electron from that molecular orbital.

The measurement of the energies of the electrons ejected by radiation incident on gaseous molecules is called *photoelectron spectroscopy*. A photoelectron spectrum of N$_2$ is shown in Figure 9.16. According to Figure 9.13, the ground-state configuration of N$_2$ is $KK(\sigma_g 2s)^2(\sigma_u 2s)^2(\pi_u 2p_x)^2(\pi_u 2p_y)^2(\sigma_g 2p_z)^2$. The peaks in the photoelectron spectrum correspond to the energies of occupied molecular orbitals. Photoelectron spectra provide striking experimental support for the molecular-orbital picture being developed here.

9–12. Molecular–Orbital Theory Also Applies to Heteronuclear Diatomic Molecules

The molecular-orbital theory we have developed can be extended to *heteronuclear diatomic molecules*, that is, diatomic molecules in which the two nuclei are different. It is important to realize that the energies of the atomic orbitals on the two atoms from

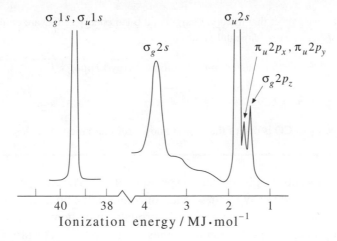

FIGURE 9.16
The photoelectron spectrum of N_2. The peaks in this plot are caused by electrons being ejected from various molecular orbitals.

which the molecular orbitals are constructed will now be different. This difference must be considered in light of the approximation made earlier that only orbitals of similar energy combine to give molecular orbitals. For small changes in atomic number, the energy difference for the same atomic orbital on the two bonded atoms is small (e.g., CO, NO). For many heteronuclear diatomic molecules (e.g., HF, HCl), however, the energies of the respective atomic orbitals can be significantly different, and we will need to rethink which atomic orbitals are involved in constructing the molecular orbitals for such molecules.

Consider the cyanide ion, CN^-. The atomic numbers of carbon (6) and nitrogen (7) differ by only one unit, so the same energy ordering may still be valid. The total number of valence electrons is 10 (carbon has 4 electrons and nitrogen has 5 electrons in the $n = 2$ shell), and the overall charge on the ion is -1. Accordingly, the ground-state electron configuration of CN^- is predicted to be

$$K K (\sigma 2s)^2 (\sigma^* 2s)^2 (\pi 2p_x)^2 (\pi 2p_y)^2 (\sigma 2p_z)^2,$$

with a bond order of three. Note that we do not use the subscripts g and u here because heteronuclear molecules do not possess inversion symmetry.

EXAMPLE 9–6
Discuss the bonding in the carbon monoxide molecule, CO.

SOLUTION: The CO molecule has a total of 10 valence electrons. Note that CO is isoelectronic with N_2. The ground-state electron configuration of CO is therefore $K K (\sigma 2s)^2 (\sigma^* 2s)^2 (\pi 2p_x)^2 (\pi 2p_y)^2 (\sigma 2p_z)^2$, so the bond order is three. Because both N_2 and CO have triple bonds and because all three atoms (N, O, C) are approximately the same size, we expect that the bond length and bond energy of CO are comparable with those of N_2. The experimental values are

	Bond length/pm	Bond energy/kJ·mol^{-1}
N_2	110	942
CO	113	1071

The bond energy of CO is one of the largest known for diatomic molecules.

Figure 9.17 presents the photoelectron spectrum of CO. The energies of the molecular orbitals are revealed nicely by these data. In addition, the photoelectron spectrum exhibits peaks characteristic of the atomic $1s$ orbitals on carbon and oxygen. Notice the high binding energy of the $1s$ atomic orbitals. This energy is a result of their being close to the nuclei and these data further verify that the $1s$ electrons do not play a significant role in the bonding of these molecules.

Now consider the diatomic molecule HF. This molecule illustrates the case in which the valence electrons on the atoms occupy different electron shells. The energies of the valence electrons in the $2s$ and $2p$ atomic orbitals on fluorine are $-1.477\,E_h$ and $-0.684\,E_h$, respectively. The energy of the valence electron in the $1s$ atomic orbital on hydrogen is $-0.500\,E_h$. Because the $2p$ atomic orbitals on fluorine are the closest in energy to the $1s$ orbital on hydrogen, a first approximation to the molecular orbital would be to consider linear combinations of these two different types of atomic orbitals. But which $2p$ atomic orbital should be used? Defining the z-axis as the internuclear axis, Figure 9.18 shows the overlap of the fluorine $2p_z$ and $2p_x$ orbitals with the

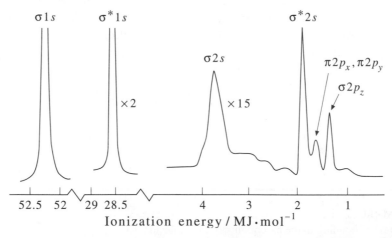

FIGURE 9.17
The photoelectron spectrum of CO. The energies associated with various molecular orbitals are identified. The $\sigma 1s$ and σ^*1s orbitals are essentially the $1s$ electrons of the oxygen and carbon atoms, respectively. The relatively large ionization energies of these electrons indicate that they are held tightly by the nuclei and play no role in bonding.

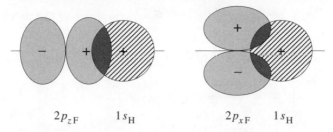

$$2p_{zF} \qquad 1s_H \qquad\qquad 2p_{xF} \qquad 1s_H$$

FIGURE 9.18
The overlap of the fluorine $2p_z$ and $2p_x$ atomic orbitals with the hydrogen $1s$ atomic orbital. Because of the change in sign of the $2p_x$ wave function, the net overlap between the $2p_x$ and hydrogen $1s$ is zero for all internuclear distance. A set of two σ molecular orbitals result from the overlap of the fluorine $2p_z$ atomic orbital and the hydrogen $1s$ atomic orbital. The bonding σ orbital, σ_b, is the one shown.

hydrogen $1s$ orbital. The fluorine $2p_y$ atomic orbital overlaps the hydrogen $1s$ atomic orbital in a similar manner as the $2p_x$ orbital except that it is directed along the y-axis instead of the x-axis. The hydrogen $1s$ and fluorine $2p_z$ orbitals overlap constructively, so we can use linear combinations of these two orbitals. However, because of the change in sign of the wave function for the $2p_x$ ($2p_y$) atomic orbital with respect to the y-z plane (x-z plane) and the constant sign of the hydrogen $1s$ atomic orbital, the net overlap between the $2p_x$ ($2p_y$) on fluorine and the $1s$ on hydrogen is zero for all internuclear distances. Thus, a first approximation to the molecular orbital would be the linear combinations of the fluorine $2p_z$ and hydrogen $1s$ atomic orbitals:

$$\psi = c_1 1s_H \pm c_2 2p_{zF} \tag{9.34}$$

The molecular orbitals given by the wave function in Equation 9.34 describe electron densities that are symmetric about the internuclear axis, so both are σ molecular orbitals (one bonding, σ_b, and one antibonding, σ_a). Figure 9.19 shows the molecular-orbital energy-level diagram for HF. (The $1s_F$ and $2s_F$ orbitals are not shown.) The eight valence electrons occupy the four lowest energy orbitals in Figure 9.19 in accord with the Pauli Exclusion Principle, so the ground-state valence electron configuration of HF is $(2s_F)^2(\sigma_b)^2(2p_{xF})^2(2p_{yF})^2$. The $2s_F$, $2p_{xF}$ and $2p_{yF}$ orbitals are non-bonding orbitals, so the bond order of HF is one.

9–13. An SCF–LCAO–MO Wave Function Is a Molecular Orbital Formed from a Linear Combination of Atomic Orbitals and Whose Coefficients Are Determined Self-Consistently

The molecular-orbital scheme we have presented thus far is the simplest possible molecular-orbital treatment. Each of the molecular orbitals in Figures 9.10 through 9.12 is formed from just one atomic orbital on each nucleus. In analogy with the atomic case, we can obtain better molecular orbitals by forming linear combinations

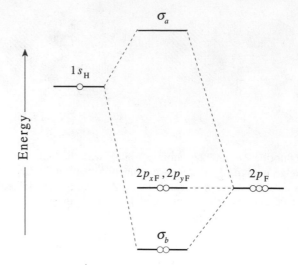

FIGURE 9.19
A molecular-orbital energy-level diagram of HF. The fluorine $1s$ and $2s$ orbitals are not shown. Note that the $2p_{xF}$ and $2p_{yF}$ orbitals are nonbonding orbitals.

of many atomic orbitals. For example, instead of using simply $\psi = c_1(1s_A + 1s_B)$, we can use a molecular-orbital trial function of the form

$$\psi = c_1(1s_A + 1s_B) + c_2(2s_A + 2s_B) + c_3(2p_{zA} + 2p_{zB}) + \cdots \qquad (9.35)$$

Note that we are including orbitals of higher energy and are achieving a more flexible trial function. We let the variational principle determine the relative importance of the various terms by yielding the relative magnitudes of the variational parameters c_1, c_2, c_3, ... Just as for atomic calculations, as we include more and more terms in Equation 9.35, we reach a limit in the calculation of the ground-state energy, and this limit is the Hartree-Fock limit. The procedure discussed in Chapter 8 for a Hartree-Fock self-consistent field calculation of atomic properties can be modified to calculate molecular properties. If we start with molecular orbitals that are linear combinations of atomic orbitals as in Equation 9.35 and determine the coefficients by a self-consistent field method, we obtain what is called an *SCF-LCAO-MO wave function*. This method was developed at the University of Chicago by Clemens Roothaan in the 1950s and is often called the Hartree-Fock-Roothaan method. Because these calculations are done using linear combinations of many atomic orbitals, molecular-orbital designations such as $\sigma 2s$ and $\pi 2p$ lose their significance, and molecular orbitals are more appropriately designated as the first σ_g orbital ($1\sigma_g$), the first σ_u orbital ($1\sigma_u$), the first π_u orbital ($1\pi_u$), and so on. The correspondence between the various notations for diatomic molecular orbitals is given in Table 9.3. Because molecular orbitals are determined computationally from linear combinations of a large set of atomic orbitals, the SCF-LCAO notation is used in the research literature.

TABLE 9.3
The correspondence between the various notations for diatomic molecular orbitals.

Simple LCAO-MO		SCF-LCAO-MO
$\sigma 1s$	$\sigma_g 1s$	$1\sigma_g$
$\sigma^* 1s$	$\sigma_u 1s$	$1\sigma_u$
$\sigma 2s$	$\sigma_g 2s$	$2\sigma_g$
$\sigma^* 2s$	$\sigma_u 2s$	$2\sigma_u$
$\pi 2p_x$	$\pi_u 2p_x$	$1\pi_u$
$\pi 2p_y$	$\pi_u 2p_y$	$1\pi_u$
$\sigma 2p_z$	$\sigma_g 2p_z$	$3\sigma_g$
$\pi^* 2p_x$	$\pi_g 2p_x$	$1\pi_g$
$\pi^* 2p_y$	$\pi_g 2p_y$	$1\pi_g$
$\sigma^* 2p_z$	$\sigma_u 2p_z$	$3\sigma_u$

We must realize that if we use only a few terms in the LCAO-MO and determine all the coefficients self-consistently, we may not achieve or reach the Hartree-Fock limit. Thus, an SCF-LCAO-MO molecular orbital is not necessarily the same as a Hartree-Fock orbital. They are the same only if the SCF-LCAO-MO molecular orbitals contain enough terms that the Hartree-Fock limit is reached.

We can use Equation 9.35 to illustrate the difference between an SCF-LCAO-MO calculation and a Hartree-Fock calculation. If $c_2 = c_3 = 0$ in Equation 9.35, and only c_1 is varied, we have the result we obtained in Section 9–4. The energy and bond length are given in the first row in Table 9.4. If we allow the nuclear charge Z in the exponent of the $1s$ orbitals to vary, then we obtain the second entry in Table 9.4. Now consider the LCAO-MO given by Equation 9.35 in which the atomic orbitals are Slater orbitals (Equation 8.12):

$$\psi_{nlm}(r, \theta, \phi) = \frac{(2\zeta)^{n+1/2}}{(2n!)^{1/2}} r^{n-1} e^{-\zeta r} Y_l^m(\theta, \phi)$$

If c_1, c_2, and c_3 are taken to be variational parameters and ζ is taken to be Z/n as in a hydrogen atomic orbital, the binding energy is $0.1321 E_h$ and the bond length is $1.40 a_0$ (the third entry in Table 9.4). If, in addition, the values of Z in the $1s$, $2s$, and $2p$ orbitals are varied independently, then the binding energy is $0.1335 E_h$ and the bond length is $1.40 a_0$ (the fourth entry in Table 9.4). This is still not the Hartree-Fock limit. If more terms (about nine terms) are included in Equation 9.35, then eventually the Hartree-Fock limit of a dissociation energy of $0.1336 E_h$ and a bond length of $1.40 a_0$ is reached. Hartree-Fock wave functions have been calculated for many diatomic molecules. Figure 9.20 shows the contour diagrams of the total electron density and for the individual molecular orbitals for the homonuclear diatomic molecules H_2 through F_2. The ground-state configurations of H_2 through F_2 are also given in Figure 9.20.

352

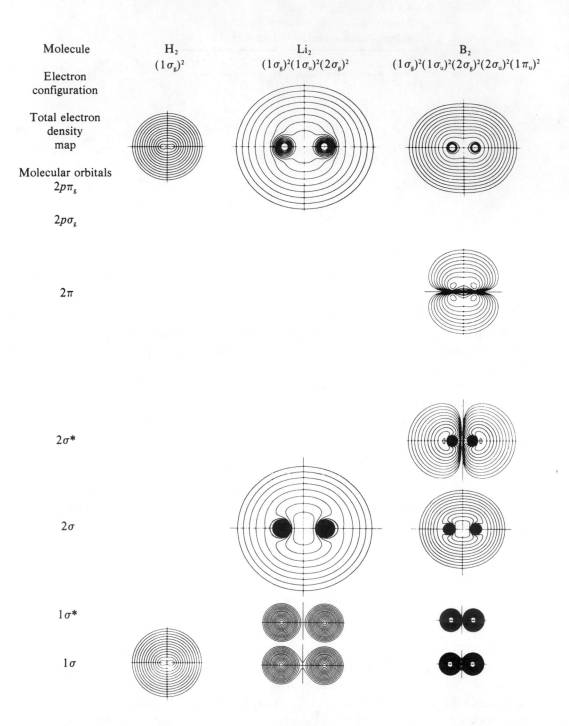

FIGURE 9.20
Contour maps of the various molecular orbitals and the total electron density of the homonuclear diatomic molecules H_2 through F_2.

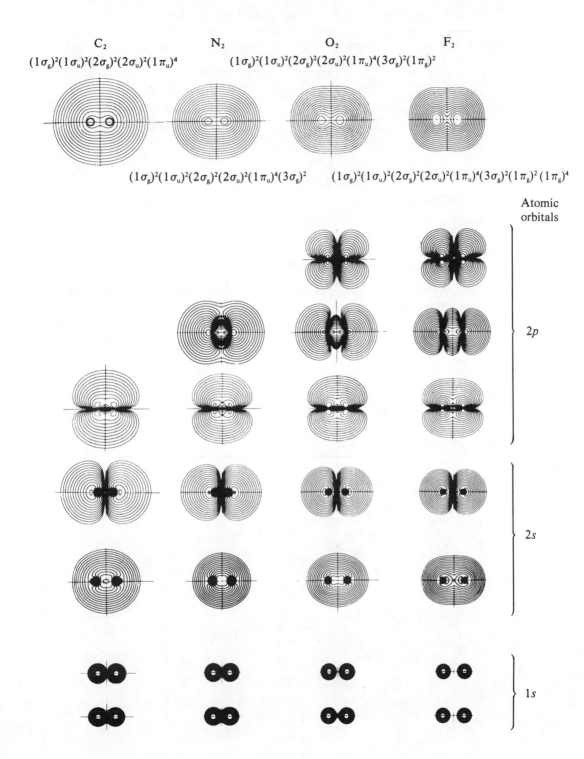

C_2

$(1\sigma_g)^2(1\sigma_u)^2(2\sigma_g)^2(2\sigma_u)^2(1\pi_u)^4$

N_2

$(1\sigma_g)^2(1\sigma_u)^2(2\sigma_g)^2(2\sigma_u)^2(1\pi_u)^4(3\sigma_g)^2$

O_2

$(1\sigma_g)^2(1\sigma_u)^2(2\sigma_g)^2(2\sigma_u)^2(1\pi_u)^4(3\sigma_g)^2(1\pi_g)^2$

F_2

$(1\sigma_g)^2(1\sigma_u)^2(2\sigma_g)^2(2\sigma_u)^2(1\pi_u)^4(3\sigma_g)^2(1\pi_g)^2(1\pi_g)^4$

Atomic orbitals

$2p$

$2s$

$1s$

TABLE 9.4
A demonstration of the convergence to the Hartree-Fock limit for H_2.

LCAO-MO	Effective nuclear charge	Total energy/E_h	Binding energy/E_h	Bond length/a_0
$1s_A + 1s_B$	1.00	-1.0990	0.0990	1.61
$1s_A + 1s_B$	1.197	-1.1282	0.1282	1.38
Equation 9.35	1.231	-1.1321	0.1321	1.40
Equation 9.35	$Z(1s) = 1.378$	-1.1335	0.1335	1.40
	$Z(2s) = 1.176$			
	$Z(2p) = 1.820$			
Hartree-Fock		-1.1336	0.1336	1.40
Experimental			0.1642	1.41

Table 9.5 shows some results of calculated bond lengths for various diatomic molecules. The agreement with experimental values is impressive. We will discuss the details of Hartree-Fock calculations of polyatomic molecules in Chapter 10, where we will see that similar impressive results can be achieved for polyatomic molecules.

TABLE 9.5
Calculated and experimental bond lengths for various diatomic molecules. The Hartree-Fock calculations used the 6-31G* basis set (to be discussed in Chapter 11) to represent the atomic orbitals.

Molecule	Calculated bond length/pm	Experimental bond length/pm
H_2	73.0	74.2
LiH	163.6	159.6
HF	91.1	91.7
NaH	191.4	188.7
HCl	126.6	127.5
LiCl	207.2	202.1
CO	111.4	112.8
N_2	107.8	107.9
ClF	161.3	162.8
Li_2	281.2	267.3
Na_2	313.0	307.8
NaCl	239.7	236.1
Cl_2	199.0	198.8
F_2	134.5	141.2

9–14. Electronic States of Molecules Are Designated by Molecular Term Symbols

In Section 8–8, the electronic states of atoms were designated by atomic term symbols. The electronic states of molecules are also designated by term symbols. Molecular term symbols happen to be easier to deduce than atomic term symbols. To determine molecular term symbols, we first calculate the possible values for the total orbital angular momentum, M_L, which is the sum of the orbital angular momenta of the electrons occupying the molecular orbitals:

$$M_L = m_{l_1} + m_{l_2} + \cdots \qquad (9.36)$$

where $m_l = 0$ for a σ orbital, $m_l = \pm 1$ for a π orbital, and so on. (Recall that m_l is the projection in units of $\hbar$ of the orbital angular momentum on the z-axis.) Different electron configurations give rise to different allowed values of M_L. The various values of $|M_L|$ are associated with capital Greek letters according to

$|M_L|$	Letter
0	Σ
1	Π
2	Δ
3	Φ

Once M_L has been determined, we then determine the possible values for the total spin angular momentum, M_S:

$$M_S = m_{s_1} + m_{s_2} + \cdots \qquad (9.37)$$

The values of M_S correspond to the projections of the total spin, S, of the molecule along the bond axis. For $S = 0$, $M_S = 0$; for $S = 1/2$, $M_S = \pm 1/2$; for $S = 1$, $M_S = \pm 1$, 0, and so on. Hence, as for atoms, the total spin S can be determined from the obtained values of M_S. For a particular set of M_L and S, the molecular term symbol is then represented by

$$^{2S+1}|M_L|$$

The superscript $2S + 1$ is the spin multiplicity and indicates the number of values of M_S for a particular value of S. Recall that the state is called a singlet if $2S + 1 = 1$, a doublet if $2S + 1 = 2$, a triplet if $2S + 1 = 3$, and so on. The determination of molecular term symbols from molecular-orbital electron configurations is best illustrated by example.

Consider the H_2 molecule first. The ground-state electron configuration of H_2 is $(1\sigma_g)^2$, so $m_l = 0$ for each electron in the occupied σ orbitals. Therefore,

$$M_L = 0 + 0 = 0$$

The spins of the two electrons must be paired to satisfy the Pauli Exclusion Principle, so

$$M_S = +\frac{1}{2} - \frac{1}{2} = 0$$

Because M_S equals 0, S must equal zero. Therefore, the term symbol for the ground-state electron configuration of H_2 is $^1\Sigma$ (a singlet sigma state).

EXAMPLE 9–7
Determine the term symbols for He_2^+ and He_2?

SOLUTION: He_2^+: The ground-state electron configuration is $(1\sigma_g)^2(1\sigma_u)^1$. We need to consider the values of m_l and m_s for all three electrons. The possible values are listed below.

$$
\begin{array}{ll}
m_{l1} = 0 & m_{s1} = +1/2 \\
m_{l2} = 0 & m_{s2} = -1/2 \\
m_{l3} = 0 & m_{s3} = \pm1/2 \\
\hline
M_L = 0 & M_S = \pm1/2
\end{array}
$$

The fact that $M_L = 0$ says that we have a Σ state. The $M_S = \pm 1/2$ corresponds to the two projections of $S = 1/2$, so the term symbol for the ground state of He_2^+ is $^2\Sigma$ (a doublet sigma state).

He_2: The ground-state electron configuration is $(1\sigma_g)^2(1\sigma_u)^2$. In this case, $M_L = 0$ and $M_S = 0$. Therefore, the term symbol for the ground state of He_2 is $^1\Sigma$.

Now consider B_2. This molecule is more complicated and illustrates the general case that needs to be considered. The ground-state electron configuration of B_2 is $(1\sigma_g)^2(1\sigma_u)^2(2\sigma_g)^2(2\sigma_u)^2(1\pi_u)^1(1\pi_u)^1$. Because the first four molecular orbitals of B_2 have $M_L = 0$ and $M_S = 0$, only the two electrons that occupy the $1\pi_u$ orbitals need be considered. The $1\pi_u$ orbital is doubly degenerate and, according to Hund's rules, each of these two electrons occupies its own $1\pi_u$ orbital and thus can have $m_l = \pm1$ and $m_s = \pm1/2$. To determine the term symbol for the molecular electronic state, we use the same approach introduced for determining atomic term symbols. For the electron configuration $(1\pi_u)^1(1\pi_u)^1$, the allowed values for M_L are 2, 0, and -2, and M_S can take on values of 1, 0, and -1. We now construct a table of all possible combinations of (m_{l1}, m_{s1}) and (m_{l2}, m_{s2}) that correspond to the possible values of M_L and M_S.

M_L	M_S		
	1	0	-1
2	~~$1^+, 1^+$~~	$1^+, 1^-$	~~$1^-, 1^-$~~
0	$1^+, -1^+$	$1^+, -1^-$; $1^-, -1^+$	$1^-, -1^-$
-2	~~$-1^+, -1^+$~~	$-1^+, -1^-$	~~$-1^-, -1^-$~~

In the entries of the above table, the superscripts $+$ and $-$ are used to designate spin quantum numbers of $m_s = +1/2$ and $m_s = -1/2$, respectively. The numbers in each entry are the corresponding m_l quantum numbers. For example, the entry $1^+, -1^+$ corresponds to $m_{l1} = 1, m_{s1} = 1/2$ and $m_{l2} = -1, m_{s2} = 1/2$, or $M_L = m_{l1} + m_{l2} = 0$ and $M_S = m_{s1} + m_{s2} = 1$. Not all the entries in the above table are allowed. The Pauli Exclusion Principle requires that no two electrons in the same orbitals have the same set of quantum numbers; hence the configurations $1^+, 1^+$; $1^-, 1^-$; $-1^+, -1^+$, and $-1^-, -1^-$ do not correspond to allowed quantum states and are crossed out. This leaves the following combinations of $(m_{l1},\ m_{s1})$ and $(m_{l2},\ m_{s2})$ from which the allowed term symbols are to be derived.

		M_S	
	1	0	-1
2		$1^+, 1^-$	
M_L 0	$1^+, -1^+$	$1^+, -1^-$; $1^-, -1^+$	$1^-, -1^-$
-2		$-1^+, -1^-$	

Looking across the middle row, we have three configurations $1^+, -1^+$; $1^+, -1^-$ (or $1^-, -1^+$); and $1^-, -1^-$ that correspond to $M_L = 0$ and $M_S = 1, 0, -1$, or a $^3\Sigma$ state. This leaves

		M_S	
	1	0	-1
2		$1^+,\ \ 1^-$	
M_L 0		$1^-, -1^+$	
-2		$-1^+, -1^-$	

Two of the remaining terms in the column ($1^+, 1^-$ and $-1^+, -1^-$) correspond to $M_L = 2$ and -2 ($|M_L| = 2$) and $M_S = 0$, or to a $^1\Delta$ state. The remaining term ($1^-, -1^+$) corresponds to $M_L = 0$ and $M_S = 0$ or to a $^1\Sigma$ state. We find that there are three possible molecular states, $^1\Delta$, $^3\Sigma$, and $^1\Sigma$, for B_2. Because Hund's rules apply to molecular electronic states as well as to atomic electronic states, the state with the largest spin multiplicity will be the ground state of B_2. Thus, we predict that the ground state of B_2 is a $^3\Sigma$ state.

EXAMPLE 9–8
Determine the term symbols for the ground states of O_2 and O_2^+?

SOLUTION: The molecule O_2 has a ground-state electron configuration of (see Example 9–5) $(1\sigma_g)^2(1\sigma_u)^2(2\sigma_g)^2(2\sigma_u)^2(3\sigma_g)^2(1\pi_u)^2(1\pi_u)^2(1\pi_g)^1(1\pi_g)^1$. The only electrons that we need to consider in determining the molecular term symbol are the two that occupy the $1\pi_g$ orbitals. This is identical to what we just discussed for the molecule B_2. Thus, we know that according to Hund's rule, the term symbol for the ground state of O_2 is $^3\Sigma$.

The ground-state electron configuration of O_2^+ is

$$(1\sigma_g)^2(1\sigma_u)^2(2\sigma_g)^2(2\sigma_u)^2(3\sigma_g)^2(1\pi_u)^2(1\pi_u)^2(1\pi_g)^1.$$

The only electron we need to consider in determining the term symbol is the one electron in the $1\pi_g$ orbital. The allowed values of m_l and m_s for an electron in a $1\pi_g$ orbital are $m_l = \pm 1$ and $m_s = \pm 1/2$. These values correspond to $|M_L| = 1$ and $M_S = 1/2$, or a term symbol of $^2\Pi$.

9–15. Molecular Term Symbols Designate the Symmetry Properties of Molecular Wave Functions

Term symbols are also used to denote symmetry properties of a molecular wave function. (We will study the symmetry properties of molecules in detail in Chapter 12.) For homonuclear diatomic molecules, inversion through the point midway between the two nuclei leaves the nuclear configuration of the molecule unchanged. This need not be the case for the molecular wave function, however. Table 9.3 summarizes this symmetry behavior of the molecular orbitals for all the homonuclear diatomic orbitals discussed in this chapter.

Because a molecular electronic wave function is a product of molecular orbitals, the symmetry of the molecular electronic wave function of a homonuclear diatomic molecule must be either gerade (g) or ungerade (u). Consider the simplest case of the product of two molecular orbitals. If both orbitals are gerade, the product is gerade. If both orbitals are ungerade, the product is also gerade because the product of two odd functions is an even function. If the two orbitals have opposite symmetry, the product is ungerade. The resultant symmetry is indicated by either a g or a u right-side subscript on the molecular term symbol. For example, the ground-state electron configuration of O_2 is $(1\sigma_g)^2(1\sigma_u)^2(2\sigma_g)^2(2\sigma_u)^2(3\sigma_g)^2(1\pi_u)^2(1\pi_u)^2(1\pi_g)^1(1\pi_g)^1$. As usual, we can ignore completely filled orbitals and focus on $(1\pi_g)^1(1\pi_g)^1$. According to Table 9.3 (or Figures 9.11 and 9.12), the symmetry of $(1\pi_g)^1(1\pi_g)^1$ is $g \cdot g = g$, so the molecular term symbol for the ground electronic state of O_2 is $^3\Sigma_g$. Similarly, that for O_2^+ is $^2\Pi_g$.

EXAMPLE 9–9
Determine the symmetry designation (g or u) for the term symbol of the ground-state electron configuration of B_2.

SOLUTION: The ground-state electron configuration is

$$(1\sigma_g)^2(1\sigma_u)^2(2\sigma_g)^2(2\sigma_u)^2(1\pi_u)^1(1\pi_u)^1,$$

corresponding to a term symbol of $^3\Sigma$. We can ignore completely occupied orbitals, so the product of the symmetry of the molecular orbitals occupied by the two unpaired electrons is $u \cdot u = g$, so the term symbol is $^3\Sigma_g$.

Finally, Σ electronic states ($M_L = 0$) are labeled with a $+$ or $-$ right-side superscript to indicate the behavior of the molecular wave function when it is reflected through a plane containing the nuclei. Because σ orbitals are symmetric about the internuclear axis, they do not change sign when they are reflected through a plane containing the two nuclei. Figure 9.21 shows that one of the doubly degenerate π_u orbitals changes sign and the other does not. Similarly, one of the doubly degenerate π_g orbitals changes sign and the other one does not (see Figure 9.12). Using these observations, we can determine whether or not a Σ electronic state is labeled with a $+$ or $-$ superscript.

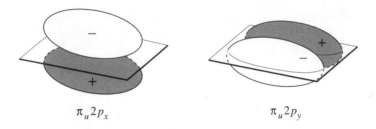

$\pi_u 2p_x$ $\qquad\qquad$ $\pi_u 2p_y$

FIGURE 9.21
The behavior of the two $1\pi_u$ orbitals with respect to a plane containing the two nuclei, which we arbitrarily choose as the y-z plane. (See Figure 9.12.)

EXAMPLE 9–10
Determine the complete molecular term symbol of the ground state of O_2.

SOLUTION: According to Example 9–8, the molecular term symbol of O_2 without the $\pm$ designation is $^3\Sigma_g$. The electron configuration is (filled orbitals)$(1\pi_g 2p_x)^1$ $(1\pi_g 2p_y)^1$, so the symmetry with respect to a reflection through the x-z plane is $(+)(-) = (-)$. Therefore, the complete molecular term symbol of O_2 is $^3\Sigma_g^-$.

EXAMPLE 9–11
Determine the sign designation $(+)$ or $(-)$ for the ground-state electron configuration of He_2^+.

SOLUTION: The ground-state electron configuration of He_2^+ is $(1\sigma_g)^2(1\sigma_u)^1$, corresponding to a term symbol of $^2\Sigma_u$. Because the $1\sigma_g$ and $1\sigma_u$ wave functions are unchanged upon reflection through a plane containing the two nuclei, the total molecular orbital wave function is unchanged. As a result, the complete term symbol of the ground state of He_2^+ is $^2\Sigma_u^+$.

Table 9.6 lists the term symbols of the ground states of a number of homonuclear diatomic molecules and Problem 9–30 involves the determination of these term symbols.

TABLE 9.6
The ground-state electron configurations and term symbols for the first and second-row homonuclear diatomic molecules.

Molecule	Electron configuration	Term symbol
H_2^+	$(1\sigma_g)^1$	$^2\Sigma_g^+$
H_2	$(1\sigma_g)^2$	$^1\Sigma_g^+$
He_2^+	$(1\sigma_g)^2(1\sigma_u)^1$	$^2\Sigma_u^+$
Li_2	$(1\sigma_g)^2(1\sigma_u)^2(2\sigma_g)^2$	$^1\Sigma_g^+$
B_2	$(1\sigma_g)^2(1\sigma_u)^2(2\sigma_g)^2(2\sigma_u)^2(1\pi_u)^1(1\pi_u)^1$	$^3\Sigma_g^-$
C_2	$(1\sigma_g)^2(1\sigma_u)^2(2\sigma_g)^2(2\sigma_u)^2(1\pi_u)^2(1\pi_u)^2$	$^1\Sigma_g^+$
N_2^+	$(1\sigma_g)^2(1\sigma_u)^2(2\sigma_g)^2(2\sigma_u)^2(1\pi_u)^2(1\pi_u)^2(3\sigma_g)^1$	$^2\Sigma_g^+$
N_2	$(1\sigma_g)^2(1\sigma_u)^2(2\sigma_g)^2(2\sigma_u)^2(1\pi_u)^2(1\pi_u)^2(3\sigma_g)^2$	$^1\Sigma_g^+$
O_2^+	$(1\sigma_g)^2(1\sigma_u)^2(2\sigma_g)^2(2\sigma_u)^2(3\sigma_g)^2(1\pi_u)^2(1\pi_u)^2(1\pi_g)^1$	$^2\Pi_g$
O_2	$(1\sigma_g)^2(1\sigma_u)^2(2\sigma_g)^2(2\sigma_u)^2(3\sigma_g)^2(1\pi_u)^2(1\pi_u)^2(1\pi_g)^1(1\pi_g)^1$	$^3\Sigma_g^-$
F_2	$(1\sigma_g)^2(1\sigma_u)^2(2\sigma_g)^2(2\sigma_u)^2(3\sigma_g)^2(1\pi_u)^2(1\pi_u)^2(1\pi_g)^2(1\pi_g)^2$	$^1\Sigma_g^+$

9–16. Most Molecules Have Excited Electronic States

So far we have considered only the ground electronic states of diatomic molecules. In this section we will consider some of the excited electronic states of molecular hydrogen. As we saw in Section 9–8, the electron configuration of the ground electronic state of H_2 is $(1\sigma_g)^2$, whose molecular term symbol is $^1\Sigma_g^+$. The first excited state has the electron configuration $(1\sigma_g)^1(1\sigma_u)^1$, which, as the following example shows, gives rise to the term symbols $^1\Sigma_u^+$ and $^3\Sigma_u^+$.

EXAMPLE 9–12
Show that the electron configuration $(1\sigma_g)^1(1\sigma_u)^1$ gives rise to the term symbols $^1\Sigma_u^+$ and $^3\Sigma_u^+$.

SOLUTION: The values of m_l are 0 for both electrons, so $M_L = 0$. The possible values of m_{s1} and m_{s2} are $m_{s1} = \pm 1/2$ and $m_{s2} = \pm 1/2$, respectively, and so

$M_S = 1, 0, -1$. We now construct a table of all possible combinations of (m_{l1}, m_{s1}) and (m_{l2}, m_{s2}) that correspond to the possible values of M_L and M_S.

		M_S		
		1	0	-1
M_L	0	$0^+, 0^+$	$0^+, 0^-; 0^-, 0^+$	$0^-, 0^-$

Looking across the middle row, we see that the entries $0^+, 0^+$; $0^+, 0^-$ (or $0^-, 0^+$) and $0^-, 0^-$ correspond to $M_L = 0$ and $M_S = 1, 0, -1$, or a $^3\Sigma$ state. The remaining entry $0^-, 0^+$ (or $0^+, 0^-$) corresponds to $M_L = 0$ and $M_S = 0$, or a $^1\Sigma$ state.

The product $1\sigma_g \times 1\sigma_u$ leads to a u state, so we have the states $^3\Sigma_u$ and $^1\Sigma_u$. Furthermore, both the $1\sigma_g$ and $1\sigma_u$ orbitals are symmetric with respect to a reflection through a plane containing the two nuclei, so the complete molecular term symbols are $^3\Sigma_u^+$ and $^1\Sigma_u^+$.

According to Hund's rule, the $^3\Sigma_u^+$ excited state has a lower energy than the $^1\Sigma_u^+$ excited state. Figure 9.22 shows the potential energy curves of the ground state and two of the excited electronic states of H_2.

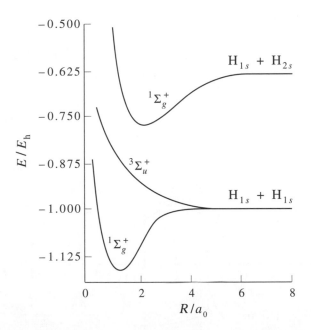

FIGURE 9.22
The internuclear potential energy curves of the ground state and two of the excited electronic states of H_2. Note that the two lowest curves go to $-1.0E_h$ at large distances, indicating two isolated ground-state hydrogen atoms. (The ground state of a hydrogen atom is $-\frac{1}{2}E_h$.) The other excited state shown dissociates into one ground-state hydrogen atom and one excited hydrogen atom with its electron in the atomic $2s$ orbital.

Note that the triplet state corresponding to an electron configuration of $(1\sigma_g)^1(1\sigma_u)^1$ (a $^3\Sigma_u^+$ state) is always repulsive. The second excited state shown in Figure 9.22 corresponds to an electron configuration of $(1\sigma_g)^1(2\sigma_g)^1$, or a term symbol of $^1\Sigma_g^+$. Like the ground state H_2 molecule, this excited state has a bond order of one. Because the $2\sigma_g$ orbital is larger than the $1\sigma_g$ orbital, however, we would predict that the bond length of H_2 is longer in this excited state than in the ground state. Experimental measurements confirm this prediction; the bond length is $\approx 35\%$ longer in this $^1\Sigma_g^+$ excited state than it is in the ground state.

Problems

9-1. Express the Hamiltonian operator for a hydrogen molecule in atomic units.

9-2. Plot the product $1s_A 1s_B$ along the internuclear axis for several values of R.

9-3. The overlap integral, Equation 9.10, and other integrals that arise in two-center systems like H_2 are called *two-center integrals*. Two-center integrals are most easily evaluated by using a coordinate system called *elliptic coordinates*. In this coordinate system (Figure 9.23), there are two fixed points separated by a distance R. A point P is given by the three coordinates

$$\lambda = \frac{r_A + r_B}{R}$$

$$\mu = \frac{r_A - r_B}{R}$$

and the angle ϕ, which is the angle that the $(r_A,\ r_B,\ R)$ triangle makes about the interfocal axis. The differential volume element in elliptic coordinates is

$$d\mathbf{r} = \frac{R^3}{8}(\lambda^2 - \mu^2)d\lambda\,d\mu\,d\phi$$

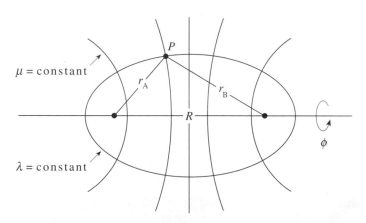

FIGURE 9.23
Elliptic coordinates are the natural coordinates for evaluating two-center integrals such as the overlap integral, Equation 9.10.

Given the above definitions of λ, μ, and ϕ, show that

$$1 \leq \lambda < \infty$$

$$-1 \leq \mu \leq 1$$

and

$$0 \leq \phi \leq 2\pi$$

Now use elliptic coordinates to evaluate the overlap integral, Equation 9.10,

$$S = \int d\mathbf{r} 1s_A 1s_B$$

$$= \frac{Z^3}{\pi} \int d\mathbf{r} e^{-Zr_A} e^{-Zr_B}$$

9-4. Determine the normalized wave function for $\psi_- = c_1(1s_A - 1s_B)$.

9-5. Repeat the calculation in Section 9–3 for $\psi_- = (1s_A - 1s_B)$.

9-6. Use the elliptic coordinate system of Problem 9–3 to derive analytic expressions for S, J, and K for the simple molecular-orbital treatment of H_2^+.

9-7. Plot ψ_b and ψ_a given by Equations 9.27 and 9.28 for several values of R along the internuclear axis.

9-8. Show that

$$H_{AA} = H_{BB} = -\frac{1}{2} + J$$

and that

$$H_{AB} = -\frac{S}{2} + K$$

in the simple molecular-orbital treatment of H_2^+. The quantities J and K are given by Equations 9.23 and 9.24, respectively.

9-9. Show explicitly that an s orbital on one hydrogen atom and a p_x orbital on another have zero overlap. Use the $2s$ and $2p_x$ wave functions given in Table 6.6 to set up the overlap integral. Take the z axis to lie along the internuclear axis. *Hint*: You need not evaluate any integrals, but simply show that the overlap integral can be separated into two parts that exactly cancel one another.

9-10. Show that $\Delta E_- = (J - K)/(1 - S)$ for the antibonding orbital ψ_- of H_2^+.

9-11. Show that ψ given by Equation 9.29 is an eigenfunction of $\hat{S}_z = \hat{S}_{z1} + \hat{S}_{z2}$ with $S_z = 0$.

9-12. Use molecular-orbital theory to explain why the dissociation energy of N_2 is greater than that of N_2^+, but the dissociation energy of O_2^+ is greater than that of O_2.

9-13. Discuss the bond properties of F_2 and F_2^+ using molecular-orbital theory.

9-14. Predict the relative stabilities of the species N_2, N_2^+, and N_2^-.

9-15. Predict the relative bond strengths and bond lengths of diatomic carbon, C_2, and its negative ion, C_2^-.

9-16. Write out the ground-state molecular-orbital electron configurations for Na_2 through Ar_2. Would you predict a stable Mg_2 molecule?

9-17. Determine the ground-state electron configurations of NO^+ and NO. Compare the bond orders of these two species.

9-18. Determine the bond order of a cyanide ion.

9-19. The force constants for the diatomic molecules B_2 through F_2 are given in the table below. Is the order what you expect? Explain.

Diatomic molecule	$k/\mathrm{N \cdot m^{-1}}$
B_2	350
C_2	930
N_2	2260
O_2	1140
F_2	450

9-20. In Section 9–7, we constructed molecular orbitals for homonuclear diatomic molecules using the $n = 2$ atomic orbitals on each of the bonded atoms. In this problem, we will consider the molecular orbitals that can be constructed from the $n = 3$ atomic orbitals. These orbitals are important in describing diatomic molecules of the first row of transition metals. Once again we choose the z-axis to lie along the molecular bond. What are the designations for the $3s_A \pm 3s_B$ and $3p_A \pm 3p_B$ molecular orbitals? The $n = 3$ shell also contains a set of five $3d$ orbitals. (The shapes of the $3d$ atomic orbitals are shown in Figure 6.7.) Given that molecular orbitals with two nodal planes that contain the internuclear axis are called δ orbitals, show that ten $3d_A \pm 3d_B$ molecular orbitals consist of a bonding σ orbital, a pair of bonding π orbitals, a pair of bonding δ orbitals, and their corresponding antibonding orbitals.

9-21. Determine the largest bond order for a first-row transition-metal homonuclear diatomic molecule (see Problem 9–20).

9-22. Figure 9.19 plots a schematic representation of the energies of the molecular orbitals of HF. How will the energy-level diagram for the diatomic OH radical differ from that of HF? What is the highest occupied molecular orbital of OH?

9-23. A common light source used in photoelectron spectroscopy is a helium discharge, which generates light at 58.4 nm. A photoelectron spectrometer measures the kinetic energy of the electrons ionized when the molecule absorbs this light. What is the largest electron binding energy that can be measured using this radiation source? Explain how a measurement of the kinetic energy of the ionized electrons can be used to determine the energy of the occupied molecular orbitals of a molecule. *Hint*: Recall the photoelectron effect discussed in Chapter 1.

9-24. Using Figure 9.19, you found that the highest occupied molecular orbital for HF is a fluorine $2p$ atomic orbital. The measured ionization energy for an electron from this nonbonding molecular orbital of HF is 1550 kJ·mol^{-1}. However, the measured ionization energy of a $2p$ electron from a fluorine atom is 1795 kJ·mol^{-1}. Why is the ionization energy of an electron from the $2p$ atomic orbital on a fluorine atom greater for the fluorine atom than for the HF molecule?

9-25. In this problem, we consider the heteronuclear diatomic molecule CO. The ionization energies of an electron from the valence atomic orbitals on the carbon atom and the oxygen atom are listed below.

Atom	Valence orbital	Ionization energy/MJ·mol^{-1}
O	$2s$	3.116
	$2p$	1.524
C	$2s$	1.872
	$2p$	1.023

Use these data to construct a molecular-orbital energy-level diagram for CO. What are the symmetry designations of the molecular orbitals of CO? What is the electron configuration of the ground state of CO? What is the bond order of CO? Is CO paramagnetic or diamagnetic?

9-26. The molecule BF is isoelectronic with CO. However, the molecular orbitals for BF are different from those for CO. Unlike CO, the energy difference between the $2s$ orbitals of boron and fluorine is so large that the $2s$ orbital of boron combines with a $2p$ orbital on fluorine to make a molecular orbital. The remaining $2p$ orbitals on fluorine combine with two of the $2p$ orbitals on B to form π orbitals. The third $2p$ orbital on B is nonbonding. The energy ordering of the molecular orbitals is $\psi(2s_B + 2p_F) < \psi(2p_B - 2p_F) < \psi(2s_B - 2p_F) < \psi(2p_B + 2p_F) < \psi(2p_B)$. What are the symmetry designations of the molecular orbitals of BF? What is the electron configuration of the ground state of BF? What is the bond order of BF? Is BF diamagnetic or paramagnetic? How do the answers to these last two questions compare with those obtained for CO (Problem 9–25)?

9-27. The photoelectron spectrum of O_2 exhibits two bands of 52.398 MJ·mol^{-1} and 52.311 MJ·mol^{-1} that correspond to the ionization of an oxygen $1s$ electron. Explain this observation.

9-28. The experimental ionization energies for a fluorine $1s$ electron from HF and F_2 are 66.981 and 67.217 MJ·mol^{-1}. Explain why these ionization energies are different even though the $1s$ electrons of the fluorine are not involved in the chemical bond.

9-29. Show that filled orbitals can be ignored in the determination of molecular term symbols.

9-30. Deduce the ground-state term symbols of all the diatomic molecules given in Table 9.6.

9-31. Determine the ground-state molecular term symbols of O_2, N_2, N_2^+, and O_2^+.

9-32. The highest occupied molecular orbitals for an excited electronic configuration of an oxygen molecule are

$$(1\pi_g)^1(3\sigma_u)^1$$

Determine the molecular term symbols for oxygen with this electronic configuration.

9-33. Determine the values for the energies of the separated hydrogen atoms shown in Figure 9.22. Determine the energy difference of the dissociated limits.

9-34. For a set of point charges $Z_i e$ that lie along a line, we define the dipole moment (μ) of the charge distribution by

$$\mu = e \sum Z_i x_i$$

where e is the protonic charge and x_i is the distance of the charge $Z_i e$ from the origin. Consider the molecule LiH. A molecular-orbital calculation of LiH reveals that the bond length of this diatomic molecule is 159 pm and that there is a net charge of $+0.76e$ on the lithium atom and a net charge of $-0.76e$ on the hydrogen atom. First, determine the location of the center-of-mass of the LiH molecule. Use the center-of-mass as the origin along the x-axis and determine the dipole moment of the LiH molecule. How does your value compare with the experimental value of 19.62×10^{-30} C·m?

9-35. Show that the value of the dipole moment μ defined in Problem 9–34 is independent of where we place the origin along the x-axis so long as the net charge of the molecule is equal to zero. Recalculate the dipole moment of LiH by placing the origin on the hydrogen atom, and compare your answer with that obtained for Problem 9–34.

9-36. What would be the value of the dipole moment of LiH if its bond were purely ionic? Estimate the amount of ionic character in LiH. (See Problem 9–34).

9-37. A dipole moment is actually a vector quantity defined by

$$\mu = e \sum_i Z_i \mathbf{r}_i$$

where $\mathbf{r}_i$ is a vector from some origin to the charge $Z_i e$. Show that $\boldsymbol{\mu}$ is independent of where we take the origin if the net charge on the molecule is zero.

9-38. The dipole moment of HCl is 3.697×10^{-30} C·m. The bond length of HCl is 127.5 pm. If HCl is modeled as two point charges separated by its bond length, then what are the net charges on the H and Cl atom?

9-39. Use the data in the table below to compute the fractional charges on the hydrogen atom and halide atom for the hydrogen halides. Is your finding in agreement with the order of the electronegativities of the halogen atoms, F > Cl > Br > I?

	R_e/pm	$\mu/10^{-30}$ C·m
HF	91.7	6.37
HCl	127.5	3.44
HBr	141.4	2.64
HI	160.9	1.40

9-40. When we built up the molecular orbitals for diatomic molecules, we combined only those orbitals with the same energy because we said that only those with similar energies mix well. This problem is meant to illustrate this idea. Consider two atomic orbitals χ_A and χ_B. Show that a linear combination of these orbitals leads to the secular determinant

$$\begin{vmatrix} \alpha_A - E & \beta - ES \\ \beta - ES & S\alpha_B - E \end{vmatrix} = 0$$

where

$$\alpha_A = \int \chi_A h^{\text{eff}} \chi_A \, d\tau$$

$$\alpha_B = \int \chi_B h^{\text{eff}} \chi_B \, d\tau$$

$$\beta = \int \chi_B h^{\text{eff}} \chi_A \, d\tau = \int \chi_A h^{\text{eff}} \chi_B \, d\tau$$

$$S = \int \chi_A \chi_B \, d\tau$$

where h^{eff} is some effective one-electron Hamiltonian operator for the electron that occupies the molecular orbital ϕ. Show that this secular determinant expands to give

$$(1 - S^2)E^2 + [2\beta S - \alpha_A - \alpha_B]E + \alpha_A \alpha_B - \beta^2 = 0$$

It is usually a satisfactory first approximation to neglect S. Doing this, show that

$$E_\pm = \frac{\alpha_A + \alpha_B \pm [(\alpha_A - \alpha_B)^2 + 4\beta^2]^{1/2}}{2}$$

Now if χ_A and χ_B have the same energy, show that $\alpha_A = \alpha_B = \alpha$ and that

$$E_\pm = \alpha \pm \beta$$

giving one level of β units below α and one level of β units above α; that is, one level of β units more stable than the isolated orbital energy and one level of β units less stable.

Now investigate the case in which $\alpha_A \neq \alpha_B$, say $\alpha_A > \alpha_B$. Show that

$$E_\pm = \frac{\alpha_A + \alpha_B}{2} \pm \frac{\alpha_A - \alpha_B}{2}\left[1 + \frac{4\beta^2}{(\alpha_A - \alpha_B)^2}\right]^{1/2}$$

$$= \frac{\alpha_A + \alpha_B}{2} \pm \frac{\alpha_A - \alpha_B}{2}\left[1 + \frac{2\beta^2}{(\alpha_A - \alpha_B)^2} - \frac{2\beta^4}{(\alpha_A - \alpha_B)^4} + \cdots\right]$$

$$E_\pm = \frac{\alpha_A + \alpha_B}{2} \pm \frac{\alpha_A - \alpha_B}{2} \pm \frac{\beta^2}{\alpha_A - \alpha_B} + \cdots$$

where we have assumed that $\beta^2 < (\alpha_A - \alpha_B)^2$ and have used the expansion

$$(1 + x)^{1/2} = 1 + \frac{x}{2} - \frac{x^2}{8} + \cdots$$

Show that

$$E_+ = \alpha_A + \frac{\beta^2}{\alpha_A - \alpha_B} + \cdots$$

$$E_- = \alpha_B - \frac{\beta^2}{\alpha_A - \alpha_B} + \cdots$$

Using this result, discuss the stabilization-destabilization of α_A and α_B versus the case above in which $\alpha_A = \alpha_B$. For simplicity, assume that $\alpha_A - \alpha_B$ is large.

9-41. In the Born-Oppenheimer approximation, we assume that because the nuclei are so much more massive than the electrons, the electrons can adjust essentially instantaneously to any nuclear motion, and hence we have a unique and well-defined energy, $E(R)$, at each internuclear separation R. Under this same approximation, $E(R)$ is the internuclear potential and so is the potential field in which the nuclei vibrate. Argue, then, that under the Born-Oppenheimer approximation, the force constant is independent of isotopic substitution. Using the above ideas, and given that the dissociation energy for H_2 is $D_0 = 432.1 \text{ kJ·mol}^{-1}$ and that the fundamental vibrational frequency ν is $1.319 \times 10^{14} \text{ s}^{-1}$, calculate D_0 and ν for deuterium, D_2. Realize that the observed dissociation energy is given by

$$D_0 = D_e - \frac{1}{2}h\nu$$

where D_e is the value of $E(R)$ at R_e.

9-42. In this problem, we evaluate the overlap integral (Equation 9.10) using spherical coordinates centered on atom A. The integral to evaluate is (Problem 9–3)

$$S(R) = \frac{1}{\pi} \int d\mathbf{r}_A e^{-r_A} e^{-r_B}$$
$$= \frac{1}{\pi} \int_0^\infty dr_A e^{-r_A} r_A^2 \int_0^{2\pi} d\phi \int_0^\pi d\theta \sin\theta e^{-r_B}$$

where r_A, r_B, and θ are shown in the figure.

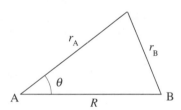

To evaluate the above integral, we must express r_B in terms of r_A, θ, and ϕ. We can do this using the law of cosines

$$r_B = (r_A^2 + R^2 - 2r_A R \cos\theta)^{1/2}$$

So the first integral we must consider is

$$I_\theta = \int_0^\pi e^{-(r_A^2 + R^2 - 2r_A R \cos\theta)^{1/2}} \sin\theta d\theta$$

Let $\cos\theta = x$ to get

$$\int_{-1}^1 e^{-(r_A^2 + R^2 - 2r_A Rx)^{1/2}} dx$$

Now let $u = (r_A^2 + R^2 - 2r_A Rx)^{1/2}$ and show that

$$dx = -\frac{u\,du}{r_A R}$$

Show that the limits of the integration over u are $u = r_A + R$ when $x = -1$ and $u = |R - r_A|$ when $x = 1$. Then show that

$$I_\theta = \frac{1}{r_A R} \left[e^{-(R-r_A)}(R + 1 - r_A) - e^{-(R+r_A)}(R + 1 + r_A) \right] \qquad r_A < R$$

$$= \frac{1}{r_A R} \left[e^{-(r_A - R)}(r_A - R + 1) - e^{-(R+r_A)}(R + 1 + r_A) \right] \qquad r_A > R$$

Now substitute this result into $S(R)$ above to get

$$S(R) = e^{-R} \left(1 + R + \frac{R^2}{3} \right)$$

Compare the length of this problem to Problem 9–3.

9-43. Let's use the method that we developed in Problem 9–42 to evaluate the Coulomb integral, J, given by Equation 9.19. Let

$$I = -\int \frac{d\mathbf{r}\, 1s_A^* 1s_A}{r_B} = -\frac{1}{\pi} \int d\mathbf{r} \frac{e^{-2r_A}}{(r_A^2 + R^2 - 2r_A R \cos\theta)^{1/2}}$$

$$= -\frac{1}{\pi} \int_0^\infty dr_A\, r_A^2 e^{-2r_A} \int_0^{2\pi} d\phi \int_0^\pi \frac{d\theta \sin\theta}{(r_A^2 + R^2 - 2r_A R \cos\theta)^{1/2}}$$

Using the approach of Problem 9–42, let $\cos\theta = x$ and $u = (r_A^2 + R^2 - 2r_A Rx)^{1/2}$ to show that

$$I = \frac{2}{R} \int_0^\infty dr_A\, r_A e^{-2r_A} \int_{R+r_A}^{|R-r_A|} du = \frac{2}{R} \int_0^\infty dr_A\, r_A e^{-2r_A}[|R - r_A| - (R + r_A)]$$

$$= e^{-2R} \left(1 + \frac{1}{R} \right) - \frac{1}{R}$$

and that the Coulomb integral, J, is given by

$$J = e^{-2R} \left(1 + \frac{1}{R} \right)$$

Hint: You need to use the integrals

$$\int xe^{ax}\,dx = e^{ax}\left(\frac{x}{a} - \frac{1}{a^2} \right)$$

and

$$\int x^2 a^{ax}\,dx = e^{ax}\left(\frac{x^2}{a} - \frac{2x}{a^2} + \frac{2}{a^3} \right)$$

Linus Pauling was born in Portland, Oregon, on February 28, 1901, and died in 1994. He received his Ph.D. in chemistry in 1925 from the California Institute of Technology for his dissertation on X-ray crystallography of organic compounds and the structure of crystals. After spending a year studying at the University of Munich, he joined the faculty at the California Institute of Technology, where he remained for almost 40 years. Pauling was a pioneer in the application of quantum mechanics to chemistry. His book *The Nature of the Chemical Bond* (1939) is one of the most influential chemistry texts of the twentieth century. In the 1930s, he became interested in biological molecules and developed a structural theory of protein molecules, work that led to elucidating that sickle cell anemia is caused by a faulty structure of hemoglobin. In the early 1950s, he proposed the alpha helix as the basic structure of proteins. Pauling was awarded the Nobel Prize for chemistry in 1954 "for his research into the nature of the chemical bond and its application to the elucidation of the structure of complex structures." During the 1950s, Pauling was in the forefront of the fight against nuclear testing, for which he was awarded the Nobel Peace Prize in 1963. From the early 1980s until his death, he was embroiled in the controversy of advocating the use of vitamin C as protection against the common cold and other serious maladies such as cancer.

CHAPTER 10

Bonding In Polyatomic Molecules

In Chapter 9, we developed molecular-orbital theory to describe the bonding in di-atomic molecules. We showed that the electrons in molecules occupy molecular or-bitals in accordance with the Pauli Exclusion Principle. These molecular orbitals are constructed by forming linear combinations of atomic orbitals on the individual bonded atoms. In this chapter, these ideas are extended to polyatomic molecules. We will see that molecular-orbital theory can be used successfully to describe bonding in large molecules. We begin the discussion by introducing localized bond orbitals in terms of hybrid orbitals. This is followed by a discussion on bonding in small molecules such as water and methane. Then we will discuss Hückel molecular-orbital theory, which is a fairly simple but useful theory to describe the π molecular orbitals of conjugated and aromatic hydrocarbons such as benzene.

10–1. Hybrid Orbitals Account for Molecular Shape

The ground-state electron configuration of a carbon atom, $1s^2 2s^2 2p_x^1 2p_y^1$, does not seem to lead to the tetrahedral bonding in methane and other saturated hydrocarbons. In fact, the electron configuration seems to imply that carbon should be divalent instead of tetravalent. You may have learned in general chemistry and organic chemistry, however, that we explain that carbon is tetravalent by promoting one of the $2s$ electrons to the $2p_z$ orbital, giving an electron configuration for carbon of $1s^2 2s^1 2p_x^1 2p_y^1 2p_z^1$, and then the four singly occupied orbitals combine to form four equivalent sp^3 *hybrid orbitals*, each of which points to a corner of a tetrahedron. In this section, we will discuss hybridization from a quantum-mechanical point of view.

Consider first the linear molecule beryllium hydride, BeH_2. The two Be–H bonds in BeH_2 are equivalent, and the bond angle between them is $180°$. The ground-state electron configuration of a beryllium atom is $1s^2 2s^2$. The question arises as to how we can describe the direction of the two bonds using the atomic orbitals on the beryllium atom. In constructing molecular orbitals, linear combinations are formed from atomic orbitals that are similar in energy. The $2s$ and $2p$ orbitals on any given atom are similar in energy and as a result we should consider the possibility that more than one atomic orbital from a given atom can contribute to a molecular orbital. In the particular case of BeH_2, the resultant orbitals on the beryllium atom must point in opposite directions to explain the linear structure of the molecule. Whereas the occupied $2s$ orbital is spherically symmetric, the desired geometry is similar to the spatial orientation of the $2p$ atomic orbital that points along the internuclear axis. Thus, it is reasonable to use the following linear combination to describe the molecular orbitals formed between the beryllium atom and the hydrogen atoms:

$$\psi_{Be-H} = c_1\psi_{Be(2s)} + c_2\psi_{Be(2p)} + c_3\psi_{H(1s)} \tag{10.1}$$

Whether these molecular orbitals correspond to bonding or antibonding orbitals depends on the signs of the coefficients. The first two terms in Equation 10.1 can be thought of as representing a new "orbital" on beryllium, given by $c_1\psi_{Be(2s)} + c_2\psi_{Be(2p)}$. Linear combinations of atomic orbitals on the same atom are called *hybrid orbitals*. The linear combination of two atomic orbitals yields two hybrid orbitals. Because these hybrid orbitals are made up of a $2s$ orbital and a $2p$ orbital, they are called *sp hybrid orbitals*.

The two *sp* hybrid orbitals are equivalent and are directed 180° from each other. The normalized *sp* hybrid orbitals are given by

$$\psi_{sp} = \frac{1}{\sqrt{2}}(2s \pm 2p_z) \tag{10.2}$$

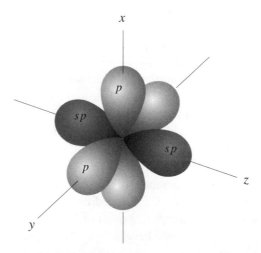

FIGURE 10.1
An illustration of the *sp* hybrid orbitals. The two *sp* hybrid orbitals are equivalent and are directed 180° from each other. The two remaining $2p$ orbitals are perpendicular to each other and to the line formed by the *sp* orbitals.

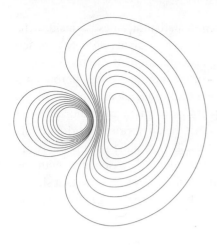

FIGURE 10.2
A contour map of one of the sp hybrid orbitals.
The two sp hybrid orbitals are equivalent and are
directed 180° from each other.

where the z-axis is arbitrarily defined to lie along the H–Be–H bonds. Figure 10.1 illustrates the sp hybrid orbitals. The remaining two $2p$ orbitals are perpendicular to the sp hybrid orbitals. Whereas each $2p$ orbital consists of two lobes that point in opposite directions, each sp hybrid orbital concentrates electron density in one direction (Figure 10.2). This is understandable since the sign of the $2p$ wave function is different in the two directions ($\pm z$) but the sign of the $2s$ wave function is everywhere positive. Examining Equation 10.2, we note that the linear combination $c_1 \psi_{Be(2s)} + c_2 \psi_{Be(2p)}$ constructively builds electron density along the $+z$ direction; the linear combination $c_1 \psi_{Be(2s)} - c_2 \psi_{Be(2p)}$ constructively builds electron density along the $-z$ direction. The beryllium-hydrogen bond orbitals result from a linear combination of each sp hybrid orbital and a hydrogen $1s$ orbital as in Equation 10.1. The bonding of BeH_2 can be depicted as in Figure 10.3.

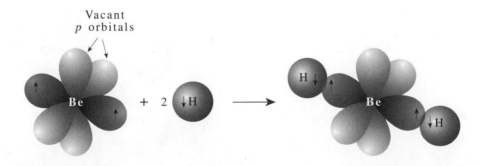

FIGURE 10.3
The formation of two equivalent localized bond orbitals in BeH_2. Each bond orbital is formed from the overlap of a beryllium sp orbital and a hydrogen $1s$ orbital. There are four valence electrons in BeH_2, two from the beryllium atom and one from each of the two hydrogen atoms. The four valence electrons occupy the two localized bond orbitals, forming the two localized beryllium-hydrogen bonds in BeH_2.

This approach can be extended to more complicated molecules. Consider the molecule BH_3. The three B–H bonds are equivalent, lie in a plane, and are directed $120°$ from each other. The ground-state electron configuration of atomic boron is $1s^2 2s^2 2p^1$. Once again, we need a set of equivalent orbitals on the central boron atom that describe the observed bonding. Because we need three equivalent hybrid orbitals, we consider hybrid orbitals that are linear combinations of three atomic orbitals on boron. The electron configuration of boron suggests that the appropriate hybrid orbitals will be constructed from the $2s$ orbital and two of the $2p$ orbitals; such orbitals are called sp^2 *hybrid orbitals*. Figure 10.4 illustrates the geometry associated with the sp^2 hybrid orbitals. A contour map of an sp^2 hybrid orbital looks similar to that shown for an sp orbital in Figure 10.2. The normalized sp^2 hybrid orbitals are given by the linear combinations

$$\psi_1 = \frac{1}{\sqrt{3}} 2s + \sqrt{\frac{2}{3}} 2p_z \tag{10.3}$$

$$\psi_2 = \frac{1}{\sqrt{3}} 2s - \frac{1}{\sqrt{6}} 2p_z + \frac{1}{\sqrt{2}} 2p_x \tag{10.4}$$

and

$$\psi_3 = \frac{1}{\sqrt{3}} 2s - \frac{1}{\sqrt{6}} 2p_z - \frac{1}{\sqrt{2}} 2p_x \tag{10.5}$$

Note that the sum of the squares of the coefficients for a particular atomic orbital for the set of hybrid orbitals is equal to one. For example, for the $2s$ atomic orbital

$$\left(\frac{1}{\sqrt{3}} \right)^2 + \left(\frac{1}{\sqrt{3}} \right)^2 + \left(\frac{1}{\sqrt{3}} \right)^2 = 1$$

Notice that the number of hybrid orbitals is equal to the number of atomic orbitals we started with.

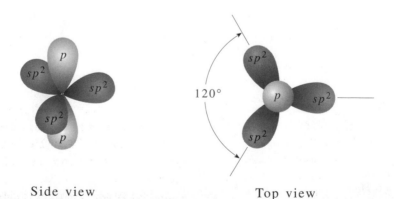

Side view   Top view

FIGURE 10.4
The geometry associated with sp^2 orbitals. The three sp^2 orbitals lie in a plane and point to the vertices of an equilateral triangle. The remaining $2p$ orbital is perpendicular to the plane formed by the three sp^2 orbitals.

EXAMPLE 10–1

Show that the three sp^2 hybrid orbitals are orthogonal to one another.

SOLUTION: If two orbitals, ψ_1 and ψ_2, are orthogonal, then

$$\int d\tau \, \psi_1^* \psi_2 = 0$$

Substituting the first two sp^2 hybrid orbitals (Equations 10.3 and 10.4) into this orthogonality integral gives

$$\int d\tau \, \psi_1^* \psi_2 = \int d\tau \left(\frac{1}{\sqrt{3}} 2s^* + \sqrt{\frac{2}{3}} 2p_z^* \right) \left(\frac{1}{\sqrt{3}} 2s - \frac{1}{\sqrt{6}} 2p_z + \frac{1}{\sqrt{2}} 2p_x \right)$$

Expanding the product in the above integral gives

$$\int d\tau \, \psi_1^* \psi_2 = \frac{1}{3} \int d\tau 2s^* 2s + \frac{\sqrt{2}}{3} \int d\tau 2p_z^* 2s - \frac{1}{\sqrt{18}} \int d\tau 2s^* 2p_z$$

$$- \frac{1}{3} \int d\tau 2p_z^* 2p_z + \frac{1}{\sqrt{6}} \int d\tau 2s^* 2p_x + \frac{1}{\sqrt{3}} \int d\tau 2p_z^* 2p_x$$

Because the atomic orbitals are orthogonal, this expression simplifies to

$$\int d\tau \, \psi_1^* \psi_2 = \frac{1}{3} \int d\tau 2s^* 2s - \frac{1}{3} \int d\tau 2p_z^* 2p_z$$

Each of the integrals on the right side of the above equation is equal to one because the atomic orbitals are normalized. Thus, we obtain

$$\int d\tau \, \psi_1^* \psi_2 = \frac{1}{3} - \frac{1}{3} = 0$$

which establishes that ψ_1 and ψ_2 are orthogonal. Likewise $\int d\tau \psi_1^* \psi_3 = 0$ and $\int d\tau \psi_2^* \psi_3 = 0$, thereby proving that the three sp^2 orbitals are mutually orthogonal.

The three sp^2 hybrid orbitals are able to account for the planar bonding and bond angles observed in BH_3, as shown schematically in Figure 10.5. In constructing the hybrid orbitals in Equations 10.3 to 10.5, we arbitrarily choose the molecule to lie in the x-z plane. In principle, any set of two of the p orbitals can be used to construct the sp^2 hybrid orbitals.

The final example we consider is the tetrahedral molecule, methane, CH_4. As indicated in the beginning of this section, the observed molecular bonding is not easily described in terms of the ground-state electron configuration of the central carbon atom. Following the approach used above, however, we can construct four equivalent hybrid orbitals that point to the corners of a tetrahedron. The four hybrid orbitals involve the linear combinations of four atomic orbitals. For carbon, this involves the $2s$ and the

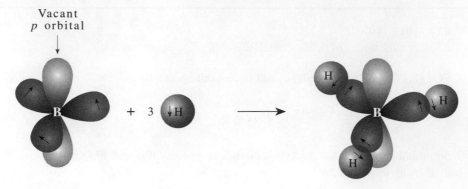

FIGURE 10.5
A schematic illustration of the bonding in BH_3. Each of the three boron-hydrogen bond orbitals is formed from the overlap of a boron sp^2 orbital and a hydrogen $1s$ orbital. The six valence electrons in BH_3 occupy the three bond orbitals to form the three boron-hydrogen bonds.

three $2p$ orbitals thereby generating sp^3 *hybrid orbitals*. The normalized sp^3 hybrids orbitals are given by

$$\psi_1 = \frac{1}{2}(2s + 2p_x + 2p_y + 2p_z) \tag{10.6}$$

$$\psi_2 = \frac{1}{2}(2s - 2p_x - 2p_y + 2p_z) \tag{10.7}$$

$$\psi_3 = \frac{1}{2}(2s + 2p_x - 2p_y - 2p_z) \tag{10.8}$$

and

$$\psi_4 = \frac{1}{2}(2s - 2p_x + 2p_y - 2p_z) \tag{10.9}$$

Figure 10.6 illustrates the geometry associated with the sp^3 hybrid orbitals.

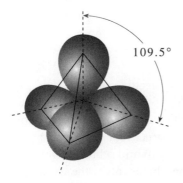

FIGURE 10.6
The geometry associated with sp^3 hybrid orbitals. The four sp^3 hybrid orbitals point toward the vertices of a tetrahedron. The angle between the center lines of any pair of sp^3 hybrid orbitals in this structure is 109.5° (Problems 10–7 and 10–9).

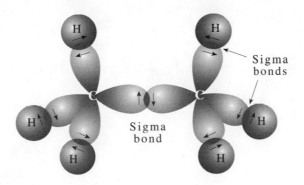

FIGURE 10.7
A schematic representation of the bonding in ethane, CH_3CH_3. Six of the seven bond orbitals in ethane result from the overlap of sp^3 orbitals on the carbon atoms and $1s$ orbitals on the hydrogen atoms. The seventh bond orbital involves the overlap of an sp^3 orbital on each carbon. There are 14 valence electrons in ethane. Each of these seven bonding orbitals is occupied by two valence electrons of opposite spins. The electrons are depicted by arrows in the above figure.

The sp^3 hybrid orbitals on the carbon atom are used to describe the bonding in saturated hydrocarbons. Figure 10.7 illustrates this by presenting an illustration of the bonding in ethane, C_2H_6.

In this section, we have shown that molecular shapes can be accounted for by using hybrid atomic orbitals. We have not illustrated the mathematical details by which equivalent hybrid orbitals are generated from linear combinations of atomic orbitals. Problem 10–6 asks you to provide these details for the sp hybrid orbitals, and Problem 10–10 addresses the sp^3 hybrid orbitals. The choice of linear combinations is greatly simplified by consideration of the molecular symmetry, a topic we will explore in Chapter 12.

EXAMPLE 10–2
Which atomic orbitals on the sulfur atom are involved in the hybrid orbitals that account for the bonding in SF_6.

SOLUTION: SF_6 is an octahedral molecule, in which the six S–F bonds are equivalent. The bond angles are all 90°. We assume that the S–F bonds lie along the x-, y-, and z-axes, with the origin of the coordinate system centered on the sulfur atom. To account for this bonding, we require six equivalent hybrid orbitals that point in directions consistent with the known molecular structure. The ground-state electron configuration of a sulfur atom is $[Ne]3s^23p^4$. Using the $3s$ and three $3p$ orbitals would limit us to a total of four hybrid orbitals, and hence only four S–F bonds. The $3d$ orbitals, however, are also similar in energy to the $3s$ and $3p$. In constructing six hybrid orbitals for the sulfur atom, we consider linear combinations of two $3d$ orbitals, the $3s$ orbital and the three $3p$ orbitals. These six atomic orbitals give rise to six hybrid orbitals, which are called d^2sp^3 *hybrid orbitals*. If we consider that the six fluorine atoms lie along the x-, y-, and z-axes, the two d orbitals used to construct the hybrid orbitals are the d_{z^2} and $d_{x^2-y^2}$ orbitals (Figure 10.8).

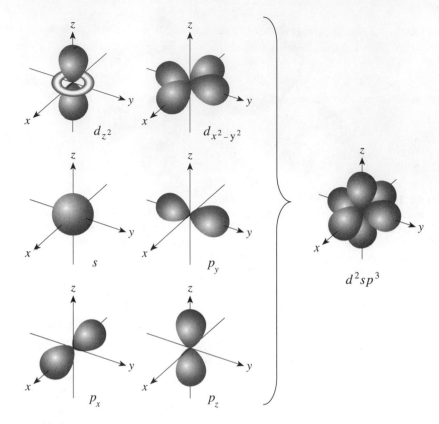

FIGURE 10.8
The six d^2sp^3 hybrid orbitals point along the x-, y-, and z-axes and are formed from linear combinations of an ns orbital, the three np orbitals, and the nd_{z^2} and $nd_{x^2-y^2}$ orbitals.

10–2. Different Hybrid Orbitals Are Used for the Bonding Electrons and the Lone Pair Electrons in Water

The valence electrons on the central atoms of the molecules discussed in the last section (beryllium for BeH_2, boron for BH_3, and carbon for CH_4) all occupy hybrid orbitals involved in bonding to a hydrogen atom. In this section, we consider the description of molecules in which the central atom has lone pairs of electrons by considering the specific case of water (H_2O).

The ground-state electron configuration of an oxygen atom is $1s^2 2s^2 2p_x^2 2p_y^1 2p_z^1$. Because the $2p_y$ and $2p_z$ orbitals contain only one electron, they can form chemical bonds with the hydrogen $1s$ electron. We could form two bond orbitals by forming the linear combinations

$$\phi_1 = c_1 1s_{H_A} + c_2 2p_{yO}$$

$$\phi_2 = c_3 1s_{H_B} + c_4 2p_{zO}$$

(10.10)

This model predicts, however, that the bond angle in water is 90°, compared with the observed bond angle of 104.5°. Even though the oxygen atom contains the needed electrons in half-filled atomic orbitals to account for the number of chemical bonds in water, the geometry of the molecule necessitates the consideration of hybrid orbitals on the central oxygen molecule. The bond angle in water (104.5°) is between that predicted using sp^2 hybrid orbitals (120°) and $2p$ orbitals (90°). This result is not surprising. In the case of BH_3, all the valence electrons on the boron atom are involved in the three equivalent B–H bonds, thus the angles between the bonds are equivalent. In water, two of the valence electrons on the oxygen atom are involved in the bonds with hydrogen atoms while four remain as two sets of lone pair electrons on the oxygen atom. We expect the pair of bonding orbitals to be equivalent and the pair of orbitals for the lone pairs to be equivalent, but there is no reason to expect the bonding orbitals to be equivalent to the lone pair orbitals. Proceeding as we did in the previous section, the general form of the hybrid orbitals on the oxygen atom is given by

$$\psi = c_1 2s + c_2 2p_y + c_3 2p_z \tag{10.11}$$

We need to determine the coefficients c_1, c_2, and c_3 such that two orthogonal orbitals are directed at an angle of 104.5°. The details of this calculation are left as an exercise (Problem 10–12). The results for the two bonding hybrid orbitals, ψ_1 and ψ_2, are given by

$$\psi_1 = 0.45 \cdot 2s + 0.71 \cdot 2p_y + 0.55 \cdot 2p_z \tag{10.12}$$

and

$$\psi_2 = 0.45 \cdot 2s - 0.71 \cdot 2p_y + 0.55 \cdot 2p_z \tag{10.13}$$

EXAMPLE 10–3
Show that the molecular orbitals ψ_1 and ψ_2 given by Equations 10.12 and 10.13 are orthogonal.

SOLUTION: The two orbitals, ψ_1 and ψ_2, are orthogonal if

$$\int d\tau \psi_1^* \psi_2 = 0$$

Substituting in Equations 10.12 and 10.13 gives

$$\int d\tau \psi_1^* \psi_2 = \int d\tau (0.45 \cdot 2s^* + 0.71 \cdot 2p_y^* + 0.55 \cdot 2p_z^*)$$
$$\times (0.45 \cdot 2s - 0.71 \cdot 2p_y + 0.55 \cdot 2p_z)$$

Multiplying out the above product gives

$$\int d\tau \psi_1^* \psi_2 = \int d\tau (0.45)^2 2s^* 2s + \int d\tau (0.45)(0.55)2s^* 2p_z$$

$$- \int d\tau (0.45)(0.71)2s^* 2p_y + \int d\tau (0.55)(0.45)2p_z^* 2s$$

$$+ \int d\tau (0.55)^2 2p_z^* 2p_z - \int d\tau (0.55)(0.71)2p_z^* 2p_y$$

$$+ \int d\tau (0.71)(0.45)2p_y^* 2s + \int d\tau (0.71)(0.55)2p_y^* 2p_z$$

$$- \int d\tau (0.71)^2 2p_y^* 2p_y$$

Because the $2s$, $2p_x$, and $2p_y$ atomic orbitals are mutually orthogonal, the above sum simplifies to

$$\int d\tau \psi_1^* \psi_2 = \int d\tau (0.55)^2 2p_z^* 2p_z - \int d\tau (0.71)^2 2p_y^* 2p_y + \int d\tau (0.45)^2 2s^* 2s$$

The atomic orbitals are normalized, giving

$$\int d\tau \psi_1^* \psi_2 = (0.55)^2 - (0.71)^2 + (0.45)^2 = 0$$

Thus, the two hybrid orbitals are orthogonal.

EXAMPLE 10–4

Show that the hybrid orbitals ψ_1 and ψ_2 given by Equations 10.12 and 10.13 are directed at an angle of 104.5° with respect to one another.

SOLUTION: Because the $2s$ orbital is spherically symmetric, the directionalities of ψ_1 and ψ_2 are determined by the relative contributions of $2p_y$ and $2p_z$. The following figure illustrates this directionality.

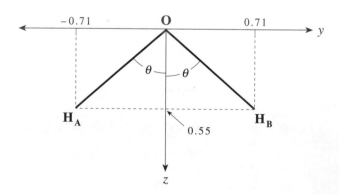

Note that ψ_1 and ψ_2 may be depicted as vectors whose components are the coefficients of $2p_y$ and $2p_z$; the $2p_y$ and $2p_z$ atomic orbitals are the (orthonormal) unit vectors. The angle θ in the above figure is given by

$$\tan \theta = \frac{0.71}{0.55} = 1.29$$

or $\theta = 52.24°$. Therefore, the bond angle is 2θ, or $104.5°$.

The hybrid orbitals given by Equations 10.12 and 10.13 are normalized. Because the $2s$, $2p_y$, and $2p_z$ orbitals are orthonormal, the normalization condition says that $c_1^2 + c_2^2 + c_3^2 = 1$. We can give a physical interpretation to this result by saying that c_1^2, the square of the coefficient of the $2s$ orbital in the hybrid orbital, is the fraction of s character in the hybrid orbital, with a corresponding interpretation for c_2^2 and c_3^2. Therefore, we can say that the hybrid orbitals on the oxygen atom in water have an s character of 0.20 (0.45^2) and a p character of 0.81 ($0.55^2 + 0.71^2$). Note that the sp^2 hybrid orbitals constructed for BH$_3$ are 0.333 s character and 0.667 p character, different from those for water. The quantitative contribution of the s and p orbitals to a particular type of hybrid atomic orbitals is a function of the bond angle and the number of valence electrons on the central atom.

The two hybrid bond orbitals we have constructed consist of the oxygen $2s$, $2p_y$, and $2p_z$ orbitals. Because we are using three atomic orbitals, there must be a third hybrid orbital ψ_3. Thus, we have an unused hybrid orbital and an unused $2p_x$ orbital to accomodate the two lone pair of electrons on the oxygen atom. Because we require that the two lone pair orbitals be equivalent, we form two such orbitals by taking linear combinations of ψ_3 and the oxygen $2p_x$ orbital (Problem 10–13).

10–3. Why is BeH$_2$ Linear and H$_2$O Bent?

We have seen that BeH$_2$ is linear and H$_2$O is bent. Although introducing hybrid orbitals can provide us an explanation of the observed geometry, the physical origin of this difference in molecular structure is not accounted for. The major difference between BeH$_2$ and H$_2$O is the number of valence electrons on the central atom (beryllium has two valence electrons, whereas oxygen has six). The effect of the number of valence electrons on the molecular structure can be quantitatively understood in terms of the shape and occupation of the molecular orbitals.

The molecular orbitals for both BeH$_2$ and H$_2$O involve linear combinations of the valence orbitals of the central atom ($2s$, $2p_x$, $2p_y$, and $2p_z$) and the $1s$ orbitals of the two hydrogen atoms. Using these six atomic orbitals, we write the LCAO-MO wave function by the general form

$$\psi = c_1 1s_{H_a} + c_2 1s_{H_b} + c_3 2s_A + c_4 2p_{xA} + c_5 2p_{yA} + c_6 2p_{zA} \qquad (10.14)$$

where the subscript A is used to indicate the central atom of the dihydride molecule. This particular LCAO-MO involves six atomic orbitals, and therefore it must generate six molecular orbitals. Once again the energies and coefficients for these six molecular orbitals are obtained by solving a secular determinantal equation. The calculated energies and corresponding wave functions clearly depend on the molecular geometry because the Hamiltonian operator explicitly depends on molecular geometry. Here, we first examine the six molecular orbitals that result from the linear combinations of atomic orbitals given in Equation 10.14 for a linear molecule AH_2. Then, we consider how these molecular orbitals change as the molecule bends.

Figure 10.9 shows the six molecular orbitals for the linear molecule AH_2 that arise from Equation 10.14. The occupied $1s$ orbital, the $1\sigma_g$, on atom A is a nonbonding orbital and is not shown in the figure. (We will use the notation for molecular orbitals introduced in Section 9–13.) The LCAO-MO given by Equation 10.14 generates two bonding orbitals ($2\sigma_g$, $1\sigma_u$), two antibonding orbitals ($3\sigma_g$, $2\sigma_u$), and a doubly degenerate set of nonbonding orbitals ($1\pi_u$). As Figure 10.9 shows, the

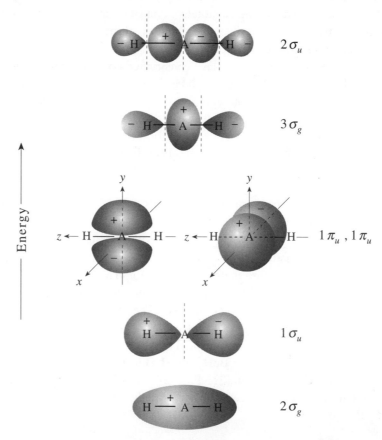

FIGURE 10.9
The six molecular orbitals for a linear AH_2 molecule that are constructed from the linear combination of atomic orbitals given by Equation 10.14.

two bonding orbitals $2\sigma_g$ and $1\sigma_u$ concentrate electron density between the central A atom and the hydrogen atoms. The two orbitals $3\sigma_g$ and $2\sigma_u$, however, have nodes between the central A atom and the hydrogen atoms and so are antibonding orbitals. The doubly degenerate $1\pi_u$ orbitals are the p_x and p_y orbitals on the central atom and so are nonbonding orbitals. The $2\sigma_g$ and $3\sigma_g$ orbitals result from linear combinations of the $2s$ orbital on the central atom with the $1s$ orbitals on the two hydrogen atoms, and the $1\sigma_u$ and $2\sigma_u$ orbitals result from linear combinations of the $2p_z$ orbital on the central atom (the z-axis is defined as the internuclear axis) with the $1s$ orbitals on the two hydrogen atoms. The molecular-orbital energy-level diagram of BeH₂ is shown in Figure 10.10. The energy ordering of these six molecular orbitals is $2\sigma_g < 1\sigma_u < 1\pi_u = 1\pi_u < 3\sigma_g < 2\sigma_u$. This ordering is independent of the number of valence electrons on the central A atom in the linear AH₂ molecule.

Now let's consider what happens to these molecular orbitals as the molecule bends. Surely the energies and the molecular orbitals depend upon the shape of the molecule. For example, for a linear AH₂ molecule there is no net overlap between the $1s$ orbitals on the hydrogen atoms and the $2p_x$ and $2p_y$ orbitals on the central atom (Figure 10.11). For a bent molecule, however, this is no longer the case. If the molecule bends in the y–z plane, a net overlap (of bonding character) results with the $2p_y$ orbital, as shown in Figure 10.11. The net overlap between the $1s$ orbitals on the hydrogen atoms and the

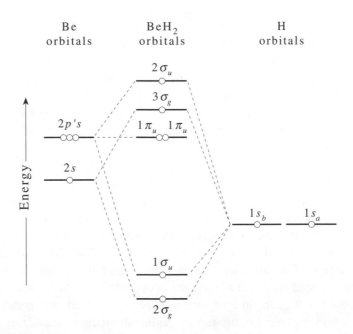

FIGURE 10.10
A molecular orbital energy level diagram for the valence electrons in the linear BeH₂ molecule. Note that the doubly degenerate $1\pi_u$ orbitals are nonbonding orbitals.

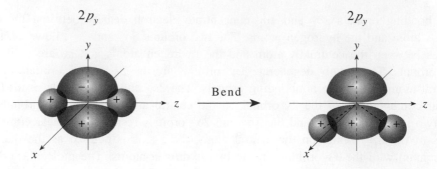

FIGURE 10.11
The net overlap of hydrogen $1s$ orbitals with a central atom $2p_y$ orbital is zero for a linear molecule. If the molecule is bent, however, a nonzero net orbital overlap of the $2p_y$ orbital on the central A atom with the $1s$ orbitals on the two hydrogen atoms results.

$2p_x$ orbital on the central atom is still zero, however. Because there is a net overlap of the $1s$ orbitals with only one of the $2p$ orbitals, the degeneracy of the π orbitals is lifted upon bending. What was once a nonbonding orbital of the linear molecule becomes a bonding orbital in the bent molecule.

The molecular orbitals for a bent triatomic molecule will be represented by different linear combinations of the atomic orbitals from those found for a linear structure. By solving for the energy as a function of all bond angles between a linear geometry and a 90° geometry, we can determine how the molecular orbitals for the linear molecule evolve into those characteristic of the 90° geometry. A plot of the energy of a molecular orbital as a function of a systematic change in molecular geometry is called a *Walsh correlation diagram*. Figure 10.12 shows the Walsh correlation diagram for a triatomic molecule, AH_2, for which the energies of all six of the molecular orbitals shown in Figure 10.9 are plotted as a function of bond angle. Note that the labels designating the orbitals for the 90° geometry are different from those used for the linear molecule. By their definitions, the σ and π designations can be used to describe only a linear molecule. The labels for a bent molecule, a_1, b_1, and b_2, reflect specific symmetry properties of the molecule and are discussed in Chapter 12. Here we will use them simply as a short-hand notation for the molecular orbitals of the bent molecule.

Whether the geometry of a molecule is linear or bent (at a particular angle) depends on which structure is lowest in energy, which can be determined using the Walsh correlation diagram. The data presented in Figure 10.12 show that bending a molecule affects the energy of the six molecular orbitals differently. The molecular geometry will therefore depend on which orbitals are occupied. The Walsh correlation diagram shows that the bonding $2\sigma_g$ and $1\sigma_u$ orbitals are destabilized with bending. Bending lifts the degeneracy of the $1\pi_u$ orbitals in the linear structure, stabilizing the $3a_1$ (see Figure 10.12), and not affecting the energy of the $1b_1$ orbital. For large bending angles, the energy of the $3a_1$ orbital drops below that of the $1b_2$ orbital.

This energy-correlation diagram can be used to predict general features of molecular geometry. Consider, for example, the molecule BeH_2, which has a total of four

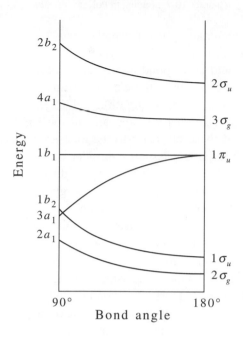

FIGURE 10.12
The Walsh correlation diagram for the valence electrons of an AH$_2$ molecule. The right side of the diagram gives the energy ordering of the molecular orbitals for an H–A–H bond angle of 180°. The left side gives the energy ordering of the molecular orbitals for an H–A–H bond angle of 90°. The solid lines tell us how the energies of the molecular orbitals depend upon H–A–H bond angles between 90° and 180°.

valence electrons. For a linear structure, this would correspond to the electron configuration of $(2\sigma_g)^2(1\sigma_u)^2$. A bent structure would have an electron configuration of $(2a_1)^2(1b_2)^2$ or $(2a_1)^2(3a_1)^2$, depending on the bond angle. Because bending destabilizes the energy of the lowest two molecular orbitals, the electron configuration $(2\sigma_g)^2(1\sigma_u)^2$ is lower in energy than either $(2a_1)^2(1b_2)^2$ or $(2a_1)^2(3a_1)^2$. The Walsh correlation diagram correctly predicts that BeH$_2$ is linear.

EXAMPLE 10–5
Predict the geometry of the ground state of BH$_2$.

SOLUTION: The species BH$_2$ has five valence electrons. A linear structure would have an electron configuration of $(2\sigma_g)^2(1\sigma_u)^2(1\pi_u)^1$. A bent structure could have the electron configuration $(2a_1)^2(3a_1)^2(1b_2)^1$ or $(2a_1)^2(1b_2)^2(3a_1)^1$, depending on the H–B–H angle. Figure 10.12 shows that the decrease in the energy of the $3a_1$ orbitals outweighs the increase in energy of the sum of the $2a_1$ and $1b_2$ orbitals as the molecule is bent from 180°. We therefore expect the molecule to be bent. However, the destabilization of the energy of the four total electrons in the $2a_1$ and $1b_2$ orbitals that accompanies bending quickly outweighs the stabilization in energy derived from the one electron in the $3a_1$ orbital. Thus, the bond angle should lie somewhere near the middle of the diagram between 180° and 90°, and we would predict a ground-state electron configuration of $(2a_1)^2(1b_2)^2(3a_1)^1$. This configuration is in agreement with the experimentally determined bond angle for BH$_2$ of 131°.

Now let's consider a water molecule, which has eight valence electrons. Each of the four lowest energy molecular orbitals given in the Walsh diagram will be populated by two electrons. A linear structure would correspond to an electron configuration of $(2\sigma_g)^2(1\sigma_u)^2(1\pi_u)^4$. A bent structure could have the electron configuration $(2a_1)^2(3a_1)^2(1b_2)^2(1b_1)^2$ or $(2a_1)^2(1b_2)^2(3a_1)^2(1b_1)^2$ depending on the H–O–H bond angle. The energy of the $1b_1$ orbital in the bent geometry is the same as the $1\pi_u$ in the linear geometry, hence the contribution of these electrons to the total energy is independent of the molecular geometry. The important issue to consider is how the stabilization in energy of one of the $1\pi_u$ orbitals with bending (corresponding to the $3a_1$ orbital in the bent molecule) compares with the destabilization in energy that accompanies generation of the $(2a_1)^2(1b_2)^2$ electron configuration from the $(2\sigma_g)^2(1\sigma_u)^2$ electron configuration of the linear molecule. As we found for BH$_2$ in Example 10–5, the stabilization of the $3a_1$ orbital is greater than the destabilization associated with formation of the $(2a_1)^2(1b_2)^2$ electronic configuration for small bending angles. Thus, the Walsh correlation diagram predicts that water is a bent molecule, which is in agreement with experimental measurements. The exact value of the H–O–H bond angle (104.5°) can be calculated using the computational techniques discussed in the next chapter. The molecular-orbital energy-level diagram for H$_2$O is shown in Figure 10.13. Water has eight valence electrons, and so Figure 10.13 shows that the ground-state electron configuration of H$_2$O is $(2a_1)^2(1b_2)^2(3a_1)^2(1b_1)^2$.

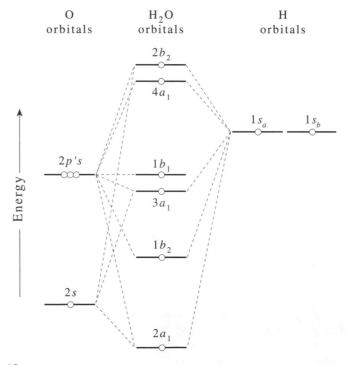

FIGURE 10.13
A molecular-orbital energy-level diagram for the valence electrons in H$_2$O. (The bond angle is 104.5°, the equilibrium value.) Note that the $1b_1$ orbital is a nonbonding orbital.

10–4. Photoelectron Spectroscopy Can Be Used to Study Molecular Orbitals

We discussed photoelectron spectroscopy in Chapter 9, where we showed photoelectron spectra of N_2 and CO. Photoelectron spectroscopy can also be used for polyatomic molecules. Figure 10.14 shows the photoelectron spectrum of H_2O vapor. The electron configuration $(2a_1)^2(1b_2)^2(3a_1)^2(1b_1)^2$ suggests that ionization from each of these occupied orbitals will be observed, and the three bands shown in Figure 10.14 correspond to ionization of electrons from the $1b_1$, $3a_1$, and $1b_2$ molecular orbitals. (The ionization from the $2a_1$ level is off the scale of the figure.) The structure evident in the $3a_1$ and $1b_2$ bands reflects ionization from the different vibrational levels associated with that state. Thus, an analysis of the photoelectron spectrum can be used to determine the energy spacing between the vibrational levels associated with various electronic states.

Figure 10.15 shows the molecular-orbital energy-level diagram of CH_4. Methane has eight valence electrons, so its ground-state valence configuration is $(2a_1)^2(1t_2)^6$ (the designation of the molecular orbitals will be explained in Chapter 12). Thus, for the valence electrons, we predict that only two bands are observed in the photelectron spectrum of CH_4, as confirmed in Figure 10.16. The $1a_1$ orbital, which corresponds to the core $1s$ orbital on carbon, is off the scale in the figure. Once again, notice the vibrational structure superimposed on the bands.

As a final example, the molecular-orbital energy-level diagram of the valence electrons in ethene, C_2H_4, is shown in Figure 10.17. The photoelectron spectrum of ethene is shown in Figure 10.18. Ethene has twelve valence electrons, so its ground-state valence electron configuration is $(2a_g)^2(2b_{1u})^2(1b_{2u})^2(1b_{3g})^2(3a_g)^2(1b_{3u})^2$. (Once again, we simply consider $2a_g$, $2b_{1u}$, etc. as a short-hand notation for the molecular

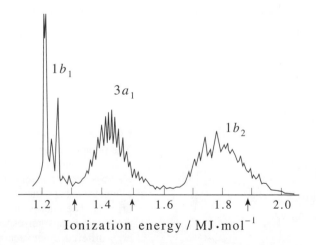

FIGURE 10.14
The photoelectron spectrum of water. The three bands shown correspond to ionization from the three highest energy occupied molecular orbitals. The fine structure on each band reflects ionization to different vibrational levels. The arrows indicate the calculated ionization energies from the $v = 0$ vibrational states (see Chapter 11).

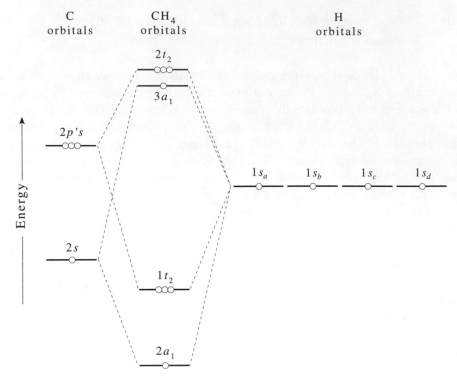

FIGURE 10.15
A molecular-orbital energy-level diagram for the valence electrons in CH_4.

orbitals.) Although not evident from the notation, the lowest five states in Figure 10.17 are associated with σ orbitals, and the sixth state, the highest occupied molecular orbital, is associated with a π orbital. Figure 10.17 shows that the energy difference between π bonding and π antibonding orbitals is less than the difference between

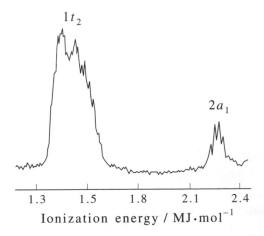

FIGURE 10.16
The photoelectron spectrum of methane. The two bands observed in the photoelectron spectrum reflect the ionization of electrons from the $1t_2$ and $2a_1$ molecular orbitals. The energy difference between these two bands corresponds to the energy difference between the $1t_2$ and $2a_1$ molecular orbitals (see Figure 10.15). The bands are broad because ionization occurs to many different vibrational levels of the molecules.

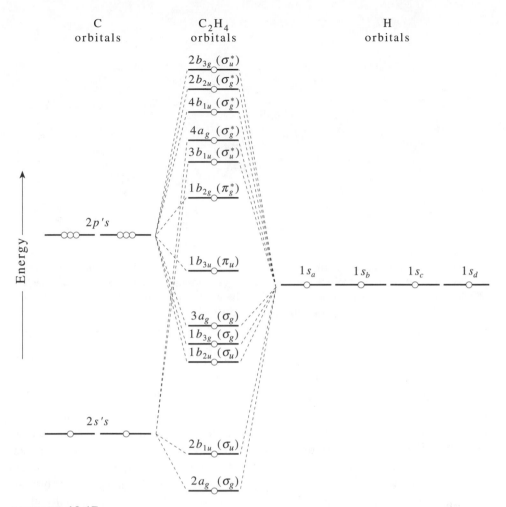

FIGURE 10.17

A molecular-orbital energy-level diagram for C_2H_4. The first five orbitals are σ orbitals, the sixth is a π orbital, the next is a π^* orbital, and the remaining five are σ^* orbitals.

σ bonding and σ antibonding orbitals, suggesting that unsaturated hydrocarbons such as ethene should absorb light at longer wavelengths (less energy) than do saturated hydrocarbons. Ethene, for example, turns out to have an ultraviolet absorption peak at $58\,500$ cm^{-1}, whereas ethane (with no π orbitals) does not begin to absorb strongly until $62\,500$ cm^{-1}. Therefore, we can develop a simplified molecular-orbital treatment of unsaturated hydrocarbons that includes only the π orbitals. In this approximation, the relatively complicated energy-level diagram in Figure 10.17 consists of simply two molecular orbitals, a π bonding orbital and a π antibonding orbital (see Figure 10.20). We will discuss this simple molecular-orbital theory in the following sections.

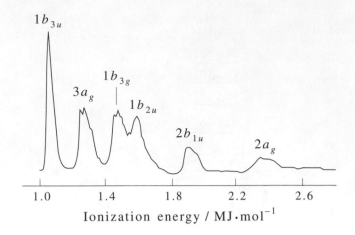

$1b_{3u}$

$3a_g$

$1b_{3g}$

$1b_{2u}$

$2b_{1u}$

$2a_g$

1.0   1.4   1.8   2.2   2.6

Ionization energy / MJ·mol^{-1}

FIGURE 10.18
The photoelectron spectrum of ethene. The bands in the photoelectron spectrum correspond to the ionization of electrons from the different molecular orbitals of the molecule. The energies of the bands in the photoelectron spectrum can be used to determine the energy spacings between the lowest six molecular orbitals in Figure 10.17.

10–5. Conjugated Hydrocarbons and Aromatic Hydrocarbons Can Be Treated by a π-Electron Approximation

In this section, we will discuss a well-known theory of bonding in unsaturated hydrocarbons. The simplest unsaturated hydrocarbon is ethene, C_2H_4. Ethene is a planar unsaturated hydrocarbon, all of whose bond angles are approximately 120°. We can describe the structure of ethene by saying that the carbon atoms form sp^2 hybrid orbitals and that each C–H bond results from an overlap of the $1s$ hydrogen orbital with an sp^2 hybrid orbital on each carbon atom. Part of the C–C bond in ethene results from the

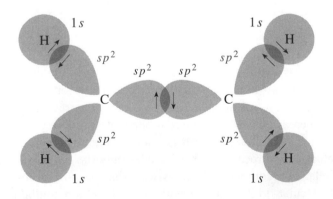

FIGURE 10.19
The σ-bond framework of an ethene molecule.

overlap of an sp^2 hybrid orbital from each carbon atom. All five bonds are σ bonds and collectively are called the σ-*bond framework* of the ethene molecule (Figure 10.19).

If this σ framework lies in the $x-y$ plane, thus implying that the $2p_x$ and $2p_y$ orbitals were used to construct the hybrid orbitals, then the $2p_z$ orbital on each carbon atom is still available for bonding. The charge distribution that results from the overlap of the $2p_z$ orbitals produces a π bond between the carbon atoms. Thus, the picture we are developing here assumes that unsaturated hydrocarbons will have both σ and π bonds. In large systems, such as conjugated polyenes and benzene, you learned in organic chemistry that the π orbitals can be delocalized over the entire molecule. In such cases, we could view the π electrons moving in some fixed, effective, electrostatic potential due to the electrons in the σ framework. This approximation is called the π-*electron approximation*. The π-electron approximation can be developed formally by starting with the Schrödinger equation, but we will simply accept it here as a physically intuitive approach to the bonding in unsaturated hydrocarbons.

We now turn our attention to describing the delocalized molecular orbitals occupied by these π electrons. You need to realize that the Hamiltonian operator we are considering contains an effective potential due to the electrons in the σ framework and that the explicit form of this effective Hamiltonian operator has not been specified in our treatment so far. With this in mind, let's return to ethene. Here, each carbon atom contributes a $2p_z$ orbital to the delocalized π orbital, and using the same approach as we used to describe the σ bond of the wave function of H_2, we would write the wave function of the π orbital of ethene, ψ_π, as

$$\psi_\pi = c_1 2p_{zA} + c_2 2p_{zB} \tag{10.15}$$

The secular determinant associated with this wave function is

$$\begin{vmatrix} H_{11} - ES_{11} & H_{12} - ES_{12} \\ H_{12} - ES_{12} & H_{22} - ES_{22} \end{vmatrix} = 0 \tag{10.16}$$

where the H_{ij} are integrals involving the effective Hamiltonian operator and the S_{ij} are overlap integrals involving $2p_z$ atomic orbitals. Because the carbon atoms in ethene are equivalent, $H_{11} = H_{22}$. The diagonal elements of the Hamiltonian operator in the secular determinant are called *Coulomb integrals*. The off-diagonal elements of the Hamiltonian operator are called *resonance integrals* or *exchange integrals*. Note that the resonance integral involves two atomic centers because it has contributions of atomic orbitals from two different carbon atoms. To determine the energies and associated molecular orbitals, we either need to specify the effective Hamiltonian operator or propose approximations for evaluating the various entries in the secular determinant. Here, we examine an approximation proposed by Erich Hückel in 1930, which along with various extensions and modifications has found wide use in organic chemistry. There are three simple assertions of *Hückel molecular-orbital theory*. First, the overlap integrals, S_{ij}, are set to zero unless $i = j$, where $S_{ii} = 1$. Second, all the Coulomb integrals (the diagonal elements of the Hamiltonian operator in the secular determinant) are assumed to be the same for all carbon atoms and are commonly denoted by α. Third, the resonance integrals involving nearest-neighbor carbon atoms are assumed to be the same and are denoted by β; the remaining resonance integrals are set equal to zero. Thus, the Hückel secular determinant for ethene (Equation 10.16) is given by

$$\begin{vmatrix} \alpha - E & \beta \\ \beta & \alpha - E \end{vmatrix} = 0 \tag{10.17}$$

The two roots of this secular determinant are $E = \alpha \pm \beta$.

 To evaluate the energy quantitatively, we would need to know the effective Hamiltonian operator. Fortunately, in Hückel theory, we do not need to worry about this because α and β are assigned values that are determined by experimental measurements. Because α approximates the energy of an electron in an isolated $2p_z$ orbital, it can be used as a reference point for the zero of energy. The quantity β has been determined from the consideration of a variety of experimental data and is given a value of approximately -75 kJ·mol^{-1}. The Hückel approximations allow us to determine the energies (in terms of α and β) and wave functions for the π molecular orbitals without explicitly specifying the Hamiltonian operator.

 There are two π electrons in ethene. In the ground state, both electrons occupy the orbital of lowest energy. Because β is negative, the lowest energy is $E = \alpha + \beta$. The π-electronic energy of ethene is $E_\pi = 2\alpha + 2\beta$. Figure 10.20 shows an energy-level diagram for the π electrons of ethene (cf. Figure 10.17). Because α is used to specify the zero of energy, the two energies found from the secular determinant, $E = \alpha \pm \beta$, must correspond to bonding and antibonding orbitals.

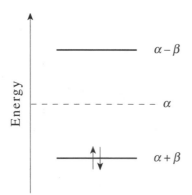

FIGURE 10.20
The ground-state electron configuration of the π electrons in ethene.

EXAMPLE 10–6
Find the bonding Hückel molecular orbitals for ethene.

SOLUTION: Recall from Section 7–2 that the secular determinantal equation originates from the following pair of linear algebraic equations for c_1 and c_2 in Equation 10.16:

$$c_1(H_{11} - ES_{11}) + c_2(H_{12} - ES_{12}) = 0$$

and

$$c_1(H_{12} - ES_{12}) + c_2(H_{22} - ES_{22}) = 0$$

Using the Hückel approximations, we rewrite these equations as

$$c_1(\alpha - E) + c_2\beta = 0$$

and

$$c_1\beta + c_2(\alpha - E) = 0$$

To find the values of c_1 and c_2 associated with each value of E, we substitute in an allowed value of E and solve for the coefficients. For $E = \alpha + \beta$, either equation yields that $c_1 = c_2$. Thus,

$$\psi_\pi = c_1(2p_{z,A} + 2p_{z,B})$$

The value of c_1 can be found by requiring that the wave function be normalized. The normalization condition on ψ_π gives

$$c_1^2(1 + 2S + 1) = 1$$

Using the Hückel assumption that $S = 0$, we find that $c_1 = 1/\sqrt{2}$. Problem 10–24 asks you to find the wave function for $E = \alpha - \beta$.

10–6. Butadiene Is Stabilized by a Delocalization Energy

The case of butadiene is more interesting than that of ethene. Although butadiene exists in both the *cis* and *trans* configurations, we will ignore that and picture the butadiene molecule as simply a linear sequence of four carbon atoms, each of which contributes a $2p_z$ orbital to the π-electron orbital (Figure 10.21). Because we are considering the linear combination of four atomic orbitals, the dimension of the secular determinant

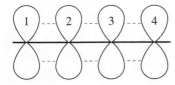

FIGURE 10.21
A schematic representation of the $2p_z$ orbitals of each of the carbon atoms in a butadiene molecule.

will be 4×4 and will give rise to four different energies and four different π molecular orbitals. We can express all the molecular orbitals, ψ_i, by the single expression

$$\psi_i = \sum_{j=1}^{4} c_{ij} 2p_{zj} \tag{10.18}$$

where the c_{ij} are the coefficients of the $2p_z$ atomic orbital on the jth atom ($2p_{zj}$) in the ith molecular orbital. The secular determinantal equation for the butadiene molecule is (Problem 10–26)

$$\begin{vmatrix} H_{11} - ES_{11} & H_{12} - ES_{12} & H_{13} - ES_{13} & H_{14} - ES_{14} \\ H_{12} - ES_{12} & H_{22} - ES_{22} & H_{23} - ES_{23} & H_{24} - ES_{24} \\ H_{13} - ES_{13} & H_{23} - ES_{23} & H_{33} - ES_{33} & H_{34} - ES_{34} \\ H_{14} - ES_{14} & H_{24} - ES_{24} & H_{34} - ES_{34} & H_{44} - ES_{44} \end{vmatrix} = 0 \tag{10.19}$$

Using the Hückel approximations, $H_{jj} = \alpha$, $S_{jj} = 1$, $S_{ij} = 0$ if $i \neq j$ and the $H_{ij} = \beta$ for neighboring carbon atoms, and $H_{ij} = 0$ for distant carbon atoms. Therefore, $H_{12} = H_{23} = H_{34} = \beta$ and $H_{13} = H_{14} = H_{24} = 0$, and the secular determinant becomes

$$\begin{vmatrix} \alpha - E & \beta & 0 & 0 \\ \beta & \alpha - E & \beta & 0 \\ 0 & \beta & \alpha - E & \beta \\ 0 & 0 & \beta & \alpha - E \end{vmatrix} = 0 \tag{10.20}$$

If we factor β from each column and let $x = (\alpha - E)/\beta$, then we can rewrite this determinantal equation as

$$\beta^4 \begin{vmatrix} x & 1 & 0 & 0 \\ 1 & x & 1 & 0 \\ 0 & 1 & x & 1 \\ 0 & 0 & 1 & x \end{vmatrix} = 0 \tag{10.21}$$

If this determinant (MathChapter E) is expanded, the secular equation is

$$x^4 - 3x^2 + 1 = 0 \tag{10.22}$$

We can solve this equation for x^2 to obtain

$$x^2 = \frac{3 \pm \sqrt{5}}{2} \tag{10.23}$$

from which we find the four roots $x = \pm 1.618, \ \pm 0.618$.

Recalling that $x = (\alpha - E)/\beta$ and that β is a negative quantity, we can construct the Hückel energy-level diagram for butadiene, as shown in Figure 10.22. There are

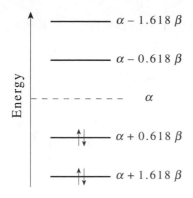

FIGURE 10.22
The ground-state electron configuration of the π electrons in butadiene.

four π electrons in butadiene and, in the ground state, these electrons occupy the two orbitals of lowest energy (Figure 10.22). The total π-electronic energy of butadiene is

$$
\begin{aligned}
E_\pi &= 2(\alpha + 1.618\beta) + 2(\alpha + 0.618\beta) \\
&= 4\alpha + 4.472\beta
\end{aligned}
\tag{10.24}
$$

We can make an interesting comparison of the energy given by Equation 10.24 with that predicted for a localized structure in which two π electrons are localized between carbon atoms 1 and 2 and between carbon atoms 3 and 4, respectively. This localized structure is equivalent to two isolated ethene molecules. We found that the energy of the π orbital in ethene is $2\alpha + 2\beta$. If we compare the energy of two ethene molecules with that obtained for the delocalized orbitals of butadiene, we see there is an energy stabilization that results from delocalization, E_{deloc}:

$$
E_{\text{deloc}} = E_\pi(\text{butadiene}) - 2E_\pi(\text{ethene}) = 0.472\beta < 0
\tag{10.25}
$$

If we use the value of -75 kJ·mol^{-1} for β, the delocalization energy of butadiene is on the order of -35 kJ·mol^{-1}. This is the energy by which butadiene is stabilized relative to two isolated double bonds, or the stability that butadiene derives because its π electrons are delocalized over the entire length of the molecule instead of being localized to the two end bonds.

Associated with each of the four molecular-orbital energies of butadiene that we have found is a molecular orbital. To specify these molecular orbitals, we need to determine the coefficients c_{ij} of Equation 10.18. The approach is the same as that carried out in Example 10–6, but the algebra is quite a bit more lengthy (see Problem 10–28). The resulting wave functions are

$$\psi_1 = 0.3717 \cdot 2p_{z1} + 0.6015 \cdot 2p_{z2} + 0.6015 \cdot 2p_{z3} + 0.3717 \cdot 2p_{z4}$$
$$E_1 = \alpha + 1.618\beta$$

$$\psi_2 = 0.6015 \cdot 2p_{z1} + 0.3717 \cdot 2p_{z2} - 0.3717 \cdot 2p_{z3} - 0.6015 \cdot 2p_{z4}$$
$$E_2 = \alpha + 0.618\beta$$

$$\psi_3 = 0.6015 \cdot 2p_{z1} - 0.3717 \cdot 2p_{z2} - 0.3717 \cdot 2p_{z3} + 0.6015 \cdot 2p_{z4}$$
$$E_3 = \alpha - 0.618\beta$$

$$\psi_4 = 0.3717 \cdot 2p_{z1} - 0.6015 \cdot 2p_{z2} + 0.6015 \cdot 2p_{z3} - 0.3717 \cdot 2p_{z4}$$
$$E_4 = \alpha - 1.618\beta$$

(10.26)

and these molecular orbitals are shown schematically in Figure 10.23. Notice that as the energy of the molecular orbital increases so do the number of nodes. This is a general result for π molecular orbitals.

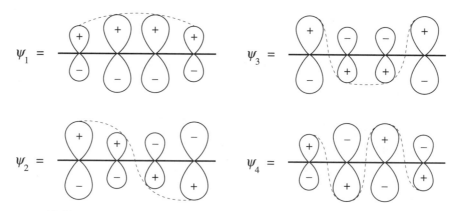

FIGURE 10.23
A schematic representation of the π molecular orbitals of butadiene. Note that the corresponding energy increases with the number of nodes.

EXAMPLE 10–7
Show that ψ_1 in Equation 10.26 is normalized and that it is orthogonal to ψ_2.

SOLUTION: We want to show first that

$$\int d\mathbf{r}\psi_1^*\psi_1 = 1$$

Using the fact that Hückel theory (as we have discussed it) sets all the overlap integrals to zero, we have

$$\int d\mathbf{r}\psi_1^*\psi_1 = (0.3717)^2 + (0.6015)^2 + (0.6015)^2 + (0.3717)^2 = 1.000$$

To show that ψ_1 is orthogonal to ψ_2, we must show that

$$\int d\mathbf{r}\psi_1^*\psi_2 = 0$$

Once again, because all the overlap integrals equal zero, we have

$$\int d\mathbf{r}\psi_1^*\psi_2 = (0.3717)(0.6015) + (0.6015)(0.3717)$$
$$-(0.6015)(0.3717) - (0.3717)(0.6015) = 0$$

It is straightforward to show that all four molecular orbitals in Equation 10.26 are normalized and that they are mutually orthogonal.

As our final example, we consider benzene. Here, we approach this problem using the basic principles of Hückel theory outlined above. In Chapter 12, we will learn that by considering the symmetry properties of the molecule, the same results can be obtained without performing much of the tedious algebra. Benzene has six carbon atoms, each contributing a $2p_z$ orbital from which the π molecular orbitals are to be constructed. Because we are considering linear combinations of six atomic orbitals, the dimension of the secular determinant will be 6×6 and will give rise to six different energies and six different π molecular orbitals. The Hückel secular determinantal equation for benzene is given by (Problem 10–29)

$$\begin{vmatrix} \alpha - E & \beta & 0 & 0 & 0 & \beta \\ \beta & \alpha - E & \beta & 0 & 0 & 0 \\ 0 & \beta & \alpha - E & \beta & 0 & 0 \\ 0 & 0 & \beta & \alpha - E & \beta & 0 \\ 0 & 0 & 0 & \beta & \alpha - E & \beta \\ \beta & 0 & 0 & 0 & \beta & \alpha - E \end{vmatrix} = 0 \quad (10.27)$$

This 6×6 secular determinant leads to a sixth-degree polynomial for E. Using the same approach as for butadiene, we let $x = (\alpha - E)/\beta$. The resulting determinant can be expanded to give

$$x^6 - 6x^4 + 9x^2 - 4 = 0 \quad (10.28)$$

The six roots to this equation are $x = \pm 1, \pm 1$, and ± 2, giving the following energies for the six molecular orbitals:

$$E_1 = \alpha + 2\beta$$

$$E_2 = E_3 = \alpha + \beta$$

$$E_4 = E_5 = \alpha - \beta \tag{10.29}$$

$$E_6 = \alpha - 2\beta$$

The Hückel energy-level diagram for benzene is given in Figure 10.24. The six π electrons are placed into the three lowest energy molecular orbitals. The total π-electron energy in benzene is given by

$$E_\pi = 2(\alpha + 2\beta) + 4(\alpha + \beta) = 6\alpha + 8\beta \tag{10.30}$$

Compared with the π-electron energy of three ethene molecules, the delocalization (or resonance) energy in benzene is 2β. Thus, Hückel molecular-orbital theory predicts that benzene is stabilized by about $150\ \text{kJ} \cdot \text{mol}^{-1}$. The resulting wave functions for the six π molecular orbitals of benzene are given by

$$\psi_1 = \tfrac{1}{\sqrt{6}}(2p_{z1} + 2p_{z2} + 2p_{z3} + 2p_{z4} + 2p_{z5} + 2p_{z6}) \qquad E_1 = \alpha + 2\beta$$

$$\psi_2 = \tfrac{1}{\sqrt{4}}(2p_{z2} + 2p_{z3} - 2p_{z5} - 2p_{z6}) \qquad E_2 = \alpha + \beta$$

$$\psi_3 = \tfrac{1}{\sqrt{3}}(2p_{z1} + \tfrac{1}{2}2p_{z2} - \tfrac{1}{2}2p_{z3} - 2p_{z4} - \tfrac{1}{2}2p_{z5} + \tfrac{1}{2}2p_{z6}) \quad E_3 = \alpha + \beta$$

$$\psi_4 = \tfrac{1}{\sqrt{4}}(2p_{z2} - 2p_{z3} + 2p_{z5} - 2p_{z6}) \qquad E_4 = \alpha - \beta \tag{10.31}$$

$$\psi_5 = \tfrac{1}{\sqrt{3}}(2p_{z1} - \tfrac{1}{2}2p_{z2} - \tfrac{1}{2}2p_{z3} + 2p_{z4} - \tfrac{1}{2}2p_{z5} - \tfrac{1}{2}2p_{z6}) \quad E_5 = \alpha - \beta$$

$$\psi_6 = \tfrac{1}{\sqrt{6}}(2p_{z1} - 2p_{z2} + 2p_{z3} - 2p_{z4} + 2p_{z5} - 2p_{z6}) \qquad E_6 = \alpha - 2\beta$$

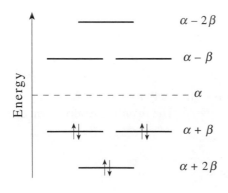

FIGURE 10.24
The ground-state electron configuration of the π electrons in benzene.

EXAMPLE 10–8

Draw the π molecular orbitals for benzene and indicate the nodal planes.

SOLUTION:

ψ_6

$E = \alpha - 2\beta$

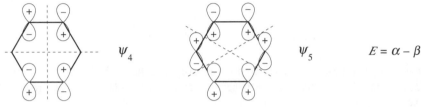

ψ_4

ψ_5

$E = \alpha - \beta$

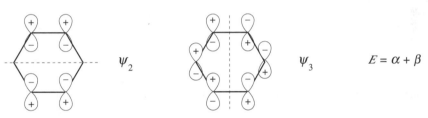

ψ_2

ψ_3

$E = \alpha + \beta$

ψ_1

$E = \alpha + 2\beta$

Note that as we found for ethene and butadiene, the energy increases with the number of nodal planes.

Problems

10-1. Show that $\psi_{sp} = \frac{1}{\sqrt{2}}(2s \pm 2p_z)$ is normalized.

10-2. Show that the three sp^2 hybrid orbitals given by Equations 10.3 through 10.5 are normalized.

10-3. Prove that the three sp^2 hybrid orbitals given by Equations 10.3 through 10.5 are directed at angles of $120°$ with respect to one another. (See Example 10–4.)

10-4. Represent the three sp^2 hybrid orbitals given by Equations 10.3 through 10.5 as vectors, where the coefficient of $2p_x$ is the x component and the coefficient of $2p_z$ is the z component. Now determine the angles between the hybrid orbitals using the formula for the dot product of two vectors. (Don't include the $2s$ orbital because it is spherically symmetric and so has no directionality.)

10-5. The following three orbitals are equivalent to the three sp^2 hybrid orbitals given by Equations 10.3 through 10.5

$$\phi_1 = \left(\frac{1}{3}\right)^{1/2} 2s - \left(\frac{1}{3}\right)^{1/2} 2p_x + \left(\frac{1}{3}\right)^{1/2} 2p_z$$

$$\phi_2 = \left(\frac{1}{3}\right)^{1/2} 2s + \frac{1}{2}(1 + 3^{-1/2})2p_x + \frac{1}{2}(1 - 3^{-1/2})2p_z$$

$$\phi_3 = \left(\frac{1}{3}\right)^{1/2} 2s + \frac{1}{2}(-1 + 3^{-1/2})2p_x - \frac{1}{2}(1 + 3^{-1/2})2p_z$$

First show that these orbitals are normalized. Now use the method introduced in Problem 10–4 to show that the angles between these orbitals are $120°$. (These orbitals are the orbitals given by Equations 10.3 through 10.5 rotated by $45°$.)

10-6. Given that one sp hybrid orbital is

$$\xi_1 = \frac{1}{\sqrt{2}}(2s + 2p_z)$$

construct a second one by requiring that it be normalized and orthogonal to ξ_1.

10-7. The relation between a tetrahedron and a cube is shown in the following figure:

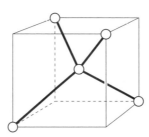

Use this figure to show that the bond angles in a regular tetrahedron are $109.47°$. (*Hint:* If we let the edge of the cube be of length a, then the diagonal on a face of the cube has a length $\sqrt{2}\,a$, by the Pythagorean theorem. The distance from the center of the cube to a face is equal to $a/2$. Using this information, determine the tetrahedral angle.)

10-8. Show that the sp^3 hybrid orbitals given by Equations 10.6 through 10.9 are orthonormal.

10-9. Using the vector approach described in Problem 10–4, show that the cosine of the angle between the sp^3 hybrid orbitals given by Equations 10.6 through 10.9 is $-1/3$. What is the angle equal to?

10-10. The sp^3 hybrid orbitals given by Equations 10.6 through 10.9 are symmetric but not unique. We construct an equivalent set in this problem. We can write the four sp^3 hybrid orbitals on the carbon atom as

$$\xi_1 = a_1 2s + b_1 2p_x + c_1 2p_y + d_1 2p_z$$

$$\xi_2 = a_2 2s + b_2 2p_x + c_2 2p_y + d_2 2p_z$$

$$\xi_3 = a_3 2s + b_3 2p_x + c_3 2p_y + d_3 2p_z \tag{1}$$

$$\xi_4 = a_4 2s + b_4 2p_x + c_4 2p_y + d_4 2p_z$$

By requiring these four hybrid orbitals to be equivalent, we have that $a_1 = a_2 = a_3 = a_4$. Because there is one $2s$ orbital distributed among four equivalent hybrid orbitals, we also say that $a_1^2 + a_2^2 + a_3^2 + a_4^2 = 1$. Thus, we have that $a_1 = a_2 = a_3 = a_4 = 1/\sqrt{4}$. Without loss of generality, we take one of the hybrid orbitals to be directed along the positive z axis. Because the $2p_x$ and $2p_y$ orbitals are directed along only the x and y axes, respectively, then b and c are zero in this orbital. If we let this orbital be ξ_1, then

$$\xi_1 = \frac{1}{\sqrt{4}} 2s + d_1 2p_z$$

By requiring that ξ_1 be normalized, show that

$$\xi_1 = \frac{1}{\sqrt{4}} 2s + \sqrt{\frac{3}{4}} 2p_z \tag{2}$$

Equation 2 is the first of our four sp^3 hybrid orbitals.

Without any loss of generality, take the second hybrid orbital to lie in the x-z plane, so that

$$\xi_2 = \frac{1}{\sqrt{4}} 2s + b_2 2p_x + d_2 2p_z \tag{3}$$

Show that if we require ξ_2 to be normalized and orthogonal to ξ_1, then

$$\xi_2 = \frac{1}{\sqrt{4}} 2s + \sqrt{\frac{2}{3}} 2p_x - \frac{1}{\sqrt{12}} 2p_z$$

Show that the angle between ξ_1 and ξ_2 is $109.47°$. Now determine ξ_3 such that it is normalized and orthogonal to ξ_1 and ξ_2. Last, determine ξ_4.

10-11. Calculate the bond angle between ψ_1 and ψ_2 in Example 10–4 using the vector approach described in Problem 10–4. (Remember not to use the $2s$ part of ψ_1 and ψ_2.)

10-12. Using the coordinate system shown below for a water molecule,

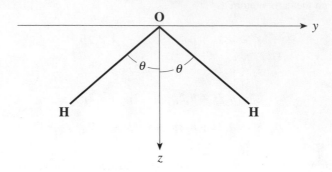

show that we can write the two bonding hybrid atomic orbitals on the oxygen atom as

$$\psi_1 = N[\gamma 2s + (\sin\theta)2p_y + (\cos\theta)2p_z]$$

and

$$\psi_2 = N[\gamma 2s - (\sin\theta)2p_y + (\cos\theta)2p_z]$$

where γ is a constant and N is the normalization constant. Now use the fact that these orbitals must be orthogonal to show that

$$\cos^2\theta - \sin^2\theta = \cos 2\theta = -\gamma^2$$

Finally, given that the H–O–H bond angle of water is 104.5°, determine the orthonormal hybrid orbitals ψ_1 and ψ_2 (see Equations 10.12 and 10.13).

10-13. In Problem 10–12, you found two bonding hybrid orbitals for the oxygen atom of a water molecule. In this problem, we will find the two equivalent lone-pair orbitals. Starting with the results of Problem 10–12, show that the third sp^2 hybrid orbital is given by

$$\psi_3 = 0.77 \cdot 2s - 0.64 \cdot 2p_z$$

At this point the lone pair orbitals are given by ψ_3 and the oxygen $2p_x$ orbital. Construct two equivalent lone pair orbitals by taking the appropriate linear combinations of ψ_3 and the $2p_x$ orbital. Which pair of orbitals, ψ_3 and the $2p_x$ orbital or your set of equivalent orbitals, is the correct description of the lone-pair orbitals for a water molecule? Explain your reasoning.

10-14. Figure 10.9 shows a schematic representation of the various molecular orbitals for a linear AH_2 molecule. We could draw similar pictures for the molecular orbitals of a linear XY_2 molecule. For example, the $3\sigma_g$ and $4\sigma_g$ molecular orbitals can be represented as

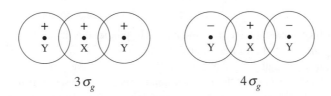

$$3\sigma_g \qquad\qquad 4\sigma_g$$

Draw a schematic representation of the $2\sigma_u$, $1\pi_u$, $2\pi_u$, and $1\pi_g$ orbitals.

10-15. Explain why the energies of the $3\sigma_g$ and $2\sigma_u$ orbitals for an XY_2 molecule are insensitive to small changes in the bond angle.

10-16. Explain why the doubly degenerate $1\pi_u$ orbitals for a linear XY_2 molecule do not remain degenerate when the molecule is bent.

10-17. Explain why the $3\sigma_u$ molecular orbital of a linear XY_2 molecule increases in energy as the molecule bends. (*Hint*: The $3\sigma_u$ molecular orbital is a linear combination of the $2p_z$ orbitals from each atom.)

10-18. Use Figure 10.25 to predict whether the following molecules are linear or bent:

    a. CO_2 **b.** CO_2^+ **c.** CF_2

10-19. Use Figure 10.25 to predict whether the following molecules are linear or bent:

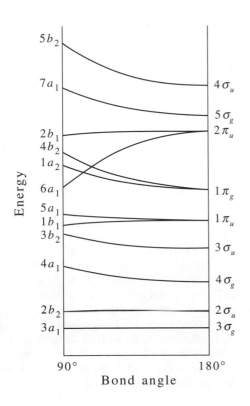

FIGURE 10.25
The Walsh correlation diagram for the valence electrons of a XY_2 molecule. The right side of the diagram gives the energy ordering of the molecular orbitals for an Y–X–Y bond angle of 180°. The left side gives the energy ordering of the molecular orbitals for an Y–X–Y bond angle of 90°. The solid lines tell us how the energies of the molecular orbitals depend upon Y–X–Y bond angles between 90° and 180°. The $1\sigma_g$, $2\sigma_g$, and $1\sigma_u$ orbitals correspond to the core $1s$ electrons on the bonded atoms and are not shown in the figure.

a. OF_2 **b.** NO_2^+ **c.** CN_2

10-20. Walsh correlation diagrams can be used to predict the shapes of polyatomic molecules that contain more than three atoms. In this and the following three problems we consider molecules that have the general formula XH_3. We will restrict our discussion to XH_3 molecules, where all the H–X–H bond angles are the same. If the molecule is planar, then the H–X–H bond angle is $120°$. A nonplanar XH_3 molecule, then, has an H–X–H bond angle that is less than $120°$. Figure 10.26 shows the Walsh correlation diagram that describes how the energies of the molecular orbitals for an XH_3 molecule change as a function of the H–X–H bond angle. Note that because XH_3 is not linear, the labels used to describe the orbitals on the two sides of the correlation diagram do not have designations such as σ and π. We see that the lowest-energy molecular orbital is insensitive to the H–X–H bond angle. Which atomic orbital(s) contribute to the lowest-energy molecular orbital? Explain why the energy of this molecular orbital is insensitive to changes in the H–X–H bond angle.

10-21. Consider the Walsh correlation diagram given in Figure 10.26. The $2a_1'$ molecular orbital of the planar XH_3 molecule is a linear combination of the $2p$ orbital on X that lies in the molecular plane and the $1s$ orbital on each hydrogen atom. Why does the energy of this molecular orbital increase as the H–X–H bond angle decreases from $120°$ to $90°$?

10-22. Orbitals designated by the letter "e" in a Walsh correlation diagram are doubly degenerate. Which atomic orbitals can contribute to the $1e'$ molecular orbitals of the planar XH_3 molecule?

10-23. Use the Walsh correlation diagram in Figure 10.26 to determine which of the following molecules are planar: (a) BH_3, (b) CH_3, (c) CH_3^-, and (d) NH_3. (Orbitals designated by the letter "e" are doubly degenerate.)

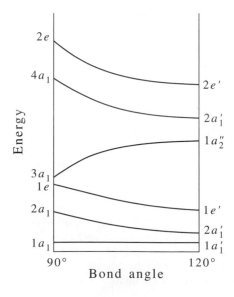

FIGURE 10.26
The Walsh correlation diagram for XH_3 molecules. The right side of the diagram gives the energy ordering of the molecular orbitals for an H–X–H bond angle of $120°$. The left side gives the energy ordering of the molecular orbitals for an H–X–H bond angle of $90°$. The solid lines tell us how the energies of the molecular orbitals depend upon H–X–H bond angles between $90°$ and $120°$.

10-24. Show that the π molecular orbital corresponding to the energy $E = \alpha - \beta$ for ethene is $\psi_\pi = \frac{1}{\sqrt{2}}(2p_{zA} - 2p_{zB})$.

10-25. Generalize our Hückel molecular-orbital treatment of ethene to include overlap of $2p_{zA}$ and $2p_{zB}$. Determine the energies and the wave functions.

10-26. Show that the four molecular orbitals for butadiene (Equation 10.18),

$$\psi_i = \sum_{j=1}^{4} c_{ij} 2p_{zj}$$

lead to the secular determinant given by Equation 10.19.

10-27. Show that

$$\begin{vmatrix} x & 1 & 0 & 0 \\ 1 & x & 1 & 0 \\ 0 & 1 & x & 1 \\ 0 & 0 & 1 & x \end{vmatrix} = 0$$

gives the algebraic equation

$$x^4 - 3x^2 + 1 = 0$$

10-28. Show that the four π molecular orbitals for butadiene are given by Equations 10.26.

10-29. Derive the Hückel theory secular determinant for benzene (see Equation 10.27).

10-30. Calculate the Hückel π-electron energies of cyclobutadiene. What do Hund's rules say about the ground state of cyclobutadiene? Compare the stability of cyclobutadiene with that of two isolated ethene molecules.

10-31. Calculate the Hückel π-electron energy of trimethylenemethane:

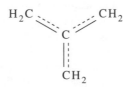

Compare the π-electron energy of trimethylenemethane with that of two isolated ethene molecules.

10-32. Calculate the π-electron energy levels and the total π-electron energy of bicyclobutadiene:

10-33. Show that the Hückel molecular orbitals of benzene given in Equations 10.31 are orthonormal.

10-34. Set up, but do not try to solve, the Hückel molecular-orbital theory determinantal equation for naphthalene, $C_{10}H_8$.

10-35. A Hückel calculation for naphthalene, $C_{10}H_8$, gives the molecular-orbital energy levels $E_i = \alpha + m_i\beta$, where the 10 values of m_i are 2.3028, 1.6180, 1.3029, 1.0000, 0.6180, -0.6180, -1.0000, -1.3029, -1.6180, and -2.3028. Calculate the ground-state π-electron energy of naphthalene.

10-36. The total π-electron energy of naphthalene (Problem 10–35) is

$$E_n = 10\alpha + 13.68\beta$$

Calculate the delocalization energy of naphthalene.

10-37. Using Hückel molecular-orbital theory, determine whether the linear state (H–H–H$^+$) or the triangular state

of H_3^+ is the more stable state. Repeat the calculation for H_3 and H_3^-.

10-38. Set up a Hückel theory secular determinant for pyridine.

10-39. The coefficients in Hückel molecular orbitals can be used to calculate charge distribution and bond orders. We will use butadiene as a concrete example. The molecular orbitals of butadiene can be expressed as

$$\psi_i = \sum_{j=1}^{4} c_{ij} 2p_{zj}$$

where the c_{ij} are determined by the set of linear algebraic equations that lead to the secular determinantal equation. The resulting molecular orbitals for butadiene are given by Equations 10.26:

$$\psi_1 = 0.3717\, 2p_{z1} + 0.6015\, 2p_{z2} + 0.6015\, 2p_{z3} + 0.3717\, 2p_{z4}$$

$$\psi_2 = 0.6015\, 2p_{z1} + 0.3717\, 2p_{z2} - 0.3717\, 2p_{z3} - 0.6015\, 2p_{z4}$$

$$\psi_3 = 0.6015\, 2p_{z1} - 0.3717\, 2p_{z2} - 0.3717\, 2p_{z3} + 0.6015\, 2p_{z4}$$

$$\psi_4 = 0.3717\, 2p_{z1} - 0.6015\, 2p_{z2} + 0.6015\, 2p_{z3} - 0.3717\, 2p_{z4}$$

These molecular orbitals are presented schematically in Figure 10.23. Because we have set $S_{ij} = \delta_{ij}$ in Equation 10.19, we have in effect assumed that the $2p_z$'s are orthonormal. Using this fact, show that the c_{ij} satisfy

$$\sum_{j=1}^{4} c_{ij}^2 = 1 \qquad i = 1,\ 2,\ 3,\ 4 \tag{1}$$

Equation 1 allows us to interpret c_{ij}^2 as the fractional π-electronic charge on the jth carbon atom due to an electron in the ith molecular orbital. Thus, the total π-electron charge on the jth carbon atom is

$$q_j = \sum_i n_i c_{ij}^2 \tag{2}$$

where n_i is the number of electrons in the ith molecular orbital. Show that

$$
\begin{aligned}
q_1 &= 2c_{11}^2 + 2c_{21}^2 + 0c_{31}^2 + 0c_{41}^2 \\
&= 2(0.3717)^2 + 2(0.6015)^2 \\
&= 1.000
\end{aligned}
$$

for butadiene. Show that the other q's are also equal to unity, indicating that the π electrons in butadiene are uniformly distributed over the molecule.

10-40. Another interesting quantity that can be defined in terms of the c_{ij} in Problem 10–39 is the π-bond order. We can interpret the product $c_{ir}c_{is}$ as the π-electron charge in the ith molecular orbital between the adjacent carbon atoms r and s. We define the π-bond order between the adjacent carbon atoms r and s by

$$P_{rs}^{\pi} = \sum_i n_i c_{ir} c_{is} \tag{1}$$

where n_i is the number of π electrons in the ith molecular orbital. Show that

$$P_{12}^{\pi} = 0.8942$$

and

$$P_{23}^{\pi} = 0.4473$$

for butadiene. Clearly, $P_{12}^{\pi} = P_{34}^{\pi}$ by symmetry. If we recall that there is a σ bond between each carbon atom, then we can define a total bond order

$$P_{rs}^{\text{total}} = 1 + P_{rs}^{\pi} \tag{2}$$

where the first term on the right side is due to the σ bond between atoms r and s. For butadiene, show that

$$P_{12}^{\text{total}} = P_{34}^{\text{total}} = 1.894$$

$$P_{23}^{\text{total}} = 1.447 \tag{3}$$

Equations 3 are in excellent agreement with the experimental observations involving the reactivity of these bonds in butadiene.

10-41. Calculate the delocalization energy, the charge on each carbon atom, and the bond orders for the allyl radical, cation, and anion. Sketch the molecular orbitals for the allyl system.

10-42. Calculate the π-electron charge on each carbon atom and the total bond orders in benzene. Comment on the result.

10-43. Because of the symmetry inherent in the Hückel theory secular determinants of linear and cyclic conjugated polyenes, we can write mathematical formulas for the energy levels for an arbitrary number of carbon atoms in the system (for present purposes, we consider cyclic polyenes with only an even number of carbon atoms). The formula for linear chains is

$$E_n = \alpha + 2\beta \cos \frac{\pi n}{N+1} \qquad n = 1, 2, \ldots, N$$

and the formula for cyclic chains with N even is

$$E_n = \alpha + 2\beta \cos \frac{2\pi n}{N} \qquad n = 0, \pm 1, \ldots, \pm \left(\frac{N}{2} - 1\right), \frac{N}{2}$$

where α and β are as defined in the text and N is the number of carbon atoms in the conjugated π system.

(a) Use these formulas to verify the results given in the chapter for butadiene and benzene.

(b) Now use these formulas to predict energy levels for linear hexatriene (C_6H_8) and octatetraene (C_8H_{10}). How does the delocalization energy of these molecules per carbon atom vary as the chains grow in length?

(c) Compare the results for hexatriene and benzene. Which molecule has a greater delocalization energy? Why?

10-44. The problem of a linear conjugated polyene of N carbon atoms can be solved in general. The energies E_j and the coefficients of the atomic orbitals in the jth molecular orbital are given by

$$E_j = \alpha + 2\beta \cos \frac{j\pi}{N+1} \qquad j = 1, 2, 3, \ldots, N$$

and

$$c_{jk} = \left(\frac{2}{N+1}\right)^{1/2} \sin \frac{jk\pi}{N+1} \qquad k = 1, 2, 3, \ldots, N$$

Derive the energy levels and the wave functions for butadiene using these formulas.

10-45. We can calculate the electronic states of a hypothetical one-dimensional solid by modeling the solid as a one-dimensional array of atoms with one orbital per atom, and using Hückel theory to calculate the allowed energies. Use the formula for E_j in Problem 10–44 to show that energies will form essentially a continuous band of width 4β. *Hint:* Calculate $E_1 - E_N$ and let N be very large so that you can use $\cos x \approx 1 - x^2/2 + \cdots$.

10-46. The band of electronic energies that we calculated in Problem 10–45 can accomodate N pairs of electrons of opposite spins, or a total of $2N$ electrons. If each atom contributes one electron (as in the case of a polyene), the band is occupied by a total of N electrons. Using some ideas you may have learned in general chemistry, would you expect such a system to be a conductor or an insulator?

10-47. The dipole moment of a polyatomic molecule is defined by

$$\boldsymbol{\mu} = e \sum_j z_j \mathbf{r}_j$$

where $z_j e$ is the magnitude of a charge located at the point given by $\mathbf{r}_j$. Show that the value of $\boldsymbol{\mu}$ is independent of the origin chosen for $\mathbf{r}_j$ if the net charge is zero. Show that $\boldsymbol{\mu} = 0$ for SO_3 (trigonal planar), CCl_4 (tetrahedral), SF_6 (octahedral), XeF_4 (square planar), and PF_5 (trigonal bipyramidal).

John Pople was born in Somerset, England, on October 31, 1925. He received his Ph.D. in mathematics from Cambridge University in 1951. He remained there until 1958 and then emigrated to the United States in 1964 to the Carnegie Institute of Technology (now Carnegie Mellon University). He then joined the faculty of Northwestern University in 1993, where he is still active. We will see in this chapter that much of modern computational quantum chemistry uses gaussian orbitals as a basis set. Although Pople was not the first person to propose the use of gaussian orbitals, he saw their computational advantages early on and over a period of years developed computational algorithms for the *ab initio* calculation of molecular properties based upon gaussian orbitals. The computer programs developed by Pople and his many collaborators have been packaged as a commercially available program called Gaussian 94, which is constantly being upgraded and extended by computational theorists employed in the private sector, as well as in academia. There are other commercially available computational quantum chemical programs, but Gaussian 94 is perhaps the most widely used. The availability of such programs has made it possible for chemistry students to calculate quantities such as molecular energies and structures, bond energies, dipole moments, vibrational frequencies, thermochemical properties, and reaction pathways. Pople shared the 1998 Nobel Prize in chemistry with Walter Kohn "for his development of computational methods in chemistry".

CHAPTER 11

Computational Quantum Chemistry

At one time, quantum-chemical calculations were the domain of professional quantum chemists using large, powerful mainframe computers. Over the years, however, computer programs have become readily available that can be used by nonexperts to calculate reliable values of molecular properties such as geometries and energies. With recent advances in computer technology, it is now possible to carry out computations on relatively large molecules without having to be a quantum chemist or a computer whiz.

Many of these programs express the atomic orbitals that make up a molecular orbital as linear combinations of Gaussian functions. We will see in this chapter that Gaussian functions are particularly convenient from a computational standpoint. Our discussion of Gaussian functions will lead us to designations such as STO-6G, 6-31G**, and other cryptic codes that are becoming part of the vocabulary of not just computational physical chemists, but also experimental physical chemists, inorganic chemists, organic chemists, and biochemists.

In the last section of this chapter, we will discuss GAUSSIAN 94, one of the most popular quantum chemical computer programs that is available today. GAUSSIAN 94 was developed by Professor John Pople and his colleagues at Carnegie-Mellon University over many years. As the name GAUSSIAN 94 implies, the latest version of this program was released in 1994, and newer versions are in development.

11–1. Gaussian Basis Sets Are Often Used in Modern Computational Chemistry

Contemporary molecular-orbital theory calculations of the properties of polyatomic molecules are done using computers. We will examine both how such calculations are carried out and the accuracy of the methods used in predicting the properties of molecules. In analogy to the discussion of multielectron atoms in Chapter 8, the wave function for a closed-shell molecule with N electrons (N must be an even number) is given by the Slater determinantal wave function

$$\psi(1, 2, \ldots, N) = \frac{1}{\sqrt{N!}} \begin{vmatrix} \psi_1(1)\alpha(1) & \psi_1(1)\beta(1) & \cdots & \psi_{N/2}(1)\alpha(1) & \psi_{N/2}(1)\beta(1) \\ \psi_1(2)\alpha(2) & \psi_1(2)\beta(2) & \cdots & \psi_{N/2}(2)\alpha(2) & \psi_{N/2}(2)\beta(2) \\ \vdots & \vdots & \vdots & \vdots & \vdots \\ \psi_1(N)\alpha(N) & \psi_1(N)\beta(N) & \cdots & \psi_{N/2}(N)\alpha(N) & \psi_{N/2}(N)\beta(N) \end{vmatrix} \quad (11.1)$$

where the individual entries are products of (one-electron) molecular orbitals and spin functions. To carry out a molecular-orbital theory calculation, we need to determine each of the molecular orbitals, ψ_i, as well as calculate the energy associated with this total wave function. The standard method for determining the molecular orbitals and their associated energies is to express the molecular orbitals as linear combinations of atomic orbitals (LCAO-MO) and then determine the coefficients in the linear combination by a self-consistent field calculation (LCAO-MO-SCF). The Hartee-Fock method, which we introduced in Section 8–7 to calculate orbitals of multi-electron atoms, is a systematic procedure for doing this. In Section 9–13 we briefly discussed Roothaan's extension of the Hartree-Fock method to the calculation of molecular orbitals in the LCAO-MO approximation. The set of resulting equations for the coefficients in the LCAO-MO are called the Roothaan-Hartree-Fock equations.

The set of atomic functions used to construct LCAO-MOs is called the *basis set*. In the case of the diatomic molecules discussed in Section 9–7, hydrogen-like atomic orbitals form the basis set. For example, $1s_{H_A}$ and $1s_{H_B}$ constitute the basis set for the molecular orbital $\sigma 1s$. The first basis set used in large-scale computational studies of polyatomic molecules consisted of Slater atomic orbitals, abbreviated STOs,

$$S_{nlm}(r, \theta, \phi) = \frac{(2\zeta)^{n+1/2}}{[(2n)!]^{1/2}} r^{n-1} e^{-\zeta r} Y_l^m(\theta, \phi) \quad (11.2)$$

introduced in Chapter 8. The difference between the STOs and the hydrogen-like atomic orbitals is that the Slater orbitals have no nodes and the *orbital exponent*, ζ (zeta), is not necessarily equal to Z/n. In principle, the orbital exponents should be chosen to minimize the energy, but this selection is still a formidable task even with modern computers. In practice, an optimal set of orbital exponents has been chosen that has turned out to be the most reliable in numerous molecular calculations. Table 11.1 lists these orbital exponents for the atoms in the first two rows of the periodic table. Note that the value of the orbital exponent for the Slater orbitals is the same for $2s$ and $2p$ orbitals. Also note that ζ increases with increasing atomic number, reflecting the contraction of the orbitals as the nuclear charge increases.

Although Slater orbitals were used for many years, they are not used directly any more because the integrals in the resulting secular determinants are difficult to evaluate. In particular, integrals involving more than one nuclear center, called *multicenter integrals*, are awkward to calculate using Slater orbitals. If we use Gaussian functions instead of Slater orbitals, however, all the multicenter integrals are very easy to evaluate

TABLE 11.1
Orbital exponents, ζ, for the Slater orbitals of the atoms of the first two rows of the periodic table.

Atom	ζ_{1s}	$\zeta_{2s} = \zeta_{2p}$
H	1.24	
He	1.69	
Li	2.69	0.80
Be	3.68	1.15
B	4.68	1.50
C	5.67	1.72
N	6.67	1.95
O	7.66	2.25
F	8.56	2.55
Ne	9.64	2.88

(Problem 11–9). Thus, it would seem desirable to use Gaussian-type orbitals of the form

$$G_{nlm}(r, \theta, \phi) = N_n r^{n-1} e^{-\alpha r^2} Y_l^m(\theta, \phi) \tag{11.3}$$

for the basis sets in molecular-orbital calculations. The problem with this idea is that Slater orbitals and Gaussian orbitals have very different behavior for small values of r. Figure 11.1 compares a normalized $S_{100} = \phi_{1s}^{STO}$ Slater orbital (Equation 11.2) with a normalized $G_{100} = \phi_{1s}^{GF}$ Gaussian orbital (Equation 11.3) for a hydrogen atom, with orbital exponents $\zeta = 1.24$ and $\alpha = 0.4166$ in ϕ_{1s}^{STO} and ϕ_{1s}^{GF}, respectively. [This value of α has been chosen to maximize the overlap between the ϕ_{1s}^{STO} orbital and the ϕ_{1s}^{GF} orbital (cf. Problem 11–10).] In carrying out calculations, we then use

$$\phi_{1s}(r) = \phi_{1s}^{STO}(r, 1.24) \tag{11.4}$$

or

$$\phi_{1s}(r) = \phi_{1s}^{GF}(r, 0.4166) \tag{11.5}$$

depending on whether we are using the Slater orbitals (STO) or Gaussian functions (GF) as our basis set. The $1s$ orbital in each of these basis sets is

$$\phi_{1s}^{STO}(r, \zeta) = S_{100}(r, \zeta) = \left(\frac{\zeta^3}{\pi}\right)^{1/2} e^{-\zeta r} \tag{11.6}$$

and

$$\phi_{1s}^{GF}(r, \alpha) = G_{100}(r, \alpha) = \left(\frac{2\alpha}{\pi}\right)^{3/4} e^{-\alpha r^2} \tag{11.7}$$

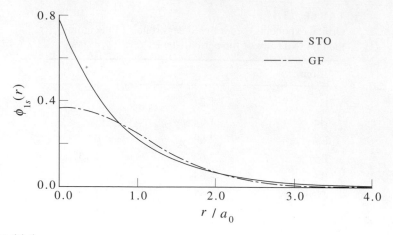

FIGURE 11.1
A comparison of the normalized Slater orbital, $S_{100} = \phi_{1s}^{STO}$ (Equation 11.6), to the Gaussian orbital (Equation 11.7), $G_{100} = \phi_{1s}^{GF}$, with orbital exponents $\zeta = 1.24$ and $\alpha = 0.4166$, respectively.

Note that the Slater orbital shown in Figure 11.1 has a cusp at $r = 0$, whereas the slope of the Gaussian orbital is zero at $r = 0$.

EXAMPLE 11–1
Show that $\phi_{1s} = \phi_{1s}^{STO}(0, 1.24) = 0.779$ and that $\phi_{1s} = \phi_{1s}^{GF}(0, 0.4166) = 0.370$ as shown in Figure 11.1.

SOLUTION: From Equation 11.6,

$$\phi_{1s}^{STO}(r, \zeta) = \left(\frac{\zeta^3}{\pi} \right)^{1/2} e^{-\zeta r}$$

Setting $r = 0$ and $\zeta = 1.24$ gives

$$\phi_{1s}^{STO}(0, 1.24) = 0.779$$

From Equation 11.7 with $\alpha = 0.4166$

$$\phi_{1s}^{GF}(r, 0.4166) = \left(\frac{(2)(0.4166)}{\pi} \right)^{3/4} e^{-0.4166 r^2}$$

and at $r = 0$

$$\phi_{1s}^{GF}(0, 0.4166) = 0.370$$

The Gaussian orbital does a reasonably good job of describing the Slater orbital for values of r greater than a_0, but it underestimates its magnitude for values of r less than a_0. These discrepancies turn out to be very significant in molecular calculations. To overcome this difficulty, a number of researchers in quantum chemistry have curve fit Slater orbitals to sums of Gaussian functions, the fit improving with N, the number of Gaussian functions used. Figure 11.2 shows this fit as a function of N. For example, for $N = 3$, the Slater orbital $\phi_{1s}^{STO}(r, 1.24) = 0.779e^{-1.24r}$ is expressed by

$$\phi_{1s}^{STO}(r) = \sum_{i=1}^{3} d_{1si}\phi_{1s}^{GF}(r, \alpha_{1si})$$

$$= 0.4446\,\phi_{1s}^{GF}(r, 0.1688) + 0.5353\,\phi_{1s}^{GF}(r, 0.6239) + 0.1543\,\phi_{1s}^{GF}(r, 3.425)$$

$$(11.8)$$

Because we are using a sum of three Gaussian functions to represent one Slater orbital, such a basis set is called the *STO-3G basis set*. In the STO-3G basis set, all atomic orbitals are described by a sum of three Gaussian functions. Although this procedure leads to a proliferation of integrals to evaluate, each one is relatively easy, so the overall procedure is quite efficient.

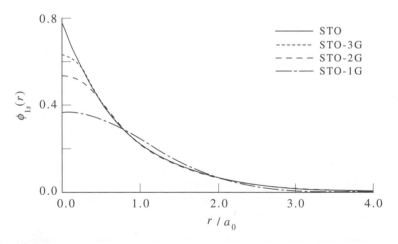

FIGURE 11.2
The Slater ϕ_{1s}^{STO} orbital is compared with $\phi_{1s}(r)$ represented by sums of different numbers of Gaussian functions.

EXAMPLE 11–2
Show that Equation 11.8 gives the value 0.628 at $r = 0$, as indicated in Figure 11.2.

SOLUTION: Using Equation 11.7, we have

$$\phi_{1s}^{GF}(r, 0.1688) = \left[\frac{(2)(0.1688)}{\pi}\right]^{3/4} e^{-0.1688r^2} = 0.1877 e^{-0.1688r^2}$$

$$\phi_{1s}^{GF}(r, 0.6239) = 0.5003 \, e^{-0.6239r^2}$$

$$\phi_{1s}^{GF}(r, 3.425) = 1.7943 \, e^{-3.425r^2}$$

Therefore,

$$\phi_{1s}(r = 0) = (0.4446)(0.1877) + (0.5353)(0.5003) + (0.1543)(1.7943)$$

$$= 0.6281$$

EXAMPLE 11–3

What is the general form of an atomic orbital in the STO-6G basis set? Use the data in Table 11.2 to write the expression for a Slater $1s$ orbital in the STO-6G basis set.

SOLUTION: The STO-6G basis set is one in which each Slater orbital is described by a sum of six Gaussian functions. Thus, for a $1s$ orbital, we have

$$\phi_{1s}^{STO}(r) = \sum_{i=1}^{6} d_{1si} \phi_{1s}^{GF}(r, \alpha_{1si})$$

where d_{1si} is the coefficient of each of the contributing Gaussian functions, and where $\phi_{1s}^{GF}(r, \alpha_{1si})$ is defined by Equation 11.7. Using the data in the Table 11.2, we see that the $1s$ orbital in the STO-6G basis set is given by the following expression:

$$\phi_{1s}^{STO}(r) = 0.1303 \, \phi_{1s}^{GF}(r, 0.1000) + 0.4165 \, \phi_{1s}^{GF}(r, 0.2431)$$

$$+ 0.3706 \, \phi_{1s}^{GF}(r, 0.6260) + 0.1685 \, \phi_{1s}^{GF}(r, 1.8222)$$

$$+ 0.0494 \, \phi_{1s}^{GF}(r, 6.5131) + 0.0092 \, \phi_{1s}^{GF}(r, 35.5231)$$

At $r = 0$, the STO-6G representation of the $1s$ orbital gives $\phi_{1s}(0) = 0.733$. The value of the STO $1s$ orbital at $r = 0$ is 0.779. Comparing Examples 11–2 and 11–3 shows that the STO-6G basis set provides a better representation of the $1s$ orbital than the STO-3G basis set for small values of r.

Each atomic orbital in the basis set is now expressed as a sum of Gaussian functions. If we let the atomic orbitals be denoted by ϕ_k, then the ith molecular orbital is

$$\psi_i = \sum_{k=1}^{M} c_{ki} \phi_k \tag{11.9}$$

where M is the number of atomic orbitals used to construct the molecular orbitals, or in other words, M is the number of atomic orbitals in the basis set. Realize that each

TABLE 11.2
Values of the expansion coefficients, d_{1si}, and exponents, α_{1si}, for a Slater $1s$ orbital with $\zeta = 1.24$ in the STO-6G basis set. These values are obtained from an "optimal" fit of a linear combination of six Gaussian functions to a $1s$ Slater orbital with $\zeta = 1.24$.

d_{1si}	α_{1si}
0.1303	0.1000
0.4165	0.2431
0.3706	0.6260
0.1685	1.8222
0.0494	6.5131
0.0092	35.5231

atomic orbital ϕ_k in Equation 11.9 is a sum of Gaussian functions, as in Equation 11.8. Our task now is to determine the c_{ki} in Equation 11.9 that minimize the energy of the molecule. This condition leads to a set of algebraic equations for the coefficients called the Roothaan equations, which can be written as

$$\sum_{j=1}^{M}(F_{ij} - E_i S_{ij})c_{ji} = 0 \tag{11.10}$$

The Roothaan equations are an extension of the Hartree-Fock equations to polyatomic molecules. The quantity F_{ij} is the ijth matrix element of the Fock operator and S_{ij} is the overlap integral between the basis functions ϕ_i and ϕ_j. Equation 11.10 will have a nontrivial solution only if

$$|F_{ij} - E_i S_{ij}| = 0 \tag{11.11}$$

Here, as for diatomic molecules (Section 9–13), the Fock operator depends on the coefficients in the molecular-orbital expansion, Equation 11.9, and so Equation 11.11 must be solved self-consistently. As we said in the introduction to this chapter, a number of user-friendly computer programs can be used to solve Equation 11.11 in order to calculate molecular properties to a high degree of accuracy. Most of these programs express the atomic orbitals in Equation 11.9 in terms of Gaussian functions.

11–2. Extended Basis Sets Accurately Account for the Size and Shape of Molecular Charge Distributions

While the STO-NG ($N = 1,\ 2,\ 3,\ \ldots$) basis sets were popular in the 1980s, they are not widely used today. Using a finite sum of Gaussian functions to describe an atomic orbital results in several inadequacies that affect the accuracy of the calculations. Here we consider one of the major limitations, and then we will learn how this problem can be overcome by modifying the linear combination of Gaussian functions used to represent an atomic orbital.

Because the atomic orbitals in the STO-NG basis sets use fixed exponents, α_{ki}, all orbitals of a given type are identical in size. For example, the p_x, p_y, and p_z atomic orbitals all have the same radial function, $r \exp(-\alpha r^2)$ (Equation 11.3) and thus are identical. However, this generally will not give an accurate picture of the electron density for a particular atom within a molecule. Consider the linear triatomic molecule HCN, where we define the bonds to lie along the z direction. In HCN, the p_z orbitals on the carbon and nitrogen atoms form a σ orbital, and the p_x and p_y orbitals form π orbitals. We learned in Chapters 9 and 10 that π orbitals are more diffuse than σ orbitals; therefore we expect the radial function for the p_z orbital to peak at a smaller value of r than the radial function for the p_x or p_y orbital does. Such a description requires one value of the orbital exponent for the $2p_z$ orbital and a different value for the $2p_x$ and $2p_y$ orbitals. Furthermore, such effects are expected to be molecule dependent, implying that different molecules would have different orbital exponents.

A similar problem arises in the inability of the STO-NG basis sets to reproduce the anisotropic charge distributions of many hydrocarbon molecules. We can visualize this problem by comparing methane and ethyne. In methane, the electron densities in the four sp^3 hybrid orbitals are equivalent. Thus, expressing the three $2p$ orbitals by functions that have the same orbital exponent is reasonable. For ethyne, however, the electron density along the bond direction is much less diffuse than that along the axes perpendicular to the bond. This leads to an anisotropic distribution of the electron density, an effect that cannot be described by the STO-NG basis set because all three $2p$ orbitals have the same r dependence.

Tabulating separate mathematical functions to describe the atomic orbitals for every molecule is clearly impractical. We need a general approach that allows the size of the orbital to be optimized as part of the Hartree-Fock calculation. This can be done if the basis set consists of functions that can adjust the shape of the atomic orbital. Computational chemists have solved this problem by expressing each atomic orbital as a sum of two Slater-type orbitals that differ only in the value of their exponent ζ. For example, the $2s$ orbital is written as

$$\phi_{2s}(r) = \phi_{2s}^{\text{STO}}(r, \zeta_1) + d\phi_{2s}^{\text{STO}}(r, \zeta_2) \tag{11.12}$$

The advantage of this approach can be seen as follows. The Slater orbitals $\phi_{2s}^{\text{STO}}(r, \zeta_1)$ and $\phi_{2s}^{\text{STO}}(r, \zeta_2)$ represent different size $2s$ orbitals. Using a linear combination of these two functions, we can construct an atomic orbital whose size can range between that specified by $\phi_{2s}^{\text{STO}}(r, \zeta_1)$ and $\phi_{2s}^{\text{STO}}(r, \zeta_2)$ by varying the constant d, as shown in Figure 11.3. Because both functions are of the same type (ϕ_{2s}^{STO} in this case), the linear combination retains the desired symmetry of the atomic orbital. Basis sets generated from a sum of two Slater orbitals with different orbital exponents are called *double-zeta*

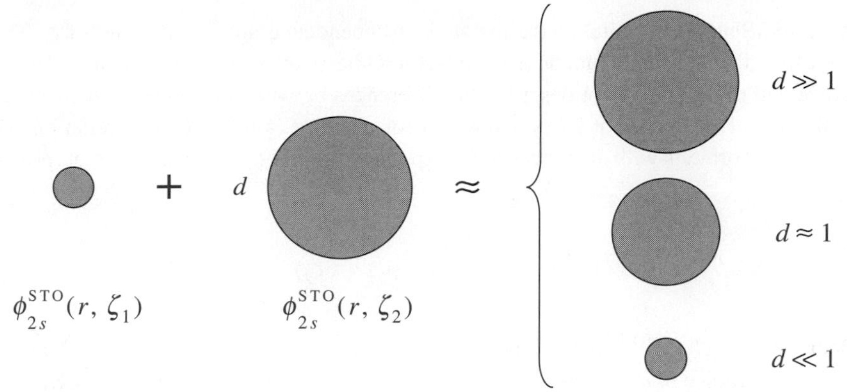

FIGURE 11.3
A linear combination of two Slater orbitals of the same type (ϕ_{2s}^{STO} in the case shown) but with different orbital exponents ζ_1 and ζ_2 can generate an atomic orbital of adjustable size by varying the constant d.

basis sets because each orbital in the basis set is the sum of two Slater orbitals that differ only in their value of the orbital exponent, ζ (zeta).

In general, only the valence orbitals are expressed by a double-zeta representation. The inner-shell electrons are still described by a single Slater orbital. For example, the electrons in the $1s$ atomic orbital on a carbon atom would be described by a single ϕ_{1s}^{STO} Slater orbital with ζ given in Table 11.1, whereas the electrons in the $2s$ atomic orbital would be described by a linear combination of two ϕ_{2s}^{STO} Slater orbitals with different values of the orbital exponent, ζ. Basis sets that describe the inner-shell electrons by a single Slater orbital and the valence-shell electrons by a sum of Slater orbitals are commonly referred to as *split-valence basis sets*.

EXAMPLE 11–4
Describe how a double-zeta basis set can be used to overcome the problems encountered in describing the $2p$ orbitals on the carbon atom in HCN by using only a STO-3G basis set.

SOLUTION: The (normalized) Slater orbitals for the $2p_x$, $2p_y$, and $2p_z$ orbitals are given by

$$\phi_{2p_x}^{STO}(r, \zeta) = \left(\frac{\zeta^5}{\pi}\right)^{1/2} xe^{-\zeta r}$$

$$\phi_{2p_y}^{STO}(r, \zeta) = \left(\frac{\zeta^5}{\pi}\right)^{1/2} ye^{-\zeta r}$$

$$\phi_{2p_z}^{STO}(r, \zeta) = \left(\frac{\zeta^5}{\pi}\right)^{1/2} ze^{-\zeta r}$$

All three Slater $2p$ orbitals have the same r dependence and same value of ζ. The corresponding Gaussian functions in an STO-3G basis set also have the same r dependence and therefore cannot describe the differences between the $2p$ orbitals involved in the σ and π orbitals of HCN. Consider, for example, a linear combination of two $\phi_{2p_x}^{STO}$ Slater orbitals with different orbital exponents, ζ_1 and ζ_2 (for discussion purposes, $\zeta_1 > \zeta_2$):

$$\phi_{2p_x}^{STO}(r, \zeta_1) + d\phi_{2p_x}^{STO}(r, \zeta_2)$$

Because $\zeta_1 > \zeta_2$, $\phi_{2p_x}^{STO}(r, \zeta_2)$ will be larger (or more diffuse) than $\phi_{2p_x}^{STO}(r, \zeta_1)$. Depending on our choice of d, we can vary the size of the $2p_x$ orbital between that represented by the two individual functions $\phi_{2p_x}^{STO}(r, \zeta_1)$ and $\phi_{2p_x}^{STO}(r, \zeta_2)$. The same procedure allows us to vary the size of the $2p_y$ and $2p_z$ orbitals.

Because the π orbitals are more diffuse than the σ orbital, the $2p_x$ and $2p_y$ orbitals on the carbon and nitrogen atoms must be larger than the $2p_z$ orbital. This difference in size can be accounted for by an appropriate choice of the coefficient d for each orbital. The larger Slater orbitals ($\phi_{2p_i}^{STO}(r, \zeta_2)$, $i = x$, y, or z), will make a greater contribution to the $2p_x$ and $2p_y$ orbitals than to the $2p_z$ orbital. The smaller Slater orbital $\phi_{2p_z}^{STO}(r, \zeta_1)$ will make a greater contribution to the $2p_z$ orbital than to the $2p_x$ and $2p_y$ orbitals.

How different are ζ_1 and ζ_2 in the double-zeta basis sets? The optimal orbital exponents for the $2p_x$, $2p_y$, and $2p_z$ orbitals of the carbon atom in HCN are 1.51, 1.51, and 2.08, respectively. (The z-axis lies along the bonds.) The standard orbital exponent in the STO-NG data set for the $2p$ orbital on a carbon atom is 1.72 (Table 11.1). Thus, the contraction of the $2p_z$ orbitals and expansion of the $2p_x$ and $2p_y$ orbitals of the carbon atom in a HCN molecule correspond to roughly a 20% change in the value of ζ. A linear combination of a smaller $2p_z$ orbital ($\zeta_1 \approx 2.08$) and larger $2p_x$ and $2p_y$ orbitals ($\zeta_2 \approx 1.51$) could be used to describe the $2p$ orbitals on the carbon atom in HCN.

Once again, to facilitate the evaluation of the secular determinant, each Slater orbital in the split-valence basis set is expressed in terms of Gaussian functions. Thus, each of the two Slater orbitals, $\phi_{2s}^{STO}(r, \zeta_1)$ and $\phi_{2s}^{STO}(r, \zeta_2)$, in Equation 11.12 is a linear combination of Gaussian functions. In principle, any number of Gaussian functions can be used to describe $\phi_{2s}^{STO}(r, \zeta_1)$ and $\phi_{2s}^{STO}(r, \zeta_2)$, giving rise to an infinite number of possible basis sets. We need a short-hand notation that tells us the number of Gaussian functions used to describe the various Slater atomic orbitals in a split-valence basis set. We will use the notation N-MPG, where N is the number of Gaussian functions used to describe the inner-shell orbitals; the hyphen indicates that we have a split-valence basis set; and the numbers M and P designate the number of Gaussian functions that are used to fit $\phi_{2s}^{STO}(r, \zeta_1)$ and $\phi_{2s}^{STO}(r, \zeta_2)$, respectively. Because $\zeta_1 > \zeta_2$ (by convention), M corresponds to the number of Gaussian functions used to express the smaller Slater orbital and P corresponds to the number of Gaussian functions used to express the larger Slater orbital. The G simply tells us that we are using Gaussian functions.

For example, one popular split-valence basis set used extensively in computational chemistry is the 6-31G basis set. Consider a carbon atom in a 6-31G basis. The 6 tells us that the $1s$ orbital on the carbon atom (the inner-shell orbital) is given by a sum of 6 Gaussian functions. The hyphen indicates a split-valence basis set, telling us that the valence $2s$ and $2p$ orbitals are each represented by a pair of Slater orbitals. One of these Slater orbitals, the smaller one, is represented by a sum of three Gaussian functions (hence the 3), and the larger orbital is represented by a single Gaussian function (hence the 1). The time required to evaluate the elements of the secular determinant depends upon the number of functions used. Computational chemists need to use computer time efficiently, thus one of the most important decisions in performing any calculation is the choice of basis set. We will explore this issue in more detail in the final two sections. First, however, we need to construct some of the more popular Gaussian basis sets that are derived from a split-valence representation of the atomic orbitals.

EXAMPLE 11–5
Describe the general procedure for constructing the functional forms of the $1s$ and $2s$ orbitals for a carbon atom in the 5-31G basis set. (See Table 11.3.)

SOLUTION: The label 5-31G tells us we are using a Gaussian basis set for which the inner-shell ($1s$) orbitals are described by a sum of 5 Gaussian functions and the valence orbitals are described by a double-zeta representation where one Slater orbital (the smaller one) is represented by a linear combination of three Gaussian functions and the other Slater orbital (the larger one) is represented by a single Gaussian function. The $1s$ orbital in the 5-31G basis set is given by the best fit of a sum of 5 Gaussian functions to a single ϕ_{1s}^{STO} Slater orbital using $\zeta = 5.67$ from Table 11.1. Thus,

$$\phi_{1s}^{STO}(r, 5.67) = \sum_{i=1}^{5} d_{1si} \phi_{1si}^{GF}(r, \alpha_{1si})$$

where the values for d_{1si} and α_{1si} for the carbon atom in the 5-31G basis set are given in Table 11.3.

TABLE 11.3
The coefficients and orbital exponents for the five Gaussian functions of the ground-state carbon atom in the 5-31G basis set. These values are determined by an "optimal" fit of a linear combination of five Gaussian functions to a Slater orbital with $\zeta = 5.67$.

α_{1si}	d_{1si}	$\alpha_{2si} = \alpha_{2pi}$	d_{2si}	α_{2s}'
1264.250	0.005473	7.942731	−0.1207731	0.158512
190.1443	0.040791	1.907238	−0.1697932	
43.12859	0.181220	0.5535774	1.149812	
11.94438	0.463485			
3.651485	0.452471			

The $2s$ orbital is described by a double-zeta basis set, or a linear combination of two ϕ_{2s}^{STO} Slater orbitals, $\phi_{2s}^{STO}(r, \zeta_1)$ and $\phi_{2s}^{STO}(r, \zeta_2)$, of different orbital exponents, $\zeta_1 > \zeta_2$. The smaller Slater orbital, $\phi_{2s}^{STO}(r, \zeta_1)$, is described by a linear combination of 3 Gaussian functions, and the larger Slater orbital, $\phi_{2s}^{STO}(r, \zeta_2)$, is described by a single Gaussian function. Thus, we have

$$\phi_{2s}^{STO}(r, \zeta_1) = \sum_{i=1}^{3} d_{2si} \phi_{2s}^{GF}(r, \alpha_{2si})$$

$$\phi_{2s}^{STO}(r, \zeta_2) = \phi_{2s}^{GF}(r, \alpha'_{2s})$$

so the $2s$ orbital in the 5-31G basis set is given by

$$\phi_{2s}(r) = \sum_{i=1}^{3} d_{2si} \phi_{2s}^{GF}(r, \alpha_{2si}) + d'_{2s} \phi_{2s}^{GF}(r, \alpha'_{2s})$$

where the values of the d_{2si} the α_{2si}, and α'_{2s} are given in Table 11.3. The value of the coefficient d'_{2s} is optimized as part of the Roothaan-Hartree-Fock procedure. The three $2p$ orbitals are given by a similar procedure.

11–3. Asterisks in the Designation of a Basis Set Denote Orbital Polarization Terms

Consider the formation of a simple $\sigma 1s$ molecular orbital in H_2. This orbital is formed from a $1s$ orbital on each hydrogen atom. Surely, however, the electron distribution about each hydrogen atom does not remain spherically symmetric as the two atoms approach each other. We can take this effect into account in the following way. If we let the internuclear axis be the z-axis, we can construct the molecular orbital from a linear combination of a $1s$ orbital and a $2p_z$ orbital on each hydrogen atom instead of from just a $1s$ orbital. In this manner, we can generally account for the fact that atomic orbitals distort as atoms are brought together. Such an effect is called a *polarization effect*. We can account for polarization by adding orbitals of higher orbital angular momentum quantum number, l, to the mathematical expression for a given atomic orbital just as we added a $2p_z$ orbital to a $1s$ hydrogen orbital above. For example, d character can be added to the description of the valence electrons in $2p$ orbitals, thereby providing a representation of the asymmetric shape of the electron density along the chemical bonds involving $2p$ orbitals. The addition of $3d$ orbitals to the $2p$ orbitals of the atoms of the second row elements in the periodic table is denoted by an asterisk *, for example, 6-31G*. A double asterisk, **, denotes that polarization is also being taken into account for the orbital descriptions on hydrogen atoms by adding $2p$ orbitals to the hydrogen $1s$ orbitals. In particular, the basis set 6-31G** is used in systems in which hydrogen bonding occurs. As you may have surmised, there are many basis sets used in modern computation chemistry (Table 11.4).

TABLE 11.4

Various split-valence Gaussian basis sets for the first- and second-row atoms in the periodic table.

Basis set[a]	Atom	Atomic orbital	Inner shell	Number of contributing Gaussian functions — Valence shell — Smaller	Larger
3-21G	H	$1s$	—	2	1
	Li–Ne	$1s$	3	—	—
		$2s$, $2p$	—	2	1
5-31G	H	$1s$	—	3	1
	Li–Ne	$1s$	5	—	—
		$2s$, $2p$	—	3	1
6-31G	H	$1s$	—	3	1
	Li–Ne	$1s$	6	—	—
		$2s$, $2p$	—	3	1

[a] Addition of a * superscript (e.g., 6-31G*) indicates that a single set of Gaussian $3d$ functions is added to the split-valence basis set description of each non-hydrogen atom. A ** superscript (e.g., 6-31G**) indicates that a single set of Gaussian $2p$ functions is also added to the split-valence basis set description of each hydrogen atom.

EXAMPLE 11–6

Table 11.8 gives the results of 6–31G** calculation of a water molecule. Describe what is meant by this notation.

SOLUTION: Let's refer to Table 11.4. The "6" in 6–31G** tells us that the $1s$ orbital (the inner-shell orbital) on the oxygen atom is represented by a linear combination of six Gaussian functions. The six Gaussian functions represent a Slater $1s$ orbital with $\zeta = 7.66$ (see Table 11.1). The "31" in 6–31G** tells us that the hydrogen $1s$ orbital and the $2s$ and $2p$ orbitals on the oxygen atom are represented by a double-zeta basis set, with the smaller orbitals represented by a linear combination of three Gaussian functions and the larger orbitals represented by a single Gaussian function. The "**" part of 6–31G** tells is two things. First, d orbital character, represented by a single set of Gaussian $3d$ functions, is added to the oxygen $2p$ orbitals and, in addition, p orbital character, represented by a single set of Gaussian $2p$ functions, is added to the hydrogen $1s$ orbitals to account for a polarization effect in each case.

Table 11.5 lists the coefficients d_{ki} and exponents α_{ki} for the split-valence 6-31G basis functions for the carbon, nitrogen, and oxygen atoms. Tabulated values can be found for many atoms. Example 11–7 shows that it is easy to write the basis set functions for the various atomic orbitals from such tables.

Even the most sophisticated basis sets we have considered here, 6-31G* and 6-31G**, have deficiencies that limit their ultimate use. Quantum chemists are currently

TABLE 11.5

The Gaussian exponents and expansion coefficients for the $n = 1$ ($1s$) and $n = 2$ ($2s$, $2p$) orbitals of atomic carbon, nitrogen, and oxygen in the 6-31G basis set.

Atom	α_{1si}	d_{1si}	$\alpha_{2si} = \alpha_{2pi}$	d_{2si}	d_{2pi}	$\alpha'_{2s} = \alpha'_{2p}$
C	3047.5	0.0018347	7.8683	−0.11933	0.068999	0.15599
	457.47	0.014037	1.8813	−0.16085	0.31642	
	103.95	0.068843	0.54425	1.1435	0.74431	
	29.210	0.23218				
	9.2867	0.46794				
	3.1639	0.36312				
N	4173.5	0.0018348	11.862	−0.11496	0.067579	0.22077
	627.46	0.013995	2.7714	−0.16912	0.32391	
	142.90	0.068587	0.7890	1.1458	0.74089	
	40.234	0.23224				
	12.820	0.46907				
	4.3904	0.36046				
O	5484.7	0.0018311	15.855	−0.11078	0.070874	0.28114
	825.23	0.039502	3.6730	−0.14803	0.33975	
	188.05	0.068445	1.0343	1.1308	0.72716	
	52.965	0.23271				
	16.898	0.47019				
	5.7996	0.35852				

exploring triple-zeta and quadruple-zeta basis sets (larger sums of Slater-type orbitals). With improvements in computational speed and computer algorithms, calculations using larger and more accurate basis sets will become practical.

EXAMPLE 11–7

Using the data in Table 11.5, determine the mathematical formulas for the $1s$, $2s$, and $2p$ orbitals for a carbon atom in the 6-31G basis set?

SOLUTION: Using the data in Table 11.5, we can write out the functional forms of the $1s$, $2s$, and $2p$ orbitals of a carbon atom in the 6-31G basis set.

$$\phi_{1s}(r) = \sum_{i=1}^{6} d_{1si} \phi_{1s}^{GF}(r, \alpha_{1si})$$

$$= 0.0018347\, \phi_{1s}^{GF}(r, 3047.5) + 0.014037\, \phi_{1s}^{GF}(r, 457.47)$$

$$+ 0.068843\, \phi_{1s}^{GF}(r, 103.95) + 0.023218\, \phi_{1s}^{GF}(r, 29.210)$$

$$+ 0.46794\, \phi_{1s}^{GF}(r, 9.2867) + 0.36231\, \phi_{1s}^{GF}(r, 3.1639)$$

$$\phi_{2s}(r) = \sum_{i=1}^{3} d_{2si}\phi_{2s}^{\text{GF}}(r, \alpha_{2si}) + d_{2s}'\phi_{2s}^{\text{GF}}(r, \alpha_{2s}')$$

$$= -0.11933\ \phi_{2s}^{\text{GF}}(r, 7.8683) - 0.16085\ \phi_{2s}^{\text{GF}}(r, 1.8813)$$

$$+ 1.1435\ \phi_{2s}^{\text{GF}}(r, 0.54425) + d_{2s}'\phi_{2s}^{\text{GF}}(r, 0.15599)$$

$$\phi_{2p}(r) = \sum_{i=1}^{3} d_{2pi}\phi_{2p}^{\text{GF}}(r, \alpha_{2pi}) + d_{2p}'\phi_{2p}^{\text{GF}}(r, \alpha_{2p}')$$

$$= 0.068999\ \phi_{2p}^{\text{GF}}(r, 7.8683) + 0.31642\ \phi_{2p}^{\text{GF}}(r, 1.8813)$$

$$+ 0.74431\ \phi_{2p}^{\text{GF}}(r, 0.54425) + d_{2p}'\phi_{2p}^{\text{GF}}(r, 0.15599)$$

The values of d_{2s}' and d_{2p}' are optimized as part of the Hartree-Fock procedure.

11–4. The Ground-State Energy of H$_2$ Can Be Calculated Essentially Exactly

For our first calculation, let's return to a discussion of the simplest diatomic molecule, H$_2$. In Chapter 9, molecular orbitals for H$_2$ were generated using the LCAO-MO approach. When only the $1s$ orbitals on the two bonded atoms are used to generate the $1\sigma_g$ molecular orbital, an energy of $-1.099\ E_{\text{h}}$ is obtained. In light of what we have learned in this chapter, however, limiting the basis set for this molecular orbital to a sum of just two atomic orbitals is clearly a severe approximation. It is worth investigating what energies are obtained for the lowest energy molecular orbital of H$_2$ when different basis sets are used. GAUSSIAN 94 calculations of the total energy of H$_2$ and the equilibrium bond length using the different basis sets considered in the last section are given by the first six entries of Table 11.6.

TABLE 11.6

Calculated energy and bond length for H$_2$ for different basis sets.

Description of wave functions	Total energy/E_{h}	R_e/pm
Molecular orbital ($1s_{\text{A}} + 1s_{\text{B}}$)	-1.099	85.0
STO-3G	-1.117	71.2
STO-6G	-1.124	71.1
3-21G	-1.123	73.5
6-31G	-1.127	73.0
6-31G**	-1.131	73.2
Best configuration interaction	-1.174	74.2
Experimental	-1.174	74.2

For all the basis sets used, the calculated energy is greater than the experimental value, in accord with the variational principle. The 6-31G** basis set affords the maximum flexibility in determining the molecular orbitals of H_2, yet the calculated total energy differs from the experimental value by 3.6%, and the calculated bond length is 1.0 pm too short. If we are to have confidence in the accuracy of calculated energies and geometries of polyatomic molecules, we should be able to determine the energy and bond length of H_2 essentially exactly.

We know from Chapter 8 that the difference between the energy calculated for the Hartree-Fock theory and the exact energy is called the correlation energy. For calculations using the 6-31G** basis set, the correlation energy of H_2 is $-0.043 E_h$. This correlation energy arises from the correlation between electrons of opposite spin and is ignored by the one-electron orbitals used in the Hartree-Fock method.

We can use the simple H_2 molecular-orbital wave function given by Equation 9.30 to discuss the limitations of the Hartree-Fock procedure. Ignoring the normalization constant, Equation 9.30 gives

$$\psi_{MO} = 1s_A(1)1s_B(2) + 1s_A(2)1s_B(1) + 1s_A(1)1s_A(2) + 1s_B(1)1s_B(2) \qquad (11.13)$$

We can interpret the four terms in this wave function. The first two terms represent one electron on each nucleus, there being two terms because of the indistinguishability of the two electrons. These two terms describe a purely covalently bonded hydrogen molecule, given by the Lewis formula H–H. The last two terms in Equation 11.13 represent both electrons being on nucleus A or both on nucleus B. These two terms describe purely ionic structures for H_2, given by the Lewis formulas:

$$H_A : \quad H_B \qquad\qquad H_A \quad : H_B$$

Thus, according to Equation 11.13, the ionic terms $[1s_A(1)1s_A(2)$ and $1s_B(1)1s_B(2)]$ carry the same weight as the covalent terms $[1s_A(1)1s_B(2)$ and $1s_B(1)1s_A(2)]$. Thus, dissociation of H_2 is just as likely to create the ions H^+ and H^- as it is to create neutral hydrogen atoms. This prediction clearly is not correct. However, this prediction is reflected in Figure 11.4, where the potential energy curves for H_2 are plotted against distance for a Hartree-Fock calculation using the 6-31G** basis set.

We see that the Hartree-Fock calculation (6-31G** basis set) gives an energy that exceeds the energy of two neutral hydrogen atoms for distances greater than $\approx 3a_0$. In addition, the calculated potential energy underestimates the bond strength at all distances. This result stems from the fact that the single determinantal wave function (Equation 11.1) overemphasizes the ionic configuration (H^+H^-). These results show that an accurate description of the molecular bonding cannot be achieved using a single Slater determinantal wave function. Several methods are used to improve upon a Hartree-Fock calculation. One method employs linear combinations of Slater determinants involving excited electronic states of the molecule instead of using just a single Slater determinant involving the lowest orbitals for the molecular wave function. This approach is called *configuration interaction*, and it is the most complete treatment of electronic structure possible for a given basis set. The details of configuration

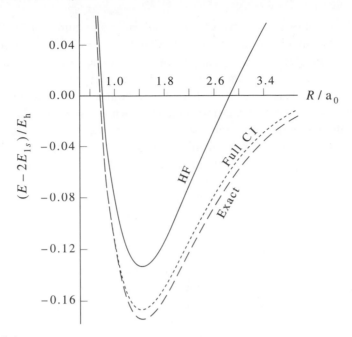

FIGURE 11.4

The energy of H_2 is plotted as a function of internuclear separation for Hartree-Fock calculations (HF) using the 6-31G** basis set with and without configuration interaction. These calculations are compared with the exact result published by Kolos and Wolniewicz in 1968. The calculation including configuration interaction is nearly identical to the exact result.

interaction are beyond the scope of this book, but you should be aware of its existence because most calculations reported today use it. Table 11.6 lists the calculated energy for H_2 obtained using Hartree-Fock theory with configuration interaction. The calculated potential energy curve using the 6-31G** basis set along with configuration interaction is also shown in Figure 11.4. The calculated curve is nearly identical to the extremely accurate results published later by Kolos and Wolniewicz in 1968 using a variational trial function containing more than 100 variational parameters that was limited to H_2. This example demonstrates the power of modern quantum-mechanical calculations. Fortunately, most of the commercially available quantum chemistry programs have routines for including configuration interaction so that calculations beyond the Hartree-Fock approximation are readily accessible.

11–5. GAUSSIAN 94 Calculations Provide Accurate Information About Molecules

We have said that a number of commercially available computer programs can be used by nonexperts to calculate molecular properties. GAUSSIAN 94, GAMESS, and SPARTAN are three of the most widely used. In this section, we will discuss

GAUSSIAN 94 and examine some of the actual input and output of GAUSSIAN 94 using water and ammonia as examples. To carry out a GAUSSIAN 94 calculation, you must provide three pieces of information: the method and level of calculation to be used, the total charge and spin multiplicity of the molecule to be considered, and the starting geometry (x, y, and z coordinates) of each atom in the molecule.

A sample input file for a water molecule is shown in Table 11.7. Line 1 specifies the name of the file (water) that will be used to store the results of the calculation. Line 2 contains a string of control words preceded by the # character. These words tell GAUSSIAN 94 various details about the desired calculation. For this example, the level of theory is a RHF (Restricted Hartree-Fock) calculation using the 6-31G* basis set. The term "restricted" means that the wave function is written in the form given by Equation 11.1; that is, all orbitals are doubly occupied, meaning that all spins are paired. The second control sequence "geom=coord" tells the computer program how the initial geometry of the molecule is specified in the input file; in this case, "coord" is a short-hand notation for Cartesian coordinates. The final control word "opt" tells the program is to optimize the molecular geometry as part of the calculation. If this phrase were deleted, the atoms would be restricted to the positions indicated in the input file, and the energy of the molecule would be minimized by the Hartree-Fock procedure for this fixed geometry. The third line of the input file is left blank to indicate the end of the sequence of control words. Line 4 is reserved for the title of the calculation. Here we have used a descriptive title so that we know the molecule, the level of calculation, and the basis set used. Line 5 is left blank to indicate the termination of the title. Line 6 contains the total charge and spin multiplicity of water. Water is a neutral molecule, hence the total charge is zero. Recall that the valence-electron configuration of water is $(2a_1)^2(1b_2)^2(3a_1)^2(1b_1)^2$ (see Figure 10.13) and therefore water has a net spin of $s = 0$, which corresponds to a total spin multiplicity $S = 2s + 1$, or $S = 1$, a singlet

TABLE 11.7

An input file for the GAUSSIAN 94 program for a water molecule.

```
Line #
    1   %chk=water
    2   #rhf/6-31G* geom=coord opt
    3
    4   Restricted Hartree Fock Calculation of Water with the
        6-31G* basis set
    5
    6   0   1
    7   H    0.754565   0.000000    0.4587771
    8   O    0.000000   0.000000   -0.1146943
    9   H   -0.754565   0.000000    0.4587771
   10
```

state. Lines 7 to 9 specify the type of atom and Cartesian coordinates (x, y, z) for each atom in the molecule. The coordinates are in units of Ångstroms (10^{-10} m or 10^2 pm). Line 10 is left blank to indicate the end of the input file.

EXAMPLE 11–8

Set up an input file for running a geometry optimized restricted Hartree-Fock calculation using the 6-31G** basis set for an ammonia molecule.

SOLUTION:

```
Line #
    1      %chk=ammonia
    2      #rhf/6-31G** geom=coord opt
    3
    4      Restricted Hartree Fock Calculation of
           Ammonia with the 6-31G** basis set
    5
    6      0    1
    7      N     0.000000      0.000000       0.000000
    8      H     0.962752      0.000000      -0.373049
    9      H    -0.450719     -0.803397      -0.466363
   10      H    -0.440104      0.826195      -0.435709
   11
```

The coordinates (in Ångstroms) given for the four atoms are our initial guess. They need only be a rough estimate because the molecular geometry will be optimized (the "opt" control word is specified in line 2) as part of the calculation.

Table 11.8 gives the energies, geometry, net atomic charges, and dipole moment for a water molecule that result from a calculation using the input file in Table 11.7. According to Koopmans' approximation (Section 8–7), the negative of the energies of the one-electron orbitals corresponds to the ionization energies. We can therefore test the accuracy of the calculated energies of the molecular orbitals by comparing the values obtained with the peaks in the photoelectron spectrum (Figure 10.14). The calculated values are indicated along the energy axis of Figure 10.14. We find excellent agreement between the calculated and experimental values for the energies of the $1b_2$, $3a_1$, and $1b_1$ molecular orbitals. Excellent agreement is also found between the calculated and experimental molecular geometries. The calculated dipole moment of water differs from the experimental value by approximately 20% (Problem 11–30). This difference can be significantly narrowed using configuration interaction but requires more computer time to carry out the calculation. Table 11.9 gives the results for an RHF/6-31G* calculation of ammonia. Figure 11.5 shows the photoelectron spectrum of ammonia. Similar to the case for water, as discussed in Chapter 10, the bands in the photoelectron spectrum are broad due to ionization from different vibrational levels of the molecule. The calculated values of the orbital energies using Koopmans'

TABLE 11.8

Calculated results for water using the input file in Table 11.7.

Restricted Hartree Fock Calculation of Water with the 6-31G*
basis set

Molecular orbital energies

Orbital	Energy/MJ · mol^{-1}	Experimental energy/MJ · mol^{-1}
1a1	-53.9	
2a1	-3.53	
1b2	-1.87	
3a1	-1.49	
1b1	-1.31	
Total	-199.70	-200.78

Bond lengths

Bond	Calculated length/pm	Experimental value/pm
O--H	94.7	95.8
H--H	150.8	

Bond angle

Angle	Calculated	Experimental
H-O-H	105.5°	104.5°

Net atomic charges

Atom	Net charge
H	+0.41
O	-0.82
H	+0.41

Dipole moment/D

Calculated	2.3
Experimental	1.85

approximation are, once again, in excellent agreement with the experimental values. As for water, the calculated and experimental molecular geometries for ammonia are in excellent agreement. The dipole moment differs by about 30% (Problem 11–31) and like that for water, the calculated dipole moment is too large. A more accurate value can be obtained using configuration interaction.

Hartree-Fock calculations have been carried out on a large number of molecules and in many cases, excellent agreement between theory and experiment is observed. In Table 11.10, optimized structural information is displayed for various molecules using different basis sets. The experimental values are also given for comparison.

TABLE 11.9

Calculated results for an amonia molecule using the input file in Example 11–8.

Restricted Hartree Fock Calculation of Ammonia with the 6-31G* basis set

Molecular orbital energies

Orbital	Energy/MJ · mol^{-1}
1a1	-40.79
2a1	-2.99
e	-1.65
e	-1.65
3a1	-1.10
Total	-147.53

Bond lengths

Bond	Calculated length/pm	Experimental value/pm
N–H	100.0	101.2
H–H	160.9	

Bond angle

Angle	Calculated	Experimental
H–N–H	107.1°	106.7°

Net atomic charges

Atom	Net charge
N	-1.11
H	+0.37

Dipole moment/D

Calculated	2.0
Experimental	1.5

Tables 11.11 and 11.12 present calculated and experimental bond lengths and bond angles for the optimized structure of large molecules. Although there are small discrepancies between the calculated and experimental values, these results clearly show that the computational technology discussed in this section is not limited to small molecules. In most cases, the agreement is improved by using advanced methods that account for electron correlation. The data given in Tables 11.10 to 11.12 reveal the power and accuracy of modern computational techniques.

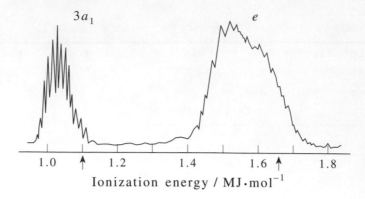

Ionization energy / MJ·mol^{-1}

FIGURE 11.5
The photoelectron spectrum of ammonia. This range corresponds to ionization from the two highest occupied molecular orbitals, e and $3a_1$. The fine structure of each band corresponds to ionization from different vibrational levels. The arrows indicate the calculated ionization energies from the $v = 0$ vibrational states at the HF/6-31G* level (Table 11.9) using Koopmans' approximation.

TABLE 11.10
Calculated and experimental bond lengths (pm) and bond angles (degrees) for small molecules for different basis sets.

Molecule	Geometrical parameter	STO-3G	3-21G	3-21G*	6-31G*	6-31G**	Expt.
H_2	r(HH)	71.2	73.5	73.5	73.0	73.2	74.2
LiH	r(LiH)	151.0	164.0	164.0	163.6	162.3	159.6
CH_4	r(CH)	108.3	108.3	108.3	108.4	108.4	109.2
NH_3	r(NH)	103.3	100.3	100.3	100.2	100.1	101.2
	∠(HNH)	104.2	112.4	112.4	107.2	107.6	106.7
H_2O	r(OH)	99.0	96.7	96.7	94.7	94.3	95.8
	∠(HOH)	100.0	107.6	107.6	105.5	105.9	104.5
HF	r(FH)	95.6	93.7	93.7	91.1	90.1	91.7
NaH	r(NaH)	165.4	192.6	193.0	191.4	191.2	188.7
SiH_4	r(SiH)	142.2	148.7	147.5	147.5	147.6	148.1
PH_3	r(PH)	137.8	142.3	140.2	140.3	140.5	142.0
	∠(HPH)	95.0	96.1	95.2	95.4	95.6	93.3
H_2S	r(SH)	132.9	135.0	132.7	132.6	132.7	133.6
	∠(HSH)	92.5	95.8	94.4	94.4	94.4	92.1
HCl	r(HCl)	131.3	129.3	126.7	126.7	126.6	127.5

TABLE 11.11
Calculated and experimental molecular bond lengths.

Bond type	Molecule	Bond length/pm				
		STO-3G	3-21G	3-21G*	6-31G*	Expt.
C≡C	Acetylene	116.8	118.8	118.8	118.5	120.3
	Propyne	117.0	118.8	118.8	118.7	120.6
C=C	Cyclopropene	127.7	128.2	228.2	127.6	130.0
	Allene	128.8	129.2	129.2	129.6	130.8
	Cyclobutene	131.4	132.6	132.6	132.2	133.2
	Ethene	130.6	131.5	131.5	131.7	133.9
	Cyclopentadiene	131.9	132.9	132.9	132.9	134.5
C–C	Acetonitrile	148.8	145.7	145.7		145.8
	Acetaldehyde	153.7	150.7	150.7	150.4	150.1
	Cyclopentadiene	152.2	151.9	151.9	150.7	150.6
	Cyclobutane	155.4	157.1	157.1	154.8	154.8
C≡N	Hydrogen cyanide	115.3	113.7	113.7	113.3	115.3
	Acetonitrile	115.4	113.9	113.9	113.3	115.7
	Hydrogen isocyanide	117.0	116.0	116.0	115.4	116.9
C=O	Carbon dioxide	118.8	115.6	115.6	114.3	116.2
	Formaldehyde	121.7	120.7	120.7	118.4	120.8
	Acetone	121.9	121.1	121.1		122.2
C–S	Methanethiol	179.8	189.5	182.3	181.7	181.9
C–Cl	Chloromethane	180.2	189.2	180.6	178.5	178.1
N=N	Diazene	126.7	123.9	123.9	121.6	125.2
O–O	Ozone	128.5	130.8	130.8	120.4	127.8
	Hydrogen peroxide	139.6	147.3	147.3	139.3	145.2

TABLE 11.12
Calculated and experimental molecular bond angles.

Angles	Molecule	Angle/degrees			
		STO-3G	3-21G	3-21G*	Expt.
C=C–C	Acrolein	122.4	120.5	120.5	119.8
	Isobutene	122.4	122.6	122.6	122.4
	Propene	125.1	124.7	124.7	124.3
C–C–C	Isobutane	110.9	110.4	110.4	110.8
	Propane	112.4	111.6	111.6	112.4
C–C=O	Acetone	122.4	122.5	122.5	121.4
	Acetic acid	126.8	127.4	127.4	126.6
C–O–C	Dimethylether	108.7	114.0	114.0	111.7
N–C=O	Formamide	124.3	125.3	125.3	124.7
O–O–O	Ozone	116.2	117.0	117.0	116.8
Cl–O–Cl	Oxygen dichloride	109.3	112.0	113.2	110.9

Problems

11-1. Show that a three-dimensional Gaussian function centered at $r_0 = x_0 i + y_0 j + z_0 k$ is a product of three one-dimensional Gaussian functions centered on x_0, y_0, and z_0.

11-2. Show that

$$\int_{-\infty}^{\infty} e^{-(x-x_0)^2} dx = \int_{-\infty}^{\infty} e^{-x^2} dx = 2\int_{0}^{\infty} e^{-x^2} dx = \pi^{1/2}$$

11-3. The Gaussian integral

$$I_0 = \int_{0}^{\infty} e^{-ax^2} dx$$

can be evaluated by a trick. First write

$$I_0^2 = \int_{0}^{\infty} dx e^{-ax^2} \int_{0}^{\infty} dy e^{-ay^2} = \int_{0}^{\infty}\int_{0}^{\infty} dx dy e^{-a(x^2+y^2)}$$

Now convert the integration variables from Cartesian coordinates to polar coordinates and show that

$$I_0 = \frac{1}{2}\left(\frac{\pi}{a}\right)^{1/2}$$

11-4. Show that the integral

$$I_{2n} = \int_{0}^{\infty} x^{2n} e^{-ax^2} dx$$

can be obtained from I_0 in Problem 11–3 by differentiating n times with respect to a. Using the result of Problem 11–3, show that

$$I_{2n} = \frac{1 \cdot 3 \cdot 5 \cdots (2n-1)}{2(2a)^n}\left(\frac{\pi}{a}\right)^{1/2}$$

11-5. Show that the Gaussian function

$$\phi(r) = \left(\frac{2\alpha}{\pi}\right)^{3/4} e^{-\alpha r^2}$$

is normalized.

11-6. Show that the product of a (not normalized) Gaussian function centered at R_A and one centered at R_B, i.e.

$$\phi_1 = e^{-\alpha|r-R_A|^2} \qquad \text{and} \qquad \phi_2 = e^{-\beta|r-R_B|^2}$$

is a Gaussian function centered at

$$R_p = \frac{\alpha R_A + \beta R_B}{\alpha + \beta}$$

For simplicity, work in one dimension and appeal to Problem 11–1 to argue that it is true in three dimensions.

11-7. Show explicitly that if

$$
\phi_{1s}(\alpha, \mathbf{r} - \mathbf{R}_A) = \left(\frac{2\alpha}{\pi}\right)^{3/4} e^{-\alpha|\mathbf{r} - \mathbf{R}_A|^2}
$$

and

$$
\phi_{1s}(\beta, \mathbf{r} - \mathbf{R}_B) = \left(\frac{2\beta}{\pi}\right)^{3/4} e^{-\beta|\mathbf{r} - \mathbf{R}_B|^2}
$$

are normalized Gaussian $1s$ functions, then

$$
\phi_{1s}(\alpha, \mathbf{r} - \mathbf{R}_A)\phi_{1s}(\beta, \mathbf{r} - \mathbf{R}_B) = K_{AB}\phi_{1s}(p, \mathbf{r} - \mathbf{R}_p)
$$

where $p = \alpha + \beta$, $\mathbf{R}_p = (\alpha\mathbf{R}_A + \beta\mathbf{R}_B)/(\alpha + \beta)$ (see Problem 11–6), and

$$
K_{AB} = \left[\frac{2\alpha\beta}{(\alpha + \beta)\pi}\right]^{3/4} e^{-\frac{\alpha\beta}{\alpha+\beta}|\mathbf{R}_A - \mathbf{R}_B|^2}
$$

11-8. Plot the product of the two (unnormalized) Gaussian functions

$$
\phi_1 = e^{-2(x-1)^2} \qquad \text{and} \qquad \phi_2 = e^{-3(x-2)^2}
$$

Interpret the result.

11-9. Using the result of Problem 11–7, show that the overlap integral of the two normalized Gaussian functions

$$
\phi_{1s} = \left(\frac{2\alpha}{\pi}\right)^{3/4} e^{-\alpha|\mathbf{r} - \mathbf{R}_A|^2} \quad \text{and} \quad \phi_{1s} = \left(\frac{2\beta}{\pi}\right)^{3/4} e^{-\beta|\mathbf{r} - \mathbf{R}_B|^2}
$$

is

$$
S(|\mathbf{R}_A - \mathbf{R}_B|) = \left[\frac{4\alpha\beta}{(\alpha + \beta)^2}\right]^{3/4} e^{-\frac{\alpha\beta|\mathbf{R}_A - \mathbf{R}_B|^2}{\alpha+\beta}}
$$

Plot this result as a function of $|\mathbf{R}_A - \mathbf{R}_B|$.

11-10. One criterion for the best possible "fit" of a Gaussian function to a Slater orbital is a fit that minimizes the integral of the square of their difference. For example, we can find the optimal value of α in $\phi_{1s}^{GF}(r, \alpha)$ by minimizing

$$
I = \int d\mathbf{r}[\phi_{1s}^{STO}(r, 1.00) - \phi_{1s}^{GF}(r, \alpha)]^2
$$

with respect to α. If the two functions $\phi_{1s}^{STO}(r, 1.00)$ and $\phi_{1s}^{GF}(r, \alpha)$ are normalized, show that minimizing I is equivalent to maximizing the overlap integral of $\phi_{1s}^{STO}(r, 1.00)$ and $\phi_{1s}^{GF}(r, \alpha)$:

$$
S = \int d\mathbf{r}\phi_{1s}^{STO}(r, 1.00)\phi_{1s}^{GF}(r, \alpha)
$$

11-11. Show that S in Problem 11–10 is given by

$$
S = 4\pi^{1/2}\left(\frac{2\alpha}{\pi}\right)^{3/4}\int_0^\infty r^2 e^{-r}e^{-\alpha r^2}\,dr
$$

Using a numerical integration computer program such as **Mathematica** or **MathCad**, show that the following results are correct:

α	S
0.10	0.8642
0.15	0.9367
0.20	0.9673
0.25	0.9776
0.30	0.9772
0.35	0.9706
0.40	0.9606

These numbers show that the maximum occurs around $\alpha = 0.25$. A more detailed calculation would show that the maximum actually occurs at $\alpha = 0.27095$. Thus, the normalized Gaussian $1s$ function $\phi_{1s}^{GF}(r, 0.2709)$ is an optimal fit to the $1s$ Slater orbital $\phi_{1s}^{STO}(r, 1.00)$.

11-12. Compare $\phi_{1s}^{STO}(r, 1.00)$ and $\phi_{1s}^{GF}(r, 0.27095)$ graphically by plotting them on the same graph.

11-13. In Problems 11–11 and 11–12, we discussed a one-term Gaussian fit to a $1s$ Slater orbital $\phi_{1s}^{STO}(r, 1.00)$. Can we use the result of Problem 11–11 to find the optimal Gaussian fit to a $1s$ Slater orbital with a different orbital exponent, $\phi_{1s}^{GF}(r, \zeta)$? The answer is "yes." To see how, start with the overlap integral of $\phi_{1s}^{STO}(r, \zeta)$ and $\phi_{1s}^{GF}(r, \beta)$:

$$S = 4\pi^{1/2} \left(\frac{2\beta}{\pi} \right)^{3/4} \zeta^{3/2} \int_0^\infty r^2 e^{-\zeta r} e^{-\beta r^2} dr$$

Now let $u = \zeta r$ to get

$$S = 4\pi^{1/2} \left(\frac{2\beta/\zeta^2}{\pi} \right)^{3/4} \int_0^\infty u^2 e^{-u} e^{-(\beta/\zeta^2)u^2} du$$

Compare this result for S with that in Problem 11–11 to show that $\beta = \alpha\zeta^2$ or, in more detailed notation,

$$\alpha(\zeta = \zeta) = \alpha(\zeta = 1.00) \times \zeta^2$$

11-14. Use the result of Problem 11–13 to verify the value of α used in Equation 11.5 and Figure 11.1.

11-15. Because of the scaling law developed in Problem 11–13, Gaussian fits are usually made with respect to a Slater orbital with $\zeta = 1.00$ and then the various Gaussian exponents are scaled according to $\alpha(\zeta = \zeta) = \alpha(\zeta = 1.00) \times \zeta^2$. Given the fit

$$\phi_{1s}^{STO-3G}(r, 1.0000) = 0.4446\phi_{1s}^{GF}(r, 0.10982)$$
$$+ 0.5353\phi_{1s}^{GF}(r, 0.40578)$$
$$+ 0.1543\phi_{1s}^{GF}(r, 2.2277)$$

verify Equation 11.8.

11-16. The Gaussian function exponents and expansion coefficients for the valence shell orbitals of chlorine are as follows:

$\alpha_{3si} = \alpha_{3pi}$	d_{3si}	$\alpha'_{3s} = \alpha'_{3p}$	d_{3p}
3.18649	−0.25183	1.42657	−0.014299
1.19427	0.061589		0.323572
0.420377	1.06018		0.743507

Write the expressions for the Gaussian functions corresponding to the $3s$ and $3p$ atomic orbitals of chlorine. Plot the function for the $3s$ orbital for several values of the expansion coefficient for the α'_{3s} term.

11-17. The input file to a computational quantum chemistry program must specify the coordinates of the atoms that comprise the molecule. Determine a set of Cartesian coordinates of the atoms in the molecule CH_4. The HCH bond angle is 109.5° and the C–H bond length is 109.1 pm. (*Hint:* Use the figure in Problem 10–7.)

11-18. The input file to a computational quantum chemistry program must specify the coordinates of the atoms that comprise the molecule. Determine a set of Cartesian coordinates of the atoms in the molecule CH_3Cl. The HCH bond angle is 110.0° and the C–H and C–Cl bond lengths are 109.6 and 178.1 pm, respectively. (*Hint:* locate the origin at the carbon atom.)

11-19. The calculated vibrational frequencies and bond lengths for three diatomic molecules are listed below.

	Calculated values (6-31G*)	
Molecule	Frequency/cm^{-1}	R_e/pm
H_2	4647	73.2
CO	2438	111.4
N_2	2763	107.9

Determine the force constants that correspond to these vibrational frequencies. How do these values compare with the data in Table 5.1? How do the calculated bond lengths compare with the experimental values (also Table 5.1)? Why do you think the bond-length calculations show a higher accuracy than the vibrational-frequency calculations?

11-20. Normalize the following Gaussian functions

a. $\phi(r) = xe^{-\alpha r^2}$

b. $\phi(r) = x^2 e^{-\alpha r^2}$

11-21. Which hydrogen atomic orbital corresponds to the following normalized Gaussian orbital?

$$G(x, y, z; \alpha) = \left(\frac{128\alpha^5}{\pi^3} \right)^{1/4} y e^{-\alpha r^2}$$

How many radial and angular nodes does the above function have? Is this result what you would expect for the corresponding hydrogenic function?

11-22. Using Equations 6.62 for the spherical harmonic components of ϕ_{2p_x} and ϕ_{2p_y}, show that the Slater orbitals for the $2p_x$, $2p_y$, and $2p_z$ orbitals are given by the formulas in Example 11–4. Recall that the $2p_x$ and $2p_y$ orbitals are given by Equations 6.62.

11-23. Consider the normalized functions

$$G_1(x, y, z; \alpha) = \left(\frac{2048\alpha^7}{9\pi^3}\right)^{1/4} x^2 e^{-\alpha r^2}$$

$$G_2(x, y, z; \alpha) = \left(\frac{2048\alpha^7}{9\pi^3}\right)^{1/4} y^2 e^{-\alpha r^2}$$

$$G_3(x, y, z; \alpha) = \left(\frac{2048\alpha^7}{9\pi^3}\right)^{1/4} z^2 e^{-\alpha r^2}$$

Which hydrogen atomic orbital corresponds to the linear combination $G_1(x, y, z; \alpha) - G_2(x, y, z; \alpha)$?

11-24. What is meant by the phrase "triple-zeta basis set"?

11-25. Part of the output of most computational programs is a list of numbers that comprise what is called Mulliken Population Analysis. This list assigns a net charge to each atom in the molecule. The value of this net charge is the difference between the charge of the isolated atom, Z, and the calculated charge on the bonded atom, q. Thus if $Z - q > 0$, the atom is assigned a net positive charge and if $Z - q < 0$, the atom is assigned a net negative charge. What would be the sum of the Mulliken Populations for the molecules H_2CO, CO_3^{2-}, and NH_4^+?

11-26. In this problem, we show that the Mulliken Populations (Problem 11–25) can be used to calculate the molecular dipole moment. Consider the formaldehyde molecule, H_2CO. The calculated bond lengths for the CO and CH bonds are 121.7 pm and 110.0 pm, respectively, and the optimized H–C–H bond angle was found to be 114.5°. Use this information along with the Mulliken Population Analysis shown below

to calculate the dipole moment of formaldehyde. The experimentally determined values for the bond lengths and bond angles are $R_{CO} = 120.8$ pm, $R_{CH} = 111.6$ pm and $\angle(HCH) = 116.5°$. What is the value of the dipole moment if you combine the experimental geometry and the calculated Mulliken Populations? How do your calculated dipole moments compare with the experimental value of 7.8×10^{-30} C·m?

11-27. The experimentally determined dipole moment of CO is 3.66×10^{-31} C·m, with the oxygen atom being positively charged. The Mulliken Populations from Hartree-Fock calculations using the STO-3G or the 6-31G* basis sets predict a dipole moment of 5.67×10^{-31} C·m and 1.30×10^{-30} C·m, respectively, and pointing in the opposite direction of the experimental results. The experimental and two calculated bond lengths are 112.8 pm, 114.6 pm, and 111.4 pm, respectively. Why do you think the bond-length calculation is significantly more accurate than the dipole-moment calculation?

11-28. The orbital energies calculated for formaldehyde using STO-3G and 3-21G basis sets are given below.

Orbital	STO-3G energy/E_h	3-21G energy/E_h
$1a_1$	−20.3127	−20.4856
$2a_1$	−11.1250	−11.2866
$3a_1$	−1.3373	−1.4117
$4a_1$	−0.8079	−0.8661
$1b_2$	−0.6329	−0.6924
$5a_1$	−0.5455	−0.6345
$1b_1$	−0.4431	−0.5234
$2b_2$	−0.3545	−0.4330
$2b_1$	0.2819	0.1486
$6a_1$	0.6291	0.2718
$3b_2$	0.7346	0.3653
$7a_1$	0.9126	0.4512

Determine the ground-state electronic configuration of formaldehyde. The photoelectron spectrum of formaldehyde is shown below.

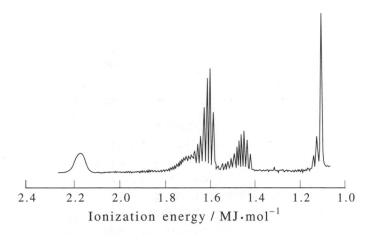

Ionization energy / $MJ \cdot mol^{-1}$

Assign the bands. Which calculated set of energies shows the best agreement with the photoelectron spectrum? Why is there such a large energy separation between the $1a_1$ and $2a_1$ orbitals? Predict the ionization energy and electron affinity of formaldehyde for each calculated set of energy levels. How do these compare with the experimental values?

11-29. The units of dipole moment given by Gaussian 94 are called debyes (D), after the Dutch-American chemist, Peter Debye, who was awarded the Nobel Prize for chemistry in 1936 for his work on dipole moments. One debye is equal to 10^{-18} esu · cm where esu (*electrostatic units*) is a non-SI unit for electric charge. Given that the protonic charge is 4.803×10^{-10} esu, show that the conversion factor between debyes and C · m (coulomb · meters) is 1 $D = 3.33 \times 10^{-30}$ C · m.

11-30. Using the geometry and the charges given in Table 11.8, verify the value of the dipole moment of water.

11-31. Using the geometry and the charges given in Table 11.9, verify the value of the dipole moment of ammonia.

MATRICES

Many physical operations such as magnification, rotation, and reflection through a plane can be represented mathematically by quantities called matrices. A matrix is simply a two-dimensional array that obeys a certain set of rules called matrix algebra. Even if matrices are entirely new to you, they are so convenient that learning some of their simpler properties is worthwhile.

Consider the lower of the two vectors shown in Figure F.1. The x and y components of the vector are given by $x_1 = r \cos \alpha$ and $y_1 = r \sin \alpha$, where r is the length of $\mathbf{r}_1$. Now let's rotate the vector counterclockwise through an angle θ, so that $x_2 = r \cos(\alpha + \theta)$ and $y_2 = r \sin(\alpha + \theta)$ (see Figure F.1). Using trigonometric formulas, we can write

$$x_2 = r \cos(\alpha + \theta) = r \cos \alpha \cos \theta - r \sin \alpha \sin \theta$$

$$y_2 = r \sin(\alpha + \theta) = r \cos \alpha \sin \theta + r \sin \alpha \cos \theta$$

or

$$x_2 = x_1 \cos \theta - y_1 \sin \theta$$

$$y_2 = x_1 \sin \theta + y_1 \cos \theta$$

(F.1)

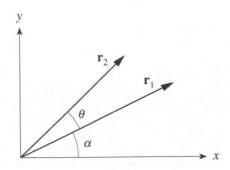

FIGURE F.1
An illustration of the rotation of a vector $\mathbf{r}_1$ through an angle θ.

We can display the set of coefficients of x_1 and y_1 in the form

$$R = \begin{pmatrix} \cos\theta & -\sin\theta \\ \sin\theta & \cos\theta \end{pmatrix} \tag{F.2}$$

We have expressed R in the form of a *matrix*, which is an array of numbers (or functions in this case) that obey a certain set of rules, called matrix algebra. We will denote a matrix by a sans serif symbol, e.g., A, B, etc. Unlike determinants (MathChapter E), matrices do not have to be square arrays. Furthermore, unlike determinants, matrices cannot be reduced to a single number. The matrix R in Equation F.2 corresponds to a rotation through an angle θ.

The entries in a matrix A are called its *matrix elements* and are denoted by a_{ij}, where, as in the case of determinants, i designates the row and j designates the column. Two matrices, A and B, are equal if and only if they are of the same dimension and $a_{ij} = b_{ij}$ for all i and j. In other words, equal matrices are identical. Matrices can be added or subtracted only if they have the same number of rows and columns, in which case the elements of the resultant matrix are given by $a_{ij} + b_{ij}$. Thus, if

$$A = \begin{pmatrix} -3 & 6 & 4 \\ 1 & 0 & 2 \end{pmatrix} \quad \text{and} \quad B = \begin{pmatrix} 2 & 1 & 1 \\ -6 & 4 & 3 \end{pmatrix}$$

then

$$C = A + B = \begin{pmatrix} -1 & 7 & 5 \\ -5 & 4 & 5 \end{pmatrix}$$

If we write

$$A + A = 2A = \begin{pmatrix} -6 & 12 & 8 \\ 2 & 0 & 4 \end{pmatrix}$$

we see that scalar multiplication of a matrix means that each element is multiplied by the scalar. Thus,

$$c\,M = \begin{pmatrix} cM_{11} & cM_{12} \\ cM_{21} & cM_{22} \end{pmatrix} \tag{F.3}$$

EXAMPLE F–1
Using the matrices A and B above, form the matrix $D = 3A - 2B$.

SOLUTION:

$$D = 3\begin{pmatrix} -3 & 6 & 4 \\ 1 & 0 & 2 \end{pmatrix} - 2\begin{pmatrix} 2 & 1 & 1 \\ -6 & 4 & 3 \end{pmatrix}$$

$$= \begin{pmatrix} -9 & 18 & 12 \\ 3 & 0 & 6 \end{pmatrix} - \begin{pmatrix} 4 & 2 & 2 \\ -12 & 8 & 6 \end{pmatrix} = \begin{pmatrix} -13 & 16 & 10 \\ 15 & -8 & 0 \end{pmatrix}$$

One of the most important aspects of matrices is matrix multiplication. For simplicity, we will discuss the multiplication of square matrices first. Consider some linear transformations of (x_1, y_1) into (x_2, y_2):

$$\begin{aligned} x_2 &= a_{11}x_1 + a_{12}y_1 \\ y_2 &= a_{21}x_1 + a_{22}y_1 \end{aligned} \tag{F.4}$$

represented by the matrix

$$A = \begin{pmatrix} a_{11} & a_{12} \\ a_{21} & a_{22} \end{pmatrix} \tag{F.5}$$

Now let's transform (x_2, y_2) into (x_3, y_3):

$$\begin{aligned} x_3 &= b_{11}x_2 + b_{12}y_2 \\ y_3 &= b_{21}x_2 + b_{22}y_2 \end{aligned} \tag{F.6}$$

represented by the matrix

$$B = \begin{pmatrix} b_{11} & b_{12} \\ b_{21} & b_{22} \end{pmatrix} \tag{F.7}$$

Let the transformation of (x_1, y_1) directly into (x_3, y_3) be given by

$$\begin{aligned} x_3 &= c_{11}x_1 + c_{12}y_1 \\ y_3 &= c_{21}x_1 + c_{22}y_1 \end{aligned} \tag{F.8}$$

represented by the matrix

$$C = \begin{pmatrix} c_{11} & c_{12} \\ c_{21} & c_{22} \end{pmatrix} \tag{F.9}$$

Symbolically, we can write that

$$C = BA$$

because C results from transforming from (x_1, y_1) to (x_2, y_2) by means of A followed by transforming (x_2, y_2) to (x_3, y_3) by means of B. Let's find the relation between the elements of C and those of A and B. Substitute Equations F.4 into F.6 to obtain

$$\begin{aligned} x_3 &= b_{11}(a_{11}x_1 + a_{12}y_1) + b_{12}(a_{21}x_1 + a_{22}y_1) \\ y_3 &= b_{21}(a_{11}x_1 + a_{12}y_1) + b_{22}(a_{21}x_1 + a_{22}y_1) \end{aligned} \tag{F.10}$$

or

$$\begin{aligned} x_3 &= (b_{11}a_{11} + b_{12}a_{21})x_1 + (b_{11}a_{12} + b_{12}a_{22})y_1 \\ y_3 &= (b_{21}a_{11} + b_{22}a_{21})x_1 + (b_{21}a_{12} + b_{22}a_{22})y_1 \end{aligned}$$

Thus, we see that

$$C = BA = \begin{pmatrix} b_{11} & b_{12} \\ b_{21} & b_{22} \end{pmatrix} \begin{pmatrix} a_{11} & a_{12} \\ a_{21} & a_{22} \end{pmatrix} = \begin{pmatrix} b_{11}a_{11} + b_{12}a_{21} & b_{11}a_{12} + b_{12}a_{22} \\ b_{21}a_{11} + b_{22}a_{21} & b_{21}a_{12} + b_{22}a_{22} \end{pmatrix} \quad \text{(F.11)}$$

This result may look complicated, but it has a nice pattern which we will illustrate two ways. Mathematically, the ijth element of C is given by the formula

$$c_{ij} = \sum_k b_{ik} a_{kj} \quad \text{(F.12)}$$

For example,

$$c_{11} = \sum_k b_{1k} a_{k1} = b_{11}a_{11} + b_{12}a_{21}$$

as in Equation F.11. A more pictorial way is to notice that any element in C can be obtained by multiplying elements in any row in B by the corresponding elements in any column in A, adding them, and then placing them in C where the row and column intersect. For example, c_{11} is obtained by multiplying the elements of row 1 of B with the elements of column 1 of A, or by the scheme

$$\rightarrow \begin{pmatrix} b_{11} & b_{12} \\ b_{21} & b_{22} \end{pmatrix} \begin{pmatrix} \downarrow \\ a_{11} & a_{12} \\ a_{21} & a_{22} \end{pmatrix} = \begin{pmatrix} b_{11}a_{11} + b_{12}a_{21} & \cdot \\ \cdot & \cdot \end{pmatrix}$$

and c_{12} by

$$\rightarrow \begin{pmatrix} b_{11} & b_{12} \\ b_{21} & b_{22} \end{pmatrix} \begin{pmatrix} a_{11} & \downarrow \\ a_{21} & a_{12} \\ & a_{22} \end{pmatrix} = \begin{pmatrix} \cdot & b_{11}a_{12} + b_{12}a_{22} \\ \cdot & \cdot \end{pmatrix}$$

EXAMPLE F–2
Find C = BA if

$$B = \begin{pmatrix} 1 & 2 & 1 \\ 3 & 0 & -1 \\ -1 & -1 & 2 \end{pmatrix} \quad \text{and} \quad A = \begin{pmatrix} -3 & 0 & -1 \\ 1 & 4 & 0 \\ 1 & 1 & 1 \end{pmatrix}$$

SOLUTION:

$$C = \begin{pmatrix} 1 & 2 & 1 \\ 3 & 0 & -1 \\ -1 & -1 & 2 \end{pmatrix} \begin{pmatrix} -3 & 0 & -1 \\ 1 & 4 & 0 \\ 1 & 1 & 1 \end{pmatrix}$$

$$= \begin{pmatrix} -3+2+1 & 0+8+1 & -1+0+1 \\ -9+0-1 & 0+0-1 & -3+0-1 \\ 3-1+2 & 0-4+2 & 1+0+2 \end{pmatrix}$$

$$= \begin{pmatrix} 0 & 9 & 0 \\ -10 & -1 & -4 \\ 4 & -2 & 3 \end{pmatrix}$$

EXAMPLE F–3

The matrix R given by Equation F.2 represents a rotation through the angle θ. Show that R^2 represents a rotation through an angle 2θ.

SOLUTION:

$$R^2 = \begin{pmatrix} \cos\theta & -\sin\theta \\ \sin\theta & \cos\theta \end{pmatrix} \begin{pmatrix} \cos\theta & -\sin\theta \\ \sin\theta & \cos\theta \end{pmatrix}$$

$$= \begin{pmatrix} \cos^2\theta - \sin^2\theta & -2\sin\theta\cos\theta \\ 2\sin\theta\cos\theta & \cos^2\theta - \sin^2\theta \end{pmatrix}$$

Using standard trigonometric identities, we get

$$R^2 = \begin{pmatrix} \cos 2\theta & -\sin 2\theta \\ \sin 2\theta & \cos 2\theta \end{pmatrix}$$

which represents rotation through an angle 2θ.

Matrices do not have to be square to be multiplied together, but either Equation F.11 or the pictorial method illustrated above suggests that the number of columns of B must be equal to the number of rows of A. When this is so, A and B are said to be *compatible*. For example, Equations F.4 can be written in matrix form as

$$\begin{pmatrix} x_2 \\ y_2 \end{pmatrix} = \begin{pmatrix} a_{11} & a_{12} \\ a_{21} & a_{22} \end{pmatrix} \begin{pmatrix} x_1 \\ y_1 \end{pmatrix} \tag{F.13}$$

An important aspect of matrix multiplication is that BA does not necessarily equal AB. For example, if

$$A = \begin{pmatrix} 0 & 1 \\ 1 & 0 \end{pmatrix} \quad \text{and} \quad B = \begin{pmatrix} 1 & 0 \\ 0 & -1 \end{pmatrix}$$

then

$$AB = \begin{pmatrix} 0 & 1 \\ 1 & 0 \end{pmatrix} \begin{pmatrix} 1 & 0 \\ 0 & -1 \end{pmatrix} = \begin{pmatrix} 0 & -1 \\ 1 & 0 \end{pmatrix}$$

and

$$BA = \begin{pmatrix} 1 & 0 \\ 0 & -1 \end{pmatrix} \begin{pmatrix} 0 & 1 \\ 1 & 0 \end{pmatrix} = \begin{pmatrix} 0 & 1 \\ -1 & 0 \end{pmatrix}$$

and so $AB = -BA$ in this case. If it does happen that $AB = BA$, then A and B are said to *commute*.

EXAMPLE F–4

Do the matrices A and B commute if

$$A = \begin{pmatrix} 2 & 1 \\ 0 & 1 \end{pmatrix} \quad \text{and} \quad B = \begin{pmatrix} 1 & 1 \\ 0 & 1 \end{pmatrix}$$

SOLUTION:

$$AB = \begin{pmatrix} 2 & 3 \\ 0 & 1 \end{pmatrix}$$

and

$$BA = \begin{pmatrix} 2 & 2 \\ 0 & 1 \end{pmatrix}$$

so they do not commute.

Another property of matrix multiplication that differs from ordinary scalar multiplication is that the equation

$$AB = O$$

where O is the zero matrix (all elements equal to zero) does not imply that A or B necessarily is a zero matrix. For example,

$$\begin{pmatrix} 1 & 1 \\ 2 & 2 \end{pmatrix} \begin{pmatrix} -1 & 1 \\ 1 & -1 \end{pmatrix} = \begin{pmatrix} 0 & 0 \\ 0 & 0 \end{pmatrix}$$

A linear transformation that leaves (x_1, y_1) unaltered is called the identity transformation, and the corresponding matrix is called the *identity matrix* or the *unit matrix*. All the elements of the identity matrix are equal to zero, except those along the diagonal, which equal one:

$$I = \begin{pmatrix} 1 & 0 & 0 & \cdots & 0 \\ 0 & 1 & 0 & \cdots & 0 \\ 0 & 0 & 1 & \cdots & 0 \\ \vdots & \vdots & \vdots & \vdots & \vdots \\ 0 & 0 & 0 & \cdots & 1 \end{pmatrix}$$

The elements of I are δ_{ij}, the Kronecker delta, which equals one when $i = j$ and zero when $i \neq j$. The unit matrix has the property that

$$IA = AI \tag{F.14}$$

The unit matrix is an example of a *diagonal matrix*. The only nonzero elements of a diagonal matrix are along its diagonal. Diagonal matrices are necessarily square matrices.

If $BA = AB = I$, then B is said to be the *inverse* of A, and is denoted by A^{-1}. Thus, A^{-1} has the property that

$$AA^{-1} = A^{-1}A = I \tag{F.15}$$

If A represents some transformation, then A^{-1} undoes that transformation and restores the original state. There are recipes for finding the inverse of a matrix, but we won't need them (see Problem F–9, however). Nevertheless, it should be clear on physical grounds that the inverse of R in Equation F.2 is

$$R^{-1} = R(-\theta) = \begin{pmatrix} \cos\theta & \sin\theta \\ -\sin\theta & \cos\theta \end{pmatrix} \tag{F.16}$$

which is obtained from R by replacing θ by $-\theta$. In other words, if $R(\theta)$ represents a rotation through an angle θ, then $R^{-1} = R(-\theta)$ and represents the reverse rotation. It is easy to show that R and R^{-1} satsify Equation F.15. Using Equations F.2 and F.16, we have

$$\begin{aligned} R^{-1}R &= \begin{pmatrix} \cos\theta & \sin\theta \\ -\sin\theta & \cos\theta \end{pmatrix} \begin{pmatrix} \cos\theta & -\sin\theta \\ \sin\theta & \cos\theta \end{pmatrix} \\ &= \begin{pmatrix} \cos^2\theta + \sin^2\theta & 0 \\ 0 & \cos^2\theta + \sin^2\theta \end{pmatrix} \\ &= \begin{pmatrix} 1 & 0 \\ 0 & 1 \end{pmatrix} \end{aligned}$$

and

$$\begin{aligned} RR^{-1} &= \begin{pmatrix} \cos\theta & -\sin\theta \\ \sin\theta & \cos\theta \end{pmatrix} \begin{pmatrix} \cos\theta & \sin\theta \\ -\sin\theta & \cos\theta \end{pmatrix} \\ &= \begin{pmatrix} \cos^2\theta + \sin^2\theta & 0 \\ 0 & \cos^2\theta + \sin^2\theta \end{pmatrix} \\ &= \begin{pmatrix} 1 & 0 \\ 0 & 1 \end{pmatrix} \end{aligned}$$

We can associate a determinant with a square matrix by writing

$$\det A = |A| = \begin{vmatrix} a_{11} & a_{12} & \cdots & a_{1n} \\ a_{21} & a_{22} & \cdots & a_{2n} \\ \vdots & \vdots & \vdots & \vdots \\ a_{n1} & a_{2n} & \cdots & a_{nn} \end{vmatrix}$$

Thus, the determinant of R is

$$\begin{vmatrix} \cos\theta & -\sin\theta \\ \sin\theta & \cos\theta \end{vmatrix} = \cos^2\theta + \sin^2\theta = 1$$

and det $R^{-1} = 1$ also. If det $A = 0$, then A is said to be a *singular matrix*. Singular matrices do not have inverses.

A quantity that arises in group theory, which we will study in the next chapter, is the sum of the diagonal elements of a matrix, called the *trace* of the matrix. Thus, the trace of the matrix

$$B = \begin{pmatrix} 1/2 & 0 & 1 \\ 0 & 2 & 1 \\ 1 & 1 & 1/2 \end{pmatrix}$$

is 3, which we write as Tr B = 3.

Problems

F-1. Given the two matrices

$$A = \begin{pmatrix} 1 & 0 & -1 \\ -1 & 2 & 0 \\ 0 & 1 & 1 \end{pmatrix} \quad \text{and} \quad B = \begin{pmatrix} -1 & 1 & 0 \\ 3 & 0 & 2 \\ 1 & 1 & 1 \end{pmatrix}$$

form the matrices C = 2A − 3B and D = 6B − A.

F-2. Given the three matrices

$$A = \frac{1}{2}\begin{pmatrix} 0 & 1 \\ 1 & 0 \end{pmatrix} \quad B = \frac{1}{2}\begin{pmatrix} 0 & -i \\ i & 0 \end{pmatrix} \quad C = \frac{1}{2}\begin{pmatrix} 1 & 0 \\ 0 & -1 \end{pmatrix}$$

show that $A^2 + B^2 + C^2 = \frac{3}{4}I$, where I is a unit matrix. Also show that

$$AB - BA = iC$$

$$BC - CB = iA$$

$$CA - AC = iB$$

F-3. Given the matrices

$$A = \frac{1}{\sqrt{2}}\begin{pmatrix} 0 & 1 & 0 \\ 1 & 0 & 1 \\ 0 & 1 & 0 \end{pmatrix} \quad B = \frac{1}{\sqrt{2}}\begin{pmatrix} 0 & -i & 0 \\ i & 0 & -i \\ 0 & i & 0 \end{pmatrix} \quad C = \begin{pmatrix} 1 & 0 & 0 \\ 0 & 0 & 0 \\ 0 & 0 & -1 \end{pmatrix}$$

show that

$$AB - BA = iC$$

$$BC - CB = iA$$

$$CA - AC = iB$$

and

$$A^2 + B^2 + C^2 = 2I$$

where I is a unit matrix.

F-4. Do you see any similarity between the results of Problems F–2 and F–3 and the commutation relations involving the components of angular momentum?

F-5. A three-dimensional rotation about the z axis can be represented by the matrix

$$R = \begin{pmatrix} \cos\theta & -\sin\theta & 0 \\ \sin\theta & \cos\theta & 0 \\ 0 & 0 & 1 \end{pmatrix}$$

Show that

$$\det R = |R| = 1$$

Also show that

$$R^{-1} = R(-\theta) = \begin{pmatrix} \cos\theta & \sin\theta & 0 \\ -\sin\theta & \cos\theta & 0 \\ 0 & 0 & 1 \end{pmatrix}$$

F-6. The *transpose* of a matrix A, which we denote by $\tilde{A}$, is formed by replacing the first row of A by its first column, its second row by its second column, etc. Show that this procedure is equivalent to the relation $\tilde{a}_{ij} = a_{ji}$. Show that the transpose of the matrix R given in Problem F–5 is

$$\tilde{R} = \begin{pmatrix} \cos\theta & \sin\theta & 0 \\ -\sin\theta & \cos\theta & 0 \\ 0 & 0 & 1 \end{pmatrix}$$

Note that $\tilde{R} = R^{-1}$. When $\tilde{R} = R^{-1}$, the matrix R is said to be *orthogonal*.

F-7. Given the matrices

$$C_3 = \begin{pmatrix} -\frac{1}{2} & -\frac{\sqrt{3}}{2} \\ \frac{\sqrt{3}}{2} & -\frac{1}{2} \end{pmatrix} \qquad \sigma_v = \begin{pmatrix} 1 & 0 \\ 0 & -1 \end{pmatrix}$$

$$\sigma_v' = \begin{pmatrix} -\frac{1}{2} & \frac{\sqrt{3}}{2} \\ \frac{\sqrt{3}}{2} & \frac{1}{2} \end{pmatrix} \qquad \sigma_v'' = \begin{pmatrix} -\frac{1}{2} & -\frac{\sqrt{3}}{2} \\ -\frac{\sqrt{3}}{2} & \frac{1}{2} \end{pmatrix}$$

show that

$$\sigma_v C_3 = \sigma_v'' \qquad C_3 \sigma_v = \sigma_v'$$

$$\sigma_v'' \sigma_v' = C_3 \qquad C_3 \sigma_v'' = \sigma_v$$

Calculate the determinant associated with each matrix. Calculate the trace of each matrix.

F-8. Which of the matrices in Problem F–7 are orthogonal (see Problem F–6)?

F-9. The inverse of a matrix A can be found by using the following procedure:

    a. Replace each element of A by its cofactor in the corresponding determinant (see MathChapter E for a definition of a cofactor).

    b. Take the transpose of the matrix obtained in step 1.

    c. Divide each element of the matrix obtained in Step 2 by the determinant of A.

For example, if

$$A = \begin{pmatrix} 1 & 2 \\ 3 & 4 \end{pmatrix}$$

then det A $= -2$ and

$$A^{-1} = -\frac{1}{2} \begin{pmatrix} 4 & -2 \\ -3 & 1 \end{pmatrix}$$

Show that $AA^{-1} = A^{-1}A = I$. Use the above procedure to find the inverse of

$$A = \begin{pmatrix} \frac{1}{2} & \frac{1}{\sqrt{2}} \\ \frac{1}{\sqrt{2}} & 0 \end{pmatrix} \quad \text{and} \quad A = \begin{pmatrix} 0 & 2 & 3 \\ 1 & 1 & 1 \\ 2 & 0 & 1 \end{pmatrix}$$

F-10. Recall that a singular matrix is one whose determinant is equal to zero. Referring to the procedure in Problem F–9, do you see why a singular matrix has no inverse?

F-11. Consider the simultaneous algebraic equations

$$x + y = 3$$
$$4x - 3y = 5$$

Show that this pair of equations can be written in the matrix form

$$Ax = c \tag{1}$$

where

$$\mathbf{x} = \begin{pmatrix} x \\ y \end{pmatrix} \qquad \mathbf{c} = \begin{pmatrix} 3 \\ 5 \end{pmatrix} \quad \text{and} \quad A = \begin{pmatrix} 1 & 1 \\ 4 & -3 \end{pmatrix}$$

Now multiply Equation 1 from the left by A^{-1} to obtain

$$\mathbf{x} = A^{-1}\mathbf{c} \tag{2}$$

Now show that

$$A^{-1} = -\frac{1}{7} \begin{pmatrix} -3 & -1 \\ -4 & 1 \end{pmatrix}$$

and that

$$\mathbf{x} = -\frac{1}{7}\begin{pmatrix} -3 & -1 \\ -4 & 1 \end{pmatrix}\begin{pmatrix} 3 \\ 5 \end{pmatrix} = \begin{pmatrix} 2 \\ 1 \end{pmatrix}$$

or that $x = 2$ and $y = 1$. Do you see how this procedure generalizes to any number of simultaneous equations?

F-12. Solve the following simultaneous algebraic equations by the matrix inverse method developed in Problem F–11:

$$x + y - z = 1$$
$$2x - 2y + z = 6$$
$$x + 3z = 0$$

First show that

$$A^{-1} = \frac{1}{13}\begin{pmatrix} 6 & 3 & 1 \\ 5 & -4 & 3 \\ -2 & -1 & 4 \end{pmatrix}$$

and evaluate $\mathbf{x} = A^{-1}\mathbf{c}$.

This photograph shows the team of researchers that discovered buckminsterfullerene (C_{60}), one of the most symmetric molecules known. Buckminsterfullerene is so symmetric that of its 174 normal modes of vibration, only four are infrared active. Kneeling in the photo from left to right are Sean O'Brien, Rick Smalley, Harry Kroto, and Jim Heath; Bob Curl is standing in the back. Smalley, Kroto, and Curl were professors of chemistry and O'Brien and Heath were graduate students at the time (1985). Smalley, Kroto, and Curl shared the Nobel Prize in chemistry in 1996 "for their discovery of fullerenes." **Richard E. Smalley** was born in Akron, Ohio, on June 6, 1943, and received his Ph.D. in chemistry from Princeton University in 1973. In 1976, he moved to Rice University, where he remains today. His current research interests lie in nanotechnology. **Robert F. Curl, Jr.** was born in Alice, Texas, on August 23, 1933, and received his Ph.D. in chemistry from the University of California at Berkeley in 1957. He joined the faculty at Rice University in 1958, where he remains today. Curl's research interests are in laser spectroscopy and chemical kinetics. **Harold Kroto** was born in Wisbech, England, on October 7, 1939, and received his Ph.D. in chemistry from the University of Sheffield in 1964. After postdoctoral research in Canada and the United States, he returned to England to the University of Sussex in Brighton, where he remains today. Kroto's research involves the spectroscopy of molecules in interstellar space.

CHAPTER 12

Group Theory:
The Exploitation of Symmetry

Group theory is perhaps the best example of a subject developed in pure mathematics and subsequently found to have wide application in the physical sciences. Many molecules have a certain degree of symmetry: methane is a tetrahedral molecule; benzene is hexagonal; sulfur hexafluoride and many inorganic ions are octahedral; and so on. By using the properties of groups, we can take advantage of molecular symmetry in a systematic way to predict the properties of molecules and to simplify molecular calculations. Group theory can be used to predict whether or not a molecule has a dipole moment, to derive selection rules for spectroscopic transitions, to determine which atomic orbitals to use to construct hybrid orbitals, to predict which molecular vibrations lead to infrared spectra, to predict which elements are necessarily equal to zero in secular determinants, to label and designate molecular orbitals, and many other useful things. In this chapter, we will develop some of the ideas of group theory and show how group theory can be used to simplify and organize molecular calculations. In the next chapter, we will apply group theory to molecular vibrations and infrared spectroscopy.

12–1. The Exploitation of the Symmetry of a Molecule Can Be Used to Significantly Simplify Numerical Calculations

In Chapter 10, we applied Hückel molecular-orbital theory to benzene. We used the $2p_z$ orbitals on each carbon atom as our atomic orbitals and found the 6×6 secular determinant

$$\begin{vmatrix} x & 1 & 0 & 0 & 0 & 1 \\ 1 & x & 1 & 0 & 0 & 0 \\ 0 & 1 & x & 1 & 0 & 0 \\ 0 & 0 & 1 & x & 1 & 0 \\ 0 & 0 & 0 & 1 & x & 1 \\ 1 & 0 & 0 & 0 & 1 & x \end{vmatrix} = 0 \qquad (12.1)$$

which gives

$$x^6 - 6x^4 + 9x^2 - 4 = 0 \qquad (12.2)$$

when expanded. This resulting secular equation is a sixth-order equation whose six roots are $x = \pm 1, \pm 1,$ and ± 2. Now, if instead of using just the $2p_z$ orbital on each carbon atom to evaluate the secular determinant, we use the following six molecular orbitals that are linear combinations of $2p_z$ orbitals, where we denote the $2p_z$ orbital centered on the jth carbon atom by ψ_j,

$$\phi_1 = \frac{1}{\sqrt{6}}(\psi_1 + \psi_2 + \psi_3 + \psi_4 + \psi_5 + \psi_6)$$

$$\phi_2 = \frac{1}{\sqrt{6}}(\psi_1 - \psi_2 + \psi_3 - \psi_4 + \psi_5 - \psi_6)$$

$$\phi_3 = \frac{1}{\sqrt{12}}(2\psi_1 + \psi_2 - \psi_3 - 2\psi_4 - \psi_5 + \psi_6)$$

$$\phi_4 = \frac{1}{\sqrt{12}}(\psi_1 + 2\psi_2 + \psi_3 - \psi_4 - 2\psi_5 - \psi_6) \qquad (12.3)$$

$$\phi_5 = \frac{1}{\sqrt{12}}(2\psi_1 - \psi_2 - \psi_3 + 2\psi_4 - \psi_5 - \psi_6)$$

$$\phi_6 = \frac{1}{\sqrt{12}}(-\psi_1 + 2\psi_2 - \psi_3 - \psi_4 + 2\psi_5 - \psi_6)$$

then the secular determinant comes out to be (Problem 12–2)

$$\begin{vmatrix} x+2 & 0 & 0 & 0 & 0 & 0 \\ 0 & x-2 & 0 & 0 & 0 & 0 \\ 0 & 0 & x+1 & \dfrac{x+1}{2} & 0 & 0 \\ 0 & 0 & \dfrac{x+1}{2} & x+1 & 0 & 0 \\ 0 & 0 & 0 & 0 & x-1 & \dfrac{1-x}{2} \\ 0 & 0 & 0 & 0 & \dfrac{1-x}{2} & x-1 \end{vmatrix} = 0 \qquad (12.4)$$

When we evaluate this determinant, we get

$$(x+2)(x-2)\begin{vmatrix} x+1 & \dfrac{x+1}{2} \\ \dfrac{x+1}{2} & x+1 \end{vmatrix}\begin{vmatrix} x-1 & \dfrac{1-x}{2} \\ \dfrac{1-x}{2} & x-1 \end{vmatrix} = 0 \qquad (12.5)$$

Notice that the *block diagonal form* of the determinant in Equation 12.4 expands into a product of smaller determinants. Equation 12.5 gives

$$\frac{9}{16}(x+2)(x-2)(x+1)^2(x-1)^2 = 0 \qquad (12.6)$$

whose six roots are ± 1, ± 1, and ± 2, which are identical to those found from Equation 12.2.

In the specific case of benzene, we see that by choosing the set of orbitals given by Equations 12.3, we obtain a secular determinant that is in a block diagonal form, and instead of having to find the six roots of a sixth-degree polynomial (Equation 12.2), we end up with Equation 12.6. Using the "judicious" choice of linear combinations given by Equations 12.3, we reduced the complexity of the resulting secular equation enormously. The obvious question is "Where do these linear combinations come from?", or "Are we able to generate them in a routine manner?". The hexagonal planar symmetry of the benzene molecule happens to lead in a natural way to Equations 12.3, and by taking advantage of this symmetry we will be able to construct these so-called *symmetry orbitals* in a straightforward manner. Furthermore, when we learn how to do this, we will also be able to deduce all the zero matrix elements in Equation 12.4 without even evaluating integrals such as

$$\int \phi_1 \hat{H} \phi_2 d\tau \qquad \text{or} \qquad \int \phi_1 \hat{H} \phi_4 d\tau$$

We will be able to identify which matrix elements must necessarily equal zero because of the symmetry of the molecule. We have chosen benzene as a specific example, but the result is general. By using the symmetry orbitals as dictated by the molecular symmetry, the resulting secular determinant will appear in a block diagonal form, thus reducing the amount of numerical labor significantly.

The procedure we will use is based upon group theory. Its applicability lies in its relation to the symmetry properties of a molecule. To develop this relation, we must first discuss what we mean by the symmetry properties of molecules. Then we will relate these properties to what in mathematics is called a group.

12–2. The Symmetry of Molecules Can Be Described by a Set of Symmetry Elements

The symmetry of a molecule can be described in terms of its *symmetry elements*. For example, a water molecule has the symmetry elements shown in Figure 12.1. The element C_2 is an *axis of symmetry* and σ_v and σ_v' are *planes of symmetry*, or *mirror planes*. Because the hydrogen atoms are indistinguishable, a rotation about the C_2 axis through $180°$ or a reflection through the σ_v plane leaves the molecule indistinguishable from its original configuration, or essentially unchanged. Furthermore, because the molecule is planar, a reflection through the σ_v' plane leaves the molecule essentially unchanged. The C_2 axis is called a two-fold axis. The subscript 2 designates a rotation

456

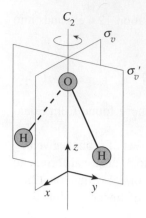

FIGURE 12.1
The C_2, σ_v, and σ_v' symmetry elements of a water molecule. The molecule lies in the mirror plane σ_v' and σ_v is perpendicular to the σ_v' plane. The C_2 axis lies along the intersection of the σ_v and σ_v' planes.

by $360°/2$; or two such rotations brings us back to the original configuration. A C_n axis is called an *n-fold rotation axis of symmetry*; a rotation by $360°/n$ leaves that molecule essentially unchanged.

There are only five types of symmetry elements we must consider, and these are listed in Table 12.1. The identity element, which may seem rather trivial, is included to make the connection with group theory later on. Of course, all molecules have an identity element. Figure 12.1 illustrates a two-fold axis and two planes of symmetry for a H_2O molecule. Sulfur hexafluoride and xenon tetrafluoride (Figure 12.2) are examples of molecules with a center of symmetry, i, and allene and methane are examples of molecules with a four-fold rotation-reflection axis, S_4 (Figure 12.3).

Symmetry elements have *symmetry operations* associated with them. The symmetry operation associated with an *n*-fold axis, C_n, is a rotation by $360°/n$ and that

TABLE 12.1
The five symmetry elements and their associated operators.

Symmetry elements		Symmetry operations	
Symbol	Description	Symbol	Description
E	Identity	$\hat{E}$	No change
C_n	*n*-Fold axis of symmetry	$\hat{C}_n$	Rotation about the axis by $360°/n$
σ	Plane of symmetry	$\hat{\sigma}$	Reflection through the plane
i	Center of symmetry	$\hat{i}$	Reflection through the center
S_n	*n*-Fold rotation-reflection axis of symmetry, also called an improper rotation	$\hat{S}_n$	Rotation about the axis by $360°/n$ followed by reflection through a plane perpendicular to the axis

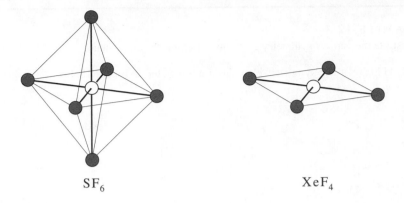

FIGURE 12.2
Sulfur hexafluoride, an octahedral molecule, and xenon tetrafluoride, a square planar molecule, are examples of molecules with a center of symmetry.

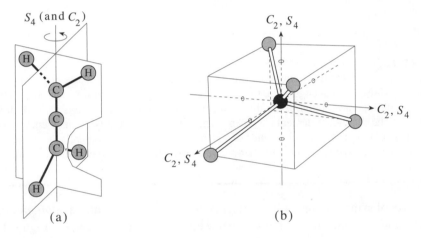

FIGURE 12.3
An illustration of the four-fold rotation-reflection axis, S_4, in (a) allene and (b) methane. We see that the S_4 axis is also a C_2 axis for both molecules.

associated with a plane of symmetry is a reflection through that plane. Note that we have distinguished between a symmetry element and a symmetry operation in Table 12.1 by placing a carat over the symbol for the operation as we do for operators. A symmetry element may have more than one symmetry operation associated with it. For example, a three-fold axis implies a counterclockwise rotation by 120°, *and* a counterclockwise rotation by 240°, which we write as $\hat{C}_3$ and $\hat{C}_3\hat{C}_3 = \hat{C}_3^2$, respectively. Similarly, a four-fold axis implies rotations by 90°, 180°, and 270°, which we write as $\hat{C}_4$, $\hat{C}_4^2$, and $\hat{C}_4^3$, respectively.

EXAMPLE 12–1
Illustrate the various symmetry elements for ammonia and ethene.

SOLUTION: The symmetry elements are illustrated in Figure 12.4.

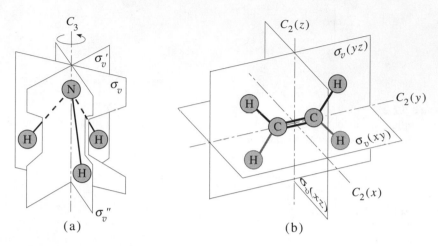

(a)   (b)

FIGURE 12.4
(a) The symmetry elements of ammonia. Each mirror plane contains an N–H bond and bisects the opposite H–N–H bond angle. The three planes are at angles of 120° to each other. The C_3 axis lies along the intersection of the three mirror planes. Of course, there is also an identity element. (b) The symmetry elements of ethene. There are three mutually perpendicular two-fold axes [$C_2(x)$, $C_2(y)$, and $C_2(z)$], a center of inversion (i), and three mutually perpendicular mirror planes [$\sigma(xy)$, $\sigma(yz)$, and $\sigma(xz)$], and, of course, the identity element.

The set of symmetry operations for a given molecule constitutes a *point group*. For example, a water molecule has the symmetry elements E, C_2, σ_v, and σ_v' (Figure 12.1). The point group consisting of these symmetry elements is designated $\mathbf{C}_{2v}$. We say that ammonia belongs to the $\mathbf{C}_{3v}$ point group, which consists of E, C_3, and three planes of symmetry (Figure 12.4a). Only about 30 point groups are of interest to chemists, and in this chapter, we will discuss only a few of them (listed in Table 12.2). Using the examples of the molecules described by each point group helps you visualize its symmetry elements and their corresponding symmetry operations.

Table 12.2 introduces some notation that needs explaining. First, if a molecule contains several symmetry axes, then the one with the largest value of n (if there is such an axis) is called the *principal axis*. Thus, the C_6 axis in benzene is the principal axis. If a plane of symmetry is parallel to a unique axis or to a principal axis, it is designated by σ_v (for *vertical*). If it is perpendicular, it is designated by σ_h (for *horizontal*). If a plane of symmetry bisects the angle between two-fold axes that are perpendicular to a principal axis, it is designated by σ_d (for *dihedral*). Figure 12.5 shows the dihedral planes in a benzene molecule. A σ_d plane is just a special type of a σ_v plane. A point group designated by $\mathbf{C}_{nv}$ has an n-fold axis and n σ_v mirror planes. A

TABLE 12.2

Examples of common point groups of interest to chemists. The number in front of a symmetry element is the number of times such a symmetry element occurs.

Point group	Symmetry elements	Molecular examples
C_{2v}	$E, C_2, 2\sigma_v$	H_2O, CH_2Cl_2, C_6H_5Cl
C_{3v}	$E, C_3, 3\sigma_v$	NH_3, CH_3Cl
C_{2h}	E, C_2, i, σ_h	*trans*-$HClC=CClH$
D_{2h}	$E, 3C_2$ (mutually perpendicular), $i, 3\sigma_v$ (mutually perpendicular)	C_2H_4 (ethene)
D_{3h}	$E, C_3, 3C_2$ (perpendicular to the C_3 axis), $\sigma_h, S_3, 3\sigma_v$	SO_3, BF_3
D_{4h}	$E, C_4, 4C_2$ (perpendicular to the C_4 axis), $i, S_4, \sigma_h, 2\sigma_v, 2\sigma_d$	XeF_4
D_{6h}	$E, C_6, 3C_2, 3C_2',$ $i, S_6, \sigma_h, 3\sigma_v, 3\sigma_d$	C_6H_6 (benzene)
D_{2d}	$E, S_4, 3C_2, 2\sigma_d$	$H_2C=C=CH_2$ (allene)
T_d	$E, 4C_3, 3C_2, 3S_4, 6\sigma_d$	CH_4

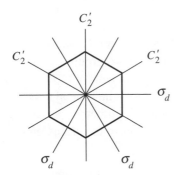

FIGURE 12.5
An illustration of the dihedral planes of symmetry (σ_d) in a benzene molecule.

point group designated by C_{nh} has an n-fold axis and a mirror plane perpendicular to the n-fold axis. Point groups that have an n-fold axis and n two-fold axes perpendicular to it are designated by D_n. If there is also a mirror plane perpendicular to the n-fold axis, then the point group is designated by D_{nh}, a common point group for planar molecules. A point group that has the elements of D_n and n dihedral mirror planes is designated by D_{nd}. Allene (Figure 12.3) is a good example of a D_{2d} molecule. The last point group listed in Table 12.2 is T_d, which stands for tetrahedral. Methane is the prototypical tetrahedral molecule.

12–3. The Symmetry Operations of a Molecule Form a Group

A group is a set of entities that satisfy certain requirements. Specifically, the set A, B, C, ... is said to form a group if

1. there is a rule for combining any two members of the group, and moreover, the result is a member of the group. This combining rule is commonly called multiplication and is denoted by AB, etc. The product AB must be one of the members of the group, and so we say that a group must be closed under multiplication.

2. the"multiplication" rule must be associative; that is,

$$A(BC) = (AB)C$$

In other words, whether we combine B and C and then multiply by A or combine A and B and then multiply the result into C makes no difference.

3. the set of entities contains an identity element, E. Whatever the combination rule, there must be an element such that

$$EA = AE, \qquad EB = BE, \qquad \text{and so on}$$

4. for every entity in the group there is an inverse that is also a member of the group. We will denote the inverse of A by A^{-1}, the inverse of B by B^{-1}, and so on. An inverse has the property that

$$AA^{-1} = A^{-1}A = E, \qquad BB^{-1} = B^{-1}B = E, \qquad \text{and so on}$$

This may seem like a lot of abstract mathematics at this point (which it is), but we will now show that the set of symmetry operations of a molecule forms a group.

Let's use the water molecule as an example. The symmetry elements of H_2O are E, C_2, σ_v, and σ_v' (Figure 12.1), and each of these symmetry elements has one symmetry operation associated with it. Thus, the $\mathbf{C_{2v}}$ point group has four symmetry operations. The number of symmetry operations in a group is called the *order* of the group, and is denoted by h.

EXAMPLE 12–2
Determine the order of the $\mathbf{D_{3h}}$ point group.

SOLUTION: According to Table 12.2, there are 10 symmetry elements in $\mathbf{D_{3h}}$. However, C_3 and S_3 each have two associated symmetry operations ($\hat{C}_3$ and $\hat{C}_3^2$, and $\hat{S}_3$ and $\hat{S}_3^2$), so the order of $\mathbf{D_{3h}}$ is 12.

The multiplication rule of the group is the sequential application of the corresponding symmetry operations. To see the effect of the various symmetry operations, we again consider the water molecule. We define a coordinate system whose origin is

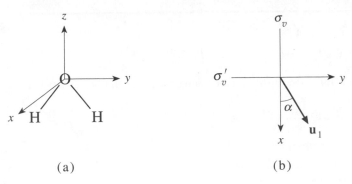

FIGURE 12.6
(a) An illustration of the coordinate axes attached to the oxygen atom in a water molecule. The y axis lies in the plane of the molecule (the σ_v' plane), the x axis is perpendicular to that plane, and the z axis bisects the H–O–H bond angle. (b) A view looking down the z axis, showing the arbitrary unit vector, $\mathbf{u}_1$.

at the oxygen atom (Figure 12.6). We now consider how an arbitrary vector $\mathbf{u}_1$ that points away from this origin is transformed by the symmetry operators of the $\mathbf{C}_{2v}$ point group. When the various symmetry operations are applied, the symmetry elements and the coordinate axes remain fixed in space, and only the vector is transformed. Because the z axis lies along both the C_2 axis and the intersection of the mirror planes, we will look down this axis to see how the vector is transformed. For example, if we apply $\hat{C}_2$ to $\mathbf{u}_1$, we obtain $\mathbf{u}_2$. Applying $\hat{\sigma}_v$ to $\mathbf{u}_2$ gives $\mathbf{u}_3$.

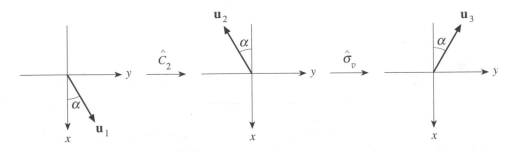

This final result, however, is what we get if we apply $\hat{\sigma}_v'$ directly to the original vector, $\mathbf{u}_1$.

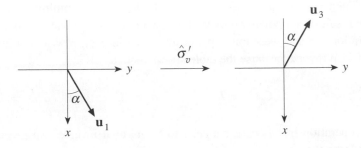

Thus, we see that $\hat{\sigma}_v \hat{C}_2 = \hat{\sigma}_v'$.

EXAMPLE 12–3
Evaluate the products $\hat{\sigma}'_v\hat{\sigma}_v$ and $\hat{\sigma}'_v\hat{C}_2$ for a water molecule.

SOLUTION: Once again, we look down the z axis in Figure 12.6 and write

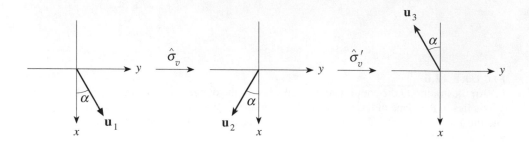

which is equivalent to applying $\hat{C}_2$ directly to the original vector $\mathbf{u}_1$. Thus, we find that $\hat{\sigma}'_v\hat{\sigma}_v = \hat{C}_2$. Similarly,

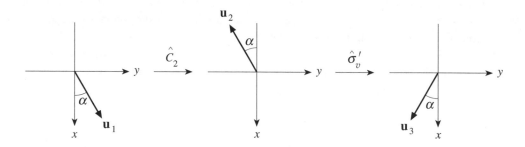

which is equivalent to applying $\hat{\sigma}_v$ directly to the original vector $\mathbf{u}_1$. Thus $\hat{\sigma}'_v\hat{C}_2 = \hat{\sigma}_v$.

We can summarize these results in a *group multiplication table* (Table 12.3), in which each entry is obtained by applying an operation in the top row followed by an operation from the first column. Thus, the result of $\hat{\sigma}'_v\hat{C}_2$ is equivalent to the operation that occurs at the intersection of the column headed by $\hat{C}_2$ and the row headed by $\hat{\sigma}'_v$, namely $\hat{\sigma}_v$. There are several things to notice in Table 12.3. The multiplication is closed because the result of applying two operations sequentially is always equivalent to one of the operations. Each operation appears just once in each row and column. Every operation has an inverse because the table shows us that

$$\hat{E}\hat{E} = \hat{E}, \qquad \hat{C}_2\hat{C}_2 = \hat{E}, \qquad \hat{\sigma}_v\hat{\sigma}_v = \hat{E}, \quad \text{and} \quad \hat{\sigma}'_v\hat{\sigma}'_v = \hat{E}$$

Thus, each operation in this case happens to be its own inverse. An inverse has the effect of undoing the operation.

TABLE 12.3

The group multiplication table of the C_{2v} point group.

Second operation	First operation			
	$\hat{E}$	$\hat{C}_2$	$\hat{\sigma}_v$	$\hat{\sigma}_v'$
$\hat{E}$	$\hat{E}$	$\hat{C}_2$	$\hat{\sigma}_v$	$\hat{\sigma}_v'$
$\hat{C}_2$	$\hat{C}_2$	$\hat{E}$	$\hat{\sigma}_v'$	$\hat{\sigma}_v$
$\hat{\sigma}_v$	$\hat{\sigma}_v$	$\hat{\sigma}_v'$	$\hat{E}$	$\hat{C}_2$
$\hat{\sigma}_v'$	$\hat{\sigma}_v'$	$\hat{\sigma}_v$	$\hat{C}_2$	$\hat{E}$

The multiplication (successive application of each operation) is also associative, and we can refer to the group multiplication table to show this. Let's see if

$$\hat{C}_2(\hat{\sigma}_v\hat{\sigma}_v') \overset{?}{=} (\hat{C}_2\hat{\sigma}_v)\hat{\sigma}_v'$$

The product in the parentheses on the left gives $\hat{C}_2$, and the one on the right gives $\hat{\sigma}_v'$. Thus, we have

$$\hat{C}_2\hat{C}_2 \overset{?}{=} \hat{\sigma}_v'\hat{\sigma}_v'$$

but we find that both of these equal $\hat{E}$. If we continue this search, we find that our multiplication rule is associative for all cases. Thus, the four symmetry operations of C_{2v} satisfy the conditions of being a group and are collectively referred to as the point group C_{2v}.

The other point group we will consider is C_{3v}, and we will use NH_3 as the example of a molecule whose symmetry properties are described by the C_{3v} point group. In this case, we must carefully distinguish between a symmetry element and a symmetry operation. In principle, either a clockwise or a counterclockwise rotation by 120° about the C_3 symmetry axis leaves the molecule unchanged. By convention, however, only counterclockwise rotations are considered. Thus, the C_3 symmetry axis has two symmetry operations associated with it, a counterclockwise rotation by 120° and one by 240°, which we write as $\hat{C}_3$ and $\hat{C}_3^2$, respectively. The geometry in this case is a little more complicated than for H_2O because of the 120° angles. To see the effect of the various group operations, we will set up a coordinate system whose origin is at the nitrogen atom and then follow the effect of the various symmetry operations on an arbitrary vector that points away from the origin. As before, the symmetry elements and the coordinate axes remain fixed in space, and the vector is transformed. Because the z-axis lies along C_3 and the intersection of the mirror planes, it is easier to look down the C_3 axis as we looked down the C_2 axis for water. Let's determine the result of $\hat{C}_3\hat{\sigma}_v'$.

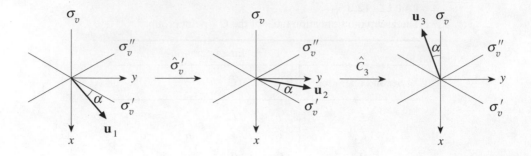

The end result here is equivalent to a reflection of $\mathbf{u}_1$ through $\hat{\sigma}_v''$, and so we see that $\hat{C}_3\hat{\sigma}_v' = \hat{\sigma}_v''$.

EXAMPLE 12–4
Evaluate $\hat{\sigma}_v''\hat{\sigma}_v'$ for the $\mathbf{C}_{3v}$ point group.

SOLUTION:

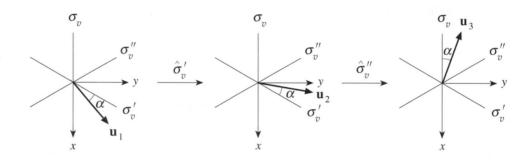

which is equivalent to applying $\hat{C}_3$ directly to $\mathbf{u}_1$.

The complete multiplication table for the $\mathbf{C}_{3v}$ point group is given in Table 12.4.

12–4. Symmetry Operations Can Be Represented by Matrices

Let's consider H_2O, which we have shown belongs to the $\mathbf{C}_{2v}$ point group. We once again construct a set of Cartesian coordinates centered on the oxygen atom (Figure 12.6) and follow the effect of each symmetry operation on an arbitrary vector, $\mathbf{u}$, where $\mathbf{u} = u_x\mathbf{i} + u_y\mathbf{j} + u_z\mathbf{k}$. If the vector is rotated by $180°$ about the z-axis, we have

$$\hat{C}_2 u_x = -u_x, \qquad \hat{C}_2 u_y = -u_y, \qquad \text{and} \qquad \hat{C}_2 u_z = u_z$$

TABLE 12.4
The group multiplication table for the C_{3v} point group.

Second operation	First operation					
	$\hat{E}$	$\hat{C}_3$	$\hat{C}_3^2$	$\hat{\sigma}_v$	$\hat{\sigma}_v'$	$\hat{\sigma}_v''$
$\hat{E}$	$\hat{E}$	$\hat{C}_3$	$\hat{C}_3^2$	$\hat{\sigma}_v$	$\hat{\sigma}_v'$	$\hat{\sigma}_v''$
$\hat{C}_3$	$\hat{C}_3$	$\hat{C}_3^2$	$\hat{E}$	$\hat{\sigma}_v'$	$\hat{\sigma}_v''$	$\hat{\sigma}_v$
$\hat{C}_3^2$	$\hat{C}_3^2$	$\hat{E}$	$\hat{C}_3$	$\hat{\sigma}_v''$	$\hat{\sigma}_v$	$\hat{\sigma}_v'$
$\hat{\sigma}_v$	$\hat{\sigma}_v$	$\hat{\sigma}_v''$	$\hat{\sigma}_v'$	$\hat{E}$	$\hat{C}_3^2$	$\hat{C}_3$
$\hat{\sigma}_v'$	$\hat{\sigma}_v'$	$\hat{\sigma}_v$	$\hat{\sigma}_v''$	$\hat{C}_3$	$\hat{E}$	$\hat{C}_3^2$
$\hat{\sigma}_v''$	$\hat{\sigma}_v''$	$\hat{\sigma}_v'$	$\hat{\sigma}_v$	$\hat{C}_3^2$	$\hat{C}_3$	$\hat{E}$

We can express this result as a matrix equation by writing

$$
\hat{C}_2 \begin{pmatrix} u_x \\ u_y \\ u_z \end{pmatrix} = C_2 \begin{pmatrix} u_x \\ u_y \\ u_z \end{pmatrix} = \begin{pmatrix} -1 & 0 & 0 \\ 0 & -1 & 0 \\ 0 & 0 & 1 \end{pmatrix} \begin{pmatrix} u_x \\ u_y \\ u_z \end{pmatrix}
$$

The above matrix

$$
C_2 = \begin{pmatrix} -1 & 0 & 0 \\ 0 & -1 & 0 \\ 0 & 0 & 1 \end{pmatrix}
$$

represents the operation $\hat{C}_2$. We will denote a matrix that represents an operator by its corresponding sans serif symbol, without the caret. Similarly, $\hat{\sigma}_v$ changes u_y to $-u_y$ and $\hat{\sigma}_v'$ changes u_x to $-u_x$, so the reflection operations can be represented by

$$
\sigma_v = \begin{pmatrix} 1 & 0 & 0 \\ 0 & -1 & 0 \\ 0 & 0 & 1 \end{pmatrix} \quad \text{and} \quad \sigma_v' = \begin{pmatrix} -1 & 0 & 0 \\ 0 & 1 & 0 \\ 0 & 0 & 1 \end{pmatrix}
$$

The identity operation is represented by a unit matrix

$$
E = \begin{pmatrix} 1 & 0 & 0 \\ 0 & 1 & 0 \\ 0 & 0 & 1 \end{pmatrix}
$$

We will now show that these four matrices multiply in the same manner as the C_{2v} group multiplication table in Table 12.3. For example,

$$
C_2 \sigma_v = \begin{pmatrix} -1 & 0 & 0 \\ 0 & -1 & 0 \\ 0 & 0 & 1 \end{pmatrix} \begin{pmatrix} 1 & 0 & 0 \\ 0 & -1 & 0 \\ 0 & 0 & 1 \end{pmatrix} = \begin{pmatrix} -1 & 0 & 0 \\ 0 & 1 & 0 \\ 0 & 0 & 1 \end{pmatrix} = \sigma_v'
$$

EXAMPLE 12–5
Show that $\sigma_v \sigma_v' = C_2$.

SOLUTION: Using the matrix representations for σ_v and σ_v', we have

$$\sigma_v \sigma_v' = \begin{pmatrix} 1 & 0 & 0 \\ 0 & -1 & 0 \\ 0 & 0 & 1 \end{pmatrix} \begin{pmatrix} -1 & 0 & 0 \\ 0 & 1 & 0 \\ 0 & 0 & 1 \end{pmatrix} = \begin{pmatrix} -1 & 0 & 0 \\ 0 & -1 & 0 \\ 0 & 0 & 1 \end{pmatrix} = C_2$$

A set of matrices that multiply together in the same manner as a group multiplication table is said to be a *representation* of that group. Thus, the above four matrices form a representation for the C_{2v} point group. In particular, it is said to be a three-dimensional representation because it consists of 3×3 matrices. This is not the only representation, however. Consider the set of 1×1 matrices

$$E = (1), \qquad C_2 = (1), \qquad \sigma_v = (1), \quad \text{and} \quad \sigma_v' = (1)$$

Certainly, these four matrices multiply in accord with the group multiplication table given in Table 12.3. For example, $C_2 \sigma_v = \sigma_v'$, $\sigma_v \sigma_v' = C_2$, etc., and so the above set of 1×1 matrices also forms a representation for the C_{2v} point group. This one-dimensional representation may seem trivial at this point, but it turns out to be one of the most important representations of any point group. Representations do not have to be diagonal matrices. For example, the 2×2 matrices

$$E = \begin{pmatrix} 1 & 0 \\ 0 & 1 \end{pmatrix}, \quad C_2 = \begin{pmatrix} -1 & 0 \\ 0 & -1 \end{pmatrix}, \quad \sigma_v = \begin{pmatrix} 0 & 1 \\ 1 & 0 \end{pmatrix}, \quad \text{and} \quad \sigma_v' = \begin{pmatrix} 0 & -1 \\ -1 & 0 \end{pmatrix}$$

form a (two-dimensional) representation that consists of nondiagonal matrices. Showing that these four matrices multiply in accord with Table 12.3 is an easy exercise. For example,

$$C_2 \sigma_v = \begin{pmatrix} -1 & 0 \\ 0 & -1 \end{pmatrix} \begin{pmatrix} 0 & 1 \\ 1 & 0 \end{pmatrix} = \begin{pmatrix} 0 & -1 \\ -1 & 0 \end{pmatrix} = \sigma_v'$$

You might ask how many representations there are, and the answer is an infinite number. A few of these, however, are special in the sense that all other representations can be expressed in terms of them. These special representations are called *irreducible representations*, and all others are called *reducible*. There are recipes to determine the irreducible representations for any point group, but fortunately this process has already been done for all the point groups.

There are four irreducible representations of the C_{2v} point group, all one-dimensional. One-dimensional irreducible representations are designated by either A or B. If they are symmetric under a rotation about the principal axis, they are designated by A; if they are antisymmetric, they are designated by B. Note that in Table 12.5 the A irreducible representations have a $+1$ under $\hat{C}_2$, and the B's have a -1. Numerical subscripts are used to distinguish irreducible representations of a similar type. The

TABLE 12.5
The irreducible representations of the C_{2v} point group.

	$\hat{E}$	$\hat{C}_2$	$\hat{\sigma}_v$	$\hat{\sigma}_v'$
A_1	(1)	(1)	(1)	(1)
A_2	(1)	(1)	(−1)	(−1)
B_1	(1)	(−1)	(1)	(−1)
B_2	(1)	(−1)	(−1)	(1)

totally symmetric representation is always designated by A_1. The four irreducible representations of C_{2v} are given in Table 12.5. Note that each one multiplies in accord with the group multiplication table. All point groups have a totally symmetric one-dimensional irreducible representation like A_1 in Table 12.5.

There is an important relation between the dimensions of the irreducible representations and the order of the point group. Recall that the order is the number of symmetry operations in the group. If we let the dimension of the jth irreducible representation be d_j and the order of the group be h, then

$$\sum_{j=1}^{N} d_j^2 = h \tag{12.7}$$

where N is the total number of irreducible representations. Because all point groups have a totally symmetric irreducible representation, A_1, d_1 always equals one in Equation 12.7. Therefore, the only solution to Equation 12.7 with $d_1 = 1$ and $h = 4$ is $d_1 = d_2 = d_3 = d_4 = 1$, so the C_{2v} point group must have four one-dimensional irreducible representations.

EXAMPLE 12–6
Use Equation 12.7 to determine the possible number and dimensions of the irreducible representations of the C_{3v} point group.

SOLUTION: There are six symmetry operations associated with C_{3v}, and so Equation 12.7 reads (with $d_1 = 1$)

$$1 + \sum_{j=2}^{N} d_j^2 = 6$$

There are only two ways to satisfy this relation. There are either six one-dimensional irreducible representations ($1^2 + 1^2 + 1^2 + 1^2 + 1^2 + 1^2 = 6$) or two one-dimensional and one two-dimensional irreducible representations ($1^2 + 1^2 + 2^2 = 6$). We shall see that the C_{3v} point group is characterized by the latter.

12–5. The C_{3v} Point Group Has a Two-Dimensional Irreducible Representation

The irreducible representations of the C_{3v} point group are given in Table 12.6. Note how the number and dimensions of these irreducible representations satisfy Equation 12.7. Note also that C_{3v} has a totally symmetric irreducible representation A_1, which can be deduced by applying the six operations of C_{3v} to the z component of the vector in Figure 12.7.

EXAMPLE 12–7
Use the irreducible representation labelled E in Table 12.6 to show that $\hat{\sigma}_v'\hat{\sigma}_v'' = \hat{C}_3^2$, in accord with the group multiplication table of the C_{3v} point group.

SOLUTION: We have

$$\sigma_v'\sigma_v'' = \begin{pmatrix} -\frac{1}{2} & \frac{\sqrt{3}}{2} \\ \frac{\sqrt{3}}{2} & \frac{1}{2} \end{pmatrix} \begin{pmatrix} -\frac{1}{2} & -\frac{\sqrt{3}}{2} \\ -\frac{\sqrt{3}}{2} & \frac{1}{2} \end{pmatrix} = \begin{pmatrix} -\frac{1}{2} & \frac{\sqrt{3}}{2} \\ -\frac{\sqrt{3}}{2} & -\frac{1}{2} \end{pmatrix} = C_3^2$$

Two-dimensional irreducible representations are designated by the letter E (not to be confused with the identity element E). The two-dimensional irreducible representations in Table 12.6 can be obtained by applying the symmetry operations to an arbitrary unit vector, $\mathbf{u}_1$, as shown in Figure 12.7. According to Figure 12.7, $\mathbf{u}_1 = u_{1x}\mathbf{i} + u_{1y}\mathbf{j} = (\cos\alpha)\mathbf{i} + (\sin\alpha)\mathbf{j}$. Therefore, if we apply $\hat{C}_3$ to $\mathbf{u}_1$, we get a new vector $\mathbf{u}_2$

$$\mathbf{u}_2 = \hat{C}_3\mathbf{u}_1 = \cos(120° + \alpha)\mathbf{i} + \sin(120° + \alpha)\mathbf{j}$$

$$= \left(-\frac{1}{2}\cos\alpha - \frac{\sqrt{3}}{2}\sin\alpha\right)\mathbf{i} + \left(\frac{\sqrt{3}}{2}\cos\alpha - \frac{1}{2}\sin\alpha\right)\mathbf{j}$$

$$= \left(-\frac{1}{2}u_{1x} - \frac{\sqrt{3}}{2}u_{1y}\right)\mathbf{i} + \left(\frac{\sqrt{3}}{2}u_{1x} - \frac{1}{2}u_{1y}\right)\mathbf{j}$$

TABLE 12.6
The irreducible representations of the C_{3v} point group.

	$\hat{E}$	$\hat{C}_3$	$\hat{C}_3^2$	$\hat{\sigma}_v$	$\hat{\sigma}_v'$	$\hat{\sigma}_v''$
A_1	(1)	(1)	(1)	(1)	(1)	(1)
A_2	(1)	(1)	(1)	(−1)	(−1)	(−1)
E	$\begin{pmatrix} 1 & 0 \\ 0 & 1 \end{pmatrix}$	$\begin{pmatrix} -\frac{1}{2} & -\frac{\sqrt{3}}{2} \\ \frac{\sqrt{3}}{2} & -\frac{1}{2} \end{pmatrix}$	$\begin{pmatrix} -\frac{1}{2} & \frac{\sqrt{3}}{2} \\ -\frac{\sqrt{3}}{2} & -\frac{1}{2} \end{pmatrix}$	$\begin{pmatrix} 1 & 0 \\ 0 & -1 \end{pmatrix}$	$\begin{pmatrix} -\frac{1}{2} & \frac{\sqrt{3}}{2} \\ \frac{\sqrt{3}}{2} & \frac{1}{2} \end{pmatrix}$	$\begin{pmatrix} -\frac{1}{2} & -\frac{\sqrt{3}}{2} \\ -\frac{\sqrt{3}}{2} & \frac{1}{2} \end{pmatrix}$

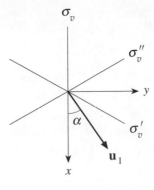

FIGURE 12.7
A view looking down the z axis of a set of coordinate axes centered on the nitrogen atom in NH_3, a $\mathbf{C}_{3v}$ molecule. The x–z plane lies in the mirror plane σ_v (see Figure 12.4).

or

$$u_{2x} = -\frac{1}{2}u_{1x} - \frac{\sqrt{3}}{2}u_{1y}$$

$$u_{2y} = \frac{\sqrt{3}}{2}u_{1x} - \frac{1}{2}u_{1y}$$

or in matrix form

$$\begin{pmatrix} u_{2x} \\ u_{2y} \end{pmatrix} = \begin{pmatrix} -\frac{1}{2} & -\frac{\sqrt{3}}{2} \\ \frac{\sqrt{3}}{2} & -\frac{1}{2} \end{pmatrix} \begin{pmatrix} u_{1x} \\ u_{1y} \end{pmatrix}$$

Thus, the matrix representation of $\hat{C}_3$ is

$$C_3 = \begin{pmatrix} -\frac{1}{2} & -\frac{\sqrt{3}}{2} \\ \frac{\sqrt{3}}{2} & -\frac{1}{2} \end{pmatrix}$$

in agreement with Table 12.6. Furthermore, note that $C_3 C_3 = C_3^2$ in Table 12.6. Note also that C_3 is given by Equation F.2 if we let $\theta = 120°$.

EXAMPLE 12–8
Deduce the 2×2 matrix corresponding to $\hat{\sigma}_v''$ in Table 12.6.

SOLUTION: The geometry is shown in Figure 12.8. We write $\mathbf{u}_1$ as

$$\mathbf{u}_1 = u_{1x}\mathbf{i} + u_{1y}\mathbf{j} = (\cos\alpha)\mathbf{i} + (\sin\alpha)\mathbf{j}$$

Figure 12.8 shows that a reflection through σ_v'' sends α to $240° - \alpha$, so $\mathbf{u}_1$ becomes the vector $\mathbf{u}_2$ where

$$\mathbf{u}_2 = u_{2x}\mathbf{i} + u_{2y}\mathbf{j} = [\cos(240° - \alpha)]\mathbf{i} + [\sin(240° - \alpha)]\mathbf{j}$$

$$= \left(-\frac{1}{2}\cos\alpha - \frac{\sqrt{3}}{2}\sin\alpha\right)\mathbf{i} + \left(-\frac{\sqrt{3}}{2}\cos\alpha + \frac{1}{2}\sin\alpha\right)\mathbf{j}$$

or

$$u_{2x} = -\frac{1}{2}u_{1x} - \frac{\sqrt{3}}{2}u_{1y}$$

$$u_{2y} = -\frac{\sqrt{3}}{2}u_{1x} + \frac{1}{2}u_{1y}$$

or in matrix form

$$\begin{pmatrix} u_{2x} \\ u_{2y} \end{pmatrix} = \sigma_v'' \begin{pmatrix} u_{1x} \\ u_{1y} \end{pmatrix} = \begin{pmatrix} -\frac{1}{2} & -\frac{\sqrt{3}}{2} \\ -\frac{\sqrt{3}}{2} & \frac{1}{2} \end{pmatrix} \begin{pmatrix} u_{1x} \\ u_{1y} \end{pmatrix}$$

in agreement with Table 12.6.

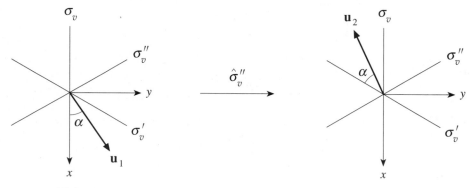

FIGURE 12.8
The result of a reflection of the unit vector $\mathbf{u}_1$ in Figure 12.7 through the mirror plane $\hat{\sigma}_v''$.

Notice that the reason the E irreducible representation is two-dimensional is that the u_x and u_y transform together so that the result of a given operation must be written as a linear combination of u_x and u_y. The components u_x and u_y are said to form a *basis* for E or to belong to E. The basis for A_1 is simply u_z because u_z is unaffected by each operation of the group. Notice that the number of bases is equal to the dimension of the irreducible representation. (Although it is not obvious, the basis for A_2 is the set of vectors that represents a rotation of the molecule about the z-axis).

EXAMPLE 12–9
Show that the basis for B_1 in Table 12.5 for the $\mathbf{C}_{2v}$ point group is u_x.

SOLUTION: Refer to just the u_x portion shown in the figures in Example 12–3, and see that

$$\hat{E}u_x = (+1)u_x, \quad \hat{C}_2u_x = (-1)u_x, \quad \hat{\sigma}_v u_x = (+1)u_x, \quad \text{and} \quad \hat{\sigma}_v' u_x = (-1)u_x$$

in accord with B_1 in Table 12.5.

12–6. The Most Important Summary of the Properties of a Point Group Is Its Character Table

Tables 12.5 and 12.6 show the irreducible representations of the point groups C_{2v} and C_{3v}, respectively. It turns out that for almost all the applications of group theory we do not need the complete matrices, only the sum of the diagonal elements. The sum of the diagonal elements of a matrix is called its *trace*, or more commonly in group theory, its *character*. Of course, the matrices must be square, but all the matrices we encounter in group theory are square. The characters of the irreducible representations of a point group are displayed in a *character table*, as shown in Tables 12.7 and 12.8 for C_{2v} and C_{3v}, respectively. The numerical entries in these tables follow immediately from Tables 12.5 and 12.6, but some new notation has been introduced. We will use character tables throughout the rest of this chapter, so let's examine the various parts of Tables 12.7 and 12.8 in detail.

As expected, the A irreducible representations in Tables 12.7 and 12.8 have $+1$ under $\hat{C}_2$ and $\hat{C}_3$, respectively, while the B's have -1. Numerical subscripts serve to distinguish irreducible representations of a similar type. By convention, the totally symmetric irreducible representation is designated by A_1 in Tables 12.7 and 12.8. Now let's look at the first of the two right columns in Tables 12.7 and 12.8. This column lists the bases for the various irreducible representations. We showed earlier that u_x forms a basis of B_1 in Table 12.7 and that u_x and u_y jointly form a basis for the two-dimensional irreducible representation E in Table 12.8. The R_j denote vectors that depict rotations about the indicated axes. We stated earlier that the basis for A_2 in Table 12.8 is the set of vectors that represent a rotation about the principal (z) axis. If we look down the z

TABLE 12.7
The character table of the C_{2v} point group.

C_{2v}	$\hat{E}$	$\hat{C}_2$	$\hat{\sigma}_v$	$\hat{\sigma}_v'$		
A_1	1	1	1	1	z	x^2, y^2, z^2
A_2	1	1	-1	-1	R_z	xy
B_1	1	-1	1	-1	x, R_y	xz
B_2	1	-1	-1	1	y, R_x	yz

TABLE 12.8
An expanded form of the character table of the C_{3v} point group.

C_{3v}	$\hat{E}$	$\hat{C}_3$	$\hat{C}_3^2$	$\hat{\sigma}_v$	$\hat{\sigma}_v'$	$\hat{\sigma}_v''$		
A_1	1	1	1	1	1	1	z	$x^2 + y^2, z^2$
A_2	1	1	1	-1	-1	-1	R_z	
E	2	-1	-1	0	0	0	$(x, y)(R_x, R_y)$	$(x^2 - y^2, xy)\,(xz, yz)$

axis of a C_{3v} molecule such as ammonia, a rotation about this axis can be depicted as follows.

To see that these vectors form a basis of A_2, we apply each symmetry operation in turn. The operations $\hat{E}$, $\hat{C}_3$, and $\hat{C}_3^2$ do not change the directions of the vectors, but $\hat{\sigma}_v$, $\hat{\sigma}_v'$, and $\hat{\sigma}_v''$ do. Symbolically, we have

$$\hat{E}(R_z) = R_z \qquad \hat{\sigma}_v(\hat{R}_z) = -R_z$$
$$\hat{C}_3(R_z) = R_z \qquad \hat{\sigma}_v'(\hat{R}_z) = -R_z$$
$$\hat{C}_3^2(R_z) = R_z \qquad \hat{\sigma}_v''(\hat{R}_z) = -R_z$$

Thus, we see from Table 12.8 that R_z is a basis for A_2 because the characters for $\hat{E}$, $\hat{C}_3$, and $\hat{C}_3^2$ are $+1$ and the characters for $\hat{\sigma}_v$, $\hat{\sigma}_v'$, and $\hat{\sigma}_v''$ are -1.

EXAMPLE 12–10
Show that R_z forms a basis for A_2 in Table 12.7 (the C_{2v} point group).

SOLUTION: We can represent a rotation about the C_2 axis as follows (looking down the axis).

The operations $\hat{E}$ and $\hat{C}_2$ do not change the directions of the arrows, but $\hat{\sigma}_v$ and $\hat{\sigma}_v'$ do. Therefore, $\hat{E}R_z = \hat{C}_2R_z = (+1)R_z$ and $\hat{\sigma}_vR_z = \hat{\sigma}_v'R_z = (-1)R_z$. Table 12.7 shows, then, that R_z is a basis for A_2.

We will use the information in the first of the right columns of a character table when we apply group theory to the vibrational spectra of polyatomic molecules in Chapter 13. The final column lists how certain products of x, y, and z transform. This column is used in discussing d orbitals and in Raman spectroscopy.

We now turn to an explanation of the title to Table 12.8, which reads "An expanded form of the character table for the C_{3v} point group." If you look up the C_{3v} character table in any book on group theory, you will find the table given in Table 12.9. To explain the difference between Tables 12.8 and 12.9, we need to introduce the idea of a *class*. Notice that the columns headed by $\hat{C}_3$ and $\hat{C}_3^2$ are the same in Table 12.8. The reason for this is that $\hat{C}_3$ (counterclockwise rotation by 120°) and $\hat{C}_3^2$ (counterclockwise rotation by 240° = clockwise rotation by 120°) are essentially the same operation physically. Whether we rotate clockwise or counterclockwise by 120° is rather arbitrary. Thus, $\hat{C}_3$ and $\hat{C}_3^2$ head identical columns in Table 12.8. Similarly, the three mirror planes are essentially equivalent. Their labeling is arbitrary, and so their corresponding symmetry operations are equivalent physically. Symmetry operations that are essentially equivalent are said to belong to the same class. There is a more mathematical definition of class, but the physical argument given here is all we need.

Table 12.9 shows that there are three classes of symmetry operations in $\mathbf{C}_{3v}$. Note from Tables 12.7 and 12.9 through 12.14 that the number of classes equals the number of irreducible representations. This is a general result; character tables are square. Note that Table 12.14 has triply degenerate representations, which are designated by the letter T. The point group $\mathbf{C}_{2h}$, which describes a *trans*-dichloroethene molecule, has a center of inversion, i, and the character table is given in Table 12.10. Irreducible representations that are symmetric under an inversion have a subscript g (for the German word, *gerade*, meaning even), while those that are antisymmetric have a subscript u (for the German word, *ungerade*, meaning odd). Recall that these labels were also used to describe the properties of the molecular orbitals of homonuclear diatomic molecules under inversion through the point halfway between the two bonded atoms. Thus, the

TABLE 12.9

The character table of the $\mathbf{C}_{3v}$ point group.

$\mathbf{C}_{3v}$	$\hat{E}$	$2\hat{C}_3$	$3\hat{\sigma}_v$		
A_1	1	1	1	z	$x^2 + y^2, z^2$
A_2	1	1	−1	R_z	
E	2	−1	0	$(x, y)\ (R_x, R_y)$	$(x^2 − y^2, xy)\ (xz, yz)$

TABLE 12.10

The character table of the $\mathbf{C}_{2h}$ point group. *trans*-dichloroethene is an example of a $\mathbf{C}_{2h}$ molecule.

$\mathbf{C}_{2h}$	$\hat{E}$	$\hat{C}_2$	$\hat{\imath}$	$\hat{\sigma}_h$		
A_g	1	1	1	1	R_z	x^2, y^2, z^2, xy
B_g	1	−1	1	−1	R_x, R_y	xz, yz
A_u	1	1	−1	−1	z	
B_u	1	−1	−1	1	x, y	

TABLE 12.11

The character table of the $\mathbf{D}_{3h}$ point group. Sulfur trioxide is an example of a $\mathbf{D}_{3h}$ molecule.

$\mathbf{D}_{3h}$	$\hat{E}$	$2\hat{C}_3$	$3\hat{C}_2$	$\hat{\sigma}_h$	$2\hat{S}_3$	$3\hat{\sigma}_v$		
A_1'	1	1	1	1	1	1		$x^2 + y^2, z^2$
A_2'	1	1	-1	1	1	-1	R_z	
E'	2	-1	0	2	-1	0	(x, y)	$(x^2 - y^2, xy)$
A_1''	1	1	1	-1	-1	-1		
A_2''	1	1	-1	-1	-1	1	z	
E''	2	-1	0	-2	1	0	(R_x, R_y)	(xz, yz)

TABLE 12.12

The character table of the $\mathbf{D}_{4h}$ point group. Xenon tetrafluoride is an example of a $\mathbf{D}_{4h}$ molecule.

$\mathbf{D}_{4h}$	$\hat{E}$	$2\hat{C}_4$	$\hat{C}_2$	$2\hat{C}_2'$	$2\hat{C}_2''$	$\hat{i}$	$2\hat{S}_4$	$\hat{\sigma}_h$	$2\hat{\sigma}_v$	$2\hat{\sigma}_d$		
A_{1g}	1	1	1	1	1	1	1	1	1	1		$x^2 + y^2, z^2$
A_{2g}	1	1	1	-1	-1	1	1	1	-1	-1	R_z	
B_{1g}	1	-1	1	1	-1	1	-1	1	1	-1		$x^2 - y^2$
B_{2g}	1	-1	1	-1	1	1	-1	1	-1	1		xy
E_g	2	0	-2	0	0	2	0	-2	0	0	(R_x, R_y)	(xz, yz)
A_{1u}	1	1	1	1	1	-1	-1	-1	-1	-1		
A_{2u}	1	1	1	-1	-1	-1	-1	-1	1	1	z	
B_{1u}	1	-1	1	1	-1	-1	1	-1	-1	1		
B_{2u}	1	-1	1	-1	1	-1	1	-1	1	-1		
E_u	2	0	-2	0	0	-2	0	2	0	0	(x, y)	

first irreducible representation in Table 12.10 is designated by A_g because the character under the operator $\hat{C}_2$ is $+1$ (hence A) and the character under the operator $\hat{i}$ is $+1$ (hence the subscript g).

12–7. Several Mathematical Relations Involve the Characters of Irreducible Representations

In this section, we will give without proof a number of the mathematical properties associated with character tables. As we said in the previous section, character tables are square; in other words, the number of irreducible representations equals the number of classes. Furthermore, because an identity operation is represented by a unit matrix and the character is the sum of the diagonal elements of this matrix, the character

TABLE 12.13

The character table of the $\mathbf{D}_{6h}$ point group. Benzene is an example of a $\mathbf{D}_{6h}$ molecule.

$\mathbf{D}_{6h}$	$\hat{E}$	$2\hat{C}_6$	$2\hat{C}_3$	$\hat{C}_2$	$3\hat{C}_2'$	$3\hat{C}_2''$	$\hat{\imath}$	$2\hat{S}_3$	$2\hat{S}_6$	$\hat{\sigma}_h$	$3\hat{\sigma}_d$	$3\hat{\sigma}_v$		
A_{1g}	1	1	1	1	1	1	1	1	1	1	1	1		x^2+y^2, z^2
A_{2g}	1	1	1	1	-1	-1	1	1	1	1	-1	-1	R_z	
B_{1g}	1	-1	1	-1	1	-1	1	-1	1	-1	1	-1		
B_{2g}	1	-1	1	-1	-1	1	1	-1	1	-1	-1	1		
E_{1g}	2	1	-1	-2	0	0	2	1	-1	-2	0	0	(R_x, R_y)	(xz, yz)
E_{2g}	2	-1	-1	2	0	0	2	-1	-1	2	0	0		(x^2-y^2, xy)
A_{1u}	1	1	1	1	1	1	-1	-1	-1	-1	-1	-1		
A_{2u}	1	1	1	1	-1	-1	-1	-1	-1	-1	1	1	z	
B_{1u}	1	-1	1	-1	1	-1	-1	1	-1	1	-1	1		
B_{2u}	1	-1	1	-1	-1	1	-1	1	-1	1	1	-1		
E_{1u}	2	1	-1	-2	0	0	-2	-1	1	2	0	0	(x, y)	
E_{2u}	2	-1	-1	2	0	0	-2	1	1	-2	0	0		

TABLE 12.14

The character table of the $\mathbf{T}_d$ point group. Methane is an example of a $\mathbf{T}_d$ molecule.

$\mathbf{T}_d$	$\hat{E}$	$8\hat{C}_3$	$3\hat{C}_2$	$6\hat{S}_4$	$6\hat{\sigma}_d$		
A_1	1	1	1	1	1		$x^2+y^2+z^2$
A_2	1	1	1	-1	-1		
E	2	-1	2	0	0		$(2z^2-x^2-y^2, x^2-y^2)$
T_1	3	0	-1	1	-1	(R_x, R_y, R_z)	
T_2	3	0	-1	-1	1	(x, y, z)	(xy, xz, yz)

of the identity operator is equal to the dimension of the irreducible representations. Tables 12.7 through 12.14 show that the character for the identity operator is 1 for A and B irreducible representations and 2 and 3 for E and T irreducible representations, respectively. We stated earlier that if d_j is the dimension of the jth irreducible representation and h is the order of the group (Section 12–4), then

$$\sum_{j=1}^{N} d_j^2 = h \tag{12.8}$$

where N is the number of irreducible representations. We can write this equation in different, but more common, notation. First, let $\hat{R}$ be any symmetry operation, and let $\chi(\hat{R})$ be the character of a matrix representation of $\hat{R}$. Furthermore, let $\chi_j(\hat{R})$ be the character of the jth irreducible representation of $\hat{R}$. Now because $\chi_j(\hat{E}) = d_j$, we can write Equation 12.8 in the more commonly seen form

$$\sum_{j=1}^{N} [\chi_j(\hat{E})]^2 = h \tag{12.9}$$

Not only is this equation valid for Tables 12.7 through 12.14, but it is valid for the character tables of all point groups.

A more thorough treatment of group theory views each row of a character table as an abstract vector. Recall from MathChapter C that the scalar product or the dot product of two vectors $\mathbf{u}$ and $\mathbf{v}$ is given by (Equation C.6)

$$\mathbf{u} \cdot \mathbf{v} = |u||v| \cos\theta \tag{12.10}$$

or by (Equation C.9)

$$\mathbf{u} \cdot \mathbf{v} = u_1 v_1 + u_2 v_2 + u_3 v_3 \tag{12.11}$$

where $|u|$ and $|v|$ are the lengths of $\mathbf{u}$ and $\mathbf{v}$ and θ is the angle between them. Therefore, if $\mathbf{u}$ and $\mathbf{v}$ are perpendicular (orthogonal), then $\cos\theta = 0$ and

$$\mathbf{u} \cdot \mathbf{v} = u_1 v_1 + u_2 v_2 + u_3 v_3 = 0 \tag{12.12}$$

Although we wrote $\mathbf{u}$ and $\mathbf{v}$ as three-dimensional vectors in Equations 12.11 and 12.12, they can be n-dimensional vectors, in which case Equation 12.11 becomes

$$\mathbf{u} \cdot \mathbf{v} = \sum_{k=1}^{n} u_k v_k \tag{12.13}$$

If the sum in Equation 12.13 equals zero, we say that $\mathbf{u}$ and $\mathbf{v}$ are *orthogonal*. Of course, if $n > 3$, it is difficult to visualize that $\mathbf{u}$ and $\mathbf{v}$ are perpendicular, but they are in a generalized sense. Now let's return to the rows in the $\mathbf{C}_{2v}$ character table (Table 12.7). Each row contains four characters, and we can think of each row as a four-dimensional vector whose components are $\chi_j(\hat{R})$. Now let's take the dot product (Equation 12.13) of any two rows of the character table. If we do this, then we find that the rows are, in fact, orthogonal, or that

$$\sum_{\hat{R}} \chi_i(\hat{R})\chi_j(\hat{R}) = 0 \qquad i \neq j \tag{12.14}$$

For example, the dot product of A_1 with B_2 in Table 12.7 is

$$\chi_{A_1}(\hat{E})\chi_{B_2}(\hat{E}) + \chi_{A_1}(\hat{C}_2)\chi_{B_2}(\hat{C}_2) + \chi_{A_1}(\hat{\sigma}_v)\chi_{B_2}(\hat{\sigma}_v) + \chi_{A_1}(\hat{\sigma}_v')\chi_{B_2}(\hat{\sigma}_v')$$

or

$$(1) \times (1) + (1) \times (-1) + (1) \times (-1) + (1) \times (1) = 0$$

EXAMPLE 12–11
Show that the row A_1 is orthogonal to the row E in the $\mathbf{C}_{3v}$ point group (Table 12.9). (Remember that the columns are headed by classes.)

SOLUTION: The dot product of the two rows A_1 and E in Table 12.8 is

$$(1) \times (2) + \underbrace{(1) \times (-1) + (1) \times (-1)}_{2\hat{C}_3} + \underbrace{(1) \times (0) + (1) \times (0) + (1) \times (0)}_{3\hat{\sigma}_v} = 0$$

Note that because we have summed over symmetry operations, we must include the products $\chi_i(\hat{R})\chi_j(\hat{R})$ for each of the two $\hat{C}_3$ rotations and the three $\hat{\sigma}_v$ reflections.

We can sum over classes rather than symmetry operations in Equation 12.14. If we let $n(\hat{R})$ be the number of symmetry operations in the class containing $\hat{R}$, then Equation 12.14 can be written as

$$\sum_{\text{classes}} n(\hat{R})\chi_i(\hat{R})\chi_j(\hat{R}) = 0 \qquad i \neq j \tag{12.15}$$

Applying Equation 12.15 to Example 12–11 gives

$$n(\hat{E})\chi_{A_1}(\hat{E})\chi_E(\hat{E}) + n(\hat{C}_3)\chi_{A_1}(\hat{C}_3)\chi_E(\hat{C}_3) + n(\hat{\sigma}_v)\chi_{A_1}(\hat{\sigma}_v)\chi_E(\hat{\sigma}_v)$$

or

$$\underbrace{(1) \times (1) \times (2)}_{1\hat{E}} + \underbrace{(2) \times (1) \times (-1)}_{2\hat{C}_3} + \underbrace{(3) \times (1) \times (0)}_{3\hat{\sigma}_v} = 0$$

If we let the i in Equation 12.14 or 12.15 stand for the totally symmetric irreducible representation, then $\chi_i(\hat{R}) = 1$ in each term in the summation, and so Equation 12.14 says that

$$\sum_{\hat{R}} \chi_j(\hat{R}) = \sum_{\text{classes}} n(\hat{R})\chi_j(\hat{R}) = 0 \qquad j \neq A_1 \tag{12.16}$$

In other words, the sum of the characters in any row (other than the first) is equal to zero. Confirm that this statement is true for Tables 12.7 through 12.14.

The square of the length of a vector $\mathbf{v}$ is given by (Example C–2)

$$\mathbf{v} \cdot \mathbf{v} = v_1^2 + v_2^2 + v_3^2$$

which we can readily generalize to n dimensions

$$\mathbf{v} \cdot \mathbf{v} = (\text{length})^2 = \sum_{k=1}^{n} v_k^2$$

Because each row of a character table may be treated as an n-dimensional vector, we may express this equation as

$$(\text{length})^2 = \sum_{\hat{R}}[\chi_j(\hat{R})]^2 \tag{12.17}$$

If we apply Equation 12.17 to any of the rows in Tables 12.7 through 12.14, then we find that

$$\sum_{\hat{R}}[\chi_j(\hat{R})]^2 = h \tag{12.18}$$

which is a general result. In terms of a sum over classes, we have

$$\sum_{\text{classes}} n(\hat{R})[\chi_j(\hat{R})]^2 = h \tag{12.19}$$

Equation 12.18 or 12.19 says that the length of the vector corresponding to any row in a character table equals the square root of the order of the group. Applying Equation 12.19 to A_2 of $\mathbf{C}_{3v}$ (Table 12.9) gives

$$\sum_{\text{classes}} n(\hat{R})[\chi_j(\hat{R})]^2 = 1 \times (1)^2 + 2 \times (1)^2 + 3 \times (-1)^2 = 6$$

We can combine Equations 12.14 through 12.19 into

$$\sum_{\hat{R}}\chi_i(\hat{R})\chi_j(\hat{R}) = \sum_{\text{classes}} n(\hat{R})\chi_i(\hat{R})\chi_j(\hat{R}) = h\delta_{ij} \tag{12.20}$$

where δ_{ij} is the Krönecker delta. Everything in this section up to this point can be summarized by Equation 12.20, which says that the rows of a character table are orthogonal and of length $h^{1/2}$.

In applying group theory to a physical problem such as a molecular orbital calculation (as in Section 12–1), we are going to apply the symmetry operations of the molecular point group to the atomic orbitals on each atom to construct a certain (usually reducible) representation. We will then reduce this reducible representation into its irreducible representations and then use this result to construct symmetry orbitals or linear combinations of the atomic orbitals that make optimum use of the symmetry of the molecule (see Equations 12.3). The one question that remains, then, is how do we reduce a given reducible representation into its component irreducible representations. As usual, we will work with the characters of the various representations. The mathematical question we want to answer is how to determine the a_j's in the expression

$$\chi(\hat{R}) = \sum_{j} a_j \chi_j(\hat{R}) \tag{12.21}$$

where $\chi(\hat{R})$ is the character of the symmetry operator $\hat{R}$ in the reducible representation, Γ. These coefficients will tell us how many times each irreducible representation is contained in Γ. Determining the a_j's using the orthogonality relation given in Equation 12.20 is actually fairly easy. Multiply Equation 12.21 by $\chi_i(\hat{R})$ and sum both sides over $\hat{R}$:

$$\sum_{\hat{R}} \chi(\hat{R})\chi_i(\hat{R}) = \sum_j a_j \sum_{\hat{R}} \chi_i(\hat{R})\chi_j(\hat{R})$$

But Equation 12.20 says that the sum over $\hat{R}$ on the right side is $h\delta_{ij}$, so we have

$$\sum_{\hat{R}} \chi(\hat{R})\chi_i(\hat{R}) = \sum_j ha_j \delta_{ij}$$

The only term that survives in the right side is the term with $i = j$ (otherwise $\delta_{ij} = 0$), so we have

$$a_i = \frac{1}{h} \sum_{\hat{R}} \chi(\hat{R})\chi_i(\hat{R}) \tag{12.22}$$

We can also write Equation 12.22 as a sum over classes:

$$a_i = \frac{1}{h} \sum_{\hat{R}} \chi(\hat{R})\chi_i(\hat{R}) = \frac{1}{h} \sum_{\text{classes}} n(\hat{R})\chi(\hat{R})\chi_i(\hat{R}) \tag{12.23}$$

We now know enough group theory to apply it to some molecular calculations.

EXAMPLE 12–12
Suppose that the characters of a reducible representation of the C_{2v} point group are $\chi(\hat{E}) = 4$, $\chi(\hat{C}_2) = 2$, $\chi(\hat{\sigma}_v) = 0$, and $\chi(\hat{\sigma}_v') = 2$. We usually express this result by writing $\Gamma = 4\,2\,0\,2$. Determine how many times each irreducible representation of C_{2v} is contained in $\Gamma = 4\,2\,0\,2$.

SOLUTION: We use Equation 12.23 with $h = 4$.

$$a_{A_1} = \frac{1}{4}[(4) \times (1) + (2) \times (1) + (0) \times (1) + (2) \times (1)] = 2$$

$$a_{A_2} = \frac{1}{4}[(4) \times (1) + (2) \times (1) + (0) \times (-1) + (2) \times (-1)] = 1$$

$$a_{B_1} = \frac{1}{4}[(4) \times (1) + (2) \times (-1) + (0) \times (1) + (2) \times (-1)] = 0$$

$$a_{B_2} = \frac{1}{4}[(4) \times (1) + (2) \times (-1) + (0) \times (-1) + (2) \times (1)] = 1$$

Thus,

$$\Gamma = 2A_1 + A_2 + B_2$$

EXAMPLE 12–13

Suppose that $\Gamma = 3\ 0\ -1$ for $\mathbf{C}_{3v}$. Determine the a_i's in Equation 12.23.

SOLUTION: We use Equation 12.23 as a sum over classes, in which case we have

$$a_{A_1} = \frac{1}{6}[(1) \times (3) \times (1) + (2) \times (0) \times (1) + (3) \times (-1) \times (1)] = 0$$

$$a_{A_2} = \frac{1}{6}[(1) \times (3) \times (1) + (2) \times (0) \times (1) + (3) \times (-1) \times (-1)] = 1$$

$$a_{E} = \frac{1}{6}[(1) \times (3) \times (2) + (2) \times (0) \times (-1) + (3) \times (-1) \times (0)] = 1$$

or $\Gamma = A_2 + E$.

12–8. We Use Symmetry Arguments to Predict Which Elements in a Secular Determinant Equal Zero

Recall from Chapters 9 and 10 that we encountered molecular integrals of the type

$$H_{ij} = \int \phi_i^* \hat{H} \phi_j d\tau \quad \text{and} \quad S_{ij} = \int \phi_i^* \phi_j d\tau \tag{12.24}$$

We will now show that integrals like these will be equal to zero if we choose ϕ_i^* and ϕ_j such that they belong to different irreducible representations. For simplicity, we will prove this only for one-dimensional irreducible representations, but the result is general. Let's start with the overlap integral

$$S_{ij} = \int \phi_i^* \phi_j d\tau \tag{12.25}$$

This integral is just some number, and its value certainly cannot depend upon how we orient the molecule. A symmetry operation of the molecule $\hat{R}$ transforms ϕ_i and ϕ_j to $\hat{R}\phi_i$ and $\hat{R}\phi_j$, respectively. The resulting (transformed) overlap integral is

$$\hat{R}S_{ij} = \int \hat{R}\phi_i^* \hat{R}\phi_j d\tau$$

Because the value of S_{ij} cannot change when we apply a symmetry operation of the point group of the molecule

$$\hat{R}S_{ij} = \int \hat{R}\phi_i^* \hat{R}\phi_j d\tau = S_{ij} = \int \phi_i^* \phi_j d\tau \tag{12.26}$$

Suppose now that ϕ_i^* and ϕ_j are bases for the (one-dimensional) irreducible representations Γ_a and Γ_b. If that is so, then

$$\hat{R}\phi_i^* = \chi_a(\hat{R})\phi_i^* \quad \text{and} \quad \hat{R}\phi_j = \chi_b(\hat{R})\phi_j \tag{12.27}$$

In fact, Equations 12.27 are exactly what we mean when we say that ϕ_i^* and ϕ_j are bases for the one-dimensional irreducible representations Γ_a and Γ_b (see Example 12–9). If we substitute Equations 12.27 into Equation 12.26, we obtain

$$S_{ij} = \chi_a(\hat{R})\chi_b(\hat{R}) \int \phi_i^*\phi_j d\tau = \chi_a(\hat{R})\chi_b(\hat{R})S_{ij} \tag{12.28}$$

Equation 12.28 requires that

$$\chi_a(\hat{R})\chi_b(\hat{R}) = 1 \quad \text{for all } \hat{R} \tag{12.29}$$

Because $\chi_i(\hat{R})$ is either 1 or -1 for any one-dimensional irreducible representation, Equation 12.29 is true only if $\chi_a(\hat{R}) = \chi_b(\hat{R})$, or if Γ_a and Γ_b are the same irreducible representation. If $\chi_a(\hat{R}) \neq \chi_b(\hat{R})$, then $\chi_a(\hat{R})\chi_b(\hat{R})$ will equal -1 for some symmetry operation $\hat{R}$, and the only way that S_{ij} can equal $-S_{ij}$ in Equation 12.28 is for S_{ij} to equal zero. Thus, we have proved (at least for one-dimensional irreducible representations) one of the most useful results of group theory; namely, that S_{ij} must necessarily be equal to zero if ϕ_i^* and ϕ_j are bases of different irreducible representations.

Let's apply this result to the H_2O molecule (which lies in the y-z plane, Figure 12.6a) and evaluate S_{ij} for a $2p_x$ orbital on the oxygen atom ($2p_{xO}$) and the sum of the $1s$ orbitals on the hydrogen atoms ($1s_{H_A} + 1s_{H_B}$). This linear combination of hydrogen $1s$ orbitals is symmetric under all four operations of the C_{2v} point group, and so transforms as A_1. We chose $1s_{H_A} + 1s_{H_B}$ rather than $1s_{H_A}$ or $1s_{H_B}$ individually for this very reason. The $2p_x$ orbital on the oxygen atom transforms as x, which transforms as B_1 according to Table 12.7. Therefore, we can say that the overlap integral of $2p_{xO}$ and $1s_{H_A} + 1s_{H_B}$ is zero by symmetry. Table 12.7 shows that the same is true for $2p_{yO}$, but not for $2p_{zO}$.

EXAMPLE 12–14
Show that the overlap integral involving $2p_{xN}$ and $1s_{H_A} + 1s_{H_B} + 1s_{H_C}$ in the NH_3 molecule (C_{3v}) is equal to zero.

SOLUTION: The linear combination $1s_{H_A} + 1s_{H_B} + 1s_{H_C}$ belongs to the totally symmetric irreducible representation A_1 and, according to Table 12.9, $2p_{xN}$ belongs to E. Therefore, the overlap integral is equal to zero.

The other integrals in a secular determinant are the H_{ij} in Equation 12.24. The molecular Hamiltonian operator is symmetric under all the group operations of the molecule, because the molecule is indistinguishable under all these operations. There-fore, $\hat{H}$ must belong to A_1. Using the same procedure we did for S_{ij}, we have

$$H_{ij} = \int \phi_i^* \hat{H} \phi_j d\tau = \hat{R} H_{ij} = \int (\hat{R}\phi_i^*)(\hat{R}\hat{H})(\hat{R}\phi_j) d\tau$$

$$= \chi_a(\hat{R}) \chi_{A_1}(\hat{R}) \chi_b(\hat{R}) H_{ij} \tag{12.30}$$

Therefore, because H_{ij} is independent of the orientation of the molecule, we must have that

$$\chi_a(\hat{R}) \chi_{A_1}(\hat{R}) \chi_b(\hat{R}) = 1 \qquad \text{for all } \hat{R} \tag{12.31}$$

But $\chi_{A_1}(\hat{R}) = 1$ for all $\hat{R}$, so Equation 12.31 is the same as Equation 12.29, which implies once again that ϕ_i^* and ϕ_j must belong to the same irreducible representation. Thus, both H_{ij} and S_{ij} will necessarily equal zero unless ϕ_i^* and ϕ_j belong to the same irreducible representation. This affords an enormous simplification in evaluating typically large secular determinants in molecular calculations.

Table 12.13 gives the character table for the $\mathbf{D}_{6h}$ point group, which describes benzene. Referring back to Equation 12.3, let's show that the symmetry orbital ϕ_1 belongs to A_{2u}. To do so, we must show that ϕ_1 transforms according to the characters of each symmetry operation of A_{2u}. Of course, $\hat{E}\phi_1 = \phi_1$. Figure 12.9 shows some of the symmetry elements of a benzene molecule, and Figure 12.10 illustrates some of the symmetry operations. A counterclockwise rotation by 60° about the principal axis gives (remember that ψ_j stands for a $2p_z$ orbital on carbon atom j)

$$\hat{C}_6\phi_1 = \hat{C}_6\psi_1 + \hat{C}_6\psi_2 + \hat{C}_6\psi_3 + \hat{C}_6\psi_4 + \hat{C}_6\psi_5 + \hat{C}_6\psi_6$$

$$= \psi_6 + \psi_1 + \psi_2 + \psi_3 + \psi_4 + \psi_5 = \phi_1$$

Similarly, (see Figure 12.10)

$$\hat{C}_2'\phi_1 = -\psi_1 - \psi_6 - \psi_5 - \psi_4 - \psi_3 - \psi_2 = -\phi_1$$

$$\hat{\sigma}_d\phi_1 = \psi_2 + \psi_1 + \psi_6 + \psi_5 + \psi_4 + \psi_3 = \phi_1$$

$$\hat{S}_3\phi_1 = \hat{\sigma}_h\hat{C}_3\phi_1 = -\psi_5 - \psi_6 - \psi_1 - \psi_2 - \psi_3 - \psi_4 = -\phi_1$$

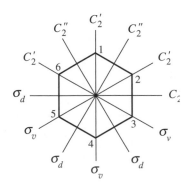

FIGURE 12.9

The symmetry elements of a benzene molecule, which belongs to the $\mathbf{D}_{6h}$ point group. The C_6 and S_6 axes are perpendicular to the plane, σ_h lies in the plane and i is at the center.

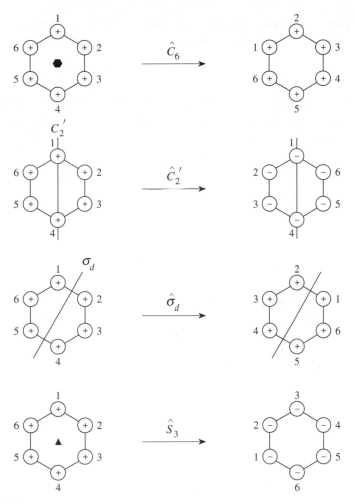

FIGURE 12.10
The effect of some of the symmetry operations of the $\mathbf{D}_{6h}$ point group on the carbon $2p_z$ orbitals, which are used to construct molecular orbitals for benzene.

and so on. In each case, ϕ_1 transforms according to the characters in A_{2u}. Similarly, we can easily show that ϕ_2 belongs to B_{2g}. For example,

$$
\begin{aligned}
\hat{C}_2'\phi_2 &= \hat{C}_2'\psi_1 - \hat{C}_2'\psi_2 + \hat{C}_2'\psi_3 - \hat{C}_2'\psi_4 + \hat{C}_2'\psi_5 - \hat{C}_2'\psi_6 \\
&= -\psi_1 + \psi_6 - \psi_5 + \psi_4 - \psi_3 + \psi_2 = -\phi_2
\end{aligned}
$$

Of the other four symmetry orbitals in Equation 12.3, ϕ_3 and ϕ_4 belong to E_{1g}, and ϕ_5 and ϕ_6 belong to E_{2u}. Because the integrals in Equation 12.24 equal zero if ϕ_i^* and ϕ_j belong to different irreducible representations, we can now see why the 6×6 benzene secular determinant (Equation 12.1) factors into two 1×1 and two 2×2 determinants when we use the linear combinations in Equations 12.3. The final question that we must face is how to find these symmetry orbitals; in other words, how to find linear combinations of atomic orbitals that act as bases for the various irreducible representations. This question is the topic of the next section.

12–9. Generating Operators Are Used to Find Linear Combinations of Atomic Orbitals That Are Bases for Irreducible Representations

There is a straightforward procedure to find linear combinations of atomic orbitals that are bases for the irreducible representations. It involves a quantity called a *generating operator*, whose formula we give without proof. The generating operator for the jth irreducible representation is

$$\hat{P}_j = \frac{d_j}{h} \sum_{\hat{R}} \chi_j(\hat{R})\hat{R} \qquad (12.32)$$

Recall that d_j is the dimensionality of the jth irreducible representation. Equation 12.32 may look formidable, but it is really easy to use. Before we use it to generate symmetry orbitals for benzene, with its relatively large $\mathbf{D}_{6h}$ character table, let's use it to generate symmetry orbitals for butadiene. Recall that we applied Hückel molecular-orbital theory to butadiene in Section 10–6. The skeletal π-electron framework of butadiene

suggests that we use the $\mathbf{C}_{2h}$ point-group elements (Table 12.10). If we denote the $2p_z$ orbital an carbon atom i by ψ_i, Equation 12.32 gives

$$\hat{P}_{A_g}\psi_1 = \tfrac{1}{4}\sum_{\hat{R}} \chi_{A_g}(\hat{R})\hat{R}\psi_1$$
$$= \tfrac{1}{4}[(1)\hat{E}\psi_1 + (1)\hat{C}_2\psi_1 + (1)\hat{i}\psi_1 + (1)\hat{\sigma}_h\psi_1]$$
$$= \tfrac{1}{4}(\psi_1 + \psi_4 - \psi_4 - \psi_1) = 0$$
$$\hat{P}_{A_g}\psi_2 = \tfrac{1}{4}(\psi_2 + \psi_3 - \psi_3 - \psi_2) = 0$$

with similar results for ψ_3 and ψ_4. Similarly, using ψ_1 and ψ_2, we get

$$\hat{P}_{B_g}\psi_1 = \tfrac{1}{4}(\psi_1 - \psi_4 - \psi_4 + \psi_1) \propto \psi_1 - \psi_4$$
$$\hat{P}_{B_g}\psi_2 = \tfrac{1}{4}(\psi_2 - \psi_3 - \psi_3 + \psi_2) \propto \psi_2 - \psi_3$$
$$\hat{P}_{A_u}\psi_1 = \tfrac{1}{4}(\psi_1 + \psi_4 + \psi_4 + \psi_1) \propto \psi_1 + \psi_4 \qquad (12.33)$$
$$\hat{P}_{A_u}\psi_2 = \tfrac{1}{4}(\psi_2 + \psi_3 + \psi_3 + \psi_2) \propto \psi_2 + \psi_3$$
$$\hat{P}_{B_u}\psi_1 = \hat{P}_{B_u}\psi_2 = 0$$

We have ignored the numerical factors in front of the various symmetry orbitals because we are interested only in their functional form. Their subsequent normalization is a simple matter.

Equations 12.33 give us four symmetry orbitals, two belonging to B_g symmetry and two belonging to A_u symmetry. Using these symmetry orbitals, the Hückel molecular-orbital theory secular determinant of butadiene factors into two 2×2 blocks. The actual form is (Problem 12–28)

$$\begin{vmatrix} x & 1 & | & 0 & 0 \\ 1 & x+1 & | & 0 & 0 \\ -- & -- & -- & -- & -- \\ 0 & 0 & | & x & 1 \\ 0 & 0 & | & 1 & x-1 \end{vmatrix} = 0$$

or

$$(x^2 + x - 1)(x^2 - x - 1) = 0 \tag{12.34}$$

or

$$x = \frac{-1 \pm \sqrt{5}}{2} \qquad \text{and} \qquad x = \frac{1 \pm \sqrt{5}}{2}$$

or $x = 0.6180$, -1.6180, 1.6180, and -0.6180. These are the very same values we obtained in Section 10–6, but there we had to deal with a quartic equation for x because the secular determinant was not in block diagonal form.

Note above that no symmetry orbitals belong to A_g or B_u. It turns out that we really did not have to apply the generating operators for A_g and B_u to learn this. Let's apply the four group operations to the four $2p_z$ orbitals in Figure 12.11. For the identity operation, $\hat{E}\psi_j = \psi_j$ for each j. We can write this result in matrix form:

$$\hat{E} \begin{pmatrix} \psi_1 \\ \psi_2 \\ \psi_3 \\ \psi_4 \end{pmatrix} = \begin{pmatrix} 1 & 0 & 0 & 0 \\ 0 & 1 & 0 & 0 \\ 0 & 0 & 1 & 0 \\ 0 & 0 & 0 & 1 \end{pmatrix} \begin{pmatrix} \psi_1 \\ \psi_2 \\ \psi_3 \\ \psi_4 \end{pmatrix}$$

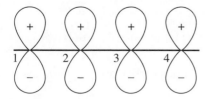

FIGURE 12.11
A schematic illustration of the four $2p_z$ orbitals used to form Hückel molecular orbitals for butadiene.

The character of this matrix is $\chi(\hat{E}) = 4$. Similarly, $\hat{C}_2\psi_1 = \psi_4$, $\hat{C}_2\psi_2 = \psi_3$, $\hat{C}_2\psi_3 = \psi_2$, and $\hat{C}_2\psi_4 = \psi_1$, or

$$\hat{C}_2 \begin{pmatrix} \psi_1 \\ \psi_2 \\ \psi_3 \\ \psi_4 \end{pmatrix} = \begin{pmatrix} 0 & 0 & 0 & 1 \\ 0 & 0 & 1 & 0 \\ 0 & 1 & 0 & 0 \\ 1 & 0 & 0 & 0 \end{pmatrix} \begin{pmatrix} \psi_1 \\ \psi_2 \\ \psi_3 \\ \psi_4 \end{pmatrix}$$

for $\chi(\hat{C}_2) = 0$. Similarly,

$$\hat{i} \begin{pmatrix} \psi_1 \\ \psi_2 \\ \psi_3 \\ \psi_4 \end{pmatrix} = \begin{pmatrix} 0 & 0 & 0 & -1 \\ 0 & 0 & -1 & 0 \\ 0 & -1 & 0 & 0 \\ -1 & 0 & 0 & 0 \end{pmatrix} \begin{pmatrix} \psi_1 \\ \psi_2 \\ \psi_3 \\ \psi_4 \end{pmatrix}$$

and

$$\hat{\sigma}_v \begin{pmatrix} \psi_1 \\ \psi_2 \\ \psi_3 \\ \psi_4 \end{pmatrix} = \begin{pmatrix} -1 & 0 & 0 & 0 \\ 0 & -1 & 0 & 0 \\ 0 & 0 & -1 & 0 \\ 0 & 0 & 0 & -1 \end{pmatrix} \begin{pmatrix} \psi_1 \\ \psi_2 \\ \psi_3 \\ \psi_4 \end{pmatrix}$$

for $\chi(\hat{i}) = 0$ and $\chi(\hat{\sigma}_v) = -4$. These results tell us that the four $2p_z$ orbitals belong to the reducible representation

	$\hat{E}$	$\hat{C}_2$	$\hat{i}$	$\hat{\sigma}_v$
Γ	4	0	0	-4

We can even write Γ without writing out all the matrices. Note that there is a 1 on a diagonal in a representation if the $2p_z$ orbital is unchanged, a -1 if it remains on the original atom but changes direction, and a 0 if it is moved to another atom. The operation $\hat{E}$ leaves all four unchanged; the operations $\hat{C}_2$ and $\hat{i}$ move all of them to different atoms; and $\hat{\sigma}_v$ changes the direction of all four but doesn't move them otherwise.

We can use Equation 12.23 to reduce Γ into irreducible representations:

$$a_{A_g} = \tfrac{1}{4}[(4) \times (1) + (0) \times (1) + (0) \times (1) + (-4) \times (1)] = 0$$

$$a_{B_g} = \tfrac{1}{4}[(4) \times (1) + (0) \times (-1) + (0) \times (1) + (-4) \times (-1)] = 2$$

$$a_{A_u} = \tfrac{1}{4}[(4) \times (1) + (0) \times (1) + (0) \times (-1) + (-4) \times (-1)] = 2$$

$$a_{B_u} = \tfrac{1}{4}[(4) \times (1) + (0) \times (-1) + (0) \times (-1) + (-4) \times (1)] = 0$$

which we write as $\Gamma = 2B_g + 2A_u$, in agreement with our earlier result that no symmetry orbitals belong to A_g or B_u.

EXAMPLE 12–15

If we carry out a molecular-orbital calculation on H_2O using the minimal basis set atomic orbitals $1s_{H_A}$, $1s_{H_B}$, $1s_O$, $2s_O$, $2p_{xO}$, $2p_{yO}$, and $2p_{zO}$ without using group theory, we would obtain a 7×7 secular determinant. How will this determinant look if we use group theory to generate symmetry orbitals?

SOLUTION: The water molecule belongs to C_{2v}. First determine the reducible representation of the seven atomic orbitals as we did above for butadiene. We picture the s orbitals as spheres on each atom and the $2p$ orbitals on the oxygen atom, much like the coordinate axes in Figure 12.6a. Certainly $\chi(\hat{E}) = 7$. The operation $\hat{C}_2$ moves the $1s_H$ orbitals to different atoms, does not alter $1s_O$, $2s_O$, and $2p_{zO}$, and changes the direction of $2p_{xO}$ and $2p_{yO}$; therefore $\chi(\hat{C}_2) = 3 - 2 = 1$. Similarly, $\hat{\sigma}_v$ moves the $1s_H$ to different atoms, does not alter $1s_O$, $2s_O$, $2p_{xO}$, and $2p_{zO}$ and changes the direction of $2p_{yO}$. Therefore, $\chi(\hat{\sigma}_v) = 4 - 1 = 3$. Last, $\chi(\hat{\sigma}'_v) = 6 - 1 = 5$, to give

	$\hat{E}$	$\hat{C}_2$	$\hat{\sigma}_v$	$\hat{\sigma}'_v$
Γ	7	1	3	5

Using Equation 12.23, we find that

$$a_{A_1} = \tfrac{1}{4}[(7) \times (1) + (1) \times (1) + (3) \times (1) + (5) \times (1)] = 4$$

$$a_{A_2} = \tfrac{1}{4}[(7) \times (1) + (1) \times (1) + (3) \times (-1) + (5) \times (-1)] = 0$$

$$a_{B_1} = \tfrac{1}{4}[(7) \times (1) + (1) \times (-1) + (3) \times (1) + (5) \times (-1)] = 1$$

$$a_{B_2} = \tfrac{1}{4}[(7) \times (1) + (1) \times (-1) + (3) \times (-1) + (5) \times (1)] = 2$$

which we write as $\Gamma = 4A_1 + B_1 + 2B_2$. So we see that four combinations belong to A_1, none to A_2, one to B_1 and two to B_2. The original 7×7 secular determinant blocks into a 1×1, a 2×2, and a 4×4 determinant. The energies obtained for the 1×1, 2×2, and 4×4 determinants correspond to the molecular orbitals of B_1, B_2, and A_1 symmetry, respectively. We can easily show, using Equation 12.32, that $1s_{H_A} + 1s_{H_B}$, $1s_O$, $2s_O$, and $2p_{zO}$ belong to A_1; that $2p_{xO}$ belongs to B_1; and that $2p_{yO}$ and $1s_{H_A} - 1s_{H_B}$ belong to B_2.

For the final topic in this chapter, we will use Equation 12.32 to derive the symmetry orbitals for benzene that are given in Equations 12.3. First we deduce the reducible representation for the six $2p_z$ orbitals that we use in the Hückel molecular-orbital treatment for benzene. The result of applying each symmetry operation to the six $2p_z$ orbitals is (see Figure 12.9)

	$\hat{E}$	$2\hat{C}_6$	$2\hat{C}_3$	$\hat{C}_2$	$3\hat{C}'_2$	$3\hat{C}''_2$	$\hat{i}$	$2\hat{S}_3$	$2\hat{S}_6$	$\hat{\sigma}_h$	$3\hat{\sigma}_d$	$3\hat{\sigma}_v$
Γ	6	0	0	0	-2	0	0	0	0	-6	0	2

Note that the symmetry operations $\hat{C}_6$, $\hat{C}_6^5$, $\hat{C}_3$, $\hat{C}_3^2$, $\hat{C}_2$, $\hat{C}''_2$, $\hat{i}$, $\hat{S}_3$, $\hat{S}_3^2$, $\hat{S}_6$, $\hat{S}_6^5$, and $\hat{\sigma}_d$ move the $2p_z$ orbitals to other atoms; that each $\hat{C}'_2$ operation leaves two $2p_z$ orbitals on their original atoms but changes their direction; that $\hat{\sigma}_h$ reverses the direction of all six $2p_z$ orbitals; and that $\hat{\sigma}_v$ leaves two $2p_z$ orbitals on their original atoms and preserves their directions. Equation 12.23 gives (Problem 12.29)

$$\Gamma = B_{2g} + E_{1g} + A_{2u} + E_{2u}$$

This result tells us that the secular determinant for benzene will contain two 1×1 blocks (for the B_{2g} and A_{2u} symmetry orbitals) and two 2×2 blocks (for the E_{1g} and E_{2u} symmetry orbitals) (see Equation 12.4). We can use Equation 12.32 to generate the symmetry orbitals. Using Equation 12.32 for A_{2u} gives (use Figures 12.8 and 12.9)

$$\hat{P}_{A_{2u}} \psi_1 = \frac{1}{24} (\underbrace{\psi_1}_{\hat{E}} + \underbrace{\psi_2 + \psi_6}_{2\hat{C}_6} + \underbrace{\psi_3 + \psi_5}_{2\hat{C}_3} + \underbrace{\psi_4}_{\hat{C}_2} + \underbrace{\psi_1 + \psi_3 + \psi_5}_{3\hat{C}_2'} + \underbrace{\psi_2 + \psi_4 + \psi_6}_{3\hat{C}_2''}$$

$$+ \underbrace{\psi_4}_{\hat{\imath}} + \underbrace{\psi_3 + \psi_5}_{2\hat{S}_3} + \underbrace{\psi_2 + \psi_6}_{2\hat{S}_6} + \underbrace{\psi_1}_{\hat{\sigma}_h} + \underbrace{\psi_2 + \psi_4 + \psi_6}_{3\hat{\sigma}_d} + \underbrace{\psi_1 + \psi_3 + \psi_5}_{3\hat{\sigma}_v})$$

$$\approx \psi_1 + \psi_2 + \psi_3 + \psi_4 + \psi_5 + \psi_6$$

We obtain the same result for $\hat{P}_{A_{2u}} \psi_j$ for $j = 2$ through 6, so this molecular orbital is the one symmetry orbital belonging to A_{2u}. Similarly, the one symmetry orbital belonging to B_{2g} is

$$\hat{P}_{B_{2g}} \psi_1 \approx (\underbrace{\psi_1}_{\hat{E}} - \underbrace{\psi_2 - \psi_6}_{2\hat{C}_6} + \underbrace{\psi_3 + \psi_5}_{2\hat{C}_3} - \underbrace{\psi_4}_{\hat{C}_2} + \underbrace{\psi_1 + \psi_3 + \psi_5}_{3\hat{C}_2'} - \underbrace{\psi_2 + \psi_4 + \psi_6}_{3\hat{C}_2''}$$

$$- \underbrace{\psi_4}_{\hat{\imath}} + \underbrace{\psi_3 + \psi_5}_{2\hat{S}_3} - \underbrace{\psi_2 - \psi_6}_{2\hat{S}_6} + \underbrace{\psi_1}_{\hat{\sigma}_h} - \underbrace{\psi_2 - \psi_4 - \psi_6}_{3\hat{\sigma}_d} + \underbrace{\psi_1 + \psi_3 + \psi_5}_{3\hat{\sigma}_v})$$

$$\approx \psi_1 - \psi_2 + \psi_3 - \psi_4 + \psi_5 - \psi_6$$

(Once again, we obtain the same symmetry orbital if we evaluate $\hat{P}_{B_{2g}} \psi_j$ for $j = 2$ through 6.) Because E_{1g} is two-dimensional, two symmetry orbitals belong to this irreducible representation. For example,

$$\hat{P}_{E_{1g}} \psi_1 \approx (2\psi_1 + \psi_2 - \psi_3 - 2\psi_4 - \psi_5 + \psi_6)$$

and

$$\hat{P}_{E_{1g}} \psi_2 \approx (\psi_1 + 2\psi_2 + \psi_3 - \psi_4 - 2\psi_5 - \psi_6)$$

Applying $\hat{P}_{E_{1g}}$ to the other $2p_z$ orbitals will give different linear combinations, but only two of them can be linearly independent. We are free to choose any two or even any linear combination of them, whatever is convenient. (See Problem 12–26.) For simplicity, we have used the two given above to evaluate the secular determinant in

Equations 12.3. Last, two linearly independent symmetry orbitals, belonging to E_{2u}, are

$$\hat{P}_{E_{2u}}\psi_1 \approx (2\psi_1 - \psi_2 - \psi_3 + 2\psi_4 - \psi_5 - \psi_6)$$

and

$$\hat{P}_{E_{2u}}\psi_2 \approx (-\psi_1 + 2\psi_2 - \psi_3 - \psi_4 + 2\psi_5 - \psi_6)$$

in accord with Equations 12.3.

As we have seen, group theory can be used to simplify molecular calculations. By choosing linear combinations of atomic orbitals that belong to irreducible representations of the molecular point group, many of the integrals involved are necessarily zero. This affords an enormous advantage in large calculations. Group theory can also be used to classify the vibrations of polyatomic molecules. Once again, by exploiting molecular symmetry we will be able to assert which vibrations lead to an infrared spectrum (are infrared active) and which do not (are infrared inactive). We will see how to do this in the next chapter.

Problems

12-1. Neglecting overlap, show that ϕ_1 and ϕ_2 given by Equations 12.3 are orthonormal to the other four molecular orbitals.

12-2. Using the six molecular orbitals given by Equations 12.3, verify that $H_{11} = \alpha + 2\beta$, $H_{22} = \alpha - 2\beta$, $H_{12} = H_{13} = H_{14} = H_{15} = H_{16} = 0$ (see Equation 12.4).

12-3. List the various symmetry elements for the trigonal planar molecule SO_3.

12-4. Verify that a methane molecule has the symmetry elements given in Table 12.2.

12-5. Verify that a benzene molecule has the symmetry elements given in Table 12.2.

12-6. Verify that a xenon tetrafluoride (square planar) molecule has the symmetry elements given in Table 12.2.

12-7. Explain why $\hat{C}_4^3 = \hat{C}_4^{-1}$.

12-8. Deduce the group multiplication table for the point group $\mathbf{C}_{2v}$ (see Table 12.3).

12-9. Determine the order of the $\mathbf{D}_{4h}$ point group (see Table 12.2).

12-10. Determine the order of the $\mathbf{D}_{6h}$ point group (see Table 12.2).

12-11. Evaluate the products $\hat{\sigma}_v\hat{\sigma}_v'$, $\hat{C}_2\hat{\sigma}_v$, and $\hat{C}_2\hat{\sigma}_v'$ for a $\mathbf{C}_{2v}$ point group (see Table 12.3).

12-12. Evaluate the products $\hat{C}_3\hat{\sigma}_v$ and $\hat{C}_3\hat{\sigma}_v''$ for a $\mathbf{C}_{3v}$ point group (see Table 12.4).

12-13. Show that Equation 12.7 is valid for the point groups given in Tables 12.9 through 12.14.

12-14. Show that the 2×2 matrices given in Table 12.6 are a representation for the $\mathbf{C}_{3v}$ point group.

12-15. In Section 12–4, we derived matrix representations for various symmetry operators. Starting with an arbitrary vector $\mathbf{u}$, where $\mathbf{u} = u_x \mathbf{i} + u_y \mathbf{j} + u_z \mathbf{k}$, show that the matrix representation for a counterclockwise rotation about the z-axis by an angle α, $\hat{C}_{360/\alpha}$, is given by

$$\begin{pmatrix} \cos \alpha & -\sin \alpha & 0 \\ \sin \alpha & \cos \alpha & 0 \\ 0 & 0 & 1 \end{pmatrix}$$

Show that the corresponding matrix for rotation-reflection $\hat{S}_{360/\alpha}$ about the z-axis by an angle α is

$$\begin{pmatrix} \cos \alpha & -\sin \alpha & 0 \\ \sin \alpha & \cos \alpha & 0 \\ 0 & 0 & -1 \end{pmatrix}$$

12-16. Show that u_x forms a basis for the irreducible representation B_1 of the point group $\mathbf{C}_{2v}$.

12-17. Show that R_x forms a basis for the irreducible representation B_2 of the point group $\mathbf{C}_{2v}$.

12-18. Show that $(u_x,\ u_y)$ forms a joint basis for the irreducible representation E of the point group $\mathbf{C}_{3v}$.

12-19. Show that the rows of the character table of $\mathbf{C}_{2h}$ satisfy Equation 12.20.

12-20. Show that the rows of the character table of $\mathbf{D}_{3h}$ satisfy Equation 12.20.

12-21. Suppose the characters of a reducible representation of the $\mathbf{T}_d$ point group are $\chi(\hat{E}) = 17$, $\chi(\hat{C}_3) = 2$, $\chi(\hat{C}_2) = 5$, $\chi(\hat{S}_4) = -3$, and $\chi(\hat{\sigma}_d) = -5$, or $\Gamma = 17\ 2\ 5\ -3\ -5$. Determine how many times each irreducible representation of $\mathbf{T}_d$ is contained in Γ.

12-22. Suppose the characters of a reducible representation of the $\mathbf{C}_{2v}$ point group are $\Gamma = 27\ -1\ 1\ 5$. Determine how many times each irreducible representation of $\mathbf{C}_{2v}$ is contained in Γ.

12-23. Suppose the characters of a reducible representation of the $\mathbf{D}_{3h}$ point group are $\Gamma = 12\ 0\ -2\ 4\ -2\ 2$. Determine how many times each irreducible representation of $\mathbf{D}_{3h}$ is contained in Γ.

12-24. In Example 12–14, we showed that the overlap integral involving $2p_{xN}$ and $1s_{H_A} + 1s_{H_B} + 1s_{H_C}$ in the NH_3 molecule is equal to zero. Is this necessarily true for the $2p_{zN}$ rather than the $2p_{xN}$ orbital?

12-25. Show that the molecular orbital ϕ_2 given by Equations 12.3 belongs to the irreducible representation B_{2g}.

12-26. Because the benzene molecular orbitals ϕ_3 and ϕ_4 in Equations 12.3 belong to a two-dimensional irreducible representation (E_{1g}), they are not unique. Any two linear combinations of ϕ_3 and ϕ_4 will also form a basis for E_{1g}. Consider

$$\phi_3' = \phi_3 = \frac{1}{\sqrt{12}}(2\psi_1 + \psi_2 - \psi_3 - 2\psi_4 - \psi_5 + \psi_6)$$

and

$$\phi_4' = \frac{1}{2}(\psi_2 + \psi_3 - \psi_5 - \psi_6)$$

First show that

$$\phi_4' = \frac{\sqrt{12}}{6}(2\phi_4 - \phi_3)$$

Now show that ϕ_4' is normalized. (Realize that ϕ_3 and ϕ_4 are not necessarily orthogonal because they are degenerate (see Problem 4–29).) Evaluate the 2×2 block of the secular determinant corresponding to E_{1g} and show that the final value of the energy is the same as that given in Equation 12.4.

12-27. Arrange the benzene molecular orbitals given by Equations 12.3 in the order of the number of nodal planes perpendicular to the plane of the molecule. Label the molecular orbitals according to the irreducible representation (see Example 10–8).

12-28. Using the symmetry orbitals for butadiene given by Equations 12.33, show that the Hückel theory secular determinant is given by Equation 12.34.

12-29. Show that if we used a $2p_z$ orbital on each carbon atom as the basis for a (reducible) representation for benzene, ($\mathbf{D}_{6h}$) then $\Gamma = 6\ 0\ 0\ 0\ -2\ 0\ 0\ 0\ 0\ -6\ 0\ 2$. Reduce Γ into its component irreducible representations. What does your answer tell you about the expected Hückel secular determinant?

12-30. Show that if we used a $2p_z$ orbital on each carbon atom as the basis for a (reducible) representation for cyclobutadiene ($\mathbf{D}_{4h}$), then $\Gamma = 4\ 0\ 0\ 0\ -2\ 0\ 0\ -4\ 0\ 2$. Reduce Γ into its component irreducible representations. What does your answer tell you about the expected Hückel secular determinant?

12-31. Consider an allyl anion, $CH_2CHCH_2^-$, which belongs to the point group $\mathbf{C}_{2v}$. Show that if we use ψ_1, ψ_2, and ψ_3 ($2p_z$ on each carbon atom) to calculate the Hückel secular determinant, then we obtain

$$\begin{vmatrix} \alpha - E & \beta & 0 \\ \beta & \alpha - E & \beta \\ 0 & \beta & \alpha - E \end{vmatrix} = \begin{vmatrix} x & 1 & 0 \\ 1 & x & 1 \\ 0 & 1 & x \end{vmatrix} = 0$$

or $x^3 - 2x = 0$, or $x = 0,\ \pm\sqrt{2}$. Now show that if we use the ψ_j as the basis for a (reducible) representation for the allyl anion, then $\Gamma = 3\ -1\ 1\ -3$. Now show that $\Gamma = A_2 + 2B_1$. What does this say about the expected Hückel secular determinant? Now use the generating operator, Equation 12.32, to derive three symmetry orbitals for the allyl anion. Normalize them and use them to calculate the Hückel secular determinantal equation and solve for the π-electron energies.

12-32. Apply the analysis we use in Example 12–15 to a minimal basis set for NH_3.

12-33. Just as we have orthogonality conditions for the characters of irreducible representations, there are also orthogonality conditions of their matrix elements. For example, if $\Gamma_i(\hat{R})_{mn}$ denotes the mn matrix element of the matrix of the ith irreducible representation, then

$$\sum_{\hat{R}} \Gamma_i(\hat{R})_{mn} \Gamma_j(\hat{R})_{m'n'} = \frac{h}{d_i} \delta_{ij} \delta_{mm'} \delta_{nn'}$$

This rather complicated looking equation is called the *great orthogonality theorem*. Show how this equation applies to the elements of the matrices in Table 12.6.

12-34.

  a. Let $i = j$, $m = n$, and $m' = n'$ in the great orthogonality theorem (Problem 12–33) and sum over n and n' to derive Equation 12.18.

  b. Let $m = n$, $m' = n'$ and sum over n and n' to derive Equation 12.14.

  c. Combine these results to derive Equation 12.20.

12-35. Consider the point group $\mathbf{C}_s$, which contains only the symmetry elements E and σ. Determine the character table for $\mathbf{C}_s$. (The molecule NOCl belongs to this point group.)

12-36. Consider the simple point group $\mathbf{C}_s$ whose character table is

$\mathbf{C}_s$	$\hat{E}$	$\hat{\sigma}$
A'	1	1
A''	1	-1

where $\hat{\sigma}$ represents reflection through the y axis in a two-dimensional x, y Cartesian coordinate system. Show that the bases for this point group are even and odd functions of x over a symmetric interval, $-a \leq x \leq a$. Now use group theory to show that

$$\int_{-a}^{a} f_{\text{even}}(x) f_{\text{odd}}(x) dx = 0$$

12-37. We calculated the π-electron energy of a trimethylenemethane molecule in Problem 10–31. Derive the symmetry orbitals for the π orbitals by applying the generating operator, Equation 12.32, to the atomic $2p_z$ orbital on each carbon atom. Identify the irreducible representation to which each resulting symmetry orbital belongs. Derive the Hückel secular determinant corresponding to these symmetry orbitals and compare it to the one that you obtained in Problem 10–31. Compare the π-electron energies.

12-38. We calculated the π-electron energy of a bicyclobutadiene molecule in Problem 10–32. Using the point group $\mathbf{C}_{2h}$, derive the symmetry orbitals for the π orbitals by applying the generating operator, Equation 12.32, to the atomic $2p_z$ orbital on each carbon atom. Identify the irreducible representation to which each resulting symmetry orbital belongs. Derive the Hückel secular determinant corresponding to these symmetry orbitals and compare it to the one that you obtained in Problem 10–32. Compare the π-electron energies.

12-39. Use the generating operator, Equation 12.32, to derive the symmetry orbitals for the π orbitals of the (bent) allyl radical ($C_3H_5\cdot$) from a basis set consisting of a $2p_z$ orbital on each carbon atom. (Assume the three carbon atoms lie in the $x-y$ plane.) Now create a set of orthonormal molecular orbitals from these symmetry orbitals. Sketch each orbital. How do your results compare with the π orbitals predicted by Hückel theory (see Problem 10–47)?

The following four problems illustrate the application of group theory to the formation of hybrid orbitals.

12-40. Consider a trigonal planar molecule XY_3 whose point group is $\mathbf{D}_{3h}$. All three bonds are unmoved by the operation of $\hat{E}$; all three are moved by the operation of $\hat{C}_3$; one is unmoved by the operation of $\hat{C}_2$; all three are unmoved by the operation of $\hat{\sigma}_h$; all are moved by the operation of $\hat{S}_3$; and one is unmoved by the operation of $\hat{\sigma}_v$. This result leads to the reducible representation $\Gamma = 3 \ 0 \ 1 \ 3 \ 0 \ 1$. Now show that $\Gamma = A_1' + E'$. Argue now that this result suggests that hybrid orbitals with $\mathbf{D}_{3h}$ symmetry can be formed from an s orbital and the p_x and p_y orbitals (or the $d_{x^2-y^2}$ and d_{z^2} orbitals) to give sp^2 (or sd^2) hybrid orbitals.

12-41. Consider a tetrahedral molecule XY_4 whose point group is $\mathbf{T}_d$. Using the procedure introduced in Problem 12–40, show that $\Gamma = 4 \ 1 \ 0 \ 0 \ 2$, which reduces to $\Gamma = A_1 + T_2$. Now argue that hybrid orbitals with $\mathbf{T}_d$ symmetry can be formed from an s orbital and the p_x, p_y and p_z orbitals (or the d_{xy}, d_{xz}, and d_{yz} orbitals) to give sp^3 (or sd^3) hybrid orbitals.

12-42. Consider a square planar molecule XY_4 whose point group is $\mathbf{D}_{4h}$. Using the procedule introduced in Problem 12–40, show that $\Gamma = 4 \ 0 \ 0 \ 2 \ 0 \ 0 \ 0 \ 4 \ 2 \ 0$, which reduced to $\Gamma = A_{1g} + B_{1g} + E_u$. Now argue that hybrid orbitals with $\mathbf{D}_{4h}$ symmetry can be formed from a s orbital, a $d_{x^2-y^2}$ orbital, and the p_x and p_y orbitals to give sdp^2 hybrid orbitals.

12-43. Consider a trigoanl bipyramidal molecule XY_5 whose point group is $\mathbf{D}_{3h}$. Using the procedure introduced in Problem 12–40, show that $\Gamma = 5 \ 2 \ 1 \ 3 \ 0 \ 3$, and that $\Gamma = 2A_1' + A_2'' + E'$. Now argue that hybrid orbitals with $\mathbf{D}_{3h}$ symmetry can be formed from an s orbital, a d_{z^2} orbital, a p_z orbital, and p_x and p_y orbitals to give sdp^3 hybrid orbitals.

Gerhard Herzberg was born in Hamburg, Germany, on December 25, 1904. He received his Doctor of Engineering Physics from the Technical University at Darmstadt in 1928, having published 12 papers in atomic and molecular physics. After spending a year with Max Born and James Franck at Göttingen, Herzberg returned to Darmstadt for a period during which he concentrated on spectroscopy. In 1935, he was dismissed from his position because his wife was Jewish. Realizing he must leave Germany, he found a position at the University of Saskatchewan as a research professor in physics. In 1945, he received a position at the Yerkes Observatory at the University of Chicago. Herzberg's early ambition had been to become an astronomer, but positions with funding were not available in Germany at that time. After three years, the Herzberg family returned to Canada. The National Research Council of Canada invited him to establish a laboratory for fundamental research in spectroscopy. Under his leadership, NRC became a world leader in spectroscopy. His three comprehensive volumes, *Molecular Spectra and Molecular Structure: I. Spectra of Diatomic Molecules; II. Infrared and Raman Spectra of Polyatomic Molecules*, and *III. Electronic Spectra and Electronic Structure of Polyatomic Molecules*, are classic books in spectroscopy. Herzberg was awarded the Nobel Prize for chemistry in 1971 "for his contributions to the knowledge of electronic structure and geometry of molecules, particularly free radicals."

CHAPTER 13

Molecular Spectroscopy

To this point, we have mainly focused on the theoretical description of atomic and molecular orbitals and molecular structure. The interaction of electromagnetic radiation with atoms and molecules, or spectroscopy, is one of the most important experimental probes for studying atomic and molecular structure. We will see in this chapter that the absorption properties of molecules in various regions of the electromagnetic spectrum yield important information about molecular structure. For example, microwave absorption spectroscopy is used to investigate the rotation of molecules and yields moments of inertia and bond lengths. Infrared absorption spectroscopy is used to study the vibrations of molecules and yields information concerning the stiffness or rigidity of chemical bonds. This information, in turn, provides how the potential energy of the molecule changes with the vibrational motion of the bonded atoms. We have already discussed the quantum-mechanical properties of a harmonic oscillator (Section 5–4) and a rigid rotator (Section 5–8), which are simple models for molecular vibrations and rotations, respectively. We will extend these models and compare the predictions of these extensions to experimental data. In this chapter, we will also introduce normal coordinates to describe the vibrational spectra of polyatomic molecules, which will give us the normal vibrational modes of a polyatomic molecule. We will then show how group theory can be used to predict which normal modes can be studied by infrared absorption spectroscopy. Next will come a discussion of the electronic spectra of diatomic molecules. Last, we use time-dependent perturbation theory to derive the various spectroscopic selection rules used throughout this chapter.

13–1. Different Regions of the Electromagnetic Spectrum Are Used to Investigate Different Molecular Processes

Molecular spectroscopy is the study of the interaction of electromagnetic radiation with molecules. Electromagnetic radiation is customarily divided into different energy regions reflecting the different types of molecular processes that can be caused by

495

such radiation. The classifications we focus on in this chapter are summarized in Table 13.1. The absorption of microwave radiation generally causes transitions between rotational energy levels; the absorption of infrared radiation generally causes transitions between vibrational levels and is accompanied by transitions between rotational energy levels; and the absorption of visible and ultraviolet radiation causes transitions between electronic energy levels, accompanied by simultaneous transitions between vibrational and rotational levels. The frequency of the radiation absorbed is given by

$$\Delta E = E_u - E_l = h\nu \tag{13.1}$$

where E_u and E_l are the energies of the upper and lower states, respectively.

EXAMPLE 13–1

Calculate ΔE for radiation of wave number, $\tilde{\nu} = 1.00 \text{ cm}^{-1}$. To what type of molecular process will absorption of this radiation correspond?

SOLUTION: Recall that wave number is given by reciprocal wavelength, or that

$$\tilde{\nu} = \frac{1}{\lambda}$$

Therefore, ΔE is related to $\tilde{\nu}$ by

$$\Delta E = h\nu = \frac{hc}{\lambda} = hc\tilde{\nu}$$

and so

$$\Delta E = hc\tilde{\nu}$$
$$= (6.626 \times 10^{-34} \text{ J·s})(2.998 \times 10^8 \text{ m·s}^{-1})(1.00 \text{ cm}^{-1})(100 \text{ cm·m}^{-1})$$
$$= 1.99 \times 10^{-23} \text{ J}$$

According to Table 13.1, absorption of this of energy corresponds to a rotational transition.

TABLE 13.1

Regions of the electromagnetic spectrum and the corresponding molecular processes.

Region	Microwave	Far infrared	Infrared	Visible and ultraviolet
Frequency/Hz	10^9—10^{11}	10^{11}—10^{13}	10^{13}—10^{14}	10^{14}—10^{16}
Wavelength/m	3×10^{-1}—3×10^{-3}	3×10^{-3}—3×10^{-5}	3×10^{-5}—6.9×10^{-7}	6.9×10^{-7}—2×10^{-7}
Wave number/cm^{-1}	0.033–3.3	3.3–330	330–14 500	14 500–50 000
Energy/J·molecule^{-1}	6.6×10^{-25}—6.6×10^{-23}	6.6×10^{-23}—6.6×10^{-21}	6.6×10^{-21}—2.9×10^{-19}	2.9×10^{-19}—1.0×10^{-18}
Molecular process	Rotation of polyatomic molecules	Rotation of small molecules	Vibrations of flexible bonds	Electronic transitions

13–2. Rotational Transitions Accompany Vibrational Transitions

The quantum-mechanical properties of a diatomic harmonic oscillator were described in Section 5–4. Recall that the allowed energies of a harmonic oscillator are given by

$$E_v = (v + \tfrac{1}{2})h\nu \qquad v = 0, \ 1, \ 2, \ \ldots \tag{13.2}$$

where

$$\nu = \frac{1}{2\pi} \left(\frac{k}{\mu} \right)^{1/2} \tag{13.3}$$

is the fundamental vibrational frequency of the oscillator. In Equation 13.3, k is the force constant and μ is the reduced mass of the molecule. Transitions among the vibrational levels that result from the absorption of radiation are subject to a selection rule that states that $\Delta v = \pm 1$ and that the dipole moment of the molecule must vary during a vibration. We will derive this and other selection rules in Sections 13–11 through 13–13. For now, we will use the result that in the case of absorption, $\Delta v = +1$ for a harmonic oscillator, so the spectrum consists of one line in the infrared region that occurs at the frequency $\nu_{\text{obs}} = (k/\mu)^{1/2}/2\pi$.

The vibrational energy of a molecule in wavenumbers is denoted by $G(v)$, where $G(v)$ is called the *vibrational term*. Currently, vibrational and electronic energies are tabulated in units of cm^{-1}, so $G(v)$ is often expressed in these units. Introducing $G(v)$ into Equation 13.2 gives

$$G(v) = \left(v + \tfrac{1}{2} \right) \tilde{\nu} \tag{13.4}$$

where $G(v) = E_v / hc$ and

$$\tilde{\nu} = \frac{1}{2\pi c} \left(\frac{k}{\mu} \right)^{1/2} \tag{13.5}$$

A tilde ($\tilde{\ }$) over a symbol emphasizes that the quantity is expressed in wavenumbers.

We discussed the quantum-mechanical properties of a diatomic rigid rotator in Section 5–8. The allowed energies of a rigid rotator are given by

$$E_J = \frac{\hbar^2}{2I} J(J + 1) \qquad J = 0, \ 1, \ 2, \ \ldots \tag{13.6}$$

where I is the moment of inertia, μR_e^2, and R_e is the bond length. The degeneracy associated with Equation 13.6 is

$$g_J = 2J + 1 \tag{13.7}$$

The rotational energy of a molecule in wavenumbers is denoted by $F(J)$, which is called the *rotational term*. Once again, $F(J)$ is commonly tabulated in units of cm^{-1}.

The transitions between the various rotational energy levels of a rigid rotator that result from the absorption of radiation are governed by a selection rule that states that $\Delta J = \pm 1$ and that the molecule must have a permanent dipole moment. As we showed in Section 5–9, the rotational spectrum of a rigid rotator consists of a series of equally spaced lines in the microwave region.

EXAMPLE 13–2

Equation 13.6 is customarily written as

$$F(J) = \tilde{B} J(J+1) \tag{13.8}$$

where $\tilde{B}$ is called the *rotational constant* of the molecule. Derive an equation for $\tilde{B}$, which has units of wave numbers.

SOLUTION: From Equation 13.6,

$$E_J = \frac{\hbar^2}{2I} J(J+1)$$

where E_J is expressed in joules. The relation between energy and the rotational term is given by $F(J) = E_J/hc$, so

$$F(J) = \frac{E_J}{hc} = \frac{\hbar^2}{2hcI} J(J+1)$$
$$= \frac{h}{8\pi^2 cI} J(J+1)$$

Comparing this result with Equation 13.8, we see that

$$\tilde{B} = \frac{h}{8\pi^2 cI} \tag{13.9}$$

Typical values of $\tilde{B}$ for diatomic molecules are of the order of 1 cm^{-1} (cf. Table 13.2).

Within the rigid rotator-harmonic oscillator approximation, the rotational and vibrational energy of a diatomic molecule is given by the sum of Equations 13.4 and 13.8

$$\tilde{E}_{v,J} = G(v) + F(J) = (v + \tfrac{1}{2})\tilde{\nu} + \tilde{B} J(J+1) \qquad \begin{array}{l} v = 0,\ 1,\ 2,\ \ldots \\ J = 0,\ 1,\ 2,\ \ldots \end{array} \tag{13.10}$$

where $\tilde{\nu}$ and $\tilde{B}$ are given by Equations 13.5 and 13.9, respectively. Typical values of $\tilde{\nu}$ and $\tilde{B}$ are on the order of 10^3 cm^{-1} and 1 cm^{-1}, respectively, so the spacing between vibrational energy levels is about 100 to 1000 times the spacing between rotational levels. This result is shown schematically in Figure 13.1.

TABLE 13.2
Spectroscopic parameters of some diatomic molecules in the ground electronic state.

Molecule	$\tilde{B}_e/\text{cm}^{-1}$	$\tilde{\alpha}_e/\text{cm}^{-1}$	$\tilde{D}/\text{cm}^{-1}$	$\tilde{\nu}_e/\text{cm}^{-1}$	$\tilde{x}_e\tilde{\nu}_e/\text{cm}^{-1}$	$R_e(v=0)/\text{pm}$	$D_0/\text{kJ}\cdot\text{mol}^{-1}$
H_2	60.8530	3.0622	4.71×10^{-2}	4401.213	121.336	74.14	432.1
$H^{19}F$	20.9557	0.798	2.15×10^{-3}	4138.32	89.88	91.68	566.2
$H^{35}Cl$	10.5934	0.3072	5.319×10^{-4}	2990.946	52.819	127.46	427.8
$H^{79}Br$	8.4649	0.2333	3.458×10^{-4}	2648.975	45.218	141.44	362.6
$H^{127}I$	6.5122	0.1689	2.069×10^{-4}	2309.014	39.644	160.92	294.7
$^{12}C^{16}O$	1.9313	0.0175	6.122×10^{-6}	2169.814	13.288	112.83	1070.2
$^{14}N^{16}O$	1.67195	0.0171	5.4×10^{-6}	1904.20	14.075	115.08	626.8
$^{14}N^{14}N$	1.9982	0.01732	5.76×10^{-6}	2358.57	14.324	109.77	941.6
$^{16}O^{16}O$	1.4456	0.0159	4.84×10^{-6}	1580.19	11.98	120.75	493.6
$^{19}F^{19}F$	0.89019	0.13847	3.3×10^{-6}	916.64	11.236	141.19	154.6
$^{35}Cl^{35}Cl$	0.2440	0.00149	1.86×10^{-7}	559.72	2.675	198.79	239.2
$^{79}Br^{79}Br$	0.0821	0.0003187	2.09×10^{-8}	325.321	1.0774	228.11	190.1
$^{127}I^{127}I$	0.03737	0.0001138	4.25×10^{-9}	214.502	0.6147	266.63	148.8
$^{35}Cl^{19}F$	0.5165	0.004358	8.77×10^{-7}	786.15	6.161	162.83	252.5
$^{23}Na^{23}Na$	0.1547	0.0008736	5.81×10^{-7}	159.125	0.7255	307.89	71.1
$^{39}K^{39}K$	0.05674	0.000165	8.63×10^{-8}	92.021	0.2829	390.51	53.5

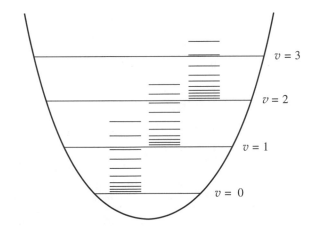

FIGURE 13.1
An energy diagram showing the rotational levels associated with each vibrational state for a diatomic molecule.

When a molecule absorbs infrared radiation, the vibrational transition is accompanied by a rotational transition. The selection rules for absorption of infrared radiation in the rigid rotator-harmonic oscillator approximation are

$$\Delta v = +1 \quad \text{(absorption)}$$
$$\Delta J = \pm 1 \tag{13.11}$$

For the case $\Delta J = +1$, Equation 13.10 gives

$$
\begin{aligned}
\tilde{\nu}_{obs}(\Delta J = +1) &= \tilde{E}_{v+1,J+1} - \tilde{E}_{v,J} \\
&= \left(v + \tfrac{3}{2}\right)\tilde{\nu} + \tilde{B}(J+1)(J+2) - \left(v + \tfrac{1}{2}\right)\tilde{\nu} - \tilde{B}J(J+1) \\
&= \tilde{\nu} + 2\tilde{B}(J+1) \qquad J = 0,\ 1,\ 2,\ \ldots
\end{aligned}
\tag{13.12}
$$

and likewise for the case $\Delta J = -1$, we have

$$
\tilde{\nu}_{obs}(\Delta J = -1) = \tilde{E}_{v+1,J-1} - \tilde{E}_{v,J} = \tilde{\nu} - 2\tilde{B}J \qquad J = 1,\ 2,\ \ldots
\tag{13.13}
$$

In both Equations 13.12 and 13.13, J is the initial rotational quantum number. Typically, $\tilde{\nu} \approx 10^3\ \mathrm{cm}^{-1}$ and $\tilde{B} \approx 1\ \mathrm{cm}^{-1}$, so the spectrum predicted by Equations 13.12 and 13.13 contains lines at $10^3\ \mathrm{cm}^{-1} \pm$ integral multiples of $\approx 1\ \mathrm{cm}^{-1}$. Notice that there is no line at $\tilde{\nu}$ because the transition $\Delta J = 0$ is forbidden. The rotational-vibrational spectrum of HBr(g) is shown in Figure 13.2. The gap centered around 2560 cm^{-1} corresponds to the missing line at $\tilde{\nu}$. On each side of the gap is a series of lines whose spacing is about 10 cm^{-1}. The series toward the high-frequency side is called the *R branch* and is due to rotational transitions with $\Delta J = +1$. The series toward the low frequencies is called the *P branch* and is due to rotational transitions with $\Delta J = -1$.

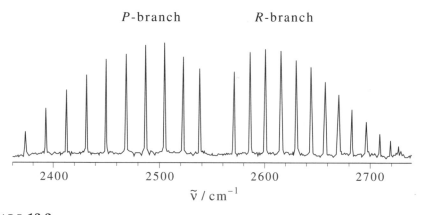

FIGURE 13.2
The rotational-vibrational spectrum of the $0 \rightarrow 1$ vibrational transition of HBr(g). The R branch and the P branch are indicated in the figure.

EXAMPLE 13–3
The bond length in $^{12}\mathrm{C}^{14}\mathrm{N}$ is 117 pm and its force constant is 1630 N·m^{-1}. Predict the vibrational-rotational spectrum of $^{12}\mathrm{C}^{14}\mathrm{N}$.

SOLUTION: First we must calculate the fundamental frequency $\tilde{\nu}$ (Equation 13.5) and the rotational constant $\tilde{B}$ (Equation 13.9). Both quantities require the reduced mass, which is

$$\mu = \frac{(12.0 \text{ amu})(14.0 \text{ amu})}{(12.0 + 14.0) \text{ amu}}(1.661 \times 10^{-27} \text{ kg·amu}^{-1}) = 1.07 \times 10^{-26} \text{ kg}$$

Using Equation 13.5 for $\tilde{\nu}$,

$$\tilde{\nu} = \frac{1}{2\pi c}\left(\frac{k}{\mu}\right)^{1/2} = \frac{1}{2\pi(2.998 \times 10^8 \text{ m})}\left(\frac{1630 \text{ N·m}^{-1}}{1.07 \times 10^{-26} \text{ kg}}\right)^{1/2}$$
$$= 2.07 \times 10^5 \text{ m}^{-1} = 2.07 \times 10^3 \text{ cm}^{-1}$$

Using Equation 13.9 for $\tilde{B}$,

$$\tilde{B} = \frac{h}{8\pi^2 cI} = \frac{h}{8\pi^2 c\mu R_e^2}$$
$$= \frac{6.626 \times 10^{-34} \text{ J·s}}{8\pi^2(2.998 \times 10^8 \text{ m·s}^{-1})(1.07 \times 10^{-26} \text{ kg})(117 \times 10^{-12} \text{ m})^2}$$
$$= 191 \text{ m}^{-1} = 1.91 \text{ cm}^{-1}$$

The vibration–rotation spectrum will consist of lines at $\tilde{\nu} \pm 2\tilde{B}J$ where $J = 1$, 2, 3, ... There will be no line at $\tilde{\nu}$, and the separation of the lines in the P and R branches will be $2\tilde{B} = 3.82 \text{ cm}^{-1}$ (cf. Figure 13.2 for HBr).

If we compare the results of Example 13–3 with experimental data, or look closely at Figure 13.2, we see several features we cannot explain. Close examination shows that the lines in the R branch are more closely spaced with increasing frequency and that the lines of the P branch become further apart with decreasing frequency. We will discuss the spacing of the lines in the R and P branches in the next section.

13–3. Vibration–Rotation Interaction Accounts for the Unequal Spacing of the Lines in the P and R Branches of a Vibration-Rotation Spectrum

The energies of a rigid rotator-harmonic oscillator are given by (Equation 13.10)

$$\tilde{E}_{v,J} = G(v) + F(J) = \tilde{\nu}(v + \tfrac{1}{2}) + \tilde{B}J(J+1)$$

where $\tilde{B} = h/8\pi^2 c\mu R_e^2$. Because the vibrational amplitude increases with the vibrational state (cf. Figure 13.1), we expect that R_e should increase slightly with v, causing $\tilde{B}$ to decrease with increasing v. We will indicate the dependence of $\tilde{B}$ upon v by using $\tilde{B}_v$ in place of $\tilde{B}$:

$$\tilde{E}_{v,J} = \tilde{\nu}(v + \tfrac{1}{2}) + \tilde{B}_v J(J+1) \tag{13.14}$$

The dependence of $\tilde{B}$ on v is called *vibration-rotation interaction*. If we consider a $v = 0 \rightarrow 1$ transition, then the frequencies of the R and P branches will be given by (Problem 13–10)

$$
\begin{aligned}
\tilde{\nu}_R(\Delta J = +1) &= \tilde{E}_{1,\,J+1} - \tilde{E}_{0,\,J} \\
&= \tfrac{3}{2}\tilde{\nu} + \tilde{B}_1(J+1)(J+2) - \tfrac{1}{2}\tilde{\nu} - \tilde{B}_0 J(J+1) \\
&= \tilde{\nu} + 2\tilde{B}_1 + (3\tilde{B}_1 - \tilde{B}_0)J + (\tilde{B}_1 - \tilde{B}_0)J^2 \quad J = 0, 1, 2, \ldots \ (13.15)
\end{aligned}
$$

and

$$
\tilde{\nu}_P(\Delta J = -1) = \tilde{E}_{1,\,J-1} - \tilde{E}_{0,\,J} = \tilde{\nu} - (\tilde{B}_1 + \tilde{B}_0)J + (\tilde{B}_1 - \tilde{B}_0)J^2 \quad J = 1, 2, 3, \ldots
$$
$$(13.16)$$

In both cases, J corresponds to the initial rotational quantum number. Note that Equations 13.15 and 13.16 reduce to Equations 13.12 and 13.13 if $\tilde{B}_1 = \tilde{B}_0$. Because the bond length increases with increasing v, $\tilde{B}_1 < \tilde{B}_0$, and therefore the spacing between the lines in the R branch decreases and the spacing between the lines in the P branch increases with increasing J. This behavior is reflected in Figure 13.2.

EXAMPLE 13–4

The lines in the R and P branches are customarily labeled by the initial value of the rotational quantum number, giving rise to the lines. Thus, the lines given by Equation 13.15 are $R(0)$, $R(1)$, $R(2)$, ..., and those given by Equation 13.16 are $P(1)$, $P(2)$, ... Given the following data for $^1\text{H}^{127}\text{I}$

Line	Frequency/cm^{-1}
$R(0)$	2242.087
$R(1)$	2254.257
$P(1)$	2216.723
$P(2)$	2203.541

calculate $\tilde{B}_0$ and $\tilde{B}_1$ and $R_e(v=0)$ and $R_e(v=1)$. Take the reduced mass of the molecule to be 1.660×10^{-27} kg.

SOLUTION: Using Equation 13.15 with $J = 0$ and 1 and Equation 13.16 with $J = 1$ and 2, we have

$$
\left.\begin{aligned}
2242.087 \text{ cm}^{-1} &= \tilde{\nu}_0 + 2\tilde{B}_1 \\
2254.257 \text{ cm}^{-1} &= \tilde{\nu}_0 + 6\tilde{B}_1 - 2\tilde{B}_0
\end{aligned}\right\} R \text{ branch}
$$

and

$$
\left.\begin{aligned}
2216.723 \text{ cm}^{-1} &= \tilde{\nu}_0 - 2\tilde{B}_0 \\
2203.541 \text{ cm}^{-1} &= \tilde{\nu}_0 + 2\tilde{B}_1 - 6\tilde{B}_0
\end{aligned}\right\} P \text{ branch}
$$

If we subtract the first line of the P branch from the second line of the R branch, we find

$$37.534 \text{ cm}^{-1} = 6\tilde{B}_1$$

or $\tilde{B}_1 = 6.256 \text{ cm}^{-1}$. If we subtract the second line of the P branch from the first line of the R branch, we find

$$38.546 \text{ cm}^{-1} = 6\tilde{B}_0$$

or $\tilde{B}_0 = 6.424 \text{ cm}^{-1}$. Using the fact that $\tilde{B}_v = h/8\pi^2 c\mu R_e^2(v)$, we obtain $R_e(v=0) = 162.0$ pm and $R_e(v=1) = 164.1$ pm.

The dependence of $\tilde{B}_v$ on v is usually expressed as

$$\tilde{B}_v = \tilde{B}_e - \tilde{\alpha}_e(v + \tfrac{1}{2}) \tag{13.17}$$

Using the values of $\tilde{B}_0$ and $\tilde{B}_1$ from the above example, we find that $\tilde{B}_e = 6.508 \text{ cm}^{-1}$ and that $\tilde{\alpha}_e = 0.168 \text{ cm}^{-1}$. Values of $\tilde{B}_e$ and $\tilde{\alpha}_e$ as well as other spectroscopic parameters are given in Table 13.2.

13–4. The Lines in a Pure Rotational Spectrum Are Not Equally Spaced

Table 13.3 lists some of the observed lines in the pure rotational spectrum (no vibrational transitions) of $H^{35}Cl$. The differences listed in the third column clearly show that the lines are not exactly equally spaced as the rigid rotator approximation predicts. The discrepancy can be resolved by realizing that a chemical bond is not truly rigid. As the molecule rotates more energetically (increasing J), the centrifugal force causes the bond to stretch slightly. This small effect can be treated by perturbation theory, and the end result is that the energy can be written as

$$F(J) = \tilde{B}J(J+1) - \tilde{D}J^2(J+1)^2 \tag{13.18}$$

where $\tilde{D}$ is called the *centrifugal distortion constant*. Rigid rotator and nonrigid rotator energy levels are sketched in Figure 13.3.

The frequencies of the absorption due to $J \rightarrow J+1$ transitions are given by

$$\tilde{\nu} = F(J+1) - F(J)$$
$$= 2\tilde{B}(J+1) - 4\tilde{D}(J+1)^3 \qquad J = 0, 1, 2, \ldots \tag{13.19}$$

The predictions of this equation are given in Table 13.3, where we obtain $\tilde{B} = 10.403 \text{ cm}^{-1}$ and $\tilde{D} = 0.00044 \text{ cm}^{-1}$ for $H^{35}Cl$ by fitting Equation 13.19 to the experimental data. These values differ slightly from those in Table 13.2 because of

TABLE 13.3
The rotational absorption spectrum of $H^{35}Cl$.

Transition	$\tilde{\nu}_{obs}/cm^{-1}$	$\Delta\tilde{\nu}_{obs}/cm^{-1}$	$\tilde{\nu}_{calc} = 2\tilde{B}(J+1)$ $\tilde{B} = 10.243\ cm^{-1}$	$\tilde{\nu}_{calc} = 2\tilde{B}(J+1) - 4\tilde{D}(J+1)^3$ $\tilde{B} = 10.403\ cm^{-1}$ $\tilde{D} = 0.00044\ cm^{-1}$
$3 \to 4$	83.03		82.72	83.11
		21.07		
$4 \to 5$	104.10		103.40	103.81
		20.20		
$5 \to 6$	124.30		124.08	124.46
		20.73		
$6 \to 7$	145.03		144.76	145.04
		20.48		
$7 \to 8$	165.51		165.44	165.55
		20.35		
$8 \to 9$	185.86		186.12	185.97
		20.52		
$9 \to 10$	206.38		206.80	206.30
		20.12		
$10 \to 11$	226.50		227.48	226.52

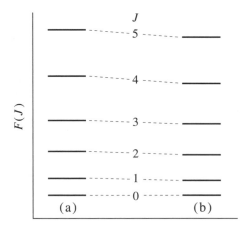

FIGURE 13.3
The rotational energy levels of (a) a rigid rotator and (b) a nonrigid rotator.

higher-order effects. The inclusion of centrifugal distortion alters the extracted value of $\tilde{B}$ (see Problem 13–18).

13–5. Overtones Are Observed in Vibrational Spectra

Thus far we have treated the vibrational motion of a diatomic molecule by means of a harmonic-oscillator model. We saw in Section 5–3, however, that the internuclear potential energy is not a simple parabola but is more like that illustrated in Figure 5.5

(cf. also Figure 13.4). The dashed line in either of these figures depicts the harmonic oscillator. Recall from Equation 5.23 that the potential energy $V(R)$ may be expanded in a Taylor series about R_e, the value of R at the minimum of $V(R)$, to give

$$V(R) - V(R_e) = \frac{1}{2!}\left(\frac{d^2 V}{d R^2}\right)_{R=R_e}(R - R_e)^2 + \frac{1}{3!}\left(\frac{d^3 V}{d R^3}\right)_{R=R_e}(R - R_e)^3 + \cdots$$

$$= \frac{k}{2}x^2 + \frac{\gamma_3}{6}x^3 + \frac{\gamma_4}{24}x^4 + \cdots \tag{13.20}$$

where x is the displacement of the nuclei from their equilibrium separation, k is the (Hooke's law) force constant, and $\gamma_j = (d^j V/d R^j)_{R=R_e}$.

The harmonic-oscillator approximation consists of keeping only the quadratic term in Equation 13.20, and it predicts that there will be only one line in the vibrational spectrum of a diatomic molecule. Experimental data show there is, indeed, one dominant line (called the *fundamental*) but also lines of weaker intensity at almost integral multiples of the fundamental. These lines are called *overtones* (Table 13.4). If the anharmonic terms in Equation 13.20 are included in the Hamiltonian operator for the vibrational motion of a diatomic molecule, the Schrödinger equation can be solved by perturbation theory to give

$$G(v) = \tilde{v}_e(v + \tfrac{1}{2}) - \tilde{x}_e\tilde{v}_e(v + \tfrac{1}{2})^2 + \cdots \qquad v = 0,\ 1,\ 2,\ \ldots \tag{13.21}$$

where $\tilde{x}_e$ is called the *anharmonicity constant*. The anharmonic correction in Equation 13.21 is much smaller than the harmonic term because $\tilde{x}_e \ll 1$ (cf. Table 13.2).

Figures 13.4 and 13.5 show the levels given by Equation 13.21. Notice that the levels are not equally spaced as they are for a harmonic oscillator and, in fact, that their separation decreases with increasing v. This is reflected by the numbers in the last

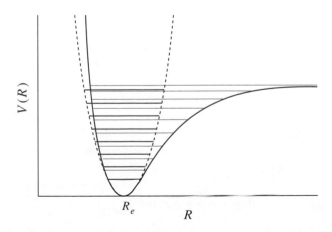

FIGURE 13.4
The energy states of a harmonic oscillator (dashed line) and an anharmonic oscillator superimposed on a harmonic-oscillator potential and a more realistic internuclear potential.

TABLE 13.4

The vibrational spectrum of $H^{35}Cl$.

| | | $\tilde{\nu}_{obs}/cm^{-1}$ | | |
| | | Harmonic oscillator | Anharmonic oscillator | |
Transition	$\tilde{\nu}_{obs}/cm^{-1}$	$\tilde{\nu} = 2885.90\nu$	$\tilde{\nu} = 2990.9\nu - 52.82\nu(\nu + 1)$	$\tilde{\nu}_{obs}(0 \to \nu)/\tilde{\nu}_{obs}(0 \to 1)$
$0 \to 1$ (fundamental)	2885.9	2885.9	2885.3	1.000
$0 \to 2$ (first overtone)	5668.0	5771.8	5665.0	1.964
$0 \to 3$ (second overtone)	8347.0	8657.7	8339.0	2.892
$0 \to 4$ (third overtone)	10 923.1	11 543.6	10 907.4	3.785
$0 \to 5$ (fourth overtone)	13 396.5	14 429.5	13 370.2	4.642

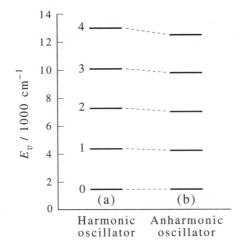

FIGURE 13.5

The vibrational energy state of $H^{35}Cl(g)$ calculated (a) in the harmonic-oscillator aproximation and (b) with a correction for anharmonicity.

column of Table 13.4. Notice from Figure 13.5 that the harmonic-oscillator approximation is best for small values of ν, which we will see are the most important values at room temperature.

The selection rule for an anharmonic oscillator is that $\Delta\nu$ can have any integral value, although the intensities of the $\Delta\nu = \pm 2, \pm 3, \ldots$ transitions are much less than for the $\Delta\nu = \pm 1$ transitions. If we recognize that most diatomic molecules are in the ground vibrational state at room temperature, the frequencies of the observed $0 \to \nu$ transitions will be given by

$$\tilde{\nu}_{obs} = G(\nu) - G(0) = \tilde{\nu}_e\nu - \tilde{x}_e\tilde{\nu}_e\nu(\nu + 1) \qquad \nu = 1, 2, \ldots \qquad (13.22)$$

The application of Equation 13.22 to the spectrum of $H^{35}Cl$ is given in Table 13.4. You can see that the agreement with experimental data is a substantial improvement over the harmonic-oscillator approximation.

EXAMPLE 13–5
Given that $\tilde{\nu}_e = 536.10$ cm^{-1} and $\tilde{x}_e\tilde{\nu}_e = 3.4$ cm^{-1} for ^{23}Na^{19}F(g), calculate the frequencies of the first and second vibrational overtone transitions.

SOLUTION: We use Equation 13.22:

$$\tilde{\nu}_{obs} = \tilde{\nu}_e v - \tilde{x}_e\tilde{\nu}_e v(v+1) \qquad v = 1,\ 2,\ \ldots$$

The fundamental is given by letting $v = 1$, and the first two overtones are given by letting $v = 2$ and 3.

Fundamental:	$\tilde{\nu}_{obs} = \tilde{\nu}_e - 2\tilde{x}_e\tilde{\nu}_e$	$= 529.3$ cm^{-1}
First overtone:	$\tilde{\nu}_{obs} = 2\tilde{\nu}_e - 6\tilde{x}_e\tilde{\nu}_e$	$= 1051.8$ cm^{-1}
Second overtone:	$\tilde{\nu}_{obs} = 3\tilde{\nu}_e - 12\tilde{x}_e\tilde{\nu}_e$	$= 1567.5$ cm^{-1}

Note that the overtones are not quite integral multiples of the fundamental frequency, and the fundamental frequency is less than the frequency for pure harmonic motion.

13–6. Electronic Spectra Contain Electronic, Vibrational, and Rotational Information

In addition to undergoing rotational and vibrational transitions as a result of absorbing microwave and infrared radiation, respectively, molecules can undergo electronic transitions. The difference in energies between electronic levels is usually such that the radiation absorbed falls in the visible or ultraviolet regions. Just as rotational transitions accompany vibrational transitions, both rotational and vibrational transitions accompany electronic transitions. Figure 13.6 shows several electronic potential energy curves of O_2 with the vibrational levels indicated on each curve. Each vibrational level has a set of rotational levels associated with it, but these are too closely spaced to be shown in the figure.

According to the Born-Oppenheimer approximation (Section 9–1), the electronic energy is independent of the vibrational-rotational energy. If we use an anharmonic oscillator-nonrigid rotator approximation, the total energy of a diatomic molecule, excluding translational energy, is given by

$$\tilde{E}_{total} = \tilde{\nu}_{el} + G(v) + F(J)$$
$$= \tilde{\nu}_{el} + \tilde{\nu}_e(v + \tfrac{1}{2}) - \tilde{x}_e\tilde{\nu}_e(v + \tfrac{1}{2})^2 + \tilde{B}J(J+1) - \tilde{D}J^2(J+1)^2 \quad (13.23)$$

where $\tilde{\nu}_{el}$ is the energy at the minimum of the electronic potential energy curve. The selection rule for *vibronic transitions* (vibrational transitions in electronic spectra) allows Δv to take on any integral value. Because rotational energies are usually much smaller than vibrational energies, we will ignore the rotational terms in Equation 13.23 and investigate only the vibrational substructure of electronic spectra.

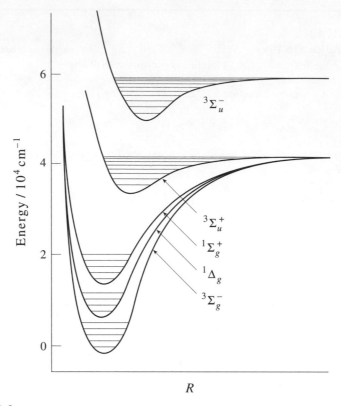

FIGURE 13.6
A potential energy diagram of O_2, showing the vibrational states associated with the various electronic states.

In electronic absorption spectroscopy, the vibronic transitions usually originate from the $v = 0$ vibrational state, because this is usually the only state appreciably populated at normal temperatures (see Section 18–4). Consequently, the predicted frequencies of an electronic transition are given by

$$\tilde{\nu}_{\text{obs}} = \tilde{T}_e + (\tfrac{1}{2}\tilde{\nu}'_e - \tfrac{1}{4}\tilde{x}'_e\tilde{\nu}'_e) - (\tfrac{1}{2}\tilde{\nu}''_e - \tfrac{1}{4}\tilde{x}''_e\tilde{\nu}''_e) + \tilde{\nu}'_e v' - \tilde{x}'_e\tilde{\nu}'_e v'(v'+1) \tag{13.24}$$

The term $\tilde{T}_e$ is the difference in energies of the minima of the two electronic potential energy curves in wave numbers, and the single primes and double primes indicate the upper and lower electronic states, respectively. The difference in energy between the minimum of the potential energy curve and the dissociated atoms is denoted by D_e. The symbol D_0 denotes the corresponding dissociation energy from the ground-vibrational level of the potential (see Figure 13.7). Consequently, $D_e = D_0 + \tfrac{1}{2}h\nu$ in the harmonic oscillator approximation, or $D_0 + \tfrac{1}{2}h(\nu_e - \tfrac{1}{2}x_e\nu_e)$ in the anharmonic oscillator approximation. The values of D_0 for various diatomic molecules are listed

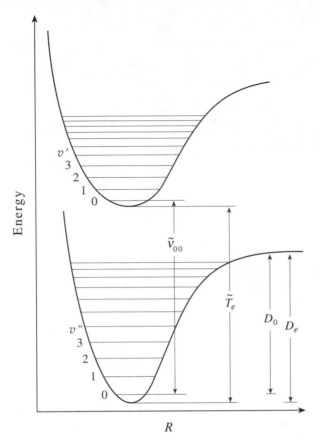

FIGURE 13.7
Two electronic states of a diatomic molecule, illustrating the two quantities $\tilde{T}_e$ and $\tilde{\nu}_{0,0}$.

in Table 13.2. Realize that $\tilde{\nu}_e$ and $\tilde{x}_e\tilde{\nu}_e$ depend on the shape of the electronic potential energy curve at its mimimum and so should differ for each electronic state.

The first two terms in parentheses in Equation 13.24 are the zero-point energies of the upper and lower states. Therefore, the quantity $\tilde{\nu}_{0,0}$ defined by

$$\tilde{\nu}_{0,0} = \tilde{T}_e + (\tfrac{1}{2}\tilde{\nu}'_e - \tfrac{1}{4}\tilde{x}'_e\tilde{\nu}'_e) - (\tfrac{1}{2}\tilde{\nu}''_e - \tfrac{1}{4}\tilde{x}''_e\tilde{\nu}''_e)$$

corresponds to the energy of the $0 \rightarrow 0$ vibronic transition. Introducing $\tilde{\nu}_{0,0}$ into Equation 13.24, we obtain

$$\tilde{\nu}_{obs} = \tilde{\nu}_{0,0} + \tilde{\nu}'_e v' - \tilde{x}'_e\tilde{\nu}'_e v'(v' + 1) \qquad v' = 0,\ 1,\ 2,\ \dots \qquad (13.25)$$

As v', the vibrational quantum number of the upper state, takes on successive values in Equation 13.25, the vibronic spacing becomes progressively smaller until the spectrum is essentially continuous as shown in Figures 13.8 and 13.9. Example 13–6 illustrates

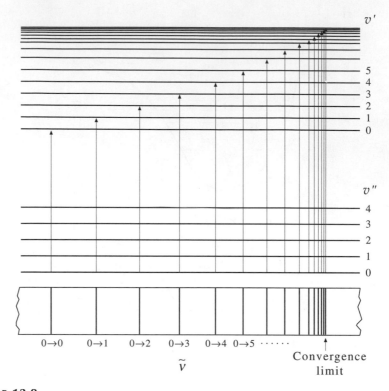

FIGURE 13.8
The electronic spectrum due to $v'' = 0$ to $v' = 0, 1, 2, \ldots$ transitions. Such a set of transitions is called an v' *progression*.

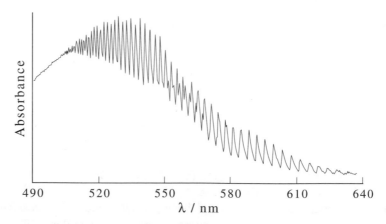

FIGURE 13.9
The absorption spectrum of $I_2(g)$ in the visible region. This spectrum is a v' progression.

how experimental data like that in Figure 13.9 can be analyzed to determine the vibrational parameters of excited electronic states. It is possible to analyze electronic spectra in even greater detail and to learn about the rotational properties of various electronic states, but we will not do that here.

EXAMPLE 13–6

The frequencies of the first few vibronic transitions to an excited electronic state of PN are:

Vibronic transition	$\tilde{\nu}_{obs}/cm^{-1}$
$0 \to 0$	39 699.10
$0 \to 1$	40 786.80
$0 \to 2$	41 858.90

Use these data to calculate $\tilde{\nu}'_e$ and $\tilde{x}'_e\tilde{\nu}'_e$ for the excited electronic state of PN.

SOLUTION: Using Equation 13.25 with $v' = 0$, 1, and 2, we have

$$39\ 699.10 = \tilde{\nu}_{0,0}$$

$$40\ 786.80 = \tilde{\nu}_{0,0} + \tilde{\nu}'_e - 2\tilde{x}'_e\tilde{\nu}'_e$$

$$41\ 858.90 = \tilde{\nu}_{0,0} + 2\tilde{\nu}'_e - 6\tilde{x}'_e\tilde{\nu}'_e$$

By subtracting the first equation from the second and third, we find

$$1087.70\ cm^{-1} = \tilde{\nu}'_e - 2\tilde{x}'_e\tilde{\nu}'_e$$

$$2159.80\ cm^{-1} = 2\tilde{\nu}'_e - 6\tilde{x}'_e\tilde{\nu}'_e$$

Solving these two equations for $\tilde{\nu}'_e$ and $\tilde{x}'_e\tilde{\nu}'_e$, we find

$$\tilde{\nu}'_e = 1103.3\ cm^{-1} \quad \text{and} \quad \tilde{x}'_e\tilde{\nu}'_e = 7.80\ cm^{-1}$$

Analysis of electronic spectra yields structural information of excited states that would be difficult to obtain otherwise.

13–7. The Franck–Condon Principle Predicts the Relative Intensities of Vibronic Transitions

Figure 13.10 shows two electronic potential energy curves with the vibrational states associated with each electronic energy state indicated. In each vibrational state, the harmonic-oscillator probability densities are plotted (cf. Figure 5.8). Notice that except for the ground vibrational state, the most likely internuclear separation occurs near the exteme of a vibration, which is called a *classical turning point* because the vibrational motion changes direction at that point. Because electrons are so much less massive than nuclei, the motion of electrons is almost instantaneous relative to the motion of nuclei.

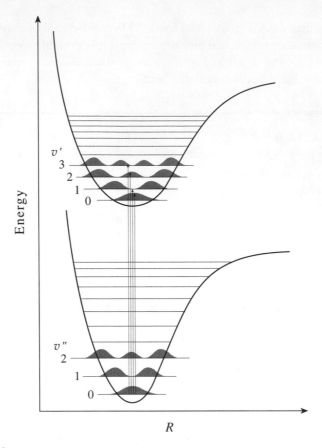

FIGURE 13.10
Two electronic potential energy curves showing the vibrational states associated with each electronic state. The minimum of the upper curve lies almost directly over the minimum of the lower curve. The shaded areas represent the harmonic-oscillator probability densities for each vibrational state. The vertical lines represent a series of $0 \rightarrow v'$ vibronic transitions.

Consequently, when an electron makes a transition from one electronic state to another, the nuclei do not move appreciably during the transition. Electronic transitions can therefore be depicted as vertical lines in a diagram such as in Figure 13.10. This argument can be made rigorous and is known as the *Franck-Condon principle*. The Franck-Condon principle allows us to estimate relative intensities of vibronic transitions. In Figure 13.10, the minima of the two electronic states lie very nearly above each other. It turns out that the relative intensity of a $0 \rightarrow v'$ transition is proportional to the product of the harmonic-oscillator wave functions in the two vibrational states. Figure 13.10 shows that the overlap of the wave functions in the upper and lower vibronic states varies with v'. Thus, we obtain a distribution of intensities like that shown in Figure 13.11a.

Figure 13.12 shows another commonly occurring case, in which the minimum of the upper potential energy curve lies at a somewhat greater value of the internuclear

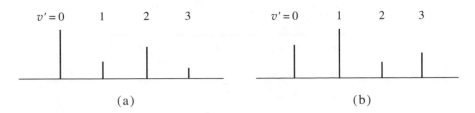

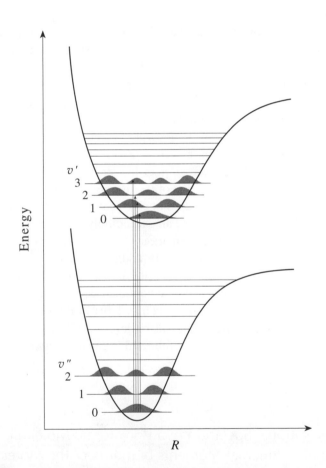

FIGURE 13.11
The distribution of intensities of the vibronic transitions for the case shown in (a) Figure 13.10 and (b) Figure 13.12.

FIGURE 13.12
Two electronic potential energy curves showing the vibrational states and the harmonic-oscillator probability densities as in Figure 13.10. In this case, however, the minimum of the upper curve occurs at a somewhat greater value of the internuclear separation than for the lower curve. The vertical lines represent the $0 \rightarrow v'$ vibronic transitions.

TABLE 13.5

The equilibrium bond lengths for various electronic states of O_2.

Electronic state	$\tilde{T}_e/\text{cm}^{-1}$	R_e/pm
$^3\Sigma_g^-$	0	120.74
$^1\Delta_g$	7 918.1	121.55
$^1\Sigma_g^+$	13 195.2	122.67
$^3\Sigma_u^+$	36 096	142
$^3\Sigma_u^-$	49 802	160

separation than for the lower curve. (See, for example, Figure 13.6 and Table 13.5 for O_2.) In this case, the $0 \rightarrow 0$ transition is not the most intense transition. The most intense transition as shown in Figure 13.12 is the $0 \rightarrow 1$ transition, and the distribution of intensities for this case is like that given in Figure 13.11b. If the excited-state potential curve is displaced to a sufficiently larger internuclear distance than that of the ground-state potential curve, then the absorption spectrum may not contain a line corresponding to $\tilde{\nu}_{0,0}$. The $^1\Sigma_g^+ \rightarrow {}^3\Pi$ absorption spectrum of $I_2(g)$, Figure 13.9, is such a case. The bond length increases from 266.6 pm to 302.5 pm as a result of this transition. The $0 \rightarrow 0$ as well as several other low-energy vibronic transitions are not observed. Performing a detailed analysis of the intensities of such vibronic transitions yields much information about the shapes of electronic potential energy curves.

Analyses of electronic spectra yield an exceedingly rich view of molecular structure. As we have seen, structural information of excited electronic states as well as ground electronic states can be obtained. In addition, the selection rules do not require the molecule to have a permanent dipole moment, nor must there be a change in dipole moment upon vibration. Thus, electronic spectroscopy yields information about homonuclear diatomic molecules (such as $I_2(g)$ in Figure 13.9) that cannot be obtained from either microwave rotational or infrared vibrational absorption spectroscopy. All the parameters given in Table 13.2 can be obtained from electronic spectra.

13–8. The Rotational Spectrum of a Polyatomic Molecule Depends Upon the Principal Moments of Inertia of the Molecule

In this section, we will model a polyatomic molecule as a rigid network of N atoms. The rotational properties of a rigid body are characterized by its *principal moments of inertia*, which are defined in the following way. Choose any set of Cartesian axes with the origin at the center of mass of the body. The moments of inertia about these three axes are

$$I_{xx} = \sum_{j=1}^{N} m_j [(y_j - y_{cm})^2 + (z_j - z_{cm})^2]$$

$$I_{yy} = \sum_{j=1}^{N} m_j [(x_j - x_{cm})^2 + (z_j - z_{cm})^2] \qquad (13.26)$$

$$I_{zz} = \sum_{j=1}^{N} m_j [(x_j - x_{cm})^2 + (y_j - y_{cm})^2]$$

where m_j is the mass of the jth atom situated at the point x_j, y_j, z_j and x_{cm}, y_{cm}, and z_{cm} are the coordinates of the center of mass of the body. Notice that the terms in square brackets in Equations 13.26 are the squares of the distances to the respective axes. In addition to these three moments of inertia, there are also products of inertia, such as

$$I_{xy} = -\sum_{j=1}^{N} m_j (x_j - x_{cm})(y_j - y_{cm})$$

with a similar equation for I_{xz} and I_{yz}. Now there is a theorem of rigid body motion that says that there always exists a particular set of Cartesian coordinates X, Y, Z, called the *principal axes*, passing through the center of mass of the body such that all the products of inertia (e.g., I_{xy}) vanish. The moments of inertia about these axes, I_{XX}, I_{YY}, and I_{ZZ} are called the *principal moments of inertia*. The principal moments of inertia are customarily denoted by I_A, I_B, and I_C. The convention is such that $I_A \leq I_B \leq I_C$.

If the molecule possesses some degree of symmetry, the principal axes are simple to find. For example, if the molecule is planar, one of the principal axes will be perpendicular to the plane. Usually an axis of symmetry of the molecule will be a principal axis. The C–H bond of $CHCl_3$ is a three-fold axis of symmetry and also a principal axis. It is rarely necessary to calculate the principal moments of inertia, however, because extensive tables of these quantities are available in the literature. They are usually given in terms of rotational constants in units of cm^{-1}, defined by

$$\tilde{A} = \frac{h}{8\pi^2 c I_A}, \qquad \tilde{B} = \frac{h}{8\pi^2 c I_B}, \qquad \text{and} \qquad \tilde{C} = \frac{h}{8\pi^2 c I_C} \qquad (13.27)$$

Because $I_A \leq I_B \leq I_C$, the rotational constants will always satisfy the relation $\tilde{A} \geq \tilde{B} \geq \tilde{C}$.

The relative magnitudes of the three principal moments of inertia are used to characterize a rigid body. If all three are equal, the body is called a *spherical top*; if only two are equal, the body is called a *symmetric top*; and if all three are different, the body is called an *asymmetric top*. The molecules CH_4 and SF_6 are examples of spherical tops; NH_3 and C_6H_6 are examples of symmetric tops; and H_2O is the classic example of an asymmetric top. Any molecule with an n-fold axis of symmetry, with $n \geq 3$, is at least a symmetric top.

The quantum-mechanical problem of a spherical top ($\tilde{A} = \tilde{B} = \tilde{C}$) can be solved exactly, and the energy levels come out to be

$$F(J) = \tilde{B}J(J+1) \qquad J = 0,\ 1,\ 2,\ \ldots \qquad (13.28a)$$

(which is the same as for a linear molecule). The degeneracy of each level, however, is

$$g_J = (2J+1)^2 \qquad (13.28b)$$

The high symmetry of spherical top molecules precludes their having a permanent dipole moment, so they do not have pure rotational spectra.

The quantum-mechanical problem of a symmetric top is also solvable in closed form. Because a symmetric top has a unique principal moment of inertia, there are two types of symmetric tops. When the unique moment of inertia is larger than the two equal ones, the molecule is called an *oblate symmetric top*. When the unique moment of inertia is smaller than the two equal ones, the molecule is called a *prolate symmetric top*. Benzene is an example of an oblate symmetric top, and chloromethane is an example of a prolate symmetric top (Figure 13.13). A less "chemical" example of an oblate symmetric top is an O-ring, and that of a prolate symmetric top is a cigar.

EXAMPLE 13–7

Classify BCl_3 and CH_3I as either an oblate or a prolate symmetric top.

SOLUTION: (a) BCl_3 is a planar molecule. The center of mass of BCl_3 sits at the central boron atom. Therefore, one of the principal axes is perpendicular to the molecular plane; the other two will lie in the plane of the molecule. The unique axis is the one perpendicular to the plane of the molecule. (Call this the z axis.) All three chlorine atoms lie as far as possible from this axis, and so the moment of inertia about this axis (I_{zz}) will be larger than the other two. The moments of inertia about the two axes lying in the plane of the molecule (the x and y axes) can be made to be equal by orienting the axes properly, and in any case the chlorine atoms will lie closer to either one of these axes than to the z axis. Therefore, $I_{xx} = I_{yy} < I_{zz}$, and so BCl_3 is an oblate symmetric top.

(b) CH_3I is a tetrahedral molecule. The center of mass of CH_3I will sit along the C–I bond somewhere between the carbon atom and the iodine atom, and the unique axis (call it the z axis) contains the C–I bond. Only the three relatively light hydrogen atoms lie off this axis, and so the moment of inertia about this (unique) axis must be smaller than the moments of inertia about the other two axes. Furthermore, these two axes (x and y) can be oriented so that $I_{xx} = I_{yy}$. All five atoms lie off these two axes, and so $I_{xx} = I_{yy} > I_{zz}$, and CH_3I is a prolate symmetric top.

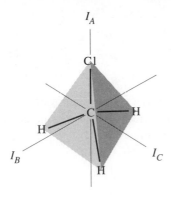 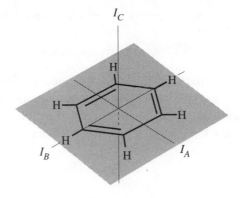

Prolate symmetric top
$$I_A < I_B = I_C$$

(a)

Oblate symmetric top
$$I_C > I_A = I_B$$

(b)

FIGURE 13.13
(a) Chloromethane is an example of a prolate symmetric top, and (b) benzene is an example of an oblate symmetric top. The assignments of the axes are in accord with the convention that $I_A \leq I_B \leq I_C$.

The energy levels of an oblate symmetric top are given by

$$F(J, K) = \tilde{B}J(J + 1) + (\tilde{C} - \tilde{B})K^2 \tag{13.29}$$

with $J = 0, 1, 2, \ldots$ and $K = 0, \pm 1, \pm 2, \ldots, \pm J$ and with a degeneracy $g_{JK} = 2J + 1$. Note that the energy levels depend upon two quantum numbers, J and K. The quantum number J is a measure of the total rotational angular momentum of the molecule, and K is a measure of the component of the rotational angular momentum along the unique axis of the symmetric top; that is, the axis having the unique moment of inertia.

The energy levels of a prolate symmetric top are given by

$$F(J, K) = \tilde{B}J(J + 1) + (\tilde{A} - \tilde{B})K^2 \tag{13.30}$$

with $J = 0, 1, 2, \ldots$ and $K = 0, \pm 1, \pm 2, \ldots, \pm J$ and a degeneracy $g_{JK} = 2J + 1$.

Not all symmetric top molecules have dipole moments (C_6H_6 and XeF_4, for example), but of those that do, most have the dipole moment directed along the symmetry axis (NH_3 and CH_3CN, for example). The selection rules for such molecules are

$$\Delta J = 0, \pm 1 \quad \Delta K = 0 \qquad \text{for } K \neq 0$$
$$\Delta J = \pm 1 \quad \Delta K = 0 \qquad \text{for } K = 0 \tag{13.31}$$

For the selection rules, $\Delta J = +1$ (absorption) and $\Delta K = 0$,

$$\tilde{\nu} = 2\tilde{B}(J+1) \tag{13.32}$$

which is the same result as for a linear molecule.

Polyatomic molecules, however, are likely to be less rigid than diatomic molecules, and so centrifugal distortion effects are more pronounced for polyatomic molecules. If we take these effects into account, Equation 13.32 becomes

$$\tilde{\nu} = 2\tilde{B}(J+1) - 2\tilde{D}_{JK}K^2(J+1) - 4\tilde{D}_J(J+1)^3 \tag{13.33}$$

where $\tilde{D}_{JK}$ and $\tilde{D}_J$ are centrifugal distortion constants. Equation 13.33 predicts that any rotational transitions will consist of a series of lines under high resolution as K takes on the values 0, 1, 2, ... Figure 13.14 shows this K dependence for the $J = 8 \rightarrow 9$ transition of CF$_3$CCH. Thus, although the rigid-rotator model of a symmetric top molecule predicts the same spectrum as that of a linear molecule, an observable distinction occurs when centrifugal distortion is taken into account.

There are no simple expressions for the energy levels of an asymmetric top molecule. Generally, their rotational spectra are fairly complicated and do not exhibit any simple pattern.

13–9. The Vibrations of Polyatomic Molecules Are Represented by Normal Coordinates

The vibrational spectra of polyatomic molecules turn out to be easily understood in terms of the harmonic-oscillator approximation. The key point is the introduction of normal coordinates, which we discuss in this section.

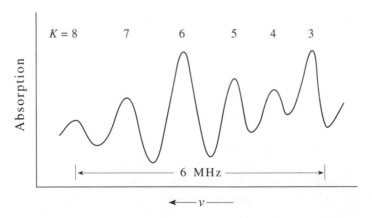

FIGURE 13.14
Part of the $J = 8 \rightarrow 9$ transition of CF$_3$CCH, showing the effect of centrifugal distortion. Notice that the transitions shown in the figure span only 6 MHz.

Consider a molecule containing N nuclei. A complete specification of this molecule in space requires $3N$ coordinates, three Cartesian coordinates for each nucleus. We say that the N-atomic molecule has a total of $3N$ *degrees of freedom*. Of these $3N$ coordinates, three can be used to specify the center of mass of the molecule. Motion along these three coordinates corresponds to translational motion of the center of mass of the molecule, and so we call these three coordinates *translational degrees of freedom*. It requires two coordinates to specify the orientation of a linear molecule about its center of mass and three coordinates to specify the orientation of a nonlinear molecule about its center of mass. Because motion along these coordinates corresponds to rotational motion, we say that a linear molecule has two *degrees of rotational freedom* and that a nonlinear molecule has three degrees of rotational freedom. The remaining coordinates ($3N - 5$ for a linear molecule and $3N - 6$ for a nonlinear molecule) specify the relative positions of the N nuclei. Because motion along these coordinates corresponds to vibrational motion, we say that a linear molecule has $3N - 5$ *vibrational degrees of freedom* and that a nonlinear molecule has $3N - 6$ vibrational degrees of freedom. These results are summarized in Table 13.6.

TABLE 13.6

The number of various degrees of freedom of a polyatomic molecule containing N atoms.

	Linear	Nonlinear
Translational degrees of freedom	3	3
Rotational degrees of freedom	2	3
Vibrational degrees of freedom	$3N - 5$	$3N - 6$

EXAMPLE 13–8

Determine the number of various degrees of freedom of HCl, CO_2, H_2O, NH_3, and CH_4.

SOLUTION:

	Total	Translational	Rotational	Vibrational
HCl	6	3	2	1
CO_2 (linear)	9	3	2	4
H_2O	9	3	3	3
NH_3	12	3	3	6
CH_4	15	3	3	9

In the absence of external fields, the energy of a molecule does not depend upon the position of its center of mass or its orientation. The potential energy of a polyatomic molecule is therefore a function of only the $3N - 5$ or $3N - 6$ vibrational coordinates. If we let the displacements about the equilibrium values of these coordinates be denoted by $q_1, q_2, \ldots, q_{N_{\text{vib}}}$, where N_{vib} is the number of vibrational degrees of freedom, then the potential energy is given by the multidimensional generalization of the one-dimensional case given by Equation 5.23:

$$
\Delta V = V(q_1, q_2, \ldots, q_{N_{\text{vib}}}) - V(0, 0, \ldots, 0) = \frac{1}{2} \sum_{i=1}^{N_{\text{vib}}} \sum_{j=1}^{N_{\text{vib}}} \left(\frac{\partial^2 V}{\partial q_i \partial q_j} \right) q_i q_j + \cdots
$$

$$
= \frac{1}{2} \sum_{i=1}^{N_{\text{vib}}} \sum_{j=1}^{N_{\text{vib}}} f_{ij} q_i q_j + \cdots \tag{13.34}
$$

In general, there are other terms that contain higher powers of q_i, but these anharmonic terms are neglected here. The presence of the cross terms in Equation 13.34 makes the solution of the corresponding Schrödinger equation very difficult to obtain. A theorem of classical mechanics, however, allows us to eliminate all the cross terms in Equation 13.34. The details are too specialized to go into here, but a straightforward procedure using matrix algebra can be used to find a new set of coordinates $\{Q_j\}$, such that

$$
\Delta V = \frac{1}{2} \sum_{j=1}^{N_{\text{vib}}} F_j Q_j^2 \tag{13.35}
$$

Note the lack of cross terms in this expression. These new coordinates are called *normal coordinates* or *normal modes*. In terms of normal coordinates, the vibrational Hamiltonian operator is

$$
\hat{H}_{\text{vib}} = -\sum_{j=1}^{N_{\text{vib}}} \frac{\hbar^2}{2\mu_j} \frac{d^2}{dQ_j^2} + \frac{1}{2} \sum_{j=1}^{N_{\text{vib}}} F_j Q_j^2 \tag{13.36}
$$

Recall from Section 3–9 that if a Hamiltonian operator can be written as a sum of independent terms, the total wave function is a product of individual wave functions and the energy is a sum of independent energies. Applying this theorem to Equation 13.36, we have

$$
\hat{H}_{\text{vib}} = \sum_{j=1}^{N_{\text{vib}}} \hat{H}_{\text{vib},j} = \sum_{j=1}^{N_{\text{vib}}} \left(-\frac{\hbar^2}{2\mu_j} \frac{d^2}{dQ_j^2} + \frac{1}{2} F_j Q_j^2 \right) \tag{13.37}
$$

$$
\psi_{\text{vib}}(Q_1, Q_2, \ldots, Q_{N_{\text{vib}}}) = \psi_{\text{vib},1}(Q_1) \psi_{\text{vib},2}(Q_2) \cdots \psi_{\text{vib}, N_{\text{vib}}}(Q_{N_{\text{vib}}})
$$

and

$$
E_{\text{vib}} = \sum_{j=1}^{N_{\text{vib}}} h\nu_j (v_j + \tfrac{1}{2}) \qquad \text{each } v_j = 0, \ 1, \ 2, \ \ldots \tag{13.38}
$$

The practical consequence of Equations 13.36 through 13.38 is that under the harmonic-oscillator approximation, the vibrational motion of a polyatomic molecule appears as N_{vib} independent harmonic oscillators. In the absence of degeneracies, each will have its own characteristic fundamental frequency v_j. The normal modes of formaldehyde (H_2CO) and chloromethane (CH_3Cl) are shown in Figure 13.15. A selection rule for vibrational absorption spectroscopy is that the dipole moment of the molecule must vary during the normal mode motion. When this is so, the normal mode is said to be *infrared active*. Otherwise, it is *infrared inactive*.

The three normal modes of H_2O are shown below.

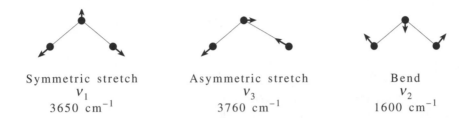

Symmetric stretch	Asymmetric stretch	Bend
v_1	v_3	v_2
$3650 \ cm^{-1}$	$3760 \ cm^{-1}$	$1600 \ cm^{-1}$

Note that the dipole moment changes during the motion of all three normal modes, so all three normal modes of H_2O are infrared active. Therefore, H_2O has three bands in its infrared spectrum. For CO_2, there are four normal modes ($3N - 5$).

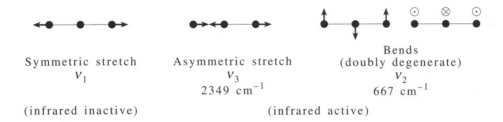

Symmetric stretch	Asymmetric stretch	Bends (doubly degenerate)
v_1	v_3	v_2
	$2349 \ cm^{-1}$	$667 \ cm^{-1}$
(infrared inactive)	(infrared active)	

There is no change in dipole moment during the symmetric stretch of CO_2, so this mode is infrared inactive. The other modes are infrared active, but the bending mode is doubly degenerate, so it leads to only one infrared band.

The two infrared active normal modes of CO_2 differ in an important respect. In the asymmetric stretch, the dipole moment oscillates parallel to the molecular axis, and in the bending mode, it oscillates perpendicular to the molecular axis. These two modes lead to quite different vibration-rotation spectra. The parallel case is similar to that of a diatomic molecule. The selection rules,

$$\Delta v = +1 \qquad \text{(absorption)}$$
$$\Delta J = \pm 1 \qquad \text{(parallel band)}$$

are the same as for a diatomic molecule and lead to a vibration-rotation spectrum consisting of a P branch and an R branch, like that shown in Figure 13.2. Such an

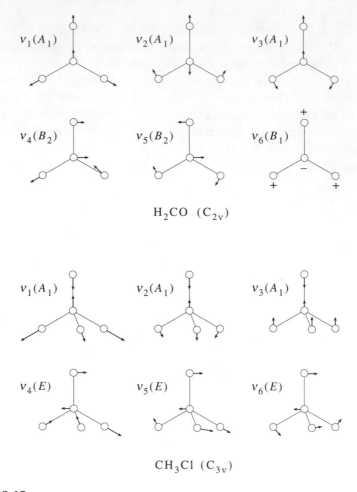

FIGURE 13.15
The normal modes of formaldehyde and chloromethane. For a given normal mode, the arrows indicate how the atoms move. Each atom oscillates about its equilibrium position with the same frequency and phase, but different atoms have different amplitudes of oscillation. Although specific molecules are indicated, the normal modes are characteristic of the symmetry of the molecules and so are more general. The designations in parentheses will be explained in the next section.

absorption band is called a *parallel band*. If the dipole moment oscillates perpendicular to the intermolecular axis, the selection rules are

$$\Delta v = +1 \qquad \text{(absorption)}$$
$$\Delta J = 0, \pm 1 \qquad \text{(perpendicular band)}$$

In this case, there is a band due to $\Delta J = 0$, called the *Q branch*, centered between the *P* and *R* branches, as shown in Figure 13.16b for the bending vibration of HCN.

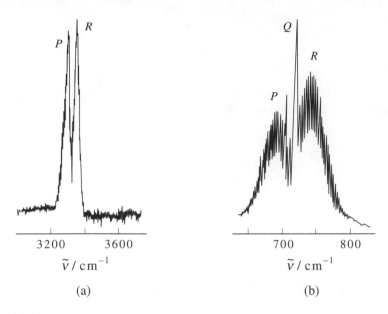

FIGURE 13.16
Two bands in the infrared spectrum of HCN. (a) A parallel band, showing the P and R branches and no Q branch. (b) A perpendicular band showing all three branches.

13–10. Normal Coordinates Belong to Irreducible Representations of Molecular Point Groups

Group theory can be used to characterize the various normal coordinates belonging to any molecule. This section uses the group-theoretic ideas presented in Chapter 12. (If you have not studied Chapter 12, then you may skip this section and go on to the next section.)

The fact that the vibrational properties of a molecule cannot change under any symmetry operation of the molecule forces the normal coordinates to transform as the irreducible representations of the group that describes the molecule. For example, let's consider the normal coordinates of a water molecule shown in the previous section. Recall from Chapter 12 that H_2O belongs to the $\mathbf{C}_{2v}$ point group, whose character table is given in Table 12.7. If we let the symmetric stretch normal coordinate be Q_{ss}, we can write (see Figure 12.1 for a reminder of the symmetry elements of the $\mathbf{C}_{2v}$ point group)

or,

$$\hat{E}Q_{ss} = Q_{ss}, \qquad \hat{C}_2 Q_{ss} = Q_{ss}, \qquad \hat{\sigma}_v Q_{ss} = Q_{ss}, \quad \text{and} \quad \hat{\sigma}'_v Q_{ss} = Q_{ss}$$

which shows that Q_{ss} belongs to the totally symmetric irreducible representation A_1. Similarly, the bending mode can be shown to belong to A_1. For the asymmetric stretch (Q_{as}), however, we have

or,

$$\hat{E}Q_{as} = Q_{as}, \qquad \hat{C}_2 Q_{as} = -Q_{as}, \qquad \hat{\sigma}_v Q_{as} = -Q_{as}, \quad \text{and} \quad \hat{\sigma}'_v Q_{as} = Q_{as}$$

which shows that Q_{as} belongs to B_2.

EXAMPLE 13–9

The normal modes in Figure 13.15 are labelled by the irreducible representations to which they belong. Verify the assignments for the six normal modes of H_2CO.

SOLUTION: The molecular point group of H_2CO is C_{2v}, whose character table is given in Table 12.7. We can express our results by the following table, whose entries are the result of acting upon the normal mode with the symmetry elements of C_{2v} (see Figure 12.1).

	$\hat{E}$	$\hat{C}_2$	$\hat{\sigma}_v$	$\hat{\sigma}'_v$
ν_1	1	1	1	1
ν_2	1	1	1	1
ν_3	1	1	1	1
ν_4	1	−1	−1	1
ν_5	1	−1	−1	1
ν_6	1	−1	1	−1

By referring to Table 12.7, we see that ν_1, ν_2, and ν_3 belong to A_1, that ν_4 and ν_5 belong to B_2, and that ν_6 belongs to B_1.

We can use group theory to determine how many normal coordinates belong to each irreducible representation. The procedure is to place an arbitrary (three-dimensional) vector onto each of the N atoms in the molecule to construct a $3N \times 3N$ reducible representation, which we then reduce using the character table. Actually, all we need are the characters of the $3N \times 3N$ reducible representation, which can be obtained fairly easily. We carried out a similar procedure in Section 12–9 when we derived symmetry orbitals for various molecules. We learned some simple rules there that allowed us to write the character for each group operation. For example, if an atom is moved under the symmetry operation, the contribution to the character is 0 for that atom. Only atoms whose positions are unchanged contribute to the character of any group operation. Because the identity operation leaves the (three-dimensional) vector on each of the N atoms unchanged, its character, $\chi(\hat{E})$, is equal to $3N$. When the operation $\hat{C}_2$ leaves an atom unmoved, it contributes -1 to $\chi(\hat{C}_2)$ because two of the axes change sign and one does not. When the operation $\hat{\sigma}$ leaves an atom unmoved, it contributes $+1$ to $\chi(\hat{\sigma})$ because two of the axes are left unchanged and one changes sign. Table 13.7 summarizes the contributions to the characters that each unmoved atom makes to the $3N$-dimensional reducible representation. The entries for the various rotation axes in Table 13.7 can be deduced by remembering that the matrix for a rotation through an angle θ about the z axis is (MathChapter F)

$$\begin{pmatrix} x' \\ y' \\ z' \end{pmatrix} = \begin{pmatrix} \cos\theta & -\sin\theta & 0 \\ \sin\theta & \cos\theta & 0 \\ 0 & 0 & 1 \end{pmatrix} \begin{pmatrix} x \\ y \\ z \end{pmatrix}$$

where $\theta = 360°/n$. For a rotation-reflection axis, the 1 in the above matrix becomes a -1.

TABLE 13.7
The contribution that each unmoved atom makes to the character of the $3N$-dimensional representation obtained by operating on arbitrary (three-dimensional) vectors attached to each of the N atoms in the molecule.

Operation, $\hat{R}$	Contribution to $\chi(\hat{R})$ per unmoved atom
$\hat{E}$	3
$\hat{\sigma}$	1
$\hat{\imath}$	-3
$\hat{C}_2$	-1
$\hat{C}_3, \hat{C}_3^2$	0
$\hat{C}_4, \hat{C}_4^3$	1
$\hat{C}_6, \hat{C}_6^5$	2
$\hat{S}_2$	-3
$\hat{S}_3, \hat{S}_3^2$	-2
$\hat{S}_4, \hat{S}_4^3$	-1
$\hat{S}_6, \hat{S}_6^5$	0

Let's apply this procedure to H_2O. The point group is C_{2v}, whose character table is given in Table 12.7. The operations $\hat{C}_2$ and $\hat{\sigma}_v$ leave only the oxygen atom unmoved, and $\hat{\sigma}_v'$ leaves all three atoms unmoved. Therefore, the nine-dimensional reducible representation is

	$\hat{E}$	$\hat{C}_2$	$\hat{\sigma}_v$	$\hat{\sigma}_v'$
Γ_{3N}	9	-1	1	3

We can use Equation 12.23 to reduce Γ_{3N} into its irreducible representations:

$$a_{A_1} = \tfrac{1}{4}[(9) \times (1) + (-1) \times (1) + (1) \times (1) + (3) \times (1)] = 3$$

$$a_{A_2} = \tfrac{1}{4}[(9) \times (1) + (-1) \times (1) + (1) \times (-1) + (3) \times (-1)] = 1$$

$$a_{B_1} = \tfrac{1}{4}[(9) \times (1) + (-1) \times (-1) + (1) \times (1) + (3) \times (-1)] = 2$$

$$a_{B_2} = \tfrac{1}{4}[(9) \times (1) + (-1) \times (-1) + (1) \times (-1) + (3) \times (1)] = 3$$

or

$$\Gamma_{3N} = 3A_1 + A_2 + 2B_1 + 3B_2$$

This sum of irreducible representations accounts for all the degrees of freedom of the molecule. The irreducible representations of x, y, and z (B_1, B_2, and A_1) correspond to the three translational degrees of freedom, which we will denote by T_x, T_y, and T_z, respectively, and those of R_x, R_y, and R_z (B_2, B_1, and A_2) correspond to the three rotational degrees of freedom (see Table 12.7). If we subtract these from Γ_{3N}, we have

$$\Gamma_{vib} = 2A_1 + B_2$$

consistent with our determination of the symmetries of the normal coordinates of H_2O using their depictions shown earlier.

EXAMPLE 13–10

Determine the symmetries of the normal coordinates of a planar XY_3 molecule.

SOLUTION: The point group is D_{3h}, whose character table is given in Table 12.11. Using Table 13.7 to determine the characters of the 12-dimensional reducible representation, we obtain

	$\hat{E}$	$2\hat{C}_3$	$3\hat{C}_2$	$\hat{\sigma}_h$	$2\hat{S}_3$	$3\hat{\sigma}_v$
Γ_{3N}	12	0	-2	4	-2	2

Using Equation 12.23, we get

$$\Gamma_{3N} = A_1' + A_2' + 3E' + 2A_2'' + E''$$

The $\mathbf{D}_{3h}$ character table shows that T_x and T_y jointly belong to E', that R_x and R_y jointly belong to E'', that T_z belongs to A_2'', and that R_z belongs to A_2'. Subtracting these from Γ_{3N} gives

$$\Gamma_{\text{vib}} = A_1' + 2E' + A_2''$$

Because E' is two-dimensional, Γ_{vib} represents the six normal modes of a $\mathbf{D}_{3h}$ XY_3 molecule such as SO_3 or BF_3.

After we learn about selection rules in the next three sections, we will return to group theory and show how it can be used to determine which normal coordinates are infrared active and which are not.

13–11. Selection Rules Are Derived from Time-Dependent Perturbation Theory

The spectroscopic selection rules determine which transitions from one state to another are possible. The very nature of transitions implies a time-dependent phenomenon, so we must use the time-dependent Schrödinger equation (Equation 4.15)

$$\hat{H}\Psi = i\hbar \frac{\partial \Psi}{\partial t} \tag{13.39}$$

We showed in Section 4–4 that if $\hat{H}$ does not depend explicitly on time, then

$$\Psi_n(\mathbf{r}, t) = \psi_n(\mathbf{r})e^{-iE_n t/\hbar}$$

where $\psi_n(\mathbf{r})$ satisfies the time-independent Schrödinger equation

$$\hat{H}\psi_n(\mathbf{r}) = E_n \psi_n(\mathbf{r})$$

Recall that $\psi_n^* \psi_n$ is independent of time and that the states described by ψ_n are called *stationary states* (cf. Section 4–4).

The idea of stationary states applies to isolated systems. Consider now a molecule interacting with electromagnetic radiation. The electromagnetic field may be written approximately as

$$\mathbf{E} = \mathbf{E}_0 \cos 2\pi \nu t \tag{13.40}$$

where ν is the frequency of the radiation and $\mathbf{E}_0$ is the electric field vector. If $\boldsymbol{\mu}$ is the dipole moment of the molecule (see Problem 10–47 for a review of molecular dipole moments), then the Hamiltonian operator for the interaction of the electric field with the molecule is (Problem 13–49)

$$\hat{H}^{(1)} = -\boldsymbol{\mu} \cdot \mathbf{E} = -\boldsymbol{\mu} \cdot \mathbf{E}_0 \cos 2\pi \nu t \tag{13.41}$$

Thus, we must solve

$$\hat{H}\Psi = i\hbar\frac{\partial \Psi}{\partial t} \tag{13.42}$$

where

$$\hat{H} = \hat{H}^{(0)} + \hat{H}^{(1)} = \hat{H}^{(0)} - \boldsymbol{\mu}\cdot\mathbf{E}_0 \cos 2\pi \nu t \tag{13.43}$$

and $\hat{H}^{(0)}$ is the Hamiltonian operator of the isolated molecule. We will see below that the time-dependent term $\hat{H}^{(1)}$ can cause transitions from one stationary state to another.

To solve Equation 13.42, we will treat the time-dependent term $\hat{H}^{(1)}$ as a small perturbation. The procedure we will use is called *time-dependent perturbation theory* and is an extension of the perturbation theory developed in Chapter 7 to time-dependent phenomena. Although an isolated molecule generally has an infinite number of stationary states, for simplicity of notation, we will consider only a two-state system. For such a system, in the absence of time-dependent perturbations

$$\hat{H}^{(0)}\psi = i\hbar\frac{\partial \psi}{\partial t} \tag{13.44}$$

where there are only two stationary states, ψ_1 and ψ_2, with

$$\Psi_1(t) = \psi_1 e^{-iE_1 t/\hbar} \quad \text{and} \quad \Psi_2(t) = \psi_2 e^{-iE_2 t/\hbar} \tag{13.45}$$

EXAMPLE 13–11

Show that Ψ_1 and Ψ_2 given by Equation 13.45 satisfy Equation 13.44.

SOLUTION: Substitute $\Psi_1(t) = \psi_1 e^{-iE_1 t/\hbar}$ into Equation 13.44 to get

$$\hat{H}^{(0)}\Psi_1 = \hat{H}^{(0)}\psi_1 e^{-iE_1 t/\hbar} = e^{-iE_1 t/\hbar}\hat{H}^{(0)}\psi_1 = E_1\psi_1 e^{-iE_1 t/\hbar}$$

and

$$i\hbar\frac{\partial \Psi_1}{\partial t} = i\hbar\psi_1\frac{d}{dt}e^{-iE_1 t/\hbar} = E_1\psi_1 e^{-iE_1 t/\hbar}$$

In the first line, we have used the fact that $\hat{H}^{(0)}$ is independent of time, and in the second line, we used the fact that ψ_1 is independent of time and that $\hat{H}^{(0)}\psi_1 = E_1\psi_1$. The proof that Ψ_2 is also a solution is similar.

Assume now that initially the system is in state 1. We let the perturbation begin at $t = 0$ and assume that $\Psi(t)$ is a linear combination of $\Psi_1(t)$ and $\Psi_2(t)$ with coefficients that depend upon time. Thus, we write

$$\Psi(t) = a_1(t)\Psi_1(t) + a_2(t)\Psi_2(t) \tag{13.46}$$

where $a_1(t)$ and $a_2(t)$ are to be determined. Recall from Chapter 4 that for such a linear combination, $a_i^* a_i$ is the probability that the molecule is in state i. We substitute Equation 13.46 into Equation 13.42 to obtain

$$a_1(t)\hat{H}^{(0)}\Psi_1 + a_2(t)\hat{H}^{(0)}\Psi_2 + a_1(t)\hat{H}^{(1)}\Psi_1 + a_2(t)\hat{H}^{(1)}\Psi_2$$
$$= a_1(t)i\hbar\frac{\partial\Psi_1}{dt} + a_2(t)i\hbar\frac{\partial\Psi_2}{dt} + i\hbar\Psi_1\frac{da_1}{dt} + i\hbar\Psi_2\frac{da_2}{dt} \quad (13.47)$$

Using the result given in Example 13–11, we can cancel the first two terms on both sides of Equation 13.47 to obtain

$$a_1(t)\hat{H}^{(1)}\Psi_1 + a_2(t)\hat{H}^{(1)}\Psi_2 = i\hbar\Psi_1\frac{da_1}{dt} + i\hbar\Psi_2\frac{da_2}{dt} \quad (13.48)$$

We now multiply Equation 13.48 by ψ_2^* and integrate over the spatial coordinates to get

$$a_1(t)\int\psi_2^*\hat{H}^{(1)}\Psi_1 d\tau + a_2(t)\int\psi_2^*\hat{H}^{(1)}\Psi_2 d\tau$$
$$= i\hbar\frac{da_1}{dt}\int\psi_2^*\Psi_1 d\tau + i\hbar\frac{da_2}{dt}\int\psi_2^*\Psi_2 d\tau \quad (13.49)$$

The first integral on the right side vanishes because $\Psi_1 = \psi_1 e^{-iE_1 t/\hbar}$ (Equation 13.45) and because ψ_2 and ψ_1 are orthogonal. Similarly, the second integral on the right side is equal to $i\hbar e^{-iE_2 t/\hbar} da_2/dt$ because $\Psi_2 = \psi_2 e^{-iE_2 t/\hbar}$ and because ψ_2 is normalized. Solving Equation 13.49 for $i\hbar da_2/dt$ gives

$$i\hbar\frac{da_2}{dt} = a_1(t)e^{iE_2 t/\hbar}\int\psi_2^*\hat{H}^{(1)}\Psi_1 d\tau + a_2(t)e^{iE_2 t/\hbar}\int\psi_2^*\hat{H}^{(1)}\Psi_2 d\tau$$

Using Equation 13.45 for Ψ_1 and Ψ_2 finally gives

$$i\hbar\frac{da_2}{dt} = a_1(t)\exp\left[\frac{-i(E_1 - E_2)t}{\hbar}\right]\int\psi_2^*\hat{H}^{(1)}\psi_1 d\tau + a_2(t)\int\psi_2^*\hat{H}^{(1)}\psi_2 d\tau$$
$$(13.50)$$

Because the system is initially in state 1,

$$a_1(0) = 1 \quad \text{and} \quad a_2(0) = 0 \quad (13.51)$$

Because $\hat{H}^{(1)}$ is considered a small perturbation, there are not enough transitions out of state 1 to cause a_1 and a_2 to differ appreciably from their initial values. Thus, as an approximation, we may replace $a_1(t)$ and $a_2(t)$ in the right side of Equation 13.51 by their initial values $[a_1(0) = 1, a_2(0) = 0]$ to get

$$i\hbar\frac{da_2}{dt} = \exp\left[\frac{-i(E_1 - E_2)t}{\hbar}\right]\int\psi_2^*\hat{H}^{(1)}\psi_1 d\tau \quad (13.52)$$

For convenience only, we will take the electric field to be in the z direction, in which case we can write

$$
\begin{aligned}
\hat{H}^{(1)} &= -\mu_z \mathrm{E}_{0z} \cos 2\pi \nu t \\
&= -\frac{\mu_z \mathrm{E}_{0z}}{2} (e^{i2\pi\nu t} + e^{-i2\pi\nu t})
\end{aligned}
$$

where μ_z is the z component of the molecular dipole moment and E_{0z} is the magnitude of the electric field along the z-axis. We substitute this expression for $\hat{H}^{(1)}$ into Equation 13.52 and obtain

$$
\frac{da_2}{dt} \propto (\mu_z)_{12} \mathrm{E}_{0z} \left\{ \exp \left[\frac{i(E_2 - E_1 + h\nu)t}{\hbar} \right] + \exp \left[\frac{i(E_2 - E_1 - h\nu)t}{\hbar} \right] \right\} \quad (13.53)
$$

where we have defined

$$
(\mu_z)_{12} = \int \psi_2^* \mu_z \psi_1 d\tau \quad (13.54)
$$

The quantity $(\mu_z)_{12}$ is the z component of the *transition dipole moment* between states 1 and 2. Note that if $(\mu_z)_{12} = 0$, then $da_2/dt = 0$ and there will be no transitions out of state 1 into state 2. The dipole transition moment is what underlies the selection rules assumed so far. Transitions occur only between states for which the transition moment is nonzero.

We will derive explicit selection rules in the next two sections, but before doing so let's integrate Equation 13.53 between 0 and t to obtain

$$
\begin{aligned}
a_2(t) \propto (\mu_z)_{12} \mathrm{E}_{0z} \\
\times \left\{ \frac{1 - \exp[i(E_2 - E_1 + h\nu)t/\hbar]}{E_2 - E_1 + h\nu} + \frac{1 - \exp[i(E_2 - E_1 - h\nu)t/\hbar]}{E_2 - E_1 - h\nu} \right\} \quad (13.55)
\end{aligned}
$$

Because we have taken $E_2 > E_1$, the so-called *resonance denominators* in Equation 13.55 cause the second term in this equation to become much larger than the first term and be of major importance in determining $a_2(t)$ when

$$
E_2 - E_1 \approx h\nu \quad (13.56)
$$

Thus, we obtain in a natural way the Bohr frequency condition we have used repeatedly. When a system makes a transition from one state to another, it absorbs (or emits) a photon whose energy is equal to the difference in the energies of the two states.

The probability of absorption or the intensity of absorption is proportional to the probability of observing the molecules in state 2, which is given by $a_2^*(t)a_2(t)$. Using only the second term in Equation 13.55, we obtain (Problem 13–40)

$$
a_2^*(t)a_2(t) \propto \frac{\sin^2[(E_2 - E_1 - \hbar\omega)t/2\hbar]}{(E_2 - E_1 - \hbar\omega)^2} \quad (13.57)
$$

Equation 13.57 is plotted in Figure 13.17. Note that the plot indicates strong absorption when $\hbar\omega = h\nu \approx E_2 - E_1$.

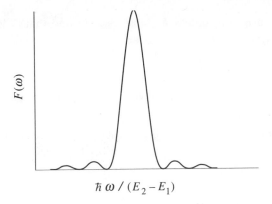

$$\hbar\omega / (E_2 - E_1)$$

FIGURE 13.17
The function $F(\omega) = \sin^2[(E_2 - E_1 - \hbar\omega)t/2\hbar]/(E_2 - E_1 - \hbar\omega)^2$, which represents the probability of making a $1 \to 2$ transition in the time interval 0 to t, plotted against frequency ω. Note that this function peaks when $E_2 - E_1 = \hbar\omega = h\nu$.

13–12. The Selection Rule in the Rigid-Rotator Approximation Is $\Delta J = \pm 1$

We can use Equation 13.54 and the properties of the spherical harmonics to derive the selection rule for a rigid rotator. Recall that the rigid-rotator wave functions are the spherical harmonics, which are developed in Section 6–2. Once again, if we assume the electric field lies along the z-axis, then the dipole transition moment between any two states in the rigid-rotator approximation is

$$(\mu_z)_{J,M;J',M'} = \int_0^{2\pi} \int_0^{\pi} Y_{J'}^{M'}(\theta, \phi)^* \mu_z Y_J^M(\theta, \phi) \sin\theta d\theta d\phi$$

Using the fact that $\mu_z = \mu \cos\theta$ gives

$$(\mu_z)_{J,M,J',M'} = \mu \int_0^{2\pi} \int_0^{\pi} Y_{J'}^{M'}(\theta, \phi)^* Y_J^M(\theta, \phi) \cos\theta \sin\theta d\theta d\phi \tag{13.58}$$

Notice that μ must be nonzero for the transition moment to be non-zero. Thus, we have now proven our earlier assertion that a molecule must have a permanent dipole moment for it to have a pure rotational spectrum, at least in the rigid-rotator approximation.

We can also determine for which values of J, M, J', and M' the integral in Equation 13.58 will be nonzero. Recall that (Equation 6.30)

$$Y_J^M(\theta, \phi) = N_{JM} P_J^{|M|}(\cos\theta)e^{iM\phi} \tag{13.59}$$

where N_{JM} is a normalization constant. Substitute Equation 13.59 into Equation 13.58 and let $x = \cos\theta$ to obtain

$$(\mu_z)_{J,M;J',M'} = \mu N_{J,M} N_{J',M'} \int_0^{2\pi} d\phi e^{i(M-M')\phi} \int_{-1}^{1} dx\, x P_{J'}^{|M'|}(x) P_J^{|M|}(x) \tag{13.60}$$

The integral over ϕ is zero unless $M = M'$, so we find that $\Delta M = 0$ is part of the rigid-rotator selection rule. Integration over ϕ for $M = M'$ gives a factor of 2π, so we have

$$(\mu_z)_{J,M;J',M'} = 2\pi \mu N_{JM} N_{J'M} \int_{-1}^{1} dx \, x \, P_{J'}^{|M|}(x) P_J^{|M|}(x) \tag{13.61}$$

We can evaluate this integral in general by using the identity (Problem 6–8)

$$(2J + 1)x \, P_J^{|M|}(x) = (J - |M| + 1) P_{J+1}^{|M|}(x) + (J + |M|) P_{J-1}^{|M|}(x) \tag{13.62}$$

By using this relation in Equation 13.61, we obtain

$$(\mu_z)_{J,M;J',M'} = 2\pi \mu N_{JM} N_{J'M} \int_{-1}^{1} dx \, P_{J'}^{|M|}(x)$$
$$\left[\frac{(J - |M| + 1)}{2J + 1} P_{J+1}^{|M|}(x) + \frac{(J + |M|)}{2J + 1} P_{J-1}^{|M|}(x) \right]$$

Using the orthogonality relation for the $P_J^M(x)$ (Equation 6.28), we find that the above integral will vanish unless $J' = J + 1$ or $J' = J - 1$. This finding leads to the selection rule $J' = J \pm 1$, or $\Delta J = \pm 1$. Thus, we have shown that the selection rule for pure rotational spectra in the rigid-rotator approximation is that the molecule must have a permanent dipole moment and that $\Delta J = \pm 1$ and $\Delta M = 0$.

EXAMPLE 13–12
Using the explicit formulas for the spherical harmonics given in Table 6.3, show that the rotational transition $J = 0 \rightarrow J = 1$ is allowed, but $J = 0 \rightarrow J = 2$ is forbidden in microwave spectroscopy (in the rigid-rotator approximation).

SOLUTION: Referring to Equation 13.58, we see that we must show that the integral

$$I_{0 \rightarrow 1} = \int_0^{2\pi} \int_0^{\pi} Y_1^M(\theta, \phi)^* Y_0^0(\theta, \phi) \cos \theta \sin \theta d\theta d\phi$$

is nonzero and that

$$I_{0 \rightarrow 2} = \int_0^{2\pi} \int_0^{\pi} Y_2^M(\theta, \phi)^* Y_0^0(\theta, \phi) \cos \theta \sin \theta d\theta d\phi$$

is equal to zero. In either case, we can easily see that the integral over ϕ will be zero unless $M = 0$, so we will concentrate only on the θ integration. For $I_{0 \rightarrow 1}$, we have

$$I_{0 \rightarrow 1} = 2\pi \int_0^{\pi} \left(\frac{3}{4\pi} \right)^{1/2} \cos \theta \left(\frac{1}{4\pi} \right)^{1/2} \cos \theta \sin \theta d\theta$$
$$= \frac{\sqrt{3}}{2} \int_{-1}^{1} dx \, x^2 = \frac{1}{\sqrt{3}} \neq 0$$

For $I_{0 \to 2}$, we have

$$
\begin{aligned}
I_{0 \to 2} &= 2\pi \int_0^\pi \left(\frac{5}{16\pi}\right)^{1/2} (3\cos^2\theta - 1)\left(\frac{1}{4\pi}\right)^{1/2} \cos\theta \sin\theta \, d\theta \\
&= \frac{\sqrt{5}}{4} \int_{-1}^1 dx(3x^3 - x) = 0
\end{aligned}
$$

because the integrand is an odd function of x.

13–13. The Harmonic-Oscillator Selection Rule Is $\Delta v = \pm 1$

Using Equation 13.54 and the fact that the harmonic-oscillator wave functions are (Equation 5.35)

$$
\psi_v(q) = N_v H_v(\alpha^{1/2}q)e^{-\alpha q^2/2} \tag{13.63}
$$

where $H_v(\alpha^{1/2}q)$ is a Hermite polynomial and where $\alpha = (k\mu/\hbar^2)^{1/2}$, we obtain the transition dipole moment when the electric field is along the z-axis as follows

$$
(\mu_z)_{v,v'} = \int_{-\infty}^\infty N_v N_{v'} H_{v'}(\alpha^{1/2}q)e^{-\alpha q^2/2}\mu_z(q)H_v(\alpha^{1/2}q)e^{-\alpha q^2/2}dq \tag{13.64}
$$

We now expand $\mu_z(q)$ about the equilibrium nuclear separation:

$$
\mu_z(q) = \mu_0 + \left(\frac{d\mu}{dq}\right)_0 q + \cdots \tag{13.65}
$$

where μ_0 is the dipole moment at the equilibrium bond length and q is the displacement from that equilibrium value. Thus when $q = 0$, $\mu_z = \mu_0$. If we substitute the first two terms of the expansion in Equation 13.65 into Equation 13.64, we have:

$$
\begin{aligned}
(\mu_z)_{v,v'} &= N_v N_{v'} \mu_0 \int_{-\infty}^\infty H_{v'}(\alpha^{1/2}q)H_v(\alpha^{1/2}q)e^{-\alpha q^2}dq \\
&\quad + N_v N_{v'}\left(\frac{d\mu}{dq}\right)_0 \int_{-\infty}^\infty H_{v'}(\alpha^{1/2}q)qH_v(\alpha^{1/2}q)e^{-\alpha q^2}dq
\end{aligned} \tag{13.66}
$$

The first integral here vanishes if $v \neq v'$ due to the orthogonality of the Hermite polynomials.

The second integral can be evaluated in general by using the Hermite polynomial identity (Problem 5–24):

$$
\xi H_v(\xi) = v H_{v-1}(\xi) + \tfrac{1}{2}H_{v+1}(\xi) \tag{13.67}
$$

If we substitute Equation 13.67 into Equation 13.66, letting $\alpha^{1/2}q = \xi$, we obtain

$$(\mu_z)_{v,v'} = \frac{N_v N_{v'}}{\alpha} \left(\frac{d\mu}{dq}\right)_0 \int_{-\infty}^{\infty} H_{v'}(\xi)\left[vH_{v-1}(\xi) + \frac{1}{2}H_{v+1}(\xi)\right]e^{-\xi^2}d\xi \qquad (13.68)$$

Using now the orthogonality property of the Hermite polynomials, we see that $(\mu_z)_{v,v'}$ vanishes unless $v' = v \pm 1$. Thus, the selection rule for vibrational transitions under the harmonic-oscillator approximation is that $\Delta v = \pm 1$. In addition, the factor $(d\mu/dq)_0$ in front of the transition moment integral reminds us that the dipole moment of the molecule must vary during a vibration (Equation 13.65), or the transition will not take place.

EXAMPLE 13–13
Using the explicit formulas for the Hermite polynomials given in Table 5.2, show that a $0 \to 1$ vibrational transition is allowed and that a $0 \to 2$ transition is forbidden for a harmonic oscillator.

SOLUTION: Letting $\xi = \alpha^{1/2}x$ in Table 5.3, we have

$$\psi_0(\xi) = \left(\frac{\alpha}{\pi}\right)^{1/4} e^{-\xi^2/2}$$

$$\psi_1(\xi) = \sqrt{2}\left(\frac{\alpha}{\pi}\right)^{1/4} \xi e^{-\xi^2/2}$$

$$\psi_2(\xi) = \frac{1}{\sqrt{2}}\left(\frac{\alpha}{\pi}\right)^{1/4} (2\xi^2 - 1)e^{-\xi^2/2}$$

The dipole transition moment is given by the integral

$$I_{0\to v} \propto \int_{-\infty}^{\infty} \psi_v(\xi)\xi\psi_0(\xi)d\xi$$

The transition is allowed if $I_{0\to v} \neq 0$ and is forbidden if $I_{0\to v} = 0$. For $v = 1$, we have

$$I_{0\to 1} \propto \left(\frac{2\alpha}{\pi}\right)^{1/2} \int_{-\infty}^{\infty} \xi^2 e^{-\xi^2}d\xi \neq 0$$

because the integrand is everywhere positive. For $v = 2$,

$$I_{0\to 2} \propto \left(\frac{\alpha}{2\pi}\right)^{1/2} \int_{-\infty}^{\infty} (2\xi^3 - \xi)e^{-\xi^2}d\xi = 0$$

because the integrand is an odd function and the limits go from $-\infty$ to $+\infty$.

13–14. Group Theory Is Used to Determine the Infrared Activity of Normal Mode Vibrations

In the previous section, we saw that a normal mode will be infrared active if the dipole moment of the molecule changes as the molecule vibrates. Thus, for example, the symmetric stretch of CO_2 will be infrared inactive, whereas the three other modes will be infrared active. We can use the vibrational selection rule and group theory to prove this. If we write Equation 13.54 in terms of normal coordinates, we see that the selection rule for the $v = 0$ to $v = 1$ vibrational state says that the integral

$$I_{0 \to 1} = \int \psi_0(Q_1, Q_2, \ldots, Q_{N_{vib}}) \begin{Bmatrix} \mu_x \\ \mu_y \\ \mu_z \end{Bmatrix} \psi_1(Q_1, Q_2, \ldots, Q_{N_{vib}}) dQ_1, dQ_2, \ldots, dQ_{N_{vib}}$$

(13.69)

must be nonzero. We have written Equation 13.69 in a form that includes all three components of the dipole moment μ_x, μ_y, and μ_z. This equation is then general for an electric field oriented along any particular direction with respect to the dipole moment of the molecule. In the harmonic oscillator approximation, $\psi_0(Q_1, Q_2, \ldots, Q_{N_{vib}})$ is the product (Equation 13.37),

$$\psi_0(Q_1, Q_2, \ldots, Q_{N_{vib}}) = ce^{-\alpha_1 Q_1^2 - \alpha_2 Q_2^2 - \cdots - \alpha_{N_{vib}} Q_{N_{vib}}^2}$$

(13.70)

where c is just a normalization constant, and $\alpha_j = (\mu_j k_j)^{1/2}/2\hbar$. The normal modes belong to the irreducible representations of the molecular point group. Therefore, for nondegenerate modes, the effect of any symmetry operation on Q_j gives $\pm Q_j$, so $\psi_0(Q_1, Q_2, \ldots, Q_{N_{vib}})$, being a quadratic function of the Q_js, is invariant under all the symmetry operations of the group. In other words, it belongs to the totally symmetric irreducible representation, A_1. (We will not prove it here, but this is also the case for degenerate vibrations.) We can express this result by the equation

$$\hat{R}\psi_0(Q_1, Q_2, \ldots, Q_{N_{vib}}) = \psi_0(Q_1, Q_2, \ldots, Q_{N_{vib}})$$

(13.71)

for all the group operations, $\hat{R}$. According to Table 5.3 and Equation 13.37, $\psi_1(Q_1, Q_2, \ldots, Q_{N_{vib}})$ for a state in which the normal coordinate Q_j is excited to the $v = 1$ level is

$$\psi_1(Q_1, Q_2, \ldots, Q_{N_{vib}}) = \psi_0(Q_1)\psi_0(Q_2) \cdots \psi_0(Q_{j-1})\psi_1(Q_j)\psi_0(Q_{j+1}) \cdots \psi_0(Q_{N_{vib}})$$

$$= c'Q_j e^{-\alpha_1 Q_1^2 - \alpha_2 Q_2^2 - \cdots - \alpha_{N_{vib}} Q_{N_{vib}}^2}$$

(13.72)

Therefore, $\psi_1(Q_1, Q_2, \ldots, Q_{N_{vib}})$ transforms as the normal coordinate Q_j. If we let the character of the operation $\hat{R}$ of the irreducible representation of Q_j be $\chi_{Q_j}(\hat{R})$, then we can write

$$\hat{R}\psi_1(Q_1, Q_2, \ldots, Q_{N_{vib}}) = \chi_{Q_j}(\hat{R})\psi_1(Q_1, Q_2, \ldots, Q_{N_{vib}})$$

(13.73)

Now let's return to the selection rule integral, Equation 13.69. Surely $I_{0\to 1}$ must be invariant under all the operations of the group, so

$$\hat{R}I_{0\to 1} = I_{0\to 1} = \int (\hat{R}\psi_0)(\hat{R}\mu_x)(\hat{R}\psi_1)dQ_1 dQ_2 \ldots dQ_{N_{\text{vib}}}$$

$$= \chi_{A_1}(\hat{R})\chi_{\mu_x}(\hat{R})\chi_{Q_j}(\hat{R})\int \psi_0 \mu_x \psi_1 dQ_1 dQ_2 \ldots dQ_{N_{\text{vib}}}$$

$$= \chi_{A_1}(\hat{R})\chi_{\mu_x}(\hat{R})\chi_{Q_j}(\hat{R})I_{0\to 1} \qquad (13.74)$$

Thus, the products of the characters here must equal 1 for all $\hat{R}$. Because $\chi_{A_1}(\hat{R}) = 1$ for all $\hat{R}$, the product $\chi_{\mu_x}(\hat{R})\chi_{Q_j}(\hat{R})$ must equal 1 for all $\hat{R}$. This can be so for one-dimensional irreducible representations only if μ_x (or x itself) and Q_j belong to the same irreducible representation, with a similar result for μ_y (or y itself) and μ_z (or z itself). In summary, then, $I_{0\to 1}$ will be nonzero only if Q_j belongs to the same irreducible representation as x, y, or z. Note that this proof is very similar to the proof we presented in Section 12–8 for the matrix elements H_{ij} and S_{ij}.

Let's apply this result to the normal modes of H_2O. The $\mathbf{C}_{2v}$ character table (Table 12.7) shows that x belongs to B_1, y to B_2, and z to A_1. But we saw in Section 13–9 that the symmetric stretch and the bending mode belong to A_1 and the asymmetric stretch belongs to B_2. Therefore, all three normal modes of H_2O are infrared active.

EXAMPLE 13–14
Determine the infrared activity (active or inactive) of the normal modes of SO_3 shown below.

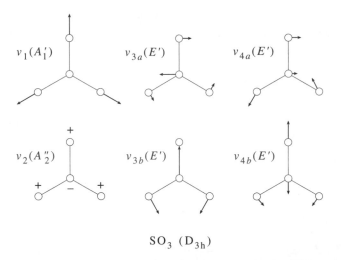

$$SO_3 \ (D_{3h})$$

SOLUTION: The $\mathbf{D}_{3h}$ character table (Table 12.11) shows that x and y belong to E' and that z belongs to A_2''. Referring to the normal modes shown above we see that the

v_1 mode (which belongs to A_1') is infrared inactive, and the others (which belong to A_2'' and E') are infrared active.

Problems

13-1. The spacing between the lines in the microwave spectrum of $H^{35}Cl$ is 6.350×10^{11} Hz. Calculate the bond length of $H^{35}Cl$.

13-2. The microwave spectrum of $^{39}K^{127}I$ consists of a series of lines whose spacing is almost constant at 3634 MHz. Calculate the bond length of $^{39}K^{127}I$.

13-3. The equilibrium internuclear distance of $H^{127}I$ is 160.4 pm. Calculate the value of B in wave numbers and megahertz.

13-4. Assuming the rotation of a diatomic molecule in the $J = 10$ state may be approximated by classical mechanics, calculate how many revolutions per second $^{23}Na^{35}Cl$ makes in the $J = 10$ rotational state. The rotational constant of $^{23}Na^{35}Cl$ is 6500 MHz.

13-5. The results we derived for a rigid rotator apply to linear polyatomic molecules as well as to diatomic molecules. Given that the moment of inertia I for $H^{12}C^{14}N$ is 1.89×10^{-46} kg·m^2 (cf. Problem 13–6), predict the microwave spectrum of $H^{12}C^{14}N$.

13-6. This problem involves the calculation of the moment of inertia of a linear triatomic molecule such as $H^{12}C^{14}N$ (see Problem 13–5). The moment of inertia of a linear molecule is

$$I = \sum_j m_j d_j^2$$

where d_j is the distance of the jth mass from the center of mass. Thus, the moment of inertia of $H^{12}C^{14}N$ is

$$I = m_H d_H^2 + m_C d_C^2 + m_N d_N^2 \tag{1}$$

Show that Equation 1 can be written as

$$I = \frac{m_H m_C R_{HC}^2 + m_H m_N R_{HN}^2 + m_C m_N R_{CN}^2}{m_H + m_C + m_N}$$

where the R's are the various internuclear distances. Given that $R_{HC} = 106.8$ pm and $R_{CN} = 115.6$ pm, calculate the value of I and compare the result with that given in Problem 13–5.

13-7. The far infrared spectrum of $^{39}K^{35}Cl$ has an intense line at 278.0 cm^{-1}. Calculate the force constant and the period of vibration of $^{39}K^{35}Cl$.

13-8. The force constant of $^{79}Br^{79}Br$ is 240 N·m^{-1}. Calculate the fundamental vibrational frequency and the zero-point energy of $^{79}Br_2$.

13-9. Prove that

$$\langle x^2 \rangle = \frac{\hbar}{2(\mu k)^{1/2}}$$

for the ground state of a harmonic oscillator. Use this equation to calculate the root-mean-square amplitude of $^{14}N_2$ in its ground state. Compare your result to the bond length. Use $k = 2260 \text{ N} \cdot \text{m}^{-1}$ for $^{14}N_2$.

13-10. Derive Equations 13.15 and 13.16.

13-11. Given that $B = 58\,000$ MHz and $\tilde{\nu} = 2160.0 \text{ cm}^{-1}$ for CO, calculate the frequencies of the first few lines of the R and P branches in the vibration-rotation spectrum of CO.

13-12. Given that $R_e = 156.0$ pm and $k = 250.0 \text{ N} \cdot \text{m}^{-1}$ for $^6Li^{19}F$, use the rigid rotator-harmonic oscillator approximation to construct to scale an energy-level diagram for the first five rotational levels in the $v = 0$ and $v = 1$ vibrational states. Indicate the allowed transitions in an absorption experiment, and calculate the frequencies of the first few lines in the R and P branches of the vibration-rotation spectrum of $^6Li^{19}F$.

13-13. Using the values of $\tilde{\nu}_e$, $\tilde{x}_e\tilde{\nu}_e$, $\tilde{B}_e$, and $\tilde{\alpha}_e$ given in Table 13.2, construct to scale an energy-level diagram for the first five rotational levels in the $v = 0$ and $v = 1$ vibrational states for $H^{35}Cl$. Indicate the allowed transitions in an absorption experiment, and calculate the frequencies of the first few lines in the R and P branches.

13-14. The following data are obtained for the vibration-rotation spectrum of $H^{79}Br$. Determine $\tilde{B}_0$, $\tilde{B}_1$, $\tilde{B}_e$, and $\tilde{\alpha}_e$ from these data.

Line	Frequency/cm^{-1}
$R(0)$	2642.60
$R(1)$	2658.36
$P(1)$	2609.67
$P(2)$	2592.51

13-15. The following lines were observed in the microwave absorption spectrum of $H^{127}I$ and $D^{127}I$ between 60 cm^{-1} and 90 cm^{-1}.

$$\tilde{\nu}/\text{cm}^{-1}$$

$H^{127}I$	64.275	77.130	89.985	
$D^{127}I$	65.070	71.577	78.084	84.591

Use the rigid-rotator approximation to determine the values of $\tilde{B}$, I, and R_e for each molecule. Do your results for the bond length agree with what you would expect based upon the Born-Oppenheimer approximation? Take the mass of ^{127}I to be 126.904 amu and the mass of D to be 2.014 amu.

13-16. The following spectroscopic constants were determined for pure samples of $^{74}Ge^{32}S$ and $^{72}Ge^{32}S$:

Molecule	B_e/MHz	α_e/MHz	D/kHz	$R_e(v = 0)$/pm
$^{74}Ge^{32}S$	5593.08	22.44	2.349	0.201 20
$^{72}Ge^{32}S$	5640.06	22.74	2.388	0.201 20

Determine the frequency of the $J = 0$ to $J = 1$ transition for $^{74}Ge^{32}S$ and $^{72}Ge^{32}S$ in their ground vibrational states. The width of a microwave absorption line is on the order of 1 kHz. Could you distinguish a pure sample of $^{74}Ge^{32}S$ from a 50/50 mixture of $^{74}Ge^{32}S$ and $^{72}Ge^{32}S$ using microwave spectroscopy?

13-17. The frequencies of the rotational transitions in the nonrigid-rotator approximation are given by Equation 13.19. Show how both $\tilde{B}$ and $\tilde{D}$ may be obtained by curve fitting $\tilde{\nu}$ to Equation 13.19. Use this method and the data in Table 13.3 to determine both $\tilde{B}$ and $\tilde{D}$ for $H^{35}Cl$.

13-18. The following data are obtained in the microwave spectrum of $^{12}C^{16}O$. Use the method of Problem 13–17 to determine the values of $\tilde{B}$ and $\tilde{D}$ from these data.

Transitions	Frequency/cm^{-1}
$0 \rightarrow 1$	3.845 40
$1 \rightarrow 2$	7.690 60
$2 \rightarrow 3$	11.535 50
$3 \rightarrow 4$	15.379 90
$4 \rightarrow 5$	19.223 80
$5 \rightarrow 6$	23.066 85

13-19. Using the parameters given in Table 13.2, calculate the frequencies (in cm^{-1}) of the $0 \rightarrow 1$, $1 \rightarrow 2$, $2 \rightarrow 3$, and $3 \rightarrow 4$ rotational transitions in the ground vibrational state of $H^{35}Cl$ in the nonrigid-rotator approximation.

13-20. The vibrational term of a diatomic molecule is given by

$$G(v) = \left(v + \tfrac{1}{2}\right)\tilde{\nu}_e - \left(v + \tfrac{1}{2}\right)^2 \tilde{x}_e\tilde{\nu}_e$$

where v is the vibrational quantum number. Show that the spacing between the adjacent levels ΔG is given by

$$\Delta G = G(v + 1) - G(v) = \tilde{\nu}_e\{1 - 2\tilde{x}_e(v + 1)\} \qquad (1)$$

The diatomic molecule dissociates in the limit that $\Delta G \rightarrow 0$. Show that the maximum vibrational quantum number, v_{max}, is given by

$$v_{max} = \frac{1}{2\tilde{x}_e} - 1$$

Use this result to show that the dissociation energy $\tilde{D}_e$ of the diatomic molecule can be written as

$$\tilde{D}_e = \frac{\tilde{\nu}_e(1 - \tilde{x}_e^2)}{4\tilde{x}_e} \approx \frac{\tilde{\nu}_e}{4\tilde{x}_e} \qquad (2)$$

Referring to Equation 1, explain how the constants $\tilde{\nu}_e$ and $\tilde{x}_e$ can be evaluated from a plot of ΔG versus $v + 1$. This type of plot is called a *Birge-Sponer plot*. Once the values of $\tilde{\nu}_e$ and $\tilde{x}_e$ are known, Equation 2 can be used to determine the dissociation energy of the molecule. Use the following experimental data for H_2 to calculate the dissociation energy, $\tilde{D}_e$.

v	$G(v)/\text{cm}^{-1}$	v	$G(v)/\text{cm}^{-1}$
0	4161.12	7	26 830.97
1	8087.11	8	29 123.93
2	11 782.35	9	31 150.19
3	15 250.36	10	32 886.85
4	18 497.92	11	34 301.83
5	21 505.65	12	35 351.01
6	24 287.83	13	35 972.97

Explain why your Birge-Sponer plot is not linear for high values of v. How does the value of $\tilde{D}_e$ obtained from the Birge-Sponer analysis compare with the experimental value of 38 269.48 cm^{-1}?

13-21. An analysis of the vibrational spectrum of the ground-state homonuclear diatomic molecule C_2 gives $\tilde{v}_e = 1854.71$ cm^{-1} and $\tilde{v}_e \tilde{x}_e = 13.34$ cm^{-1}. Suggest an experimental method that can be used to determine these spectroscopic parameters. Use the expression derived in Problem 13–20 to determine the number of bound vibrational levels for the ground state of C_2.

13-22. A simple function that is a good representation of an internuclear potential is the Morse potential,

$$U(q) = D_e(1 - e^{-\beta q})^2$$

where q is $R - R_e$. Show that the force constant calculated for a Morse potential is given by

$$k = 2D_e \beta^2$$

Given that $D_e = 7.31 \times 10^{-19}$ J·molecule^{-1} and $\beta = 1.83 \times 10^{10}$ m^{-1} for HCl, calculate the value of k.

13-23. The Morse potential is presented in Problem 13–22. Given that $D_e = 8.19 \times 10^{-19}$ J·molecule^{-1}, $\tilde{v}_e = 1580.0$ cm^{-1}, and $R_e = 121$ pm for $^{16}O_2$, plot a Morse potential for $^{16}O_2$. Plot the corresponding harmonic-oscillator potential on the same graph.

13-24. The fundamental line in the infrared spectrum of $^{12}C^{16}O$ occurs at 2143.0 cm^{-1}, and the first overtone occurs at 4260.0 cm^{-1}. Calculate the values of $\tilde{v}_e$ and $\tilde{x}_e \tilde{v}_e$ for $^{12}C^{16}O$.

13-25. Using the parameters given in Table 13.2, calculate the fundamental and the first three overtones of $H^{79}Br$.

13-26. The frequencies of the vibrational transitions in the anharmonic-oscillator approximation are given by Equation 13.22. Show how the values of both $\tilde{v}_e$ and $\tilde{x}_e \tilde{v}_e$ may be obtained by plotting $\tilde{v}_{obs}/v$ versus $(v + 1)$. Use this method and the data in Table 13.4 to determine the values $\tilde{v}_e$ and $\tilde{x}_e \tilde{v}_e$ for $H^{35}Cl$.

13-27. The following data are obtained from the infrared spectrum of $^{127}I^{35}Cl$. Using the method of Problem 13–26, determine the values of $\tilde{\nu}_e$ and $\tilde{x}_e\tilde{\nu}_e$ from these data.

Transitions	Frequency/cm^{-1}
$0 \rightarrow 1$	381.20
$0 \rightarrow 2$	759.60
$0 \rightarrow 3$	1135.00
$0 \rightarrow 4$	1507.40
$0 \rightarrow 5$	1877.00

13-28. The values of $\tilde{\nu}_e$ and $\tilde{x}_e\tilde{\nu}_e$ of $^{12}C^{16}O$ are 2169.81 cm^{-1} and 13.29 cm^{-1} in the ground electronic state and 1514.10 cm^{-1} and 17.40 cm^{-1} in the first excited electronic state. If the $0 \rightarrow 0$ vibronic transition occurs at $6.475\,15 \times 10^4$ cm^{-1}, calculate the value of $\tilde{T}_e = \tilde{\nu}'_{el} - \tilde{\nu}''_{el}$, the energy difference between the minima of the potential curves of the two electronic states.

13-29. Given the following parameters for $^{12}C^{16}O$: $\tilde{T}_e = 6.508\,043 \times 10^4$ cm^{-1}, $\tilde{\nu}'_e = 1514.10$ cm^{-1}, $\tilde{x}'_e\tilde{\nu}'_e = 17.40$ cm^{-1}, $\tilde{\nu}''_e = 2169.81$ cm^{-1}, and $\tilde{x}''_e\tilde{\nu}''_e = 13.29$ cm^{-1}, construct to scale an energy-level diagram of the first two electronic states, showing the first four vibrational states in each electronic state. Indicate the allowed transitions from $v'' = 0$, and calculate the frequencies of these transitions. Also, calculate the zero-point vibrational energy in each electronic state.

13-30. An analysis of the rotational spectrum of $^{12}C^{32}S$ gives the following results:

v	0	1	2	3
$\tilde{B}_v$/cm^{-1}	0.817 08	0.811 16	0.805 24	0.799 32

Determine the values of $\tilde{B}_e$ and $\tilde{\alpha}_e$ from these data.

13-31. The frequencies of the first few vibronic transitions to an excited state of BeO are as follows:

Vibronic transitions	$0 \rightarrow 2$	$0 \rightarrow 3$	$0 \rightarrow 4$	$0 \rightarrow 5$
$\tilde{\nu}_{obs}$/cm^{-1}	12 569.95	13 648.43	14 710.85	15 757.50

Use these data to calculate the values of $\tilde{\nu}_e$ and $\tilde{x}_e\tilde{\nu}_e$ for the excited state of BeO.

13-32. The frequencies of the first few vibronic transitions to an excited state of $^7Li_2$ are as follows:

Vibronic transitions	$0 \rightarrow 0$	$0 \rightarrow 1$	$0 \rightarrow 2$	$0 \rightarrow 3$	$0 \rightarrow 4$	$0 \rightarrow 5$
$\tilde{\nu}_{obs}$/cm^{-1}	14 020	14 279	14 541	14 805	15 074	15 345

Use these data to calculate the values of $\tilde{\nu}_e$ and $\tilde{x}_e\tilde{\nu}_e$ for the excited state of $^7Li_2$.

13-33. Determine the number of translational, rotational, and vibrational degrees of freedom in

    a. CH_3Cl **b.** OCS

    c. C_6H_6 **d.** H_2CO

13-34. Determine which of the following molecules will exhibit a microwave rotational absorption spectrum: H_2, HCl, CH_4, CH_3I, H_2O, and SF_6.

13-35. Classify each of the following molecules as a spherical, a symmetric, or an asymmetric top: CH_3Cl, CCl_4, SO_2, and SiH_4.

13-36. Classify each of the following molecules as either a prolate or an oblate symmetric top: FCH_3, $HCCl_3$, PF_3, and CH_3CCH.

13-37. Show that the components of the moment of inertia of the trigonal planar molecule shown below are $I_{xx} = I_{yy} = 3m/2$ and $I_{zz} = 3m$ if all the masses are m units, all the bond lengths are unit length, and all the bond angles are $120°$.

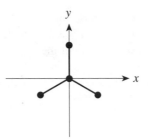

13-38. This problem illustrates how the principal moments of inertia can be obtained as an eigenvalue problem. We will work in two dimensions for simplicity. Consider the "molecule" represented below,

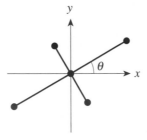

where all the masses are unit masses and the long and short bond lengths are 2 and 1, respectively. Show that

$$I_{xx} = 2\cos^2\theta + 8\sin^2\theta$$

$$I_{yy} = 8\cos^2\theta + 2\sin^2\theta$$

$$I_{xy} = -6\cos\theta\sin\theta$$

The fact that $I_{xy} \neq 0$ indicates that these I_{ij} are not the principal moments of inertia. Now solve the secular determinantal equation for λ

$$\begin{vmatrix} I_{xx} - \lambda & I_{xy} \\ I_{xy} & I_{yy} - \lambda \end{vmatrix} = 0$$

and compare your result with the values of I_{xx} and I_{yy} that you would obtain if you align the "molecule" and the coordinate system such that $\theta = 90°$. What does this comparison tell you? What are the values of I_{xx} and I_{yy} if $\theta = 0°$?

13-39. Sketch an energy-level diagram for a prolate symmetric top and an oblate symmetric top. How do they differ? Indicate some of the allowed transitions in each case.

13-40. Derive Equation 13.57 from Equation 13.55.

13-41. Show that the first few associated Legendre functions satisfy the recursion formula given by Equation 13.62.

13-42. Calculate the ratio of the dipole transition moments for the $0 \rightarrow 1$ and $1 \rightarrow 2$ rotational transitions in the rigid-rotator approximation.

13-43. Calculate the ratio of the dipole transition moments for the $0 \rightarrow 1$ and $1 \rightarrow 2$ vibrational transitions in the harmonic-oscillator approximation.

13-44. Use Table 13.7 to determine the 12-dimensional reducible representation for the vibrational motion of NH_3. Use this result to determine the symmetries and the infrared activity of the normal coordinates of NH_3.

13-45. Use Table 13.7 to determine the 15-dimensional reducible representation for the vibrational motion of CH_2Cl_2. Use this result to determine the symmetries and the infrared activity of the normal coordinates of CH_2Cl_2.

13-46. Use Table 13.7 to determine the 18-dimensional reducible representation for the vibrational motion of *trans*-dichloroethene. Use this result to determine the symmetries and the infrared activity of the normal coordinates of *trans*-dichloroethene.

13-47. Use Table 13.7 to determine the 15-dimensional reducible representation for the vibrational motion of XeF_4 (square planar). Use this result to determine the symmetries and the infrared activity of the normal coordinates of XeF_4.

13-48. Use Table 13.7 to determine the 15-dimensional reducible representation for the vibrational motion of CH_4. Use this result to determine the symmetries and the infrared activity of the normal coordinates of CH_4.

13-49. Consider a molecule with a dipole moment μ in an electric field E. We picture the dipole moment as a positive charge and a negative charge of magnitude q separated by a vector $\mathbf{l}$.

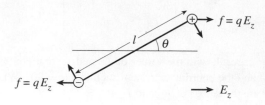

The field E causes the dipole to rotate into a direction parallel to E. Therefore, work is required to rotate the dipole to an angle θ to E. The force causing the molecule to rotate is actually a torque (torque is the angular analog of force) and is given by $l/2$ times the force perpendicular to $\mathbf{l}$ at each end of the vector $\mathbf{l}$. Show that this torque is equal to $\mu E \sin \theta$ and that the energy required to rotate the dipole from some initial angle θ_0 to some arbitrary angle θ is

$$V = \int_{\theta_0}^{\theta} \mu E \sin \theta' d\theta'$$

Given that θ_0 is customarily taken to be $\pi/2$, show that

$$V = -\mu E \cos \theta = -\mu \cdot E$$

The magnetic analog of this result will be given by Equation 14.10.

13-50. The observed vibrational-rotational lines for the $v = 0$ to $v = 1$ transition of $^{12}C^{16}O(g)$ are listed below. Determine $\tilde{B}_0$, $\tilde{B}_1$, $\tilde{B}_e$, $\tilde{\alpha}_e$, $\tilde{I}_e$, and r_e.

2238.89	2215.66	2189.84	2161.83	2127.61	2094.69	2059.79
2236.06	2212.46	2186.47	2158.13	2123.62	2090.56	2055.31
2233.34	2209.31	2183.14	2154.44	2119.64	2086.27	2050.72
2230.49	2206.19	2179.57	2150.83	2115.56	2081.95	2046.14
2227.55	2202.96	2176.12	2147.05	2111.48	2077.57	
2224.63	2199.77	2172.63	2139.32	2107.33	2073.19	
2221.56	2196.53	2169.05	2135.48	2103.12	2068.69	
2218.67	2193.19	2165.44	2131.49	2099.01	2064.34	

[*Hint:* Recall that the transition $(v'' = 0, J'' = 0) \rightarrow (v'' = 1, J'' = 0)$ is forbidden.]

13-51. This problem is a three-dimensional version of Problem 13–41. The rotational spectrum of a polyatomic molecule can be predicted once the values of $\tilde{A}$, $\tilde{B}$, and $\tilde{C}$ are known. These, in turn, can be calculated from the principal moments of inertia I_A, I_B, and I_C. In this problem, we show how I_A, I_B, and I_C can be determined from the molecular geometry. We set up an arbitrarily oriented coordinate system, whose origin sits at the center-of-mass of the molecule, and determine the moments of inertia I_{xx}, I_{xy}, I_{xz}, I_{yy}, I_{yz}, and I_{zz}. The principal moments of inertia are the solution to the secular determinantal equation

$$
\begin{vmatrix}
I_{xx} - \lambda & I_{xy} & I_{xz} \\
I_{xy} & I_{yy} - \lambda & I_{yz} \\
I_{xz} & I_{yz} & I_{zz} - \lambda
\end{vmatrix} = 0
$$

The assignment for the subscripts A, B, and C to the three roots of this determinant are done according to the convention $I_A \leq I_B \leq I_C$. Use this approach to find the principal moments of inertia for the planar formate radical, HCO_2, given the following geometry:

The H–C bond length is 109.7 pm, the C=O bond length is 120.2 pm, and the C–O bond length is 134.3 pm.

Richard R. Ernst was born in Winterthur, Switzerland, on August 14, 1933. In 1962, he received his Ph.D. in technical sciences from the Swiss Federal Institute of Technology (ETH) in Zurich; his dissertation was on high-resolution nuclear magnetic resonance (NMR). Ernst then joined Varian Associates in Palo Alto, California, from 1963 to 1968. In 1968, he returned to ETH in Zurich, where he became director of research and professor of physical chemistry, a position he still holds. While at Varian, Ernst developed Fourier transform NMR spectroscopy. The basic idea of FTNMR is to perturb many different nuclei at once and to sort out their individual responses by a mathematical technique called Fourier transformation. This procedure leads to greatly enhanced sensitivity of NMR signals. While in Switzerland, Ernst pioneered the development of two-dimensional NMR, which turns out to be enormously important for determining the structure of biological molecules. In his youth, he played the cello in a small orchestra but never became very serious with it, although he toyed with the idea of becoming a composer. Over the years, he has become interested in Asian art, particularly Tibetan art. Ernst was awarded the Nobel Prize for chemistry in 1991 "for his contribution to the development of the methodology of high resolution NMR spectroscopy."

CHAPTER 14

Nuclear Magnetic Resonance Spectroscopy

Certainly one of the most important spectroscopic techniques is nuclear magnetic resonance (NMR) spectroscopy, particularly to organic chemists and biochemists. Hardly a chemical laboratory in the world does not have at least one NMR spectrometer. You may have learned about the application of NMR to the determination of the structures of organic molecules in your course in organic chemistry. In this chapter, we will study NMR spectroscopy in a fairly quantitative manner, using the principles of quantum mechanics developed in previous chapters. NMR spectroscopy involves transitions of the orientations of nuclear spins in magnetic fields. Consequently, in this chapter we will examine the quantum-mechanical states of nuclear spins interacting with magnetic fields, and learn how we can induce transitions between these states when we irradiate the nuclei with electromagnetic radiation. We will focus exclusively on magnetic resonance that results from transitions involving the protons in hydrogen atoms. First, we will discuss the properties of isolated nuclei in magnetic fields, and then we will show how the chemical or the electronic environment in a molecule can affect the energies of hydrogen nuclei (protons) in external magnetic fields. This discussion will lead us to simple NMR spectra in which hydrogen nuclei in different chemical or electronic environments yield characteristic absorption frequencies in NMR experiments. Last, we will see how these spectra are modifed under high resolution to give information not only about the electronic environment about a given nucleus, but also about the arrangements of its neighboring hydrogen atoms.

14–1. Nuclei Have Intrinsic Spin Angular Momenta

We learned in Section 8–4 that an electron has an intrinsic spin angular momentum, whose z components are equal to $\pm\hbar/2$, or that it has a spin of 1/2, with z components $\pm 1/2$. We defined two spin functions, $\alpha(\sigma)$ and $\beta(\sigma)$, where σ is a spin variable, which satisfy the eigenvalue equations

$$\hat{S}^2\alpha = \tfrac{1}{2}\left(\tfrac{1}{2}+1\right)\hbar^2\alpha \qquad \hat{S}^2\beta = \tfrac{1}{2}\left(\tfrac{1}{2}+1\right)\hbar^2\beta$$
$$\hat{S}_z\alpha = \tfrac{1}{2}\hbar\alpha \qquad\qquad \hat{S}_z\beta = -\tfrac{1}{2}\hbar\beta \tag{14.1}$$

We associated α with $s_z = \hbar/2$ and β with $s_z = -\hbar/2$. We expressed the orthonormality of α and β formally by the equations

$$\int \alpha^*(\sigma)\alpha(\sigma)d\sigma = \int \beta^*(\sigma)\beta(\sigma)d\sigma = 1$$
$$\int \alpha^*(\sigma)\beta(\sigma)d\sigma = \int \alpha(\sigma)\beta^*(\sigma)d\sigma = 0 \tag{14.2}$$

Because an electron is charged, the intrinsic spin confers the property of a magnetic dipole to an electron. In other words, because of its spin, an electron acts like a magnet when it is placed in a magnetic field. Nuclei also have an intrinsic spin angular momentum (which we designate by I) and an associated magnetic dipole. Unlike electrons, the spins of nuclei are not restricted to 1/2. The commonly occurring nuclei ^{12}C and ^{16}O have a spin of 0, a proton (^1H) and ^{19}F have a spin of 1/2, and a deuteron (^2H) and ^{14}N have a spin of 1. Table 14.1 lists the spins and other properties of some nuclei important in NMR spectroscopy. Because essentially all organic compounds contain hydrogen, for simplicity, we will focus almost exclusively on protons, with their spin of 1/2, in this chapter. The nuclear spin eigenvalue equations for protons analogous to Equations 14.1 for electrons are

$$\hat{I}^2\alpha = \tfrac{1}{2}\left(\tfrac{1}{2}+1\right)\hbar^2\alpha \qquad \hat{I}^2\beta = \tfrac{1}{2}\left(\tfrac{1}{2}+1\right)\hbar^2\beta \tag{14.3a}$$
$$\hat{I}_z\alpha = \tfrac{1}{2}\hbar\alpha \qquad\qquad \hat{I}_z\beta = -\tfrac{1}{2}\hbar\beta \tag{14.3b}$$

TABLE 14.1
The properties of some common nuclei used in NMR experiments.

Nucleus	Spin	Nuclear g factor	Magnetic moment (in nuclear magnetons)	Magnetogyric ratio $\gamma/10^7$ rad·T^{-1}·s^{-1}
^1H	1/2	5.5854	2.7928	26.7522
^2H	1	0.8574	0.8574	4.1066
^{13}C	1/2	1.4042	0.7021	6.7283
^{14}N	1	0.4036	0.4036	1.9338
^{31}P	1/2	2.2610	1.1305	10.841

and the nuclear spin functions satisfy orthonormality conditions equivalent to Equations 14.2

As we stated above, a charged particle with a nonzero spin acts as a magnetic dipole and consequently will interact with a magnetic field. Let's look into this concept more closely. Recall from your course in physics that the motion of an electric charge around a closed loop produces a magnetic dipole, μ, (Figure 14.1) whose magnitude is given by

$$\mu = iA \tag{14.4}$$

where i is the current in amperes (coulombs per second) and A is the area of the loop in square meters. Note that the SI units of magnetic dipole are ampere·meter^2 (A·m^2). If we consider a circular loop for simplicity, then

$$i = \frac{qv}{2\pi r} \tag{14.5}$$

where v is the velocity of the charge q and r is the radius of the circle. Substituting Equation 14.5 and $A = \pi r^2$ into Equation 14.4 gives

$$\mu = \frac{qrv}{2} \tag{14.6}$$

More generally, if the orbit is not circular, then Equation 14.6 becomes (see Math-Chapter C)

$$\mu = \frac{q(\mathbf{r} \times \mathbf{v})}{2} \tag{14.7}$$

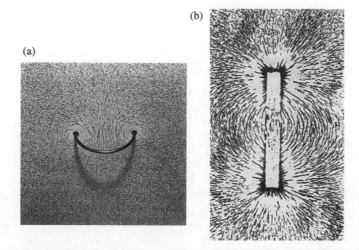

FIGURE 14.1
(a) Iron filings sprinkled around a loop carrying an electric current show the spatial distribution of the magnetic field produced by the current loop. This field is very similar to the field produced by a bar magnet (b).

Equation 14.7 says that μ is perpendicular to the plane formed by **r** and **v** (the plane of motion). Problem 14–1 has you show that Equation 14.7 reduces to Equation 14.6 for the case of a circular orbit.

We can express μ in terms of angular momentum by using the fact that $\mathbf{L} = \mathbf{r} \times \mathbf{p}$ and that $\mathbf{p} = m\mathbf{v}$, so that Equation 14.7 becomes

$$\mu = \frac{q}{2m}\mathbf{L} \tag{14.8}$$

Equation 14.8 says that the magnetic moment μ is proportional to the angular momentum **L**.

Of course, a nucleus is not a circular current-carrying loop, but Equation 14.8 can still be applied to a nucleus by replacing the classical angular momentum **L** by the spin angular momentum **I** and writing

$$\mu = g_{\mathrm{N}}\frac{q}{2m_{\mathrm{N}}}\mathbf{I} = g_{\mathrm{N}}\beta_{\mathrm{N}}\mathbf{I} = \gamma\mathbf{I} \tag{14.9}$$

where g_{N} is the *nuclear g factor*, β_{N} is the *nuclear magneton* $(q/2m_{\mathrm{N}})$, m_{N} is the mass of the nucleus, and $\gamma = g_{\mathrm{N}}\beta_{\mathrm{N}}$ is the *magnetogyric ratio*. The nuclear g factor is a unitless constant whose magnitude is on the order of unity and is characteristic of each nucleus. The magnetogyric ratio is also a characteristic quantity for each nucleus. The detection sensitivity of a particular type of nucleus in an NMR experiment depends upon the value of γ. The larger the value of γ, the easier it is to observe the nucleus. Some nuclear g factors and magnetogyric ratios are given in Table 14.1.

14–2. Magnetic Moments Interact with Magnetic Fields

A magnetic dipole will tend to align itself in a magnetic field, and its potential energy will be given by (see Problem 13–49)

$$V = -\mu \cdot \mathbf{B} \tag{14.10}$$

where **B** is the strength of the magnetic field. The quantity **B** is defined through the equation

$$\mathbf{F} = q(\mathbf{v} \times \mathbf{B}) \tag{14.11}$$

where **F** is the force acting upon a charge q moving with a velocity **v** in a magnetic field of strength **B**. The SI unit of magnetic field strength is tesla (T). From Equation 14.11, we see that one tesla is equal to one newton/ampere·meter $(\mathrm{N \cdot A^{-1} \cdot m^{-1}})$.

EXAMPLE 14–1
Show that an ampere·meter^2 $(\mathrm{A \cdot m^2})$ is equal to a joule·tesla^{-1} $(\mathrm{J \cdot T^{-1}})$.

SOLUTION: Equation 14.4 shows that the units of a magnetic dipole moment are

$$\mu = A \cdot m^2$$

Equation 14.10 gives the units

$$J = (A \cdot m^2) \cdot T$$

where T stands for tesla, the unit of the intensity of a magnetic field. Therefore, we see that

$$A \cdot m^2 = J \cdot T^{-1}$$

or that the units of μ are either $A \cdot m^2$ or $J \cdot T^{-1}$.

Although tesla is the SI unit of magnetic field strength, another unit, called a *gauss*, is so commonly used in NMR that we will use it frequently. The relation between a gauss (G) and a tesla (T) is $1\ G = 10^{-4}\ T$. Table 14.2 lists some magnetic field strengths to give an idea of some typical values found in nature and in the laboratory.

If, as usual, we take the magnetic field to be in the z direction, Equation 14.10 becomes

$$V = -\mu_z B_z \tag{14.12}$$

Using Equation 14.9 for μ_z, we have

$$V = -\gamma B_z I_z \tag{14.13}$$

If we replace I_z by its operator equivalent $\hat{I}_z$, then Equation 14.13 gives the Hamiltonian operator that accounts for the interaction of the nucleus with the external magnetic field. Thus, we write the spin Hamiltonian operator of a single isolated nucleus as

$$\hat{H} = -\gamma B_z \hat{I}_z \tag{14.14}$$

The corresponding Schrödinger equation for the nuclear spin is

$$\hat{H}\psi = -\gamma B_z \hat{I}_z \psi = E\psi \tag{14.15}$$

The wave functions in this case are the spin eigenfunctions, so $\hat{I}_z \psi_I = \hbar m_I \psi_I$, where as usual $m_I = I,\ I-1,\ \ldots,\ -I$. Therefore, Equation 14.15 gives

$$E = -\hbar \gamma m_I B_z \tag{14.16}$$

We can use Equation 14.16 to calculate the difference in energy between a proton aligned with a magnetic field and one aligned against it. The energy of a proton aligned with or against a magnetic field is given by Equation 14.16 with $m_I = +1/2$ or $m_I = -1/2$, respectively. Thus, the energy difference is given by

$$\Delta E = E(m_I = -1/2) - E(m_I = 1/2) = \hbar \gamma B_z \tag{14.17}$$

TABLE 14.2

Some approximate magnetic field strengths in units of teslas and gauss.

Source	B/T	B/G
Surface of a pulsar	10^8	10^{12}
Maximum achieved in laboratory		
Transient	10^3	10^7
Steady	30	300 000
Superconducting magnet	15	150 000
Electromagnet	2	20 000
Small bar magnet	0.01	100
Near household wiring	10^{-4}	1
Surface of the Earth	5×10^{-5}	0.5

Note that ΔE depends linearly on the strength of the magnetic field. Figure 14.2 shows ΔE as a function of B_z for a spin 1/2 nucleus. If a proton aligned with an applied magnetic field is irradiated with electromagnetic radiation of frequency given by $\Delta E = \hbar \gamma B_z = h\nu = \hbar\omega$, the radiation will cause the proton to make a transition from the lower energy state ($m_I = 1/2$) to the higher energy state ($m_I = -1/2$). For a field of 21 100 gauss (2.11 T), the energy difference for protons is

$$\Delta E = (1.054 \times 10^{-34} \text{ J·s·rad}^{-1})(26.7522 \times 10^7 \text{ rad·T}^{-1}\text{·s}^{-1})(2.11 \text{ T})$$
$$= 5.95 \times 10^{-26} \text{ J}$$

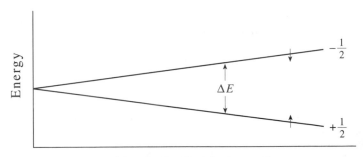

FIGURE 14.2

The relative energies of a spin 1/2 nucleus in a magnetic field. In the state of lower energy, the nucleus is aligned with the field ($m_I = +1/2$), and in the state of higher energy, it is aligned against the field ($m_I = -1/2$). The magnitude of the energy difference depends upon the strength of the magnetic field.

Using the relation $\Delta E = h\nu$, this result corresponds to a frequency of 90 MHz, which is in the radiofrequency region. Generally, the frequency associated with a transition of one aligned nuclear spin state to another for a spin $+1/2$ nucleus is given by

$$\nu = \frac{\gamma B_z}{2\pi} \qquad \text{(Hz)} \qquad (14.18)$$

or

$$\omega = \gamma B_z \qquad (\text{rad} \cdot \text{s}^{-1}) \qquad (14.19)$$

EXAMPLE 14–2

What magnetic field strength must be applied to a free proton for spin transitions to occur at 60.0 MHz?

SOLUTION: According to Table 14.1, $\gamma = 26.7522 \times 10^7 \text{ rad} \cdot \text{T}^{-1} \cdot \text{s}^{-1}$ for ^1H, so

$$B_z = \frac{2\pi\nu}{\gamma} = \frac{\omega}{\gamma} = \frac{(2\pi \text{ rad})(60.0 \times 10^6 \text{ s}^{-1})}{26.7522 \times 10^7 \text{ rad} \cdot \text{T}^{-1} \cdot \text{s}^{-1}}$$
$$= 1.41 \text{ T} = 14\,100 \text{ G}$$

Figure 14.3 gives the frequency ν as a function of magnetic field strength B_z for (free) proton spin transitions.

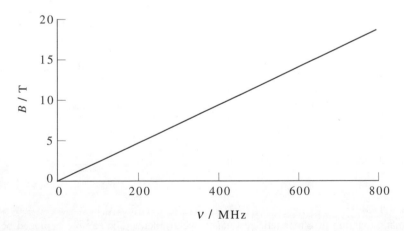

FIGURE 14.3
The frequencies that induce (free) proton spin transitions as a function of the strength of the magnetic field, according to Equation 14.18. Commercial NMR spectrometers operate at frequencies of 60 MHz, 90 MHz, 250 MHz, 270 MHz, 300 MHz, 500 MHz, 600 MHz, and 750 MHz.

14–3. Proton NMR Spectrometers Operate at Frequencies Between 60 MHz and 750 MHz

According to Equation 14.18, the *resonance frequency* of a proton (the frequency at which a spin-state transition will occur) in a magnetic field is directly proportional to the strength of the magnetic field. Thus, for a fixed magnetic field strength, we can vary the frequency of the electromagnetic radiation until absorption occurs, or conversely, we can fix the frequency of the radiation and vary the strength of the magnetic field. Early spectrometers used magnets that produced fields at 14 100 G (1.41 T), which sets the frequency of proton transitions at about 60 MHz. (See Example 14–2.) Newer spectrometers, however, using superconducting magnets, operate at frequencies as high as 750 MHz. We will see later that higher operating frequencies (or higher field strengths) give greater resolution than lower operating frequencies and hence greatly simplify the interpretation of NMR spectra.

The basic elements of a proton NMR spectrometer are diagrammed in Figure 14.4. A compound containing hydrogen is placed between the poles of a strong electromagnet whose field strength can be varied by varying the current through the wires wrapped around the electromagnet. The sample is irradiated by radio-frequency radiation, and the amount absorbed by the sample is detected and recorded. When the magnetic field strength is such that the energy difference between the two nuclear spin states is the same as the energy of the radio frequency radiation, then protons make transitions from

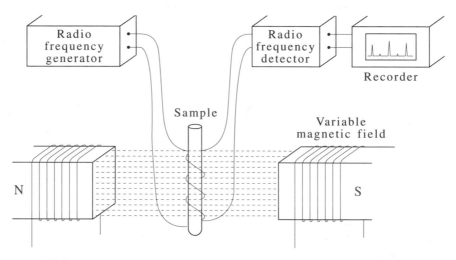

FIGURE 14.4
A schematic diagram of a magnetic resonance spectrometer. The sample is placed between the poles of an electromagnet, whose strength can be varied by varying the electric current through the coils wrapped around the magnet. The sample is irradiated by a fixed frequency of radio-frequency radiation. The amount of radiation absorbed by the sample is measured by a radio-frequency detector, whose output is fed into a recorder. The magnetic field strength is varied, and the radio-frequency radiation absorbed by the sample is measured and plotted versus the magnetic field strength by the recorder. The result is an NMR spectrum.

one spin state to another, and the radiation is absorbed by the sample as illustrated in Figure 14.5.

Although NMR spectra may be recorded by either varying the magnetic field strength at fixed frequency or by varying the frequency at fixed magnetic field strength, the spectra obtained are indistinguishable. It is standard practice to calibrate NMR spectra in hertz (Hz), as if the frequency had been varied at constant magnetic field strength, and furthermore to present spectra with the strength of the magnetic field increasing from left to right. The NMR spectrum of iodomethane (CH_3I) is shown in Figure 14.6. The strong peak, or signal, in this spectrum reflects absorption by the three equivalent hydrogen nuclei in iodomethane. We will discuss both the top and bottom scales in this and other NMR spectra in the following section, but note that the top scale is calibrated in Hz and the bottom scale is unitless.

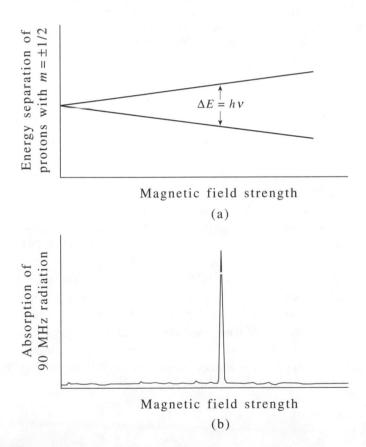

FIGURE 14.5

The energy separation of a proton aligned with or against an applied magnetic field increases with the strength of the magnetic field as shown in (a). When the strength of the magnetic field is such that the separation matches the energy of the radio frequency radiation (say 90 MHz) that irradiates the sample, the sample will absorb the radiation and give the NMR spectrum shown in (b). The condition for absorption, or resonance, is $\Delta E = \hbar \gamma B_z = h\nu$.

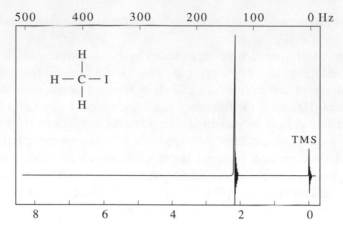

FIGURE 14.6
An NMR spectrum of iodomethane taken on a 60-MHz spectrometer. There is a strong signal at 130 Hz on the top horizontal axis and 2.16 on the lower horizontal axis. This signal reflects the absorption by the three equivalent hydrogen nuclei in iodomethane. The small signal at 0 on the horizontal axis is a reference peak and can be ignored for now.

14–4. The Magnetic Field Acting upon Nuclei in Molecules Is Shielded

In Section 14–3, we showed that the two spin states of a spin 1/2 nucleus such as a proton have different energies in a magnetic field, and that the frequency associated with a transition from one state to another is given by Equation 14.18, $v = \gamma B_z/2\pi$. According to this equation, all the hydrogen nuclei in a molecule absorb at the same frequency. If this were the case, then NMR spectroscopy would be little more than an expensive technique for testing for the presence of hydrogen.

The quantity B_z in Equation 14.18 is the magnetic field the nucleus experiences. For an isolated or bare nucleus, this field is just the external magnetic field. But a nucleus in a molecule is surrounded by electrons, and the applied magnetic field causes a circulatory motion of these electrons, which in turn generates an additional small magnetic field at the nucleus. For most substances, this electronically generated magnetic field, B_{elec}, opposes the applied magnetic field. It turns out that the magnitude of B_{elec} is proportional to the applied field, so we can write

$$B_{elec} = -\sigma B_0 \tag{14.20}$$

where B_0 is the applied magnetic field (assumed to be in the z direction) and σ is a (unitless) proportionality constant. The negative sign in Equation 14.20 accounts for the fact that B_{elec} opposes B_0. The electrons effectively shield the nucleus from B_0, so σ is called a *shielding constant*. Typical values of shielding constants for hydrogen nuclei in organic compounds are approximately 10^{-5}.

An important property of a shielding constant is that its value depends upon the electronic or chemical environment around the nucleus. Therefore, the two sets of chemically equivalent hydrogen nuclei in a molecule such as methyl formate ($HCOOCH_3$) experience different local fields.

The total magnetic field that any nucleus experiences is the sum of the applied, external field, B_0, and the shielding field, $B_{elec} = -\sigma B_0$, so that the total field is given by $B_z = (1 - \sigma)B_0$. If we substitute this expression into Equation 14.18 or 14.19, we see that the frequency (at fixed magnetic field strength) or the field strength (at fixed frequency) at which a nucleus will undergo a spin transition is given by

$$B_0 = \frac{2\pi v}{\gamma(1 - \sigma)} = \frac{\omega}{\gamma(1 - \sigma)} \tag{14.21}$$

Equation 14.21 shows that the field strength at which a nuclear spin transition will take place depends upon σ, which in turn depends upon the chemical environment of the nucleus. Thus, in iodomethane, with its three chemically equivalent hydrogen nuclei, the NMR spectrum (Figure 14.6) has only one absorption line, whereas in methyl formate, with hydrogen nuclei in two different chemical or electronic environments, there are two lines (Figure 14.7).

The spectra in Figures 14.6 and 14.7 show a relatively small peak at the right side, at the zero position on both the top and the bottom scales. This peak at zero is due to a small amount of added tetramethylsilane, $Si(CH_3)_4$ (TMS), which is used as an internal reference or standard. Tetramethylsilane is used because it has 12 equivalent hydrogen atoms and is relatively nonreactive. Furthermore, the hydrogen atoms in most organic compounds absorb at fields smaller than does TMS, or downfield from TMS, and so the

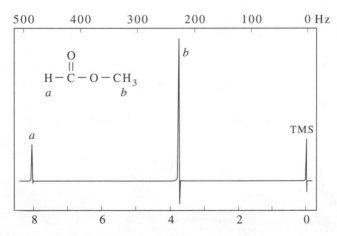

FIGURE 14.7
An NMR spectrum of methyl formate taken on a 60-MHz spectrometer. The small signal at 0 on the horizontal scale is simply a reference signal and can be ignored for now. The signals at 3.6 and 8.1 on the lower scale reflect the two sets of equivalent hydrogen nuclei in methyl formate. Note that the signal due to the three hydrogen nuclei labelled b is about three times greater than the signal due to the single hydrogen nucleus labelled a.

absorption due to TMS will appear at the right edge of the spectrum. (Recall that NMR spectra are conventionally presented with the strength of the magnetic field increasing from left to right.) The top and bottom scales in the spectra in Figures 14.6 and 14.7 indicate the absorption lines relative to this TMS standard.

The top scale in both Figures 14.6 and 14.7 is in hertz (Hz), and runs from 0 Hz (on the right) to 500 Hz (on the left). The bottom scale in both figures is a derived scale related to the top scale and is defined in the following way. From Equation 14.21, the resonance frequency of a hydrogen nucleus ν_H is

$$\nu_H = \frac{\gamma B_0}{2\pi}(1 - \sigma_H) \tag{14.22}$$

Equation 14.22 shows that the resonance frequency is proportional to the strength of the magnetic field generated by the spectrometer. Different NMR spectrometers, then, will record different resonance frequencies relative to TMS for similar hydrogen nuclei in similar compounds. For example, the hydrogen nuclei in CH_3I will absorb at 130 Hz in a 60-MHz spectrometer, at 195 Hz ($130 \times 90/60$ Hz) in a 90-MHz spectrometer, and at 585 Hz ($130 \times 270/60$ Hz) in a 270-MHz spectrometer.

To avoid this complication and to be able to compare spectra taken by different spectrometers, we standardize the measured resonance frequency (relative to TMS) by dividing it by the frequency of the spectrometer. This procedure yields a spectrometer-independent number, the *chemical shift* (δ_H), defined by

$$\delta_H = \frac{\text{resonance frequency of nucleus H relative to TMS}}{\text{spectrometer frequency}} \times 10^6$$

$$= \left(\frac{\nu_H - \nu_{TMS}}{\nu_{\text{spectrometer}}}\right) \times 10^6 \tag{14.23}$$

From Figure 14.6, we find that $\nu_H - \nu_{TMS} = 130$ Hz, so Equation 14.23 gives $\delta = 2.16$ ppm, which is given on the bottom scale in Figure 14.6. Because the numerator of Equation 14.23 is measured in Hz and the denominator is measured in MHz, the factor of 10^6 in Equation 14.23 yields values of δ (in ppm) usually between 0 and 10 for hydrogen nuclei in organic compounds.

Consider two different hydrogen nuclei that absorb at frequencies ν_1 and ν_2. Using Equation 14.22, we can write

$$\nu_1 = \frac{\gamma B_0}{2\pi}(1 - \sigma_1)$$

and

$$\nu_2 = \frac{\gamma B_0}{2\pi}(1 - \sigma_2)$$

Therefore,

$$\delta_1 - \delta_2 = \left(\frac{\nu_1 - \nu_2}{\nu_{\text{spectrometer}}}\right) \times 10^6 = \frac{\gamma B_0}{2\pi \nu_{\text{spectrometer}}}(\sigma_2 - \sigma_1) \times 10^6 \tag{14.24}$$

Realizing that typical values of σ are of the order of 10^{-5}, we can neglect σ compared to 1 in Equation 14.21 and replace $\nu_{\text{spectrometer}}$ with $\gamma B_0 / 2\pi$ in Equation 14.24 to get

$$\delta_1 - \delta_2 = (\sigma_2 - \sigma_1) \times 10^6 \qquad (14.25)$$

Notice that the separation between the two absorption lines for the chemical shift scale is independent of the applied magnetic field.

EXAMPLE 14–3

Show that the top and bottom scales in Figure 14.7 are consistent for the two signals labelled a and b. Estimate the difference in chemical shifts for the hydrogen nuclei labelled a and b. What would be the separation between the two signals on a 270-MHz spectrometer?

SOLUTION: The a signal occurs at approximately 480 Hz, so using Equation 14.23, we find that

$$\delta_a = \left(\frac{480 \text{ Hz}}{60 \text{ MHz}} \right) \times 10^6 = 8.0 \text{ ppm}$$

Similarly, the b signal at 230 Hz corresponds to $\delta_b = 3.8$ ppm.

The separation between the two signal is $\delta_a - \delta_b = 8.0 \text{ ppm} - 3.8 \text{ ppm} = 4.2 \text{ ppm}$, so Equation 14.25 gives

$$\sigma_b - \sigma_a = 4.2 \times 10^{-6}$$

On a 270-MHz spectrometer, the separation in the frequency between the two signals would be (Equation 14.24)

$$\begin{aligned} \nu_a - \nu_b &= \nu_{\text{spectrometer}}(\delta_a - \delta_b) \times 10^{-6} \\ &= (270 \text{ MHz})(4.2 \text{ ppm}) \times 10^{-6} \\ &= 1130 \text{ Hz} \end{aligned}$$

The separation is different on the hertz scale but remains the same on the chemical shift scale.

EXAMPLE 14–4

Show that

$$\delta_{\text{H}} = (\sigma_{\text{TMS}} - \sigma_{\text{H}}) \times 10^6$$

and interpret this result.

SOLUTION: Simply let $\delta_1 = \delta_{\text{H}}$, $\delta_2 = 0$, and $\sigma_2 - \sigma_1 = \sigma_{\text{TMS}} - \sigma_{\text{H}}$ in Equation 14.25. This results says that the chemical shift of a given proton decreases with an increase in shielding constant (so long as δ_{H} remains positive).

14–5. Chemical Shifts Depend upon the Chemical Environment of the Nucleus

Because the shielding of a nucleus is caused by the enhanced electronic currents set up in the molecule by the external applied magnetic field, we can expect that the degree of shielding increases with increasing electron density around the nucleus. As Equation 14.21 shows, the larger the shielding constant, the greater is the external magnetic field required to produce resonance. Thus, according to Example 14.4, we expect that the greater the electron density, the smaller the chemical shift and the more upfield (toward the right in the spectrum) the resonance will occur. Table 14.3 lists typical hydrogen chemical shifts in organic molecules. (Note that all the chemical shifts are positive; this is another reason that TMS is used as a standard.) Table 14.3 shows that the hydrogen nuclei in alkanes absorb at relatively high fields, or have relatively small chemical shifts ($\delta = 0.8$ to 1.7). The hydrogen nuclei in alkanes are relatively well shielded by the surrounding electrons. If we substitute an electron-withdrawing group on an alkane carbon atom, as in CH_3Cl, the hydrogen nuclei will be less shielded, the value of σ will be smaller, and the hydrogen nuclei will absorb at lower field strengths, or at higher chemical shifts. This downfield shift increases with the number of electron-withdrawing groups attached, as the following data show:

$$CH_4 \qquad CH_3Cl \qquad CH_2Cl_2 \qquad CHCl_3$$
$$\delta = 0.23 \qquad \delta = 3.05 \qquad \delta = 5.33 \qquad \delta = 7.26$$

TABLE 14.3
Chemical shifts for hydrogen nuclei in various chemical or electronic environments.

Type of compound	Type of proton	Example	δ
	Tetramethylsilane	$(CH_3)_4Si$	0
Alkane	$RC\underline{H}_3$	$CH_3CH_2CH_3$	0.8–1.0
Alkane	$R_2C\underline{H}_2$	$(CH_3)_2C\underline{H}_2$	1.2–1.4
Alkane	$R_3C\underline{H}$	$(CH_3)_3C\underline{H}$	1.4–1.6
Aromatic	$Ar\underline{H}$	Benzene	6.0–8.5
Aromatic	$ArC\underline{H}_3$	p-Xylene	2.2–2.5
Chloroalkane	$RC\underline{H}_2Cl$	$CH_3C\underline{H}_2Cl$	3.4–3.8
Bromoalcane	$RC\underline{H}_2Br$	$CH_3C\underline{H}_2Br$	3.3–3.6
Iodoalkane	$RC\underline{H}_2I$	$CH_3C\underline{H}_2I$	3.1–3.3
Ether	$ROC\underline{H}_2R$	$CH_3OC\underline{H}_2CH_3$	3.3–3.9
Ester	$RCOOC\underline{H}_2R$	$CH_3COOC\underline{H}_2CH_3$	3.7–4.1
Ester	$RC\underline{H}_2COOR$	$CH_3C\underline{H}_2COOCH_3$	2.0–2.2
Ketone	$RCOC\underline{H}_3$	$CH_3COC\underline{H}_3$	2.1–2.6

Also, as you might expect, there is a correlation between electronegativity and chemical shift; the greater the electronegativity of X in CH_3X, the greater the chemical shift:

$$CH_3I \qquad CH_3Br \qquad CH_3Cl \qquad CH_3F$$
$$\delta = 2.16 \quad \delta = 2.68 \quad \delta = 3.05 \quad \delta = 4.26$$

The electronegativity effect also can be transmitted through adjacent carbon atoms:

$$CH_3Cl \qquad CH_3{-}CH_2Cl \qquad CH_3{-}CH_2{-}CH_2Cl$$
$$\delta = 3.05 \qquad \delta = 1.42 \qquad \delta = 1.04$$

In the NMR spectrum of methyl formate (Figure 14.7), the two signals (not counting the reference signal from TMS) are due to the hydrogen nuclei as indicated in the figure. From Table 14.3, we see that the smaller signal arises from the hydrogen labelled a in Figure 14.7 and that the larger signal arises from the methyl hydrogens.

The relative areas of the two peaks in Figure 14.7 reflect the number of equivalent hydrogen atoms in each set. Each hydrogen atom in a set contributes to the observed signal, so the area of a signal peak is proportional to the number of hydrogen atoms generating that signal. The relative signal areas in Figure 14.7 are in the ratio 3:1, in quantitative agreement with the numbers of equivalent hydrogen atoms in the two sets of hydrogen atoms in methyl formate. The relative areas are often difficult to determine visually, but the areas are measured electronically by an NMR spectrometer. In many spectrometers, the chemical shift and relative areas of each peak are printed digitally right on the spectrum.

EXAMPLE 14–5

Suppose we have a compound that we know to be either methyl acetate (CH_3COOCH_3) or ethyl formate ($HCOOCH_2CH_3$). Both substances have the same molecular formula, $C_3H_6O_2$. Given that the NMR spectrum of the compound is as shown below,

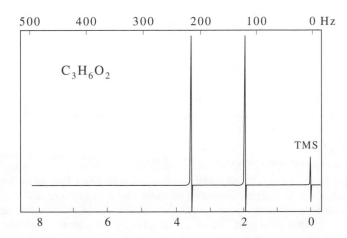

determine which substance the compound is.

SOLUTION: From the Lewis formula of methyl acetate, we see that methyl acetate has two methyl groups. One methyl group is attached to an oxygen atom, and the other is attached to a carbon atom. Consequently, these two methyl groups are not equivalent, so methyl acetate has two sets of equivalent hydrogen nuclei, each set containing three hydrogen atoms. The Lewis formula of ethyl formate suggests there are three different sets of hydrogen atoms in this molecule. Because only two signals are observed in the NMR spectrum, we conclude the unknown compound must be methyl acetate. For added assurance, note that the positions of the two signals are in agreement with the values given in Table 14.3, and the relative areas are 1:1.

14–6. Spin–Spin Coupling Can Lead to Multiplets in NMR Spectra

There is an important feature of NMR spectra we have not discussed yet. To see this feature, let's consider 1,1,2-trichloroethane.There are two types of hydrogen atoms in this molecule. One set contains one hydrogen atom, and the other contains two structurally equivalent hydrogen atoms. Consequently, we predict that the NMR spectrum will contain two signals whose areas are in the ratio 1:2. The NMR spectrum of 1,1,2-trichloroethane is shown in Figure 14.8. It appears to be more complicated than we predicted. Instead of just two single peaks, we have two groups of closely spaced peaks. One group consists of three closely spaced peaks (labelled a) and the other consists of two closely spaced peaks (labelled b). The signals due to the two sets of

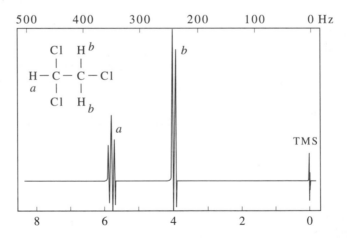

FIGURE 14.8
The NMR spectrum of 1,1,2-trichloroethane taken on a 60-MHz spectrometer. There are two sets of hydrogen atoms, labelled a and b in the Lewis formula. Instead of two single peaks in the spectrum, there is a signal consisting of three closely spaced peaks (a triplet) and a signal consisting of two closely spaced peaks (a doublet). The signals due to the two sets of hydrogen atoms are said to be split. The observed splitting gives information about the number of protons adjacent to each set of equivalent hydrogen atoms. The relative areas under the two multiplets are 1:2, in accord with the number of hydrogen atoms in the two equivalent sets.

hydrogen atoms in 1,1,2-trichloroethane are said to be *split*. The three peaks labelled a in Figure 14.8 are collectively called a *triplet*, and the two peaks labelled b are called a *doublet*.

The areas of the doublet and triplet in Figure 14.8 are in the ratio 2:1, as we would have predicted, but why does the splitting occur? Recall that protons behave as tiny magnets and so create their own magnetic fields. Therefore, any given hydrogen nucleus will be acted upon not only by the externally applied field and by the magnetic field generated by the motion of its nearby electrons, but also by the magnetic field due to the magnetic dipoles of its neighboring hydrogen nuclei on adjacent carbon atoms. The effect of the neighboring hydrogen nuclei is to split the signal of the given hydrogen nucleus into multiplets. The interaction between nuclear spins is called *spin-spin interaction*.

Now we will consider this splitting into multiplets due to spin-spin interaction in a quantitative manner. For simplicity, let's consider a molecule with just two hydrogen atoms in different electronic environments. In the absence of spin-spin interaction, the spin Hamiltonian operator of such a molecule consists of two terms similar to Equation 14.14 but with B_z replaced by $B_0(1 - \sigma_j)$, where σ_j is the chemical shift of the jth hydrogen nucleus. Thus, we can write $\hat{H}$ as

$$\hat{H} = -\gamma B_0(1 - \sigma_1)\hat{I}_{z1} - \gamma B_0(1 - \sigma_2)\hat{I}_{z2} \tag{14.26}$$

This Hamiltonian operator has no terms that account for the interaction between spins on neighboring hydrogen nuclei. The classical expression for the interaction between two magnetic dipole moments involves a factor $\boldsymbol{\mu}_1 \cdot \boldsymbol{\mu}_2$, where $\boldsymbol{\mu}_1$ and $\boldsymbol{\mu}_2$ are the magnetic dipole moments. Quantum mechanically, $\boldsymbol{\mu}$ is proportional to the spin $\mathbf{I}$ (Equation 14.9), so we can account for the effect of spin-spin coupling by including a term proportional to $\hat{\mathbf{I}}_1 \cdot \hat{\mathbf{I}}_2$ in the Hamiltonian operator. (See a similar interaction term in Equation 8.55.) We let the proportionality constant be J_{12}, so we write the spin Hamiltonian operator for an interacting two-spin system as

$$\hat{H} = -\gamma B_0(1 - \sigma_1)\hat{I}_{z1} - \gamma B_0(1 - \sigma_2)\hat{I}_{z2} + \frac{hJ_{12}}{\hbar^2}\hat{\mathbf{I}}_1 \cdot \hat{\mathbf{I}}_2 \tag{14.27}$$

The factor of $h/\hbar^2$ is included in the spin-spin interaction term to give J_{12} units of hertz. The quantity J_{12} is called the *spin-spin coupling constant*.

In this section, we will assume the spin-spin interaction term can be treated by first-order perturbation theory. The unperturbed spin Hamiltonian operator and the perturbation term are

$$\hat{H}^{(0)} = -\gamma B_0(1 - \sigma_1)\hat{I}_{z1} - \gamma B_0(1 - \sigma_2)\hat{I}_{z2} \tag{14.28}$$

and

$$\hat{H}^{(1)} = \frac{hJ_{12}}{\hbar^2}\hat{\mathbf{I}}_1 \cdot \hat{\mathbf{I}}_2 \tag{14.29}$$

The unperturbed wave functions are the four spin-function products for a two-spin system:

$$\psi_1 = \alpha(1)\alpha(2) \qquad \psi_2 = \beta(1)\alpha(2)$$
$$\psi_3 = \alpha(1)\beta(2) \qquad \psi_4 = \beta(1)\beta(2) \tag{14.30}$$

Recall from Section 7–4 that the energy through first-order is given by (Equations 7.47 and 7.48)

$$E_j = E_j^{(0)} + \int d\tau_1 d\tau_2 \psi_j^* \hat{H}^{(1)} \psi_j \tag{14.31}$$

where τ_1 and τ_2 are spin variables. (We don't use σ here for the spin variable as we did in Section 8–4 and Equation 14.2 to avoid confusion with the notation for shielding constants.) The $E_j^{(0)}$ are given by

$$\hat{H}^{(0)} \psi_j = E_j^{(0)} \psi_j \tag{14.32}$$

where the ψ_j are given in Equation 14.30. For example, $E_1^{(0)}$ can be determined readily by using the fact that $\hat{I}_{zj}\alpha(j) = \dfrac{\hbar}{2}\alpha(j)$ for $j = 1$ and 2:

$$
\begin{aligned}
\hat{H}^{(0)} \psi_1 &= \hat{H}^{(0)} \alpha(1)\alpha(2) \\
&= -\gamma B_0 (1-\sigma_1)\hat{I}_{z1}\alpha(1)\alpha(2) - \gamma B_0 (1-\sigma_2)\hat{I}_{z2}\alpha(1)\alpha(2) \\
&= -\frac{\hbar\gamma B_0 (1-\sigma_1)}{2}\alpha(1)\alpha(2) - \frac{\hbar\gamma B_0 (1-\sigma_2)}{2}\alpha(1)\alpha(2) \\
&= E_1^{(0)}\alpha(1)\alpha(2) = E_1^{(0)}\psi_1
\end{aligned}
\tag{14.33}
$$

so that

$$E_1^{(0)} = -\hbar\gamma B_0 \left(1 - \frac{\sigma_1 + \sigma_2}{2}\right) \tag{14.34}$$

EXAMPLE 14–6
Show that $E_3^{(0)} = \hbar\gamma B_0 (\sigma_1 - \sigma_2)/2$.

SOLUTION: To determine $E_3^{(0)}$, we use

$$\hat{H}^{(0)} \psi_3 = E_3^{(0)} \psi_3$$

Therefore, we have

$$
\begin{aligned}
\hat{H}^{(0)} \psi_3 &= \hat{H}^{(0)} \alpha(1)\beta(2) \\
&= -\gamma B_0 (1-\sigma_1)\hat{I}_{z1}\alpha(1)\beta(2) - \gamma B_0 (1-\sigma_2)\hat{I}_{z2}\alpha(1)\beta(2) \\
&= -\frac{\hbar\gamma B_0 (1-\sigma_1)}{2}\alpha(1)\beta(2) + \frac{\hbar\gamma B_0 (1-\sigma_2)}{2}\alpha(1)\beta(2)
\end{aligned}
$$

$$= \frac{\hbar \gamma B_0}{2}(\sigma_1 - \sigma_2)\alpha(1)\beta(2)$$

so that

$$E_3^{(0)} = \frac{\hbar \gamma B_0}{2}(\sigma_1 - \sigma_2) \tag{14.35}$$

Similarly (Problem 14–16),

$$E_2^{(0)} = -\frac{\hbar \gamma B_0}{2}(\sigma_1 - \sigma_2) \tag{14.36}$$

and

$$E_4^{(0)} = \hbar \gamma B_0 \left(1 - \frac{\sigma_1 + \sigma_2}{2}\right) \tag{14.37}$$

To calculate the first-order corrections, we must evaluate integrals of the type

$$H_{ii} = \frac{h J_{12}}{\hbar^2} \int d\tau_1 d\tau_2 \psi_i^* \hat{\mathbf{I}}_1 \cdot \hat{\mathbf{I}}_2 \psi_i \tag{14.38}$$

The dot product of $\hat{\mathbf{I}}_1 \cdot \hat{\mathbf{I}}_2$ is (MathChapter C)

$$\hat{\mathbf{I}}_1 \cdot \hat{\mathbf{I}}_2 = \hat{I}_{x1}\hat{I}_{x2} + \hat{I}_{y1}\hat{I}_{y2} + \hat{I}_{z1}\hat{I}_{z2} \tag{14.39}$$

The integral involving $\hat{I}_{z1}\hat{I}_{z2}$ is fairly easy to evaluate because of Equation 14.3b. If we take $\psi_1 = \alpha(1)\alpha(2)$ as an example, we obtain

$$\hat{I}_{z1}\hat{I}_{z2}\alpha(1)\alpha(2) = [\hat{I}_{z1}\alpha(1)][\hat{I}_{z2}\alpha(2)]$$
$$= \frac{\hbar}{2}\alpha(1)\frac{\hbar}{2}\alpha(2) = \frac{\hbar^2}{4}\alpha(1)\alpha(2)$$

and so

$$H_{z,11} = \frac{h J_{12}}{\hbar^2} \int d\tau_1 d\tau_2 \alpha^*(1)\alpha^*(2)\hat{I}_{z1}\hat{I}_{z2}\alpha(1)\alpha(2)$$

$$= \frac{h J_{12}}{\hbar^2}\frac{\hbar^2}{4} \int d\tau_1 \alpha^*(1)\alpha(1) \int d\tau_2 \alpha^*(2)\alpha(2)$$

$$= \frac{h J_{12}}{4} \tag{14.40}$$

Similarly (Problem 14–17), we find that

$$H_{z,22} = H_{z,33} = -\frac{h J_{12}}{4} \tag{14.41}$$

and

$$H_{z,44} = \frac{h J_{12}}{4} \tag{14.42}$$

The integrals involving $\hat{I}_{x1}\hat{I}_{x2}$ and $\hat{I}_{y1}\hat{I}_{y2}$ are not as easy to evaluate. Problems 14–18 through 14–21 lead you through the proof that

$$\hat{I}_x\alpha = \frac{\hbar}{2}\beta \qquad \hat{I}_y\alpha = \frac{i\hbar}{2}\beta$$

$$\hat{I}_x\beta = \frac{\hbar}{2}\alpha \qquad \hat{I}_y\beta = -\frac{i\hbar}{2}\alpha \tag{14.43}$$

These equations along with Equation 14.3b are listed in Table 14.4 for convenience. Using these relations, we see, for example, that

$$\hat{I}_{x1}\hat{I}_{x2}\alpha(1)\alpha(2) = [\hat{I}_{x1}\alpha(1)][\hat{I}_{x2}\alpha(2)]$$

$$= \frac{\hbar}{2}\beta(1)\frac{\hbar}{2}\beta(2) = \frac{\hbar^2}{4}\beta(1)\beta(2)$$

and so

$$H_{x,11} = \frac{h J_{12}}{\hbar^2} \int\int d\tau_1 d\tau_2 \alpha^*(1)\alpha^*(2)\hat{I}_{x1}\hat{I}_{x2}\alpha(1)\alpha(2)$$

$$= \frac{h J_{12}}{\hbar^2} \int\int d\tau_1 d\tau_2 \alpha^*(1)\alpha^*(2)\frac{\hbar^2}{4}\beta(1)\beta(2)$$

$$= \frac{h J_{12}}{4} \int d\tau_1 \alpha^*(1)\beta(1) \int d\tau_2 \alpha^*(2)\beta(2) = 0$$

where we have used the orthogonality of the α and β functions. Similarly, we can show

TABLE 14.4
A summary of the results of $\hat{I}_x$, $\hat{I}_y$, and $\hat{I}_z$ operating on α and β.

$\hat{I}_x\alpha = \frac{\hbar}{2}\beta$	$\hat{I}_y\alpha = \frac{i\hbar}{2}\beta$	$\hat{I}_z\alpha = \frac{\hbar}{2}\alpha$
$\hat{I}_x\beta = \frac{\hbar}{2}\alpha$	$\hat{I}_y\beta = -\frac{i\hbar}{2}\alpha$	$\hat{I}_z\beta = -\frac{\hbar}{2}\beta$

that the x and y terms in $\hat{\mathbf{I}}_1 \cdot \hat{\mathbf{I}}_2$ do not contribute to any of the first-order energies in this case (Problem 14–22), and so the energy of each level to first-order is

$$E_1 = -h\nu_0 \left(1 - \frac{\sigma_1 + \sigma_2}{2}\right) + \frac{hJ_{12}}{4}$$

$$E_2 = -\frac{h\nu_0}{2}(\sigma_1 - \sigma_2) - \frac{hJ_{12}}{4}$$

$$E_3 = \frac{h\nu_0}{2}(\sigma_1 - \sigma_2) - \frac{hJ_{12}}{4}$$

$$E_4 = h\nu_0 \left(1 - \frac{\sigma_1 + \sigma_2}{2}\right) + \frac{hJ_{12}}{4}$$

$$(14.44)$$

where

$$\nu_0 = \frac{\gamma B_0}{2\pi} \qquad (14.45)$$

The energy levels given by Equations 14.44 (along with the allowed transitions) are sketched in Figure 14.9. The selection rule for transitions between nuclear spin

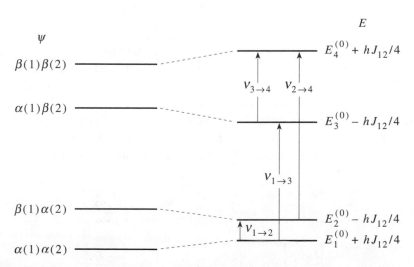

FIGURE 14.9
The energy levels of a two-spin system calculated by first-order perturbation theory. The allowed transitions are indicated by the vertical arrows.

states says that only one type of nucleus at a time can undergo a transition. Thus, the allowed transitions for absorption (as indicated in Figure 14.9) are

$$\alpha(1)\alpha(2) \longrightarrow \beta(1)\alpha(2) \qquad (1 \to 2)$$

$$\alpha(1)\alpha(2) \longrightarrow \alpha(1)\beta(2) \qquad (1 \to 3)$$

$$\beta(1)\alpha(2) \longrightarrow \beta(1)\beta(2) \qquad (2 \to 4)$$

$$\alpha(1)\beta(2) \longrightarrow \beta(1)\beta(2) \qquad (3 \to 4)$$

The frequencies associated with the allowed transitions are

$$\nu_{1\to 2} = \nu_0(1 - \sigma_1) - \frac{J_{12}}{2}$$

$$\nu_{1\to 3} = \nu_0(1 - \sigma_2) - \frac{J_{12}}{2}$$

$$\nu_{2\to 4} = \nu_0(1 - \sigma_2) + \frac{J_{12}}{2} \qquad (14.46)$$

$$\nu_{3\to 4} = \nu_0(1 - \sigma_1) + \frac{J_{12}}{2}$$

We can express the above four resonance frequencies by

$$\nu_1^{\pm} = \nu_0(1 - \sigma_1) \pm \frac{J_{12}}{2}$$

$$\qquad\qquad (14.47)$$

$$\nu_2^{\pm} = \nu_0(1 - \sigma_2) \pm \frac{J_{12}}{2}$$

Realizing that J_{12} is small enough to use first-order perturbation theory, we know the four resonant frequencies occur as a pair of two closely spaced lines, or two doublets, as sketched in Figure 14.10. The centers of the doublets are separated by $\nu_0|\sigma_1 - \sigma_2|$, and the separations of the peaks *within* the two doublets is J_{12}. A molecule in which the two hydrogen nuclei are in very different chemical environments so that $\nu_0|\sigma_1 - \sigma_2| \gg J_{12}$ is called an *AX system*. Figure 14.11 is a sketch of the NMR spectrum of an AX spin system taken at 90 MHz and 200 MHz. In the 90-MHz spectrum, one doublet is

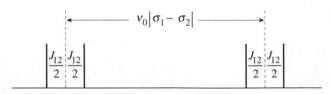

FIGURE 14.10
The splitting pattern in the first-order spectrum of an AX spin system. The centers of the doublets are separated by $\nu_0|\sigma_1 - \sigma_2|$, and the separation within each doublet is J_{12}.

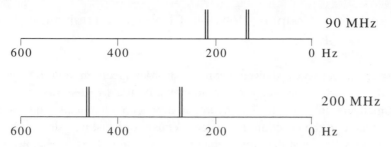

FIGURE 14.11
An idealized spectrum of an AX spin system taken at 90 MHz (top) and 200 MHz (bottom), illustrating that the spacing between the centers of the doublets increases with increasing spectrometer frequency but that the spacing within the doublets is independent of spectrometer frequency.

centered at 130 Hz and the other at 210 Hz. The separation within the doublet is 6.5 Hz. In the 200-MHz spectrum, v_0 is now 200 MHz, so the two doublets are centered at

$$\text{center of doublet 1} = (130 \text{ Hz}) \left(\frac{200 \text{ MHz}}{90 \text{ MHz}} \right) = 289 \text{ Hz}$$

and the other at

$$\text{center of doublet 2} = (210 \text{ Hz}) \left(\frac{200 \text{ MHz}}{90 \text{ MHz}} \right) = 467 \text{ Hz}$$

so their separation increases from 80 Hz at 90 MHz to 178 Hz at 200 MHz. The spacing within the doublets, however, is still 6.5 Hz because J_{12} is independent of the frequency of the spectrometer.

The condition for the use of first-order perturbation theory is that $J_{12} \ll v_0 |\sigma_1 - \sigma_2|$, and this is the condition that leads to two separated doublets. Such a spectrum is called a *first-order spectrum*. Typical values of coupling constants are around 5 Hz, so a first-order spectrum will result if the separation between multiplets is 100 Hz or so. For example, in Figure 14.8, $J = 6$ Hz and $v_0 |\sigma_1 - \sigma_2| = 110$ Hz. We will see in Section 14–9 that the resultant spectrum will not consist of two separated doublets of equal intensity unless $J_{12} \ll v_0 |\sigma_1 - \sigma_2|$.

The designation "AX" to describe the two-spin system discussed above comes from a notation commonly used in NMR studies. For any molecule, each nonequivalent hydrogen atom is given a letter, A, B, C, and so forth. If there is more than one hydrogen atom of a type, we use a subscript to denote the number as in A_3 or B_2. Hydrogen atoms whose chemical shifts are relatively similar in magnitude are assigned letters that are close to each other in the alphabet, such as AB. Hydrogen atoms whose chemical shifts differ by a relatively large amount are assigned letters that are far apart in the alphabet, such as AX. Thus, a two-spin system with $J_{12} \ll v_0 |\sigma_1 - \sigma_2|$ is an AX system. One with $J_{12} \approx v_0 |\sigma_1 - \sigma_2|$ is an AB system. Figure 14.8 shows that 1,1,2-trichloroethane is an example of an A_2X system when measured by a 60-MHz spectrometer.

14–7. Spin–Spin Coupling Between Chemically Equivalent Protons Is Not Observed

In the previous section we showed that the first-order spectrum of an AX system leads to an NMR spectrum consisting of two doublets. Figure 14.12 shows the spectrum of dichloromethane, in which the two hydrogen atoms are chemically equivalent (an A_2 system). Note that the spectrum in this case consists of just one singlet. Not only do the two protons absorb at the same frequency, as you might expect, but no splitting of this signal due to spin-spin coupling is observed, as you might not expect.

The spin Hamiltonian operator of an A_2 system is (see Equation 14.27)

$$\hat{H} = -\gamma B_0(1 - \sigma_A)\hat{I}_{z1} - \gamma B_0(1 - \sigma_A)\hat{I}_{z2} + \frac{hJ_{AA}}{\hbar^2}\hat{\mathbf{I}}_1 \cdot \hat{\mathbf{I}}_2 \tag{14.48}$$

The spin Hamiltonian operator for an A_2 system is similar to that of an AX system, except that in this case the two shielding constants are equal. As we did in the previous section, we will use perturbation theory to determine the first-order spectrum. We will use

$$\hat{H}^{(0)} = -\gamma B_0(1 - \sigma_A)(\hat{I}_{z1} + \hat{I}_{z2}) \tag{14.49}$$

as the unperturbed Hamiltonian operator and

$$\hat{H}^{(1)} = \frac{hJ_{AA}}{\hbar^2}\hat{\mathbf{I}}_1 \cdot \hat{\mathbf{I}}_2 \tag{14.50}$$

as the perturbation term. A primary difference between the first-order perturbation theory treatment of an AX system and an A_2 system is in the form of the unperturbed

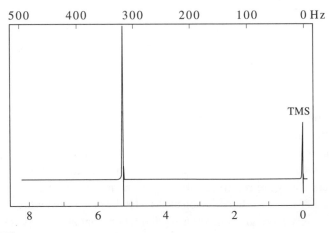

FIGURE 14.12
An NMR spectrum of dichloromethane taken on a 60-MHz spectrometer.

spin wave functions. Because the two nuclei are equivalent, and so indistinguishable, in the A_2 case, we must use combinations of the α's and β's that are either symmetric or antisymmetric, as we did when we developed the spin wave functions of the two electrons in a helium atom in Section 8–5. The four acceptable combinations are

$$\phi_1 = \alpha(1)\alpha(2) \qquad\qquad \phi_2 = \frac{1}{\sqrt{2}}[\alpha(1)\beta(2) - \beta(1)\alpha(2)]$$

$$\phi_3 = \frac{1}{\sqrt{2}}[\alpha(1)\beta(2) + \beta(1)\alpha(2)] \quad \phi_4 = \beta(1)\beta(2) \tag{14.51}$$

We can now use Equation 14.31 to calculate the four first-order energies. For example,

$$
\begin{aligned}
E_1 &= E_1^{(0)} + E_1^{(1)} \\
&= \int\!\!\int d\tau_1 d\tau_2 \alpha^*(1)\alpha^*(2)\left[-\gamma B_0(1-\sigma_A)(\hat{I}_{z1} + \hat{I}_{z2})\right]\alpha(1)\alpha(2) \\
&+ \int\!\!\int d\tau_1 d\tau_2 \alpha^*(1)\alpha^*(2)\frac{hJ_{AA}}{\hbar^2}(\hat{I}_{x1}\hat{I}_{x2} + \hat{I}_{y1}\hat{I}_{y2} + \hat{I}_{z1}\hat{I}_{z2})\alpha(1)\alpha(2) \quad (14.52)
\end{aligned}
$$

The first integral in Equation 14.52 is evaluated readily using the fact that (Equations 14.3b)

$$(\hat{I}_{z1} + \hat{I}_{z2})\alpha(1)\alpha(2) = \left(\frac{\hbar}{2} + \frac{\hbar}{2}\right)\alpha(1)\alpha(2) = \hbar\alpha(1)\alpha(2)$$

The second integral is evaluated using the relations in Table 14.4:

$$
\begin{aligned}
(\hat{I}_{x1}\hat{I}_{x2} &+ \hat{I}_{y1}\hat{I}_{y2} + \hat{I}_{z1}\hat{I}_{z2})\alpha(1)\alpha(2) \\
&= \frac{\hbar^2}{4}\beta(1)\beta(2) - \frac{\hbar^2}{4}\beta(1)\beta(2) + \frac{\hbar^2}{4}\alpha(1)\alpha(2)
\end{aligned}
$$

To evaluate E_1, we multiply this relation by $\alpha^*(1)\alpha^*(2)$ and integrate over the spin coordinates. The first and second terms here will vanish due to the orthogonality of the spin functions α and β, and so we have

$$
\begin{aligned}
E_1 &= -\hbar\gamma B_0(1-\sigma_A)\int d\tau_1 \alpha^*(1)\alpha(1)\int d\tau_2 \alpha^*(2)\alpha(2) \\
&+ \frac{hJ_{AA}}{\hbar^2}\frac{\hbar^2}{4}\int d\tau_1 \alpha^*(1)\alpha(1)\int d\tau_2 \alpha^*(2)\alpha(2) \\
&= -\hbar\gamma B_0(1-\sigma_A) + \frac{hJ_{AA}}{4} \tag{14.53}
\end{aligned}
$$

EXAMPLE 14–7
Evaluate E_2 through first order.

SOLUTION: The value of E_2 through first order is given by

$$E_2 = E_2^{(0)} + E_2^{(1)}$$

$$= \iint d\tau_1 d\tau_2 \phi_2^* \hat{H}^{(0)} \phi_2 + \iint d\tau_1 d\tau_2 \phi_2^* \hat{H}^{(1)} \phi_2 \tag{1}$$

The first integral in Equation 1 requires that we evaluate

$$(\hat{I}_{z1} + \hat{I}_{z2})\phi_2 = \frac{1}{\sqrt{2}}(\hat{I}_{z1} + \hat{I}_{z2})[\alpha(1)\beta(2) - \beta(1)\alpha(2)]$$

$$= \frac{1}{\sqrt{2}}\left[\left(\frac{\hbar}{2} - \frac{\hbar}{2}\right) - \left(\frac{\hbar}{2} - \frac{\hbar}{2}\right)\right][\alpha(1)\beta(2) - \beta(1)\alpha(2)] = 0$$

Substituting this result into Equation 1 shows that $E_2^{(0)} = 0$. The second integral involves evaluating the term

$$\frac{1}{\sqrt{2}}(\hat{I}_{x1}\hat{I}_{x2} + \hat{I}_{y1}\hat{I}_{y2} + \hat{I}_{z1}\hat{I}_{z2})[\alpha(1)\beta(2) - \beta(1)\alpha(2)]$$

Using the relations in Table 14.4, we see that

$$\hat{I}_{x1}\hat{I}_{x2}\alpha(1)\beta(2) = \frac{\hbar^2}{4}\beta(1)\alpha(2)$$

$$\hat{I}_{x1}\hat{I}_{x2}\beta(1)\alpha(2) = \frac{\hbar^2}{4}\alpha(1)\beta(2)$$

$$\hat{I}_{y1}\hat{I}_{y2}\alpha(1)\beta(2) = \frac{\hbar^2}{4}\beta(1)\alpha(2)$$

$$\hat{I}_{y1}\hat{I}_{y2}\beta(1)\alpha(2) = \frac{\hbar^2}{4}\alpha(1)\beta(2)$$

$$\hat{I}_{z1}\hat{I}_{z2}\alpha(1)\beta(2) = -\frac{\hbar^2}{4}\alpha(1)\beta(2)$$

$$\hat{I}_{z1}\hat{I}_{z2}\beta(1)\alpha(2) = -\frac{\hbar^2}{4}\beta(1)\alpha(2)$$

Putting this all together gives

$$(\hat{I}_{x1}\hat{I}_{x2} + \hat{I}_{y1}\hat{I}_{y2} + \hat{I}_{z1}\hat{I}_{z2})[\alpha(1)\beta(2) - \beta(1)\alpha(2)] = -\frac{3\hbar^2}{4}[\alpha(1)\beta(2) - \beta(1)\alpha(2)]$$

and substituting this result into Equation 1 gives

$$E_2 = E_2^{(1)} = -\frac{3h J_{AA}}{4} \tag{14.54}$$

Similarly, we find that (Problem 14–27)

$$E_3 = \frac{h J_{AA}}{4} \tag{14.55}$$

and

$$E_4 = \hbar \gamma B_0 (1 - \sigma_A) + \frac{h J_{AA}}{4} \tag{14.56}$$

These four energy levels are sketched in Figure 14.13. The selection rules state that not only does one spin at a time undergo a transition, but that only transitions between states of the same spin symmetry are allowed (Problem 14–39). Thus, the allowed transitions are $1 \rightarrow 3$ and $3 \rightarrow 4$. The frequencies corresponding to these transitions are

$$\nu_{1 \rightarrow 3} = \nu_{3 \rightarrow 4} = \frac{E_3 - E_1}{h} = \frac{\gamma B_0 (1 - \sigma_A)}{2\pi} = \nu_0 (1 - \sigma_A) \tag{14.57}$$

Therefore, although the spin-spin coupling between equivalent protons alters the energy levels, the selection rules are such that the spin-spin coupling constant effect cancels in the transition frequencies, so only a single proton resonance is observed in a molecule such as dichloromethane (Figure 14.12).

14–8. The $n + 1$ Rule Applies Only to First-Order Spectra

The splitting observed for 1,1,2-trichloroethane in Figure 14.8 shows a doublet and a triplet, and that for chloroethane in Figure 14.14 shows a triplet and a quartet. The

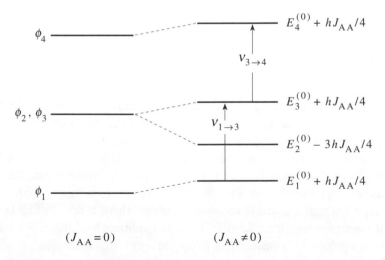

FIGURE 14.13
The energy levels of an A_2 system calculated by first-order perturbation theory. The two allowed transitions, indicated by vertical arrows, have the same frequency (Equation 14.57). The wave functions are defined by Equations 14.51.

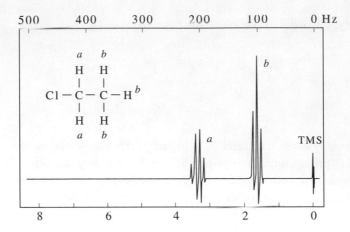

FIGURE 14.14
An NMR spectrum of chloroethane taken on a 60-MHz spectrometer. The equivalent sets of hydrogen atoms are labelled a and b.

splitting in each case can be predicted by a simple rule called the $n + 1$ *rule*. The $n + 1$ rule states that if a proton has n equivalent neighboring protons, then its NMR signal will be split into $n + 1$ closely spaced peaks. Each proton senses the number of equivalent protons on the carbon atoms adjacent to the one to which it is bonded.

To illustrate the $n + 1$ rule, consider 1,1,2-trichloroethane, whose NMR spectrum is shown in Figure 14.8. The two equivalent hydrogen atoms labelled b are neighbors of hydrogen atom a.

$$
\begin{array}{ccc}
 & \text{Cl} & \text{H}^{b} \\
 & | & | \\
{}^{a}\text{H}-\text{C}- & \text{C} & -\text{Cl} \\
 & | & | \\
 & \text{Cl} & \text{H}_{b}
\end{array}
$$

The two sets of hydrogen atoms are bonded to adjacent carbon atoms. Consequently, the nuclei of the two hydrogen atoms labelled b split the signal due to the nucleus of hydrogen atom a into a triplet ($n + 1 = 2 + 1 = 3$). The nucleus of the hydrogen atom labelled a in turn splits the signal due to the nuclei of the hydrogen atoms labelled b into a doublet ($n + 1 = 1 + 1 = 2$). The ratio of the area under the doublet signal to that under the triplet signal, however, is still equal to 2:1, in accord with the number of equivalent hydrogen atoms in each set. Note that there is no splitting between the nuclei of the hydrogen atoms labelled b. Signal splitting and the $n + 1$ rule applies only between groups of nonequivalent hydrogen atoms in a molecule.

Chloroethane (Figure 14.14) has two sets of equivalent hydrogen atoms, so there are two signals in the NMR spectrum. The two hydrogen atoms labelled a have three equivalent neighboring hydrogen atoms (labelled b). Therefore, the signal due to the a hydrogen nuclei is split into a quartet ($n + 1 = 3 + 1 = 4$) by the three neighboring

b hydrogen nuclei. The signal due to the three hydrogen nuclei labelled b is split by the two neighboring hydrogen nuclei a into a triplet ($n + 1 = 2 + 1 = 3$). The relative areas of the quartet and triplet are in the ratio 2:3, in accord with the number of hydrogen atoms in each equivalent set.

In the case of 1,1-dichloroethane (Figure 14.15), there are two sets of equivalent hydrogen atoms, containing one and three hydrogen atoms. Consequently, the NMR spectrum shows two main signals, one a doublet and the other a quartet, and the ratio of the area under the doublet to that under the quartet is 3:1.

To explain the basis of the $n + 1$ rule qualitatively, consider the resonance due to the hydrogen atoms labelled b in the spectrum of 1,1,2-trichloroethane (Figure 14.8). Each of these hydrogen atoms is acted upon by the magnetic field due to the nucleus of hydrogen atom a. This nucleus can be aligned in one of only two orientations ($\pm 1/2$) with respect to the externally applied magnetic field. These two possible orientations produce slightly different magnetic fields, so the nuclei of the b hydrogen atoms are acted upon by two slightly different magnetic field strengths. Consequently, the nuclei labelled b absorb at two slightly different positions in the NMR spectrum, leading to a doublet. Thus, we see that a set containing only one equivalent neighboring hydrogen atom yields a doublet.

Now consider the hydrogen atom labelled a in Figure 14.8. The nucleus in this case has two equivalent neighboring hydrogen atoms. Each of these nuclei must be aligned in one of only two orientations. This requirement leads to four possibilities:

$$\uparrow\downarrow$$

$$\downarrow\downarrow \qquad \downarrow\uparrow \qquad \uparrow\uparrow$$

Because one equivalent proton cannot be distinquished from another, the middle two combinations ($\uparrow\downarrow$ and $\downarrow\uparrow$) produce the same field, albeit twice as likely as the other

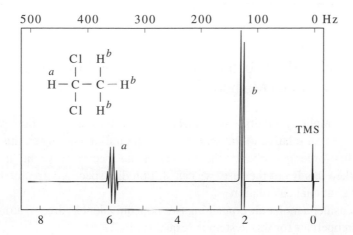

FIGURE 14.15
The NMR spectrum of 1,1-dichloroethane taken on a 60-MHz spectrometer. The equivalent sets of hydrogen atoms are labelled a and b.

two combinations ($\uparrow\uparrow$ or $\downarrow\downarrow$). Thus, the signal due to hydrogen atom a in 1,1,2-trichloroethane is split into a triplet, with the middle peak of the triplet being twice the size of the other two peaks. This leads to the 1:2:1 triplet pattern shown in Figure 14.8.

We can use chloroethane to illustrate the splitting caused by a set of three equivalent neighboring hydrogen atoms. Using the same argument as before, we write the following possibilites:

$$\uparrow\downarrow\downarrow \qquad \uparrow\uparrow\downarrow$$

$$\downarrow\uparrow\downarrow \qquad \uparrow\downarrow\uparrow$$

$$\downarrow\downarrow\downarrow \qquad \downarrow\downarrow\uparrow \qquad \downarrow\uparrow\uparrow \qquad \uparrow\uparrow\uparrow$$

This pattern of proton spin combinations leads to a quartet of intensities 1:3:3:1, as shown for the hydrogen atoms labelled a in the spectrum of chloroethane in Figure 14.14. Table 14.5 summarizes the observed multiplet splitting in first-order spectra.

TABLE 14.5
The observed multiplet splitting in first-order spectra.

Number of closely spaced lines	1	2	3	4
Name	Singlet	Doublet	Triplet	Quartet
Relative peak size	1	1:1	1:2:1	1:3:3:1
Idealized intensity pattern														

14–9. Second-Order Spectra Can Be Calculated Exactly Using the Variational Method

The relative simplicity of first-order spectra occurs because the spin-spin coupling constants are small relative to the separation of the multiplets. When this is the case, we can use first-order perturbation theory to calculate spectra, as we did in the previous sections. When this is not the case, we can still predict spectra correctly, but we must resort to a variational calculation.

Let's consider a molecule containing two nonequivalent hydrogen atoms. The spin Hamiltonian operator for this system is (Equation 14.27)

$$\hat{H} = -\gamma B_0(1 - \sigma_1)\hat{I}_{z1} - \gamma B_0(1 - \sigma_2)\hat{I}_{z2} + \frac{hJ_{12}}{\hbar^2}\hat{\mathbf{I}}_1 \cdot \hat{\mathbf{I}}_2 \qquad (14.58)$$

There are a total of four possible spin wave functions for this system:

$$\phi_1 = \alpha(1)\alpha(2) \qquad \phi_2 = \alpha(1)\beta(2)$$

$$\phi_3 = \beta(1)\alpha(2) \qquad \phi_4 = \beta(1)\beta(2)$$
(14.59)

We can calculate the energy levels of this spin system *exactly* by using a linear combination of Equations 14.59,

$$\psi = c_1\phi_1 + c_2\phi_2 + c_3\phi_3 + c_4\phi_4$$
(14.60)

as a trial function in a variational calculation. In other words, we use c_1, c_2, c_3, and c_4 as variational parameters and minimize

$$E = \frac{\iint d\tau_1 d\tau_2 \psi^* \hat{H} \psi}{\iint d\tau_1 d\tau_2 \psi^* \psi}$$
(14.61)

Ordinarily, variational calculations are not exact, but for this case Equation 14.60 represents *all* the possible two-proton spin functions, so the resultant ψ is as general as possible, and a variational calculation will yield an exact result. When Equation 14.61 is minimized with respect to the c_j's, we obtain a 4×4 secular determinantal equation (Section 7–2),

$$\begin{vmatrix} H_{11} - E & H_{12} & H_{13} & H_{14} \\ H_{12} & H_{22} - E & H_{23} & H_{24} \\ H_{13} & H_{23} & H_{33} - E & H_{34} \\ H_{14} & H_{24} & H_{34} & H_{44} - E \end{vmatrix} = 0$$
(14.62)

where

$$H_{ij} = \iint d\tau_1 d\tau_2 \phi_i^* \hat{H} \phi_j$$
(14.63)

When the determinant in Equation 14.62 is expanded, we obtain a fourth-degree polynomial in E, giving the four allowed energy levels of a two-spin system. We have evaluated integrals similar to the H_{ij} when we calculated the first-order corrections to the energies in the previous sections. All of them are fairly easy to evaluate using the relations given in Table 14.4.

EXAMPLE 14–8
Using the relations in Table 14.4, evaluate H_{12}.

SOLUTION: We must evaluate the integral

$$H_{12} = \iint d\tau_1 d\tau_2 \phi_1^* \hat{H} \phi_2$$

$$= \iint d\tau_1 d\tau_2 \alpha^*(1)\alpha^*(2)\hat{H}\alpha(1)\beta(2)$$

where $\hat{H}$ is given by Equation 14.58. Therefore, we must first evaluate terms such as

$$\hat{I}_{z1}\alpha(1)\beta(2) = \frac{\hbar}{2}\alpha(1)\beta(2)$$

$$\hat{I}_{z2}\alpha(1)\beta(2) = -\frac{\hbar}{2}\alpha(1)\beta(2)$$

$$\hat{I}_{x1}\hat{I}_{x2}\alpha(1)\beta(2) = \frac{\hbar^2}{4}\beta(1)\alpha(2)$$

$$\hat{I}_{y1}\hat{I}_{y2}\alpha(1)\beta(2) = \frac{\hbar^2}{4}\beta(1)\alpha(2)$$

$$\hat{I}_{z1}\hat{I}_{z2}\alpha(1)\beta(2) = -\frac{\hbar^2}{4}\alpha(1)\beta(2)$$

Therefore,

$$\hat{H}\phi_2 = \hat{H}\alpha(1)\beta(2) = -\left[h\nu_0(1-\sigma_1) + h\nu_0(1-\sigma_2) - \frac{hJ_{12}}{4}\right]\alpha(1)\beta(2)$$
$$+\frac{hJ_{12}}{2}\beta(1)\alpha(2)$$

and

$$H_{12} = -\left[h\nu_0(1-\sigma_1) + h\nu_0(1-\sigma_2) - \frac{hJ_{12}}{4}\right]\iint d\tau_1 d\tau_2 \alpha^*(1)\alpha^*(2)\alpha(1)\beta(2)$$
$$+\frac{hJ_{12}}{2}\iint d\tau_1 d\tau_2 \alpha^*(1)\alpha^*(2)\beta(1)\alpha(2)$$

But both of these integrals equal zero; the first because of the integration over $d\tau_2$ and the second because of the integration over $d\tau_1$.

Not only does $H_{12} = 0$, but most of the nondiagonal H_{ij} for this two-spin system are equal to zero. The only nonzero H_{ij} with $i \neq j$ is $H_{23} = H_{32}$. Using the result for $\hat{H}\phi_2 = \hat{H}\alpha(1)\beta(2)$ given in Example 14–8, we find that

$$H_{23} = \frac{hJ_{12}}{2}$$

When all the H_{ij} are evaluated (Problems 14–28 through 14–30), Equation 14.62 becomes

$$\begin{vmatrix} -d_1 - d_2 + \frac{hJ}{4} - E & 0 & 0 & 0 \\ 0 & -d_1 + d_2 - \frac{hJ}{4} - E & \frac{hJ}{2} & 0 \\ 0 & \frac{hJ}{2} & d_1 - d_2 - \frac{hJ}{4} - E & 0 \\ 0 & 0 & 0 & d_1 + d_2 + \frac{hJ}{4} - E \end{vmatrix} = 0 \quad (14.64)$$

where $d_1 = \frac{1}{2}h\nu_0(1 - \sigma_1)$, $d_2 = \frac{1}{2}h\nu_0(1 - \sigma_2)$, and we have dropped the 12 subscript on J for convenience.

When the secular determinant is expanded, we obtain two first-degree equations and one second-degree equation for the E's. These give

$$E_1 = -h\nu_0\left(1 - \frac{\sigma_1 + \sigma_2}{2}\right) + \frac{hJ}{4}$$

$$E_2 = -\frac{hJ}{4} - \frac{h}{2}[\nu_0^2(\sigma_1 - \sigma_2)^2 + J^2]^{1/2}$$

$$E_3 = -\frac{hJ}{4} + \frac{h}{2}[\nu_0^2(\sigma_1 - \sigma_2)^2 + J^2]^{1/2} \qquad (14.65)$$

$$E_4 = h\nu_0\left(1 - \frac{\sigma_1 + \sigma_2}{2}\right) + \frac{hJ}{4}$$

Note that Equations 14.65 reduce to the results given in Section 14.7 for two equivalent protons when $\sigma_1 = \sigma_2$.

The energies given by Equations 14.65 are sketched in Figure 14.16 for a two-spin system. The selection rules from Sections 14.6 and 14.7 give the allowed transitions $1 \rightarrow 2$, $1 \rightarrow 3$, $2 \rightarrow 4$, and $3 \rightarrow 4$ for non-equivalent protons (Section 14.6) and $1 \rightarrow 3$, $3 \rightarrow 4$ for equivalent protons (Section 14.7), as shown in Figure 14.16.

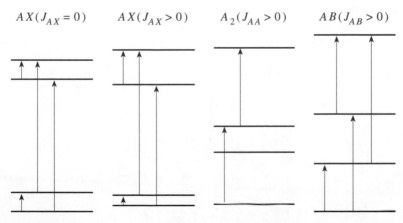

FIGURE 14.16
The four energy levels of a two-spin system for various relative values of $\nu_0|\sigma_1 - \sigma_2|$ and J. For the AX case, $\nu_0|\sigma_1 - \sigma_2| \gg J$; for A_2, $\nu_0|\sigma_1 - \sigma_2| = 0$; and for AB, $\nu_0|\sigma_1 - \sigma_2| \approx J$. The selection rules differ for non-equivalent and equivalent protons.

EXAMPLE 14–9

Determine the resonance frequency of the $1 \rightarrow 3$ transition for a two-spin system.

SOLUTION: Using Equations 14.65, we have

$$E_3 - E_1 = h\nu_{1\rightarrow3} = -\frac{hJ}{4} + \frac{h}{2}[\nu_0^2(\sigma_1 - \sigma_2)^2 + J^2]^{1/2}$$

$$+\frac{h\nu_0}{2}(2 - \sigma_1 - \sigma_2) - \frac{hJ}{4}$$

or

$$\nu_{1\rightarrow3} = \frac{\nu_0}{2}(2 - \sigma_1 - \sigma_2) - \frac{J}{2} + \frac{1}{2}[\nu_0^2(\sigma_1 - \sigma_2)^2 + J^2]^{1/2}$$

All four allowed resonance frequencies and their relative intensities are given in Table 14.6.

The observed spectra associated with the resonance frequencies and relative intensities given in Table 14.6 depend upon the relative values of $\nu_0|\sigma_1 - \sigma_2|$ and J, as shown in Figure 14.16. Note that for $J = 0$, there are just two separate singlets, as you would expect for two distinct hydrogen atoms with no coupling. At the other extreme, when $\sigma_1 = \sigma_2$, we have two chemically equivalent hydrogen atoms, with one signal, as in the case of dichloromethane (Figure 14.12). Note that even though $J \neq 0$ in this case, no coupling between chemically equivalent hydrogen atoms is observed.

TABLE 14.6

The four resonance frequencies and their relative intensities for a two-spin system.

Frequency	Relative intensity[a]
$\nu_{1\rightarrow2} = \dfrac{\nu_0}{2}(2 - \sigma_1 - \sigma_2) - \dfrac{J}{2} - \dfrac{1}{2}[\nu_0^2(\sigma_1 - \sigma_2)^2 + J^2]^{1/2}$	$(r - 1)^2/(r + 1)^2$
$\nu_{1\rightarrow3} = \dfrac{\nu_0}{2}(2 - \sigma_1 - \sigma_2) - \dfrac{J}{2} + \dfrac{1}{2}[\nu_0^2(\sigma_1 - \sigma_2)^2 + J^2]^{1/2}$	1
$\nu_{2\rightarrow4} = \dfrac{\nu_0}{2}(2 - \sigma_1 - \sigma_2) + \dfrac{J}{2} + \dfrac{1}{2}[\nu_0^2(\sigma_1 - \sigma_2)^2 + J^2]^{1/2}$	$(r - 1)^2/(r + 1)^2$
$\nu_{3\rightarrow4} = \dfrac{\nu_0}{2}(2 - \sigma_1 - \sigma_2) + \dfrac{J}{2} - \dfrac{1}{2}[\nu_0^2(\sigma_1 - \sigma_2)^2 + J^2]^{1/2}$	1

[a] where $r = \left[\dfrac{(\Delta^2 + J^2)^{1/2} + \Delta}{(\Delta^2 + J^2)^{1/2} - \Delta}\right]^{1/2}$ and $\Delta = \nu_0(\sigma_1 - \sigma_2)$

EXAMPLE 14–10

Show that there is only one singlet signal in the spectrum of an A_2 spin system.

SOLUTION: If $\sigma_1 = \sigma_2$, then $\Delta = 0$, and $r = 1$. Therefore, there is no signal (zero intensity) for the $1 \rightarrow 2$ and $2 \rightarrow 4$ transitions in Table 14.6. Furthermore, when $\sigma_1 = \sigma_2 = \sigma$,

$$\nu_{1\rightarrow 3} = \nu_{3\rightarrow 4} = \nu_0(1 - \sigma)$$

so there is only one singlet signal in the system.

For cases that are intermediate between the two extremes, $J = 0$ and $\sigma_1 = \sigma_2$, the spectrum can vary considerably (Figure 14.17). Such spectra are called *second-order spectra* and the $n + 1$ rule does not apply to such systems. Only for the case in which $J \ll \nu_0|\sigma_1 - \sigma_2|$ does the $n + 1$ rule apply, and the spectrum consists of two separated doublets of equal intensity, as shown in Figure 14.17. When $J \ll \nu_0|\sigma_1 - \sigma_2|$, we can

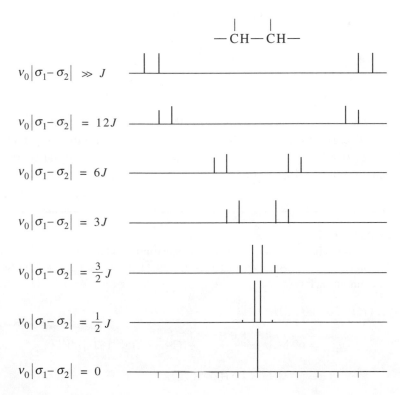

FIGURE 14.17

The splitting pattern of a two-spin system $-\overset{|}{C}H-\overset{|}{C}H-$ for various values of J and $\nu_0|\sigma_1 - \sigma_2|$.

write the square root term in Table 14.6 in the form

$$[\nu_0^2(\sigma_1 - \sigma_2)^2 + J^2]^{1/2} = \nu_0(\sigma_1 - \sigma_2)\left[1 + \frac{J^2}{\nu_0^2(\sigma_1 - \sigma_2)^2}\right]^{1/2}$$

and then use the fact that $J^2/\nu_0^2(\sigma_1 - \sigma_2)^2 \ll 1$. In this case, we can use the expansion

$$(1 + x)^{1/2} = 1 + \frac{x}{2} - \frac{x^2}{8} + \cdots$$

and write

$$[\nu_0^2(\sigma_1 - \sigma_2)^2 + J^2]^{1/2} = \nu_0(\sigma_1 - \sigma_2)\left[1 + \frac{J^2}{2\nu_0^2(\sigma_1 - \sigma_2)^2} + \cdots\right]$$

$$= \nu_0(\sigma_1 - \sigma_2) + \cdots$$

Therefore, keeping only terms linear in J, we have from Table 14.6

$$\nu_{1 \to 2} = \nu_0(1 - \sigma_1) - \frac{J}{2}$$

$$\nu_{3 \to 4} = \nu_0(1 - \sigma_1) + \frac{J}{2}$$

$$\nu_{1 \to 3} = \nu_0(1 - \sigma_2) - \frac{J}{2} \tag{14.66}$$

$$\nu_{2 \to 4} = \nu_0(1 - \sigma_2) + \frac{J}{2}$$

in agreement with our first-order perturbation theory treatment of an AX system (Equation 14.46). The general case, in which $\nu_0|\sigma_1 - \sigma_2|$ and J are of comparable magnitudes, must be handled by a computer. Fortunately, computer programs are available for analyzing second-order spectra.

EXAMPLE 14–11

Using the results in Table 14.6, compute the spectrum of a two-spin system for $\nu_0 = 60\,\text{MHz}$ and $\nu_0 = 270\,\text{MHz}$, given that $\sigma_1 - \sigma_2 = 0.24 \times 10^{-6}$ and $J = 8.0\,\text{Hz}$. Sketch the spectrum in each case.

SOLUTION: At 60 MHz,

$$\nu_{1 \to 2} = 60\,\text{MHz} - \frac{8.0\,\text{Hz}}{2} - \frac{1}{2}[(14.4\,\text{Hz})^2 + (8.0\,\text{Hz})^2]^{1/2}$$

$$= 60\,\text{MHz} - 4.0\,\text{Hz} - 8.2\,\text{Hz}$$

$$= 60\,\text{MHz} - 12.2\,\text{Hz}$$

$$\nu_{1 \to 3} = 60\,\text{MHz} - \frac{8.0\,\text{Hz}}{2} + 8.2\,\text{Hz}$$

$$= 60 \text{ MHz} + 4.2 \text{ Hz}$$

$$\nu_{2 \to 4} = 60 \text{ MHz} + \frac{8.0 \text{ Hz}}{2} + 8.2 \text{ Hz}$$

$$= 60 \text{ MHz} + 12.2 \text{ Hz}$$

$$\nu_{3 \to 4} = 60 \text{ MHz} - 4.2 \text{ Hz}$$

To calculate the relative intensities, we first must calculate r:

$$r = \left\{ \frac{[(14.4 \text{ Hz})^2 + (8.0 \text{ Hz})^2]^{1/2} + 14.4 \text{ Hz}}{[(14.4 \text{ Hz})^2 + (8.0 \text{ Hz})^2]^{1/2} - 14.4 \text{ Hz}} \right\}^{1/2} = 3.86$$

so the relative intensities are $(r-1)^2/(r+1)^2 = 0.35$ to 1. At 60 MHz, the ideal spectrum looks like this:

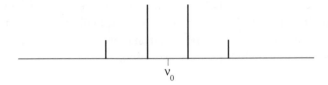

At 270 MHz,

$$\nu_{1 \to 2} = 270 \text{ MHz} - 4.0 \text{ Hz} - \frac{1}{2}[(64.8 \text{ Hz})^2 + (8.0 \text{ Hz})^2]^{1/2}$$

$$= 270 \text{ MHz} - 4.0 \text{ Hz} - 32.6 \text{ Hz}$$

$$= 270 \text{ MHz} - 36.6 \text{ Hz}$$

$$\nu_{1 \to 3} = 270 \text{ MHz} + 28.6 \text{ Hz}$$

$$\nu_{2 \to 4} = 270 \text{ MHz} + 36.6 \text{ Hz}$$

$$\nu_{3 \to 4} = 270 \text{ MHz} - 28.6 \text{ Hz}$$

For the intensities at 270 MHz,

$$r = \left\{ \frac{[(64.8 \text{ Hz})^2 + (8.0 \text{ Hz})^2]^{1/2} + 64.8 \text{ Hz}}{[(64.8 \text{ Hz})^2 + (8.0 \text{ Hz})^2]^{1/2} - 64.8 \text{ Hz}} \right\}^{1/2} = 16.3$$

so the relative intensities are $(r-1)^2/(r+1)^2 = 0.78$ to 1. The idealized 270-MHz spectrum (on the same scale as the 60-MHz spectrum) looks like this:

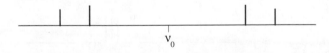

Note that the 270-MHz spectrum looks like a first-order spectrum consisting of two doublets with internal spacing $J = 8.0$ Hz, but that the 60-MHz spectrum looks like a second-order spectrum. The $n + 1$ rule works for the 270-MHz spectrum but not for the 60-MHz spectrum.

Figure 14.18 shows the calculated spectra for a three-spin system of the type $-CH_2-\overset{|}{C}H-$. In this case, the secular determinant is 8×8 because there is a total of 8 ($2 \times 2 \times 2 = 8$) spin wave functions. The calculations of the spectra are very similar to those for the two-spin system, except that the algebra is more involved. Notice from Figure 14.18 that if $J \ll v_0|\sigma_1 - \sigma_2|$, the $n + 1$ rule applies, and the spectrum consists of a separated doublet and a triplet, as in the case of 1,1,2-trichloroethane (Figure 14.8).

Figure 14.19a shows a 60-MHz spectrum of 1,2,3-trichlorobenzene. The chemical shifts of the two sets of chemically equivalent hydrogen atoms are similar enough that 1,2,3-trichlorobenzene must be treated as an AB_2 molecule. A comparison of Figures 14.18 and 14.19a suggests that $v_0|\sigma_1 - \sigma_2|/J$ is about 1.5. Figure 14.19b shows a 270-MHz spectrum of the same compound. In this case, $v_0|\sigma_1 - \sigma_2|$ is now large enough that the spectrum appears to be first order, with a separated doublet and triplet, as in the top entry in Figure 14.18. The use of an instrument of higher frequency or higher field strength is advantageous in this case, avoiding the complications of a second-order spectrum. Modern NMR spectrometers work at frequencies as high as 750 MHz, resulting in greatly improved resolution.

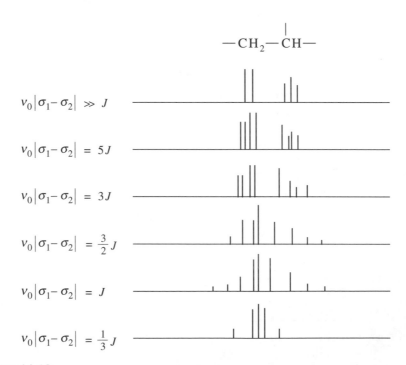

FIGURE 14.18

The splitting pattern of a three-spin system of the type $-CH_2-\overset{|}{C}H-$ for various relative values of J and $v_0|\sigma_1 - \sigma_2|$.

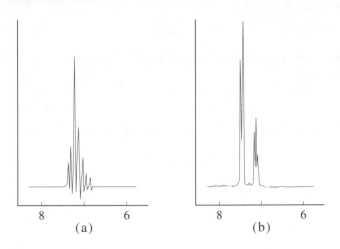

FIGURE 14.19
(a) A 60-MHz spectrum and (b) a 270-MHz spectrum of 1,2,3-trichlorobenzene. The 60-MHz spectrum is a second-order spectrum and the 270-MHz spectrum is first-order.

Problems

14-1. Show how Equation 14.7 reduces to Equation 14.6 for a circular orbit.

14-2. What magnetic field strength must be applied for C–13 spin transitions to occur at 90.0 MHz?

14-3. What magnetic field strength must be applied for proton spin transitions to occur at 270.0 MHz?

14-4. Calculate the magnetic field strength necessary to observe resonances of the nuclei given in Table 14.1 using a 300-MHz NMR spectrometer.

14-5. It turns out that a proton chemical shift of 2.2 ppm corresponds to a frequency range of 1100 Hz on a certain NMR instrument. Determine the magnetic field strength of this instrument.

14-6. Show that a chemical shift range of 8.0 ppm corresponds to a frequency range of 480 Hz on a 60-MHz instrument. What is the frequency range on a 270-MHz instrument?

14-7. Show that the top and bottom scales in Figure 14.6 are consistent.

14-8. Use Equation 14.21 to show that $B_{\text{TMS}} - B_{\text{H}}$ is directly proportional to δ_{H}, in analogy with Equation 14.23. Interpret this result.

14-9. Make a rough sketch of what you think the NMR spectrum of methyl acetate looks like.

14-10. Make rough sketches of what you think the NMR spectra of the two isomers dimethyl ether and ethanol look like and compare the two.

14-11. Make a rough sketch of what you think the NMR spectrum of diethyl ether looks like.

14-12. Make a rough sketch of what you think the NMR spectrum of 3-pentanone looks like.

14-13. Make a rough sketch of what you think the NMR spectrum of methyl propanoate looks like.

14-14. Make a rough sketch of what you think the NMR spectrum of ethyl acetate looks like.

14-15. Show that Equation 14.27 has units of joules.

14-16. Verify Equations 14.36 and 14.37.

14-17. Verify Equations 14.41 and 14.42.

14-18. The nuclear spin operators, $\hat{I}_x$, $\hat{I}_y$, and $\hat{I}_z$, like all angular momentum operators, obey the commutation relations (Problem 6–13)

$$[\hat{I}_x, \hat{I}_y] = i\hbar\hat{I}_z, \quad [\hat{I}_y, \hat{I}_z] = i\hbar\hat{I}_x, \quad \text{and} \quad [\hat{I}_z, \hat{I}_x] = i\hbar\hat{I}_y$$

Define the (non-Hermitian) operators

$$\hat{I}_+ = \hat{I}_x + i\hat{I}_y \quad \text{and} \quad \hat{I}_- = \hat{I}_x - i\hat{I}_y \tag{1}$$

and show that

$$\hat{I}_z\hat{I}_+ = \hat{I}_+\hat{I}_z + \hbar\hat{I}_+ \tag{2}$$

and

$$\hat{I}_z\hat{I}_- = \hat{I}_-\hat{I}_z - \hbar\hat{I}_- \tag{3}$$

14-19. Using the definitions of $\hat{I}_+$ and $\hat{I}_-$ from the previous problem, show that

$$\hat{I}_+\hat{I}_- = \hat{I}^2 - \hat{I}_z^2 + \hbar\hat{I}_z$$

and that

$$\hat{I}_-\hat{I}_+ = \hat{I}^2 - \hat{I}_z^2 - \hbar\hat{I}_z$$

where

$$\hat{I}^2 = \hat{I}_x^2 + \hat{I}_y^2 + \hat{I}_z^2$$

14-20. Use Equation 2 from Problem 14–18 and the fact that $\hat{I}_z\beta = -\frac{\hbar}{2}\beta$ to show that

$$\hat{I}_z\hat{I}_+\beta = \hat{I}_+\left(-\frac{\hbar}{2}\beta + \hbar\beta\right) = \frac{\hbar}{2}\hat{I}_+\beta$$

Because $\hat{I}_z\alpha = \frac{\hbar}{2}\alpha$, this result shows that

$$\hat{I}_+\beta \propto \alpha = c\alpha$$

where c is a proportionality constant. The following problem shows that $c = \hbar$, so we have

$$\hat{I}_+\beta = \hbar\alpha \tag{1}$$

Now use Equation 3 from Problem 14–18 and the fact that $\hat{I}_z\alpha = \frac{\hbar}{2}\alpha$ to show that

$$\hat{I}_-\alpha = c\beta$$

where c is a proportionality constant. The following problem shows that $c = \hbar$, so we have

$$\hat{I}_-\alpha = \hbar\beta \tag{2}$$

Notice that $\hat{I}_+$ "raises" the spin function from β to α, whereas $\hat{I}_-$ "lowers" the spin function from α to β. The two operators $\hat{I}_+$ and $\hat{I}_-$ are called raising and lowering operators, respectively.

Now argue that a consequence of the raising and lowering properties of $\hat{I}_+$ and $\hat{I}_-$ is that

$$\hat{I}_+\alpha = 0 \quad \text{and} \quad \hat{I}_-\beta = 0 \tag{3}$$

Now use Equations 1, 2, and 3 to show that

$$\hat{I}_x\alpha = \frac{\hbar}{2}\beta \qquad \hat{I}_y\alpha = \frac{i\hbar}{2}\beta$$

$$\hat{I}_x\beta = \frac{\hbar}{2}\alpha \qquad \hat{I}_y\beta = -\frac{i\hbar}{2}\alpha$$

14-21. This problem shows that the proportionality constant c in

$$\hat{I}_+\beta = c\alpha \quad \text{or} \quad \hat{I}_-\alpha = c\beta$$

is equal to $\hbar$. Start with

$$\int \alpha^*\alpha\,d\tau = 1 = \frac{1}{c^2}\int (\hat{I}_+\beta)^*(\hat{I}_+\beta)d\tau$$

Let $\hat{I}_+ = \hat{I}_x + i\hat{I}_y$ in the second factor in the above integral and use the fact that $\hat{I}_x$ and $\hat{I}_y$ are Hermitian to get

$$\int (\hat{I}_x\hat{I}_+\beta)^*\beta\,d\tau + i\int (\hat{I}_y\hat{I}_+\beta)^*\beta\,d\tau = c^2$$

Now take the complex conjugate of both sides to get

$$\int \beta^*\hat{I}_x\hat{I}_+\beta\,d\tau - i\int \beta^*\hat{I}_y\hat{I}_+\beta\,d\tau = c^2$$

$$= \int \beta^*\hat{I}_-\hat{I}_+\beta\,d\tau$$

Now use the result in Problem 14–19 to show that

$$c^2 = \int \beta^*\hat{I}_-\hat{I}_+\beta\,d\tau = \int \beta^*(\hat{I}^2 - \hat{I}_z^2 - \hbar\hat{I}_z)\beta\,d\tau$$

$$= \int \beta^*\left(\frac{3}{4}\hbar^2 - \frac{1}{4}\hbar^2 + \frac{\hbar^2}{2}\right)\beta\,d\tau = \hbar^2$$

or that $c = \hbar$.

14-22. Show that

$$
H_{y,11} = \frac{h J_{12}}{\hbar^2} \int\!\!\int d\tau_1 d\tau_2 \alpha^*(1)\alpha^*(2)\hat{I}_{y1}\hat{I}_{y2}\alpha(1)\alpha(2)
$$
$$
= 0
$$

and more generally that

$$
H_{x,jj} = H_{y,jj} = 0 \qquad j = 1,\ 2,\ 3,\ 4
$$

where $j = 1,\ 2,\ 3,\ 4$ refer to the four spin functions given by Equations 14.30.

14-23. Verify Equations 14.44.

14-24. Verify Equations 14.46.

14-25. Make a sketch like Figure 14.11 for a spectrum taken at 500 MHz.

14-26. For a first-order spectrum with (Equations 14.47)

$$
v_1^{\pm} = v_0(1 - \sigma_1) \pm \frac{J_{12}}{2}
$$

and

$$
v_2^{\pm} = v_0(1 - \sigma_2) \pm \frac{J_{12}}{2}
$$

show that the centers of the doublets are separated by $v_0|\sigma_1 - \sigma_2|$ and that the separations of the peaks within the two doublets is J_{12}.

14-27. Verify Equations 14.55 and 14.56.

14-28. Prove that

$$
H_{13} = \int\!\!\int d\tau_1 d\tau_2 \alpha^*(1)\alpha^*(2)\hat{H}\beta(1)\alpha(2) = 0
$$

with $\hat{H}$ given by Equation 14.58.

14-29. Prove that

$$
H_{11} = \int\!\!\int d\tau_1 d\tau_2 \alpha^*(1)\alpha^*(2)\hat{H}\alpha(1)\alpha(2)
$$
$$
= -\frac{1}{2}hv_0(1 - \sigma_1) - \frac{1}{2}hv_0(1 - \sigma_2) + \frac{h J_{12}}{4}
$$

with $\hat{H}$ given by Equation 14.58.

14-30. Prove that

$$
H_{44} = \int\!\!\int d\tau_1 d\tau_2 \beta^*(1)\beta^*(2)\hat{H}\beta(1)\beta(2)
$$
$$
= \frac{1}{2}hv_0(1 - \sigma_1) + \frac{1}{2}hv_0(1 - \sigma_2) + \frac{h J_{12}}{4}
$$

with $\hat{H}$ given by Equation 14.58.

14-31. Show that Equation 14.64 leads to Equation 14.65.

14-32. Sketch the splitting pattern of a two-spin system $-\overset{|}{C}H-\overset{|}{C}H-$ for $\nu_0|\sigma_1 - \sigma_2|/J = 20$, 10, 5, 2, 1, 0.10, and 0.01.

14-33. Show that a two-spin system with $J = 0$ consists of just two peaks with frequencies $\nu_0(1 - \sigma_1)$ and $\nu_0(1 - \sigma_2)$.

14-34. Show that

$$\nu_{1\to 2} = \frac{\nu_0}{2}(2 - \sigma_1 - \sigma_2) - \frac{J}{2} - \frac{1}{2}[\nu_0^2(\sigma_1 - \sigma_2)^2 + J^2]^{1/2}$$

for a general two-spin system (see Table 14.6).

14-35. Show that the frequencies given in Table 14.6 reduce to Equations 14.66 (and also Equations 14.46) when $J \ll \nu_0(\sigma_1 - \sigma_2)$.

14-36. Using the results in Table 14.6, compute the spectrum of a two-spin system for $\nu_0 = 60$ MHz and 500 MHz given that $\sigma_1 - \sigma_2 = 0.12 \times 10^{-6}$ and $J = 8.0$ Hz.

14-37. In Chapter 13, we learned that selection rules for a transition from state i to state j are governed by an integral of the form (Equation 13.52)

$$\int \psi_j^* \hat{H}^{(1)} \psi_i \, d\tau$$

where $\hat{H}^{(1)}$ is the Hamiltonian operator that causes the transitions from one state to another. In NMR spectroscopy, there are two magnetic fields to consider. There is the static field **B** that is produced by the magnets and aligns the nuclear spins of the sample. We customarily take this field to be in the z direction, and the nuclear (proton) spin states α and β are defined with respect to this field. Nuclear spin transitions occur when the spin system is irradiated with a radio-frequency field $\mathbf{B}_1 = \mathbf{B}_1^0 \cos 2\pi\nu t$. In this case,

$$\hat{H}^{(1)} = -\hat{\mu} \cdot \mathbf{B}_1 = -\gamma \hat{\mathbf{I}} \cdot \mathbf{B}_1$$

Show that the NMR selection rules are governed by integrals of the form

$$P_x = \int \psi_j^* \hat{I}_x \psi_i \, d\tau$$

with similar integrals involving $\hat{I}_y$ and $\hat{I}_z$. Now show that $P_x \neq 0$, $P_y \neq 0$, and $P_z = 0$, indicating that the radio-frequency field must be perpendicular to the static magnetic field.

14-38. Consider the two-spin system discussed in Section 14–6. In this case, the selection rule is governed by

$$P_x = \int d\tau_1 d\tau_2 \psi_j^* (\hat{I}_{x1} + \hat{I}_{x2}) \psi_i$$

with a similar equation for P_y. Using the notation given by Equations 14.30, show that the only allowed transitions are for $1 \to 2$, $1 \to 3$, $2 \to 4$, and $3 \to 4$.

14-39. Using the spin functions given by Equations 14.51, show that the only allowed transitions are $1 \to 3$ and $3 \to 4$.

Richard N. Zare was born in Cleveland Ohio, on November 19, 1939. In 1964, he received his Ph.D. in chemical physics from Harvard University, where his dissertation was on the angular distribution that would be expected when molecules are photodissociated by a beam of polarized light, and the calculation of Franck-Condon factors from potential energy curves constructed from spectroscopic constants. After teaching in the physics and chemistry departments at The Massachusetts Institute of Technology, the University of Colorado, and Columbia University, he joined Stanford University in 1977, where he is currently the Marguerite Blake Wilbur Professor of Chemistry. His research group is currently studying a variety of topics that range from the basic understanding of chemical reaction dynamics to the nature of the chemical contents of single cells. Zare has been a pioneer in using laser techniques to investigate a wide range of chemical problems. In 1988, he wrote *Angular Momentum: Understanding Spatial Aspects in Chemistry and Physics*, which has now become a classic text on angular momentum in quantum mechanics. Zare has received many awards and honors, but the most prestigious to date was his National Medal of Science in 1983. In 1996, he was elected chairman of the National Science Board, which reviews and approves the National Science Foundation's overall program plans and annual budget.

CHAPTER 15

Lasers, Laser Spectroscopy, and Photochemistry

The word *laser* is an acronym for *l*ight *a*mplification by *s*timulated *e*mission of *r*adiation. Lasers are used in a variety of devices and applications such as supermarket scanners, optical disk storage drives, compact disc players, ophthalmic and angioplastic surgery, and military targeting. Lasers have also revolutionized research in physical chemistry. Their impact on the field of spectroscopy and light-initiated reactions, or *photochemistry*, has been tremendous. Using lasers, chemists can measure the spectra and photochemical dynamics of molecules with high spectral or time resolution. Furthermore, the techniques are so sensitive that a single molecule can be studied. Every chemist today should know how lasers work and understand the unique properties of the light they generate.

To understand how a laser works, we first must learn about the various pathways by which an electronically excited atom or molecule can decay back to its ground state. The generation of laser light depends on the rates at which these excited atoms or molecules decay back to their ground states. Therefore, we will discuss a rate-equation model developed by Einstein that describes the dynamics of spectroscopic transitions between atomic energy levels. We will see that before we can even consider building a laser, we must understand transitions between more than two atomic energy levels. We will then discuss the general principles of laser design and describe some of the lasers used in research chemistry laboratories. In particular, we will illustrate the specifics of how a laser works by examining the helium-neon laser in detail. Using the laser spectroscopy of iodine chloride, $ICl(g)$, as an example, we will see that spectral features can be resolved by lasers that cannot be observed using conventional lamp-based spectrometers. We will then examine a photochemical reaction, the light-induced dissociation, or *photodissociation*, of $ICN(g)$. We will learn that the time required for the I–CN bond to break after absorption to a dissociative electronic state can be measured using lasers with outputs of femtosecond (10^{-15} s) light pulses.

591

15–1. Electronically Excited Molecules Can Relax by a Number of Processes

A molecule will not remain in an excited state indefinitely. After an excitation to an excited electronic state, a molecule invariably will relax back to its electronic ground state. Although we will consider a diatomic molecule to illustrate the mechanisms by which an electronically excited molecule can relax back to its ground state, our discussion also applies to polyatomic molecules. We will assume that the ground electronic state of the diatomic molecule is a singlet state, which we denote by S_0. Figure 15.1 shows a plot of the potential energy curves for the ground electronic state and the first excited singlet state, S_1, and the first excited triplet state, T_1. (Recall from Section 9–16 that the energy of the triplet state is less than that of the singlet state.) So that the various processes can be viewed easily, we assume that the equilibrium bond length in these three electronic states increases in the order $R_e(S_0) < R_e(S_1) < R_e(T_1)$. The vibrational levels of the ground and electronically excited states are indicated by the horizontal solid lines. The spacing between rotational levels is small compared with that between vibrational levels, so there are discrete rotational levels (not indicated in the figure) that lie between the indicated vibrational levels.

Absorption to S_1 produces a molecule that is in an excited vibrational (and possibly rotational) state, and from our study of the Franck-Condon principle in Section 13–7,

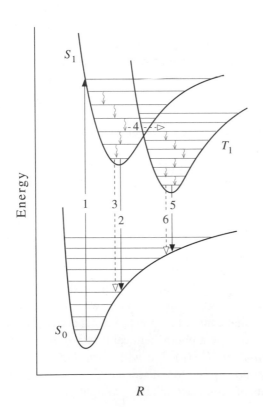

FIGURE 15.1

A schematic illustration of the absorption and the subsequent radiative and nonradiative decay pathways for an electronically excited diatomic molecule: 1, absorption from S_0 to S_1; 2, fluorescence (a radiative transition from S_1 to S_0); 3, internal conversion (a nonradiative transition from S_1 to S_0); 4, intersystem crossing (a nonradiative transition from S_1 to T_1); 5, phosphorescence (a radiative transition from T_1 to S_0); and 6, intersystem crossing (a nonradiative transition from T_1 to S_0). The wavy arrows between adjacent vibrational states illustrate the nonradiative process of vibrational relaxation.

we know that electronic transitions are depicted by vertical lines in a diagram such as in Figure 15.1. An excited-state molecule can relax by many different mechanisms. Transitions between energy levels that involve either the absorption or the emission of radiation are called *radiative transitions*. Transitions between energy levels that occur without the absorption or emission of radiation are called *nonradiative transitions*.

The various types of arrows in Figure 15.1 indicate the types of relaxation processes that can occur between the different energy levels shown. Solid arrows are used to depict radiative transitions, wavy arrows to indicate nonradiative transitions within a single electronic state, and dashed lines with an unfilled arrow tip to indicate nonradiative transitions between two electronic states. In the absence of collisions, an excited molecule can undergo only processes that conserve energy, and therefore the return to the ground state must involve the emission of a photon. Collisions between an excited molecule and other molecules in the sample, however, can result in an exchange of energy that removes some of the excess vibrational energy. This process is called *vibrational relaxation*. Because of vibrational relaxation, an excited molecule quickly relaxes to the lowest vibrational state of S_1. Once the molecule reaches the lowest vibrational state of S_1, it can relax to the ground state S_0 by either emitting a photon (a radiative process) or by exchanging energy in a collision such that it makes a nonradiative transition from the excited electronic state to one of the rotational-vibrational energy levels of the ground electronic state of the molecule. The radiative decay process involves a transition between states of the same spin multiplicity and is called *fluorescence*. The nonradiative decay process also involves the transition between states of the same spin multiplicity and is called *internal conversion*.

Notice that some of the vibrational and rotational states of the S_1 electronic state overlap the vibrational and rotational states of the T_1 electronic state in Figure 15.1. When such an overlap occurs, the molecule may undergo a nonradiative transition between states of different spin multiplicity, a process called *intersystem crossing*. Because intersystem crossing requires a change in the spin of an electron, it is usually a slower process than internal conversion. If intersystem crossing produces a molecule in the T_1 state with an excess of vibrational energy, then vibrational relaxation can occur in the T_1 state until the molecule reaches the $v = 0$ level of this state. Once the molecule reaches the lowest vibrational level in the T_1 state, it can relax to the ground electronic state by either emitting a photon (a radiative process) or by exchanging energy in a collision such that it makes a nonradiative transition from the excited electronic state to one of the rotational-vibrational energy levels of the ground electronic state. The radiative decay process involves a transition between states of different spin multiplicity ($T_1 \rightarrow S_0$) and is called *phosphorescence*. The nonradiative decay process also involves a transition between states of different spin multiplicity and is therefore another example of intersystem crossing. Because phosphorescence requires a change in the spin of an electron, it is usually a slower process than fluorescence. Because the T_1 state in Figure 15.1 is lower in energy than the S_1 state, phosphorescence occurs at a lower energy than fluorescence. Figure 15.1 and Table 15.1 summarize the various relaxation processes we have described.

Consider the absorption and fluorescence spectra of a diatomic molecule when $R_e(S_0) = R_e(S_1)$. In this case, the minimum of the potential curve for the S_1 state

TABLE 15.1

Typical time scales of the various processes by which a molecule in an excited electronic state can relax.

Process	Transition	Change in spin multiplicity	Time scale
Fluorescence	Radiative $S_1 \rightarrow S_0$	0	10^{-9} s
Internal conversion	Collisional $S_1 \rightarrow S_0$	0	10^{-7}–10^{-12} s
Vibrational relaxation	Collisional		10^{-14} s
Intersystem crossing	$S_1 \rightarrow T_1$	1	10^{-12}–10^{-6} s
Phosphoresence	$T_1 \rightarrow S_0$	1	10^{-7}–10^{-5} s
Intersystem crossing	$T_1 \rightarrow S_0$	1	10^{-8}–10^{-3} s

shown in Figure 15.1 would sit directly above that of the S_0 ground state. Figure 15.2 shows a plot of both the absorption and the fluorescence spectra in this case. The energy levels that give rise to the observed transitions are indicated above the spectra. The vibrational quantum numbers for the ground electronic state and excited electronic states are denoted by v'' and v', respectively. The absorption spectrum consists of a series of lines reflecting the transitions from the $v'' = 0$ level of the ground electronic state to the $v' = 0$, 1, 2, 3, ... levels of the excited electronic state. The data in Table 15.1 show that vibrational relaxation occurs more rapidly than electronic relaxation. Therefore, we can reasonably assume that the excited molecule relaxes to the lowest vibrational state of the S_1 electronic state before any fluorescence occurs. The fluorescence spectrum will then consist of a series of lines reflecting the transitions from the $v' = 0$ level of the excited electronic state to the $v'' = 0$, 1, 2, ... levels of the ground electronic

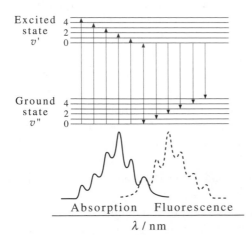

FIGURE 15.2
An illustration of fluorescence transitions of a diatomic molecule. Both the absorption and fluoresence spectra have peaks corresponding to a transition between $v'' = 0$ and $v' = 0$. The spacing between lines in the absorption spectrum is determined by the energy gap between vibrational states in the excited electronic state. The spacing between lines in the fluoresence spectrum is determined by the energy gap between vibrational states in the ground electronic state.

state. Note that both the absorption and fluorescence spectra will contain a transition between the $v'' = 0$ and $v' = 0$ levels, which is called the 0,0-transition. The remaining absorption bands occur at higher energy than the 0,0-transition, and the remaining fluorescence bands occur at lower energy than the 0,0-transition. The spacing between the lines in the absorption spectrum depends on the energy gaps between the vibrational levels of the excited electronic state. The spacing between the lines in the fluorescence spectrum depends on the energy gaps between the vibrational levels of the ground electronic state. If the vibrational frequencies of the ground electronic state and the excited electronic state are the same, the absorption and fluorescence spectra appear to be mirror images of one another, as shown in Figure 15.2. The relative intensities of the absorption and fluorescence lines can be determined using the Franck-Condon principle (see Section 13–7).

15–2. The Dynamics of Spectroscopic Transitions Between the Electronic States of Atoms Can Be Modeled by Rate Equations

To understand how lasers work, we need to learn about the rate at which atoms and molecules undergo radiative transitions. To illustrate the concepts of radiative decay, we will focus our discussion on atoms, so that we need consider only electronic states. Molecules can be treated in a similar way, but the mathematical equations become more complicated because of the need to include transitions among the various vibrational and rotational levels in addition to electronic levels. Actually, many lasers are based on the radiative properties of atomic transitions. A phenomenological approach that describes the rates of the various transitions between electronic states was proposed by Einstein at the beginning of this century. Einstein's approach is based on a few simple assumptions that account for how atoms absorb and emit photons. (His assumptions can be justified using time-dependent quantum mechanics.) The elegance of Einstein's approach is that no quantum mechanics is required except that the energy levels of the atom are assumed to be quantized.

Consider the interaction of light with a sample of N_{total} identical atoms. We will assume for simplicity that each atom has only two electronic levels, a ground level (with energy E_1) and an excited level (with energy E_2). Furthermore, we will assume that each level is nondegenerate and therefore, each level represents a single state of the system. The number of atoms in each state is designated by N_1 and N_2, respectively (see Figure 15.3). Because there are only two possible states that the atom can occupy,

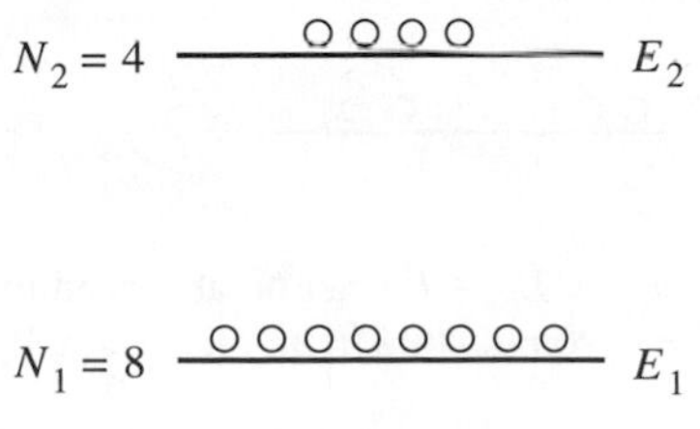

$N_2 = 4$ ⎯⎯ ○ ○ ○ ○ ⎯⎯ E_2

FIGURE 15.3
A schematic representation of a two-level energy diagram. Both levels are nondegenerate and therefore each one indicates a single state of the system. The circles represent the number of atoms in each state, eight in the ground state and four in the excited state.

$N_1 = 8$ ⎯⎯ ○ ○ ○ ○ ○ ○ ○ ○ ⎯⎯ E_1

$N_{\text{total}} = N_1 + N_2$. We will learn in Chapters 17 and 18 that the average energy of an atom or a molecule depends upon the kelvin temperature, T, and is of the order of $k_B T$, where k_B is the Boltzmann constant. Consequently, for atoms in which $E_2 - E_1$ is much greater than the thermal energy $k_B T$, the atoms do not have sufficient (thermal) energy to make a transition from state 1 to 2. Therefore, essentially all the atoms in the sample will be in the ground state, so $N_1 = N_{\text{total}}$. If we expose the sample to electromagnetic radiation of frequency v_{12}, where $hv_{12} = E_2 - E_1$, some of the atoms will absorb light and make a transition from state 1 to state 2, (Figure 15.4).

The energy density of the light is described by two related quantities. The *radiant energy density*, ρ, is defined as the radiant energy per unit volume and has units of $J \cdot m^{-3}$. The *spectral radiant energy density*, ρ_v, is a measure of the radiant energy density per unit frequency, $\rho_v = d\rho/dv$, and has units of $J \cdot m^{-3} \cdot s$. Because the transition between states 1 and 2 occurs only if light at $v = v_{12}$ is provided, we will be interested in the spectral radiant energy density at v_{12}, $\rho_v(v_{12})$, of the incident light source.

Einstein proposed that the rate of excitation from the ground electronic state to the excited electronic state is proportional to $\rho_v(v_{12})$ and to N_1, the number of molecules present in the ground state at time t. The rate of excitation from the ground electronic state to the excited electronic states is given by $-dN_1(t)/dt$, where the negative sign indicates that $N_1(t)$ decreases with increasing time. Because $-dN_1(t)/dt$ is proportional to both $\rho_v(v_{12})$ and $N_1(t)$, we can write

$$\text{rate} = -\frac{dN_1(t)}{dt} \propto \rho_v(v_{12})N_1(t)$$

or

$$\text{rate} = -\frac{dN_1(t)}{dt} = B_{12}\rho_v(v_{12})N_1(t) \tag{15.1}$$

where B_{12} is a proportionality constant called an *Einstein coefficient*. The "12" subscript of the B coefficient refers to the order of the states involved for the particular transition being discussed ($1 \rightarrow 2$). In the absence of any decay mechanism, the rate of growth of the excited-state population must be the negative of the rate of depletion of the

Absorption

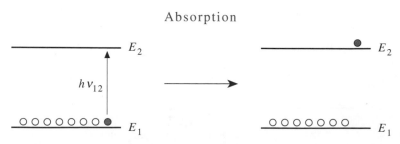

FIGURE 15.4
An illustration of the absorption process. Light of energy $hv_{12} = E_2 - E_1$ can be absorbed by an atom, which causes the atom to make a transition from the ground state to an electronically excited state.

ground-state population, $-dN_1(t)/dt = dN_2(t)/dt$, because $N_1(t) + N_2(t) = N_{total} =$ constant. Therefore,

$$-\frac{dN_1(t)}{dt} = \frac{dN_2(t)}{dt} = B_{12}\rho_\nu(\nu_{12})N_1(t) \qquad \text{(absorption only)} \qquad (15.2)$$

Note that $N_1(t)$ decreases and $N_2(t)$ increases with increasing time.

EXAMPLE 15–1

The output of a light source is usually given by a measure of its intensity. The intensity I is defined as the radiant energy per time that passes through a cross-sectional area perpendicular to the direction of propagation of the light. Show that

$$I = \rho c$$

where ρ is the radiant energy density and c is the speed of light.

SOLUTION: Consider a light beam of radiant energy dQ that passes through a cross-sectional area dA during a time period dt. The intensity is then defined as

$$I = \frac{dQ}{dt\,dA} \qquad (1)$$

and has units of $J \cdot s^{-1} \cdot m^{-2}$. During the time dt, the light beam travels a distance $c\,dt$. Therefore, dQ, the radiant energy that has passed through the cross-sectional area dA in the time dt is now contained in the volume $c\,dA\,dt$. The radiant energy density is then

$$\rho = \frac{dQ}{c\,dA\,dt} \qquad (2)$$

Solving Equation 1 for dQ and substituting the result into Equation 2 gives us $I = \rho c$. Note that we can also define a spectral intensity $I_\nu(\nu) = dI/d\nu$, which has units of $J \cdot m^{-2}$. Because $I = \rho c$, we see that the spectral intensity is related to the spectral radiant energy density by $I_\nu(\nu) = \rho_\nu(\nu)c$, where $\rho_\nu(\nu) = d\rho/d\nu$.

The above discussion accounts for only the absorption process. Atoms do not remain in excited states indefinitely, however. After a brief time, an atom emits energy and returns to the ground electronic state. Einstein's treatment proposes two pathways by which atoms relax back to the ground electronic state: *spontaneous emission* and *stimulated emission*. Spontaneous emission accounts for the process by which atoms spontaneously emit a photon of energy $h\nu_{12} = E_2 - E_1$ at some time after excitation (Figure 15.5). The rate at which spontaneous emission occurs can be described by $-dN_2(t)/dt$, where the negative sign indicates that $N_2(t)$ decreases with increasing time. We assume that the rate of spontaneous emission is simply proportional to the number of atoms in the excited state, $N_2(t)$, at time t. The proportionality constant

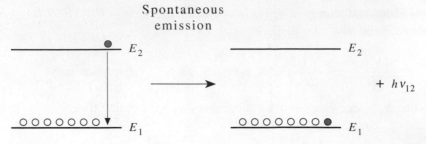

FIGURE 15.5
The spontaneous-emission process. Light of energy $h\nu_{12} = E_2 - E_1$ is emitted by an excited atom when the atom makes a transition from the electronically excited state to the ground state.

relating the rate of excited state decay, $-dN_2(t)/dt$, to the population of that level, $N_2(t)$, is given by another Einstein coefficient, A_{21}:

$$-\frac{dN_2(t)}{dt} = A_{21}N_2(t) \qquad \text{(spontaneous emission only)} \qquad (15.3)$$

In addition to spontaneous emission, Einstein proposed that the exposure of an atom in an excited electronic state to electromagnetic radiation of energy $h\nu_{12} = E_2 - E_1$ could stimulate the emission of a photon and thereby regenerate the ground-state atom (Figure 15.6). As for spontaneous emission, the rate of stimulated emission also depends on the number of excited molecules. Unlike spontaneous emission, however, the rate of stimulated emission is proportional to the spectral radiant energy density, $\rho_\nu(\nu_{12})$, in addition to $N_2(t)$, the number of atoms in state 2 at time t. The proportionality constant relating the rate of stimulated emission, $-dN_2(t)/dt$, to $\rho_\nu(\nu_{12})$ and $N_2(t)$, is given by a third Einstein coefficient, B_{21}. The order of the subscript of B is 21, indicating that

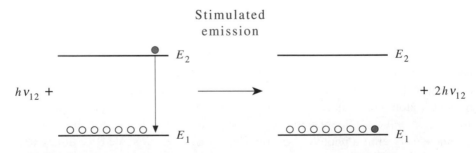

FIGURE 15.6
The stimulated-emission process. Incident light of energy $h\nu_{12} = E_2 - E_1$ stimulates an atom in an excited electronic state to emit a photon of energy $h\nu_{12}$ and thereby causes the atom to make a transition from the excited electronic state to the ground electronic state.

the transition takes place from the excited state (level 2) to the ground state $(2 \rightarrow 1)$. The rate of decay of $N_2(t)$ as a result of stimulated emission is given by

$$-\frac{dN_2(t)}{dt} = B_{21}\rho_\nu(\nu_{12})N_2(t) \qquad \text{(stimulated emission only)} \qquad (15.4)$$

Notice that the stimulated-emission process amplifies light intensity; one photon at frequency ν_{12} stimulates an atom to emit another, thus generating a second photon at frequency ν_{12}. In a large sample of atoms, this process can occur many times, resulting in a substantial amplification of an incident light beam at frequency ν_{12}. Lasers are devices that exploit the amplification of light through stimulated emission (recall that the word, laser, stands for *light amplification by stimulated emission of radiation*).

Upon exposure to light, a sample of atoms simultaneously undergoes all three processes, absorption, spontaneous emission, and stimulated emission. Thus, the rate of change in the population of either the ground electronic state or the excited electronic state must be the sum of the rates of these three individual processes, Equations 15.2, 15.3, and 15.4:

$$-\frac{dN_1(t)}{dt} = \frac{dN_2(t)}{dt} = B_{12}\rho_\nu(\nu_{12})N_1(t) - A_{21}N_2(t) - B_{21}\rho_\nu(\nu_{12})N_2(t) \qquad (15.5)$$

EXAMPLE 15–2
What are the units of the Einstein A and B coefficients?

SOLUTION: We can use Equations 15.3 and 15.4 to determine the units of the Einstein A and B coefficients. First, consider Equation 15.3:

$$-\frac{dN_2(t)}{dt} = A_{21}N_2(t)$$

Solving this equation for A_{21} gives us

$$A_{21} = -\left(\frac{1}{N_2(t)}\right)\left(\frac{dN_2(t)}{dt}\right)$$

The units of $1/N_2(t)$ and $dN_2(t)/dt$ are number^{-1} and number·s^{-1}, respectively, and so the units of A_{21} are s^{-1}.

Now consider Equation 15.4:

$$-\frac{dN_2(t)}{dt} = B_{21}\rho_\nu(\nu_{12})N_2(t)$$

Solving this equation for B_{21} gives us

$$B_{21} = -\left(\frac{1}{\rho_\nu(\nu_{12})N_2(t)}\right)\left(\frac{dN_2(t)}{dt}\right)$$

The units of the spectral radiant energy density of the electromagnetic radiation, $\rho_\nu(\nu_{12})$, are $J \cdot m^{-3} \cdot s$. Thus, the units of B_{21} are

$$(J^{-1} \cdot m^3 \cdot s^{-1} \cdot number^{-1})(number \cdot s^{-1}) = J^{-1} \cdot m^3 \cdot s^{-2} = kg^{-1} \cdot m$$

A comparison of Equations 15.2 and 15.4 shows that B_{12} and B_{21} have identical units.

The three Einstein coefficients (B_{12}, B_{21}, and A_{21}) turn out to be related to each other. We can see this relation by considering the limit at which the two energy states are in thermal equilibrium, in which case N_1 and N_2 no longer vary with time,

$$-\frac{dN_1(t)}{dt} = \frac{dN_2(t)}{dt} = 0 \tag{15.6}$$

and $\rho_\nu(\nu_{12})$ is the equilibrium spectral radiant energy density, which we can assume comes from a thermal blackbody radiation source. Recall that this quantity is given by Planck's blackbody distribution law (Equation 1.2),

$$\rho_\nu(\nu_{12}) = \frac{8\pi h}{c^3} \frac{\nu_{12}^3}{e^{h\nu_{12}/k_B T} - 1} \tag{15.7}$$

Now if we let $dN_1(t)/dt = 0$ in Equation 15.5 and solve for $\rho_\nu(\nu_{12})$, we obtain

$$\rho_\nu(\nu_{12}) = \frac{A_{21}}{(N_1/N_2)B_{12} - B_{21}} \tag{15.8}$$

We will learn in Chapter 17 that for a system in equilibrium at a temperature T, the number of atoms or molecules in the state j with energy E_j is given by

$$N_j = ce^{-E_j/k_B T} \tag{15.9}$$

where c is a proportionality constant. Using Equation 15.9 for states 1 and 2 gives us that

$$\frac{N_2}{N_1} = e^{-(E_2 - E_1)/k_B T} = e^{-h\nu_{12}/k_B T} \tag{15.10}$$

at equilibrium. If we use Equation 15.10 in Equation 15.8, we obtain

$$\rho_\nu(\nu_{12}) = \frac{A_{21}}{B_{12}e^{h\nu_{12}/k_B T} - B_{21}} \tag{15.11}$$

Equations 15.7 and 15.11 are equivalent only if (Problem 15–4)

$$B_{12} = B_{21} \tag{15.12}$$

and

$$A_{21} = \frac{8h\pi v_{12}^3}{c^3} B_{21} \tag{15.13}$$

Note that we had to include the stimulated-emission process to have consistency between the Einstein theory and Planck's blackbody radiation law.

15–3. A Two-Level System Cannot Achieve a Population Inversion

Lasers are designed to amplify light by the stimulated emission of radiation. For this amplification to occur, a photon that passes through the sample of atoms must have a greater probability of stimulating emission from an electronically excited atom than of being absorbed by an atom in its ground state. This condition requires that the rate of stimulated emission be greater than the rate of absorption, or that (see Equations 15.2 and 15.4)

$$B_{21}\rho_v(v_{12})N_2 > B_{12}\rho_v(v_{12})N_1 \tag{15.14}$$

Because $B_{21} = B_{12}$ (Equation 15.12), stimulated emission is more probable than absorption only when $N_2 > N_1$, or when the population of the excited state is greater than that of the lower state. Such a situation is called *population inversion*. According to Equation 15.10, N_2 must be less than N_1 because $hv_{12}/k_B T$ is a positive quantity. Consequently, a population inversion, for which $N_2 > N_1$, is a nonequilibrium situation. Thus, before we can expect light amplification, a population inversion between the upper and lower levels must be generated. Let's see if a population inversion can be achieved for the two-level system discussed in Section 15–2.

The rate equation for a nondegenerate two-level system is given by Equation 15.5:

$$-\frac{dN_1(t)}{dt} = \frac{dN_2(t)}{dt} = B\rho_v(v_{12})\{N_1(t) - N_2(t)\} - AN_2(t) \tag{15.15}$$

where we have dropped the indices for the Einstein coefficients because $B_{12} = B_{21}$ and the spontaneous emission process, A, occurs only from state 2 to state 1 in a two-level system. If we assume that all the atoms are in the ground state at time $t = 0$ so that $N_1 = N_{total}$ and $N_2 = 0$, Equation 15.15 gives us (Problem 15–5):

$$N_2(t) = \frac{B\rho_v(v_{12})N_{total}}{A + 2B\rho_v(v_{12})}\{1 - e^{-[A+2B\rho_v(v_{12})]t}\} \tag{15.16}$$

Figure 15.7 shows a plot of N_2/N_{total} as a function of time. The value of the excited-state population reaches a steady state as $t \to \infty$. If we let $t \to \infty$ in Equation 15.16, we find that

$$\frac{N_2(t \to \infty)}{N_{total}} = \frac{B\rho_v(v_{12})}{A + 2B\rho_v(v_{12})} \tag{15.17}$$

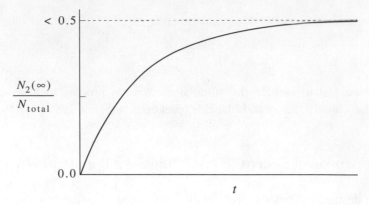

FIGURE 15.7
The ratio of the number of atoms in electronically excited states to the total number of atoms, N_2/N_{total}, is plotted as a function of time for a two-level system. The number of atoms in the excited state is always less than that in the ground state in a two-level system. Therefore, a two-level system can never achieve a population inversion.

Because $A > 0$, Equation 15.17 says that for all time t,

$$\frac{N_2}{N_{total}} = \frac{N_2}{N_1 + N_2} < \frac{1}{2} \tag{15.18}$$

Equation 15.18 reveals that the number of atoms in the excited state can never exceed the number of atoms in the ground state (Problem 15–8). Thus, a population inversion cannot occur in a two-level system.

EXAMPLE 15–3
Consider a two-level system. An incident light beam of energy $h\nu_{12} = E_2 - E_1$ is turned on for a while and then turned off. Describe how the system relaxes to equilibrium once the incident light source is turned off.

SOLUTION: Once the light source is turned off, the only pathway by which an excited atom can return to its ground state is by spontaneous emission. Because $\rho_\nu(\nu_{12}) = 0$, the rate equation (Equation 15.5) simplifies to

$$\frac{dN_2(t)}{dt} = -AN_2(t)$$

This equation can be integrated to give

$$N_2(t) = N_2(0)e^{-At}$$

The reciprocal of A is denoted by τ_R and is called the *fluorescence lifetime* or *radiative lifetime*.

15–4. Population Inversion Can Be Achieved in a Three-Level System

The ideas presented in Section 15–3 can be generalized to multilevel systems, and we will demonstrate here that a population inversion can be achieved in a three-level system. A schematic diagram of a three-level system is shown in Figure 15.8. Each level is once again assumed to be nondegenerate and therefore represents a single state of the system. In this figure, the ground state is labeled as 1 and has an energy E_1. We have drawn two excited states, labeled 2 and 3, which have energies E_2 and E_3, respectively. We will show that under certain conditions, a population inversion can be achieved between the two excited states (in other words, that $N_3 > N_2$). Once prepared, such a system provides a medium for the amplification of light of energy $h\nu_{32} = E_3 - E_2$ and is said to be able to lase.

Many rate processes are labeled in Figure 15.8. The double-headed arrows indicate that both absorption and stimulated emission occur between the two states. A single B coefficient is used for absorption and stimulated emission between a set of two states because we know that $B_{ij} = B_{ji}$. Initially, all atoms are in the ground state, so that $N_1(0) = N_{\text{total}}$. We consider the case in which this three-level system is exposed to an incident light beam of spectral radiant energy density, $\rho_\nu(\nu_{31})$ (where $h\nu_{31} = E_3 - E_1$), which excites atoms from level 1 to level 3. A light beam such as this one that is used to create excited-state populations is referred to as a *pump source*. The pump source is assumed to have no spectral radiant energy density at $h\nu_{12} = E_2 - E_1$, and as a result no atoms are excited to state 2. Once an atom populates state 3, it can decay by stimulated emission back to state 1 (induced by the pump source) or by spontaneous emission to either state 2 or state 1. The rates of spontaneous emission to state 2 and state 1 can be different; thus, we must include subscripts on the A coefficients to indicate explicitly the two states involved in the transition. An atom that relaxes from state 3 to state 2 can in turn relax back to the ground state by spontaneous emission. If light of frequency ν_{32} ($h\nu_{32} = E_3 - E_2$) is available, both absorption and stimulated emission can occur between states 3 and 2. Light of this energy is inevitably

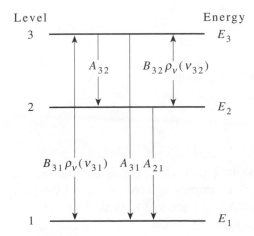

Level   Energy

3 ——————————— E_3

   A_{32} $B_{32}\rho_v(\nu_{32})$

2 ——————————— E_2

$B_{31}\rho_v(\nu_{31})$ $A_{31}\ A_{21}$

1 ——————————— E_1

FIGURE 15.8

A three-level energy diagram. Pump light of frequency given by $h\nu_{13} = E_3 - E_1$ excites an atom from the ground state (state 1) to state 3. Once populated, this excited state can relax by spontaneous emission to states 2 or 1 or by stimulated emission back to the ground state. Those excited-state atoms that relax by spontaneous emission to state 2 will also undergo spontaneous emission to state 1. If light of energy $h\nu_{32} = E_3 - E_2$ is incident on the system, absorption and stimulated emission can occur between the excited states 3 and 2.

available because it is generated by the spontaneous-emission process between these two levels.

For a three-level system, the sum of the populations of the individual energy levels is equal to the total number of atoms:

$$N_{total} = N_1(t) + N_2(t) + N_3(t) \tag{15.19}$$

The processes indicated in Figure 15.8 give rise to rate equations for each of the three energy levels (Problem 15–11). Because each level is nondegenerate, these rate equations apply to the populations of states 1, 2, and 3. When the system achieves equilibrium, the population of each level will remain constant, so $dN_1(t)/dt = 0$, $dN_2(t)/dt = 0$, and $dN_3(t)/dt = 0$. Although the three rate equations can be written and solved exactly to generate expressions for the time-dependent and equilibrium values of N_1, N_2, and N_3, we can learn an important result by considering only the rate equation for state 2. The population of state 2, N_2, is a balance between spontaneous emission from state 3 to state 2 ($A_{32}N_3$), spontaneous emission from state 2 to state 1 ($A_{21}N_2$), stimulated emission from state 3 to state 2 $[\rho_v(v_{32})B_{32}N_3]$, and absorption from state 2 to state 3 $[\rho_v(v_{32})B_{32}N_2]$ (Figure 15.8). When equilibrium is achieved, $dN_2(t)/dt = 0$ and

$$\frac{dN_2(t)}{dt} = 0 = A_{32}N_3 - A_{21}N_2 + \rho_v(v_{32})B_{32}N_3 - \rho_v(v_{32})B_{32}N_2 \tag{15.20}$$

Equation 15.20 can be rearranged to become

$$N_3[A_{32} + B_{32}\rho_v(v_{32})] = N_2[A_{21} + B_{32}\rho_v(v_{32})] \tag{15.21}$$

or

$$\frac{N_3}{N_2} = \frac{A_{21} + B_{32}\rho_v(v_{32})}{A_{32} + B_{32}\rho_v(v_{32})} \tag{15.22}$$

Notice that N_3 can be larger than N_2 if $A_{21} > A_{32}$. Therefore, a population inversion is possible between states 3 and 2 when the atoms excited to state 3 decay relatively slowly to state 2 and those in state 2 decay rapidly back to the ground state. If this is the case, a population of state 3 can be built up, and a system of atoms that satisfies this condition may lase. Such a system is called a *gain medium*.

15–5. What is Inside a Laser?

Lasers are composed of three essential elements (Figure 15.9): (1) a gain medium that amplifies light at the desired wavelength, (2) a pumping source that excites the gain medium, and (3) mirrors that direct the light beam back and forth through the gain medium. We will discuss each of these components in turn.

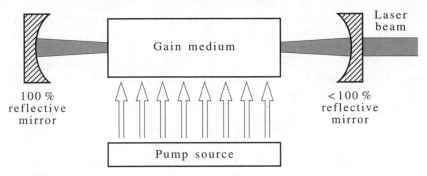

FIGURE 15.9
A diagram of the insides of a laser. The gain medium is placed between two mirrors; the arrangement of these components is called the laser cavity. A pump source excites the atoms, molecules, or ions that constitute the gain medium. The radiation that is emitted by the excited-state atoms is directed back and forth through the gain medium using the mirrors. One of the mirrors has a reflectivity that is less than 100%, which allows light to escape the cavity. This output light is the laser beam.

Gain medium:

The gain medium of a laser can be a solid-state material, a liquid solution, or a gas mixture. Since the report of the first laser in 1960, many different media have been used. In the following paragraphs, we discuss only a few of the materials currently used as laser gain media.

The first laser used a solid ruby rod as a lasing medium. Ruby is a crystal of corundum, Al_2O_3, in which some of the Al^{3+} ions are replaced by impurity Cr^{3+} ions. The impurity Cr^{3+} ions are the source of the laser light because the photophysical properties of the electronic energy levels of Cr^{3+} in the Al_2O_3 host crystal are suitable for achieving a population inversion. Naturally occurring ruby is unsuitable as a laser gain medium because of its strains and crystal defects, so ruby lasers use synthetic rods grown from molten mixtures of Cr_2O_3 and Al_2O_3. A typical chromium doping level is about 0.05% by mass. There are many solid-state gain media, like ruby, in which the active ion (Cr^{3+} for ruby) is embedded in a host material (Al_2O_3 for ruby). Examples of different solid-state gain media are given in Table 15.2 along with the wavelength of the laser light produced. Many commercially available lasers use Nd^{3+} as the gain medium. Note from the information given in Table 15.2 for various Nd^{3+} lasers that the host material can affect the wavelength of the laser light produced. Laser output can be a continuous light beam or a short burst of light. We see in Table 15.2 that solid-state lasers can produce both continuous and pulsed laser output.

EXAMPLE 15–4
A Nd^{3+}:YAG (YAG stands for yttrium-aluminum garnet) laser produces pulses at a repetition rate of 1 kHz. If each pulse is 150 ps in duration and has a radiant energy of

1.25×10^{-3} J, calculate P, the radiant power of each laser pulse, and $\langle P \rangle$, the average radiant power of the laser. Also, calculate the number of photons in a single pulse.

SOLUTION: Radiant power is a measure of radiant energy per unit time and has units of watts $(1 \text{ W} = 1 \text{ J}\cdot\text{s}^{-1})$. Therefore the radiant power P of an individual laser pulse from the described laser is

$$P = \frac{1.25 \times 10^{-3} \text{ J}}{150 \times 10^{-12} \text{ s}} = 8.3 \times 10^6 \text{ W} = 8.3 \text{ MW}$$

The average radiant power of the laser is a measure of the total power emitted per second by the laser, or

$$\langle P \rangle = (1000 \text{ pulses}\cdot\text{s}^{-1})(1.25 \times 10^{-3} \text{ J}\cdot\text{pulse}^{-1}) = 1.25 \text{ W}$$

A Nd^{3+}:YAG laser produces light at $\lambda = 1064.1$ nm (Table 15.2). The radiant energy of a 1064.1-nm photon, Q_p, is

$$Q_p = h\nu = \frac{hc}{\lambda} = \frac{(6.626 \times 10^{-34} \text{ J}\cdot\text{s})(2.998 \times 10^8 \text{ m}\cdot\text{s}^{-1})}{1064.1 \times 10^{-9} \text{ m}}$$
$$= 1.867 \times 10^{-19} \text{ J}$$

The pulse radiant energy, Q, is given by $Q = nQ_p$, where n is the number of photons in the laser pulse. Therefore, the number of 1064.1-nm photons in a 1.25×10^{-3} J laser pulse is

$$n = \frac{Q}{Q_p} = \frac{1.25 \times 10^{-3} \text{ J}}{1.867 \times 10^{-19} \text{ J}} = 6.70 \times 10^{15}$$

Examples of different gas-phase gain media and the wavelengths produced by the lasers that use them are listed in Table 15.3. The active element in a gas-phase laser can be a noble-gas atom (e.g., the He–Ne laser, see Section 15–6), a positive ion (e.g., Ar^+ laser, K^+ laser), a metal atom (e.g., He–Cd laser, Cu vapor laser), a neutral molecule (e.g., N_2 laser, CO_2 laser), or an unstable complex created by the pumping process (e.g.,

TABLE 15.2
The gain medium (active ion and host) and laser wavelength of various solid-state lasers.

Active ion	Host	Wavelength/nm	Output[a]	Duration
Cr^{3+}	Al_2O_3	694.3	Pulsed	10 ps
Nd^{3+}	$Y_3Al_5O_{15}$ (YAG)	1064.1	Both	10–150 ps
Nd^{3+}	$Y_3Li_xF_y$ (YLF)	1054.3	Both	10–100 ps
Nd^{3+}	Glass	1059	Pulsed	1 ps
Ti^{3+}	Al_2O_3 (sapphire)	780	Both	10 fs–5 ps

[a] The term "both" refers to both continuous and pulsed outputs.

TABLE 15.3
The gain medium and laser wavelength of various gas-phase lasers.

Gain medium	Wavelength/nm	Output	Pulse duration
$He(g)$, $Ne(g)$	3391, 1152, 632, 544	Continuous	Continuous
$N_2(g)$	337	Pulsed	1 ns
$Ar^+(g)$	488, 515	Continuous	Continuous
$K^+(g)$	647	Continuous	Continuous
$CO_2(g)$, $He(g)$, $N_2(g)$	Line tunable around 10 000	Pulsed	≥ 100 ns
$Cu(g)$	510	Pulsed	30 ns
$He(g)$, $Cd(g)$	441, 325	Continuous	Continuous

XeCl*). The wavelength data given in Table 15.3 reveal that gas-phase lasers produce light in the ultraviolet, visible, and infrared regions of the spectrum. Some of these lasers are capable of generating light at several different frequencies. For example, the CO_2 laser involves population inversion (and therefore lasing) between different rotational-vibrational levels of the electronic ground state. Laser light can be generated in small, discrete frequency steps dictated by the energy separation of the rotational levels of CO_2. Figure 15.10 shows the frequencies of laser light that can be generated from a population inversion between the first excited asymmetric stretch and both the first excited symmetric stretch and the second excited state of the bending mode of $CO_2(g)$.

Because the energy of laser light must correspond to an energy difference between two quantized stationary states of the gain medium, the laser light must be monochromatic (single color). The electric field of a monochromatic light source can be expressed as $E = A\cos(\omega t + \phi)$, where A is the amplitude, ω is the angular frequency of the light ($\omega = 2\pi\nu$), and ϕ is the phase angle, which serves to reference the field to some fixed point in time. The phases of light waves from a lamp vary randomly ($0 \leq \phi \leq 2\pi$). In contrast, the stimulated-emission process requires that the phases of the incident light wave and stimulated light wave have the same phase. Thus, the light waves emitted from a laser are all in phase. This property of laser light is called *coherence*. Many modern spectroscopic techniques take advantage of the coherence of laser light. We will not discuss these techniques in this text, but you should be aware of this unique property of laser light.

Pumping sources:

There are two common approaches for pumping the gain medium: optical excitation and electrical excitation. In optical excitation, a high-intensity light source is used to excite the gain medium. Devices that use continuous lamps, flashlamps, and lasers as pumping sources are commercially available. Figure 15.11 shows the optical pumping arrangement used for the first ruby laser (a solid-state gain medium of Cr^{3+} doped into Al_2O_3). The ruby rod was surrounded by a high-intensity helical flashlamp.

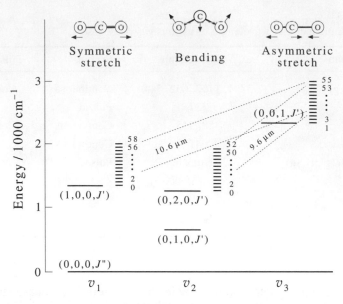

FIGURE 15.10

The energies of several of the low vibrational-rotational states of CO_2 (v_1, v_2, v_3, J). Laser light is generated by stimulating the emission between a pair of quantum states. The approximate wavelengths for the transitions between the $(0,0,1,J')$ upper state and the $(1,0,0,J')$ and $(0,2,0,J')$ lower states are indicated. The exact wavelength depends upon the rotational quantum numbers of the upper and lower states. Because the rotational states have discrete energies, the CO_2 laser is not continuously tunable.

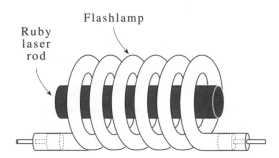

FIGURE 15.11

The arrangement for optically pumping a ruby rod, the gain medium of a ruby laser. The ruby rod is placed inside a helical flashlamp. The flashlamp emits a short burst of high-intensity light. This light excites the Cr^{3+} atoms in the ruby rod, a process that creates a population inversion between two electronic states of the Cr^{3+} atoms.

It is interesting to consider how efficiently lasers convert the energy of the pumping source into laser light. The energy of the pumping source places an upper limit on the energy output of a laser. Because of the discrete energy levels of the gain medium, a large fraction of the pump light is generally not absorbed by the gain medium. Thus, lasers tend to be inefficient devices. For example, gas lasers convert only between 0.001% and 0.1% of the input energy into laser light. Solid-state lasers are more efficient, approaching values of a few percent. A few lasers such as the CO_2 laser and some semiconductor lasers exhibit conversion efficiencies as high as 50% to 70%. If the pump source populated only the upper lasing level, improved efficiency could be achieved. This fact has prompted the use of lasers themselves as pump sources for other lasers.

Electrical excitation involves using intense electrical discharges to excite the gain material. This approach is commonly used in gas lasers. The discharge of a large current through the gain medium can be done continuously or in short pulses. Collisions between the high-energy electrons created in the discharge and the atoms or molecules in the gas container produce atoms, molecules, or ions in excited states. One of the ways that electrical pumping produces a population inversion is examined in detail in our discussion of the helium-neon (He-Ne) laser in the next section.

Laser cavity design:

Combining a gain medium with a pumping source does not make a laser. Once a population inversion is achieved, light of a specific frequency can be amplified as we demonstrated in Section 15–4. Unfortunately, a single pass of light through the gain medium generally does not produce much amplification in the light intensity. To generate high-intensity outputs, the light must be directed back and forth through the gain medium. Lasers accomplish this feat by having the gain medium placed inside an optical cavity called a *resonator*, which usually includes a pair of mirrors that direct the light back and forth through the gain medium. Only the light that travels back and forth along the path defined by the gain medium and the cavity mirrors can be amplified. If both mirrors were 100% reflective, the device would not create any output. In a laser, one of the mirrors is 100% reflective, and the other is less than 100% reflective, thereby allowing some of the light to escape from the resonator.

15–6. The Helium-Neon Laser Is an Electrical-Discharge Pumped, Continuous-Wave, Gas-Phase Laser

In 1961, the first continuous-wave laser was reported. This device, using a mixture of gaseous helium and neon as the gain medium and a direct current power supply as the pumping source, produced light in the infrared region at 1152.3 nm. In 1962, it was demonstrated that by proper choice of resonator mirrors, a He-Ne laser could also generate light at 632.8 nm (red light). Today, commercial He-Ne lasers are available that produce light at 3391.3 nm, 1152.3 nm, 632.8 nm, or 543.5 nm. The He-Ne laser is a low-power laser, producing an output radiant power on the order of a few milliwatts. The He-Ne laser is a widely produced laser; it is used in such devices as supermarket scanners, range finders, and Fourier-transform spectrometers.

The design of the He-Ne laser is shown in Figure 15.12. A glass cell containing the gas mixture serves as the gain medium. (A typical gas pressure in the cell is about 1.0 torr of helium and 0.1 torr of neon, or about an order of magnitude more helium

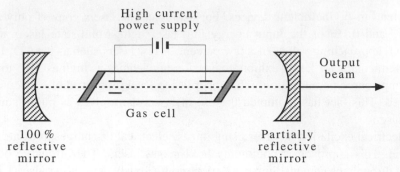

FIGURE 15.12
A schematic diagram of the helium-neon laser. A mixture of helium and neon gases is contained in a gas cell. A pair of electrodes located inside the gas cell is connected to an external high-current power supply. This circuit causes electrons to flow between the two electrodes in the cell. Collisions between the electrons and the gas atoms excite the atoms, which creates the population inversion required for lasing.

than neon.) Near each end of the gas cell is a mirror. One of the mirrors is 100% reflective at the desired lasing frequency; the other is partially reflective. Inside the gas cell are electrodes that are connected to the positive and negative poles of a direct current electrical power supply. The completed circuit causes high-energy electrons to travel through the gas cell from the cathode to the anode. While traversing the cell, these high-energy electrons collide with the gas-phase atoms. The transfer of energy during these collisions generates atoms in excited states. Because the power supply provides a steady flow of electrons, the excitation process occurs continuously. The collisions between the electrons and the atoms serve as the pump source for this laser. To understand how this excitation process ends up generating laser light, we need to examine the energy levels of helium and neon, which are diagrammed in Figure 15.13.

Because the concentration of helium in the glass cell is an order of magnitude greater than that of neon, the energetic electrons provided by the electric current have a much greater probability of colliding with helium atoms than with neon atoms. These collisions generate helium atoms in a variety of excited electronic states. We will be concerned with two relatively long-lived excited states, the $2s^3S_1$ and $2s^1S_0$ states (the $2s$ in front of $^3S_1$ and $^1S_0$ indicates that one of the electrons in the helium atom is excited to the $2s$ state), which lie $159\,809$ cm^{-1} and $166\,527$ cm^{-1} above the ground state (see Figure 15.13). The lifetimes of the $2s\ ^3S_1$ and $2s\ ^1S_0$ states are 10^{-4} s and 5×10^{-6} s, respectively. For a typical gas pressure in the cell of about 1.0 torr of helium and 0.1 torr of neon, the time between collisions of a helium atom with a neon atom is about 10^{-7} s, which is less than the lifetimes of the $2s^3S_1$ and $2s^1S_0$ states of helium.

Figure 15.13 reveals a fortuitous near equivalence of the energies of the set of excited states corresponding to the electron configurations $2p^54s$ and $2p^55s$ of neon with the energies of the $2s^3S_1$ and $2s^1S_0$ excited states of helium. Consequently, there is

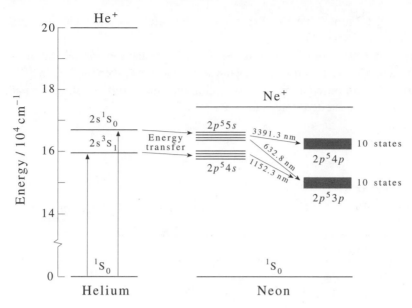

FIGURE 15.13
The electron configurations and the energies of several electronic excited states of helium and neon. An electrical discharge produces helium atoms in the $2s\,^3S_1$ and $2s\,^1S_0$ excited states. The energics of these states are similar to the sets of four states ($^3P_2$, $^3P_1$, $^3P_0$, $^1P_1$) corrresponding to the $2p^54s$ and $2p^55s$ electron configurations of neon, and so nonradiative energy transfer from helium to neon occurs readily upon collision. The lifetimes of the excited states of neon produced by these collisions are such that population inversion can be achieved. Several of the transitions used to generate laser light are indicated in the figure.

a high probability that a nonradiative energy transfer occurs during a collision between an excited helium atom, He*, and a ground-state neon atom:

$$He^*(2s^3S_1) + Ne(g) \longrightarrow He(g) + Ne^*(2p^54s)$$

$$He^*(2s^1S_0) + Ne(g) \longrightarrow He(g) + Ne^*(2p^55s)$$

As Figure 15.13 shows, there is a set of excited states of neon associated with the $2p^53p$ electron configuration that lie at a lower energy than those associated with the $2p^54s$ and $2p^55s$ electron configurations. The lifetime of these respective states are such that a population inversion can be set up and maintained. There are four states ($^3P_2$, $^3P_1$, $^3P_0$, and $^1P_1$) associated with the $2p^5ns$ electron configurations and ten states ($^3D_3$, $^3D_2$, $^3D_1$, $^1D_2$, $^3P_2$, $^3P_1$, $^3P_0$, $^1P_1$, $^3S_1$, and $^1S_0$) associated with the $2p^5np$ electron configuration. Several lasing transitions are indicated by solid arrows on the right side of Figure 15.13. The output wavelength of the first He-Ne laser was 1152.3 nm, corresponding to the $4s^1P_1 \rightarrow 3p^3P_2$ transition of neon (Figure 15.13). The most widely used He-Ne laser produces light at 632.8 nm (red light) by amplifying the emission from the $5s^1P_1 \rightarrow 3p^3P_2$ transition. A He-Ne gas mixture can also be made to lase at 3391.3 nm. Table 15.4 lists some of the data associated with neon transitions that have been observed to lase.

You may have noticed that the $4s\,{}^1P_1 \rightarrow 3p\,{}^3P_2$ transition and several of the transitions shown in Table 15.4 do not obey the selection rules given in Section 8–11. Those selection rules are based upon the assumption that spin-orbit coupling is small, which is not so for these excited states of neon.

TABLE 15.4
The wavelength and Einstein A coefficient for several of the transitions of a neon atom.

Transition	λ/nm	$A/10^6\ \text{s}^{-1}$	Relative intensity
$5s\,{}^1P_1 \rightarrow 3p\,{}^1S_0$	730.5	0.48	30
$5s\,{}^1P_1 \rightarrow 3p\,{}^3P_1$	640.1	0.60	100
$5s\,{}^1P_1 \rightarrow 3p\,{}^3P_0$	635.2	0.70	100
$5s\,{}^1P_1 \rightarrow 3p\,{}^3P_2$	632.8	6.56	300
$5s\,{}^1P_1 \rightarrow 3p\,{}^1P_1$	629.4	1.35	100
$5s\,{}^1P_1 \rightarrow 3p\,{}^1D_2$	611.8	1.28	100
$5s\,{}^1P_1 \rightarrow 3p\,{}^3D_1$	604.6	0.68	50
$5s\,{}^1P_1 \rightarrow 3p\,{}^3D_2$	593.9	0.56	50

EXAMPLE 15–5
The atomic term symbols of the four states corresponding to the $2p^5 3s$ electron configuration are ${}^3P_2$, ${}^3P_1$, ${}^3P_0$, and ${}^1P_1$ in order of increasing energy. To indicate that these states arise from an electron promoted to a $3s$ orbital, we designate these states as $3s\,{}^3P_2$, etc. Show that these four states account for all the states associated with the $2p^5 3s$ electron configuration.

SOLUTION: There are two spin orbitals associated with the $3s$ orbital and six associated with the (empty) $2p$ orbital. Thus, there are $2 \times 6 = 12$ microstates in a table of microstates corresponding to $2p^5 3s$. Using the fact that each term symbol (without the J subscript) accounts for $(2S + 1)(2L + 1)$ microstates, we have

$$3 \times 3 + 1 \times 3 = 12$$
$${}^3P {}^1P$$

15–7. High-Resolution Laser Spectroscopy Can Resolve Absorption Lines that Cannot Be Distinguished by Conventional Spectrometers

Figure 15.14 shows part of the absorption spectrum of ICl(g) measured by a conventional absorption spectrometer. The displayed spectrum consists of two absorption lines in the vicinity of $17\,299$ cm^{-1}. The separation between the absorption lines is approximately 0.2 cm^{-1}. These two absorption lines correspond to transitions from the $v'' = 0$, $J'' = 2$ state of the ground electronic state to two different rotational levels in the $v' = 32$ level of the first electronic excited state (see Problem 15–27).

Spectrometers have an inherent limit in their ability to resolve light of different wavenumbers. The range over which a spectrometer cannot distinguish a difference in wavenumbers is called the *spectral resolution* of the spectrometer. In the visible region of the electromagnetic spectrum, the spectral resolution of a spectrometer that uses a lamp as the light source is about 0.03 cm^{-1}. As a result, a lamp-based spectrometer will be unable to distinguish two absorption bands that differ by 0.03 cm^{-1}. We learned in Section 15–5 that lasers generate monochromatic radiation. Actually, there is an inherent linewidth associated with the "monochromatic" light from a laser. For lasers that emit visible light, the output beam can have a spectral width as small as 3.0×10^{-5} cm^{-1}.

We now ask whether or not the width of the absorption lines shown in Figure 15.14 are an intrinsic property of the ICl(g) molecule or are limited by the spectral resolution of the lamp-based instrument. The spectrum of ICl(g) recorded using a tunable laser with a linewidth of $\approx 3.0 \times 10^{-5}$ cm^{-1} is shown in Figure 15.15. With this higher resolution, each absorption band shown in Figure 15.14 is found to consist of a set of closely spaced absorption lines. The expanded region of part of this spectrum shows that the individual lines are separated by energies as small as 0.002 cm^{-1}. These features could not be resolved by the conventional spectrometer because the spectral resolution of the lamp-based device was insufficient to distinguish between the frequencies of the different absorption bands. Clearly new information can be observed using laser light sources instead of lamps. The lines in the high-resolution absorption spectrum shown

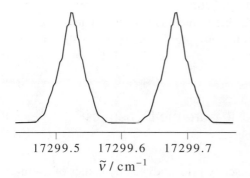

FIGURE 15.14
The absorption spectrum of ICl(g) in the vicinity of $17\,299.6$ cm^{-1} recorded using an absorption spectrometer with a spectral resolution of about 0.03 cm^{-1}. These two bands correspond to absorption from the ($v'' = 0$, $J'' = 2$) level of the electronic ground state to different rotational states of the $v' = 32$ level of the first electronic excited state.

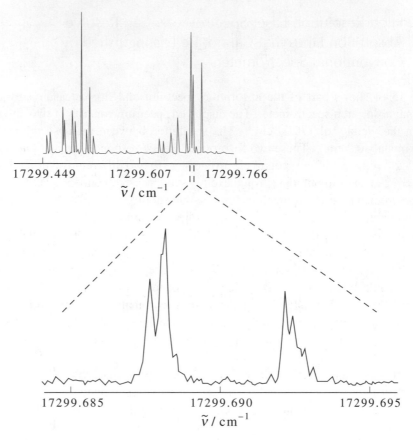

FIGURE 15.15
The absorption spectrum of ICl(g) in the vicinity of 17 299.600 cm^{-1} recorded using a high-resolution laser spectrometer. With this type of spectrometer, the single absorption lines shown in Figure 15.14 are found to consist of multiple absorption bands. A portion of this high-resolution spectrum is expanded to show the detailed features of the absorption spectrum that can be resolved.

in Figure 15.15 correspond to small changes in the energies of the rotational state of ICl(g) that are caused by the interaction of the electron spins with the nuclear spins, an effect called *hyperfine interaction*. It is possible to include hyperfine interactions in the Hamiltonian operator for a molecule and thereby predict the spacings that would be observed in its high-resolution spectrum.

15–8. Pulsed Lasers Can Be Used to Measure the Dynamics of Photochemical Processes

One application of time-resolved laser spectroscopy is to study the dynamics of chemical reactions initiated by the absorption of light. Chemical reactions that result from the absorption of light are called *photochemical reactions*. The following equations illustrate some of the many types of photochemical reactions that can occur.

$$O_3(g) + (\lambda = 300 \text{ nm}) \longrightarrow O_2(g) + O(g) \qquad \text{(Photodissociation)}$$

$$\textit{trans}\text{--butadiene} + (\lambda = 250 \text{ nm}) \longrightarrow \textit{cis}\text{--butadiene} \quad \text{(Photoisomerization)}$$

(Photodimerization)

We define the *quantum yield* Φ for a photochemical reaction by

$$\Phi = \frac{\text{number of molecules that undergo reaction}}{\text{number of photons absorbed}} \qquad (15.23)$$

The values of quantum yields vary over a wide range. For example, $\Phi \approx 10^{-3}$ for the bleaching of certain dyes, $\Phi \approx 1$ for the photodissociation of ozone, and $\Phi = 10^6$ for the reaction between $H_2(g)$ and $Cl_2(g)$.

EXAMPLE 15–6

Upon absorption of 313 nm light, acetone photodissociates according to the chemical equation

$$(CH_3)_2CO(g) + (\lambda = 313 \text{ nm}) \longrightarrow C_2H_6(g) + CO(g)$$

Exposure of a gaseous sample of acetone to a radiant power of 1.71×10^{-2} W at 313 nm for a period of 1.15×10^4 s results in the photodissociation of 8.68×10^{-5} mol of acetone. Determine the quantum yield for this photodissociation reaction. (Assume the sample absorbs all the light.)

SOLUTION: The number of molecules of acetone that photodissociates is

$$(8.68 \times 10^{-5} \text{ mol})(6.022 \times 10^{23} \text{ molecule} \cdot \text{mol}^{-1}) = 5.23 \times 10^{19} \text{ molecule}$$

The gas sample was exposed to a total radiant energy Q of

$$Q = (1.71 \times 10^{-2} \text{ W})(1.15 \times 10^4 \text{ s}) = 1.97 \times 10^2 \text{ J}$$

The number of photons is given by

$$\frac{Q}{Q_p} = \frac{Q\lambda}{hc} = \frac{(1.97 \times 10^2 \text{ J})(313 \times 10^{-9} \text{ m})}{(6.626 \times 10^{-34} \text{ J} \cdot \text{s})(2.998 \times 10^8 \text{ m} \cdot \text{s}^{-1})}$$

$$= 3.10 \times 10^{20}$$

The quantum yield for the photodissociation of acetone is given by (Equation 15.23)

$$\Phi = \frac{\text{number of molecules that undergo reaction}}{\text{number of photons absorbed}}$$

$$= \frac{5.23 \times 10^{19}}{3.10 \times 10^{20}} = 0.17$$

Figure 15.16 shows a schematic diagram of an apparatus designed to carry out time-resolved laser studies of photochemical reactions. The light source is a pulsed laser, which generates light pulses of short duration. Light pulses as short as 1×10^{-14} s, or 10 fs (femtoseconds) are routinely generated in laser laboratories. The laser output is split into two parts by a partially reflecting mirror, or beam splitter. The two pulses leave the beam splitter at the same time but then travel in different directions. The path that each pulse travels is determined using mirrors. The paths are designed so that they eventually cross one another inside the sample of interest. Consider the path traveled by each beam from the beam splitter to the sample. If the lengths of the two paths are the same, both pulses arrive at the crossing point in the sample at the same time. In this case, we say there is no time delay between the two light pulses. If the pathlengths differ, the two pulses will arrive at the sample at different times. In a time-resolved laser experiment, some property of the sample is measured as a function of the delay time between these two laser pulses. The laser pulse that initiates the photochemistry is called the *pump pulse*. The laser pulse that is used to record changes in the sample since the pump pulse arrived is called the *probe pulse*. Depending on the type of

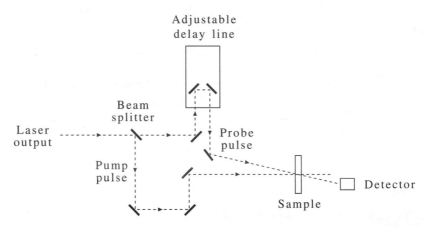

FIGURE 15.16
An illustration of the apparatus designed to carry out time-resolved laser experiments. The laser pulse is split into two pulses, a pump pulse and a probe pulse, using a beam splitter. The paths cross one another in the sample. The pump pulse is used to initiate a photochemical process in the sample, and the probe pulse is used to record how the sample changes in response to the pump pulse. A change in the pathlength of the two pulses affects the relative arrival times of the pump pulse and the probe pulse at the sample. In this manner, the sample can be probed as a function of time following the excitation by the pump pulse.

experiment being performed, the pump and probe pulses can be at the same or different wavelengths.

EXAMPLE 15–7
The probe laser pulse travels a pathlength that is 10.00 cm longer than that traveled by the pump laser pulse. Calculate the difference in arrival times of the two pulses of light at the sample.

SOLUTION: The difference in arrival times of the two laser pulses at the sample is the time required for the probe pulse to travel the extra 10.00 cm that the pump pulse did not travel. The time required for light to travel 10.00 cm is

$$t = \frac{(10.00 \text{ cm})(1 \text{ m}/100 \text{ cm})}{2.998 \times 10^8 \text{ m·s}^{-1}}$$
$$= 3.335 \times 10^{-10} \text{ s}$$

or 333.5 picoseconds.

We now illustrate the use of time-resolved laser spectroscopy by examining experimental data for the photodissociation reaction of ICN(g),

$$\text{ICN(g)} + h\nu \longrightarrow \text{I(g)} + \text{CN(g)} \tag{15.24}$$

For reasons we will see shortly, the pump and probe laser pulses have different wavelengths; the pump wavelength is set at 306 nm and the probe wavelength is set at 388 nm. Figure 15.17 shows how the energies of the ground state and one of the excited electronic states of ICN(g) depend on the distance between the iodine and carbon atoms, or the I–CN bond length. The ground state is a bound electronic state and the excited state is dissociative. Once the I–CN bond length reaches 400 pm, the bond is broken and I(g) and ground-state CN(g) radicals are produced.

EXAMPLE 15–8
The quantum yield for the photodissociation of ICN(g) into I(g) and CN(g) by a 306 nm pump pulse is 1.00. If the radiant energy of the pump pulse is 1.55×10^{-4} J, determine the number of CN(g) radicals created per pulse if only 0.100% of the incident light is absorbed by the ICN(g) sample.

SOLUTION: The radiant energy of a 306 nm photon is

$$Q_p = \frac{hc}{\lambda} = \frac{(6.626 \times 10^{-34} \text{ J·s})(2.998 \times 10^8 \text{ m·s}^{-1})}{306 \times 10^{-9} \text{ m}}$$
$$= 6.49 \times 10^{-19} \text{ J}$$

Therefore, the number of photons in a 1.55×10^{-4} J pulse is

$$\text{number of photons} = \frac{Q}{Q_p} = \frac{1.55 \times 10^{-4} \text{ J}}{6.49 \times 10^{-19} \text{ J}} = 2.39 \times 10^{14}$$

The number of CN(g) radicals produced by each laser pulse is given by (Equation 15.23)

$$\text{number of CN(g) produced} = (0.100\%)(\text{number of photons})\Phi$$
$$= (0.00100)(2.39 \times 10^{14})(1.00)$$
$$= 2.39 \times 10^{11}$$

The idea behind a time-resolved experiment is as follows. The ICN(g) sample is excited from the ground state to the excited state (Figure 15.17) using a laser pulse that is short compared with the time required for the molecule to dissociate. In that case, when the light pulse is over, the molecules find themselves on the dissociative curve with a bond length equal to the equilibrium bond length in the ground electronic state, as required by the Franck-Condon principle (see Section 13–7). These excited-state molecules respond to the repulsive force of the excited-state potential and therefore dissociate. Now suppose that at some time after the photoexcitation of ICN(g), a probe pulse of short duration interacts with the sample. The wavelength of the probe pulse

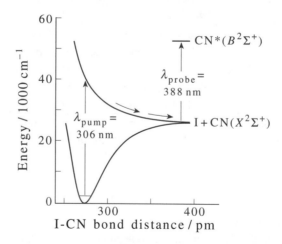

FIGURE 15.17
The potential energy curves of the ground state and first excited state of ICN(g) are plotted as a function of the I–CN bond distance. Dissociation of the I–CN bond from both states produces the CN(g) radical in its $X^2\Sigma^+$ ground state. The diagram also indicates the energy of the $B^2\Sigma^+$ excited state of CN(g). In the time-resolved study of the photodissociation of ICN(g), excitation by a pump pulse at 306 nm promotes molecules from the ground state to the excited dissociative state. A time-delayed probe pulse at 388 nm, which is on resonance with the $X^2\Sigma^+ \rightarrow B^2\Sigma^+$ transition of CN(g), is used to excite the CN(g) in the sample. The subsequent fluorescence from CN*(g) is then recorded. The intensity of the fluorescence caused by the probe pulse is a measure of the concentration of CN(g) present in the sample.

is not designed to excite the ICN(g) molecule but is tuned to the $X^2\Sigma^+ \to B^2\Sigma^+$ transition of the CN(g) radical at 388 nm (see Figure 15.17). Thus, if there are CN(g) radicals present in the sample, the light will be absorbed. The excited CN(g) molecules relax back to the ground state by fluorescence. The intensity of the fluorescence signal is a measure of the number of CN(g) radicals present in the sample at the time the probe pulse arrived at the sample. The only source of these CN(g) radicals is via the dissociation of excited ICN(g) molecules. If we monitor the intensity of the CN(g) fluorescence as a function of delay time between the pump and probe laser pulses, we thereby record the number of CN(g) molecules formed as a function of time following the initial excitation of ICN(g) to its dissociative state. This form of detection is called *laser-induced fluorescence* because we are using a laser to cause the product molecules to fluoresce.

Figure 15.18 shows a plot of the fluorescence intensity as a function of the delay time between the pump laser pulse and the probe laser pulse. At negative time delays the probe pulse arrives at the sample before the pump pulse. The ground state ICN(g) molecules do not absorb the probe pulse, so the signal intensity is zero. When the two pulses arrive at the sample at the same time ($t = 0$), we observe a small amount of CN(g) fluorescence. This result tells us that some of the excited molecules dissociate quickly after the absorption of the pump pulse. The signal continues to grow as the delay between the pump and probe pulses is increased. A constant signal level is observed for probe pulses delayed by 600 fs or longer. This observation tells us that no additional CN(g) radicals are produced for $t > 600$ fs. In other words, all the excited ICN(g) molecules have undergone photodissociation during the first 600 fs after the excitation. The solid line in Figure 15.18 is a curve fit of the experimental data to a function of the form $1 - \exp(-t/\tau)$, where τ is a constant called the *reaction half-life*. For this reaction, we find that $\tau = 205 \pm 30$ fs.

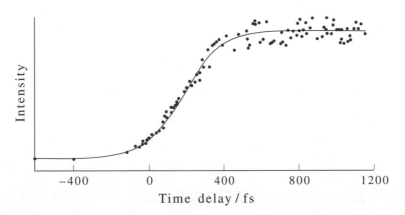

FIGURE 15.18
Experimental time-resolved data for the reaction ICN(g) $\to$ I(g) + CN(g). The intensity of the CN(g) fluoresence caused by the 388-nm probe pulse is plotted as a function of the delay time between the arrival of the probe pulse and the 306 nm pump pulse at the sample. Analysis of these data indicates that after excitation at 306 nm, the I–CN bond breaks with a reaction half-life of $\tau = 205 \pm 30$ fs. The solid line is a curve fit of the data to the function $I = 1 - \exp(-t/\tau)$.

Problems

15-1. The ground-state term symbol for O_2^+ is $^2\Pi_g$. The first electronic excited state has an energy of $38\,795$ cm^{-1} above that of the ground state and has a term symbol of $^2\Pi_u$. Is the radiative $^2\Pi_u \rightarrow\, ^2\Pi_g$ decay of the O_2^+ molecule an example of fluorescence or phosphorescence?

15-2. Consider the absorption and fluorescence spectrum of a diatomic molecule for the specific case in which $R_e(S_1) > R_e(S_0)$. Using the potential energy curves shown in Figure 15.1, draw the expected absorption and fluorescence spectra of the molecule. You can assume that the molecule relaxes to $v' = 0$ before it fluoresces. Do your spectra look like the spectra in Figure 15.2? Explain.

15-3. In Section 15–2, the spectral radiant energy density was expressed in terms of the frequency of the electromagnetic radiation. We could have chosen to express the spectral radiant energy density in terms of the wave number or wavelength of the electromagnetic radiation. Recall that the units of $\rho_\nu(\nu)$ are J·m^{-3}·s. Show that the units of $\rho_{\tilde{\nu}}(\tilde{\nu})$, the spectral radiant energy density in terms of wave numbers, are J·m^{-2} and that the units of $\rho_\lambda(\lambda)$, the spectral radiant energy density in terms of wavelength, are J·m^{-4}. What are the units of the Einstein B coefficient if we use $\rho_{\tilde{\nu}}(\tilde{\nu})$ to describe the spectral radiant energy density? What are the units of the Einstein B coefficient if we use $\rho_\lambda(\lambda)$ to describe the spectral radiant energy density?

15-4. Show that Equations 15.7 and 15.11 are equivalent only if $B_{12} = B_{21}$ and $A_{21} = (8h\pi\nu_{12}^3/c^3)B_{21}$.

15-5. Substitute Equation 15.16 into Equation 15.15 to prove that it is a solution to Equation 15.15.

15-6. Use the fact that $N_1(t) + N_2(t) = N_{\text{total}}$ to write Equation 15.15 as

$$\frac{dN_2}{B\rho_\nu(\nu_{12})N_{\text{total}} - [A + 2B\rho_\nu(\nu_{12})]N_2} = dt$$

Now show that the integral of this equation gives Equation 15.16.

15-7. Prove that Equation 15.17 implies that N_2/N_{total} is less than 1/2 because $A > 0$.

15-8. Prove that the inequality

$$\frac{N_2}{N_{\text{total}}} < \frac{1}{2}$$

implies that N_2/N_1 is less than 1. (*Hint:* Use the fact that $1/a > 1/b$ if $a < b$.)

15-9. The Einstein coefficients can also be derived using quantum mechanics. If the ground state and the excited state have a degeneracy of g_1 and g_2, respectively, the Einstein A coefficient is given by

$$A = \frac{16\pi^3\nu^3 g_1}{3\varepsilon_0 hc^3 g_2}|\mu|^2$$

where $|\mu|$ is the transition dipole moment (see Section 13–11). Now consider the $1s \rightarrow 2p$ absorption of H(g), which is observed at 121.8 nm. The radiative lifetime (see Example 15–3) of the triply degenerate excited $2p$ state of H(g) is 1.6×10^{-9} s. Determine the value of the transition dipole moment for this transition.

15-10. Use the equation given in Problem 15–9 and Equation 15.13 to derive the quantum mechanical expression for the Einstein B coefficient. Consider the $5s\,{}^1P_1 \rightarrow 3p\,{}^3P_2$ transition of neon at 632.8 nm, which is the lasing transition of most commercially available helium-neon lasers. Table 15.4 gives the Einstein A coefficient for this transition to be 6.56×10^6 s^{-1}. Determine the values of the Einstein B coefficient and the transition moment dipole for this transition. ($g_1 = g_2 = 1$.)

15-11. Derive (but do not try to solve) rate equations for $N_1(t)$, $N_2(t)$, and $N_3(t)$ for the three-level system described by Figure 15.8.

15-12. Consider the nondegenerate three-level system shown in Figure 15.8. Suppose that an incident light beam of energy $h\nu = E_3 - E_1$ is turned on for a while and then turned off. Show that the subsequent decay of the E_3 level is given by

$$N_3(t) = N_3^0 e^{-(A_{32}+A_{31})t}$$

where N_3^0 is the number of atoms in state 3 at the instant the light source is turned off. What will be the observed radiative lifetime of this excited state?

15-13. In this problem, we will generalize the result of Problem 15–12. Consider a system that has N nondegenerate levels of energy, E_1, E_2, $\ldots$, E_N such that $E_1 < E_2 < \cdots < E_N$. Suppose that all the atoms are initially in the level of energy E_1. The system is then exposed to light of energy $h\nu = E_N - E_1$. Defining $t = 0$ to be the instant the light source is turned off, show that the decay of p_N, the population in state N, is given by

$$p_N(t) = p_N^0 e^{-\sum_{i=1}^{N-1} A_{Ni} t}$$

where p_N^0 is the population of level N at $t = 0$. Show that the radiative lifetime of level N is given by $1/\sum_{i=1}^{N-1} A_{Ni}$. Use this result and the data in Table 15.4 to evaluate the radiative lifetime of the $5s\,{}^1P_1$ level of neon, assuming the only radiative decay channels are to the eight levels tabulated in Table 15.4.

15-14. The excited states of helium shown in Figure 15.13 have the electron configuration $1s2s$. Show that this electron configuration leads to a ${}^3S_1$ and a ${}^1S_0$ state. Which state has the lower energy?

15-15. According to Table 8.2, the ground-state energy of a helium atom is -2.904 hartrees. Use this value and the fact that the energy of a helium ion is given by $E = -Z^2/2n^2$ (in hartrees) to verify the energy of He$^+$ in Figure 15.13.

15-16. The 3391.3 nm line in a He-Ne laser is due to the $5s\,{}^1P_1 \rightarrow 3p\,{}^3P_2$ transition. According to the *Table of Atomic Energy Levels* by Charlotte Moore, the energies of these levels are $166\,658.484$ cm^{-1} and $163\,710.581$ cm^{-1}, respectively. Calculate the wavelength of this transition. Why does your answer not come out to be 3391.3 nm? (See Example 8–10.)

15-17. Using the method explained in Section 8–9, show that the states associated with a $2p^5ns$ electron configuration are ${}^3P_2$, ${}^3P_1$, ${}^3P_0$, and ${}^1P_1$.

15-18. Consider the excited-state electron configuration $2p^5np$, with $n \geq 3$. How many microstates are associated with this electron configuration? The term symbols that correspond to $2p^5np$ are $^3D_3$, $^3D_2$, $^3D_1$, $^1D_2$, $^3P_2$, $^3P_1$, $^3P_0$, $^1P_1$, $^3S_1$, and $^1S_0$. Show that these term symbols account for all the microstates of the electron configuration $2p^5np$, $n \geq 3$.

15-19. A titanium sapphire laser operating at 780 nm produces pulses at a repetition rate of 100 MHz. If each pulse is 25 fs in duration and the average radiant power of the laser is 1.4 W, calculate the radiant power of each laser pulse. How many photons are produced by this laser in one second?

15-20. A typical chromium doping level of a ruby rod is 0.050% by mass. How many chromium atoms are there in a ruby rod of diameter 1.15 cm and length 15.2 cm? The density of corundum (Al_2O_3) is 4.05 g·cm^{-3}, and you can assume that the doping with chromium has no effect on the density of the solid. Now suppose all the chromium atoms are in the upper lasing level. If a laser pulse of 100 ps is generated by the simultaneous stimulated emission of all the chromium atoms, determine the radiant power of the laser pulse. (See Table 15.2.)

15-21. Which laser pulse contains more photons, a 10-ns, 1.60-mJ pulse at 760 nm or a 500-ms, 1.60-mJ pulse at 532 nm?

15-22. Consider a flashlamp-pumped Nd^{3+}:YAG laser operating at a repetition rate of 10 Hz. Suppose the average radiant power of the flashlamp is 100 W. Determine the maximum number of photons that each laser pulse can contain using this pump source. The actual number of photons per laser pulse is 6.96×10^{17}. Determine the efficiency for converting the flashlamp output into laser output. (See Table 15.2.)

15-23. Chemical lasers are devices that create population inversions by a chemical reaction. One example is the HF gas laser, in which HF(g) is generated by the reaction

$$F(g) + H_2(g) \longrightarrow HF(g) + H(g)$$

The major product of this reaction is HF(g) in the excited $v = 3$ vibrational state. The reaction creates a population inversion in which $N(v)$, the number of molecules in each vibrational state, is such that $N(3) > N(v)$ for $v = 0$, 1, and 2. The output of the HF(g) laser corresponds to transitions between rotational lines of the $v = 3 \rightarrow v = 2$ ($\lambda = 2.7$–$3.2 \, \mu$m) transition. Why is there no lasing action from $v = 3 \rightarrow v = 1$ and $v = 3 \rightarrow v = 0$ even though there is a population inversion between these pairs of levels?

15-24. A CO_2 laser operating at 9.6 μm uses an electrical power of 5.00 kW. If this laser produces 100-ns pulses at a repetition rate of 10 Hz and has an efficiency of 27%, how many photons are in each laser pulse?

15-25. Figure 15.10 displays the energy levels of the CO_2 laser. Given the following spectroscopic data for CO_2(g), calculate the spacing between the $J' = 1 \rightarrow 0$ and $J' = 2 \rightarrow 1$ laser lines for the $001 \rightarrow 100$ vibrational transition.

$$\text{Fundamental frequency}(J' = 0 \rightarrow 0) \; 100 \rightarrow 001 = 960.80 \text{ cm}^{-1}$$

$$\tilde{B}(001) = 0.3871 \text{ cm}^{-1} \qquad \tilde{B}(100) = 0.3902 \text{ cm}^{-1}$$

Why is no lasing observed at the fundamental frequency of 960.80 cm^{-1}?

15-26. The upper level of the $H_2(g)$ laser is the lowest excited state of the molecule, the $B^1\Sigma_u^+$ state, and the lower level is the $X^1\Sigma_g^+$ ground state. The lasing occurs between the $v' = 5$ level of the excited state and the $v'' = 12$ level of the ground state. Use the following spectroscopic data to determine the wavelength of the laser light from the $H_2(g)$ laser.

State	$\tilde{T}_e/\text{cm}^{-1}$	$\tilde{\nu}_e/\text{cm}^{-1}$	$\tilde{\nu}_e\tilde{x}_e/\text{cm}^{-1}$
$B^1\Sigma_u^+$	91 689.9	1356.9	19.93
$X^1\Sigma_g^+$	0	4401.2	121.34

A 1.0 ns pulse can be generated with a pulse radiant power of 100 kW. Calculate the radiant energy of such a laser pulse. How many photons are there in this pulse?

15-27. In this problem, we will determine the excited-state rotational quantum numbers for the $X \to A$ absorption bands of $ICl(g)$ that are shown in Figure 15.14. The transition is from the $v'' = 0$ state of the X state to a highly excited vibrational level of the A state ($v' = 32$). To accurately calculate the vibrational term $G(v)$ for the excited A state, we will need to include a second-order anharmonic correction to take into account the shape of the potential curve. First-order corrections will be sufficient for the ground electronic state. Extending the approach discussed in Chapter 13, we would write

$$G(v) = \tilde{\nu}_e(v + \tfrac{1}{2}) - \tilde{\nu}_e\tilde{x}_e(v + \tfrac{1}{2})^2 + \tilde{\nu}_e\tilde{y}_e(v + \tfrac{1}{2})^3$$

Some of the spectroscopic constants for the X ground state and the A excited state of $ICl(g)$ are tabulated below.

State	$\tilde{T}_e/\text{cm}^{-1}$	$\tilde{\nu}_e/\text{cm}^{-1}$	$\tilde{\nu}_e\tilde{x}_e/\text{cm}^{-1}$	$\tilde{\nu}_e\tilde{y}_e/\text{cm}^{-1}$	$\tilde{B}_e/\text{cm}^{-1}$	$\tilde{\alpha}_e/\text{cm}^{-1}$
A	13 745.6	212.30	1.927	−0.03257	0.08389	0.00038
X	0	384.18	1.46			

Determine the value of $\tilde{\nu}$ corresponding to the transition $X(v'' = 0, J'' = 0) \to A(v' = 32, J' = 0)$. Given that the ground state for the lines shown in Figure 15.14 is the $v'' = 0, J'' = 2$ level of the X state and that the rotational term for this level is $F(2) = 0.65\,\text{cm}^{-1}$, determine the closest value of J', the rotational number of the $v'' = 32$ level of the excited A state that gives the two observed spectral lines. Using your results, do you think that the individual lines between 17 299.45 and 17 299.55 cm^{-1} in Figure 15.15 can be attributed to transitions to different excited rotational states from the $X(v'' = 0, J'' = 2)$ ground state?

15-28. Hydrogen iodide decomposes to hydrogen and iodine when it is irradiated with radiation of frequency 1.45×10^{15} Hz. When 2.31 J of energy is absorbed by $HI(g)$, 0.153 mg of $HI(g)$ is decomposed. Calculate the quantum yield for this reaction.

15-29. Ozone decomposes to $O_2(g)$ and $O(g)$ with a quantum yield of 1.0 when it is irradiated with radiation of wavelength 300 nm. If ozone is irradiated with a power of 100 W, how long will it take for 0.020 mol of $O_3(g)$ to decompose?

15-30. The quantum yield for the photosubstitution reaction

$$Cr(CO)_6 + NH_3 + h\nu \longrightarrow Cr(CO)_5NH_3 + CO$$

in octane solution at room temperature is 0.71 for a photolysis wavelength of 308 nm. How many $Cr(CO)_6$ molecules are destroyed per second when the solution is irradiated by a continuous laser with an output radiant power of 1.00 mW at 308 nm? If you wanted to produce one mole of $Cr(CO)_5NH_3$ per minute of exposure, what would the output radiant power of the laser need to be? (For both questions, assume the sample is sufficiently concentrated so that all the incident light is absorbed.)

15-31. A mole of photons is called an *einstein*. Calculate the radiant energy of an einstein if the photons have a wavelength of 608.7 nm.

15-32. The width of the duration of an electromagnetic pulse, Δt, and the width of the frequency distribution of the pulse, $\Delta \nu$, are related by $\Delta t \Delta \nu = 1/2\pi$. Compute the width of the frequency distribution of a 10-fs laser pulse and a 1-ms laser pulse. Can you record high-resolution spectra like that shown in Figure 15.15 for ICl(g) using a tunable femtosecond laser?

15-33. In Section 15–8, we found that in the photodissociation reaction of ICN(g), 205 fs is required for the I(g) and CN(g) photofragments to separate by 400 pm (Figure 15.18). Calculate the relative velocity of the two photofragments. (*Hint*: The equilibrium bond length in the ground state is 275 pm.)

15-34. In the photolysis of ICN(g), the CN(g) fragment can be generated in several different vibrational and rotational states. At what wavelength would you set your probe laser to excite the $v'' = 0$, $J'' = 3$ of the $X^2\Sigma^+$ ground state to the $v' = 0$, $J' = 3$ level of the $B^2\Sigma^+$ excited state? Use the following spectroscopic data.

State	$\tilde{T}_e/cm^{-1}$	$\tilde{\nu}_e/cm^{-1}$	$\tilde{\nu}_e\tilde{x}_e/cm^{-1}$	$\tilde{B}_e/cm^{-1}$	$\tilde{\alpha}_e/cm^{-1}$
$B^2\Sigma^+$	25 751.8	2164.13	20.25	1.970	0.0222
$X^2\Sigma^+$	0	2068.71	13.14	1.899	0.0174

Calculate the energy-level spacing between the $v'' = 0$, $J'' = 3$ and the $v'' = 0$, $J'' = 4$ levels. Can the formation dynamics of a single vibrational-rotational state of CN(g) be monitored by a femtosecond pump-probe experiment? (*Hint*: See Problem 15–32.)

15-35. The $X^1A_1 \rightarrow \tilde{A}$ electronic excitation of $CH_3I(g)$ at 260 nm results in the following two competing photodissociation reactions:

$$CH_3I(g) + h\nu \longrightarrow CH_3(g) + I(g)(^2P_{3/2})$$
$$\longrightarrow CH_3(g) + I^*(g)(^2P_{1/2})$$

The energy difference between the excited $^2P_{1/2}$ state and the ground $^2P_{3/2}$ state of I(g) is 7603 cm^{-1}. The total quantum yield for dissociation is 1.00 with 31% of the excited molecules producing $I^*(g)$. Assuming that $I^*(g)$ relaxes by only radiative decay, calculate the number of photons emitted per second by a $CH_3I(g)$ sample that absorbs 10% of the light generated by a 1.00-mW 260-nm laser.

15-36. The frequency of laser light can be converted using nonlinear optical materials. The most common form of frequency conversion is second harmonic generation, whereby laser light of frequency ν is converted to light at frequency 2ν. Calculate the wavelength of the second harmonic light from a Nd^{3+}:YAG laser. If the output pulse of a Nd^{3+}:YAG laser at 1064.1 nm has a radiant energy of 150.0 mJ, how many photons are contained in this pulse?

Calculate the maximum number of photons that can be generated at the second harmonic. (*Hint*: Energy must be conserved.)

15-37. There are nonlinear optical materials that can sum two laser beams at frequencies ν_1 and ν_2 and thereby generate light at frequency $\nu_3 = \nu_1 + \nu_2$. Suppose that part of the output from a krypton ion laser operating at 647.1 nm is used to pump a rhodamine 700 dye laser that produces laser light at 803.3 nm. The dye laser beam is then combined with the remaining output from the krypton ion laser in a nonlinear optical material that sums the two laser beams. Calculate the wavelength of the light created by the nonlinear optical material.

The following four problems examine how the intensity of absorption lines are quantified.

15-38. The *decadic absorbance* A of a sample is defined by $A = \log(I_0/I)$, where I_0 is the light intensity incident on the sample and I is the intensity of the light after it has passed through the sample. The decadic absorbance is proportional to c, the molar concentration of the sample, and l, the path length of the sample in meters, or in an equation

$$A = \varepsilon cl$$

where the proportionality factor ε is called the *molar absorption coefficient*. This expression is called the *Beer-Lambert law*. What are the units of A and ε? If the intensity of the transmitted light is 25.0% of that of the incident light, then what is the decadic absorbance of the sample? At 200 nm, a 1.42×10^{-3} M solution of benzene has decadic absorbance of 1.08. If the pathlength of the sample cell is 1.21×10^{-3} m, what is the value of ε? What percentage of the incident light is transmitted through this benzene sample? (It is common to express ε in the non SI units $L \cdot mol^{-1} \cdot cm^{-1}$ because l and c are commonly expressed in cm and $mol \cdot L^{-1}$, respectively. This difference in units leads to annoying factors of 10 that you need to be aware of.)

15-39. The Beer-Lambert law (Problem 15–38) can also be written as

$$I = I_0 e^{-\sigma Nl}$$

where N is the number of molecules per cubic meter and l is the pathlength of the cell in units of meters. What are the units of σ? The constant σ in this equation is called the *absorption cross section*. Derive an expression relating σ to ε, the molar absorption coefficient introduced in Problem 15–38. Determine σ for the benzene solution described in Problem 15–38.

15-40. The Beer-Lambert law (Problem 15–38) can also be written in terms of the natural logarithm instead of the base ten logarithm:

$$A_e = \ln \frac{I_0}{I} = \kappa cl$$

In this form, the constant κ is called the *molar napierian absorption coefficient*, and A_e is called the *napierian absorbance*. What are the units of κ? Derive a relationship between κ and ε (see Problem 15–38). Determine κ for the benzene solution described in Problem 15–38.

15-41. A re-examination of the spectra in Chapter 13 reveals that the transitions observed have a line width. We define A, the integrated absorption intensity to be

$$A = \int_{-\infty}^{\infty} \kappa(\tilde{\nu}) d\tilde{\nu}$$

where $\kappa(\tilde{\nu})$ is the molar napierian absorption coefficient in terms of wavenumbers, $\tilde{\nu}$ (see Problem 15–40). What are the units of A? Now suppose that the absorption line has a Gaussian line shape, or that

$$\kappa(\tilde{\nu}) = \kappa(\tilde{\nu}_{max}) e^{-\alpha(\tilde{\nu}-\tilde{\nu}_{max})^2}$$

where α is a constant and $\tilde{\nu}_{max}$ is the maximum frequency of absorption. Plot $\kappa(\tilde{\nu})$. How is α related to $\Delta\tilde{\nu}_{1/2}$, the width of the absorption line at half of its maximum intensity? Now show that

$$A = 1.07\kappa(\tilde{\nu}_{max})\Delta\tilde{\nu}_{1/2}$$

(*Hint:* $\int_0^{\infty} e^{-\beta x^2} dx = (\pi/4\beta)^{1/2}$.)

NUMERICAL METHODS

You learned in high school that a quadratic equation $ax^2 + bx + c = 0$ has two roots, given by the so-called quadratic formula:

$$x = \frac{-b \pm \sqrt{b^2 - 4ac}}{2a}$$

Thus, the two values of x (called roots) that satisfy the equation $x^2 + 3x - 2 = 0$ are

$$x = \frac{-3 \pm \sqrt{17}}{2}$$

Although there are general formulas for the roots of cubic and quartic equations, they are very inconvenient to use, and furthermore, there are no formulas for equations of the fifth degree or higher. Unfortunately, in practice we encounter such equations frequently and must learn to deal with them. Fortunately, with the advent of hand calculators and personal computers, the numerical solution of polynomial equations and other types of equations, such as $x - \cos x = 0$, is routine. Although these and other equations can be solved by "brute force" trial and error, much more organized procedures can arrive at an answer to almost any desired degree of accuracy. Perhaps the most widely known procedure is the Newton-Raphson method, which is best illustrated by a figure. Figure G.1 shows a function $f(x)$ plotted against x. The solution to the equation $f(x) = 0$ is denoted by x_*. The idea behind the Newton-Raphson method is to guess an initial value of x (call it x_0) "sufficiently close" to x_*, and draw the tangent to the curve $f(x)$ at x_0, as shown in Figure G.1. Very often, the extension of the tangent line through the horizontal axis will lie closer to x_* than does x_0. We denote this value of x by x_1 and repeat the process using x_1 to get a new value of x_2, which will lie even closer to x_*. By repeating this process (called iteration) we can approach x_* to essentially any desired degree of accuracy.

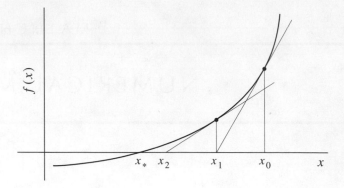

FIGURE G.1
A graphical illustration of the Newton-Raphson method.

We can use Figure G.1 to derive a convenient formula for the iterative values of x. The slope of $f(x)$ at x_n, $f'(x_n)$, is given by

$$f'(x_n) = \frac{f(x_n) - 0}{x_n - x_{n+1}}$$

Solving this equation for x_{n+1} gives

$$x_{n+1} = x_n - \frac{f(x_n)}{f'(x_n)} \tag{G.1}$$

which is the iterative formula for the Newton-Raphson method. As an application of this formula, consider the chemical equation

$$2\,NOCl(g) \rightleftharpoons 2\,NO(g) + Cl_2(g)$$

whose related equilibrium constant is 2.18 at a certain temperature. (Chemical equilibrium is discussed in Chapter 26, but we're simply using the algebraic equation below as an example at this point.) If 1.00 atm of $NOCl(g)$ is introduced into a reaction vessel, then at equilibrium $P_{NOCl} = 1.00 - 2x$, $P_{NO} = 2x$, and $P_{Cl_2} = x$; these pressures satisfy the equilibrium-constant expression

$$\frac{P_{NO}^2 P_{Cl_2}}{P_{NOCl}^2} = \frac{(2x)^2 x}{(1.00 - 2x)^2} = 2.18$$

which we write as

$$f(x) = 4x^3 - 8.72x^2 + 8.72x - 2.18 = 0$$

Because of the stoichiometry of the reaction equation, the value of x we are seeking must be between 0 and 0.5, so let's choose 0.250 as our initial guess (x_0). Table G.1 shows the results of using Equation G.1. Notice that we have converged to three significant figures in just three steps.

TABLE G.1

The results of the application of the Newton-Raphson method to the solution of the equation $f(x) = 4x^3 - 8.72x^2 + 8.72x - 2.18 = 0$.

n	x_n	$f(x_n)$	$f'(x_n)$
0	0.2500	-4.825×10^{-1}	5.110
1	0.3442	-4.855×10^{-2}	4.139
2	0.3559	-6.281×10^{-4}	4.033
3	0.3561	-1.704×10^{-5}	4.031
4	0.3561		

EXAMPLE G–1

In Chapter 16, we will solve the cubic equation

$$x^3 + 3x^2 + 3x - 1 = 0$$

Use the Newton-Raphson method to find the real root of this equation to five significant figures.

SOLUTION: We write the equation as

$$f(x) = x^3 + 3x^2 + 3x - 1 = 0$$

By inspection, a solution lies between 0 and 1. Using $x_0 = 0.5$ results in the following table:

n	x_n	$f(x_n)$	$f'(x_n)$
0	0.500000	1.37500	6.7500
1	0.296300	0.178294	5.04118
2	0.260930	0.004809	4.76983
3	0.259920	-0.000005	4.76220
4	0.259920		

The answer to five significant figures is $x = 0.25992$. Note that $f(x_n)$ is significantly smaller at each step, as it should be as we approach the value of x that satisfies $f(x) = 0$, but that $f'(x_n)$ does not vary appreciably. The same behavior can be seen in Table G.1.

As powerful as it is, the Newton-Raphson method does not always work; when it does work, it is obvious the method is working, and when it doesn't work, it may be even more obvious. A spectacular failure is provided by the equation $f(x) = x^{1/3} = 0$, for which $x_* = 0$. If we begin with $x_0 = 1$, we will obtain $x_1 = -2, x_2 = +4, x_3 = -8$, and so on. Figure G.2 shows why the method is failing to converge. The message here is that you should always plot $f(x)$ first to get an idea of where the relevant roots

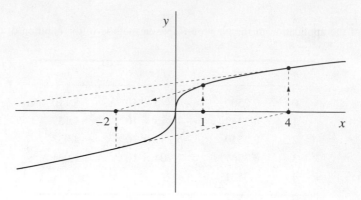

FIGURE G.2
A plot of $y = x^{1/3}$, illustrating that the Newton-Raphson method fails in this case.

lie and to see that the function does not have any peculiar properties. You should do Problems G–1 to G–9 to become proficient with the Newton-Raphson method.

There are also numerical methods to evaluate integrals. You learned in calculus that an integral is the area between a curve and the horizontal axis (area under a curve) between the integration limits, so that the value of

$$I = \int_a^b f(u)\,du \tag{G.2}$$

is given by the shaded area in Figure G.3. Recall a fundamental theorem of calculus, which says that if

$$F(x) = \int_a^x f(u)\,du$$

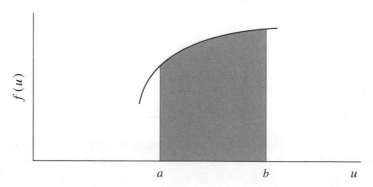

FIGURE G.3
The integral of $f(u)$ from a to b is given by the shaded area.

then

$$\frac{dF}{dx} = f(x)$$

The function $F(x)$ is sometimes called the antiderivative of $f(x)$. If there is no elementary function $F(x)$ whose derivative is $f(x)$, we say that the integral of $f(x)$ cannot be evaluated analytically. By elementary function, we mean a function that can be expressed as a finite combination of polynomial, trigonometric, exponential, and logarithmic functions.

It turns out that numerous integrals cannot be evaluated analytically. A particularly important example of an integral that cannot be evaluated in terms of elementary functions is

$$\phi(x) = \int_0^x e^{-u^2} du \qquad (G.3)$$

Equation G.3 serves to define the (nonelementary) function $\phi(x)$. The value of $\phi(x)$ for any value of x is given by the area under the curve $f(u) = e^{-u^2}$ from $u = 0$ to $u = x$.

Let's consider the more general case given by Equation G.2 or the shaded area in Figure G.3. We can approximate this area in a number of ways. First divide the interval (a,b) into n equally spaced subintervals $u_1 - u_0$, $u_2 - u_1$, ..., $u_n - u_{n-1}$ with $u_0 = a$ and $u_n = b$. We will let $h = u_{j+1} - u_j$ for $j = 0, 1, \ldots, n - 1$. Figure G.4 shows a magnification of one of the subintervals, say the u_j, u_{j+1} subinterval. One way to approximate the area under the curve is to connect the points $f(u_j)$ and $f(u_{j+1})$ by a straight line as shown in Figure G.4. The area under the straight line approximation to $f(u)$ in the interval is the sum of the area of the rectangle $[hf(u_j)]$ and the area of the triangle $\{\frac{1}{2}h[f(u_{j+1}) - f(u_j)]\}$. Using this approximation for all intervals, the total area under the curve from $u = a$ to $u = b$ is given by the sum

$$I \approx I_n = hf(u_0) + \frac{h}{2}[f(u_1) - f(u_0)]$$
$$+ hf(u_1) + \frac{h}{2}[f(u_2) - f(u_1)]$$
$$\vdots$$
$$+ hf(u_{n-2}) + \frac{h}{2}[f(u_{n-1}) - f(u_{n-2})]$$
$$+ hf(u_{n-1}) + \frac{h}{2}[f(u_n) - f(u_{n-1})]$$
$$= \frac{h}{2}[f(u_0) + 2f(u_1) + 2f(u_2) + \cdots + 2f(u_{n-1}) + f(u_n)] \qquad (G.4)$$

Note that the coefficients in Equation G.4 go as 1, 2, 2, ..., 2, 1. Equation G.4 is easy to implement on a hand calculator for $n = 10$ or so and on a personal computer using a spreadsheet for larger values of n. The approximation to the integral given by Equation G.4 is called the *trapezoidal approximation*. [The error goes as Ah^2, where A is a constant that depends upon the nature of the function $f(u)$. In fact, if M is the

632

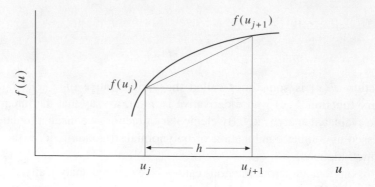

FIGURE G.4
An illustration of the area of the $j + 1$st subinterval for the trapezoidal approximation.

largest value of $|f''(u)|$ in the interval (a, b), then the error is *at most* $M(b - a)h^2/12$.]
Table G.2 shows the values of

$$\phi(1) = \int_0^1 e^{-u^2} du \tag{G.5}$$

for $n = 10 \, (h = 0.1)$, $n = 100 \, (h = 0.01)$, and $n = 1000 \, (h = 0.001)$. The "accepted" value (using more sophisticated numerical integration methods) is 0.74682413, to eight decimal places.

We can develop a more accurate numerical integration routine by approximating $f(u)$ in Figure G.4 by something other than a straight line. If we approximate $f(u)$ by a quadratic function, we have *Simpson's rule*, whose formula is

$$I_{2n} = \frac{h}{3}[f(u_0) + 4f(u_1) + 2f(u_2) + 4f(u_3) + 2f(u_4) + \cdots$$
$$+ 2f(u_{2n-2}) + 4f(u_{2n-1}) + f(u_{2n})] \tag{G.6}$$

Note that the coefficients go as 1, 4, 2, 4, 2, 4, ..., 4, 2, 4, 1. We write I_{2n} in Equation G.6 because Simpson's rule requires that there be an even number of intervals. Table G.2 shows the values of $\phi(1)$ in Equation G.5 for $n = 10, 100$, and 1000. Note that

TABLE G.2
The application of the trapezoidal approximation (Equation G.4) and Simpson's rule (Equation G.6) to the evaluation of $\phi(1)$ given by Equation G.5. The exact value to eight decimal places is 0.74682413.

n	h	I_n (trapezoidal)	I_{2n} (Simpson's tule)
10	0.1	0.74621800	0.74682494
100	0.01	0.74681800	0.74682414
1000	0.001	0.74682407	0.74682413

with $n = 100$, the result for Simpson's rule differs from the "accepted" value by only one unit in the eighth decimal place. The error for Simpson's rule goes as h^4 compared with h^2 for the trapezoidal approximation. In fact, if M is the largest value of $|f^{(4)}(u)|$ in the interval (a, b), then the error is *at most* $M(b - a)h^4/180$. Problems G–10 to G–13 illustrate the use of the trapezoidal approximation and Simpson's rule.

EXAMPLE G–2
One theory (from Debye) of the molar heat capacity of a monatomic crystal gives

$$\overline{C}_V = 9R\left(\frac{T}{\Theta_D}\right)^3 \int_0^{\Theta_D/T} \frac{x^4 e^x}{(e^x - 1)^2}\, dx$$

where R is the molar gas constant ($8.314\,\text{J}\cdot\text{K}^{-1}\cdot\text{mol}^{-1}$) and Θ_D, the Debye temperature, is a parameter characteristic of the crystalline substance. Given that $\Theta_D = 309$ K for copper, calculate the molar heat capacity of copper at $T = 103$ K.

SOLUTION: At $T = 103$ K, the basic integral to evaluate numerically is

$$I = \int_0^3 \frac{x^4 e^x}{(e^x - 1)^2}\, dx$$

Using the trapezoidal approximation (Equation G.5) and Simpson's rule (Equation G.6), we find the following values of I:

n	h	I_n (trapezoidal)	I_{2n} (Simpson's rule)
10	0.3	5.9725	5.9648
100	0.03	5.9649	5.9648
1000	0.003	5.9648	5.9648

The molar heat capacity at 103 K is given by

$$\overline{C}_V = 9R\left(\frac{1}{3}\right)^3 I$$

or $\overline{C}_V = 16.5\,\text{J}\cdot\text{mol}^{-1}\cdot\text{K}^{-1}$, in agreement with the experimental value.

Although the Newton-Raphson method and Simpson's rule can be implemented easily on a spreadsheet, there are a number of easy-to-use numerical software packages such as *MathCad*, *Kaleidagraph*, *Mathematica*, or *Maple* that can be used to evaluate the roots of algebraic equations and integrals by even more sophisticated numerical methods.

Problems

G-1. Solve the equation $x^5 + 2x^4 + 4x = 5$ to four significant figures for the root that lies between 0 and 1.

G-2. Use the Newton-Raphson method to derive the iterative formula

$$x_{n+1} = \frac{1}{2}\left(x_n + \frac{A}{x_n}\right)$$

for the value of $\sqrt{A}$. This formula was discovered by a Babylonian mathematician more than 2000 years ago. Use this formula to evaluate $\sqrt{2}$ to five significant figures.

G-3. Use the Newton-Raphson method to solve the equation $e^{-x} + (x/5) = 1$ to four significant figures. This equation occurs in Problem 1–5.

G-4. Consider the chemical reaction described by the equation

$$CH_4(g) + H_2O(g) \rightleftharpoons CO(g) + 3\,H_2(g)$$

at 300 K. If 1.00 atm of $CH_4(g)$ and $H_2O(g)$ are introduced into a reaction vessel, the pressures at equilibrium obey the equation

$$\frac{P_{CO}\,P_{H_2}^3}{P_{CH_4}\,P_{H_2O}} = \frac{(x)(3x)^3}{(1-x)(1-x)} = 26$$

Solve this equation for x.

G-5. In Chapter 16, we will solve the cubic equation

$$64x^3 + 6x^2 + 12x - 1 = 0$$

Use the Newton-Raphson method to find the only real root of this equation to five significant figures.

G-6. Solve the equation $x^3 - 3x + 1 = 0$ for all three of its roots to four decimal places.

G-7. In Example 16–3 we will solve the cubic equation

$$\overline{V}^3 - 0.1231\overline{V}^2 + 0.02056\overline{V} - 0.001271 = 0$$

Use the Newton-Raphson method to find the root to this equation that is near $\overline{V} = 0.1$.

G-8. In Section 16–3 we will solve the cubic equation

$$\overline{V}^3 - 0.3664\overline{V}^2 + 0.03802\overline{V} - 0.001210 = 0$$

Use the Newton-Raphson method to show that the three roots to this equation are 0.07073, 0.07897, and 0.2167.

G-9. The Newton-Raphson method is not limited to polynomial equations. For example, in Problem 4–38 we solved the equation

$$\varepsilon^{1/2} \tan \varepsilon^{1/2} = (12 - \varepsilon)^{1/2}$$

for ε by plotting $\varepsilon^{1/2} \tan \varepsilon^{1/2}$ and $(12 - \varepsilon)^{1/2}$ versus ε on the same graph and noting the intersections of the two curves. We found that $\varepsilon = 1.47$ and 11.37. Solve the above equation using the Newton-Raphson method and obtain the same values of ε.

G-10. Use the trapezoidal approximation and Simpson's rule to evaluate

$$I = \int_0^1 \frac{dx}{1 + x^2}$$

This integral can be evaluated analytically; it is given by $\tan^{-1}(1)$, which is equal to $\pi/4$, so $I = 0.78539816$ to eight decimal places.

G-11. Evaluate $\ln 2$ to six decimal places by evaluating

$$\ln 2 = \int_1^2 \frac{dx}{x}$$

What must n be to assure six-digit accuracy?

G-12. Use Simpson's rule to evaluate

$$I = \int_0^\infty e^{-x^2} dx$$

and compare your result with the exact value, $\sqrt{\pi}/2$.

G-13. The integral

$$I = \int_0^\infty \frac{x^3 dx}{e^x - 1}$$

occurs in Problem 1–42, where we use its exact value $\pi^4/15$. Use Simpson's rule to evaluate I to six decimal places.

G-14. Use a numerical software package such as *MathCad*, *Kaleidagraph*, or *Mathematica* to evaluate the integral

$$S = 4\pi^{1/2} \left(\frac{2\alpha}{\pi} \right)^{3/4} \int_0^\infty r^2 e^{-r} e^{-\alpha r^2} dr$$

for values of α between 0.200 and 0.300 and show that S has a maximum value at $\alpha = 0.271$ (see Problem 11–11).

Johannes Diderik van der Waals was born in Leiden, the Netherlands, on November 23, 1837, and died in 1923. Because he had not learned Latin and Greek, he was at first not able to continue with university studies and so worked as a school teacher in a secondary school. After passage of new legislation, however, van der Waals obtained an exemption from the university requirements in classical languages and defended his doctoral dissertation at Leiden University in 1873. In his dissertation, he proposed an explanation of the continuity of the gaseous and liquid phases and the phenomenon of the critical point, as well as a derivation of a new equation of state of gases, now called the van der Waals equation. A few years later, he proposed the law of corresponding states, which reduces the properties of all gases to one common denominator. Although his dissertation was written in Dutch, his work quickly came to the attention of Maxwell, who published a review of it in English in the British journal *Nature* in 1875 and so brought the work to the attention of a much broader audience. In 1876, van der Waals was appointed the first Professor of Physics at the newly created University of Amsterdam. The University became a center for both theoretical and experimental research on fluids, largely through van der Waals' influence. Van der Waals was awarded the Nobel Prize for physics in 1910 "for the work on the equation of state for gases and liquids."

CHAPTER 16

The Properties of Gases

To this point, we have learned about the properties of individual atoms and molecules. For most of the rest of the book, we will study systems consisting of large numbers of atoms and molecules. In particular, we will explore the relations between the macroscopic properties of systems and the dependence these properties have upon the properties of the constituent atoms and molecules. We begin our study with the properties of gases. First, we will discuss the ideal-gas equation and then some extensions of this equation, of which the van der Waals equation is the most famous. Although the van der Waals equation accounts in part for deviations from ideal-gas behavior, a more systematic and accurate approach is to use a so-called virial expansion, which is an expression for the pressure of a gas as a polynomial in the density. We will relate the coefficients in this polynomial to the energy of interaction between the molecules of the gas. This relation will take us into a discussion of how molecules interact with one another. We will see that deviations from ideal-gas behavior teach us a great deal about molecular interactions.

16–1. All Gases Behave Ideally If They Are Sufficiently Dilute

If a gas is sufficiently dilute that its constituent molecules are so far apart from each other on the average that we can ignore their interactions, it obeys the equation of state

$$PV = nRT \tag{16.1a}$$

If we divide both sides of this equation by n, we obtain

$$P\overline{V} = RT \tag{16.1b}$$

where $\overline{V} = V/n$ is the molar volume. We will always indicate a molar quantity by drawing a line above the symbol. Either of Equations 16.1, familiar even to high school students, is called the *ideal-gas equation of state*. Equations 16.1 are called an equation of state because they serve as a relation between the pressure, volume, and temperature of the gas for a given quantity of gas. A gas that obeys Equations 16.1 is called an ideal gas, or the gas is said to behave ideally.

The distinction between V and $\overline{V}$ illustrates an important character of the quantities or the variables used to describe macroscopic systems. These quantities are of two types, called extensive quantities and intensive quantities. *Extensive quantities*, or *extensive variables*, are directly proportional to the size of a system. Volume, mass, and energy are examples of extensive quantities. *Intensive quantities*, or *intensive variables*, do not depend upon the size of the system. Pressure, temperature, and density are examples of intensive quantities. If we divide an extensive quantity by the number of particles or the number of moles in a system, we obtain an intensive quantity. For example, V (dm^3) is an extensive quantity but $\overline{V}$ ($dm^3 \cdot mol^{-1}$) is an intensive quantity. Distinguishing between extensive and intensive quantities is often important in describing the properties of chemical systems.

The reason Equations 16.1 are encountered so frequently in chemistry courses is that *all* gases obey Equations 16.1, as long as they are sufficiently dilute. Any individual characteristics of the gas, such as the shape or size of its molecules or how the molecules interact with each other, are lost in Equations 16.1. In a sense, these equations are a common denominator for all gases. Experimentally, most gases satisfy Equations 16.1 to approximately 1% at one atm and 0°C.

Equations 16.1 require us to discuss the system of units (SI) adopted by the International Union of Pure and Applied Chemistry (IUPAC). For example, although the SI unit of volume is m^3 (meters cubed), the unit L (liter), which is defined as exactly 1 dm^3 (decimeters cubed), is an acceptable unit of volume in the IUPAC system. The SI unit of pressure is a pascal (Pa), which is equal to one newton per square meter ($Pa = N \cdot m^{-2} = kg \cdot m^{-1} \cdot s^{-2}$). Recall that a newton is the SI unit of force, so we see that pressure is a force per unit area. Pressure can be measured experimentally by observing how high a column of liquid (usually mercury) is supported by the gas. If m is the mass of the liquid and g is the gravitational acceleration constant, the pressure is given by

$$P = \frac{F}{A} = \frac{mg}{A} = \frac{\rho h A g}{A} = \rho h g \qquad (16.2)$$

where A is the base area of the column, ρ is the density of the fluid, and h is the height of the column. The gravitational acceleration constant is equal to 9.8067 $m \cdot s^{-2}$, or 980.67 $cm \cdot s^{-2}$. Note that the area cancels out in Equation 16.2.

EXAMPLE 16–1

Calculate the pressure exerted by a 76.000-cm column of mercury. Take the density of mercury to be 13.596 $g \cdot cm^{-3}$.

SOLUTION: $P = (13.596 \text{ g} \cdot \text{cm}^{-3})(76.000 \text{ cm})(980.67 \text{ cm} \cdot \text{s}^{-2})$

$\qquad\qquad = 1.0133 \times 10^6 \text{ g} \cdot \text{cm}^{-1} \cdot \text{s}^{-2}$

A pascal is equal to $N \cdot m^{-2}$ or $kg \cdot m^{-1} \cdot s^{-2}$, so the pressure in pascals is

$$P = (1.0133 \times 10^6 \text{ g} \cdot \text{cm}^{-1} \cdot \text{s}^{-2})(10^{-3} \text{ kg} \cdot \text{g}^{-1})(100 \text{ cm} \cdot \text{m}^{-1})$$
$$= 1.0133 \times 10^5 \text{ Pa} = 101.33 \text{ kPa}$$

Strictly speaking, new textbooks should use the IUPAC-suggested SI units, but the units of pressure are particularly problematic. Although a pascal is the SI unit of pressure and will probably see increasing use, the atmosphere will undoubtedly continue to be widely used. One *atmosphere* (atm) is defined as 1.01325×10^5 Pa $= 101.325$ kPa. [One atmosphere used to be defined as the pressure that supports a 76.0-cm column of mercury (see Example 16.1).] Note that one kPa is approximately 1% of an atmosphere. One atmosphere used to be the standard of pressure, in the sense that tabulated properties of substances were presented at one atm. With the change to SI units, the standard is now one *bar*, which is equal to 10^5 Pa, or 0.1 MPa. The relation between bars and atmospheres is 1 atm $= 1.01325$ bar. One other commonly used unit of pressure is a *torr*, which is the pressure that supports a 1.00-mm column of mercury. Thus 1 torr $= (1/760)$ atm. Because we are experiencing a transition period between the widespread use of atm and torr on the one hand and the future use of bar and kPa on the other hand, students of physical chemistry must be proficient in both sets of pressure units. The relations between the various units of pressure are collected in Table 16.1

Of the three quantities, volume, pressure, and temperature, temperature is the most difficult to conceptualize. We will present a molecular interpretation of temperature later, but here we will give an operational definition. The fundamental temperature scale is based upon the ideal-gas law, Equations 16.1. Specifically, we define T to be

$$T = \lim_{P \to 0} \frac{P\overline{V}}{R} \tag{16.3}$$

because all gases behave ideally in the limit of $P \to 0$. The unit of temperature is the kelvin, which is denoted K. Note that we do not use a degree symbol when the temperature is expressed in kelvin. Because P and $\overline{V}$ cannot take on negative values, the lowest possible value of the temperature is 0 K. Temperatures as low as 1×10^{-7} K have

TABLE 16.1
Various units for expressing pressure.

1 pascal (Pa)	=	$1 \text{ N} \cdot \text{m}^{-2} = 1 \text{ kg} \cdot \text{m}^{-1} \cdot \text{s}^{-2}$
1 atmosphere (atm)	=	1.01325×10^5 Pa
	=	1.01325 bar
	=	101.325 kPa
	=	1013.25 mbar
	=	760 torr
1 bar	=	10^5 Pa $= 0.1$ MPa

been achieved in the laboratory. The temperature of absolute zero (0 K) corresponds to a substance that has no thermal energy. There is no fundamental limit to the maximum value of T. There are, of course, practical limitations, and the highest value of T achieved in the laboratory is around 100 million (10^8) K, which has been generated inside a magnetic confinement in nuclear fusion research facilities.

To establish the unit of kelvin, the triple point of water has been assigned the temperature of 273.16 K. (We will learn about the properties of a "triple point" in Chapter 23. For our present purposes, it is sufficient to know that the triple point of a substance corresponds to an equilibrium system that contains gas, liquid, and solid.) We now have a definition for 0 K and 273.16 K. A kelvin is then defined as 1/273.16 of the temperature of the triple point of water. These definitions of 0 K and 273.16 K generate a linear temperature scale.

Figure 16.1 plots experimental $\overline{V}$ versus T for Ar(g) at different pressures. As expected from our definition of the temperature scale, the extraplotation of these data shows that $T \rightarrow 0$ as $\overline{V} \rightarrow 0$.

The kelvin scale is related to the commonly used Celsius scale by

$$t/^\circ C = T/K - 273.15 \tag{16.4}$$

We will use the lower case t for $^\circ$C and the upper case T for K. Note also that the degree symbol ($^\circ$) is associated with values of the temperature in the Celsius scale. Equation 16.4 tells us that 0 K $= -273.15^\circ$C, or that 0°C $= 273.15$ K. Because of the general use of $^\circ$C in laboratories, a significant amount of thermodynamic data are tabulated for substances at 0°C (273.15 K) and 25°C (298.15 K); this latter value is commonly called "room temperature."

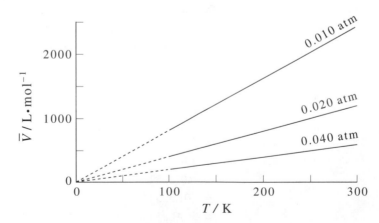

FIGURE 16.1
Experimental molar volumes (solid lines) of Ar(g) are plotted as a function of T/K at 0.040 atm, 0.020 atm, and 0.010 atm. All three pressures extrapolate to the origin (dashed lines).

If we measure $P\overline{V}$ at 273.15 K for any gas at a sufficiently low pressure that its behavior is ideal, then

$$P\overline{V} = R(273.15 \text{ K})$$

Figure 16.2 shows $P\overline{V}$ data plotted against P for several gases at $T = 273.15$ K. All the data plotted extrapolate to $P\overline{V} = 22.414$ L·atm·mol^{-1} as $P \to 0$, where the gases certainly behave ideally. Therefore, we can write

$$R = \frac{P\overline{V}}{T} = \frac{22.414 \text{ L·atm·mol}^{-1}}{273.15 \text{ K}} = 0.082058 \text{ L·atm·mol}^{-1}\text{·K}^{-1}$$

Using the fact that 1 atm = 1.01325×10^5 Pa and that 1 L = 10^{-3} m^3, we have

$$R = (0.082058 \text{ L·atm·mol}^{-1}\text{·K}^{-1})(1.01325 \times 10^5 \text{ Pa·atm}^{-1})(10^{-3} \text{ m}^3\text{·L}^{-1})$$

$$= 8.3145 \text{ Pa·m}^3\text{·mol}^{-1}\text{·K}^{-1}$$

$$= 8.3145 \text{ J·mol}^{-1}\text{·K}^{-1}$$

where we have used the fact that 1 Pa·m^3 = 1 N·m = 1 J. Because of the change of the standard of pressure from atmospheres to bars, it is also convenient to know the value of R in units of L·bar·mol^{-1}·K^{-1}. Using the fact that 1 atm = 1.01325 bar, we see that

$$R = (0.082058 \text{ L·atm·mol}^{-1}\text{·K}^{-1})(1.01325 \text{ bar·atm}^{-1})$$

$$= 0.083145 \text{ L·bar·mol}^{-1}\text{·K}^{-1} = 0.083145 \text{ dm}^3\text{·bar·mol}^{-1}\text{·K}^{-1}$$

Table 16.2 gives the value of R in various units.

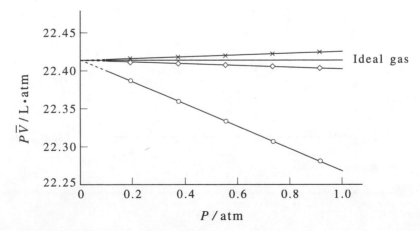

FIGURE 16.2
A plot of experimental values of $P\overline{V}$ versus P for H_2(g) (crosses), N_2(g) (diamonds), and CO_2(g) (circles) at $T = 273.15$ K. The data for all three gases extrapolate to a value of $P\overline{V} = 22.414$ L·atm as $P \to 0$ (ideal behavior).

TABLE 16.2
The values of the molar gas constant R in various units.

$$
\begin{aligned}
R \; &= \; 8.3145 \ \text{J} \cdot \text{mol}^{-1} \cdot \text{K}^{-1} \\
&= \; 0.083145 \ \text{dm}^3 \cdot \text{bar} \cdot \text{mol}^{-1} \cdot \text{K}^{-1} \\
&= \; 83.145 \ \text{cm}^3 \cdot \text{bar} \cdot \text{mol}^{-1} \cdot \text{K}^{-1} \\
&= \; 0.082058 \ \text{L} \cdot \text{atm} \cdot \text{mol}^{-1} \cdot \text{K}^{-1} \\
&= \; 82.058 \ \text{cm}^3 \cdot \text{atm} \cdot \text{mol}^{-1} \cdot \text{K}^{-1}
\end{aligned}
$$

16–2. The van der Waals Equation and the Redlich–Kwong Equation Are Examples of Two-Parameter Equations of State

The ideal-gas equation is valid for all gases at sufficiently low pressures. As the pressure on a given quantity of gas is increased, however, deviations from the ideal-gas equation appear. These deviations can be displayed graphically by plotting $P\overline{V}/RT$ as a function of pressure, as shown in Figure 16.3. The quantity $P\overline{V}/RT$ is called the *compressibility factor* and is denoted by Z. Note that $Z = 1$ under all conditions for an ideal gas. For real gases, $Z = 1$ at low pressures, but deviations from ideal behavior ($Z \neq 1$) are seen as the pressure increases. The extent of the deviations from ideal behavior at a given pressure depends upon the temperature and the nature of the gas. The closer the gas is to the point at which it begins to liquefy, the larger the deviations from ideal behavior will be. Figure 16.4 shows Z plotted against P for methane at various temperatures. Note that Z dips below unity at lower temperatures but lies above unity at higher temperatures. At lower temperatures the molecules are moving less rapidly, and so

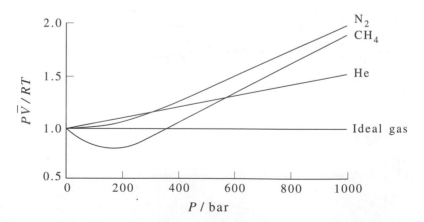

FIGURE 16.3
A plot of $P\overline{V}/RT$ versus P for one mole of helium, nitrogen, and methane at 300 K. This figure shows that the ideal-gas equation, for which $P\overline{V}/RT = 1$, is not valid at high pressure.

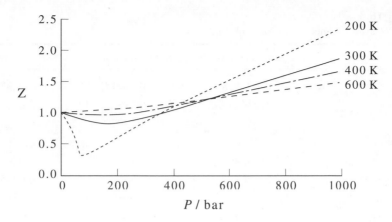

FIGURE 16.4
The compressibility factor of methane versus pressure at various temperatures. This figure shows that the effect of molecular attraction becomes less important at higher temperatures.

are more influenced by their attractive forces. Because of these attractive forces, the molecules are drawn together, thus making $\overline{V}_{real}$ less than $\overline{V}_{ideal}$, which in turn causes Z to be less than unity. A similar effect can be seen in Figure 16.3: the order of the curves shows that the effect of molecular attractions are in the order $CH_4 > N_2 > He$ at 300 K. At higher temperatures, the molecules are moving rapidly enough that their attraction is much smaller than $k_B T$ (which we will see in Chapter 18 is a measure of their thermal energy). The molecules are influenced primarily by their repulsive forces at higher temperatures, which tend to make $\overline{V} > \overline{V}_{ideal}$, and so $Z > 1$.

Our picture of an ideal gas views the molecules as moving independently of each other, not experiencing any intermolecular interactions. Figures 16.3 and 16.4 show that this picture fails at high pressures, and that the attractive and repulsive intermolecular interactions must be taken into account. Many equations extend the ideal-gas equation to account for the intermolecular interactions. Perhaps the most well known is the *van der Waals equation*,

$$\left(P + \frac{a}{\overline{V}^2} \right) (\overline{V} - b) = RT \tag{16.5}$$

where $\overline{V}$ designates molar volume. Notice that Equation 16.5 reduces to the ideal-gas equation when $\overline{V}$ is large, as it must. The constants a and b in Equation 16.5 are called *van der Waals constants*, whose values depend upon the particular gas (Table 16.3). We will see in Section 16–7 that the value of a reflects how strongly the molecules of a gas attract each other and the value of b reflects the size of the molecules.

Let's use Equation 16.5 to calculate the pressure (in bars) exerted by 1.00 mol of $CH_4(g)$ that occupies a 250-mL container at 0°C. From Table 16.3, we find that

TABLE 16.3

van der Waals constants for various substances.

Species	$a/\text{dm}^6 \cdot \text{bar} \cdot \text{mol}^{-2}$	$a/\text{dm}^6 \cdot \text{atm} \cdot \text{mol}^{-2}$	$b/\text{dm}^3 \cdot \text{mol}^{-1}$
Helium	0.034598	0.034145	0.023733
Neon	0.21666	0.21382	0.017383
Argon	1.3483	1.3307	0.031830
Krypton	2.2836	2.2537	0.038650
Hydrogen	0.24646	0.24324	0.026665
Nitrogen	1.3661	1.3483	0.038577
Oxygen	1.3820	1.3639	0.031860
Carbon monoxide	1.4734	1.4541	0.039523
Carbon dioxide	3.6551	3.6073	0.042816
Ammonia	4.3044	4.2481	0.037847
Methane	2.3026	2.2725	0.043067
Ethane	5.5818	5.5088	0.065144
Ethene	4.6112	4.5509	0.058199
Propane	9.3919	9.2691	0.090494
Butane	13.888	13.706	0.11641
2-Methyl propane	13.328	13.153	0.11645
Pentane	19.124	18.874	0.14510
Benzene	18.876	18.629	0.11974

$a = 2.3026$ dm$^6 \cdot$bar$\cdot$mol^{-2} and $b = 0.043067$ dm$^3 \cdot$mol^{-1} for methane. If we divide Equation 16.5 by $\overline{V} - b$ and solve for P, we obtain

$$
\begin{aligned}
P &= \frac{RT}{\overline{V} - b} - \frac{a}{\overline{V}^2} \\
&= \frac{(0.083145 \text{ dm}^3 \cdot \text{bar} \cdot \text{mol}^{-1} \cdot \text{K}^{-1})(273.15 \text{ K})}{(0.250 \text{ dm}^3 \cdot \text{mol}^{-1} - 0.043067 \text{ dm}^3 \cdot \text{mol}^{-1})} - \frac{2.3026 \text{ dm}^6 \cdot \text{bar} \cdot \text{mol}^{-2}}{(0.250 \text{ dm}^3 \cdot \text{mol}^{-1})^2} \\
&= 72.9 \text{ bar}
\end{aligned}
$$

By comparison, the ideal-gas equation predicts that $P = 90.8$ bar. The prediction of the van der Waals equation is in much better agreement with the experimental value of 78.6 bar than is the ideal-gas equation.

The van der Waals equation qualitatively gives the behavior shown in Figures 16.3 and 16.4. We can rewrite Equation 16.5 in the form

$$
Z = \frac{P\overline{V}}{RT} = \frac{\overline{V}}{\overline{V} - b} - \frac{a}{RT\overline{V}} \tag{16.6}
$$

At high pressures, the first term in Equation 16.6 dominates because $\overline{V} - b$ becomes small, and at low pressures the second term dominates.

EXAMPLE 16–2

Use the van der Waals equation to calculate the molar volume of ethane at 300 K and 200 atm.

SOLUTION: When we try to solve the van der Waals equation for $\overline{V}$, we obtain a cubic equation,

$$\overline{V}^3 - \left(b + \frac{RT}{P}\right)\overline{V}^2 + \frac{a}{P}\overline{V} - \frac{ab}{P} = 0$$

which we must solve numerically using the Newton-Raphson method (MathChapter G). Using the values of a and b from Table 16.3, we have

$$\overline{V}^3 - (0.188 \text{ L·mol}^{-1})\overline{V}^2 + (0.0275 \text{ L}^2\text{·mol}^{-1})\overline{V} - 0.00179 \text{ L}^3\text{·mol}^{-3} = 0$$

The Newton-Raphson method gives us

$$\overline{V}_{n+1} = \overline{V}_n - \frac{\overline{V}_n^3 - 0.188\overline{V}_n^2 + 0.0275\overline{V}_n - 0.00179}{3\overline{V}_n^2 - 0.376\overline{V}_n + 0.0275}$$

where we have suppressed the units for convenience. The ideal-gas value of $\overline{V}$ is $\overline{V}_{\text{ideal}} = RT/P = 0.123 \text{ L·mol}^{-1}$, so let's use 0.10 L·mol^{-1} as our initial guess. In this case, we obtain

n	$\overline{V}_n/\text{L·mol}^{-1}$	$f(\overline{V}_n)/\text{L}^3\text{·mol}^{-3}$	$f'(\overline{V}_n)/\text{L}^2\text{·mol}^{-2}$
0	0.100	8.00×10^{-5}	2.00×10^{-2}
1	0.096	2.53×10^{-6}	1.90×10^{-2}
2	0.096		

The experimental value is 0.071 L·mol^{-1}. The calculation of pressure preceding this example and the calculation of the volume in this example show that the van der Waals equation, while more accurate than the ideal-gas equation, is not particularly accurate. We will learn shortly that there are more accurate equations of state.

Two other relatively simple equations of state that are much more accurate and hence more useful than the van der Waals equation are the *Redlich-Kwong equation*

$$P = \frac{RT}{\overline{V} - B} - \frac{A}{T^{1/2}\overline{V}(\overline{V} + B)} \tag{16.7}$$

and the *Peng-Robinson equation*

$$P = \frac{RT}{\overline{V} - \beta} - \frac{\alpha}{\overline{V}(\overline{V} + \beta) + \beta(\overline{V} - \beta)} \tag{16.8}$$

TABLE 16.4
The Redlich-Kwong equation parameters for various substances.

Species	$A/\mathrm{dm^6 \cdot bar \cdot mol^{-2} \cdot K^{1/2}}$	$A/\mathrm{dm^6 \cdot atm \cdot mol^{-2} \cdot K^{1/2}}$	$B/\mathrm{dm^3 \cdot mol^{-1}}$
Helium	0.079905	0.078860	0.016450
Neon	1.4631	1.4439	0.012049
Argon	16.786	16.566	0.022062
Krypton	33.576	33.137	0.026789
Hydrogen	1.4333	1.4145	0.018482
Nitrogen	15.551	15.348	0.026738
Oxygen	17.411	17.183	0.022082
Carbon monoxide	17.208	16.983	0.027394
Carbon dioxide	64.597	63.752	0.029677
Ammonia	87.808	86.660	0.026232
Methane	32.205	31.784	0.029850
Ethane	98.831	97.539	0.045153
Ethene	78.512	77.486	0.040339
Propane	183.02	180.63	0.062723
Butane	290.16	286.37	0.08068
2-Methyl propane	272.73	269.17	0.080715
Pentane	419.97	414.48	0.10057
Benzene	453.32	447.39	0.082996

where A, B, α, and β, are parameters that depend upon the gas. The values of A and B in the Redlich-Kwong equation are listed in Table 16.4 for a variety of substances. The parameter α in the Peng-Robinson equation is a somewhat complicated function of temperature, so we will not tabulate values of α and β. Equations 16.7 and 16.8, like the van der Waals equation (Example 16–2), can be written as cubic equations in $\overline{V}$. For example, the Redlich-Kwong equation becomes (Problem 16–26)

$$\overline{V}^3 - \frac{RT}{P}\overline{V}^2 - \left(B^2 + \frac{BRT}{P} - \frac{A}{T^{1/2}P}\right)\overline{V} - \frac{AB}{T^{1/2}P} = 0 \qquad (16.9)$$

Problem 16–28 has you show that the Peng-Robinson equation of state is also a cubic equation in $\overline{V}$.

EXAMPLE 16–3
Use the Redlich-Kwong equation to calculate the molar volume of ethane at 300 K and 200 atm.

SOLUTION: Substitute $T = 300\,\text{K}$, $P = 200\,\text{atm}$, $A = 97.539\,\text{dm}^6 \cdot \text{atm} \cdot \text{mol}^{-1} \cdot \text{K}^{1/2}$, and $B = 0.045153\,\text{dm}^3 \cdot \text{mol}^{-1}$ into Equation 16.9, to obtain

$$\overline{V}^3 - 0.1231\overline{V}^2 + 0.02056\overline{V} - 0.001271 = 0$$

where we have suppressed the units for convenience. Solving this equation by the Newton-Raphson method gives $\overline{V} = 0.0750\,\text{dm}^3 \cdot \text{mol}^{-1}$, compared with the van der Waals result of $\overline{V} = 0.096\,\text{dm}^3 \cdot \text{mol}^{-1}$ and the experimental result of $0.071\,\text{dm}^3 \cdot \text{mol}^{-1}$ (see Example 16–2). The prediction of the Redlich-Kwong equation is nearly quantitative, unlike the van der Waals equation, which predicts a value of $\overline{V}$ that is about 30% too large.

Figure 16.5 compares experimental pressure versus density data for ethane at 400 K with the predictions of the various equations of state introduced in this chapter. Note that the Redlich-Kwong and Peng-Robinson equations are nearly quantitative, whereas the van der Waals equation fails completely at pressures greater than 200 bar. One of the impressive features of the Redlich-Kwong and Peng-Robinson equations is that they are nearly quantitative in regions where the gas liquefies. For example, Figure 16.6 shows pressure versus density data for ethane at 305.33 K, where it liquefies at around 40 bar. The horizontal region in the figure represents liquid and vapor in equilibrium with each other. Note that the Peng-Robinson equation is better in the liquid-vapor region but that the Redlich-Kwong equation is better at high pressures. The van der Waals equation is not shown because it gives negative values of the pressure under these conditions.

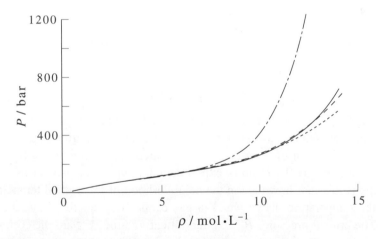

FIGURE 16.5
Experimental pressure versus density data for ethane at 400 K (solid line) is compared with the predictions of the van der Waals equation (dot-dashed line), the Redlich-Kwong equation (long dashed line), and the Peng-Robinson equation (short dashed line).

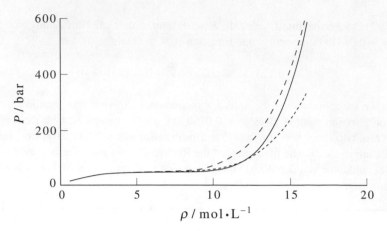

FIGURE 16.6
The experimental pressure versus density data (solid line) for ethane at 305.33 K is compared with the predictions of the Redlich-Kwong equation (long dashed line) and the Peng-Robinson equation (short dashed line). The liquid and vapor phases are in equilibrium in the horizontal region.

Although Figures 16.5 and 16.6 show comparisons only for ethane, the conclusions as to the relative accuracies of the equations are general. In general, the Redlich-Kwong equation is superior at high pressures, whereas the Peng-Robinson equation is superior in the liquid-vapor region. In fact, these two equations of state have been "constructed" so that this is so. There are more sophisticated equations of state (some containing more than 10 parameters!) that can reproduce the experimental data to a high degree of accuracy over a large range of pressure, density, and temperature.

16–3. A Cubic Equation of State Can Describe Both the Gaseous and Liquid States

A remarkable feature of equations of state that can be written as cubic equations in $\overline{V}$ is that they describe both the gaseous *and* the liquid regions of a substance. To understand this feature, we start by discussing some experimentally determined plots of P as a function of $\overline{V}$ at constant T, which are commonly called *isotherms* (*iso* = constant). Figure 16.7 shows experimental P versus $\overline{V}$ isotherms for carbon dioxide. The isotherms shown are in the neighborhood of the critical temperature, T_c, which is the temperature above which a gas cannot be liquefied, regardless of the pressure. The critical pressure, P_c, and the critical volume, $\overline{V}_c$, are the corresponding pressure and the molar volume at the *critical point*. For example, for carbon dioxide, $T_c = 304.14$ K (30.99°C), $P_c = 72.9$ atm, and $\overline{V}_c = 0.094$ L·mol^{-1}. Note that the isotherms in Figure 16.7 flatten out as $T \to T_c$ from above and that there are horizontal regions when T is less than T_c. In the horizontal regions, gas and liquid coexist in equilibrium with each other. The dashed curve connecting the ends of the horizontal

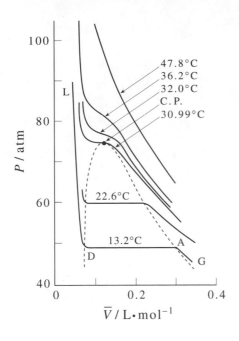

FIGURE 16.7
Experimental pressure-volume isotherms of carbon dioxide around its critical temperature, 30.99°C. Points G, A, D, and L are discussed in the text.

lines in Figure 16.7 is called the *coexistence curve*, because any point within this curve corresponds to liquid and gas coexisting in equilibrium with each other. At any point on or outside this curve, only one phase is present. For example, at point G in the figure we have only a gas phase. If we now start at G and compress the gas along the 13.2°C isotherm, liquid will first appear when we reach the horizontal line at point A. The pressure will remain constant as we condense the gas at molar volume 0.3 L·mol^{-1} (point A) to liquid at molar volume of approximately 0.07 L·mol^{-1} (point D). After reaching point D, the pressure increases sharply with a further decrease in volume, because we now have all liquid and the volume of a liquid changes very little with pressure.

Note that as the temperature increases toward the critical temperature, the horizontal lines shorten and disappear at the critical temperature. At this point, the meniscus between the liquid and its vapor disappears and there is no distinction between liquid and gas; the surface tension disappears and the gas and liquid phases both have the same (critical) density. We will discuss the critical point in more detail in Chapter 23.

Figure 16.8 shows similar isotherms for the van der Waals equation and the Redlich-Kwong equation. Notice that the two equations of state give fairly similar plots. The spurious loops obtained for $T < T_c$ result from the approximate nature of these equations of state. Figure 16.9 shows a single van der Waals or Redlich-Kwong isotherm for $T < T_c$. The curve GADL is the curve that would be observed experimentally upon compressing the gas. The horizontal line DA is drawn so that the areas of the loop both above and below DA are equal. (This so-called *Maxwell equal-area construction* will be justified in Chapter 23.) The line GA represents compression of the gas. Along the line AD, liquid and vapor are in equilibrium with each other. The point A represents

650

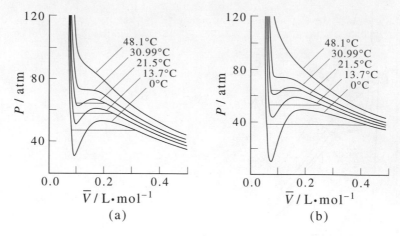

FIGURE 16.8
Pressure-volume isotherms of carbon dioxide around its critical temperature, as calculated from (a) the van der Waals equation (Equation 16.5) and (b) the Redlich-Kwong equation (Equation 16.7).

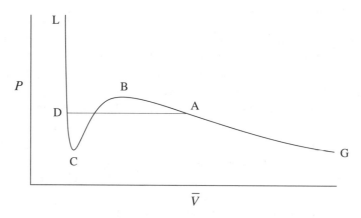

FIGURE 16.9
A typical van der Waals pressure-volume isotherm at a temperature less than the critical temperature. The horizontal line has been drawn so that areas of the loop above and below are equal.

the coexisting vapor and the point D represents the liquid. The line DL represents the change of volume of the liquid with increasing pressure. The steepness of this line results from the relative incompressibility of the liquid. The segment AB is a metastable region corresponding to the superheated vapor, and the segment CD corresponds to the supercooled liquid. The segment BC is a region in which $(\partial P/\partial \overline{V})_T > 0$. This condition signifies an unstable region, which is not observed for equilibrium systems.

Figure 16.9 shows that we can obtain three values of the volume along the line DA for a given pressure if the temperature is less than the critical temperature. This result is consistent with the fact that the van der Waals equation can be written as a cubic polynomial in the (molar) volume (see Example 16–2). The volume corresponding to point D is the molar volume of the liquid, the volume corresponding to point A is the molar volume of the vapor in equilibrium with the liquid, and the third root, lying between A and D is spurious.

At 142.69 K and 35.00 atm, argon exists as two phases in equilibrium with each other, and the densities of the liquid and vapor phases are 22.491 mol·L^{-1} and 5.291 mol·L^{-1}, respectively. Let's see what the van der Waals equation predicts in this case. As we saw in Example 16–2, we can write the van der Waals equation as

$$\overline{V}^3 - \left(b + \frac{RT}{P}\right)\overline{V}^2 + \frac{a}{P}\overline{V} - \frac{ab}{P} = 0 \tag{16.10}$$

Using the values of a and b from Table 16.3, $T = 142.69$ K, and $P = 35.00$ atm, Equation 16.10 becomes

$$\overline{V}^3 - 0.3664\overline{V}^2 + 0.03802\overline{V} - 0.001210 = 0$$

where, for convenience, we have suppressed the units of the coefficients. The three roots of this equation are (Problem 16–22) 0.07073 L·mol^{-1}, 0.07897 L·mol^{-1}, and 0.2167 L·mol^{-1}. The smallest root represents the molar volume of liquid argon, and the largest represents the molar volume of the vapor. The corresponding densities are 14.14 mol·L^{-1} and 4.615 mol·L^{-1}, which are in poor agreement with the experimental values (22.491 mol·L^{-1} and 5.291 mol·L^{-1}). The Redlich-Kwong equation gives 20.13 mol·L^{-1} and 5.147 mol·L^{-1}, and the Peng-Robinson equation gives 23.61 mol·L^{-1} and 5.564 mol·L^{-1} (Problem 16–23). Both the Redlich-Kwong and the Peng-Robinson equations are fairly accurate, and the Peng-Robinson equation is about 10% more accurate in this liquid region.

The point C.P. in Figure 16.7 is the critical point, where $T = T_c$, $P = P_c$, and $\overline{V} = \overline{V}_c$. The point C.P. is an inflection point, and so

$$\left(\frac{\partial P}{\partial \overline{V}}\right)_T = 0 \qquad \text{and} \qquad \left(\frac{\partial^2 P}{\partial \overline{V}^2}\right)_T = 0 \qquad \text{at C.P.}$$

We can use these two conditions to determine the critical constants in terms of a and b (Problem 16–25). An easier way to do this, however, is to write the van der Waals equation as a cubic equation in $\overline{V}$, Equation 16.10.

$$\overline{V}^3 - \left(b + \frac{RT}{P}\right)\overline{V}^2 + \frac{a}{P}\overline{V} - \frac{ab}{P} = 0$$

Being a cubic equation, it has three roots. For $T > T_c$, only one of these roots is real (the other two are complex), and for $T < T_c$ and $P \approx P_c$, all three roots are real. At $T = T_c$, these three roots merge into one, and so we can write Equation 16.10 as $(\overline{V} - \overline{V}_c)^3 = 0$, or

$$\overline{V}^3 - 3\overline{V}_c\overline{V}^2 + 3\overline{V}_c^2\overline{V} - \overline{V}_c^3 = 0 \tag{16.11}$$

If we compare this equation with Equation 16.10 at the critical point, we have

$$3\overline{V}_c = b + \frac{RT_c}{P_c}, \qquad 3\overline{V}_c^2 = \frac{a}{P_c}, \qquad \text{and} \qquad \overline{V}_c^3 = \frac{ab}{P_c} \qquad (16.12)$$

Eliminate P_c between the second two of these to obtain

$$\overline{V}_c = 3b \qquad (16.13a)$$

and then substitute this result into the third of Equations 16.12 to obtain

$$P_c = \frac{a}{27b^2} \qquad (16.13b)$$

and last, substitute Equations 16.13a and 16.13b into the first of Equations 16.12 to obtain

$$T_c = \frac{8a}{27bR} \qquad (16.13c)$$

The critical constants of a number of substances are given in Table 16.5.

The values of the critical constants in terms of the parameters A and B of the Redlich-Kwong equation can be determined in a similar fashion. The mathematics is a bit more involved, and the results are (Problem 16–27)

$$\overline{V}_c = 3.8473B, \qquad P_c = 0.029894\frac{A^{2/3}R^{1/3}}{B^{5/3}}, \qquad \text{and} \qquad T_c = 0.34504\left(\frac{A}{BR}\right)^{2/3} \qquad (16.14)$$

The following example shows that the van der Waals equation and the Redlich-Kwong equation make an interesting prediction about the value of $P_c\overline{V}_c/RT_c$.

EXAMPLE 16–4

Calculate the ratio $P_c\overline{V}_c/RT_c$ for the van der Waals equation and the Redlich-Kwong equation.

SOLUTION: Multiplying Equation 16.13b by 16.13a and dividing by R times Equation 16.13c gives

$$\frac{P_c\overline{V}_c}{RT_c} = \frac{1}{R}\left(\frac{a}{27b^2}\right)(3b)\left(\frac{27bR}{8a}\right) = \frac{3}{8} = 0.375 \qquad (16.15)$$

Similarly, the Redlich-Kwong equation gives

$$\frac{P_c\overline{V}_c}{RT_c} = \frac{1}{R}\left(\frac{0.029894A^{2/3}R^{1/3}}{B^{5/3}}\right)(3.8473B)\left(\frac{(BR)^{2/3}}{0.34504A^{2/3}}\right) = 0.33333 \qquad (16.16)$$

Equations 16.15 and 16.16 predict that $P_c\overline{V}_c/RT_c$ should be the same value for all substances but that the numerical values differ slightly for the two approximate

TABLE 16.5
The experimental critical constants of various substances.

Species	T_c/K	P_c/bar	P_c/atm	$\overline{V}_c$/L·mol^{-1}	$P_c\overline{V}_c/RT_c$
Helium	5.1950	2.2750	2.2452	0.05780	0.30443
Neon	44.415	26.555	26.208	0.04170	0.29986
Argon	150.95	49.288	48.643	0.07530	0.29571
Krypton	210.55	56.618	55.878	0.09220	0.29819
Hydrogen	32.938	12.838	12.670	0.06500	0.30470
Nitrogen	126.20	34.000	33.555	0.09010	0.29195
Oxygen	154.58	50.427	50.768	0.07640	0.29975
Carbon monoxide	132.85	34.935	34.478	0.09310	0.29445
Chlorine	416.9	79.91	78.87	0.1237	0.28517
Carbon dioxide	304.14	73.843	72.877	0.09400	0.27443
Water	647.126	220.55	217.66	0.05595	0.2295
Ammonia	405.30	111.30	109.84	0.07250	0.23945
Methane	190.53	45.980	45.379	0.09900	0.28735
Ethane	305.34	48.714	48.077	0.1480	0.28399
Ethene	282.35	50.422	49.763	0.1290	0.27707
Propane	369.85	42.477	41.922	0.2030	0.28041
Butane	425.16	37.960	37.464	0.2550	0.27383
2-Methylpropane	407.85	36.400	35.924	0.2630	0.28231
Pentane	469.69	33.643	33.203	0.3040	0.26189
Benzene	561.75	48.758	48.120	0.2560	0.26724

equations of state. The experimental values of $P_c\overline{V}_c/RT_c$ given in Table 16.5 show that neither equation of state is quantitative. The corresponding value for $P_c\overline{V}_c/RT_c$ for the Peng-Robinson equation is 0.30740 (Problem 16–28), which is closer to the experimental values than either of the values given by the van der Waals equation or the Redlich-Kwong equation. Note, however, that all three equations of state do predict a constant value for $P_c\overline{V}_c/RT_c$, and the experimental data in Table 16.5 show that this value is indeed fairly constant. This observation is an example of the law of corresponding states, which says that the properties of all gases are the same if we compare them under the same conditions relative to their critical point. We will discuss the law of corresponding states more thoroughly in the next section.

Although we have written $\overline{V}_c$, P_c, and T_c in terms of a and b in Equations 16.13 or in terms of A and B in Equations 16.14, in practice these constants are usually evaluated in terms of experimental critical constants. Because there are three critical constants and only two constants for each equation of state, there is some ambiguity in doing so. For example, we could use Equations 16.13a and 16.13b to evaluate a and b

in terms of $\overline{V}_c$ and P_c, or use another pair of equations. Because P_c and T_c are known more accurately, we use Equations 16.13b and 16.13c to obtain

$$a = \frac{27(RT_c)^2}{64P_c} \quad \text{and} \quad b = \frac{RT_c}{8P_c} \tag{16.17}$$

Likewise, from Equations 16.14, we obtain the Redlich-Kwong constants,

$$A = 0.42748 \frac{R^2 T_c^{5/2}}{P_c} \quad \text{and} \quad B = 0.086640 \frac{RT_c}{P_c} \tag{16.18}$$

The van der Waals and Redlich-Kwong constants in Tables 16.3 and 16.4 have been obtained in this way.

EXAMPLE 16–5

Use the critical-constant data in Table 16.5 to evaluate the van der Waals constants for ethane.

SOLUTION:

$$a = \frac{27(0.083145 \text{ dm}^3 \cdot \text{bar} \cdot \text{mol}^{-1} \text{K}^{-1})^2 (305.34 \text{ K})^2}{64(48.714 \text{ bar})}$$

$$= 5.5817 \text{ dm}^6 \cdot \text{bar} \cdot \text{mol}^{-2} = 5.5088 \text{ dm}^6 \cdot \text{atm} \cdot \text{mol}^{-2}$$

and

$$b = \frac{(0.083145 \text{ dm}^3 \cdot \text{bar} \cdot \text{mol}^{-1} \text{K}^{-1})(305.34 \text{ K})}{8(48.714 \text{ bar})}$$

$$= 0.065144 \text{ dm}^3 \cdot \text{mol}^{-1}$$

EXAMPLE 16–6

Use the critical-constant data in Table 16.5 to evaluate A and B, the Redlich-Kwong constants for ethane.

SOLUTION:

$$A = 0.42748 \frac{(0.083145 \text{ dm}^3 \cdot \text{bar} \cdot \text{mol}^{-1} \cdot \text{K}^{-1})^2 (305.34 \text{ K})^{5/2}}{48.714 \text{ bar}}$$

$$= 98.831 \text{ dm}^6 \cdot \text{bar} \cdot \text{mol}^{-2} \cdot \text{K}^{1/2} = 97.539 \text{ dm}^6 \cdot \text{atm} \cdot \text{mol}^{-2} \cdot \text{K}^{1/2}$$

and

$$B = 0.086640 \frac{(0.083145 \text{ dm}^3 \cdot \text{bar} \cdot \text{mol}^{-1} \cdot \text{K}^{-1})(305.34 \text{ K})}{48.714 \text{ bar}}$$

$$= 0.045153 \text{ dm}^3 \cdot \text{mol}^{-1}$$

16–4. The van der Waals Equation and the Redlich–Kwong Equation Obey the Law of Corresponding States

Let's start with the van der Waals equation, which we can write in an interesting and practical form by substituting the second of Equations 16.12 for a and Equation 16.13a for b into Equation 16.5:

$$\left(P + \frac{3P_c \overline{V}_c^2}{\overline{V}^2}\right)\left(\overline{V} - \frac{1}{3}\overline{V}_c\right) = RT$$

Divide through by P_c and $\overline{V}_c$ to get

$$\left(\frac{P}{P_c} + \frac{3\overline{V}_c^2}{\overline{V}^2}\right)\left(\frac{\overline{V}}{\overline{V}_c} - \frac{1}{3}\right) = \frac{RT}{P_c \overline{V}_c} = \frac{RT}{\frac{3}{8}RT_c} = \frac{8}{3}\frac{T}{T_c}$$

where we have used Equation 16.15 for $P_c \overline{V}_c$. Now introduce the *reduced quantities* $P_R = P/P_c$, $\overline{V}_R = \overline{V}/\overline{V}_c$, and $T_R = T/T_c$ to obtain the van der Waals equation written in terms of reduced quantities:

$$\left(P_R + \frac{3}{\overline{V}_R^2}\right)\left(\overline{V}_R - \frac{1}{3}\right) = \frac{8}{3}T_R \tag{16.19}$$

Equation 16.19 is remarkable in that there are no quantities in this equation that are characteristic of any particular gas; it is a universal equation for *all* gases. It says, for example, that the value of P_R will be the same for all gases at the same values of $\overline{V}_R$ and T_R. Let's consider $CO_2(g)$ and $N_2(g)$ for $\overline{V}_R = 20$ and $T_R = 1.5$. According to Equation 16.19, $P_R = 0.196$ when $\overline{V}_R = 20.0$ and $T_R = 1.5$. Using the values of the critical constants given in Table 16.5, we find that the reduced quantities $P_R = 0.196$, $\overline{V}_R = 20.0$, and $T_R = 1.5$ correspond to $P_{CO_2} = 14.3$ atm $= 14.5$ bar, $\overline{V}_{CO_2} = 1.9$ L·mol^{-1}, and $T_{CO_2} = 456$ K and to $P_{N_2} = 6.58$ atm $= 6.66$ bar, $\overline{V}_{N_2} = 1.8$ L·mol^{-1}, and $T_{N_2} = 189$ K. These two gases under these conditions are said to be at corresponding states (same values of P_R, $\overline{V}_R$, and T_R). According to the van der Waals equation, these quantities are related by Equation 16.19, so Equation 16.19 is an example of the *law of corresponding states*, that all gases have the same properties if they are compared at corresponding conditions (same values of P_R, $\overline{V}_R$, and T_R).

EXAMPLE 16–7
Express the Redlich-Kwong equation in terms of reduced quantities.

SOLUTION: Equations 16.18 show that

$$A = 0.42748\frac{R^2 T_c^{5/2}}{P_c} \quad \text{and} \quad B = 0.086640\frac{RT_c}{P_c}$$

Substituting these equivalencies into Equation 16.7 gives

$$P = \frac{RT}{\overline{V} - 0.086640 \dfrac{RT_c}{P_c}} - \frac{0.42748 R^2 T_c^{5/2}/P_c}{T^{1/2}\overline{V}\left(\overline{V} + 0.086640 \dfrac{RT_c}{P_c}\right)}$$

Divide the numerator and the denominator of the first term on the right side by $\overline{V}_c$ and the second by $\overline{V}_c^2$ to get

$$P = \frac{RT/\overline{V}_c}{\overline{V}_R - 0.086640 \dfrac{RT_c}{P_c\overline{V}_c}} - \frac{0.42748 R^2 T_c^2/P_c\overline{V}_c^2}{T_R^{1/2}\overline{V}_R\left(\overline{V}_R + 0.086640 \dfrac{RT_c}{P_c\overline{V}_c}\right)}$$

Divide both sides by P_c and use the fact that $P_c\overline{V}_c/RT_c = 1/3$ in the second term to get

$$P_R = \frac{RT/P_c\overline{V}_c}{\overline{V}_R - 0.25992} - \frac{3.8473}{T_R^{1/2}\overline{V}_R(\overline{V}_R + 0.25992)}$$

Finally, multiply and divide the numerator of the first term on the right side by T_c to obtain

$$P_R = \frac{3T_R}{\overline{V}_R - 0.25992} - \frac{3.8473}{T_R^{1/2}\overline{V}_R(\overline{V}_R + 0.25992)}$$

Thus, we see that the Redlich-Kwong equation also obeys a law of corresponding states.

The compressibility factor, Z, associated with the van der Waals equation also obeys the law of corresponding states. To demonstrate this point, we start with Equation 16.6 and substitute the second of Equations 16.12 for a and Equation 16.13b for b to get

$$Z = \frac{P\overline{V}}{RT} = \frac{\overline{V}}{\overline{V} - \frac{1}{3}\overline{V}_c} - \frac{3P_c\overline{V}_c^2}{RT\overline{V}}$$

Now use Equation 16.15 for $P_c\overline{V}_c$ in the second term and introduce reduced variables to get

$$Z = \frac{\overline{V}_R}{\overline{V}_R - \frac{1}{3}} - \frac{9}{8\overline{V}_R T_R} \tag{16.20}$$

Similarly, the compressibility factor for the Redlich-Kwong equation is (Problem 16-30)

$$Z = \frac{\overline{V}_R}{\overline{V}_R - 0.25992} - \frac{1.2824}{T_R^{3/2}(\overline{V}_R + 0.25992)} \tag{16.21}$$

Equations 16.20 and 16.21 express Z as a universal function of $\overline{V}_R$ and T_R, or of any other two reduced quantities, such as P_R and T_R. Although these equations can be used to illustrate the law of corresponding states, they are based on approximate equations of state. Nevertheless, the law of corresponding states is valid for a great variety of gases. Figure 16.10 shows experimental data for Z plotted against P_R at various values of T_R for 10 gases. Note that the data for all 10 gases fall on the same curves, thus illustrating the law of corresponding states in a more general way than either Equation 16.20 or 16.21. Much more extensive graphs are available, particularly in the engineering literature, and are of great use in practical applications.

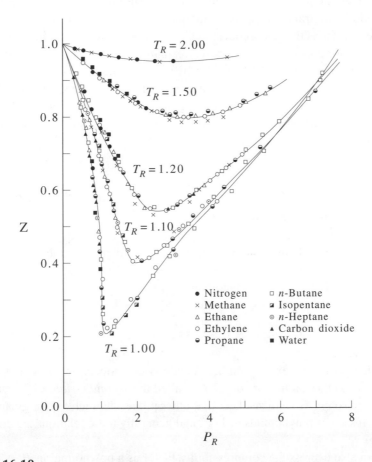

FIGURE 16.10
An illustration of the law of corresponding states. The compressibility factor, Z, is plotted against the reduced pressure, P_R, of each of the 10 indicated gases. Each curve represents a given reduced temperature. Note that for a given reduced temperature, all 10 gases fall on the same curve because reduced quantities are used.

EXAMPLE 16–8
Use Figure 16.10 to estimate the molar volume of ammonia at 215°C and 400 bar.

SOLUTION: Using the critical-constant data in Table 16.5, we find that $T_R = 1.20$ and $P_R = 3.59$. Figure 16.10 shows that $Z \approx 0.60$ under these conditions. The molar volume is

$$\overline{V} \approx \frac{RTZ}{P} = \frac{(0.08314 \text{ L·bar·mol}^{-1}·\text{K}^{-1})(488 \text{ K})(0.60)}{400 \text{ bar}}$$
$$\approx 0.061 \text{ L·mol}^{-1} = 61 \text{ cm}^3·\text{mol}^{-1}$$

The law of corresponding states has a nice physical interpretation. Any temperature scale we use to describe a gas is necessarily arbitrary. Even the Kelvin scale, with its fundamental zero temperature, is arbitrary in the sense that the size of a degree on the Kelvin scale is arbitrary. Thus, the numerical value we assign to the temperature is meaningless as far as the gas is concerned. A gas does "know" its critical temperature, and therefore is "aware" of its temperature *relative* to its critical temperature or its reduced temperature, $T_R = T/T_c$. Similarly, pressure and volume scales are imposed by us, but the reduced pressure and the reduced volume are quantities that are of significance to a particular gas. Thus, any gas that has a certain reduced temperature, pressure, and volume will behave in the same manner as another gas under the same conditions.

16–5. Second Virial Coefficients Can Be Used to Determine Intermolecular Potentials

The most fundamental equation of state, in the sense that it has the most sound theoretical foundation, is the *virial equation of state*. The virial equation of state expresses the compressibility factor as a polynomial in $1/\overline{V}$:

$$Z = \frac{P\overline{V}}{RT} = 1 + \frac{B_{2V}(T)}{\overline{V}} + \frac{B_{3V}(T)}{\overline{V}^2} + \cdots \tag{16.22}$$

The coefficients in this expression are functions of temperature only and are called *virial coefficients*. In particular, $B_{2V}(T)$ is called the *second virial coefficient*, $B_{3V}(T)$ the third, and so on. We will see later that other properties such as energy and entropy can be expressed as polynomials in $1/\overline{V}$, and generally these relations are called *virial expansions*.

We can also express the compressibility factor as a polynomial in P

$$Z = \frac{P\overline{V}}{RT} = 1 + B_{2P}(T)P + B_{3P}(T)P^2 + \cdots \tag{16.23}$$

Equation 16.23 is also called a virial expansion or a virial equation of state. The virial coefficients $B_{2V}(T)$ and $B_{2P}(T)$ are related by (Problem 16–36)

$$B_{2V}(T) = RT B_{2P}(T) \tag{16.24}$$

Note in Equation 16.22 or 16.23 that $Z \to 1$ as $\overline{V}$ becomes large or as P becomes small, just as it should. Table 16.6 gives an idea of the magnitudes of the terms in Equation 16.22 as a function of pressure for argon at 25°C. Notice that even at 100 bar the first three terms are sufficient for calculating Z.

TABLE 16.6

The contribution of the first few terms in the virial expansion of Z, Equation 16.22, for argon at 25°C.

P/bar	$Z = P\overline{V}/RT$
	$1 + \dfrac{B_{2V}(T)}{\overline{V}} + \dfrac{B_{3V}(T)}{\overline{V}^2} +$ remaining terms
1	$1 - 0.00064 + 0.00000 + (+0.00000)$
10	$1 - 0.00648 + 0.00020 + (-0.00007)$
100	$1 - 0.06754 + 0.02127 + (-0.00036)$
1000	$1 - 0.38404 + 0.08788 + (+0.37232)$

The second virial coefficient is the most important virial coefficient because it reflects the first deviation from ideality as the pressure of the gas is increased (or the volume is decreased). As such, it is the most easily measured virial coefficient and is well tabulated for many gases. According to Equation 16.23, it can be determined experimentally from the slope of a plot of Z against P, as shown in Figure 16.11. Figure 16.12 shows $B_{2V}(T)$ plotted against temperature for helium, nitrogen, methane, and carbon dioxide. Note that $B_{2V}(T)$ is negative at low temperatures and increases with temperature, eventually going through a shallow maximum (observable only for helium in Figure 16.12). The temperature at which $B_{2V}(T) = 0$ is called the *Boyle temperature*. At the Boyle temperature, the repulsive and attractive parts of the intermolecular interactions cancel each other, and the gas appears to behave ideally (neglecting any effect of virial coefficients beyond the second).

Not only are Equations 16.22 and 16.23 used to summarize experimental P–V–T data, but they also allow us to derive exact relations between the virial coefficients and the intermolecular interactions. Consider two interacting molecules as shown in Figure 16.13. The interaction of the two molecules depends upon the distance between their centers, r, and upon their orientations. Because the molecules are rotating, their orientations partially average out, so for simplicity we assume that the interaction depends only upon r. This approximation turns out to be satisfactory for

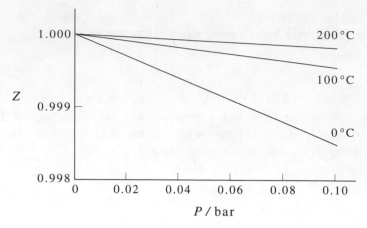

FIGURE 16.11

A plot of Z versus P at low pressures for $NH_3(g)$ at 0°C, 100°C, and 200°C. The slopes of the lines are equal to $B_{2V}(T)/RT$ according to Equations 16.23 and 16.24. The respective slopes give $B_{2V}(0°C) = -0.345$ $dm^3 \cdot mol^{-1}$, $B_{2V}(100°C) = -0.142$ $dm^3 \cdot mol^{-1}$, and $B_{2V}(200°C) = -0.075$ $dm^3 \cdot mol^{-1}$.

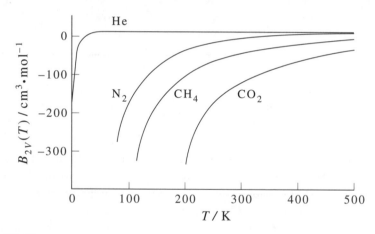

FIGURE 16.12

The second virial coefficient $B_{2V}(T)$ of several gases plotted against temperature. Note that $B_{2V}(T)$ is negative at low temperatures and increases with temperature up to a point, where it passes through a shallow maximum (observable here only for helium).

many molecules, especially if they are not very polar. If we let $u(r)$ be the potential energy of two molecules separated by a distance r, the relation between the second virial coefficient $B_{2V}(T)$ and $u(r)$ is given by

$$B_{2V}(T) = -2\pi N_A \int_0^\infty [e^{-u(r)/k_B T} - 1]r^2 dr \qquad (16.25)$$

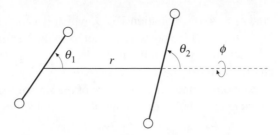

FIGURE 16.13
Two interacting linear molecules. Generally, the intermolecular interaction between two molecules depends upon the distance between their centers (r) and upon their orientations (θ_1, θ_2, and ϕ).

where N_A is the Avogadro constant and k_B is the Boltzmann constant, which is equal to the molar gas constant R divided by the Avogadro constant. Note that $B_{2V}(T) = 0$ if $u(r) = 0$; in other words, there are no deviations from ideal behavior if there are no intermolecular interactions.

Equation 16.25 shows that once $u(r)$ is known, it is a simple matter to calculate $B_{2V}(T)$ as a function of temperature, or conversely, to determine $u(r)$ if $B_{2V}(T)$ is known. In principle, $u(r)$ can be calculated from quantum mechanics, but this is a difficult computational problem. We can show from perturbation theory, however, that

$$u(r) \longrightarrow -\frac{c_6}{r^6} \tag{16.26}$$

for large values of r. In this expression, c_6 is a constant whose value depends upon the particular interacting molecules. The negative sign in Expression 16.26 indicates that the two molecules attract each other. This attraction is what causes substances to condense at sufficiently low temperatures. There is no known exact expression like 16.26 for small distances, but it must be of a form that reflects the repulsion that occurs when two molecules approach closely. Usually, we assume that

$$u(r) \longrightarrow \frac{c_n}{r^n} \tag{16.27}$$

for small values of r. In Equation 16.27, n is an integer, often taken to be 12, and c_n is a constant whose value depends upon the two molecules.

An intermolecular potential that embodies the long-range (attractive) behavior of Equation 16.26 and the short-range (repulsive) behavior of Equation 16.27 is simply the sum of the two. If we take n to be 12, then

$$u(r) = \frac{c_{12}}{r^{12}} - \frac{c_6}{r^6} \tag{16.28}$$

Equation 16.28 is usually written in the form

$$u(r) = 4\varepsilon \left[\left(\frac{\sigma}{r} \right)^{12} - \left(\frac{\sigma}{r} \right)^{6} \right] \tag{16.29}$$

where $c_{12} = 4\varepsilon\sigma^{12}$ and $c_6 = 4\varepsilon\sigma^6$. Equation 16.29, which is called the *Lennard-Jones potential*, is plotted in Figure 16.14. The two parameters in the Lennard-Jones potential have the following physical interpretation: ε is the depth of the potential well and σ is the distance at which $u(r) = 0$ (Figure 16.14). As such, ε is a measure of how strongly the molecules attract each other, and σ is a measure of the size of the molecules. These *Lennard-Jones parameters* are tabulated for a number of molecules in Table 16.7.

EXAMPLE 16–9

Show that the minimum of the Lennard-Jones potential occurs at $r_{min} = 2^{1/6}\sigma = 1.12\sigma$. Evaluate $u(r)$ at r_{min}.

SOLUTION: To find r_{min}, we differentiate Equation 16.29:

$$\frac{du}{dr} = 4\varepsilon\left[-\frac{12\sigma^{12}}{r^{13}} + \frac{6\sigma^6}{r^7}\right] = 0$$

which gives $r_{min}^6 = 2\sigma^6$, or $r_{min} = 2^{1/6}\sigma$. Therefore,

$$u(r_{min}) = 4\varepsilon\left[\left(\frac{\sigma}{2^{1/6}\sigma}\right)^{12} - \left(\frac{\sigma}{2^{1/6}\sigma}\right)^{6}\right] = 4\varepsilon\left(\frac{1}{4} - \frac{1}{2}\right) = -\varepsilon$$

Thus ε is the depth of the potential well, relative to the infinite separation.

If we substitute the Lennard-Jones potential into Equation 16.25, we obtain

$$B_{2V}(T) = -2\pi N_A \int_0^\infty \left[\exp\left\{-\frac{4\varepsilon}{k_B T}\left[\left(\frac{\sigma}{r}\right)^{12} - \left(\frac{\sigma}{r}\right)^{6}\right]\right\} - 1\right] r^2 dr \qquad (16.30)$$

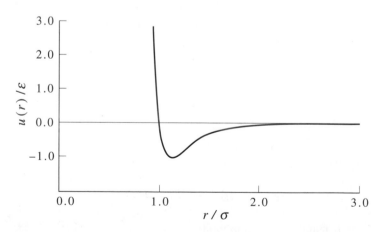

FIGURE 16.14

A plot of $u(r)/\varepsilon = 4\left[\left(\frac{\sigma}{r}\right)^{12} - \left(\frac{\sigma}{r}\right)^{6}\right]$ versus r/σ for the Lennard-Jones potential. Note that the depth of the potential well is ε and that $u(r) = 0$ at $r/\sigma = 1$.

TABLE 16.7
Lennard-Jones parameters, ε and σ, for various substances.

Species	(ε/k_B)/K	σ/pm	$(2\pi\sigma^3 N_A/3)$/cm$^3 \cdot$mol^{-1}
He	10.22	256	21.2
Ne	35.6	275	26.2
Ar	120	341	50.0
Kr	164	383	70.9
Xe	229	406	86.9
H_2	37.0	293	31.7
N_2	95.1	370	63.9
O_2	118	358	57.9
CO	100	376	67.0
CO_2	189	449	114.2
CF_4	152	470	131.0
CH_4	149	378	68.1
C_2H_4	199	452	116.5
C_2H_6	243	395	77.7
C_3H_8	242	564	226.3
$C(CH_3)_4$	232	744	519.4

Equation 16.30 may look complicated, but it can be simplified. We first define a reduced temperature T^* by $T^* = k_B T/\varepsilon$ and let $r/\sigma = x$ to get

$$B_{2V}(T^*) = -2\pi\sigma^3 N_A \int_0^\infty \left[\exp\left\{-\frac{4}{T^*}(x^{-12} - x^{-6})\right\} - 1\right] x^2 dx$$

We then divide both sides by $2\pi\sigma^3 N_A/3$ to get

$$B_{2V}^*(T^*) = -3 \int_0^\infty \left[\exp\left\{-\frac{4}{T^*}(x^{-12} - x^{-6})\right\} - 1\right] x^2 dx \qquad (16.31)$$

where $B_{2V}^*(T^*) = B_{2V}(T^*)/(2\pi\sigma^3 N_A/3)$. Equation 16.31 shows that the reduced second virial coefficient, $B_{2V}^*(T^*)$, depends upon only the reduced temperature, T^*. The integral in Equation 16.31 must be evaluated numerically (MathChapter G) for each value of T^*. Extensive tables of $B_{2V}^*(T^*)$ versus T^* are available.

Equation 16.31 is another example of the law of corresponding states. If we take experimental values of $B_{2V}(T)$, divide them by $2\pi\sigma^3 N_A/3$, and then plot the data versus $T^* = k_B T/\varepsilon$, the result for *all* gases will fall on one curve. Figure 16.15 shows such a plot for six gases. Conversely, a plot such as the one in Figure 16.15 (or better yet, numerical tables) can be used to evaluate $B_{2V}(T)$ for any gas.

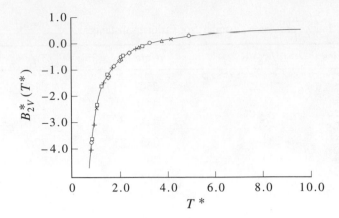

FIGURE 16.15
A plot of the reduced second virial coefficient $B_{2V}^*(T^*) = B_{2V}(T^*)/(2\pi\sigma^3 N_A/3)$ (solid line) against the reduced temperature $T^* = k_B T/\varepsilon$. Experimental data of six gases (argon, nitrogen, oxygen, carbon dioxide, methane, and sulfur hexafluoride) are also plotted. This plot is another illustration of the law of corresponding states.

EXAMPLE 16–10
Estimate $B_{2V}(T)$ for $N_2(g)$ at $0°C$.

SOLUTION: Table 16.7 gives $\varepsilon/k_B = 95.1$ K and $2\pi\sigma^3 N_A/3 = 63.9$ cm^3·mol^{-1} for $N_2(g)$. Thus, $T^* = 2.87$, and Figure 16.15 gives $B_{2V}^*(T^*) \approx -0.2$. Therefore,

$$B_{2V}(T) \approx (63.9 \text{ cm}^3 \cdot \text{mol}^{-1})(-0.2)$$

$$\approx -10 \text{ cm}^3 \cdot \text{mol}^{-1}$$

If we had used numerical tables for $B_{2V}^*(T^*)$ instead of Figure 16.15, we would have obtained $B_{2V}^*(T^*) = -0.16$, or $B_{2V}(T) = -10$ cm^3·mol^{-1}.

The value of $B_{2V}(T)$ has a simple interpretation. Consider Equation 16.23 under conditions where we can ignore the terms in P^2 and higher

$$\frac{P\overline{V}}{RT} = 1 + B_{2P}(T)P = 1 + \frac{B_{2V}(T)}{RT}P$$

By multiplying through by RT/P and using $\overline{V}_{\text{ideal}} = RT/P$, we can rewrite this equation in the form

$$\overline{V} = \overline{V}_{\text{ideal}} + B_{2V}(T)$$

or

$$B_{2V}(T) = \overline{V} - \overline{V}_{\text{ideal}} \tag{16.32}$$

Thus, we see that $B_{2V}(T)$ represents the difference between the actual value of $\overline{V}$

and the ideal-gas value $\overline{V}_{\text{ideal}}$ at pressures such that the contribution of the third virial coefficient is negligible.

EXAMPLE 16–11

The molar volume of isobutane at 300.0 K and one bar is 24.31 dm$^3 \cdot$mol^{-1}. Estimate the value of B_{2V} for isobutane at 300.0 K.

SOLUTION: The ideal-gas molar volume at 300.0 K and one bar is

$$\overline{V}_{\text{ideal}} = \frac{RT}{P} = \frac{(0.083145 \text{ dm}^3 \cdot \text{bar} \cdot \text{K}^{-1} \cdot \text{mol}^{-1})(300.0 \text{ K})}{1 \text{ bar}}$$

$$= 24.94 \text{ dm}^3 \cdot \text{mol}^{-1}$$

Therefore, using Equation 16.32,

$$B_{2V} = \overline{V} - \overline{V}_{\text{ideal}} = 24.31 \text{ dm}^3 \cdot \text{mol}^{-1} - 24.94 \text{ dm}^3 \cdot \text{mol}^{-1}$$

$$= -0.63 \text{ dm}^3 \cdot \text{mol}^{-1} = -630 \text{ cm}^3 \cdot \text{mol}^{-1}$$

Although we have been discussing calculating $B_{2V}(T)$ in terms of the Lennard-Jones potential, in practice it's the other way around: Lennard-Jones parameters are usually determined from experimental values of $B_{2V}(T)$. This determination is usually made through trial and error using tables of $B_{2V}^*(T^*)$. The values of the Lennard-Jones parameters in Table 16.7 were determined from experimental second virial coefficient data. Because the second virial coefficient reflects the initial deviations from ideal behavior, which are caused by intermolecular interactions, experimental P–V–T data turn out to be a rich source of information concerning intermolecular interactions. Once Lennard-Jones parameters have been determined, they can be used to calculate many other fluid properties such as viscosity, thermal conductivity, heats of vaporization, and various crystal properties.

16–6. London Dispersion Forces Are Often the Largest Contribution to the r^{-6} Term in the Lennard-Jones Potential

In the previous section, we used the Lennard-Jones potential (Equation 16.29) to represent the intermolecular potential between molecules. The r^{-12} term accounts for the repulsion at short distances, and the r^{-6} term accounts for the attraction at larger distances. The actual form of the repulsive term is not well established, but the r^{-6} dependence of the attractive term is. In this section, we will discuss three contributions to the r^{-6} attraction and compare their relative importance.

Consider two dipolar molecules, whose dipole moments are μ_1 and μ_2. The interaction of these dipoles depends upon how they are oriented with respect to each other. The energy will vary from repulsive, when they are oriented head-to-head as shown in

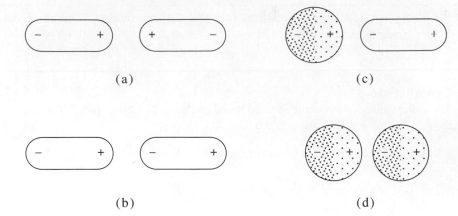

FIGURE 16.16
Two permanent dipoles oriented (a) head-to-head and (b) head-to-tail. The head-to-tail orientation is energetically favorable. (c) A molecule with a permanent dipole moment will induce a dipole moment in a neighboring molecule. (d) The instantaneous dipole-dipole correlation shown here is what leads to a London attraction between all atoms and molecules.

Figure 16.16a to attractive, when they are oriented head-to-tail (Figure 16.16b). Both molecules rotate in the gas phase, and if we were to average both dipoles randomly over their orientations, the dipole-dipole interactions would average out to zero. Because different orientations have different energies, they do not occur to equal extents. Clearly, the lower-energy head-to-tail orientation is favored over the repulsive head-to-head orientation. If we take into account the energy of the orientation, then the overall average interaction between the two molecules results in an attractive r^{-6} term of the form

$$u_{d.d}(r) = -\frac{2\mu_1^2\mu_2^2}{(4\pi\varepsilon_0)^2(3k_\text{B}T)}\frac{1}{r^6} \tag{16.33}$$

EXAMPLE 16–12
Show that the units of the right side of Equation 16.33 are energy.

SOLUTION: The units of μ are C·m (charge × separation), and so we have

$$u_{d.d}(r) \sim \frac{(\text{C·m})^4}{(\text{C}^2\cdot\text{s}^2\cdot\text{kg}^{-1}\cdot\text{m}^{-3})^2\text{J m}^6}$$

$$\sim \frac{\text{kg}^2\cdot\text{m}^4\cdot\text{s}^{-4}}{\text{J}} = \text{J}$$

EXAMPLE 16–13

Calculate the value of the coefficient of r^{-6} in Equation 16.33 at 300 K for two HCl(g) molecules. Table 16.8 lists the dipole moments of various molecules.

SOLUTION: According to Table 16.8, $\mu_1 = \mu_2 = 3.44 \times 10^{-30}$ C·m. Therefore,

$$-r^6 u_{d.d}(r)$$
$$= \frac{(2)(3.44 \times 10^{-30} \text{ C·m})^4}{(3)\left(\dfrac{8.314 \text{ J·mol}^{-1}\text{·K}^{-1}}{6.022 \times 10^{23} \text{ mol}^{-1}}\right)(300 \text{ K})(1.113 \times 10^{-10} \text{ C}^2\text{·s}^2\text{·kg}^{-1}\text{·m}^{-3})^2}$$
$$= 1.82 \times 10^{-78} \text{ J·m}^6$$

This numerical result may seem exceedingly small, but remember that we are calculating $-r^6 u_{d.d}(r)$. At a separation of 300 pm, $u_{d.d}(r)$ is equal to -2.5×10^{-21} J, compared with a thermal energy ($k_B T$) of 4.1×10^{-21} J at 300 K.

Equation 16.33 requires that both molecules have a permanent dipole moment. Even if one molecule does not have a permanent dipole moment, the one without a permanent dipole moment will have a dipole moment induced by the other. A dipole moment can be induced in a molecule that does not have a permanent dipole moment because all atoms and molecules are *polarizable*. When an atom or a molecule interacts with an electric field, the (negative) electrons are displaced in one direction and the (positive) nuclei are displaced in the opposite direction, as illustrated in Figure 16.16c. This charge separation with its associated dipole moment, is proportional to the strength of the electric field, and if we designate the induced dipole moment by μ_{induced} and the electric field by E, we have that $\mu_{\text{induced}} \propto E$. The proportionality constant, which we denote by α, is called the *polarizability*, so we have the defining expression

$$\mu_{\text{induced}} = \alpha E \tag{16.34}$$

The units of E are V·m^{-1}, so the units of α in Equation 16.34 are C·m/V·m^{-1} = C·m^2·V^{-1}. We can put α into more transparent units by using the fact that energy = (charge)$^2/4\pi\varepsilon_0$ (distance), which in SI units gives

$$\text{joule} \sim \frac{\text{C}^2}{(4\pi\varepsilon_0)\text{m}} = \text{C}^2\text{·m}^{-1}/4\pi\varepsilon_0$$

Similarly, from electrostatics, we have that

$$\text{joule} = \text{coulomb} \times \text{volt} = \text{C·V}$$

Equating these two expressions for joules gives C·V = C^2·m$^{-1}/4\pi\varepsilon_0$, or C·V^{-1} = $(4\pi\varepsilon_0)$ m. Now we substitute this result into the above units for α (C·m^2·V^{-1}) to get

$$\alpha \sim (4\pi\varepsilon_0)\text{m}^3$$

Thus, we see that $\alpha/4\pi\varepsilon_0$ has units of m^3. The quantity $\alpha/4\pi\varepsilon_0$, which is sometimes referred to as the *polarizability volume*, has units of volume. The easier it is for the electric field to deform the atomic or molecular charge distribution, the greater is the polarizability. The polarizability of an atom or a molecule is proportional to its size (note the units of $\alpha/4\pi\varepsilon_0$), or to its number of electrons. This trend can be seen in Table 16.8, which lists the polarizability volumes of some atoms and molecules.

TABLE 16.8
The dipole moment (μ), the polarizability volume ($\alpha/4\pi\varepsilon_0$), and the ionization energies (I) of various atoms and molecules.

Species	$\mu/10^{-30}$ C·m	$(\alpha/4\pi\varepsilon_0)/10^{-30}$ m^3	$I/10^{-18}$ J
He	0	0.21	3.939
Ne	0	0.39	3.454
Ar	0	1.63	2.525
Kr	0	2.48	2.243
Xe	0	4.01	1.943
N_2	0	1.77	2.496
CH_4	0	2.60	2.004
C_2H_6	0	4.43	1.846
C_3H_8	0.03	6.31	1.754
CO	0.40	1.97	2.244
CO_2	0	2.63	2.206
HCl	3.44	2.63	2.043
HI	1.47	5.42	1.664
NH_3	5.00	2.23	1.628
H_2O	6.14	1.47	2.020

We now return to the dipole-induced dipole interaction shown in Figure 16.16c. Because the induced dipole moment is always in a head-to-tail orientation with respect to the permanent dipole moment, the interaction is always attractive and is given by

$$u_{\text{induced}}(r) = -\frac{\mu_1^2\alpha_2}{(4\pi\varepsilon_0)^2 r^6} - \frac{\mu_2^2\alpha_1}{(4\pi\varepsilon_0)^2 r^6} \tag{16.35}$$

The first term represents a permanent dipole moment in molecule 1 and an induced dipole moment in molecule 2, and the second represents the opposite situation.

EXAMPLE 16–14
Calculate the value of the coefficient of r^{-6} for $u_{\text{induced}}(r)$ for two HCl(g) molecules.

SOLUTION: The two terms in Equation 16.35 are the same for identical molecules. Using the data in Table 16.8,

$$
\begin{aligned}
-r^6 u_{\text{induced}}(r) &= \frac{2\mu^2 (\alpha/4\pi\varepsilon_0)}{4\pi\varepsilon_0} \\
&= \frac{(2)(3.44 \times 10^{-30}\ \text{C}\cdot\text{m})^2 (2.63 \times 10^{-30}\ \text{m}^3)}{1.113 \times 10^{-10}\ \text{C}^2\cdot\text{s}^2\cdot\text{kg}^{-1}\cdot\text{m}^{-3}} \\
&= 5.59 \times 10^{-79}\ \text{J}\cdot\text{m}^6
\end{aligned}
$$

Note that this result is about 30% of the result we obtained in Example 16–13 for $-r^6 u_{d.d}(r)$.

Both Equations 16.33 and 16.35 equal zero when neither molecule has a permanent dipole moment. The third contribution to the r^{-6} term in Equation 16.29 is nonzero even if both molecules are nonpolar. This contribution was first calculated by the German scientist Fritz London in 1930 using quantum mechanics and is now called a *London dispersion attraction*. Although this attraction is a strictly quantum-mechanical effect, it lends itself to the following commonly used classical picture. Consider two atoms as shown in Figure 16.16d separated by a distance r. The electrons on one atom do not completely shield the high positive charge on the nucleus from the electrons on the other atom. Because the molecule is polarizable, the electronic wave function can distort a bit to further lower the interaction energy. If we average this electronic attraction quantum mechanically, we obtain an attractive term that varies as r^{-6}. The exact quantum-mechanical calculation is somewhat complicated, but an approximate form of the final result is

$$
u_{\text{disp}}(r) = -\frac{3}{2}\left(\frac{I_1 I_2}{I_1 + I_2}\right)\frac{\alpha_1 \alpha_2}{(4\pi\varepsilon_0)^2}\frac{1}{r^6} \tag{16.36}
$$

where I_j is the ionization energy of atom or molecule j. Note that Equation 16.36 does not involve a permanent dipole moment and that the interaction energy is proportional to the product of the polarizability volumes. Thus, the importance of $u_{\text{disp}}(r)$ increases with the sizes of the atoms or molecules, and, in fact, is often the dominant contribution to the r^{-6} term in Equation 16.29.

EXAMPLE 16–15

Calculate the value of the coefficient of r^{-6} for $u_{\text{disp}}(r)$ for two HCl(g) molecules.

SOLUTION: Using the data in Table 16.8, we have

$$
\begin{aligned}
-r^6 u_{\text{disp}}(r) &= \frac{3}{2}\left(\frac{2.043 \times 10^{-18}\ \text{J}}{2}\right)(2.63 \times 10^{-30}\ \text{m}^3)^2 \\
&= 1.06 \times 10^{-77}\ \text{J}\cdot\text{m}^6
\end{aligned}
$$

This quantity is about six times greater than $-r^6 u_{d.d}(r)$ and 20 times greater than $-r^6 u_{\text{induced}}(r)$. Similar calculations show that the dispersion term is significantly larger than either the dipole-dipole term or the dipole-induced dipole term except for very polar molecules such as NH_3, H_2O, and HCN.

The total contribution to the r^{-6} term in the Lennard-Jones potential is given by the sum of Equations 16.33, 16.35, and 16.36, giving

$$u(r) = \frac{c_{12}}{r^{12}} - \frac{c_6}{r^6}$$

with (Problem 16–53)

$$c_6 = \frac{2\mu^4}{3(4\pi\varepsilon_0)^2 k_B T} + \frac{2\alpha\mu^2}{(4\pi\varepsilon_0)^2} + \frac{3}{4}\frac{I\alpha^2}{(4\pi\varepsilon_0)^2} \tag{16.37}$$

for identical atoms or molecules.

16–7. The van der Waals Constants Can Be Written in Terms of Molecular Parameters

Although the Lennard-Jones potential is fairly realistic, it is also difficult to use. For example, the second virial coefficient (Example 16–10) must be evaluated numerically and one must resort to numerical tables to calculate the properties of gases. Consequently, intermolecular potentials that can be evaluated analytically are often used to estimate the properties of gases. The simplest of these potentials is the so-called *hard-sphere potential* (Figure 16.17a), whose mathematical form is

$$u(r) = \begin{array}{cc} \infty & r < \sigma \\ 0 & r > \sigma \end{array} \tag{16.38}$$

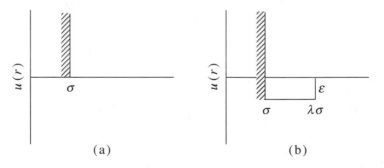

(a)   (b)

FIGURE 16.17
(a) A schematic illustration of a hard-sphere potential and (b) a square-well potential. The parameter σ is the diameter of the molecules, ε is the depth of the attractive well, and $(\lambda - 1)\sigma$ is the width of the well.

This potential represents hard spheres of diameter σ. Equation 16.38 depicts the repulsive region as varying infinitely steeply rather than as r^{-12}. As simplistic as this potential may seem, it does account for the finite size of molecules, which turns out to be the dominating feature in determining the structure of liquids and solids. Its obvious deficiency is the lack of any attractive term. At high temperatures, however, meaning high with respect to ε/k_B in the Lennard-Jones potential, the molecules are traveling with enough energy that the attractive potential is significantly "washed out," so the hard-sphere potential is useful under these conditions.

The second virial coefficient is easy to evaluate for the hard sphere potential. Substituting Equation 16.38 into Equation 16.25 gives

$$
\begin{aligned}
B_{2V}(T) &= -2\pi N_A \int_0^\infty [e^{-u(r)/k_B T} - 1] r^2 dr \\
&= -2\pi N_A \int_0^\sigma [0 - 1] r^2 dr - 2\pi N_A \int_\sigma^\infty [e^0 - 1] r^2 dr \\
&= \frac{2\pi \sigma^3 N_A}{3}
\end{aligned}
\tag{16.39}
$$

which is equal to four times the volume of N_A spheres. (Remember that σ is the diameter of the spheres.) Thus, the hard-sphere second virial coefficient is independent of temperature. Note that the high-temperature limit of the second virial coefficients shown in Figures 16.12 and 16.15 is fairly constant. The curves actually go through a slight maximum because molecules are not really "hard."

Another simple potential used fairly often is the *square-well potential* (Figure 16.17b):

$$
u(r) = \begin{array}{ll}
\infty & r < \sigma \\
-\varepsilon & \sigma < r < \lambda\sigma \\
0 & r > \lambda\sigma
\end{array}
\tag{16.40}
$$

The parameter ε is the depth of the well and $(\lambda - 1)\sigma$ is its width. This potential provides an attractive region, as crude as it is. The second virial coefficient can be evaluated analytically for the square-well potential

$$
\begin{aligned}
B_{2V}(T) &= -2\pi N_A \int_0^\sigma [0 - 1] r^2 dr - 2\pi N_A \int_\sigma^{\lambda\sigma} [e^{\varepsilon/k_B T} - 1] r^2 dr \\
&\quad - 2\pi N_A \int_{\lambda\sigma}^\infty [e^0 - 1] r^2 dr \\
&= \frac{2\pi \sigma^3 N_A}{3} - \frac{2\pi \sigma^3 N_A}{3}(\lambda^3 - 1)(e^{\varepsilon/k_B T} - 1) \\
&= \frac{2\pi \sigma^3 N_A}{3}[1 - (\lambda^3 - 1)(e^{\varepsilon/k_B T} - 1)]
\end{aligned}
\tag{16.41}
$$

Note that Equation 16.41 reduces to Equation 16.39 when $\lambda = 1$ or $\varepsilon = 0$, there being no attractive well in either case. Figure 16.18 shows Equation 16.41 compared with

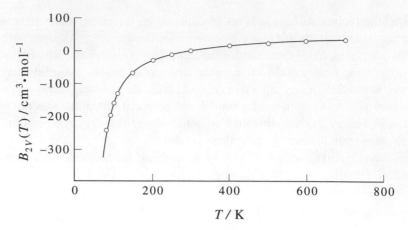

FIGURE 16.18
A comparison of the square-well second virial coefficient for nitrogen. The square-well parameters for nitrogen are $\sigma = 327.7$ pm, $\varepsilon/k_B = 95.2$ K, and $\lambda = 1.58$. The solid circles represent experimental data.

experimental data for nitrogen. The agreement is amazingly good, but the square-well potential does have three adjustable parameters.

We will finish this chapter with a discussion of the second virial coefficients for the three cubic equations of state introduced in Section 16–2. First, we write the van der Waals equation in the form

$$
\begin{aligned}
P &= \frac{RT}{\overline{V} - b} - \frac{a}{\overline{V}^2} \\
&= \frac{RT}{\overline{V}} \frac{1}{(1 - b/\overline{V})} - \frac{a}{\overline{V}^2}
\end{aligned}
\tag{16.42}
$$

We now use the binomial expansion of $1/(1-x)$ (MathChapter I),

$$
\frac{1}{1-x} = 1 + x + x^2 + \cdots
$$

to write Equation 16.42 as (letting $x = b/\overline{V}$)

$$
\begin{aligned}
P &= \frac{RT}{\overline{V}} \left[1 + \frac{b}{\overline{V}} + \frac{b^2}{\overline{V}^2} + \cdots \right] - \frac{a}{\overline{V}^2} \\
&= \frac{RT}{\overline{V}} + (RTb - a)\frac{1}{\overline{V}^2} + \frac{RTb^2}{\overline{V}^3} + \cdots
\end{aligned}
$$

or

$$
Z = \frac{P\overline{V}}{RT} = 1 + \left(b - \frac{a}{RT} \right) \frac{1}{\overline{V}} + \frac{b^2}{\overline{V}^2} + \cdots
$$

Comparing this result with Equation 16.22, we see that

$$B_{2V}(T) = b - \frac{a}{RT} \tag{16.43}$$

for the van der Waals equation. We will now derive a similar result from Equation 16.25 and interpret a and b in terms of molecular parameters. The intermolecular potential that we will use is a hybrid of the hard-sphere potential and the Lennard-Jones potential

$$u(r) = \begin{matrix} \infty & r < \sigma \\ -\dfrac{c_6}{r^6} & r > \sigma \end{matrix} \tag{16.44}$$

We substitute this potential into Equation 16.25 to obtain

$$B_{2V}(T) = -2\pi N_A \int_0^\sigma (-1)r^2 dr - 2\pi N_A \int_\sigma^\infty [e^{c_6/k_B T r^6} - 1]r^2 dr$$

In the second integral, we assume that $c_6/k_B T r^6 \ll 1$ and use the expansion for e^x (MathChapter I)

$$e^x = 1 + x + \frac{x^2}{2!} + \cdots$$

and keep only the first two terms to obtain

$$\begin{aligned} B_{2V}(T) &= \frac{2\pi\sigma^3 N_A}{3} - \frac{2\pi N_A c_6}{k_B T} \int_\sigma^\infty \frac{r^2 dr}{r^6} \\ &= \frac{2\pi\sigma^3 N_A}{3} - \frac{2\pi N_A c_6}{3k_B T \sigma^3} \end{aligned} \tag{16.45}$$

Comparing this result with Equation 16.43 gives

$$a = \frac{2\pi N_A^2 c_6}{3\sigma^3} \quad \text{and} \quad b = \frac{2\pi\sigma^3 N_A}{3}$$

Thus, we see that a is directly proportional to c_6, the coefficient of r^{-6} in the intermolecular potential, and that b is equal to four times the volume of the molecules. From a molecular point of view, the van der Waals equation is based on an intermolecular potential that is a hard-sphere potential at small distances and a weak attractive potential (such that $c_6/k_B T r^6 \ll 1$) at larger distances.

In a similar fashion (Problem 16–55), the second virial coefficient for the Redlich-Kwong equation is

$$B_{2V}(T) = B - \frac{A}{RT^{3/2}} \tag{16.46}$$

and the second virial coefficient for the Peng-Robinson equation is (Problem 16–56)

$$B_{2V}(T) = \beta - \frac{\alpha}{RT} \tag{16.47}$$

The second virial coefficient from the van der Waals equation and the Peng-Robinson equation have the same functional form, but they have different numerical values because the values of the constants are different. Also, the parameter α is a function of temperature in the Peng-Robinson equation.

Problems

16-1. In an issue of the journal *Science* a few years ago, a research group discussed experiments in which they determined the structure of cesium iodide crystals at a pressure of 302 gigapascals (GPa). How many atmospheres and bars is this pressure?

16-2. In meteorology, pressures are expressed in units of millibars (mbar). Convert 985 mbar to torr and to atmospheres.

16-3. Calculate the value of the pressure (in atm) exerted by a 33.9-foot column of water. Take the density of water to be 1.00 g·mL^{-1}.

16-4. At which temperature are the Celsius and Farenheit temperature scales equal?

16-5. A travel guide says that to convert Celsius temperatures to Farenheit temperatures, double the Celsius temperature and add 30. Comment on this recipe.

16-6. Research in surface science is carried out using ultra-high vacuum chambers that can sustain pressures as low as 10^{-12} torr. How many molecules are there in a 1.00-cm^3 volume inside such an apparatus at 298 K? What is the corresponding molar volume $\overline{V}$ at this pressure and temperature?

16-7. Use the following data for an unknown gas at 300 K to determine the molecular mass of the gas.

P/bar	0.1000	0.5000	1.000	1.01325	2.000
$\rho/\text{g·L}^{-1}$	0.1771	0.8909	1.796	1.820	3.652

16-8. Recall from general chemistry that Dalton's law of partial pressures says that each gas in a mixture of ideal gases acts as if the other gases were not present. Use this fact to show that the partial pressure exerted by each gas is given by

$$P_j = \left(\frac{n_j}{\sum n_j} \right) P_{\text{total}} = y_j P_{\text{total}}$$

where P_j is the partial pressure of the jth gas and y_j is its mole fraction.

16-9. A mixture of $H_2(g)$ and $N_2(g)$ has a density of 0.216 g·L^{-1} at 300 K and 500 torr. What is the mole fraction composition of the mixture?

16-10. One liter of $N_2(g)$ at 2.1 bar and two liters of $Ar(g)$ at 3.4 bar are mixed in a 4.0-L flask to form an ideal-gas mixture. Calculate the value of the final pressure of the mixture if the initial and final temperature of the gases are the same. Repeat this calculation if the initial temperatures of the $N_2(g)$ and $Ar(g)$ are 304 K and 402 K, respectively, and the final temperature of the mixture is 377 K. (Assume ideal-gas behavior.)

16-11. It takes 0.3625 g of nitrogen to fill a glass container at 298.2 K and 0.0100 bar pressure. It takes 0.9175 g of an unknown homonuclear diatomic gas to fill the same bulb under the same conditions. What is this gas?

16-12. Calculate the value of the molar gas constant in units of $dm^3 \cdot torr \cdot K^{-1} \cdot mol^{-1}$.

16-13. Use the van der Waals equation to plot the compressibility factor, Z, against P for methane for $T = 180$ K, 189 K, 190 K, 200 K, and 250 K. *Hint*: Calculate Z as a function of $\overline{V}$ and P as a function of $\overline{V}$, and then plot Z versus P.

16-14. Use the Redlich-Kwong equation to plot the compressibility factor, Z, against P for methane for $T = 180$ K, 189 K, 190 K, 200 K, and 250 K. *Hint*: Calculate Z as a function of $\overline{V}$ and P as a function of $\overline{V}$, and then plot Z versus P.

16-15. Use both the van der Waals and the Redlich-Kwong equations to calculate the molar volume of CO at 200 K and 1000 bar. Compare your result to the result you would get using the ideal-gas equation. The experimental value is $0.04009 \text{ L} \cdot mol^{-1}$.

16-16. Compare the pressures given by (a) the ideal-gas equation, (b) the van der Waals equation, (c) the Redlich-Kwong equation, and (d) the Peng-Robinson equation for propane at 400 K and $\rho = 10.62 \text{ mol} \cdot dm^{-3}$. The experimental value is 400 bar. Take $\alpha = 9.6938 \text{ L}^2 \cdot bar \cdot mol^{-2}$ and $\beta = 0.05632 \text{ L} \cdot mol^{-1}$ for the Peng-Robinson equation.

16-17. Use the van der Waals equation and the Redlich-Kwong equation to calculate the value of the pressure of one mole of ethane at 400.0 K confined to a volume of 83.26 cm^3. The experimental value is 400 bar.

16-18. Use the van der Waals equation and the Redlich-Kwong equation to calculate the molar density of one mole of methane at 500 K and 500 bar. The experimental value is $10.06 \text{ mol} \cdot L^{-1}$.

16-19. Use the Redlich-Kwong equation to calculate the pressure of methane at 200 K and a density of $27.41 \text{ mol} \cdot L^{-1}$. The experimental value is 1600 bar. What does the van der Waals equation give?

16-20. The pressure of propane versus density at 400 K can be fit by the expression

$$P/\text{bar} = 33.258(\rho/\text{mol} \cdot L^{-1}) - 7.5884(\rho/\text{mol} \cdot L^{-1})^2$$
$$+1.0306(\rho/\text{mol} \cdot L^{-1})^3 - 0.058757(\rho/\text{mol} \cdot L^{-1})^4$$
$$-0.0033566(\rho/\text{mol} \cdot L^{-1})^5 + 0.00060696(\rho/\text{mol} \cdot L^{-1})^6$$

for $0 \leq \rho/\text{mol} \cdot L^{-1} \leq 12.3$. Use the van der Waals equation and the Redlich-Kwong equation to calculate the pressure for $\rho = 0 \text{ mol} \cdot L^{-1}$ up to $12.3 \text{ mol} \cdot L^{-1}$. Plot your results. How do they compare to the above expression?

16-21. The Peng-Robinson equation is often superior to the Redlich-Kwong equation for temperatures near the critical temperature. Use these two equations to calculate the pressure of $CO_2(g)$ at a density of $22.0 \text{ mol} \cdot L^{-1}$ at 280 K [the critical temperature of $CO_2(g)$ is 304.2 K]. Use $\alpha = 4.192 \text{ bar} \cdot L^2 \cdot mol^{-2}$ and $\beta = 0.02665 \text{ L} \cdot mol^{-1}$ for the Peng-Robinson equation.

16-22. Show that the van der Waals equation for argon at $T = 142.69$ K and $P = 35.00$ atm can be written as

$$\overline{V}^3 - 0.3664\,\overline{V}^2 + 0.03802\,\overline{V} - 0.001210 = 0$$

where, for convenience, we have supressed the units in the coefficients. Use the Newton-Raphson method (MathChapter G) to find the three roots to this equation, and calculate the values of the density of liquid and vapor in equilibrium with each other under these conditions.

16-23. Use the Redlich-Kwong equation and the Peng-Robinson equation to calculate the densities of the coexisting argon liquid and vapor phases at 142.69 K and 35.00 atm. Use the Redlich-Kwong constants given in Table 16.4 and take $\alpha = 1.4915$ atm·L^2·mol^{-2} and $\beta = 0.01981$ L·mol^{-1} for the Peng-Robinson equation.

16-24. Butane liquid and vapor coexist at 370.0 K and 14.35 bar. The densities of the liquid and vapor phases are 8.128 mol·L^{-1} and 0.6313 mol·L^{-1}, respectively. Use the van der Waals equation, the Redlich-Kwong equation, and the Peng-Robinson equation to calculate these densities. Take $\alpha = 16.44$ bar·L^2·mol^{-2} and $\beta = 0.07245$ L·mol^{-1} for the Peng-Robinson equation.

16-25. Another way to obtain expressions for the van der Waals constants in terms of critical parameters is to set $(\partial P/\partial \overline{V})_T$ and $(\partial^2 P/\partial \overline{V}^2)_T$ equal to zero at the critical point. Why are these quantities equal to zero at the critical point? Show that this procedure leads to Equations 16.12 and 16.13.

16-26. Show that the Redlich-Kwong equation can be written in the form

$$\overline{V}^3 - \frac{RT}{P}\overline{V}^2 - \left(B^2 + \frac{BRT}{P} - \frac{A}{PT^{1/2}}\right)\overline{V} - \frac{AB}{PT^{1/2}} = 0$$

Now compare this equation with $(\overline{V} - \overline{V}_c)^3 = 0$ to get

$$3\overline{V}_c = \frac{RT_c}{P_c} \tag{1}$$

$$3\overline{V}_c^2 = \frac{A}{P_c T_c^{1/2}} - \frac{BRT_c}{P_c} - B^2 \tag{2}$$

and

$$\overline{V}_c^3 = \frac{AB}{P_c T_c^{1/2}} \tag{3}$$

Note that Equation 1 gives

$$\frac{P_c \overline{V}_c}{RT_c} = \frac{1}{3} \tag{4}$$

Now solve Equation 3 for A and substitute the result and Equation 4 into Equation 2 to obtain

$$B^3 + 3\overline{V}_c B^2 + 3\overline{V}_c^2 B - \overline{V}_c^3 = 0 \tag{5}$$

Divide this equation by $\overline{V}_c^3$ and let $B/\overline{V}_c = x$ to get

$$x^3 + 3x^2 + 3x - 1 = 0$$

Solve this cubic equation by the Newton-Raphson method (MathChapter G) to obtain $x = 0.25992$, or

$$B = 0.25992\overline{V}_c \tag{6}$$

Now substitute this result and Equation 4 into Equation 3 to obtain

$$A = 0.42748\frac{R^2 T_c^{5/2}}{P_c}$$

16-27. Use the results of the previous problem to derive Equations 16.14.

16-28. Write the Peng-Robinson equation as a cubic polynomial equation in $\overline{V}$ (with the coefficient of $\overline{V}^3$ equal to one), and compare it with $(\overline{V} - \overline{V}_c)^3 = 0$ at the critical point to obtain

$$\frac{RT_c}{P_c} - \beta = 3\overline{V}_c \tag{1}$$

$$\frac{\alpha_c}{P_c} - 3\beta^2 - 2\beta\frac{RT_c}{P_c} = 3\overline{V}_c^2 \tag{2}$$

and

$$\frac{\alpha_c\beta}{P_c} - \beta^2\frac{RT_c}{P_c} - \beta^3 = \overline{V}_c^3 \tag{3}$$

(We write α_c because α depends upon the temperature.) Now eliminate α_c/P_c between Equations 2 and 3, and then use Equation 1 for $\overline{V}_c$ to obtain

$$64\beta^3 + 6\beta^2\frac{RT_c}{P_c} + 12\beta\left(\frac{RT_c}{P_c}\right)^2 - \left(\frac{RT_c}{P_c}\right)^3 = 0$$

Let $\beta/(RT_c/P_c) = x$ and get

$$64x^3 + 6x^2 + 12x - 1 = 0$$

Solve this equation using the Newton-Raphson method to obtain

$$\beta = 0.077796\frac{RT_c}{P_c}$$

Substitute this result and Equation 1 into Equation 2 to obtain

$$\alpha_c = 0.45724\frac{(RT_c)^2}{P_c}$$

Last, use Equation 1 to show that

$$\frac{P_c\overline{V}_c}{RT_c} = 0.30740$$

16-29. Look up the boiling points of the gases listed in Table 16.5 and plot these values versus the critical temperatures T_c. Is there any correlation? Propose a reason to justify your conclusions from the plot.

16-30. Show that the compressibility factor Z for the Redlich-Kwong equation can be written as in Equation 16.21.

16-31. Use the following data for ethane and argon at $T_R = 1.64$ to illustrate the law of corresponding states by plotting Z against $\overline{V}_R$.

Ethane ($T = 500$ K)		Argon ($T = 247$ K)	
P/bar	$\overline{V}$/L·mol^{-1}	P/atm	$\overline{V}$/L·mol^{-1}
0.500	83.076	0.500	40.506
2.00	20.723	2.00	10.106
10.00	4.105	10.00	1.999
20.00	2.028	20.00	0.9857
40.00	0.9907	40.00	0.4795
60.00	0.6461	60.00	0.3114
80.00	0.4750	80.00	0.2279
100.0	0.3734	100.0	0.1785
120.0	0.3068	120.0	0.1462
160.0	0.2265	160.0	0.1076
200.0	0.1819	200.0	0.08630
240.0	0.1548	240.0	0.07348
300.0	0.1303	300.0	0.06208
350.0	0.1175	350.0	0.05626
400.0	0.1085	400.0	0.05219
450.0	0.1019	450.0	0.04919
500.0	0.09676	500.0	0.04687
600.0	0.08937	600.0	0.04348
700.0	0.08421	700.0	0.04108

16-32. Use the data in Problem 16–31 to illustrate the law of corresponding states by plotting Z against P_R.

16-33. Use the data in Problem 16.31 to test the quantitative reliability of the van der Waals equation by comparing a plot of Z versus $\overline{V}_R$ from Equation 16.20 to a similar plot of the data.

16-34. Use the data in Problem 16.31 to test the quantitative reliability of the Redlich-Kwong equation by comparing a plot of Z versus $\overline{V}_R$ from Equation 16.21 to a similar plot of the data.

16-35. Use Figure 16.10 to estimate the molar volume of CO at 200 K and 180 bar. An accurate experimental value is 78.3 cm^3·mol^{-1}.

16-36. Show that $B_{2V}(T) = RT B_{2P}(T)$ (see Equation 16.24).

16-37. Use the following data for $NH_3(g)$ at 273 K to determine $B_{2P}(T)$ at 273 K.

P/bar	0.10	0.20	0.30	0.40	0.50	0.60	0.70
$(Z-1)/10^{-4}$	1.519	3.038	4.557	6.071	7.583	9.002	10.551

16-38. The density of oxygen as a function of pressure at 273.15 K is listed below.

P/atm	0.2500	0.5000	0.7500	1.0000
ρ/g·dm^{-3}	0.356985	0.714154	1.071485	1.428962

Use the data to determine $B_{2V}(T)$ of oxygen. Take the atomic mass of oxygen to be 15.9994 and the value of the molar gas constant to be 8.31451 J·K^{-1}·mol^{-1} = 0.0820578 dm^3·atm·K^{-1}·mol^{-1}.

16-39. Show that the Lennard-Jones potential can be written as

$$u(r) = \varepsilon \left(\frac{r^*}{r}\right)^{12} - 2\varepsilon \left(\frac{r^*}{r}\right)^{6}$$

where r^* is the value of r at which $u(r)$ is a minimum.

16-40. Using the Lennard-Jones parameters given in Table 16.7, compare the depth of a typical Lennard-Jones potential to the strength of a covalent bond.

16-41. Compare the Lennard-Jones potentials of $H_2(g)$ and $O_2(g)$ by plotting both on the same graph.

16-42. Use the data in Tables 16.5 and 16.7 to show that *roughly* $\epsilon/k_B = 0.75 \, T_c$ and $b_0 = 0.7 \, \overline{V}_c$. Thus, critical constants can be used as rough, first estimates of ϵ and b_0 ($= 2\pi N_A \sigma^3/3$).

16-43. Prove that the second virial coefficient calculated from a general intermolecular potential of the form

$$u(r) = \text{(energy parameter)} \times f\left(\frac{r}{\text{distance parameter}}\right)$$

rigorously obeys the law of corresponding states. Does the Lennard-Jones potential satisfy this condition?

16-44. Use the following data for argon at 300.0 K to determine the value of B_{2V}. The accepted value is $-15.05 \text{ cm}^3 \cdot \text{mol}^{-1}$.

P/atm	$\rho/\text{mol}\cdot\text{L}^{-1}$	P/atm	$\rho/\text{mol}\cdot\text{L}^{-1}$
0.01000	0.000406200	0.4000	0.0162535
0.02000	0.000812500	0.6000	0.0243833
0.04000	0.00162500	0.8000	0.0325150
0.06000	0.00243750	1.000	0.0406487
0.08000	0.00325000	1.500	0.0609916
0.1000	0.00406260	2.000	0.0813469
0.2000	0.00812580	3.000	0.122094

16-45. Using Figure 16.15 and the Lennard-Jones parameters given in Table 16.7, estimate $B_{2V}(T)$ for $CH_4(g)$ at 0°C.

16-46. Show that $B_{2V}(T)$ obeys the law of corresponding states for a square-well potential with a *fixed* value of λ (in other words, if all molecules had the same value of λ).

16-47. Using the Lennard-Jones parameters in Table 16.7, show that the following second virial cofficient data satisfy the law of corresponding states.

Argon		Nitrogen		Ethane	
T/K	$B_{2V}(T)$ $/10^{-3} \text{ dm}^3 \cdot \text{mol}^{-1}$	T/K	$B_{2V}(T)$ $/10^{-3} \text{ dm}^3 \cdot \text{mol}^{-1}$	T/K	$B_{2V}(T)$ $/10^{-3} \text{ dm}^3 \cdot \text{mol}^{-1}$
173	−64.3	143	−79.8	311	−164.9
223	−37.8	173	−51.9	344	−132.5
273	−22.1	223	−26.4	378	−110.0
323	−11.0	273	−10.3	411	−90.4
423	+1.2	323	−0.3	444	−74.2
473	4.7	373	+6.1	478	−59.9
573	11.2	423	11.5	511	−47.4
673	15.3	473	15.3		
		573	20.6		
		673	23.5		

16-48. In Section 16–4, we expressed the van der Waals equation in reduced units by dividing $P, \overline{V}$, and T by their critical values. This suggests we can write the second virial coefficient in reduced form by dividing $B_{2V}(T)$ by $\overline{V}_c$ and T by T_c (instead of $2\pi N_A \sigma^3/3$ and ε/k as we did in Section 16–5). Reduce the second virial coefficient data given in the previous problem by using the values of $\overline{V}_c$ and T_c in Table 16.5 and show that the reduced data satisfy the law of corresponding states.

16-49. Listed below are experimental second virial coefficient data for argon, krypton, and xenon.

$$B_{2V}(T)/10^{-3}\text{dm}^3 \cdot \text{mol}^{-1}$$

T/K	Argon	Krypton	Xenon
173.16	−63.82		
223.16	−36.79		
273.16	−22.10	−62.70	−154.75
298.16	−16.06		−130.12
323.16	−11.17	−42.78	−110.62
348.16	−7.37		−95.04
373.16	−4.14	−29.28	−82.13
398.16	−0.96		
423.16	+1.46	−18.13	−62.10
473.16	4.99	−10.75	−46.74
573.16	10.77	+0.42	−25.06
673.16	15.72	7.42	−9.56
773.16	17.76	12.70	−0.13
873.16	19.48	17.19	+7.95
973.16			14.22

Use the Lennard-Jones parameters in Table 16.7 to plot $B_{2V}^*(T^*)$, the reduced second virial coefficient, versus T^*, the reduced temperature, to illustrate the law of corresponding states.

16-50. Use the critical temperatures and the critical molar volumes of argon, krypton, and xenon to illustrate the law of corresponding states with the data given in Problem 16–49.

16-51. Evaluate $B_{2V}^*(T^*)$ in Equation 16.31 numerically from $T^* = 1.00$ to 10.0 using a packaged numerical integration program such as *MathCad* or *Mathematica*. Compare the reduced second virial coefficient data from Problem 16–49 and $B_{2V}^*(T^*)$ by plotting them all on the same graph.

16-52. Show that the units of the right side of Equation 16.35 are energy.

16-53. Show that the sum of Equations 16.33, 16.35, and 16.36 gives Equation 16.37.

16-54. Compare the values of the coefficient of r^{-6} for $N_2(g)$ using Equation 16.37 and the Lennard-Jones parameters given in Table 16.7.

16-55. Show that

$$B_{2V}(T) = B - \frac{A}{RT^{3/2}}$$

and

$$B_{3V}(T) = B^2 + \frac{AB}{RT^{3/2}}$$

for the Redlich-Kwong equation.

16-56. Show that the second and third virial coefficients of the Peng-Robinson equation are

$$B_{2V}(T) = \beta - \frac{\alpha}{RT}$$

and

$$B_{3V}(T) = \beta^2 + \frac{2\alpha\beta}{RT}$$

16-57. The square-well parameters for krypton are $\varepsilon/k_B = 136.5$ K, $\sigma = 327.8$ pm, and $\lambda = 1.68$. Plot $B_{2V}(T)$ against T and compare your results with the data given in Problem 16–49.

16-58. The coefficient of thermal expansion α is defined as

$$\alpha = \frac{1}{\overline{V}} \left(\frac{\partial \overline{V}}{\partial T} \right)_P$$

Show that

$$\alpha = \frac{1}{T}$$

for an ideal gas.

16-59. The isothermal compressibility κ is defined as

$$\kappa = -\frac{1}{\overline{V}} \left(\frac{\partial \overline{V}}{\partial P} \right)_T$$

Show that

$$\kappa = \frac{1}{P}$$

for an ideal gas.

PARTIAL DIFFERENTIATION

Recall from your course in calculus that the derivative of a function $y(x)$ at some point x is defined as

$$\frac{dy}{dx} = \lim_{\Delta x \to 0} \frac{y(x + \Delta x) - y(x)}{\Delta x} \tag{H.1}$$

Physically, dy/dx expresses the variation of y when x is varied. Much of your calculus course was spent in starting with Equation H.1 to derive formulas for the derivatives of the commonly occurring functions. The function y in Equation H.1 depends upon only one variable, x. For the function $y(x)$, x is called the independent variable and y, whose value depends upon the value of x, is called the dependent variable.

Functions can depend upon more than one variable. For example, we know that the pressure of an ideal gas depends upon the temperature, volume, and number of moles through the equation

$$P = \frac{nRT}{V} \tag{H.2}$$

In this case, there are three independent variables; the temperature, volume, and amount of gas can be varied independently. The pressure is the dependent variable. We can emphasize this dependency by writing

$$P = P(n, T, V)$$

Experimentally, we may wish to vary only one of the independent variables at a time (say the temperature) to produce a change in pressure with two of the independent variables fixed (fixed volume and fixed number of moles). To form the derivative of P with respect to T with n and V held constant, we simply refer to Equation H.1 and write

$$\left(\frac{\partial P}{\partial T}\right)_{n,V} = \lim_{\Delta T \to 0} \frac{P(n, T + \Delta T, V) - P(n, T, V)}{\Delta T} \tag{H.3}$$

683

We call $(\partial P/\partial T)_{n,V}$ the partial derivative of P with respect to T, with n and V held constant. To actually evaluate this partial derivative, we simply differentiate P with respect to T in Equation H.2, treating n and V as if they were constants. Thus, for an ideal gas

$$\left(\frac{\partial P}{\partial T}\right)_{n,V} = \frac{nR}{V}$$

We can also have

$$\left(\frac{\partial P}{\partial n}\right)_{T,V} = \frac{RT}{V}$$

and

$$\left(\frac{\partial P}{\partial V}\right)_{n,T} = -\frac{nRT}{V^2}$$

EXAMPLE H–1
Evaluate the two first partial derivatives of P for the van der Waals equation

$$P = \frac{RT}{\overline{V} - b} - \frac{a}{\overline{V}^2} \tag{H.4}$$

SOLUTION: In this case, P depends upon T and $\overline{V}$, so we have $P = P(T, \overline{V})$. The two first partial derivatives of P are

$$\left(\frac{\partial P}{\partial T}\right)_{\overline{V}} = \frac{R}{\overline{V} - b} \tag{H.5}$$

and

$$\left(\frac{\partial P}{\partial \overline{V}}\right)_T = -\frac{RT}{(\overline{V} - b)^2} + \frac{2a}{\overline{V}^3} \tag{H.6}$$

The partial derivatives given by Equations H.5 and H.6 are themselves functions of T and $\overline{V}$, so we can form second partial derivatives by differentiating Equations H.5 and H.6:

$$\left(\frac{\partial^2 P}{\partial T^2}\right)_{\overline{V}} = 0$$

and

$$\left(\frac{\partial^2 P}{\partial \overline{V}^2}\right)_T = \frac{2RT}{(\overline{V} - b)^3} - \frac{6a}{\overline{V}^4}$$

We can also form another type of second derivative, however. For example, we can form

$$\left[\frac{\partial}{\partial \overline{V}}\left(\frac{\partial P}{\partial T}\right)_{\overline{V}}\right]_T = \left[\frac{\partial}{\partial \overline{V}}\left(\frac{R}{\overline{V} - b}\right)\right]_T$$

$$= -\frac{R}{(\overline{V} - b)^2} \qquad\qquad (H.7)$$

and we can also form

$$\left[\frac{\partial}{\partial T}\left(\frac{\partial P}{\partial \overline{V}}\right)_T\right]_{\overline{V}} = \left[\frac{\partial}{\partial T}\left(-\frac{RT}{(\overline{V} - b)^2} + \frac{2a}{\overline{V}^3}\right)\right]_{\overline{V}}$$

$$= -\frac{R}{(\overline{V} - b)^2} \qquad\qquad (H.8)$$

The above two second derivatives are called cross derivatives, mixed derivatives, or second cross partial derivatives. These derivatives are commonly written as

$$\left(\frac{\partial^2 P}{\partial \overline{V}\partial T}\right) \quad \text{or} \quad \left(\frac{\partial^2 P}{\partial T \partial \overline{V}}\right)$$

We don't indicate which variable is held constant because they differ with each differentiation. Notice that these two cross derivatives are equal (see Equations H.7 and H.8), so that

$$\left(\frac{\partial^2 P}{\partial \overline{V}\partial T}\right) = \left(\frac{\partial^2 P}{\partial T \partial \overline{V}}\right) \qquad\qquad (H.9)$$

Thus, the order in which we take the two partial derivatives of P makes no difference in this case. It turns out that cross derivatives are generally equal.

EXAMPLE H–2
Suppose that

$$S = -\left(\frac{\partial A}{\partial T}\right)_V \quad \text{and} \quad P = -\left(\frac{\partial A}{\partial V}\right)_T$$

where A, S, and P are functions of T and V. Prove that

$$\left(\frac{\partial S}{\partial V}\right)_T = \left(\frac{\partial P}{\partial T}\right)_V$$

SOLUTION: Take the partial derivative of S with respect to V at constant T:

$$\left(\frac{\partial S}{\partial V}\right)_T = -\left(\frac{\partial^2 A}{\partial V \partial T}\right)$$

and the partial derivative of P with respect to T at constant V:

$$\left(\frac{\partial P}{\partial T}\right)_V = -\left(\frac{\partial^2 A}{\partial T \partial V}\right)$$

and equate the two cross derivatives of A to obtain

$$\left(\frac{\partial S}{\partial V}\right)_T = \left(\frac{\partial P}{\partial T}\right)_V$$

The partial derivatives given in Equations H.5 and H.6 indicate how P changes with one independent variable, keeping the other one fixed. We often want to know how a dependent variable changes with a change in the values of both (or more) of its independent variables. Using the example $P = P(T, \overline{V})$ (for one mole), we write

$$\Delta P = P(T + \Delta T, \overline{V} + \Delta \overline{V}) - P(T, \overline{V})$$

If we add and subtract $P(T, \overline{V} + \Delta \overline{V})$ to this equation, we obtain

$$\Delta P = [P(T + \Delta T, \overline{V} + \Delta \overline{V}) - P(T, \overline{V} + \Delta \overline{V})] \\ + [P(T, \overline{V} + \Delta \overline{V}) - P(T, \overline{V})]$$

Multiply the first two terms in brackets by $\Delta T/\Delta T$ and the second two terms by $\Delta \overline{V}/\Delta \overline{V}$ to get

$$\Delta P = \left[\frac{P(T + \Delta T, \overline{V} + \Delta \overline{V}) - P(T, \overline{V} + \Delta \overline{V})}{\Delta T}\right] \Delta T \\ + \left[\frac{P(T, \overline{V} + \Delta \overline{V}) - P(T, \overline{V})}{\Delta \overline{V}}\right] \Delta \overline{V}$$

Now let $\Delta T \to 0$ and $\Delta \overline{V} \to 0$, in which case we have

$$dP = \lim_{\Delta T \to 0} \left[\frac{P(T + \Delta T, \overline{V}) - P(T, \overline{V})}{\Delta T}\right] \Delta T \\ + \lim_{\Delta \overline{V} \to 0} \left[\frac{P(T, \overline{V} + \Delta \overline{V}) - P(T, \overline{V})}{\Delta \overline{V}}\right] \Delta \overline{V} \tag{H.10}$$

The first limit gives $(\partial P/\partial T)_{\overline{V}}$ (by definition) and the second gives $(\partial P/\partial \overline{V})_T$, so that Equation H.10 gives our desired result:

$$dP = \left(\frac{\partial P}{\partial T}\right)_{\overline{V}} dT + \left(\frac{\partial P}{\partial \overline{V}}\right)_T d\overline{V} \tag{H.11}$$

Equation H.11 is called the total derivative of P. It simply says that the change in P is given by how P changes with T (keeping $\overline{V}$ constant) times the infinitesimal change in T plus how P changes with $\overline{V}$ (at constant T) times the infinitesimal change in $\overline{V}$.

EXAMPLE H–3

We can use Equation H.11 to estimate the change in pressure when both the temperature and the molar volume are changed slightly. To this end, for finite ΔT and $\Delta \overline{V}$, we write Equation H.11 as

$$\Delta P \approx \left(\frac{\partial P}{\partial T}\right)_{\overline{V}} \Delta T + \left(\frac{\partial P}{\partial \overline{V}}\right)_T \Delta \overline{V}$$

Use this equation to estimate the change in pressure of one mole of an ideal gas if the temperature is changed from 273.15 K to 274.00 K and the volume is changed from 10.00 L to 9.90 L.

SOLUTION: We first need

$$\left(\frac{\partial P}{\partial T}\right)_{\overline{V}} = \left[\frac{\partial}{\partial T}\left(\frac{RT}{\overline{V}}\right)\right]_{\overline{V}} = \frac{R}{\overline{V}}$$

and

$$\left(\frac{\partial P}{\partial \overline{V}}\right)_T = \left[\frac{\partial}{\partial \overline{V}}\left(\frac{RT}{\overline{V}}\right)\right]_{\overline{V}} = -\frac{RT}{\overline{V}^2}$$

so that

$$\Delta P \approx \frac{R}{\overline{V}}\Delta T - \frac{RT}{\overline{V}^2}\Delta \overline{V}$$

$$\approx \frac{(8.314 \text{ J·K}^{-1}\text{·mol}^{-1})}{(10.00 \text{ L·mol}^{-1})}(0.85 \text{ K})$$

$$-\frac{(8.314 \text{ J·K}^{-1}\text{·mol}^{-1})(273.15 \text{ K})}{(10.00 \text{ L·mol}^{-1})^2}(-0.10 \text{ L·mol}^{-1})$$

$$\approx 3.0 \text{ J·L}^{-1}$$

$$\approx 3.0 \times 10^3 \text{ J·m}^{-3} = 3.0 \times 10^3 \text{ Pa} = 0.030 \text{ bar}$$

Incidently, in this particularly simple case, we calculate the exact change in P from

$$\Delta P = \frac{RT_2}{\overline{V}_2} - \frac{RT_1}{\overline{V}_1}$$

$$= (8.314 \text{ J·K}^{-1}\text{·mol}^{-1})\left(\frac{274.00 \text{ K}}{9.90 \text{ L·mol}^{-1}} - \frac{273.15 \text{ K}}{10.00 \text{ L·mol}^{-1}}\right)$$

$$= 3.0 \text{ J·L}^{-1} = 3.0 \text{ J·dm}^{-3} = 0.030 \text{ bar}$$

Equation H.4 gives P as a function of T and $\overline{V}$, or $P = P(T, \overline{V})$. We can form the total derivative of P by differentiating the right side of Equation H.4 with respect to T and $\overline{V}$ to obtain

$$dP = \frac{R}{\overline{V} - b}dT - \frac{RT}{(\overline{V} - b)^2}d\overline{V} + \frac{2a}{\overline{V}^3}d\overline{V}$$

$$= \frac{R}{\overline{V} - b}dT + \left[-\frac{RT}{(\overline{V} - b)^2} + \frac{2a}{\overline{V}^3}\right]d\overline{V} \qquad \text{(H.12)}$$

We can see from Example H–1 that Equation H.12 is just Equation H.11 written for the van der Waals equation. Suppose, however, that we are given an arbitrary expression for dP, say

$$dP = \frac{RT}{\overline{V} - b}dT + \left[\frac{RT}{(\overline{V} - b)^2} - \frac{a}{T\overline{V}^2}\right]d\overline{V} \qquad \text{(H.13)}$$

and are asked to determine the equation of state $P = P(T, \overline{V})$ that leads to Equation H.13. In fact, a simpler question is to ask if there even is a function $P(T, \overline{V})$ whose total derivative is given by Equation H.13. How can we tell? If there is such a function $P(T, \overline{V})$, then its total derivative is (Equation H.11)

$$dP = \left(\frac{\partial P}{\partial T}\right)_{\overline{V}} dT + \left(\frac{\partial P}{\partial \overline{V}}\right)_T d\overline{V}$$

Furthermore, according to Equation H.9, the cross derivatives of a function $P(T, \overline{V})$,

$$\left(\frac{\partial^2 P}{\partial \overline{V}\partial T}\right) = \left[\frac{\partial}{\partial \overline{V}}\left(\frac{\partial P}{\partial T}\right)_{\overline{V}}\right]_T$$

and

$$\left(\frac{\partial^2 P}{\partial T\partial \overline{V}}\right) = \left[\frac{\partial}{\partial T}\left(\frac{\partial P}{\partial \overline{V}}\right)_T\right]_{\overline{V}}$$

must be equal. If we apply this requirement to Equation H.13, we find that

$$\frac{\partial}{\partial T}\left[\frac{RT}{(\overline{V} - b)^2} - \frac{a}{T\overline{V}^2}\right] = \frac{R}{(\overline{V} - b)^2} + \frac{a}{T^2\overline{V}^2}$$

and

$$\frac{\partial}{\partial \overline{V}}\left(\frac{RT}{\overline{V} - b}\right) = -\frac{RT}{(\overline{V} - b)^2}$$

Thus, we see that the cross-derivatives are not equal, so the expression given by Equation H.13 is not the derivative of any function $P(T, \overline{V})$. The differential given by Equation H.13 is called an *inexact differential*.

We can obtain an example of an *exact differential* simply by explicitly differentiating any function $P(T, \overline{V})$, such as we did for the van der Waals equation to obtain

Equation H.12. Equations H.7 and H.8 show that the cross derivatives are equal, as they must be for an exact differential.

EXAMPLE H–4

Is

$$dP = \left[\frac{R}{\overline{V} - B} + \frac{A}{2T^{3/2}\overline{V}(\overline{V} + B)} \right] dT$$

$$+ \left[-\frac{RT}{(\overline{V} - B)^2} + \frac{A(2\overline{V} + B)}{T^{1/2}\overline{V}^2(\overline{V} + B)^2} \right] d\overline{V} \qquad \text{(H.14)}$$

an exact differential?

SOLUTION: We evaluate the two derivatives

$$\left[\frac{\partial}{\partial \overline{V}} \left\{ \frac{R}{\overline{V} - B} + \frac{A}{2T^{3/2}\overline{V}(\overline{V} + B)} \right\} \right]_T = -\frac{R}{(\overline{V} - B)^2} - \frac{A(2\overline{V} + B)}{2T^{3/2}\overline{V}^2(\overline{V} + B)^2}$$

and

$$\left[\frac{\partial}{\partial T} \left\{ -\frac{RT}{(\overline{V} - B)^2} + \frac{A(2\overline{V} + B)}{T^{1/2}\overline{V}^2(\overline{V} + B)^2} \right\} \right]_{\overline{V}} = -\frac{R}{(\overline{V} - B)^2} - \frac{A(2\overline{V} + B)}{2T^{3/2}\overline{V}^2(\overline{V} + B)^2}$$

These derivatives are equal and so Equation H.14 represents an exact differential. Equation H.14 is the total derivative of P for the Redlich-Kwong equation of state.

Exact and inexact differentials play a significant role in physical chemistry. If dy is an exact differential, then

$$\int_1^2 dy = y_2 - y_1 \qquad \text{(exact differential)}$$

so the integral depends only upon the end points (1 and 2) and not upon the path from 1 to 2. This statement is not true for an inexact differential, however, so

$$\int_1^2 dy \neq y_2 - y_1 \qquad \text{(inexact differential)}$$

The integral in this case depends not only upon the end points but also upon the path from 1 to 2.

Problems

H-1. The isothermal compressibility, κ_T, of a substance is defined as

$$\kappa_T = -\frac{1}{V} \left(\frac{\partial V}{\partial P} \right)_T$$

Obtain an expression for the isothermal compressibility of an ideal gas.

H-2. The coefficient of thermal expansion, α, of a substance is defined as

$$\alpha = \frac{1}{V}\left(\frac{\partial V}{\partial T}\right)_P$$

Obtain an expression for the coefficient of thermal expansion of an ideal gas.

H-3. Prove that

$$\left(\frac{\partial P}{\partial V}\right)_{n,T} = \frac{1}{\left(\dfrac{\partial V}{\partial P}\right)_{n,T}}$$

for an ideal gas and for a gas whose equation of state is $P = nRT/(V - nb)$, where b is a constant. This relation is generally true and is called the reciprocal identity. Notice that the same variables must be held fixed on both sides of the identity.

H-4. Given that

$$U = kT^2\left(\frac{\partial \ln Q}{\partial T}\right)_{N,V}$$

where

$$Q(N, V, T) = \frac{1}{N!}\left(\frac{2\pi m k_B T}{h^2}\right)^{3N/2} V^N$$

and k_B, m, and h are constants, determine U as a function of T.

H-5. Show that the total derivative of P for the Redlich-Kwong equation,

$$P = \frac{RT}{\overline{V} - B} - \frac{A}{T^{1/2}\overline{V}(\overline{V} + B)}$$

is given by Equation H.14.

H-6. Show explicitly that

$$\left(\frac{\partial^2 P}{\partial \overline{V}\partial T}\right) = \left(\frac{\partial^2 P}{\partial T\partial \overline{V}}\right)$$

for the Redlich-Kwong equation (Problem H–5).

H-7. We will derive the following equation in Chapter 19:

$$\left(\frac{\partial U}{\partial V}\right)_T = T\left(\frac{\partial P}{\partial T}\right)_V - P$$

Evaluate $(\partial U/\partial V)_T$ for an ideal gas, for a van der Waals gas (Equation H.4), and for a Redlich-Kwong gas (Problem H–5).

H-8. Given that the heat capacity at constant volume is defined by

$$C_V = \left(\frac{\partial U}{\partial T}\right)_V$$

and given the expression in Problem H–7, derive the equation

$$\left(\frac{\partial C_V}{\partial V}\right)_T = T\left(\frac{\partial^2 P}{\partial T^2}\right)_V$$

H-9. Use the expression in Problem H–8 to determine $(\partial C_V/\partial V)_T$ for an ideal gas, a van der Waals gas (Equation H.4), and a Redlich-Kwong gas (see Problem H–5).

H-10. Is

$$dV = \pi r^2 dh + 2\pi r h dr$$

an exact or inexact differential?

H-11. Is

$$dx = C_V(T)dT + \frac{nRT}{V}dV$$

an exact or inexact differential? The quantity $C_V(T)$ is simply an arbitrary function of T. What about dx/T?

H-12. Prove that

$$\frac{1}{Y}\left(\frac{\partial Y}{\partial P}\right)_{T,n} = \frac{1}{\overline{Y}}\left(\frac{\partial \overline{Y}}{\partial P}\right)_T$$

and that

$$\left(\frac{\partial P}{\partial \overline{Y}}\right)_T = n\left(\frac{\partial P}{\partial Y}\right)_{T,n}$$

where $Y = Y(P, T, n)$ is an extensive variable.

H-13. Equation 16.5 gives P for the van der Waals equation as a function of $\overline{V}$ and T. Show that P expressed as a function of V, T, and n is

$$P = \frac{nRT}{V - nb} - \frac{n^2 a}{V^2} \tag{1}$$

Now evaluate $(\partial P/\partial \overline{V})_T$ from Equation 16.5 and $(\partial P/\partial V)_{T,n}$ from Equation 1 above and show that (see Problem H–12)

$$\left(\frac{\partial P}{\partial \overline{V}}\right)_T = n\left(\frac{\partial P}{\partial V}\right)_{T,n}$$

H-14. Referring to Problem H–13, show that

$$\left(\frac{\partial P}{\partial T}\right)_{\overline{V}} = \left(\frac{\partial P}{\partial T}\right)_{V,n}$$

and generally that

$$\left[\frac{\partial y(x, \overline{V})}{\partial x}\right]_{\overline{V}} = \left[\frac{\partial y(x, n, V)}{\partial x}\right]_{V,n}$$

where y and x are intensive variables and $y(x, n, V)$ can be written as $y(x, V/n)$.

Ludwig Boltzmann was born in Vienna, Austria, on February 20, 1844, and died in 1906. In 1867, he received his doctorate from the University of Vienna, where he studied with Stefan (of the Stefan-Boltzmann equation). He worked on the kinetic theory of gases and did experimental work on gases and radiation during his stay there. Although known for his theoretical work, he was an able experimentalist but was handicapped by poor vision. He was an early proponent of the atomic theory, and much of his work involved a study of the atomic theory of matter. In 1869, Boltzmann extended Maxwell's theory of the distribution of energy among colliding gas molecules and gave a new expression for this distribution, now known as the Boltzmann factor. In addition, the distribution of the speeds and the energies of gas molecules is now called the Maxwell-Boltzmann distribution. In 1877, he published his famous equation, $S = k_B \ln W$, which expresses the relation between entropy and probability. At the time, the atomic nature of matter was not generally accepted, and Boltzmann's work was criticized by a number of eminent scientists. Unfortunately, Boltzmann did not live to see the atomic theory and his work corroborated. He had always suffered from depression and committed suicide in 1906 by drowning.

CHAPTER 17

The Boltzmann Factor and Partition Functions

In previous chapters, we learned that the energy states of atoms and molecules, and for all systems in fact, are quantized. These allowed energy states are found by solving the Schrödinger equation. A practical question that arises is how the molecules are distributed over these energy states at a given temperature. For example, we may ask what fraction of the molecules are to be found in the ground vibrational state, the first excited vibrational state, and so on. You may have an intuitive feel that the populations of excited states increase with increasing temperature, and we will see in this chapter that this is the case. Two central themes of this chapter are the Boltzmann factor and the partition function. The Boltzmann factor is one of the most fundamental and useful quantities of physical chemistry. The Boltzmann factor tells us that if a system has states with energies E_1, E_2, E_3, ..., the probability p_j that the system will be in the state with energy E_j depends exponentially on the energy of that state, or

$$p_j \propto e^{-E_j/k_B T}$$

where k_B is the Boltzmann constant and T is the kelvin temperature. We will derive this result in Section 17–2 and then discuss its implications and applications in the remainder of the chapter.

The sum of the probabilities must equal 1, so the normalization constant for the above probability is $1/Q$ where

$$Q = \sum_j e^{-E_j/k_B T}$$

The quantity Q is called a partition function, and we will see that partition functions play a central role in calculating the properties of any system. For example, we will show that we can calculate the energy, heat capacity, and pressure of a system in terms of Q. In Chapter 18, we will use partition functions to calculate the heat capacities of monatomic and polyatomic ideal gases.

17–1. The Boltzmann Factor Is One of the Most Important Quantities in the Physical Sciences

Consider some macroscopic system such as a liter of gas, a liter of water, or a kilogram of some solid. From a mechanical point of view, such a system can be described by specifying the number of particles, N, the volume, V, and the forces between the particles. Even though the system contains on the order of Avogadro's number of particles, we can still consider its Hamiltonian operator and its associated wave functions, which will depend upon the coordinates of all the particles. The Schrödinger equation for this N-body system is

$$\hat{H}_N \Psi_j = E_j \Psi_j \qquad j = 1, 2, 3, \ldots \tag{17.1}$$

where the energies depend upon both N and V, which we will emphasize by writing $E_j(N, V)$.

For the special case of an ideal gas, the total energy $E_j(N, V)$ will simply be a sum of the individual molecular energies,

$$E_j(N, V) = \epsilon_1 + \epsilon_2 + \cdots + \epsilon_N \tag{17.2}$$

because the molecules of an ideal gas are independent of each other. For example, for a monatomic ideal gas in a cubic container with sides of length a, if we ignore the electronic states and focus only on the translational states, then the ϵ_js are just the translational energies given by (Equation 3.60)

$$\epsilon_{n_x n_y n_z} = \frac{h^2}{8ma^2} \left(n_x^2 + n_y^2 + n_z^2 \right) \tag{17.3}$$

Note that $E_j(N, V)$ depends upon N through the number of terms in Equation 17.2 and upon V through the fact that $a = V^{1/3}$ in Equation 17.3.

For a more general system in which the particles interact with each other, the $E_j(N, V)$ cannot be written as a sum of individual particle energies, but we can still consider the set of allowed macroscopic energies $\{E_j(N, V)\}$, at least in principle.

What we want to do now is to determine the probability that a system will be in the state j with energy $E_j(N, V)$. To do this, we consider a huge collection of such systems in thermal contact with an essentially infinite heat bath (called a heat reservoir) at a temperature T. Each system has the same values of N, V, and T but is likely to be in a different quantum state, consistent with the values of N and V. Such a collection of systems is called an *ensemble* (Figure 17.1). We will denote the number of systems in the state j with energy $E_j(N, V)$ by a_j and the total number of systems in the ensemble by $\mathcal{A}$.

We now ask for the relative number of systems of the ensemble that would be found in each state. As an example, let's focus on two particular states, 1 and 2, with

695

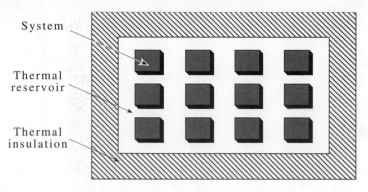

FIGURE 17.1
An ensemble, or collection, of (macroscopic) systems in thermal equilibrium with a heat reservoir. The number of systems in the state j [with energy $E_j(N, V)$] is a_j, and the total number of systems in the ensemble is $\mathcal{A}$. Because the ensemble is a conceptual construction, we may consider $\mathcal{A}$ to be as large as we want.

energies $E_1(N, V)$ and $E_2(N, V)$. The relative number of systems in the states with energies E_1 and E_2 must depend upon E_1 and E_2, so we write

$$\frac{a_2}{a_1} = f(E_1, E_2) \tag{17.4}$$

where a_1 and a_2 are the number of systems in the ensemble in states 1 and 2 and where the functional form of f is to be determined. Now, because energy is a quantity that must always be referred to a zero of energy, the dependence on E_1 and E_2 in Equation 17.4 must be of the form

$$f\left(E_1, E_2\right) = f\left(E_1 - E_2\right) \tag{17.5}$$

In this way, any arbitrary zero of energy associated with E_1 and E_2 will cancel. Thus, we have so far

$$\frac{a_2}{a_1} = f\left(E_1 - E_2\right) \tag{17.6}$$

Equation 17.6 must be true for any two energy states, so we can also write

$$\frac{a_3}{a_2} = f\left(E_2 - E_3\right) \quad \text{and} \quad \frac{a_3}{a_1} = f\left(E_1 - E_3\right) \tag{17.7}$$

But

$$\frac{a_3}{a_1} = \frac{a_2}{a_1} \cdot \frac{a_3}{a_2}$$

so using Equations 17.6 and 17.7, we find that the function f must satisfy

$$f\left(E_1 - E_3\right) = f\left(E_1 - E_2\right) f\left(E_2 - E_3\right) \tag{17.8}$$

The form of the function f that satisfies this equation may not be obvious as first sight, but if you remember that

$$e^{x+y} = e^x e^y$$

then we can see that

$$f(E) = e^{\beta E}$$

where β is an arbitrary constant (see also Problem 17–2). To verify that this form for f does indeed satisfy Equation 17.8, we substitute this functional form of $f(E)$ into Equation 17.8:

$$e^{\beta(E_1 - E_3)} = e^{\beta(E_1 - E_2)} e^{\beta(E_2 - E_3)} = e^{\beta(E_1 - E_3)}$$

Thus, we find from Equation 17.6 that

$$\frac{a_2}{a_1} = e^{\beta(E_1 - E_2)} \tag{17.9}$$

There is nothing special about the states 1 and 2, so we can write Equation 17.9 more generally as

$$\frac{a_n}{a_m} = e^{\beta(E_m - E_n)} \tag{17.10}$$

The form of this equation implies that both a_m and a_n are given by

$$a_j = C e^{-\beta E_j} \tag{17.11}$$

where j represents either state m or n and C is a constant.

17–2. The Probability That a System in an Ensemble Is in the State j with Energy $E_j(N, V)$ Is Proportional to $e^{-E_j(N,V)/k_B T}$

Equation 17.11 has two quantities, C and β, that we must determine. Determining C is fairly easy. We sum both sides of Equation 17.11 over j to obtain

$$\sum_j a_j = C \sum_j e^{-\beta E_j}$$

But the summation over a_j must equal $\mathcal{A}$, the total number of systems in the ensemble. Therefore, we have

$$C = \frac{\sum_j a_j}{\sum_j e^{-\beta E_j}} = \frac{\mathcal{A}}{\sum_j e^{-\beta E_j}}$$

If we substitute this result back into Equation 17.11, we obtain

$$\frac{a_j}{\mathcal{A}} = \frac{e^{-\beta E_j}}{\sum_j e^{-\beta E_j}} \tag{17.12}$$

The ratio $a_j/\mathcal{A}$ is the fraction of systems in our ensemble that will be found in the state j with energy E_j. In the limit of large $\mathcal{A}$, which we are certainly able to take because we can make our ensemble as large as we want, $a_j/\mathcal{A}$ becomes a probability (MathChapter B), so Equation 17.12 can be written as

$$p_j = \frac{e^{-\beta E_j}}{\sum_i e^{-\beta E_i}} \tag{17.13}$$

where p_j is the probability that a randomly chosen system will be in state j with energy $E_j(N, V)$.

Equation 17.13 is a central result of physical chemistry. We customarily let the denominator in this expression be denoted by Q, and if we specifically include the dependence of E_j on N and V, then we write

$$Q(N, V, \beta) = \sum_i e^{-\beta E_i(N, V)} \tag{17.14}$$

Equation 17.13 becomes

$$p_j(N, V, \beta) = \frac{e^{-\beta E_j(N, V)}}{Q(N, V, \beta)} \tag{17.15}$$

We are not quite ready to determine β at this point, but later we will present several different arguments to show that

$$\beta = \frac{1}{k_B T} \tag{17.16}$$

where k_B is the Boltzmann constant and T is the kelvin temperature. Thus, we can write Equation 17.15 as

$$p_j(N, V, T) = \frac{e^{-E_j(N, V)/k_B T}}{Q(N, V, T)} \tag{17.17}$$

We will use Equations 17.15 and 17.17 interchangeably. Equation 17.15 is just as acceptable as Equation 17.17. From a theoretical point of view, β, or $1/k_B T$, often happens to be a more convenient quantity to use than T itself.

The quantity $Q(N, V, \beta)$, or $Q(N, V, T)$, is called *the partition function* of the system, and we will see in the next few chapters that we can express all the macroscopic properties of a system in terms of $Q(N, V, \beta)$. At this point, it may not seem possible to determine all the energy states $\{E_j(N, V)\}$ never mind $Q(N, V, \beta)$, but you will learn that we can determine $Q(N, V, \beta)$ for a number of interesting and important systems.

17–3. We Postulate That the Average Ensemble Energy Is Equal to the Observed Energy of a System

Using Equation 17.15, we can calculate the average energy of a system in an ensemble of systems. If we denote the average energy by $\langle E \rangle$, then (see MathChapter B)

$$\langle E \rangle = \sum_j p_j(N, V, \beta) E_j(N, V) = \sum_j \frac{E_j(N, V) e^{-\beta E_j(N,V)}}{Q(N, V, \beta)} \tag{17.18}$$

Note that $\langle E \rangle$ is a function of N, V, and β. We can express Equation 17.18 entirely in terms of $Q(N, V, \beta)$. First we differentiate $\ln Q(N, V, \beta)$ with respect to β, with N and V held constant:

$$\left(\frac{\partial \ln Q(N, V, \beta)}{\partial \beta} \right)_{N,V} = \frac{1}{Q(N, V, \beta)} \left(\frac{\partial \sum e^{-\beta E_j(N,V)}}{\partial \beta} \right)_{N,V}$$

$$= \frac{1}{Q(N, V, \beta)} \sum_j [-E_j(N, V)] e^{-\beta E_j(N,V)}$$

$$= -\sum_j \frac{E_j(N, V) e^{-\beta E_j(N,V)}}{Q(N, V, \beta)} \tag{17.19}$$

If we compare Equation 17.19 with 17.18, we see that

$$\langle E \rangle = - \left(\frac{\partial \ln Q}{\partial \beta} \right)_{N,V} \tag{17.20}$$

We can also express Equation 17.20 as a temperature derivative rather than a β derivative. If we use the chain rule of differentiation, we can write for any function f that

$$\frac{\partial f}{\partial T} = \frac{\partial f}{\partial \beta} \cdot \frac{\partial \beta}{\partial T} = \frac{\partial f}{\partial \beta} \cdot \frac{d(1/k_B T)}{dT} = -\frac{1}{k_B T^2} \frac{\partial f}{\partial \beta}$$

or

$$\frac{\partial f}{\partial \beta} = -k_B T^2 \frac{\partial f}{\partial T}$$

Applying this result to Equation 17.20 with $f = \ln Q$ gives us the alternate form

$$\langle E \rangle = k_B T^2 \left(\frac{\partial \ln Q}{\partial T} \right)_{N,V} \tag{17.21}$$

Equation 17.20 is often the easier one to use, however.

EXAMPLE 17–1
Derive an equation for $\langle E \rangle$ for the simple system of a (bare) proton in a magnetic field B_z.

SOLUTION: According to Equation 14.16, the energy can take on one of the two values

$$E_{\pm\frac{1}{2}} = \mp\frac{1}{2}\hbar\gamma B_z$$

where γ is the magnetogyric ratio. The partition function consists of just two terms:

$$Q(T, B_z) = e^{\beta\hbar\gamma B_z/2} + e^{-\beta\hbar\gamma B_z/2}$$
$$= e^{\hbar\gamma B_z/2k_B T} + e^{-\hbar\gamma B_z/2k_B T}$$

The average energy is obtained from either Equation 17.20 or 17.21:

$$\langle E \rangle = -\left(\frac{\partial \ln Q}{\partial \beta}\right)_{B_z} = -\frac{1}{Q(\beta, B_z)}\left(\frac{\partial Q}{\partial \beta}\right)_{B_z}$$
$$= -\frac{\hbar\gamma B_z}{2}\left(\frac{e^{\beta\hbar\gamma B_z/2} - e^{-\beta\hbar\gamma B_z/2}}{e^{\beta\hbar\gamma B_z/2} + e^{-\beta\hbar\gamma B_z/2}}\right)$$
$$= -\frac{\hbar\gamma B_z}{2}\left(\frac{e^{\hbar\gamma B_z/2k_B T} - e^{-\hbar\gamma B_z/2k_B T}}{e^{\hbar\gamma B_z/2k_B T} + e^{-\hbar\gamma B_z/2k_B T}}\right)$$

This expression for $\langle E \rangle$ (in units of $\hbar\gamma B_z/2$) is plotted against T (in units of $\hbar\gamma B_z/2k_B$) in Figure 17.2. Note that $\langle E \rangle \to -\hbar\gamma B_z/2$ as $T \to 0$ and that $\langle E \rangle \to 0$ as $T \to \infty$. As $T \to 0$, there is no thermal energy, so the proton orients itself parallel to the magnetic field with certainty. As $T \to \infty$, however, the thermal energy of the proton increases to such an extent that the proton is equally likely to point in either direction.

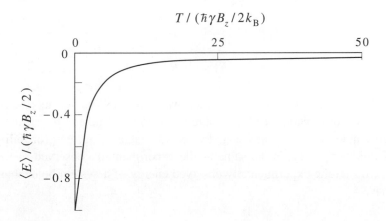

FIGURE 17.2
The average energy of a (bare) proton in a magnetic field plotted against the temperature (see Example 17–1).

We will learn in Chapter 18 that for a monatomic ideal gas,

$$Q(N, V, \beta) = \frac{[q(V, \beta)]^N}{N!} \tag{17.22}$$

where

$$q(V, \beta) = \left(\frac{2\pi m}{h^2 \beta}\right)^{3/2} V \tag{17.23}$$

For a monatomic ideal gas in its electronic ground state, the energy of the system is only in the translational degrees of freedom. Before we substitute Equation 17.22 into Equation 17.20, we write $\ln Q$ for convenience as a sum of terms that involve β and terms that are independent of β:

$$\ln Q = N \ln q - \ln N!$$
$$= -\frac{3N}{2} \ln \beta + \frac{3N}{2} \ln \left(\frac{2\pi m}{h^2}\right) + N \ln V - \ln N!$$
$$= -\frac{3N}{2} \ln \beta + \text{terms involving only } N \text{ and } V$$

Now we can see more easily that

$$\left(\frac{\partial \ln Q}{\partial \beta}\right)_{N,V} = -\frac{3N}{2} \frac{d \ln \beta}{d\beta} = -\frac{3N}{2\beta} = -\frac{3}{2} N k_B T$$

and that (Equation 17.20)

$$\langle E \rangle = \tfrac{3}{2} N k_B T$$

For n moles, $N = n N_A$ and $k_B N_A = R$, so

$$\langle E \rangle = \tfrac{3}{2} n R T$$

We will arrive at this same result when we study the kinetic theory of gases in Chapter 27. This observation leads us to a fundamental postulate of physical chemistry that the ensemble average of any quantity, as calculated using the probability distribution of Equation 17.17, is the same as the experimentally observed value of that quantity. If we let the experimentally observed energy of a system be denoted by U, then we have

$$\overline{U} = \langle \overline{E} \rangle = \tfrac{3}{2} R T$$

for one mole of a monatomic ideal gas. (We indicate a molar quantity by an overbar.)

EXAMPLE 17–2
We will learn in the next chapter that for the rigid rotator-harmonic oscillator model of an ideal diatomic gas, the partition function is given by

$$Q(N, V, \beta) = \frac{[q(V, \beta)]^N}{N!}$$

where

$$q(V, \beta) = \left(\frac{2\pi m}{h^2 \beta}\right)^{3/2} V \cdot \frac{8\pi^2 I}{h^2 \beta} \cdot \frac{e^{-\beta h \nu/2}}{1 - e^{-\beta h \nu}}$$

In this expression, I is the moment of inertia and ν is the fundamental vibrational frequency of the diatomic molecule. Note that $q(V, \beta)$ for a diatomic molecule is the same as the expression for $q(V, \beta)$ for a monatomic gas (Equation 17.23, a translational term), except that it is multiplied by a rotational term, $8\pi^2 I/h^2 \beta$, and a vibrational term, $e^{-\beta h \nu/2}/(1 - e^{-\beta h \nu})$. The reason for this difference will become apparent in Section 17–8. Use this partition function to calculate the average energy of one mole of a diatomic ideal gas.

SOLUTION: Once again, for convenience we write $\ln Q$ as the sum of terms that involve β and terms that are independent of β:

$$\ln Q = N \ln q - \ln N!$$
$$= -\frac{3N}{2} \ln \beta - N \ln \beta - \frac{N\beta h \nu}{2} - N \ln(1 - e^{-\beta h \nu})$$
$$+ \text{ terms not involving } \beta$$

Now

$$\left(\frac{\partial \ln Q}{\partial \beta}\right)_{N,V} = -\frac{3N}{2}\frac{d \ln \beta}{d\beta} - N\frac{d \ln \beta}{d\beta} - \frac{Nh\nu}{2} - N\frac{d \ln(1 - e^{-\beta h \nu})}{d\beta}$$
$$= -\frac{3N}{2\beta} - \frac{N}{\beta} - \frac{Nh\nu}{2} - \frac{Nh\nu e^{-\beta h \nu}}{1 - e^{-\beta h \nu}}$$

or

$$U = \langle E \rangle = \frac{3}{2}Nk_{\mathrm{B}}T + Nk_{\mathrm{B}}T + \frac{Nh\nu}{2} + \frac{Nh\nu e^{-\beta h \nu}}{1 - e^{-\beta h \nu}}$$

For one mole, $N = N_{\mathrm{A}}$ and $N_{\mathrm{A}}k_{\mathrm{B}} = R$, so

$$\overline{U} = \frac{3}{2}RT + RT + \frac{N_{\mathrm{A}}h\nu}{2} + \frac{N_{\mathrm{A}}h\nu e^{-\beta h \nu}}{1 - e^{-\beta h \nu}} \tag{17.24}$$

Equation 17.24 has a nice physical interpretation. The first term represents the average translational energy, the second term represents the average rotational energy, the third term represents the zero-point vibrational energy, and the fourth term represents the average vibrational energy. The fourth term is negligible at low temperatures for most gases but increases with increasing temperature as the excited vibrational states become populated.

17–4. The Heat Capacity at Constant Volume Is the Temperature Derivative of the Average Energy

The constant-volume heat capacity, C_V, of a system is defined as

$$C_V = \left(\frac{\partial \langle E \rangle}{\partial T} \right)_{N,V} = \left(\frac{\partial U}{\partial T} \right)_{N,V} \tag{17.25}$$

The heat capacity C_V is then a measure of how the energy of the system changes with temperature at constant amount and volume. Consequently, C_V can be expressed in terms of $Q(N, V, T)$ through Equation 17.21. We have seen that $\overline{U} = 3RT/2$ for one mole of a monatomic ideal gas, so

$$\overline{C}_V = \tfrac{3}{2} R \qquad \left(\begin{array}{c} \text{monatomic} \\ \text{ideal gas} \end{array} \right) \tag{17.26}$$

For a diatomic ideal gas, we obtain from Equation 17.24

$$
\begin{aligned}
\overline{C}_V &= \frac{5}{2} R + N_A h\nu \frac{\partial}{\partial T} \left(\frac{e^{-\beta h\nu}}{1 - e^{-\beta h\nu}} \right) \\
&= \frac{5}{2} R - \frac{N_A h\nu}{k_B T^2} \frac{\partial}{\partial \beta} \left(\frac{e^{-\beta h\nu}}{1 - e^{-\beta h\nu}} \right) \\
&= \frac{5}{2} R + R \left(\frac{h\nu}{k_B T} \right)^2 \frac{e^{-h\nu/k_B T}}{(1 - e^{-h\nu/k_B T})^2} \qquad \left(\begin{array}{c} \text{diatomic} \\ \text{ideal gas} \end{array} \right)
\end{aligned}
\tag{17.27}
$$

Figure 17.3 shows the theoretical (Equation 17.27) versus the experimental molar heat capacity of $O_2(g)$ as a function of temperature. The agreement between the two is seen to be excellent.

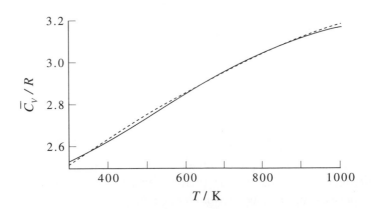

FIGURE 17.3
The experimental and theoretical (Equation 17.27) molar heat capacity of $O_2(g)$ from 300 K to 1000 K. The theoretical curve (solid curve) is calculated using $h\nu/k_B = 2240$ K.

EXAMPLE 17–3

In 1905, Einstein proposed a simple model for an atomic crystal that can be used to calculate the molar heat capacity. He pictured an atomic crystal as N atoms situated at lattice sites, with each atom vibrating as a three-dimensional harmonic oscillator. Because all the lattice sites are identical, he further assumed that each atom vibrated with the same frequency. The partition function associated with this model is (Problem 17–20)

$$Q = e^{-\beta U_0} \left(\frac{e^{-\beta h\nu/2}}{1 - e^{-\beta h\nu}} \right)^{3N} \tag{17.28}$$

where ν, which is characteristic of the particular crystal, is the frequency with which the atoms vibrate about their lattice positions and U_0 is the sublimation energy at 0 K, or the energy needed to separate all the atoms from one another at 0 K. Calculate the molar heat capacity of an atomic crystal from this partition function.

SOLUTION: The average energy is given by (Equation 17.20)

$$
\begin{aligned}
U &= -\left(\frac{\partial \ln Q}{\partial \beta} \right)_{N,V} \\
&= -\left(\frac{\partial}{\partial \beta} \left[-\beta U_0 - \frac{3N}{2} \beta h\nu - 3N \ln(1 - e^{-\beta h\nu}) \right] \right)_{N,V} \\
&= U_0 + \frac{3Nh\nu}{2} + \frac{3Nh\nu e^{-\beta h\nu}}{1 - e^{-\beta h\nu}}
\end{aligned}
$$

Note that U consists of three terms: U_0, the sublimation energy at 0 K; $3Nh\nu/2$, the zero-point energy of N three-dimensional harmonic oscillators; and a term that represents the increase in vibrational energy as the temperature increases.

The heat capacity at constant volume is given by

$$
\begin{aligned}
C_V &= \left(\frac{\partial U}{\partial T} \right)_{N,V} = -\frac{1}{k_B T^2} \left(\frac{\partial U}{\partial \beta} \right)_{N,V} \\
&= -\frac{3Nh\nu}{k_B T^2} \left[-\frac{h\nu e^{-\beta h\nu}}{1 - e^{-\beta h\nu}} - \frac{h\nu e^{-2\beta h\nu}}{(1 - e^{-\beta h\nu})^2} \right]
\end{aligned}
$$

or

$$\overline{C}_V = 3R \left(\frac{h\nu}{k_B T} \right)^2 \frac{e^{-h\nu/k_B T}}{(1 - e^{-h\nu/k_B T})^2} \tag{17.29}$$

where we have used the fact that $N = N_A$ and $N_A k_B = R$ for one mole.

Equation 17.29 contains one adjustable parameter, the vibrational frequency ν. Figure 17.4 shows the molar heat capacity of diamond as a function of temperature calculated with $\nu = 2.75 \times 10^{13}$ s^{-1}. The agreement with experiment is seen to be fairly good considering the simplicity of the model.

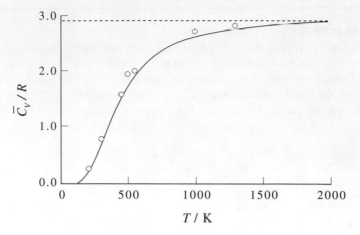

FIGURE 17.4
The observed and theoretical (Einstein model) molar heat capacity of diamond as a function of temperaure. The solid curve is calculated using Equation 17.29, and the circles represent experimental data.

It is interesting to look at the high-temperature limit of Equation 17.29. At high temperatures, $h\nu/k_B T$ is small, so we can use the fact that $e^x \approx 1 + x$ for small x (MathChapter I). Thus, Equation 17.29 becomes

$$\overline{C}_V \approx 3R \left(\frac{h\nu}{k_B T} \right)^2 \frac{1 - \dfrac{h\nu}{k_B T} + \cdots}{\left(1 - 1 + \dfrac{h\nu}{k_B T} + \cdots \right)^2}$$

$$\approx 3R \left(\frac{h\nu}{k_B T} \right)^2 \frac{1}{\left(\dfrac{h\nu}{k_B T} \right)^2} = 3R$$

This result predicts that the molar heat capacities of atomic crystals should level off at a value of $3R = 24.9 \text{ J} \cdot \text{K}^{-1} \cdot \text{mol}^{-1}$ at high temperatures. This prediction is known as the law of Dulong and Petit, which played an important role in the determination of atomic masses in the 1800s. This prediction is in good agreement with the data shown in Figure 17.4.

17–5. We Can Express the Pressure in Terms of a Partition Function

We will show in Section 19–6 that the pressure of a macroscopic system is given by

$$P_j(N, V) = - \left(\frac{\partial E_j}{\partial V} \right)_N \tag{17.30}$$

Using the fact that the average pressure is given by

$$\langle P \rangle = \sum_j p_j(N, V, \beta) P_j(N, V)$$

we can write

$$\langle P \rangle = \sum_j p_j(N, V, \beta) \left(-\frac{\partial E_j}{\partial V} \right)_N = \sum_j \left(-\frac{\partial E_j}{\partial V} \right)_N \frac{e^{-\beta E_j(N,V)}}{Q(N, V, \beta)} \qquad (17.31)$$

This expression can be written in a more compact form. Let's start with

$$Q(N, V, \beta) = \sum_j e^{-\beta E_j(N,V)}$$

and differentiate it with respect to V keeping N and β fixed:

$$\left(\frac{\partial Q}{\partial V} \right)_{N,\beta} = -\beta \sum_j \left(\frac{\partial E_j}{\partial V} \right)_N e^{-\beta E_j(N,V)}$$

Comparing this result with the second equality of Equation 17.31 shows that

$$\langle P \rangle = \frac{k_B T}{Q(N, V, \beta)} \left(\frac{\partial Q}{\partial V} \right)_{N,\beta}$$

or equivalently,

$$\langle P \rangle = k_B T \left(\frac{\partial \ln Q}{\partial V} \right)_{N,\beta} \qquad (17.32)$$

Just as we equated the ensemble average of the energy with the observed energy, we equate the ensemble average pressure with the observed pressure, $P = \langle P \rangle$. Thus, we see that we can calculate the observed pressure if we know $Q(N, V, \beta)$.

We can use this result to derive the ideal-gas equation of state. First, consider a monatomic ideal gas. Recall from Equation 17.22 that $Q(N, V, \beta)$ for a monatomic ideal gas is given by

$$Q(N, V, \beta) = \frac{[q(V, \beta)]^N}{N!}$$

where

$$q(V, \beta) = \left(\frac{2\pi m}{h^2 \beta} \right)^{3/2} V$$

Let's use this result to calculate the pressure of a monatomic ideal gas. To evaluate Equation 17.32, we write out $\ln Q$ first for convenience:

$$\ln Q = N \ln q - \ln N!$$
$$= \frac{3N}{2} \ln \left(\frac{2\pi m}{h^2 \beta} \right) + N \ln V - \ln N!$$

Because N and β are fixed in Equation 17.32, we write $\ln Q$ as

$$\ln Q = N \ln V + \text{terms in } N \text{ and } \beta \text{ only}$$

Therefore,

$$\left(\frac{\partial \ln Q}{\partial V} \right)_{N,\beta} = \frac{N}{V}$$

and substituting this result into Equation 17.32 gives us

$$P = \frac{Nk_{\mathrm{B}}T}{V}$$

as you might have expected.

Notice that the ideal-gas equation results from the fact that $\ln Q = N \ln V +$ terms in N and β, which comes from the fact that $q(V, T)$ is directly proportional to V in Equation 17.22. Example 17–2 shows that $q(V, T)$ is directly proportional to V for a diatomic ideal gas also, and so $PV = Nk_{\mathrm{B}}T$ for a diatomic ideal gas. This is the case for a polyatomic ideal gas as well, so the ideal-gas equation of state results for any ideal gas, monatomic, diatomic, or polyatomic.

EXAMPLE 17–4

Calculate the equation of state associated with the partition function

$$Q(N, V, \beta) = \frac{1}{N!} \left(\frac{2\pi m}{h^2 \beta} \right)^{3N/2} (V - Nb)^N e^{\beta a N^2 / V}$$

where a and b are constants. Can you identify the resulting equation of state?

SOLUTION: We use Equation 17.32 to calculate the equation of state. First, we evaluate $\ln Q$, which gives

$$\ln Q = N \ln(V - Nb) + \frac{\beta a N^2}{V} + \text{terms in } N \text{ and } \beta \text{ only}$$

We now differentiate with respect to V, keeping N and β constant, to get

$$\left(\frac{\partial \ln Q}{\partial V} \right)_{N,\beta} = \frac{N}{V - Nb} - \frac{\beta a N^2}{V^2}$$

and so

$$P = \frac{Nk_{\mathrm{B}}T}{V - Nb} - \frac{aN^2}{V^2}$$

Bringing the last term to the left side and multiplying by $V - Nb$ gives us

$$\left(P + \frac{aN^2}{V^2} \right)(V - Nb) = Nk_{\mathrm{B}}T$$

which is the van der Waals equation.

17–6. The Partition Function of a System of Independent, Distinguishable Molecules Is the Product of Molecular Partition Functions

The general results we have derived up to now are valid for arbitrary systems. To apply these equations, we need to have the set of eigenvalues $\{E_j(N, V)\}$ for the N-body Schrödinger equation. In general, this is an impossible task. For many important physical systems, however, writing the total energy of the system as a sum of individual energies is a good approximation. (See Section 3–9.) This procedure leads to a great simplification of the partition function and allows us to apply the results with relative ease.

First, let's consider a system that consists of independent, distinguishable particles. Although atoms and molecules are certainly not distinguishable in general, they can be treated as such in a number of cases. An excellent example is that of a perfect crystal. In a perfect crystal, each atom is confined to one and only one lattice site, which we could, at least in principle, identify by a set of three coordinates. Because each particle, then, is confined to a lattice site and the lattice sites are distinguishable, the particles themselves are distinguishable. We can treat the vibration of each particle about its lattice site as independent to a fairly good approximation, just as we did for normal modes of polyatomic molecules.

We will denote the individual particle energies by $\{\varepsilon_j^a\}$, where the superscript denotes the particle (they are distinguishable), and the subscript denotes the energy state of the particle. In this case, the total energy of the system $E_l(N, V)$ can be written as

$$E_l(N, V) = \underbrace{\varepsilon_i^a(V) + \varepsilon_j^b(V) + \varepsilon_k^c(V) + \cdots}_{N \text{ terms}}$$

and the system partition function becomes

$$Q(N, V, T) = \sum_l e^{-\beta E_l} = \sum_{i,j,k,\ldots} e^{-\beta(\varepsilon_i^a + \varepsilon_j^b + \varepsilon_k^c + \cdots)}$$

Because the particles are distinguishable and independent, we can sum over $i, j, k, \ldots$ independently, in which case $Q(N, V, T)$ can be written as a product of individual summations (Problem 17–21):

$$Q(N, V, T) = \sum_i e^{-\beta \varepsilon_i^a} \sum_j e^{-\beta \varepsilon_j^b} \sum_k e^{-\beta \varepsilon_k^c} \cdots$$

$$= q_a(V, T) q_b(V, T) q_c(V, T) \cdots \tag{17.33}$$

where each of the $q(V, T)$ is given by

$$q(V, T) = \sum_i e^{-\beta \varepsilon_i} = \sum_i e^{-\varepsilon_i / k_B T} \tag{17.34}$$

In many cases, the $\{\varepsilon_i\}$ is a set of molecular energies; thus $q(V, T)$ is called a *molecular partition function*.

Equation 17.33 is an important result. It shows that if we can write the total energy as a sum of individual, independent terms, and if the atoms or molecules are *distinguishable*, then the system partition function $Q(N, V, T)$ reduces to a product of molecular partition functions $q(V, T)$. Because $q(V, T)$ requires a knowledge of the allowed energies of only individual atoms or molecules, its evaluation is often feasible, as we will see for a number of cases in Chapter 18.

If the energy states of all the atoms or molecules are the same (as for an atomic crystal), then Equation 17.33 becomes

$$Q(N, V, T) = [q(V, T)]^N \qquad \binom{\text{independent, distinguishable}}{\text{atoms or molecules}} \qquad (17.35)$$

where

$$q(V, T) = \sum_j e^{-\varepsilon_j / k_B T}$$

The Einstein model of atomic crystals (Example 17–3) considers the atoms to be fixed at lattice sites, so Equation 17.35 should be applicable to that model. Notice that the partition function of that model (Equation 17.28) can be written in the form of Equation 17.35 if we let $u_0 = U_0/N$ be the sublimation energy per atom at 0 K, in which case we have

$$Q = \left[e^{-\beta u_0} \left(\frac{e^{-\beta h\nu/2}}{1 - e^{-\beta h\nu}} \right)^3 \right]^N \qquad (17.36)$$

17–7. The Partition Function of a System of Independent, Indistinguishable Atoms or Molecules Can Usually Be Written as $[q(V, T)]^N / N!$

Equation 17.35 is an attractive result, but atoms and molecules are, in general, not distinguishable; thus the utility of Equation 17.35 is severely limited. The reduction of a system partition function $Q(N, V, T)$ to molecular partition functions $q(V, T)$ becomes somewhat more complicated when the inherent indistinguishability of atoms and molecules cannot be ignored. For indistinguishable particles, the total energy is

$$E_{ijk\ldots} = \underbrace{\varepsilon_i + \varepsilon_j + \varepsilon_k + \cdots}_{N \text{ terms}}$$

(note the lack of distinguishing superscripts, as in Equation 17.33) and the system partition function is

$$Q(N, V, T) = \sum_{i,j,k,\ldots} e^{-\beta(\varepsilon_i + \varepsilon_j + \varepsilon_k + \cdots)} \tag{17.37}$$

Because the particles are indistinguishable, we cannot sum over i, j, k, ... separately as we did in Equation 17.33. To see why, we must consider a fundamental property of all particles.

We learned in Chapter 8 that a consequence of the Pauli Exclusion Principle is that electronic wave functions must be antisymmetric under the interchange of two electrons and that no two electrons in an atom or a molecule can occupy the same energy state. The Pauli Exclusion Principle, as we have applied it to electrons, is part of a more general principle of nature that applies to all particles. All known particles fall into one of two classes: those whose wave functions must be symmetric under the interchange of two identical particles, and those whose wave functions must be antisymmetric under such an interchange. Particles of the first type are called *bosons*, and particles of the second type are called *fermions*. Experimentally, particles of integer spin are bosons and particles of half-integer spin are fermions. Thus, electrons, which have spin 1/2, behave as fermions and their wave functions must satisfy the antisymmetry requirement. Other examples of fermions are protons (spin 1/2) and neutrons (spin 1/2). Examples of bosons are photons (spin 1) and deuterons (spin 0). Although no two identical fermions can occupy the same single-particle energy state, there is no such restriction on bosons. These restrictions are important to recognize when we attempt to carry out the summation in Equation 17.37.

Let's go back now to the summation in Equation 17.37 for the case of fermions. Because no two identical fermions can occupy the same single-particle energy state, terms in which two or more indices are the same cannot be included in the summation. Therefore, the indices i, j, k, ... are *not* independent of one another, so a direct evaluation of $Q(N, V, T)$ by means of Equation 17.37 poses problems for fermions.

EXAMPLE 17–5
Consider a system of two noninteracting identical fermions, each of which has states with energies ε_1, ε_2, ε_3, and ε_4. Enumerate the allowed total energies in the summation in Equation 17.37.

SOLUTION: For this system

$$Q(2, V, T) = \sum_{i,j=1}^{4} e^{-\beta(\varepsilon_i + \varepsilon_j)}$$

Of the 16 terms that would occur in an unrestricted evaluation of Q, only six are allowed for two identical fermions; these are the terms with energies

$$\begin{array}{cc} \varepsilon_1 + \varepsilon_2 & \varepsilon_2 + \varepsilon_3 \\ \varepsilon_1 + \varepsilon_3 & \varepsilon_2 + \varepsilon_4 \\ \varepsilon_1 + \varepsilon_4 & \varepsilon_3 + \varepsilon_4 \end{array}$$

The six terms in which the ε_j are written in reverse order are the same as those above (because the particles are indistinguishable), and the four terms in which the ε_j are the same are not allowed (because the particles are fermions).

Bosons do not have the restriction that no two of the same type can occupy the same single-particle state, but the summation in Equation 17.37 is still complicated. To see why, consider a term in Equation 17.37 in which all the indices are the same except for one; for example, a term like

$$E = \underbrace{\varepsilon_2 + \varepsilon_{10} + \varepsilon_{10} + \varepsilon_{10} + \cdots + \varepsilon_{10}}_{N \text{ particles, } N \text{ terms}}$$

(in reality, these indices might be enormous numbers). Because the particles are indistinguishable, the position of the term ε_2 is not important, and we could just as easily have $\varepsilon_{10} + \varepsilon_2 + \varepsilon_{10} + \varepsilon_{10} + \cdots + \varepsilon_{10}$ or $\varepsilon_{10} + \varepsilon_{10} + \varepsilon_2 + \varepsilon_{10} + \cdots + \varepsilon_{10}$ and so on. Because these terms all represent the same state, such a state should be included only once in Equation 17.37, but an unrestricted summation over all the indices (summing over i, j, k, ... independently) in Equation 17.37 would produce N terms of this type (the ε_2 can be located in any of the N positions).

Now consider the other extreme in which all the N particles are in different molecular states; that is, for example, a system with energy $\varepsilon_1 + \varepsilon_2 + \varepsilon_3 + \varepsilon_4 + \cdots + \varepsilon_N$. Because the particles are indistinguishable, all $N!$ arrangements obtained by permuting these N terms are identical and should occur only once in Equation 17.37. Yet such terms will appear $N!$ times in an unrestricted summation. Consequently, a direct evaluation of $Q(N, V, T)$ by means of Equation 17.37 poses problems for bosons as well as fermions.

EXAMPLE 17–6
Redo Example 17–5 for bosons instead of fermions.

SOLUTION: In this case there are 10 allowed terms: the six that are allowed in Example 17–5 and the four in which the ε_j are the same (bosons do not have the restriction that no two can occupy the same state).

Note that in every case, the terms in Equation 17.37 that cause difficulty are those in which two or more indices are the same. If it were not for such terms, we could carry out the summation in Equation 17.37 in an unrestricted manner (obtaining $[q(V, T)]^N$

as in Section 17–6) and then divide by $N!$ (to obtain $[q(V, T)]^N / N!$) to account for the over-counting. For example, if we could ignore terms like $\varepsilon_1 + \varepsilon_1$, $\varepsilon_2 + \varepsilon_2$, etc. in the evaluation of $Q(2, V, T)$, there would be a total of 12 terms, the six enumerated in Example 17–5 and the six in which the energies are written in reverse order. By dividing by 2!, we would obtain the correct number of allowed terms.

Certainly, if the number of quantum states available to any particle is much greater than the number of particles, it would be unlikely for any two particles to be in the same state. Although most of the quantum-mechanical systems we have studied have an infinite number of energy states, at any given temperature many of these will not be readily accessible because the energies of these states are much larger than $k_B T$, which is roughly the average energy of a molecule. If, however, the number of quantum states with energies less than roughly $k_B T$ is much larger than the number of particles, then essentially all the terms in Equation 17.37 will contain ε's with different indices, and so we can evaluate $Q(N, V, T)$ to a good approximation by summing over $i, j, k, \ldots$ independently in Equation 17.37 and then dividing by $N!$ to get

$$Q(N, V, T) = \frac{[q(V, T)]^N}{N!} \quad \left(\begin{array}{c}\text{independent, indistinguishable}\\ \text{atoms or molecules}\end{array}\right) \quad (17.38)$$

where

$$q(V, T) = \sum_j e^{-\varepsilon_j/k_B T} \quad (17.39)$$

The number of translational states alone is usually sufficient to guarantee that the number of energy states available to any atom or molecule is greater than the number of particles in the system. Therefore, this procedure yields an excellent approximation in many cases. The criterion that the number of available states exceeds the number of particles so that Equation 17.38 can be used is

$$\frac{N}{V}\left(\frac{h^2}{8mk_B T}\right)^{3/2} \ll 1 \quad (17.40)$$

Notice that this criterion is favored by large particle mass, high temperature, and low density.

Although our discussion at this point is limited to ideal gases (independent, indistinguishable particles), we show the values of $(N/V)(h^2/8mk_B T)^{3/2}$ in Table 17.1 even for some liquids at their boiling points, just to show that Inequality 17.40 is easily satisfied in most cases. Note that the exceptional systems include liquid helium and liquid hydrogen (because of their small masses and low temperatures) and electrons in metals (because of their very small mass). These systems are the prototype examples of quantum systems that must be treated by special methods (which we will not discuss).

When Equation 17.38 is valid, that is, when the number of available molecular states is much greater than the number of particles, we say that the particles obey *Boltzmann statistics*. As Inequality 17.40 indicates, Boltzmann statistics becomes increasingly

TABLE 17.1
The quantity $(N/V)(h^2/8mk_B T)^{3/2}$ at a pressure of one bar for a number of simple systems.

System	T/K	$\dfrac{N}{V}\left(\dfrac{h^2}{8mk_B T}\right)^{3/2}$
Liquid helium	4	1.5
Gaseous helium	4	0.11
Gaseous helium	20	1.8×10^{-3}
Gaseous helium	100	3.3×10^{-5}
Liquid hydrogen	20	0.29
Gaseous hydrogen	20	5.1×10^{-3}
Gaseous hydrogen	100	9.4×10^{-5}
Liquid neon	27	1.0×10^{-2}
Gaseous neon	27	7.8×10^{-5}
Liquid krypton	127	5.1×10^{-5}
Electrons in metals (Na)	300	1400

valid with increasing temperature. Let's test Inequality 17.40 for $N_2(g)$ at 20°C and one bar. Under these conditions,

$$\frac{N}{V} = \frac{P}{k_B T} = \frac{10^5 \text{ Pa}}{(1.381 \times 10^{-23} \text{ J}\cdot\text{K}^{-1})(293.2 \text{ K})}$$
$$= 2.470 \times 10^{25} \text{ m}^{-3}$$

and

$$\frac{h^2}{8mk_B T} = \frac{(6.626 \times 10^{-34} \text{ J}\cdot\text{s})^2}{(8)(4.653 \times 10^{-26} \text{ kg})(1.381 \times 10^{-23} \text{ J}\cdot\text{K}^{-1})(293.2 \text{ K})}$$
$$= 2.913 \times 10^{-22} \text{ m}^2$$

and so

$$\frac{N}{V}\left(\frac{h^2}{8mk_B T}\right)^{3/2} = (2.470 \times 10^{25} \text{ m}^{-3})(2.913 \times 10^{-22} \text{ m}^2)^{3/2}$$
$$= 1.23 \times 10^{-7}$$

which is much less than unity.

Let's test Inequality 17.40 for liquid nitrogen at its boiling point, −195.8°C. Experimentally, the density of $N_2(l)$ is 0.808 g·mL^{-1} at its boiling point. Therefore,

$$\frac{N}{V} = (0.808 \text{ g}\cdot\text{mL}^{-1})\left(\frac{1 \text{ mol } N_2}{28.02 \text{ g } N_2}\right)\left(\frac{6.022 \times 10^{23}}{1 \text{ mol}}\right)\left(\frac{10^6 \text{ mL}}{1 \text{ m}^3}\right)$$
$$= 1.737 \times 10^{28} \text{ m}^{-3}$$

and

$$\frac{N}{V}\left(\frac{h^2}{8mk_BT}\right)^{3/2} = (1.737 \times 10^{28} \text{ m}^{-3})(1.104 \times 10^{-21} \text{ m}^2)^{3/2}$$

$$= 6.37 \times 10^{-4}$$

Thus, Equation 17.38 is valid, even for liquid nitrogen at its boiling point.

17–8. A Molecular Partition Function Can Be Decomposed into Partition Functions for Each Degree of Freedom

In this section, we will explore the similarity between a system partition function, Equation 17.14, and a molecular partition function, Equation 17.39. We will start by substituting Equation 17.38 into Equation 17.21:

$$\langle E \rangle = k_B T^2 \left(\frac{\partial \ln Q}{\partial T}\right)_{N,V}$$

$$= Nk_B T^2 \left(\frac{\partial \ln q}{\partial T}\right)_V$$

$$= N \sum_j \varepsilon_j \frac{e^{-\varepsilon_j/k_B T}}{q(V,T)} \tag{17.41}$$

But Equation 17.38 is valid only for independent particles, so

$$\langle E \rangle = N \langle \varepsilon \rangle \tag{17.42}$$

where $\langle \varepsilon \rangle$ is the average energy of any one molecule. If we compare Equations 17.41 and 17.42, we see that

$$\langle \varepsilon \rangle = \sum_j \varepsilon_j \frac{e^{-\varepsilon_j/k_B T}}{q(V,T)} \tag{17.43}$$

We can conclude from this equation that the probability that a molecule is in its jth molecular energy state, π_j, is given by

$$\pi_j = \frac{e^{-\varepsilon_j/k_B T}}{q(V,T)} = \frac{e^{-\varepsilon_j/k_B T}}{\sum_j e^{-\varepsilon_j/k_B T}} \tag{17.44}$$

Note how similar this equation is to Equation 17.13.

Equation 17.44 can be reduced even further if we assume that the energy of a molecule can be written as

$$\varepsilon = \varepsilon_i^{\text{trans}} + \varepsilon_j^{\text{rot}} + \varepsilon_k^{\text{vib}} + \varepsilon_l^{\text{elec}} \tag{17.45}$$

Because the various energy terms are distinguishable here, we can apply the reasoning behind Equation 17.33 and write

$$q(V, T) = q_{trans}q_{rot}q_{vib}q_{elec} \tag{17.46}$$

where, for example

$$q_{trans} = \sum_j e^{-\varepsilon_j^{trans}/k_B T} \tag{17.47}$$

Note that the partition function for a diatomic molecule we used in Example 17–2 was expressed as

$$q(V, \beta) = q_{trans}(V, T)q_{rot}(T)q_{vib}(T)$$

where

$$q_{trans}(V, T) = \left(\frac{2\pi m}{h^2 \beta}\right)^{3/2} V$$

$$q_{rot}(T) = \frac{8\pi^2 I}{h^2 \beta}$$

and

$$q_{vib}(T) = \frac{e^{-\beta h\nu/2}}{1 - e^{-\beta h\nu}}$$

If we substitute Equations 17.45 and 17.46 into Equation 17.44, we obtain

$$\pi_{ijkl} = \frac{e^{-\varepsilon_i^{trans}/k_B T} e^{-\varepsilon_j^{rot}/k_B T} e^{-\varepsilon_k^{vib}/k_B T} e^{-\varepsilon_l^{elec}/k_B T}}{q_{trans}q_{rot}q_{vib}q_{elec}} \tag{17.48}$$

where π_{ijkl} is the probability that a molecule is in the ith translational state, the jth rotational state, the kth vibrational state, and the lth electronic state. Now if we sum Equation 17.48 over i (all translational states), j (all rotational states), and l (all electronic states), we obtain

$$\pi_k^{vib} = \sum_{i,j,l} \pi_{ijkl} = \frac{\left(\sum_i e^{-\varepsilon_i^{trans}/k_B T}\right)\left(\sum_j e^{-\varepsilon_j^{rot}/k_B T}\right)\left(\sum_l e^{-\varepsilon_l^{elec}/k_B T}\right) e^{-\varepsilon_k^{vib}/k_B T}}{q_{trans}q_{rot}q_{vib}q_{elec}}$$

$$= \frac{e^{-\varepsilon_k^{vib}/k_B T}}{q_{vib}} = \frac{e^{-\varepsilon_k^{vib}/k_B T}}{\sum_k e^{-\varepsilon_k^{vib}/k_B T}} \tag{17.49}$$

where, as the notation suggests, π_k^{vib} is the probability that a molecule is in its kth vibrational state. Furthermore, the average vibrational energy of a molecule is given by

$$\langle \varepsilon^{\text{vib}} \rangle = \sum_k \varepsilon_k^{\text{vib}} \frac{e^{-\varepsilon_k^{\text{vib}}/k_{\text{B}}T}}{q_{\text{vib}}}$$

$$= k_{\text{B}}T^2 \frac{\partial \ln q_{\text{vib}}}{\partial T} = -\frac{\partial \ln q_{\text{vib}}}{\partial \beta} \tag{17.50}$$

Again, note the similarity with Equation 17.21. Of course, we also have the relations

$$\langle \varepsilon^{\text{trans}} \rangle = k_{\text{B}}T^2 \left(\frac{\partial \ln q_{\text{trans}}}{\partial T} \right)_V = -\left(\frac{\partial \ln q_{\text{trans}}}{\partial \beta} \right)_V \tag{17.51}$$

and

$$\langle \varepsilon^{\text{rot}} \rangle = k_{\text{B}}T^2 \frac{\partial \ln q_{\text{rot}}}{\partial T} = -\frac{\partial \ln q_{\text{rot}}}{\partial \beta} \tag{17.52}$$

EXAMPLE 17–7

Use the partition function for a diatomic molecule given in Example 17–2 to calculate $\langle \varepsilon^{\text{vib}} \rangle$.

SOLUTION: According to Example 17–2, we can write

$$q_{\text{vib}}(T) = \frac{e^{-\beta h\nu/2}}{1 - e^{-\beta h\nu}}$$

and so

$$\langle \varepsilon^{\text{vib}} \rangle = -\left(\frac{\partial \ln q_{\text{vib}}}{\partial \beta} \right)$$

$$= \frac{h\nu}{2} + \frac{h\nu e^{-\beta h\nu}}{1 - e^{-\beta h\nu}}$$

in agreement with Equation 17.24.

To this point, we have written partition functions as summations over energy *states*. Each state is represented by a wave function with an associated energy. Thus, we write

$$q(V, T) = \sum_{\substack{j \\ (\text{states})}} e^{-\varepsilon_j/k_{\text{B}}T} \tag{17.53}$$

We will call sets of states that have the same energy, *levels*. We can write $q(V, T)$ as a summation over levels by including the degeneracy, g_j, of the level:

$$q(V, T) = \sum_{\substack{j \\ \text{(levels)}}} g_j e^{-\varepsilon_j / k_B T} \tag{17.54}$$

In the notation of Equation 17.53, the terms representing a degenerate level are repeated g_j times, whereas in Equation 17.54, they are written once and multiplied by g_j. For example, we learned in Section 5–8 (Equation 5.57) that the energy and degeneracy for a rigid rotator are

$$\varepsilon_J = \frac{\hbar^2}{2I} J(J + 1)$$

and

$$g_J = 2J + 1$$

Thus, we can write the rotational partition function by summing over levels:

$$q_{\text{rot}}(T) = \sum_{J=0}^{\infty} (2J + 1) e^{-\hbar^2 J(J+1)/2I k_B T} \tag{17.55}$$

Including degeneracies explicitly as in Equation 17.54 is usually more convenient, so we will use Equation 17.54 rather than Equation 17.53 in later chapters.

Problems

17-1. How would you describe an ensemble whose systems are one-liter containers of water at 25°C?

17-2. Show that Equation 17.8 is equivalent to $f(x + y) = f(x) f(y)$. In this problem, we will prove that $f(x) \propto e^{ax}$. First, take the logarithm of the above equation to obtain

$$\ln f(x + y) = \ln f(x) + \ln f(y)$$

Differentiate both sides with respect to x (keeping y fixed) to get

$$\left[\frac{\partial \ln f(x + y)}{\partial x} \right]_y = \frac{d \ln f(x + y)}{d(x + y)} \left[\frac{\partial (x + y)}{\partial x} \right]_y = \frac{d \ln f(x + y)}{d(x + y)}$$
$$= \frac{d \ln f(x)}{dx}$$

Now differentiate with respect to y (keeping x fixed) and show that

$$\frac{d \ln f(x)}{dx} = \frac{d \ln f(y)}{dy}$$

For this relation to be true for all x and y, each side must equal a constant, say a. Show that

$$f(x) \propto e^{ax} \qquad \text{and} \qquad f(y) \propto e^{ay}$$

17-3. Show that $a_l/a_i = e^{\beta(E_i - E_l)}$ implies that $a_j = Ce^{-\beta E_j}$.

17-4. Prove to yourself that $\sum_i e^{-\beta E_i} = \sum_j e^{-\beta E_j}$.

17-5. Show that the partition function in Example 17–1 can be written as

$$Q(\beta, B_z) = 2\cosh\left(\frac{\beta \hbar \gamma B_z}{2}\right) = 2\cosh\left(\frac{\hbar \gamma B_z}{2k_B T}\right)$$

Use the fact that $d\cosh x / dx = \sinh x$ to show that

$$\langle E \rangle = -\frac{\hbar \gamma B_z}{2} \tanh \frac{\beta \hbar \gamma B_z}{2} = -\frac{\hbar \gamma B_z}{2} \tanh \frac{\hbar \gamma B_z}{2k_B T}$$

17-6. Use either the expression for $\langle E \rangle$ in Example 17–1 or the one in Problem 17–5 to show that

$$\langle E \rangle \longrightarrow -\frac{\hbar \gamma B_z}{2} \qquad \text{as} \qquad T \longrightarrow 0$$

and that

$$\langle E \rangle \longrightarrow 0 \qquad \text{as} \qquad T \longrightarrow \infty$$

17-7. Generalize the results of Example 17–1 to the case of a spin-1 nucleus. Determine the low-temperature and high-temperature limits of $\langle E \rangle$.

17-8. If N_w is the number of protons aligned with a magnetic field B_z and N_o is the number of protons opposed to the field, show that

$$\frac{N_o}{N_w} = e^{-\hbar \gamma B_z / k_B T}$$

Given that $\gamma = 26.7522 \times 10^7$ rad·T^{-1}·s^{-1} for a proton, calculate N_o/N_w as a function of temperature for a field strength of 5.0 T. At what temperature is $N_o = N_w$? Interpret this result physically.

17-9. In Section 17–3, we derived an expression for $\langle E \rangle$ for a monatomic ideal gas by applying Equation 17.20 to $Q(N, V, T)$ given by Equation 17.22. Apply Equation 17.21 to

$$Q(N, V, T) = \frac{1}{N!}\left(\frac{2\pi m k_B T}{h^2}\right)^{3N/2} V^N$$

to derive the same result. Note that this expression for $Q(N, V, T)$ is simply Equation 17.22 with β replaced by $1/k_B T$.

17-10. A gas absorbed on a surface can sometimes be modelled as a two-dimensional ideal gas. We will learn in Chapter 18 that the partition function of a two-dimensional ideal gas is

$$Q(N, A, T) = \frac{1}{N!}\left(\frac{2\pi m k_B T}{h^2}\right)^N A^N$$

where A is the area of the surface. Derive an expression for $\langle E \rangle$ and compare your result with the three-dimensional result.

17-11. Although we will not do so in this book, it is possible to derive the partition function for a monatomic van der Waals gas.

$$Q(N, V, T) = \frac{1}{N!} \left(\frac{2\pi m k_B T}{h^2} \right)^{3N/2} (V - Nb)^N e^{aN^2/Vk_B T}$$

where a and b are the van der Waals constants. Derive an expression for the energy of a monatomic van der Waals gas.

17-12. An approximate partition function for a gas of hard spheres can be obtained from the partition function of a monatomic gas by replacing V in Equation 17.22 (and the following equation) by $V - b$, where b is related to the volume of the N hard spheres. Derive expressions for the energy and the pressure of this system.

17-13. Use the partition function in Problem 17–10 to calculate the heat capacity of a two-dimensional ideal gas.

17-14. Use the partition function for a monatomic van der Waals gas given in Problem 17–11 to calculate the heat capacity of a monatomic van der Waals gas. Compare your result with that of a monatomic ideal gas.

17-15. Using the partition function given in Example 17–2, show that the pressure of an ideal diatomic gas obeys $PV = Nk_B T$, just as it does for a monatomic ideal gas.

17-16. Show that if a partition function is of the form

$$Q(N, V, T) = \frac{[q(V, T)]^N}{N!}$$

and if $q(V, T) = f(T)V$ [as it does for a monatomic ideal gas (Equation 17.22) and a diatomic ideal gas (Example 17–2)], then the ideal-gas equation of state results.

17-17. Use Equation 17.27 and the value of $\tilde{\nu}$ for O_2 given in Table 5.1 to calculate the value of the molar heat capacity of $O_2(g)$ from 300 K to 1000 K (see Figure 17.3).

17-18. Show that the heat capacity given by Equation 17.29 in Example 17–3 obeys a law of corresponding states.

17-19. Consider a system of independent, distinguishable particles that have only two quantum states with energy ε_0 (let $\varepsilon_0 = 0$) and ε_1. Show that the molar heat capacity of such a system is given by

$$\overline{C}_V = R(\beta\varepsilon)^2 \frac{e^{-\beta\varepsilon}}{(1 + e^{-\beta\varepsilon})^2}$$

and that $\overline{C}_V$ plotted against $\beta\varepsilon$ passes through a maximum value at $\beta\varepsilon$, given by the solution to $\beta\varepsilon/2 = \coth \beta\varepsilon/2$. Use a table of values of $\coth x$ (for example, the *CRC Standard Mathematical Tables*) to show that $\beta\varepsilon = 2.40$.

17-20. Deriving the partition function for an Einstein crystal is not difficult (see Example 17–3). Each of the N atoms of the crystal is assumed to vibrate independently about its lattice position, so that the crystal is pictured as N independent harmonic oscillators, each vibrating in three directions. The partition function of a harmonic oscillator is

$$q_{ho}(T) = \sum_{v=0}^{\infty} e^{-\beta(v+\frac{1}{2})h\nu}$$

$$= e^{-\beta h\nu/2} \sum_{v=0}^{\infty} e^{-\beta v h\nu}$$

This summation is easy to evaluate if you recognize it as the so-called geometric series (MathChapter I)

$$\sum_{v=0}^{\infty} x^v = \frac{1}{1-x}$$

Show that

$$q_{ho}(T) = \frac{e^{-\beta h\nu/2}}{1 - e^{-\beta h\nu}}$$

and that

$$Q = e^{-\beta U_0} \left(\frac{e^{-\beta h\nu/2}}{1 - e^{-\beta h\nu}} \right)^{3N}$$

where U_0 simply represents the zero-of-energy, where all N atoms are infinitely separated.

17-21. Show that

$$S = \sum_{i=1}^{2} \sum_{j=0}^{1} x^i y^j = x(1+y) + x^2(1+y) = (x+x^2)(1+y)$$

by summing over j first and then over i. Now obtain the same result by writing S as a product of two separate summations.

17-22. Evaluate

$$S = \sum_{i=0}^{2} \sum_{j=0}^{1} x^{i+j}$$

by summing over j first and then over i. Now obtain the same result by writing S as a product of two separate summations.

17-23. How many terms are there in the following summations?

a. $S = \sum_{i=1}^{3} \sum_{j=1}^{2} x^i y^j$ **b.** $S = \sum_{i=1}^{3} \sum_{j=0}^{2} x^i y^j$ **c.** $S = \sum_{i=1}^{3} \sum_{j=1}^{2} \sum_{k=1}^{2} x^i y^j z^k$

17-24. Consider a system of two noninteracting identical fermions, each of which has states with energies ε_1, ε_2, and ε_3. How many terms are there in the unrestricted evaluation of $Q(2, V, T)$? Enumerate the allowed total energies in the summation in Equation 17.37 (see Example 17–5). How many terms occur in $Q(2, V, T)$ when the fermion restriction is taken into account?

17-25. Redo Problem 17–24 for the case of bosons instead of fermions.

17-26. Consider a system of three noninteracting identical fermions, each of which has states with energies ε_1, ε_2, and ε_3. How many terms are there in the unrestricted evaluation of $Q(3, V, T)$? Enumerate the allowed total energies in the summation of Equation 17.37 (see Example 17–5). How many terms occur in $Q(3, V, T)$ when the fermion restriction is taken into account?

17-27. Redo Problem 17–26 for the case of bosons instead of fermions.

17-28. Evaluate $(N/V)(h^2/8mk_BT)^{3/2}$ (see Table 17.1) for $O_2(g)$ at its normal boiling point, 90.20 K. Use the ideal-gas equation of state to calculate the density of $O_2(g)$ at 90.20 K.

17-29. Evaluate $(N/V)(h^2/8mk_BT)^{3/2}$ (see Table 17.1) for He(g) at its normal boiling point 4.22 K. Use the ideal-gas equation of state to calculate the density of He(g) at 4.22 K.

17-30. Evaluate $(N/V)(h^2/8mk_BT)^{3/2}$ for the electrons in sodium metal at 298 K. Take the density of sodium to 0.97 g·mL^{-1}. Compare your result with the value given in Table 17.1.

17-31. Evaluate $(N/V)(h^2/8mk_BT)^{3/2}$ (see Table 17.1) for liquid hydrogen at its normal boiling point 20.3 K. The density of $H_2(l)$ at its boiling point is 0.067 g·mL^{-1}.

17-32. Because the molecules in an ideal gas are independent, the partition function of a mixture of monatomic ideal gases is of the form

$$Q(N_1, N_2, V, T) = \frac{[q_1(V, T)]^{N_1}}{N_1!} \frac{[q_2(V, T)]^{N_2}}{N_2!}$$

where

$$q_j(V, T) = \left(\frac{2\pi m_j k_B T}{h^2}\right)^{3/2} V \qquad j = 1, 2$$

Show that

$$\langle E \rangle = \frac{3}{2}(N_1 + N_2)k_B T$$

and that

$$PV = (N_1 + N_2)k_B T$$

for a mixture of monatomic ideal gases.

17-33. We will learn in Chapter 18 that the rotational partition function of an asymmetric top molecule is given by

$$q_{rot}(T) = \frac{\pi^{1/2}}{\sigma}\left(\frac{8\pi^2 I_A k_B T}{h^2}\right)^{1/2}\left(\frac{8\pi^2 I_B k_B T}{h^2}\right)^{1/2}\left(\frac{8\pi^2 I_C k_B T}{h^2}\right)^{1/2}$$

where σ is a constant and I_A, I_B, and I_C are the three (distinct) moments of inertia. Show that the rotational contribution to the molar heat capacity is $\overline{C}_{V,rot} = \frac{3}{2}R$.

17-34. The allowed energies of a harmonic oscillator are given by $\varepsilon_v = (v + \frac{1}{2})h\nu$. The corresponding partition function is given by

$$q_{\text{vib}}(T) = \sum_{v=0}^{\infty} e^{-(v+\frac{1}{2})h\nu/k_B T}$$

Let $x = e^{-h\nu/k_B T}$ and use the formula for the summation of a geometric series (Problem 17–20) to show that

$$q_{\text{vib}}(T) = \frac{e^{-h\nu/2k_B T}}{1 - e^{-h\nu/k_B T}}$$

17-35. Derive an expression for the probability that a harmonic oscillator will be found in the vth state. Calculate the probability that the first few vibrational states are occupied for HCl(g) at 300 K. (See Table 5–1 and Problem 17–34.)

17-36. Show that the fraction of harmonic oscillators in the ground vibrational state is given by

$$f_0 = 1 - e^{-h\nu/k_B T}$$

Calculate f_0 for N_2(g) at 300 K, 600 K, and 1000 K (see Table 5.1).

17-37. Use Equation 17.55 to show that the fraction of rigid rotators in the Jth rotational level is given by

$$f_J = \frac{(2J + 1)e^{-\hbar^2 J(J+1)/2Ik_B T}}{q_{\text{rot}}(T)}$$

Plot the fraction in the Jth level relative to the $J = 0$ level (f_J/f_0) against J for HCl(g) at 300 K. Take $\tilde{B} = 10.44$ cm^{-1}.

17-38. Equations 17.20 and 17.21 give the ensemble average of E, which we assert is the same as the experimentally observed value. In this problem, we will explore the standard deviation about $\langle E \rangle$ (MathChapter B). We start with either Equation 17.20 or 17.21:

$$\langle E \rangle = U = -\left(\frac{\partial \ln Q}{\partial \beta}\right)_{N,V} = k_B T^2 \left(\frac{\partial \ln Q}{\partial T}\right)_{N,V}$$

Differentiate again with respect to β or T to show that (MathChapter B)

$$\sigma_E^2 = \langle E^2 \rangle - \langle E \rangle^2 = k_B T^2 C_V$$

where C_V is the heat capacity. To explore the relative magnitude of the spread about $\langle E \rangle$, consider

$$\frac{\sigma_E}{\langle E \rangle} = \frac{(k_B T^2 C_V)^{1/2}}{\langle E \rangle}$$

To get an idea of the size of this ratio, use the values of $\langle E \rangle$ and C_V for a (monatomic) ideal gas, namely, $\frac{3}{2} N k_B T$ and $\frac{3}{2} N k_B$, respectively, and show that $\sigma_E/\langle E \rangle$ goes as $N^{-1/2}$. What does this trend say about the likely observed deviations from the average macroscopic energy?

17-39. Following Problem 17–38, show that the variance about the average values of a *molecular* energy is given by

$$\sigma_\varepsilon^2 = \langle \varepsilon^2 \rangle - \langle \varepsilon \rangle^2 = \frac{k_B T^2 C_V}{N}$$

and that $\sigma_\varepsilon / \langle \varepsilon \rangle$ goes as order unity. What does this result say about the deviations from the average molecular energy?

17-40. Use the result of Problem 17–38 to show that C_V is never negative.

17-41. The lowest electronic states of Na(g) are tabulated below.

Term symbol	Energy/cm^{-1}	Degeneracy
$^2S_{1/2}$	0.000	2
$^2P_{1/2}$	16 956.183	2
$^2P_{3/2}$	16 973.379	4
$^2S_{1/2}$	25 739.86	2

Calculate the fraction of the atoms in each of these electronic states in a sample of Na(g) at 1000 K. Repeat this calculation for a temperature of 2500 K.

17-42. The vibrational frequency of NaCl(g) is 159.23 cm^{-1}. Calculate the molar heat capacity, $\overline{C}_V$, at 1000 K. (See Equation 17.27.)

17-43. The energies and degeneracies of the two lowest electronic states of atomic iodine are listed below.

Energy/cm^{-1}	Degeneracy
0	4
7603.2	2

What temperature is required so that 2% of the atoms are in the excited state?

SERIES AND LIMITS

Frequently, we need to investigate the behavior of an equation for small values (or perhaps large values) of one of the variables in the equation. For example, we might want to know the low-frequency behavior of the Planck distribution law for blackbody radiation (Equation 1.2):

$$\rho_\nu(T)d\nu = \frac{8\pi h}{c^3}\frac{\nu^3 d\nu}{e^{\beta h\nu}-1} \tag{I.1}$$

To do this, we first have to use the fact that e^x can be written as the infinite series (i.e., a series containing an unending number of terms)

$$e^x = \sum_{n=0}^{\infty}\frac{x^n}{n!} = 1 + x + \frac{x^2}{2!} + \frac{x^3}{3!} + \cdots \tag{I.2}$$

and then realize that if x is small, then x^2, x^3, etc. are even smaller. We can express this result by writing

$$e^x = 1 + x + O(x^2)$$

where $O(x^2)$ is a bookkeeping symbol that reminds us we are neglecting terms involving x^2 and higher powers of x. If we apply this result to Equation I.1, we have

$$\begin{aligned}\rho_\nu(T)d\nu &= \frac{8\pi h}{c^3}\frac{\nu^3 d\nu}{1 + \beta h\nu + O[(\beta h\nu)^2] - 1}\\ &\approx \frac{8\pi h}{c^3}\frac{\nu^3 d\nu}{\beta h\nu}\\ &= \frac{8\pi k_B T}{c^3}\nu^2 d\nu\end{aligned}$$

Thus, we see that $\rho_\nu(T)$ goes as ν^2 for small values of ν. In this MathChapter, we will review some useful series and apply them to some physical problems.

723

One of the most useful series we will use is the geometric series:

$$\frac{1}{1-x} = \sum_{n=0}^{\infty} x^n = 1 + x + x^2 + x^3 + \cdots \qquad |x| < 1 \qquad (\text{I.3})$$

This result can be derived by algebraically dividing 1 by $1 - x$, or by the following trick. Consider the finite series (i.e., a series with a finite number of terms)

$$S_N = 1 + x + x^2 + \cdots + x^N$$

Now multiply S_N by x:

$$xS_N = x + x^2 + \cdots + x^{N+1}$$

Now notice that

$$S_N - xS_N = 1 - x^{N+1}$$

or that

$$S_N = \frac{1 - x^{N+1}}{1 - x} \qquad (\text{I.4})$$

If $|x| < 1$, then $x^{N+1} \to 0$ as $N \to \infty$, so we recover Equation I.3.

Recovering Equation I.3 from Equation I.4 brings us to an important point regarding infinite series: Equation I.3 is valid only if $|x| < 1$. It makes no sense at all if $|x| \geq 1$. We say that the infinite series in Equation I.3 converges for $|x| < 1$ and diverges for $|x| \geq 1$. How can we tell whether a given infinite series converges or diverges? There are a number of so-called convergence tests, but one simple and useful one is the *ratio test*. To apply the ratio test, we form the ratio of the $(n + 1)$th term, u_{n+1}, to the nth term, u_n, and then let n become very large:

$$r = \lim_{n \to \infty} \left| \frac{u_{n+1}}{u_n} \right| \qquad (\text{I.5})$$

If $r < 1$, the series converges; if $r > 1$, the series diverges; and if $r = 1$, the test is inconclusive. Let's apply this test to the geometric series (Equation I.3). In this case, $u_{n+1} = x^{n+1}$ and $u_n = x^n$, so

$$r = \lim_{n \to \infty} \left| \frac{x^{n+1}}{x^n} \right| = |x|$$

Thus, we see that the series converges if $|x| < 1$ and diverges if $|x| > 1$. It actually diverges at $x = 1$, but the ratio test does not tell us that. We would have to use a more sophisticated convergence test to determine the behavior at $x = 1$.

For the exponential series (Equation I.2), we have

$$r = \lim_{n \to \infty} \left| \frac{x^{n+1}/(n+1)!}{x^n/n!} \right| = \lim_{n \to \infty} \left| \frac{x}{n+1} \right|$$

Thus, we conclude that the exponential series converges for all values of x.

In Chapter 18, we encounter the summation

$$S = \sum_{v=0}^{\infty} e^{-vh\nu/k_B T} \tag{I.6}$$

where ν represents the vibrational frequency of a diatomic molecule and the other symbols have their usual meanings. We can sum this series by letting

$$x = e^{-h\nu/k_B T}$$

in which case we have

$$S = \sum_{v=0}^{\infty} x^v$$

The quantity x is less than 1, and according to Equation I.3, $S = 1/(1-x)$, or

$$S = \frac{1}{1 - e^{-h\nu/k_B T}} \tag{I.7}$$

We say that S has been evaluated in closed form because its numerical evaluation requires only a finite number of steps, in contrast to Equation I.6, which would require an infinite number of steps.

A practical question that arises is how do we find the infinite series that corresponds to a given function. For example, how do we derive Equation I.2? First, assume that the function $f(x)$ can be expressed as a power series in x:

$$f(x) = c_0 + c_1 x + c_2 x^2 + c_3 x^3 + \cdots$$

where the c_j are to be determined. Then let $x = 0$ and find that $c_0 = f(0)$. Now differentiate once with respect to x

$$\frac{df}{dx} = c_1 + 2c_2 x + 3c_3 x_2 + \cdots$$

and let $x = 0$ to find that $c_1 = (df/dx)_{x=0}$. Differentiate again,

$$\frac{d^2 f}{dx^2} = 2c_2 + 3 \cdot 2c_3 x + \cdots$$

and let $x = 0$ to get $c_2 = (d^2 f/dx^2)_{x=0}/2$. Differentiate once more,

$$\frac{d^3 f}{dx^3} = 3 \cdot 2c_3 + 4 \cdot 3 \cdot 2x + \cdots$$

and let $x = 0$ to get $c_3 = (d^3 f/dx^3)_{x=0}/3!$. The general result is

$$c_n = \frac{1}{n!} \left(\frac{d^n f}{dx^n} \right)_{x=0} \tag{I.8}$$

so we can write

$$f(x) = f(0) + \left(\frac{df}{dx}\right)_{x=0} x + \frac{1}{2!}\left(\frac{d^2 f}{dx^2}\right)_{x=0} x^2 + \frac{1}{3!}\left(\frac{d^3 f}{dx^3}\right)_{x=0} x^3 + \cdots \quad \text{(I.9)}$$

Equation I.9 is called the Maclaurin series of $f(x)$. If we apply Equation I.9 to $f(x) = e^x$, we find that

$$\left(\frac{d^n e^x}{dx^n}\right)_{x=0} = 1$$

so

$$e^x = 1 + x + \frac{x^2}{2!} + \frac{x^3}{3!} + \cdots$$

Some other important Maclaurin series, which can be obtained from a straightforward application of Equation I.9 (Problem I–7) are

$$\sin x = x - \frac{x^3}{3!} + \frac{x^5}{5!} - \frac{x^7}{7!} + \cdots \quad \text{(I.10)}$$

$$\cos x = 1 - \frac{x^2}{2!} + \frac{x^4}{4!} - \frac{x^6}{6!} + \cdots \quad \text{(I.11)}$$

$$\ln(1 + x) = x - \frac{x^2}{2} + \frac{x^3}{3} - \frac{x^4}{4} + \cdots \quad -1 < x \le 1 \quad \text{(I.12)}$$

and

$$(1 + x)^n = 1 + nx + \frac{n(n-1)}{2!}x^2 + \frac{n(n-1)(n-2)}{3!}x^3 + \cdots \quad x^2 < 1 \quad \text{(I.13)}$$

Series I.10 and I.11 converge for all values of x, but as indicated, Series I.12 converges only for $-1 < x \le 1$ and Series I.13 converges only for $x^2 < 1$. Note that if n is a positive integer in Series I.13, the series truncates. For example, if $n = 2$ or 3, we have

$$(1 + x)^2 = 1 + 2x + x^2$$

and

$$(1 + x)^3 = 1 + 3x + 3x^2 + x^3$$

Equation I.13 for a positive integer is called the binomial expansion. If n is not a positive integer, the series continues indefinitely, and Equation I.13 is called the binomial series. Any handbook of mathematical tables will have the Maclaurin series for many functions. Problem I–13 discusses a Taylor series, which is an extension of a Maclaurin series.

We can use the series presented here to derive a number of results used throughout the book. For example, the limit

$$\lim_{x \to 0} \frac{\sin x}{x}$$

occurs several times. Because this limit gives 0/0, we could use l'Hôpital's rule, which tells us that

$$\lim_{x \to 0} \frac{\sin x}{x} = \lim_{x \to 0} \frac{\dfrac{d \sin x}{dx}}{\dfrac{dx}{dx}} = \lim_{x \to 0} \cos x = 1$$

We could derive the same result by dividing Equation I.10 by x and then letting $x \to 0$. (These two methods are really equivalent. See Problem I–14.)

We will do one final example involving series and limits. According to a theory by Debye, the temperature dependence of the molar heat capacity of a crystal is given by

$$\overline{C}_V(T) = 9R \left(\frac{T}{\Theta_D} \right)^3 \int_0^{\Theta_D/T} \frac{x^4 e^x dx}{(e^x - 1)^2} \tag{I.14}$$

In this equation, T is the kelvin temperature, R is the molar gas constant, and Θ_D is a parameter characteristic of the particular crystal. The parameter Θ_D has units of temperature and is called the Debye temperature of the crystal. We want to determine both the low-temperature and the high-temperature limits of $\overline{C}_V(T)$. In the low-temperature limit, the upper limit of the integral becomes very large. For large values of x, we can neglect 1 compared with e^x in the denominator of the integrand, showing that the integrand goes as $x^4 e^{-x}$ for large x. But $x^4 e^{-x} \to 0$ as $x \to \infty$, so the upper limit of the integral can safely be set to ∞, giving

$$\lim_{T \to 0} \overline{C}_V(T) = 9R \left(\frac{T}{\Theta_D} \right)^3 \int_0^\infty \frac{x^4 e^x dx}{(e^x - 1)^2}$$

Whatever the value of the integral here, it is just a constant, so we see that

$$\overline{C}_V(T) \to \text{constant} \times T^3 \quad \text{as} \quad T \to 0$$

This famous result for the low-temperature heat capacity of a crystal is called the T^3 law. The low-temperature heat capacity goes to zero as T^3. We will use the T^3 law in Chapter 21.

Now let's look at the high-temperature limit. For high temperatures, the upper limit of the integral in Equation I.14 becomes very small. Consequently, during the integration from 0 to Θ_D/T, x is always small. Therefore, we can use Equation I.2 for e^x, giving

$$\lim_{T \to \infty} \overline{C}_V(T) = 9R \left(\frac{T}{\Theta_D} \right)^3 \int_0^{\Theta_D/T} \frac{x^4 [1 + x + O(x^2)] dx}{[1 + x + O(x^2) - 1]^2}$$

$$= 9R \left(\frac{T}{\Theta_D} \right)^3 \int_0^{\Theta_D/T} x^2 dx$$

$$= 9R \left(\frac{T}{\Theta_D} \right)^3 \cdot \frac{1}{3} \left(\frac{\Theta_D}{T} \right)^3 = 3R$$

This result is called the Law of Dulong and Petit; the molar heat capacity of a crystal becomes $3R = 24.9$ J·K^{-1}·mol^{-1} for monatomic crystals at high temperatures. By "high temperatures", we actually mean that $T \gg \Theta_D$, which for many substances is less than 1000 K.

Problems

I-1. Calculate the percentage difference between e^x and $1 + x$ for $x = 0.0050$, 0.0100, $0.0150, \ldots, 0.1000$.

I-2. Calculate the percentage difference between $\ln(1 + x)$ and x for $x = 0.0050$, 0.0100, $0.0150, \ldots, 0.1000$.

I-3. Write out the expansion of $(1 + x)^{1/2}$ through the quadratic term.

I-4. Evaluate the series

$$S = \sum_{v=0}^{\infty} e^{-(v+\frac{1}{2})\beta h v}$$

I-5. Show that

$$\frac{1}{(1-x)^2} = 1 + 2x + 3x^2 + 4x^3 + \cdots$$

I-6. Evaluate the series

$$S = \frac{1}{2} + \frac{1}{4} + \frac{1}{8} + \frac{1}{16} + \cdots$$

I-7. Use Equation I.9 to derive Equations I.10 and I.11.

I-8. Show that Equations I.2, I.10, and I.11 are consistent with the relation $e^{ix} = \cos x + i \sin x$.

I-9. In Example 17–3, we derived a simple formula for the molar heat capacity of a solid based on a model by Einstein:

$$\overline{C}_V = 3R \left(\frac{\Theta_E}{T}\right)^2 \frac{e^{-\Theta_E/T}}{(1 - e^{-\Theta_E/T})^2}$$

where R is the molar gas constant and $\Theta_E = h v / k_B$ is a constant, called the Einstein constant, that is characteristic of the solid. Show that this equation gives the Dulong and Petit limit ($\overline{C}_V \to 3R$) at high temperatures.

I-10. Evaluate the limit of

$$f(x) = \frac{e^{-x} \sin^2 x}{x^2}$$

as $x \to 0$.

I-11. Evaluate the integral

$$I = \int_0^a x^2 e^{-x} \cos^2 x \, dx$$

for small values of a by expanding I in powers of a through quadratic terms.

I-12. Prove that the series for $\sin x$ converges for all values of x.

I-13. A Maclaurin series is an expansion about the point $x = 0$. A series of the form

$$f(x) = c_0 + c_1(x - x_0) + c_2(x - x_0)^2 + \cdots$$

is an expansion about the point x_0 and is called a Taylor series. First show that $c_0 = f(x_0)$. Now differentiate both sides of the above expansion with respect to x and then let $x = x_0$ to show that $c_1 = (df/dx)_{x=x_0}$. Now show that

$$c_n = \frac{1}{n!} \left(\frac{d^n f}{dx^n} \right)_{x=x_0}$$

and so

$$f(x) = f(x_0) + \left(\frac{df}{dx} \right)_{x_0} (x - x_0) + \frac{1}{2} \left(\frac{d^2 f}{dx^2} \right)_{x_0} (x - x_0)^2 + \cdots$$

I-14. Show that l'Hôpital's rule amounts to forming a Taylor expansion of both the numerator and the denominator. Evaluate the limit

$$\lim_{x \to 0} \frac{\ln(1 + x) - x}{x^2}$$

both ways.

I-15. In Problem 18–45, we will need to sum the series

$$s_1 = \sum_{v=0}^{\infty} v x^v$$

and

$$s_2 = \sum_{v=0}^{\infty} v^2 x^v$$

To sum the first one, start with (Equation I.3)

$$s_0 = \sum_{v=0}^{\infty} x^v = \frac{1}{1 - x}$$

Differentiate with respect to x and then multiply by x to obtain

$$s_1 = \sum_{v=0}^{\infty} v x^v = x \frac{ds_0}{dx} = x \frac{d}{dx} \left(\frac{1}{1 - x} \right) = \frac{x}{(1 - x)^2}$$

Using the same approach, show that

$$s_2 = \sum_{v=0}^{\infty} v^2 x^v = \frac{x + x^2}{(1 - x)^3}$$

William Francis Giauque was born on May 12, 1895, in Niagara Falls, Ontario, Canada, to American parents and died in 1982. After working for two years in the laboratory at Hooker Electro-Chemical Company in Niagara Falls, he entered the University of California at Berkeley with the intent of becoming a chemical engineer. He decided to study chemistry, however, and remained at Berkeley to receive his Ph.D. in chemistry with a minor in physics in 1922. His dissertation was on the behavior of materials at very low temperatures. Upon receiving his Ph.D., Giauque accepted a faculty position in the College of Chemistry at Berkeley and remained there for the rest of his life. He made exhaustive and meticulous thermochemical studies that explored the Third Law of Thermodynamics. In particular, his very low temperature studies of the entropies of substances validated the Third Law. Giauque developed the technique of adiabatic demagnetization to achieve low temperatures. He achieved a temperature of 0.25 K, and other research groups subsequently reached temperatures as low as 0.0014 K using Giauque's technique. Together with his graduate student Herrick Johnston, he spectroscopically identified the two hitherto unknown oxygen isotopes 17 and 18 in 1929. He was awarded the Nobel Prize for chemistry in 1949 "for his contributions in the field of chemical thermodynamics, particularly concerning the behavior of substances at extremely low temperatures."

CHAPTER 18

Partition Functions and Ideal Gases

In this chapter, we will apply the general results of the preceding chapter to calculate the partition functions and heat capacities of ideal gases. We have shown in Section 17–7 that if the number of available quantum states is much greater than the number of particles, we can write the partition function of the entire system in terms of the individual atomic or molecular partition functions:

$$Q(N, V, T) = \frac{[q(V, T)]^N}{N!}$$

This equation is particularly applicable to ideal gases because the molecules are independent and the densities of gases that behave ideally are low enough that the inequality given by Equation 17.40 is satisfied. We will discuss a monatomic ideal gas first and then diatomic and polyatomic ideal gases.

18–1. The Translational Partition Function of an Atom in a Monatomic Ideal Gas is $(2\pi m k_B T / h^2)^{3/2} V$

The energy of an atom in an ideal monatomic gas can be written as the sum of its translational energy and its electronic energy

$$\varepsilon_{atomic} = \varepsilon_{trans} + \varepsilon_{elec}$$

so the atomic partition function can be written as

$$q(V, T) = q_{trans}(V, T) q_{elec}(T) \tag{18.1}$$

We will evaluate the translational partition function first.

731

The translational energy states in a cubic container are given by (Section 3–9)

$$\varepsilon_{n_x n_y n_z} = \frac{h^2}{8ma^2} \left(n_x^2 + n_y^2 + n_z^2\right) \quad n_x, n_y, n_z = 1, \ 2, \ \ldots \tag{18.2}$$

We substitute Equation 18.2 into q_{trans} (Equation 17.47) to get

$$q_{\text{trans}} = \sum_{n_x, n_y, n_z = 1}^{\infty} e^{-\beta \varepsilon_{n_x n_y n_z}} = \sum_{n_x=1}^{\infty} \sum_{n_y=1}^{\infty} \sum_{n_z=1}^{\infty} \exp\left[-\frac{\beta h^2}{8ma^2} \left(n_x^2 + n_y^2 + n_z^2\right)\right] \tag{18.3}$$

Because $e^{a+b+c} = e^a e^b e^c$, we can write the triple summation as a product of three single summations:

$$q_{\text{trans}} = \sum_{n_x=1}^{\infty} \exp\left(-\frac{\beta h^2 n_x^2}{8ma^2}\right) \sum_{n_y=1}^{\infty} \exp\left(-\frac{\beta h^2 n_y^2}{8ma^2}\right) \sum_{n_z=1}^{\infty} \exp\left(-\frac{\beta h^2 n_z^2}{8ma^2}\right)$$

Now, each of these three single summations is alike, because each one is simply

$$\sum_{n=1}^{\infty} \exp\left(-\frac{\beta h^2 n^2}{8ma^2}\right) = e^{-\beta h^2/8ma^2} + e^{-4\beta h^2/8ma^2} + e^{-9\beta h^2/8ma^2} + \cdots$$

Thus, we can write Equation 18.3 as

$$q_{\text{trans}}(V, T) = \left[\sum_{n=1}^{\infty} \exp\left(-\frac{\beta h^2 n^2}{8ma^2}\right)\right]^3 \tag{18.4}$$

This summation cannot be expressed in terms of any simple analytic function. This situation does not present any difficulty, however, for the following reason. Graphically, a summation such as $\sum_{n=1}^{\infty} f_n$ is equal to the sum of the areas under rectangles of unit width centered at 1, 2, 3, ... and of height f_1, f_2, f_3, ... as shown in Figure 18.1. If the heights of successive rectangles differ by a very small amount, the area of the rectangles is essentially equal to the area under the continuous curve obtained by letting the summation index n be a continuous variable (Figure 18.1). Problem 18–2 helps you prove that the successive terms in the summation in Equation 18.4 do indeed differ very little from each other under most conditions.

Thus, it is an excellent approximation to replace the summation in Equation 18.4 by an integration:

$$q_{\text{trans}}(V, T) = \left(\int_0^{\infty} e^{-\beta h^2 n^2/8ma^2} \, dn\right)^3 \tag{18.5}$$

Note that the integral starts at $n = 0$, whereas the summation in Equation 18.4 starts at $n = 1$. For the small values of $\beta h^2/8ma^2$ we are considering here, the difference is negligible (Problem 18–41). If we denote $\beta h^2/8ma^2$ by α, the above integral becomes (see MathChapter B)

$$\int_0^{\infty} e^{-\alpha n^2} \, dn = \left(\frac{\pi}{4\alpha}\right)^{1/2}$$

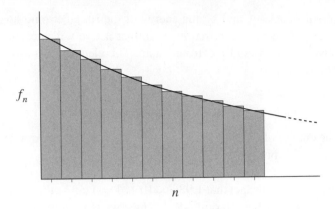

FIGURE 18.1
An illustration of the approximation of a summation $\sum_{n=1}^{\infty} f_n$ by an integral. The summation is equal to the areas of the rectangles and the integral is equal to the area under the curve obtained by letting n be a continuous variable.

so we have that

$$q_{\text{trans}}(V, T) = \left(\frac{2\pi m k_{\text{B}} T}{h^2} \right)^{3/2} V \tag{18.6}$$

where we have written V for a^3. Note that q_{trans} is a function of V and T.

We can calculate the average translational energy of an ideal-gas atom from this partition function by using Equation 17.51:

$$
\begin{aligned}
\langle \varepsilon_{\text{trans}} \rangle &= k_{\text{B}} T^2 \left(\frac{\partial \ln q_{\text{trans}}}{\partial T} \right)_V \\
&= k_{\text{B}} T^2 \left(\frac{\partial}{\partial T} \left[\frac{3}{2} \ln T + \text{terms independent of } T \right] \right)_V \\
&= \frac{3}{2} k_{\text{B}} T
\end{aligned}
\tag{18.7}
$$

in agreement with what we found in Section 17–3.

18–2. Most Atoms Are in the Ground Electronic State at Room Temperature

In this section, we will investigate the electronic contributions to $q(V, T)$. It is more convenient to write the electronic partition function as a sum over levels rather than a sum over states (Section 17–8), so we write

$$q_{\text{elec}} = \sum_i g_{ei} e^{-\beta \varepsilon_{ei}} \tag{18.8}$$

where g_{ei} is the degeneracy, and ε_{ei} the energy of the ith electronic level. We first fix the arbitrary zero of energy such that $\varepsilon_{e1} = 0$; that is, we will measure all electronic energies relative to the ground electronic state. The electronic contribution to q can then be written as

$$q_{\text{elec}}(T) = g_{e1} + g_{e2}e^{-\beta\varepsilon_{e2}} + \cdots \tag{18.9}$$

where ε_{ej} is the energy of the jth electronic level relative to the ground state. Note that q_{elec} is a function of T but not of V.

As we have seen in Chapter 8, these ε's are typically of the order of tens of thousands of wave numbers. Using the fact that 1.986×10^{-23} J $= 1$ cm^{-1}, the Boltzmann constant in wave numbers is $k_B = 0.6950$ cm$^{-1}\cdot$K^{-1}. Thus, we see that typically

$$\beta\varepsilon_{\text{elec}} \approx \frac{40\,000 \text{ cm}^{-1}}{0.6950 \text{ cm}^{-1}\cdot\text{K}^{-1}}\frac{1}{T} \approx \frac{10^4 \text{ K}}{T}$$

which is equal to 10 even for $T = 1000$ K. Therefore, $e^{-\beta\varepsilon_{e2}}$ in Equation 18.9 typically is around 10^{-5} for most atoms at ordinary temperatures, so only the first term in the summation for q_{elec} is significantly different from zero. There are some cases, however, such as the halogen atoms, for which the first excited state lies only a few hundred wave numbers above the ground state, so that several terms in q_{elec} are necessary. Even in these cases, the sum in Equation 18.9 converges very rapidly.

As we learned in Chapter 8, the electronic energies of atoms and ions are determined by atomic spectroscopy and are well tabulated. The standard reference, "Moore's tables," lists the energy levels and energies of many atoms and ions. Table 18.1 lists the first few levels for H, He, Li, and F.

We can make some general observations from tables like Table 18.1. The noble gas atoms have a $^1S_0$ ground state with the first excited state of order of 10^5 cm^{-1} or higher; alkali metal atoms have a $^2S_{1/2}$ ground state with the next state of order of 10^4 cm^{-1} or higher; halogen atoms have a $^2P_{3/2}$ ground state with the next state, a $^2P_{1/2}$ state, only of order of 10^2 cm^{-1} higher. Thus, at ordinary temperatures, the electronic partition function of noble gas atoms is essentially unity and that of alkali metal atoms is two, while those for halogen atoms consist of two terms.

Using the data in Table 18.1, we can now calculate the fraction of helium atoms in the first triplet state $^3S_1$. This fraction is given by

$$\begin{aligned}
f_2 &= \frac{g_{e2}e^{-\beta\varepsilon_{e2}}}{q_{\text{elec}}(T)} \\
&= \frac{g_{e2}e^{-\beta\varepsilon_{e2}}}{g_{e1} + g_{e2}e^{-\beta\varepsilon_{e2}} + g_{e3}e^{-\beta\varepsilon_{e3}} + \cdots} \\
&= \frac{3e^{-\beta\varepsilon_{e2}}}{1 + 3e^{-\beta\varepsilon_{e2}} + e^{-\beta\varepsilon_{e3}} + \cdots}
\end{aligned} \tag{18.10}$$

At 300 K, $\beta\varepsilon_{e2} = 770$, so $f_2 \approx 10^{-334}$. Even at 3000 K, $f_2 \approx 10^{-33}$. This is typical of the noble gases. The energy separation between the ground and excited levels must

TABLE 18.1

Some atomic energy levels.[a]

Atom	Electron configuration	Term symbol	Degeneracy $g_e = 2J + 1$	energy/cm^{-1}
H	$1s$	$^2S_{1/2}$	2	0.
	$2p$	$^2P_{1/2}$	2	82 258.907
	$2s$	$^2S_{1/2}$	2	82 258.942
	$2p$	$^2P_{3/2}$	4	82 259.272
He	$1s^2$	$^1S_0$	1	0.
	$1s2s$	$^3S_1$	3	159 850.318
		$^1S_0$	1	166 271.70
Li	$1s^22s$	$^2S_{1/2}$	2	0.
	$1s^22p$	$^2P_{1/2}$	2	14 903.66
		$^2P_{3/2}$	4	14 904.00
	$1s^23s$	$^2S_{1/2}$	2	27 206.12
F	$1s^22s^22p^5$	$^2P_{3/2}$	4	0.
		$^2P_{1/2}$	2	404.0
	$1s^22s^22p^43s$	$^4P_{5/2}$	6	102 406.50
		$^4P_{3/2}$	4	102 681.24
		$^4P_{1/2}$	2	102 841.20
		$^2P_{3/2}$	4	104 731.86
		$^2P_{1/2}$	2	105 057.10

[a] From C.E. Moore, "Atomic Energy Levels" *Natl. Bur. Std, Circ.* 1 467, U.S. Government Printing Office, Washington D.C., 1949

be less than a few hundred cm^{-1} or so before any population of the excited level is significant.

EXAMPLE 18–1

Using the data in Table 18.1, calculate the fraction of fluorine atoms in the first excited state at 300 K, 1000 K, and 2000 K.

SOLUTION: Using the second line of Equation 18.10 with $g_{e1} = 4$, $g_{e2} = 2$, and $g_{e3} = 6$, we have

$$f_2 = \frac{2e^{-\beta\varepsilon_{e2}}}{4 + 2e^{-\beta\varepsilon_{e2}} + 6e^{-\beta\varepsilon_{e3}} + \cdots}$$

with $\varepsilon_{e2} = 404.0$ cm^{-1} and $\varepsilon_{e3} = 102\,406.50$ cm^{-1}. We also have

$$\beta\varepsilon_{e2} = \frac{404.0 \text{ cm}^{-1}}{(0.6950 \text{ cm}^{-1}\cdot\text{K}^{-1})T} = \frac{581.3 \text{ K}}{T}$$

and

$$\beta\varepsilon_{e3} = \frac{147\,300 \text{ K}}{T}$$

Clearly, we can neglect the third term in the denominator of f_2.
   The value of f_2 for the various temperatures is

$$f_2(T = 300 \text{ K}) = \frac{2e^{-581/300}}{4 + 2e^{-581/300}} = 0.0672$$

$$f_2(T = 1000 \text{ K}) = \frac{2e^{-581/1000}}{4 + 2e^{-581/1000}} = 0.219$$

$$f_2(T = 2000 \text{ K}) = 0.272$$

Thus, the population of the first excited state is significant at these temperatures and so the first two terms of the summation in Equation 18.9 must be evaluated in determining $q_{\text{elec}}(T)$.

For most atoms and molecules, the first two terms of the electronic partition function are sufficient, or

$$q_{\text{elec}}(T) \approx g_{e1} + g_{e2}e^{-\beta\varepsilon_{e2}} \tag{18.11}$$

At temperatures at which the second term is not negligible with respect to the first term, we must check the possible contribution of higher terms as well.
   This completes our discussion of the partition function of monatomic ideal gases. In summary, we have

$$Q(N, V, T) = \frac{(q_{\text{trans}}q_{\text{elec}})^N}{N!} \tag{18.12}$$

where

$$q_{\text{trans}}(V, T) = \left(\frac{2\pi m k_{\text{B}} T}{h^2}\right)^{3/2} V \tag{18.13}$$

$$q_{\text{elec}}(T) = g_{e1} + g_{e2}e^{-\beta\varepsilon_{e2}} + \cdots$$

We can now calculate some of the properties of a monatomic ideal gas. The average energy is

$$U = k_{\text{B}}T^2\left(\frac{\partial \ln Q}{\partial T}\right)_{N,V} = Nk_{\text{B}}T^2\left(\frac{\partial \ln q}{\partial T}\right)_V = \frac{3}{2}Nk_{\text{B}}T + \frac{Ng_{e2}\varepsilon_{e2}e^{-\beta\varepsilon_{e2}}}{q_{\text{elec}}} + \cdots \tag{18.14}$$

The first term represents the average kinetic energy, and the second term represents the average electronic energy (in excess of the ground-state energy). The contribution of the electronic degrees of freedom to the average energy is small at ordinary temperatures. If we ignore the very small contribution from the electronic degrees of freedom, the molar heat capacity at constant volume is given by

$$\overline{C}_V = \left(\frac{d\overline{U}}{dT}\right)_{N,V} = \frac{3}{2}R$$

The pressure is

$$P = k_{\mathrm{B}}T\left(\frac{\partial \ln Q}{\partial V}\right)_{N,T} = Nk_{\mathrm{B}}T\left(\frac{\partial \ln q}{\partial V}\right)_T$$

$$= Nk_{\mathrm{B}}T\left[\frac{\partial}{\partial V}(\ln V + \text{terms not involving } V)\right]_T$$

$$= \frac{Nk_{\mathrm{B}}T}{V} \tag{18.15}$$

which is the ideal gas equation of state. Note that Equation 18.15 results because $q(V, T)$ is of the form $f(T)V$, and only the translational energy of the atoms contributes to the pressure. This is expected intuitively, because the pressure is due to bombardment of the walls of the container by the atoms and molecules of the gas.

In the next few sections, we will treat a diatomic ideal gas. In addition to translational and electronic degrees of freedom, diatomic molecules also possess vibrational and rotational degrees of freedom. The general procedure would be to set up the Schrödinger equation for two nuclei and n electrons and to solve this equation for the set of eigenvalues of the diatomic molecule. Fortunately, a series of very good approximations can be used to reduce this complicated two-nuclei, n-electron problem to a set of simpler problems. The simplest of these approximations is the rigid rotator-harmonic oscillator approximation, which we described in Chapters 5 and 13. We will set up this approximation in the next section and then discuss the vibrational and rotational partition functions within this approximation in Sections 18–4 and 18–5.

18–3. The Energy of a Diatomic Molecule Can Be Approximated as a Sum of Separate Terms

When treating diatomic or polyatomic molecules, we use the rigid rotator-harmonic oscillator approximation (Section 13–2). In this case, we can write the total energy of the molecule as a sum of its translational, rotational, vibrational, and electronic energies:

$$\varepsilon = \varepsilon_{\mathrm{trans}} + \varepsilon_{\mathrm{rot}} + \varepsilon_{\mathrm{vib}} + \varepsilon_{\mathrm{elec}} \tag{18.16}$$

737

As for a monatomic ideal gas, the inequality given by Equation 17.40 is easily satisfied at normal temperatures, and so we can write

$$Q(N, V, T) = \frac{[q(V, T)]^N}{N!} \tag{18.17}$$

Furthermore, Equation 18.16 allows us to write

$$q(V, T) = q_{\text{trans}} q_{\text{rot}} q_{\text{vib}} q_{\text{elec}} \tag{18.18}$$

so the partition function of a molecular ideal gas is given by

$$Q(N, V, T) = \frac{\left(q_{\text{trans}} q_{\text{rot}} q_{\text{vib}} q_{\text{elec}}\right)^N}{N!} \tag{18.19}$$

The translational partition function of a diatomic molecule is similar to the result we found in Section 18–1 for an atom:

$$q_{\text{trans}}(V, T) = \left[\frac{2\pi (m_1 + m_2) k_B T}{h^2}\right]^{3/2} V \tag{18.20}$$

Note that Equation 18.20 is essentially the same as Equation 18.6. The electronic partition function will be similar to Equation 18.9. We will discuss the vibrational and rotational contributions to the partition function in the next two sections. Although Equation 18.19 is not exact, it is often a good approximation, particularly for small molecules.

Before we consider q_{rot} and q_{vib}, we must choose a zero of energy for the rotational, vibrational, and electronic states. The natural choice for the zero of rotational energy is the $J = 0$ state, where the rotational energy is zero. In the vibrational case, however, we have two sensible choices. One is to take the zero of vibrational energy to be that of the ground state, and the other is take the zero to be the bottom of the internuclear potential well. In the first case, the energy of the ground vibrational state is zero, and in the second case it is $h\nu/2$. We will choose the zero of vibrational energy to be the bottom of the internuclear potential well of the lowest electronic state, so the energy of the ground vibrational state will be $h\nu/2$.

Last, we take the zero of the electronic energy to be the separated atoms at rest in their ground electronic states (see Figure 18.2). Recall that the depth of the ground electronic state potential well is denoted by D_e (D_e is a positive number; see Section 13–6), and so the energy of the ground electronic state is $\varepsilon_{e1} = -D_e$, and the electronic partition function is

$$q_{\text{elec}} = g_{e1} e^{D_e/k_B T} + g_{e2} e^{-\varepsilon_{e2}/k_B T} \tag{18.21}$$

where D_e and ε_{e2} are shown in Figure 18.2. We also introduced in Section 13–6 a quantity D_0 that is equal to $D_e - \frac{1}{2}h\nu$. As Figure 18.2 shows, D_0 is the energy difference between the lowest vibrational state and the dissociated molecule. The quantity D_0 can be measured spectroscopically, and values of D_0 and D_e for several diatomic molecules are given in Table 18.2.

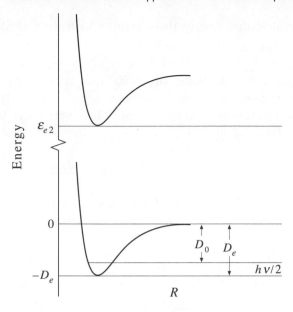

FIGURE 18.2
The ground and first excited electronic states as a function of the internuclear separation, illustrating the quantities D_e and D_0 of the ground state and ε_{e2}. The quantities D_e and D_0 are related by $D_e = D_0 + h\nu/2$ as shown in the figure.

TABLE 18.2
Molecular constants for several diatomic molecules. These parameters were obtained from a variety of sources and do not represent the most accurate values because they were obtained under the rigid rotator-harmonic oscillator approximation.

Molecule	Electronic state	Θ_{vib}/K	Θ_{rot}/K	$D_0/kJ\cdot mol^{-1}$	$D_e/kJ\cdot mol^{-1}$
H_2	$^1\Sigma_g^+$	6332	85.3	432.1	457.6
D_2	$^1\Sigma_g^+$	4394	42.7	435.6	453.9
Cl_2	$^1\Sigma_g^+$	805	0.351	239.2	242.3
Br_2	$^1\Sigma_g^+$	463	0.116	190.1	191.9
I_2	$^1\Sigma_g^+$	308	0.0537	148.8	150.3
O_2	$^3\Sigma_g^-$	2256	2.07	493.6	503.0
N_2	$^1\Sigma_g^+$	3374	2.88	941.6	953.0
CO	$^1\Sigma^+$	3103	2.77	1070	1085
NO	$^2\Pi_{1/2}$	2719	2.39	626.8	638.1
HCl	$^1\Sigma^+$	4227	15.02	427.8	445.2
HBr	$^1\Sigma^+$	3787	12.02	362.6	377.7
HI	$^1\Sigma^+$	3266	9.25	294.7	308.6
Na_2	$^1\Sigma_g^+$	229	0.221	71.1	72.1
K_2	$^1\Sigma_g^+$	133	0.081	53.5	54.1

18–4. Most Molecules Are in the Ground Vibrational State at Room Temperature

In this section, we will evaluate the vibrational part of the partition function of a diatomic molecule under the harmonic-oscillator approximation. If we measure the vibrational energy levels relative to the bottom of the internuclear potential well, the energies are given by (Section 5–4)

$$\varepsilon_v = \left(v + \tfrac{1}{2}\right) h\nu \qquad\qquad v = 0,\ 1,\ 2,\ \ldots \qquad\qquad (18.22)$$

with $\nu = (k/\mu)^{1/2}/2\pi$, where k is the force constant of the molecule and μ is its reduced mass. The vibrational partition function q_{vib} becomes

$$q_{\mathrm{vib}}(T) = \sum_v e^{-\beta\varepsilon_v} = \sum_{v=0}^{\infty} e^{-\beta\left(v+\frac{1}{2}\right)h\nu}$$

$$= e^{-\beta h\nu/2} \sum_{v=0}^{\infty} e^{-\beta h\nu v}$$

This summation can be evaluated easily by recognizing it to be a geometric series (MathChapter I):

$$\sum_{n=0}^{\infty} x^n = \frac{1}{1-x}$$

with $x = e^{-\beta h\nu} < 1$. Thus we can write

$$\sum_{v=0}^{\infty} e^{-\beta h\nu v} = \sum_{v=0}^{\infty} \left(e^{-\beta h\nu}\right)^v = \frac{1}{1 - e^{-\beta h\nu}}$$

so $q_{\mathrm{vib}}(T)$ becomes

$$q_{\mathrm{vib}}(T) = \frac{e^{-\beta h\nu/2}}{1 - e^{-\beta h\nu}} \qquad\qquad (18.23)$$

Note that this is the vibrational term encountered in Example 17–2, which presented the partition function for the rigid rotator-harmonic oscillator model of an ideal diatomic gas. If we introduce a quantity, $\Theta_{\mathrm{vib}} = h\nu/k_{\mathrm{B}}$, called the *vibrational temperature*, $q_{\mathrm{vib}}(T)$ can be written as

$$q_{\mathrm{vib}}(T) = \frac{e^{-\Theta_{\mathrm{vib}}/2T}}{1 - e^{-\Theta_{\mathrm{vib}}/T}} \qquad\qquad (18.24)$$

This is one of the rare cases in which q can be summed directly without having to approximate it by an integral, as we did for the translational case in Section 18–1 and will do shortly for the rotational case in Section 18–5.

We can calculate the average vibrational energy from $q_{vib}(T)$

$$\langle E_{vib} \rangle = N k_B T^2 \frac{d \ln q_{vib}}{dT} = N k_B \left(\frac{\Theta_{vib}}{2} + \frac{\Theta_{vib}}{e^{\Theta_{vib}/T} - 1} \right) \qquad (18.25)$$

Table 18.2 gives Θ_{vib} for several diatomic molecules. The vibrational contribution to the molar heat capacity is

$$\overline{C}_{V,vib} = \frac{d \langle \overline{E}_{vib} \rangle}{dT} = R \left(\frac{\Theta_{vib}}{T} \right)^2 \frac{e^{-\Theta_{vib}/T}}{\left(1 - e^{-\Theta_{vib}/T} \right)^2} \qquad (18.26)$$

Figure 18.3 shows the vibrational contribution of an ideal diatomic gas to the molar heat capacity as a function of temperature. The high temperature limit of $\overline{C}_{V,vib}$ is R, and $\overline{C}_{V,vib}$ is one-half of this value at $T/\Theta_{vib} = 0.34$.

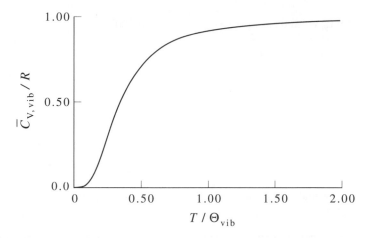

FIGURE 18.3
The vibrational contribution to the molar heat capacity of an ideal diatomic gas as a function of reduced temperature, T/Θ_{vib}.

EXAMPLE 18–2
Calculate the vibrational contribution to the molar heat capacity of $N_2(g)$ at 1000 K. The experimental value is 3.43 $J \cdot K^{-1} \cdot mol^{-1}$.

SOLUTION: We use Equation 18.26 with $\Theta_{vib} = 3374$ K (Table 18.2). Thus, $\Theta_{vib}/T = 3.374$ and

$$\frac{\overline{C}_{V,vib}}{R} = (3.374)^2 \frac{e^{-3.374}}{(1 - e^{-3.374})^2} = 0.418$$

or

$$\overline{C}_{V,vib} = (0.418)(8.314 \; J \cdot K^{-1} \cdot mol^{-1}) = 3.48 \; J \cdot K^{-1} \cdot mol^{-1}$$

The agreement with the experimental value is quite good.

An interesting quantity to calculate is the fraction of molecules in various vibrational states. The fraction of molecules in the vth vibrational state is

$$f_v = \frac{e^{-\beta h \nu (v + \frac{1}{2})}}{q_{\text{vib}}} \tag{18.27}$$

If we substitute Equation 18.23 into this equation, we obtain

$$f_v = \left(1 - e^{-\beta h \nu}\right) e^{-\beta h \nu v} = \left(1 - e^{-\Theta_{\text{vib}}/T}\right) e^{-v\Theta_{\text{vib}}/T} \tag{18.28}$$

The following example illustrates the use of this equation.

EXAMPLE 18–3
Use Equation 18.28 to calculate the fraction of $N_2(g)$ molecules in the $v = 0$ and $v = 1$ vibrational states at 300 K.

SOLUTION: We first calculate $\exp(-\Theta_{\text{vib}}/T)$ for 300 K:

$$e^{-\Theta_{\text{vib}}/T} = e^{-3374 \text{ K}/300 \text{ K}} = e^{-11.25} = 1.31 \times 10^{-5}$$

Therefore,

$$f_0 = 1 - e^{-\Theta_{\text{vib}}/T} \approx 1$$

and

$$f_1 = (1 - e^{-\Theta_{\text{vib}}/T})e^{-\Theta_{\text{vib}}/T} \approx 1.31 \times 10^{-5}$$

Notice that essentially all the nitrogen molecules are in the ground vibrational state at 300 K.

Figure 18.4 shows the population of vibrational levels of $Br_2(g)$ at 300 K. Notice that most molecules are in the ground vibrational state and that the population of the higher vibrational states decreases exponentially. Bromine has a smaller force constant and a larger mass (and hence a smaller value of Θ_{vib}) than most diatomic molecules, however (*cf.* Table 18.2), so the population of excited vibrational states of $Br_2(g)$ at a given temperature is greater than most other molecules.

We can use Equation 18.28 to calculate the fraction of molecules in all excited vibrational states. This quantity is given by $\sum_{v=1}^{\infty} f_v$ but because $\sum_{v=0}^{\infty} f_v = 1$, we can write

$$f_{v>0} = \sum_{v=1}^{\infty} f_v = 1 - f_0 = 1 - (1 - e^{-\Theta_{\text{vib}}/T})$$

or simply

$$f_{v>0} = e^{-\Theta_{\text{vib}}/T} = e^{-\beta h \nu} \tag{18.29}$$

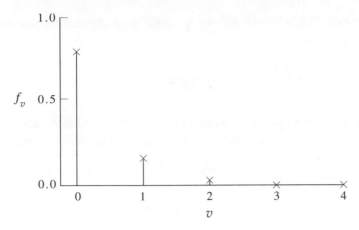

FIGURE 18.4
The population of the vibrational levels of $Br_2(g)$ at 300 K.

Table 18.3 gives the fraction of molecules in excited vibrational states for several diatomic molecules.

TABLE 18.3
The fraction of molecules in excited vibrational states at 300 K and 1000 K.

Gas	Θ_{vib}/K	$f_{v>0}$ ($T = 300$ K)	$f_{v>0}$ ($T = 1000$ K)
H_2	6215	1.01×10^{-9}	2.00×10^{-3}
HCl	4227	7.59×10^{-7}	1.46×10^{-2}
N_2	3374	1.30×10^{-5}	3.43×10^{-2}
CO	3103	3.22×10^{-5}	4.49×10^{-2}
Cl_2	805	6.82×10^{-2}	4.47×10^{-1}
I_2	308	3.58×10^{-1}	7.35×10^{-1}

18–5. Most Molecules Are in Excited Rotational States at Ordinary Temperatures

The energy levels of a rigid rotator are given by (Section 5–8)

$$\varepsilon_J = \frac{\hbar^2 J(J+1)}{2I} \qquad J = 0,\ 1,\ 2,\ \ldots \qquad (18.30a)$$

where I is the moment of inertia of the rotator. Each energy level has a degeneracy of

$$g_J = 2J + 1 \qquad (18.30b)$$

Using Equations 18.30a and 18.30b, we can write the rotational partition function of a rigid rotator as

$$q_{rot}(T) = \sum_{J=0}^{\infty}(2J+1)e^{-\beta\hbar^2 J(J+1)/2I} \tag{18.31}$$

where we sum over levels rather than states by including the degeneracy explicitly. For convenience, we introduce a quantity that has units of temperature and is called the *rotational temperature*, Θ_{rot}:

$$\Theta_{rot} = \frac{\hbar^2}{2Ik_B} = \frac{hB}{k_B} \tag{18.32}$$

where $B = h/8\pi^2 I$ (Equation 5.62). Substituting Equation 18.32 into Equation 18.31 gives

$$q_{rot}(T) = \sum_{J=0}^{\infty}(2J+1)e^{-\Theta_{rot}J(J+1)/T} \tag{18.33}$$

Unlike the harmonic-oscillator partition function, the summation in Equation 18.33 cannot be written in closed form. However, as the data in Table 18.2 will verify, the value of Θ_{rot}/T is quite small at ordinary temperatures for diatomic molecules that do not contain hydrogen atoms. For example, Θ_{rot} for CO(g) is 2.77 K, so Θ_{rot}/T is about 10^{-2} at room temperature. Just as we were able to approximate the summation in Equation 18.4 very well by an integral because $\alpha = \beta h^2/8ma^2$ is typically small at normal temperatures, we are able to approximate the summation in Equation 18.33 by an integral because Θ_{rot}/T is small for most molecules at ordinary temperatures. Therefore, it is an excellent approximation to write $q_{rot}(T)$ as

$$q_{rot}(T) = \int_0^{\infty}(2J+1)e^{-\Theta_{rot}J(J+1)/T}dJ$$

This integral is easy to evaluate because if we let $x = J(J+1)$, then $dx = (2J+1)dJ$ and $q_{rot}(T)$ becomes

$$q_{rot}(T) = \int_0^{\infty} e^{-\Theta_{rot}x/T}dx$$

$$= \frac{T}{\Theta_{rot}} = \frac{8\pi^2 Ik_B T}{h^2} \qquad \Theta_{rot} \ll T \tag{18.34}$$

Note that this is the rotational term encountered in Example 17–2, which presented the partition function for the rigid rotator-harmonic oscillator model of an ideal diatomic gas. This approximation improves as the temperature increases and is called the high-temperature limit. For low temperatures or for molecules with large values of Θ_{rot}, say H_2(g) with $\Theta_{rot} = 85.3$ K, we can use Equation 18.33 directly. For example, the first four terms of Equation 18.33 are sufficient to calculate $q_{rot}(T)$ to within 0.1%

for $T < 3\Theta_{rot}$. For simplicity, we will use only the high-temperature limit, because $\Theta_{rot} \ll T$ for most molecules at room temperature. (See Table 18.2.)

The average rotational energy is

$$\langle E_{rot} \rangle = N k_B T^2 \left(\frac{d \ln q_{rot}}{dT} \right) = N k_B T$$

and the rotational contribution to the molar heat capacity is

$$\overline{C}_{V,rot} = R$$

A diatomic molecule has two rotational degrees of freedom, and each one contributes $R/2$ to $\overline{C}_{V,rot}$.

We can also calculate the fraction of molecules in the Jth rotational level:

$$f_J = \frac{(2J + 1)e^{-\Theta_{rot}J(J+1)/T}}{q_{rot}}$$

$$= (2J + 1)(\Theta_{rot}/T)e^{-\Theta_{rot}J(J+1)/T} \tag{18.35}$$

EXAMPLE 18–4

Use Equation 18.35 to calculate the population of the rotational levels of CO at 300 K.

SOLUTION: Using $\Theta_{rot} = 2.77$ K from Table 18.2, we have that $\Theta_{rot}/T = 0.00923$ at 300 K. Therefore,

$$f_J = (2J + 1)(0.00923)e^{-0.00923J(J+1)}$$

We can present our results in the form of a table:

J	f_J
0	0.00923
2	0.0437
4	0.0691
6	0.0814
8	0.0807
10	0.0702
12	0.0547
16	0.0247
18	0.0145

These results are plotted in Figure 18.5.

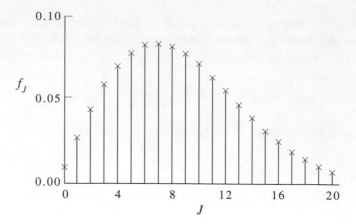

FIGURE 18.5
The fraction of molecules in the Jth rotational level for CO at 300 K.

Contrary to the case for vibrational levels, most molecules are in the excited rotational levels at ordinary temperatures. We can estimate the most probable value of J by treating Equation 18.35 as if J were continuous and by setting the derivative with respect to J equal to zero to obtain (Problem 18–18)

$$J_{mp} \approx \left(\frac{T}{2\Theta_{rot}} \right)^{1/2} - \frac{1}{2} \tag{18.36}$$

This equation gives a value of 7 for CO at 300 K (in agreement with Figure 18.5).

We can also use Equation 18.35 to rationalize the observed intensities of the lines in the P and R branches of the vibrational-rotational spectrum of a diatomic molecule (Figure 13.2). Note that the envelope of the lines in Figure 18.5 is similar to the lines in the P and R branches in Figure 13.2. The reason the two figures are similar is that the intensities of the rotational lines are proportional to the number of molecules in the rotational level from which the transition occurs. Thus, we see that the shape of the P and R branches reflects the thermal population of rotational energy levels.

18–6. Rotational Partition Functions Contain a Symmetry Number

Although it is not apparent from our derivation of $q_{rot}(T)$, Equations 18.33 and 18.34 apply only to heteronuclear diatomic molecules. The underlying reason is that the wave function of a homonuclear diatomic molecule must possess a certain symmetry with respect to the interchange of the two identical nuclei in the molecule. In particular, if the two nuclei have integral spins (bosons), the molecular wave function must be symmetric with respect to an interchange of the two nuclei; if the nuclei have half odd integer spin (fermions), the molecular wave function must be antisymmetric. This symmetry requirement has a profound effect on the population of the rotational energy levels of a homonuclear diatomic molecule, which can be understood only by a careful analysis of the general symmetry properties of the wave function of a diatomic molecule. This analysis is somewhat involved and will not be done here, but we need the final result. At temperatures such that $\Theta_{rot} \ll T$, which we have seen applies to most molecules at ordinary temperatures, q_{rot} for a homonuclear diatomic molecule is

$$q_{rot}(T) = \frac{T}{2\Theta_{rot}} \tag{18.37}$$

Note that this equation is the same as Equation 18.34 for a heteronuclear diatomic molecule except for the factor of 2 in the denominator. This factor comes from the additional symmetry of the homonuclear diatomic molecule; in particular, a homonuclear diatomic molecule has two indistinguishable orientations. There is a two-fold axis of symmetry perpendicular to the internuclear axis.

Equations 18.34 and 18.37 can be written as one equation by writing q_{rot} as

$$q_{rot}(T) = \frac{T}{\sigma\,\Theta_{rot}} \tag{18.38}$$

where $\sigma = 1$ for a heteronuclear diatomic molecule and 2 for a homonuclear diatomic molecule. The factor σ is called the *symmetry number* of the molecule and represents the number of indistinguishable orientations of the molecule.

Having studied each contribution to the molecular partition function of a diatomic molecule, we can now include the rigid rotator-harmonic oscillator approximation in the partition function of a diatomic molecule to obtain

$$
\begin{aligned}
q(V, T) &= q_{trans}q_{rot}q_{vib}q_{elec} \\
&= \left(\frac{2\pi M k_B T}{h^2}\right)^{3/2} V \cdot \frac{T}{\sigma\,\Theta_{rot}} \cdot \frac{e^{-\Theta_{vib}/2T}}{1 - e^{-\Theta_{vib}/T}} \cdot g_{e1} e^{D_e/k_B T} \tag{18.39}
\end{aligned}
$$

Remember that this expression requires that $\Theta_{rot} \ll T$, that only the ground electronic state is populated, that the zero of the electronic energy is taken to be the separated atoms at rest in their ground electronic states, and that the zero of energy for the vibrational energy is that at the bottom of the internuclear potential well of the lowest electronic state. Note that only q_{trans} is a function of V, and that this function is of the form $f(T)V$, which, as we have seen before, is responsible for the ideal-gas equation of state.

EXAMPLE 18–5
Derive an expression for the molar energy $\overline{U}$ of a diatomic ideal gas from Equation 18.39. Identify each of the terms.

SOLUTION: We start with

$$Q(N, V, T) = \frac{[q(V, T)]^N}{N!}$$

and

$$U = k_B T^2 \left(\frac{\partial \ln Q}{\partial T} \right)_{N,V} = N k_B T^2 \left(\frac{\partial \ln q}{\partial T} \right)_V$$

Using Equation 18.39 for $q(V, T)$, we have

$$\ln q = \frac{3}{2} \ln T + \ln T - \frac{\Theta_{vib}}{2T} - \ln(1 - e^{-\Theta_{vib}/T}) + \frac{D_e}{k_B T}$$

$$+ \text{ terms not containing } T$$

Therefore,

$$\left(\frac{\partial \ln q}{\partial T} \right)_V = \frac{3}{2T} + \frac{1}{T} + \frac{\Theta_{vib}}{2T^2} + \frac{(\Theta_{vib}/T^2)e^{-\Theta_{vib}/T}}{1 - e^{-\Theta_{vib}/T}} - \frac{D_e}{k_B T^2}$$

and letting $N = N_A$ and $N_A k_B = R$ for one mole,

$$\overline{U} = \frac{3}{2} RT + RT + R \frac{\Theta_{vib}}{2} + R \frac{\Theta_{vib} e^{-\Theta_{vib}/T}}{1 - e^{-\Theta_{vib}/T}} - N_A D_e \qquad (18.40)$$

The first term represents the average translational energy ($RT/2$ for each of the three translational degrees of freedom), the second term represents the average rotational energy ($RT/2$ for each of the two rotational degrees of freedom), the third term represents the zero-point vibrational energy, the fourth term represents the average vibrational energy in excess of the zero-point vibrational energy, and the last term reflects the electronic energy relative to the zero of electronic energy that we have chosen, namely the two separated atoms at rest in their ground electronic states.

The heat capacity is obtained by differentiating Equation 18.40 with respect to T:

$$\frac{\overline{C}_V}{R} = \frac{5}{2} + \left(\frac{\Theta_{vib}}{T} \right)^2 \frac{e^{-\Theta_{vib}/T}}{(1 - e^{-\Theta_{vib}/T})^2} \qquad (18.41)$$

Figure 17.3 presents a comparison of Equation 18.41 with experimental data for oxygen. The agreement is good and is typical of that found for other properties. The agreement can be improved considerably by including the first corrections to the rigid rotator-harmonic oscillator model. These include effects such as centrifugal distortion and anharmonicity. The consideration of these effects introduces a new set of molecular constants, all of which can be determined spectroscopically and are well tabulated. The use of such additional parameters from spectroscopic data can give calculated values of the heat capacity that are actually more accurate than calorimetric ones.

18–7. The Vibrational Partition Function of a Polyatomic Molecule Is a Product of Harmonic Oscillator Partition Functions for Each Normal Coordinate

The discussion in Section 18–3 for diatomic molecules applies equally well to polyatomic molecules, and so

$$Q(N, V, T) = \frac{[q(V, T)]^N}{N!}$$

As before, the number of translational energy states alone is sufficient to guarantee that the number of energy states available to any molecule is much greater than the number of molecules in the system.

As for diatomic molecules, we use a rigid rotator-harmonic oscillator approximation. This allows us to separate the rotational motion from the vibrational motion of the molecule, so that we can treat each one separately. Both problems are somewhat more complicated for polyatomic molecules than for diatomic molecules. Nevertheless, we can write the polyatomic analog of Equation 18.19:

$$Q(N, V, T) = \frac{(q_{\text{trans}} q_{\text{rot}} q_{\text{vib}} q_{\text{elec}})^N}{N!} \tag{18.42}$$

In Equation 18.42, q_{trans} is given by

$$q_{\text{trans}}(V, T) = \left[\frac{2\pi M k_{\text{B}} T}{h^2} \right]^{3/2} V \tag{18.43}$$

where M is the total mass of the molecule. We choose as the zero of energy the n atoms completely separated in their ground electronic states. Thus, the energy of the ground electronic state is $-D_e$, and then the electronic partition function is

$$q_{\text{elec}} = g_{e1} e^{D_e/k_{\text{B}} T} + \cdots \tag{18.44}$$

To calculate $Q(N, V, T)$ we must investigate q_{rot} and q_{vib}.

We learned in Section 13–9 that the vibrational motion of a polyatomic molecule can be expressed in terms of normal coordinates. By introducing normal coordinates, the vibrational motion of a polyatomic molecule can be expressed as a set of *independent* harmonic oscillators. Consequently, the vibrational energy of a polyatomic molecule can be written as

$$\varepsilon_{\text{vib}} = \sum_{j=1}^{\alpha} \left(v_j + \tfrac{1}{2} \right) h \nu_j \qquad v_j = 0, 1, 2, \ldots \tag{18.45}$$

where ν_j is the vibrational frequency associated with the jth normal mode and α is the number of vibrational degrees of freedom ($3n - 5$ for a linear molecule and $3n - 6$ for a nonlinear molecule, where n is the number of atoms in the molecule). Because the normal modes are independent,

$$q_{vib} = \prod_{j=1}^{\alpha} \frac{e^{-\Theta_{vib,j}/2T}}{\left(1 - e^{-\Theta_{vib,j}/T}\right)} \tag{18.46}$$

$$E_{vib} = Nk_B \sum_{j=1}^{\alpha} \left(\frac{\Theta_{vib,j}}{2} + \frac{\Theta_{vib,j} e^{-\Theta_{vib,j}/T}}{1 - e^{-\Theta_{vib,j}/T}} \right) \tag{18.47}$$

and

$$C_{V,vib} = Nk_B \sum_{j=1}^{\alpha} \left[\left(\frac{\Theta_{vib,j}}{T} \right)^2 \frac{e^{-\Theta_{vib,j}/T}}{(1 - e^{-\Theta_{vib,j}/T})^2} \right] \tag{18.48}$$

where $\Theta_{vib,j}$ is a characteristic vibrational temperature defined by

$$\Theta_{vib,j} = \frac{h\nu_j}{k_B} \tag{18.49}$$

Table 18.4 contains values of $\Theta_{vib,j}$ for several polyatomic molecules.

TABLE 18.4

Values of the characteristic rotational temperatures, the characteristic vibrational temperatures, D_0 for the ground state, and the symmetry number, σ, for some polyatomic molecules. The numbers in parentheses indicate the degeneracy of that mode.

Molecule	Θ_{rot}/K	$\Theta_{vib,j}/K$	$D_0/kJ \cdot mol^{-1}$	σ
CO_2	0.561	3360, 954(2), 1890	1596	2
H_2O	40.1, 20.9, 13.4	5360, 5160, 2290	917.6	2
NH_3	13.6, 13.6, 8.92	4800, 1360, 4880(2), 2330(2)	1158	3
ClO_2	2.50, 0.478, 0.400	1360, 640, 1600	378	2
SO_2	2.92, 0.495, 0.422	1660, 750, 1960	1063	2
N_2O	0.603	3200, 850(2), 1840	1104	1
NO_2	11.5, 0.624, 0.590	1900, 1080, 2330	928.0	2
CH_4	7.54, 7.54, 7.54	4170, 2180(2), 4320(3), 1870(3)	1642	12
CH_3Cl	7.32, 0.637, 0.637	4270, 1950, 1050, 4380(2) 2140(2), 1460(2)	1551	3
CCl_4	0.0823, 0.0823, 0.0823	660, 310(2), 1120(3), 450(3)	1292	12

EXAMPLE 18–6

Calculate the contribution of each normal mode to the vibrational heat capacity of CO_2 at 400 K.

SOLUTION: The values of $\Theta_{\mathrm{vib},j}$ are given in Table 18.4. Note that the $\Theta_{\mathrm{vib}} = 954$ K mode (bending mode) is doubly degenerate. For $\Theta_{\mathrm{vib},j} = 954$ K (the doubly degerate bending mode),

$$\frac{\overline{C}_{V,j}}{R} = \left(\frac{954}{400}\right)^2 \frac{e^{-954/400}}{(1 - e^{-954/400})^2} = 0.635$$

For $\Theta_{\mathrm{vib},j} = 1890$ K (the asymmetric stretch),

$$\frac{\overline{C}_{V,j}}{R} = \left(\frac{1890}{400}\right)^2 \frac{e^{-1890/400}}{(1 - e^{-1890/400})^2} = 0.202$$

For $\Theta_{\mathrm{vib},j} = 3360$ K (the symmetric stretch),

$$\frac{\overline{C}_{V,j}}{R} = \left(\frac{3360}{400}\right)^2 \frac{e^{-3360/400}}{(1 - e^{-3360/400})^2} = 0.016$$

The total vibrational heat capacity at 400 K is

$$\frac{\overline{C}_{V,\mathrm{vib}}}{R} = 2(0.635) + 0.202 + 0.016 = 1.488$$

Note that the contribution from each mode decreases as $\Theta_{\mathrm{vib},j}$ increases. Because $\Theta_{\mathrm{vib},j}$ is proportional to the frequency of the mode, it requires higher temperatures to excite modes with larger values of $\Theta_{\mathrm{vib},j}$. The molar vibrational heat capacity from 200 K to 2000 K contributed by each mode is shown in Figure 18.6.

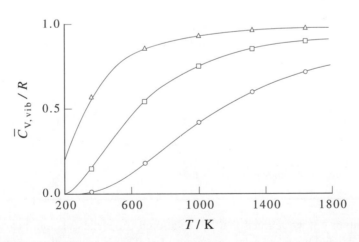

FIGURE 18.6
The contribution of each normal mode to the molar vibrational heat capacity of CO_2. The curve indicated by triangles corresponds to $\Theta_{\mathrm{vib},j} = 954$ K; the curve indicated by squares to $\Theta_{\mathrm{vib},j} = 1890$ K; and the curve indicated by circles to $\Theta_{\mathrm{vib},j} = 3360$ K. Note that modes with smaller values of $\Theta_{\mathrm{vib},j}$, or ν_j, contribute more at a given temperature.

18–8. The Form of the Rotational Partition Function of a Polyatomic Molecule Depends Upon the Shape of the Molecule

In this section, we will discuss the rotational partition functions of polyatomic molecules. Let's consider a linear polyatomic molecule first. In the rigid-rotator approximation, the energies and degeneracies of a linear polyatomic molecule are the same as for a diatomic molecule, $\varepsilon_J = J(J+1)h^2/8\pi^2 I$ with $J = 0, 1, 2, \ldots$ and $g_J = 2J+1$. In this case, the moment of inertia I is

$$I = \sum_{j=1}^{n} m_j d_j^2$$

where d_j is the distance of the jth nucleus from the center of mass of the molecule. Consequently, the rotational partition function of a linear polyatomic molecule is the same as that of a diatomic molecule, namely,

$$q_{rot} = \frac{8\pi^2 I k_B T}{\sigma h^2} = \frac{T}{\sigma \Theta_{rot}} \tag{18.50}$$

As before, we have introduced a symmetry number, which is unity for unsymmetrical molecules such as N_2O and COS and equal to two for symmetrical molecules such as CO_2 and C_2H_2. Recall that the symmetry number is the number of different ways the molecule can be rotated into a configuration indistinguishable from the original.

EXAMPLE 18–7
What is the symmetry number of ammonia, NH_3?

SOLUTION: Ammonia is a trigonal pyramidal molecule and has the three indistinguishable orientations shown below looking down the three-fold axis of symmetry.

Therefore, the symmetry number is three.

In Chapter 13, we learned that the rotational properties of nonlinear polyatomic molecules depend upon the relative magnitudes of their principal moments of inertia. If all three principal moments of inertia are equal, the molecule is called a *spherical top*. If two of the three are equal, the molecule is called a *symmetric top*. If all three are different, the molecule is called an *asymmetric top*. Just as we defined a characteristic rotational temperature of a diatomic molecule by Equation 18.32, $\Theta_{rot} = \hbar^2/2Ik_B$, we define three characteristic rotational temperatures in terms of the three principal moments of inertia according to

$$\Theta_{rot,j} = \frac{\hbar^2}{2I_j k_B} \qquad j = A, \ B, \ C \qquad (18.51)$$

Thus, we have the various cases

$$\Theta_{rot,A} = \Theta_{rot,B} = \Theta_{rot,C} \qquad \text{spherical top}$$

$$\Theta_{rot,A} = \Theta_{rot,B} \neq \Theta_{rot,C} \qquad \text{symmetric top}$$

$$\Theta_{rot,A} \neq \Theta_{rot,B} \neq \Theta_{rot,C} \qquad \text{asymmetric top}$$

The quantum-mechanical problem of a spherical top can be solved exactly to give

$$\varepsilon_J = \frac{J(J+1)\hbar^2}{2I} \qquad (18.52)$$

$$g_J = (2J+1)^2 \qquad J = 0, \ 1, \ 2, \ \ldots$$

The rotational partition function is

$$q_{rot}(T) = \sum_{J=0}^{\infty} (2J+1)^2 e^{-\hbar^2 J(J+1)/2I k_B T} \qquad (18.53)$$

For almost all spherical top molecules $\Theta_{rot} \ll T$ at ordinary temperatures, so we convert the sum in Equation 18.53 to an integral:

$$q_{rot}(T) = \frac{1}{\sigma} \int_0^{\infty} (2J+1)^2 e^{-\Theta_{rot} J(J+1)/T} dJ$$

Note that we have included the symmetry number σ. For $\Theta_{rot} \ll T$, the most important values of J are large (Problem 18–26), and so we may neglect 1 compared with J in the integrand of the above expression for q_{rot} to obtain

$$q_{rot}(T) = \frac{1}{\sigma} \int_0^{\infty} 4J^2 e^{-\Theta_{rot} J^2/T} dJ$$

If we let $\Theta_{rot}/T = a$, we can write

$$q_{rot}(T) = \frac{4}{\sigma} \int_0^{\infty} x^2 e^{-ax^2} dx$$

$$= \frac{4}{\sigma} \cdot \frac{1}{4a} \left(\frac{\pi}{a}\right)^{1/2}$$

or, upon substituting Θ_{rot}/T for a,

$$q_{rot}(T) = \frac{\pi^{1/2}}{\sigma} \left(\frac{T}{\Theta_{rot}}\right)^{3/2} \qquad \text{spherical top} \qquad (18.54)$$

The corresponding expressions for a symmetric top and an asymmetric top are

$$q_{rot}(T) = \frac{\pi^{1/2}}{\sigma}\left(\frac{T}{\Theta_{rot,A}}\right)\left(\frac{T}{\Theta_{rot,C}}\right)^{1/2} \qquad \text{symmetric top} \qquad (18.55)$$

and

$$q_{rot}(T) = \frac{\pi^{1/2}}{\sigma}\left(\frac{T^3}{\Theta_{rot,A}\Theta_{rot,B}\Theta_{rot,C}}\right)^{1/2} \qquad \text{asymmetric top} \qquad (18.56)$$

Notice how Equation 18.56 reduces to Equation 18.55 when $\Theta_{rot,A} = \Theta_{rot,B}$ and how both Equations 18.55 and 18.56 reduce to Equation 18.54 when $\Theta_{rot,A} = \Theta_{rot,B} = \Theta_{rot,C}$. Table 18.4 contains values of $\Theta_{rot,A}$, $\Theta_{rot,B}$, and $\Theta_{rot,C}$ for several polyatomic molecules. The average molar rotational energy of a nonlinear polyatomic molecule is

$$\overline{U}_{rot} = N_A k_B T^2 \left(\frac{d \ln q_{rot}(T)}{dT}\right)$$

$$= RT^2\left(\frac{d \ln T^{3/2}}{dT}\right) = \frac{3RT}{2}$$

or $RT/2$ for each rotational degree of freedom, and $\overline{C}_{V,rot} = 3R/2$.

18–9. Calculated Molar Heat Capacities Are in Very Good Agreement with Experimental Data

We can now use the results of Sections 18–7 and 18–8 to construct $q(V, T)$ for polyatomic molecules. For an ideal gas of linear polyatomic molecules, $q(V, T)$ is the product of Equations 18.43, 18.44, 18.46, and 18.50:

$$q(V, T) = \left(\frac{2\pi M k_B T}{h^2}\right)^{3/2} V \cdot \frac{T}{\sigma \Theta_{rot}} \cdot \left(\prod_{j=1}^{3n-5}\frac{e^{-\Theta_{vib,j}/2T}}{1 - e^{-\Theta_{vib,j}/T}}\right) \cdot g_{e1}e^{D_e/k_B T} \qquad (18.57)$$

The energy is

$$\frac{U}{N k_B T} = \frac{3}{2} + \frac{2}{2} + \sum_{j=1}^{3n-5}\left(\frac{\Theta_{vib,j}}{2T} + \frac{\Theta_{vib,j}/T}{e^{\Theta_{vib,j}/T} - 1}\right) - \frac{D_e}{k_B T} \qquad (18.58)$$

and the heat capacity is

$$\frac{C_V}{N k_B} = \frac{3}{2} + \frac{2}{2} + \sum_{j=1}^{3n-5}\left(\frac{\Theta_{vib,j}}{T}\right)^2 \frac{e^{-\Theta_{vib,j}/T}}{(1 - e^{-\Theta_{vib,j}/T})^2} \qquad (18.59)$$

For an ideal gas of nonlinear polyatomic molecules,

$$q(V, T) = \left(\frac{2\pi M k_B T}{h^2}\right)^{3/2} V \cdot \frac{\pi^{1/2}}{\sigma} \left(\frac{T^3}{\Theta_{rot,A}\Theta_{rot,B}\Theta_{rot,C}}\right)^{1/2}$$

$$\times \left[\prod_{j=1}^{3n-6} \frac{e^{-\Theta_{vib,j}/2T}}{(1 - e^{-\Theta_{vib,j}/T})}\right] \cdot g_{e1} e^{D_e/k_B T} \qquad (18.60)$$

$$\frac{U}{Nk_B T} = \frac{3}{2} + \frac{3}{2} + \sum_{j=1}^{3n-6} \left(\frac{\Theta_{vib,j}}{2T} + \frac{\Theta_{vib,j}/T}{e^{\Theta_{vib,j}/T} - 1}\right) - \frac{D_e}{k_B T} \qquad (18.61)$$

and

$$\frac{C_V}{Nk_B} = \frac{3}{2} + \frac{3}{2} + \sum_{j=1}^{3n-6} \left(\frac{\Theta_{vib,j}}{T}\right)^2 \frac{e^{-\Theta_{vib,j}/T}}{(1 - e^{-\Theta_{vib,j}/T})^2} \qquad (18.62)$$

EXAMPLE 18–8

Calculate the molar heat capacity of gaseous water at 300 K.

SOLUTION: We use Equation 18.62 with $\Theta_{vib,j} = 2290$ K, 5160 K, and 5360 K (Table 18.4). For $\Theta_{vib,j} = 2290$ K,

$$\frac{\overline{C}_{V,j}}{R} = \left(\frac{2290}{300}\right)^2 \frac{e^{2290/300}}{(e^{2290/300} - 1)^2} = 0.0282$$

Similarly $\overline{C}_{V,j}/R = 1.00 \times 10^{-5}$ for $\Theta_{vib,j} = 5160$ K and 5.56×10^{-6} for $\Theta_{vib,j} = 5360$ K. The total molar heat capacity of water at 300 K is

$$\frac{\overline{C}_V}{R} = 3.000 + 0.0282 + 1.00 \times 10^{-5} + 5.56 \times 10^{-6} = 3.028$$

The experimental value is 3.011. Notice that the vibrational degrees of freedom contribute very little to the heat capacity of water at 300 K. The calculated and experimental values at 1000 K are 3.948 and 3.952, respectively. Figure 18.7 shows the molar heat capacity of water from 300 K to 1200 K.

Table 18.5 gives the vibrational contribution to the molar heat capacity at 300 K for a variety of molecules of different shapes. It can be seen that the vibrational contributions are far from their high-temperature limits and that the agreement between the calculated and experimental values of $\overline{C}_V/R$ is good. A calculation for more complicated molecules would show similar agreement between the calculated values and the experimental data.

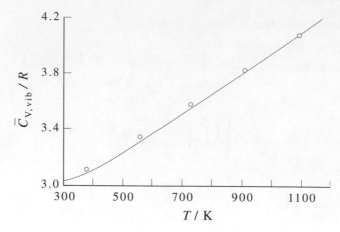

FIGURE 18.7
A comparison of the molar heat capacity of water vapor calculated from Equation 18.62 and the experimental value. The experimental data are indicated by the circles.

TABLE 18.5
Vibrational contributions to the molar heat capacity of some polyatomic molecules at 300 K.

Molecule	Θ_{vib}/K	Degeneracy	Vibrational Contribution to $\overline{C}_V$	$\overline{C}_{V,vib}/R$	Total $\overline{C}_V/R$ (calc)	Total $\overline{C}_V/R$ (exptl)
CO_2	1890	1	0.073			
	3360	1	0.000			
	954	2	0.458	0.99	3.49	3.46
N_2O	1840	1	0.082			
	3200	1	0.003			
	850	2	0.533	1.15	2.65	
NH_3	4800	1	0.000			
	1360	1	0.226			
	4880	2	0.000			
	2330	2	0.026	0.28	3.28	
CH_4	4170	1	0.000			
	2180	2	0.037			
	4320	3	0.000			
	1870	3	0.077	0.30	3.30	3.29
H_2O	2290	1	0.028			
	5160	1	0.000			
	5360	1	0.000	0.03	3.03	3.01

Problems

18-1. Equation 18.7 shows that $\langle \varepsilon_{trans} \rangle = \frac{3}{2} k_B T$ in three dimensions, and Problem 18–3 shows that $\langle \varepsilon_{trans} \rangle = \frac{1}{2} k_B T$ in one dimension and $\frac{2}{2} k_B T$ in two dimensions. Show that typical values of translational quantum numbers at room temperature are $O(10^9)$ for $m = 10^{-26}$ kg, $a = 1$ dm, and $T = 300$ K.

18-2. Show that the difference between the successive terms in the summation in Equation 18.4 is very small for $m = 10^{-26}$ kg, $a = 1$ dm, and $T = 300$ K. Recall from Problem 18–1 that typical values of n are $O(10^9)$.

18-3. Show that

$$q_{trans}(a, T) = \left(\frac{2\pi m k_B T}{h^2} \right)^{1/2} a$$

in one dimension and that

$$q_{trans}(a, T) = \left(\frac{2\pi m k_B T}{h^2} \right) a^2$$

in two dimensions. Use these results to show that $\langle \varepsilon_{trans} \rangle$ has a contribution of $k_B T/2$ to its total value for each dimension.

18-4. Using the data in Table 8.6, calculate the fraction of sodium atoms in the first excited state at 300 K, 1000 K, and 2000 K.

18-5. Using the data in Table 18.1, evaluate the fraction of lithium atoms in the first excited state at 300 K, 1000 K, and 2000 K.

18-6. Show that each dimension contributes $R/2$ to the molar translational heat capacity.

18-7. Using the values of Θ_{vib} and D_0 in Table 18.2, calculate the vaues of D_e for CO, NO, and K_2.

18-8. Calculate the characteristic vibrational temperature Θ_{vib} for $H_2(g)$ and $D_2(g)$ ($\tilde{\nu}_{H_2} = 4401$ cm^{-1} and $\tilde{\nu}_{D_2} = 3112$ cm^{-1}).

18-9. Plot the vibrational contribution to the molar heat capacity of Cl_2 (g) from 250 K to 1000 K.

18-10. Plot the fraction of HCl(g) molecules in the first few vibrational states at 300 K and 1000 K.

18-11. Calculate the fraction of molecules in the ground vibrational state and in all the excited states at 300 K for each of the molecules in Table 18.2.

18-12. Calculate the value of the characteristic rotational temperature Θ_{rot} for $H_2(g)$ and $D_2(g)$. (The bond lengths of H_2 and D_2 are 74.16 pm.) The atomic mass of deuterium is 2.014.

18-13. The average molar rotational energy of a diatomic molecule is RT. Show that typical values of J are given by $J(J + 1) = T/\Theta_{rot}$. What are typical values of J for $N_2(g)$ at 300 K?

18-14. There is a mathematical procedure to calculate the error in replacing a summation by an integral as we do for the translational and rotational partition functions. The formula is called the Euler-Maclaurin summation formula and goes as follows:

$$\sum_{n=a}^{b} f(n) = \int_a^b f(n)dn + \frac{1}{2}\{f(b) + f(a)\} - \frac{1}{12}\left\{\left.\frac{df}{dn}\right|_{n=a} - \left.\frac{df}{dn}\right|_{n=b}\right\}$$
$$+ \frac{1}{720}\left\{\left.\frac{d^3 f}{dn^3}\right|_{n=a} - \left.\frac{d^3 f}{dn^3}\right|_{n=b}\right\} + \cdots$$

Apply this formula to Equation 18.33 to obtain

$$q_{rot}(T) = \frac{T}{\Theta_{rot}}\left\{1 + \frac{1}{3}\left(\frac{\Theta_{rot}}{T}\right) + \frac{1}{15}\left(\frac{\Theta_{rot}}{T}\right)^2 + O\left[\left(\frac{\Theta_{rot}}{T}\right)^3\right]\right\}$$

Calculate the correction to replacing Equation 18.33 by an integral for $N_2(g)$ at 300 K; $H_2(g)$ at 300 K (being so light, H_2 is an extreme example).

18-15. Apply the Euler-Maclaurin summation formula (Problem 18–14) to the one-dimensional version of Equation 18.4 to obtain

$$q_{trans}(a, T) = \left(\frac{2\pi m k_B T}{h^2}\right)^{1/2} a + \left[\frac{1}{2} + \frac{h^2}{48ma^2 k_B T}\right] e^{-h^2/8ma^2 k_B T}$$

Show that the correction amounts to about $10^{-8}\%$ for $m = 10^{-26}$ kg, $a = 1$ dm, and $T = 300$ K.

18-16. We were able to evaluate the vibrational partition function for a harmonic oscillator exactly by recognizing the summation as a geometric series. Apply the Euler-Maclaurin summation formula (Problem 18–14) to this case and show that

$$\sum_{v=0}^{\infty} e^{-\beta(v+\frac{1}{2})h\nu} = e^{-\Theta_{vib}/2T} \sum_{v=0}^{\infty} e^{-v\Theta_{vib}/T}$$
$$= e^{-\Theta_{vib}/2T}\left[\frac{T}{\Theta_{vib}} + \frac{1}{2} + \frac{\Theta_{vib}}{12T} + \cdots\right]$$

Show that the corrections to replacing the summation by an integration are very large for $O_2(g)$ at 300 K. Fortunately, we don't need to replace the summation by an integration in this case.

18-17. Plot the fraction of NO(g) molecules in the various rotational levels at 300 K and at 1000 K.

18-18. Show that the values of J at the maximum of a plot of f_J versus J (Equation 18.35) is given by

$$J_{max} \approx \left(\frac{T}{2\Theta_{rot}}\right)^{1/2} - \frac{1}{2}$$

Hint: Treat J as a continuous variable. Use this result to verify the values of J at the maxima in the plots in Problem 18–17.

18-19. The experimental heat capacity of $N_2(g)$ can be fit to the empirical formula

$$\overline{C}_V(T)/R = 2.283 + (6.291 \times 10^{-4} \text{ K}^{-1})T - (5.0 \times 10^{-10} \text{ K}^{-2})T^2$$

over the temperature range 300 K $< T <$ 1500 K. Plot $\overline{C}_V(T)/R$ versus T over this range using Equation 18.41, and compare your results with the experimental curve.

18-20. The experimental heat capacity of $CO(g)$ can be fit to the empirical formula

$$\overline{C}_V(T)/R = 2.192 + (9.240 \times 10^{-4} \text{ K}^{-1})T - (1.41 \times 10^{-7} \text{ K}^{-2})T^2$$

over the temperature range 300 K $< T <$ 1500 K. Plot $\overline{C}_V(T)/R$ versus T over this range using Equation 18.41, and compare your results with the experimental curve.

18-21. Calculate the contribution of each normal mode to the molar vibrational heat capacity of $H_2O(g)$ at 600 K.

18-22. In analogy to the characteristic vibrational temperature, we can define a characteristic electronic temperature by

$$\Theta_{\text{elec},j} = \frac{\varepsilon_{ej}}{k_B}$$

where ε_{ej} is the energy of the jth excited electronic state relative to the ground state. Show that if we define the ground state to be the zero of energy, then

$$q_{\text{elec}} = g_0 + g_1 e^{-\Theta_{\text{elec},1}/T} + g_2 e^{-\Theta_{\text{elec},2}/T} + \cdots$$

The first and second excited electronic states of $O(g)$ lie 158.2 cm^{-1} and 226.5 cm^{-1} above the ground electronic state. Given $g_0 = 5$, $g_1 = 3$, and $g_2 = 1$, calculate the values of $\Theta_{\text{elec},1}$, $\Theta_{\text{elec},2}$, and q_{elec} (ignoring any higher states) for $O(g)$ at 5000 K.

18-23. Determine the symmetry numbers for H_2O, HOD, CH_4, SF_6, C_2H_2, and C_2H_4.

18-24. The HCN(g) molecule is a linear molecule, and the following constants determined spectroscopically are $I = 18.816 \times 10^{-47}$ kg·m^2, $\tilde{\nu}_1 = 2096.7$ cm^{-1} (HC–N stretch), $\tilde{\nu}_2 = 713.46$ cm^{-1} (H–C–N bend, two-fold degeneracy), and $\tilde{\nu}_3 = 3311.47$ cm^{-1} (H–C stretch). Calculate the values of Θ_{rot} and Θ_{vib} and $\overline{C}_V$ at 3000 K.

18-25. The acetylene molecule is linear, the C≡C bond length is 120.3 pm, and the C–H bond length is 106.0 pm. What is the symmetry number of acetylene? Determine the moment of inertia (Section 13–8) of acetylene and calculate the value of Θ_{rot}. The fundamental frequencies of the normal modes are $\tilde{\nu}_1 = 1975$ cm^{-1}, $\tilde{\nu}_2 = 3370$ cm^{-1}, $\tilde{\nu}_3 = 3277$ cm^{-1}, $\tilde{\nu}_4 = 729$ cm^{-1}, and $\tilde{\nu}_5 = 600$ cm^{-1}. The normal modes $\tilde{\nu}_4$ and $\tilde{\nu}_5$ are doubly degenerate. All the other modes are nondegenerate. Calculate $\Theta_{\text{vib},j}$ and $\overline{C}_V$ at 300 K.

18-26. Plot the summand in Equation 18.53 versus J, and show that the most important values of J are large for $T \gg \Theta_{\text{rot}}$. We use this fact in going from Equation 18.53 to Equation 18.54.

18-27. Use the Euler-Maclaurin summation formula (Problem 18–14) to show that

$$q_{\text{rot}}(T) = \frac{\pi^{1/2}}{\sigma}\left(\frac{T}{\Theta_{\text{rot}}}\right)^{3/2} + \frac{1}{6} + O\left(\frac{\Theta_{\text{rot}}}{T}\right)$$

for a spherical top molecule. Show that the correction to replacing Equation 18.53 by an integral is about 1% for CH_4 and 0.001% for CCl_4 at 300 K.

18-28. The N–N and N–O bond lengths in the (linear) molecule N_2O are 109.8 pm and 121.8 pm, respectively. Calculate the center of mass and the moment of inertia of $^{14}N^{14}N^{16}O$. Compare your answer with the value obtained from Θ_{rot} in Table 18.4.

18-29. $NO_2(g)$ is a bent triatomic molecule. The following data determined from spectroscopic measurements are $\tilde{\nu}_1 = 1319.7$ cm^{-1}, $\tilde{\nu}_2 = 749.8$ cm^{-1}, $\tilde{\nu}_3 = 1617.75$ cm^{-1}, $\tilde{A}_0 = 8.0012$ cm^{-1}, $\tilde{B}_0 = 0.43304$ cm^{-1}, and $\tilde{C}_0 = 0.41040$ cm^{-1}. Determine the three characteristic vibrational temperatures and the characteristic rotational temperatures for each of the principle axes of $NO_2(g)$ at 1000 K. Calculate the value of $\overline{C}_V$ at 1000 K.

18-30. The experimental heat capacity of $NH_3(g)$ can be fit to the empirical formula

$$\overline{C}_V(T)/R = 2.115 + (3.919 \times 10^{-3} \text{ K}^{-1})T - (3.66 \times 10^{-7} \text{ K}^{-2})T^2$$

over the temperature range 300 K $< T <$ 1500 K. Plot $\overline{C}_V(T)/R$ versus T over this range using Equation 18.62 and the molecular parameters in Table 18.4, and compare your results with the experimental curve.

18-31. The experimental heat capacity of $SO_2(g)$ can be fit to the empirical formula

$$\overline{C}_V(T)/R = 6.8711 - \frac{1454.62 \text{ K}}{T} + \frac{160\,351 \text{ K}^2}{T^2}$$

over the temperature range 300 K $< T <$ 1500 K. Plot $\overline{C}_V(T)/R$ versus T over this range using Equation 18.62 and the molecular parameters in Table 18.4, and compare your results with the experimental curve.

18-32. The experimental heat capacity of $CH_4(g)$ can be fit to the empirical formula

$$\overline{C}_V(T)/R = 1.099 + (7.27 \times 10^{-3} \text{ K}^{-1})T + (1.34 \times 10^{-7} \text{ K}^{-2})T^2$$
$$-(8.67 \times 10^{-10} \text{ K}^{-3})T^3$$

over the temperature range 300 K $< T <$ 1500 K. Plot $\overline{C}_V(T)/R$ versus T over this range using Equation 18.62 and the molecular parameters in Table 18.4, and compare your results with the experimental curve.

18-33. Show that the moment of inertia of a diatomic molecule is μR_e^2, where μ is the reduced mass, and R_e is the equilibrium bond length.

18-34. Given that the values of Θ_{rot} and Θ_{vib} for H_2 are 85.3 K and 6332 K, respectively, calculate these quantities for HD and D_2. *Hint*: Use the Born-Oppenheimer approximation.

18-35. Using the result for $q_{rot}(T)$ obtained in Problem 18–14, derive corrections to the expressions $\langle E_{rot} \rangle = RT$ and $C_{V,rot} = R$ given in Section 18–5. Express your result in terms of powers of Θ_{rot}/T.

18-36. Show that the thermodynamic quantities P and C_V are independent of the choice of a zero of energy.

18-37. Molecular nitrogen is heated in an electric arc. The spectroscopically determined relative populations of excited vibrational levels are listed below.

v	0	1	2	3	4	$\cdots$
$\dfrac{f_v}{f_0}$	1.000	0.200	0.040	0.008	0.002	$\cdots$

Is the nitrogen in thermodynamic equilibrium with respect to vibrational energy? What is the vibrational temperature of the gas? Is this value necessarily the same as the translational temperature? Why or why not?

18-38. Consider a system of independent diatomic molecules constrained to move in a plane, that is, a two-dimensional ideal diatomic gas. How many degrees of freedom does a two-dimensional diatomic molecule have? Given that the energy eigenvalues of a two-dimensional rigid rotator are

$$\varepsilon_J = \frac{\hbar^2 J^2}{2I} \qquad J = 0,\ 1,\ 2,\ \ldots$$

(where I is the moment of inertia of the molecule) with a degeneracy $g_J = 2$ for all J except $J = 0$, derive an expression for the rotational partition function. The vibrational partition function is the same as for a three-dimensional diatomic gas. Write out

$$q(T) = q_{\text{trans}}(T)q_{\text{rot}}(T)q_{\text{vib}}(T)$$

and derive an expression for the average energy of this two-dimensional ideal diatomic gas.

18-39. What molar constant-volume heat capacities would you expect under classical conditions for the following gases: (a) Ne, (b) O_2, (c) H_2O, (d) CO_2, and (e) $CHCl_3$?

18-40. In Chapter 13, we learned that the harmonic-oscillator model can be corrected to include anharmonicity. The energy of an anharmonic oscillator was given as (Equation 13.21)

$$\tilde{\varepsilon}_v = \left(v + \tfrac{1}{2}\right)\tilde{\nu}_e - \tilde{x}_e\tilde{\nu}_e\left(v + \tfrac{1}{2}\right)^2 + \cdots$$

where the frequency $\tilde{\nu}_e$ is expressed in cm^{-1}. Substitute this expression for $\tilde{\varepsilon}_v$ into the summation for the vibrational partition function to obtain

$$q_{\text{vib}}(T) = \sum_{v=0}^{\infty} e^{-\beta\tilde{\nu}_e(v+\frac{1}{2})}e^{\beta\tilde{x}_e\tilde{\nu}_e(v+\frac{1}{2})^2}$$

Now expand the second factor in the summand, keeping only the linear term in $\tilde{x}_e\tilde{\nu}_e$, to obtain

$$q_{\text{vib}}(T) = \frac{e^{-\Theta_{\text{vib}}/2T}}{1 - e^{-\Theta_{\text{vib}}/T}} + \beta\tilde{x}_e\tilde{\nu}_e e^{-\Theta_{\text{vib}}/2T}\sum_{v=0}^{\infty}\left(v + \tfrac{1}{2}\right)^2 e^{-\Theta_{\text{vib}}v/T} + \cdots$$

where $\Theta_{\text{vib}}/T = \beta\tilde{\nu}_e$. Given that (Problem I–15)

$$\sum_{v=0}^{\infty} vx^v = \frac{x}{(1-x)^2}$$

and

$$\sum_{v=0}^{\infty} v^2 x^v = \frac{x^2 + x}{(1-x)^3}$$

show that

$$q_{\text{vib}}(T) = q_{\text{vib,ho}}(T) \left[1 + \beta \tilde{x}_e \tilde{\nu}_e \left(\tfrac{1}{4} + 2 q_{\text{vib,ho}}^2(T) \right) + \cdots \right]$$

where $q_{\text{vib,ho}}(T)$ is the harmonic-oscillator partition function. Estimate the magnitude of the correction for $\text{Cl}_2(\text{g})$ at 300 K, for which $\Theta_{\text{vib}} = 805$ K and $\tilde{x}_e \tilde{\nu}_e = 2.675$ cm^{-1}.

18-41. Prove that

$$\int_0^{\infty} e^{-\alpha n^2} dn \approx \int_1^{\infty} e^{-\alpha n^2} dn$$

if α is very small. *Hint*: Prove that

$$\int_0^1 e^{-\alpha n^2} dn \ll \int_0^{\infty} e^{-\alpha n^2} dn$$

by expanding the exponential in the first integral.

18-42. In this problem, we will derive an expression for the number of translational energy states with (translational) energy between ε and $\varepsilon + d\varepsilon$. This expression is essentially the degeneracy of the state whose energy is

$$\varepsilon_{n_x n_y n_z} = \frac{h^2}{8ma^2}(n_x^2 + n_y^2 + n_z^2) \qquad n_x,\ n_y,\ n_z = 1,\ 2,\ 3,\ \ldots \tag{1}$$

The degeneracy is given by the number of ways the integer $M = 8ma^2\varepsilon/h^2$ can be written as the sum of the squares of three positive integers. In general, this is an erratic and discontinuous function of M (the number of ways will be zero for many values of M), but it becomes smooth for large M, and we can derive a simple expression for it. Consider a three-dimensional space spanned by n_x, n_y, and n_z. There is a one-to-one correspondence between energy states given by Equation 1 and the points in this n_x, n_y, n_z space with coordinates given by positive integers. Figure 18.8 shows a two-dimensional version of this space. Equation 1 is an equation for a sphere of radius $R = (8ma^2\varepsilon/h^2)^{1/2}$ in this space

$$n_x^2 + n_y^2 + n_z^2 = \frac{8ma^2\varepsilon}{h^2} = R^2$$

We want to calculate the number of lattice points that lie at some fixed distance from the origin in this space. In general, this is very difficult, but for large R we can proceed as follows. We treat R, or ε, as a continuous variable and ask for the number of lattice points between ε and $\varepsilon + \Delta\varepsilon$. To calculate this quantity, it is convenient to first calculate the number of lattice points consistent with an energy $\leq \varepsilon$. For large ε, an excellent approximation can be made by equating the number of lattice points consistent with an energy $\leq \varepsilon$ with the volume of one octant of a sphere of radius R. We take only one octant because n_x, n_y, and

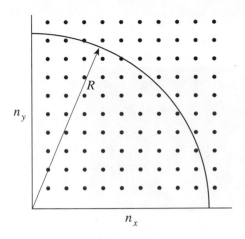

FIGURE 18.8
A two-dimensional version of the (n_x, n_y, n_z) space, the space with the quantum numbers n_x, n_y, and n_z as axes. Each point corresponds to an energy of a particle in a (two-dimensional) box.

n_z are restricted to be positive integers. If we denote the number of such states by $\Phi(\varepsilon)$, we can write

$$\Phi(\varepsilon) = \frac{1}{8}\left(\frac{4\pi R^3}{3}\right) = \frac{\pi}{6}\left(\frac{8ma^2\varepsilon}{h^2}\right)^{3/2}$$

The number of states with energy between ε and $\varepsilon + \Delta\varepsilon$ ($\Delta\varepsilon/\varepsilon \ll 1$) is

$$\omega(\varepsilon, \Delta\varepsilon) = \Phi(\varepsilon + \Delta\varepsilon) - \Phi(\varepsilon)$$

Show that

$$\omega(\varepsilon, \Delta\varepsilon) = \frac{\pi}{4}\left(\frac{8ma^2}{h^2}\right)^{3/2}\varepsilon^{1/2}\Delta\varepsilon + O[(\Delta\varepsilon)^2]$$

Show that if we take $\varepsilon = 3k_BT/2$, $T = 300$ K, $m = 10^{-25}$ kg, $a = 1$ dm, and $\Delta\varepsilon$ to be 0.010ε (in other words 1% of ε), then $\omega(\varepsilon, \Delta\varepsilon)$ is $O(10^{28})$. So, even for a system as simple as a single particle in a box, the degeneracy can be very large at room temperature.

18-43. The translational partition function can be written as a single integral over the energy ε if we include the degeneracy

$$q_{\text{trans}}(V, T) = \int_0^\infty \omega(\varepsilon)e^{-\varepsilon/k_BT}d\varepsilon$$

where $\omega(\varepsilon)d\varepsilon$ is the number of states with energy between ε and $\varepsilon + d\varepsilon$. Using the result from the previous problem, show that $q_{\text{trans}}(V, T)$ is the same as that given by Equation 18.6.

James Prescott Joule was born in Salford, near Manchester, England, on December 24, 1818, and died in 1889. He and his elder brother were tutored at home by John Dalton, then in his 70s. Joule's father was a wealthy brewer, which allowed Joule freedom from having to seek employment. Joule conducted his pioneering experiments in laboratories he built at his own expense in his home or in his father's brewery. From 1837 to 1847, he carried out a series of experiments that led to the general law of energy conservation and to the mechanical equivalent of heat. Joule announced all his measurements in a public lecture at St. Ann's Church in Manchester, England and, because his earlier reports had been rejected by the British Association, later had his lecture published in the *Manchester Courier*, a newspaper for which his brother wrote musical critiques. In 1847, he presented his results to the British Association meeting in Oxford, where the 22-year-old William Thomson (later Lord Kelvin) immediately appreciated the importance of Joule's work. Thomson later asked Joule to carry out experiments on the expansion of gases. This work led to the discovery of the Joule-Thomson effect, which demonstrated that a nonideal gas cools when undergoing a free expansion. Joule was elected to the Royal Society in 1850. Later in life, he suffered severe financial losses, and in 1878 friends obtained a pension for him from the government. The SI unit of energy is named in his honor.

CHAPTER 19

The First Law of Thermodynamics

Thermodynamics is the study of the various properties and, particularly, the relations between the various properties of systems in equilibrium. It is primarily an experimental science that was developed in the 1800s and still is of great practical value in many fields, such as chemistry, biology, geology, physics, and engineering. For example, we will use thermodynamics to show the quantitative relationship between the vapor pressure of a liquid and its heat of vaporization, or to show that if a gas obeys the equation of state $P\overline{V} = RT$, then its energy depends only upon its temperature. One of the most important and fruitful applications of thermodynamics is the analysis of chemical equilibria, where thermodynamics can be used to determine the temperature and pressure that optimize the products of a given chemical reaction. No industrial process would ever be undertaken without a thorough thermodynamic analysis of the chemical reactions involved.

All the results of thermodynamics are based on three fundamental laws. These laws summarize an enormous body of experimental data, and there are absolutely no known exceptions. In fact, Einstein said of thermodynamics:

> A theory is the more impressive the greater the simplicity of its premises is, the more different kinds of things it relates, and the more extended is its area of applicability. Therefore, the deep impression which classical thermodynamics made upon me. It is the only physical theory of universal content concerning which I am convinced that, within the framework of the applicability of its basic concepts, it will never be overthrown.[1]

Einstein's assessment is worth comment. Realize that thermodynamics was developed in the 1800s before the atomic theory of matter was generally accepted. The laws and results of thermodynamics are not based on any atomic or molecular theory; they

[1]From "Albert Einstein: Philosopher-Scientist", edited by P.A. Schlipp, Open Court Publishing company, La Salle, IL (1973).

are independent of atomic and molecular models. The development of thermodynamics along these lines is called *classical thermodynamics*. This character of classical thermodynamics is both a strength and a weakness. We can be assured that classical thermodynamic results will never need to be modified as our knowledge of atomic and molecular structure improves, but classical thermodynamics gives us only a limited insight at the molecular level.

With the development of atomic and molecular theories in the late 1800s and early 1900s, thermodynamics was given a molecular interpretation, or a molecular basis. This field is called *statistical thermodynamics* because it relates averages of molecular properties to macroscopic thermodynamic properties such as temperature or pressure. The material in Chapters 17 and 18 is actually an elementary treatment of statistical thermodynamics. Many of the results of statistical thermodynamics depend upon the molecular models used, so these results are not as solidly based as those of classical thermodynamics. Nevertheless, the intuitive advantage of having a molecular picture of certain quantities or processes is very convenient. Consequently, in our development of thermodynamics in this and the following chapters, we will use a mixture of classical and statistical thermodynamics, even though this approach will cost us some of the rigor of the results.

The First Law of Thermodynamics is the law of conservation of energy applied to macroscopic systems. To present the first law, we must introduce the concepts of work and heat as they are used in thermodynamics. As we will see in the next section, work and heat are modes of energy transfer between a system and its surroundings.

19–1. A Common Type of Work Is Pressure–Volume Work

The concepts of work and heat play important roles in thermodynamics. Both work and heat refer to the manner in which energy is transferred between some system of interest and its surroundings. By *system* we mean that part of the world we are investigating and by *surroundings* we mean everything else. We define *heat*, q, to be the manner of energy transfer that results from a temperature difference between the system and its surroundings. Heat input to a system is considered a positive quantity; heat evolved by a system is considered a negative quantity. We define *work*, w, to be the transfer of energy between the system of interest and its surroundings as a result of the existence of unbalanced forces between the two. If the energy of the system is increased by the work, we say that work is done *on* the system by the surroundings, and we take it to be a positive quantity. On the other hand, if the energy of the system is decreased by the work, we say that the system does work on the surroundings, or that work is done *by* the system, and we take it to be a negative quantity. A common example of work in physical chemistry occurs during the expansion or compression of a gas as a result of the difference in pressures exerted by the gas and on the gas.

An important aspect of work is that it can always be related to the raising or lowering of a mass in the surroundings. To see the consequences of this statement, consider the situation in Figure 19.1, where a gas is confined to a cylinder that exerts a force Mg on the gas. In Figure 19.1a, the initial pressure of the gas, P_i, is sufficient

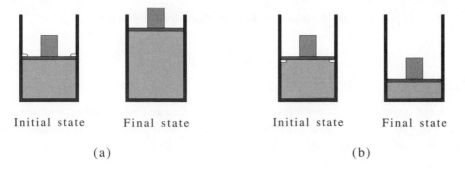

Initial state	Final state		Initial state	Final state
(a)			(b)	

FIGURE 19.1
The effect of work is equivalent to the raising or lowering of a mass in the surroundings. In (a) work is done *by* the system because the mass is raised, and in (b) work is done *on* the system because the mass is lowered. (The system is defined as the gas inside the piston.)

to push the piston upward, so there are pins holding it in position. Now, remove the pins and allow the gas to lift the mass upward to the new position shown, and let the pressure of the gas now be P_f. In this process, the mass M has been raised a distance h, so the work done by the system is

$$w = -Mgh$$

The negative sign here is in accord with our convention that work done *by* a system is taken to be a negative quantity. If we divide Mg by A, the area of the piston, and multiply h by A, then we have

$$w = -\frac{Mg}{A} \cdot Ah$$

But Mg/A is the external pressure exerted on the gas and Ah is the change in volume experienced by the gas, so we have

$$w = -P_{ext}\Delta V \tag{19.1}$$

Note that $\Delta V > 0$ in an expansion, so $w < 0$. Clearly, the external pressure must be less than the pressure of the inital state of the gas in order that the expansion occur. After the expansion, $P_{ext} = P_f$.

Now consider the situation in Figure 19.1b, where the initial pressure of the gas is less than the external pressure $P_{ext} = Mg/A$, so the gas is compressed when the pins are removed. In this case, the mass M is lowered a distance h, and the work is given by

$$w = -Mgh = -\frac{Mg}{A}(Ah) = -P_{ext}\Delta V$$

But now $\Delta V < 0$, so $w > 0$. After the compression, we have $P_{ext} = P_f$. The work is positive because work is done *on* the gas when it is compressed.

If P_{ext} is not constant during the expansion, the work is given by

$$w = -\int_{V_i}^{V_f} P_{ext} dV \qquad (19.2)$$

where the limits on the integral indicate an initial state and a final state; we must have knowledge of how P_{ext} varies with V along the path connecting these two states so we can carry out the integration in Equation 19.2. Equation 19.2 is applicable to either expansion or compression. If P_{ext} is constant, Equation 19.2 gives Equation 19.1

$$w = -P_{ext}(V_f - V_i) = -P_{ext}\Delta V$$

EXAMPLE 19–1

Consider an ideal gas that occupies 1.00 dm^3 at a pressure of 2.00 bar. If the gas is compressed isothermally at a constant external pressure, P_{ext}, so that the final volume is 0.500 dm^3, what is the smallest value P_{ext} can have? Calculate the work involved using this value of P_{ext}.

SOLUTION: For a compression to occur, the value of P_{ext} must be at least as large as the final pressure of the gas. Given the inital pressure and volume, and the final volume, we can determine the final pressure. The final pressure of the gas is

$$P_f = \frac{P_i V_i}{V_f} = \frac{(2.00 \text{ bar})(1.00 \text{ dm}^3)}{0.500 \text{ dm}^3} = 4.00 \text{ bar}$$

This is the smallest value P_{ext} can be to compress the gas isothermally from 1.00 dm^3 to 0.500 dm^3. The work involved using this value of P_{ext} is

$$w = -P_{ext}\Delta V = -(4.00 \text{ bar})(-0.500 \text{ dm}^3) = 2.00 \text{ dm}^3 \cdot \text{bar}$$
$$= (2.00 \text{ dm}^3 \cdot \text{bar})(10^{-3} \text{ m}^3 \cdot \text{dm}^{-3})(10^5 \text{ Pa} \cdot \text{bar}^{-1}) = 200 \text{ Pa} \cdot \text{m}^3 = 200 \text{ J}$$

Of course, P_{ext} can be any value greater than 4.00 bar, so 200 J represents the smallest value of w for the isothermal compression at constant pressure from a volume of 1.00 dm^3 to 0.500 dm^3.

Figure 19.2 illustrates the work involved in Example 19–1. As Equation 19.2 implies, the work is the area under the curve of P_{ext} versus V. The smooth curve is an isotherm (P versus V at constant T) of an ideal gas; Figure 19.2a shows a constant-pressure compression at an external pressure equal to P_f, the final pressure of the gas; and Figure 19.2b shows one at an external pressure greater than P_f. We see that the work is different for different values of P_{ext}.

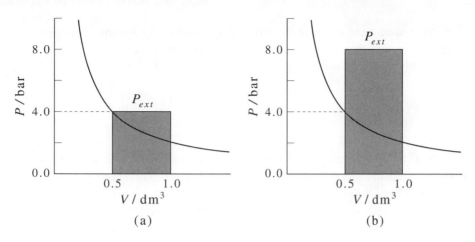

FIGURE 19.2
An illustration of the work involved in an isothermal constant-pressure compression from $V_i = 1.00$ dm^3 to $V_f = 0.500$ dm^3 at different values of P_{ext}. The smooth curve is an isotherm (P vs. V at constant T) for an ideal gas. In (a) P_{ext} is equal to P_f, the final pressure of the gas, and in (b) P_{ext} is larger than $P_f = 4.00$ bar, and pins must be used to stop the compression at $V_f = 0.500$ dm^3. Otherwise the gas would be compressed further, until it reaches the volume that corresponds to P_{ext} on the isotherm. The work is equal to the area of the P_{ext}-V rectangles.

19–2. Work and Heat Are Not State Functions, but Energy Is a State Function

Work and heat have a property that makes them quite different from energy. To appreciate this difference, we must first discuss what we mean by the state of a system. We say that a system is in a definite state when all the variables needed to describe the system completely are defined. For example, the state of one mole of an ideal gas can be described completely by specifying P, $\overline{V}$, and T. In fact, because P, $\overline{V}$, and T are related by $P\overline{V} = RT$, any two of these three variables will suffice to specify the state of the gas. Other systems may require more variables, but usually only a few will suffice. A *state function* is a property that depends only upon the state of the system, and not upon how the system was brought to that state, or upon the history of the system. Energy is an example of a state function. An important mathematical property of a state function is that its differential can be integrated in a normal way:

$$\int_1^2 dU = U_2 - U_1 = \Delta U \tag{19.3}$$

As the notation suggests, the value of ΔU is *independent* of the path taken between the initial and final states 1 and 2; it depends only upon the initial and final states through $\Delta U = U_2 - U_1$.

Work and heat are *not* state functions. For example, the external pressure used to compress a gas can have any value as long as it is large enough to compress the gas. Consequently, the work done on the gas,

$$w = -\int_1^2 P_{ext} dV$$

will depend upon the pressure used to compress the gas. The value of P_{ext} must exceed the pressure of the gas to compress it. The minimum work required occurs when P_{ext} is just infinitesimally greater than the pressure of the gas at every stage of the compression, which means that the gas is essentially in equilibrium during the entire compression. In this special but important case, we can replace P_{ext} by the pressure of the gas (P) in Equation 19.2. When P_{ext} and P differ only infinitesimally, the process is called a *reversible process* because the process could be reversed (from compression to expansion) by decreasing the external pressure infinitesimally. Necessarily, a strictly reversible process would require an infinite time to carry out because the process must be adjusted by an infinitesimal amount at each stage. Nevertheless, a reversible process serves as a useful idealized limit.

Figure 19.3 shows that a reversible, isothermal compression of a gas requires the minimum possible amount of work. Let w_{rev} denote the reversible work. To calculate w_{rev} for the compression of an ideal gas isothermally from V_1 to V_2, we use Equation 19.2

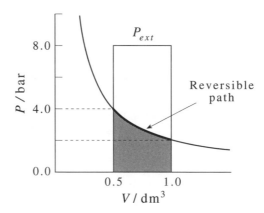

FIGURE 19.3
The work of isothermal compression is the area under the P_{ext} versus V curves shown in the figure. The external pressure must exceed the pressure of the gas in order to compress it. The minimum amount of work occurs when the expansion is carried out reversibly; that is, when P_{ext} is just infinitesimally greater than the pressure of the gas at every stage of the compression. The gray area is the minimum work needed to compress the gas from $V_1 = 1.00$ dm^3 to $V_2 = 0.500$ dm^3. The constant-pressure compression curves are the same as those in Figure 19.2.

with P_{ext} replaced by the equilibrium value of the pressure of the gas, which is nRT/V for an ideal gas. Therefore,

$$w_{rev} = -\int_1^2 P_{gas} dV = -\int_1^2 \frac{nRT}{V} dV = -nRT \int_1^2 \frac{dV}{V}$$

$$= -nRT \ln \frac{V_2}{V_1} \qquad (19.4)$$

Because $V_2 < V_1$ for compression, we see that $w_{rev} > 0$ as it should be; in other words, we have done work on the gas.

EXAMPLE 19–2

Consider an ideal gas that occupies 1.00 dm^3 at 2.00 bar. Calculate the work required to compress the gas isothermally to a volume of 0.667 dm^3 at a constant pressure of 3.00 bar followed by another isothermal compression to 0.500 dm^3 at a constant pressure of 4.00 bar (Figure 19.4). Compare the result with the work of compressing the gas isothermally and reversibly from 1.00 dm^3 to 0.500 dm^3. Compare both results to the one obtained in Example 19–1.

SOLUTION: In the two-stage compression, $\Delta V = -(1.00 - 0.667)$ dm^3 in the first step and $-(0.667 - 0.500)$ dm^3 in the second step. Therefore,

$$w = -(3.00 \text{ bar})(-0.333 \text{ dm}^3) - (4.00 \text{ bar})(-0.167 \text{ dm}^3)$$

$$= 1.67 \text{ dm}^3 \cdot \text{bar} = 167 \text{ J}$$

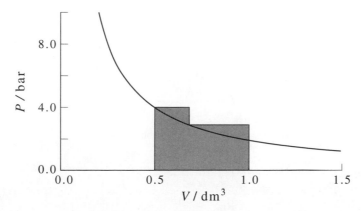

FIGURE 19.4
An illustration of the constant-pressure compression of a gas as described in Example 19–2. The work required is given by the areas under the two rectangles.

We use Equation 19.4 for the reversible process

$$w_{rev} = -nRT \ln \frac{V_2}{V_1} = -nRT \ln \frac{0.500 \text{ dm}^3}{1.00 \text{ dm}^3}$$

Because the gas is ideal and the process is isothermal, nRT is equal to either $P_1 V_1$ or $P_2 V_2$, both of which equal 2.00 dm$^3 \cdot$ bar, and so

$$w_{rev} = -(2.00 \text{ dm}^3 \cdot \text{bar}) \ln 0.500 = 1.39 \text{ dm}^3 \cdot \text{bar} = 139 \text{ J}$$

Note that w_{rev} is less than that for the two-stage process and that the work for that process is less than the work required in Example 19–1 (200 J). (Compare Figures 19.2, 19.3, and 19.4.)

Just as the reversible isothermal compression of a gas requires the minimum amount of work to be done on the gas, a reversible isothermal expansion requires the gas to do a maximum amount of work in the process. In a reversible expansion, the external pressure is infinitesimally less than the pressure of the gas at each stage. If P_{ext} were any larger, the expansion would not occur. The work involved in the reversible isothermal expansion of an ideal gas is also given by Equation 19.4. Because $V_2 > V_1$ for expansion, we see that $w_{rev} < 0$; the gas has done work on the surroundings, in fact, the maximum possible.

EXAMPLE 19–3
Derive an expression for the reversible isothermal work of an expansion of a van der Waals gas.

SOLUTION: The expression for the reversible work is

$$w_{rev} = -\int_1^2 P dV$$

where

$$P = \frac{nRT}{V - nb} - \frac{an^2}{V^2}$$

We substitute this expression for P into w_{rev} to obtain

$$w_{rev} = -nRT \int_1^2 \frac{dV}{V - nb} - an^2 \int_1^2 \frac{dV}{V^2}$$
$$= -nRT \ln \frac{V_2 - nb}{V_1 - nb} + an^2 \left(\frac{1}{V_2} - \frac{1}{V_1} \right)$$

Note that this equation reduces to Equation 19.4 when $a = b = 0$.

19–3. The First Law of Thermodynamics Says the Energy Is a State Function

Because the work involved in a process depends upon how the process is carried out, work is *not* a state function. Thus, we write

$$\int_1^2 \delta w = w \qquad \text{(not } \Delta w \text{ or } w_2 - w_1) \tag{19.5}$$

It makes no sense at all to write w_2, w_1, $w_2 - w_1$, or Δw. The value of w obtained in Equation 19.5 depends upon the *path* from state 1 to 2, so work is called a *path function*. Mathematically, δw in Equation 19.5 is called an *inexact differential*, as opposed to an *exact differential* like dU, which can be integrated in the normal way to obtain $U_2 - U_1$ (see MathChapter H).

Work and heat are defined only for processes in which energy is transferred between a system and its surroundings. Both work and heat are path functions. Although a system in a given state has a certain amount of energy, it does not possess work or heat. The difference between energy and work and heat can be summarized by writing

$$\int_1^2 dU = U_2 - U_1 = \Delta U \qquad (U \text{ is a state function}) \tag{19.6}$$

$$\int_1^2 \delta w = w \quad \text{(not } w_2 - w_1) \qquad \text{(path function)} \tag{19.7}$$

and

$$\int_1^2 \delta q = q \quad \text{(not } q_2 - q_1) \qquad \text{(path function)} \tag{19.8}$$

For a process in which energy is transferred both as work and heat, the law of conservation of energy says that the energy of the system obeys the equation

$$dU = \delta q + \delta w \tag{19.9}$$

in differential form, or

$$\Delta U = q + w \tag{19.10}$$

in integrated form. Equations 19.9 and 19.10 are statements of the *First Law of Thermodynamics*. The First Law of Thermodynamics, which is essentially a statement of the law of conservation of energy, also says that even though δq and δw are separately path functions or inexact differentials, their sum is a state function or an exact differential. All state functions are exact differentials.

19–4. An Adiabatic Process Is a Process in Which No Energy as Heat Is Transferred

Not only are work and heat not state functions, but we can prove that even reversible work and reversible heat are not state functions by a direct calculation. Consider the three paths, depicted in Figure 19.5, that occur between the same initial and final states, P_1, V_1, T_1 and P_2, V_2, T_1. Path A involves a reversible isothermal expansion of an ideal gas from P_1, V_1, T_1 to P_2, V_2, T_1. Because the energy of an ideal gas depends upon only the temperature (see Equation 18.40, for example),

$$\Delta U_{\text{A}} = 0 \tag{19.11}$$

so

$$\delta w_{\text{rev,A}} = -\delta q_{\text{rev,A}}$$

for an isothermal process involving an ideal gas. Furthermore, because the process is reversible,

$$\delta w_{\text{rev,A}} = -\delta q_{\text{rev,A}} = -\frac{RT_1}{V}dV \tag{19.12}$$

so

$$w_{\text{rev,A}} = -q_{\text{rev,A}} = -RT_1 \int_{V_1}^{V_2} \frac{dV}{V} = -RT_1 \ln \frac{V_2}{V_1} \tag{19.13}$$

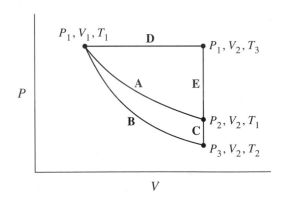

FIGURE 19.5
An illustration of three different pathways (A, B + C, and D + E) to take an ideal gas from P_1, V_1, T_1 to P_2, V_2, T_1. In each case, the value of ΔU is the same (ΔU is a state function), but the values of q and w are different (q and w are path functions).

Note that w_{rev} is negative ($V_2 > V_1$) because work is done by the gas. Furthermore, q_{rev} is positive because energy as heat entered the system to maintain the temperature constant as the system used its energy to do the work.

Another path (B + C) in Figure 19.5 consists of two parts. The first part (B) involves a reversible expansion from P_1, V_1, T_1 to P_3, V_2, T_2 and is carried out such that no energy as heat is transferred between the system and its surroundings. A process in which no energy as heat is transferred is called an *adiabatic process*. For an adiabatic process, $q = 0$, so

$$dU = \delta w \tag{19.14}$$

Path C of the path B + C involves heating the gas reversibly at constant volume from P_3, V_2, T_2 to P_2, V_2, T_1. As stated above, ΔU depends upon only temperature; it is independent of P and V for an ideal gas. To calculate ΔU for a change from state 1 of temperature T_1 to state 2 of temperature T_2, recall that the constant volume heat capacity is defined as (Equation 17.25)

$$C_V(T) = \left(\frac{\partial U}{\partial T}\right)_V$$

and therefore, for an ideal gas

$$\frac{dU}{dT} = \left(\frac{\partial U}{\partial T}\right)_V = C_V(T)$$

or $dU = C_V(T)dT$, which can be integrated to give

$$\Delta U = \int_{T_1}^{T_2} C_V(T)dT$$

We can now calculate the total work involved in the path B + C. Because process B is adiabatic,

$$q_{rev,B} = 0 \tag{19.15}$$

so

$$w_{rev,B} = \Delta U_B = \int_{T_1}^{T_2} \left(\frac{\partial U}{\partial T}\right)_V dT = \int_{T_1}^{T_2} C_V(T)dT \tag{19.16}$$

For process C, no pressure-volume work is involved (it is a constant-volume process), so

$$q_{rev,C} = \Delta U_C = \int_{T_2}^{T_1} C_V(T)dT \tag{19.17}$$

For the total path B + C, then

$$q_{rev,B+C} = q_{rev,B} + q_{rev,C} = 0 + \int_{T_2}^{T_1} C_V(T)dT$$

$$= \int_{T_2}^{T_1} C_V(T)dT \tag{19.18}$$

and

$$w_{\text{rev,B+C}} = w_{\text{rev,B}} + w_{\text{rev,C}} = \int_{T_1}^{T_2} C_V(T)dT + 0$$

$$= \int_{T_1}^{T_2} C_V(T)dT \tag{19.19}$$

Note that

$$\Delta U_{\text{B+C}} = \Delta U_{\text{B}} + \Delta U_{\text{C}} = \int_{T_1}^{T_2} C_V(T)dT + \int_{T_2}^{T_1} C_V(T)dT = 0$$

which is the same as in path A, because the energy U is a state function. However, $w_{\text{rev,A}} \neq w_{\text{rev,B+C}}$ and $q_{\text{rev,A}} \neq q_{\text{rev,B+C}}$, because both work and heat are path functions.

EXAMPLE 19–4

Calculate ΔU, w_{rev}, and q_{rev} for the paths D + E in Figure 19.5, where D represents a reversible constant-pressure (P_1) expansion of an ideal gas from V_1, T_1 to V_2, T_3 and E represents cooling the gas reversibly from T_3 to T_1 at a constant volume V_2.

SOLUTION: For path D,

$$\Delta U_{\text{D}} = \int_{T_1}^{T_3} C_V(T)dT$$

$$w_{\text{rev,D}} = -P_1(V_2 - V_1)$$

and

$$q_{\text{rev,D}} = \Delta U_{\text{D}} - w_{\text{rev,D}} = \int_{T_1}^{T_3} C_V(T)dT + P_1(V_2 - V_1)$$

For path E,

$$\Delta U_{\text{E}} = \int_{T_3}^{T_1} C_V(T)dT$$

$$w_{\text{rev,E}} = 0$$

and

$$q_{\text{rev,E}} = \Delta U_{\text{E}} = \int_{T_3}^{T_1} C_V(T)dT$$

Therefore, for the overall process,

$$\Delta U_{\text{D+E}} = \Delta U_{\text{D}} + \Delta U_{\text{E}} = \int_{T_1}^{T_3} C_V(T)dT + \int_{T_3}^{T_1} C_V(T)dT = 0$$

$$w_{\text{rev,D+E}} = w_{\text{rev,D}} + w_{\text{rev,E}} = -P_1(V_2 - V_1)$$

and

$$q_{rev,D+E} = q_{rev,D} + q_{rev,E} = P_1(V_2 - V_1)$$

Note that $\Delta U = 0$ for all three processes indicated in Figure 19.5, but that w_{rev} and q_{rev} are different for each one.

19–5. The Temperature of a Gas Decreases in a Reversible Adiabatic Expansion

Path B in Figure 19.5 represents the reversible adiabatic expansion of an ideal gas from T_1, V_1 to T_2, V_2. As the figure suggests, $T_2 < T_1$, which means that the gas cools during a (reversible) adiabatic expansion. We can determine the final temperature T_2 for this process. For an adiabatic process, $q = 0$, and so

$$dU = \delta w = dw$$

Note that the above expression tells us that $\delta w = dw$ is an exact differential when $\delta q = 0$. Likewise, $\delta q = dq$ is an exact differential if $\delta w = 0$. The work done by the gas (the system) in the expansion is "paid for" by a decrease in the energy of the gas, which amounts to a decrease in the temperature of the gas. Because the work involved in a reversible expansion is maximum, the gas must suffer a maximum drop in temperature in a reversible adiabatic expansion. Recall that for an ideal gas, U depends only upon the temperature and $dU = C_V(T)dT = n\overline{C}_V(T)dT$, where $\overline{C}_V(T)$ is the molar constant-volume heat capacity. Using the fact that $dw = -PdV = -nRTdV/V$ for a reversible expansion, the relation $dU = dw$ gives

$$C_V(T)dT = -\frac{nRT}{V}dV \tag{19.20}$$

We divide both sides by T and n and integrate to obtain

$$\int_{T_1}^{T_2} \frac{\overline{C}_V(T)}{T} dT = -R \int_{V_1}^{V_2} \frac{dV}{V} = -R \ln \frac{V_2}{V_1} \tag{19.21}$$

We learned in Section 18–2 that $\overline{C}_V = 3R/2$ for a monatomic ideal gas, so Equation 19.21 becomes

$$\frac{3R}{2} \int_{T_1}^{T_2} \frac{dT}{T} = \frac{3R}{2} \ln \frac{T_2}{T_1} = -R \ln \frac{V_2}{V_1}$$

or

$$\frac{3}{2} \ln \frac{T_2}{T_1} = -\ln \frac{V_2}{V_1} = \ln \frac{V_1}{V_2}$$

or

$$\left(\frac{T_2}{T_1}\right)^{3/2} = \frac{V_1}{V_2} \qquad \left(\begin{array}{c}\text{monatomic} \\ \text{ideal gas}\end{array}\right) \tag{19.22}$$

Thus, the gas cools in a reversible adiabatic expansion ($V_2 > V_1$).

EXAMPLE 19–5
Calculate the final temperature if argon (assumed to be ideal) at an initial temperature of 300 K expands reversibly and adiabatically from a volume of 50.0 L to 200 L.

SOLUTION: First solve Equation 19.22 for T_2/T_1,

$$\frac{T_2}{T_1} = \left(\frac{V_1}{V_2}\right)^{2/3}$$

and then let $T_1 = 300$ K, $V_1 = 50.0$ L, and $V_2 = 200$ L to obtain

$$T_2 = (300\text{K}) \left(\frac{50.0 \text{ L}}{200 \text{ L}}\right)^{2/3} = 119 \text{ K}$$

We can express Equation 19.22 in terms of pressure and volume by using $PV = nRT$ to eliminate T_1 and T_2:

$$\left(\frac{P_2 V_2}{P_1 V_1}\right)^{3/2} = \frac{V_1}{V_2}$$

Upon taking both sides to the 2/3 power and rearranging, we obtain

$$P_1 V_1^{5/3} = P_2 V_2^{5/3} \quad \text{(monatomic ideal gas)} \tag{19.23}$$

This equation shows how the pressure and volume are related in a reversible, adiabatic process for an ideal monatomic gas. Compare this result to Boyle's law, which says that

$$P_1 V_1 = P_2 V_2$$

for an isothermal process.

EXAMPLE 19–6
Derive the analogs of Equations 19.22 and 19.23 for an ideal diatomic gas. Assume the temperature is such that the vibrational contribution to the heat capacity can be ignored.

SOLUTION: Assuming that $\overline{C}_{V,\text{vib}} \approx 0$, we have from Equation 18.41 that $\overline{C}_V = 5R/2$. Equation 19.20 for a diatomic ideal gas becomes

$$\frac{5R}{2}\int_{T_1}^{T_2}\frac{dT}{T} = \frac{5R}{2}\ln\frac{T_2}{T_1} = -R\ln\frac{V_2}{V_1}$$

so

$$\left(\frac{T_2}{T_1}\right)^{5/2} = \frac{V_1}{V_2} \qquad \text{(diatomic ideal gas)}$$

Substituting $T = PV/nR$ into the above equation gives

$$\left(\frac{P_2 V_2}{P_1 V_1}\right)^{5/2} = \frac{V_1}{V_2}$$

or

$$P_1 V_1^{7/5} = P_2 V_2^{7/5} \qquad \text{(diatomic ideal gas)}$$

19–6. Work and Heat Have a Simple Molecular Interpretation

Let's go back to Equation 17.18 for the average energy of a macroscopic system,

$$U = \sum_j p_j(N, V, \beta) E_j(N, V) \tag{19.24}$$

with

$$p_j(N, V, \beta) = \frac{e^{-\beta E_j(N,V)}}{Q(N, V, \beta)} \tag{19.25}$$

Equation 19.24 represents the average energy of an equilibrium system that has the variables N, V, and T fixed. If we differentiate Equation 19.24, we obtain

$$dU = \sum_j p_j dE_j + \sum_j E_j dp_j \tag{19.26}$$

Because $E_j = E_j(N, V)$, we can view dE_j as the change in E_j due to a small change in the volume, dV, keeping N fixed. Therefore, substituting $dE_j = (\partial E_j/\partial V)_N dV$ into Equation 19.26 gives

$$dU = \sum_j p_j \left(\frac{\partial E_j}{\partial V}\right)_N dV + \sum_j E_j dp_j$$

This result suggests we can interpret the first term in Equation 19.26 to be the average change in energy of the system caused by a small change in its volume, in other words, the average work.

Furthermore, if this change is done reversibly, so that the system remains essentially in equilibrium at each stage, then the p_j in Equation 19.26 will be given by Equation 19.25 thoughout the entire process. We can emphasize this by writing

$$dU = \sum_j p_j(N, V, \beta) \left(\frac{\partial E_j}{\partial V}\right)_N dV + \sum_j E_j(N, V)dp_j(N, V, \beta) \qquad (19.27)$$

If we compare this result with the macroscopic equation (Equation 19.9)

$$dU = \delta w_{\text{rev}} + \delta q_{\text{rev}} \qquad (19.28)$$

we see that

$$\delta w_{\text{rev}} = \sum_j p_j(N, V, \beta) \left(\frac{\partial E_j}{\partial V}\right)_N dV \qquad (19.29)$$

and

$$\delta q_{\text{rev}} = \sum_j E_j(N, V)dp_j(N, V, \beta) \qquad (19.30)$$

Thus, we see that reversible work, δw_{rev}, results from an infinitesimal change in the allowed energies of a system, without changing the probability distribution of its states. Reversible heat, on the other hand, results from a change in the probability distribution of the states of a system, without changing the allowed energies.

If we compare Equation 19.29 with

$$\delta w_{\text{rev}} = -PdV$$

we see that we can identify the pressure of the gas with

$$P = -\sum_j p_j(N, V, \beta) \left(\frac{\partial E_j}{\partial V}\right)_N = -\left\langle \left(\frac{\partial E}{\partial V}\right)_N \right\rangle \qquad (19.31)$$

Recall that we used this equation without proof in Section 17–5 to show that $PV = RT$ for one mole of an ideal gas.

19–7. The Enthalpy Change Is Equal to the Energy Transferred as Heat in a Constant-Pressure Process Involving Only P–V Work

For a reversible process in which the only work involved is pressure-volume work, the first law tells us that

$$\Delta U = q + w = q - \int_{V_1}^{V_2} PdV \qquad (19.32)$$

If the process is carried out at constant volume, then $V_1 = V_2$ and

$$\Delta U = q_V \tag{19.33}$$

where the subscript V on q emphasizes that Equation 19.33 applies to a constant-volume process. Thus, we see that ΔU can be measured experimentally by measuring the energy as heat (by means of a calorimeter) associated with a constant-volume process (in a rigid closed container).

Many processes, particularly chemical reactions, are carried out at constant pressure (open to the atmosphere). The energy as heat associated with a constant-pressure process, q_P, is not equal to ΔU. It would be convenient to have a state function analogous to U so that we could write an expression like that in Equation 19.33. To this end, let P be constant in Equation 19.32 so that

$$q_P = \Delta U + P_{\text{ext}} \int_{V_1}^{V_2} dV = \Delta U + P\Delta V \tag{19.34}$$

where we have used the subscript P on q_P to emphasize that this is a constant-pressure process. This equation suggests that we define a new state function by

$$H = U + PV \tag{19.35}$$

At constant pressure,

$$\Delta H = \Delta U + P\Delta V \qquad \text{(constant pressure)} \tag{19.36}$$

Equation 19.34 shows that

$$q_P = \Delta H \tag{19.37}$$

Thus, this new state function H, called the *enthalpy*, plays the same role in a constant-pressure process that U plays in a constant-volume process. The value of ΔH can be determined experimentally by measuring the energy as heat associated with a constant-pressure process, or conversely, q_P can be determined from ΔH. Because most chemical reactions take place at constant pressure, the enthalpy is a practical and important thermodynamic function.

Let's apply these results to the melting of ice at 0°C and one atm. For this process, $q_P = 6.01 \text{ kJ} \cdot \text{mol}^{-1}$. Using Equation 19.37, we find that

$$\Delta \overline{H} = q_P = 6.01 \text{ kJ} \cdot \text{mol}^{-1}$$

where the overbar on H signifies that $\Delta \overline{H}$ is a molar quantity. We can also calculate the value of $\Delta \overline{U}$ using Equation 19.36 and the fact that the molar volume of ice $(\overline{V}_s)$ is 0.0196 L·mol^{-1} and that of water $(\overline{V}_l)$ is 0.0180 L·mol^{-1}:

$$\Delta \overline{U} = \Delta \overline{H} - P\Delta \overline{V}$$

$$= 6.01 \text{ kJ·mol}^{-1} - (1 \text{ atm})(0.0180 \text{ L·mol}^{-1} - 0.0196 \text{ L·mol}^{-1})$$

$$= 6.01 \text{ kJ·mol}^{-1} - (1.60 \times 10^{-3} \text{ L·atm·mol}^{-1}) \left(\frac{8.314 \text{ J}}{0.08206 \text{ L·atm}} \right) \left(\frac{1 \text{ kJ}}{10^3 \text{ J}} \right)$$

$$\approx 6.01 \text{ kJ·mol}^{-1}$$

Thus, in this case, there is essentially no difference between $\Delta \overline{H}$ and $\Delta \overline{U}$.

Let's look at the vaporization of water at 100°C and one atm. For this process, $q_P = 40.7 \text{ kJ·mol}^{-1}$, $\overline{V}_l = 0.0180 \text{ L·mol}^{-1}$, and $\overline{V}_g = 30.6 \text{ L·mol}^{-1}$. Therefore,

$$\Delta \overline{H} = q_P = 40.7 \text{ kJ·mol}^{-1}$$

But

$$\Delta \overline{V} = 30.6 \text{ L·mol}^{-1} - 0.0180 \text{ L·mol}^{-1} = 30.6 \text{ L·mol}^{-1}$$

so

$$\Delta \overline{U} = \Delta \overline{H} - P\Delta \overline{V}$$

$$= 40.7 \text{ kJ·mol}^{-1} - (1 \text{ atm})(30.6 \text{ L·mol}^{-1}) \left(\frac{8.314 \text{ J}}{0.08206 \text{ L·atm}} \right)$$

$$= 37.6 \text{ kJ·mol}^{-1}$$

Notice that the numerical values of $\Delta \overline{H}$ and $\Delta \overline{U}$ are significantly different ($\approx 8\%$) in this case because $\Delta \overline{V}$ for this process is fairly large. We can give a physical interpretation of these results. Of the 40.7 kJ that are absorbed at constant pressure, 37.6 kJ ($q_V = \Delta \overline{U}$) are used to overcome the intermolecular forces holding the water molecules in the liquid state (hydrogen bonds) and 3.1 kJ (40.7 kJ − 37.6 kJ) are used to increase the volume of the system against the atmospheric pressure.

EXAMPLE 19–7
The value of ΔH at 298 K and one bar for the reaction described by

$$2 \text{ H}_2(g) + \text{O}_2(g) \longrightarrow 2 \text{ H}_2\text{O}(l)$$

is −572 kJ. Calculate ΔU for this reaction as written.

SOLUTION: Because the reaction is carried out at a constant pressure of 1.00 bar, $\Delta H = q_P = -572 \text{ kJ}$. To calculate ΔU, we must first calculate ΔV. Initially, we have three moles of gas at 298 K and 1.00 bar, and so

$$V = \frac{nRT}{P} = \frac{(3 \text{ mol})(0.08314 \text{ L·bar·K}^{-1}\text{·mol}^{-1})(298 \text{ K})}{1.00 \text{ bar}}$$

$$= 74.3 \text{ L}$$

Afterward, we have two moles of liquid water, whose volume is about 36 mL, which is negligible compared with 74.3 L. Thus, $\Delta V = -74.3$ L and

$$\Delta U = \Delta H - P\Delta V$$
$$= -572 \text{ kJ} + (1.00 \text{ bar})(73.4 \text{ L}) \left(\frac{1 \text{ kJ}}{10 \text{ bar·L}} \right) = -572 \text{ kJ} + 7.43 \text{ kJ}$$
$$= -565 \text{ kJ}$$

The numerical difference between ΔH and ΔU in this case is about 1%.

Example 19–7 is a special case of a general result for reactions or processes that involve ideal gases, which says that

$$\Delta H = \Delta U + RT \Delta n_{\text{gas}} \tag{19.38}$$

where

$$\Delta n_{\text{gas}} = \left(\begin{array}{c} \text{number of moles of} \\ \text{gaseous products} \end{array} \right) - \left(\begin{array}{c} \text{number of moles of} \\ \text{gaseous reactants} \end{array} \right)$$

As Example 19–7 implies, the numerical difference between ΔH and ΔU is usually small.

19–8. Heat Capacity Is a Path Function

Recall that heat capacity is defined as the energy as heat required to raise the temperature of a substance by one kelvin. The heat capacity also depends upon the temperature T. Because the energy required to raise the temperature of a substance by one kelvin depends upon the amount of substance, heat capacity is an *extensive quantity*. Heat capacity is also a path function; for example, its value depends upon whether we heat the substance at constant volume or at constant pressure. If the substance is heated at constant volume, the added energy as heat is q_V and the heat capacity is denoted by C_V. Because $\Delta U = q_V$, C_V is given by

$$C_V = \left(\frac{\partial U}{\partial T} \right)_V \approx \frac{\Delta U}{\Delta T} = \frac{q_V}{\Delta T} \tag{19.39}$$

If the substance is heated at constant pressure, the added energy as heat is q_P and the heat capacity is denoted by C_P. Because $\Delta H = q_P$, C_P is given by

$$C_P = \left(\frac{\partial H}{\partial T} \right)_P \approx \frac{\Delta H}{\Delta T} = \frac{q_P}{\Delta T} \tag{19.40}$$

We expect that C_P is larger than C_V because not only do we increase the temperature when we add energy as heat in a constant-pressure process, but we also do work against atmospheric pressure as the substance expands as it is heated. Calculating the difference between C_P and C_V for an ideal gas is easy. We start with $H = U + PV$ and replace PV by nRT to obtain

$$H = U + nRT \qquad \text{(ideal gas)} \qquad (19.41)$$

Notice that because U depends only upon the temperature (at constant n) for an ideal gas, H also depends only upon temperature. Thus, we can differentiate Equation 19.41 with respect to temperature to obtain

$$\frac{dH}{dT} = \frac{dU}{dT} + nR \qquad (19.42)$$

But

$$\frac{dH}{dT} = \left(\frac{\partial H}{\partial T}\right)_P = C_P \qquad \text{(ideal gas)}$$

and

$$\frac{dU}{dT} = \left(\frac{\partial U}{\partial T}\right)_V = C_V \qquad \text{(ideal gas)}$$

so Equation 19.42 becomes

$$C_P - C_V = nR \qquad \text{(ideal gas)} \qquad (19.43)$$

Recall from Chapter 17 that C_V is $3R/2$ for one mole of a monatomic ideal gas and is approximately $3R$ for one mole of a nonlinear polyatomic ideal gas at room temperature. Therefore, the difference between $\overline{C}_P$ and $\overline{C}_V$ is significant for gases. For solids and liquids, however, the difference is small.

EXAMPLE 19–8
We will prove generally that (Section 22–3)

$$\overline{C}_P - \overline{C}_V = T \left(\frac{\partial P}{\partial T}\right)_{\overline{V}} \left(\frac{\partial \overline{V}}{\partial T}\right)_P$$

First, use this result to show that $\overline{C}_P - \overline{C}_V = R$ for an ideal gas and then derive an expression for $\overline{C}_P - \overline{C}_V$ for a gas that obeys the equation of state

$$P\overline{V} = RT + B(T)P$$

SOLUTION: For an ideal gas, $P\overline{V} = RT$, so

$$\left(\frac{\partial P}{\partial T}\right)_{\overline{V}} = \frac{R}{\overline{V}} \qquad \text{and} \qquad \left(\frac{\partial \overline{V}}{\partial T}\right)_P = \frac{R}{P}$$

and so

$$\overline{C}_P - \overline{C}_V = T\left(\frac{R}{\overline{V}}\right)\left(\frac{R}{P}\right) = R\left(\frac{RT}{P\overline{V}}\right) = R$$

To determine $(\partial P/\partial T)_{\overline{V}}$ for a gas that obeys the equation of state, $P\overline{V} = RT + B(T)P$, we first solve for P.

$$P = \frac{RT}{\overline{V} - B(T)}$$

and then differentiate with respect to temperature:

$$\left(\frac{\partial P}{\partial T}\right)_{\overline{V}} = \frac{R}{\overline{V} - B(T)} + \frac{RT}{[\overline{V} - B(T)]^2}\frac{dB}{dT}$$

$$= \frac{P}{T} + \frac{P}{\overline{V} - B(T)}\frac{dB}{dT}$$

Similarly,

$$\overline{V} = \frac{RT}{P} + B(T)$$

and

$$\left(\frac{\partial \overline{V}}{\partial T}\right)_P = \frac{R}{P} + \frac{dB}{dT}$$

Therefore, using the equation for $\overline{C}_P - \overline{C}_V$ given in the statement of this example,

$$\overline{C}_P - \overline{C}_V = T\left(\frac{\partial P}{\partial T}\right)_{\overline{V}}\left(\frac{\partial \overline{V}}{\partial T}\right)_P$$

$$= T\left[\frac{P}{T} + \frac{P}{\overline{V} - B(T)}\frac{dB}{dT}\right]\left[\frac{R}{P} + \frac{dB}{dT}\right]$$

$$= R + \left[\frac{RT}{\overline{V} - B(T)} + P\right]\frac{dB}{dT} + \frac{PT}{\overline{V} - B(T)}\left(\frac{dB}{dT}\right)^2$$

$$= R + 2\left(\frac{dB}{dT}\right)P + \frac{1}{R}\left(\frac{dB}{dT}\right)^2 P^2$$

where we have used the fact that $P = RT/[\overline{V} - B(T)]$ in going from the third line to the last line. Notice that this expression is the same as that for an ideal gas if $B(T)$ is a constant.

19–9. Relative Enthalpies Can Be Determined from Heat Capacity Data and Heats of Transition

By integrating Equation 19.40, we can calculate the difference in the enthalpy of a substance that does not change phase between two temperatures:

$$H(T_2) - H(T_1) = \int_{T_1}^{T_2} C_P(T)dT \tag{19.44}$$

If we let $T_1 = 0$ K, we have

$$H(T) - H(0) = \int_0^T C_P(T')dT' \tag{19.45}$$

[Notice that we have written the integration variable in Equation 19.45 with a prime, which is standard mathematical notation used to distinguish an integration limit (T in this case) from the integration variable, T'.] It would appear from Equation 19.44 that if we had heat-capacity data from 0 K to any other temperature, T, we could calculate $H(T)$ relative to $H(0)$. That is not entirely true, however. Equation 19.45 is applicable to a temperature range in which no phase transitions occur. If there is a phase transition, we must add the enthalpy change for that transition because heat is absorbed without a change in T for a phase transition. For example, if T in Equation 19.45 is in the liquid region of a substance and the only phase change between 0 K and T is a solid-liquid transition, then

$$H(T) - H(0) = \int_0^{T_{\text{fus}}} C_P^s(T)dT + \Delta_{\text{fus}}H + \int_{T_{\text{fus}}}^T C_P^l(T')dT' \tag{19.46}$$

where $C_P^s(T)$ and $C_P^l(T)$ stand for the heat capacity of the solid and liquid phases, respectively, T_{fus} stands for the melting temperature, and $\Delta_{\text{fus}}H$ is the enthalpy change upon melting (the heat of fusion):

$$\Delta_{\text{fus}}H = H^l(T_{\text{fus}}) - H^s(T_{\text{fus}})$$

Figure 19.6 shows the molar heat capacity of benzene as a function of temperature. Notice that the plot of C_P versus T is not continuous, but has jump discontinuities at the temperatures corresponding to phase transitions. The melting point and boiling point of benzene at one atm are 278.7 K and 353.2 K, respectively. As Equation 19.45 implies, the area under the curve in Figure 19.6 from 0 K to $T \le 278.7$ K gives the molar enthalpy of solid benzene [relative to $\overline{H}(0)$]. To calculate the molar enthalpy of liquid benzene, say, at 300 K and one atm, we take the area under the curve in Figure 19.6 from 0 K to 300 K and add the molar enthalpy of fusion, which is 9.95 kJ·mol^{-1}. Figure 19.7 shows the molar enthalpy of benzene as a function of temperature. Notice that $\overline{H}(T) - \overline{H}(0)$ increases smoothly within a phase and that there is a jump at a phase transition.

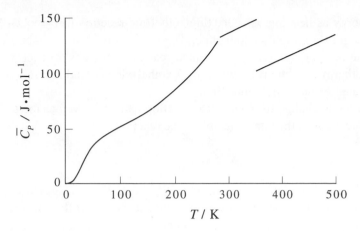

FIGURE 19.6
The constant-pressure molar heat capacity of benzene from 0 K to 500 K. The melting point and boiling point of benzene at one atm are 278.7 K and 353.2 K, respectively

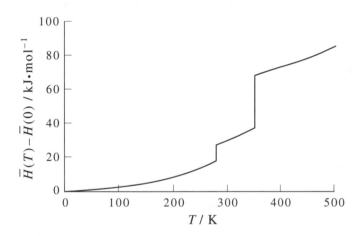

FIGURE 19.7
The molar enthalpy of benzene [relative to $\overline{H}(0)$] from 0 K to 500 K.

19–10. Enthalpy Changes for Chemical Equations Are Additive

Because most chemical reactions take place at constant pressure (open to the atmosphere), the enthalpy change associated with chemical reactions, $\Delta_r H$, (the subscript r indicates that the enthalpy change is for a chemical reaction) plays a central role in *thermochemistry*, which is the branch of thermodynamics that concerns the measurement of the evolution or absorption of energy as heat associated with chemical reactions. For example, the combustion of methane,

$$CH_4(g) + 2\,O_2(g) \longrightarrow CO_2(g) + 2\,H_2O(l)$$

releases energy as heat and is called an exothermic reaction (*exo* = out). Most combustion reactions are highly exothermic. The heat evolved in a combustion reaction is called the *heat of combustion*. Chemical reactions that absorb energy as heat are called endothermic reactions (*endo* = in). Exothermic and endothermic reactions are illustrated schematically in Figure 19.8.

The enthalpy change for a chemical reaction can be viewed as the total enthalpy of the products minus the total enthalpy of the reactants:

$$\Delta_r H = H_{prod} - H_{react} \qquad (19.47)$$

For an exothermic reaction, H_{prod} is less than H_{react}, so $\Delta_r H < 0$. Figure 19.8a represents an exothermic reaction; the enthalpy of the reactants is greater than the enthalpy of the products, so $q_P = \Delta_r H < 0$, and energy as heat is evolved as the reaction proceeds. For an endothermic reaction, H_{prod} is greater than H_{react}, so $\Delta_r H > 0$. Figure 19.8b represents an endothermic reaction; the enthalpy of the reactants is less than the enthalpy of the products, so $q_P = \Delta_r H > 0$, and energy as heat must be supplied to drive the reaction up the enthalpy "hill."

Let's consider several examples of chemical reactions carried out at one bar. For the combustion of one mole of methane to form one mole of $CO_2(g)$ and two moles of $H_2O(l)$, the value of $\Delta_r H$ is -890.36 kJ at 298 K. The negative value of $\Delta_r H$ tells us that the reaction gives off energy as heat and is therefore exothermic.

An example of an endothermic reaction is the water-gas reaction:

$$C(s) + H_2O(g) \longrightarrow CO(g) + H_2(g)$$

For this reaction, $\Delta_r H = +131$ kJ at 298 K, so energy as heat must be supplied to drive the reaction from left to right.

An important and useful property of $\Delta_r H$ for chemical equations is additivity. This property of $\Delta_r H$ follows directly from the fact that the enthalpy is a state function. If we add two chemical equations to obtain a third chemical equation, the value of

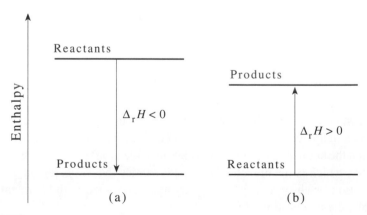

FIGURE 19.8
An enthalpy diagram for (a) an exothermic reaction and (b) and endothermic reaction.

$\Delta_r H$ for the resulting equation is equal to the sum of the $\Delta_r H$ for the two equations being added together. The additivity of $\Delta_r H$ is best illustrated by example. Consider the following two chemical equations.

$$C(s) + \tfrac{1}{2}O_2(g) \longrightarrow CO(g) \qquad \Delta_r H(1) = -110.5 \text{ kJ} \qquad (1)$$

$$CO(g) + \tfrac{1}{2}O_2(g) \longrightarrow CO_2(g) \qquad \Delta_r H(2) = -283.0 \text{ kJ} \qquad (2)$$

If we add these two chemical equations as if they were algebraic equations, we get

$$C(s) + O_2(g) \longrightarrow CO_2(g) \qquad (3)$$

The additive property of $\Delta_r H$ tells us that $\Delta_r H$ for Equation 3 is simply

$$\Delta_r H(3) = \Delta_r H(1) + \Delta_r H(2)$$
$$= -110.5 \text{ kJ} + (-283.0 \text{ kJ}) = -393.5 \text{ kJ}$$

In effect, we can imagine Equations 1 and 2 as representing a two-step process with the same initial and final states as Equation 3. The total enthalpy change for the two equations together must, therefore, be the same as if the reaction proceeded in a single step.

The additivity property of $\Delta_r H$ values is known as *Hess's Law*. Thus, if the values of $\Delta_r H(1)$ and $\Delta_r H(2)$ are known, we need not independently determine the experimental value of $\Delta_r H(3)$ because its value is equal to the sum $\Delta_r H(1) + \Delta_r H(2)$.

Now let's consider the following combination of chemical equations.

$$SO_2(g) \longrightarrow S(s) + O_2(g) \qquad (1)$$

$$S(s) + O_2(g) \longrightarrow SO_2(g) \qquad (2)$$

Because Equation 2 is simply the reverse of Equation 1, we conclude from Hess's Law that

$$\Delta_r H(\text{reverse}) = -\Delta_r H(\text{forward}) \qquad (19.48)$$

As an example of the application of Hess's Law, consider the use of

$$2\,P(s) + 3\,Cl_2(g) \longrightarrow 2\,PCl_3(l) \qquad \Delta_r H(1) = -640 \text{ kJ} \qquad (1)$$

and

$$2\,P(s) + 5\,Cl_2(g) \longrightarrow 2\,PCl_5(s) \qquad \Delta_r H(2) = -887 \text{ kJ} \qquad (2)$$

to calculate the value of $\Delta_r H$ for the equation

$$PCl_3(l) + Cl_2(g) \longrightarrow PCl_5(s) \qquad (3)$$

In this case, we add Equation 2 to the reverse of Equation 1 to obtain Equation 4:

$$2\,PCl_3(l) + 2\,Cl_2(g) \longrightarrow 2\,PCl_5(s) \tag{4}$$

Thus, from Hess's law, we obtain

$$\Delta_r H(4) = \Delta_r H(2) - \Delta_r H(1)$$
$$= -887\,kJ + 640\,kJ = -247\,kJ$$

We now multiply Equation 4 through by 1/2 to obtain Equation 3:

$$PCl_3(l) + Cl_2(g) \longrightarrow PCl_5(s) \tag{3}$$

and so

$$\Delta_r H(3) = \frac{1}{2}\Delta_r H(4) = \frac{-247\,kJ}{2} = -124\,kJ$$

EXAMPLE 19–9

The molar enthalpies of combustion of isobutane and n-butane are $-2869\,kJ\cdot mol^{-1}$ and $-2877\,kJ\cdot mol^{-1}$, respectively at 298K and one atm. Calculate $\Delta_r H$ for the conversion of one mole of n-butane to one mole of isobutane.

SOLUTION: The equations for the two combustion reactions are

$$n\text{–}C_4H_{10}(g) + \tfrac{13}{2}O_2(g) \longrightarrow 4\,CO_2(g) + 5\,H_2O(l) \tag{1}$$

$$\Delta_r H(1) = -2877\,kJ\cdot mol^{-1}$$

and

$$i\text{–}C_4H_{10}(g) + \tfrac{13}{2}O_2(g) \longrightarrow 4\,CO_2(g) + 5\,H_2O(l) \tag{2}$$

$$\Delta_r H(2) = -2869\,kJ\cdot mol^{-1}$$

If we reverse the second equation and add the result to the first equation, then we obtain the desired equation

$$n\text{–}C_4H_{10}(g) \longrightarrow i\text{–}C_4H_{10}(g) \tag{3}$$

$$\Delta_r H(3) = \Delta_r H(1) - \Delta_r H(2)$$
$$= -2877\,kJ\cdot mol^{-1} - (-2869\,kJ\cdot mol^{-1}) = -8\,kJ\cdot mol^{-1}$$

The heat of this reaction cannot be measured directly because competing reactions occur.

19–11. Heats of Reactions Can Be Calculated from Tabulated Heats of Formation

The enthalpy change of a chemical reaction, $\Delta_r H$, depends upon the number of moles of the reactants. Recently, the physical chemistry division of the International Union of Pure and Applied Chemistry (IUPAC) has proposed a systematic procedure for tabulating reaction enthalpies. The *standard reaction enthalpy* of a chemical reaction is denoted by $\Delta_r H°$ and refers to the enthalpy change associated with one mole of a specified reagent when all reactants and products are in their standard states, which for a gas is the equivalent hypothetical ideal gas at a pressure of one bar at the temperature of interest.

For example, consider the combustion of carbon to form carbon dioxide $CO_2(g)$. (The standard state of a solid is the pure crystalline substance at one bar pressure at the temperature of interest.) The balanced reaction can be written in many ways, including

$$C(s) + O_2(g) \longrightarrow CO_2(g) \tag{19.49}$$

and

$$2\,C(s) + 2\,O_2(g) \longrightarrow 2\,CO_2(g) \tag{19.50}$$

The quantity $\Delta_r H°$ implies Equation 19.49 because only one mole of the (specified) reactant C(s) is combusted. The value of $\Delta_r H°$ for this reaction at 298 K is $\Delta_r H° = -393.5$ kJ·mol^{-1}. The corresponding reaction enthalpy for Equation 19.50 is

$$\Delta_r H = 2\Delta_r H° = -787.0 \text{ kJ}$$

We see that $\Delta_r H$ is an extensive quantity, whereas $\Delta_r H°$ is an intensive quantity. The advantage of the terminology is that it removes the ambiguity of how the balanced reaction corresponding to an enthalpy change is written.

Certain subscripts are used in place of r to indicate specific types of processes. For example, the subscript "c" is used for a combustion reaction and "vap" is used for vaporization [e.g., $H_2O(l) \to H_2O(g)$]. Table 19.1 lists many of the subscripts you will encounter.

The *standard molar enthalpy of formation*, $\Delta_f H°$, is a particularly useful quantity. This intensive quantity is the standard reaction enthalpy for the formation of one mole of a molecule from its constituent elements. The degree superscript tells us that all reactants and products are in their standard states. The value of $\Delta_f H°$ of $H_2O(l)$ is -285.8 kJ·mol^{-1} at 298.15 K. This quantity implies that the balanced reaction is written as

$$H_2(g) + \tfrac{1}{2}O_2(g) \longrightarrow H_2O(l)$$

because $\Delta_f H°$ refers to the heat of formation of one mole of $H_2O(l)$. (The standard state for a liquid is the normal state of the liquid at one bar at the temperature of interest.) A value of $\Delta_f H°$ for $H_2O(l)$ equal to -285.8 kJ·mol^{-1} tells us that one mole

TABLE 19.1
Common subscripts for the enthalpy changes of processes.

Subscript	Reaction
vap	Vaporization, evaporation
sub	Sublimation
fus	Melting, fusion
trs	Transition between phases in general
mix	Mixing
ads	Adsorption
c	Combustion
f	Formation

of $H_2O(l)$ lies 285.8 kJ "downhill" on the enthalpy scale relative to its constituent elements (Figure 19.9b) when the reactants and products are in their standard states.

Most compounds cannot be formed directly from their elements. For example, an attempt to make the hydrocarbon acetylene (C_2H_2) by the direct reaction of carbon with hydrogen

$$2\,C(s) + H_2(g) \longrightarrow C_2H_2(g) \tag{19.51}$$

yields not just C_2H_2 but a complex mixture of various hydrocarbons such as C_2H_4 and C_2H_6, among others. Nevertheless, we can determine the value of $\Delta_f H^\circ$ for acetylene

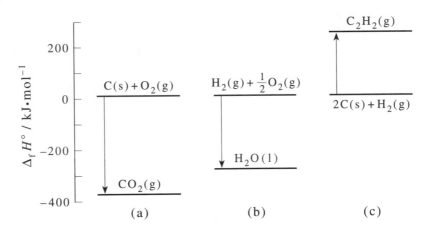

FIGURE 19.9
Standard enthalpy changes involved in the formation of $CO_2(g)$, $H_2O(l)$, and $C_2H_2(g)$ from their elements, based upon the convention that $\Delta_f H^\circ = 0$ for a pure element in its stable form at one bar and at the temperature of interest.

by using Hess's Law, together with the available $\Delta_c H°$ data on combustion reactions. All three species in Equation 19.51 burn in oxygen, and at 298 K we have

$$C(s) + O_2(g) \longrightarrow CO_2(g) \quad \Delta_c H°(1) = -393.5 \text{ kJ·mol}^{-1} \tag{1}$$

$$H_2(g) + \tfrac{1}{2}O_2(g) \longrightarrow H_2O(l) \quad \Delta_c H°(2) = -285.8 \text{ kJ·mol}^{-1} \tag{2}$$

$$C_2H_2(g) + \tfrac{5}{2}O_2(g) \longrightarrow 2\,CO_2(g) + H_2O(l) \quad \Delta_c H°(3) = -1299.6 \text{ kJ·mol}^{-1} \tag{3}$$

If we multiply Equation 1 by 2, reverse Equation 3, and add the results to Equation 2, we obtain

$$2\,C(s) + H_2(g) \longrightarrow C_2H_2(g) \tag{4}$$

with

$$\Delta_r H°(4) = 2\,\Delta_c H°(1) + \Delta_c H°(2) - \Delta_c H°(3)$$
$$= (2)(-393.5 \text{ kJ·mol}^{-1}) + (-285.8 \text{ kJ·mol}^{-1}) - (-1299.5 \text{ kJ·mol}^{-1})$$
$$= +226.7 \text{ kJ·mol}^{-1}$$

Note that the stoichiometric coefficients have no units in the IUPAC convention. Because Equation 4 represents the formation of one mole of $C_2H_2(g)$ from its elements, $\Delta_f H°[C_2H_2(g)] = +226.7 \text{ kJ·mol}^{-1}$ at 298 K (Figure 19.9c). Thus, we see that we can obtain values of $\Delta_f H°$ even if the compound cannot be formed directly from its elements.

EXAMPLE 19–10
Given that the standard enthalpies of combustion of $C(s)$, $H_2(g)$, and $CH_4(g)$ are $-393.51 \text{ kJ·mol}^{-1}$, $-285.83 \text{ kJ·mol}^{-1}$, and $-890.36 \text{ kJ·mol}^{-1}$, respectively, at 298 K, calculate the standard enthalpy of formation of methane, $CH_4(g)$.

SOLUTION: The chemical equations for the three combustion reactions are as follows:

$$C(s) + O_2(g) \longrightarrow CO_2(g) \quad \Delta_c H°(1) = -393.51 \text{ kJ·mol}^{-1} \tag{1}$$

$$H_2(g) + \tfrac{1}{2}O_2(g) \longrightarrow H_2O(l) \quad \Delta_c H°(2) = -285.83 \text{ kJ·mol}^{-1} \tag{2}$$

$$CH_4(g) + 2\,O_2(g) \longrightarrow CO_2(g) + 2\,H_2O(l) \quad \Delta_c H°(3) = -890.36 \text{ kJ·mol}^{-1} \tag{3}$$

If we reverse Equation 3, multiply Equation 2 by 2, and add the results to Equation 1, we obtain the equation for the formation of $CH_4(g)$ from its elements.

$$C(s) + 2\,H_2(g) \longrightarrow CH_4(g) \tag{4}$$

along with

$$\Delta_r H°(4) = \Delta_c H°(1) + 2\,\Delta_c H°(2) - \Delta_c H°(3)$$
$$= (-393.51 \text{ kJ·mol}^{-1}) + (2)(-285.83 \text{ kJ·mol}^{-1}) - (-890.36 \text{ kJ·mol}^{-1})$$
$$= -74.81 \text{ kJ·mol}^{-1}$$

Because Equation (4) represents the formation of one mole of $CH_4(g)$ directly from its elements, we have $\Delta_f H°[CH_4(g)] = -74.81 \text{ kJ·mol}^{-1}$ at 298 K.

As suggested by Figure 19.9, we can set up a table of $\Delta_f H°$ values for compounds by setting the values of $\Delta_f H°$ for the elements equal to zero. That is, for each pure element in its stable form at one bar at the temperature of interest, we set $\Delta_f H°$ equal to zero. Thus, standard enthalpies of formation of compounds are given relative to the elements in their normal physical states at one bar. Table 19.2 lists values of $\Delta_f H°$ at 25°C for a number of substances. If you look at Table 19.2, you will see that $\Delta_f H°[C(\text{diamond})] = +1.897 \text{ kJ·mol}^{-1}$, $\Delta_f H°[Br_2(g)] = +30.907 \text{ kJ·mol}^{-1}$, and $\Delta_f H°[I_2(g)] = +62.438 \text{ kJ·mol}^{-1}$. The values of $\Delta_f H°$ for these forms of the elements are not equal to zero because $C(\text{diamond})$, $Br_2(g)$, and $I_2(g)$ are not the normal physical states of these elements at 25°C and one bar. The normal physical states of these elements at 25°C and one bar are $C(\text{graphite})$, $Br_2(l)$, and $I_2(s)$.

EXAMPLE 19–11
Use Table 19.2 to calculate the molar enthalpy of vaporization $\Delta_{vap} H°$ of bromine at 25°C.

SOLUTION: The equation that represents the vaporization of one mole of bromine is

$$Br_2(l) \longrightarrow Br_2(g)$$

Therefore,

$$\Delta_{vap} H° = \Delta_f H°[Br_2(g)] - \Delta_f H°[Br_2(l)]$$
$$= 30.907 \text{ kJ·mol}^{-1}$$

Note that this result is not the value of $\Delta_{vap} H°$ at its normal boiling point of 58.8°C. The value of $\Delta_{vap} H°$ at 58.8°C is 29.96 kJ·mol^{-1}. (We will learn how to calculate the temperature variation of ΔH in the next section.)

We can use Hess's law to understand how enthaplies of formation are used to calculate enthalpy changes. Consider the general chemical equation

$$aA + bB \longrightarrow yY + zZ$$

TABLE 19.2

Standard molar enthalpies of formation, $\Delta_f H^\circ$, for various substances at 25°C and one bar.

Substance	Formula	$\Delta_f H^\circ/\text{kJ}\cdot\text{mol}^{-1}$
Acetylene	$C_2H_2(g)$	+226.73
Ammonia	$NH_3(g)$	−46.11
Benzene	$C_6H_6(l)$	+49.03
Bromine	$Br_2(g)$	+30.907
Butane	$C_4H_{10}(g)$	−125.6
Carbon(diamond)	$C(s)$	+1.897
Carbon(graphite)	$C(s)$	0
Carbon dioxide	$CO_2(g)$	−393.509
Carbon monoxide	$CO(g)$	−110.5
Cyclohexane	$C_6H_{12}(l)$	−156.4
Ethane	$C_2H_6(g)$	−84.68
Ethanol	$C_2H_5OH(l)$	−277.69
Ethene	$C_2H_4(g)$	+52.28
Glucose	$C_6H_{12}O_6(s)$	−1260
Hexane	$C_6H_{14}(l)$	−198.7
Hydrazine	$N_2H_4(l)$	+50.6
	$N_2H_4(g)$	+95.40
Hydrogen bromide	$HBr(g)$	−36.3
Hydrogen chloride	$HCl(g)$	−92.31
Hydrogen fluoride	$HF(g)$	−273.3
Hydrogen iodide	$HI(g)$	+26.5
Hydrogen peroxide	$H_2O_2(l)$	−187.8
Iodine	$I_2(g)$	+62.438
Methane	$CH_4(g)$	−74.81
Methanol	$CH_3OH(l)$	−239.1
	$CH_3OH(g)$	−201.5
Nitrogen oxide	$NO(g)$	+90.37
Nitrogen dioxide	$NO_2(g)$	+33.85
Dinitrogen tetraoxide	$N_2O_4(g)$	+9.66
	$N_2O_4(l)$	−19.5
Octane	$C_8H_{18}(l)$	−250.1
Pentane	$C_5H_{12}(l)$	−173.5
Propane	$C_3H_8(g)$	−103.8
Sucrose	$C_{12}H_{22}O_{11}(s)$	−2220
Sulfur dioxide	$SO_2(g)$	−296.8
Sulfur trioxide	$SO_3(g)$	−395.7
Tetrachloromethane	$CCl_4(l)$	−135.44
	$CCl_4(g)$	−102.9
Water	$H_2O(l)$	−285.83
	$H_2O(g)$	−241.8

where a, b, y, and z are the number of moles of the respective species. We can calculate $\Delta_r H$ in two steps, as shown in the following diagram:

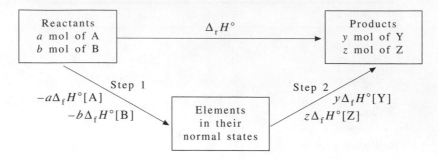

First, we decompose compounds A and B into their constituent elements (step 1); and then we combine the elements to form the compounds Y and Z (step 2). In the first step, we have

$$\Delta_r H(1) = -a\Delta_f H^\circ[A] - b\Delta H_f^\circ[B]$$

We have omitted the degree superscript on the $\Delta_r H$ because this value is not necessarily referenced to one mole of a particular reagent. The minus signs occur here because the reaction involved is the reverse of the formation of the compounds from their elements; we are forming the elements from the compounds. In the second step, we have

$$\Delta_r H(2) = y\Delta_f H^\circ[Y] + z\Delta_f H^\circ[Z]$$

The sum of $\Delta_r H(1)$ and $\Delta_r H(2)$ gives $\Delta_r H$ for the general equation:

$$\Delta_r H = y\Delta_f H^\circ[Y] + z\Delta_f H^\circ[Z] - a\Delta_f H^\circ[A] - b\Delta_f H^\circ[B] \qquad (19.52)$$

Note that the right side of Equation 19.52 is the total enthalpy of the products minus the total enthalpy of the reactants (see Equation 19.47).

When using Equation 19.52, you need to specify whether each substance is a gas, liquid, or solid because the value of $\Delta_f H^\circ$ depends upon the physical state of the substance. Using Equation 19.52, we determine $\Delta_r H$ for the reaction

$$C_2H_2(g) + \tfrac{5}{2}O_2(g) \longrightarrow 2CO_2(g) + H_2O(l)$$

at 298 K to be

$$\Delta_r H = (2)\Delta_f H^\circ[CO_2(g)] + (1)\Delta_f H^\circ[H_2O(l)]$$
$$-(1)\Delta_f H^\circ[C_2H_2(g)] - (\tfrac{5}{2})\Delta_f H^\circ[O_2(g)]$$

Using the data in Table 19.2, we obtain

$$\Delta_r H = (2)(-393.509 \text{ kJ·mol}^{-1}) + (1)(-285.83 \text{ kJ·mol}^{-1})$$
$$-(1)(+226.73 \text{ kJ·mol}^{-1}) - (\tfrac{5}{2})(0 \text{ kJ·mol}^{-1})$$
$$= -1299.58 \text{ kJ·mol}^{-1}$$

Note that $\Delta_f H^\circ[O_2(g)] = 0$ because the $\Delta_f H^\circ$ value for any element in its stable state at 298 K and one bar is zero. To determine $\Delta_r H$ for

$$2\,C_2H_2(g) + 5\,O_2(g) \longrightarrow 4\,CO_2(g) + 2\,H_2O(l)$$

we multiply $\Delta_r H = -1299.58$ kJ·mol^{-1} by 2 mol to obtain $\Delta_r H = -2599.16$ kJ.

EXAMPLE 19–12
Use the $\Delta_f H^\circ$ data in Table 19.2 to calculate the value of $\Delta_r H^\circ$ for the combustion of liquid ethanol, $C_2H_5OH(l)$, at 25°C:

$$C_2H_5OH(l) + 3\,O_2(g) \longrightarrow 2\,CO_2(g) + 3\,H_2O(l)$$

SOLUTION: Referring to Table 19.2, we find that $\Delta_f H^\circ[CO_2(g)] = -393.509$ kJ·mol^{-1}; $\quad \Delta_f H^\circ[H_2O(l)] = -285.83$ kJ·mol^{-1}; $\quad \Delta_f H^\circ[O_2(g)] = 0$; and $\Delta_f H^\circ[C_2H_5OH(l)] = -277.69$ kJ·mol^{-1}. Application of Equation 19.52 yields

$$\Delta_r H^\circ = (2)\Delta_f H^\circ[CO_2(g)] + (3)\Delta_f H^\circ[H_2O(l)]$$

$$-(1)\Delta_f H^\circ[C_2H_5OH(l)] - (3)\Delta_f H^\circ[O_2(g)]$$

$$= (2)(-393.509 \text{ kJ·mol}^{-1}) + (3)(-285.83 \text{ kJ·mol}^{-1})$$

$$-(1)(-277.69 \text{ kJ·mol}^{-1}) - (3)(0)$$

$$= -1366.82 \text{ kJ·mol}^{-1}$$

19–12. The Temperature Dependence of $\Delta_r H$ Is Given in Terms of the Heat Capacities of the Reactants and Products

Up to now, we have calculated reaction enthalpies at 25°C. We will see in this section that we can calculate $\Delta_r H$ at other temperatures if we have sufficient heat-capacity data. Consider the general reaction

$$a\text{A} + b\text{B} \longrightarrow y\text{Y} + z\text{Z}$$

We can express $\Delta_r H$ at a temperature T_2 in the form

$$\Delta_r H(T_2) = y[H_Y(T_2) - H_Y(0)] + z[H_Z(T_2) - H_Z(0)]$$

$$-a[H_A(T_2) - H_A(0)] - b[H_B(T_2) - H_B(0)] \qquad (19.53)$$

where, from Equation 19.45,

$$H_Y(T_2) - H_Y(0) = \int_0^{T_2} C_{P,Y}(T)dT \qquad (19.54)$$

etc. Similarly, $\Delta_r H(T_1)$ is given by

$$\Delta_r H(T_1) = y[H_Y(T_1) - H_Y(0)] + z[H_Z(T_1) - H_Z(0)]$$
$$-a[H_A(T_1) - H_A(0)] - b[H_B(T_1) - H_B(0)] \qquad (19.55)$$

with

$$H_Y(T_1) - H_Y(0) = \int_0^{T_1} C_{P,Y}(T)dT \qquad (19.56)$$

etc. If we substitute Equation 19.54 into Equation 19.53 and Equation 19.56 into Equation 19.55, and then subtract the resultant $\Delta_r H(T_1)$ from $\Delta_r H(T_2)$, we obtain

$$\Delta_r H(T_2) = \Delta_r H(T_1) + \int_{T_1}^{T_2} \Delta C_P(T)dT \qquad (19.57)$$

where, as the notation suggests,

$$\Delta C_P(T) = yC_{P,Y}(T) + zC_{P,Z}(T) - aC_{P,A}(T) - bC_{P,B}(T) \qquad (19.58)$$

Thus, if we know $\Delta_r H$ at T_1, say 25°C, we can calculate $\Delta_r H$ at any other temperature using Equation 19.57. In writing Equation 19.57 we have assumed there are no phase transitions between T_1 and T_2.

Equation 19.57 has a simple physical interpretation given by Figure 19.10. To calculate the value of $\Delta_r H$ at some temperature T_2 given the value of $\Delta_r H$ at T_1, we can follow the path 1—2—3 in Figure 19.10. This pathway involves taking the

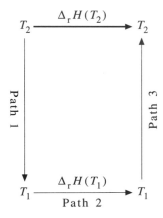

FIGURE 19.10
An illustration of Equation 19.57. Along path 1 we take the reactants from T_2 to T_1. Along path 2 we let the reaction occur at T_1. Then along path 3, we bring the products from T_1 back to T_2. Because ΔH is a state function, we have that $\Delta H(T_2) = \Delta H_1 + \Delta H_2 + \Delta H_3$.

reactants from temperature T_2 to T_1, letting the reaction occur at T_1, and then taking the products from T_1 back to T_2. The mathematical expressions for ΔH for each step are

$$\Delta H_1 = \int_{T_2}^{T_1} C_P(\text{reactants})dT = -\int_{T_1}^{T_2} C_P(\text{reactants})dT$$

$$\Delta H_2 = \Delta_r H(T_1)$$

$$\Delta H_3 = \int_{T_1}^{T_2} C_P(\text{products})dT$$

and so

$$\Delta H(T_2) = \Delta H_1 + \Delta H_2 + \Delta H_3$$

$$= \Delta_r H(T_1) + \int_{T_1}^{T_2} \left[C_P(\text{products}) - C_P(\text{reactants}) \right] dT$$

As a simple application of Equation 19.57, consider

$$H_2O(s) \longrightarrow H_2O(l)$$

Let's calculate $\Delta_{fus} H^\circ$ of water at $-10°C$ and one bar given that $\Delta_{fus} H^\circ(0°C) = 6.01 \text{ kJ·mol}^{-1}$, $C_P^\circ(s) = 37.7 \text{ J·K}^{-1}·\text{mol}^{-1}$ and $C_P^\circ(l) = 75.3 \text{ J·K}^{-1}·\text{mol}^{-1}$. Because the equation is written in terms of one mole of reactant and the reactants and products are in their standard states, we use a superscript $\circ$ on calculated thermodynamic quantities. Therefore,

$$\Delta C_P^\circ = C_P^\circ(l) - C_P^\circ(s) = 37.6 \text{ J·K}^{-1}·\text{mol}^{-1}$$

and

$$\Delta_{fus} H^\circ(-10°C) = \Delta_{fus} H^\circ(0°C) + \int_{0°C}^{-10°C} \left(37.6 \text{ J·K}^{-1}·\text{mol}^{-1} \right) dT$$

$$= 6.01 \text{ kJ·mol}^{-1} - 376 \text{ J·mol}^{-1} = 5.64 \text{ kJ·mol}^{-1}$$

EXAMPLE 19–13
The standard molar enthaply of formation, $\Delta_f H^\circ$, of $NH_3(g)$ is $-46.11 \text{ kJ·mol}^{-1}$ at 25°C. Using the heat capacity data given below, calculate the standard molar heat of formation of $NH_3(g)$ at 1000 K.

$$C_P^\circ(H_2)/\text{J·K}^{-1}·\text{mol}^{-1} = 29.07 - (0.837 \times 10^{-3} \text{ K}^{-1})T + (2.012 \times 10^{-6} \text{ K}^{-2})T^2$$

$$C_P^\circ(N_2)/\text{J·K}^{-1}·\text{mol}^{-1} = 26.98 + (5.912 \times 10^{-3} \text{ K}^{-1})T - (0.3376 \times 10^{-6} \text{ K}^{-2})T^2$$

$$C_P^\circ(NH_3)/\text{J·K}^{-1}·\text{mol}^{-1} = 25.89 + (32.58 \times 10^{-3} \text{ K}^{-1})T - (3.046 \times 10^{-6} \text{ K}^{-2})T^2$$

where $298 \text{ K} < T < 1500 \text{ K}$

SOLUTION: We use the equation

$$\Delta_f H^\circ(1000 \text{ K}) = \Delta_f H^\circ(298 \text{ K}) + \int_{298 \text{ K}}^{1000 \text{ K}} \Delta C_P^\circ(T) dT$$

The relevant chemical equation for the formation of one mole of $NH_3(g)$ from its elements is

$$\tfrac{1}{2} N_2(g) + \tfrac{3}{2} H_2(g) \longrightarrow NH_3(g)$$

and so

$$\Delta C_P^\circ(T)/\text{J·K}^{-1}\text{·mol}^{-1} = (1) \, C_P^\circ(NH_3) - (\tfrac{1}{2}) \, C_P^\circ(N_2) - (\tfrac{3}{2}) \, C_P^\circ(H_2)$$

$$= -31.21 + (30.88 \times 10^{-3} \text{ K}^{-1})T - (5.895 \times 10^{-6} \text{ K}^{-2})T^2$$

The integral of $\Delta C_P(T)$ is

$$\int_{298 \text{ K}}^{1000 \text{ K}} \left[-31.21 + (30.88 \times 10^{-3} \text{ K}^{-1})T - (5.895 \times 10^{-6} \text{ K}^{-2})T^2 \right] dT$$

$$= (-21.91 + 14.07 - 1.913) \text{ kJ·mol}^{-1}$$

$$= -9.75 \text{ kJ·mol}^{-1}$$

so

$$\Delta_f H^\circ(1000 \text{ K}) = \Delta_f H^\circ(25^\circ\text{C}) - 9.75 \text{ kJ·mol}^{-1}$$

$$= -46.11 \text{ kJ·mol}^{-1} - 9.75 \text{ kJ·mol}^{-1}$$

$$= -55.86 \text{ kJ·mol}^{-1}$$

The pressure dependence of $\Delta_r H$ (which we will study in Chapter 22) is usually much smaller than its temperature dependence.

Problems

19-1. Suppose that a 10-kg mass of iron at 20°C is dropped from a height of 100 meters. What is the kinetic energy of the mass just before it hits the ground? What is its speed? What would be the final temperature of the mass if all its kinetic energy at impact is transformed into internal energy? Take the molar heat capacity of iron to be $\overline{C}_P = 25.1$ J·mol^{-1}·K^{-1} and the gravitational acceleration constant to be 9.80 m·s^{-2}.

19-2. Consider an ideal gas that occupies 2.50 dm^3 at a pressure of 3.00 bar. If the gas is compressed isothermally at a constant external pressure, P_{ext}, so that the final volume is 0.500 dm^3, calculate the smallest value P_{ext} can have. Calculate the work involved using this value of P_{ext}.

19-3. A one-mole sample of $CO_2(g)$ occupies 2.00 dm^3 at a temperature of 300 K. If the gas is compressed isothermally at a constant external pressure, P_{ext}, so that the final volume is

0.750 dm³, calculate the smallest value P_{ext} can have, assuming that $CO_2(g)$ satisfies the van der Waals equation of state under these conditions. Calculate the work involved using this value of P_{ext}.

19-4. Calculate the work involved when one mole of an ideal gas is compressed reversibly from 1.00 bar to 5.00 bar at a constant temperature of 300 K.

19-5. Calculate the work involved when one mole of an ideal gas is expanded reversibly from 20.0 dm³ to 40.0 dm³ at a constant temperature of 300 K.

19-6. Calculate the minimum amount of work required to compress 5.00 moles of an ideal gas isothermally at 300 K from a volume of 100 dm³ to 40.0 dm³.

19-7. Consider an ideal gas that occupies 2.25 L at 1.33 bar. Calculate the work required to compress the gas isothermally to a volume of 1.50 L at a constant pressure of 2.00 bar followed by another isothermal compression to 0.800 L at a constant pressure of 3.75 bar (Figure 19.4). Compare the result with the work of compressing the gas isothermally and reversibly from 2.25 L to 0.800 L.

19-8. Show that for an isothermal reversible expansion from a molar volume $\overline{V}_1$ to a final molar volume $\overline{V}_2$, the work is given by

$$w = -RT \ln \left(\frac{\overline{V}_2 - B}{\overline{V}_1 - B} \right) - \frac{A}{BT^{1/2}} \ln \left[\frac{(\overline{V}_2 + B)\overline{V}_1}{(\overline{V}_1 + B)\overline{V}_2} \right]$$

for the Redlich-Kwong equation.

19-9. Use the result of Problem 19–8 to calculate the work involved in the isothermal reversible expansion of one mole of $CH_4(g)$ from a volume of $1.00 \ dm^3 \cdot mol^{-1}$ to $5.00 \ dm^3 \cdot mol^{-1}$ at 300 K. (See Table 16.4 for the values of A and B.)

19-10. Repeat the calculation in Problem 19-9 for a van der Waals gas.

19-11. Derive an expression for the reversible isothermal work of an expansion of a gas that obeys the Peng-Robinson equation of state.

19-12. One mole of a monatomic ideal gas initially at a pressure of 2.00 bar and a temperature of 273 K is taken to a final pressure of 4.00 bar by the reversible path defined by $P/V =$ constant. Calculate the values of ΔU, ΔH, q, and w for this process. Take $\overline{C}_V$ to be equal to $12.5 \ J \cdot mol^{-1} \cdot K^{-1}$.

19-13. The isothermal compressibility of a substance is given by

$$\beta = -\frac{1}{V} \left(\frac{\partial V}{\partial P} \right)_T \tag{1}$$

For an ideal gas, $\beta = 1/P$, but for a liquid, β is fairly constant over a moderate pressure range. If β is constant, show that

$$\frac{V}{V_0} = e^{-\beta(P-P_0)} \tag{2}$$

where V_0 is the volume at a pressure P_0. Use this result to show that the reversible isothermal work of compressing a liquid from a volume V_0 (at a pressure P_0) to a volume V (at a pressure P) is given by

$$w = -P_0(V - V_0) + \beta^{-1}V_0\left(\frac{V}{V_0}\ln\frac{V}{V_0} - \frac{V}{V_0} + 1\right)$$

$$= -P_0V_0[e^{-\beta(P-P_0)} - 1] + \beta^{-1}V_0\{1 - [1 + \beta(P - P_0)]e^{-\beta(P-P_0)}\} \qquad (3)$$

(You need to use the fact that $\int \ln x\,dx = x\ln x - x$.)

The fact that liquids are incompressible is reflected by β being small, so that $\beta(P - P_0) \ll 1$ for moderate pressures. Show that

$$w = \beta P_0 V_0(P - P_0) + \frac{\beta V_0(P - P_0)^2}{2} + O(\beta^2)$$

$$= \frac{\beta V_0}{2}(P^2 - P_0^2) + O(\beta^2) \qquad (4)$$

Calculate the work required to compress one mole of toluene reversibly and isothermally from 10 bar to 100 bar at 20°C. Take the value of β to be 8.95×10^{-5} bar^{-1} and the molar volume to be 0.106 L·mol^{-1} at 20°C.

19-14. In the previous problem, you derived an expression for the reversible, isothermal work done when a liquid is compressed. Given that β is typically $O(10^{-4})$ bar^{-1}, show that $V/V_0 \approx 1$ for pressures up to about 100 bar. This result, of course, reflects the fact that liquids are not very compressible. We can exploit this result by substituting $dV = -\beta V dP$ from the defining equation of β into $w = -\int P dV$ and then treating V as a constant. Show that this approximation gives Equation 4 of Problem 19–13.

19-15. Show that

$$\frac{T_2}{T_1} = \left(\frac{V_1}{V_2}\right)^{R/\overline{C}_V}$$

for a reversible adiabatic expansion of an ideal gas.

19-16. Show that

$$\left(\frac{T_2}{T_1}\right)^{3/2} = \frac{\overline{V}_1 - b}{\overline{V}_2 - b}$$

for a reversible, adiabatic expansion of a monatomic gas that obeys the equation of state $P(\overline{V} - b) = RT$. Extend this result to the case of a diatomic gas.

19-17. Show that

$$\frac{T_2}{T_1} = \left(\frac{P_2}{P_1}\right)^{R/\overline{C}_P}$$

for a reversible adiabatic expansion of an ideal gas.

19-18. Show that

$$P_1 V_1^{(\overline{C}_V + R)/\overline{C}_V} = P_2 V_2^{(\overline{C}_V + R)/\overline{C}_V}$$

for an adiabatic expansion of an ideal gas. Show that this formula reduces to Equation 19.23 for a monatomic gas.

19-19. Calculate the work involved when one mole of a monatomic ideal gas at 298 K expands reversibly and adiabatically from a pressure of 10.00 bar to a pressure of 5.00 bar.

19-20. A quantity of $N_2(g)$ at 298 K is compressed reversibly and adiabatically from a volume of 20.0 dm^3 to 5.00 dm^3. Assuming ideal behavior, calculate the final temperature of the $N_2(g)$. Take $\overline{C}_V = 5R/2$.

19-21. A quantity of $CH_4(g)$ at 298 K is compressed reversibly and adiabatically from 50.0 bar to 200 bar. Assuming ideal behavior, calculate the final temperature of the $CH_4(g)$. Take $\overline{C}_V = 3R$.

19-22. One mole of ethane at 25°C and one atm is heated to 1200°C at constant pressure. Assuming ideal behavior, calculate the values of w, q, ΔU, and ΔH given that the molar heat capacity of ethane is given by

$$\overline{C}_P/R = 0.06436 + (2.137 \times 10^{-2} \text{ K}^{-1})T$$
$$- (8.263 \times 10^{-6} \text{ K}^{-2})T^2 + (1.024 \times 10^{-9} \text{ K}^{-3})T^3$$

over the above temperature range. Repeat the calculation for a constant-volume process.

19-23. The value of $\Delta_r H°$ at 25°C and one bar is +290.8 kJ for the reaction

$$2 \text{ ZnO(s)} + 2 \text{ S(s)} \longrightarrow 2 \text{ ZnS(s)} + O_2(g)$$

Assuming ideal behavior, calculate the value of $\Delta_r U°$ for this reaction.

19-24. Liquid sodium is being considered as an engine coolant. How many grams of sodium are needed to absorb 1.0 MJ of heat if the temperature of the sodium is not to increase by more than 10°C. Take $\overline{C}_P = 30.8 \text{ J·K}^{-1}\cdot\text{mol}^{-1}$ for Na(l) and 75.2 J·K^{-1}·mol^{-1} for $H_2O(l)$.

19-25. A 25.0-g sample of copper at 363 K is placed in 100.0 g of water at 293 K. The copper and water quickly come to the same temperature by the process of heat transfer from copper to water. Calculate the final temperature of the water. The molar heat capacity of copper is 24.5 J·K^{-1}·mol^{-1} and that of water is 75.2 J·K^{-1}·mol^{-1}.

19-26. A 10.0-kg sample of liquid water is used to cool an engine. Calculate the heat removed (in joules) from the engine when the temperature of the water is raised from 293 K to 373 K. Take $\overline{C}_P = 75.2 \text{ J·K}^{-1}\cdot\text{mol}^{-1}$ for $H_2O(l)$.

19-27. In this problem, we will derive a general relation between C_P and C_V. Start with $U = U(P, T)$ and write

$$dU = \left(\frac{\partial U}{\partial P}\right)_T dP + \left(\frac{\partial U}{\partial T}\right)_P dT \tag{1}$$

We could also consider V and T to be the independent variables of U and write

$$dU = \left(\frac{\partial U}{\partial V}\right)_T dV + \left(\frac{\partial U}{\partial T}\right)_V dT \tag{2}$$

Now take $V = V(P, T)$ and substitute its expression for dV into Equation 2 to obtain

$$dU = \left(\frac{\partial U}{\partial V}\right)_T \left(\frac{\partial V}{\partial P}\right)_T dP + \left[\left(\frac{\partial U}{\partial V}\right)_T \left(\frac{\partial V}{\partial T}\right)_P + \left(\frac{\partial U}{\partial T}\right)_V\right] dT$$

Compare this result with Equation 1 to obtain

$$\left(\frac{\partial U}{\partial P}\right)_T = \left(\frac{\partial U}{\partial V}\right)_T \left(\frac{\partial V}{\partial P}\right)_T \tag{3}$$

and

$$\left(\frac{\partial U}{\partial T}\right)_P = \left(\frac{\partial U}{\partial V}\right)_T \left(\frac{\partial V}{\partial T}\right)_P + \left(\frac{\partial U}{\partial T}\right)_V \tag{4}$$

Last, substitute $U = H - PV$ into the left side of Equation (4) and use the definitions of C_P and C_V to obtain

$$C_P - C_V = \left[P + \left(\frac{\partial U}{\partial V}\right)_T\right] \left(\frac{\partial V}{\partial T}\right)_P$$

Show that $C_P - C_V = nR$ if $(\partial U/\partial V)_T = 0$, as it is for an ideal gas.

19-28. Following Problem 19–27, show that

$$C_P - C_V = \left[V - \left(\frac{\partial H}{\partial P}\right)_T\right] \left(\frac{\partial P}{\partial T}\right)_V$$

19-29. Starting with $H = U + PV$, show that

$$\left(\frac{\partial U}{\partial T}\right)_P = C_P - P\left(\frac{\partial V}{\partial T}\right)_P$$

Interpret this result physically.

19-30. Given that $(\partial U/\partial V)_T = 0$ for an ideal gas, prove that $(\partial H/\partial V)_T = 0$ for an ideal gas.

19-31. Given that $(\partial U/\partial V)_T = 0$ for an ideal gas, prove that $(\partial C_V/\partial V)_T = 0$ for an ideal gas.

19-32. Show that $C_P - C_V = nR$ if $(\partial H/\partial P)_T = 0$, as is true for an ideal gas.

19-33. Differentiate $H = U + PV$ with respect to V at constant temperature to show that $(\partial H/\partial V)_T = 0$ for an ideal gas.

19-34. Given the following data for sodium, plot $\overline{H}(T) - \overline{H}(0)$ against T for sodium: melting point, 361 K; boiling point, 1156 K; $\Delta_{fus}H^\circ = 2.60$ kJ·mol^{-1}; $\Delta_{vap}H^\circ = 97.4$ kJ·mol^{-1}; $\overline{C}_P(s) = 28.2$ J·mol^{-1}·K^{-1}; $\overline{C}_P(l) = 32.7$ J·mol^{-1}·K^{-1}; $\overline{C}_P(g) = 20.8$ J·mol^{-1}·K^{-1}.

19-35. The $\Delta_r H^\circ$ values for the following equations are

$$2\,\text{Fe(s)} + \tfrac{3}{2}\text{O}_2(\text{g}) \rightarrow \text{Fe}_2\text{O}_3(\text{s}) \qquad \Delta_r H^\circ = -206 \text{ kJ·mol}^{-1}$$

$$3\,\text{Fe(s)} + 2\,\text{O}_2(\text{g}) \rightarrow \text{Fe}_3\text{O}_4(\text{s}) \qquad \Delta_r H^\circ = -136 \text{ kJ·mol}^{-1}$$

Use these data to calculate the value of $\Delta_r H$ for the reaction described by

$$4\,\text{Fe}_2\text{O}_3(\text{s}) + \text{Fe(s)} \longrightarrow 3\,\text{Fe}_3\text{O}_4(\text{s})$$

19-36. Given the following data,

$$\tfrac{1}{2} H_2(g) + \tfrac{1}{2} F_2(g) \rightarrow HF(g) \quad \Delta_r H^\circ = -273.3 \text{ kJ} \cdot \text{mol}^{-1}$$

$$H_2(g) + \tfrac{1}{2} O_2(g) \rightarrow H_2O(l) \quad \Delta_r H^\cup = -285.8 \text{ kJ} \cdot \text{mol}^{-1}$$

calculate the value of $\Delta_r H$ for the reaction described by

$$2 F_2(g) + 2 H_2O(l) \longrightarrow 4 HF(g) + O_2(g)$$

19-37. The standard molar heats of combustion of the isomers m-xylene and p-xylene are -4553.9 kJ·mol^{-1} and -4556.8 kJ·mol^{-1}, respectively. Use these data, together with Hess's Law, to calculate the value of $\Delta_r H^\circ$ for the reaction described by

$$m\text{-xylene} \longrightarrow p\text{-xylene}$$

19-38. Given that $\Delta_r H^\circ = -2826.7$ kJ for the combustion of 1.00 mol of fructose at 298.15 K,

$$C_6H_{12}O_6(s) + 6 O_2(g) \longrightarrow 6 CO_2(g) + 6 H_2O(l)$$

and the $\Delta_f H^\circ$ data in Table 19.2, calculate the value of $\Delta_f H^\circ$ for fructose at 298.15 K.

19-39. Use the $\Delta_f H^\circ$ data in Table 19.2 to calculate the value of $\Delta_c H^\circ$ for the combustion reactions described by the equations:

a. $CH_3OH(l) + \tfrac{3}{2} O_2(g) \longrightarrow CO_2(g) + 2 H_2O(l)$
b. $N_2H_4(l) + O_2(g) \longrightarrow N_2(g) + 2 H_2O(l)$

Compare the heat of combustion per gram of the fuels $CH_3OH(l)$ and $N_2H_4(l)$.

19-40. Using Table 19.2, calculate the heat required to vaporize 1.00 mol of $CCl_4(l)$ at 298 K.

19-41. Using the $\Delta_f H^\circ$ data in Table 19.2, calculate the values of $\Delta_r H^\circ$ for the following:

a. $C_2H_4(g) + H_2O(l) \longrightarrow C_2H_5OH(l)$
b. $CH_4(g) + 4 Cl_2(g) \longrightarrow CCl_4(l) + 4 HCl(g)$

In each case, state whether the reaction is endothermic or exothermic.

19-42. Use the following data to calculate the value of $\Delta_{vap} H^\circ$ of water at 298 K and compare your answer to the one you obtain from Table 19.2: $\Delta_{vap} H^\circ$ at 373 K = 40.7 kJ·mol^{-1}; $\overline{C}_P(l) = 75.2$ J·mol^{-1}·K^{-1}; $\overline{C}_P(g) = 33.6$ J·mol^{-1}·K^{-1}.

19-43. Use the following data and the data in Table 19.2 to calculate the standard reaction enthalpy of the water-gas reaction at 1273 K. Assume that the gases behave ideally under these conditions.

$$C(s) + H_2O(g) \longrightarrow CO(g) + H_2(g)$$

$$C_P^\circ[CO(g)]/R = 3.231 + (8.379 \times 10^{-4} \text{ K}^{-1})T - (9.86 \times 10^{-8} \text{ K}^{-2})T^2$$

$$C_P^\circ[H_2(g)]/R = 3.496 + (1.006 \times 10^{-4} \text{ K}^{-1})T + (2.42 \times 10^{-7} \text{ K}^{-2})T^2$$

$$C_P^\circ[H_2O(g)]/R = 3.652 + (1.156 \times 10^{-3} \text{ K}^{-1})T + (1.42 \times 10^{-7} \text{ K}^{-2})T^2$$

$$C_P^\circ[C(s)]/R = -0.6366 + (7.049 \times 10^{-3} \text{ K}^{-1})T - (5.20 \times 10^{-6} \text{ K}^{-2})T^2$$
$$+ (1.38 \times 10^{-9} \text{ K}^{-3})T^3$$

19-44. The standard molar enthalpy of formation of $CO_2(g)$ at 298 K is -393.509 kJ·mol^{-1}. Use the following data to calculate the value of $\Delta_f H^\circ$ at 1000 K. Assume the gases behave ideally under these conditions.

$$C_P^\circ[CO_2(g)]/R = 2.593 + (7.661 \times 10^{-3} \text{ K}^{-1})T - (4.78 \times 10^{-6} \text{ K}^{-2})T^2$$
$$+ (1.16 \times 10^{-9} \text{ K}^{-3})T^3$$

$$C_P^\circ[O_2(g)]/R = 3.094 + (1.561 \times 10^{-3} \text{ K}^{-1})T - (4.65 \times 10^{-7} \text{ K}^{-2})T^2$$

$$C_P^\circ[C(s)]/R = -0.6366 + (7.049 \times 10^{-3} \text{ K}^{-1})T - (5.20 \times 10^{-6} \text{ K}^{-2})T^2$$
$$+ (1.38 \times 10^{-9} \text{ K}^{-3})T^3$$

19-45. The value of the standard reaction enthalpy for

$$CH_4(g) + 2O_2(g) \longrightarrow CO_2(g) + 2H_2O(g)$$

is -802.2 kJ at 298 K. Use the heat-capacity data in Problems 19–43 and 19–44 in addition to

$$C_P^\circ[CH_4(g)]/R = 2.099 + (7.272 \times 10^{-3} \text{ K}^{-1})T + (1.34 \times 10^{-7} \text{ K}^{-2})T^2$$
$$- (8.66 \times 10^{-10} \text{ K}^{-3})T^3$$

to derive a general equation for the value of $\Delta_r H^\circ$ at any temperature between 300 K and 1500 K. Plot $\Delta_r H^\circ$ versus T. Assume that the gases behave ideally under these conditions.

19-46. In all the calculations thus far, we have assumed the reaction takes place at constant temperature, so that any energy evolved as heat is absorbed by the surroundings. Suppose, however, that the reaction takes place under adiabatic conditions, so that all the energy released as heat stays within the system. In this case, the temperature of the system will increase, and the final temperature is called the *adiabatic flame temperature*. One relatively simple way to estimate this temperature is to suppose the reaction occurs at the initial temperature of the reactants and then determine to what temperature the products can be raised by the quantity $\Delta_r H^\circ$. Calculate the adiabatic flame temperature if one mole of $CH_4(g)$ is burned in two moles of $O_2(g)$ at an initial temperature of 298 K. Use the results of the previous problem.

19-47. Explain why the adiabatic flame temperature defined in the previous problem is also called the maximum flame temperature.

19-48. How much energy as heat is required to raise the temperature of 2.00 moles of $O_2(g)$ from 298 K to 1273 K at 1.00 bar? Take

$$\overline{C}_P[O_2(g)]/R = 3.094 + (1.561 \times 10^{-3} \text{ K}^{-1})T - (4.65 \times 10^{-7} \text{ K}^{-2})T^2$$

19-49. When one mole of an ideal gas is compressed adiabatically to one-half of its original volume, the temperature of the gas increases from 273 K to 433 K. Assuming that $\overline{C}_V$ is independent of temperature, calculate the value of $\overline{C}_V$ for this gas.

19-50. Use the van der Waals equation to calculate the minimum work required to expand one mole of $CO_2(g)$ isothermally from a volume of 0.100 dm^3 to a volume of 100 dm^3 at 273 K. Compare your result with that which you calculate assuming ideal behavior.

19-51. Show that the work involved in a reversible, adiabatic pressure change of one mole of an ideal gas is given by

$$w = \overline{C}_V T_1 \left[\left(\frac{P_2}{P_1} \right)^{R/\overline{C}_P} - 1 \right]$$

where T_1 is the initial temperature and P_1 and P_2 are the initial and final pressures, respectively.

19-52. In this problem, we will discuss a famous experiment called the *Joule-Thomson experiment*. In the first half of the 19th century, Joule tried to measure the temperature change when a gas is expanded into a vacuum. The experimental setup was not sensitive enough, however, and he found that there was no temperature change, within the limits of his error. Soon afterward, Joule and Thomson devised a much more sensitive method for measuring the temperature change upon expansion. In their experiments (see Figure 19.11), a constant applied pressure P_1 causes a quantity of gas to flow slowly from one chamber to another through a porous plug of silk or cotton. If a volume, V_1, of gas is pushed through the porous plug, the work done on the gas is $P_1 V_1$. The pressure on the other side of the plug

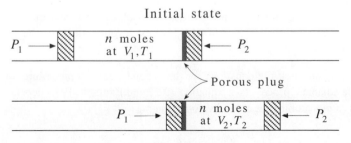

FIGURE 19.11
A schematic description of the Joule-Thomson experiment.

is maintained at P_2, so if a volume V_2 enters the right-side chamber, then the net work is given by

$$w = P_1 V_1 - P_2 V_2$$

The apparatus is constructed so that the entire process is adiabatic, so $q = 0$. Use the First Law of Thermodynamics to show that

$$U_2 + P_2 V_2 = U_1 + P_1 V_1$$

or that $\Delta H = 0$ for a Joule-Thomson expansion.

Starting with

$$dH = \left(\frac{\partial H}{\partial P}\right)_T dP + \left(\frac{\partial H}{\partial T}\right)_P dT$$

show that

$$\left(\frac{\partial T}{\partial P}\right)_H = -\frac{1}{C_P}\left(\frac{\partial H}{\partial P}\right)_T$$

Interpret physically the derivative on the left side of this equation. This quantity is called the *Joule-Thomson coefficient* and is denoted by μ_{JT}. In Problem 19–54 you will show that it equals zero for an ideal gas. Nonzero values of $(\partial T/\partial P)_H$ directly reflect intermolecular interactions. Most gases cool upon expansion [a positive value of $(\partial T/\partial P)_H$] and a Joule-Thomson expansion is used to liquefy gases.

19-53. The Joule-Thomson coefficient (Problem 19–52) depends upon the temperature and pressure, but assuming an average constant value of $0.15 \text{ K} \cdot \text{bar}^{-1}$ for $N_2(g)$, calculate the drop in temperature if $N_2(g)$ undergoes a drop in pressure of 200 bar.

19-54. Show that the Joule-Thomson coefficient (Problem 19–52) can be written as

$$\mu_{JT} = \left(\frac{\partial T}{\partial P}\right)_H = -\frac{1}{C_P}\left[\left(\frac{\partial U}{\partial V}\right)_T \left(\frac{\partial V}{\partial P}\right)_T + \left(\frac{\partial (PV)}{\partial P}\right)_T\right]$$

Show that $(\partial T/\partial P)_H = 0$ for an ideal gas.

19-55. Use the rigid rotator-harmonic oscillator model and the data in Table 18.2 to plot $\overline{C}_P(T)$ for CO(g) from 300 K to 1000 K. Compare your result with the expression given in Problem 19–43.

19-56. Use the rigid rotator-harmonic oscillator model and the data in Table 18.4 to plot $\overline{C}_P(T)$ for $CH_4(g)$ from 300 K to 1000 K. Compare your result with the expression given in Problem 19–45.

19-57. Why do you think the equations for the dependence of temperature on volume for a reversible adiabatic process (see Equation 19.22 and Example 19.6) depend upon whether the gas is a monatomic gas or a polyatomic gas?

THE BINOMIAL DISTRIBUTION AND STIRLING'S APPROXIMATION

In the next chapter, we will learn about entropy, a thermodynamic state function that has a molecular interpretation of being a measure of the disorder of a system. In doing so, we will have to put the idea of the disorder of a system on a quantitative basis. A problem we will encounter is that of determining how many ways we can arrange N distinguishable objects such that there are n_1 objects in the first group, n_2 objects in the second group, and so on, such that

$$n_1 + n_2 + n_3 + \cdots = N$$

that is, such that all the objects are accounted for. This problem is actually a fairly standard one in statistics.

Let's solve the problem of dividing the N distinguishable objects into two groups first and then generalize our results to any number of groups. First, we calculate the number of permutations of N distinguishable objects, that is, the number of possible different arrangements or ways to order N distinguishable objects. Let's choose one of the N objects and place it in the first position, one of the $N - 1$ remaining objects and place it in the second postion, and so on until all N objects are ordered. Clearly, there are N choices for the first position, $N - 1$ choices for the second position, and so on until finally there is only one object left for the Nth position. The total number of ways of doing this ordering is the product of all the choices:

$$N(N - 1)(N - 2) \ldots (2)(1) = N!$$

Next, we calculate the number of ways of dividing N distinguishable objects into two groups, one containing N_1 objects and the other containing the $N - N_1 = N_2$ remaining objects. There are

$$\underbrace{N(N - 1) \ldots (N - N_1 + 1)}_{N_1 \text{ terms}}$$

809

ways to form the first group. This product can be written more conveniently as

$$N(N - 1)(N - 2) \dots (N - N_1 + 1) = \frac{N!}{(N - N_1)!} \tag{J.1}$$

as can be seen by noting that

$$N! = (N)(N - 1) \dots (N - N_1 + 1) \times (N - N_1)!$$

The number of ways of forming the second group is $N_2! = (N - N_1)!$. You might think that the total number of arrangements is the product of the two factors, $N!/(N - N_1)!$ and $N_2!$, but this product drastically overcounts the situation because the order in which we arrange the N_1 objects in the first group and the N_2 objects in the second group is immaterial to the problem stated. All $N_1!$ orders of the first group and $N_2!$ orders of the second group correspond to just one division of N distinguishable objects into two groups containing N_1 and N_2 objects. Therefore, we divide the product of $N!/(N - N_1)!$ and $N_2!$ by $N_1!$ and $N_2!$ to obtain

$$W(N_1, N_2) = \frac{N!}{(N - N_1)!N_1!} = \frac{N!}{N_1!N_2!} \tag{J.2}$$

where we let $W(N_1, N_2)$ denote the result. (Problem J–12 shows that $0! = 1$.)

EXAMPLE J–1

Use Equation J.2 to calculate the number of ways of arranging four distinguishable objects into two groups, containing three objects and one object. Verify your result with an explicit enumeration.

SOLUTION: We have $N = 4$, $N_1 = 3$, and $N_2 = 1$, and so Equation J.2 gives

$$W(3, 1) = \frac{4!}{3!1!} = 4$$

If we let a, b, c, and d be the four distinguishable objects, the four arrangements are $abc : d$, $abd : c$, $acd : b$, and $bcd : a$. There are no others.

The combinatorial factor in Equation J.2 is called a binomial coefficient because the expansion of the binomial $(x + y)^N$ is given by

$$(x + y)^N = \sum_{N_1=0}^{N} \frac{N!}{N_1!(N - N_1)!} x^{N_1} y^{N-N_1} \tag{J.3}$$

For example,

$$(x + y)^2 = x^2 + 2xy + y^2 = \sum_{N_1=0}^{2} \frac{2!}{N_1!(2 - N_1)!} x^{N_1} y^{2-N_1}$$

and

$$(x + y)^3 = x^3 + 3x^2y + 3xy^2 + y^3 = \sum_{N_1=0}^{3} \frac{3!}{N_1!(3 - N_1)!} x^{N_1} y^{3-N_1}$$

Equation J.3 may be written in a more symmetric form:

$$(x + y)^N = \sum_{N_1=0}^{N} \sum_{N_2=0}^{N} {}^* \frac{N!}{N_1!N_2!} x^{N_1} y^{N_2} \tag{J.4}$$

where the asterisk on the summation signs indicates that only terms with $N_1 + N_2 = N$ are included. This symmetric form of the binomial expansion suggests the form of the multinomial expansion given below in Equation J.6. Simple numerical examples verify that Equations J.3 and J.4 are equivalent.

The generalization of Equation J.2 to the division of N distinguishable objects into r groups, the first containing N_1, the second containing N_2, and so on, is

$$W(N_1, N_2, \ldots, N_r) = \frac{N!}{N_1!N_2! \cdots N_r!} \tag{J.5}$$

with $N_1 + N_2 + \cdots + N_r = N$. This quantity is called a multinomial coefficient because it occurs in the multinomial expansion:

$$(x_1 + x_2 + \cdots + x_r)^N = \sum_{N_1=0}^{N} \sum_{N_2=0}^{N} \cdots \sum_{N_r=0}^{N} {}^* \frac{N!}{N_1!N_2! \cdots N_r!} x_1^{N_1} x_2^{N_2} \ldots x_r^{N_r} \tag{J.6}$$

where the asterisk indicates that only terms such that $N_1 + N_2 + \cdots + N_r = N$ are included. Note how Equation J.6 is a straightforward generalization of Equation J.4.

EXAMPLE J–2
Calculate the number of ways of dividing 10 distinguishable objects into three groups containing 2, 5, and 3 objects.

SOLUTION: We use Equation J.5:

$$W(2, 5, 3) = \frac{10!}{2!5!3!} = 2520$$

If we use Equation J.5 to calculate something like the number of ways of distributing Avogadro's number of particles over their energy states, then we are forced to deal with factorials of huge numbers. Even the evaluation of 100! would be a chore, never mind $10^{23}!$, unless we have a good approximation for $N!$. We shall see that there is an approximation for $N!$ that actually improves as N gets larger. Such an approximation is called an asymptotic approximation, that is, an approximation to a function that gets better as the argument of the function increases.

Because $N!$ is a product, it is convenient to deal with $\ln N!$ because the latter is a sum. The asymptotic expansion to $\ln N!$ is called Stirling's approximation and is given by

$$\ln N! = N \ln N - N \tag{J.7}$$

which is surely a lot easier to use than calculating $N!$ and then taking its logarithm. Table J.1 shows the value of $\ln N!$ versus Stirling's approximation for a number of values of N. Note that the agreement, which we express in terms of relative error, improves markedly with increasing N.

EXAMPLE J–3

A more refined version of Stirling's approximation (one we will *not* have to use in the next chapter) says that

$$\ln N! = N \ln N - N + \ln(2\pi N)^{1/2}$$

Use this version of Stirling's approximation to calculate $\ln N!$ for $N = 10$ and compare the relative error with that in Table J.1.

SOLUTION: For $N = 10$,

$$\ln N! = N \ln N - N + \ln(2\pi N)^{1/2} = 15.096$$

and using the value of $\ln 10!$ from Table J.1, we see that

$$\text{relative error} = \frac{15.104 - 15.096}{15.104} = 0.0005$$

The relative error is significantly smaller than that in Table J.1. The relative errors for the other entries in Table J.1 are essentially zero for this extended version of Stirling's approximation.

TABLE J.1

A numerical comparison of $\ln N!$ with Stirling's approximation.

N	$\ln N!$	$N \ln N - N$	Relative error[a]
10	15.104	13.026	0.1376
50	148.48	145.60	0.0194
100	363.74	360.52	0.0089
500	2611.3	2607.3	0.0015
1000	5912.1	5907.7	0.0007

[a]relative error $= (\ln N! - N \ln N + N)/\ln N!$

The proof of Stirling's approximation is not difficult. Because $N!$ is given by $N! = N(N-1)(N-2)\ldots(2)(1)$, $\ln N!$ is given by

$$\ln N! = \sum_{n=1}^{N} \ln n \tag{J.8}$$

Figure J.1 shows $\ln x$ plotted versus x for integer values of x. According to Equation J.8, the sum of the areas under the rectangles up to N in Figure J.1 is $\ln N!$. Figure J.1 also shows the continuous curve $\ln x$ plotted on the same graph. Thus, $\ln x$ is seen to form an envelope to the rectangles, and this envelope becomes a steadily smoother approximation to the rectangles as x increases. Therefore, we can approximate the area under these rectangles by the integral of $\ln x$. The area under $\ln x$ will poorly approximate the rectangles only in the beginning. If N is large enough (we are deriving an asymptotic expansion), this area will make a negligible contribution to the total area. We may write, then,

$$\ln N! = \sum_{n=1}^{N} \ln n \approx \int_{1}^{N} \ln x\, dx = N \ln N - N \quad (N \text{ large}) \tag{J.9}$$

which is Stirling's approximation to $\ln N!$. The lower limit could just as well have been taken as 0 in Equation J.9, because N is large. (Remember that $x \ln x \to 0$ as $x \to 0$.) We will use Stirling's approximation frequently in the next few chapters.

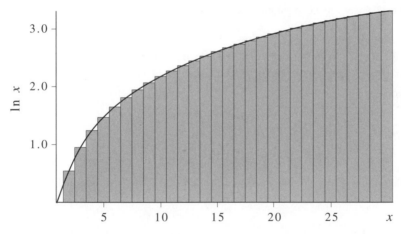

FIGURE J.1
A plot of $\ln x$ versus x. The sum of the areas under the rectangles up to N is $\ln N!$.

Problems

J-1. Use Equation J.3 to write the expansion of $(1 + x)^5$. Use Equation J.4 to do the same thing.

J-2. Use Equation J.6 to write out the expression for $(x + y + z)^2$. Compare your result to the one that you obtain by multiplying $(x + y + z)$ by $(x + y + z)$.

J-3. Use Equation J.6 to write out the expression for $(x + y + z)^4$. Compare your result to the one that you obtain by multiplying $(x + y + z)^2$ from Problem J–2 by itself.

J-4. How many permutations of the letters a, b, c are there?

J-5. The coefficients of the expansion of $(1 + x)^n$ can be arranged in the following form:

n								
0					1			
1				1		1		
2			1		2		1	
3		1		3		3		1
4	1		4		6		4		1

Do you see a pattern in going from one row to the next? The triangular arrangement here is called Pascal's triangle.

J-6. In how many ways can a committee of three be chosen from nine people?

J-7. Calculate the relative error for $N = 50$ using the formula for Stirling's approximation given in Example J–3, and compare your result with that given in Table J.1 using Equation J.7. Take $\ln N!$ to be 148.47776 (*CRC Handbook of Chemistry and Physics*).

J-8. Prove that $x \ln x \to 0$ as $x \to 0$.

J-9. Prove that the maximum value of $W(N, N_1) = N!/(N - N_1)!N_1!$ is given by $N_1 = N/2$. (*Hint*: Treat N_1 as a continuous variable.)

J-10. Prove that the maximum value of $W(N_1, N_2, \ldots, N_r)$ in Equation J.5 is given by $N_1 = N_2 = \cdots = N_r = N/r$.

J-11. Prove that

$$\sum_{k=0}^{N} \frac{N!}{k!(N - k)!} = 2^N$$

J-12. The quantity $n!$ as we have defined it is defined only for positive integer values of n. Consider now the function of x *defined* by

$$\Gamma(x) = \int_0^{\infty} t^{x-1}e^{-t}dt \tag{1}$$

Integrate by parts (letting $u = t^{x-1}$ and $dv = e^{-t}dt$) to get

$$\Gamma(x) = (x-1)\int_0^\infty t^{x-2}e^{-t}dt = (x-1)\Gamma(x-1) \tag{2}$$

Now use Equation 2 to show that $\Gamma(x) = (x-1)!$ if x is a positive integer. Although Equation 2 provides us with a general function that is equal to $(n-1)!$ when x takes on integer values, it is defined just as well for non-integer values. For example, show that $\Gamma(3/2)$, which in a sense is $(\frac{1}{2})!$, is equal to $\pi^{1/2}/2$. Equation 1 can also be used to explain why $0! = 1$. Let $x = 1$ in Equation 1 to show that $\Gamma(1)$, which we can write as $0!$, is equal to 1. The function $\Gamma(x)$ defined by Equation 1 is called the *gamma function* and was introduced by Euler to generalize the idea of a factorial to general values of n. The gamma function arises in many problems in chemistry and physics.

Rudolf Clausius was born in Köslin, Prussia (now Koszalin, Poland), on January 2, 1822, and died in 1888. Although Clausius was initially attracted to history, he eventually received his Ph.D. in mathematical physics from the University of Halle in 1847. He held a position for several years at the University of Zurich but returned to Germany and in 1871 settled at the University of Bonn, where he remained for the rest of his life. Clausius is credited with creating the early foundations of thermodynamics. In 1850, he published his first great paper on the theory of heat, in which he rejected the then-current caloric theory and argued that the energy of a system is a thermodynamic state function. In 1865, he published his second landmark paper, in which he introduced another new thermodynamic state function, which he called entropy, and expressed the Second Law of Thermodynamics in terms of the entropy. Clausius also studied the kinetic theory of gases and made important contributions to it. He was chauvinistic and strongly defended German achievements against what he considered the infringements of others. Most of Clausius' work was done before 1870 because of two events in his life. In 1870, he was wounded while serving in an ambulance corps in the Franco-Prussian War and suffered life-long pain from his injury. More tragically, his wife died in childbirth, and he assumed the responsibility of raising six young children.

CHAPTER 20

Entropy and the Second Law of Thermodynamics

In this chapter, we will introduce and develop the concept of entropy. We will see that energy considerations alone are not sufficient to predict in which direction a process or a chemical reaction can occur spontaneously. We will demonstrate that isolated systems that are not in equilibrium will evolve in a direction that increases their disorder, and then we will introduce a thermodynamic state function called entropy that gives a quantitative measure of the disorder of a system. One statement of the Second Law of Thermodynamics, which governs the direction in which systems evolve to their equilibrium states, is that the entropy of an isolated system always increases as a result of any spontaneous (irreversible) process. In the second half of this chapter, we will give a quantitative molecular definition of entropy in terms of partition function.

20–1. The Change of Energy Alone Is Not Sufficient to Determine the Direction of a Spontaneous Process

For years, scientists wondered why some reactions or processes proceed spontaneously and others do not. We all know that under the right conditions iron rusts, and that objects do not spontaneously unrust. We all know that hydrogen and oxygen react explosively to form water but that an input of energy by means of electrolysis is required to decompose water into hydrogen and oxygen. At one time scientists believed that a criterion for a reaction or a process to proceed spontaneously was that it should be exothermic, or evolve energy. This belief was motivated by the fact that the products of an exothermic reaction lie at a lower energy or enthalpy than the reactants. After all, balls *do* roll downhill and opposite charges *do* attract each other. In fact, the variational principle of quantum mechanics (Section 7–1) is based upon the fact that a system will always seek its state of lowest energy. Mechanical systems evolve in such a way as to minimize their energy.

817

FIGURE 20.1
Two bulbs connected by a stopcock. Initially, one bulb contains a colored gas such as bromine and the other one is evacuated. When the two bulbs are connected by opening the stopcock, the bromine occupies both bulbs at a uniform pressure as seen by the uniform color.

Now consider the situation in Figure 20.1, however, where one bulb contains a gas at some low pressure at which it may be considered to behave ideally, and the other bulb is evacuated. When the two bulbs are connected by opening the stopcock between them, the gas will expand into the evacuated bulb until the pressures in the two bulbs are equal, at which time the system will be in equilibrium. Yet a careful determination of the thermal processes of this experiment shows that both ΔU and ΔH are essentially zero. Furthermore, the unaided reverse process has never been observed. Gases do not spontaneously occupy only part of a container, leaving the other part as a vacuum.

Another example of a spontaneous process that is not exothermic is depicted in Figure 20.2, where two pure gases are separated by a stopcock. When the stopcock is opened, the two gases will mix, and both will eventually become evenly distributed between the two bulbs, in which case the system will be in equilibrium. Yet once again, the value of ΔU or ΔH for this process is essentially zero. Furthermore, the reverse process has never been observed. Mixtures of gases do not spontaneously unmix.

There are many spontaneous endothermic processes. A simple example of a spontaneous endothermic reaction is the melting of ice at a temperature above $0°C$. This spontaneous process has a value of $\Delta_{fus} H°$ equal to $+6.0 \text{ kJ·mol}^{-1}$ when the temperature is around $0°C$. An especially interesting endothermic chemical reaction is the reation of solid barium hydroxide, $Ba(OH)_2(s)$, with solid ammonium nitrate, $NH_4NO_3(s)$:

$$Ba(OH)_2(s) + 2 NH_4NO_3(s) \longrightarrow Ba(NO_3)_2(s) + 2 H_2O(l) + 2 NH_3(aq)$$

FIGURE 20.2
Two bulbs connected by a stopcock. Initially, each bulb is occupied by a pure gas, say bromine and nitrogen. When the two bulbs are connected by opening the stopcock, the two gases mix uniformly, so each bulb contains the same uniform mixture.

The energy absorbed by mixing stoichiometric amounts of these two reagents in a test tube can cool the system to below $-20°C$.

These and numerous other examples indicate that spontaneous processes have a direction that cannot be explained by the First Law of Thermodynamics. Of course, each of these processes obeys the First Law of Thermodynamics, but using this law, we cannot tell why one direction occurs spontaneously and its reverse does not. Although mechanical systems tend to achieve their state of lowest energy, clearly some other factor is involved that we have not yet discussed.

20–2. Nonequilibrium Isolated Systems Evolve in a Direction That Increases Their Disorder

If we examine the above processes from a microscopic or molecular point of view, we see that each one involves an increase in disorder or randomness of the system. For example, in Figure 20.1, the gas molecules in the final state are able to move over a volume that is twice as large as in the initial state. In a sense, locating any gas molecule in the final state is twice as difficult as in the initial state. Recall that we found that the number of accessible translational states increases with the volume of the container, Problem 18–42. A similar argument applies to the mixing of two gases. Not only is each gas spread over a larger volume, but they are also mixed together. Clearly the final (mixed) state is more disordered than the initial (separated) state. The melting of ice at a temperature greater than $0°C$ also involves an increase in disorder. Our molecular picture of a solid being an ordered lattice array of its constituent particles and a liquid being a more random arrangement directly implies that the melting of ice involves an increase in disorder.

These examples suggest that not only do systems evolve spontaneously in a direction that lowers their energy but that they also seek to increase their disorder. There is a competition between the tendency to minimize energy and to maximize disorder. If disorder is not a factor, as is the case for a simple mechanical system, then energy is the key factor and the direction of any spontaneous process is that which minimizes the energy. If energy is not a factor, however, as is the case when mixing two gases, then disorder is the key factor and the direction of any spontaneous process is that which maximizes the disorder. In general, some compromise between decreasing energy and increasing disorder must be met.

What we need is to devise some particular property that puts this idea of disorder on a useful, quantitative basis. Like energy, we want this property to be a state function because then it will be a property of the state of the system, and not of its previous history. Thus, we will rule out heat, although the transfer of energy as heat to a system certainly does increase its disorder. To try to get an idea of what an appropriate function might be, let's consider, for simplicity, the heat transfer associated with a reversible, small change in the temperature and volume of an ideal gas. From the First Law (Equation 19.9), we have

$$\delta q_{rev} = dU - \delta w_{rev} = C_V(T)dT + PdV$$

$$= C_V(T)dT + \frac{nRT}{V}dV \tag{20.1}$$

Example 19–4 showed us that δq_{rev} is not a state function. In mathematical terms, this means that the right side of Equation 20.1 is not an exact differential; in other words, it can not be written as the derivative of some function of T and V (see MathChapter H). The first term, however, can be written as the derivative of a function of T because C_V is a function of only temperature for an ideal gas , so $C_V(T)dT$ can be written as

$$C_V(T)dT = d\left[\int C_V(T)dT + \text{constant}\right]$$

The fact that the second term cannot be written as a derivative means that

$$\frac{nRT}{V}dV \neq d\left(\int \frac{nRT}{V}dV + \text{constant}\right)$$

because T depends upon V. It is really a work term, so the evaluation of w_{rev} depends upon the path. If we divide Equation 20.1 by T, however, we get a very interesting result:

$$\frac{\delta q_{rev}}{T} = \frac{C_V(T)dT}{T} + \frac{nR}{V}dV \tag{20.2}$$

Notice now that $\delta q_{rev}/T$ is an exact differential. The right side can be written in the form

$$d\left[\int \frac{C_V(T)}{T}dT + nR\int \frac{dV}{V} + \text{constant}\right]$$

so $\delta q_{rev}/T$ is the derivative of a state function that is a function of T and V (see also MathChapter H). If we let this state function be denoted by S, Equation 20.2 reads

$$dS = \frac{\delta q_{rev}}{T} \tag{20.3}$$

Notice that the inexact differential δq_{rev} has been converted to an exact differential by multiplying it by $1/T$. In mathematical terms, we say that $1/T$ is an *integrating factor* of δq_{rev}.

The state function S that we have described here is called the *entropy*. Because entropy is a state function, $\Delta S = 0$ for a cyclic process; that is, a process in which the final state is the same as the initial state. We can indicate this concept mathematically by writing

$$\oint dS = 0 \tag{20.4}$$

where the circle on the integral sign indicates a cyclic process. From Equation 20.3, we can also write

$$\oint \frac{\delta q_{rev}}{T} = 0 \tag{20.5}$$

Equation 20.5 is a statement of the fact that $\delta q_{rev}/T$ is the derivative of a state function. Although we proved Equation 20.5 only for the case of an ideal gas, it is generally true (Problem 20–5).

20–3. Unlike q_{rev}, Entropy Is a State Function

In the previous chapter, we calculated the reversible work and reversible heat for two processes that take place between the same initial and final states (Figure 20.3). The first process involved a reversible isothermal expansion of an ideal gas from P_1, V_1, T_1 to P_2, V_2, T_1 (path A). For this process (cf. Equations 19.12 and 19.13),

$$\delta q_{rev,A} = \frac{nRT_1}{V}dV \tag{20.6}$$

and so

$$q_{rev,A} = nRT_1 \ln \frac{V_2}{V_1}$$

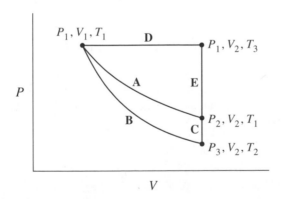

FIGURE 20.3
An illustration of three different paths (A, B+C, and D+E) from an initial state P_1, V_1, T_1 to a final state P_2, V_2, T_1 of an ideal gas. Path A represents a reversible isothermal expansion from P_1, V_1 to P_2, V_2. Path B+C represents a reversible adiabatic expansion (B) from P_1, V_1, T_1 to P_3, V_2, T_2 followed by reversibly heating the gas at constant volume (C) from P_3, V_2, T_2 to P_2, V_2, T_1. Path D+E represents a reversible expansion at constant pressure P_1 (D) from P_1, V_1, T_1 to P_1, V_2, T_3 followed by a reversible cooling at constant volume V_2 (E) from P_1, V_2, T_3 to P_2, V_2, T_1.

The other process involved a reversible adiabatic expansion of an ideal gas from P_1, V_1, T_1 to P_3, V_2, T_2 (path B), followed by heating the gas reversibly at constant volume from P_3, V_2, T_2 to P_2, V_2, T_1 (path C). For this process (cf. Equations 19.15 and 19.17),

$$\delta q_{rev,B} = 0$$
$$\delta q_{rev,C} = C_V(T)dT \tag{20.7}$$

and

$$q_{rev,B+C} = \int_{T_2}^{T_1} C_V(T)dT$$

where T_2 is given by (cf. Equation 19.21)

$$\int_{T_1}^{T_2} \frac{C_V(T)}{T}dT = -nR\ln\frac{V_2}{V_1} \tag{20.8}$$

The point here is that q_{rev} differs for the two paths, A and B + C, indicating that q_{rev} is not a state function.

Now let's evaluate

$$\Delta S = \int_1^2 \frac{\delta q_{rev}}{T}$$

for these two paths. For path A from P_1, T_1, V_1 to P_2, V_2, T_1, we have, using Equation 20.6,

$$\Delta S_A = \int_1^2 \frac{\delta q_{rev,A}}{T_1} = \int_{V_1}^{V_2} \frac{1}{T_1}\frac{nRT_1}{V}dV$$
$$= nR \int_{V_1}^{V_2} \frac{dV}{V} = nR\ln\frac{V_2}{V_1} \tag{20.9}$$

For the reversible adiabatic expansion from P_1, V_1, T_1 to P_3, V_2, T_2 (path B) followed by a reversible heating at constant volume from P_3, V_2, T_2 to P_2, V_2, T_1 (path C), we have, using Equation 20.7,

$$\Delta S_B = \int_1^2 \frac{\delta q_{rev,B}}{T} = 0$$

and

$$\Delta S_C = \int_2^1 \frac{\delta q_{rev,C}}{T} = \int_{T_2}^{T_1} \frac{C_V(T)}{T}dT = -\int_{T_1}^{T_2} \frac{C_V(T)}{T}dT$$

But using Equation 20.8, ΔS_C turns out to be

$$\Delta S_C = nR\ln\frac{V_2}{V_1}$$

and so

$$\Delta S_{B+C} = \Delta S_B + \Delta S_C = 0 + nR \ln \frac{V_2}{V_1} = nR \ln \frac{V_2}{V_1} \qquad (20.10)$$

Thus, we see that the ΔS_A (Equation 20.9) is equal to ΔS_{B+C} (Equation 20.10) and that the value of ΔS is independent of the path.

EXAMPLE 20–1

Calculate q_{rev} and ΔS for a reversible expansion of an ideal gas at constant pressure P_1 from T_1, V_1 to T_3, V_2 (path D in Figure 20.3) followed by a reversible cooling of the gas at constant volume V_2 from P_1, T_3 to P_2, T_1 (path E).

SOLUTION: For path D (cf. Example 19–4),

$$\delta q_{rev,D} = dU_D - \delta w_{rev,D} = C_V(T)dT + P_1 dV \qquad (20.11)$$

and so

$$q_{rev,D} = \int_{T_1}^{T_3} C_V(T)dT + P_1(V_2 - V_1)$$

For path E, $\delta w_{rev} = 0$, and so

$$\delta q_{rev,E} = dU_E = C_V(T)dT \qquad (20.12)$$

and

$$q_{rev,E} = \int_{T_3}^{T_1} C_V(T)dT$$

For the complete process (paths D + E),

$$q_{rev,D+E} = q_{rev,D} + q_{rev,E} = P_1(V_2 - V_1)$$

To calculate ΔS for path D, we use Equation 20.11 to write

$$\Delta S_D = \int \frac{\delta q_{rev,D}}{T}$$

$$= \int_{T_1}^{T_3} \frac{C_V(T)}{T}dT + P_1 \int_{V_1}^{V_2} \frac{dV}{T}$$

To evaluate the second integral here, we must know how T varies with V for this process. But this is given by $P_1 V = nRT$, so

$$\Delta S_D = \int_{T_1}^{T_3} \frac{C_V(T)}{T}dT + nR \int_{V_1}^{V_2} \frac{dV}{V}$$

$$= \int_{T_1}^{T_3} \frac{C_V(T)}{T}dT + nR \ln \frac{V_2}{V_1}$$

For path E, $\delta w_{rev} = 0$, and using Equation 20.12 for $\delta q_{rev,E}$ gives

$$\Delta S_E = \int \frac{\delta q_{rev,E}}{T} = \int_{T_3}^{T_1} \frac{C_V(T)}{T} dT$$

The value of ΔS for the complete process (paths D + E) is

$$\Delta S_{D+E} = \Delta S_D + \Delta S_E = nR \ln \frac{V_2}{V_1}$$

Notice that this is the very same result we obtained for paths A and B + C, once again suggesting that S is a state function.

EXAMPLE 20–2

We shall prove in Example 22–4 that similar to that found for an ideal gas, U is a function of only the temperature for a gas that obeys the equation of state

$$P = \frac{RT}{\overline{V} - b}$$

where b is a constant that reflects the size of the molecules. Calculate q_{rev} and ΔS for both the paths A and B + C in Figure 20.3 for one mole of such a gas.

SOLUTION: Path A represents an isothermal expansion, so $dU_A = 0$ because U depends only upon the temperature. Therefore,

$$\delta q_{rev,A} = -\delta w_{rev,A} = P d\overline{V} = \frac{RT}{\overline{V} - b} d\overline{V}$$

and

$$q_{rev,A} = \int_{\overline{V}_1}^{\overline{V}_2} \frac{RT d\overline{V}}{\overline{V} - b} = RT \int_{\overline{V}_1}^{\overline{V}_2} \frac{d\overline{V}}{\overline{V} - b} = RT \ln \frac{\overline{V}_2 - b}{\overline{V}_1 - b}$$

The entropy change is given by

$$\Delta S_A = \int_1^2 \frac{\delta q_{rev,A}}{T} = R \int_{\overline{V}_1}^{\overline{V}_2} \frac{d\overline{V}}{\overline{V} - b} = R \ln \frac{\overline{V}_2 - b}{\overline{V}_1 - b}$$

For path B, a reversible adiabatic expansion, $q_{rev,B} = 0$, so

$$\Delta S_B = 0$$

For path C, $\delta w_{rev,C} = 0$, and

$$\delta q_{rev,C} = dU_C = C_V(T) dT$$

and

$$q_{rev,C} = \int_{T_2}^{T_1} C_V(T) dT$$

The molar entropy change is given by

$$\Delta \overline{S}_C = \int_{T_2}^{T_1} \frac{\delta q_{rev,C}}{T} = \int_{T_2}^{T_1} \frac{\overline{C}_V(T)}{T} dT = -\int_{T_1}^{T_2} \frac{\overline{C}_V(T)}{T} dT$$

and so

$$\Delta \overline{S}_{B+C} = \Delta \overline{S}_B + \Delta \overline{S}_C = -\int_{T_1}^{T_2} \frac{\overline{C}_V(T)}{T} dT$$

But T_2, the temperature at the end of the reversible adiabatic expansion, can be found from

$$dU = \delta q_{rev} + \delta w_{rev}$$

Using the fact that $d\overline{U} = \overline{C}_V(T) dT$ and $\delta q_{rev} = 0$ gives

$$\overline{C}_V(T) dT = -Pd\overline{V} = -\frac{RT}{\overline{V} - b} d\overline{V}$$

Divide through by T and integrate from the initial state to the final state to get

$$\int_{T_1}^{T_2} \frac{\overline{C}_V(T)}{T} dT = -R \int_{\overline{V}_1}^{\overline{V}_2} \frac{d\overline{V}}{\overline{V} - b} = -R \ln \frac{\overline{V}_2 - b}{\overline{V}_1 - b}$$

Substituting this result into the above expression for $\Delta \overline{S}_{B+C}$ gives

$$\Delta \overline{S}_{B+C} = R \ln \frac{\overline{V}_2 - b}{\overline{V}_1 - b}$$

Therefore, we see that even though $q_{rev,A} \neq q_{rev,B+C}$, nevertheless,

$$\Delta \overline{S}_A = \Delta \overline{S}_{B+C}$$

We will show several times in the following sections that the entropy is related to the disorder of a system, but for now, notice that if we add energy as heat to a system, then its entropy increases because its thermal disorder increases. Furthermore, notice that because $dS = \delta q_{rev}/T$, energy delivered as heat at a lower temperature contributes more to an entropy (disorder) increase than at a higher temperature. The lower the temperature, the lower the disorder, so the energy added as heat has proportionally more "order" to convert to "disorder."

20–4. The Second Law of Thermodynamics States That the Entropy of an Isolated System Increases as a Result of a Spontaneous Process

We all know that energy as heat will flow spontaneously from a region of high temperature to a region of low temperature. Let's investigate the role entropy plays in this process. Consider the two-compartment system shown in Figure 20.4, where parts A and B are large one-component systems. Both systems are at equilibrium, but they are not at equilibrium with each other. Let the temperatures of these two systems be T_A and T_B. The two systems are separated from each other by a rigid, heat-conducting wall so that energy as heat can flow from one system to the other, but the two-compartment system itself is isolated. When we call a system *isolated*, we mean that the system is separated from its surroundings by rigid walls that do not allow matter or energy to pass through them. We may picture the walls as rigid, totally non-heat conducting, and impervious to matter. Consequently, the system can do no work nor can work be done on the system, nor can it exhange energy as heat with the surroundings. The two-compartment system is described by the equations

$$U_A + U_B = \text{constant}$$

$$V_A = \text{constant} \qquad V_B = \text{constant} \qquad (20.13)$$

$$S = S_A + S_B$$

Because V_A and V_B are fixed, we have for each separate system

$$dU_A = \delta q_{rev} + \delta w_{rev} = T_A dS_A \quad (dV_A = 0)$$

$$dU_B = \delta q_{rev} + \delta w_{rev} = T_B dS_B \quad (dV_B = 0) \qquad (20.14)$$

The entropy change of the two-compartment system is given by

$$dS = dS_A + dS_B$$
$$= \frac{dU_A}{T_A} + \frac{dU_B}{T_B} \qquad (20.15)$$

But $dU_A = -dU_B$ because the two-compartment system is isolated, so we have

$$dS = dU_B \left(\frac{1}{T_B} - \frac{1}{T_A} \right) \qquad (20.16)$$

Experimentally, we know that if $T_B > T_A$, then $dU_B < 0$ (energy as heat flows from system B to system A), in which case $dS > 0$. Similarly, $dS > 0$ if $T_B < T_A$

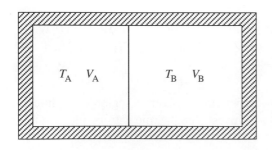

FIGURE 20.4
A two-compartment system in which A and B are large, one-component systems. Each system is at equilibrium, but they are not at equilibrium with each other. The two systems are separated from each other by a rigid, heat-conducting wall. The total two-compartment system itself is isolated.

because $dU_B > 0$ in this case (energy as heat flows from system A to system B). We may interpret this result by saying that the spontaneous flow of energy as heat from a body at a higher temperature to a body at a lower temperature is governed by the condition $dS > 0$. If $T_A = T_B$, then the two-compartment system is in equilibrium and $dS = 0$.

We can generalize this result by investigating the role entropy plays in governing the direction of any spontaneous process. To be able to focus on the entropy alone, we will consider an infinitesimal spontaneous change in an isolated system. We choose an isolated system because the energy remains constant in an isolated system, and we wish to separate the effect due to a change in energy from the effect due to a change in entropy. Because the energy remains constant, the driving force for any spontaneous process in an isolated system must be due to an increase in entropy, which we can express mathematically by $dS > 0$. Because the system is isolated, this increase in entropy must be created within the system itself. Unlike energy, entropy is not necessarily conserved; it increases whenever a spontaneous process takes place. In fact, the entropy of an isolated system will continue to increase until no more spontaneous processes occur, in which case the system will be in equilibrium (Figure 20.5). Thus, we conclude that the entropy of an isolated system is a maximum when the system is in equilibrium. Consequently, $dS = 0$ at equilibrium. Furthermore, not only is $dS = 0$ in an isolated system at equilibrium, but $dS = 0$ for any reversible process in an isolated system because, by definition, a reversible process is one in which the system remains essentially in equilibrium during the entire process. To summarize our conclusions thus far, then, we write

$$dS > 0 \quad \text{(spontaneous process in an isolated system)}$$

$$dS = 0 \quad \text{(reversible process in an isolated system)}$$

(20.17)

Because we have considered an isolated system, no energy as heat can flow in or out of the system. For other types of systems, however, energy as heat can flow in or out, and it is convenient to view dS in any spontaneous infinitesimal process as

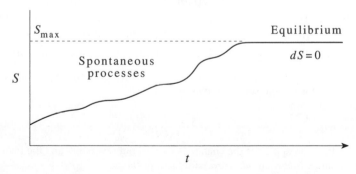

FIGURE 20.5
A schematic plot of entropy versus time for an isolated system. The entropy increases ($dS > 0$) until no more spontaneous processes occur, in which case the system is in equilibrium, and $dS = 0$.

consisting of two parts. One part of dS is the entropy created by the irreversible process itself, and the other part is the entropy due to the energy as heat exchanged between the system and its surroundings. These two contributions account for the entire change in entropy. We will denote the part of dS that is created by the irreversible process by dS_{prod} because it is *produced* by the system. This quantity is always positive. We will denote the part of dS that is due to the exchange of energy as heat with the surroundings by dS_{exch} because it is due to *exchange*. This quantity is given by $\delta q/T$, and it can be positive, negative, or zero. Note that δq need not be δq_{rev}. The quantity δq will be δq_{rev} if the exchange is reversible and δq_{irr} if the exchange is irreversible. Thus, we write for *any* process

$$dS = dS_{prod} + dS_{exch}$$

$$= dS_{prod} + \frac{\delta q}{T} \tag{20.18}$$

For a reversible process, $\delta q = \delta q_{rev}$, $dS_{prod} = 0$, so

$$dS = \frac{\delta q_{rev}}{T} \tag{20.19}$$

in agreement with Equation 20.3. For an irreversible or spontaneous process, $dS_{prod} > 0$, $dS_{exch} = \delta q_{irr}/T$, and so

$$dS > \frac{\delta q_{irr}}{T} \tag{20.20}$$

Equations 20.19 and 20.20 can be written as one equation,

$$dS \geq \frac{\delta q}{T} \tag{20.21}$$

or

$$\Delta S \geq \int \frac{\delta q}{T} \tag{20.22}$$

where the equality sign holds for a reversible process and the inequality sign holds for an irreversible process. Equation 20.22 is one of a number of ways of expressing the Second Law of Thermodynamics and is called the *Inequality of Clausius*.

A formal statement of the Second Law of Thermodynamics is as follows:

There is a thermodynamic state function of a system called the entropy, S, such that for any change in the thermodynamic state of the system,

$$dS \geq \frac{\delta q}{T}$$

where the equality sign applies if the change is carried out reversibly and the inequality sign applies if the change is carried out irreversibly at any stage.

We can use Equation 20.22 to prove quite generally that the entropy of an isolated system always increases during a spontaneous (irreversible) process or that $\Delta S > 0$.

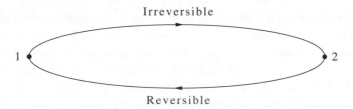

FIGURE 20.6
A cyclic process in which the system is first isolated and undergoes an irreversible process from state 1 to state 2. Then the system is allowed to interact with its surroundings and is brought back to state 1 by some reversible path. Because entropy is a state function, $\Delta S = 0$ for a cyclic process.

Consider a cyclic process (Figure 20.6) in which a system is first isolated and undergoes an irreversible process from state 1 to state 2. Now let the system interact with its surroundings and return to state 1 by any reversible path. Because S is a state function, $\Delta S = 0$ for this cyclic process, so according to Equation 20.22,

$$\Delta S = 0 > \int_1^2 \frac{\delta q_{\text{irr}}}{T} + \int_2^1 \frac{\delta q_{\text{rev}}}{T}$$

The inequality applies because the cyclic process is irreversible from 1 to 2. The first integral here equals zero because the system is isolated, i.e., $\delta q_{\text{irr}} = 0$. The second integral is by definition equal to $S_1 - S_2$, so we have $0 > S_1 - S_2$. Because the final state is state 2 and the initial state is state 1,

$$\Delta S = S_2 - S_1 > 0$$

Thus we see that the entropy increases when the isolated system goes from state 1 to state 2 by a general irreversible process.

Because the universe itself may be considered to be an isolated system and all naturally occurring processes are irreversible, one statement of the Second Law of Thermodynamics says that the entropy of the universe is constantly increasing. In fact, Clausius summarized the first two laws of thermodynamics by

The energy of the Universe is constant;

the entropy is tending to a maximum.

20–5. The Most Famous Equation of Statistical Thermodynamics Is $S = k_{\text{B}} \ln W$

In this section, we will discuss the molecular interpretation of entropy more quantitatively than we have up to now. We have shown that entropy is a state function that is related to the disorder of a system. Disorder can be expressed in a number of ways, but the way that has turned out to be the most useful is the following. Consider an ensemble of $\mathcal{A}$ isolated systems, each with energy E, volume V, and number of particles N. Realize that whatever the value of E, it must be an eigenvalue of the Schrödinger equation for the system. As we discussed in Chapter 17, the energy is a function of N and V, so we can write $E = E(N, V)$ (see, for example, Equations 17.2 and 17.3). Although all the systems have the same energy, they may be in different quantum states because of degeneracy. Let the degeneracy associated with the energy E be $\Omega(E)$, so that we can label the $\Omega(E)$ degenerate quantum states by $j = 1, 2, \ldots, \Omega(E)$. (The degeneracies of systems that consist of N particles turn out to be enormous; they are numbers of the order of e^N for energies not too close to the ground-state energy.) Now, let a_j be the number of systems in the ensemble that are in the state j. Because the $\mathcal{A}$ systems of the ensemble are distinguishable, the number of ways of having a_1 systems in state 1, a_2 systems in state 2, etc. is given by (MathChapter J)

$$W(a_1, a_2, a_3, \ldots) = \frac{\mathcal{A}!}{a_1! a_2! a_3! \ldots} = \frac{\mathcal{A}!}{\prod_j a_j!} \tag{20.23}$$

with

$$\sum_j a_j = \mathcal{A}$$

If all $\mathcal{A}$ systems are in one particular state (a totally ordered arrangement), say state 1, then $a_1 = \mathcal{A}$, $a_2 = a_3 = \cdots = 0$ and $W = 1$, which is the smallest value W can have. In the other extreme, when all the a_j are equal (a disordered arrangement), W takes on its largest value (Problem J–10). Therefore, W can be taken to be a quantitative measure of the disorder of a system. We will not set the entropy proportional to W, however, but to $\ln W$ according to

$$S = k_B \ln W \tag{20.24}$$

where k_B is the Boltzmann constant. Note that $S = 0$ for a completely ordered system ($a_1 = 1$, $a_2 = a_3 = \cdots 0$) and achieves a maximum value for a completely disordered system ($a_1 = a_2 = a_3 = \cdots$). Equation 20.24 was formulated by Boltzmann and is the most famous equation of statistical thermodynamics. In fact, this equation is the only inscription on a monument to Boltzmann in the central cemetary in Vienna. It gives us a quantitative relation between the thermodynamic quantity, entropy, and the statistical quantity, W.

We set S equal to $\ln W$ rather than W for the following reason. We want S to be such that the total entropy of a system that is made up of two parts (say A and B) is given by

$$S_{\text{total}} = S_A + S_B$$

In other words, we want S to be an extensive state function. Now if W_A is the value of W for system A and W_B is the value of W for system B, W_{AB} for the composite system is given by

$$W_{AB} = W_A W_B$$

The entropy of the composite system is

$$S_{AB} = k_B \ln W_{AB} = k_B \ln W_A W_B = k_B \ln W_A + k_B \ln W_B$$
$$= S_A + S_B$$

An alternate form of Equation 20.24 expresses S in terms of the degeneracy Ω. We can determine this expression in the following way. Given no other information, there is no reason to choose one of the Ω degenerate quantum states over any other; each one should occur in an ensemble with equal probability (this concept is actually one of the postulates of statistical thermodynamics). Consequently, we expect that the ensemble of isolated systems should contain equal numbers of systems in each quantum state.

Because S is a maximum for an isolated system at equilibrium, W must also be a maximum. The value of W is maximized when all the a_j are equal (Problem J–10). Let the total number of systems in the ensemble be $\mathcal{A} = n\Omega$ and let each $a_j = n$, so that the set of Ω degenerate quantum states is replicated n times in the ensemble. (We will never need the value of n.) Using Stirling's approximation (MathChapter J) in Equation 20.23, we get

$$S_{ensemble} = k_B \ln W = k_B[\mathcal{A} \ln \mathcal{A} - \sum_{j=1}^{\Omega} a_j \ln a_j]$$

$$= k_B[n\Omega \ln(n\Omega) - \sum_{j=1}^{\Omega}(n \ln n)] = k_B[n\Omega \ln(n\Omega) - \Omega(n \ln n)]$$

$$= k_B(n\Omega \ln \Omega)$$

The entropy of a typical system in the ensemble is given by $S_{ensemble} = \mathcal{A} S_{system} = n\Omega S_{system}$, and so

$$S = k_B \ln \Omega \qquad (20.25)$$

where we have dropped the subscript, system. Equation 20.25 is an alternate form of Equation 20.24 and relates entropy to disorder. As a concrete example, consider a system of N (distinguishable) spins (or dipoles) that can be oriented in one of two possible directions with equal probability. Then, each spin has a degeneracy of 2 associated with it, and the degeneracy of the N spins is 2^N. The entropy of this system is $Nk_B \ln 2$. We will use this result when we discuss the entropy of carbon monoxide at 0 K in Section 21–8.

As another example of the use of Equation 20.25, Problem 20–23 has you show that

$$\Omega(E) = c(N)f(E)V^N$$

for an ideal gas of N particles, where $c(N)$ is a function of N and $f(E)$ is a function of the energy. Now let's determine ΔS for an isothermal expansion of one mole of an ideal gas from a volume V_1 to V_2.

$$\Delta S = k_B \ln \Omega_2 - k_B \ln \Omega_1$$

$$= k_B \ln \frac{\Omega_2}{\Omega_1} = k_B \ln \frac{c(N)f(E_2)V_2^N}{c(N)f(E_1)V_1^N}$$

But $f(E_1) = f(E_2)$ in this case because we are considering an isothermal expansion of an ideal gas, so $E_2 = E_1$. Therefore, we have for one mole

$$\Delta \overline{S} = N_A k_B \ln \frac{V_2}{V_1} = R \ln \frac{V_2}{V_1}$$

in agreement with Equation 20.9.

EXAMPLE 20–3
Use the fact that

$$\Omega(E) = c(N)f(E)V^N$$

for an ideal gas to show that the change in entropy (per mole) when two gases are mixed isothermally is given by

$$\Delta_{mix}\overline{S}/R = -y_1 \ln y_1 - y_2 \ln y_2 \tag{20.26}$$

where y_1 and y_2 are the mole fractions of the two gases.

SOLUTION: Consider the process depicted in Figure 20.2. Then $\Delta_{mix}S$ is given by

$$\Delta_{mix}S = S_{mixture} - S_1 - S_2$$

$$= k_B \ln \frac{\Omega_{mixture}}{\Omega_1 \Omega_2}$$

where 1 and 2 refer to $N_2(g)$ and $Br_2(g)$, respectively. The quantities Ω_1 and Ω_2 are given by

$$\Omega_1 = c(N_1)f(E_1)V_1^{N_1} \quad \text{and} \quad \Omega_2 = c(N_2)f(E_2)V_2^{N_2}$$

Because the molecules in a mixture of ideal gases are independent of each other,

$$\Omega_{mixture} = c(N_1)f(E_1)(V_1 + V_2)^{N_1} \times c(N_2)f(E_2)(V_1 + V_2)^{N_2}$$

Substitute these expressions for Ω_{N_2}, Ω_{Br_2}, and $\Omega_{mixture}$ into the above equation for $\Delta_{mix}S$ to get

$$\Delta_{mix}S = k_B \ln \frac{(V_1 + V_2)^{N_1}}{V_1^{N_1}} \cdot \frac{(V_1 + V_2)^{N_2}}{V_2^{N_2}}$$

$$= -k_B N_1 \ln \left(\frac{V_1}{V_1 + V_2} \right) - k_B N_2 \ln \left(\frac{V_2}{V_1 + V_2} \right)$$

Because V is proportional to n for an ideal gas,

$$\frac{V_1}{V_1 + V_2} = \frac{n_1}{n_1 + n_2} = y_1 \quad \text{and} \quad \frac{V_2}{V_1 + V_2} = \frac{n_2}{n_1 + n_2} = y_2$$

so we have

$$\Delta_{mix} S = -k_B N_1 \ln y_1 - k_B N_2 \ln y_2$$
$$= -R n_1 \ln y_1 - R n_2 \ln y_2$$

Now, finally divide by $n_1 + n_2$ and R to get

$$\Delta_{mix} \overline{S}/R = -y_1 \ln y_1 - y_2 \ln y_2$$

Note that $\Delta_{mix} \overline{S}$ is always a positive quantity because y_1 and y_2, being mole fractions, are always less than one. Thus, the isothermal mixing of two (ideal) gases is a spontaneous process. We will derive Equation 20.26 using classical thermodynamics in the next section.

20–6. We Must Always Devise a Reversible Process to Calculate Entropy Changes

The discussion so far has been fairly abstract, and it will be helpful at this point to illustrate the change of entropy in a spontaneous process by means of some calculations involving an ideal gas for simplicity. First, let's consider the situation in Figure 20.1, in which an ideal gas at T and V_1 is allowed to expand into a vacuum to a total volume of V_2. We use Equation 20.19 even though this is *not* a reversible process. Remember that because the entropy is a state function, it depends only upon the initial and final states and not upon the path between them. Equation 20.19 tells us that we can calculate ΔS by integrating $\delta q_{rev}/T$ over a reversible path,

$$\Delta S = \int_1^2 \frac{\delta q_{rev}}{T} \tag{20.27}$$

regardless of whether the process is reversible or not. Even though the irreversible process occurs adiabatically, we use a reversible path to calculate the entropy change from the state T, V_1 to T, V_2. This path will not represent the actual adiabatic process, which does not matter because we are interested in only the entropy change between the initial state and the final state. To calculate ΔS, then, we start with

$$\delta q_{rev} = dU - \delta w_{rev}$$

But $dU = 0$ for the expansion of an ideal gas into a vacuum because U depends upon only temperature and is independent of volume for an ideal gas. Therefore, we have $\delta q_{rev} = -\delta w_{rev}$. The reversible work is given by

$$\delta w_{rev} = -PdV = -\frac{nRT}{V}dV$$

so

$$\Delta S = \int_1^2 \frac{\delta q_{rev}}{T} = -\int_1^2 \frac{\delta w_{rev}}{T} = nR \int_{V_1}^{V_2} \frac{dV}{V} = nR \ln \frac{V_2}{V_1} \qquad (20.28)$$

Note that $\Delta S > 0$ because $V_2 > V_1$. Thus, we see that the entropy increases in the expansion of an ideal gas into a vacuum.

Because Equation 20.19 tells us to calculate ΔS by expanding the gas reversibly and isothermally from V_1 to V_2, Equation 20.28 holds for the reversible isothermal expansion. Because S is a state function, however, the value of ΔS obtained from Equation 20.28 is the same as the value of ΔS for the irreversible isothermal expansion from V_1 to V_2. How, then, do a reversible and an irreversible isothermal expansion differ? The answer lies in the value of ΔS for the surroundings. (Remember that the condition $\Delta S \geq 0$ applies to an isolated system. If the system is not isolated, then the condition $\Delta S \geq 0$ applies to the sum of the entropy changes in the system and its surroundings, in other words, the entire universe.)

Let's look at the entropy change of the surroundings, ΔS_{surr}, for both a reversible and an irreversible isothermal expansion. During the reversible expansion, $\Delta U = 0$ (the process is isothermal and the gas is ideal) and the gas absorbs a quantity of energy as heat, $q_{rev} = -w_{rev} = nRT \ln V_2/V_1$, from its surroundings. The entropy of the surroundings, therefore, decreases according to

$$\Delta S_{surr} = -\frac{q_{rev}}{T} = -nR \ln \frac{V_2}{V_1}$$

The total entropy change is given by

$$\Delta S_{total} = \Delta S_{sys} + \Delta S_{surr} = nR \ln \frac{V_2}{V_1} - nR \ln \frac{V_2}{V_1} = 0$$

as it should be because the entire process is carried out reversibly.

In the irreversible expansion, $\Delta U = 0$ (the process is isothermal and the gas is ideal). No work is done in the expansion, so $w_{irr} = 0$ and therefore, $q_{irr} = 0$. No energy as heat is delivered to the system by the surroundings and so

$$\Delta S_{surr} = 0$$

Thus, the total entropy change is given by

$$\Delta S_{total} = \Delta S_{sys} + \Delta S_{surr} = nR \ln \frac{V_2}{V_1} + 0 = nR \ln \frac{V_2}{V_1}$$

and so $\Delta S > 0$ as we expect for an irreversible process.

Did we use $q_{irr} = 0$ to calculate ΔS_{surr} in this process? We actually did because no work was done by the process. In the general case of an isothermal process in which no work is done ($\delta w = 0$), the process is one of pure heat transfer and $dU = \delta q = dq$, where dq is an exact differential because U is a state function. Therefore, q is path independent and so we can use q_{irr} to calculate the entropy is this particular case.

EXAMPLE 20–4

In Example 20–2 we stated that U is a function of only the temperature for a gas that obeys the equation of state

$$P = \frac{RT}{\overline{V} - b}$$

where b is a constant that reflects the size of the molecules. Calculate $\Delta \overline{S}$ when one mole of such a gas at T and $\overline{V}_1$ is allowed to expand into a vacuum to a total volume of $\overline{V}_2$.

SOLUTION: We start with

$$\delta q_{rev} = dU - \delta w_{rev}$$

Because U is a function of only the temperature, and hence is independent of the volume, $dU = 0$ for the expansion. Therefore,

$$\delta q_{rev} = -\delta w_{rev} = Pd\overline{V} = \frac{RT}{\overline{V} - b} d\overline{V}$$

and

$$\Delta \overline{S} = \int_1^2 \frac{\delta q_{rev}}{T} = R \int_{\overline{V}_1}^{\overline{V}_2} \frac{d\overline{V}}{\overline{V} - b} = R \ln \frac{\overline{V}_2 - b}{\overline{V}_1 - b}$$

Once again, the entropy increases when a gas expands into a vacuum.

Let's look at the mixing of two ideal gases, as depicted in Figure 20.2. Because the two gases are ideal, each acts independently of the other. Thus, we can consider each gas separately to expand from $V_{initial}$ to V_{final}. For nitrogen, we have (using Equation 20.28)

$$\Delta S_{N_2} = n_{N_2} R \ln \frac{V_{N_2} + V_{Br_2}}{V_{N_2}} = -n_{N_2} R \ln \frac{V_{N_2}}{V_{N_2} + V_{Br_2}}$$

and for bromine,

$$\Delta S_{Br_2} = n_{Br_2} R \ln \frac{V_{N_2} + V_{Br_2}}{V_{Br_2}} = -n_{Br_2} R \ln \frac{V_{Br_2}}{V_{N_2} + V_{Br_2}}$$

The total entropy change is

$$\Delta S = \Delta S_{N_2} + \Delta S_{Br_2}$$

$$= -n_{N_2} R \ln \frac{V_{N_2}}{V_{N_2} + V_{Br_2}} - n_{Br_2} R \ln \frac{V_{Br_2}}{V_{N_2} + V_{Br_2}}$$

Because V is proportional to n for an ideal gas, we can write the above equation as

$$\Delta S = -n_{N_2} R \ln \frac{n_{N_2}}{n_{N_2} + n_{Br_2}} - n_{Br_2} R \ln \frac{n_{Br_2}}{n_{N_2} + n_{Br_2}} \qquad (20.29)$$

If we divide both sides by the total number of moles, $n_{total} = n_{N_2} + n_{Br_2}$ and introduce mole fractions

$$y_{N_2} = \frac{n_{N_2}}{n_{total}} \quad \text{and} \quad y_{Br_2} = \frac{n_{Br_2}}{n_{total}}$$

then Equation 20.29 becomes

$$\Delta_{mix} \overline{S}/R = -y_{N_2} \ln y_{N_2} - y_{Br_2} \ln y_{Br_2}$$

More generally, $\Delta_{mix} \overline{S}$ for the isothermal mixing of N ideal gases is given by

$$\Delta_{mix} \overline{S} = -R \sum_{j=1}^{N} y_j \ln y_j \qquad (20.30)$$

in agreement with Equation 20.26. Equation 20.30 says that $\Delta_{mix} \overline{S} > 0$ because the arguments of the logarithms are less than unity. Thus, Equation 20.30 shows that there is an increase in entropy whenever ideal gases mix isothermally.

Last, let's consider ΔS when two equal sized pieces of the same metal at different temperatures, T_h and T_c, are brought into thermal contact and then isolated from their surroundings. Clearly, the two pieces of metal will come to the same final temperature, T, which can be calculated by

heat lost by hotter piece = heat gained by colder piece

$$C_V(T_h - T) = C_V(T - T_c)$$

Solving for T gives

$$T = \frac{T_h + T_c}{2}$$

We now will calculate the entropy change for each piece of metal. Remember that we must calculate ΔS along a reversible path, even though the actual process is irreversible. As usual, we use Equation 20.19,

$$dS = \frac{\delta q_{rev}}{T}$$

There is essentially no work done, so $\delta q_{rev} = dU = C_V dT$. Therefore,

$$\Delta S = \int_{T_1}^{T_2} \frac{C_V dT}{T}$$

If we take C_V to be constant from T_1 to T_2, then

$$\Delta S = C_V \ln \frac{T_2}{T_1} \tag{20.31}$$

Now, for the initially hotter piece, $T_1 = T_h$ and $T_2 = (T_h + T_c)/2$, and so

$$\Delta S_h = C_V \ln \frac{T_h + T_c}{2T_h}$$

Similarly,

$$\Delta S_c = C_V \ln \frac{T_h + T_c}{2T_c}$$

The total change in entropy is given by

$$\Delta S = \Delta S_h + \Delta S_c$$
$$= C_V \ln \frac{(T_h + T_c)^2}{4T_h T_c} \tag{20.32}$$

We will now prove that $(T_h + T_c)^2 > 4T_h T_c$, and that $\Delta S > 0$. Start with

$$(T_h - T_c)^2 = T_h^2 - 2T_h T_c + T_c^2 > 0$$

Add $4T_h T_c$ to both sides and obtain

$$T_h^2 + 2T_h T_c + T_c^2 = (T_h + T_c)^2 > 4T_h T_c$$

Therefore, the value of the argument of the logarithm in Equation 20.32 is greater than one, so we see that $\Delta S > 0$ in this irreversible process.

EXAMPLE 20–5

The constant-pressure molar heat capacity of $O_2(g)$ from 300 K to 1200 K is given by

$$\overline{C}_p(T)/J \cdot K^{-1} \cdot mol^{-1} = 25.72 + (12.98 \times 10^{-3} \ K^{-1})T - (38.62 \times 10^{-7} \ K^{-2})T^2$$

where T is in kelvins. Calculate the value of $\Delta \overline{S}$ when one mole of $O_2(g)$ is heated at constant pressure from 300 K to 1200 K.

SOLUTION: As usual, we start with Equation 20.19

$$dS = \frac{\delta q_{rev}}{T}$$

In this case, $\delta q_{rev} = \overline{C}_P(T)dT$, so

$$\Delta \overline{S} = \int_{T_1}^{T_2} \frac{\overline{C}_P(T)}{T}dT$$

Using the given expression for $\overline{C}_P(T)$, we have

$$\Delta \overline{S}/\text{J·K}^{-1}\text{·mol}^{-1} = \int_{300 \text{ K}}^{1200 \text{ K}} \frac{25.72}{T}dT + \int_{300 \text{ K}}^{1200 \text{ K}} (12.98 \times 10^{-3} \text{ K}^{-1})dT$$

$$- \int_{300 \text{ K}}^{1200 \text{ K}} (38.62 \times 10^{-7} \text{ K}^{-2})TdT$$

$$= 25.72 \ln \frac{1200 \text{ K}}{300 \text{ K}} + (12.98 \times 10^{-3} \text{ K}^{-1})(900 \text{ K})$$

$$- (38.62 \times 10^{-7} \text{ K}^{-2})\left[(1200 \text{ K})^2 - (300 \text{ K})^2\right]/2$$

$$= 35.66 + 11.68 - 2.61 = 44.73$$

Note the increase in entropy due to the increased thermal disorder.

20–7. Thermodynamics Gives Us Insight into the Conversion of Heat into Work

The concept of entropy and the Second Law of Thermodynamics was first developed by a French engineer named Sadi Carnot in the 1820s in a study of the efficiency of the newly developed steam engines and other types of heat engines. Although primarily of historical interest to chemists, the result of Carnot's analysis is still worth knowing. Basically, a steam engine works in a cyclic manner; in each cycle, it withdraws energy as heat from some high-temperature thermal reservoir, uses some of this energy to do work, and then discharges the rest of the energy as heat to a lower-temperature thermal reservoir. A schematic representation of a heat engine is shown in Figure 20.7. The maximum amount of work will be obtained if the cyclic process is carried out reversibly. Of course, the maximum amount of work cannot be acheived in practice because the reversible path is an idealized process, but the results will give us a measure of the maximum efficiency that can be expected. Because the process is cyclic and reversible,

$$\Delta U_{\text{engine}} = w + q_{\text{rev,h}} + q_{\text{rev,c}} = 0 \qquad (20.33)$$

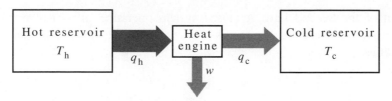

FIGURE 20.7
A highly schematic illustration of a heat engine. Energy as heat (q_h) is withdrawn from a high-temperature thermal reservoir at temperature T_h. The engine does work (w) and delivers an amount of energy as heat (q_c) to the lower-temperature reservoir at temperature T_c.

and

$$\Delta S_{engine} = \frac{\delta q_{rev,h}}{T_h} + \frac{\delta q_{rev,c}}{T_c} = 0 \tag{20.34}$$

where $\delta q_{rev,h}$ is the energy withdrawn reversibly as heat from the high-temperature reservoir at temperature T_h, and $\delta q_{rev,c}$ is the energy discharged reversibly as heat to the lower-temperature reservoir at temperature T_c. Note that the sign convention for energy transferred as heat means that $\delta q_{rev,h}$ is a positive quantity and that $\delta q_{rev,c}$ is a negative quantity. From Equation 20.33, we have that the work done by the engine is

$$-w = q_{rev,h} + q_{rev,c}$$

The work done by the engine is a negative quantity, so $-w$ is a positive quantity. We can define the efficiency of the process by the ratio of the work done by the engine divided by the amount of energy withdrawn as heat from the hot reservoir, or

$$\text{maximum efficiency} = \frac{-w}{q_{rev,h}} = \frac{q_{rev,h} + q_{rev,c}}{q_{rev,h}}$$

Equation 20.34 says that $q_{rev,c} = -q_{rev,h}(T_c/T_h)$, so the efficiency can be written as

$$\text{maximum efficiency} = 1 - \frac{T_c}{T_h} = \frac{T_h - T_c}{T_h} \tag{20.35}$$

Equation 20.35 is really a remarkable result because it is independent of the specific design of the engine or of the working substance. For a heat engine working between 373 K and 573 K, the maximum possible efficiency is

$$\text{maximum efficiency} = \frac{200}{573} = 35\%$$

In practice, the efficiency would be less due to factors such as friction. Equation 20.35 indicates that a greater efficiency is obtained by engines working with a higher value of T_h or a lower value of T_c.

Note that the efficiency equals zero if $T_h = T_c$, which says that no net work can be obtained from an isothermal cyclic process. This conclusion is known as Kelvin's statement of the Second Law. A closed system operating in an isothermal cyclic manner cannot convert heat into work without some accompanying change in the surroundings.

20–8. Entropy Can Be Expressed in Terms of a Partition Function

We presented the equation $S = k_B \ln W$ in Section 20–5. This equation can be used as the starting point to derive most of the important results of statistical thermodynamics. For example, we can use it to derive an expression for the entropy in terms of the system partition function, $Q(N, V, \beta)$, as we have for the energy and the pressure:

$$U = k_B T^2 \left(\frac{\partial \ln Q}{\partial T} \right)_{N,V} = - \left(\frac{\partial \ln Q}{\partial \beta} \right)_{N,V} \tag{20.36}$$

and

$$P = k_B T \left(\frac{\partial \ln Q}{\partial V} \right)_{N,T} \tag{20.37}$$

Substitute Equation 20.23 into Equation 20.24 and then use Stirling's approximation for the factorials (MathChapter J) to get

$$
\begin{aligned}
S_{\text{ensemble}} &= k_B \ln \frac{\mathcal{A}!}{\prod_j a_j!} = k_B \ln \mathcal{A}! - k_B \sum_j \ln a_j! \\
&= k_B \mathcal{A} \ln \mathcal{A} - k_B \mathcal{A} - k_B \sum_j a_j \ln a_j + k_B \sum_j a_j \\
&= k_B \mathcal{A} \ln \mathcal{A} - k_B \sum_j a_j \ln a_j
\end{aligned} \tag{20.38}
$$

where we have used the fact that $\sum a_j = \mathcal{A}$ and have subscripted S with "ensemble" to emphasize that it is the entropy of the entire ensemble of $\mathcal{A}$ systems. The entropy of a typical system is given by $S_{\text{system}} = S_{\text{ensemble}}/\mathcal{A}$. If we use the fact that the probability of finding a system in the jth quantum state is given by

$$p_j = \frac{a_j}{\mathcal{A}}$$

and then substitute $a_j = \mathcal{A} p_j$ into Equation 20.38, we obtain

$$
\begin{aligned}
S_{\text{ensemble}} &= k_B \mathcal{A} \ln \mathcal{A} - k_B \sum_j p_j \mathcal{A} \ln p_j \mathcal{A} \\
&= k_B \mathcal{A} \ln \mathcal{A} - k_B \sum_j p_j \mathcal{A} \ln p_j - k_B \sum_j p_j \mathcal{A} \ln \mathcal{A}
\end{aligned} \tag{20.39}
$$

But the last term here cancels with the first because

$$\sum_j p_j \mathcal{A} \ln \mathcal{A} = \mathcal{A} \ln \mathcal{A} \sum_j p_j = \mathcal{A} \ln \mathcal{A}$$

where we have used the facts that $\mathcal{A} \ln \mathcal{A}$ is a constant and $\sum_j p_j = 1$. If we furthermore divide Equation 20.39 through by $\mathcal{A}$, we obtain

$$S_{\text{system}} = -k_{\text{B}} \sum_j p_j \ln p_j \tag{20.40}$$

Note that if all the p_j's are zero except for one (which must equal unity because $\sum_j p_j = 1$), the system is completely ordered and $S = 0$. Therefore, we see that according to our molecular picture of entropy, $S = 0$ for a perfectly ordered system. Problem 20–39 asks you to show that S is a maximum when all the p_j's are equal, in which case the system is maximally disordered.

To derive an expression for S in terms of $Q(N, V, T)$, we substitute

$$p_j(N, V, \beta) = \frac{e^{-\beta E_j(N,V)}}{Q(N, V, \beta)} \tag{20.41}$$

into Equation 20.40 to obtain

$$
\begin{aligned}
S &= -k_{\text{B}} \sum_j p_j \ln p_j \\
&= -k_{\text{B}} \sum_j \frac{e^{-\beta E_j}}{Q} \left(-\beta E_j - \ln Q \right) \\
&= \beta k_{\text{B}} \sum_j \frac{E_j e^{-\beta E_j}}{Q} + \frac{k_{\text{B}} \ln Q}{Q} \sum_j e^{-\beta E_j} \\
&= \frac{U}{T} + k_{\text{B}} \ln Q
\end{aligned}
\tag{20.42}
$$

We used the fact that $\beta k_{\text{B}} = 1/T$ to go from the third line to the last line. Using Equation 20.36 for U gives S in terms of the partition function, $Q(N, V, T)$.

$$S = k_{\text{B}} T \left(\frac{\partial \ln Q}{\partial T} \right)_{N,V} + k_{\text{B}} \ln Q \tag{20.43}$$

Recall from Chapter 18 that

$$Q(N, V, T) = \frac{1}{N!} \left(\frac{2\pi m k_{\text{B}} T}{h^2} \right)^{3N/2} V^N g_{el}$$

for a monatomic ideal gas where all the atoms are in their ground electronic state. Using Equation 20.43, we obtain for the molar entropy of one mole of a monatomic ideal gas,

$$\overline{S} = \frac{3}{2}R + R\ln\left[\left(\frac{2\pi m k_B T}{h^2}\right)^{3/2}\overline{V}g_{e1}\right] - k_B\ln N_A! \tag{20.44}$$

Applying Stirling's approximation to the last term gives

$$-k_B\ln N_A! = -k_B N_A\ln N_A + k_B N_A = -R\ln N_A + R$$

Therefore,

$$\overline{S} = \frac{5}{2}R + R\ln\left[\left(\frac{2\pi m k_B T}{h^2}\right)^{3/2}\frac{\overline{V}g_{e1}}{N_A}\right] \tag{20.45}$$

EXAMPLE 20–6
Use Equation 20.45 to calculate the molar entropy of argon at 298.2 K and one bar, and compare your result with the experimental value of 154.8 J·K⁻¹·mol⁻¹.

SOLUTION: At 298.2 K and one bar,

$$\frac{N_A}{\overline{V}} = \frac{N_A P}{RT}$$

$$= \frac{(6.022 \times 10^{23}\ \text{mol}^{-1})(1\ \text{bar})}{(0.08314\ \text{L·bar·K}^{-1}\text{·mol}^{-1})(298.2\ \text{K})}$$

$$= 2.429 \times 10^{22}\ \text{L}^{-1} = 2.429 \times 10^{25}\ \text{m}^{-3}$$

and

$$\left(\frac{2\pi m k_B T}{h^2}\right)^{3/2} = \left[\frac{2\pi(0.03995\ \text{kg·mol}^{-1})(1.3806 \times 10^{-23}\ \text{J·K}^{-1})(298.2\ \text{K})}{(6.022 \times 10^{23}\ \text{mol}^{-1})(6.626 \times 10^{-34}\ \text{J·s})^2}\right]^{3/2}$$

$$= (3.909 \times 10^{21}\ \text{m}^{-2})^{3/2}$$

$$= 2.444 \times 10^{32}\ \text{m}^{-3}$$

Therefore

$$\frac{\overline{S}}{R} = \frac{5}{2} + \ln\left[\frac{2.444 \times 10^{32}\ \text{m}^{-3}}{2.429 \times 10^{25}\ \text{m}^{-3}}\right]$$

$$= 18.62$$

or

$$\overline{S} = (18.62)(8.314\ \text{J·K}^{-1}\text{·mol}^{-1}) = 154.8\ \text{J·K}^{-1}\text{·mol}^{-1}$$

This value of $\overline{S}$ agrees exactly with the experimentally determined value.

EXAMPLE 20–7
Show that Equation 20.45 gives Equation 20.26 for the molar entropy of mixing nitrogen and bromine as ideal gases.

SOLUTION: First we write Equation 20.45 as

$$S = Nk_B \ln V + \text{terms not involving } V$$

The initial state is given by

$$S_1 = S_{1,N_2} + S_{1,Br_2}$$
$$= n_{N_2} R \ln V_{N_2} + n_{Br_2} R \ln V_{Br_2} + \text{terms not involving } V$$

where we have written $Nk_B = nR$. The final state is given by

$$S_2 = S_{2,N_2} + S_{2,Br_2}$$
$$= n_{N_2} R \ln(V_{N_2} + V_{Br_2}) + n_{Br_2} R \ln(V_{N_2} + V_{Br_2}) + \text{terms not involving } V$$

Therefore

$$\Delta_{mix} S = S_2 - S_1 = n_{N_2} R \ln \frac{V_{N_2} + V_{Br_2}}{V_{N_2}} + n_{Br_2} R \ln \frac{V_{N_2} + V_{Br_2}}{V_{Br_2}}$$

Because V is proportional to n for an ideal gas, we have

$$\Delta_{mix} S = -n_{N_2} R \ln \frac{n_{N_2}}{n_{N_2} + n_{Br_2}} - n_{Br_2} R \ln \frac{n_{Br_2}}{n_{N_2} + n_{Br_2}}$$

If we divide this result through by $n_{N_2} + n_{Br_2}$, then we obtain Equation 20.26.

20–9. The Molecular Formula $S = k_B \ln W$ Is Analogous to the Thermodynamic Formula $dS = \delta q_{rev}/T$

In this last section, we will show that Equation 20.24, or its equivalent, Equation 20.40, is consistent with our thermodynamic definition of the entropy. As a bonus, we will finally prove that $\beta = 1/k_B T$.

If we differentiate Equation 20.40 with respect to p_j, we get

$$dS = -k_B \sum_j \left(dp_j + \ln p_j dp_j \right)$$

But $\sum dp_j = 0$ because $\sum p_j = 1$, so

$$dS = -k_B \sum_j \ln p_j dp_j \tag{20.46}$$

Now substitute Equation 20.41 into the $\ln p_j$ term in Equation 20.46 to obtain

$$dS = -k_B \sum_j \left[-\beta E_j(N, V) - \ln Q\right] dp_j$$

The term involving $\ln Q$ drops out because

$$\sum_j \ln Q \, dp_j = \ln Q \sum_j dp_j = 0$$

and so

$$dS = \beta k_B \sum_j E_j(N, V) dp_j(N, V, \beta) \qquad (20.47)$$

But we showed in Section 19–4 that $\sum_j E_j(N, V) dp_j(N, V, \beta)$ is the energy as heat that a system gains or loses in a reversible process, so Equation 20.47 becomes

$$dS = \beta k_B \delta q_{\text{rev}} \qquad (20.48)$$

Equation 20.48 shows, furthermore, that βk_B is an integrating factor of δq_{rev}, or $\beta k_B = 1/T$, or $\beta = 1/k_B T$. Thus, we have finally proved that $\beta = 1/k_B T$.

In the next chapter, we will discuss the experimental determination of the entropies of substances.

Problems

20-1. Show that

$$\oint dY = 0$$

if Y is a state function.

20-2. Let $z = z(x, y)$ and $dz = xydx + y^2dy$. Although dz is not an exact differential (why not?), what combination of dz and x and/or y is an exact differential?

20-3. Use the criterion developed in MathChapter H to prove that δq_{rev} in Equation 20.1 is not an exact differential (see also Problem H–11).

20-4. Use the criterion developed in MathChapter H to prove that $\delta q_{\text{rev}}/T$ in Equation 20.1 is an exact differential.

20-5. In this problem, we will prove that Equation 20.5 is valid for an arbitrary system. To do this, consider an isolated system made up of two equilibrium subsystems, A and B, which are in thermal contact with each other; in other words, they can exchange energy as heat between themselves. Let subsystem A be an ideal gas and let subsystem B be arbitrary. Suppose now that an infinitesimal reversible process occurs in A accompanied by an exchange of energy as heat δq_{rev} (ideal). Simultaneously, another infinitesimal reversible process takes place in B accompanied by an exchange of energy as heat δq_{rev} (arbitrary). Because the composite system is isolated, the First Law requires that

$$\delta q_{rev}(\text{ideal}) = -\delta q_{rev}(\text{arbitrary})$$

Now use Equation 20.4 to prove that

$$\oint \frac{\delta q_{rev}(\text{arbitrary})}{T} = 0$$

Therefore, we can say that the definition given by Equation 20.4 holds for any system.

20-6. Calculate q_{rev} and ΔS for a reversible cooling of one mole of an ideal gas at a constant volume V_1 from P_1, V_1, T_1 to P_2, V_1, T_4 followed by a reversible expansion at constant pressure P_2 from P_2, V_1, T_4 to P_2, V_2, T_1 (the final state for all the processes shown in Figure 20.3). Compare your result for ΔS with those for paths A, B + C, and D + E in Figure 20.3.

20-7. Derive Equation 20.8 without referring to Chapter 19.

20-8. Calculate the value of ΔS if one mole of an ideal gas is expanded reversibly and isothermally from 10.0 dm^3 to 20.0 dm^3. Explain the sign of ΔS.

20-9. Calculate the value of ΔS if one mole of an ideal gas is expanded reversibly and isothermally from 1.00 bar to 0.100 bar. Explain the sign of ΔS.

20-10. Calculate the values of q_{rev} and ΔS along the path D + E in Figure 20.3 for one mole of a gas whose equation of state is given in Example 20–2. Compare your result with that obtained in Example 20–2.

20-11. Show that ΔS_{D+E} is equal to ΔS_A and ΔS_{B+C} for the equation of state given in Example 20–2.

20-12. Calculate the values of q_{rev} and ΔS along the path described in Problem 20–6 for one mole of a gas whose equation of state is given in Example 20–2. Compare your result with that obtained in Example 20–2.

20-13. Show that

$$\Delta S = C_P \ln \frac{T_2}{T_1}$$

for a constant-pressure process if C_P is independent of temperature. Calculate the change in entropy of 2.00 moles of $H_2O(l)$ ($\overline{C}_P = 75.2$ J·K^{-1}·mol^{-1}) if it is heated from 10°C to 90°C.

20-14. Show that

$$\Delta \overline{S} = \overline{C}_V \ln \frac{T_2}{T_1} + R \ln \frac{V_2}{V_1}$$

if one mole of an ideal gas is taken from T_1, V_1 to T_2, V_2, assuming that $\overline{C}_V$ is independent of temperature. Calculate the value of $\Delta \overline{S}$ if one mole of $N_2(g)$ is expanded from 20.0 dm^3 at 273 K to 300 dm^3 at 400 K. Take $\overline{C}_P = 29.4$ J·K^{-1}·mol^{-1}.

20-15. In this problem, we will consider a two-compartment system like that in Figure 20.4, except that the two subsystems have the same temperature but different pressures and the wall that separates them is flexible rather than rigid. Show that in this case,

$$dS = \frac{dV_B}{T}(P_B - P_A)$$

Interpret this result with regard to the sign of dV_B when $P_B > P_A$ and when $P_B < P_A$.

20-16. In this problem, we will illustrate the condition $dS_{prod} \geq 0$ with a concrete example. Consider the two-component system shown in Figure 20.8. Each compartment is in equilibrium with a heat reservoir at different temperatures T_1 and T_2, and the two compartments are separated by a rigid heat-conducting wall. The total change of energy as heat of compartment 1 is

$$dq_1 = d_e q_1 + d_i q_1$$

where $d_e q_1$ is the energy as heat exchanged with the reservoir and $d_i q_1$ is the energy as heat exchanged with compartment 2. Similarly,

$$dq_2 = d_e q_2 + d_i q_2$$

Clearly,

$$d_i q_1 = -d_i q_2$$

Show that the entropy change for the two-compartment system is given by

$$dS = \frac{d_e q_1}{T_1} + \frac{d_e q_2}{T_2} + d_i q_1 \left(\frac{1}{T_1} - \frac{1}{T_2} \right)$$
$$= dS_{exchange} + dS_{prod}$$

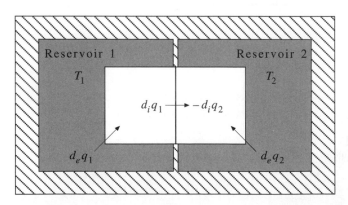

FIGURE 20.8
A two-compartment system with each compartment in contact with an (essentially infinite) heat reservoir, one at temperature T_1 and the other at temperature T_2. The two compartments are separated by a rigid heat-conducting wall.

where

$$dS_{\text{exchange}} = \frac{d_e q_1}{T_1} + \frac{d_e q_2}{T_2}$$

is the entropy *exchanged* with the reservoirs (surroundings) and

$$dS_{\text{prod}} = d_i q_1 \left(\frac{1}{T_1} - \frac{1}{T_2} \right)$$

is the entropy *produced* within the two-compartment system. Now show that the condition $dS_{\text{prod}} \geq 0$ implies that energy as heat flows spontaneously from a higher temperature to a lower temperature. The value of dS_{exchange}, however, has no restriction and can be positive, negative, or zero.

20-17. Show that

$$\Delta S \geq \frac{q}{T}$$

for an isothermal process. What does this equation say about the sign of ΔS? Can ΔS decrease in a reversible isothermal process? Calculate the entropy change when one mole of an ideal gas is compressed reversibly and isothermally from a volume of 100 dm^3 to 50.0 dm^3 at 300 K.

20-18. Vaporization at the normal boiling point (T_{vap}) of a substance (the boiling point at one atm) can be regarded as a reversible process because if the temperature is decreased infinitesimally below T_{vap}, all the vapor will condense to liquid, whereas if it is increased infinitesimally above T_{vap}, all the liquid will vaporize. Calculate the entropy change when two moles of water vaporize at 100.0°C. The value of $\Delta_{\text{vap}} \overline{H}$ is 40.65 kJ·mol^{-1}. Comment on the sign of $\Delta_{\text{vap}} S$.

20-19. Melting at the normal melting point (T_{fus}) of a substance (the melting point at one atm) can be regarded as a reversible process because if the temperature is changed infinitesimally from exactly T_{fus}, then the substance will either melt or freeze. Calculate the change in entropy when two moles of water melt at 0°C. The value of $\Delta_{\text{fus}} \overline{H}$ is 6.01 kJ·mol^{-1}. Compare your answer with the one you obtained in Problem 20–18. Why is $\Delta_{\text{vap}} S$ much larger than $\Delta_{\text{fus}} S$?

20-20. Consider a simple example of Equation 20.23 in which there are only two states, 1 and 2. Show that $W(a_1, a_2)$ is a maximum when $a_1 = a_2$. *Hint*: Consider ln W, use Stirling's approximation, and treat a_1 and a_2 as continuous variables.

20-21. Extend Problem 20–20 to the case of three states. Do you see how to generalize it to any number of states?

20-22. Show that the system partition function can be written as a summation over levels by writing

$$Q(N, V, T) = \sum_E \Omega(N, V, E) e^{-E/k_B T}$$

Now consider the case of an isolated system, for which there is only one term in $Q(N, V, T)$. Now substitute this special case for Q into Equation 20.43 to derive the equation $S = k_B \ln \Omega$.

20-23. In this problem, we will show that $\Omega = c(N) f(E) V^N$ for an ideal gas (Example 20–3). In Problem 18–42 we showed that the number of translational energy states between ε and $\varepsilon + \Delta\varepsilon$ for a particle in a box can be calculated by considering a sphere in n_x, n_y, n_z space,

$$n_x^2 + n_y^2 + n_z^2 = \frac{8ma^2\varepsilon}{h^2} = R^2$$

Show that for an N-particle system, the analogous expression is

$$\sum_{j=1}^{N}(n_{xj}^2 + n_{yj}^2 + n_{zj}^2) = \frac{8ma^2 E}{h^2} = R^2$$

or, in more convenient notation

$$\sum_{j=1}^{3N} n_j^2 = \frac{8ma^2 E}{h^2} = R^2$$

Thus, instead of dealing with a three-dimensional sphere as we did in Problem 18–42, here we must deal with a $3N$-dimensional sphere. Whatever the formula for the volume of a $3N$-dimensional sphere is (it is known), we can at least say that it is proportional to R^{3N}. Show that this proportionality leads to the following expression for $\Phi(E)$, the number of states with energy $\leq E$,

$$\Phi(E) \propto \left(\frac{8ma^2 E}{h^2}\right)^{3N/2} = c(N) E^{3N/2} V^N$$

where $c(N)$ is a constant whose value depends upon N and $V = a^3$. Now, following the argument developed in Problem 18–42, show that the number of states between E and $E + \Delta E$ (which is essentially Ω) is given by

$$\Omega = c(N) f(E) V^N \Delta E$$

where $f(E) = E^{\frac{3N}{2}-1}$.

20-24. Show that if a process involves only an isothermal transfer of energy as heat (*pure heat transfer*), then

$$dS_{sys} = \frac{dq}{T} \qquad \text{(pure heat transfer)}$$

20-25. Calculate the change in entropy of the system and of the surroundings and the total change in entropy if one mole of an ideal gas is expanded isothermally and reversibly from a pressure of 10.0 bar to 2.00 bar at 300 K.

20-26. Redo Problem 20–25 for an expansion into a vacuum, with an initial pressure of 10.0 bar and a final pressure of 2.00 bar.

20-27. The molar heat capacity of 1-butene can be expressed as

$$\overline{C}_P(T)/R = 0.05641 + (0.04635\,\text{K}^{-1})T - (2.392 \times 10^{-5}\,\text{K}^{-2})T^2 + (4.80 \times 10^{-9}\,\text{K}^{-3})T^3$$

over the temperature range 300 K $< T <$ 1500 K. Calculate the change in entropy when one mole of 1-butene is heated from 300 K to 1000 K at constant pressure.

20-28. Plot $\Delta_{mix}\overline{S}$ against y_1 for the mixing of two ideal gases. At what value of y_1 is $\Delta_{mix}\overline{S}$ a maximum? Can you give a physical interpretation of this result?

20-29. Calculate the entropy of mixing if two moles of $N_2(g)$ are mixed with one mole $O_2(g)$ at the same temperature and pressure. Assume ideal behavior.

20-30. Show that $\Delta_{mix}\overline{S} = R \ln 2$ if equal volumes of any two ideal gases under the same conditions are mixed.

20-31. Derive the equation $dU = TdS - PdV$. Show that

$$d\overline{S} = \overline{C}_V \frac{dT}{T} + R\frac{d\overline{V}}{\overline{V}}$$

for one mole of an ideal gas. Assuming that $\overline{C}_V$ is independent of temperature, show that

$$\Delta\overline{S} = \overline{C}_V \ln\frac{T_2}{T_1} + R\ln\frac{\overline{V}_2}{\overline{V}_1}$$

for the change from $T_1, \overline{V}_1$ to $T_2, \overline{V}_2$. Note that this equation is a combination of Equations 20.28 and 20.31.

20-32. Derive the equation $dH = TdS + VdP$. Show that

$$\Delta\overline{S} = \overline{C}_P \ln\frac{T_2}{T_1} - R\ln\frac{P_2}{P_1}$$

for the change of one mole of an ideal gas from T_1, P_1 to T_2, P_2, assuming that $\overline{C}_P$ is independent of temperature.

20-33. Calculate the change in entropy if one mole of $SO_2(g)$ at 300 K and 1.00 bar is heated to 1000 K and its pressure is decreased to 0.010 bar. Take the molar heat capacity of $SO_2(g)$ to be

$$\overline{C}_P(T)/R = 7.871 - \frac{1454.6 \text{ K}}{T} + \frac{160\,351 \text{ K}^2}{T^2}$$

20-34. In the derivation of Equation 20.32, argue that $\Delta S_c > 0$ and $\Delta S_h < 0$. Now show that

$$\Delta S = \Delta S_c + \Delta S_h > 0$$

by showing that

$$\Delta S_c - |\Delta S_h| > 0$$

20-35. We can use the equation $S = k_B \ln W$ to derive Equation 20.28. First, argue that the probability that an ideal-gas molecule is found in a subvolume V_s of some larger volume V is V_s/V. Because the molecules of an ideal gas are independent, the probability that

N ideal-gas molecules are found in V_s is $(V_s/V)^N$. Now show that the change in entropy when the volume of one mole of an ideal gas changes isothermally from V_1 to V_2 is

$$\Delta S = R \ln \frac{V_2}{V_1}$$

20-36. The relation $n_j \propto e^{-\varepsilon_j/k_B T}$ can be derived by starting with $S = k_B \ln W$. Consider a gas with n_0 molecules in the ground state and n_j in the jth state. Now add an energy $\varepsilon_j - \varepsilon_0$ to this system so that a molecule is promoted from the ground state to the jth state. If the volume of the gas is kept constant, then no work is done, so $dU = dq$,

$$dS = \frac{dq}{T} = \frac{dU}{T} = \frac{\varepsilon_j - \varepsilon_0}{T}$$

Now, assuming that n_0 and n_j are large, show that

$$dS = k_B \ln \left\{ \frac{N!}{(n_0 - 1)! n_1! \cdots (n_j + 1)! \cdots} \right\} - k_B \ln \left\{ \frac{N!}{(n_0! n_1! \cdots n_j! \cdots)} \right\}$$

$$= k_B \ln \left\{ \frac{n_j!}{(n_j + 1)!} \frac{n_0!}{(n_0 - 1)!} \right\} = k_B \ln \frac{n_0}{n_j}$$

Equating the two expressions for dS, show that

$$\frac{n_j}{n_0} = e^{-(\varepsilon_j - \varepsilon_0)/k_B T}$$

20-37. We can use Equation 20.24 to calculate the probability of observing fluctuations from the equilibrium state. Show that

$$\frac{W}{W_{eq}} = e^{-\Delta S/k_B}$$

where W represents the nonequilibrium state and ΔS is the entropy difference between the two states. We can interpret the ratio W/W_{eq} as the probability of observing the nonequilibrium state. Given that the entropy of one mole of oxygen is $205.0 \text{ J} \cdot \text{K}^{-1} \cdot \text{mol}^{-1}$ at $25°C$ and one bar, calculate the probability of observing a decrease in entropy that is one millionth of a percent of this amount.

20-38. Consider one mole of an ideal gas confined to a volume V. Calculate the probability that all the N_A molecules of this ideal gas will be found to occupy one half of this volume, leaving the other half empty.

20-39. Show that S_{system} given by Equation 20.40 is a maximum when all the p_j are equal. Remember that $\sum p_j = 1$, so that

$$\sum_j p_j \ln p_j = p_1 \ln p_1 + p_2 \ln p_2 + \cdots + p_{n-1} \ln p_{n-1}$$

$$+ (1 - p_1 - p_2 - \cdots - p_{n-1}) \ln(1 - p_1 - p_2 - \cdots - p_{n-1})$$

See also Problem J–10.

20-40. Use Equation 20.45 to calculate the molar entropy of krypton at 298.2 K and one bar, and compare your result with the experimental value of 164.1 $J \cdot K^{-1} \cdot mol^{-1}$.

20-41. Use Equation 18.39 and the data in Table 18.2 to calculate the entropy of nitrogen at 298.2 K and one bar. Compare your result with the experimental value of 191.6 $J \cdot K^{-1} \cdot mol^{-1}$.

20-42. Use Equation 18.57 and the data in Table 18.4 to calculate the entropy of $CO_2(g)$ at 298.2 K and one bar. Compare your result with the experimental value of 213.8 $J \cdot K^{-1} \cdot mol^{-1}$.

20-43. Use Equation 18.60 and the data in Table 18.4 to calculate the entropy of $NH_3(g)$ at 298.2 K and one bar. Compare your result with the experimental value of 192.8 $J \cdot K^{-1} \cdot mol^{-1}$.

20-44. Derive Equation 20.35.

20-45. The boiling point of water at a pressure of 25 atm is 223°C. Compare the theoretical efficiencies of a steam engine operating between 20°C and the boiling point of water at 1 atm and at 25 atm.

Walther Nernst was born in Briessen, Prussia (now Wabrzezno, Poland), on June 25, 1864, and died in 1941. Although he aspired to be a poet, his chemistry teacher kindled his interest in science. Between 1883 and 1887, Nernst studied physics with von Helmholtz, Boltzmann, and Kohlrausch. He received his doctorate in physics at the University of Würzburg in 1887. Nernst was Ostwald's assistant at the University of Leibzig from 1887 to 1891, after which he went to the University of Göttingen, where he established the Kaiser Wilhelm Institute for Physical Chemistry and Electrochemistry in 1894. Upon moving to the University of Berlin in 1905, Nernst began his studies of the behavior of substances at very low temperatures. He proposed one of the early versions of the Third Law of Thermodynamics, which says that the physical activities of substances tend to vanish as the temperature approaches absolute zero. The Third Law made it possible to calculate thermodynamic quantities such as equilibrium constants from thermal data. He was awarded the Nobel Prize for chemistry in 1920 "in recognition of his work in thermochemistry." He was an early automoblile enthusiast and served during World War I as a driver. Nernst lost both of his sons in World War I. His anti-Nazi stance in the 1930s led to increasing isolation, so he retired to his country home, where he died in 1941.

CHAPTER 21

Entropy and the Third Law
of Thermodynamics

In the previous chapter, we introduced the concept of entropy. We showed that entropy is created or generated whenever a spontaneous or irreversible process occurs in an isolated system. We also showed that the entropy of an isolated system that is not in equilibrium will increase until the system reaches equilibrium, from which time the entropy will remain constant. We expressed this condition mathematically by writing $dS \geq 0$ for a process that occurs at constant U and V. Although we calculated the *change* in entropy for a few processes, we did not attempt to calculate absolute values of the entropy of substances. (See Example 20–6 and Problems 20–41 through 20–43, however.) In this chapter, we will introduce the Third Law of Thermodynamics, so that we can calculate absolute values of the entropy of substances.

21–1. Entropy Increases with Increasing Temperature

We start with the First Law of Thermodynamics for a reversible process:

$$dU = \delta q_{\text{rev}} + \delta w_{\text{rev}}$$

Using the fact that $\delta q_{\text{rev}} = T dS$ and $\delta w_{\text{rev}} = -P dV$, we obtain a combination of the First and Second Laws of Thermodynamics:

$$dU = T dS - P dV \tag{21.1}$$

We can derive a number of relationships between thermodynamic quantities using the laws of thermodynamics and the fact that state functions are exact differentials. Example 21–1 derives the following two important relationships

$$\left(\frac{\partial S}{\partial T} \right)_V = \frac{C_V}{T} \tag{21.2}$$

853

and

$$\left(\frac{\partial S}{\partial V}\right)_T = \frac{1}{T}\left[P + \left(\frac{\partial U}{\partial V}\right)_T\right] \tag{21.3}$$

EXAMPLE 21–1
Express U as a function of V and T and then use this result and Equation 21.1 to derive Equations 21.2 and 21.3.

SOLUTION: If we treat U as a function of V and T, its total derivative is (Math-Chapter H)

$$dU = \left(\frac{\partial U}{\partial T}\right)_V dT + \left(\frac{\partial U}{\partial V}\right)_T dV \tag{21.4}$$

We substitute Equation 21.4 into Equation 21.1 and solve for dS to obtain

$$dS = \frac{1}{T}\left(\frac{\partial U}{\partial T}\right)_V dT + \frac{1}{T}\left[P + \left(\frac{\partial U}{\partial V}\right)_T\right]dV$$

Using the definition that $(\partial U/\partial T)_V = C_V$, we obtain

$$dS = \frac{C_V dT}{T} + \frac{1}{T}\left[P + \left(\frac{\partial U}{\partial V}\right)_T\right]dV$$

If we compare this equation for dS with the total derivative of $S = S(T, V)$,

$$dS = \left(\frac{\partial S}{\partial T}\right)_V dT + \left(\frac{\partial S}{\partial V}\right)_T dV$$

we see that

$$\left(\frac{\partial S}{\partial T}\right)_V = \frac{C_V}{T} \quad \text{and} \quad \left(\frac{\partial S}{\partial V}\right)_T = \frac{1}{T}\left[P + \left(\frac{\partial U}{\partial V}\right)_T\right]$$

Equation 21.2 tells us how S varies with temperature at constant volume. If we integrate with respect to T (keeping V constant), we obtain

$$\Delta S = S(T_2) - S(T_1) = \int_{T_1}^{T_2} \frac{C_V(T)dT}{T} \qquad \text{(constant } V) \tag{21.5}$$

Thus, if we know $C_V(T)$ as a function of T, we can calculate ΔS. Note that because C_V is always positive, the entropy increases with increasing temperature.

Equation 21.5 is restricted to constant volume. To derive a similar equation for constant pressure, we start with

$$dH = d(U + PV) = dU + PdV + VdP$$

and substitute Equation 21.1 for dU to obtain

$$dH = TdS + VdP \tag{21.6}$$

Proceeding in a similar manner as in Example 21–1 (Problem 21–1), we obtain

$$\left(\frac{\partial S}{\partial T}\right)_P = \frac{C_P(T)}{T} \tag{21.7}$$

and

$$\left(\frac{\partial S}{\partial P}\right)_T = \frac{1}{T}\left[\left(\frac{\partial H}{\partial P}\right)_T - V\right] \tag{21.8}$$

From Equation 21.7, we get

$$\Delta S = S(T_2) - S(T_1) = \int_{T_1}^{T_2} \frac{C_P(T)dT}{T} \qquad \text{(constant } P\text{)} \tag{21.9}$$

Thus, if we know C_P as a function of T, we can calculate ΔS. Most processes we will consider occur at constant pressure, so we will usually use Equation 21.9 to calculate ΔS.

If we let $T_1 = 0$ K in Equation 21.9, then we have

$$S(T) = S(0\text{ K}) + \int_0^T \frac{C_P(T')dT'}{T'} \qquad \text{(constant } P\text{)} \tag{21.10}$$

Equation 21.10 tells us that we can calculate the entropy of a substance if we know $S(0\text{ K})$ and $C_P(T)$ from $T = 0$ K to the temperature of interest. (Notice once again that we use a prime on the variable of integration to distinguish it from an integration limit.)

21–2. The Third Law of Thermodynamics Says That the Entropy of a Perfect Crystal Is Zero at 0 K

Let's discuss $S(0\text{ K})$ first. Around the turn of the century, the German chemist Walther Nernst, after studying numerous chemical reactions, postulated that $\Delta_r S \to 0$ as $T \to 0$. Nernst did not make any statement concerning the entropy of any particular substance at 0 K, only that all pure crystalline substances have the same entropy at 0 K. We have added the "pure crystalline" condition here to avoid some apparent exceptions to Nernst's postulate that we will resolve later. In 1911, Planck, who incidentally did a great deal of research in thermodynamics (including his doctoral thesis), extended Nernst's postulate by postulating that the entropy of a pure substance approaches zero at 0 K. Planck's postulate is consistent with Nernst's but takes it further. There are several equivalent statements of what is now called the *Third Law of Thermodynamics*, but the one we will use is

Every substance has a finite positive entropy, but at zero kelvin the entropy may become zero, and does so in the case of a perfectly crystalline substance.

The Third Law of Thermodynamics is unlike the first two laws in that it introduces no new state function. The first law gives us the energy and the second law gives us the entropy; the third law provides a numerical scale for entropy.

Although the Third Law was formulated before the full development of the quantum theory, it is much more plausible and intuitive if we think of it in terms of molecular quantum states or levels. One of our molecular formulas for the entropy is (Equation 20.24)

$$S = k_B \ln W \tag{21.11}$$

where W is the number of ways the total energy of a system may be distributed over its various energy states. At 0 K, we expect that the system will be in its lowest energy state. Therefore, $W = 1$ and $S = 0$. Another way to see this result is to start with Equation 20.40 for S:

$$S = -k_B \sum_j p_j \ln p_j \tag{21.12}$$

where p_j is the probability of finding the system in the jth quantum state with energy E_j. At 0 K, there is no thermal energy, so we expect the system to be in the ground state; thus, $p_0 = 1$ and all the other p_j's equal zero. Therefore, S in Equation 21.12 equals zero. Even if the ground state has a degeneracy of n, say, then each of the n quantum states with energy E_0 would have a probability of $1/n$, and S in Equation 21.12 would be

$$S(0 \text{ K}) = -k_B \sum_{j=1}^{n} \frac{1}{n} \ln \frac{1}{n} = k_B \ln n \tag{21.13}$$

Even if the degeneracy of the ground state were as large as the Avogadro constant, $\overline{S}$ would be equal to only 7.56×10^{-22} J·K^{-1}·mol^{-1}, which is well below a measurable value of $\overline{S}$.

Because the Third Law of Thermodynamics asserts that $S(0 \text{ K}) = 0$, we can write Equation 21.10 as

$$S(T) = \int_0^T \frac{C_P(T')dT'}{T'} \tag{21.14}$$

21–3. $\Delta_{trs}S = \Delta_{trs}H/T_{trs}$ at a Phase Transition

We made a tacit assumption when we wrote Equation 21.14; we assumed that there is no phase transition between 0 and T. Suppose there is such a transition at T_{trs} between 0 and T. We can calculate the entropy change upon the phase transition, $\Delta_{trs}S$, by using the equation

$$\Delta_{trs}S = \frac{q_{rev}}{T_{trs}} \qquad (21.15)$$

A phase transition is a good example of a reversible process. A phase transition can be reversed by changing the temperature ever so slightly. In the melting of ice, for example, at one atm, the system will be all ice if T is just slightly less than 273.15 K and all liquid if T is just slightly greater than 273.15 K. Furthermore, a phase transition takes place at a fixed temperature, so Equation 21.15 becomes (recall that $\Delta H = q_P$ for a phase transition)

$$\Delta_{trs}S = \frac{\Delta_{trs}H}{T_{trs}} \qquad (21.16)$$

EXAMPLE 21–2
Calculate the molar entropy change upon melting and upon vaporization at one atm for H_2O. Use $\Delta_{fus}\overline{H} = 6.01 \text{ kJ} \cdot \text{mol}^{-1}$ at 273.15 K and $\Delta_{vap}\overline{H} = 40.7 \text{ kJ} \cdot \text{mol}^{-1}$ at 373.15 K.

SOLUTION: Using Equation 21.16, we have

$$\Delta_{fus}\overline{S} = \frac{6.01 \text{ kJ} \cdot \text{mol}^{-1}}{273.15 \text{ K}} = 22.0 \text{ J} \cdot \text{K}^{-1} \cdot \text{mol}^{-1}$$

and

$$\Delta_{vap}\overline{S} = \frac{40.7 \text{ kJ} \cdot \text{mol}^{-1}}{373.15 \text{ K}} = 109 \text{ J} \cdot \text{K}^{-1} \cdot \text{mol}^{-1}$$

Note that $\Delta_{vap}\overline{S}$ is much larger than $\Delta_{fus}\overline{S}$. This makes sense molecularly because the difference in disorder between a gas and a liquid phase is much greater than the difference in disorder between a liquid and a solid phase.

To calculate $S(T)$, we integrate $C_P(T)/T$ up to the first phase transition temperature, add a $\Delta_{trs}H/T_{trs}$ term for the phase transition, and then integrate $C_P(T)/T$ from the first phase transition temperature to the second, and so on. For example, if the substance has no solid-solid phase transition, we would have, for T greater than the boiling point,

$$S(T) = \int_0^{T_{fus}} \frac{C_P^s(T)dT}{T} + \frac{\Delta_{fus}H}{T_{fus}} + \int_{T_{fus}}^{T_{vap}} \frac{C_P^l(T)dT}{T}$$
$$+ \frac{\Delta H_{vap}}{T_{vap}} + \int_{T_{vap}}^{T} \frac{C_P^g(T')dT'}{T'} \qquad (21.17)$$

where T_{fus} is the melting point, $C_P^s(T)$ is the heat capacity of the solid phase, T_{vap} is the boiling point, $C_P^l(T)$ is the heat capacity of the liquid phase, $C_P^g(T)$ is the heat capacity of the gaseous phase, and $\Delta_{\text{fus}}H$ and ΔH_{vap} are the enthalpies of fusion and vaporization, respectively.

21–4. The Third Law of Thermodynamics Asserts That $C_P \rightarrow 0$ as $T \rightarrow 0$

It has been shown experimentally and theoretically that $C_P^s(T) \rightarrow T^3$ as $T \rightarrow 0$ for most nonmetallic crystals (C_P^s for metallic crystals goes as $aT + bT^3$ as $T \rightarrow 0$, where a and b are constants). This T^3 temperature dependence is valid from 0 K to about 15 K and is called the *Debye T^3 law*, after the Dutch chemist Peter Debye, who first showed theoretically that $C_P^s(T) \rightarrow T^3$ as $T \rightarrow 0$ for nonmetallic solids.

EXAMPLE 21–3
According to the Debye theory, the low-temperature molar heat capacity of nonmetallic solids goes as

$$\overline{C}_P(T) = \frac{12\pi^4}{5} R \left(\frac{T}{\Theta_D} \right)^3 \qquad 0 < T \leq T_{\text{low}}$$

where T_{low} depends upon the particular solid, but is about 10 K to 20 K for most solids, and Θ_D is a constant characteristic of the solid. The parameter Θ_D has units of temperature and is called the *Debye temperature* of the solid. Show that if $\overline{C}_P$ is given by the above expression, the low-temperature contribution to the molar entropy is given by

$$\overline{S}(T) = \frac{\overline{C}_P(T)}{3} \qquad 0 < T \leq T_{\text{low}}$$

SOLUTION: Substitute the given expression for $\overline{C}_P(T)$ into Equation 21.14 to get

$$\overline{S}(T) = \int_0^T \frac{\overline{C}_P(T')dT'}{T'} = \frac{12\pi^4 R}{5\Theta_D^3} \int_0^T T'^2 dT'$$

$$= \frac{12\pi^4 R}{5\Theta_D^3} \frac{T^3}{3} = \frac{\overline{C}_P(T)}{3} \qquad (21.18)$$

EXAMPLE 21–4
Given that the molar heat capacity of solid chlorine is 3.39 J·K^{-1}·mol^{-1} at 14 K and obeys the Debye T^3 law below 14 K, calculate the molar entropy of solid chlorine at 14 K.

SOLUTION: We use Equation 21.18 and get

$$\overline{S}(\text{at } 14 \text{ K}) = \frac{\overline{C}_p(\text{at } 14 \text{ K})}{3}$$

$$= \frac{3.39 \text{ J} \cdot \text{K}^{-1} \cdot \text{mol}^{-1}}{3} = 1.13 \text{ J} \cdot \text{K}^{-1} \cdot \text{mol}^{-1}$$

21–5. Practical Absolute Entropies Can Be Determined Calorimetrically

Given suitable heat capacity data and enthalpies of transition and transition temperatures, we can use Equation 21.17 to calculate entropies based on the convention of setting $S(0 \text{ K}) = 0$. Such entropies are called third-law entropies, or practical absolute entropies. Table 21.1 gives the entropy of $N_2(g)$ at 298.15 K. The entropy at 10.00 K was determined by using Equation 21.18 with $\overline{C}_p = 6.15 \text{ J} \cdot \text{K}^{-1} \cdot \text{mol}^{-1}$. At 35.61 K, the solid undergoes a phase change in crystalline structure with $\Delta_{\text{trs}}\overline{H} = 0.2289 \text{ kJ} \cdot \text{mol}^{-1}$, so $\Delta_{\text{trs}}\overline{S} = 6.43 \text{ J} \cdot \text{K}^{-1} \cdot \text{mol}^{-1}$. At 63.15 K, $N_2(s)$ melts with $\Delta_{\text{fus}}\overline{H} = 0.71 \text{ kJ} \cdot \text{mol}^{-1}$, so $\Delta_{\text{fus}}\overline{S} = 11.2 \text{ J} \cdot \text{K}^{-1} \cdot \text{mol}^{-1}$. Finally, $N_2(l)$ at one atm boils at 77.36 K with $\Delta_{\text{vap}}\overline{H} = 5.57 \text{ kJ} \cdot \text{mol}^{-1}$, giving $\Delta_{\text{vap}}\overline{S} = 72.0 \text{ J} \cdot \text{K}^{-1} \cdot \text{mol}^{-1}$. For the regions between the phase transitions, $\overline{C}_p(T)/T$ data were integrated numerically (Problem 21–14). According to Equation 21.17, the molar entropy is given by the area under the curve of $\overline{C}_p(T)/T$ plotted against the temperature.

The small correction at the end of Table 21.1 needs explaining. The values of entropies of gases presented in the literature are called *standard entropies*, which by

TABLE 21.1
The standard molar entropy of nitrogen at 298.15 K.

Process	$\overline{S}^{\circ}/\text{J} \cdot \text{K}^{-1} \cdot \text{mol}^{-1}$
0 to 10.00 K	2.05
10.00 to 35.61 K	25.79
Transition	6.43
35.61 to 63.15 K	23.41
Fusion	11.2
63.15 to 77.36 K	11.46
Vaporization	72.0
77.36 K to 298.15 K	39.25
Correction for nonideality	0.02
Total	191.6

convention are corrected for the nonideality of the gas at one bar. We will learn how to make this correction in Section 22–6. Recall that the standard state of a (real) gas at any temperature is that of the corresponding (hypothetical) ideal gas at one bar.

Figure 21.1 shows the molar entropy of nitrogen plotted against temperature from 0 K to 400 K. Note that the molar entropy increases smoothly with temperature between phase transitions and that there are discontinuous jumps at each phase transition. Note also that the jump at the vaporization transition is much larger than the jump at the melting point. Figure 21.2 shows a similar plot for benzene. Note that benzene does not undergo any solid-solid phase transitions.

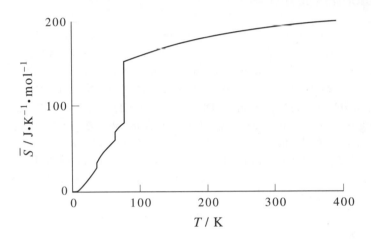

FIGURE 21.1
The molar entropy of nitrogen plotted against temperature from 0 K to 400 K.

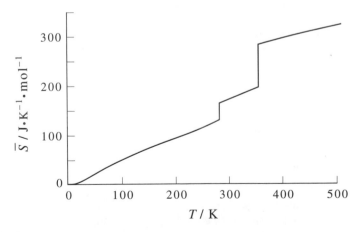

FIGURE 21.2
The molar entropy of benzene plotted against temperature from 0 K to 500 K.

21–6. Practical Absolute Entropies of Gases Can Be Calculated from Partition Functions

Recall from Section 20–8 that the entropy can be written as (Equation 20.43)

$$S = k_B \ln Q + k_B T \left(\frac{\partial \ln Q}{\partial T} \right)_{N,V} \tag{21.19}$$

where $Q(N, V, T)$ is the system partition function

$$Q(N, V, T) = \sum_j e^{-E_j(N,V)/k_B T} \tag{21.20}$$

Equation 21.19 is consistent with the Third Law of Thermodynamics. Let's write Equation 21.19 for S more explicitly by substituting Equation 21.20 into it:

$$S = k_B \ln \sum_j e^{-E_j/k_B T} + \frac{1}{T} \frac{\sum_j E_j e^{-E_j/k_B T}}{\sum_j e^{-E_j/k_B T}} \tag{21.21}$$

We want to study the behavior of this equation as $T \to 0$. Assume for generality that the first n states have the same energy $E_1 = E_2 = \cdots = E_n$ (in other words, that the ground state is n-fold degenerate) and that the next m states have the same energy $E_{n+1} = E_{n+2} = \cdots = E_{n+m}$ (the first excited state is m-fold degenerate), and so on.

Let's look at the summations in Equations 21.21 as $T \to 0$. Writing out Equation 21.20 explicitly gives

$$\sum_j e^{-E_j/k_B T} = n e^{-E_1/k_B T} + m e^{-E_{n+1}/k_B T} + \cdots$$

If we factor out $e^{-E_1/k_B T}$, then

$$\sum_j e^{-E_j/k_B T} = e^{-E_1/k_B T} \left[n + m e^{-(E_{n+1}-E_1)/k_B T} + \cdots \right]$$

But $E_{n+1} - E_1 > 0$ (essentially by definition), so

$$e^{-(E_{n+1}-E_1)/k_B T} \to 0 \quad \text{as} \quad T \to 0$$

Therefore, as $T \to 0$,

$$\sum_j e^{E_j/k_B T} \to n e^{-E_1/k_B T}$$

In the limit of small T, then, the first terms in each summation in Equation 21.21 dominate, and we have

$$S = k_B \ln(ne^{-E_1/k_B T}) + \frac{1}{T} \frac{n E_1 e^{-E_1/k_B T}}{ne^{-E_1/k_B T}}$$

$$= k_B \ln n - \frac{E_1}{T} + \frac{E_1}{T} = k_B \ln n$$

Thus, as $T \to 0$, S is proportional to the logarithm of the degeneracy of the ground state (see Equation 21.13). As we argued in Section 21–2, even if n were as large as the Avogadro constant, S itself would be completely negligible.

We learned in Chapter 17 (Equation 17.38) that

$$Q(N, V, T) = \frac{[q(V, T)]^N}{N!} \tag{21.22}$$

for an ideal gas. Furthermore, we learned in Chapter 18 that for a
(1) monatomic ideal gas (Equation 18.13):

$$q(V, T) = \left(\frac{2\pi m k_B T}{h^2} \right)^{3/2} V \cdot g_{e1} \tag{21.23}$$

(2) diatomic ideal gas (Equation 18.39):

$$q(V, T) = \left(\frac{2\pi M k_B T}{h^2} \right)^{3/2} V \cdot \frac{T}{\sigma \Theta_{rot}} \cdot \frac{e^{-\Theta_{vib}/2T}}{1 - e^{-\Theta_{vib}/T}} \cdot g_{e1} e^{D_e/k_B T} \tag{21.24}$$

(3) linear polyatomic ideal gas (Equation 18.57):

$$q(V, T) = \left(\frac{2\pi M k_B T}{h^2} \right)^{3/2} V \cdot \frac{T}{\sigma \Theta_{rot}} \cdot \left[\prod_{j=1}^{3n-5} \frac{e^{-\Theta_{vib, j}/2T}}{1 - e^{-\Theta_{vib, j}/T}} \right] \cdot g_{e1} e^{D_e/k_B T} \tag{21.25}$$

(4) nonlinear polyatomic ideal gas (Equation 18.60):

$$q(V, T) = \left(\frac{2\pi M k_B T}{h^2} \right)^{3/2} V \cdot \frac{\pi^{1/2}}{\sigma} \left(\frac{T^3}{\Theta_A \Theta_B \Theta_C} \right)^{1/2} \left[\prod_{j=1}^{3n-6} \frac{e^{-\Theta_{vib, j}/2T}}{1 - e^{-\Theta_{vib, j}/T}} \right] g_{e1} e^{D_e/k_B T}$$
$$\tag{21.26}$$

The various quantities in these equations are defined and discussed in Chapter 18.

If we substitute Equation 21.22 into Equation 21.19, then we obtain

$$S = N k_B \ln q - k_B \ln N! + N k_B T \left(\frac{\partial \ln q}{\partial T} \right)_V$$

If we use Stirling's approximation for $\ln N! (= N \ln N - N)$, then (Problem 21-27)

$$S = N k_B + N k_B \ln \left[\frac{q(V, T)}{N} \right] + N k_B T \left(\frac{\partial \ln q}{\partial T} \right)_V \tag{21.27}$$

Let's use Equations 21.27 and 21.24 to calculate the standard molar entropy of $N_2(g)$ at 298.15 K and compare the result with the value in Table 21.1 obtained from heat capacity data. If we substitute Equation 21.24 into Equation 21.27, we obtain

$$\frac{\overline{S}^{\circ}}{R} = \ln\left[\left(\frac{2\pi M k_B T}{h^2}\right)^{3/2} \frac{\overline{V} e^{5/2}}{N_A}\right] + \ln\frac{Te}{2\Theta_{rot}} - \ln(1 - e^{-\Theta_{vib}/T})$$
$$+ \frac{\Theta_{vib}/T}{e^{\Theta_{vib}/T} - 1} + \ln g_{e1} \qquad (21.28)$$

The first term represents the translational contribution to S, the second represents the rotational contribution, the third and fourth represent the vibrational contribution, and the last term represents the electronic contribution to S. The necessary parameters are $\Theta_{rot} = 2.88$ K, $\Theta_{vib} = 3374$ K, and $g_{e1} = 1$. At 298.15 K and one bar, the various factors are

$$\left(\frac{2\pi M k_B T}{h^2}\right)^{3/2} = \left[\frac{2\pi(4.653 \times 10^{-26} \text{ kg})(1.3807 \times 10^{-23} \text{ J·K}^{-1})(298.15 \text{ K})}{(6.626 \times 10^{-34} \text{ J·s})^2}\right]^{3/2}$$
$$= 1.436 \times 10^{32} \text{ m}^{-3}$$

$$\frac{\overline{V}}{N_A} = \frac{RT}{N_A P} = \frac{(0.08314 \text{ L·bar·mol}^{-1}\text{·K}^{-1})(298.15 \text{ K})}{(6.022 \times 10^{23} \text{ mol}^{-1})(1 \text{ bar})}$$
$$= 4.117 \times 10^{-23} \text{ L} = 4.117 \times 10^{-26} \text{ m}^{-3}$$

$$\frac{Te}{2\Theta_{rot}} = \frac{(298.15 \text{ K})(2.71828)}{2(2.88 \text{ K})} = 140.7$$

$$1 - e^{-\Theta_{vib}/T} = 1 - e^{-11.31} \approx 1.000$$

$$\frac{\Theta_{vib}/T}{e^{\Theta_{vib}/T} - 1} = \frac{11.31}{e^{11.31} - 1} = 1.380 \times 10^{-4}$$

Thus, the standard molar entropy $\overline{S}^{\circ}$ at 298.15 K is

$$\overline{S}^{\circ} = \overline{S}_{trans}^{\circ} + \overline{S}_{rot}^{\circ} + \overline{S}_{vib}^{\circ} + \overline{S}_{elec}^{\circ}$$
$$= (150.4 + 41.13 + 1.15 \times 10^{-3} + 0) \text{ J·K}^{-1}\text{·mol}^{-1}$$
$$= 191.5 \text{ J·K}^{-1}\text{·mol}^{-1}$$

compared with the value of 191.6 $\text{J·K}^{-1}\text{·mol}^{-1}$ given in Table 21.1. The two values agree essentially exactly. This type of agreement is quite common, and in many cases the statistical thermodynamic value is more accurate than the calorimetric value. Table 21.2 gives standard molar entropies for several substances. The accepted literature values are often a combination of statistical thermodynamic and calorimetric values.

EXAMPLE 21–5
Use the equations of this section to calculate the standard molar entropy of carbon dioxide at 298.15 K and compare the result with the value in Table 21.2.

TABLE 21.2
Standard molar entropies ($\overline{S}^{\circ}$) of various substances at 298.15 K.

Substance	$\overline{S}^{\circ}/\text{J}\cdot\text{K}^{-1}\cdot\text{mol}^{-1}$	Substance	$\overline{S}^{\circ}/\text{J}\cdot\text{K}^{-1}\cdot\text{mol}^{-1}$
Ag(s)	42.55	HCl(g)	186.9
Ar(g)	154.8	HCN(g)	201.8
Br_2(g)	245.5	HI(g)	206.6
Br_2(l)	152.2	H_2O(g)	188.8
C(s)(diamond)	2.38	H_2O(l)	70.0
C(s)(graphite)	5.74	Hg(l)	75.9
CH_4(g)	186.3	I_2(s)	116.1
C_2H_2(g)	200.9	I_2(g)	260.7
C_2H_4(g)	219.6	K(s)	64.7
C_2H_6(g)	229.6	N_2(g)	191.6
CH_3OH(l)	126.8	Na(s)	51.3
CH_3Cl(g)	234.6	NH_3(g)	192.8
CO(g)	197.7	NO(g)	210.8
CO_2(g)	213.8	NO_2(g)	240.1
Cl_2(g)	223.1	O_2(g)	205.2
H_2(g)	130.7	O_3(g)	238.9
HBr(g)	198.7	SO_2(g)	248.2

SOLUTION: Carbon dioxide is a symmetric linear molecule with four vibrational degrees of freedom. We substitute Equation 21.25 into Equation 21.27 to obtain

$$\frac{\overline{S}^{\circ}}{R} = 1 + \ln\left[\left(\frac{2\pi Mk_B T}{h^2}\right)^{3/2}\frac{\overline{V}}{N_A}\right] + \ln\left(\frac{T}{\sigma\Theta_{\text{rot}}}\right)$$

$$-\sum_{j=1}^{4}\frac{\Theta_{\text{vib},j}}{2T} - \sum_{j=1}^{4}\ln\left(1 - e^{-\Theta_{\text{vib},j}/T}\right) + \ln g_{e1} + \frac{D_e}{k_B T}$$

$$+ T\left[\frac{3}{2T} + \frac{1}{T} + \sum_{j=1}^{4}\frac{\Theta_{\text{vib},j}}{2T^2} + \sum_{j=1}^{4}\frac{(\Theta_{\text{vib},j}/T^2)e^{-\Theta_{\text{vib},j}/T}}{1 - e^{-\Theta_{\text{vib},j}/T}} - \frac{D_e}{k_B T^2}\right]$$

or

$$\frac{\overline{S}^{\circ}}{R} = \frac{7}{2} + \ln\left[\left(\frac{2\pi Mk_B T}{h^2}\right)^{3/2}\frac{\overline{V}}{N_A}\right] + \ln\left(\frac{T}{\sigma\Theta_{\text{rot}}}\right)$$

$$+ \sum_{j=1}^{4}\left[\frac{(\Theta_{\text{vib},j}/T)e^{-\Theta_{\text{vib},j}/T}}{1 - e^{-\Theta_{\text{vib},j}/T}} - \ln(1 - e^{-\Theta_{\text{vib},j}/T})\right] + \ln g_{e1}$$

Paralleling the calculation for N_2(g), we find that $(2\pi Mk_B T/h^2)^{3/2} = 2.826 \times 10^{32}$ m^{-3} and $\overline{V}/N_A = 4.117 \times 10^{-26}$ m^{-3}. Using the value of $\Theta_{\text{rot}} = 0.561$ K from Table 18.4, we find that $T/2\Theta_{\text{rot}} = 265.8$. Similarly, we use Table 18.4 to show that the four values of $\Theta_{\text{vib},j}/T$ are 3.199 (twice), 6.338, and 11.27. Last, $g_{e1} = 1$. Putting all this together, we find that

$$\frac{\overline{S}^\circ}{R} = \frac{7}{2} + \ln\left[(2.826 \times 10^{32} \text{ m}^{-3})(4.117 \times 10^{-26} \text{ m}^{-3})\right] + \ln 265.8$$

$$+ 2\left[\frac{3.199e^{-3.199}}{1 - e^{-3.199}} - \ln(1 - e^{-3.199})\right] + \left[\frac{6.338e^{-6.338}}{1 - e^{-6.338}} - \ln(1 - e^{-6.338})\right]$$

$$+ \left[\frac{11.27e^{-11.27}}{1 - e^{-11.27}} - \ln(1 - e^{-11.27})\right]$$

$$= 3.5 + 16.27 + 5.58 + 2(0.178) + 0.01 + O(10^{-4})$$

$$= 25.71$$

or

$$\overline{S}^\circ = 25.71R = 213.8 \text{ J}\cdot\text{K}^{-1}\cdot\text{mol}^{-1}$$

which is in excellent agreement with the value in Table 21.2.

21-7. The Values of Standard Molar Entropies Depend Upon Molecular Mass and Molecular Structure

Let's look at the standard molar entropy values in Table 21.2 and try to determine some trends. First, notice that the standard molar entropies of the gaseous substances are the largest, and the standard molar entropies of the solid substances are the smallest. These values reflect the fact that solids are more ordered than liquids and gases.

Now consider the standard molar entropies of the noble gases given in Table 21.3. The increase in standard molar entropy of the noble gases is a consequence of their increasing mass as we move down the periodic table. Thus, an increase in mass leads to an increase in thermal disorder (more translational energy levels are available) and a greater entropy. We know from quantum theory that the greater the molecular mass, the

TABLE 21.3
Standard molar entropies $(\overline{S}^\circ)$ for the noble gases, the gaseous halogens, and the hydrogen halides at 298.15 K.

Noble gas	$\overline{S}^\circ$/J·K^{-1}·mol^{-1}	Halogen	$\overline{S}^\circ$/J·K^{-1}·mol^{-1}	Hydrogen halide	$\overline{S}^\circ$/J·K^{-1}·mol^{-1}
He(g)	126.2	F_2(g)	202.8	HF(g)	173.8
Ne(g)	146.3	Cl_2(g)	223.1	HCl(g)	186.9
Ar(g)	154.8	Br_2(g)	245.5	HBr(g)	198.7
Kr(g)	164.1	I_2(g)	260.7	HI(g)	206.6
Xe(g)	169.7				

more closely spaced are the energy levels. The same trend can be seen by comparing the standard molar entropies at 298.15 K of the gaseous halogens and hydrogen halides (see Table 21.3 and Figure 21.3).

Generally speaking, the more atoms of a given type in a molecule, the greater is the capacity of the molecule to take up energy and thus the greater is its entropy (the greater the number of atoms, the more different ways in which the molecules can vibrate.) This trend is illustrated by the series $C_2H_2(g)$, $C_2H_4(g)$, and $C_2H_6(g)$, whose standard molar entropies in joules per kelvin per mole at 298.15 K are 201, 220, and 230, respectively. For molecules with the same geometry and number of atoms, the standard molar entropy increases with increasing molecular mass.

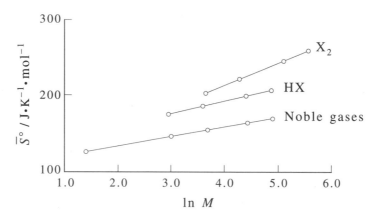

FIGURE 21.3
Standard molar entropies $(\overline{S}^{\,\circ})$ for the noble gases, the gaseous halogens, and the hydrogen halides at 298.15 K plotted against $\ln M$, where M is the molecular mass.

EXAMPLE 21–6
Arrange the following molecules in order of increasing standard molar entropy: $CH_2Cl_2(g)$; $CHCl_3(g)$; $CH_3Cl(g)$.

SOLUTION: The number of atoms is the same in each case, but chlorine has a greater mass than hydrogen. Thus, we predict that

$$\overline{S}^{\,\circ}[CH_3Cl(g)] < \overline{S}^{\,\circ}[CH_2Cl_2(g)] < \overline{S}^{\,\circ}[CHCl_3(g)]$$

This ordering is in agreement with the values of the standard molar entropies at 298.15 K, which are in units of joules per kelvin per mole, 234.6, 270.2, and 295.7, respectively.

An interesting comparison is given by the isomers acetone and trimethylene oxide (whose molecular structures are shown below), whose standard molar entropies for the gaseous forms at 298.15 K are 298 J·K^{-1}·mol^{-1} and 274 J·K^{-1}·mol^{-1}, respectively. The entropy of acetone is higher than the entropy of trimethylene oxide because of the free rotation of the methyl groups about the carbon-carbon bonds in the acetone molecule. The relatively rigid ring structure of the trimethylene oxide molecule restricts the movement of the ring atoms. This restriction gives rise to a lower molar entropy because the capacity of the rigid isomer to take up energy is less than that of the more flexible acetone molecule, which has more possibilities for intermolecular motion. For molecules with approximately the same molecular masses, the more compact the molecule is, the smaller is its entropy.

Note that Table 21.2 gives $\overline{S}^{\,\circ} = 245.5$ J·K^{-1}·mol^{-1} for Br$_2$(g) at 298.15 K and one bar. But bromine is a liquid at 298.15 K and one bar, so where does such a value come from? Even though bromine is a liquid under these conditions, we can calculate $\overline{S}^{\,\circ}[\text{Br}_2(\text{g})]$ according to the scheme outlined in Figure 21.4. Therefore, we need the values of the molar heat capacity of Br$_2$(l) (75.69 J·K^{-1}·mol^{-1}), the molar heat capacity of Br$_2$(g) (36.02 J·K^{-1}·mol^{-1}), the normal boiling point of Br$_2$(l) (332.0 K), and the molar enthalpy of vaporization at 332.0 K (29.54 kJ·mol^{-1}) . We start with Br$_2$(l)

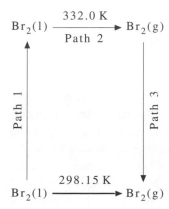

FIGURE 21.4
The scheme used to calculate $\overline{S}^{\,\circ}[\text{Br}_2(\text{g})]$ at 298.15 K. In Path 1, Br$_2$(l) is heated to its boiling point, 332.0 K. Then Br$_2$(l) is vaporized to Br$_2$(g) at 332.0 K (Path 2), and finally Br$_2$(g) is cooled from 332.0 K back to 298.15 K (Path 3).

at 298.15 K and heat it to its boiling point. The value of $\Delta\overline{S}$ for this first step is (Equation 21.7)

$$\Delta\overline{S}_1 = \overline{S}^{\,l}(332.0 \text{ K}) - \overline{S}^{\,l}(298.15 \text{ K}) = \overline{C}_P^{\,l}\ln\frac{T_2}{T_1}$$

$$= (75.69 \text{ J}\cdot\text{K}^{-1}\cdot\text{mol}^{-1})\ln\frac{332.0 \text{ K}}{298.15 \text{ K}} = 8.140 \text{ J}\cdot\text{K}^{-1}\cdot\text{mol}^{-1}$$

Now vaporize the bromine at its normal boiling point (step 2 in Figure 21.4):

$$\Delta\overline{S}_2 = \overline{S}^{\,g}(332.0 \text{ K}) - \overline{S}^{\,l}(332.0 \text{ K}) = \frac{\Delta_{\text{vap}}\overline{H}}{T_{\text{vap}}} = \frac{29.54 \text{ kJ}\cdot\text{mol}^{-1}}{332.0 \text{ K}}$$

$$= 88.98 \text{ J}\cdot\text{K}^{-1}\cdot\text{mol}^{-1}$$

Last, cool the gas from 332.0 K back to 298.15 K (step 3):

$$\Delta\overline{S}_3 = \overline{S}^{\,g}(298.15 \text{ K}) - \overline{S}^{\,g}(332.0 \text{ K}) = \overline{C}_P^{\,g}\ln\frac{298.15 \text{ K}}{332.0 \text{ K}}$$

$$= (36.02 \text{ J}\cdot\text{K}^{-1}\cdot\text{mol}^{-1})\ln\frac{298.15}{332.0} = -3.87 \text{ J}\cdot\text{K}^{-1}\cdot\text{mol}^{-1}$$

If we add these three steps and then add the results to $S_{298}^{\circ}[\text{Br}_2(l)] = 152.2 \text{ J}\cdot\text{K}^{-1}\cdot\text{mol}^{-1}$ (Table 21.2), we obtain

$$\overline{S}_{298}^{\circ}[\text{Br}_2(g)] = \overline{S}_{298}^{\circ}[\text{Br}_2(l)] + \Delta\overline{S}_1 + \Delta\overline{S}_2 + \Delta\overline{S}_3$$

$$= 152.2 \text{ J}\cdot\text{K}^{-1}\cdot\text{mol}^{-1} + 8.14 \text{ J}\cdot\text{K}^{-1}\cdot\text{mol}^{-1}$$

$$+ 88.98 \text{ J}\cdot\text{K}^{-1}\cdot\text{mol}^{-1} - 3.87 \text{ J}\cdot\text{K}^{-1}\cdot\text{mol}^{-1}$$

$$= 245.5 \text{ J}\cdot\text{K}^{-1}\cdot\text{mol}^{-1}$$

in agreement with the value of $\text{Br}_2(g)$ in Table 21.2. Incidentally, the spectroscopic value of $\overline{S}^{\,\circ}[\text{Br}_2(g)]$, using Equation 21.24 and the data in Chapter 18 is 245.5 J·K^{-1}·mol^{-1} (Problem 21–33).

21–8. The Spectroscopic Entropies of a Few Substances Do Not Agree with the Calorimetric Entropies

Table 21.4 compares calculated values of the molar entropies of several polyatomic gases with those measured calorimetrically. Note again, that the agreement with experiment is quite good. In fact, calculated values of the entropy are often more accurate than measured values, provided sophisticated enough spectroscopic models are used.

There is, however, a class of molecules for which the type of agreement in Table 21.4 is not found. For example, for carbon monoxide, $\overline{S}_{\text{calc}} = 160.3 \text{ J}\cdot\text{K}^{-1}\cdot\text{mol}^{-1}$ and $\overline{S}_{\text{exp}} = 155.6 \text{ J}\cdot\text{K}^{-1}\cdot\text{mol}^{-1}$ at its boiling point (81.6 K), for a discrepancy of 4.7 J·K^{-1}·mol^{-1}. Other such discrepancies are found, and in all cases $\overline{S}_{\text{calc}} > \overline{S}_{\text{exp}}$.

TABLE 21.4

The standard molar entropies of several polyatomic gases at 298.15 K and one bar.

Gas	$\overline{S}^{\circ}$(calc)/J·K^{-1}·mol^{-1}	$\overline{S}^{\circ}$(exp)/J·K^{-1}·mol^{-1}
CO_2	213.8	213.7
NH_3	192.8	192.6
NO_2	240.1	240.2
CH_4	186.3	186.3
C_2H_2	200.9	200.8
C_2H_4	219.6	219.6
C_2H_6	229.6	229.5

The difference $\overline{S}_{calc} - \overline{S}_{exp}$ is often referred to as *residual entropy*. The explanation of these cases is the following. Carbon monoxide has a very small dipole moment ($\approx 4 \times 10^{-31}$ C·m), so when carbon monoxide is crystallized, the molecules do not have a strong tendency to line up in an energetically favorable way. The resultant crystal, then, is a random mixture of the two possible orientations, CO and OC. As the crystal is cooled down toward 0 K, each molecule gets locked into its orientation and cannot realize the state of lowest energy with $W = 1$, that is, all the molecules oriented in the same direction. Instead, the number of configurations W of the crystal is 2^N, because each of the N molecules exists equally likely (almost equally likely because the dipole moment is so small) in two states. Thus, the molar entropy of the crystal at 0 K is $S = R \ln 2$ instead of zero. If $R \ln 2 = 5.7$ J·K^{-1}·mol^{-1} is added to the experimental entropy, the agreement in the case of carbon monoxide becomes satisfactory. If it were possible to obtain carbon monoxide in its true equilibrium state at $T = 0$ K, this discrepancy would not occur. A similar situation occurs with dinitrogen oxide, which is a linear molecule with the structure NNO. For H$_3$CD, the residual entropy is 11.7 J·K^{-1}·mol^{-1}, which is explained by realizing that each molecule of monodeuterated methane can assume four different orientations in the low-temperature crystal, so $\overline{S}_{residual} = R \ln 4 = 11.5$ J·K^{-1}·mol^{-1}, in very close agreement with the experimental value.

21–9. Standard Entropies Can Be Used to Calculate Entropy Changes of Chemical Reactions

One of the most important uses of tables of standard molar entropies is for the calculation of entropy changes of chemical reactions. These changes are calculated in much the same way we calculated standard enthalpy changes of reactions from standard molar enthalpies of formation in Chapter 19. For the general reaction

$$a\,A + b\,B \longrightarrow y\,Y + z\,Z$$

the standard entropy change is given by

$$\Delta_r S^\circ = y S^\circ[\text{Y}] + z S^\circ[\text{Z}] - a S^\circ[\text{A}] - b S^\circ[\text{B}]$$

For example, using the values of S° given in Table 21.2 for the substances in the reaction described by the chemical equation

$$\text{H}_2(\text{g}) + \tfrac{1}{2}\text{O}_2(\text{g}) \rightleftharpoons \text{H}_2\text{O}(\text{l})$$

$$\begin{aligned}
\Delta_r S^\circ &= (1)S^\circ[\text{H}_2\text{O}(\text{l})] - (1)S^\circ[\text{H}_2(\text{g})] - (\tfrac{1}{2})S^\circ[\text{O}_2(\text{g})] \\
&= (1)(70.0 \text{ J}\cdot\text{K}^{-1}\cdot\text{mol}^{-1}) - (1)(130.7 \text{ J}\cdot\text{K}^{-1}\cdot\text{mol}^{-1}) - (\tfrac{1}{2})(205.2 \text{ J}\cdot\text{K}^{-1}\cdot\text{mol}^{-1}) \\
&= -163.3 \text{ J}\cdot\text{K}^{-1}\cdot\text{mol}^{-1}
\end{aligned}$$

This value of $\Delta_r S^\circ$ represents the value of $\Delta_r S$ for the combustion of one mole of $\text{H}_2(\text{g})$ or the formation of one mole of $\text{H}_2\text{O}(\text{l})$, when all the reactants and products are in their standard states. The large negative value of $\Delta_r S^\circ$ reflects the loss of gaseous reactants to produce a condensed phase, an ordering process.

We will use tables of standard enthalpies of formation and standard entropies to calculate equilibrium constants of chemical reactions in Chapter 26.

Problems

21-1. Form the total derivative of H as a function of T and P and equate the result to dH in Equation 21.6 to derive Equations 21.7 and 21.8.

21-2. The molar heat capacity of $\text{H}_2\text{O}(\text{l})$ has an approximately constant value of $\overline{C}_P = 75.4 \text{ J}\cdot\text{K}^{-1}\cdot\text{mol}^{-1}$ from 0°C to 100°C. Calculate ΔS if two moles of $\text{H}_2\text{O}(\text{l})$ are heated from 10°C to 90°C at constant pressure.

21-3. The molar heat capacity of butane can be expressed by

$$\overline{C}_P/R = 0.05641 + (0.04631 \text{ K}^{-1})T - (2.392 \times 10^{-5} \text{ K}^{-2})T^2 + (4.807 \times 10^{-9} \text{ K}^{-3})T^3$$

over the temperature range $300 \text{ K} \leq T \leq 1500 \text{ K}$. Calculate ΔS if one mole of butane is heated from 300 K to 1000 K at constant pressure.

21-4. The molar heat capacity of $\text{C}_2\text{H}_4(\text{g})$ can be expressed by

$$\overline{C}_V(T)/R = 16.4105 - \frac{6085.929 \text{ K}}{T} + \frac{822\,826 \text{ K}^2}{T^2}$$

over the temperature range $300 \text{ K} < T < 1000 \text{ K}$. Calculate ΔS if one mole of ethene is heated from 300 K to 600 K at constant volume.

21-5. Use the data in Problem 21–4 to calculate ΔS if one mole of ethene is heated from 300 K to 600 K at constant pressure. Assume ethene behaves ideally.

21-6. We can calculate the difference in the results of Problems 21–4 and 21–5 in the following way. First, show that because $\overline{C}_P - \overline{C}_V = R$ for an ideal gas,

$$\Delta \overline{S}_P = \Delta \overline{S}_V + R \ln \frac{T_2}{T_1}$$

Check to see numerically that your answers to Problems 21–4 and 21–5 differ by $R \ln 2 = 0.693 R = 5.76 \text{ J} \cdot \text{K}^{-1} \cdot \text{mol}^{-1}$.

21-7. The results of Problems 21–4 and 21–5 must be connected in the following way. Show that the two processes can be represented by the diagram

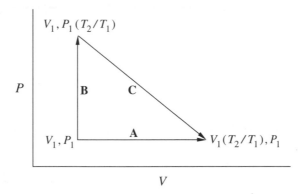

where paths A and B represent the processes in Problems 21–5 and 21–4, respectively. Now, path A is equivalent to the sum of paths B and C. Show that ΔS_C is given by

$$\Delta S_C = nR \ln \frac{V_1 \left(\dfrac{T_2}{T_1} \right)}{V_1} = nR \ln \frac{P_1 \left(\dfrac{T_2}{T_1} \right)}{P_1} = nR \ln \frac{T_2}{T_1}$$

and that the result given in Problem 21–6 follows.

21-8. Use Equations 20.23 and 20.24 to show that $S = 0$ at 0 K, where every system will be in its ground state.

21-9. Prove that $S = -k_B \sum p_j \ln p_j = 0$ when $p_1 = 1$ and all the other $p_j = 0$. In other words, prove that $x \ln x \to 0$ as $x \to 0$.

21-10. It has been found experimentally that $\Delta_{vap}\overline{S} \approx 88$ J·K^{-1}·mol^{-1} for many nonassociated liquids. This rough rule of thumb is called *Trouton's rule*. Use the following data to test the validity of Trouton's rule.

Substance	$t_{fus}/°C$	$t_{vap}/°C$	$\Delta_{fus}\overline{H}/kJ·mol^{-1}$	$\Delta_{vap}\overline{H}/kJ·mol^{-1}$
Pentane	−129.7	36.06	8.42	25.79
Hexane	−95.3	68.73	13.08	28.85
Heptane	−90.6	98.5	14.16	31.77
Ethylene oxide	−111.7	10.6	5.17	25.52
Benzene	5.53	80.09	9.95	30.72
Diethyl ether	−116.3	34.5	7.27	26.52
Tetrachloromethane	−23	76.8	3.28	29.82
Mercury	−38.83	356.7	2.29	59.11
Bromine	−7.2	58.8	10.57	29.96

21-11. Use the data in Problem 21–10 to calculate the value of $\Delta_{fus}\overline{S}$ for each substance.

21-12. Why is $\Delta_{vap}\overline{S} > \Delta_{fus}\overline{S}$?

21-13. Show that if $C_P^s(T) \rightarrow T^\alpha$ as $T \rightarrow 0$, where α is a positive constant, then $S(T) \rightarrow 0$ as $T \rightarrow 0$.

21-14. Use the following data to calculate the standard molar entropy of $N_2(g)$ at 298.15 K.

$$C_P^\circ[N_2(s_1)]/R = -0.03165 + (0.05460 \text{ K}^{-1})T + (3.520 \times 10^{-3} \text{ K}^{-2})T^2$$
$$- (2.064 \times 10^{-5} \text{ K}^{-3})T^3$$
$$10 \text{ K} \leq T \leq 35.61 \text{ K}$$

$$C_P^\circ[N_2(s_2)]/R = -0.1696 + (0.2379 \text{ K}^{-1})T - (4.214 \times 10^{-3} \text{ K}^{-2})T^2$$
$$+ (3.036 \times 10^{-5} \text{ K}^{-3})T^3$$
$$35.61 \text{ K} \leq T \leq 63.15 \text{ K}$$

$$C_P^\circ[N_2(l)]/R = -18.44 + (1.053 \text{ K}^{-1})T - (0.0148 \text{ K}^{-2})T^2$$
$$+ (7.064 \times 10^{-5} \text{ K}^{-3})T^3$$
$$63.15 \text{ K} \leq T \leq 77.36 \text{ K}$$

$C_P^\circ[N_2(g)]/R = 3.500$ from 77.36 K $\leq T \leq 1000$ K, $\overline{C}_P(T = 10.0 \text{ K}) = 6.15$ J·K^{-1}·mol^{-1}, $T_{trs} = 35.61$ K, $\Delta_{trs}\overline{H} = 0.2289$ kJ·mol^{-1}, $T_{fus} = 63.15$ K, $\Delta_{fus}\overline{H} = 0.71$ kJ·mol^{-1}, $T_{vap} = 77.36$ K, and $\Delta_{vap}\overline{H} = 5.57$ kJ·mol^{-1}. The correction for nonideality (Problem 22–20) $= 0.02$ J·K^{-1}·mol^{-1}.

21-15. Use the data in Problem 21–14 and $\overline{C}_P[N_2(g)]/R = 3.307 + (6.29 \times 10^{-4} \text{ K}^{-1})T$ for $T \geq 77.36$ K to plot the standard molar entropy of nitrogen as a function of temperature from 0 K to 1000 K.

21-16. The molar heat capacities of solid, liquid, and gaseous chlorine can be expressed as

$$C_P^\circ[Cl_2(s)]/R = -1.545 + (0.1502 \text{ K}^{-1})T - (1.179 \times 10^{-3} \text{ K}^{-2})T^2$$
$$+ (3.441 \times 10^{-6} \text{ K}^{-3})T^3$$
$$15 \text{ K} \leq T \leq 172.12 \text{ K}$$

$$C_P^\circ[Cl_2(l)]/R = 7.689 + (5.582 \times 10^{-3} \text{ K}^{-1})T - (1.954 \times 10^{-5} \text{ K}^{-2})T^2$$
$$172.12 \text{ K} \leq T \leq 239.0 \text{ K}$$

$$C_P^\circ[Cl_2(g)]/R = 3.812 + (1.220 \times 10^{-3} \text{ K}^{-1})T - (4.856 \times 10^{-7} \text{ K}^{-2})T^2$$
$$239.0 \text{ K} \leq T \leq 1000 \text{ K}$$

Use the above molar heat capacities and $T_{fus} = 172.12$ K, $\Delta_{fus}\overline{H} = 6.406$ kJ·mol^{-1}, $T_{vap} = 239.0$ K, $\Delta_{vap}\overline{H} = 20.40$ kJ·mol^{-1}, $\Theta_D = 116$ K and the correction for nonideality $= 0.502$ J·K^{-1}·mol^{-1} to calculate the standard molar entropy of chlorine at 298.15 K. Compare your result with the value given in Table 21.2.

21-17. Use the data in Problem 21–16 to plot the standard molar entropy of chlorine as a function of temperature from 0 K to 1000 K.

21-18. Use the following data to calculate the standard molar entropy of cyclopropane at 298.15 K.

$$C_P^\circ[C_3H_6(s)]/R = -1.921 + (0.1508 \text{ K}^{-1})T - (9.670 \times 10^{-4} \text{ K}^{-2})T^2$$
$$+ (2.694 \times 10^{-6} \text{ K}^{-3})T^3$$
$$15 \text{ K} \leq T \leq 145.5 \text{ K}$$

$$C_P^\circ[C_3H_6(l)]/R = 5.624 + (4.493 \times 10^{-2} \text{ K}^{-1})T - (1.340 \times 10^{-4} \text{ K}^{-2})T^2$$
$$145.5 \text{ K} \leq T \leq 240.3 \text{ K}$$

$$C_P^\circ[C_3H_6(g)]/R = -1.793 + (3.277 \times 10^{-2} \text{ K}^{-1})T - (1.326 \times 10^{-5} \text{ K}^{-2})T^2$$
$$240.3 \text{ K} \leq T \leq 1000 \text{ K}$$

$T_{fus} = 145.5$ K, $T_{vap} = 240.3$ K, $\Delta_{fus}\overline{H} = 5.44$ kJ·mol^{-1}, $\Delta_{vap}\overline{H} = 20.05$ kJ·mol^{-1}, and $\Theta_D = 130$ K. The correction for nonideality $= 0.54$ J·K^{-1}·mol^{-1}.

21-19. Use the data in Problem 21–18 to plot the standard molar entropy of cyclopropane from 0 K to 1000 K.

21-20. The constant-pressure molar heat capacity of N_2O as a function of temperature is tabulated below. Dinitrogen oxide melts at 182.26 K with $\Delta_{fus}\overline{H} = 6.54$ kJ·mol^{-1}, and boils at 184.67 K with $\Delta_{vap}\overline{H} = 16.53$ kJ·mol^{-1} at one bar. Assuming the heat capacity of solid dinitrogen oxide can be described by the Debye theory up to 15 K, calculate the molar entropy of $N_2O(g)$ at its boiling point.

T/K	$\overline{C}_P/J \cdot K^{-1} \cdot mol^{-1}$	T/K	$\overline{C}_P/J \cdot K^{-1} \cdot mol^{-1}$
15.17	2.90	120.29	45.10
19.95	6.19	130.44	47.32
25.81	10.89	141.07	48.91
33.38	16.98	154.71	52.17
42.61	23.13	164.82	54.02
52.02	28.56	174.90	56.99
57.35	30.75	180.75	58.83
68.05	34.18	182.26	Melting point
76.67	36.57	183.55	77.70
87.06	38.87	183.71	77.45
98.34	41.13	184.67	Boiling point
109.12	42.84		

21-21. Methylammonium chloride occurs as three crystalline forms, called β, γ, and α, between 0 K and 298.15 K. The constant-pressure molar heat capacity of methylammonium chloride as a function of temperature is tabulated below. The $\beta \rightarrow \gamma$ transition occurs at 220.4 K with $\Delta_{trs}\overline{H} = 1.779$ kJ·mol^{-1} and the $\gamma \rightarrow \alpha$ transition occurs at 264.5 K with $\Delta_{trs}\overline{H} = 2.818$ kJ·mol^{-1}. Assuming the heat capacity of solid methylammonium chloride can be described by the Debye theory up to 12 K, calculate the molar entropy of methylammonium chloride at 298.15 K.

T/K	$\overline{C}_P/J \cdot K^{-1} \cdot mol^{-1}$	T/K	$\overline{C}_P/J \cdot K^{-1} \cdot mol^{-1}$
12	0.837	180	73.72
15	1.59	200	77.95
20	3.92	210	79.71
30	10.53	220.4	$\beta \rightarrow \gamma$ transition
40	18.28	222	82.01
50	25.92	230	82.84
60	32.76	240	84.27
70	38.95	260	87.03
80	44.35	264.5	$\gamma \rightarrow \alpha$ transition
90	49.08	270	88.16
100	53.18	280	89.20
120	59.50	290	90.16
140	64.81	295	90.63
160	69.45		

21-22. The constant-pressure molar heat capacity of chloroethane as a function of temperature is tabulated below. Chloroethane melts at 134.4 K with $\Delta_{fus}\overline{H} = 4.45$ kJ·mol^{-1}, and boils at 286.2 K with $\Delta_{vap}\overline{H} = 24.65$ kJ·mol^{-1} at one bar. Furthermore, the heat capacity of solid chloroethane can be described by the Debye theory up to 15 K. Use these data to calculate the molar entropy of chloroethane at its boiling point.

T/K	$\overline{C}_P/J \cdot K^{-1} \cdot mol^{-1}$	T/K	$\overline{C}_P/J \cdot K^{-1} \cdot mol^{-1}$
15	5.65	130	84.60
20	11.42	134.4	90.83 (solid)
25	16.53		97.19 (liquid)
30	21.21	140	96.86
35	25.52	150	96.40
40	29.62	160	96.02
50	36.53	180	95.65
60	42.47	200	95.77
70	47.53	220	96.04
80	52.63	240	97.78
90	55.23	260	99.79
100	59.66	280	102.09
110	65.48	286.2	102.13
120	73.55		

21-23. The constant-pressure molar heat capacity of nitromethane as a function of temperature is tabulated below. Nitromethane melts at 244.60 K with $\Delta_{fus}\overline{H} = 9.70 \text{ kJ} \cdot mol^{-1}$, and boils at 374.34 K at one bar with $\Delta_{vap}H = 38.27 \text{ kJ} \cdot mol^{-1}$. Furthermore, the heat capacity of solid nitromethane can be described by the Debye theory up to 15 K. Use these data to calculate the molar entropy of nitromethane at 298.15 K and one bar. The vapor pressure of nitromethane is 36.66 torr at 298.15 K. (Be sure to take into account ΔS for the isothermal compression of nitromethane from its vapor pressure to one bar at 298.15 K).

T/K	$\overline{C}_P/J \cdot K^{-1} \cdot mol^{-1}$	T/K	$\overline{C}_P/J \cdot K^{-1} \cdot mol^{-1}$
15	3.72	200	71.46
20	8.66	220	75.23
30	19.20	240	78.99
40	28.87	244.60	melting point
60	40.84	250	104.43
80	47.99	260	104.64
100	52.80	270	104.93
120	56.74	280	105.31
140	60.46	290	105.69
160	64.06	300	106.06
180	67.74		

21-24. Use the following data to calculate the standard molar entropy of CO(g) at its normal boiling point. Carbon monoxide undergoes a solid-solid phase transition at 61.6 K. Compare your result with the calculated value of 160.3 $J \cdot K^{-1} \cdot mol^{-1}$. Why is there a discrepancy between the calculated value and the experimental value?

$$\overline{C}_P[CO(s_1)]/R = -2.820 + (0.3317 \ K^{-1})T - (6.408 \times 10^{-3} \ K^{-2})T^2$$
$$+ (6.002 \times 10^{-5} \ K^{-3})T^3$$
$$10 \ K \leq T \leq 61.6 \ K$$

$$\overline{C}_P[CO(s_2)]/R = 2.436 + (0.05694 \ K^{-1})T$$
$$61.6 \ K \leq T \leq 68.1 \ K$$

$$\overline{C}_P[CO(l)]/R = 5.967 + (0.0330 \ K^{-1})T - (2.088 \times 10^{-4} \ K^{-2})T^2$$
$$68.1 \ K \leq T \leq 81.6 \ K$$

and $T_{trs}(s_1 \rightarrow s_2) = 61.6$ K, $T_{fus} = 68.1$ K, $T_{vap} = 81.6$ K, $\Delta_{fus}\overline{H} = 0.836$ kJ·mol^{-1}, $\Delta_{trs}\overline{H} = 0.633$ kJ·mol^{-1}, $\Delta_{vap}\overline{H} = 6.04$ kJ·mol^{-1}, $\Theta_D = 79.5$ K, and the correction for nonideality= 0.879 J·K^{-1}·mol^{-1}.

21-25. The molar heat capacities of solid and liquid water can be expressed by

$$\overline{C}_P[H_2O(s)]/R = -0.2985 + (2.896 \times 10^{-2} \ K^{-1})T - (8.6714 \times 10^{-5} \ K^{-2})T^2$$
$$+ (1.703 \times 10^{-7} \ K^{-3})T^3$$
$$10 \ K \leq T \leq 273.15 \ K$$

$$\overline{C}_P[H_2O(l)]/R = 22.447 - (0.11639 \ K^{-1})T + (3.3312 \times 10^{-4} \ K^{-2})T^2$$
$$-(3.1314 \times 10^{-7} \ K^{-3})T^3$$
$$273.15 \ K \leq T \leq 298.15 \ K$$

and $T_{fus} = 273.15$ K, $\Delta_{fus}\overline{H} = 6.007$ kJ·mol^{-1}, $\Delta_{vap}\overline{H}(T = 298.15 \ K) = 43.93$ kJ·mol^{-1}, $\Theta_D = 192$ K, the correction for nonideality = 0.32 J·K^{-1}·mol^{-1}, and the vapor pressure of H_2O at 298.15 K = 23.8 torr. Use these data to calculate the standard molar entropy of $H_2O(g)$ at 298.15 K. You need the vapor pressure of water at 298.15 K because that is the equilibrium pressure of $H_2O(g)$ when it is vaporized at 298.15 K. You must include the value of ΔS that results when you compress the $H_2O(g)$ from 23.8 torr to its standard value of one bar. Your answer should come out to be 185.6 J·K^{-1}·mol^{-1}, which does not agree exactly with the value in Table 21.2. There is a residual entropy associated with ice, which a detailed analysis of the structure of ice gives as $\Delta S_{residual} = R \ln(3/2) = 3.4$ J·K^{-1}·mol^{-1}, which is in good agreement with $\overline{S}_{calc} - \overline{S}_{exp}$.

21-26. Use the data in Problem 21–25 and the empirical expression

$$\overline{C}_P[H_2O(g)]/R = 3.652 + (1.156 \times 10^{-3} \ K^{-1})T - (1.424 \times 10^{-7} \ K^{-2})T^2$$
$$300 \ K \leq T \leq 1000 \ K$$

to plot the standard molar entropy of water from 0 K to 500 K.

21-27. Show for an ideal gas that

$$\overline{S} = R \ln \frac{qe}{N_A} + RT \left(\frac{\partial \ln q}{\partial T} \right)_V$$

21-28. Show that Equations 17.21 and 21.19 are consistent with Equations 21.2 and 21.3.

21-29. Substitute Equation 21.23 into Equation 21.19 and derive the equation (Problem 20–31)

$$\Delta \overline{S} = \overline{C}_V \ln \frac{T_2}{T_1} + R \ln \frac{V_2}{V_1}$$

for one mole of a monatomic ideal gas.

21-30. Use Equation 21.24 and the data in Chapter 18 to calculate the standard molar entropy of $Cl_2(g)$ at 298.15 K. Compare your answer with the experimental value of $223.1 \; J \cdot K^{-1} \cdot mol^{-1}$.

21-31. Use Equation 21.24 and the data in Chapter 18 to calculate the standard molar entropy of $CO(g)$ at its standard boiling point, 81.6 K. Compare your answer with the experimental value of $155.6 \; J \cdot K^{-1} \cdot mol^{-1}$. Why is there a discrepancy of about $5 \; J \cdot K^{-1} \cdot mol^{-1}$?

21-32. Use Equation 21.26 and the data in Chapter 18 to calculate the standard molar entropy of $NH_3(g)$ at 298.15 K. Compare your answer with the experimental value of $192.8 \; J \cdot K^{-1} \cdot mol^{-1}$.

21-33. Use Equation 21.24 and the data in Chapter 18 to calculate the standard molar entropy of $Br_2(g)$ at 298.15 K. Compare your answer with the experimental value of $245.5 \; J \cdot K^{-1} \cdot mol^{-1}$.

21-34. The vibrational and rotational constants for $HF(g)$ within the harmonic oscillator-rigid rotator model are $\tilde{\nu}_0 = 3959 \; cm^{-1}$ and $\tilde{B}_0 = 20.56 \; cm^{-1}$. Calculate the standard molar entropy of $HF(g)$ at 298.15 K. How does this value compare with that in Table 21.3?

21-35. Calculate the standard molar entropy of $H_2(g)$ and $D_2(g)$ at 298.15 K given that the bond length of both diatomic molecules is 74.16 pm and the vibrational temperatures of $H_2(g)$ and $D_2(g)$ are 6215 K and 4394 K, respectively. Calculate the standard molar entropy of $HD(g)$ at 298.15 K ($R_e = 74.13$ pm and $\Theta_{vib} = 5496$ K).

21-36. Calculate the standard molar entropy of $HCN(g)$ at 1000 K given that $I = 1.8816 \times 10^{-46} \; kg \cdot m^2$, $\tilde{\nu}_1 = 2096.70 \; cm^{-1}$, $\tilde{\nu}_2 = 713.46 \; cm^{-1}$, and $\tilde{\nu}_3 = 3311.47 \; cm^{-1}$. Recall that $HCN(g)$ is a linear triatomic molecule and therefore the bending mode, ν_2, is doubly degenerate.

21-37. Given that $\tilde{\nu}_1 = 1321.3 \; cm^{-1}$, $\tilde{\nu}_2 = 750.8 \; cm^{-1}$, $\tilde{\nu}_3 = 1620.3 \; cm^{-1}$, $\tilde{A}_0 = 7.9971 \; cm^{-1}$, $\tilde{B}_0 = 0.4339 \; cm^{-1}$, and $\tilde{C}_0 = 0.4103 \; cm^{-1}$, calculate the standard molar entropy of $NO_2(g)$ at 298.15 K. (Note that $NO_2(g)$ is a bent triatomic molecule.) How does your value compare with that in Table 21.2?

21-38. In Problem 21–48, you are asked to calculate the value of $\Delta_r S^\circ$ at 298.15 K using the data in Table 21.2 for the reaction described by

$$2 \, CO(g) + O_2(g) \longrightarrow 2 \, CO_2(g)$$

Use the data in Table 18.2 to calculate the standard molar entropy of each of the substances in this reaction [see Example 21–5 for the calculation of the standard molar entropy of $CO_2(g)$]. Then use these results to calculate the standard entropy change for the above reaction. How does your answer compare with what you obtained in Problem 21–48?

21-39. Calculate the value of $\Delta_r S^\circ$ for the reaction described by

$$H_2(g) + \tfrac{1}{2} O_2(g) \longrightarrow H_2O(g)$$

at 500 K using the data in Tables 18.2 and 18.4.

21-40. In each case below, predict which molecule of the pair has the greater molar entropy under the same conditions (assume gaseous species).

a. CO   CO$_2$

b. CH$_3$CH$_2$CH$_3$
$$\begin{array}{c} H_2C \!-\! CH_2 \\ \diagdown \diagup \\ CH_2 \end{array}$$

c. CH$_3$CH$_2$CH$_2$CH$_2$CH$_3$
$$H_3C \!-\! \overset{\displaystyle CH_3}{\underset{\displaystyle CH_3}{\overset{|}{\underset{|}{C}}}} \!-\! CH_3$$

21-41. In each case below, predict which molecule of the pair has the greater molar entropy under the same conditions (assume gaseous species).

a. H$_2$O   D$_2$O

b. CH$_3$CH$_2$OH
$$\begin{array}{c} H_2C \!-\! CH_2 \\ \diagdown \diagup \\ O \end{array}$$

c. CH$_3$CH$_2$CH$_2$CH$_2$NH$_2$
$$\begin{array}{c} \overset{\displaystyle H}{\overset{|}{N}} \\ H_2C \diagup \diagdown CH_2 \\ | \qquad | \\ H_2C \!-\!-\! CH_2 \end{array}$$

21-42. Arrange the following reactions according to increasing values of $\Delta_r S^\circ$ (do not consult any references).

a. $S(s) + O_2(g) \longrightarrow SO_2(g)$
b. $H_2(g) + O_2(g) \longrightarrow H_2O_2(l)$
c. $CO(g) + 3\,H_2(g) \longrightarrow CH_4(g) + H_2O(l)$
d. $C(s) + H_2O(g) \longrightarrow CO(g) + H_2(g)$

21-43. Arrange the following reactions according to increasing values of $\Delta_r S^\circ$ (do not consult any references).

a. $2\,H_2(g) + O_2(g) \longrightarrow 2\,H_2O(l)$
b. $NH_3(g) + HCl(g) \longrightarrow NH_4Cl(s)$
c. $K(s) + O_2(g) \longrightarrow KO_2(s)$
d. $N_2(g) + 3\,H_2(g) \longrightarrow 2\,NH_3(g)$

21-44. In Problem 21-40, you are asked to predict which molecule, CO(g) or CO$_2$(g), has the greater molar entropy. Use the data in Tables 18.2 and 18.4 to calculate the standard molar entropy of CO(g) and CO$_2$(g) at 298.15 K. Does this calculation confirm your intuition? Which degree of freedom makes the dominant contribution to the molar entropy of CO? Of CO$_2$?

21-45. Table 21.2 gives $\overline{S}\,^{\circ}[CH_3OH(l)] = 126.8$ J·K^{-1}·mol^{-1} at 298.15 K. Given that $T_{vap} = 337.7$ K, $\Delta_{vap}\overline{H}(T_b) = 36.5$kJ·mol^{-1}, $\overline{C}_P[CH_3OH(l)] = 81.12$J·K^{-1}·mol^{-1}, and $\overline{C}_P[CH_3OH(g)] = 43.8$ J·K^{-1}·mol^{-1}, calculate the value of $\overline{S}\,^{\circ}[CH_3OH(g)]$ at 298.15 K and compare your answer with the experimental value of 239.8 J·K^{-1}·mol^{-1}.

21-46. Given the following data, $T_{fus} = 373.15$ K, $\Delta\overline{H}_{vap}(T_{vap}) = 40.65$ kJ·mol^{-1}, $\overline{C}_P[H_2O(l)] = 75.3$ J·K^{-1}·mol^{-1}, and $\overline{C}_P[H_2O(g)] = 33.8$ J·K^{-1}·mol^{-1}, show that the values of $\overline{S}\,^{\circ}[H_2O(l)]$ and $\overline{S}\,^{\circ}[H_2O(g)]$ in Table 21.2 are consistent.

21-47. Use the data in Table 21.2 to calculate the value of $\Delta_r S°$ for the following reactions at 25°C and one bar.

a. $C(s, graphite) + O_2(g) \longrightarrow CO_2(g)$
b. $CH_4(g) + 2\,O_2(g) \longrightarrow CO_2(g) + 2\,H_2O(l)$
c. $C_2H_2(g) + H_2(g) \longrightarrow C_2H_4(g)$

21-48. Use the data in Table 21.2 to calculate the value of $\Delta_r S°$ for the following reactions at 25°C and one bar.

a. $CO(g) + 2\,H_2(g) \longrightarrow CH_3OH(l)$
b. $C(s, graphite) + H_2O(l) \longrightarrow CO(g) + H_2(g)$
c. $2\,CO(g) + O_2(g) \longrightarrow 2\,CO_2(g)$

Hermann von Helmholtz was born in Potsdam, Germany, on August 31, 1821, and died in 1894. Although he wanted to study physics, his family could not afford to send him to the University, so he studied medicine in Berlin because he could obtain state financial aid. He was, however, required to repay his stipend by service as a surgeon in the army for eight years. He was later appointed professor at the University of Königsberg, and he also held positions at Bonn, Heidelberg, and Berlin. In 1885, in recognition of his position as the foremost scientist in Germany, he was appointed president of the newly founded Physico-Technical Institute in Berlin, an institution devoted to purely scientific research. Helmholtz was one of the greatest scientists of the 19th century, making important discoveries in physiology, optics, acoustics, electromagnetic theory, and thermodynamics. His work in physiology showed that physiological phenomena are based upon the laws of physics and not on some vague "vital force." In thermodynamics, he derived the equation now known as the Gibbs-Helmholtz equation, which we will discuss in this chapter. Helmholtz was always generous with his students and other scientists, but unfortunately he was a barely intelligible lecturer, even to the likes of Planck, who was a student in several of his classes. Helmholtz's great influence in German science was recognized by the Kaiser, who bestowed him with the title "von."

CHAPTER 22

Helmholtz and Gibbs Energies

The criterion that $dS > 0$ for a spontaneous process and that $dS = 0$ for a reversible process applies only to an isolated system. Consequently, in the various processes we discussed in Chapter 20, we had to consider the entropy change of both the system *and* its surroundings to determine the sign of ΔS_{total} and establish whether a process is spontaneous or not. Although of great fundamental and theoretical importance, the criterion that $dS \geq 0$ in an isolated system is too restrictive for practical applications. In this chapter, we will introduce two new state functions that can be used to determine the direction of a spontaneous process in systems that are not isolated.

22–1. The Sign of the Helmholtz Energy Change Determines the Direction of a Spontaneous Process in a System at Constant Volume and Temperature

Let's consider a system with its volume and temperature held constant. The criterion that $dS \geq 0$ does not apply to a system at constant temperature and volume because the system is not isolated; a system must be in thermal contact with a thermal reservoir to be at constant temperature. If the criterion $dS \geq 0$ does not apply, then what is the criterion for a spontaneous process that we can use for a system at constant temperature and volume? Let's start with the expression of the First Law of Thermodynamics, Equation 19.9,

$$dU = \delta q + \delta w \tag{22.1}$$

Because $\delta w = -P_{ext}dV$ and $dV = 0$ (constant volume), then $\delta w = 0$. If we substitute Equation 20.3, $dS \geq \delta q / T$, and $\delta w = 0$ into Equation 22.1, we obtain

$$dU \leq TdS \qquad \text{(constant } V) \tag{22.2}$$

881

The equality holds for a reversible process and the inequality for an irreversible process. Note that if the system is isolated, then $dU = 0$ and we have $dS \geq 0$ as in Chapter 20. We can write Equation 22.2 as

$$dU - TdS \leq 0$$

If T and V are held constant, we can write this expression as

$$d(U - TS) \leq 0 \qquad \text{(constant } T \text{ and } V) \tag{22.3}$$

Equation 22.3 prompts us to define a new thermodynamic state function by

$$A = U - TS \tag{22.4}$$

so Equation 22.3 becomes

$$dA \leq 0 \qquad \text{(constant } T \text{ and } V) \tag{22.5}$$

The quantity A is called the *Helmholtz energy*. In a system held at constant T and V, the Helmholtz energy will decrease until all the possible spontaneous processes have occurred, at which time the system will be in equilibrium and A will be a minimum. At equilibrium, $dA = 0$ (see Figure 22.1). Note that Equation 22.5 is the analog of the criterion that $dS \geq 0$ to occur in an isolated system (cf. Figures 20.5 and 22.1).

For an isothermal change from one state to another, Equation 22.4 gives

$$\Delta A = \Delta U - T\Delta S \tag{22.6}$$

Using Equation 22.5, we see that

$$\Delta A = \Delta U - T\Delta S \leq 0 \qquad \text{(constant } T \text{ and } V) \tag{22.7}$$

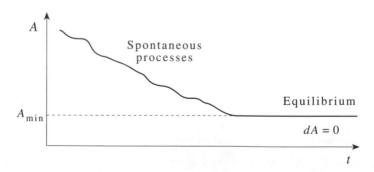

FIGURE 22.1
The Helmholtz energy, A, of a system will decrease during any spontaneous processes that occur at constant T and V and will achieve its minimum value at equilibrium.

where the equality holds for a reversible change and the inequality holds for an irreversible, spontaneous change. A process for which $\Delta A > 0$ cannot take place spontaneously in a system at constant T and V. Consequently, something, such as work, must be done on the system to effect the change.

Notice that if $\Delta U < 0$ and $\Delta S > 0$ in Equation 22.6, then both the energy change and the entropy change contribute to ΔA being negative. But if they have the same sign, some sort of compromise must be reached and the value of ΔA is a quantitative measure of whether a process is spontaneous or not. The Helmholtz energy represents this compromise between the tendency of a system to decrease its energy and to increase its entropy. Because ΔS is multiplied by T, we see that the sign of ΔU is more important at low temperatures but the sign of ΔS is more important at high temperatures.

We can apply the criterion that $\Delta A < 0$ for an irreversible (spontaneous) process in a system at constant T and V to the mixing of two ideal gases, which we discussed in Section 20–6. For that process, $\Delta U = 0$ and $\Delta \overline{S} = -y_1 R \ln y_1 - y_2 R \ln y_2$. Therefore, for the mixing of two ideal gases at constant T and V, $\Delta \overline{A} = RT(y_1 \ln y_1 + y_2 \ln y_2)$, which is a negative quantity because y_1 and y_2 are less than one. Thus, we see once again that the isothermal mixing of two ideal gases is a spontaneous process.

In addition to serving as our criterion for spontaneity in a system at constant temperature and volume, the Helmholtz energy has an important physical interpretation. Let's start with Equation 22.6

$$\Delta A = \Delta U - T \Delta S \tag{22.8}$$

for a spontaneous (irreversible) process, so that $\Delta A < 0$. In this process, the initial and final states are well-defined equilibrium states, and there is no fundamental reason we have to follow an irreversible path to get from one state to the other. In fact, we can gain some considerable insight into the process if we look at any reversible path connecting these two states. For a reversible path we can replace ΔS by q_{rev}/T, giving

$$\Delta A = \Delta U - q_{rev}$$

But according to the first law, $\Delta U - q_{rev}$ is equal to w_{rev}, so we have

$$\Delta A = w_{rev} \qquad \text{(isothermal, reversible)} \tag{22.9}$$

If $\Delta A < 0$, the process will occur spontaneously and w_{rev} represents the work that can be done *by* the system if this change is carried out reversibly. This quantity is the maximum work that could be obtained. If any irreversible process such as friction occurs, then the quantity of work that can be obtained will be less than w_{rev}. If $\Delta A > 0$, the process will not occur spontaneously and w_{rev} represents the work that must be done *on* the system to produce the change in a reversible manner. If there is any irreversibility in the process, the quantity of work required will be even greater than w_{rev}.

22–2. The Gibbs Energy Determines the Direction of a Spontaneous Process for a System at Constant Pressure and Temperature

Most reactions occur at constant pressure rather than at constant volume because they are open to the atmosphere. Let's see what the criterion of spontaneity is for a system at constant temperature and pressure. Once again, we start with Equation 22.1, but now we substitute $dS \geq \delta q/T$ and $\delta w = -PdV$ to obtain

$$dU \leq TdS - PdV$$

or

$$dU - TdS + PdV \leq 0$$

Because both T and P are constant, we can write this expression as

$$d(U - TS + PV) \leq 0 \qquad \text{(constant } T \text{ and } P) \tag{22.10}$$

We now define a new thermodynamic state function by

$$G = U - TS + PV \tag{22.11}$$

so Equation 22.5 becomes

$$dG \leq 0 \qquad \text{(constant } T \text{ and } P) \tag{22.12}$$

Note that Equation 22.11 is the analog of Equation 22.4.

The quantity G is called the *Gibbs energy*. In a system at constant T and P, the Gibbs energy will decrease as the result of any spontaneous processes until the system reaches equilibrium, where $dG = 0$. A plot of G versus time for a system at constant T and P would be similar to the plot of A versus time for a system at constant T and V (Figure 22.1). Thus, we see that the Gibbs energy, G, is the analog of the Helmoltz energy, A, for a process that takes place at constant temperature and pressure.

Equation 22.11 can also be written as

$$G = H - TS \tag{22.13}$$

where $H = U + PV$ is the enthalpy. Note that the enthalpy plays the same role in a constant T and P process that the energy U plays in a constant T and V process (cf. Equation 22.4). Note also that G can be written as

$$G = A + PV \tag{22.14}$$

thus relating the Gibbs energy and the Helmholtz energy in the same manner that H and U are related.

The analog of Equation 22.7 is

$$\Delta G = \Delta H - T\Delta S \leq 0 \qquad \text{(constant } T \text{ and } P\text{)} \qquad (22.15)$$

The equality holds for a reversible process, whereas the inequality holds for an irreversible (spontaneous) process. If $\Delta H < 0$ and $\Delta S > 0$ in Equation 22.15, both terms in Equation 22.15 contribute to ΔG being negative. But if ΔH and ΔS have the same sign, then $\Delta G = \Delta H - T\Delta S$ represents the compromise between the tendency of a system to decrease its enthalpy and to increase its entropy in a constant T and P process. Because of the factor of T multiplying ΔS in Equation 22.15, the ΔH term can dominate at low temperatures, whereas the $T\Delta S$ term can dominate at high temperatures. Of course if $\Delta H > 0$ and $\Delta S < 0$, then $\Delta G > 0$ at all temperatures and the process is never spontaneous.

An example of a reaction favored by its value of $\Delta_r H$ but disfavored by its value of $\Delta_r S$ is

$$NH_3(g) + HCl(g) \longrightarrow NH_4Cl(s)$$

The value of $\Delta_r H$ for this reaction at 298.15 K and one bar is -176.2 kJ, whereas the corresponding value of $\Delta_r S$ is -0.285 kJ$\cdot$K^{-1}, giving $\Delta_r G = \Delta_r H - T\Delta_r S = -91.21$ kJ at 298.15 K. Therefore, this reaction proceeds spontaneously at 298.15 K and one bar.

A process for which the sign of ΔG changes with a small change in temperature is the vaporization of a liquid at its normal boiling point. We represent this process by

$$H_2O(l) \longrightarrow H_2O(g)$$

The expression for the molar Gibbs energy of vaporization, $\Delta_{vap}\overline{G}$, for this process is

$$\Delta_{vap}\overline{G} = \overline{G}[H_2O(g)] - \overline{G}[H_2O(l)]$$
$$= \Delta_{vap}\overline{H} - T\Delta_{vap}\overline{S}$$

The molar enthalpy of vaporization of water at one atm near 100°C, $\Delta_{vap}\overline{H}$, is equal to 40.65 kJ$\cdot$mol^{-1} and $\Delta_{vap}\overline{S} = 108.9$ J$\cdot$K$^{-1}\cdot$mol^{-1}. Thus, we can write $\Delta_{vap}\overline{G}$ as

$$\Delta_{vap}\overline{G} = 40.65 \text{ kJ}\cdot\text{mol}^{-1} - T(108.9 \text{ J}\cdot\text{K}^{-1}\cdot\text{mol}^{-1})$$

At $T = 373.15$ K,

$$\Delta_{vap}\overline{G} = 40.65 \text{ kJ}\cdot\text{mol}^{-1} - (373.15 \text{ K})(108.9 \text{ J}\cdot\text{K}^{-1}\cdot\text{mol}^{-1})$$
$$= 40.65 \text{ kJ}\cdot\text{mol}^{-1} - 40.65 \text{ kJ}\cdot\text{mol}^{-1} = 0$$

The fact that $\Delta_{vap}\overline{G} = 0$ means that liquid and vapor water are in equilibrium with each other at one atm and 373.15 K. The molar Gibbs energy of liquid water is equal to the molar Gibbs energy of water vapor at 373.15 K and one atm. The transfer of one mole of liquid water to water vapor under these conditions is a reversible process, and so $\Delta_{vap}\overline{G} = 0$.

Now let's consider a temperature less than the normal boiling point, say 363.15 K. At this temperature, $\Delta_{vap}\overline{G} = +1.10$ kJ·mol^{-1}. The positive sign means that the formation of one mole of water vapor at one atm from one mole of liquid water at one atm and 363.15 K is not a spontaneous process. If the temperature is above the normal boiling point, however, say 383.15 K, then $\Delta_{vap}\overline{G} = -1.08$ kJ·mol^{-1}. The negative sign means that the formation of one mole of water vapor from one mole of liquid water at one atm and 383.15 K is a spontaneous process.

EXAMPLE 22–1

The molar enthalpy of fusion of ice at 273.15 K and one atm is $\Delta_{fus}\overline{H} = 6.01$ kJ·mol^{-1}, and the molar entropy of fusion under the same conditions is $\Delta_{fus}\overline{S} = 22.0$ J·K^{-1}·mol^{-1}. Show that $\Delta_{fus}\overline{G} = 0$ at 273.15 K and one atm, that $\Delta_{fus}\overline{G} < 0$ when the temperature is greater than 273.15 K, and that $\Delta_{fus}\overline{G} > 0$ when the temperature is less than 273.15 K.

SOLUTION: Assuming that $\Delta_{fus}\overline{H}$ and $\Delta_{fus}\overline{S}$ do not vary appreciably around 273.15 K, we can write

$$\Delta_{fus}\overline{G} = 6010 \text{ J·mol}^{-1} - T(22.0 \text{ J·K}^{-1}\text{·mol}^{-1})$$

If $T = 273.15$ K, then $\Delta_{fus}\overline{G} = 0$, indicating that ice and liquid water are in equilibrium with each other at 273.15 K and one atm. If $T < 273.15$ K, then $\Delta_{fus}\overline{G} > 0$, indicating that ice will not spontaneously melt under these conditions. If $T > 273.15$ K, then $\Delta_{fus}\overline{G} < 0$, indicating that ice will melt under these conditions.

The value of ΔG can be related to the maximum work that can be obtained from a process carried out at constant T and P. To show this, we start by differentiating $G = U - TS + PV$ to get

$$dG = dU - TdS - SdT + PdV + VdP$$

and substitute $dU = TdS + \delta w_{rev}$ for dU to get

$$dG = -SdT + VdP + \delta w_{rev} + PdV$$

Because the reversible pressure-volume (P-V) work is $-PdV$, the quantity $\delta w_{rev} + PdV$ is the reversible work other than P-V work (such as electrical work). Therefore, we can write dG as

$$dG = -SdT + VdP + \delta w_{nonPV}$$

where δw_{nonPV} represents the total work exclusive of P-V work. For a reversible process taking place at constant T and P, $dG = \delta w_{nonPV}$, or

$$\Delta G = w_{nonPV} \qquad \text{(reversible, constant } T \text{ and } P) \qquad (22.16)$$

If $\Delta G < 0$, the process will occur spontaneously, and w_{nonPV} is the work exclusive of P-V work that can be done by the system if the change is carried out reversibly. This is the maximum work that can be obtained from the process. If any irreversibility occurs in the process, the quantity of work obtained will be less than the maximum. If $\Delta G > 0$, the process will not occur spontaneously and w_{nonPV} is the minimum work, exclusive of P-V work, that must be done on the system to make the process occur. For example, it is known experimentally that ΔG for the formation of one mole of $H_2O(l)$ at 298.15 K and one bar from $H_2(g)$ and $O_2(g)$ at 298.15 K and one bar is -237.1 kJ·mol^{-1}. Thus, a maximum of 237.1 kJ·mol^{-1} of useful work (that is, work exclusive of P-V work) can be obtained from the spontaneous reaction

$$H_2(g, \text{ 1 bar, 298.15 K}) + \tfrac{1}{2}O_2(g, \text{ 1 bar, 298.15 K}) \longrightarrow H_2O(l, \text{ 1 bar, 298.15 K})$$

Conversely, it would require at least 237.1 kJ·mol^{-1} of energy to drive the (nonspontaneous) reaction

$$H_2O(l, \text{ 1 bar, 298.15 K}) \longrightarrow H_2(g, \text{ 1 bar, 298.15 K}) + \tfrac{1}{2}O_2(g, \text{ 1 bar, 298.15 K})$$

EXAMPLE 22–2
The value of $\Delta\overline{G}$ for the decomposition of one mole of $H_2O(l)$ to $H_2(g)$ and $O_2(g)$ at one bar and 298.15 K is $+237.1$ kJ·mol^{-1}. Calculate the minimum voltage required to decompose one mole of $H_2O(l)$ to $H_2(g)$ and $O_2(g)$ at one bar and 298.15 K by electrolysis.

SOLUTION: Electrolysis represents the non-P-V work required to carry out the decomposition, so we write

$$\Delta\overline{G} = w_{nonPV} = +237.1 \text{ kJ·mol}^{-1}$$

You might remember from physics that electrical work is given by charge × voltage. The charge involved in electrolyzing one mole of $H_2O(l)$ can be determined from the chemical equation of the reaction

$$H_2O(l) \longrightarrow H_2(g) + \tfrac{1}{2}O_2(g)$$

The oxidation state of hydrogen goes from $+1$ to 0 and that of oxygen goes from -2 to 0. Thus two electrons are transferred per $H_2O(l)$ molecule, or two times the Avogadro constant of electrons per mole. The total charge of two moles of electrons is

$$\text{total charge} = (1.602 \times 10^{-19} \text{ C·e}^{-1})(12.044 \times 10^{23} \text{ e}) = 1.929 \times 10^5 \text{ C}$$

The minimum voltage, $\mathcal{E}$, required to decompose one mole is given by

$$\mathcal{E} = \frac{\Delta\overline{G}}{1.929 \times 10^5 \text{ C}} = \frac{237.1 \times 10^3 \text{ J·mol}^{-1}}{1.929 \times 10^5 \text{ C}} = 1.23 \text{ volts}$$

where we have used the fact that one joule is a coulomb times a volt ($1J = 1CV$).

22–3. Maxwell Relations Provide Several Useful Thermodynamic Formulas

A number of the thermodynamic functions we have defined cannot be measured directly. Consequently, we need to be able to express these quantities in terms of others that can be experimentally determined. To do so, we start with the definitions of A and G, Equations 22.4 and 22.11. Differentiate Equation 22.4 to obtain

$$dA = dU - TdS - SdT$$

For a reversible process, $dU = TdS - PdV$, so

$$dA = -PdV - SdT \qquad (22.17)$$

By comparing Equation 22.17 with the formal total derivative of $A = A(V, T)$,

$$dA = \left(\frac{\partial A}{\partial V}\right)_T dV + \left(\frac{\partial A}{\partial T}\right)_V dT$$

we see that

$$\left(\frac{\partial A}{\partial V}\right)_T = -P \quad \text{and} \quad \left(\frac{\partial A}{\partial T}\right)_V = -S \qquad (22.18a, b)$$

Now if we use the fact that the cross derivatives of A are equal (MathChapter H),

$$\left(\frac{\partial^2 A}{\partial T \partial V}\right) = \left(\frac{\partial^2 A}{\partial V \partial T}\right)$$

we find that

$$\left(\frac{\partial P}{\partial T}\right)_V = \left(\frac{\partial S}{\partial V}\right)_T \qquad (22.19)$$

Equation 22.19, which is obtained by equating the second cross partial derivatives of A, is called a *Maxwell relation*. There are many useful Maxwell relations involving various thermodynamic quantities. Equation 22.19 is particularly useful because it allows us to determine how the entropy of a substance changes with volume if we know its equation of state. Integrating Equation 22.19 at constant T, we have

$$\Delta S = \int_{V_1}^{V_2} \left(\frac{\partial P}{\partial T}\right)_V dV \qquad \text{(constant } T\text{)} \qquad (22.20)$$

We have applied the condition of constant T to Equation 22.20 because we have integrated $(\partial S/\partial V)_T$; in other words, T is held constant in the derivative, so T must be held constant when we integrate.

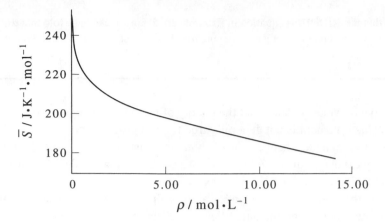

FIGURE 22.2
The molar entropy of ethane at 400 K plotted against density ($\rho = 1/\overline{V}$). The value of $\overline{S}^{id}$ at 400 K is 246.45 J·mol^{-1}·K^{-1}.

Equation 22.20 allows us to determine the entropy of a substance as a function of volume or density (recall that $\rho = 1/\overline{V}$) from P-V-T data. If we let V_1 in Equation 22.20 be very large, where the gas is sure to behave ideally, then Equation 22.20 becomes

$$S(T, V) - S^{id} = \int_{V^{id}}^{V} \left(\frac{\partial P}{\partial T}\right)_V dV'$$

Figure 22.2 plots the molar entropy of ethane at 400 K versus density. (Problem 22–3 involves calculating the molar entropy as a function of density using the van der Waals equation and the Redlich-Kwong equation.)

We can also use Equation 22.20 to derive an equation we derived earlier in Section 20–3 by another method. For an ideal gas, $(\partial P/\partial T)_V = nR/V$, so

$$\Delta S = nR \int_{V_1}^{V_2} \frac{dV}{V} = nR \ln \frac{V_2}{V_1} \qquad \text{(isothermal process)} \qquad (22.21)$$

EXAMPLE 22–3
Calculate $\Delta \overline{S}$ for an isothermal expansion from $\overline{V}_1$ to $\overline{V}_2$ for a gas that obeys the equation of state

$$P(\overline{V} - b) = RT$$

SOLUTION: We use Equation 22.20 to obtain

$$\Delta \overline{S} = \int_{\overline{V}_1}^{\overline{V}_2} \left(\frac{\partial P}{\partial T}\right)_{\overline{V}} d\overline{V} = R \int_{\overline{V}_1}^{\overline{V}_2} \frac{d\overline{V}}{\overline{V} - b} = R \ln \frac{\overline{V}_2 - b}{\overline{V}_1 - b}$$

Note that we derived this equation in Example 20–2, but we had to be told that $dU = 0$ in an isothermal process for a gas obeying the above equation of state. We did not need this information to derive our result here.

We have previously stated that the energy of an ideal gas depends only upon temperature. This statement is not generally true for real gases. Suppose we want to know how the energy of a gas changes with volume at constant temperature. Unfortunately, this quantity cannot be measured directly. We can use Equation 22.19, however, to derive a practical equation for $(\partial U/\partial V)_T$; in other words, we can derive an equation that tells us how the energy of a substance varies with its volume at constant temperature in terms of readily measurable quantities. We differentiate Equation 22.4 with respect to V at constant temperature to obtain

$$\left(\frac{\partial A}{\partial V}\right)_T = \left(\frac{\partial U}{\partial V}\right)_T - T\left(\frac{\partial S}{\partial V}\right)_T$$

Substituting Equation 22.18a for $(\partial A/\partial V)_T$ and Equation 22.19 for $(\partial S/\partial V)_T$ gives

$$\left(\frac{\partial U}{\partial V}\right)_T = -P + T\left(\frac{\partial P}{\partial T}\right)_V \tag{22.22}$$

Equation 22.22 gives $(\partial U/\partial V)_T$ in terms of P-V-T data. Equations like Equation 22.22 that relate thermodynamic functions to functions of P, V, and T are sometimes called thermodynamic equations of state.

We can integrate Equation 22.22 with respect to V to determine U relative to the ideal gas value,

$$U(T, V) - U^{\mathrm{id}} = \int_{V^{\mathrm{id}}}^{V'} \left[T\left(\frac{\partial P}{\partial T}\right)_V - P\right] dV' \qquad \text{(constant } T\text{)}$$

where V^{id} is a large volume, where the gas is sure to behave ideally. This equation along with the P-V-T data gives us U as a function of pressure. Figure 22.3 shows $\overline{U}$ plotted against pressure for ethane at 400 K. Problem 22–4 involves calculating $\overline{U}$ as a function of volume for the van der Waals equation and the Redlich-Kwong equation. We can also use Equation 22.22 to show that the energy of an ideal gas is independent of the volume at constant temperature. For an ideal gas, $(\partial P/\partial T)_V = nR/V$, so

$$\left(\frac{\partial U}{\partial V}\right)_T = -P + T\frac{nR}{V} = -P + P = 0$$

which proves that the energy of an ideal gas depends only upon temperature.

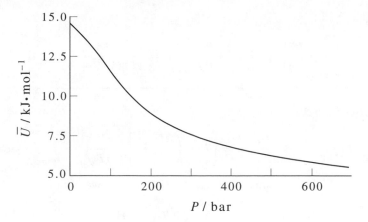

FIGURE 22.3
The molar energy $\overline{U}$ plotted against pressure for ethane at 400 K. The value of $\overline{U}^{\text{id}}$ is equal to 14.55 kJ·mol^{-1}.

EXAMPLE 22–4
In Example 20–2, we stated we would prove later that the energy of a gas that obeys the equation of state

$$P(\overline{V} - b) = RT$$

is independent of the volume. Use Equation 22.22 to prove this.

SOLUTION: For $P(\overline{V} - b) = RT$,

$$\left(\frac{\partial P}{\partial T}\right)_{\overline{V}} = \frac{R}{\overline{V} - b}$$

and so

$$\left(\frac{\partial \overline{U}}{\partial \overline{V}}\right)_T = -P + \frac{RT}{\overline{V} - b} = -P + P = 0$$

EXAMPLE 22–5
Evaluate $(\partial \overline{U}/\partial \overline{V})_T$ for one mole of a Redlich-Kwong gas.

SOLUTION: Recall that the Redlich-Kwong equation (Equation 16.7) is

$$P = \frac{RT}{\overline{V} - B} - \frac{A}{T^{1/2}\overline{V}(\overline{V} + B)}$$

Therefore,

$$\left(\frac{\partial P}{\partial T}\right)_{\overline{V}} = \frac{R}{\overline{V} - B} + \frac{A}{2T^{3/2}\overline{V}(\overline{V} + B)}$$

and so

$$\left(\frac{\partial \overline{U}}{\partial \overline{V}}\right)_T = T\left(\frac{\partial P}{\partial T}\right)_{\overline{V}} - P = \frac{3A}{2T^{1/2}\overline{V}(\overline{V} + B)}$$

We derived the equation

$$C_P - C_V = \left[P + \left(\frac{\partial U}{\partial V}\right)_T\right]\left(\frac{\partial V}{\partial T}\right)_P$$

in Problem 19–27. Using Equation 22.22 for $(\partial U/\partial V)_T$, we obtain

$$C_P - C_V = T\left(\frac{\partial P}{\partial T}\right)_V \left(\frac{\partial V}{\partial T}\right)_P \tag{22.23}$$

For an ideal gas $(\partial P/\partial T)_V = nR/V$ and $(\partial V/\partial T)_P = nR/P$, and so $C_P - C_V = nR$, in agreement with Equation 19.43.

An alternative equation for $C_P - C_V$ that is more convenient than Equation 22.23 for solids and liquids is (Problem 22–11)

$$C_P - C_V = -T\left(\frac{\partial V}{\partial T}\right)_P^2 \left(\frac{\partial P}{\partial V}\right)_T \tag{22.24}$$

Each of the partial derivatives here can be expressed in terms of familiar tabulated physical quantities. The isothermal compressibility of a substance is defined as

$$\kappa = -\frac{1}{V}\left(\frac{\partial V}{\partial P}\right)_T \tag{22.25}$$

and the coefficient of thermal expansion is defined as

$$\alpha = \frac{1}{V}\left(\frac{\partial V}{\partial T}\right)_P \tag{22.26}$$

Using these definitions, Equation 22.24 becomes

$$C_P - C_V = \frac{\alpha^2 TV}{\kappa} \tag{22.27}$$

EXAMPLE 22–6

The coefficient of thermal expansion, α, of copper at 298 K is 5.00×10^{-5} K^{-1}, and its isothermal compressibility, κ, is 7.85×10^{-7} atm^{-1}. Given that the density of copper is 8.92 $g \cdot cm^{-3}$ at 298 K, calculate the value of $\overline{C}_P - \overline{C}_V$ for copper.

SOLUTION: For copper, the molar volume, $\overline{V}$, is given by

$$\overline{V} = \frac{63.54 \text{ g} \cdot \text{mol}^{-1}}{8.92 \text{ g} \cdot \text{cm}^{-3}}$$

$$= 7.12 \text{ cm}^3 \cdot \text{mol}^{-1} = 7.12 \times 10^{-3} \text{ L} \cdot \text{mol}^{-1}$$

and

$$\overline{C}_P - \overline{C}_V = \frac{(5.00 \times 10^{-5} \text{ K}^{-1})^2 (298 \text{ K})(7.12 \times 10^{-3} \text{ L·mol}^{-1})}{7.85 \times 10^{-7} \text{ atm}^{-1}}$$

$$= 6.76 \times 10^{-3} \text{ L·atm·K}^{-1}\text{·mol}^{-1}$$

$$= 0.684 \text{ J·K}^{-1}\text{·mol}^{-1}$$

The experimental value of $\overline{C}_P$ is 24.43 J·K^{-1}·mol^{-1}. Note that $\overline{C}_P - \overline{C}_V$ is small compared with $\overline{C}_P$ (or $\overline{C}_V$) and is also much smaller for solids than for gases, as you might expect.

22–4. The Enthalpy of an Ideal Gas Is Independent of Pressure

Equation 22.18a can be used directly to give the volume dependence of the Helmholtz energy. By integrating at constant temperature, we have

$$\Delta A = -\int_{V_1}^{V_2} P \, dV \qquad \text{(constant } T) \tag{22.28}$$

For the case of an ideal gas, we have

$$\Delta A = -nRT \int_{V_1}^{V_2} \frac{dV}{V} = -nRT \ln \frac{V_2}{V_1} \qquad \text{(constant } T) \tag{22.29}$$

Notice that this result is $-T$ times Equation 22.21 for ΔS. This result must be so because $\Delta U = 0$ for an ideal gas at constant T, so $\Delta A = -T \Delta S$.

If we differentiate Equation 22.11, $G = U - TS + PV$, and substitute $dU = TdS - PdV$, we get

$$dG = -SdT + VdP \tag{22.30}$$

By comparing Equation 22.30 with

$$dG = \left(\frac{\partial G}{\partial T} \right)_P dT + \left(\frac{\partial G}{\partial P} \right)_T dP$$

we see that

$$\left(\frac{\partial G}{\partial T} \right)_P = -S \quad \text{and} \quad \left(\frac{\partial G}{\partial P} \right)_T = V \tag{22.31a, b}$$

Note that Equation 22.31a says that G decreases with increasing temperature (because $S \geq 0$) and that Equation 22.31b says that G increases with increasing pressure (because $V > 0$).

If we now take cross derivatives of G as we did for A in the previous section, we find that

$$-\left(\frac{\partial S}{\partial P}\right)_T = \left(\frac{\partial V}{\partial T}\right)_P \qquad (22.32)$$

This Maxwell relation gives us an equation we can use to calculate the pressure dependence of S. We integrate Equation 22.32 with T constant to get

$$\Delta S = -\int_{P_1}^{P_2} \left(\frac{\partial V}{\partial T}\right)_P dP \qquad \text{(constant } T) \qquad (22.33)$$

Equation 22.33 can be used to obtain the molar entropy as a function of pressure by integrating $(\partial V/\partial T)_P$ data from some low pressure, where the gas is sure to behave ideally, to some arbitrary pressure. Figure 22.4 shows the molar entropy of ethane at 400 K obtained in this way plotted against pressure.

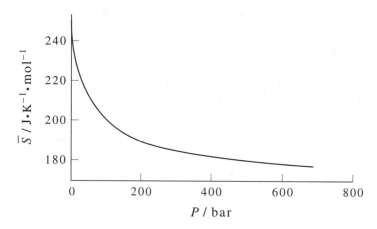

FIGURE 22.4
The molar entropy of ethane at 400 K plotted against pressure. The value of $\overline{S}^{\text{id}}$ at 400 K is 246.45 J·mol^{-1}·K^{-1}.

For an ideal gas, $(\partial V/\partial T)_P = nR/P$, so Equation 22.33 gives us

$$\Delta S = -nR \int_{P_1}^{P_2} \frac{dP}{P} = -nR \ln \frac{P_2}{P_1}$$

This results is not really a new one for us because if we let $P_2 = nRT/V_2$ and $P_1 = nRT/V_1$, we obtain Equation 22.21.

EXAMPLE 22–7

Use the virial expansion in the pressure

$$Z = 1 + B_{2P}P + B_{3P}P^2 + \cdots$$

to derive a virial expansion for $\Delta\overline{S}$ for a reversible isothermal change in pressure.

SOLUTION: Solve the above equation for $\overline{V}$:

$$\overline{V} = \frac{RT}{P} + RTB_{2P} + RTB_{3P}P + \cdots$$

and write

$$\left(\frac{\partial\overline{V}}{\partial T}\right)_P = \frac{R}{P} + R\left(B_{2P} + T\frac{dB_{2P}}{dT}\right) + R\left(B_{3P} + T\frac{dB_{3P}}{dT}\right)P + \cdots$$

Substitute this result into Equation 22.33 and integrate from P_1 to P_2 to obtain

$$\Delta\overline{S} = -\ln\frac{P_2}{P_1} - R\left(B_{2P} + T\frac{dB_{2P}}{dT}\right)P - \frac{R}{2}\left(B_{3P} + T\frac{dB_{3P}}{dT}\right)P^2 + \cdots$$

We can also use Equations 22.31 to show that the enthalpy of an ideal gas is independent of the pressure, just as its energy is independent of the volume. First, we differentiate Equation 22.13 with respect to P at constant T to obtain

$$\left(\frac{\partial G}{\partial P}\right)_T = \left(\frac{\partial H}{\partial P}\right)_T - T\left(\frac{\partial S}{\partial P}\right)_T$$

Now use Equation 22.31b for $(\partial G/\partial P)_T$ and Equation 22.32 for $(\partial S/\partial P)_T$ to obtain

$$\left(\frac{\partial H}{\partial P}\right)_T = V - T\left(\frac{\partial V}{\partial T}\right)_P \tag{22.34}$$

Note that Equation 22.34 is the analog of Equation 22.22. Equation 22.34 is also called a thermodynamic equation of state. It allows us to calculate the pressure dependence of H from P-V-T data (Such data for ethane at 400 K are shown in Figure 22.5). For an ideal gas, $(\partial V/\partial T)_P = nR/P$, so $(\partial H/\partial P)_T = 0$.

EXAMPLE 22–8

Evaluate $(\partial\overline{H}/\partial P)_T$ for a gas whose equation of state is

$$P\overline{V} = RT + B(T)P$$

SOLUTION: We have

$$\left(\frac{\partial\overline{V}}{\partial T}\right)_P = \frac{R}{P} + \frac{dB}{dT}$$

so Equation 22.34 gives us

$$\left(\frac{\partial \overline{H}}{\partial P}\right)_T = \overline{V} - T\left(\frac{\partial \overline{V}}{\partial T}\right)_P = \frac{RT}{P} + B(T) - \frac{RT}{P} - T\frac{dB}{dT}$$

or

$$\left(\frac{\partial \overline{H}}{\partial P}\right)_T = B(T) - T\frac{dB}{dT}$$

Note that $(\partial \overline{H}/\partial P)_T = 0$ when $B(T) = 0$.

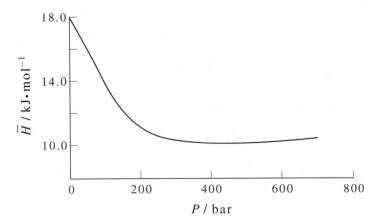

FIGURE 22.5
The molar enthalpy of ethane at 400 K plotted against pressure. The value of $\overline{H}^{\text{id}}$ at 400 K is
17.867 kJ·mol^{-1}.

22–5. The Various Thermodynamic Functions Have Natural Independent Variables

We may seem to be deriving a lot of equations in this chapter, but they can be organized neatly by recognizing that the energy, enthalpy, entropy, Helmholtz energy, and Gibbs energy depend upon natural sets of variables. For example, Equation 21.1 summarizes the First and Second Laws of Thermodynamics by

$$dU = TdS - PdV \tag{22.35}$$

Note that when S and V are considered to be the independent variables of U, then the total derivative of U,

$$dU = \left(\frac{\partial U}{\partial S}\right)_V dS + \left(\frac{\partial U}{\partial V}\right)_S dV \tag{22.36}$$

takes on a simple form, in the sense that the coefficients of dS and dV are simple thermodynamic functions. Consequently, we say that the natural variables of U are S and V, and we have

$$\left(\frac{\partial U}{\partial S}\right)_V = T \quad \text{and} \quad \left(\frac{\partial U}{\partial V}\right)_S = -P \tag{22.37}$$

This concept of natural variables is particularly clear if we consider V and T instead of S and V to be the independent variables of U, in which case we get (cf. Equation 22.22)

$$dU = \left[T\left(\frac{\partial P}{\partial T}\right)_V - P\right]dV + C_V dT \tag{22.38}$$

Certainly U can be considered to be a function of V and T, but its total derivative is not as simple as if it were considered to be a function of S and V (cf. Equation 22.36). Equation 22.35 also gives us that a criterion for a spontaneous process is that $dU < 0$ for a system at constant S and V.

We can write Equation 22.35 in terms of dS rather than dU to get

$$dS = \frac{1}{T}dU + \frac{P}{T}dV \tag{22.39}$$

which suggests that the natural variables of S are U and V. Furthermore, the criterion for a spontaneous process is that $dS > 0$ at constant U and V (Equation 22.2 for an isolated system). Equation 22.39 gives us

$$\left(\frac{\partial S}{\partial U}\right)_V = \frac{1}{T} \quad \text{and} \quad \left(\frac{\partial S}{\partial V}\right)_U = \frac{P}{T} \tag{22.40}$$

The total derivative of the enthalpy is given by (Equation 21.6)

$$dH = TdS + VdP \tag{22.41}$$

which suggests that the natural variables of H are S and P. The criterion of spontaneity involving H is that $dH < 0$ at constant S and P.

The total derivative of the Helmholtz energy is

$$dA = -SdT - PdV \tag{22.42}$$

from which we obtain

$$\left(\frac{\partial A}{\partial T}\right)_V = -S \quad \text{and} \quad \left(\frac{\partial A}{\partial V}\right)_T = -P \tag{22.43}$$

Equation 22.42, plus the spontaneity criterion that $dA < 0$ at constant T and V, suggest that T and V are the natural variables of A. The Maxwell relations obtained from Equation 22.43 are useful because the variables held constant are more experimentally controllable than are S and V, as in Equations 22.37, or U and V, as in Equations 22.40. The Maxwell relation from Equations 22.43 is

$$\left(\frac{\partial S}{\partial V}\right)_T = \left(\frac{\partial P}{\partial T}\right)_V \tag{22.44}$$

which allows us to calculate the volume dependence of S in terms of P-V-T data (see Figure 22.2).

Last, let's consider the Gibbs energy, whose total derivative is

$$dG = -SdT + VdP \tag{22.45}$$

Equation 22.45, plus the spontaneity criterion $dG < 0$ for a system at constant T and P, tell us that the natural variables of G are T and P. Equation 22.45 gives us

$$\left(\frac{\partial G}{\partial T}\right)_P = -S \quad \text{and} \quad \left(\frac{\partial G}{\partial P}\right)_T = V \tag{22.46}$$

The Maxwell relation we obtain from Equations 22.46 is

$$\left(\frac{\partial S}{\partial P}\right)_T = -\left(\frac{\partial V}{\partial T}\right)_P \tag{22.47}$$

which we can use to calculate the pressure dependence of S in terms of P-V-T data (Figure 22.4).

This section is meant to provide both a summary of many of the equations we have derived so far and a way to bring some order to them. You do not need to memorize these equations because they can all be obtained from Equation 22.35:

$$dU = TdS - PdV \tag{22.48}$$

which is nothing more than the First and Second Laws of Thermodynamics expressed as one equation. If we add $d(PV)$ to both sides of this equation, we obtain

$$d(U + PV) = TdS - PdV + VdP + PdV$$

or

$$dH = TdS + VdP \tag{22.49}$$

If we subtract $d(TS)$ from both sides of Equation 22.48, we have

$$d(U - TS) = TdS - PdV - TdS - SdT$$

or

$$dA = -SdT - PdV \tag{22.50}$$

If we add $d(PV)$ and subtract $d(TS)$ from Equation 22.48, or subtract $d(TS)$ from Equation 22.49, or add $d(PV)$ to Equation 22.50, we get

$$dG = -SdT + VdP \tag{22.51}$$

The other equations of this section follow by comparing the total derivative of each function in terms of its natural variables to the above equations for dU, dH, dA, and dG. Table 22.1 summarizes some of the principal equations we have derived in this and previous chapters.

TABLE 22.1
The four principal thermodynamic energies, their differential expressions, and the corresponding Maxwell relations.

Thermodynamic energy	Differential expression	Corresponding Maxwell relations
U	$dU = TdS - PdV$	$\left(\dfrac{\partial T}{\partial V}\right)_S = -\left(\dfrac{\partial P}{\partial S}\right)_V$
H	$dH = TdS + VdP$	$\left(\dfrac{\partial T}{\partial P}\right)_S = \left(\dfrac{\partial V}{\partial S}\right)_P$
A	$dA = -SdT - PdV$	$\left(\dfrac{\partial S}{\partial V}\right)_T = \left(\dfrac{\partial P}{\partial T}\right)_V$
G	$dG = -SdT + VdP$	$\left(\dfrac{\partial S}{\partial P}\right)_T = -\left(\dfrac{\partial V}{\partial T}\right)_P$

22–6. The Standard State for a Gas at Any Temperature Is the Hypothetical Ideal Gas at One Bar

One of the most important applications of Equation 22.33 involves the correction for nonideality that we make to obtain the standard molar entropies of gases. The standard molar entropies of gases tabulated in the literature are expressed in terms of a hypothetical ideal gas at one bar and at the same temperature. This correction is usually small and is obtained in the following two-step procedure (Figure 22.6). We first take our real gas from its pressure of one bar to some very low pressure P^{id}, where it is sure to behave ideally. We use Equation 22.33 to do this, giving

$$\overline{S}(P^{\text{id}}) - \overline{S}(1 \text{ bar}) = -\int_{1 \text{ bar}}^{P^{\text{id}}} \left(\frac{\partial \overline{V}}{\partial T}\right)_P dP$$

$$= \int_{P^{\text{id}}}^{1 \text{ bar}} \left(\frac{\partial \overline{V}}{\partial T}\right)_P dP \quad \text{(constant } T\text{)} \tag{22.52}$$

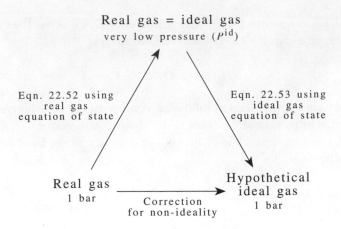

FIGURE 22.6
The scheme to bring the experimental entropies of gases to the standard state of a (hypothetical) ideal gas at the same temperature.

The superscript "id" on P emphasizes that this value is for conditions for which the gas behaves ideally. The quantity $(\partial \overline{V}/\partial T)_P$ can be determined from the equation of state of the actual gas. We now calculate the change in entropy as we increase the pressure back to one bar, but *taking the gas to be ideal*. We use Equation 22.52 for this process, but with $(\partial \overline{V}/\partial T)_P = R/P$, giving

$$S^\circ(1 \text{ bar}) - \overline{S}(P^{\text{id}}) = -\int_{P^{\text{id}}}^{1 \text{ bar}} \frac{R}{P} dP \tag{22.53}$$

The superscript $\circ$ of $S^\circ(1 \text{ bar})$ emphasizes that this is the standard molar entropy of the gas. We add Equations 22.52 and 22.53 to get

$$S^\circ(\text{at } 1 \text{ bar}) - \overline{S}(\text{at } 1 \text{ bar}) = \int_{P^{\text{id}}}^{1 \text{ bar}} \left[\left(\frac{\partial \overline{V}}{\partial T} \right)_P - \frac{R}{P} \right] dP \tag{22.54}$$

In Equation 22.54, $\overline{S}$ is the molar entropy we calculate from heat-capacity data and heats of transitions (Section 21–3), and S° is the molar entropy of the corresponding hypothetical ideal gas at one bar.

Equation 22.54 tells us that we can calculate the necessary correction to obtain the standard entropy if we know the equation of state. Because the pressures involved are around one bar, we can use the virial expansion using only the second virial coefficient. Using Equation 16.22,

$$\frac{P\overline{V}}{RT} = 1 + \frac{B_{2V}(T)}{RT} P + \cdots \tag{22.55}$$

we have

$$\left(\frac{\partial \overline{V}}{\partial T}\right)_P = \frac{R}{P} + \frac{d B_{2V}}{dT} + \cdots$$

Substituting this result into Equation 22.54 gives

$$S^\circ(\text{at 1 bar}) = \overline{S}(\text{at 1 bar}) + \frac{d B_{2V}}{dT} \times (1 \text{ bar}) + \cdots \qquad (22.56)$$

where we have neglected P^{id} with respect to one bar. The second term on the right side of Equation 22.56 represents the correction that we add to $\overline{S}$ to get S°.

We can use Equation 22.56 to calculate the nonideality correction to the entropy of $N_2(g)$ at 298.15 K that we used in Table 21.1. The experimental value of $d B_{2V}/dT$ for $N_2(g)$ at 298.15 K and one bar is $0.192 \text{ cm}^3 \cdot \text{mol}^{-1} \cdot \text{K}^{-1}$. Therefore, the correction for nonideality is given by

$$\text{correction for nonideality} = (0.192 \text{ cm}^3 \cdot \text{mol}^{-1} \cdot \text{K}^{-1})(1 \text{ bar})$$

$$= 0.192 \text{ cm}^3 \cdot \text{bar} \cdot \text{mol}^{-1} \cdot \text{K}^{-1}$$

$$= (0.192 \text{ cm}^3 \cdot \text{bar} \cdot \text{mol}^{-1} \cdot \text{K}^{-1})$$

$$\times \left(\frac{1 \text{ dm}^3}{10 \text{ cm}}\right)^3 \left(\frac{8.314 \text{ J} \cdot \text{mol}^{-1} \cdot \text{K}^{-1}}{0.08314 \text{ dm}^3 \cdot \text{bar} \cdot \text{mol}^{-1} \cdot \text{K}^{-1}}\right)$$

$$= 0.02 \text{ J} \cdot \text{K}^{-1} \cdot \text{mol}^{-1}$$

which is what was used in Table 21.1. The correction in this case is rather small, but that is not always so. If second virial coefficent data are not available, then an approximate equation of state can be used (Problems 22–20 through 22–22).

22–7. The Gibbs–Helmholtz Equation Describes the Temperature Dependence of the Gibbs Energy

Both of Equations 22.31 are useful because they tell us how the Gibbs energy varies with pressure and with temperature. Let's look at Equation 22.31b first. We can use Equation 22.31b to calculate the pressure dependence of the Gibbs energy:

$$\Delta G = \int_{P_1}^{P_2} V d P \qquad (\text{constant } T) \qquad (22.57)$$

For one mole of an ideal gas, we have

$$\Delta \overline{G} = RT \int_{P_1}^{P_2} \frac{dP}{P} = RT \ln \frac{P_2}{P_1} \qquad (22.58)$$

We could have obtained the same result by using

$$\Delta \overline{G} = \Delta \overline{H} - T\Delta \overline{S} \qquad \text{(isothermal)}$$

For an isothermal change in an ideal gas, $\Delta \overline{H} = 0$ and $\Delta \overline{S}$ is given by Equation 22.21.

It is customary to let $P_1 = 1$ bar (exactly) in Equation 22.58 and to write it in the form

$$\overline{G}(T, P) = G°(T) + RT \ln(P/1 \text{ bar}) \qquad (22.59)$$

where $G°(T)$ is called the standard molar Gibbs energy. The standard molar Gibbs energy in this case is the Gibbs energy of one mole of the ideal gas at a pressure of one bar. Note that $G°(T)$ depends upon only the temperature. Equation 22.59 gives the Gibbs energy of an ideal gas relative to the standard Gibbs energy. According to Equation 22.59, $\overline{G}(T, P) - G°(T)$ increases logarithmically with P, which we have seen is entirely an entropic effect for an ideal gas (because H is independent of P for an ideal gas). We will see in Chapter 24 that Equation 22.59 plays a central role in chemical equilibria involving gas-phase reactions.

EXAMPLE 22–9

Solids and liquids are fairly incompressible, so V in Equation 22.57 may be taken to be constant to a good approximation in this case. Derive an expression for $\overline{G}(T, P)$ analogous to Equation 22.59 for a solid or a liquid.

SOLUTION: We integrate Equation 22.57 at constant T to get

$$\overline{G}(P_2, T) - \overline{G}(P_1, T) = \overline{V}(P_2 - P_1)$$

We let $P_1 = 1$ bar and $\overline{G}(P_1 = 1 \text{ bar}, T) = G°(T)$ to get

$$\overline{G}(T, P) = G°(T) + \overline{V}(P - 1)$$

where P must be expressed in bars. In this case, $\overline{G}(T, P)$ increases linearly with P, but because the volume of a condensed phase is much smaller than that of a gas, the slope of $\overline{G}(T, P)$ versus P, or $(\partial \overline{G}/\partial P)_T = \overline{V}$, is very small. Consequently, at ordinary pressures $\overline{G}(T, P)$ is almost independent of pressure and is approximately equal to $G°(T)$ for a condensed phase.

Equation 22.31a determines the temperature dependence of the Gibbs energy. We can derive a useful equation for the temperature dependence of G by starting with Equation 22.31a (Problem 22–24), but an easier way is to start with $G = H - TS$ and divide by T to obtain

$$\frac{G}{T} = \frac{H}{T} - S$$

Now differentiate partially with respect to T keeping P fixed:

$$\left(\frac{\partial G/T}{\partial T}\right)_P = -\frac{H}{T^2} + \frac{1}{T}\left(\frac{\partial H}{\partial T}\right)_P - \left(\frac{\partial S}{\partial T}\right)_P$$

These last two terms cancel because of the relation $(\partial S/\partial T)_P = C_P(T)/T$ (Equation 21.7), so we have

$$\left(\frac{\partial G/T}{\partial T}\right)_P = -\frac{H}{T^2} \tag{22.60}$$

Equation 22.60 is called the *Gibbs-Helmholtz equation*. This equation can be directly applied to any process, in which case it becomes

$$\left(\frac{\partial \Delta G/T}{\partial T}\right)_P = -\frac{\Delta H}{T^2} \tag{22.61}$$

This equation is simply another form of the Gibbs-Helmholtz equation. We will use Equations 22.60 and 22.61 a number of times in the following chapters. For example, Equation 22.61 is used in Chapter 26 to derive an equation for the temperature dependence of an equilibrium constant.

We can determine the Gibbs energy as a function of temperature directly from equations we derived in Chapters 19 and 21. In Chapter 19, we learned how to calculate the enthalpy of a substance as a function of temperature in terms of its heat capacity and its various heats of transition. For example, if there exists only one solid phase, so that there are no solid-solid phase transitions between $T = 0$ K and its melting point, then (Equation 19.46)

$$H(T) - H(0) = \int_0^{T_{\text{fus}}} C_P^s(T)dT + \Delta_{\text{fus}}H$$

$$+ \int_{T_{\text{fus}}}^{T_{\text{vap}}} C_P^l(T)dT + \Delta_{\text{vap}}H \tag{22.62}$$

$$+ \int_{T_{\text{vap}}}^{T} C_P^g(T')dT'$$

for a temperature above the boiling point, where Figure 19.7 shows $\overline{H}(T) - \overline{H}(0)$ versus T for benzene. We calculate $H(T)$ relative to $H(0)$ because it is not possible to calculate an absolute enthalpy; $H(0)$ is essentially our zero of energy.

In Chapter 21, we learned to calculate absolute entropies according to Equation 21.17),

$$S(T) = \int_0^{T_{fus}} \frac{C_P^s(T)}{T} dT + \frac{\Delta_{fus}H}{T_{fus}}$$
$$+ \int_{T_{fus}}^{T_{vap}} \frac{C_P^l(T)}{T} dT + \frac{\Delta_{vap}H}{T_{vap}} \qquad (22.63)$$
$$+ \int_{T_{vap}}^{T} \frac{C_P^g(T')}{T'} dT'$$

Figure 21.2 shows $\overline{S}(T)$ versus T for benzene. We can use Equations 22.62 and 22.63 to calculate $\overline{G}(T) - \overline{H}(0)$ because

$$\overline{G}(T) - \overline{H}(0) = \overline{H}(T) - \overline{H}(0) - T\overline{S}(T)$$

Figure 22.7 shows $\overline{G}(T) - \overline{H}(0)$ versus T for benzene. There are several features of Figure 22.7 to appreciate. First note that $\overline{G}(T) - \overline{H}(0)$ decreases with increasing T. Furthermore, $\overline{G}(T) - \overline{H}(0)$ is a continuous function of temperature, even at a phase transition. To see that this is so, consider the equation (Equation 21.16)

$$\Delta_{trs}S = \frac{\Delta_{trs}H}{T_{trs}}$$

Because $\Delta_{trs}G = \Delta_{trs}H - T_{trs}\Delta_{trs}S$, we see that $\Delta_{trs}G = 0$, indicating that the two phases are in equilibrium with each other. Two phases in equilibrium with each other have the same value of G, so $G(T)$ is continuous at a phase transition. Figure 22.7

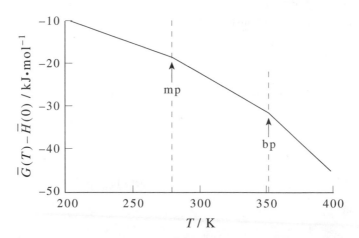

FIGURE 22.7
A plot of $\overline{G}(T) - \overline{H}(0)$ versus T for benzene. Note that $\overline{G}(T) - \overline{H}(0)$ is continuous but its derivative (the slope of the curve) is discontinuous at the phase transitions.

also shows that there is a discontinuity in the slope at each phase transition. (Benzene melts at 278.7 K and boils at 353.2 K at one atm.) We can understand why there is a discontinuity in the slope of $G(T)$ versus T at each phase transition by looking at Equation 22.31a

$$\left(\frac{\partial G}{\partial T}\right)_P = -S$$

Because entropy is an intrinsically positive quantity, the slope of $G(T)$ versus T is negative. Furthermore, because $S(\text{gas}) > S(\text{liquid}) > S(\text{solid})$, the slopes within each single phase region increase in going from solid to liquid to gas, so the slope, $(\partial G/\partial T)_P$, is discontinuous in passing from one phase to another.

The values of $H°(T) - H°(0)$, $S°(T)$, and $G°(T) - H°(0)$ are tabulated for a variety of substances. We will use these values to calculate equilibrium constants in Chapter 26.

22–8. Fugacity Is a Measure of the Nonideality of a Gas

In the previous section, we showed that the molar Gibbs energy of an ideal gas is given by

$$\overline{G}(T, P) = G°(T) + RT \ln \frac{P}{P°} \tag{22.64}$$

The pressure $P°$ is equal to one bar and $G°(T)$ is called the standard molar Gibbs energy. Recall that this equation is derived by starting with

$$\left(\frac{\partial \overline{G}}{\partial P}\right)_T = \overline{V} \tag{22.65}$$

and then integrating, using the ideal gas expression, RT/P, for $\overline{V}$. Let's now generalize Equation 22.64 to the case of a real gas.

We could start with the virial expansion,

$$\frac{P\overline{V}}{RT} = 1 + B_{2P}(T)P + B_{3P}(T)P^2 + \cdots$$

and substitute this into Equation 22.65 to obtain a virial expansion for the molar Gibbs energy,

$$\int_{P^{\text{id}}}^{P} d\overline{G} = RT \int_{P^{\text{id}}}^{P} \frac{dP'}{P'} + RT B_{2P}(T) \int_{P^{\text{id}}}^{P} dP' + RT B_{3P}(T) \int_{P^{\text{id}}}^{P} P' dP'$$

where we are integrating from some low pressure, say P^{id}, where the gas is sure to behave ideally, to some arbitrary pressure P. The result of the integration is

$$\overline{G}(T, P) = \overline{G}(T, P^{id}) + RT \ln \frac{P}{P^{id}} + RT B_{2P}(T) P + \frac{RT B_{3P}(T) P^2}{2} + \cdots \quad (22.66)$$

Now according to Equation 22.64, $\overline{G}(T, P^{id}) = G°(T) + RT \ln P^{id}/P°$, where $G°(T)$ is the molar Gibbs energy of an ideal gas at a pressure of $P° = 1$ bar. Therefore, Equation 22.66 can be written as

$$\overline{G}(T, P) = G°(T) + RT \ln \frac{P}{P°} + RT B_{2P}(T) P + \frac{RT B_{3P}(T) P^2}{2} + \cdots \quad (22.67)$$

Equation 22.67 is the generalization of Equation 22.64 to any real gas. Although Equation 22.67 is exact, it differs for each gas, depending upon the values of $B_{2P}(T)$, $B_{3P}(T)$, and so on. It turns out to be much more convenient, particularly for calculations involving chemical equilibria, as we will see in Chapter 24, to maintain the form of Equation 22.64 by defining a new thermodynamic function, $f(P, T)$, called *fugacity*, by the equation

$$\overline{G}(T, P) = G°(T) + RT \ln \frac{f(P, T)}{f°} \quad (22.68)$$

The nonideality is buried in $f(P, T)$. Because all gases behave ideally as $P \rightarrow 0$, fugacity must have the property that

$$f(P, T) \rightarrow P \qquad \text{as} \qquad P \rightarrow 0$$

so that Equation 22.68 reduces to Equation 22.64.

Equations 22.67 and 22.68 are equivalent if

$$\frac{f(P, T)}{f°} = \frac{P}{P°} \exp \left[B_{2P}(T) P + B_{3P}(T) P^2 + \cdots \right] \quad (22.69)$$

It might seem at this point that we are just going in circles, but by incorporating the nonideality of a gas through its fugacity, we can preserve the thermodynamic equations we have derived for ideal gases and write those corresponding to a real gas by simply replacing $P/P°$ by $f/f°$. All we need at this stage is a straightforward way to determine the fugacity of a gas at any pressure and temperature. Before looking into this, however, we must discuss the choice of the standard state in Equation 22.68. Being a type of energy, the Gibbs energy must always be taken relative to some chosen standard state.

Note that the standard molar Gibbs energy $G°(T)$ is taken to be the same quantity in Equations 22.64 and 22.68. The standard state in Equation 22.64 is the ideal gas at one bar, so this must be the standard state in Equation 22.68 as well. Thus, the standard state of the real gas in Equation 22.68 is taken to be the corresponding ideal gas at one bar; in other words, the standard state of the real gas is one bar after it has been adjusted to ideal behavior. In an equation, we have that $f° = P°$. Note that this choice is also suggested by Equation 22.69, because otherwise $f(P, T)$ would not reduce to P when $B_{2P}(T) = B_{3P}(T) = 0$.

This choice of standard state not only allows all gases to be brought to a single common state, but also leads to a procedure to calculate $f(P, T)$ at any pressure and temperature. To do so, consider the scheme in Figure 22.8, which depicts the difference in molar Gibbs energy between a real gas at P and T and an ideal gas at P and T. We can calculate this difference by starting with the real gas at P and T and then calculating the change in Gibbs energy when the pressure is reduced to essentially zero (step 2), where the gas is certain to behave ideally. Then we calculate the change in Gibbs energy as we compress the gas back to pressure P, but taking the gas to behave ideally (step 3). The sum of steps 2 and 3, then, will be the difference in Gibbs energy of an ideal gas at P and T and the real gas at P and T (step 1). In an equation, we have

$$\Delta \overline{G}_1 = \overline{G}^{\text{id}}(T, P) - \overline{G}(T, P) \tag{22.70}$$

Substituting Equations 22.64 and 22.68 into Equation 22.70, we have

$$\Delta \overline{G}_1 = RT \ln \frac{P}{P^\circ} - RT \ln \frac{f}{f^\circ}$$

But the standard state of the real gas has been chosen such that $f^\circ = P^\circ = 1$ bar, so

$$\Delta \overline{G}_1 = RT \ln \frac{P}{f} \tag{22.71}$$

We now use Equation 22.65 to calculate the change in the Gibbs energy along steps 2 and 3:

$$\Delta \overline{G}_2 = \int_P^{P \to 0} \left(\frac{\partial G}{\partial P} \right)_T dG = \int_P^{P \to 0} \overline{V} dP'$$

$$\Delta \overline{G}_3 = \int_{P \to 0}^P \overline{V}^{\text{id}} dP' = \int_{P \to 0}^P \frac{RT}{P'} dP'$$

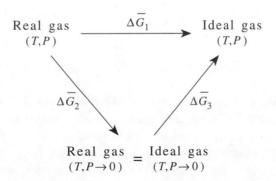

FIGURE 22.8
An illustration of the scheme used to relate the fugacity of a gas to its standard state, which is a (hypothetical) ideal gas at $P = 1$ bar and the temperature T of interest.

The sum of $\Delta \overline{G}_2$ and $\Delta \overline{G}_3$ gives another expression for $\Delta \overline{G}_1$

$$\Delta \overline{G}_1 = \Delta \overline{G}_2 + \Delta \overline{G}_3 = \int_{P \to 0}^{P} \left(\frac{RT}{P'} - \overline{V} \right) dP'$$

Equating this expression for $\Delta \overline{G}_1$ to $\Delta \overline{G}_1$ in Equation 22.71, we have

$$\ln \frac{P}{f} = \int_0^P \left(\frac{1}{P'} - \frac{\overline{V}}{RT} \right) dP'$$

or

$$\ln \frac{f}{P} = \int_0^P \left(\frac{\overline{V}}{RT} - \frac{1}{P'} \right) dP' \tag{22.72}$$

Given either P-V-T data or the equation of state of the real gas, Equation 22.72 allows us to calculate the ratio of the fugacity to the pressure of a gas at any pressure and temperature. Note that if the gas behaves ideally under the conditions of interest (in other words, if $\overline{V} = \overline{V}^{\text{id}}$ in Equation 22.72), then $\ln f/P = 0$, or $f = P$. Therefore, the extent of the deviation of f/P from unity is a direct indication of the extent of the deviation of the gas from ideal behavior. The ratio f/P is called the *fugacity coefficient*, γ,

$$\gamma = \frac{f}{P} \tag{22.73}$$

For an ideal gas, $\gamma = 1$.

By introducing the compressibility factor, $Z = P\overline{V}/RT$, Equation 22.72 can be written as

$$\ln \gamma = \int_0^P \frac{Z - 1}{P'} dP' \tag{22.74}$$

Even though the lower limit here is $P = 0$, the integrand is finite (Problem 22–27). Furthermore, $(Z - 1)/P = 0$ for an ideal gas (Problem 22–27), and hence $\ln \gamma = 0$ and $f = P$. Figure 22.9 shows $(Z - 1)/P$ plotted against P for $CO(g)$ at 200 K. According to Equation 22.74, the area under this curve from 0 to P is equal to $\ln \gamma$ at the pressure P. Figure 22.10 shows the resulting values of $\gamma = f/P$ plotted against the pressure for $CO(g)$ at 200 K.

We can also calculate the fugacity if we know the equation of state of the gas.

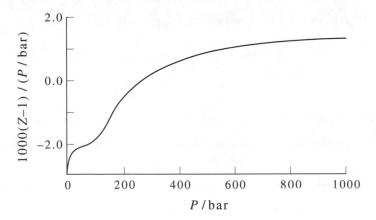

FIGURE 22.9
A plot of $(Z - 1)/P$ versus P for CO(g) at 200 K. The area under this curve from $P = 0$ to P gives $\ln \gamma$ at the pressure P.

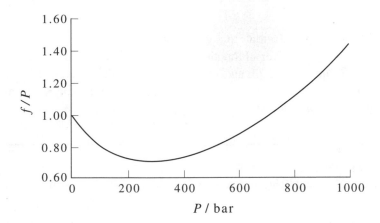

FIGURE 22.10
A plot of $\gamma = f/P$ against P for CO(g) at 200 K. These values of f/P were obtained from a numerical integration of $(Z - 1)/P$ shown in Figure 22.9.

EXAMPLE 22–10
Derive an expression for the fugacity of a gas that obeys the equation of state

$$P(\overline{V} - b) = RT$$

where b is a constant.

SOLUTION: We solve for $\overline{V}$ and substitute into Equation 22.72 to get

$$\ln \gamma = \int_0^P \frac{b}{RT} dP = \frac{bP}{RT}$$

or

$$\gamma = e^{bP/RT}$$

Problems 22–33 through 22–38 derive expressions for $\ln \gamma$ for the van der Waals equation and the Redlich-Kwong equation.

We can write Equation 22.74 in a form that shows that the fugacity coefficient is a function of the reduced pressure and the reduced temperature. If we change the integration variable to $P_R = P/P_c$, where P_c is the critical pressure of the gas, then Equation 22.74 takes the form

$$\ln \gamma = \int_0^{P_R} \left(\frac{Z - 1}{P_R'} \right) dP_R' \tag{22.75}$$

Now recall from Chapter 16 that, to a good approximation for most gases, the compressibility factor Z is a universal function of P_R and T_R (see Figure 16.10). Therefore, the right side of Equation 22.75, and so $\ln \gamma$ itself, is also a universal function of P_R and T_R. Figure 22.11 shows the experimental values of γ for many gases as a family of curves of constant T_R plotted against P_R.

EXAMPLE 22–11
Use Figure 22.11 and Table 16.5 to estimate the fugacity of nitrogen at 623 K and 1000 atm.

SOLUTION: We find from Table 16.5 that $T_c = 126.2$ K and $P_c = 33.6$ atm for $N_2(g)$. Therefore $T_R = 4.94$ at 623 K and $P_R = 29.8$ at 1000 atm. Reading from the curves in Figure 22.11, we find that $\gamma \approx 1.7$. At 1000 atm and 623 K, the fugacity of nitrogen is 1700 atm.

Problems

22-1. The molar enthalpy of vaporization of benzene at its normal boiling point (80.09°C) is 30.72 kJ·mol^{-1}. Assuming that $\Delta_{vap}\overline{H}$ and $\Delta_{vap}\overline{S}$ stay constant at their values at 80.09°C, calculate the value of $\Delta_{vap}\overline{G}$ at 75.0°C, 80.09°C, and 85.0°C. Interpret these results physically.

22-2. Redo Problem 22–1 without assuming that $\Delta_{vap}\overline{H}$ and $\Delta_{vap}\overline{S}$ do not vary with temperature. Take the molar heat capacities of liquid and gaseous benzene to be 136.3 J·K^{-1}·mol^{-1}

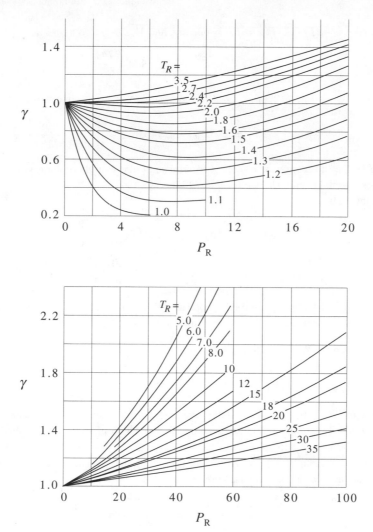

FIGURE 22.11
The fugacity coefficients of gases plotted against the reduced pressure, P/P_c, for various values of the reduced temperature, T/T_c.

and 82.4 $J \cdot K^{-1} \cdot mol^{-1}$, respectively. Compare your results with those you obtained in Problem 22–1. Are any of your physcial interpretations different?

22-3. Substitute $(\partial P/\partial T)_{\overline{V}}$ from the van der Waals equation into Equation 22.19 and integrate from $\overline{V}^{id}$ to $\overline{V}$ to obtain

$$\overline{S}(T, \overline{V}) - \overline{S}^{id}(T) = R \ln \frac{\overline{V} - b}{\overline{V}^{id} - b}$$

Now let $\overline{V}^{\,id} = RT/P^{\,id}$, $P^{\,id} = P^{\circ} =$ one bar, and $\overline{V}^{\,id} \gg b$ to obtain

$$\overline{S}(T, \overline{V}) - \overline{S}^{\,id}(T) = -R \ln \frac{RT/P^{\circ}}{\overline{V} - b}$$

Given that $\overline{S}^{\,id} = 246.35$ J·mol^{-1}·K^{-1} for ethane at 400 K, show that

$$\overline{S}(\overline{V})/\text{J·mol}^{-1}\cdot\text{K}^{-1} = 246.35 - 8.3145 \ln \frac{33.258 \text{ L·mol}^{-1}}{\overline{V} - 0.065144 \text{ L·mol}^{-1}}$$

Calculate $\overline{S}$ as a function of $\rho = 1/\overline{V}$ for ethane at 400 K and compare your results with the experimental results shown in Figure 22.2.

Show that

$$\overline{S}(\overline{V})/\text{J·mol}^{-1}\cdot\text{K}^{-1} = 246.35 - 8.3145 \ln \frac{33.258 \text{ L·mol}^{-1}}{\overline{V} - 0.045153 \text{ L·mol}^{-1}}$$

$$- 13.68 \ln \frac{\overline{V} + 0.045153 \text{ L·mol}^{-1}}{\overline{V}}$$

for the Redlich-Kwong equation for ethane at 400 K. Calculate $\overline{S}$ as a function of $\rho = 1/\overline{V}$ and compare your results with the experimental results shown in Figure 22.2.

22-4. Use the van der Waals equation to derive

$$\overline{U}(T, \overline{V}) - \overline{U}^{\,id}(T) = -\frac{a}{\overline{V}}$$

Use this result along with the van der Waals equation to calculate the value of $\overline{U}$ as a function of $\overline{V}$ for ethane at 400 K, given that $\overline{U}^{\,id} = 14.55$ kJ·mol^{-1}. To do this, specify $\overline{V}$ (from 0.0700 L·mol^{-1} to 7.00 L·mol^{-1}, see Figure 22.2), calculate both $\overline{U}(\overline{V})$ and $P(\overline{V})$, and plot $\overline{U}(\overline{V})$ versus $P(\overline{V})$. Compare your result with the experimental data in Figure 22.3. Use the Redlich-Kwong equation to derive

$$\overline{U}(T, \overline{V}) - \overline{U}^{\,id}(T) = -\frac{3A}{2BT^{1/2}} \ln \frac{\overline{V} + B}{\overline{V}}$$

Repeat the above calculation for ethane at 400 K.

22-5. Show that $(\partial U/\partial V)_T = 0$ for a gas that obeys an equation of state of the form $Pf(V) = RT$. Give two examples of such equations of state that appear in the text.

22-6. Show that

$$\left(\frac{\partial \overline{U}}{\partial \overline{V}}\right)_T = \frac{RT^2}{\overline{V}^2} \frac{dB_{2V}}{dT} + \frac{RT^2}{\overline{V}^3} \frac{dB_{3V}}{dT} + \cdots$$

22-7. Use the result of the previous problem to show that

$$\Delta \overline{U} = -T \frac{dB_{2V}}{dT}(P_2 - P_1) + \cdots$$

Use Equation 16.41 for the square-well potential to show that

$$\Delta \overline{U} = -\frac{2\pi\sigma^3 N_A}{3}(\lambda^3 - 1)\frac{\varepsilon}{k_B T}e^{\varepsilon/k_B T}(P_2 - P_1) + \cdots$$

Given that $\sigma = 327.7$ pm, $\varepsilon/k_B = 95.2$ K, and $\lambda = 1.58$ for $N_2(g)$, calculate the value of $\Delta \overline{U}$ for a pressure increase from 1.00 bar to 10.0 bar at 300 K.

22-8. Determine $\overline{C}_P - \overline{C}_V$ for a gas that obeys the equation of state $P(\overline{V} - b) = RT$.

22-9. The coefficient of thermal expansion of water at 25°C is 2.572×10^{-4} K^{-1}, and its isothermal compressibility is 4.525×10^{-5} bar^{-1}. Calculate the value of $C_P - C_V$ for one mole of water at 25°C. The density of water at 25°C is 0.99705 g·mL^{-1}.

22-10. Use Equation 22.22 to show that

$$\left(\frac{\partial C_V}{\partial V}\right)_T = T\left(\frac{\partial^2 P}{\partial T^2}\right)_V$$

Show that $(\partial C_V/\partial V)_T = 0$ for an ideal gas and a van der Waals gas, and that

$$\left(\frac{\partial C_V}{\partial V}\right)_T = -\frac{3A}{4T^{3/2}\overline{V}(\overline{V} + B)}$$

for a Redlich-Kwong gas.

22-11. In this problem you will derive the equation (Equation 22.24)

$$C_P - C_V = -T\left(\frac{\partial V}{\partial T}\right)_P^2 \left(\frac{\partial P}{\partial V}\right)_T$$

To start, consider V to be a function of T and P and write out dV. Now divide through by dT at constant volume ($dV = 0$) and then substitute the expression for $(\partial P/\partial T)_V$ that you obtain into Equation 22.23 to get the above expression.

22-12. The quantity $(\partial U/\partial V)_T$ has units of pressure and is called the *internal pressure*, which is a measure of the intermolecular forces within the body of a substance. It is equal to zero for an ideal gas, is nonzero but relatively small for dense gases, and is relatively large for liquids, particularly those whose molecular interactions are strong. Use the following data to calculate the internal pressure of ethane as a function of pressure at 280 K. Compare your values with the values you obtain from the van der Waals equation and the Redlich-Kwong equation.

P/bar	$(\partial P/\partial T)_V$/bar·K^{-1}	$\overline{V}$/dm^3·mol^{-1}	P/bar	$(\partial P/\partial T)_V$/bar·K^{-1}	$\overline{V}$/dm^3·mol^{-1}
4.458	0.01740	5.000	307.14	6.9933	0.06410
47.343	4.1673	0.07526	437.40	7.9029	0.06173
98.790	4.9840	0.07143	545.33	8.5653	0.06024
157.45	5.6736	0.06849	672.92	9.2770	0.05882

22-13. Show that

$$\left(\frac{\partial \overline{H}}{\partial P}\right)_T = -RT^2\left(\frac{dB_{2P}}{dT} + \frac{dB_{3P}}{dT}P + \cdots\right)$$

$$= B_{2V}(T) - T\frac{dB_{2V}}{dT} + O(P)$$

Use Equation 16.41 for the square-well potential to obtain

$$\left(\frac{\partial \overline{H}}{\partial P} \right)_T = \frac{2\pi \sigma^3 N_A}{3} \left[\lambda^3 - (\lambda^3 - 1) \left(1 + \frac{\varepsilon}{k_B T} \right) e^{\varepsilon / k_B T} \right]$$

Given that $\sigma = 327.7$ pm, $\varepsilon / k_B = 95.2$ K, and $\lambda = 1.58$ for $N_2(g)$, calculate the value of $(\partial \overline{H} / \partial P)_T$ at 300 K. Evaluate $\Delta \overline{H} = \overline{H}(P = 10.0 \text{ bar}) - \overline{H}(P = 1.0 \text{ bar})$. Compare your result with 8.724 kJ·mol^{-1}, the value of $\overline{H}(T) - \overline{H}(0)$ for nitrogen at 300 K.

22-14. Show that the enthalpy is a function of only the temperature for a gas that obeys the equation of state $P(\overline{V} - bT) = RT$, where b is a constant.

22-15. Use your results for the van der Waals equation and the Redlich-Kwong equation in Problem 22–4 to calculate $\overline{H}(T, \overline{V})$ as a function of volume for ethane at 400 K. In each case, use the equation $\overline{H} = \overline{U} + P\overline{V}$. Compare your results with the experimental data shown in Figure 22.5.

22-16. Use Equation 22.34 to show that

$$\left(\frac{\partial C_P}{\partial P} \right)_T = -T \left(\frac{\partial^2 V}{\partial T^2} \right)_P$$

Use a virial expansion in P to show that

$$\left(\frac{\partial \overline{C}_P}{\partial P} \right)_T = -T \frac{d^2 B_{2V}}{dT^2} + O(P)$$

Use the square-well second virial coefficient (Equation 16.41) and the parameters given in Problem 22–13 to calculate the value of $(\partial \overline{C}_P / \partial P)_T$ for $N_2(g)$ at 0°C. Now calculate $\overline{C}_P$ at 100 atm and 0°C, using $\overline{C}_P^{\text{id}} = 5R/2$.

22-17. Show that the molar enthalpy of a substance at pressure P relative to its value at one bar is given by

$$\overline{H}(T, P) = \overline{H}(T, P = 1 \text{ bar}) + \int_1^P \left[\overline{V} - T \left(\frac{\partial \overline{V}}{\partial T} \right)_P \right] dP'$$

Calculate the value of $\overline{H}(T, P) - \overline{H}(T, P = 1 \text{ bar})$ at 0°C and 100 bar for mercury given that the molar volume of mercury varies with temperature according to

$$\overline{V}(t) = (14.75 \text{ mL·mol}^{-1})(1 + 0.182 \times 10^{-3}t + 2.95 \times 10^{-9}t^2 + 1.15 \times 10^{-10}t^3)$$

where t is the Celsius temperature. Assume that $\overline{V}(0)$ does not vary with pressure over this range and express your answer in units of kJ·mol^{-1}.

22-18. Show that

$$dH = \left[V - T \left(\frac{\partial V}{\partial T} \right)_P \right] dP + C_P dT$$

What does this equation tell you about the natural variables of H?

22-19. What are the natural variables of the entropy?

22-20. Experimentally determined entropies are commonly adjusted for nonideality by using an equation of state called the (modified) Berthelot equation:

$$\frac{P\overline{V}}{RT} = 1 + \frac{9}{128}\frac{PT_c}{P_cT}\left(1 - 6\frac{T_c^2}{T^2}\right)$$

Show that this equation leads to the correction

$$S°(\text{at one bar}) = \overline{S}(\text{at one bar}) + \frac{27}{32}\frac{RT_c^3}{P_cT^3}(1\ \text{bar})$$

This result needs only the critical data for the substance. Use this equation along with the critical data in Table 16.5 to calculate the nonideality correction for $N_2(g)$ at 298.15 K. Compare your result with the value used in Table 21.1.

22-21. Use the result of Problem 22–20 along with the critical data in Table 16.5 to determine the nonideality correction for $CO(g)$ at its normal boiling point, 81.6 K. Compare your result with the value used in Problem 21–24.

22-22. Use the result of Problem 22–20 along with the critical data in Table 16.5 to determine the nonideality correction for $Cl_2(g)$ at its normal boiling point, 239 K. Compare your result with the value used in Problem 21–16.

22-23. Derive the equation

$$\left(\frac{\partial(A/T)}{\partial T}\right)_V = -\frac{U}{T^2}$$

which is a Gibbs-Helmholtz equation for A.

22-24. We can derive the Gibbs-Helmholtz equation directly from Equation 22.31a in the following way. Start with $(\partial G/\partial T)_P = -S$ and substitue for S from $G = H - TS$ to obtain

$$\frac{1}{T}\left(\frac{\partial G}{\partial T}\right)_P - \frac{G}{T^2} = -\frac{H}{T^2}$$

Now show that the left side is equal to $(\partial[G/T]/\partial T)_P$ to get the Gibbs-Helmholtz equation.

22-25. Use the following data for benzene to plot $\overline{G}(T) - \overline{H}(0)$ versus T. [In this case we will ignore the (usually small) corrections due to nonideality of the gas phase.]

$$\overline{C}_P^s(T)/R = \frac{12\pi^4}{5}\left(\frac{T}{\Theta_D}\right)^3 \qquad \Theta_D = 130.5\ \text{K} \qquad 0\ \text{K} < T < 13\ \text{K}$$

$$\overline{C}_P^s(T)/R = -0.6077 + (0.1088\ \text{K}^{-1})T - (5.345 \times 10^{-4}\ \text{K}^{-2})T^2 + (1.275 \times 10^{-6}\ \text{K}^{-3})T^3$$

$$13\ \text{K} < T < 278.6\ \text{K}$$

$$\overline{C}_P^l(T)/R = 12.713 + (1.974 \times 10^{-3}\ \text{K}^{-1})T - (4.766 \times 10^{-5}\ \text{K}^{-2})T^2$$

$$278.6\ \text{K} < T < 353.2\ \text{K}$$

$$\overline{C}_P^g(T)/R = -4.077 + (0.05676 \text{ K}^{-1})T - (3.588 \times 10^{-5} \text{ K}^{-2})T^2 + (8.520 \times 10^{-9} \text{ K}^{-3})T^3$$

$$353.2 \text{ K} < T < 1000 \text{ K}$$

$$T_{\text{fus}} = 278.68 \text{ K} \qquad \Delta_{\text{fus}}\overline{H} = 9.95 \text{ kJ·mol}^{-1}$$

$$T_{\text{vap}} = 353.24 \text{ K} \qquad \Delta_{\text{vap}}\overline{H} = 30.72 \text{ kJ·mol}^{-1}$$

22-26. Use the following data for propene to plot $\overline{G}(T) - \overline{H}(0)$ versus T. [In this case we will ignore the (usually small) corrections due to nonideality of the gas phase.]

$$\overline{C}_P^s(T)/R = \frac{12\pi^4}{5}\left(\frac{T}{\Theta_D}\right)^3 \qquad \Theta_D = 100 \text{ K} \qquad 0 \text{ K} < T < 15 \text{ K}$$

$$\overline{C}_P^s(T)/R = -1.663 + (0.001112 \text{ K}^{-1})T - (9.791 \times 10^{-4} \text{ K}^{-2})T^2 + (3.740 \times 10^{-6} \text{ K}^{-3})T^3$$

$$15 \text{ K} < T < 87.90 \text{ K}$$

$$\overline{C}_P^l(T)/R = 15.935 - (0.08677 \text{ K}^{-1})T + (4.294 \times 10^{-4} \text{ K}^{-2})T^2 - (6.276 \times 10^{-7} \text{ K}^{-3})T^3$$

$$87.90 \text{ K} < T < 225.46 \text{ K}$$

$$\overline{C}_P^g(T)/R = 1.4970 + (2.266 \times 10^{-2} \text{ K}^{-1})T - (5.725 \times 10^{-6} \text{ K}^{-2})T^2$$

$$225.46 \text{ K} < T < 1000 \text{ K}$$

$$T_{\text{fus}} = 87.90 \text{ K} \qquad \Delta_{\text{fus}}\overline{H} = 3.00 \text{ kJ·mol}^{-1}$$

$$T_{\text{vap}} = 225.46 \text{ K} \qquad \Delta_{\text{vap}}\overline{H} = 18.42 \text{ kJ·mol}^{-1}$$

22-27. Use a virial expansion for Z to prove (a) that the integrand in Equation 22.74 is finite as $P \to 0$, and (b) that $(Z - 1)/P = 0$ for an ideal gas.

22-28. Derive a virial expansion in the pressure for $\ln \gamma$.

22-29. The compressibility factor for ethane at 600 K can be fit to the expression

$$Z = 1.0000 - 0.000612(P/\text{bar}) + 2.661 \times 10^{-6}(P/\text{bar})^2$$
$$- 1.390 \times 10^{-9}(P/\text{bar})^3 - 1.077 \times 10^{-13}(P/\text{bar})^4$$

for $0 \le P/\text{bar} \le 600$. Use this expression to determine the fugacity coefficient of ethane as a function of pressure at 600 K.

22-30. Use Figure 22.11 and the data in Table 16.5 to estimate the fugacity of ethane at 360 K and 1000 atm.

22-31. Use the following data for ethane at 360 K to plot the fugacity coefficient against pressure.

$\rho/\text{mol}\cdot\text{dm}^3$	P/bar	$\rho/\text{mol}\cdot\text{dm}^{-3}$	P/bar	$\rho/\text{mol}\cdot\text{dm}^{-3}$	P/bar
1.20	31.031	6.00	97.767	10.80	197.643
2.40	53.940	7.20	112.115	12.00	266.858
3.60	71.099	8.40	130.149	13.00	381.344
4.80	84.892	9.60	156.078	14.40	566.335

Compare your result with the result you obtained in Problem 22–30.

22-32. Use the following data for $N_2(g)$ at $0°C$ to plot the fugacity coefficient as a function of pressure.

P/atm	$Z = P\overline{V}/RT$	P/atm	$Z = P\overline{V}/RT$	P/atm	$Z = P\overline{V}/RT$
200	1.0390	1000	2.0700	1800	3.0861
400	1.2570	1200	2.3352	2000	3.3270
600	1.5260	1400	2.5942	2200	3.5640
800	1.8016	1600	2.8456	2400	3.8004

22-33. It might appear that we can't use Equation 22.72 to determine the fugacity of a van der Waals gas because the van der Waals equation is a cubic equation in $\overline{V}$, so we can't solve it analytically for $\overline{V}$ to carry out the integration in Equation 22.72. We can get around this problem, however, by integrating Equation 22.72 by parts. First show that

$$RT \ln \gamma = P\overline{V} - RT - \int_{\overline{V}^{\,\text{id}}}^{\overline{V}} P d\overline{V}' - RT \ln \frac{P}{P^{\text{id}}}$$

where $P^{\text{id}} \to 0$, $\overline{V}^{\,\text{id}} \to \infty$, and $P^{\text{id}}\overline{V}^{\,\text{id}} \to RT$. Substitute P from the van der Waals equation into the first term and the integral on the right side of the above equation and integrate to obtain

$$RT \ln \gamma = \frac{RT\overline{V}}{\overline{V} - b} - \frac{a}{\overline{V}} - RT - RT \ln \frac{\overline{V} - b}{\overline{V}^{\,\text{id}} - b} - \frac{a}{\overline{V}} - RT \ln \frac{P}{P^{\text{id}}}$$

Now use the fact that $\overline{V}^{\,\text{id}} \to \infty$ and that $P^{\text{id}}\overline{V}^{\,\text{id}} = RT$ to show that

$$\ln \gamma = -\ln \left[1 - \frac{a(\overline{V} - b)}{RT\overline{V}^2} \right] + \frac{b}{\overline{V} - b} - \frac{2a}{RT\overline{V}}$$

This equation gives the fugacity coefficient of a van der Waals gas as a function of $\overline{V}$. You can use the van der Waals equation itself to calculate P from $\overline{V}$, so the above equation, in conjunction with the van der Waals equation, gives $\ln \gamma$ as a function of pressure.

22-34. Use the final equation in Problem 22–33 along with the van der Waals equation to plot $\ln \gamma$ against pressure for $CO(g)$ at 200 K. Compare your result with Figure 22.10.

22-35. Show that the expression for $\ln \gamma$ for the van der Waals equation (Problem 22–33) can be written in the reduced form

$$\ln \gamma = \frac{1}{3V_R - 1} - \frac{9}{4V_R T_R} - \ln\left[1 - \frac{3(3V_R - 1)}{8T_R V_R^2}\right]$$

Use this equation along with the van der Waals equation in reduced form (Equation 16.19) to plot γ against P_R for $T_R = 1.00$ and 2.00 and compare your results with Figure 22.11.

22-36. Use the method outlined in Problem 22–33 to show that

$$\ln \gamma = \frac{B}{\overline{V} - B} - \frac{A}{RT^{3/2}(\overline{V} + B)} - \frac{A}{BRT^{3/2}} \ln \frac{\overline{V} + B}{\overline{V}}$$
$$- \ln\left[1 - \frac{A(\overline{V} - B)}{RT^{3/2}\overline{V}(\overline{V} + B)}\right]$$

for the Redlich-Kwong equation. You need to use the standard integral

$$\int \frac{dx}{x(a + bx)} = -\frac{1}{a} \ln \frac{a + bx}{x}$$

22-37. Show that $\ln \gamma$ for the Redlich-Kwong equation (see Problem 22–36) can be written in the reduced form

$$\ln \gamma = \frac{0.25992}{\overline{V}_R - 0.25992} - \frac{1.2824}{T_R^{3/2}(\overline{V}_R + 0.25992)}$$
$$- \frac{4.9340}{T_R^{3/2}} \ln \frac{\overline{V}_R + 0.25992}{\overline{V}_R} - \ln\left[1 - \frac{1.2824(\overline{V}_R - 0.25992)}{T_R^{3/2} \overline{V}_R(\overline{V}_R + 0.25992)}\right]$$

22-38. Use the expression for $\ln \gamma$ in reduced form given in Problem 22–37 along with the Redlich-Kwong equation in reduced form (Example 16–7) to plot $\ln \gamma$ versus P_R for $T_R = 1.00$ and 2.00 and compare your results with those you obtained in Problem 22–35 for the van der Waals equation.

22-39. Compare $\ln \gamma$ for the van der Waals equation (Problem 22–33) with the values of $\ln \gamma$ for ethane at 600 K (Problem 22–29).

22-40. Compare $\ln \gamma$ for the Redlich-Kwong equation (Problem 22–36) with the values of $\ln \gamma$ for ethane at 600 K (Problem 22–29).

22-41. We can use the equation $(\partial S/\partial U)_V = 1/T$ to illustrate the consequence of the fact that entropy always increases during an irreversible adiabatic process. Consider a two-compartment system enclosed by rigid adiabatic walls, and let the two compartments be separated by a rigid heat-conducting wall. We assume that each compartment is at equilibrium but that they are not in equilibrium with each other. Because no work can be done by this two-compartment system (rigid walls) and no energy as heat can be exchanged with the surroundings (adiabatic walls),

$$U = U_1 + U_2 = \text{constant}$$

Show that

$$dS = \left(\frac{\partial S_1}{\partial U_1}\right) dU_1 + \left(\frac{\partial S_2}{\partial U_2}\right) dU_2$$

because the entropy of each compartment can change only as a result of a change in energy. Now show that

$$dS = dU_1 \left(\frac{1}{T_1} - \frac{1}{T_2}\right) \geq 0$$

Use this result to discuss the direction of the flow of energy as heat from one temperature to another.

22-42. Modify the argument in Problem 22–41 to the case in which the two compartments are separated by a nonrigid, insulating wall. Derive the result

$$dS = \left(\frac{P_1}{T_1} - \frac{P_2}{T_2}\right) dV_1$$

Use this result to discuss the direction of a volume change under an isothermal pressure difference.

22-43. In this problem, we will derive virial expansions for $\overline{U}$, $\overline{H}$, $\overline{S}$, $\overline{A}$, and $\overline{G}$. Substitute

$$Z = 1 + B_{2P} P + B_{3P} P^2 + \cdots$$

into Equation 22.65 and integrate from a small pressure, P^{id}, to P to obtain

$$\overline{G}(T, P) - \overline{G}(T, P^{\text{id}}) = RT \ln \frac{P}{P^{\text{id}}} + RT B_{2P} P + \frac{RT B_{3P}}{2} P^2 + \cdots$$

Now use Equation 22.64 (realize that $P = P^{\text{id}}$ in Equation 22.64) to get

$$\overline{G}(T, P) - G^\circ(T) = RT \ln P + RT B_{2P} P + \frac{RT B_{3P}}{2} P^2 + \cdots \tag{1}$$

at $P^\circ = 1$ bar. Now use Equation 22.31a to get

$$\overline{S}(T, P) - S^\circ(T) = -R \ln P - \frac{d(RT B_{2P})}{dT} P - \frac{1}{2} \frac{d(RT B_{3P})}{dT} P^2 + \cdots \tag{2}$$

at $P^\circ = 1$ bar. Now use $\overline{G} = \overline{H} - T\overline{S}$ to get

$$\overline{H}(T, P) - H^\circ(T) = -RT^2 \frac{d B_{2P}}{dT} P - \frac{RT^2}{2} \frac{d B_{3P}}{dT} P^2 + \cdots \tag{3}$$

Now use the fact that $\overline{C}_P = (\partial \overline{H}/\partial T)_P$ to get

$$\overline{C}_P(T, P) - C_P^\circ(T) = -RT \left[2\frac{d B_{2P}}{dT} + T\frac{d^2 B_{2P}}{dT^2}\right] P - \frac{RT}{2} \left[2\frac{d B_{3P}}{dT} + T\frac{d^2 B_{3P}}{dT^2}\right] P^2 + \cdots \tag{4}$$

We can obtain expansions for $\overline{U}$ and $\overline{A}$ by using the equation $\overline{H} = \overline{U} + P\overline{V} = \overline{U} + RTZ$ and $\overline{G} = \overline{A} + P\overline{V} = \overline{A} + RTZ$. Show that

$$\overline{U} - U^\circ = -RT \left(B_{2P} + T\frac{d B_{2P}}{dT}\right) P - RT \left(B_{3P} + \frac{T}{2}\frac{d B_{3P}}{dT}\right) P^2 + \cdots \tag{5}$$

and

$$\overline{A} - A^\circ = RT \ln P - \frac{RTB_{3p}}{2} P^2 + \cdots \tag{6}$$

at $P^\circ = 1$ bar.

22-44. In this problem, we will derive the equation

$$\overline{H}(T, P) - H^\circ(T) = RT(Z - 1) + \int_{\overline{V}^{id}}^{\overline{V}} \left[T \left(\frac{\partial P}{\partial T} \right)_V - P \right] d\overline{V'}$$

where $\overline{V}^{id}$ is a very large (molar) volume, where the gas is sure to behave ideally. Start with $dH = TdS + VdP$ to derive

$$\left(\frac{\partial H}{\partial V} \right)_T = T \left(\frac{\partial S}{\partial V} \right)_T + V \left(\frac{\partial P}{\partial V} \right)_T$$

and use one of the Maxwell relations for $(\partial S / \partial V)_T$ to obtain

$$\left(\frac{\partial H}{\partial V} \right)_T = T \left(\frac{\partial P}{\partial T} \right)_V + V \left(\frac{\partial P}{\partial V} \right)_T$$

Now integrate by parts from an ideal-gas limit to an arbitrary limit to obtain the desired equation.

22-45. Using the result of Problem 22–44, show that H is independent of volume for an ideal gas. What about a gas whose equation of state is $P(\overline{V} - b) = RT$? Does U depend upon volume for this equation of state? Account for any difference.

22-46. Using the result of Problem 22–44, show that

$$\overline{H} - H^\circ = \frac{RTb}{\overline{V} - b} - \frac{2a}{\overline{V}}$$

for the van der Waals equation.

22-47. Using the result of Problem 22–44, show that

$$\overline{H} - H^\circ = \frac{RTB}{\overline{V} - B} - \frac{A}{T^{1/2}(\overline{V} + B)} - \frac{3A}{2BT^{1/2}} \ln \frac{\overline{V} + B}{\overline{V}}$$

for the Redlich-Kwong equation.

22-48. We introduced the Joule-Thomson effect and the Joule-Thomson coefficient in Problems 19–52 through 19–54. The Joule-Thomson coefficient is defined by

$$\mu_{JT} = \left(\frac{\partial T}{\partial P} \right)_H = -\frac{1}{C_P} \left(\frac{\partial H}{\partial P} \right)_T$$

and is a direct measure of the expected temperature change when a gas is expanded through a throttle. We can use one of the equations derived in this chapter to obtain a convenient working equation for μ_{JT}. Show that

$$\mu_{JT} = \frac{1}{C_P}\left[T\left(\frac{\partial V}{\partial T}\right)_P - V \right]$$

Use this result to show that $\mu_{JT} = 0$ for an ideal gas.

22-49. Use the virial equation of state of the form

$$\frac{P\overline{V}}{RT} = 1 + \frac{B_{2V}(T)}{RT}P + \cdots$$

to show that

$$\mu_{JT} = \frac{1}{C_P^{id}}\left[T\frac{dB_{2V}}{dT} - B_{2V} \right] + O(P)$$

It so happens that B_{2V} is negative and dB_{2V}/dT is positive for $T^* < 3.5$ (see Figure 16.15) so that μ_{JT} is positive for low temperatures. Therefore, the gas will cool upon expansion under these conditions. (See Problem 22–48.)

22-50. Show that

$$\mu_{JT} = -\frac{b}{C_P}$$

for a gas that obeys the equation of state $P(\overline{V} - b) = RT$. (See Problem 22–48).

22-51. The second virial coefficient for a square-well potential is (Equation 16.41)

$$B_{2V}(T) = b_0[1 - (\lambda^3 - 1)(e^{\varepsilon/k_B T} - 1)]$$

Show that

$$\mu_{JT} = \frac{b_0}{C_P}\left[(\lambda^3 - 1)\left(1 + \frac{\varepsilon}{k_B T}\right)e^{\varepsilon/k_B T} - \lambda^3 \right]$$

where $b_0 = 2\pi\sigma^3 N_A/3$. Given the following square-well parameters, calculate μ_{JT} at 0°C and compare your values with the given experimental values. Take $C_P = 5R/2$ for Ar and $7R/2$ for N_2 and CO_2.

Gas	$b_0/cm^3 \cdot mol^{-1}$	λ	ε/k_B	$\mu_{JT}(exptl)/K \cdot atm^{-1}$
Ar	39.87	1.85	69.4	0.43
N_2	45.29	1.87	53.7	0.26
CO_2	75.79	1.83	119	1.3

22-52. The temperature at which the Joule-Thomson coefficient changes sign is called the *Joule-Thomson inversion temperature*, T_i. The low-pressure Joule-Thomson inversion temperature for the square-well potential is obtained by setting $\mu_{JT} = 0$ in Problem 22–51. This procedure leads to an equation for $k_B T/\varepsilon$ in terms of λ^3 that cannot be solved analytically. Solve the equation numerically to calculate T_i for the three gases given in the previous problem. The experimental values are 794 K, 621 K, and 1500 K for Ar, N_2, and CO_2, respectively.

22-53. Use the data in Problem 22–51 to estimate the temperature drop when each of the gases undergoes an expansion for 100 atm to one atm.

22-54. When a rubber band is stretched, it exerts a restoring force, f, which is a function of its length L and its temperature T. The work involved is given by

$$w = \int f(L, T)dL \tag{1}$$

Why is there no negative sign in front of the integral, as there is in Equation 19.2 for P-V work? Given that the volume change upon stretching a rubber band is negligible, show that

$$dU = TdS + fdL \tag{2}$$

and that

$$\left(\frac{\partial U}{\partial L}\right)_T = T\left(\frac{\partial S}{\partial L}\right)_T + f \tag{3}$$

Using the definition $A = U - TS$, show that Equation 2 becomes

$$dA = -SdT + fdL \tag{4}$$

and derive the Maxwell relation

$$\left(\frac{\partial f}{\partial T}\right)_L = -\left(\frac{\partial S}{\partial L}\right)_T \tag{5}$$

Substitute Equation 5 into Equation 3 to obtain the analog of Equation 22.22

$$\left(\frac{\partial U}{\partial L}\right)_T = f - T\left(\frac{\partial f}{\partial T}\right)_L$$

For many elastic systems, the observed temperature-dependence of the force is linear. We define an *ideal rubber band* by

$$f = T\phi(L) \qquad \text{(ideal rubber band)} \tag{6}$$

Show that $(\partial U/\partial L)_T = 0$ for an ideal rubber band. Compare this result with $(\partial U/\partial V)_T = 0$ for an ideal gas.

Now let's consider what happens when we stretch a rubber band quickly (and, hence, adiabatically). In this case, $dU = dw = fdL$. Use the fact that U depends upon only the temperature for an ideal rubber band to show that

$$dU = \left(\frac{\partial U}{\partial T}\right)_L dT = fdL \tag{7}$$

The quantity $(\partial U/\partial T)_L$ is a heat capacity, so Equation 7 becomes

$$C_L dT = fdL \tag{8}$$

Argue now that if a rubber band is suddenly stretched, then its temperature will rise. Verify this result by holding a rubber band against your upper lip and stretching it quickly.

22-55. Derive an expression for ΔS for the reversible, isothermal change of one mole of a gas that obeys van der Waals equation. Use your result to calculate ΔS for the isothermal compression of one mole of ethane from $10.0\,\text{dm}^3$ to $1.00\,\text{dm}^3$ at 400 K. Compare your result to what you would get using the ideal-gas equation.

22-56. Derive an expression for ΔS for the reversible, isothermal change of one mole of a gas that obeys the Redlich-Kwong equation (Equation 16.7). Use your result to calculate ΔS for the isothermal compression of one mole of ethane from $10.0\,\text{dm}^3$ to $1.00\,\text{dm}^3$ at 400 K. Compare your result with the result you would get using the ideal-gas equation.

Josiah Willard Gibbs was born in New Haven, Connecticut, on February 11, 1839, and died there in 1903. He received his Ph.D. in engineering from Yale University in 1863, the second doctorate in science and the first in engineering awarded in the United States. He stayed on at Yale, for years without salary, and remained there for the rest of his life. In 1878, Gibbs published a long, original treatise on thermodynamics titled "On the Equilibrium of Heterogeneous Substances" in the *Transactions of the Connecticut Academy of Sciences*. In addition to introducing the concept of chemical potential, Gibbs introduced what is now called the Gibbs phase rule, which relates the number of components (C) and the number of phases (P) in a system to the number of degrees of freedom (F, the number of variables such as temperature and pressure that can be varied independently) by the equation $F = C + 2 - P$. Between its austere writing style and the obscurity of the journal in which it was published, however, this important work was not as widely appreciated as it deserved. Fortunately, Gibbs sent copies to a number of prominent European scientists. Maxwell and van der Waals immediately appreciated the signifance of the work and made it known in Europe. Eventually, Gibbs received the recognition that was his due, and Yale finally offered him a salaried position in 1880. Gibbs was an unassuming, modest person, living in New Haven in his family home his entire life.

CHAPTER 23

Phase Equilibria

The relation between all the phases of a substance at various temperatures and pressures can be concisely represented by a phase diagram. In this chapter, we will study the information presented by phase diagrams and the thermodynamic consequences of this information. In particular, we will analyze the temperature and pressure dependence of a substance in terms of its Gibbs energy, particularly using the fact that a phase with the lower Gibbs energy will always be the more stable one.

Many thermodynamic systems of interest consist of two or more phases in equilibrium with each other. For example, both the solid and liquid phases of a substance are in equilibrium with each other at its melting point. Thus, an analysis of such a system as a function of temperature and pressure gives the pressure-dependence of the melting point. One of the many unusual properties of water is that the melting point of ice decreases with increasing pressure. We will see in this chapter that this property is a direct consequence of the fact that water expands upon freezing, or that the molar volume of liquid water is less than that of ice. We will also derive an expression that allows us to calculate the vapor pressure of a liquid as a function of temperature from a knowledge of its enthalpy of vaporization. These results can all be understood using a quantity called the chemical potential, which is one of the most useful functions of chemical thermodynamics. We will see that chemical potential is analogous to electric potential. Just as electric current flows from a region of high electric potential to a region of low electric potential, matter flows from a region of high chemical potential to a region of low chemical potential. In the last section of the chapter, we will derive a statistical thermodynamic expression for the chemical potential and show how to calculate it in terms of molecular or spectroscopic quantities.

23–1. A Phase Diagram Summarizes the Solid–Liquid–Gas Behavior of a Substance

You might recall from general chemistry that we can summarize the solid-liquid-gas behavior of a substance by means of a phase diagram, which indicates under what conditions of pressure and temperature the various states of matter of a substance exist in equilibrium. Figure 23.1 shows the phase diagram of benzene, a typical substance. Note that there are three principal regions in this phase diagram. Any point within one of these regions specifies a pressure and a temperature at which the single phase exists in equilibrium. For example, according to Figure 23.1, benzene exists as a solid at 60 torr and 260 K (point A), and as a gas at 60 torr and 300 K (point B).

The lines that separate the three regions indicate pressures and temperatures at which two phases can coexist at equilibrium. For example, at all points along the line that separates the solid and gas regions (line CF), benzene exists as a solid and a gas in equilibrium with each other. This line is called the solid-gas coexistence curve. As such, it specifies the vapor pressure of solid benzene as a function of temperature. Similarly, the line that separates the liquid and gas regions (line FD) gives the vapor pressure of liquid benzene as a function of temperature, and the line that separates the solid and liquid regions (line FE) gives the melting point of benzene as a function of pressure. Notice that the three lines in the phase diagram intersect at one point (point F), at which solid, liquid, and gaseous benzene coexist at equilibrium. This point is called the *triple point*, which occurs at 278.7 K (5.5°C) and 36.1 torr for benzene.

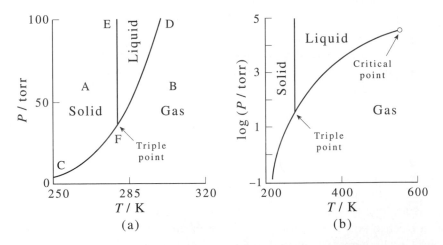

FIGURE 23.1
The phase diagram of benzene, (a) displayed as P against T, and (b) log P versus T. The log P versus T display condenses the vertical axis.

EXAMPLE 23–1
Experimentally, the vapor pressure of liquid benzene is given by

$$\ln(P/\text{torr}) = -\frac{4110 \text{ K}}{T} + 18.33 \qquad 273 \text{ K} < T < 300 \text{ K}$$

and the vapor pressure of solid benzene is given by

$$\ln(P/\text{torr}) = -\frac{5319 \text{ K}}{T} + 22.67 \qquad 250 \text{ K} < T < 280 \text{ K}$$

Calculate the pressure and the temperature at the triple point of benzene.

SOLUTION: Solid, liquid, and gaseous benzene coexist at the triple point. Therefore, at the triple point, these two equations for the vapor pressure must give the same value. Setting the two expressions above for ln P equal to each other gives

$$-\frac{4110 \text{ K}}{T} + 18.33 = -\frac{5319 \text{ K}}{T} + 22.67$$

or $T = 278.7$ K, or $5.5°$C. The pressure at the triple point is given by $\ln(P/\text{torr}) = 3.58$ or $P = 36.1$ torr.

Within a single-phase region, both the pressure and the temperature must be specified, and we say that there are two degrees of freedom within a single-phase region of a pure substance. Along any of the coexistence curves, either the pressure or the temperature alone is sufficient to specify a point on the curve, so we say that there is one degree of freedom. The triple point is a fixed point, so there are no degrees of freedom there. If we think of P and T as degrees of freedom of the system, then the number of degrees of freedom, f, at any point in a phase diagram of a pure substance is given by $f = 3 - p$, where p is the number of phases that coexist at equilibrium at that point.

If we start on the pressure axis at 760 torr (2.88 on the vertical axis in Figure 23.1b) and move horizontally to the right in the phase diagram of benzene, we can see how benzene behaves with increasing temperature at a constant pressure of 760 torr (one atmosphere). For temperatures below 278.7 K, benzene exists as a solid. At 278.7 K (5.5°C), we reach the solid-liquid coexistence curve, and benzene melts at this point. This point is called the *normal melting point*. (The melting point at a pressure of one bar is called the *standard melting point*.) Then for temperatures between 278.7 K and 353.2 K (80.1°C), benzene exists as a liquid. At the liquid-gas coexistence curve (353.2 K), benzene boils and then exists as a gas at temperatures higher than 353.2 K. Note that if we were to start at a pressure less than 760 torr (but above the triple point), the melting point is about the same as it is at 760 torr (because the solid-liquid coexistence curve is so steep), but the boiling point is lower than 353.2 K. Similarly, at a pressure greater than 760 torr, the melting point is about the same as it is at 760 torr, but the boiling point is greater than 353.2 K. Thus, the liquid-gas coexistence curve may also be interpreted as the boiling point of benzene as a function of pressure and the solid-liquid coexistence curve as the melting point as a function of pressure. Figure 23.2 shows the melting point of benzene plotted against pressure up to 10 000 atm. The slope of this curve is 0.0293 °C·atm^{-1} around 760 torr, which shows that the melting point is fairly insensitive to pressure. The melting point of benzene increases by about one degree in going from a pressure of 1 atm to 34 atm. By contrast, Figure 23.3 is a plot of the boiling point of benzene as a function of pressure; it shows that the boiling point depends strongly upon pressure. For example, the normal atmospheric pressure at an elevation of 10 000 feet (3100 meters) is 500 torr, so according to Figure 23.3, benzene boils at 67°C at this elevation. (Recall that the boiling point is defined to be that temperature at which the vapor pressure equals the atmospheric pressure.) The boiling point at exactly one atm is called the normal boiling point. The boiling point at exactly one bar is called the standard boiling point.

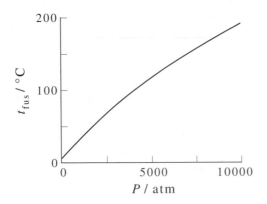

FIGURE 23.2
A plot of the melting point of benzene as a function of pressure. Notice that the melting point increases slowly with pressure. (Note that the scales of the horizontal axes in Figures 23.2 and 23.3 are very different.)

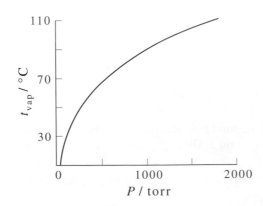

FIGURE 23.3
A plot of the boiling point of benzene as a function of pressure. Notice that the boiling point depends strongly on the pressure. (Note that the scales of the horizontal axes in Figures 23.2 and 23.3 are different.)

EXAMPLE 23–2

The vapor pressure of benzene can be expressed by the empirical formula

$$\ln(P/\text{torr}) = -\frac{3884 \text{ K}}{T} + 17.63$$

Use this formula to show that benzene boils at 67°C when the atmospheric pressure is 500 torr.

SOLUTION: Benzene boils when its vapor pressure is equal to the atmospheric pressure. Therefore $P = 500$ torr, so we have

$$\ln 500 = -\frac{3884 \text{ K}}{T} + 17.63$$

or $T = 340.2$ K, or 67.1°C.

Example 23–1 shows that the pressure at the triple point of benzene is 36.1 torr. Note from Figure 23.1 that if the pressure is less than 36.1 torr, benzene does not melt as we increase the temperature, but rather *sublimes*; that is, it passes directly from the solid phase to the gaseous phase. If the pressure at the triple point happens to be greater than one atm for a substance, it will sublime rather than melt at one atm. A noted substance with this property is carbon dioxide, whose solid phase is called dry ice because it doesn't liquefy at atmospheric pressure. Figure 23.4 shows the phase diagram for carbon dioxide. The triple point pressure of CO_2 is 5.11 atm, and so we see that CO_2 sublimes at one atm. The normal sublimation temperature of CO_2 is 195 K (-78°C).

Figure 23.5 shows the phase diagram for water. Water has the unusual property that its melting point decreases with increasing pressure (Figure 23.6). This behavior is reflected in the phase diagram of water by the slope of the solid-liquid coexistence curve. Although it is difficult to see in the phase diagram because the slope of the solid-liquid coexistence curve is so large, it does point upward to the left (has a negative

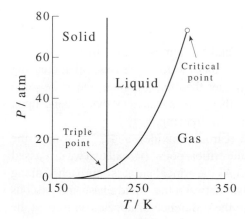

FIGURE 23.4
The phase diagram of carbon dioxide. Note that the triple point pressure of carbon dioxide is greater than one atm. Consequently, carbon dioxide sublimes at atmospheric pressure.

930

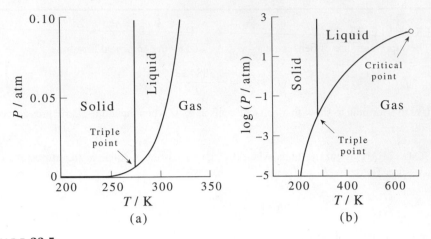

FIGURE 23.5
The phase diagram of water, (a) displayed as P against T, and (b) log P versus T. The log P versus T display condenses the vertical axis. Although it is difficult to discern because of the scale of the figure, the melting point of water decreases with increasing pressure.

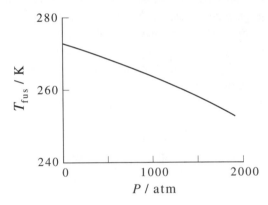

FIGURE 23.6
A plot of the melting point of water versus pressure. The melting point of water decreases with increasing pressure.

slope). Numerically, the slope of the curve around one atm is $-130 \ \text{atm}\cdot\text{K}^{-1}$. We will see in Section 23–3 that the reason the melting point of water decreases with increasing pressure is that the molar volume of ice is greater than that of water under the same conditions. Antimony and bismuth are two other such substances that expand upon freezing. Most substances, however, contract upon freezing.

In each of Figures 23.1 (benzene), 23.4 (carbon dioxide), and 23.5 (water), the liquid-gas coexistence curve ends abruptly at the critical point. (Recall that we discussed the critical behavior of gases in Section 16–3.) As the critical point is approached along the liquid-gas coexistence curve, the difference between the liquid phase and gaseous phase becomes increasingly less distinct until the difference disappears entirely at the critical point. For example, if we plot the densities of the liquid and vapor phases in equilibrium with each other along the liquid-vapor coexistence curve (such densities are called *orthobaric densities*), we see that these densities approach each other and become equal at the critical point (Figure 23.7). The liquid phase and vapor phase simply merge into a single fluid phase. Similarly, the molar enthalpy of vaporization decreases along this curve.

Figure 23.8 shows experimental values of the molar enthalpy of vaporization of benzene plotted against temperature. Notice that the value of $\Delta_{vap}\overline{H}$ decreases with increasing temperature and drops to zero at the critical temperature (289°C for benzene).

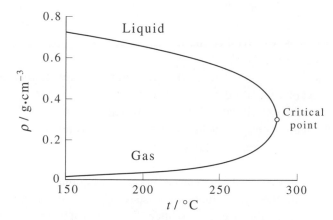

FIGURE 23.7
A plot of the orthobaric densities of the liquid and vapor phases of benzene in equilibrium along the liquid-vapor coexistence curve. Notice that the densities of the liquid and vapor phases approach one another and become equal at the critical point (289°C).

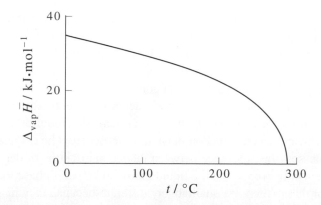

FIGURE 23.8
Experimental values of the molar enthalpy of vaporization of benzene plotted against temperature. The value of $\Delta_{vap}\overline{H}$ decreases with increasing temperature and drops to zero at the critical temperature, 289°C.

The data in Figure 23.8 reflect the fact that the difference between a liquid and its vapor decreases as the critical point is approached. Because the two phases become less and less distinct as the critical point is approached and then merge into one phase at the critical point, $\Delta_{vap}S = S(\text{gas}) - S(\text{liquid})$ becomes zero at the critical point. Therefore $\Delta_{vap}H = T\Delta_{vap}S$ also becomes zero there. Above the critical point, there is no distinction between a liquid and a gas, and a gas cannot be liquefied no matter how great the pressure.

A nice lecture demonstration illustrates the idea of the critical temperature. First, fill a glass tube with a liquid such as sulfur hexafluoride. (The critical temperature of sulfur hexafluoride is 45.5°C, which is a convenient temperature to achieve.) After evacuating all the air so that the tube contains only pure sulfur hexafluoride, seal off the tube. Below 45.5°C, the tube will contain two layers, the liquid phase and the gas phase separated by a meniscus. Now, as the tube and its contents are warmed, the meniscus becomes less distinct and just as the critical temperature is reached, the meniscus disappears entirely and the tube becomes transparent [$SF_6(g)$ is colorless]. When the tube and its contents are cooled, the liquid phase and the meniscus suddenly appear at the critical temperature.

A fluid very near its critical point constantly changes from a liquid to a vapor state, causing fluctuations in the density from one region to another. These fluctuations scatter light very strongly (somewhat like a finely dispersed fog) and the system appears milky. This effect is known as *critical opalescence*. These fluctuations are difficult to study experimentally because gravity causes the density fluctuations to be distorted. To overcome the effect of gravity, a team of scientists, engineers, and technicians designed an experiment to measure the laser light scattered by xenon at its critical point on board the Columbia space shuttle. After several preliminary experiments, they were able to measure the details of the fluctuations to within microkelvins of the critical temperature of xenon (289.72 K) on the March 1996 flight of Columbia. No other microgravity experiment has logged as many hours as this one, and the results will provide us with a detailed understanding of the liquid-vapor phase transition and the liquid-vapor interface.

Because of the existence of a critical point, a gas can be transformed into a liquid without ever passing through a two-phase state. Simply start in the gas region of the phase diagram and go into the liquid region by traveling out around the critical point. The gas passes gradually and continuously into the liquid state without a two-phase region appearing and without any apparent condensation.

You might wonder if the solid-liquid coexistence curve ends abruptly as the liquid-gas coexistence curve does. The very nature of a critical point requires that we pass from one phase to the other in a gradual, continual manner. Because the gas and liquid phases are both fluid phases, the difference between them is purely one of degree rather than actual structure. On the other hand, a liquid phase and a solid phase, or two different solid phases for that matter, are qualitatively different because they have intrinsically different structures. It is not possible to pass from one phase to the other in a gradual, continual manner. A critical point, therefore, cannot exist for such phases, and the coexistence curve separating these phases must continue indefinitely or intersect the coexistence curves of other phases. In fact, many substances exhibit a variety of solid

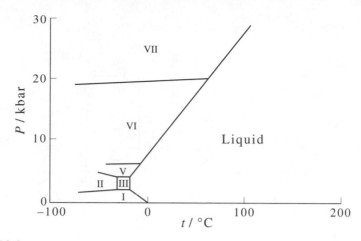

FIGURE 23.9
The phase diagram for water at high pressures showing six stable phases of ice.

phases at high pressures, and Figure 23.9 shows the high-pressure phase diagram of water, showing various distinct solid phases. Ice (I) is the "normal" ice that occurs at one atm, and the other ices are different crystalline forms of solid H_2O that are stable at very high pressures. Note, for example, that ice (VII) is stable at temperatures well above 0°C, and even above 100°C, but it is formed only at high pressures.

23–2. The Gibbs Energy of a Substance Has a Close Connection to Its Phase Diagram

Recall Figure 22.7 where the molar Gibbs energy of benzene is plotted against temperature. As the figure shows, the molar Gibbs energy is a continuous function of temperature, but there is a discontinuity in the slope of $\overline{G}(T)$ versus T at each phase transition. Figure 23.10a is a magnification of a plot of $\overline{G}(T)$ versus T in the region around the melting point of benzene (279 K). The dashed extensions represent the Gibbs energy of the supercooled liquid and the (hypothetical) superheated solid. Picture moving along the curve of $\overline{G}(T)$ versus T in Figure 23.10a with increasing temperature. Along the solid-phase branch, $\overline{G}(T)$ decreases with a slope $(\partial \overline{G}/\partial T)_P = -\overline{S}^s$. When we reach the melting point, we switch to the liquid branch because the Gibbs energy of the liquid phase is lower than that of the solid phase. The slope of the liquid branch is steeper than that of the solid branch because $(\partial \overline{G}/\partial T)_P = -\overline{S}^l$ and $\overline{S}^l > \overline{S}^s$. Therefore, the molar Gibbs energy of the liquid phase must be lower than that of the solid phase at higher temperatures. The dashed extension of the solid branch represents the (hypothetical) superheated solid, and even if it were to occur, it would be unstable relative to the liquid and would convert to liquid. The dashed lines represent what are called metastable states. Figure 23.10b shows the transition from liquid to gas at the normal boiling point (353 K) of benzene. The boiling point occurs when the liquid and

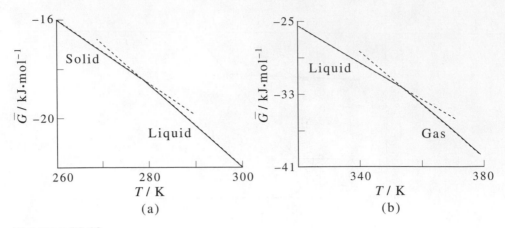

FIGURE 23.10
A plot of $\overline{G}(T)$ versus T for benzene in the region around (a) its melting point (279 K) and (b) its boiling point (353 K).

gas branches of the $\overline{G}(T)$ versus T curves intersect. The slope of the gas branch is steeper than that of the liquid branch because $\overline{S}^{\,g} > \overline{S}^{\,l}$, and so the molar Gibbs energy of the gas must be lower than that of the liquid at higher temperatures.

We can see from the equation $G = H - TS$ why the solid phase is favored at low temperatures whereas the gaseous phase is favored at high temperatures. At low temperatures, the TS term is small compared with H; thus, a solid phase is favored at low temperatures because it has the lowest enthalpy of the three phases. At high temperatures, on the other hand, H is small compared with the TS term, so we see that the gas phase with its relatively large entropy is favored at high temperatures. The liquid phase, which is intermediate in both energy and disorder to the solid and gaseous phases, is favored at intermediate temperatures.

It is also instructive to look at the molar Gibbs energy as a function of pressure at a fixed temperature. Recall that $(\partial \overline{G}/\partial P)_T = \overline{V}$, so that the slope of G versus P is always positive. For most substances, $\overline{V}^{\,g} \gg \overline{V}^{\,l} > \overline{V}^{\,s}$, so the slope of the gas branch is much greater than that of a liquid branch, which in turn is greater than that of a solid branch. Figure 23.11a sketches a plot of $\overline{G}(P)$ against P showing the gas, liquid, and solid branches at a temperature just greater than the triple-point temperature. As we increase the pressure, we move along the gas branch of $\overline{G}(P)$ until we hit the liquid branch, at which point the gas condenses to a liquid. As we continue to increase the pressure, we reach the solid branch, which necessarily lies lower than that of the liquid branch. The path we have just followed in Figure 23.11a corresponds to moving up along a vertical line that lies just to the right of the triple point in the phase diagram of a "normal" substance like benzene. For a substance such as water, however, $\overline{V}^{\,s} > \overline{V}^{\,l}$ at least for moderate pressures, so a plot of $\overline{G}(P)$ against P looks like that given in Figure 23.11b. Tracing along the curve for $\overline{G}(P)$ for increasing pressures in Figure 23.11b corresponds to moving up a vertical line just to the left of the triple point in the phase diagram of water.

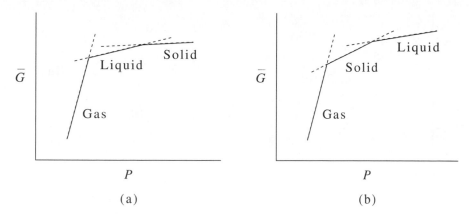

FIGURE 23.11
A plot of $\overline{G}(P)$ against P showing the gas, liquid, and solid branches at a temperature near the triple point. (a) A "normal" substance ($\overline{V}^{\,s} < \overline{V}^{\,l}$) is depicted at a temperature above the triple-point temperature, where we see a gas-liquid-solid progression with increasing pressure. (b) A substance like water ($\overline{V}^{\,s} > \overline{V}^{\,l}$) is depicted at a temperature lower than the triple point temperature, where we see a gas-solid-liquid progression.

Figure 23.12 shows the behavior of $\overline{G}(P)$ versus P at a number of temperatures for a normal substance such as benzene. Part (a) shows $\overline{G}(P)$ versus P for a temperature less than the triple-point temperature in Figure 23.1. In this case, we go directly from the gas phase to the solid phase as we increase the pressure. The molar Gibbs energy of the liquid phase at these temperatures lies higher than that of either the solid or gas phase and does not enter the picture. Part (b) shows the molar Gibbs energy situation at the triple-point temperature. At the triple point, the curves for the Gibbs energies of each of the three phases intersect, and for a "normal" substance like benzene, the Gibbs energy of the solid phase lies lower than that of the liquid phase for pressures above the triple-point pressure. Part (c) shows the Gibbs energies at a temperature slightly less than the critical temperature. Notice that the slopes of the gas and liquid branches are almost the same at the point of intersection. The reason for this similarity is that the slopes of the curves, $(\partial \overline{G}/\partial P)_T$, are equal to the molar volumes of the two phases, which are approaching each other as the critical point is approached. Part (d) shows the Gibbs energies at a temperature greater than the critical temperature. In this case, $\overline{G}(P)$ varies smoothly with pressure. There is no discontinuity in the slope in this case because only a single fluid phase is involved.

23–3. The Chemical Potentials of a Pure Substance in Two Phases in Equilibrium Are Equal

Consider a system consisting of two phases of a pure substance in equilibrium with each other. For example, we might have water vapor in equilibrium with liquid water. The Gibbs energy of this system is given by $G = G^l + G^g$, where G^l and G^g are the

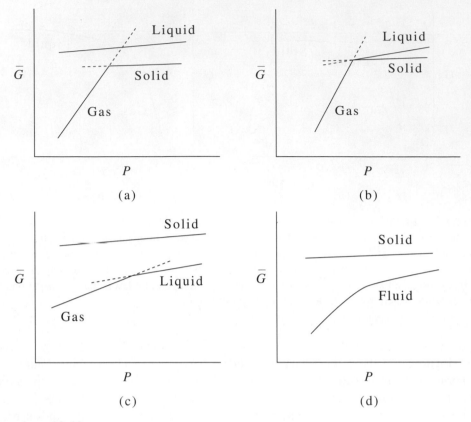

FIGURE 23.12
A plot of $\overline{G}(P)$ against P at a number of temperatures for a "normal" substance like benzene. In (a) the temperature is less than the triple-point temperature; in (b) the temperature is equal to the triple-point temperature; in (c) the temperature is a little less than the critical temperature; and in (d) the temperature is greater than the critical temperature.

Gibbs energies of the liquid phase and the gas phase, respectively. Now, suppose that dn moles are transferred from the liquid phase to the vapor phase, while T and P are kept constant. The infinitesimal change in Gibbs energy for this process is

$$dG = \left(\frac{\partial G^g}{\partial n^g}\right)_{P,T} dn^g + \left(\frac{\partial G^l}{\partial n^l}\right)_{P,T} dn^l \qquad (23.1)$$

But $dn^l = -dn^g$ for the transfer of dn moles from the liquid phase to the vapor phase, so Equation 23.1 becomes

$$dG = \left[\left(\frac{\partial G^g}{\partial n^g}\right)_{P,T} - \left(\frac{\partial G^l}{\partial n^l}\right)_{P,T}\right] dn^g \qquad (23.2)$$

The partial derivatives in Equation 23.2 are central quantities in the treatment of equilibria. They are called *chemical potentials* and are denoted by μ^g and μ^l:

$$\mu^g = \left(\frac{\partial G^g}{\partial n^g}\right)_{P,T} \quad \text{and} \quad \mu^l = \left(\frac{\partial G^l}{\partial n^l}\right)_{P,T} \tag{23.3}$$

In terms of chemical potentials, then, Equation 23.2 reads

$$dG = (\mu^g - \mu^l)dn^g \qquad \text{(constant } T \text{ and } P) \tag{23.4}$$

If the two phases are in equilibrium with each other, then $dG = 0$, and because $dn^g \neq 0$, we find that $\mu^g = \mu^l$. Thus, we find that if two phases of a single substance are in equilibrium with each other, then the chemical potentials of that substance in the two phases are equal.

If the two phases are not in equilibrium with each other, a spontaneous transfer of matter from one phase to the other will occur in the direction such that $dG < 0$. If $\mu^g > \mu^l$, the term in parentheses in Equation 23.4 is positive, so dn^g must be negative in order that $dG < 0$. In other words, matter will transfer from the vapor phase to the liquid phase, or from the phase with higher chemical potential to the phase with lower chemical potential. If, on the other hand, $\mu^g < \mu^l$, then dn^g will be positive, meaning that matter will transfer from the liquid phase to the vapor phase. Once again, the transfer occurs from the phase with higher chemical potential to the phase with lower chemical potential. Notice that chemical potential is analogous to electric potential. Just as electric current flows from a higher electric potential to a lower electric potential, matter "flows" from a higher chemical potential to a lower chemical potential (see Problem 23–19).

Although we have defined chemical potential quite generally in Equation 23.3, it takes on a simple, familiar form for a pure substance. Because G, like U, H, and S, is an extensive thermodynamic function, G is proportional to the size of a system, or $G \propto n$. We can express this proportionality by the equation $G = n\mu(T, P)$. Note that this equation is consistent with the definition of $\mu(T, P)$ because

$$\mu = \left(\frac{\partial G}{\partial n}\right)_{P,T} = \left(\frac{\partial n\mu(T, P)}{\partial n}\right)_{T,P} = \mu(T, P) \tag{23.5}$$

Therefore, for a single, pure substance, μ is the same quantity as the molar Gibbs energy and $\mu(T, P)$ is an intensive quantity like temperature and pressure.

We can use the fact that the chemical potentials of a single substance in two phases in equilibrium are equal to derive an expression for the variation of equilibrium pressure with temperature for any two phases of a given pure substance. Let the two phases be α and β, so that

$$\mu^\alpha(T, P) = \mu^\beta(T, P) \qquad \text{(equilibrium between phases)} \tag{23.6}$$

Now take the total derivative of both sides of Equation 23.6

$$\left(\frac{\partial \mu^{\alpha}}{\partial P}\right)_{T} dP + \left(\frac{\partial \mu^{\alpha}}{\partial T}\right)_{P} dT = \left(\frac{\partial \mu^{\beta}}{\partial P}\right)_{T} dP + \left(\frac{\partial \mu^{\beta}}{\partial T}\right)_{P} dT \qquad (23.7)$$

Because μ is simply the molar Gibbs energy for a single substance, we have in analogy with Equations 22.31

$$\left(\frac{\partial G}{\partial P}\right)_{T} = V \quad \text{and} \quad \left(\frac{\partial G}{\partial T}\right)_{P} = -S$$

that

$$\left(\frac{\partial \mu}{\partial P}\right)_{T} = \left(\frac{\partial \overline{G}}{\partial P}\right)_{T} = \overline{V} \quad \text{and} \quad \left(\frac{\partial \mu}{\partial T}\right)_{P} = \left(\frac{\partial \overline{G}}{\partial T}\right)_{P} = -\overline{S} \qquad (23.8)$$

where $\overline{V}$ and $\overline{S}$ are the molar volume and the molar entropy, respectively. We substitute this result into Equation 23.7 to obtain

$$\overline{V}^{\alpha} dP - \overline{S}^{\alpha} dT = \overline{V}^{\beta} dP - \overline{S}^{\beta} dT$$

Solving for dP/dT gives

$$\frac{dP}{dT} = \frac{\overline{S}^{\beta} - \overline{S}^{\alpha}}{\overline{V}^{\beta} - \overline{V}^{\alpha}} = \frac{\Delta_{trs}\overline{S}}{\Delta_{trs}\overline{V}} \qquad (23.9)$$

Equation 23.9 applies to two phases in equilibrium with each other, so we may use the fact that $\Delta_{trs}\overline{S} = \Delta_{trs}\overline{H}/T$ and write

$$\frac{dP}{dT} = \frac{\Delta_{trs}\overline{H}}{T\Delta_{trs}\overline{V}} \qquad (23.10)$$

Equation 23.10 is called the *Clapeyron equation*, and relates the slope of the two-phase boundary line in a phase diagram with the values of $\Delta_{trs}\overline{H}$ and $\Delta_{trs}\overline{V}$ for a transition between these two phases.

Let's use Equation 23.10 to calculate the slope of the solid-liquid coexistence curve for benzene around one atm (Figure 23.1). The molar enthalpy of fusion of benzene at its normal melting point (278.7 K) is 9.95 kJ·mol^{-1}, and $\Delta_{fus}\overline{V}$ under the same conditions is 10.3 cm^3·mol^{-1}. Thus, dP/dT at the normal melting point of benzene is

$$\frac{dP}{dT} = \frac{9950 \text{ J·mol}^{-1}}{(278.68 \text{ K})(10.3 \text{ cm}^3\cdot\text{mol}^{-1})} \left(\frac{10 \text{ cm}}{1 \text{ dm}}\right)^3 \left(\frac{0.08206 \text{ dm}^3\cdot\text{atm}\cdot\text{mol}^{-1}\cdot\text{K}^{-1}}{8.314 \text{ J·mol}^{-1}\cdot\text{K}^{-1}}\right)$$

$$= 34.2 \text{ atm·K}^{-1}$$

We can take the reciprocal of this result to obtain

$$\frac{dT}{dP} = 0.0292 \ \mathrm{K \cdot atm^{-1}}$$

Thus, we see that the melting point of benzene increases by 0.0292 K per atmosphere of pressure around one atm. If $\Delta_{fus}\overline{H}$ and $\Delta_{fus}\overline{V}$ were independent of pressure, we could use this result to predict that the melting point of benzene at 1000 atm is 29.2 K higher than it is at one atm, or 307.9 K. The experimental value is 306.4 K, so our assumption of constant $\Delta_{fus}\overline{H}$ and $\Delta_{fus}\overline{V}$ is fairly satisfactory. Figure 23.2 shows the experimental melting point of benzene versus pressure up to 10 000 atm. You can see from the figure that the slope is not quite constant.

EXAMPLE 23–3

Determine the value of dT/dP for ice at its normal melting point. The molar enthalpy of fusion of ice at 273.15 K and one atm is 6010 J·mol^{-1}, and $\Delta_{fus}\overline{V}$ under the same conditions is -1.63 cm^3·mol^{-1} (recall that unlike most substances water expands upon freezing, so that $\Delta_{fus}\overline{V} = \overline{V}^l - \overline{V}^s < 0$.) Estimate the melting point of ice at 1000 atm.

SOLUTION: We use the reciprocal of Equation 23.10:

$$\frac{dT}{dP} = \frac{T\Delta_{fus}\overline{V}}{\Delta_{fus}\overline{H}}$$

$$= \frac{(273.2 \ \mathrm{K})(-1.63 \ \mathrm{cm^3 \cdot mol^{-1}})}{6010 \ \mathrm{J \cdot mol^{-1}}} \left(\frac{10 \ \mathrm{cm}}{1 \ \mathrm{dm}}\right)^3$$

$$\times \left(\frac{8.314 \ \mathrm{J \cdot mol^{-1} \cdot K^{-1}}}{0.08206 \ \mathrm{dm^3 \cdot atm \cdot mol^{-1} \cdot K^{-1}}}\right)$$

$$= -0.00751 \ \mathrm{K \cdot atm^{-1}}$$

Assuming that dT/dP is constant up to 1000 atm, we find that $\Delta T = -7.51$ K, or that the melting point of ice at 1000 atm is 265.6 K. The experimental value is 263.7 K. The discrepancy arises from our assumption that the values of $\Delta_{fus}\overline{V}$ and $\Delta_{fus}\overline{H}$ are independent of pressure. Figure 23.6 shows the experimental melting point of ice versus pressure up to 2000 atm.

Notice that the melting point of ice decreases with increasing pressure, so that the slope of the solid-liquid equilibrium curve in the pressure-temperature phase diagram of water has a negative slope. Equation 23.10 shows that this slope is a direct result of the fact that $\Delta_{fus}\overline{V}$ is negative for this case.

Equation 23.10 can be used to estimate the molar volume of a liquid at its boiling point.

EXAMPLE 23–4
The vapor pressure of benzene is found to obey the empirical equation

$$\ln(P/\text{torr}) = 16.725 - \frac{3229.86 \text{ K}}{T} - \frac{118345 \text{ K}^2}{T^2}$$

from 298.2 K to its normal boiling point 353.24 K. Given that the molar enthalpy of vaporization at 353.24 K is 30.8 kJ·mol^{-1} and that the molar volume of liquid benzene at 353.24 K is 96.0 cm^3·mol^{-1}, use the above equation to determine the molar volume of the vapor at its equilibrium pressure at 353.24 K and compare this value with the ideal-gas value.

SOLUTION: We start with Equation 23.10, which we solve for $\Delta_{\text{vap}}\overline{V}$

$$\Delta_{\text{vap}}\overline{V} = \frac{\Delta_{\text{vap}}\overline{H}}{T(dP/dT)}$$

Using the above empirical vapor pressure equation at $T = 353.24$ K,

$$\frac{dP}{dT} = P\left(\frac{3229.86 \text{ K}}{T^2} + \frac{236690 \text{ K}^2}{T^3}\right)$$
$$= (760 \text{ torr})(0.0312 \text{ K}^{-1}) = 23.75 \text{ torr·K}^{-1} = 0.0312 \text{ atm·K}^{-1}$$

Therefore,

$$\Delta_{\text{vap}}\overline{V} = \frac{30800 \text{ J·mol}^{-1}}{(353.24 \text{ K})(0.0312 \text{ atm·K}^{-1})}$$
$$= (2790 \text{ J·atm}^{-1}\text{·mol}^{-1})\left(\frac{0.08206 \text{ L·atm}}{8.314 \text{ J}}\right)$$
$$= 27.6 \text{ L·mol}^{-1}$$

The molar volume of the vapor is

$$\overline{V}^{\text{g}} = \Delta_{\text{vap}}\overline{V} + \overline{V}^{\text{l}} = 27.5 \text{ L·mol}^{-1} + 0.0960 \text{ L·mol}^{-1}$$
$$= 27.7 \text{ L·mol}^{-1}$$

The corresponding value from the ideal gas equation is

$$\overline{V}^{\text{g}} = \frac{RT}{P}$$
$$= \frac{(0.08206 \text{ L·atm·K}^{-1}\text{·mol}^{-1})(353.24 \text{ K})}{1 \text{ atm}}$$
$$= 29.0 \text{ L·mol}^{-1}$$

which is slightly larger than the actual value.

23–4. The Clausius–Clapeyron Equation Gives the Vapor Pressure of a Substance As a Function of Temperature

When we used Equation 23.10 to calculate the variation of the melting points of ice (Example 23–3) and benzene, we assumed that $\Delta_{trs}\overline{H}$ and $\Delta_{trs}\overline{V}$ do not vary appreciably with pressure. Although this approximation is fairly satisfactory for solid-liquid and solid-solid transitions over a small ΔT, it is not satisfactory for liquid-gas and solid-gas transitions because the molar volume of a gas varies strongly with pressure. If the temperature is not too near the critical point, however, Equation 23.10 can be cast into a very useful form for condensed phase-gas phase transitions.

Let's apply Equation 23.10 to a liquid-vapor equilibrium. In this case, we have

$$\frac{dP}{dT} = \frac{\Delta_{vap}\overline{H}}{T(\overline{V}^g - \overline{V}^l)} \tag{23.11}$$

Equation 23.11 gives the slope of the liquid-vapor equilibrium line in the phase diagram of the substance. As long as we are not too near the critical point, $\overline{V}^g \gg \overline{V}^l$, so that we can neglect $\overline{V}^l$ compared with $\overline{V}^g$ in the denominator of Equation 23.11. Furthermore, if the vapor pressure is not too high (once again, if we are not too close to the critical point), we can assume the vapor is ideal and replace $\overline{V}^g$ by RT/P, so that Equation 23.11 becomes

$$\frac{1}{P}\frac{dP}{dT} = \frac{d\ln P}{dT} = \frac{\Delta_{vap}\overline{H}}{RT^2} \tag{23.12}$$

This equation, which was first derived by Clausius in 1850, is known as the *Clausius-Clapeyron equation*. Remember that we have neglected the molar volume of the liquid compared with the molar volume of the gas and that we assumed the vapor can be treated as an ideal gas. Nevertheless, Equation 23.12 has the advantage of being more convenient to use than Equation 23.10. As might be expected, however, Equation 23.10 is more accurate than Equation 23.12.

The real advantage of Equation 23.12 is that it can be readily integrated. If we assume $\Delta_{vap}\overline{H}$ does not vary with temperature over the integration limits of T, Equation 23.12 becomes

$$\ln\frac{P_2}{P_1} = -\frac{\Delta_{vap}\overline{H}}{R}\left(\frac{1}{T_2} - \frac{1}{T_1}\right) = \frac{\Delta_{vap}\overline{H}}{R}\left(\frac{T_2 - T_1}{T_1 T_2}\right) \tag{23.13}$$

Equation 23.13 can be used to calculate the vapor pressure at some temperature given the molar enthalpy of vaporization and the vapor pressure at some other temperature. For example, the normal boiling point of benzene is 353.2 K and $\Delta_{vap}\overline{H} = 30.8$ kJ·mol^{-1}. Assuming $\Delta_{vap}\overline{H}$ does not vary with temperature, let's calculate the vapor pressure of benzene at 373.2 K. We substitute $P_1 = 760$ torr, $T_1 = 353.2$ K, and $T_2 = 373.2$ K into Equation 23.13 to obtain

$$\ln \frac{P}{760} = \left(\frac{30800 \text{ J} \cdot \text{mol}^{-1}}{8.314 \text{ J} \cdot \text{K}^{-1} \cdot \text{mol}^{-1}} \right) \left(\frac{19.8 \text{ K}}{(353.2 \text{ K})(373.2 \text{ K})} \right)$$
$$= 0.556$$

or $P = 1330$ torr. The experimental value is 1360 torr.

EXAMPLE 23–5
The vapor pressure of water at 363.2 K is 529 torr. Use Equation 23.13 to determine the average value of $\Delta_{\text{vap}} \overline{H}$ of water between 363.2 K and 373.2 K.

SOLUTION: We use the fact that the normal boiling point of water is 373.2 K ($P = 760$ torr) and write

$$\ln \frac{760}{529} = \frac{\Delta_{\text{vap}} \overline{H}}{8.314 \text{ J} \cdot \text{K}^{-1} \cdot \text{mol}^{-1}} \frac{10.0 \text{ K}}{(363.2 \text{ K})(373.2 \text{ K})}$$

or

$$\Delta_{\text{vap}} \overline{H} = 40.8 \text{ kJ} \cdot \text{mol}^{-1}$$

The value of $\Delta_{\text{vap}} \overline{H}$ for water at its normal boiling point is 40.65 kJ·mol⁻¹.

If we integrate Equation 23.12 indefinitely rather than between definite limits, we obtain (assuming $\Delta_{\text{vap}} \overline{H}$ is constant)

$$\ln P = -\frac{\Delta_{\text{vap}} \overline{H}}{RT} + \text{constant} \tag{23.14}$$

Equation 23.14 says that a plot of the logarithm of the vapor pressure against the reciprocal of the kelvin temperature should be a straight line with a slope of $-\Delta_{\text{vap}} \overline{H}/R$. Figure 23.13 shows such a plot for benzene over the temperature range 313 K to 353 K. The slope of the line gives $\Delta_{\text{vap}} \overline{H} = 32.3$ kJ·mol⁻¹. This value represents an average value of $\Delta_{\text{vap}} \overline{H}$ over the given temperature interval. The value of $\Delta_{\text{vap}} \overline{H}$ at the normal boiling point (353 K) is 30.8 kJ·mol⁻¹.

We can recognize that $\Delta_{\text{vap}} \overline{H}$ varies with temperature by writing $\Delta_{\text{vap}} \overline{H}$ in the form

$$\Delta_{\text{vap}} \overline{H} = A + BT + CT^2 + \cdots$$

where A, B, C, ... are constants. If this equation is substituted into Equation 23.12, then integration gives

$$\ln P = -\frac{A}{RT} + \frac{B}{R} \ln T + \frac{C}{R} T + k + O(T^2) \tag{23.15}$$

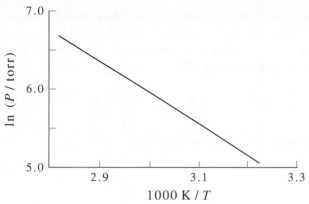

FIGURE 23.13
A plot of the logarithm of the vapor pressure of liquid benzene against the reciprocal kelvin temperature over a temperature range of 313 K to 353 K .

where k is an integration constant. Equation 23.15 expresses the variation of vapor pressure over a larger temperature range than Equation 23.14. Thus, a plot of $\ln P$ against $1/T$ will not be exactly linear, in agreement with the experimental data for most liquids and solids over an extended temperature range. For example, the vapor pressure of solid ammonia in torr is found to obey the equation

$$\ln(P/\text{torr}) = -\frac{4124.4 \text{ K}}{T} - 1.81630 \ln(T/K) + 34.4834 \qquad (23.16)$$

from 146 K to 195 K.

EXAMPLE 23–6
Use the Clausius-Clapeyron equation and Equation 23.16 to determine the molar enthalpy of sublimation of ammonia from 146 K to 195 K.

SOLUTION: According to Equation 23.12

$$\frac{d\ln P}{dT} = \frac{\Delta_{sub}\overline{H}}{RT^2}$$

Using Equation 23.16 for $\ln P$ gives us

$$\frac{\Delta_{sub}\overline{H}}{RT^2} = \frac{4124.4 K^2}{T^2} - \frac{1.8163 K}{T}$$

or

$$\Delta_{sub}\overline{H} = (4124.4 \text{ K})R - (1.8163)RT$$

$$= 34.29 \text{ kJ·mol}^{-1} - (0.0151 \text{ kJ·mol}^{-1}\text{·K}^{-1})T$$

$$146 \text{ K} < T < 195 \text{ K}$$

The Clausius-Clapeyron equation can be used to show that the slope of the solid-gas coexistence curve must be greater than the slope of the liquid-gas coexistence curve near the triple point, where these two curves meet. According to Equation 23.12, the slope of the solid-gas curve is given by

$$\frac{dP^s}{dT} = P^s \frac{\Delta_{sub}\overline{H}}{RT^2} \tag{23.17}$$

and the slope of the liquid-gas curve is given by

$$\frac{dP^l}{dT} = P^l \frac{\Delta_{vap}\overline{H}}{RT^2} \tag{23.18}$$

At the triple point, P^s and P^l, the vapor pressures of the solid and liquid, respectively, are equal, so the ratio of the slopes from Equations 23.17 and 23.18 is

$$\frac{dP^s/dT}{dP^l/dT} = \frac{\Delta_{sub}\overline{H}}{\Delta_{vap}\overline{H}} \tag{23.19}$$

at the triple point. Because enthalpy is a state function, the enthalpy change in going directly from the solid phase to the gas phase is the same as first going from the solid phase to the liquid phase and then going from the liquid phase to the gas phase. In an equation, we have

$$\Delta_{sub}\overline{H} = \Delta_{fus}\overline{H} + \Delta_{vap}\overline{H} \tag{23.20}$$

where the three $\Delta\overline{H}$s must all be evaluated at the same temperature. If we substitute Equation 23.20 into Equation 23.19, we see that

$$\frac{dP^s/dT}{dP^l/dT} = 1 + \frac{\Delta_{fus}\overline{H}}{\Delta_{vap}\overline{H}}$$

Thus, we see that the slope of the solid-gas curve is greater than that of the liquid-gas curve at the triple point.

EXAMPLE 23–7

The vapor pressures of solid and liquid ammonia near the triple point are given by

$$\log(P^s/\text{torr}) = 10.0 - \frac{1630 \text{ K}}{T}$$

and

$$\log(P^l/\text{torr}) = 8.46 - \frac{1330 \text{ K}}{T}$$

Calculate the ratio of the slopes of the solid-gas curve and the liquid-gas curve at the triple point.

SOLUTION: The derivatives of both expressions at the triple point are

$$\frac{dP^s}{dT} = (2.303 P_{tp}) \left(\frac{1630 \text{ K}}{T_{tp}^2} \right) = 4.31 \text{ torr} \cdot \text{K}^{-1}$$

and

$$\frac{dP^l}{dT} = (2.303 P_{tp}) \left(\frac{1330 \text{ K}}{T_{tp}^2} \right) = 3.52 \text{ torr} \cdot \text{K}^{-1}$$

so the ratio of the slopes is $4.31/3.52 = 1.22$.

23–5. Chemical Potential Can Be Evaluated From a Partition Function

In this section, we will derive a convenient formula for the chemical potential in terms of a partition function. Recall that the corresponding formulas for the energy and entropy are (see Equations 17.21 and 20.43)

$$U = k_B T^2 \left(\frac{\partial \ln Q}{\partial T} \right)_{N,V} \tag{23.21}$$

and

$$S = k_B T \left(\frac{\partial \ln Q}{\partial T} \right)_{N,V} + k_B \ln Q \tag{23.22}$$

Using the fact that the Helmholtz energy A is equal to $U - TS$, Equations 23.21 and 23.22 give

$$A = -k_B T \ln Q \tag{23.23}$$

Let's now include N in our discussion of natural variables, and write

$$dA = \left(\frac{\partial A}{\partial T} \right)_{N,V} dT + \left(\frac{\partial A}{\partial V} \right)_{N,T} dV + \left(\frac{\partial A}{\partial N} \right)_{T,V} dN$$
$$= -SdT - PdV + \left(\frac{\partial A}{\partial N} \right)_{T,V} dN \tag{23.24}$$

The last term in Equation 23.24 is expressed in terms of N, the number of molecules in the system. It is more conventional to express this quantity in terms of n, the number of moles in the system. We can do this by noting that

$$\left(\frac{\partial A}{\partial N} \right)_{T,V} dN = \left(\frac{\partial A}{\partial n} \right)_{T,V} dn$$

because n and N differ by a constant factor of the Avogadro constant. Therefore, we may write Equation 23.24 in the form

$$dA = -SdT - PdV + \left(\frac{\partial A}{\partial n}\right)_{T,V} dn \tag{23.25}$$

We'll now show that $(\partial A/\partial n)_{T,V}$ is just another way of writing the chemical potential, μ. If we add $d(PV)$ to both sides of Equation 23.25 and use the equation $G = A + PV$, we get

$$dG = dA + d(PV) = -SdT + VdP + \left(\frac{\partial A}{\partial n}\right)_{T,V} dn$$

But if we compare this result to the total derivative of $G = G(T, P, n)$,

$$dG = \left(\frac{\partial G}{\partial T}\right)_{P,N} dT + \left(\frac{\partial G}{\partial P}\right)_{T,N} dP + \left(\frac{\partial G}{\partial n}\right)_{T,P} dn$$
$$= -SdT + VdP + \mu dn$$

we see that

$$\mu = \left(\frac{\partial G}{\partial n}\right)_{T,P} = \left(\frac{\partial A}{\partial n}\right)_{T,V} \tag{23.26}$$

Thus, we can use either G or A to determine μ as long as we keep the natural variables of each one fixed when we take the partial derivative with respect to n.

We can now substitute Equation 23.23 into Equation 23.26 to obtain

$$\mu = -k_B T \left(\frac{\partial \ln Q}{\partial n}\right)_{V,T} = -RT \left(\frac{\partial \ln Q}{\partial N}\right)_{V,T} \tag{23.27}$$

We have gone from the second term to the third term by multipying k_B and n by the Avogadro constant. Equation 23.27 takes on a fairly simple form for an ideal gas. If we substitute the ideal-gas expression

$$Q(N, V, T) = \frac{[q(V, T)]^N}{N!}$$

into $\ln Q$, we can write

$$\ln Q = N \ln q - N \ln N + N$$

where we have used Stirling's approximation for $\ln N!$. If we substitute this result into Equation 23.27, we obtain

$$\mu = -RT(\ln q - \ln N - 1 + 1)$$
$$= -RT \ln \frac{q(V, T)}{N} \qquad \text{(ideal gas)} \tag{23.28}$$

Recall now that $q(V, T) \propto V$ for an ideal gas, and so we can write Equation 23.28 as

$$\mu = -RT \ln\left[\left(\frac{q}{V}\right)\frac{V}{N}\right] \qquad (23.29)$$

where $q(V, T)/V$ is a function of temperature only. Equation 23.29 also gives us an equation for G because $G = n\mu$. We can make Equation 23.29 look exactly like Equation 22.59 if we substitute $k_B T/P$ for V/N:

$$\mu = -RT \ln\left[\left(\frac{q}{V}\right)\frac{k_B T}{P}\right]$$

$$= -RT \ln\left[\left(\frac{q}{V}\right)k_B T\right] + RT \ln P \qquad (23.30)$$

If we compare this equation with

$$\mu(T, P) = \mu^\circ(T) + RT \ln P \qquad (23.31)$$

we see that

$$\mu^\circ(T) = -RT \ln\left[\left(\frac{q}{V}\right)k_B T\right] \qquad (23.32)$$

Once again, recall that q/V is a function of T only for an ideal gas.

To calculate $\mu^\circ(T)$, we must remember that P is expressed relative to the standard state pressure P°, which is equal to one bar or 10^5 Pa. We emphasize this convention by writing Equation 23.31 as

$$\mu(T, P) = \mu^\circ(T) + RT \ln \frac{P}{P^\circ} \qquad (23.33)$$

If we compare Equation 23.33 with Equation 23.30, we see that

$$\mu^\circ(T) = -RT \ln\left[\left(\frac{q}{V}\right)k_B T\right] + RT \ln P^\circ$$

$$= -RT \ln\left[\left(\frac{q}{V}\right)\frac{k_B T}{P^\circ}\right] \qquad (23.34)$$

The argument of the logarithm in Equation 23.34 is unitless, as it must be. Equation 23.34 gives us a molecular formula to calculate $\mu^\circ(T)$, or $G^\circ(T)$. For example, for Ar(g) at 298.15 K:

$$\frac{q(V, T)}{V} = \left(\frac{2\pi m k_B T}{h^2}\right)^{3/2}$$

$$= \left[\frac{(2\pi)(0.03995 \text{ kg·mol}^{-1})(1.3806 \times 10^{-23} \text{ J·K}^{-1})(298.15 \text{ K})}{(6.022 \times 10^{23} \text{ mol}^{-1})(6.626 \times 10^{-34} \text{ J·s})^2}\right]^{3/2}$$

$$= 2.444 \times 10^{32} \text{ m}^{-3}$$

$$\frac{k_B T}{P^\circ} = \frac{RT}{N_A P^\circ} = \frac{(8.314 \text{ J·mol}^{-1} \cdot \text{K}^{-1})(298.15 \text{ K})}{(6.022 \times 10^{23} \text{ mol}^{-1})(1.00 \times 10^5 \text{ Pa})}$$

$$= 4.116 \times 10^{-26} \text{ m}^3$$

and

$$RT = (8.314 \text{ J·K}^{-1} \cdot \text{mol}^{-1})(298.15 \text{ K}) = 2479 \text{ J·mol}^{-1}$$

and so

$$\mu^\circ(298.15 \text{ K}) = -(2479 \text{ J·mol}^{-1}) \ln\left[(2.444 \times 10^{32} \text{ m}^{-3})(4.116 \times 10^{-26} \text{ m}^3)\right]$$

$$= -(2479 \text{ J·mol}^{-1}) \ln[1.006 \times 10^7]$$

$$= -3.997 \times 10^4 \text{ J·mol}^{-1} = -39.97 \text{ kJ·mol}^{-1}$$

This result is in excellent agreement with the experimental value of $-39.97 \text{ kJ·mol}^{-1}$.

Being essentially an energy, the value of the chemical potential must be based upon some choice of a zero of energy. The chemical potential we have just calculated is based upon the ground state of the atom being zero. For diatomic molecules, we have chosen the ground-state energy (vibrational and electronic) to be $-D_0$, as illustrated in Figure 18.2. In tabulating values of $\mu^\circ(T)$, it is customary to take the ground-state energy of the molecule rather than the separated atoms as in Figure 18.2 to be the zero of energy. To see how this definition of the zero of energy changes the form of the partition function, write

$$q(V, T) = \sum_j e^{-\varepsilon_j / k_B T}$$

$$= e^{-\varepsilon_0 / k_B T} + e^{-\varepsilon_1 / k_B T} + \cdots$$

If we factor out $e^{-\varepsilon_0 / k_B T}$, we have

$$q(V, T) = e^{-\varepsilon_0 / k_B T}[1 + e^{-(\varepsilon_1 - \varepsilon_0)/k_B T} + e^{-(\varepsilon_2 - \varepsilon_0)/k_B T} + \cdots]$$

$$= e^{-\varepsilon_0 / k_B T} q^0(V, T) \tag{23.35}$$

where we have written $q^0(V, T)$ to emphasize that the ground-state energy of the molecule is taken to be zero. Substituting this result into Equation 23.34 gives

$$\mu^\circ(T) - E_0 = -RT \ln\left[\left(\frac{q^0}{V}\right)\frac{k_B T}{P^\circ}\right]$$

$$= -RT \ln\left[\left(\frac{q^0}{V}\right)\frac{RT}{N_A P^\circ}\right] \tag{23.36}$$

where $E_0 = N_A \varepsilon_0$ and $P^\circ = 1 \text{ bar} = 10^5 \text{ Pa}$.

The partition function $q^0(V, T)$ for a diatomic molecule is

$$q^0(V, T) = \left(\frac{2\pi m k_B T}{h^2}\right)^{3/2} V \cdot \frac{T}{\sigma \Theta_{rot}} \cdot \frac{1}{1 - e^{-\Theta_{vib}/T}} \cdot g_{e1} \qquad (23.37)$$

Notice that this expression is the same as Equation 18.39 except for the factor of $e^{-h\nu/2k_B T} e^{D_e/k_B T} = e^{D_0/k_B T}$ in Equation 18.39, which accounts for the ground-state energy being taken to be $-D_0$. The ground-state energy associated with $q^0(V, T)$ given by Equation 23.37 is zero. Let's use Equation 23.36 along with Equation 23.37 to calculate $\mu^\circ - E_0$ for HI(g) at 298.15 K in the harmonic oscillator-rigid rotator approximation, with $\Theta_{rot} = 9.25$ K and $\Theta_{vib} = 3266$ K (Table 18.2). Therefore,

$$\frac{q^0(V, T)}{V} = \left[\frac{(2\pi)(0.1279 \text{ kg·mol}^{-1})(1.3806 \times 10^{-23} \text{ J·K}^{-1})(298.15 \text{ K})}{(6.022 \times 10^{23} \text{ mol}^{-1})(6.626 \times 10^{-34} \text{ J·s})^2}\right]^{3/2}$$

$$\times \left(\frac{298.15 \text{ K}}{9.25 \text{ K}}\right) \frac{1}{1 - e^{-3266 \text{ K}/298.15 \text{ K}}}$$

$$= 4.51 \times 10^{34} \text{ m}^{-3}$$

$$\frac{RT}{N_A P^\circ} = \frac{(8.314 \text{ J·mol}^{-1}\cdot\text{K}^{-1})(298.15 \text{ K})}{(6.022 \times 10^{23} \text{ mol}^{-1})(10^5 \text{ Pa})}$$

$$= 4.116 \times 10^{-26} \text{ m}^3$$

and

$$\mu^\circ(298.15 \text{ K}) - E_0 = -(8.314 \text{ J·mol}^{-1}\cdot\text{K}^{-1})(298.15 \text{ K}) \ln(1.86 \times 10^9)$$

$$= -52.90 \text{ kJ·mol}^{-1}$$

The literature value, which includes anharmonic and nonrigid rotator effects, is -52.94 kJ·mol^{-1}. We will use values of $\mu^\circ(T) - E_0$ when we discuss chemical equilibria in Chapter 24.

Problems

23-1. Sketch the phase diagram for oxygen using the following data: triple point, 54.3 K and 1.14 torr; critical point, 154.6 K and 37 828 torr; normal melting point, −218.4°C; and normal boiling point, −182.9°C. Does oxygen melt under an applied pressure as water does?

23-2. Sketch the phase diagram for I_2 given the following data: triple point, 113°C and 0.12 atm; critical point, 512°C and 116 atm; normal melting point, 114°C; normal boiling point, 184°C; and density of liquid > density of solid.

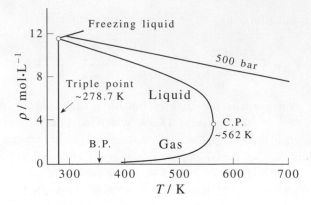

FIGURE 23.14
A density-temperature phase diagram of benzene.

23-3. Figure 23.14 shows a density-temperature phase diagram for benzene. Using the following data for the triple point and the critical point, interpret this phase diagram. Why is the triple point indicated by a line in this type of phase diagram?

	T/K	P/bar	$\rho/mol \cdot L^{-1}$ Vapor	Liquid
Triple point	278.680	0.04785	0.002074	11.4766
Critical point	561.75	48.7575	3.90	3.90
Normal freezing point	278.68	1.01325		
Normal boiling point	353.240	1.01325	0.035687	10.4075

23-4. The vapor pressures of solid and liquid chlorine are given by

$$\ln(P^s/\text{torr}) = 24.320 - \frac{3777 \text{ K}}{T}$$

$$\ln(P^l/\text{torr}) = 17.892 - \frac{2669 \text{ K}}{T}$$

where T is the absolute temperature. Calculate the temperature and pressure at the triple point of chlorine.

23-5. The pressure along the melting curve from the triple-point temperature to an arbitrary temperature can be fit empirically by the Simon equation, which is

$$(P - P_{tp})/\text{bar} = a \left[\left(\frac{T}{T_{tp}} \right)^\alpha - 1 \right]$$

where a and α are constants whose values depend upon the substance. Given that $P_{tp} = 0.04785$ bar, $T_{tp} = 278.68$ K, $a = 4237$, and $\alpha = 2.3$ for benzene, plot P against T and compare your result with that given in Figure 23.2.

23-6. The slope of the melting curve of methane is given by

$$\frac{dP}{dT} = (0.08446 \text{ bar} \cdot \text{K}^{-1.85}) T^{0.85}$$

from the triple point to arbitrary temperatures. Using the fact that the temperature and pressure at the triple point are 90.68 K and 0.1174 bar, calculate the melting pressure of methane at 300 K.

23-7. The vapor pressure of methanol along the entire liquid-vapor coexistence curve can be expressed very accurately by the empirical equation

$$\ln(P/\text{bar}) = -\frac{10.752849}{x} + 16.758207 - 3.603425x$$
$$+ 4.373232x^2 - 2.381377x^3 + 4.572199(1-x)^{1.70}$$

where $x = T/T_c$, and $T_c = 512.60$ K. Use this formula to show that the normal boiling point of methanol is 337.67 K.

23-8. The standard boiling point of a liquid is the temperature at which the vapor pressure is exactly one bar. Use the empirical formula given in the previous problem to show that the standard boiling point of methanol is 337.33 K.

23-9. The vapor pressure of benzene along the liquid-vapor coexistence curve can be accurately expressed by the empirical expression

$$\ln(P/\text{bar}) = -\frac{10.655375}{x} + 23.941912 - 22.388714x$$
$$+ 20.2085593x^2 - 7.219556x^3 + 4.84728(1-x)^{1.70}$$

where $x = T/T_c$, and $T_c = 561.75$ K. Use this formula to show that the normal boiling point of benzene is 353.24 K. Use the above expression to calculate the standard boiling point of benzene.

23-10. Plot the following data for the densities of liquid and gaseous ethane in equilibrium with each other as a function of temperature, and determine the critical temperature of ethane.

T/K	$\rho^l/\text{mol}\cdot\text{dm}^{-3}$	$\rho^g/\text{mol}\cdot\text{dm}^{-3}$	T/K	$\rho^l/\text{mol}\cdot\text{dm}^{-3}$	$\rho^g/\text{mol}\cdot\text{dm}^{-3}$
100.00	21.341	1.336×10^{-3}	283.15	12.458	2.067
140.00	19.857	0.03303	293.15	11.297	2.880
180.00	18.279	0.05413	298.15	10.499	3.502
220.00	16.499	0.2999	302.15	9.544	4.307
240.00	15.464	0.5799	304.15	8.737	5.030
260.00	14.261	1.051	304.65	8.387	5.328
270.00	13.549	1.401	305.15	7.830	5.866

23-11. Use the data in the preceding problem to plot $(\rho^l + \rho^g)/2$ against $T_c - T$, with $T_c = 305.4$ K. The resulting straight line is an empirical law called the *law of rectilinear diameters*. If this curve is plotted on the same figure as in the preceding problem, the intersection of the two curves gives the critical density, ρ_c.

23-12. Use the data in Problem 23–10 to plot $(\rho^l - \rho^g)$ against $(T_c - T)^{1/3}$ with $T_c = 305.4$ K. What does this plot tell you?

23-13. The densities of the coexisting liquid and vapor phases of methanol from the triple point to the critical point are accurately given by the empirical expressions

$$\frac{\rho^l}{\rho_c} - 1 = 2.51709(1 - x)^{0.350} + 2.466694(1 - x)$$

$$- 3.066818(1 - x^2) + 1.325077(1 - x^3)$$

and

$$\ln \frac{\rho^g}{\rho_c} = -10.619689 \frac{1 - x}{x} - 2.556682(1 - x)^{0.350}$$

$$+ 3.881454(1 - x) + 4.795568(1 - x)^2$$

where $\rho_c = 8.40$ mol·L^{-1} and $x = T/T_c$, where $T_c = 512.60$ K. Use these expressions to plot ρ^l and ρ^g against temperature, as in Figure 23.7. Now plot $(\rho^l + \rho^g)/2$ against T. Show that this line intersects the ρ^l and ρ^g curves at $T = T_c$.

23-14. Use the expressions given in the previous problem to plot $(\rho^l - \rho^g)/2$ against $(T_c - T)^{1/3}$. Do you get a reasonably straight line? If not, determine the value of the exponent of $(T_c - T)$ that gives the best straight line.

23-15. The molar enthalpy of vaporization of ethane can be expressed as

$$\Delta_{vap}\overline{H}(T)/\text{kJ·mol}^{-1} = \sum_{j=1}^{6} A_j x^j$$

where $A_1 = 12.857$, $A_2 = 5.409$, $A_3 = 33.835$, $A_4 = -97.520$, $A_5 = 100.849$, $A_6 = -37.933$, and $x = (T_c - T)^{1/3}/(T_c - T_{tp})^{1/3}$ where the critical temperature $T_c = 305.4$ K and the triple point temperature $T_{tp} = 90.35$ K. Plot $\Delta_{vap}\overline{H}(T)$ versus T and show that the curve is similar to that of Figure 23.8.

23-16. Fit the following data for argon to a cubic polynomial in T. Use your result to determine the critical temperature. Repeat using a fifth-degree polynomial.

T/K	$\Delta_{vap}\overline{H}/\text{J·mol}^{-1}$	T/K	$\Delta_{vap}\overline{H}/\text{J·mol}^{-1}$
83.80	6573.8	122.0	4928.7
86.0	6508.4	126.0	4665.0
90.0	6381.8	130.0	4367.7
94.0	6245.2	134.0	4024.7
98.0	6097.7	138.0	3618.8
102.0	5938.8	142.0	3118.2
106.0	5767.6	146.0	2436.3
110.0	5583.0	148.0	1944.5
114.0	5383.5	149.0	1610.2
118.0	5166.5	150.0	1131.5

23-17. Use the following data for methanol at one atm to plot $\overline{G}$ versus T around the normal boiling point (337.668 K). What is the value of $\Delta_{vap}\overline{H}$?

T/K	$\overline{H}/\text{kJ}\cdot\text{mol}^{-1}$	$\overline{S}/\text{J}\cdot\text{mol}^{-1}\cdot\text{K}^{-1}$
240	4.7183	112.259
280	7.7071	123.870
300	9.3082	129.375
320	10.9933	134.756
330	11.8671	137.412
337.668	12.5509	139.437
337.668	47.8100	243.856
350	48.5113	245.937
360	49.0631	247.492
380	50.1458	250.419
400	51.2257	253.189

23-18. In this problem, we will sketch $\overline{G}$ versus P for the solid, liquid, and gaseous phases for a generic ideal substance as in Figure 23.11. Let $\overline{V}^{\,s} = 0.600$, $\overline{V}^{\,l} = 0.850$, and $RT = 2.5$, in arbitrary units. Now show that

$$\overline{G}^{\,s} = 0.600(P - P_0) + \overline{G}_0^{\,s}$$
$$\overline{G}^{\,l} = 0.850(P - P_0) + \overline{G}_0^{\,l}$$

and

$$\overline{G}^{\,g} = 2.5\ln(P/P_0) + \overline{G}_0^{\,g}$$

where $P_0 = 1$ and $\overline{G}_0^{\,s}$, $\overline{G}_0^{\,l}$, and $\overline{G}_0^{\,g}$ are the respective zeros of energy. Show that if we (arbitrarily) choose the solid and liquid phases to be in equilibrium at $P = 2.00$ and the liquid and gaseous phases to be in equilibrium at $P = 1.00$, then we obtain

$$\overline{G}_0^{\,s} - \overline{G}_0^{\,l} = 0.250$$

and

$$\overline{G}_0^{\,l} = \overline{G}_0^{\,g}$$

from which we obtain (by adding these two results)

$$\overline{G}_0^{\,s} - \overline{G}_0^{\,g} = 0.250$$

Now we can express $\overline{G}^{\,s}$, $\overline{G}^{\,l}$, and $\overline{G}^{\,g}$ in terms of a common zero of energy, $\overline{G}_0^{\,g}$, which we must do to compare them with each other and to plot them on the same graph. Show that

$$\overline{G}^{\,s} - \overline{G}_0^{\,g} = 0.600(P - 1) + 0.250$$
$$\overline{G}^{\,l} - \overline{G}_0^{\,g} = 0.850(P - 1)$$
$$\overline{G}^{\,g} - \overline{G}_0^{\,g} = 2.5\ln P$$

Plot these on the same graph from $P = 0.100$ to 3.00 and compare your result with Figure 23.11.

23-19. In this problem, we will demonstrate that entropy always increases when there is a material flow from a region of higher concentration to one of lower concentration. (Compare with Problems 22–41 and 22–42.) Consider a two-compartment system enclosed by rigid, impermeable, adiabatic walls, and let the two compartments be separated by a rigid, insulating, but permeable wall. We assume that the two compartments are in equilibrium but that they are not in equilibirum with each other. Show that

$$U_1 = \text{constant}, \qquad U_2 = \text{constant}, \qquad V_1 = \text{constant}, \qquad V_2 = \text{constant},$$

and

$$n_1 + n_2 = \text{constant}$$

for this system. Now show that

$$dS = \frac{dU}{T} + \frac{P}{T}dV - \frac{\mu}{T}dn$$

in general, and that

$$dS = \left(\frac{\partial S_1}{\partial n_1}\right)dn_1 + \left(\frac{\partial S_2}{\partial n_2}\right)dn_2$$
$$= dn_1\left(\frac{\mu_2}{T} - \frac{\mu_1}{T}\right) \geq 0$$

for this system. Use this result to discuss the direction of a (isothermal) material flow under a chemical potential difference.

23-20. Determine the value of dT/dP for water at its normal boiling point of 373.15 K given that the molar enthalpy of vaporization is 40.65 kJ·mol^{-1}, and the densities of the liquid and vapor are 0.9584 g·mL^{-1} and 0.6010 g·L^{-1}, respectively. Estimate the boiling point of water at 2 atm.

23-21. The orthobaric densities of liquid and gaseous ethyl acetate are 0.826 g·mL^{-1} and 0.00319 g·mL^{-1}, respectively, at its normal boiling point (77.11°C). The rate of change of vapor pressure with temperature is 23.0 torr·K^{-1} at the normal boiling point. Estimate the molar enthalpy of vaporization of ethyl acetate at its normal boiling point.

23-22. The vapor pressure of mercury from 400°C to 1300°C can be expressed by

$$\ln(P/\text{torr}) = -\frac{7060.7 \text{ K}}{T} + 17.85$$

The density of the vapor at its normal boiling point is 3.82 g·L^{-1} and that of the liquid is 12.7 g·mL^{-1}. Estimate the molar enthalpy of vaporization of mercury at its normal boiling point.

23-23. The pressures at the solid-liquid coexistence boundary of propane are given by the empirical equation

$$P = -718 + 2.38565T^{1.283}$$

where P is in bars and T is in kelvins. Given that $T_{fus} = 85.46$ K and $\Delta_{fus}\overline{H} = 3.53$ kJ·mol^{-1}, calculate $\Delta_{fus}\overline{V}$ at 85.46 K.

23-24. Use the vapor pressure data given in Problem 23–7 and the density data given in Problem 23–13 to calculate $\Delta_{vap}\overline{H}$ for methanol from the triple point (175.6 K) to the critical point (512.6 K). Plot your result.

23-25. Use the result of the previous problem to plot $\Delta_{vap}\overline{S}$ of methanol from the triple point to the critical point.

23-26. Use the vapor pressure data for methanol given in Problem 23–7 to plot $\ln P$ against $1/T$. Using your calculations from Problem 23–24, over what temperature range do you think the Clausius-Clapeyron equation will be valid?

23-27. The molar enthalpy of vaporization of water is 40.65 kJ·mol^{-1} at its normal boiling point. Use the Clausius-Clapeyron equation to calculate the vapor pressure of water at 110°C. The experimental value is 1075 torr.

23-28. The vapor pressure of benzaldehyde is 400 torr at 154°C and its normal boiling point is 179°C. Estimate its molar enthalpy of vaporization. The experimental value is 42.50 kJ·mol^{-1}.

23-29. Use the following data to estimate the normal boiling point and the molar enthalpy of vaporization of lead.

$T/$K	1500	1600	1700	1800	1900
$P/$torr	19.72	48.48	107.2	217.7	408.2

23-30. The vapor pressure of solid iodine is given by

$$\ln(P/\text{atm}) = -\frac{8090.0 \text{ K}}{T} - 2.013 \ln(T/\text{K}) + 32.908$$

Use this equation to calculate the normal sublimation temperature and the molar enthalpy of sublimation of $I_2(s)$ at 25°C. The experimental value of $\Delta_{sub}\overline{H}$ is 62.23 kJ·mol^{-1}.

23-31. Fit the following vapor pressure data of ice to an equation of the form

$$\ln P = -\frac{a}{T} + b \ln T + cT$$

where T is temperature in kelvins. Use your result to determine the molar enthalpy of sublimation of ice at 0°C.

$t/°C$	$P/torr$	$t/°C$	$P/torr$
−10.0	1.950	−4.8	3.065
−9.6	2.021	−4.4	3.171
−9.2	2.093	−4.0	3.280
−8.8	2.168	−3.6	3.393
−8.4	2.246	−3.2	3.509
−8.0	2.326	−2.8	3.630
−7.6	2.408	−2.4	3.753
−7.2	2.493	−2.0	3.880
−6.8	2.581	−1.6	4.012
−6.4	2.672	−1.2	4.147
−6.0	2.765	−0.8	4.287
−5.6	2.862	−0.4	4.431
−5.2	2.962	0.0	4.579

23-32. The following table gives the vapor pressure data for liquid palladium as a function of temperature:

T/K	P/bar
1587	1.002×10^{-9}
1624	2.152×10^{-9}
1841	7.499×10^{-8}

Estimate the molar enthalpy of vaporization of palladium.

23-33. The sublimation pressure of CO_2 at 138.85 K and 158.75 K is 1.33×10^{-3} bar and 2.66×10^{-2} bar, respectively. Estimate the molar enthalpy of sublimation of CO_2.

23-34. The vapor pressures of solid and liquid hydrogen iodide can be expressed empirically as

$$\ln(P^s/torr) = -\frac{2906.2 \text{ K}}{T} + 19.020$$

and

$$\ln(P^l/torr) = -\frac{2595.7 \text{ K}}{T} + 17.572$$

Calculate the ratio of the slopes of the solid-gas curve and the liquid-gas curve at the triple point.

23-35. Given that the normal melting point, the critical temperature, and the critical pressure of hydrogen iodide are 222 K, 424 K and 82.0 atm, respectively, use the data in the previous problem to sketch the phase diagram of hydrogen iodide.

23-36. Consider the phase change

$$C(\text{graphite}) \rightleftharpoons C(\text{diamond})$$

Given that $\Delta_r G^\circ / \text{J} \cdot \text{mol}^{-1} = 1895 + 3.363 T$, calculate $\Delta_r H^\circ$ and $\Delta_r S^\circ$. Calculate the pressure at which diamond and graphite are in equilibrium with each other at 25°C. Take the density of diamond and graphite to be 3.51 g·cm^{-3} and 2.25 g·cm^{-3}, respectively. Assume that both diamond and graphite are incompressible.

23-37. Use Equation 23.36 to calculate $\mu^\circ - E_0$ for Kr(g) at 298.15 K. The literature value is -42.72 kJ·mol^{-1}.

23-38. Show that Equations 23.30 and 23.32 for $\mu(T, P)$ for a monatomic ideal gas are equivalent to using the relation $\overline{G} = \overline{H} - T\overline{S}$ with $\overline{H} = 5RT/2$ and S given by Equation 20.45.

23-39. Use Equation 23.37 and the molecular parameters in Table 18.2 to calculate $\mu^\circ - E_0$ for N$_2$(g) at 298.15 K. The literature value is -48.46 kJ·mol^{-1}.

23-40. Use Equation 23.37 and the molecular parameters in Table 18.2 to calculate $\mu^\circ - E_0$ for CO(g) at 298.15 K. The literature value is -50.26 kJ·mol^{-1}.

23-41. Use Equation 18.60 [without the factor of $\exp(D_e/k_B T)$] and the molecular parameters in Table 18.4 to calculate $\mu^\circ - E_0$ for CH$_4$(g) at 298.15 K. The literature value is -45.51 kJ·mol^{-1}.

23-42. When we refer to the equilibrium vapor pressure of a liquid, we tacitly assume that some of the liquid has evaporated into a vacuum and that equilibrium is then achieved. Suppose, however, that we are able by some means to exert an additional pressure on the surface of the liquid. One way to do this is to introduce an insoluble, inert gas into the space above the liquid. In this problem, we will investigate how the equilibrium vapor pressure of a liquid depends upon the total pressure exerted on it.

Consider a liquid and a vapor in equilibrium with each other, so that $\mu^l = \mu^g$. Show that

$$\overline{V}^l dP^l = \overline{V}^g dP^g$$

because the two phases are at the same temperature. Assuming that the vapor may be treated as an ideal gas and that $\overline{V}^l$ does not vary appreciably with pressure, show that

$$\ln \frac{P^g(\text{at } P^l = P)}{P^g(\text{at } P^l = 0)} = \frac{\overline{V}^l P^l}{RT}$$

Use this equation to calculate the vapor pressure of water at a total pressure of 10.0 atm at 25°C. Take P^g (at $P^l = 0$) = 0.0313 atm.

23-43. Using the fact that the vapor pressure of a liquid does not vary appreciably with the total pressure, show that the final result of the previous problem can be written as

$$\frac{\Delta P^g}{P^g} = \frac{\overline{V}^l P^l}{RT}$$

Hint: Let $P^g(\text{at } P = P^l) = P^g(\text{at } P = 0) + \Delta P$ and use the fact that ΔP is small. Calculate ΔP for water at a total pressure of 10.0 atm at 25°C. Compare your answer with the one you obtained in the previous problem.

23-44. In this problem, we will show that the vapor pressure of a droplet is not the same as the vapor pressure of a relatively large body of liquid. Consider a spherical droplet of liquid of radius r in equilibrium with a vapor at a pressure P, and a flat surface of the same liquid in equilibrium with a vapor at a pressure P_0. Show that the change in Gibbs energy for the isothermal transfer of dn moles of the liquid from the flat surface to the droplet is

$$dG = dn\, RT \ln \frac{P}{P_0}$$

This change in Gibbs energy is due to the change in surface energy of the droplet (the change in surface energy of the large, flat surface is negligible). Show that

$$dn\, RT \ln \frac{P}{P_0} = \gamma \, dA$$

where γ is the surface tension of the liquid and dA is the change in the surface area of a droplet. Assuming the droplet is spherical, show that

$$dn = \frac{4\pi r^2 dr}{\overline{V}^{\,1}}$$

$$dA = 8\pi r\, dr$$

and finally that

$$\ln \frac{P}{P_0} = \frac{2\gamma \overline{V}^{\,1}}{rRT} \tag{1}$$

Because the right side is positive, we see that the vapor pressure of a droplet is greater than that of a planar surface. What if $r \to \infty$?

23-45. Use Equation 1 of Problem 23–44 to calculate the vapor pressure at 25°C of droplets of water of radius 1.0×10^{-5} cm. Take the surface tension of water to be 7.20×10^{-4} J·m^{-2}.

23-46. Figure 23.15 shows reduced pressure, P_R, plotted against reduced volume, $\overline{V}_R$, for the van der Waals equation at a reduced temperature, T_R, of 0.85. The so-called van der Waals loop apparent in the figure will occur for any reduced temperature less than unity and is a consequence of the simplified form of the van der Waals equation. It turns out that any analytic equation of state (one that can be written as a Maclaurin expansion in the reduced density, $1/\overline{V}_R$) will give loops for subcritical temperatures ($T_R < 1$). The correct behavior as the pressure is increased is given by the path abdfg in Figure 23.15. The horizontal region bdf, not given by the van der Waals equation, represents the condensation of the gas to a liquid at a fixed pressure. We can draw the horizontal line (called a *tie line*) at the correct position by recognizing that the chemical potentials of the liquid and the vapor must be equal at the points b and f. Using this requirement, Maxwell showed that the horizontal line representing condensation should be drawn such that the areas of the loops above and below the line must be equal. To prove *Maxwell's equal-area construction rule*, integrate $(\partial \mu / \partial P)_T = \overline{V}$ by parts along the path bcdef and use the fact that μ^1 (the value of μ at point f) $= \mu^g$ (the value of μ at point b) to obtain

$$\mu^1 - \mu^g = P_0(\overline{V}^1 - \overline{V}^g) - \int_{bcdef} P\, d\overline{V}$$

$$= \int_{bcdef} (P_0 - P)\, d\overline{V}$$

where P_0 is the pressure corresponding to the tie line. Interpret this result.

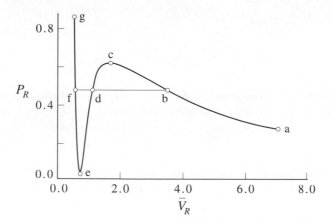

FIGURE 23.15
A plot of reduced pressure, P_R, versus reduced volume, $\overline{V}_R$, for the van der Waals equation at a reduced temperature, T_R, of 0.85.

23-47. The isothermal compressibility, κ_T, is defined by

$$\kappa_T = -\frac{1}{V}\left(\frac{\partial V}{\partial P}\right)_T$$

Because $(\partial P/\partial V)_T = 0$ at the critical point, κ_T diverges there. A question that has generated a great deal of experimental and theoretical research is the question of the manner in which κ_T diverges as T approaches T_c. Does it diverge as $\ln(T - T_c)$ or perhaps as $(T - T_c)^{-\gamma}$ where γ is some *critical exponent*? An early theory of the behavior of thermodynamic functions such as κ_T very near the critical point was proposed by van der Waals, who predicted that κ_T diverges as $(T - T_c)^{-1}$. To see how van der Waals arrived at this prediction, we consider the (double) Taylor expansion of the pressure $P(\overline{V}, T)$ about T_c and $\overline{V}_c$:

$$P(\overline{V}, T) = P(\overline{V}_c, T_c) + (T - T_c)\left(\frac{\partial P}{\partial T}\right)_c + \frac{1}{2}(T - T_c)^2\left(\frac{\partial^2 P}{\partial T^2}\right)_c$$
$$+ (T - T_c)(\overline{V} - \overline{V}_c)\left(\frac{\partial^2 P}{\partial V \partial T}\right)_c + \frac{1}{6}(\overline{V} - \overline{V}_c)^3\left(\frac{\partial^3 P}{\partial \overline{V}^3}\right)_c + \cdots$$

Why are there no terms in $(\overline{V} - \overline{V}_c)$ or $(\overline{V} - \overline{V}_c)^2$? Write this Taylor series as

$$P = P_c + a(T - T_c) + b(T - T_c)^2 + c(T - T_c)(\overline{V} - \overline{V}_c) + d(\overline{V} - \overline{V}_c)^3 + \cdots$$

Now show that

$$\left(\frac{\partial P}{\partial \overline{V}}\right)_T = c(T - T_c) + 3d(\overline{V} - \overline{V}_c)^2 + \cdots \qquad \left(\begin{array}{c} T \to T_c \\ V \to \overline{V}_c \end{array}\right)$$

and that

$$\kappa_T = \frac{-1/\overline{V}}{c(T - T_c) + 3d(\overline{V} - \overline{V}_c)^2 + \cdots}$$

Now let $\overline{V} = \overline{V}_c$ to obtain

$$\kappa_T \propto \frac{1}{T - T_c} \qquad T \to (T_c)$$

Accurate experimental measurements of κ_T as $T \to T_c$ suggest that κ_T diverges a little more strongly than $(T - T_c)^{-1}$. In particular, it is found that $\kappa_T \to (T - T_c)^{-\gamma}$ where $\gamma = 1.24$. Thus, the theory of van der Waals, although qualitatively correct, is not quantitatively correct.

23-48. We can use the ideas of the previous problem to predict how the difference in the densities (ρ^l and ρ^g) of the coexisting liquid and vapor states (*orthobaric densities*) behave as $T \to T_c$. Substitute

$$P = P_c + a(T - T_c) + b(T - T_c)^2 + c(T - T_c)(\overline{V} - \overline{V}_c) + d(\overline{V} - \overline{V}_c)^3 + \cdots \quad (1)$$

into the Maxwell equal-area construction (Problem 23–46) to get

$$P_0 = P_c + a(T - T_c) + b(T - T_c)^2 + \frac{c}{2}(T - T_c)(\overline{V}^l + \overline{V}^g - 2\overline{V}_c)$$

$$+ \frac{d}{4}[(\overline{V}^g - \overline{V}_c)^2 + (\overline{V}^l - \overline{V}_c)^2](\overline{V}^l + \overline{V}^g - 2\overline{V}_c) + \cdots \quad (2)$$

For $P < P_c$, Equation 1 gives loops and so has three roots, $\overline{V}^l$, $\overline{V}_c$, and $\overline{V}^g$ for $P = P_0$. We can obtain a first approximation to these roots by assuming that $\overline{V}_c \approx \frac{1}{2}(\overline{V}^l + \overline{V}^g)$ in Equation 2 and writing

$$P_0 = P_c + a(T - T_c) + b(T - T_c)^2$$

To this approximation, the three roots to Equation 1 are obtained from

$$d(\overline{V} - \overline{V}_c)^3 + c(T - T_c)(\overline{V} - \overline{V}_c) = 0$$

Show that the three roots are

$$\overline{V}_1 = \overline{V}^l = \overline{V}_c - \left(\frac{c}{d}\right)^{1/2}(T_c - T)^{1/2}$$

$$\overline{V}_2 = \overline{V}_c$$

$$\overline{V}_3 = \overline{V}^g = \overline{V}_c + \left(\frac{c}{d}\right)^{1/2}(T_c - T)^{1/2}$$

Now show that

$$\overline{V}^g - \overline{V}^l = 2\left(\frac{c}{d}\right)^{1/2}(T_c - T)^{1/2} \qquad \left(\begin{array}{c} T < T_c \\ T \to T_c \end{array}\right)$$

and that this equation is equivalent to

$$\rho^l - \rho^g \longrightarrow A(T_c - T)^{1/2} \qquad \left(\begin{array}{c} T < T_c \\ T \to T_c \end{array}\right)$$

Thus, the van der Waals theory predicts that the critical exponent in this case is 1/2. It has been shown experimentally that

$$\rho^l - \rho^g \longrightarrow A(T_c - T)^\beta$$

where $\beta = 0.324$. Thus, as in the previous problem, although qualitatively correct, the van der Waals theory is not quantitatively correct.

23-49. The following data give the temperature, the vapor pressure, and the density of the coexisting vapor phase of butane. Use the van der Waals equation and the Redlich-Kwong equation to calculate the vapor pressure and compare your result with the experimental values given below.

T/K	P/bar	$\rho^g/\text{mol·L}^{-1}$
200	0.0195	0.00117
210	0.0405	0.00233
220	0.0781	0.00430
230	0.1410	0.00746
240	0.2408	0.01225
250	0.3915	0.01924
260	0.6099	0.02905
270	0.9155	0.04239
280	1.330	0.06008

23-50. The following data give the temperature, the vapor pressure, and the density of the coexisting vapor phase of benzene. Use the van der Waals equation and the Redlich-Kwong equation to calculate the vapor pressure and compare your result with the experimental values given below. Use Equations 16.17 and 16.18 with $T_c = 561.75$ K and $P_c = 48.7575$ bar to calculate the van der Waals parameters and the Redlich-Kwong parameters.

T/K	P/bar	$\rho^g/\text{mol·L}^{-1}$
290.0	0.0860	0.00359
300.0	0.1381	0.00558
310.0	0.2139	0.00839
320.0	0.3205	0.01223
330.0	0.4666	0.01734
340.0	0.6615	0.02399
350.0	0.9161	0.03248

Joel Hildebrand was born in Camden, NJ, on November 16, 1881, and died in 1983. He received his Ph.D. in chemistry from the University of Pennsylvania in 1906. After spending a year at the University of Berlin with Nernst, he returned to the University of Pennsylvania as an instructor. In 1913, he joined the Department of Chemistry at the University of California at Berkeley, where he stayed for the remainder of his life. Although he officially retired in 1952, he remained professionally active until his death, publishing his last paper in 1981. Hildebrand made significant contributions to the fields of liquids and nonelectrolyte solutions. He retained a long interest in deviations from ideal solutions (Raoult's law) and the theory of regular solutions. His books, *The Solubility of Nonelectrolytes* and *Regular Solutions*, published with Robert Scott, were standard references in the field. Hildebrand was a famed, excellent teacher of general chemistry at Berkeley. His general chemistry text, *Principles of Chemistry*, influenced other schools to place greater emphasis on principles and less on the memorization of specific material in the teaching of general chemistry. Hildebrand was a great lover of the outdoors and especially enjoyed skiing and camping. He managed the U.S. Olympic Ski Team in 1936, was President of the Sierra Club from 1937 to 1940, and wrote a book on camping with his daughter Louise, *Camp Catering* or, *How to rustle grub for hikers, campers, mountaineers, canoeists, hunters, skiers, and fishermen.*

CHAPTER 24

Solutions I: Liquid-Liquid Solutions

In this and the next chapter, we will apply our thermodynamic principles to solutions. This chapter focuses on solutions that consist of two volatile liquids, such as alcohol–water solutions. We will first discuss partial molar quantities, which provide the most convenient set of thermodynamic variables to describe solutions. This discussion will lead to the Gibbs–Duhem equation, which gives us a relation between the change in the properties of one component of a solution in terms of the change in the properties of the other component. The simplest model of a solution is an ideal solution, in which both components obey Raoult's law over the entire composition range. Although a few solutions behave almost ideally, most solutions are not ideal. Just as nonideal gases can be described in terms of fugacity, nonideal solutions can be described in terms of a quantity called activity. Activity must be calculated with respect to a specific standard state, and in Section 24–8 we introduce two commonly-used standard states: a solvent, or Raoult's law standard state, and a solute, or Henry's law standard state.

24–1. Partial Molar Quantities Are Important Thermodynamic Properties of Solutions

Up to this point, we have discussed the thermodynamics of only one-component systems. We will now discuss the thermodynamics of multicomponent systems, although, for simplicity, we will discuss only systems of two components. Most of the concepts and results we will develop are applicable to multicomponent systems. Let's consider a solution consisting of n_1 moles of component 1 and n_2 moles of component 2. The Gibbs energy of this solution is a function of T and P and the two mole numbers n_1 and n_2. We emphasize this dependence of G on these variables by writing $G = G(T, P, n_1, n_2)$. The total derivative of G is given by

$$dG = \left(\frac{\partial G}{\partial T}\right)_{P,n_1,n_2} dT + \left(\frac{\partial G}{\partial P}\right)_{T,n_1,n_2} dP$$

$$+ \left(\frac{\partial G}{\partial n_1}\right)_{P,T,n_2} dn_1 + \left(\frac{\partial G}{\partial n_2}\right)_{P,T,n_1} dn_2 \tag{24.1}$$

963

If the composition of the solution is fixed, so that $dn_1 = dn_2 = 0$, then Equation 24.1 is the same as Equation 22.30, and we have

$$\left(\frac{\partial G}{\partial T}\right)_{P,n_1,n_2} = -S(P, T, n_1, n_2)$$

and

$$\left(\frac{\partial G}{\partial P}\right)_{T,n_1,n_2} = V(P, T, n_1, n_2)$$

As in the previous chapter, the partial derivatives of G with respect to mole numbers are called chemical potentials, or partial molar Gibbs energies. The standard notation for chemical potential is μ, so we can write Equation 24.1 as

$$dG = -SdT + VdP + \mu_1 dn_1 + \mu_2 dn_2 \tag{24.2}$$

where

$$\mu_j = \mu_j(T, P, n_1, n_2) = \left(\frac{\partial G}{\partial n_j}\right)_{T,P,n_{i \neq j}} = \overline{G}_j \tag{24.3}$$

We will see that the chemical potential of each component in the solution plays a central role in determining the thermodynamic properties of the solution.

Other extensive thermodynamic variables have associated partial molar values, although only the partial molar Gibbs energy is given a special symbol and name. For example, $(\partial S/\partial n_j)_{T,P,n_{i \neq j}}$ is called the partial molar entropy and is denoted by $\overline{S}_j$, and $(\partial V/\partial n_j)_{T,P,n_{i \neq j}}$ is called the partial molar volume and is denoted by $\overline{V}_j$. Generally, if $Y = Y(T, P, n_1, n_2)$ is some extensive thermodynamic property, then its associated partial molar quantity, denoted by $\overline{Y}_j$, is by definition

$$\overline{Y}_j = \overline{Y}_j(T, P, n_1, n_2) = \left(\frac{\partial Y}{\partial n_j}\right)_{T,P,n_{i \neq j}} \tag{24.4}$$

Physically, the partial molar quantity $\overline{Y}_j$ is a measure of how Y changes when n_j is changed while keeping T, P, and the other mole numbers fixed.

Partial molar quantities are intensive thermodynamic quantities. In fact, for a pure system, the chemical potential is just the Gibbs energy per mole. We can use the intensive property of partial molar quantities to derive one of the most important relations for solutions. As a concrete example, we will consider a *binary solution*, that is, one composed of two different liquids. The Gibbs energy of a binary solution (Equation 24.2) is

$$dG = -SdT + VdP + \mu_1 dn_1 + \mu_2 dn_2$$

At constant T and P, we have

$$dG = \mu_1 dn_1 + \mu_2 dn_2 \tag{24.5}$$

Now, imagine that we increase the size of the system uniformly by means of a scale parameter λ such that $dn_1 = n_1 d\lambda$ and $dn_2 = n_2 d\lambda$. Note that as we vary λ from 0 to 1, the number of moles of components 1 and 2 varies from 0 to n_1 and 0 to n_2, respectively. Because G depends extensively on n_1 and n_2, we must have that $dG = G d\lambda$. Therefore, the total Gibbs energy varies from 0 to some final value G as λ is varied. Introducing $d\lambda$ into Equation 24.5 gives

$$\int_0^1 G d\lambda = \int_0^1 n_1 \mu_1 d\lambda + \int_0^1 n_2 \mu_2 d\lambda$$

Because G, n_1, and n_2 are final values (and so do not depend upon λ) and μ_1 and μ_2 are intensive variables (and so do not depend upon the size parameter λ), we can write the above equation as

$$G \int_0^1 d\lambda = n_1 \mu_1 \int_0^1 d\lambda + n_2 \mu_2 \int_0^1 d\lambda$$

or, upon integration,

$$G(T, P, n_1, n_2) = \mu_1 n_1 + \mu_2 n_2 \tag{24.6}$$

Note that $G = \mu n$ for a one-component system, which shows once again that μ is the Gibbs energy per mole for a pure system, or more generally, that the partial molar quantity of any extensive thermodynamic quantity of a pure substance is its molar value.

Partial molar quantities have a particularly nice physical interpretation in terms of volume, for which the equivalent equation to Equation 24.6 would be

$$V(T, P, n_1, n_2) = \overline{V}_1 n_1 + \overline{V}_2 n_2 \tag{24.7}$$

Now, when 1-propanol and water are mixed, the final volume of the solution is not equal to the sum of the volumes of pure 1-propanol and water. We can use Equation 24.7 to calculate the final volume of a solution of any composition if we know the partial molar volumes of 1-propanol and water at that composition. Figure 24.1 shows the partial molar volumes of 1-propanol and water as a function of the mole fraction of 1-propanol in 1-propanol/water solutions at 20°C. We can use this figure to estimate the final volume of solution when 100 mL of 1-propanol is mixed with 100 mL of water at 20°C. The densities of 1-propanol and water at 20°C are 0.803 g·mL^{-1} and 0.998 g·mL^{-1}, respectively. Using these densities, we see that 100 mL each of 1-propanol and water corresponds to a mole fraction of 1-propanol of 0.194. Referring

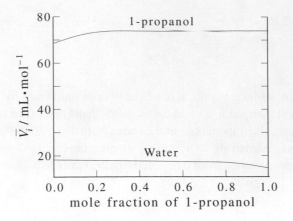

FIGURE 24.1
The partial molar volumes of 1-propanol and water in a 1-propanol/water solution at 20°C plotted against the mole fraction of 1-propanol in the solution.

to Figure 24.1, we see that this corresponds to roughly $\overline{V}_{\text{1-propanol}} = 72 \text{ mL·mol}^{-1}$ and $\overline{V}_{\text{water}} = 18 \text{ mL·mol}^{-1}$. Thus, the final volume of the solution is

$$V = n_1 \overline{V}_{\text{1-propanol}} + n_2 \overline{V}_{\text{water}}$$

$$= \left(\frac{80.3 \text{ g}}{60.09 \text{ g·mol}^{-1}} \right) (72 \text{ mL·mol}^{-1}) + \left(\frac{99.8 \text{ g}}{18.02 \text{ g·mol}^{-1}} \right) (18 \text{ mL·mol}^{-1})$$

$$= 196 \text{ mL}$$

compared with a total volume of 200 mL before mixing. Problems 24–8 through 24–12 involve the determination of partial molar volumes from solution data.

24–2. The Gibbs–Duhem Equation Relates the Change in the Chemical Potential of One Component of a Solution to the Change in the Chemical Potential of the Other

Most of our thermodynamic formulas for single-component systems (pure substances) have analogous formulas in terms of partial molar quantities. For example, if we start with $G = H - TS$ and differentiate with respect to n_j keeping T, P, and $n_{i \neq j}$ fixed, we obtain

$$\left(\frac{\partial G}{\partial n_j} \right)_{T,P,n_{i \neq j}} = \left(\frac{\partial H}{\partial n_j} \right)_{T,P,n_{i \neq j}} - T \left(\frac{\partial S}{\partial n_j} \right)_{T,P,n_{i \neq j}}$$

or

$$\mu_j = \overline{G}_j = \overline{H}_j - T\overline{S}_j \tag{24.8}$$

Furthermore, by using the fact that cross second partial derivatives are equal, we get

$$\overline{S}_j = \left(\frac{\partial S}{\partial n_j}\right)_{T,P,n_{i\neq j}} = \frac{\partial}{\partial n_j}\left(-\frac{\partial G}{\partial T}\right)_{P,n_i} = -\frac{\partial}{\partial T}\left(\frac{\partial G}{\partial n_j}\right)_{T,P,n_{i\neq j}} = -\left(\frac{\partial \mu_j}{\partial T}\right)_{P,n_i}$$

and

$$\overline{V}_j = \left(\frac{\partial V}{\partial n_j}\right)_{T,P,n_{i\neq j}} = \frac{\partial}{\partial n_j}\left(\frac{\partial G}{\partial P}\right)_{T,n_i} = \frac{\partial}{\partial P}\left(\frac{\partial G}{\partial n_j}\right)_{T,P,n_{i\neq j}} = \left(\frac{\partial \mu_j}{\partial P}\right)_{T,n_i}$$

If we substitute these two results into

$$d\mu_j = \left(\frac{\partial \mu_j}{\partial T}\right)_{P,n_i} dT + \left(\frac{\partial \mu_j}{\partial P}\right)_{T,n_i} dP$$

we obtain

$$d\mu_j = -\overline{S}_j dT + \overline{V}_j dP \tag{24.9}$$

which is an extension of Equation 22.30 to multicomponent systems.

EXAMPLE 24–1
Derive an equation for the temperature dependence of $\mu_j(T, P)$ in analogy with the Gibbs–Helmholtz equation (Equation 22.60).

SOLUTION: The Gibbs–Helmholtz equation is (Equation 22.60)

$$\left(\frac{\partial G/T}{\partial T}\right)_{P,n_i} = -\frac{H}{T^2}$$

Now differentiate with respect to n_j and interchange the order of differentiation on the left side to get

$$\left(\frac{\partial \mu_j/T}{\partial T}\right)_P = -\frac{\overline{H}_j}{T^2}$$

where $\overline{H}_j$ is the partial molar enthalpy of component j.

We will now derive one of the most useful equations involving partial molar quantities. First we differentiate Equation 24.6

$$dG = \mu_1 dn_1 + \mu_2 dn_2 + n_1 d\mu_1 + n_2 d\mu_2$$

and subtract Equation 24.5 to get

$$n_1 d\mu_1 + n_2 d\mu_2 = 0 \qquad \text{(constant } T \text{ and } P\text{)} \tag{24.10}$$

If we divide both sides by $n_1 + n_2$, we have

$$x_1 d\mu_1 + x_2 d\mu_2 = 0 \qquad \text{(constant } T \text{ and } P) \qquad (24.11)$$

where x_1 and x_2 are mole fractions. Either of Equations 24.10 or 24.11 is called the *Gibbs–Duhem equation*. The Gibbs–Duhem equation tells us that if we know the chemical potential of one component as a function of composition, we can determine the other. For example, suppose we were to know that

$$\mu_2 = \mu_2^* + RT \ln x_2 \qquad 0 \le x_2 \le 1$$

over the whole range of x_2 (0 to 1). A superscript * is the IUPAC notation for a property of a pure substance, so in this equation, $\mu_2^* = \mu_2(x_2 = 1)$ is the chemical potential of pure component 2. We can differentiate μ_2 with respect to x_2 and substitute into Equation 24.11 to get

$$d\mu_1 = -\frac{x_2}{x_1} d\mu_2 = -RT \frac{x_2}{x_1} d\ln x_2$$

$$= -RT \frac{x_2}{x_1} \frac{dx_2}{x_2} = -RT \frac{dx_2}{x_1} \qquad (0 \le x_2 \le 1)$$

But $dx_2 = -dx_1$ (because $x_1 + x_2 = 1$), so

$$d\mu_1 = RT \frac{dx_1}{x_1} \qquad (0 \le x_1 \le 1)$$

where $0 \le x_1 \le 1$ because $0 \le x_2 \le 1$. Now integrate both sides from $x_1 = 1$ (pure component 1) to arbitrary x_1 to get

$$\mu_1 = \mu_1^* + RT \ln x_1 \qquad (0 \le x_1 \le 1)$$

where $\mu_1^* = \mu_1(x_1 = 1)$. We will see later in this chapter that this result says that if one component of a binary solution obeys Raoult's law over the complete concentration range, the other component does also.

EXAMPLE 24–2
Derive a Gibbs–Duhem type of equation for the volume of a binary solution.

SOLUTION: We start with Equation 24.7, which is the analog of Equation 24.6

$$V(T, P, n_1, n_2) = n_1 \overline{V}_1 + n_2 \overline{V}_2$$

and differentiate (at constant T and P) to obtain

$$dV = n_1 d\overline{V}_1 + \overline{V}_1 dn_1 + n_2 d\overline{V}_2 + \overline{V}_2 dn_2$$

Subtract the analog of Equation 24.5

$$dV = \overline{V}_1 dn_1 + \overline{V}_2 dn_2 \qquad \text{(constant } T \text{ and } P)$$

to obtain

$$n_1 d\overline{V}_1 + n_2 d\overline{V}_2 = 0 \qquad \text{(constant } T \text{ and } P)$$

This equation says that if we know the change in the partial molar volume of one component of a binary system over a range of composition, we can determine the change in the partial molar volume of the other component over the same range.

24–3. At Equilibrium, the Chemical Potential of Each Component Has the Same Value in Each Phase in Which the Component Appears

Consider a binary solution of two liquids that is in equilibrium with its vapor phase, which contains both components. Examples are a solution of 1-propanol and water or a solution of benzene and toluene, each in equilibrium with its vapor. We wish to generalize our treatment in the previous chapter, in which we treated a pure liquid in equilibrium with its vapor phase, and develop the criterion for equilibrium in a binary solution. The Gibbs energy of the solution and its vapor is

$$G = G^{\text{sln}} + G^{\text{vap}}$$

Let n_1^{sln}, n_2^{sln} and n_1^{vap}, n_2^{vap} be the mole numbers of each component in each phase. For generality, let j denote either component 1 or 2, so n_j denotes the number of moles of component j. Now suppose that dn_j moles of component j are transferred from the solution to the vapor at constant T and P, so that $dn_j^{\text{vap}} = +dn_j$ and $dn_j^{\text{sln}} = -dn_j$. The accompanying change in the Gibbs energy is

$$
\begin{aligned}
dG &= dG^{\text{sln}} + dG^{\text{vap}} \\
&= \left(\frac{\partial G^{\text{sln}}}{\partial n_j^{\text{sln}}} \right)_{T,P,n_{i \neq j}} dn_j^{\text{sln}} + \left(\frac{\partial G^{\text{vap}}}{\partial n_j^{\text{vap}}} \right)_{T,P,n_{i \neq j}} dn_j^{\text{vap}} \\
&= \mu_j^{\text{sln}} dn_j^{\text{sln}} + \mu_j^{\text{vap}} dn_j^{\text{vap}} = (\mu_j^{\text{vap}} - \mu_j^{\text{sln}}) dn_j^{\text{vap}}
\end{aligned}
$$

If the transfer from the solution to the vapor occurs spontaneously, then $dG < 0$. Furthermore, $dn_j^{\text{vap}} > 0$, so μ_j^{vap} must be less than μ_j^{sln} in order that $dG < 0$. Therefore, molecules of component j move spontaneously from the phase of higher chemical potential (solution) to that of lower chemical potential (vapor). Similarly, if $\mu_j^{\text{vap}} > \mu_j^{\text{sln}}$, then molecules of component j move spontaneously from the vapor phase to the solution phase ($dn_j^{\text{vap}} < 0$). At equilibrium, where $dG = 0$, we have that

$$\mu_j^{\text{vap}} = \mu_j^{\text{sln}} \tag{24.12}$$

Equation 24.12 holds for each component. Although we have discussed a solution in equilibrium with its vapor phase, our choice of phases was arbitrary, so Equation 24.12 is valid for the equilibrium between any two phases in which component j occurs.

The important result here is that Equation 24.12 says that the chemical potential of each component in the liquid solution phase can be measured by the chemical potential of that component in the vapor phase. If the pressure of the vapor phase is low enough that we can consider it to be ideal, then Equation 24.12 becomes

$$\mu_j^{\text{sln}} = \mu_j^{\text{vap}} = \mu_j^{\circ}(T) + RT \ln P_j \qquad (24.13)$$

where the standard state is taken to be $P_j^{\circ} = 1$ bar. For *pure* component j, Equation 24.13 becomes

$$\mu_j^*(l) = \mu_j^*(\text{vap}) = \mu_j^{\circ}(T) + RT \ln P_j^* \qquad (24.14)$$

where the superscript $*$ represents pure (liquid) component j. Thus, for example, $\mu_j^*(l)$ is the chemical potential and P_j^* is the vapor pressure of pure j. If we subtract Equation 24.14 from Equation 24.13, we obtain

$$\mu_j^{\text{sln}} = \mu_j^*(l) + RT \ln \frac{P_j}{P_j^*} \qquad (24.15)$$

Equation 24.15 is a central equation in the study of binary solutions. Note that $\mu_j^{\text{sln}} \to \mu_j^*$ as $P_j \to P_j^*$. Strictly speaking, we should use fugacities (Section 22–8) instead of pressures in Equation 24.15, but usually the magnitudes of vapor pressures are such that pressures are quite adequate. For example, the vapor pressure of water at 293.15 K is 17.4 torr, or 0.0232 bar.

24–4. The Components of an Ideal Solution Obey Raoult's Law for All Concentrations

A few solutions have the property that the partial vapor pressure of each component is given by the simple equation

$$P_j = x_j P_j^* \qquad (24.16)$$

Equation 24.16 is called *Raoult's law*, and a solution that obeys Raoult's law over the entire composition range is said to be an *ideal solution*.

The molecular picture behind an ideal binary solution is that the two types of molecules are randomly distributed throughout the solution. Such a distribution will occur if (1) the molecules are roughly the same size and shape, and (2) the intermolecular forces in the pure liquids 1 and 2 and in a mixture of 1 and 2 are all similar. We expect ideal-solution behavior only when the molecules of the two components are similar. For example, benzene and toluene, *o*-xylene and *p*-xylene, hexane and heptane, and bromoethane and iodoethane form essentially ideal solutions. Figure 24.2 depicts an ideal solution, in which the two types of molecules are randomly distributed. The mole fraction x_j reflects the fraction of the solution surface that is occupied by

FIGURE 24.2
A molecular depiction of an ideal solution. The two types of molecules are distributed throughout the solution in a random manner.

j molecules. Because the j molecules on the surface are the molecules that can escape into the vapor phase, the partial pressure P_j is just $x_j P_j^*$.

According to Raoult's law (Equation 24.16) and Equation 24.15, the chemical potential of component j in the solution is given by

$$\mu_j^{\text{sln}} = \mu_j^*(l) + RT \ln x_j \tag{24.17}$$

Equation 24.17 also serves to define an ideal solution if it is valid for all values of x_j ($0 \leq x_j \leq 1$). Furthermore, we showed in Section 24–2 that if one component obeys Equation 24.17 from $x_j = 0$ to $x_j = 1$, then so does the other.

The total vapor pressure over an ideal solution is given by

$$\begin{aligned} P_{\text{total}} &= P_1 + P_2 = x_1 P_1^* + x_2 P_2^* = (1 - x_2) P_1^* + x_2 P_2^* \\ &= P_1^* + x_2 (P_2^* - P_1^*) \end{aligned} \tag{24.18}$$

Therefore, a plot of P_{total} against x_2 (or x_1) will be a straight line as shown in Figure 24.3.

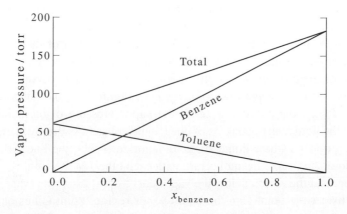

FIGURE 24.3
A plot of P_{total} against x_{benzene} for a solution of benzene and toluene at 40°C. This plot shows that a benzene/toluene solution is essentially ideal.

EXAMPLE 24-3
1-propanol and 2-propanol form essentially an ideal solution at all concentrations at 25°C. Letting the subscripts 1 and 2 denote 1-propanol and 2-propanol, respectively, and given that $P_1^* = 20.9$ torr and $P_2^* = 45.2$ torr at 25°C, calculate the total vapor pressure and the composition of the vapor phase at $x_2 = 0.75$.

SOLUTION: We use Equation 24.18:

$$P_{total}(x_2 = 0.75) = x_1 P_1^* + x_2 P_2^*$$

$$= (0.25)(20.9 \text{ torr}) + (0.75)(45.2 \text{ torr})$$

$$= 39.1 \text{ torr}$$

Let y_j denote the mole fraction of each component in the vapor phase. Then, by Dalton's law of partial pressures,

$$y_1 = \frac{P_1}{P_{total}} = \frac{x_1 P_1^*}{P_{total}} = \frac{(0.25)(20.9 \text{ torr})}{39.1 \text{ torr}} = 0.13$$

Similarly,

$$y_2 = \frac{P_2}{P_{total}} = \frac{x_2 P_2^*}{P_{total}} = \frac{(0.75)(45.2 \text{ torr})}{39.1 \text{ torr}} = 0.87$$

Note that $y_1 + y_2 = 1$. Also note that the vapor is richer than the solution in the more volatile component.

Problem 24–15 has you expand Example 24–3 by calculating P_{total} as a function of x_2 (the mole fraction of 2-propanol in the liquid phase) and as a function of y_2 (the mole fraction of 2-propanol in the vapor phase), and then plotting P_{total} against x_2 and y_2. The resulting plot, which is shown in Figure 24.4, is called a *pressure-composition diagram*. The upper curve shows the total vapor pressure as a function of the composition of the liquid phase (the liquid curve), and the lower curve shows the total vapor pressure as a function of the composition of the vapor phase (the vapor curve). Now let's see what happens when you start at the point P_a, x_a in Figure 24.4 and lower the pressure. At the point P_a, x_a, the pressure exceeds the vapor pressure of the solution, so the region above the liquid curve consists of one (liquid) phase. As the pressure is lowered, we reach the point A, where liquid starts to vaporize. Along the line AB, the system consists of liquid and vapor in equilibrium with each other. At the point B, all the liquid has vaporized, and the region below the vapor curve consists of one (vapor) phase.

Let's consider the point C in the liquid-vapor region. Point C lies on a line connecting the composition of liquid ($x_2 = 0.75$) and vapor ($y_2 = 0.87$) phases that we calculated in Example 24–3. Such a line is called a *tie line*. The overall composition of the two-phase (liquid-vapor) system is x_a. We can determine the relative amounts

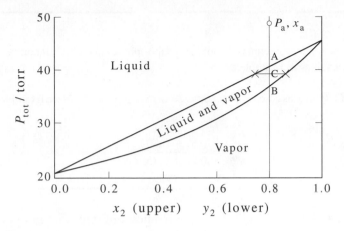

FIGURE 24.4
A pressure-composition diagram for a 1-propanol/2-propanol solution, which forms an essentially ideal solution at 25°C. This figure can be calculated using the approach in Example 24–3. The upper curve (called the liquid curve) represents P_{total} versus x_2, the mole fraction of 2-propanol in the liquid phase, and the lower curve (called the vapor curve) represents P_{total} versus y_2, the mole fraction of 2-propanol in the vapor phase. The two points marked by × represent the values of x_2 and y_2 from Example 24–3.

of liquid and vapor phase in the following way. The mole fractions in the liquid and vapor phases are

$$x_2 = \frac{n_2^l}{n_1^l + n_2^l} = \frac{n_2^l}{n^l} \quad \text{and} \quad y_2 = \frac{n_2^{vap}}{n_1^{vap} + n_2^{vap}} = \frac{n_2^{vap}}{n^{vap}}$$

where n^{vap} and n^l are the total number of moles in the vapor and liquid phases, respectively. The overall mole fraction at x_a is given by the total number of moles of component 2 divided by the total number of moles

$$x_a = \frac{n_2^l + n_2^{vap}}{n^l + n^{vap}}$$

Using a material balance of the number of moles of component 2, we have

$$x_a(n^l + n^{vap}) = x_2 n^l + y_2 n^{vap}$$

or

$$\frac{n^l}{n^{vap}} = \frac{y_2 - x_a}{x_a - x_2} \tag{24.19}$$

This equation represents what is called the *lever rule* because $n^{vap}(y_2 - x_a) = n^l(x_a - x_2)$ can be interpreted as a balance of each value of "n" times the distance from each curve to the point C in Figure 24.4. Note that $n^l = 0$ when $x_a = y_2$ (vapor curve) and that $n^{vap} = 0$ when $x_a = x_2$ (liquid curve).

EXAMPLE 24–4
Calculate the relative amounts of liquid and vapor phases at an overall composition of 0.80 for the values in Example 24–3.

SOLUTION: In this case, $x_a = 0.80$, $x_2 = 0.75$, and $y_2 = 0.87$ (see Example 24–3), so

$$\frac{n^l}{n^{\text{vap}}} = \frac{0.87 - 0.80}{0.80 - 0.75} = 1.6$$

According to Example 24–3, the mole fraction of 2-propanol in the vapor phase in equilibrium with a 1-propanol/2-propanol solution is greater than the mole fraction of 2-propanol in the solution. We can display the composition of the solution and vapor phases at various temperatures by a diagram called a *temperature-composition diagram*. To construct such a diagram, we choose some total ambient pressure such as 760 torr and write

$$760 \text{ torr} = x_1 P_1^* + x_2 P_2^* = x_1 P_1^* + (1 - x_1) P_2^*$$
$$= P_2^* - x_1 (P_2^* - P_1^*)$$

or

$$x_1 = \frac{P_2^* - 760 \text{ torr}}{P_2^* - P_1^*}$$

We then choose some temperature between the boiling points of the two components and solve the above equation for x_1, the compositon of the solution that will give a total pressure of 760 torr. A plot of temperature against x_1 shows the boiling temperature (at $P_{\text{total}} = 760$ torr) of a solution as a function of its composition (x_1). Such a curve, labeled the solution curve, is shown in Figure 24.5. For example, at $t = 90°C$, P_1^* (the vapor pressure of 1-propanol) = 575 torr and P_2^* (the vapor pressure of 2-propanol) = 1027 torr. Therefore,

$$x_1 = \frac{P_2^* - 760 \text{ torr}}{P_2^* - P_1^*} = \frac{1027 \text{ torr} - 760 \text{ torr}}{1027 \text{ torr} - 575 \text{ torr}} = 0.59$$

The point corresponding to $t = 90°C$ and $x_1 = 0.59$ is labeled by point a in Figure 24.5. We can also calculate the corresponding composition of the vapor phase as a function of temperature. The mole fraction of component 1 in the vapor phase is given by Dalton's law

$$y_1 = \frac{P_1}{760 \text{ torr}} = \frac{x_1 P_1^*}{760 \text{ torr}}$$

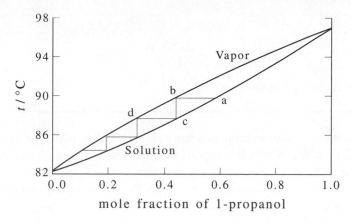

FIGURE 24.5
A temperature-composition diagram of a 1-propanol/2-propanol solution, which is essentially an ideal solution. The boiling point of 1-propanol is 97.2°C and that of 2-propanol is 82.3°C.

because the total pressure is taken (arbitrarily) to be 760 torr. We saw above that $x_1 = 0.59$ at 90°C, so we have that

$$y_1 = (0.59)(575 \text{ torr})/(760 \text{ torr}) = 0.45$$

which is labelled by point b in Figure 24.5.

EXAMPLE 24–5

The vapor pressures (in torr) of 1-propanol and 2-propanol as a function of the Celsius temperature, t, are given by the empirical formulas

$$\ln P_1^* = 18.0699 - \frac{3452.06}{t + 204.64}$$

and

$$\ln P_2^* = 18.6919 - \frac{3640.25}{t + 219.61}$$

Use these formulas to calculate x_1 and y_1 at 93.0°C, and compare your results with the values given in Figure 24.5.

SOLUTION: At 93.0°C,

$$\ln P_1^* = 18.0699 - \frac{3452.06}{93.0 + 204.64} = 6.472$$

or $P_1^* = 647$ torr. Similarly, $P_2^* = 1150$ torr. Therefore,

$$x_1 = \frac{P_2^* - 760 \text{ torr}}{P_2^* - P_1^*} = \frac{1150 \text{ torr} - 760 \text{ torr}}{1150 \text{ torr} - 647 \text{ torr}} = 0.77$$

and

$$y_1 = \frac{x_1 P_1^*}{760 \text{ torr}} = \frac{(0.77)(647 \text{ torr})}{760 \text{ torr}} = 0.65$$

in agreement with the values shown in Figure 24.5.

The temperature-composition diagram can be used to illustrate the process of fractional distillation, in which a vapor is condensed and then re-evaporated many times (Figure 24.6). If we were to start with a 1-propanol/2-propanol solution that has a mole fraction of 0.59 in 1-propanol (point a in Figure 24.5), the mole fraction of 1-propanol in the vapor will be 0.45 (point b). If this vapor is condensed (point c) and then re-evaporated, then the mole fraction of 1-propanol in the vapor phase will be about 0.30 (point d). As this process is continued, the vapor becomes increasingly richer in 2-propanol, eventually resulting in pure 2-propanol. A fractional distillation column differs from an ordinary distillation column in that the former is packed with glass beads, which provide a large surface area for the repeated condensation-evaporation process.

We can calculate the change in thermodynamic properties upon forming an ideal solution from its pure components. Let's take the Gibbs energy as an example. We define the Gibbs energy of mixing by

$$\Delta_{\text{mix}} G = G^{\text{sln}}(T, P, n_1, n_2) - G_1^*(T, P, n_1) - G_2^*(T, P, n_2) \tag{24.20}$$

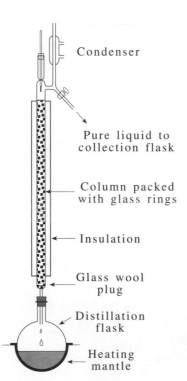

Condenser

Pure liquid to collection flask

Column packed with glass rings

Insulation

Glass wool plug

Distillation flask

Heating mantle

FIGURE 24.6
A simple fractional distillation column. Because repeated condensation and re-evaporation occur along the entire column, the vapor becomes progresively richer in the more volatile component as it moves up the column.

where G_1^* and G_2^* are the Gibbs energies of the pure components. Using Equation 24.17 for an ideal solution gives

$$\Delta_{mix} G^{id} = n_1 \mu_1^{sln} + n_2 \mu_2^{sln} - n_1 \mu_1^* - n_2 \mu_2^*$$
$$= RT(n_1 \ln x_1 + n_2 \ln x_2) \tag{24.21}$$

This quantity is always negative because x_1 and x_2 are less than one. In other words, an ideal solution will always form spontaneously from its separate components. The entropy of mixing of an ideal solution is given by

$$\Delta_{mix} S^{id} = -\left(\frac{\partial \Delta_{mix} G^{id}}{\partial T}\right)_{P, n_1, n_2} = -R(n_1 \ln x_1 + n_2 \ln x_2) \tag{24.22}$$

Note that this result for an ideal solution is the same as Equation 20.26 for the mixing of ideal gases. This similarity is due to the fact that in both cases the molecules in the final solution are randomly mixed. Nevertheless, you should realize that an ideal solution and a mixture of ideal gases differ markedly in the interactions involved. Although the molecules do not interact in a mixture of ideal gases, they interact strongly in an ideal solution. In an ideal solution, the interactions in the mixture and those in the pure liquids are essentially identical.

The volume change upon mixing of an ideal solution is given by

$$\Delta_{mix} V^{id} = \left(\frac{\partial \Delta_{mix} G^{id}}{\partial P}\right)_{T, n_1, n_2} = 0 \tag{24.23}$$

and the enthalpy of mixing is (see Equations 24.21 and 24.22)

$$\Delta_{mix} H^{id} = \Delta_{mix} G^{id} + T \Delta_{mix} S^{id} = 0 \tag{24.24}$$

Therefore, there is no volume change upon mixing, nor is there any energy as heat absorbed or evolved when an ideal solution is formed from its pure components. Both Equations 24.23 and 24.24 result from the facts that the molecules are roughly the same size and shape (hence $\Delta_{mix} V^{id} = 0$) and that the various interaction energies are the same (hence $\Delta_{mix} H^{id} = 0$). Equations 24.23 and 24.24 are indeed observed to be true experimentally for ideal solutions. For most solutions, however, $\Delta_{mix} H$ and $\Delta_{mix} V$ do not equal zero.

24–5. Most Solutions Are Not Ideal

Ideal solutions are not very common. Figures 24.7 and 24.8 show vapor pressure diagrams for carbon disulfide/dimethoxymethane $[(CH_3O)_2CH_2]$ solutions and tri-chloromethane/acetone solutions, respectively. The behavior in Figure 24.7 shows so-called positive deviations from Raoult's law because the partial vapor pressures

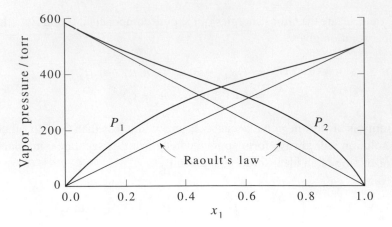

FIGURE 24.7
The vapor pressure diagram of a carbon disulfide/dimethoxymethane solution at 25°C. This system shows positive deviations from ideal, or Raoult's law, behavior.

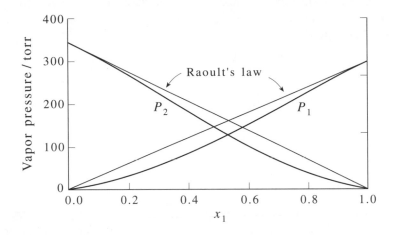

FIGURE 24.8
The vapor pressure diagram of a trichloromethane/acetone solution at 25°C. This system shows negative deviations from ideal, or Raoult's law, behavior.

of carbon disulfide and dimethoxymethane are greater than predicted on the basis of Raoult's law. Physically, positive deviations occur because carbon disulfide–dimethoxymethane interactions are more repulsive than either carbon disulfide–carbon disulfide or dimethoxymethane–dimethoxymethane interactions. Negative deviations, on the other hand, like those shown in Figure 24.8 for a trichloromethane/acetone solution, are due to stronger unlike-molecule interactions than like-molecule interactions. Problem 24–36 asks you to show that if one component of a binary solution exhibits positive deviations from ideal behavior, then the other component must do likewise.

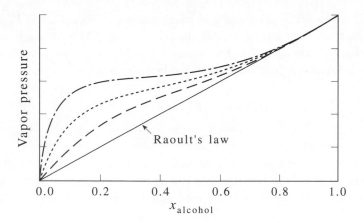

FIGURE 24.9
The vapor pressure diagram of alcohol/water solutions as a function of the number of carbon atoms in the alcohols, showing increasing deviation from ideal behavior. The dashed line corresponds to methanol, the dotted line to ethanol, and the dashed-dotted line to 1-propanol.

Figure 24.9 shows plots of methanol, ethanol, and 1-propanol vapor pressures in alcohol/water solutions. Note that the positive deviation from ideal behavior increases with the size of the hydrocarbon part of the alcohol. This behavior occurs because the water–hydrocarbon (repulsive) interactions become increasingly prevalent as the size of the hydrocarbon chain increases.

There are some important features to notice in Figures 24.7 and 24.8. Let's focus on component 1. The vapor pressure of component 1 approaches its Raoult's law value as x_1 approaches 1. In an equation, we have that

$$P_1 \longrightarrow x_1 P_1^* \quad \text{as} \quad x_1 \longrightarrow 1 \tag{24.25}$$

Although we deduced Equation 24.25 from Figures 24.7 and 24.8, it is generally true. Physically, this behavior may be attributed to the fact that there are so few component-2 molecules that most component-1 molecules see only other component-1 molecules, so that the solution behaves ideally. Raoult's law behavior is *not* observed for component 1 as $x_1 \rightarrow 0$ in Figures 24.7 and 24.8, however. Although not easily seen in Figures 24.7 and 24.8, the vapor pressure of component 1 as $x_1 \rightarrow 0$ is linear in x_1, but the slope is not equal to P_1^* as in Equation 24.25. We emphasize this behavior by writing

$$P_1 \longrightarrow k_{H,1} x_1 \quad \text{as} \quad x_1 \longrightarrow 0 \tag{24.26}$$

In the special case of an ideal solution, $k_{H,1} = P_1^*$, but ordinarily $k_{H,1} \neq P_1^*$. Equation 24.26 is called *Henry's law*, and $k_{H,1}$ is called the *Henry's law constant* of component 1. As $x_1 \rightarrow 0$, the component-1 molecules are completely surrounded by component-2 molecules, and the value of $k_{H,1}$ reflects the intermolecular interactions between the two components. As $x_1 \rightarrow 1$, on the other hand, the component-1 molecules are completely surrounded by component-1 molecules, and P_1^* is what reflects the intermolecular interactions in the pure liquid. Although we have focussed our discussion on component 1 in Figures 24.7 and 24.8, the same situtation holds for component 2. Equations 24.25 and 24.26 can be written as

$$P_j \longrightarrow x_j P_j^* \quad \text{as } x_j \longrightarrow 1$$
$$P_j \longrightarrow x_j k_{\mathrm{H},j} \quad \text{as } x_j \longrightarrow 0$$

(24.27)

Thus, in a vapor pressure diagram of a solution of two volatile liquids, the vapor pressure of each component approaches Raoult's law as the mole fraction of that component approaches one and Henry's law as the mole fraction approaches zero.

EXAMPLE 24–6

The vapor pressure (in torr) of component 1 over a binary solution is given by

$$P_1 = 180 x_1 e^{x_2^2 + \frac{1}{2} x_2^3} \qquad 0 \le x_1 \le 1$$

Determine the vapor pressure (P_1^*) and the Henry's law constant ($k_{\mathrm{H},1}$) of pure component 1.

SOLUTION: In the limit that $x_1 \to 1$, the exponential factor $\to 1$ because $x_2 \to 0$ as $x_1 \to 1$. Therefore,

$$P_1 \longrightarrow 180 x_1 \quad \text{as} \quad x_1 \longrightarrow 1$$

so $P_1^* = 180$ torr. As $x_1 \to 0$, on the other hand, the exponential factor approaches $e^{3/2}$ because $x_2 \to 1$ as $x_1 \to 0$. Thus, we have

$$P_1 \longrightarrow 180 e^{3/2} x_1 = 807 x_1 \quad \text{as} \quad x_1 \longrightarrow 0$$

and $k_{\mathrm{H},1} = 807$ torr.

We will now show that the Henry's law behavior of component 2 as $x_2 \to 0$ is a thermodynamic consequence of the Raoult's law behavior of component 1 as $x_1 \to 1$. To prove this connection, we will start with the Gibbs–Duhem equation (Equation 24.11)

$$x_1 d\mu_1 + x_2 d\mu_2 = 0 \qquad \text{(constant } T \text{ and } P)$$

Now, assuming that the vapor phase may be treated as an ideal gas, both chemical potentials can be expressed as

$$\mu_j(T, P) = \mu_j^\circ(T) + RT \ln P_j$$

(Recall that the argument of the logarithm is actually P_j/P°, where P° is one bar.) Now this form of $\mu_j(T, P)$ allows us to write

$$d\mu_1 = RT \left(\frac{\partial \ln P_1}{\partial x_1} \right)_{T,P} dx_1$$

and

$$d\mu_2 = RT \left(\frac{\partial \ln P_2}{\partial x_2} \right)_{T,P} dx_2$$

Substitute these two expressions into the Gibbs–Duhem equation to get

$$x_1 \left(\frac{\partial \ln P_1}{\partial x_1} \right)_{T,P} dx_1 + x_2 \left(\frac{\partial \ln P_2}{\partial x_2} \right)_{T,P} dx_2 = 0 \qquad (24.28)$$

But $dx_1 = -dx_2$ (because $x_1 + x_2 = 1$), so Equation 24.28 becomes

$$x_1 \left(\frac{\partial \ln P_1}{\partial x_1} \right)_{T,P} = x_2 \left(\frac{\partial \ln P_2}{\partial x_2} \right)_{T,P} \qquad (24.29)$$

which is another form of the Gibbs–Duhem equation. If component 1 obeys Raoult's law as $x_1 \to 1$, then $P_1 \to x_1 P_1^*$ and $(\partial \ln P_1/\partial x_1)_{T,P} = 1/x_1$, so the left side of Equation 24.29 becomes unity. Thus, we have the condition

$$x_2 \left(\frac{\partial \ln P_2}{\partial x_2} \right)_{T,P} = 1 \quad \text{as} \quad x_1 \to 1 \text{ or } x_2 \to 0$$

We now integrate this expression indefinitely to get

$$\ln P_2 = \ln x_2 + \text{constant} \quad \text{as} \quad x_1 \to 1 \text{ or } x_2 \to 0$$

or

$$P_2 = k_{H,2} x_2 \quad \text{as} \quad x_2 \to 0$$

Thus, we see that if component 1 obeys Raoult's law as $x_1 \to 1$, then component 2 must obey Henry's law as $x_2 \to 0$. Problem 24–32 has you prove the converse: if component 2 obeys Henry's law as $x_2 \to 0$, then component 1 must obey Raoult's law as $x_1 \to 1$.

24–6. The Gibbs–Duhem Equation Relates the Vapor Pressures of the Two Components of a Volatile Binary Solution

The following example shows that if we know the vapor pressure curve of one of the components over the entire composition range, we can calculate the vapor pressure of the other component.

EXAMPLE 24–7

The vapor pressure curve of one of the components (say component 1) of a nonideal binary solution can often be represented empirically by (see Figure 24.10)

$$P_1 = x_1 P_1^* e^{\alpha x_2^2 + \beta x_2^3} \qquad 0 \leq x_1 \leq 1$$

where α and β are parameters that are used to fit the data. Show that the vapor pressure of component 2 is necessarily given by

$$P_2 = x_2 P_2^* e^{\gamma x_1^2 + \delta x_1^3} \qquad 0 \leq x_2 \leq 1$$

where $\gamma = \alpha + 3\beta/2$ and $\delta = -\beta$. Notice that the parameters α and β must in some manner reflect the extent of the nonideality of the solution because both P_1 and P_2 reduce to the ideal solution expressions when $\alpha = \beta = 0$. Furthermore, note that $P_1 \to x_1 P_1^* e^{\alpha + \beta}$ as $x_1 \to 0$ $(x_2 \to 1)$, so the Henry's law constant of component 1 is $k_{H,1} = P_1^* e^{\alpha + \beta}$. Similarly, we find that $k_{H,2} = P_2^* e^{\alpha + \beta/2}$.

SOLUTION: We use the Gibbs–Duhem equation

$$d\mu_2 = -\frac{x_1}{x_2} d\mu_1$$

along with (Equation 24.13)

$$\mu_1 = \mu_1^\circ + RT \ln P_1$$
$$= \mu_1^\circ + RT \ln P_1^* + RT \ln x_1$$
$$\qquad + \alpha RT (1 - x_1)^2 + \beta RT (1 - x_1)^3$$

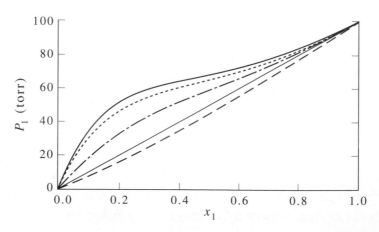

FIGURE 24.10
A plot of $P_1 = x_1 P_1^* e^{\alpha x_2^2 + \beta x_2^3}$ for $P_1^* = 100$ torr and various values of α and β. The values of α and β for the five curves, top to bottom, are 1.0, 0.60; 0.80, 0.60; 0.60, 0.20; 0,0 (ideal solution); and -0.80, 0.60.

Differentiate this equation with respect to x_1 and substitute the result into the above Gibbs–Duhem equation to obtain

$$d\mu_2 = -\frac{x_1}{x_2}RT\left[\frac{dx_1}{x_1} - 2\alpha(1 - x_1)dx_1 - 3\beta(1 - x_1)^2 dx_1\right]$$

$$= RT\left[-\frac{dx_1}{x_2} + 2\alpha x_1 dx_1 + 3\beta x_1(1 - x_1)dx_1\right]$$

Now change variables from x_1 to x_2

$$d\mu_2 = RT\left[\frac{dx_2}{x_2} - 2\alpha(1 - x_2)dx_2 - 3\beta x_2(1 - x_2)dx_2\right]$$

and integrate from $x_2 = 1$ to arbitrary x_2 and use the fact that $\mu_2 = \mu_2^*$ when $x_2 = 1$ to get

$$\mu_2 - \mu_2^* = RT\left[\ln x_2 + \alpha(1 - x_2)^2 - \frac{3\beta}{2}(x_2^2 - 1) + \beta(x_2^3 - 1)\right]$$

$$= RT\left[\ln x_2 + \alpha x_1^2 + \frac{3\beta}{2}x_1^2 - \beta x_1^3\right]$$

Using the fact that $\mu_2 = \mu_2^\circ + RT\ln P_2$ and that $\mu_2^* = \mu_2^\circ + RT\ln P_2^*$, we see that

$$\ln P_2 = \ln P_2^* + \ln x_2 + \alpha x_1^2 + \frac{3\beta}{2}x_1^2 - \beta x_1^3$$

or

$$P_2 = x_2 P_2^* e^{(\alpha + 3\beta/2)x_1^2 - \beta x_1^3}$$

We could also have used Equation 24.29 to do this problem (Problem 24–33).

Figure 24.11 shows the boiling-point diagram of a benzene/ethanol system, in which the boiling points of benzene/ethanol solutions (at one atm) are plotted against the mole fraction of ethanol. Figure 24.11 shows that if you were to start with a solution with an ethanol mole fraction of 0.2, for example, then repeated evaporation–condensation would lead to a mixture consisting of a mole fraction of about 0.4 that cannot be separated by further fractional distillation.

Such a mixture, for which there is no change in composition upon boiling, is called an *azeotrope*. Thus, it is not possible to achieve a separation of a benzene/ethanol solution by distillation into pure benzene and pure ethanol. If we start out at an ethanol mole fraction of 0.2, we would obtain a separation of pure benzene and the azeotrope. Similarly, if we started out with an ethanol mole fraction of 0.8, we would achieve a separation of pure ethanol and the benzene/ethanol azeotrope.

As our final topic in this section on nonideal solutions, let's consider the case in which the positive deviations from ideal behavior become increasingly large, as often occurs as the temperature is lowered. Figure 24.12 illustrates typical vapor pressure behavior for a series of temperatures, where $T_3 > T_c > T_2 > T_1$. The vertical axis is P_2/P_2^*, so each curve is "normalized" by the vapor pressure of pure component 2 at each

984

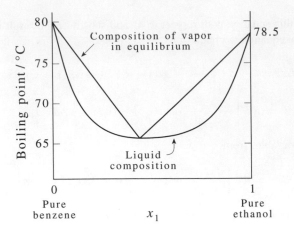

FIGURE 24.11
The boiling-point diagram of a benzene/ethanol solution, showing the occurrence of an azeotropic solution at an ethanol mole fraction of about 0.4. The quantity x_1 is the mole fraction of ethanol.

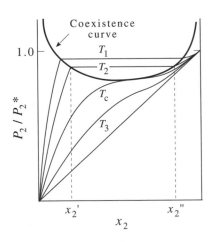

FIGURE 24.12
An illustration of the critical behavior of a binary solution as a function of temperature $(T_3 > T_c > T_2 > T_1)$.

temperature. Therefore, all the curves meet at $P_2/P_2^* = 1$ at $x_2 = 1$. For temperature T_3, which is greater than T_c, the slope of the P_2 versus x_2 curve is everywhere positive. At T_c, the curve has an inflection point, where $\partial P_2/\partial x_2 = 0$ and $\partial^2 P_2/\partial x_2^2 = 0$. For the temperatures T_1 and T_2, which are less than T_c, the curves have a horizontal portion that becomes wider as the temperature is lowered. The temperature T_c is called the *critical temperature* or *consulate temperature*, and as we will now discuss, the consulate temperature is the temperature below which the two liquids are not miscible in all proportions.

Let's follow the T_2 curve in Figure 24.12 as we start with pure component 1 $(x_2 = 0)$ and add component 2. Up to the point x_2', the added component 2 simply dissolves in component 1 to form a single solution phase. Above the concentration x_2', however, two separate or immiscible solution phases form, one of composition x_2' and one of composition x_2''. As x_2 is increased from x_2' to x_2'', the two phases must maintain a constant mole fraction of component 2 (x_2' and x_2'') and therefore, the relative proportions of the two phases change, with the volume of the phase of composition x_2'' increasing and the volume of the phase of composition x_2' decreasing. The overall composition of the two phases together is given by the value of x_2. When $x_2 > x_2''$, we obtain a single solution phase.

We can derive a lever rule to calculate the relative amounts of the two phases in the following way. Consider some overall composition x_2, which lies between x_2' and x_2''. Let n_1', n_2' and n_1'', n_2'' be the number of moles of the two components in the phases of composition x_2' and x_2'', respectively. Then, the mole fraction of component 2 in each phase is

$$x_2' = \frac{n_2'}{n_1' + n_2'} \quad \text{and} \quad x_2'' = \frac{n_2''}{n_1'' + n_2''}$$

and the overall mole fraction of component-2 is

$$x_2 = \frac{n_2' + n_2''}{n_1' + n_1'' + n_2' + n_2''}$$

Using material balance of the number of moles of component 2 allows us to write

$$x_2(n_1' + n_1'' + n_2' + n_2'') = x_2'(n_1' + n_2') + x_2''(n_1'' + n_2'')$$

We can rearrange this material balance equation to give

$$\frac{n'}{n''} = \frac{n_1' + n_2'}{n_1'' + n_2''} = \frac{x_2'' - x_2}{x_2 - x_2'} \tag{24.30}$$

Equation 24.30 gives the relative total number of moles in each phase. Note that if $x_2 = x_2''$, then $n' = 0$, and if $x_2 = x_2'$, then $n'' = 0$. As Equation 24.30 shows, when x_2 reaches x_2'', the phase of composition x_2' disappears, and there is a single solution phase of composition $x_2 = x_2''$. For $x_2 \geq x_2''$, there is a single solution phase of composition x_2. Thus, at a temperature T_2, the two liquids are immiscible when x_2 is between x_2' and x_2'' but are miscible for $x_2 < x_2'$ and $x_2 > x_2''$. Similar behavior occurs at other temperatures less than T_c, and Figure 24.12 summarizes this behavior. The heavy curve in Figure 24.12 is called a *coexistence curve*. Points inside the coexistence curve represent two solution phases, whereas points below the coexistence curve represent one solution phase. Problem 24–43 has you determine the coexistence curve for a simple model system.

We can display the results illustrated by Figure 24.12 in a temperature-composition diagram (Figure 24.13*a*). The curve separating the one-phase region from the two-phase region is the coexistence curve. The temperature T_c, the temperature above which the two liquids are totally miscible, is the consulate temperature. The coexistence curve in

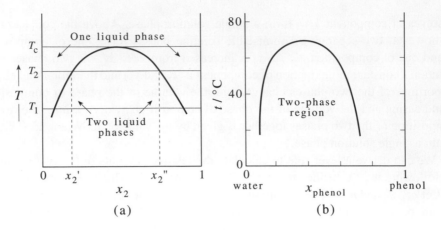

FIGURE 24.13
(a) A temperature-composition diagram for the system illustrated in Figure 24.12.
(b) A temperature-composition diagram for a water/phenol system.

Figure 24.13a looks "upside down" compared with the one in Figure 24.12, but note that the temperature decreases as you go up in Figure 24.12, whereas they decrease as you go down in Figure 24.13. Figure 24.13b shows a coexistence curve for a water/phenol system.

24–7. The Central Thermodynamic Quantity for Nonideal Solutions Is the Activity

The chemical potential of component j in a liquid solution is given by (Equation 24.15)

$$\mu_j^{\text{sln}} = \mu_j^* + RT \ln \frac{P_j}{P_j^*} \tag{24.31}$$

if we assume, as usual, that the vapor pressures involved are low enough that the vapors can be considered to behave ideally (otherwise, we replace the partial pressures by partial fugacities). An ideal solution is one in which $P_j = x_j P_j^*$ for all concentrations, so that Equation 24.31 becomes

$$\mu_j^{\text{sln}} = \mu_j^* + RT \ln x_j \quad \text{(ideal solution)} \tag{24.32}$$

Equation 24.31 is still valid for a nonideal solution, but the relation between P_j/P_j^* and composition is more complicated than simply $P_j = x_j P_j^*$. For example, we saw in Example 24–7 that partial vapor pressure data are often fit by an expression like

$$P_1 = x_1 P_1^* \exp(\alpha x_2^2 + \beta x_2^3 + \cdots) \tag{24.33}$$

The exponential factor here accounts for the nonideality of the system. The chemical potential of component 1 in this case is given by

$$\mu_1 = \mu_1^* + RT \ln x_1 + \alpha RT x_2^2 + \beta RT x_2^3 + \cdots \tag{24.34}$$

In Section 22–8, we introduced the idea of fugacity to preserve the form of the thermodynamic equations we had derived for ideal gases. We will follow a similar procedure for solutions, using an ideal solution as our standard.

To carry over the form of Equation 24.32 to nonideal solutions, we define a quantity called the *activity* by the equation

$$\mu_j^{\text{sln}} = \mu_j^* + RT \ln a_j \tag{24.35}$$

where μ_j^* is the chemical potential, or the molar Gibbs energy, of the pure liquid. Equation 24.35 is the generalization of Equation 24.32 to nonideal solutions. The first of Equations 24.27 says that $P_j = x_j P_j^*$, as $x_j \rightarrow 1$. If we substitute this result into Equation 24.31, we obtain

$$\mu_j^{\text{sln}} = \mu_j^* + RT \ln x_j \qquad (\text{as } x_j \rightarrow 1)$$

If we compare this equation with Equation 24.35, which is valid at all concentrations, we can define the activity of component j by

$$a_j = \frac{P_j}{P_j^*} \qquad (\text{ideal vapor}) \tag{24.36}$$

such that $a_j \rightarrow x_j$ as $x_j \rightarrow 1$. In other words, the activity of a pure liquid is unity (at a total pressure of one bar and at the temperature of interest). For an ideal solution, $P_j = x_j P_j^*$ for all concentrations, and so the activity of component i in an ideal solution is given by $a_j = x_j$. In a nonideal solution, a_j still is equal to P_j/P_j^*, but this ratio is no longer equal to x_j, although $a_j \rightarrow x_j$ as $x_j \rightarrow 1$.

According to Equations 24.33 and 24.36, the activity of component 1 can be represented empirically by

$$a_1 = x_1 e^{\alpha x_2^2 + \beta x_2^3 + \cdots}$$

Note that $a_1 \rightarrow 1$ as $x_1 \rightarrow 1$ ($x_2 \rightarrow 0$). The ratio a_j/x_j can be used as a measure of the deviation of the solution from ideality. This ratio is called the *activity coefficient* of component j and is denoted by γ_j:

$$\gamma_j = \frac{a_j}{x_j} \tag{24.37}$$

If $\gamma_j = 1$ for all concentrations, the solution is ideal. If $\gamma_j \neq 1$, the solution is not ideal. For example, the partial vapor pressures of chlorobenzene in equilibrium with a chlorobenzene/1-nitropropane solution at 75°C are listed below:

x_1	0.119	0.289	0.460	0.691	1.00
P_1/torr	19.0	41.9	62.4	86.4	119

According to these data, the vapor pressure of pure chlorobenzene at 75°C is 119 torr, so the activities and activity coefficients are as follows:

x_1	0.119	0.289	0.460	0.691	1.00
$a_1(= P_1/P_1^*)$	0.160	0.352	0.524	0.726	1.00
$\gamma_1(= a_1/x_1)$	1.34	1.22	1.14	1.05	1.00

Figure 24.14 shows the activity coefficient of chlorobenzene in 1-nitropropane at 75°C plotted against the mole fraction of chlorobenzene.

Activity is really just another way of expressing chemical potential because the two quantities are directly related to each other through $\mu_j = \mu_j^* + RT \ln a_j$. Therefore, just as the chemical potential of one component of a binary solution is related to the chemical potential of the other component by way of the Gibbs-Duhem equation, the activities are related to each other by

$$x_1 d \ln a_1 + x_2 d \ln a_2 = 0 \tag{24.38}$$

For example, if $a_1 = x_1$ over the entire composition range, meaning that component 1 obeys Raoult's law over the entire composition range, then

$$d \ln a_2 = -\frac{x_1}{x_2} \frac{dx_1}{x_1} = -\frac{dx_1}{x_2} = \frac{dx_2}{x_2}$$

Integrate from $x_2 = 1$ to arbitrary x_2 and use the fact that $a_2 \to 1$ as $x_2 \to 1$ to get

$$\ln a_2 = \ln x_2$$

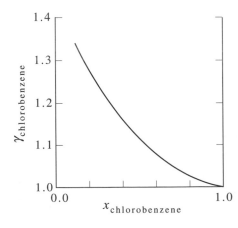

FIGURE 24.14
The activity coefficient of chlorobenzene in 1-nitropropane at 75°C plotted against the mole fraction of chlorobenzene.

or $a_2 = x_2$. Thus, we see once again that if one component obeys Raoult's law over the entire composition range, the other component will also.

EXAMPLE 24–8
Show that if

$$a_1 = x_1 e^{\alpha x_2^2}$$

then

$$a_2 = x_2 e^{\alpha x_1^2}$$

SOLUTION: We first differentiate $\ln a_1$ with respect to x_1:

$$d \ln a_1 = \frac{dx_1}{x_1} - 2\alpha(1 - x_1)dx_1$$

and substitute into Equation 24.38 to obtain

$$d \ln a_2 = -\frac{x_1}{x_2}\left(\frac{dx_1}{x_1} - 2\alpha x_2 dx_1\right)$$

$$= -\frac{dx_1}{x_2} + 2\alpha x_1 dx_1$$

Now change the integration variable from x_1 to x_2:

$$d \ln a_2 = \frac{dx_2}{x_2} - 2\alpha(1 - x_2)dx_2$$

and integrate from $x_2 = 1$ (where $a_2 = 1$) to arbitrary x_2:

$$\ln a_2 = \ln x_2 + \alpha(1 - x_2)^2$$

or

$$a_2 = x_2 e^{\alpha x_1^2}$$

24–8. Activities Must Be Calculated with Respect to Standard States

In one sense, there are two types of binary solutions, those in which the two components are miscible in all proportions and those in which they are not. Only in the latter case are the designations "solvent" and "solute" unambiguous. As we will see in this section, the different nature of these two types of solutions leads us to define different standard states.

Although we have not said so explicitly, we have tacitly assumed both components of the solutions we have considered thus far exist as pure liquids at the temperatures of the solutions. We have defined the activity of each component by (Equation 24.36)

$$a_j = \frac{P_j}{P_j^*} \qquad \text{(ideal vapor)} \tag{24.39}$$

so that $a_j \to x_j$ as $x_j \to 1$ and $a_j = 1$ when $P_j = P_j^*$. An activity defined by Equation 24.39 is said to be based upon a solvent, or Raoult's law standard state. Because of the relation (Equation 24.35) $\mu_j = \mu_j^* + RT \ln a_j$, the chemical potential of component j is also based upon a solvent, or Raoult's law, standard state. You need to realize that activities or chemical potentials are meaningless unless it is clear just what has been used as the standard state. If the two liquids are miscible in all proportions, there is no distinction between solvent and solute and a solvent standard state is normally used. If, on the other hand, one component is sparingly soluble in the other, then picking a standard state based upon Henry's law instead of Raoult's law is more convenient. To see how we define the activity in this case, we start with Equation 24.31

$$\mu_j^{\text{sln}} = \mu_j^* + RT \ln \frac{P_j}{P_j^*} \tag{24.40}$$

Because component j is sparingly soluble, we use the second of Equations 24.27, which says that $P_j \to x_j k_{\text{H},j}$ as $x_j \to 0$, where $k_{\text{H},j}$ is the Henry's law constant of component j. If we substitute the limiting value $x_j k_{\text{H},j}$ into Equation 24.40 for P_j, we obtain

$$\mu_j^{\text{sln}} = \mu_j^* + RT \ln \frac{x_j k_{\text{H},j}}{P_j^*} \qquad (x_j \to 0)$$

$$= \mu_j^* + RT \ln \frac{k_{\text{H},j}}{P_j^*} + RT \ln x_j \qquad (x_j \to 0) \tag{24.41}$$

We define the activity of component j by

$$\mu_j^{\text{sln}} = \mu_j^* + RT \ln \frac{k_{\text{H},j}}{P_j^*} + RT \ln a_j \tag{24.42}$$

so that $a_j \to x_j$ as $x_j \to 0$, as can be seen by comparing Equations 24.41 and 24.42. Equation 24.42 becomes equivalent to Equation 24.35 if we define a_j by

$$a_j = \frac{P_j}{k_{\text{H},j}} \qquad \text{(ideal vapor)} \tag{24.43}$$

and choose the standard state such that

$$\mu_j^* = \mu_j^* + RT \ln \frac{k_{\text{H},j}}{P_j^*}$$

or such that $k_{H,j} = P_j^*$. The standard state in this case requires that $k_{H,j} = P_j^*$. This standard state may not exist in practice, so it is called a hypothetical standard state. Nevertheless, the definition of activity involving Henry's law for dilute components given by Equation 24.43 is natural and useful.

The numerical value of an activity or an activity coefficient depends upon the choice of standard state. Table 24.1 lists vapor pressure data for carbon disulfide/di-methoxymethane solutions at 35.2°C, and these data are plotted in Figure 24.15. Notice that both curves approach Raoult's law as their corresponding mole fractions approach unity. The dashed lines in the figure represent the linear regions as the corresponding mole fractions approach zero. The slopes of these lines give the Henry's law constant for each component. The values come out to be $k_{H,CS_2} = 1130$ torr and $k_{H,dimeth} = 1500$ torr. We can use these values and the values of the vapor pressures of the pure components to calculate activities and activity coefficients based upon each standard state. For example, Table 24.1 gives $P_{CS_2} = 407.0$ torr and $P_{dimeth} = 277.8$ torr at $x_{CS_2} = 0.6827$. Therefore,

$$a_{CS_2}^{(R)} = \frac{P_{CS_2}}{P_{CS_2}^*} = \frac{407.0 \text{ torr}}{514.5 \text{ torr}} = 0.7911$$

TABLE 24.1
Vapor pressure data of carbon disulfide/dimethoxymethane solutions at 35.2°C

x_{CS_2}	P_{CS_2}/torr	P_{dimeth}/torr
0.0000	0.000	587.7
0.0489	54.5	558.3
0.1030	109.3	529.1
0.1640	159.5	500.4
0.2710	234.8	451.2
0.3470	277.6	412.7
0.4536	324.8	378.0
0.4946	340.2	360.8
0.5393	357.2	342.2
0.6071	381.9	313.3
0.6827	407.0	277.8
0.7377	424.3	250.1
0.7950	442.3	217.4
0.8445	458.1	184.9
0.9108	481.8	124.2
0.9554	501.0	65.1
1.0000	514.5	0.000

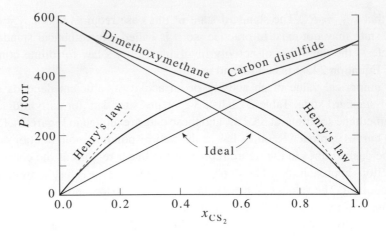

FIGURE 24.15
Vapor pressures of carbon disulfide and dimethoxymethane over their solutions at 35.2°C. The solid straight lines represent ideal behavior, and the dashed lines represent the Henry's law behavior for each component as the corresponding mole fractions approach zero.

and

$$a_{\text{dimeth}}^{(R)} = \frac{P_{\text{dimeth}}}{P_{\text{dimeth}}^*} = \frac{277.8 \text{ torr}}{587.7 \text{ torr}} = 0.4727$$

with

$$\gamma_{\text{CS}_2}^{(R)} = \frac{a_{\text{CS}_2}^{(R)}}{x_{\text{CS}_2}} = \frac{0.7911}{0.6827} = 1.159$$

and

$$\gamma_{\text{dimeth}}^{(R)} = \frac{a_{\text{dimeth}}^{(R)}}{x_{\text{dimeth}}} = \frac{0.4727}{0.3173} = 1.490$$

where the superscript (R) simply emphasizes that these values are based upon a Raoult's law, or solvent, standard state.

Similarly,

$$a_{\text{CS}_2}^{(H)} = \frac{P_{\text{CS}_2}}{k_{\text{H,CS}_2}} = \frac{407.0 \text{ torr}}{1130 \text{ torr}} = 0.360$$

$$a_{\text{dimeth}}^{(H)} = \frac{P_{\text{dimeth}}}{k_{\text{H,dimeth}}} = \frac{277.8 \text{ torr}}{1500 \text{ torr}} = 0.185$$

$$\gamma_{\text{CS}_2}^{(H)} = \frac{a_{\text{CS}_2}^{(H)}}{x_{\text{CS}_2}} = \frac{0.360}{0.6827} = 0.527$$

and

$$\gamma_{\text{dimeth}}^{(H)} = \frac{a_{\text{dimeth}}^{(H)}}{x_{\text{dimeth}}} = \frac{0.185}{0.3173} = 0.583$$

where the superscript (H) simply emphasizes that these values are based upon a Henry's law, or solute, standard state. Figure 24.16a shows the Raoult's law, or solvent-based, activities, and Figure 24.16b shows the Henry's law, or solute-based, activities plotted against the mole fraction of carbon disulfide. We will see in the next chapter that a solute, or Henry's law, standard state is particularly appropriate for a substance that does not exist as a liquid at one bar and at the temperature of the solution under study.

The activity coefficients based upon the Raoult's law standard state (which is the usual standard state for miscible liquids) are plotted in Figure 24.17. Notice that $\gamma_{\text{CS}_2} \to 1$ as $x_{\text{CS}_2} \to 1$ and that it goes to 2.2 as $x_{\text{CS}_2} \to 0$. Both of these limiting values may be deduced from the definition of γ_j (Equation 24.37)

$$\gamma_j = \frac{a_j}{x_j} = \frac{P_j}{x_j P_j^*}$$

Now $P_j \to P_j^*$ as $x_j \to 1$, and so $\gamma_j \to 1$ as $x_j \to 1$. At the other limit, however, $P_j \to x_j k_{\text{H},j}$ as $x_j \to 0$, so we see that $\gamma_j \to k_{\text{H},j}/P_j^*$ as $x_j \to 0$. The value of k_{H} for $\text{CS}_2(\text{l})$ is 1130 torr, so $\gamma_{\text{CS}_2} \to k_{\text{H,CS}_2}/P_{\text{CS}_2}^* = (1130 \text{ torr}/514.5 \text{ torr}) = 2.2$, in agreement with Figure 24.17. The activity coefficient of dimethoxymethane approaches 2.5 as $x_{\text{dimeth}} \to 0$ ($x_{\text{CS}_2} \to 1$), in agreement with $\gamma_{\text{dimeth}} \to k_{\text{H,dimeth}}/P_{\text{dimeth}}^* = (1500 \text{ torr}/587.7 \text{ torr}) = 2.5$.

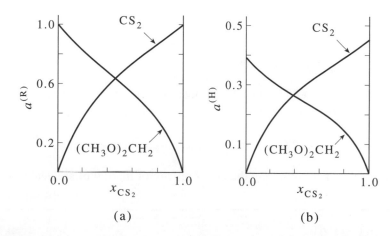

(a) (b)

FIGURE 24.16
(a) The Raoult's law activities of carbon disulfide and dimethoxymethane in carbon disulfide/dimethoxymethane solutions at 35.2°C plotted against the mole fraction of carbon disulfide. (b) The Henry's law activities for the same system.

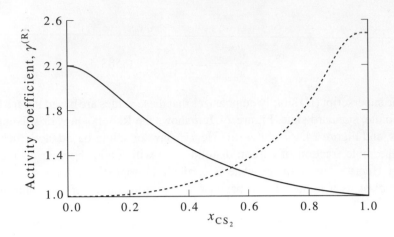

FIGURE 24.17
The Raoult's law activity coefficients of carbon disulfide (solid line) and dimethoxymethane (dashed line) plotted against x_{CS_2} for carbon disulfide/dimethoxymethane solutions at 35.2°C.

24–9. We Can Calculate the Gibbs Energy of Mixing of Binary Solutions in Terms of the Activity Coefficients

Recall from Equation 24.21 that

$$\Delta_{mix} G = n_1 \mu_1^{sln} + n_2 \mu_2^{sln} - n_1 \mu_1^* - n_2 \mu_2^*$$

But, according to Equations 24.35 and 24.37,

$$\mu_j^{sln} = \mu_j^* + RT \ln a_j = \mu_j^* + RT \ln x_j + RT \ln \gamma_j \tag{24.44}$$

so

$$\Delta_{mix} G/RT = n_1 \ln x_1 + n_2 \ln x_2 + n_1 \ln \gamma_1 + n_2 \ln \gamma_2 \tag{24.45}$$

If we divide Equation 24.45 by the total number of moles, $n_1 + n_2$, we obtain the *molar Gibbs energy of mixing*, $\Delta_{mix} \overline{G}$.

$$\Delta_{mix} \overline{G}/RT = x_1 \ln x_1 + x_2 \ln x_2 + x_1 \ln \gamma_1 + x_2 \ln \gamma_2 \tag{24.46}$$

The first two terms here represent the Gibbs energy of mixing of an ideal solution.

EXAMPLE 24–9
Use Equation 24.46 to derive a formula for $\Delta_{mix} \overline{G}$ for a binary solution in which the vapor pressure can be expressed by

$$P_1 = x_1 P_1^* e^{\alpha x_2^2} \quad \text{and} \quad P_2 = x_2 P_2^* e^{\alpha x_1^2}$$

SOLUTION: According to the above expressions for P_1 and P_2,

$$\gamma_1 = \frac{P_1}{x_1 P_1^*} = e^{\alpha x_2^2} \quad \text{and} \quad \gamma_2 = \frac{P_2}{x_2 P_2^*} = e^{\alpha x_1^2}$$

Substitute these expressions into Equation 24.46 to obtain

$$\Delta_{\text{mix}}\overline{G}/RT = x_1 \ln x_1 + x_2 \ln x_2 + \alpha x_1 x_2^2 + \alpha x_2 x_1^2$$

But

$$x_1 x_2^2 + x_2 x_1^2 = x_1 x_2 (x_1 + x_2) = x_1 x_2$$

so

$$\Delta_{\text{mix}}\overline{G}/RT = x_1 \ln x_1 + x_2 \ln x_2 + \alpha x_1 x_2 \tag{24.47}$$

Molecular theories of binary solutions show that the parameter α, which is unitless, has the form of an energy divided by RT. Therefore, we will write α as w/RT, where w is a constant whose value we will not need. With this substitution, Equation 24.47 can be written as

$$\frac{\Delta_{\text{mix}}\overline{G}}{w} = \frac{RT}{w}(x_1 \ln x_1 + x_2 \ln x_2) + x_1 x_2 \tag{24.48}$$

Figure 24.18 shows plots of $\Delta_{\text{mix}}\overline{G}/w$ for several values of RT/w. Note that the slopes of all the curves equal zero at the midpoint, $x_1 = x_2 = 1/2$. The curve for $RT/w = 0.50$ is special in the sense that curves for values of RT/w greater than 0.50

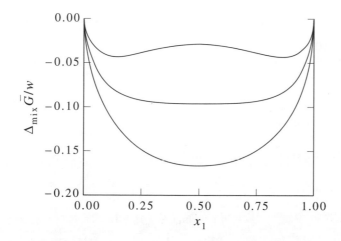

FIGURE 24.18
Plots of $\Delta_{\text{mix}}\overline{G}/w$ for $RT/w = 0.60$ (bottom curve), $RT/w = 0.50$ (middle curve), and $RT/w = 0.40$ (top curve).

are concave upward for all values of x_1, whereas curves for values of RT/w less than 0.50 are concave downward at $x_1 = 1/2$. In mathematical terms, $\partial^2(\Delta_{mix}\overline{G}/w)/\partial x_1^2$ is positive (a minimum) at $x_1 = x_2 = 1/2$ for the curves that lie below the curve with $RT/w = 0.50$, whereas $\partial^2(\Delta_{mix}\overline{G}/w)/\partial x_1^2$ is negative (a maximum) at $x_1 = x_2 = 1/2$ for curves that lie above it. The region where $\partial^2(\Delta_{mix}\overline{G}/w)/\partial x_1^2$ is negative is similar to the loops of the van der Waals equation or the Redlich–Kwong equation when $T < T_c$ (Figure 16.8), and in this case corresponds to a region in which the two liquids are not miscible. The critical value $RT/w = 0.50$ corresponds to a solution critical temperature, T_c, where the two liquids are miscible in all proportions at temperatures above $T_c = 0.50w/R$ and immiscible at temperatures below $T_c = 0.50w/R$.

Let's consider the curve with $RT/w = 0.40$ in Figure 24.18. The two minima represent two immiscible solutions in equilibrium with each other. The compositions of these two solutions are given by the values of x_1 at each minimum. Using Equation 24.47, we have

$$\frac{\partial(\Delta_{mix}\overline{G}/w)}{\partial x_1} = \frac{RT}{w}[\ln x_1 - \ln(1 - x_1)] + (1 - 2x_1) = 0 \qquad (24.49)$$

as the condition for the extrema of $\Delta_{mix}\overline{G}/w$. First note that $x_1 = 1/2$ solves Equation 24.49 for any value of RT/w, which accounts for the fact that all the curves in Figure 24.18 have either a maximum or a minimum at $x_1 = 1/2$. By plotting $(RT/w)[\ln x_1 - \ln(1 - x_1)] + (1 - 2x_1)$ against x_1 for various values of RT/w, you can see that only $x_1 = 1/2$ satisfies Equation 24.49 for $RT/w \geq 0.50$, whereas two other roots occur for $RT/w < 0.50$. The two roots give the composition of the two miscible solutions in equilibrium with each other. For the case in which $RT/w = 0.40$, the two values of x_1 are 0.145 and 0.855. Figure 24.19 shows the mole fraction of component 1 in each of the two immiscible solutions as a function of temperature (RT/w). Note that Figure 24.19 is similar to Figure 24.13.

EXAMPLE 24–10

Use Equation 24.49 to calculate the composition of the two immiscible solutions in equilibrium with each other at a temperature given by $RT/w = 0.40$.

SOLUTION: We use the Newton–Raphson method that we introduced in Math-Chapter G. The function $f(x)$ of Equation G.1 is

$$f(x) = \frac{RT}{w}[\ln x - \ln(1 - x)] + 1 - 2x$$

Equation G.1 becomes

$$x_{n+1} = x_n - \frac{\dfrac{RT}{w}[\ln x_n - \ln(1 - x_n)] + 1 - 2x_n}{\dfrac{RT}{w}\left[\dfrac{1}{x_n(1 - x_n)}\right] - 2}$$

with $RT/w = 0.40$. For one of the solutions, we start with $x_0 = 0.100$ and get

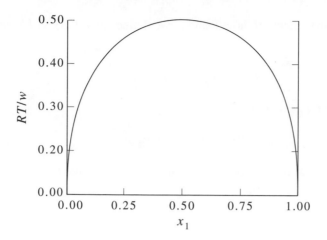

FIGURE 24.19

A temperature-composition diagram for a binary system for which $\Delta_{mix}\overline{G}/w = (RT/w)(x_1 \ln x_1 + x_2 \ln x_2) + x_1 x_2$ (Equation 24.48). The curve gives the compositions of the two immiscible solutions as a function of temperature. There is only one homogeneous phase in the region above the curve, and there are two immiscible solutions in equilibrium with each other in the region below the curve.

n	x_n	$f(x_n)$	$f'(x_n)$
0	0.100	−0.07889	2.4444
1	0.132	−0.01695	1.4851
2	0.144	−0.001370	1.2509
3	0.145	−0.000017	1.2305
4	0.145		

For the other solution, we start with $x_0 = 0.900$ and get

n	x_n	$f(x_n)$	$f'(x_n)$
0	0.900	0.07889	2.4444
1	0.868	0.01695	1.4851
2	0.856	0.00137	1.2509
3	0.855	0.000017	1.2305
4	0.855		

in agreement with Figure 24.19.

Many solutions can be described by the Equation 24.47, and such solutions are called *regular solutions*. Problems 24–37 through 24–45 involve regular solutions.

To focus on the effect of nonideality, we define an *excess Gibbs energy of mixing*, G^E:

$$G^E = \Delta_{mix}G - \Delta_{mix}G^{id} \tag{24.50}$$

We see from Equation 24.45 that

$$G^E/RT = n_1 \ln \gamma_1 + n_2 \ln \gamma_2$$

If we divide by the total number of moles $n_1 + n_2$, we obtain the *molar excess Gibbs energy of mixing*, $\overline{G}^E$:

$$\overline{G}^E/RT = x_1 \ln \gamma_1 + x_2 \ln \gamma_2 \tag{24.51}$$

For $\Delta_{mix}\overline{G}$ given by Equation 24.47,

$$\overline{G}^E/RT = \alpha x_1 x_2 \tag{24.52}$$

According to Equation 24.52, a plot of $\overline{G}^E$ against x_1 is a parabola that is symmetric about the vertical line at $x_1 = 1/2$.

We can use γ_{CS_2} and γ_{dimeth} that we calculated for Figure 24.17 to calculate the value of $\overline{G}^E$ for a carbon disulfide/dimethoxymethane solution at 35.2°C, which is shown in Figure 24.20. Note that the plot of $\overline{G}^E$ versus x_{CS_2} is not symmetric about $x_{CS_2} = 1/2$. This asymmetry implies that $\beta \neq 0$ in the empirical vapor pressure formula (Equation 24.33).

We will continue our discussion of solutions in the next chapter, where we focus on solutions in which the two components are not soluble in all proportions. In particular,

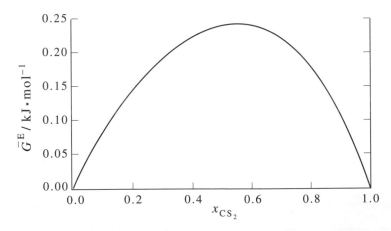

FIGURE 24.20
The molar excess Gibbs energy of mixing of carbon disulfide/dimethoxymethane solutions at 35.2°C plotted against the mole fraction of carbon disulfide.

we will discuss solutions of solids in liquids, where the terms *solute* and *solvent* are meaningful.

Problems

24-1. In the text, we went from Equation 24.5 to 24.6 using a physical argument involving varying the size of the system while keeping T and P fixed. We could also have used a mathematical process called *Euler's theorem*. Before we can learn about Euler's theorem, we must first define a *homogeneous function*. A function $f(z_1, z_2, \ldots, z_N)$ is said to be homogeneous if

$$f(\lambda z_1, \lambda z_2, \ldots, \lambda z_N) = \lambda f(z_1, z_2, \ldots, z_N)$$

Argue that extensive thermodynamic quantities are homogeneous functions of their extensive variables.

24-2. Euler's theorem says that if $f(z_1, z_2, \ldots, z_N)$ is homogeneous, then

$$f(z_1, z_2, \ldots, z_N) = z_1 \frac{\partial f}{\partial z_1} + z_2 \frac{\partial f}{\partial z_2} + \cdots + z_N \frac{\partial f}{\partial z_N}$$

Prove Euler's theorem by differentiating the equation in Problem 24–1 with respect to λ and then setting $\lambda = 1$.

Apply Euler's theorem to $G = G(n_1, n_2, T, P)$ to derive Equation 24.6. (*Hint*: Because T and P are intensive variables, they are simply irrevelant variables in this case.)

24-3. Use Euler's theorem (Problem 24–2) to prove that

$$Y(n_1, n_2, \ldots, T, P) = \sum n_j \overline{Y}_j$$

for any extensive quantity Y.

24-4. Apply Euler's theorem to $U = U(S, V, n)$. Do you recognize the resulting equation?

24-5. Apply Euler's theorem to $A = A(T, V, n)$. Do you recognize the resulting equation?

24-6. Apply Euler's theorem to $V = V(T, P, n_1, n_2)$ to derive Equation 24.7.

24-7. The properties of many solutions are given as a function of the mass percent of the components. If we let the mass percent of component 2 be A_2, derive a relation between A_2 and the mole fractions, x_1 and x_2.

24-8. The *CRC Handbook of Chemistry and Physics* gives the densities of many aqueous solutions as a function of the mass percentage of solute. If we denote the density by ρ and the mass percentage of component 2 by A_2, the *Handbook* gives $\rho = \rho(A_2)$ (in g·mL^{-1}). Show that the quantity $V = (n_1 M_1 + n_2 M_2)/\rho(A_2)$ is the volume of the solution containing n_1 moles of component 1 and n_2 moles of component 2, where M_j is the molar mass of component j. Now show that

$$\overline{V}_1 = \frac{M_1}{\rho(A_2)} \left[1 + \frac{A_2}{\rho(A_2)} \frac{d\rho(A_2)}{dA_2} \right]$$

and

$$\overline{V}_2 = \frac{M_2}{\rho(A_2)}\left[1 + \frac{(A_2 - 100)}{\rho(A_2)}\frac{d\rho(A_2)}{dA_2}\right]$$

Show that

$$V = n_1\overline{V}_1 + n_2\overline{V}_2$$

in agreement with Equation 24.7.

24-9. The density (in $g \cdot ml^{-1}$) of a 1-propanol/water solution at 20°C as a function of A_2, the mass percentage of 1-propanol, can be expressed as

$$\rho(A_2) = \sum_{j=0}^{7} \alpha_j A_2^j$$

where

$$\alpha_0 = 0.99823 \qquad\qquad \alpha_4 = 1.5312 \times 10^{-7}$$

$$\alpha_1 = -0.0020577 \qquad\qquad \alpha_5 = -2.0365 \times 10^{-9}$$

$$\alpha_2 = 1.0021 \times 10^{-4} \qquad\qquad \alpha_6 = 1.3741 \times 10^{-11}$$

$$\alpha_3 = -5.9518 \times 10^{-6} \qquad\qquad \alpha_7 = -3.7278 \times 10^{-14}$$

Use this expression to plot $\overline{V}_{H_2O}$ and $\overline{V}_{1\text{-propanol}}$ versus A_2, and compare your values with those in Figure 24.1.

24-10. Given the density of a binary solution as a function of the mole fraction of component 2 $[\rho = \rho(x_2)]$, show that the volume of the solution containing n_1 moles of component 1 and n_2 moles of component 2 is given by $V = (n_1 M_1 + n_2 M_2)/\rho(x_2)$, where M_j is the molar mass of component j. Now show that

$$\overline{V}_1 = \frac{M_1}{\rho(x_2)}\left[1 + \left(\frac{x_2(M_2 - M_1) + M_1}{M_1}\right)\frac{x_2}{\rho(x_2)}\frac{d\rho(x_2)}{dx_2}\right]$$

and

$$\overline{V}_2 = \frac{M_2}{\rho(x_2)}\left[1 - \left(\frac{x_2(M_2 - M_1) + M_1}{M_2}\right)\frac{1 - x_2}{\rho(x_2)}\frac{d\rho(x_2)}{dx_2}\right]$$

Show that

$$V = n_1\overline{V}_1 + n_2\overline{V}_2$$

in agreement with Equation 24.7.

24-11. The density (in $g \cdot mL^{-1}$) of a 1-propanol/water solution at 20°C as a function of x_2, the mole fraction of 1-propanol, can be expressed as

$$\rho(x_2) = \sum_{j=0}^{4} \alpha_j x_2^j$$

where

$$\alpha_0 = 0.99823 \qquad \alpha_3 = -0.17163$$

$$\alpha_1 = -0.48503 \qquad \alpha_4 = -0.01387$$

$$\alpha_2 = 0.47518$$

Use this expression to calculate the values of $\overline{V}_{H_2O}$ and $\overline{V}_{1\text{-propanol}}$ as a function of x_2 according to the equation in Problem 24–10.

24-12. Use the data in the *CRC Handbook of Chemistry and Physics* to curve fit the density of a water/glycerol solution to a fifth-order polynomial in the mole fraction of glycerol, and then determine the partial molar volumes of water and glycerol as a function of mole fraction. Plot your result.

24-13. Just before Example 24–2, we showed that if one component of a binary solution obeys Raoult's law over the entire composition range, the other component does also. Now show that if $\mu_2 = \mu_2' + RT \ln x_2$ for $x_{2,min} \leq x_2 \leq 1$, then $\mu_1 = \mu_1' + RT \ln x_1$ for $0 \leq x_1 < 1 - x_{2,min}$. Notice that for the range over which μ_2 obeys the simple form given, μ_1 obeys a similarly simple form. If we let $x_{2,min} = 0$, we obtain $\mu_1 = \mu_1^* + RT \ln x_1$ ($0 \leq x_1 \leq 1$).

24-14. Continue the calculations in Example 24–3 to obtain y_2 as a function of x_2 by varying x_2 from 0 to 1. Plot your result.

24-15. Use your results from Problem 24–14 to construct the pressure-composition diagram in Figure 24.4.

24-16. Calculate the relative amounts of liquid and vapor phases at an overall composition of 0.50 for one of the pair of values, $x_2 = 0.38$ and $y_2 = 0.57$, that you obtained in Problem 24–14.

24-17. In this problem, we will derive analytic expressions for the pressure-composition curves in Figure 24.4. The liquid (upper) curve is just

$$P_{total} = x_1 P_1^* + x_2 P_2^* = (1 - x_2) P_1^* + x_2 P_2^* = P_1^* + x_2 (P_2^* - P_1^*) \tag{1}$$

which is a straight line, as seen in Figure 24.4. Solve the equation

$$y_2 = \frac{x_2 P_2^*}{P_{total}} = \frac{x_2 P_2^*}{P_1^* + x_2 (P_2^* - P_1^*)}$$

for x_2 in terms of y_2 and substitute into Equation (1) to obtain

$$P_{total} = \frac{P_1^* P_2^*}{P_2^* - y_2 (P_2^* - P_1^*)}$$

Plot this result versus y_2 and show that it gives the vapor (lower) curve in Figure 24.4.

24-18. Prove that $y_2 > x_2$ if $P_2^* > P_1^*$ and that $y_2 < x_2$ if $P_2^* < P_1^*$. Interpret this result physically.

24-19. Tetrachloromethane and trichloroethylene form essentially ideal solutions at 40°C at all concentrations. Given that the vapor pressure of tetrachloromethane and trichloroethylene at 40°C are 214 torr and 138 torr, respectively, plot the pressure-composition diagram for this system (see Problem 24–17).

24-20. The vapor pressures of tetrachloromethane (1) and trichloroethylene (2) between 76.8°C and 87.2°C can be expressed empirically by the formulas

$$\ln(P_1^*/\text{torr}) = 15.8401 - \frac{2790.78}{t + 226.4}$$

and

$$\ln(P_2^*/\text{torr}) = 15.0124 - \frac{2345.4}{t + 192.7}$$

where t is the Celsius temperature. Assuming that tetrachloromethane and trichloroethylene form an ideal solution at all compositions, calculate the values of x_1 and y_1 at 82.0°C (at an ambient pressure of 760 torr).

24-21. Use the data in Problem 24–20 to construct the entire temperature-composition diagram of a tetrachloromethane/trichlororethylene solution.

24-22. The vapor pressures of benzene and toluene between 80°C and 110°C as a function of the Kelvin temperature are given by the empirical formulas

$$\ln(P_{\text{benz}}^*/\text{torr}) = -\frac{3856.6 \text{ K}}{T} + 17.551$$

and

$$\ln(P_{\text{tol}}^*/\text{torr}) = -\frac{4514.6 \text{ K}}{T} + 18.397$$

Assuming that benzene and toluene form an ideal solution, use these formulas to construct a temperature-composition diagram of this system at an ambient pressure of 760 torr.

24-23. Construct the temperature-composition diagram for 1-propanol and 2-propanol in Figure 24.5 by varying t from 82.3°C (the boiling point of 2-propanol) to 97.2°C (the boiling point of 1-propanol), calculating the values of (1) P_1^* and P_2^* at each temperature (see Example 24–5), (2) x_1 according to $x_1 = (P_2^* - 760)/(P_2^* - P_1^*)$, and (3) y_1 according to $y_1 = x_1 P_1^*/760$. Now plot t versus x_1 and y_1 on the same graph to obtain the temperature-composition diagram.

24-24. Prove that $\overline{V}_j = \overline{V}_j^*$ for an ideal solution, where $\overline{V}_j^*$ is the molar volume of pure component j.

24-25. The volume of mixing of miscible liquids is defined as the volume of the solution minus the volume of the individual pure components. Show that

$$\Delta_{\text{mix}}\overline{V} = \sum x_i(\overline{V}_i - \overline{V}_i^*)$$

at constant P and T, where $\overline{V}_i^*$ is the molar volume of pure component i. Show that $\Delta_{\text{mix}}\overline{V} = 0$ for an ideal solution (see Problem 24–24).

24-26. Suppose the vapor pressures of the two components of a binary solution are given by

$$P_1 = x_1 P_1^* e^{x_2^2/2}$$

and

$$P_2 = x_2 P_2^* e^{x_1^2/2}$$

Given that $P_1^* = 75.0$ torr and $P_2^* = 160$ torr, calculate the total vapor pressure and the composition of the vapor phase at $x_1 = 0.40$.

24-27. Plot y_1 versus x_1 for the system described in the previous problem. Why does the curve lie below the straight line connecting the origin with the point $x_1 = 1$, $y_1 = 1$? Describe a system for which the curve would lie above the diagonal line.

24-28. Use the expressions for P_1 and P_2 given in Problem 24–26 to construct a pressure-composition diagram.

24-29. The vapor pressure (in torr) of the two components in a binary solution are given by

$$P_1 = 120x_1 e^{0.20x_2^2 + 0.10x_2^3}$$

and

$$P_2 = 140x_2 e^{0.35x_1^2 - 0.10x_1^3}$$

Determine the values of P_1^*, P_2^*, $k_{H,1}$, and $k_{H,2}$.

24-30. Suppose the vapor pressure of the two components of a binary solution are given by

$$P_1 = x_1 P_1^* e^{\alpha x_2^2 + \beta x_2^3}$$

and

$$P_2 = x_2 P_2^* e^{(\alpha + 3\beta/2)x_1^2 - \beta x_1^3}$$

Show that $k_{H,1} = P_1^* e^{\alpha + \beta}$ and $k_{H,2} = P_2^* e^{\alpha + \beta/2}$.

24-31. The empirical expression for the vapor pressure that we used in Examples 24–6 and 24–7, for example,

$$P_1 = x_1 P_1^* e^{\alpha x_2^2 + \beta x_2^3 + \cdots}$$

is sometimes called the *Margules equation*. Use Equation 24.29 to prove that there can be no linear term in the exponential factor in P_1, for otherwise P_2 will not satisfy Henry's law as $x_2 \to 0$.

24-32. In the text, we showed that the Henry's law behavior of component 2 as $x_2 \to 0$ is a direct consequence of the Raoult's law behavior of component 1 as $x_1 \to 1$. In this problem, we will prove the converse: the Raoult's law behavior of component 1 as $x_1 \to 1$ is a direct consequence of the Henry's law behavior of component 2 as $x_2 \to 0$. Show that the chemical potential of component 2 as $x_2 \to 0$ is

$$\mu_2(T, P) = \mu_2^\circ(T) + RT \ln k_{H,2} + RT \ln x_2 \qquad x_2 \to 0$$

Differentiate μ_2 with respect to x_2 and substitute the result into the Gibbs–Duhem equation to obtain

$$d\mu_1 = RT\frac{dx_1}{x_1} \qquad x_2 \longrightarrow 0$$

Integrate this expression from $x_1 = 1$ to $x_1 \approx 1$ and use the fact that $\mu_1(x_1 = 1) = \mu_1^*$ to obtain

$$\mu_1(T, P) = \mu_1^*(T) + RT \ln x_1 \qquad x_1 \to 1$$

which is the Raoult's law expression for chemical potential.

24-33. In Example 24–7, we saw that if

$$P_1 = x_1 P_1^* e^{\alpha x_2^2 + \beta x_2^3}$$

then

$$P_2 = x_2 P_2^* e^{(\alpha + 3\beta/2)x_1^2 - \beta x_1^3}$$

Show that this result follows directly from Equation 24.29.

24-34. Suppose we express the vapor pressures of the components of a binary solution by

$$P_1 = x_1 P_1^* e^{\alpha x_2^2}$$

and

$$P_2 = x_2 P_2^* e^{\beta x_1^2}$$

Use the Gibbs–Duhem equation or Equation 24.29 to prove that α must equal β.

24-35. Use Equation 24.29 to show that if one component of a binary solution obeys Raoult's law for all concentrations, then the other component also obeys Raoult's law for all concentrations.

24-36. Use Equation 24.29 to show that if one component of a binary solution has positive deviations from Raoult's law, then the other component must also.

The following nine problems develop the idea of a regular solution.

24-37. If the vapor pressures of the two components in a binary solution are given by

$$P_1 = x_1 P_1^* e^{wx_2^2/RT} \qquad \text{and} \qquad P_2 = x_2 P_2^* e^{wx_1^2/RT}$$

show that

$$\Delta_{mix}\overline{G}/w = \Delta_{mix}G/(n_1 + n_2)w = \frac{RT}{w}[x_1 \ln x_1 + x_2 \ln x_2] + x_1 x_2$$

$$\Delta_{mix}\overline{S}/R = \Delta_{mix}S/(n_1 + n_2)R = -(x_1 \ln x_1 + x_2 \ln x_2)$$

and

$$\Delta_{mix}\overline{H}/w = \Delta_{mix}H/(n_1 + n_2)w = x_1 x_2$$

A solution that satisfies these equations is called a *regular solution*. A statistical thermodynamic model of binary solutions shows that w is given by

$$w = N_A(\varepsilon_{11} + \varepsilon_{22} - 2\varepsilon_{12})$$

where ε_{ij} is the interaction energy between molecules of components i and j. Note that $w = 0$ if $\varepsilon_{12} = (\varepsilon_{11} + \varepsilon_{22})/2$, which means that energetically, molecules of components 1 and 2 "like" the opposite molecules as well as their own.

24-38. Prove that $\Delta_{mix}\overline{G}$, $\Delta_{mix}\overline{S}$, and $\Delta_{mix}\overline{H}$ in the previous problem are symmetric about the point $x_1 = x_2 = 1/2$.

24-39. Plot $P_1/P_1^* = x_1 e^{wx_2^2/RT}$ versus x_1 for $RT/w = 0.60, 0.50, 0.45, 0.40,$ and 0.35. Note that some of the curves have regions where the slope is negative. The following problem has you show that this behavior occurs when $RT/w < 0.50$. These regions are similar to the loops of the van der Waals equation or the Redlich–Kwong equation when $T < T_c$ (Figure 16.8), and in this case correspond to regions in which the two liquids are not miscible. The critical value $RT/w = 0.50$ corresponds to a solution critical temperature.

24-40. Differentiate $P_1 = x_1 P_1^* e^{w(1-x_1)^2/RT}$ with respect to x_1 to prove that P_1 has a maximum or a minimum at the points $x_1 = \frac{1}{2} \pm \frac{1}{2}(1 - \frac{2RT}{w})^{1/2}$. Show that $RT/w < 0.50$ for either a maximum or a minimum to occur. Do the positions of these extrema when $RT/w = 0.35$ correspond to the plot you obtained in the previous problem?

24-41. Plot $\Delta_{mix}\overline{G}/w$ in Problem 24–37 versus x_1 for $RT/w = 0.60, 0.50, 0.45, 0.40,$ and 0.35. Note that some of the curves have regions where $\partial^2 \Delta_{mix}\overline{G}/\partial x_1^2 < 0$. These regions correspond to regions in which the two liquids are not miscible. Show that $RT/w = 0.50$ is a critical value, in the sense that unstable regions occur only when $RT/w < 0.50$. (See the previous problem.)

24-42. Plot both $P_1/P_1^* = x_1 e^{\alpha x_2^2}$ and $P_2/P_2^* = x_2 e^{\alpha x_1^2}$ for $RT/w = 1/\alpha = 0.60, 0.50, 0.45, 0.40,$ and 0.35. Prove that the loops occur for values of $RT/w < 0.50$.

24-43. Plot both $P_1/P_1^* = x_1 e^{\alpha x_2^2}$ and $P_2/P_2^* = x_2 e^{\alpha x_1^2}$ for $RT/w = 1/\alpha = 0.40$. The loops indicate regions in which the two liquids are not miscible, as explained in Problem 24–39. Draw a horizontal line connecting the left-side and the right-side intersections of the two curves. This line, which connects states in which the vapor pressure (or chemical potential) of each component is the same in the two solutions of different composition, corresponds to one of the horizontal lines in Figure 24.12. Now set $P_1/P_1^* = x_1 e^{\alpha x_2^2}$ equal to $P_2/P_2^* = x_2 e^{\alpha x_1^2}$ and solve for α in terms of x_1. Plot $RT/w = 1/\alpha$ against x_1 and obtain a coexistence curve like the one in Figure 24.19.

24-44. The molar enthalpies of mixing of solutions of tetrachloromethane (1) and cyclohexane (2) at 25°C are listed below.

x_1	$\Delta_{mix}\overline{H}/\text{J}\cdot\text{mol}^{-1}$
0.0657	37.8
0.2335	107.9
0.3495	134.9
0.4745	146.7
0.5955	141.6
0.7213	118.6
0.8529	73.6

Plot $\Delta_{mix}\overline{H}/x_2$ against x_1 according to Problem 24–37. Do tetrachloromethane and cyclohexane form a regular solution?

24-45. The molar enthalpies of mixing of solutions of tetrahydrofuran and trichloromethane at 25°C are listed below.

x_{THF}	$\Delta_{mix}\overline{H}/\text{J}\cdot\text{mol}^{-1}$
0.0568	−0.469
0.1802	−1.374
0.3301	−2.118
0.4508	−2.398
0.5702	−2.383
0.7432	−1.888
0.8231	−1.465
0.9162	−0.802

Do tetrahydrofuran and trichloromethane form a regular solution?

24-46. Derive the equation

$$x_1 d\ln\gamma_1 + x_2 d\ln\gamma_2 = 0$$

by starting with Equation 24.11. Use this equation to obtain the same result as in Example 24–8.

24-47. The vapor pressure data for carbon disulfide in Table 24.1 can be curve fit by

$$P_1 = x_1(514.5 \text{ torr})e^{1.4967x_2^2 - 0.68175x_2^3}$$

Using the results of Example 24–7, show that the vapor pressure of dimethoxymethane is given by

$$P_2 = x_2(587.7 \text{ torr})e^{0.4741x_1^2 + 0.68175x_1^3}$$

Now plot P_2 versus x_2 and compare the result with the data in Table 24.1. Plot $\overline{G}^E$ against x_1. Is the plot symmetric about a vertical line at $x_1 = 1/2$? Do carbon disulfide and dimethoxymethane form a regular solution at 35.2°C?

24-48. A mixture of trichloromethane and acetone with $x_{acet} = 0.713$ has a total vapor pressure of 220.5 torr at 28.2°C, and the mole fraction of acetone in the vapor is $y_{acet} = 0.818$. Given that the vapor pressure of pure trichloromethane at 28.2°C is 221.8 torr, calculate the activity and the activity coefficient (based upon a Raoult's law standard state) of trichloromethane in the mixture. Assume the vapor behaves ideally.

24-49. Consider a binary solution for which the vapor pressure (in torr) of one of the components (say component 1) is given empirically by

$$P_1 = 78.8 x_1 e^{0.65 x_2^2 + 0.18 x_2^3}$$

Calculate the activity and the activity coefficient of component 1 when $x_1 = 0.25$ based on a solvent and a solute standard state.

24-50. Some vapor pressure data for ethanol/water solutions at 25°C are listed below.

$x_{ethanol}$	$P_{ethanol}$/torr	P_{water}/torr
0.00	0.00	23.78
0.02	4.28	23.31
0.05	9.96	22.67
0.08	14.84	22.07
0.10	17.65	21.70
0.20	27.02	20.25
0.30	31.23	19.34
0.40	33.93	18.50
0.50	36.86	17.29
0.60	40.23	15.53
0.70	43.94	13.16
0.80	48.24	9.89
0.90	53.45	5.38
0.93	55.14	3.83
0.96	56.87	2.23
0.98	58.02	1.13
1.00	59.20	0.00

Plot these data to determine the Henry's law constant for ethanol in water and for water in ethanol at 25°C.

24-51. Using the data in Problem 24–50, plot the activity coefficients (based upon Raoult's law) of both ethanol and water against the mole fraction of ethanol.

24-52. Using the data in Problem 24–50, plot $\overline{G}^E/RT$ against x_{H_2O}. Is a water/ethanol solution at 25°C a regular solution?

24-53. Some vapor pressure data for a 2-propanol/benzene solution at 25°C are

$x_{2\text{-propanol}}$	$P_{2\text{-propanol}}/\text{torr}$	$P_{\text{total}}/\text{torr}$
0.000	0.0	94.4
0.059	12.9	104.5
0.146	22.4	109.0
0.362	27.6	108.4
0.521	30.4	105.8
0.700	36.4	99.8
0.836	39.5	84.0
0.924	42.2	66.4
1.000	44.0	44.0

Plot the activities and the activity coefficients of 2-propanol and benzene relative to a Raoult's law standard state versus the mole fraction of 2-propanol.

24-54. Using the data in Problem 24–53, plot $\overline{G}^{\text{E}}/RT$ versus $x_{2\text{-propanol}}$.

24-55. *Excess thermodynamic quantities* are defined relative to the values the quantities would have if the pure components formed an ideal solution at the same given temperature and pressure. For example, we saw that (Equation 24.47)

$$\frac{G^{\text{E}}}{(n_1 + n_2)RT} = x_1 \ln \gamma_1 + x_2 \ln \gamma_2$$

Show that

$$\frac{S^{\text{E}}}{(n_1 + n_2)R} = -(x_1 \ln \gamma_1 + x_2 \ln \gamma_2) - T\left(x_1 \frac{\partial \ln \gamma_1}{\partial T} + x_2 \frac{\partial \ln \gamma_2}{\partial T}\right)$$

24-56. Show that

$$\frac{G^{\text{E}}}{(n_1 + n_2)} = wx_1 x_2$$

$$\frac{S^{\text{E}}}{(n_1 + n_2)R} = 0$$

and

$$\frac{H^{\text{E}}}{(n_1 + n_2)} = wx_1 x_2$$

for a regular solution (see Problem 24–37).

24-57. Example 24–7 expresses the vapor pressures of the two components of a binary solution as

$$P_1 = x_1 P_1^* e^{\alpha x_2^2 + \beta x_2^3}$$

and

$$P_2 = x_2 P_2^* e^{(\alpha + 3\beta/2)x_1^2 - \beta x_1^3}$$

Show that these expressions are equivalent to

$$\gamma_1 = e^{\alpha x_2^2 + \beta x_2^3} \quad \text{and} \quad \gamma_2 = e^{(\alpha + 3\beta/2)x_1^2 - \beta x_1^3}$$

Using these expressions for the activity coefficients, derive an expression for $\overline{G}^{\mathrm{E}}$ in terms of α and β. Show that your expression reduces to that for $\overline{G}^{\mathrm{E}}$ for a regular solution.

24-58. Prove that the maxima or minima of $\Delta_{\mathrm{mix}}\overline{G}$ defined in Problem 24–37 occur at $x_1 = x_2 = 1/2$ for any value of RT/w. Now prove that

$$\frac{\partial^2 \Delta_{\mathrm{mix}}\overline{G}}{\partial x_1^2} \begin{array}{ll} > 0 & \text{for } RT/w > 0.50 \\ = 0 & \text{for } RT/w = 0.50 \\ < 0 & \text{for } RT/w < 0.50 \end{array}$$

at $x_1 = x_2 = 1/2$. Is this result consistent with the graphs you obtained in Problem 24–41?

24-59. Use the data in Table 24.1 to plot Figures 24.15 through 24.17.

Peter Debye (left) was born in Maastricht, the Netherlands, on March 24, 1884 and died in 1966. Debye was originally trained as an electrical engineer but turned his attention to physics, receiving his Ph.D. from the University of Munich in 1908. After holding positions in Switzerland, the Netherlands, and Germany, he moved to the University of Berlin in the early 1930s. Although he had been assured that he would be able retain his Dutch citizenship, Debye found that he would be unable to continue his work in Berlin unless he became a German citizen. He refused and left Germany in 1939 for Cornell University, where he remained for the rest of his life, becoming an American citizen in 1946. Debye was awarded the Nobel Prize for chemistry in 1936 "for his contributions to our knowledge of molecular structure through his investigations on dipole moments and on the diffraction of X rays and electrons in gases."
Erich Hückel (right) was born in Göttingen, Germany, on August 19, 1896 and died in 1980. He received his Ph.D. in physics from the University of Göttingen in 1921. He later worked with Peter Debye in Zurich, and together they developed a theory for the thermodynamic properties of solutions of strong electrolytes that is now known as the Debye–Hückel theory. Hückel also developed Hückel molecular orbital theory, which we learned in Chapter 10 applies to conjugated and aromatic molecules. Hückel was appointed professor of theoretical physics at the University of Marburg in 1937, where he remained until his retirement.

CHAPTER 25

Solutions II: Solid–Liquid Solutions

In the previous chapter, we studied binary solutions, such as ethanol/water solutions, in which the two components were miscible in all proportions. In such solutions, either component can be treated as a solvent. In this chapter, we will study solutions in which one of the components is present at much smaller concentrations than the other, so that the terms "solute" and "solvent" are meaningful. We will introduce a solute standard state based upon Henry's law such that the activity of the solute becomes equal to its concentration as its concentration goes to zero. In the first few sections, we will study solutions of nonelectrolytes, and then solutions of electrolytes. Unlike for solutions of nonelectrolytes, we will be able to present exact expressions for the activities and activity coefficients in dilute solutions of electrolytes. In Sections 25–3 and 25–4, we will discuss the colligative properties of solutions, such as osmotic pressure, as well as the depression of the freezing point and elevation of the boiling point of a solvent by the addition of solute.

25–1. We Use a Raoult's Law Standard State for the Solvent and a Henry's Law Standard State for the Solute for Solutions of Solids Dissolved in Liquids

In Section 24–8, we considered solutions in which one of the components is only sparingly soluble in the other. In cases such as these, we use the terms *solute* for the sparingly soluble component and *solvent* for the component in excess. We customarily denote solvent quantities by a subscript 1 and solute quantities by a subscript 2. The activities we defined for the solvent and solute are such that $a_1 \to x_1$ as $x_1 \to 1$ and $a_2 \to x_2$ as $x_2 \to 0$. Recall that a_1 is defined with respect to a Raoult's law standard state (Equation 24.39)

$$a_1 = \frac{P_1}{P_1^*} \qquad \text{(Raoult's law standard state)} \qquad (25.1)$$

and that a_2 is defined with respect to a Henry's law standard state (Equation 24.43)

$$a_{2x} = \frac{P_2}{k_{H,x}} \qquad \text{(Henry's law standard state)} \qquad (25.2)$$

where the subscript x emphasizes that a_{2x} and $k_{H,x}$ are based on a mole fraction scale ($P_2 = k_{H,x} x_2$). Even if the solute does not have a measurable vapor pressure, defining the activity by Equation 25.2 is nevertheless convenient because the ratio is still meaningful; even though P_2 and $k_{H,2}$ may be exceedingly small, the ratio $P_2 / k_{H,2}$ is finite.

Although we have defined the activities of the solvent and solute in terms of mole fractions, the use of mole fractions to express the concentration of a solute in a dilute solution is not numerically convenient. A more convenient unit is *molality* (m), which is defined as the number of moles of solute per 1000 grams of solvent. In an equation, we have

$$m = \frac{n_2}{1000 \text{ g solvent}} \qquad (25.3)$$

where n_2 is the number of moles of solute (subscript 2). Note that the units of molality are $\text{mol} \cdot \text{kg}^{-1}$. We say that a solution containing 2.00 moles of NaCl in 1.00 kg of water is 2.00 molal, or that it is a 2.00 $\text{mol} \cdot \text{kg}^{-1}$ NaCl(aq) solution. The relation between the mole fraction of solute (x_2) and molality (m) is

$$x_2 = \frac{n_2}{n_1 + n_2} = \frac{m}{\dfrac{1000 \text{ g} \cdot \text{kg}^{-1}}{M_1} + m} \qquad (25.4)$$

where M_1 is the molar mass ($\text{g} \cdot \text{mol}^{-1}$) of the solvent. The term $1000 \text{ g} \cdot \text{kg}^{-1}/M_1$ is the number of moles of solvent (n_1) in 1000 g of solvent and m, by definition, is the number of moles of solute in 1000 g of solvent. In the case of water, $1000 \text{ g} \cdot \text{kg}^{-1}/M_1$ is equal to $55.506 \text{ mol} \cdot \text{kg}^{-1}$, so Equation 25.4 becomes

$$x_2 = \frac{m}{55.506 \text{ mol} \cdot \text{kg}^{-1} + m} \qquad (25.5)$$

Note that x_2 and m are directly proportional to each other if $m \ll 55.506 \text{ mol} \cdot \text{kg}^{-1}$, which is the case for dilute solutions.

EXAMPLE 25–1
Calculate the mole fraction of a 0.200 $\text{mol} \cdot \text{kg}^{-1}$ $C_{12}H_{22}O_{11}$(aq) solution.

SOLUTION: The solution contains 0.200 moles of sucrose per 1000.0 g of water. The mole fraction of sucrose is

$$x_2 = \frac{n_2}{n_1 + n_2} = \frac{0.200 \text{ mol}}{\dfrac{1000.0 \text{ g}}{18.02 \text{ g} \cdot \text{mol}^{-1}} + 0.200 \text{ mol}} = 0.000359$$

We define the solute activity in terms of molality by requiring that

$$a_{2m} \longrightarrow m \quad \text{as} \quad m \longrightarrow 0 \qquad (25.6)$$

where the subscript m emphasizes that a_{2m} is based on a molality scale. We can express Henry's law in terms of the molality rather than the mole fraction by $P_2 = k_{\text{H},m}m$, where once again the subscript m emphasizes that $k_{\text{H},m}$ is based on a molality scale. In terms of $k_{\text{H},m}$, the activity of the solute is defined by

$$a_{2m} = \frac{P_2}{k_{\text{H},m}} \qquad (25.7)$$

Another common concentration unit is *molarity* (c), which is the number of moles of solute per 1000 mL of solution. In an equation,

$$c = \frac{n_2}{1000 \text{ mL solution}} \qquad (25.8)$$

Note that molarity has units of mol·L^{-1}. We say that a solution containing 2.00 moles of NaCl in 1.00 liter of solution is a 2.00-molar solution, or that it is a 2.00 mol·L^{-1} NaCl(aq) solution.

We define the solute activity in terms of molarity by requiring that

$$a_{2c} \longrightarrow c \quad \text{as} \quad c \longrightarrow 0 \qquad (25.9)$$

where the subscript c emphasizes that a_{2c} is based on a molarity scale. We can express Henry's law in terms of the molarity rather than the mole fraction of solute by $P_2 = k_{\text{H},c}c$, where once again the subscript c emphasizes that $k_{\text{H},c}$ is based on a molarity scale. In terms of $k_{\text{H},c}$, the activity of the solute is defined by

$$a_{2c} = \frac{P_2}{k_{\text{H},c}} \qquad (25.10)$$

Converting from molarity to molality is easy if we know the density of the solution, which is available for many solutions in handbooks. For example, the density of a 2.450 mol·L^{-1} aqueous sucrose solution at 20°C is 1.3103 g·mL^{-1}. Thus, there are 838.6 g of sucrose in 1000 mL of solution, which has a total mass of 1310.3 g. Of these 1310.3 g, 838.6 g are due to sucrose, so 1310.3 g − 838.6 g = 471.7 g are due to water. The molality then is given by

$$m = \frac{2.450 \text{ mol sucrose}}{471.7 \text{ g H}_2\text{O}} \times \frac{1000 \text{ g H}_2\text{O}}{\text{kg H}_2\text{O}} = 5.194 \text{ mol·kg}^{-1}$$

EXAMPLE 25–2
The density (in g·mL^{-1}) of an aqueous sucrose solution can be expressed as

$$\rho/\text{g·mL}^{-1} = 0.9982 + (0.1160 \text{ kg·mol}^{-1})m - (0.0156 \text{ kg}^2\cdot\text{mol}^{-2})m^2$$
$$+ (0.0011 \text{ kg}^3\cdot\text{mol}^{-3})m^3 \qquad 0 \le m \le 6 \text{ mol·kg}^{-1}$$

Calculate the molarity of a 2.00-molal aqueous sucrose solution.

SOLUTION: A 2.00-molal aqueous sucrose solution contains 2.00 moles (684.6 g) of sucrose per 1000 g of H_2O, or 2.00 moles of sucrose in 1684.6 g of solution. The density of the solution is given by

$$\rho/g \cdot mL^{-1} = 0.9982 + (0.1160 \text{ kg} \cdot mol^{-1})(2.00 \text{ mol} \cdot kg^{-1})$$
$$- (0.0156 \text{ kg}^2 \cdot mol^{-2})(4.00 \text{ mol}^2 \cdot kg^{-2})$$
$$+ (0.0011 \text{ kg}^3 \cdot mol^{-3})(8.00 \text{ mol}^3 \cdot kg^{-3})$$
$$= 1.177$$

so the volume of the solution is

$$V = \frac{\text{mass}}{\text{density}} = \frac{1684.6 \text{ g}}{1.177 \text{ g} \cdot mL^{-1}} = 1432 \text{ mL}$$

Therefore, the molarity of the solution is

$$c = \frac{2.00 \text{ mol sucrose}}{1.432 \text{ L}} = 1.40 \text{ mol} \cdot L^{-1}$$

Problem 25–5 asks you to derive a general relation between c and m.

EXAMPLE 25–3

Given the density (ρ) of the solution in $g \cdot mL^{-1}$, derive a general relation between x_2 and c.

SOLUTION: Consider exactly a one liter sample of the solution. In this case, $c = n_2$, the number of moles of solute in the one-liter sample. The mass of the solution is given by

$$\text{mass of the solution per liter} = (1000 \text{ mL} \cdot L^{-1})\rho$$

so the mass of the solvent is

$$\text{mass of the solvent per liter} = \text{mass of the solution} - \text{mass of the solute}$$
$$= (1000 \text{ mL} \cdot L^{-1})\rho - cM_2$$

where M_2 is the molar mass ($g \cdot mol^{-1}$) of the solute. Therefore, n_1, the number of moles of solvent, is

$$n_1 = \frac{(1000 \text{ mL} \cdot L^{-1})\rho - cM_2}{M_1}$$

so

$$x_2 = \frac{n_2}{n_1 + n_2} = \frac{c}{\dfrac{(1000 \text{ mL} \cdot L^{-1})\rho - cM_2}{M_1} + c}$$

$$= \frac{cM_1}{(1000 \text{ mL} \cdot L^{-1})\rho + c(M_1 - M_2)} \tag{25.11}$$

Table 25.1 summarizes the equations for the activities we have defined for the various concentration scales. In each case, the activity coefficient γ is defined by dividing the activity by the appropriate concentration. Thus, for example, $\gamma_m = a_{2m}/m$. Problem 25–12 asks you to derive a relation between the various solute activity coefficients in Table 25.1.

25–2. The Activity of a Nonvolatile Solute Can Be Obtained from the Vapor Pressure of the Solvent

The equations for the solute activities in Table 25.1 are applicable to nonvolatile as well as volatile solutes. The vapor pressure of a nonvolatile solute is so low, however, that these equations are not practical to use. Fortunately, the Gibbs–Duhem equation provides us with a way to determine the activity of a nonvolatile solute from a measurement of the activity of the solvent. We will illustrate this procedure using an

TABLE 25.1

A summary of the equations for the activities used for the various concentration scales for dilute solutions.

Solvent—Raoult's law standard state

$$a_1 = \frac{P_1}{P_1^*}$$

$a_1 \to x_1$ as $x_1 \to 1$

$$\gamma_1 = \frac{a_1}{x_1}$$

$P_1 \to P_1^* x_1$ as $x_1 \to 1$ (Raoult's law)

Solute—Henry's law standard state

Mole fraction scale

$$a_{2x} = \frac{P_2}{k_{H,x}}$$

$a_{2x} \to x_2$ as $x_2 \to 0$

$$\gamma_{2x} = \frac{a_{2x}}{x_2}$$

$P_2 \to k_{H,x} x_2$ as $x_2 \to 0$ (Henry's law)

Molality scale

$$a_{2m} = \frac{P_2}{k_{H,m}}$$

$a_{2m} \to m$ as $m \to 0$

$$\gamma_{2m} = \frac{a_{2m}}{m}$$

$P_2 \to k_{H,m} m$ as $m \to 0$ (Henry's law)

Molarity scale

$$a_{2c} = \frac{P_2}{k_{H,c}}$$

$a_{2c} \to c$ as $c \to 0$

$$\gamma_{2c} = \frac{a_{2c}}{c}$$

$P_2 \to k_{H,c} c$ as $c \to 0$ (Henry's law)

---

aqueous solution of sucrose. According to a Raoult's law standard state, the activity of the water is given by P_1/P_1^*. Now let's consider a dilute solution, in which case $a_1 = x_1$. We now want to relate a_1 to the molality of the solute, m. For a dilute solution, $m \ll 55.506 \text{ mol·kg}^{-1}$, so we can neglect m compared with $55.506 \text{ mol·kg}^{-1}$ in the denominator of Equation 25.5 and write

$$x_2 \approx \frac{m}{55.506 \text{ mol·kg}^{-1}}$$

Therefore, for small concentrations,

$$\ln a_1 = \ln x_1 = \ln(1 - x_2) \approx -x_2 \approx -\frac{m}{55.506 \text{ mol·kg}^{-1}} \tag{25.12}$$

where we have used the fact that $\ln(1 - x_2) \approx -x_2$ for small values of x_2.

Table 25.2 and Figure 25.1 give experimental data for the vapor pressure of water in equilibrium with an aqueous sucrose solution at 25°C as a function of molality and mole fraction, respectively. The equilibrium vapor pressure of pure water at 25°C is 23.756 torr, so $a_1 = P_1/P_1^* = P_1/23.756$ is given in the third column of Table 25.2.

Equation 25.12 relates a_1 to the molality m for only a dilute solution. For example, Table 25.2 shows that $a_1 = 0.93276$ at 3.00 molal, whereas Equation 25.12 gives $\ln a_1 = -0.054048$, or $a_1 = 0.9474$. To account for this discrepancy, we now define a quantity ϕ, called the *osmotic coefficient*, by

$$\ln a_1 = -\frac{m\phi}{55.506 \text{ mol·kg}^{-1}} \tag{25.13}$$

Note that $\phi = 1$ if the solution behaves as an ideal dilute solution. Thus, the deviation of ϕ from unity is a measure of the nonideality of the solution.

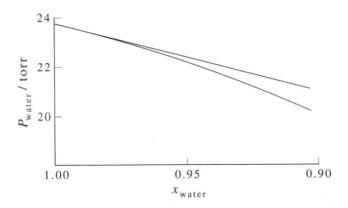

FIGURE 25.1
The vapor pressure of water in equilibrium with an aqueous sucrose solution at 25°C plotted against the mole fraction of water. Note that Raoult's law (the straight line in the figure) holds from $x_{\text{water}} = 1.00$ to about 0.97, but that deviations occur at lower values of x_{water}.

TABLE 25.2
The vapor pressure of water (P_1) in equilibrium with an aqueous sucrose solution at 25°C as a function of molality (m). Additional data are the activity of the water (a_1), the osmotic coefficient (ϕ), and the activity coefficient (γ_{2m}) of the sucrose.

$m/\text{mol·kg}^{-1}$	P_1/torr	a_1	ϕ	γ_{2m}	$\ln \gamma_{2m}$
0.00	23.756	1.00000	1.0000	1.000	0.0000
0.10	23.713	0.99819	1.0056	1.017	0.0169
0.20	23.669	0.99634	1.0176	1.034	0.0334
0.30	23.625	0.99448	1.0241	1.051	0.0497
0.40	23.580	0.99258	1.0335	1.068	0.0658
0.50	23.534	0.99067	1.0406	1.085	0.0816
0.60	23.488	0.98872	1.0494	1.105	0.0998
0.70	23.441	0.98672	1.0601	1.125	0.1178
0.80	23.393	0.98472	1.0683	1.144	0.1345
0.90	23.344	0.98267	1.0782	1.165	0.1527
1.00	23.295	0.98059	1.0880	1.185	0.1723
1.20	23.194	0.97634	1.1075	1.233	0.2095
1.40	23.089	0.97193	1.1288	1.283	0.2492
1.60	22.982	0.96740	1.1498	1.335	0.2889
1.80	22.872	0.96280	1.1690	1.387	0.3271
2.00	22.760	0.95807	1.1888	1.442	0.3660
2.50	22.466	0.94569	1.2398	1.590	0.4637
3.00	22.159	0.93276	1.2879	1.751	0.5602
3.50	21.840	0.91933	1.3339	1.924	0.6544
4.00	21.515	0.90567	1.3749	2.101	0.7424
4.50	21.183	0.89170	1.4139	2.310	0.8372
5.00	20.848	0.87760	1.4494	2.481	0.9087
5.50	20.511	0.86340	1.4823	2.680	0.9858
6.00	20.176	0.84930	1.5111	3.878	1.3553

EXAMPLE 25–4
Using the data in Table 25.2, calculate the value of ψ at 1.00 mol·kg^{-1}.

SOLUTION: We simply use Equation 25.13 and find that

$$\phi = -\frac{(55.506 \text{ mol·kg}^{-1}) \ln(0.98059)}{1.00 \text{ mol·kg}^{-1}} = 1.0880$$

in agreement with the entry in Table 25.2.

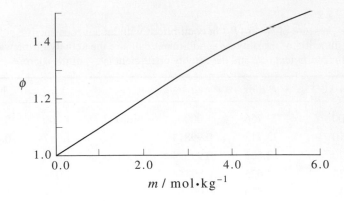

FIGURE 25.2
The osmotic coefficient (ϕ) of an aqueous sucrose solution at 25°C plotted against the molality (m). The magnitutde of the deviation of the value of ϕ from unity is a measure of the nonideality of the solution.

Figure 25.2 shows ϕ for an aqueous sucrose solution at 25°C plotted against m. Note that the solution becomes increasingly nonideal as m increases.

The fifth column in Table 25.2 gives the activity coefficient of the sucrose calculated from the activity of the water, or from the osmotic coefficient, by means of the Gibbs–Duhem equation,

$$n_1 d \ln a_1 + n_2 d \ln a_2 = 0$$

In terms of molality, m, $n_1 = 55.506$ mol and $n_2 = m$, so the Gibbs–Duhem equation becomes

$$(55.506 \text{ mol·kg}^{-1}) d \ln a_1 + m d \ln a_2 = 0 \tag{25.14}$$

Using Equation 25.13, we see that $(55.506 \text{ mol·kg}^{-1}) d \ln a_1 = -d(m\phi)$. If we substitute this result and $a_{2m} = \gamma_{2m} m$ (Table 25.1) into Equation 25.14, we obtain

$$d(m\phi) = m d \ln(\gamma_{2m} m)$$

or

$$m d\phi + \phi dm = m(d \ln \gamma_{2m} + d \ln m)$$

We can rewrite this equation as

$$d \ln \gamma_{2m} = d\phi + \frac{\phi - 1}{m} dm$$

We now integrate from $m = 0$ (where $\gamma_{2m} = \phi = 1$) to arbitrary m to get

$$\ln \gamma_{2m} = \phi - 1 + \int_0^m \left(\frac{\phi - 1}{m'} \right) dm' \tag{25.15}$$

Equation 25.15 allows us to calculate the activity coefficient of the solute from the data on the vapor pressure of the solvent. The vapor pressure of the solvent gives us the activity of the solvent from Equation 25.1; then the osmotic coefficient ϕ is calculated from Equation 25.13, and $\ln \gamma_{2m}$ is determined from Equation 25.15.

The data for ϕ in Table 25.2 can be fit with a polynomial in the molality. If we choose (arbitrarily) a 5th-degree polynomial, we find that (Problem 25–18)

$$\phi = 1.00000 + (0.07349 \text{ kg·mol}^{-1})m + (0.019783 \text{ kg}^2 \cdot \text{mol}^{-2})m^2$$
$$- (0.005688 \text{ kg}^3 \cdot \text{mol}^{-3})m^3 + (6.036 \times 10^{-4} \text{ kg}^4 \cdot \text{mol}^{-4})m^4$$
$$- (2.517 \times 10^{-5} \text{ kg}^5 \cdot \text{mol}^{-5})m^5 \qquad 0 \leq m \leq 6 \text{ mol·kg}^{-1}$$

We can substitute this expression into Equation 25.15 to obtain $\ln \gamma_{2m}$.

EXAMPLE 25–5
Use the above polynomial fit for ϕ and Equation 25.15 to calculate the value of γ_{2m} for a 1.00-molal aqueous sucrose solution.

SOLUTION: First, we need to evaluate the integral in Equation 25.15 (neglecting to write the units in the coefficients of the powers of m):

$$\int_0^1 \left(\frac{\phi - 1}{m} \right) dm = \int_0^1 [0.07349 + 0.019783m - 0.005688m^2$$
$$+ 6.036 \times 10^{-4}m^3 - 2.517 \times 10^{-5}m^4]dm$$
$$= 0.07349 + \frac{0.019783}{2} - \frac{0.005688}{3}$$
$$+ \frac{6.036 \times 10^{-4}}{4} - \frac{2.517 \times 10^{-5}}{5}$$
$$= 0.08163$$

so

$$\ln \gamma_{2m} = \phi - 1 + \int_0^1 \left(\frac{\phi - 1}{m} \right) dm$$
$$= 0.08816 + 0.08163 = 0.1698$$

or $\gamma_{2m} = 1.185$, in agreement with the entry in Table 25.2.

The values of $\ln \gamma_{2m}$ and γ_{2m} given in Table 25.2 have been calculated using the procedure in Example 25–5. Figure 25.3 shows $\ln \gamma_{2m}$ plotted against m for an aqueous sucrose solution at 25°C.

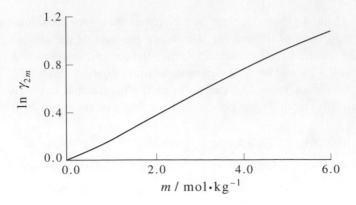

FIGURE 25.3
The logarithm of the activity coefficient ($\ln \gamma_{2m}$) of sucrose in an aqueous sucrose solution at 25°C plotted against the molality (m).

25–3. Colligative Properties Are Solution Properties That Depend Only Upon the Number Density of Solute Particles

A number of solution properties, called *colligative properties*, depend, at least in dilute solution, upon only the number of solute particles, and not upon their kind. Colligative properties include the lowering of the vapor pressure of a solvent by the addition of a solute, the elevation of the boiling point of a solution by a nonvolatile solute, the depression of the freezing point of a solution by a solute, and osmotic pressure. We will discuss only freezing-point depression and osmotic pressure.

At the freezing point of a solution, solid solvent is in equilibrium with the solvent in solution. The thermodynamic condition of this equilibrium is that

$$\mu_1^s(T_{\text{fus}}) = \mu_1^{\text{sln}}(T_{\text{fus}})$$

where as usual the subscript 1 denotes solvent and T_{fus} is the freezing point of the solution. We use Equation 24.35 for μ_1 to obtain

$$\mu_1^s = \mu_1^* + RT \ln a_1 = \mu_1^l + RT \ln a_1$$

We have written μ_1^l for μ_1^* simply to compare it with μ_1^s. Solving for $\ln a_1$, we get

$$\ln a_1 = \frac{\mu_1^s - \mu_1^l}{RT} \tag{25.16}$$

Now differentiate with respect to temperature and use the Gibbs–Helmholtz equation (Example 24–1),

$$\left[\frac{\partial(\mu_1/T)}{\partial T} \right]_{P,x_1} = -\frac{\overline{H}_1}{T^2}$$

to obtain

$$\left(\frac{\partial \ln a_1}{\partial T}\right)_{P,x_1} = \frac{\overline{H}_1^{\text{l}} - \overline{H}_1^{\text{s}}}{RT^2} = \frac{\Delta_{\text{fus}}\overline{H}}{RT^2} \tag{25.17}$$

where we have used the fact that $\overline{H}_1^{\text{l}} - \overline{H}_1^{\text{s}} = \Delta_{\text{fus}}\overline{H}$ for the pure solvent. If we integrate Equation 25.17 from pure solvent, where $a_1 = 1$, $T = T_{\text{fus}}^*$, to a solution with arbitrary values of a_1 and T_{fus}, we obtain

$$\ln a_1 = \int_{T_{\text{fus}}^*}^{T_{\text{fus}}} \frac{\Delta_{\text{fus}}\overline{H}}{RT^2}\,dT \tag{25.18}$$

Equation 25.18 can be used to determine the activity of the solvent in a solution (Problem 25–20).

You may have calculated freezing-point depressions in general chemistry using the formula

$$\Delta T_{\text{fus}} = K_{\text{f}}m \tag{25.19}$$

where K_{f} is a constant, called the *freezing-point depression constant*, whose value depends upon the solvent. We can derive Equation 25.19 from Equation 25.18 by making a few approximations appropriate to dilute solutions. If the solution is sufficiently dilute, then $\ln a_1 = \ln x_1 = \ln(1 - x_2) \approx -x_2$, and if we assume that $\Delta_{\text{fus}}\overline{H}$ is independent of temperature over the temperature range $(T_{\text{fus}}, T_{\text{fus}}^*)$, we obtain

$$-x_2 = \frac{\Delta_{\text{fus}}\overline{H}}{R} \int_{T_{\text{fus}}^*}^{T_{\text{fus}}} \frac{dT}{T^2} = \frac{\Delta_{\text{fus}}\overline{H}}{R}\left(\frac{1}{T_{\text{fus}}^*} - \frac{1}{T_{\text{fus}}}\right)$$
$$= \frac{\Delta_{\text{fus}}\overline{H}}{R}\left(\frac{T_{\text{fus}} - T_{\text{fus}}^*}{T_{\text{fus}}T_{\text{fus}}^*}\right) \tag{25.20}$$

Because x_2 and $\Delta_{\text{fus}}\overline{H}$ are positive quantities, we see immediately that $T_{\text{fus}} - T_{\text{fus}}^* < 0$, or that $T_{\text{fus}} < T_{\text{fus}}^*$. Thus, we find that the addition of a solute will lower the freezing point of a solution. We can express x_2 in terms of molality by using Equation 25.4,

$$x_2 = \frac{m}{\dfrac{1000\ \text{g}\cdot\text{kg}^{-1}}{M_1} + m} \approx \frac{M_1 m}{1000\ \text{g}\cdot\text{kg}^{-1}} \tag{25.21}$$

for small values of m (dilute solution). Furthermore, because $T_{\text{fus}}^* - T_{\text{fus}}$ is usually only a few degrees (dilute solution once again), we can replace T_{fus} in the denominator of Equation 25.20 by T_{fus}^* to a good approximation to get finally (Problem 25–23)

$$\Delta T_{\text{fus}} = T_{\text{fus}}^* - T_{\text{fus}} = K_{\text{f}}m \tag{25.22}$$

where

$$K_f = \frac{M_1}{1000 \text{ g·kg}^{-1}} \frac{R(T^*_{fus})^2}{\Delta_{fus}\overline{H}} \tag{25.23}$$

We can calculate the value of K_f for water.

$$K_f = \left(\frac{18.02 \text{ g·mol}^{-1}}{1000 \text{ g·kg}^{-1}}\right) \frac{(8.314 \text{ J·K}^{-1}\text{·mol}^{-1})(273.2 \text{ K})^2}{6.01 \text{ kJ·mol}^{-1}}$$

$$= 1.86 \text{ K·kg·mol}^{-1}$$

Equation 25.22 tells us that the freezing point of a 0.20-molal solution of sucrose in water is $-(1.86 \text{ K·kg·mol}^{-1})(0.20 \text{ mol·kg}^{-1}) = -0.37 \text{ K}$.

EXAMPLE 25–6

Calculate the value of K_f for cyclohexane, whose freezing point is 279.6 K and molar enthalpy of fusion is 2.68 kJ·mol^{-1}.

SOLUTION: We use Equation 25-23 with $M_1 = 84.16 \text{ g·mol}^{-1}$ and the above values of T^*_{fus} and $\Delta_{fus}\overline{H}$.

$$K_f = \left(\frac{84.16 \text{ g·mol}^{-1}}{1000 \text{ g·kg}^{-1}}\right) \frac{(8.314 \text{ J·K}^{-1}\text{·mol}^{-1})(279.6 \text{ K})^2}{2680 \text{ J·mol}^{-1}}$$

$$= 20.4 \text{ K·kg·mol}^{-1}$$

Thus, the freezing point of a 0.20-molal solution of hexane in cyclohexane is 4.1 K lower than the freezing point of pure cyclohexane, or $T_{fus} = 275.5 \text{ K}$.

We can derive an expression for the boiling-point elevation of a solution containing a nonvolatile solute. The analog of Equation 25.22 is (Problem 25–25)

$$\Delta T_{vap} = T_{vap} - T^*_{vap} = K_b m \tag{25.24}$$

where the *boiling-point elevation constant* is given by

$$K_b = \frac{M_1}{1000 \text{ g·kg}^{-1}} \frac{R(T^*_{vap})^2}{\Delta_{vap}\overline{H}} \tag{25.25}$$

The value of K_b for water is only 0.512 K·kg·mol^{-1}, so the boiling point elevation is a rather small effect for aqueous solutions.

25–4. Osmotic Pressure Can Be Used to Determine the Molecular Masses of Polymers

Figure 25.4 illustrates the development of osmotic pressure. In the initial state, we have pure water on the left and an aqueous sucrose solution on the right. The two liquids are separated by a membrane containing pores that allow water molecules but not solute molecules to pass through. Such a membrane is called a *semipermeable membrane*. (Many biological cells are surrounded by membranes semipermeable to water.) The levels of the two liquids in Figure 25.4 are initially the same, but water will pass through the semipermeable membrane until the chemical potentials of the water on the two sides of the membrane are equal. This process results in the situation shown in the equilibrium state, where the two liquid levels are no longer equal. The hydrostatic pressure head that is built up is called *osmotic pressure*.

Because the water is free to pass through the semipermeable membrane, the chemical potential of the water must be the same on the two sides of the membrane at equilibrium. In other words, the chemical potential of the pure water at a pressure P must equal the chemical potential of the water in the solution at a pressure $P + \Pi$ and an activity a_1. In an equation,

$$
\begin{aligned}
\mu_1^*(T, P) &= \mu_1^{\text{sln}}(T, P + \Pi, a_1) \\
&= \mu_1^*(T, P + \Pi) + RT \ln a_1
\end{aligned}
\tag{25.26}
$$

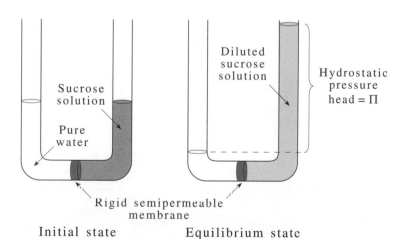

FIGURE 25.4

Passage of water through a rigid, semipermeable membrane separating pure water from an aqueous sucrose solution. The water passes through the membrane until the chemical potential of the water in the aqueous sucrose solution equals that of the pure water. The chemical potential of water in the sucrose solution increases as the hydrostatic pressure above the solution increases.

where $a_1 = P_1/P_1^*$. We can rewrite Equation 25.26. as

$$\mu_1^*(T, P + \Pi) - \mu_1^*(T, P) + RT \ln a_1 = 0 \qquad (25.27)$$

The first two terms in Equation 25.27 are the difference in the chemical potential of the pure solvent at two different pressures. Equation 23.8

$$\left(\frac{\partial \mu_1^*}{\partial P}\right)_T = \overline{V}_1^* \qquad (23.8)$$

where $\overline{V}_1^*$ is the molar volume of the pure solvent, tells us how the chemical potential varies with pressure. We can use Equation 23.8 to evaluate $\mu_1^*(T, P + \Pi) - \mu_1^*(T, P)$ by integrating both sides from P to $P + \Pi$ to get

$$\mu_1^*(T, P + \Pi) - \mu_1^*(T, P) = \int_P^{P+\Pi} \left(\frac{\partial \mu_1^*}{\partial P'}\right)_T dP' = \int_P^{P+\Pi} \overline{V}_1^* dP' \qquad (25.28)$$

If we substitute Equation 25.28 into Equation 25.27, we obtain

$$\int_P^{P+\Pi} \overline{V}^* dP' + RT \ln a_1 = 0 \qquad (25.29)$$

Assuming $\overline{V}_1^*$ does not vary with applied pressure, we can write Equation 25.29 as

$$\Pi \overline{V}_1^* + RT \ln a_1 = 0 \qquad (25.30)$$

Furthermore, if the solution is dilute, then $a_1 \approx x_1 = 1 - x_2$, with x_2 small. Therefore, we can write $\ln a_1$ as $\ln(1 - x_2) \approx -x_2$, so that Equation 25.30 becomes

$$\Pi \overline{V}_1^* = RT x_2$$

Furthermore, because x_2 is small, $n_2 \ll n_1$ and

$$x_2 = \frac{n_2}{n_1 + n_2} \approx \frac{n_2}{n_1}$$

Substitute this into the above equation to get

$$\Pi = \frac{n_2 RT}{n_1 \overline{V}_1^*} \approx \frac{n_2 RT}{V}$$

where we have replaced $n_1 \overline{V}_1^*$ by the total volume of the solution, V (dilute solution). The above equation is usually written as

$$\Pi = cRT \qquad (25.31)$$

where c is the molarity, n_2/V, of the solution. Equation 25.31 is called the van't Hoff equation for osmotic pressure. Using this equation, we calculate the osmotic pressure of a 0.100-molar aqueous solution of sucrose at 20°C to be

$$\Pi = (0.100 \text{ mol} \cdot \text{L}^{-1})(0.08206 \text{ L} \cdot \text{atm} \cdot \text{K}^{-1} \cdot \text{mol}^{-1})(293.2 \text{ K})$$

$$= 2.40 \text{ atm}$$

Thus, we see that osmotic pressure is a large effect. Because of this, osmotic pressure can be used to determine molecular masses of solutes, particularly solutes with large molecular masses such as polymers and proteins.

EXAMPLE 25–7

It is found that 2.20 g of a certain polymer dissolved in enough water to make 300 mL of solution has an osmotic pressure of 7.45 torr at 20°C. Determine the molecular mass of the polymer.

SOLUTION: The molarity of the solution is given by

$$c = \frac{\Pi}{RT} = \frac{7.45 \text{ torr}/760 \text{ torr} \cdot \text{atm}^{-1}}{(0.08206 \text{ L} \cdot \text{atm} \cdot \text{K}^{-1} \cdot \text{mol}^{-1})(293.2 \text{ K})}$$

$$= 4.07 \times 10^{-4} \text{ mol} \cdot \text{L}^{-1}$$

Therefore, there are 4.07×10^{-4} moles of polymer per liter of solution, or $(0.300)(4.07 \times 10^{-4}) = 1.22 \times 10^{-4}$ moles per 300 mL of solution. Thus, we find that 1.22×10^{-4} moles corresponds to 2.20 g, or that the molecular mass is 18,000.

If a pressure in excess of 26 atm is applied to seawater at 15°C, the chemical potential of the water in the seawater will exceed that of pure water. Consequently, pure water can be obtained from seawater by using a rigid semipermeable membrane and an applied pressure in excess of the osmotic pressure of 26 atm. This process is known as *reverse osmosis*. Reverse osmosis units are commercially available and are used to obtain fresh water from salt water using a variety of semipermeable membranes, the most common of which is cellulose acetate.

25–5. Solutions of Electrolytes Are Nonideal at Relatively Low Concentrations

When sodium chloride dissolves in water, the solution contains sodium ions and chloride ions and essentially no undissociated sodium chloride. The ions interact with each other through a coulombic potential, which varies as $1/r$. We should compare this interaction with the one between neutral solute molecules (nonelectrolytes) such as sucrose, where the interaction varies as something like $1/r^6$. Thus, the interaction between ions in solution is effective over a much greater distance than the interaction between neutral solute particles, so solutions of electrolytes deviate from ideal behavior more strongly and at lower concentrations than do solutions of nonelectrolytes. Figure 25.5 shows $\ln \gamma_{2m}$ for sucrose, sodium chloride, and calcium chloride plotted versus molality. Note that $CaCl_2(aq)$ appears to behave more nonideally than $NaCl(aq)$, which in turn behaves more nonideally than sucrose. The charge of $+2$ on the calcium ion leads to a stronger coulombic interaction and hence a stronger deviation from ideality than for NaCl. At $0.100 \ \text{mol}\cdot\text{kg}^{-1}$, the activity coefficient of sucrose is 0.998, whereas that of $CaCl_2(aq)$ is 0.518 and that of $NaCl(aq)$ is 0.778.

Before we discuss the determination of activity coefficients for electrolytes, we must first introduce notation needed to describe the thermodynamic properties of solutions of electrolytes. Consider the general salt $C_{\nu_+} A_{\nu_-}$, which dissociates into ν_+ cations and ν_- anions per formula unit as in

$$C_{\nu_+} A_{\nu_-} (s) \xrightarrow{H_2O(l)} \nu_+ C^{z+}(aq) + \nu_- A^{z-}(aq)$$

where $\nu_+ z_+ + \nu_- z_- = 0$ by electroneutrality. For example, $\nu_+ = 1$ and $\nu_- = 2$ for $CaCl_2$ and $\nu_+ = 2$ and $\nu_- = 1$ for Na_2SO_4. Therefore, $CaCl_2$ is called a 1–2 electrolyte and Na_2SO_4 is called a 2–1 electrolyte. We write the chemical potential of the salt in terms of the chemical potentials of its constituent ions according to

$$\mu_2 = \nu_+ \mu_+ + \nu_- \mu_- \tag{25.32}$$

where

$$\mu_2 = \mu_2^\circ + RT \ln a_2 \tag{25.33}$$

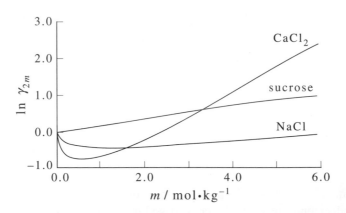

FIGURE 25.5
The logarithm of the activity coefficient ($\ln \gamma_{2m}$) of aqueous solutions of sucrose, sodium chloride, and calcium chloride plotted against molality (m) at 25°C. Note that the electrolyte solutions deviate from ideality ($\ln \gamma_{2m} = 0$) much more strongly than does sucrose at small concentrations.

and

$$\mu_+ = \mu_+^\circ + RT \ln a_+$$
$$\mu_- = \mu_-^\circ + RT \ln a_- \tag{25.34}$$

The superscript zeros here represent the chosen standard state, which we can leave unspecified at this point but is usually taken to be the solute or Henry's law standard state. If we substitute Equations 25.34 into Equation 25.32 and equate the result to Equation 25.33, we obtain

$$\nu_+ \ln a_+ + \nu_- \ln a_- = \ln a_2$$

where we have used the relation $\mu_2^\circ = \nu_+ \mu_+^\circ + \nu_- \mu_-^\circ$ in analogy with Equation 25.32. We can rewrite the above equation as

$$a_2 = a_+^{\nu_+} a_-^{\nu_-} \tag{25.35}$$

For many of the formulas that occur in the thermodynamics of solutions of electrolytes, it is convenient to define a quantity $a_\pm$, called the *mean ionic activity*, by

$$a_2 = a_\pm^\nu = a_+^{\nu_+} a_-^{\nu_-} \tag{25.36}$$

where $\nu = \nu_+ + \nu_-$. Note that $a_\pm$ is raised to the same power as the sum of the exponents in the last term in Equation 25.36, For example, we write

$$a_{\text{NaCl}} = a_\pm^2 = a_+ a_-$$

and

$$a_{\text{CaCl}_2} = a_\pm^3 = a_+ a_-^2$$

Even though we cannot determine activities of single ions, we can still *define* single-ion activity coefficients by

$$a_+ = m_+ \gamma_+ \quad \text{and} \quad a_- = m_- \gamma_-$$

where m_+ and m_- are the molalities of the individual ions, which are given by $m_+ = \nu_+ m$ and $m_- = \nu_- m$. If we substitute these expressions for a_+ and a_- into Equation 25.36, we get

$$a_2 = a_\pm^\nu = (m_+^{\nu_+} m_-^{\nu_-})(\gamma_+^{\nu_+} \gamma_-^{\nu_-}) \tag{25.37}$$

In analogy with the definition of the *mean ionic activity* $a_\pm$ in Equation 25.36, we define a mean ionic molality $m_\pm$ by

$$m_\pm^\nu = m_+^{\nu_+} m_-^{\nu_-} \tag{25.38}$$

and a *mean ionic activity coefficient* $\gamma_\pm$ by

$$\gamma_\pm^\nu = \gamma_+^{\nu_+} \gamma_-^{\nu_-} \tag{25.39}$$

Again, notice that the sum of the exponents on both sides of Equations 25.38 and 25.39 are the same. Given these definitions, we can now write Equation 25.37 as

$$a_2 = a_\pm^\nu = m_\pm^\nu \gamma_\pm^\nu \tag{25.40}$$

EXAMPLE 25–8

Write out Equation 25.40 explicitly for $CaCl_2$.

SOLUTION: In this case, $\nu_+ = 1$ and $\nu_- = 2$. Furthermore, according to the equation

$$CaCl_2(s) \xrightarrow{\text{H}_2\text{O(l)}} Ca^{2+}(aq) + 2\,Cl^-(aq)$$

we see that $m_+ = m$ and $m_- = 2m$. Thus,

$$a_2 = a_\pm^3 = (m)(2m)^2 \gamma_\pm^3 = 4m^3 \gamma_\pm^3$$

The relations between a_2, m, and $\gamma_\pm$ for other types of electrolytes are given in Table 25.3.

TABLE 25.3

The relations between the activity of a strong electrolyte, its molality, and its mean ionic activity coefficient for various types of strong electrolytes.

Type	
1–1	
KCl(aq)	$a_2 = a_+ a_- = a_\pm^2 = m_\pm^2 \gamma_\pm^2 = (m_+)(m_-)\gamma_\pm^2 = m^2 \gamma_\pm^2$
1–2	
$CaCl_2$(aq)	$a_2 = a_+ a_-^2 = a_\pm^3 = m_\pm^3 \gamma_\pm^3 = (m_+)(m_-)^2 \gamma_\pm^3 = (m)(2m)^2 \gamma_\pm^3 = 4m^3 \gamma_\pm^3$
1–3	
$LaCl_3$(aq)	$a_2 = a_+ a_-^3 = a_\pm^4 = m_\pm^4 \gamma_\pm^4 = (m_+)(m_-)^3 \gamma_\pm^4 = (m)(3m)^3 \gamma_\pm^4 = 27m^4 \gamma_\pm^4$
2–1	
Na_2SO_4(aq)	$a_2 = a_+^2 a_- = a_\pm^3 = (m_+)^2 (m_-)\gamma_\pm^3 = (2m)^2 (m)\gamma_\pm^3 = 4m^3 \gamma_\pm^3$
2–2	
$ZnSO_4$(aq)	$a_2 = a_+ a_- = a_\pm^2 = m_\pm^2 \gamma_\pm^2 = (m_+)(m_-)\gamma_\pm^2 = m^2 \gamma_\pm^2$
3–1	
$Na_3Fe(CN)_6$(aq)	$a_2 = a_+^3 a_- = a_\pm^4 = m_\pm^4 \gamma_\pm^4 = (m_+)^3 (m_-)\gamma_\pm^4 = (3m)^3 (m)\gamma_\pm^4 = 27m^4 \gamma_\pm^4$

Mean ionic activity coefficients can be determined experimentally by the same methods used for the activity coefficients of nonelectrolytes. We will illustrate their determination from the measurement of the vapor pressure of the solvent as we did for an aqueous sucrose solution in Section 25–2. In analogy with Equation 25.13, we define an osmotic coefficient for aqueous electrolyte solutions by

$$\ln a_1 = -\frac{\nu m \phi}{55.506 \text{ mol} \cdot \text{kg}^{-1}} \tag{25.41}$$

Notice that this equation differs from Equation 25.13 by the inclusion of a factor of ν here. Equation 25.41 reduces to Equation 25.13 for nonelectrolyte solutions because $\nu = 1$ in that case. Problem 25–34 asks you to show that with this factor of ν, $\phi \to 1$ as $m \to 0$ for solutions of electrolytes or nonelectrolytes. Starting with Equation 25.41 and the Gibbs–Duhem equation, you can derive the analog of Equation 25.15 straightforwardly:

$$\ln \gamma_\pm = \phi - 1 + \int_0^m \left(\frac{\phi - 1}{m'}\right) dm' \tag{25.42}$$

Table 25.4 gives the vapor pressure of an aqueous solution of NaCl as a function of molality. Also included in the table are activities of the water (calculated from

TABLE 25.4
The vapor pressure (P_{H_2O}), activity of the water (a_w), osmotic coefficient (ϕ), and logarithm of the mean ionic activity coefficient ($\ln \gamma_\pm$) of the NaCl in an aqueous solution of NaCl at 25°C as a function of molality (m).

$m/\text{mol}\cdot\text{kg}^{-1}$	P_{H_2O}/torr	a_w	ϕ	$\ln \gamma_\pm$
0.000	23.76	1.0000	1.0000	0.0000
0.200	23.60	0.9934	0.9245	−0.3079
0.400	23.44	0.9868	0.9205	−0.3685
0.600	23.29	0.9802	0.9227	−0.3977
0.800	23.13	0.9736	0.9285	−0.4143
1.000	22.97	0.9669	0.9353	−0.4234
1.400	22.64	0.9532	0.9502	−0.4267
1.800	22.30	0.9389	0.9721	−0.4166
2.200	21.96	0.9242	0.9944	−0.3972
2.600	21.59	0.9089	1.0196	−0.3709
3.000	21.22	0.8932	1.0449	−0.3396
3.400	20.83	0.8769	1.0723	−0.3046
3.800	20.43	0.8600	1.1015	0.2666
4.400	19.81	0.8339	1.1457	−0.2053
5.000	19.17	0.8068	1.1916	−0.1389

$a_1 = P_1/P_1^*$), osmotic coefficients (calculated from Equation 25.41), and mean ionic activity coefficients (calculated from Equation 25.42).

For sucrose in Section 25–2, we curve fit ϕ to a polynomial in m and then used that polynomial to calculate the value of γ_{2m}. As we will see in Section 25–6, the osmotic coefficient of electrolytes is better described by an expression of the form (a polynomial in $m^{1/2}$)

$$\phi = 1 + am^{1/2} + bm + cm^{3/2} + \cdots$$

The osmotic coefficient data for sodium chloride given in Table 25.4 can be fit by

$$\phi = 1 - (0.3920 \text{ kg}^{1/2}\cdot\text{mol}^{-1/2})m^{1/2} + (0.7780 \text{ kg}\cdot\text{mol}^{-1})m$$
$$- (0.8374 \text{ kg}^{3/2}\cdot\text{mol}^{-3/2})m^{3/2} + (0.5326 \text{ kg}^2\cdot\text{mol}^{-2})m^2$$
$$- (0.1673 \text{ kg}^{5/2}\cdot\text{mol}^{-5/2})m^{5/2} + (0.0206 \text{ kg}^3\cdot\text{mol}^{-3})m^3$$
$$0 \le m \le 5.0 \text{ mol}\cdot\text{kg}^{-1} \qquad (25.43)$$

This expression for ϕ along with Equation 25.42 were used to calculate the values of $\ln \gamma_{\pm}$ given in Table 25.4.

EXAMPLE 25–9

Verify the entry for $\ln \gamma_{\pm}$ at 1.00 molal in Table 25.4.

SOLUTION: We first write (neglecting the units in the coefficients of the powers of m in Equation 25.43)

$$\int_0^m \left(\frac{\phi - 1}{m'}\right) dm' = -(0.3920)(2m^{1/2}) + 0.7780m - (0.8374)\frac{2m^{3/2}}{3}$$
$$+ (0.5326)\frac{m^2}{2} - (0.1673)\frac{2m^{5/2}}{5} + (0.0206)\frac{m^3}{3}$$

and add this result to $\phi - 1$ to obtain

$$\ln \gamma_{\pm} = -(0.3920)(3m^{1/2}) + (0.7780)(2m) - (0.8374)\frac{5m^{3/2}}{3}$$
$$+ (0.5326)\frac{3m^2}{2} - (0.1673)\frac{7m^{5/2}}{5} + (0.0206)\frac{4m^3}{3}$$

Thus, at 1.00 molal, $\ln \gamma_{\pm} = -0.4234$, or $\gamma_{\pm} = 0.655$.

The formulas we derived in Section 25–3 for the colligative properties of solutions of nonelectrolytes take on a slightly different form for solutions of electrolytes. The difference lies in Equation 25.21 for x_2. For a strong electrolyte that dissociates into ν_+ cations and ν_- anions per formula unit, the mole fraction of solute particles is given by

$$x_2 = \frac{\nu m}{\dfrac{1000 \text{ g·kg}^{-1}}{M_1} + \nu m} \approx \frac{\nu m M_1}{1000 \text{ g·kg}^{-1}} \tag{25.44}$$

Note that the right side here contains a factor of ν. If this expression for x_2 is carried through in derivations of the formulas for the colligative effects, we obtain

$$\Delta T_{\text{fus}} = \nu K_{\text{f}} m \tag{25.45}$$

$$\Delta T_{\text{vap}} = \nu K_{\text{b}} m \tag{25.46}$$

and

$$\Pi - \nu c R T \tag{25.47}$$

EXAMPLE 25–10

A 0.050-molal aqueous solution of $K_3Fe(CN)_6$ has a freezing point of $-0.36°C$. How many ions are formed per formula unit of $K_3Fe(CN)_6$?

SOLUTION: We can solve Equation 25.45 for ν to obtain

$$\nu = \frac{\Delta T_{\text{fus}}}{K_{\text{f}} m} = \frac{0.36°C}{(1.86 \ °C·kg·mol^{-1})(0.050 \ mol·kg^{-1})} = 3.9$$

Thus, the dissolution process of $K_3Fe(CN)_6$ can be written as

$$K_3Fe(CN)_6 \xrightarrow{H_2O(l)} 3 \ K^+(aq) + Fe(CN)_6^{3-}(aq)$$

25–6. The Debye–Hückel Theory Gives an Exact Expression for $\ln \gamma_\pm$ for Very Dilute Solutions

In the previous section, we expressed the osmotic coefficient for solutions of electrolytes in the form $\phi = 1 + am^{1/2} + bm + \cdots$ rather than as a simple polynomial in m as we did for sucrose in Section 25–2. The reason we did so is that in 1925, Peter Debye and Erich Hückel showed theoretically that at low concentrations, the logarithm of the activity coefficient of ion j is given by

$$\ln \gamma_j = -\frac{\kappa q_j^2}{8\pi \varepsilon_0 \varepsilon_r k_B T} \tag{25.48}$$

and that the logarithm of the mean ionic activity coefficient is given by (see Problems 25–50 through 25–58)

$$\ln \gamma_{\pm} = -|q_+ q_-| \frac{\kappa}{8\pi \varepsilon_0 \varepsilon_r k_B T} \tag{25.49}$$

where $q_+ = z_+ e$ and $q_- = z_- e$ are the charges on the cations and anions, ε_r is the (unitless) relative permittivity of the solvent, and κ is given by

$$\kappa^2 = \sum_{j=1}^{s} \frac{q_j^2}{\varepsilon_0 \varepsilon_r k_B T} \left(\frac{N_j}{V} \right) \tag{25.50}$$

where s is the number of ionic species and N_j/V is the number density of species j. If we convert N_j/V to molarity, Equation 25.50 becomes

$$\kappa^2 = N_A (1000 \text{ L} \cdot \text{m}^{-3}) \sum_{j=1}^{s} \frac{q_j^2 c_j}{\varepsilon_0 \varepsilon_r k_B T} \tag{25.51}$$

It is customary to define a quantity I_c, called the *ionic strength*, by

$$I_c = \frac{1}{2} \sum_{j=1}^{s} z_j^2 c_j \tag{25.52}$$

where c_j is the molarity of the jth ionic species, in which case (Problem 25–46)

$$\kappa^2 = \frac{2e^2 N_A (1000 \text{ L} \cdot \text{m}^{-3})}{\varepsilon_0 \varepsilon_r k_B T} \left(I_c / \text{mol} \cdot \text{L}^{-1} \right) \tag{25.53}$$

EXAMPLE 25–11
First show that κ has units of m^{-1} and then show that $\ln \gamma_{\pm}$ in Equation 25.49 is unitless, as it must be.

SOLUTION: We start with Equation 25.50. The units of q_j are C, ε_0 are $\text{C}^2 \cdot \text{s}^2 \cdot \text{kg}^{-1} \cdot \text{m}^{-3}$, k_B are $\text{J} \cdot \text{K}^{-1} = \text{kg} \cdot \text{m}^2 \cdot \text{s}^{-2} \cdot \text{K}^{-1}$, T are K, and N_j/V are m^{-3}. Therefore, the units of κ^2 are

$$\kappa^2 \sim \frac{(\text{C}^2)(\text{m}^{-3})}{(\text{C}^2 \cdot \text{s}^2 \cdot \text{kg}^{-1} \cdot \text{m}^{-3})(\text{kg} \cdot \text{m}^2 \cdot \text{s}^{-2} \cdot \text{K}^{-1})(\text{K})} = m^{-2}$$

or

$$\kappa \sim \text{m}^{-1}$$

Using Equation 25.49 for $\ln \gamma_{\pm}$,

$$\ln \gamma_{\pm} \sim \frac{(\text{C}^2)(\text{m}^{-1})}{(\text{C}^2 \cdot \text{s}^2 \cdot \text{kg}^{-1} \cdot \text{m}^{-3})(\text{kg} \cdot \text{m}^2 \cdot \text{s}^{-2} \cdot \text{K}^{-1})(\text{K})} = \text{unitless}$$

Equation 25.49 is called the Debye–Hückel limiting law because it is the exact form that $\ln \gamma_\pm$ takes on for all electrolyte solutions for sufficiently low concentrations. Just what is meant by "sufficiently low concentrations" depends upon the system. Note that $\ln \gamma_\pm$ goes as κ in Equation 25.49, that κ goes as $I_c^{1/2}$ in Equation 25.53, and that $I_c^{1/2}$ goes as $c^{1/2}$ in Equation 25.52. Consequently, $\ln \gamma_\pm$ varies as $c^{1/2}$. This $c^{1/2}$ dependence is typical for electrolyte solutions, so when we curve fit ϕ in Section 25–5, we fit it to a polynomial in $c^{1/2}$ (or $m^{1/2}$) instead of c (or m).

Most of the experimental data for $\ln \gamma_\pm$ are given in terms of molality rather than molarity. In Figure 25.6, we plot $\ln \gamma_\pm$ versus $m^{1/2}$ for a number of 1–1 electrolytes. Note that all the curves merge into a single straight line at small concentrations, in accord with the limiting law nature of Equation 25.49. At small concentrations where the limiting law is valid, the molality and molarity scales differ by only a multiplicative constant, so a linear plot in $c^{1/2}$ is also linear in $m^{1/2}$ (Problem 25–5).

The quantity κ in Equation 25.50 is a central quantity in the Debye–Hückel theory and has the following physical interpretation. Consider an ion with charge q_i situated at the origin of a spherical coordinate system. According to Debye and Hückel (see also Problem 25–51), the net charge in a spherical shell of radius r and thickness dr surrounding this central ion is

$$p_i(r)dr = -q_i \kappa^2 r e^{-\kappa r} dr \tag{25.54}$$

If we integrate this expression from 0 to ∞, we obtain

$$\int_0^\infty p_i(r)dr = -q_i \kappa^2 \int_0^\infty r e^{-\kappa r} dr = -q_i$$

This result simply says that the total charge surrounding an ion of charge q_i is equal and of the opposite sign to q_i. In other words, it expresses the electroneutrality of the

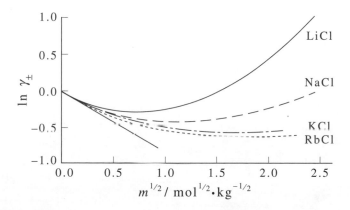

FIGURE 25.6
Values of $\ln \gamma_\pm$ versus $m^{1/2}$ for aqueous alkali halide solutions at 25°C. Note that even though the four curves are different, they all merge into one, the Debye–Hückel limiting law (Equation 25.49) at small concentrations.

solution. Equation 25.54, which is plotted in Figure 25.7, shows that there is a diffuse shell of net charge of opposite sign surrounding any given ion in solution. We say that Equation 25.54 describes an *ionic atmosphere* about the central ion. Furthermore, the maximum in the curve in Figure 25.7 occurs at $r = \kappa^{-1}$, so we say that κ^{-1}, which Example 25–11 shows has units of m, is a measure of the thickness of the ionic atmosphere.

For a 1–1 electrolyte in aqueous solution at 25°C, a handy formula for κ is (Problem 25–53)

$$\frac{1}{\kappa} = \frac{304 \text{ pm}}{(c/\text{mol·L}^{-1})^{1/2}} \tag{25.55}$$

where c is the molarity of the solution. The thickness of the ionic atmosphere in a 0.010 molar solution is approximately 3000 pm, or about 10 times the size of a typical ion.

For an aqueous solution at 25°C, Equation 25.49 becomes (Problem 25–59)

$$\ln \gamma_{\pm} = -1.173|z_+ z_-|(I_c/\text{mol·L}^{-1})^{1/2} \tag{25.56}$$

According to Equation 25.52, I_c is related to the concentration, but the relation itself depends upon the type of electrolyte. For example, for a 1–1 electrolyte, $z_+ = 1$, $z_- = -1$, $c_+ = c$, and $c_- = c$, so $I = c$. For a 1–2 electrolyte such as $CaCl_2$, $z_+ = 2$, $z_- = -1$, $c_+ = c$, and $c_- = 2c$, so $I_c = \frac{1}{2}(4c + 2c) = 3c$. Generally, I_c is equal to some numerical factor times c, where the value of the numerical factor depends upon the type of salt. Therefore, Equation 25.56 says that a plot of $\ln \gamma_{\pm}$ versus $c^{1/2}$ should be a straight line and that the slope of the line should depend upon the type of electrolyte. The slope will be -1.173 for a 1–1 electrolyte and $-(1.173)(2)(3^{1/2}) = -4.06$ for

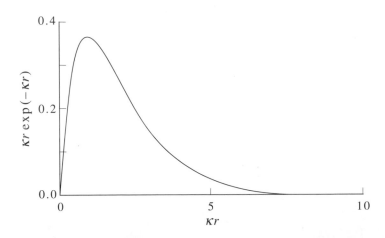

FIGURE 25.7
A plot of the net charge in a spherical shell of radius r and thickness dr surrounding a central ion of charge q_i. This plot illustrates the ionic atmosphere that surrounds each ion in solution. The maximum here corresponds to $r = \kappa^{-1}$.

a 1–2 electrolyte. Figure 25.8 shows a plot of $\ln \gamma_{\pm}$ versus $c^{1/2}$ for NaCl(aq) and CaCl$_2$(aq). Notice that the plots are indeed linear for small concentrations and that deviations from linear behavior occur at higher concentrations [$c^{1/2} \approx 0.05$ mol·L^{-1} or $c = 0.003$ mol·L^{-1} for CaCl$_2$(aq) and $c^{1/2} \approx 0.15$ mol·L^{-1} or $c = 0.02$ mol·L^{-1} for NaCl(aq)]. The slopes of the two linear portions are in the ratio of 4.06 to 1.17.

25–7. The Mean Spherical Approximation Is an Extension of the Debye–Hückel Theory to Higher Concentrations

The Debye–Hückel theory assumes that the ions are simply point ions (zero radii) and that they interact with a purely coulombic potential [$U(r) = z_{+}z_{-}e^2/4\pi\varepsilon_0\varepsilon_r r$]. In addition, the solvent is considered a continuous medium with a uniform relative permittivity ε_r (78.54 for water at 25°C). Although the assumptions of point ions and a continuum solvent may seem crude, they are quite satisfactory when the ions are far apart from each other on the average, as they are in very dilute solutions. Consequently, the Debye–Hückel expression for $\ln \gamma_{\pm}$ given by Equation 25.49 is exact in the limit of small concentrations. There is no corresponding theory for solutions of nonelectrolytes because, being neutral species, nonelectrolyte molecules do not interact with each other to any significant extent until they approach each other relatively closely, where the solvent can hardly be assumed to be a continuous medium.

Figure 25.8 emphasizes that the Debye–Hückel theory is a limiting law. It should not be considered a quantitative theory with which to calculate activity coefficients except at very low concentrations. Nevertheless, the Debye–Hückel theory has played an invaluable role as a strict limiting law that all electrolyte solutions obey. In addition, any theory that attempts to describe solutions at higher concentrations must reduce to Equation 25.49 for small concentrations. Many attempts have been made to construct

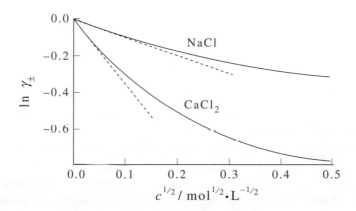

FIGURE 25.8
A plot of the logarithm of the mean ionic activity coefficient ($\ln \gamma_{\pm}$) for NaCl(aq) and CaCl$_2$(aq) at 25°C versus $c^{1/2}$. Note that both curves approach the Debye–Hückel limiting law (the straight lines) as the molarity goes to zero.

theories for more concentrated electrolyte solutions, but most have met with only limited success. One early attempt is called the Extended Debye–Hückel theory, in which Equation 25.49 is modified to be

$$\ln \gamma_\pm = -\frac{1.173|z_+ z_-|(I_c/\text{mol·L}^{-1})^{1/2}}{1 + (I_c/\text{mol·L}^{-1})^{1/2}} \tag{25.57}$$

This expression becomes Equation 25.49 in the limit of small concentrations because $I_c^{1/2}$ becomes negligible compared with unity in the denominator of Equation 25.57 in this limit.

EXAMPLE 25–12
Use Equation 25.57 to calculate $\ln \gamma_\pm$ for 0.050 molar LiCl(aq), and compare the result with that obtained from Equation 25.49. The accepted experimental value is -0.191.

SOLUTION: For a 1–1 salt such as LiCl, $I_c = c$, so

$$\ln \gamma_\pm = -1.173(0.050)^{1/2} = -0.262$$

and

$$\ln \gamma_\pm = -\frac{1.173(0.050)^{1/2}}{1 + (0.050)^{1/2}} = -0.214$$

Although Equation 25.57 provides some improvement over the Debye–Hückel limiting law, it is not very accurate even at 0.050 molar. At 0.200 molar, Equation 25.57 gives -0.362 for $\ln \gamma_\pm$ versus the experimental value of -0.274.

Another semiempirical expression for $\ln \gamma_\pm$ that has been widely used to fit experimental data is

$$\ln \gamma_\pm = -\frac{1.173|z_+ z_-|(I_c/\text{mol·L}^{-1})^{1/2}}{1 + (I_c/\text{mol·L}^{-1})^{1/2}} + Cm \tag{25.58}$$

where C is a parameter whose value depends upon the electrolyte. Although Equation 25.58 can be used to fit experimental $\ln \gamma_\pm$ data up to one molar or so, C is still strictly an adjustable parameter.

In the 1970s, significant advances were made in the theory of electrolyte solutions. Most of the work on these theories is based on a model called the *primitive model*, in which the ions are considered hard spheres with charges at their centers and the solvent is considered a continuous medium with a uniform relative permittivity. In spite of the obvious deficiencies of this model, it addresses the long-range coulombic interactions between the ions and their short-range repulsion. These turn out to be major considerations, and as we will see, the primitive model can give quite satisfactory agreement with experimental data over a fairly large concentration range.

Most of these theories that have been developed require numerical solutions to fairly complicated equations, but one is notable in that it provides analytic expressions for the various thermodynamic properties of electrolyte solutions. The name of this theory, the mean spherical approximation (MSA), derives from its original formulation, and the theory can be viewed as a Debye–Hückel theory in which the finite (nonzero) size of the ions is accounted for in a fairly rigorous manner. A central result of the mean spherical approximation is that

$$\ln \gamma_{\pm} = \ln \gamma_{\pm}^{el} + \ln \gamma^{HS} \tag{25.59}$$

where $\ln \gamma_{\pm}^{el}$ is an electrostatic (coulombic) contribution to $\ln \gamma_{\pm}$ and $\ln \gamma^{HS}$ is a hard-sphere (finite-size) contribution. For solutions of 1–1 electrolytes, $\ln \gamma_{\pm}^{el}$ is given by

$$\ln \gamma_{\pm}^{el} = \frac{x(1 + 2x)^{1/2} - x - x^2}{4\pi \rho d^3} \tag{25.60}$$

where ρ is the number density of charged particles, d is the sum of the radius of a cation and an anion, and $x = \kappa d$, where κ is given by Equation 25.53. Although it is not obvious by casual inspection, Equation 25.59 reduces to the Debye–Hückel limiting law, Equation 25.49, in the limit of small concentrations (Problem 25–60). The hard sphere contribution to $\ln \gamma_{\pm}$ is given by

$$\ln \gamma^{HS} = \frac{4y - \dfrac{9}{4}y^2 + \dfrac{3}{8}y^3}{\left(1 - \dfrac{y}{2}\right)^3} \tag{25.61}$$

where $y = \pi \rho d^3/6$.

In spite of the fact that Equations 25.60 and 25.61 are somewhat lengthy, they are easy to use because once d has been chosen, they give $\ln \gamma_{\pm}$ in terms of the molarity c. Figure 25.9 shows experimental values of $\ln \gamma_{\pm}$ for NaCl(aq) at 25°C and $\ln \gamma_{\pm}$ as calculated from Equation 25.59 with $d = 320$ pm.

Given essentially one adjustable parameter (the sum of the ionic radii), the agreement is seen to be quite good. We also show the results for the more commonly seen Equation 25.57 in Figure 25.9.

Problems

25-1. The density of a glycerol/water solution that is 40.0% glycerol by mass is 1.101 g·mL^{-1} at 20°C. Calculate the molality and the molarity of glycerol in the solution at 20°C. Calculate the molality at 0°C.

25-2. Concentrated sulfuric acid is sold as a solution that is 98.0% sulfuric acid and 2.0% water by mass. Given that the density is 1.84 g·mL^{-1}, calculate the molarity of concentrated sulfuric acid.

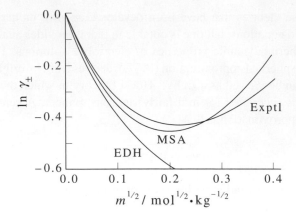

FIGURE 25.9
A comparison of $\ln \gamma_\pm$ from the mean spherical approximation (Equation 25.59) with experimental data for NaCl(aq) at 25°C. The line labelled EDH is the extended Debye–Hückel theory result, Equation 25.57. The value of d, the sum of the radii of the cation and anion, is taken to be 320 pm.

25-3. Concentrated phosphoric acid is sold as a solution that is 85% phosphoric acid and 15% water by mass. Given that the molarity is 15 mol·L⁻¹, calculate the density of concentrated phosphoric acid.

25-4. Calculate the mole fraction of glucose in an aqueous solution that is 0.500 molal in glucose.

25-5. Show that the relation between molarity and molality for a solution with a single solute is

$$c = \frac{(1000 \text{ mL·L}^{-1})\rho m}{1000 \text{ g·kg}^{-1} + mM_2}$$

where c is the molarity, m is the molality, ρ is the density of the solution in g·mL⁻¹, and M_2 is the molar mass (g·mol⁻¹) of the solute.

25-6. The *CRC Handbook of Chemistry and Physics* has tables of "concentrative properties of aqueous solutions" for many solutions. Some entries for CsCl(s) are

$A/\%$	$\rho/\text{g·mL}^{-1}$	$c/\text{mol·L}^{-1}$
1.00	1.0058	0.060
5.00	1.0374	0.308
10.00	1.0798	0.641
20.00	1.1756	1.396
40.00	1.4226	3.380

where A is the mass percent of the solute, ρ is the density of the solution, and c is the molarity. Using these data, calculate the molality at each concentration.

25-7. Derive a relation between the mass percentage (A) of a solute in a solution and its molality (m). Calculate the molality of an aqueous sucrose solution that is 18% sucrose by mass.

25-8. Derive a relation between the mole fraction of the solvent and the molality of a solution.

25-9. The volume of an aqueous sodium chloride solution at 25°C can be expressed as

$$V/\text{mL} = 1001.70 + (17.298 \text{ kg·mol}^{-1})m + (0.9777 \text{ kg}^2\text{·mol}^{-2})m^2$$

$$- (0.0569 \text{ kg}^3\text{·mol}^{-3})m^3$$

$$0 \le m \le 6 \text{ mol·kg}^{-1}$$

where m is the molality. Calculate the molarity of a solution that is 3.00 molal in sodium chloride.

25-10. If x_2^∞, m^∞, and c^∞ are the mole fraction, molality, and molarity, respectively, of a solute at infinite dilution, show that

$$x_2^\infty = \frac{m^\infty M_1}{1000 \text{ g·kg}^{-1}} = \frac{c^\infty M_1}{(1000 \text{ mL·L}^{-1})\rho_1}$$

where M_1 is the molar mass (g·mol^{-1}) and ρ_1 is the density (g·mL^{-1}) of the solvent. Note that mole fraction, molality, and molarity are all directly proportional to each other at low concentrations.

25-11. Consider two solutions whose solute activities are a_2' and a_2'', referred to the same standard state. Show that the difference in the chemical potentials of these two solutions is independent of the standard state and depends only upon the ratio a_2'/a_2''. Now choose one of these solutions to be at an arbitrary concentration and the other at a very dilute concentration (essentially infinitely dilute) and argue that

$$\frac{a_2'}{a_2''} = \frac{\gamma_{2x} x_2}{x_2^\infty} = \frac{\gamma_{2m} m}{m^\infty} = \frac{\gamma_{2c} c}{c^\infty}$$

25-12. Use Equations 25.4, 25.11, and the results of the previous two problems to show that

$$\gamma_{2x} = \gamma_{2m} \left(1 + \frac{mM_1}{1000 \text{ g·kg}^{-1}} \right) = \gamma_{2c} \left(\frac{\rho}{\rho_1} + \frac{c[M_1 - M_2]}{\rho_1[1000 \text{ mL·L}^{-1}]} \right)$$

where ρ is the density of the solution. Thus, we see that the three different activity coefficients are related to one another.

25-13. Use Equations 25.4, 25.11, and the results of Problem 25–12 to derive

$$\gamma_{2m} = \gamma_{2c} \left(\frac{\rho}{\rho_1} - \frac{cM_2}{\rho_1[1000 \text{ mL·L}^{-1}]} \right)$$

Given that the density of an aqueous citric acid ($M_2 = 192.12$ g·mol^{-1}) solution at 20°C is given by

$$\rho/\text{g·mL}^{-1} = 0.99823 + (0.077102 \text{ L·mol}^{-1})c$$

$$0 \le c < 1.772 \text{ mol·L}^{-1}$$

plot γ_{2m}/γ_{2c} versus c. Up to what concentration do γ_{2m} and γ_{2c} differ by 2%?

25-14. The *CRC Handbook of Chemistry and Physics* gives a table of mass percent of sucrose in an aqueous solution and its corresponding molarity at 25°C. Use these data to plot molality versus molarity for an aqueous sucrose solution.

25-15. Using the data in Table 25.2, calculate the activity coefficient of water (on a mole fraction basis) at a sucrose concentration of 3.00 molal.

25-16. Using the data in Table 25.2, plot the activity coefficient of water against the mole fraction of water.

25-17. Using the data in Table 25.2, calculate the value of ϕ at each value of m and reproduce Figure 25.2.

25-18. Fit the data for the osmotic coefficient of sucrose in Table 25.2 to a 4th-degree polynomial and calculate the value of γ_{2m} for a 1.00-molal solution. Compare your result with the one obtained in Example 25–5.

25-19. Using the data for sucrose given in Table 25.2, determine the value of $\ln \gamma_{2m}$ at 3.00 molal by plotting $(\phi - 1)/m$ versus m and determining the area under the curve by numerical integration (MathChapter G) rather than by curve fitting ϕ first. Compare your result with the value given in Table 25.2.

25-20. Equation 25.18 can be used to determine the activity of the solvent at its freezing point. Assuming that ΔC_P^* is independent of temperature, show that

$$\Delta_{\text{fus}} \overline{H}(T) = \Delta_{\text{fus}} \overline{H}(T_{\text{fus}}^*) + \Delta \overline{C}_P^*(T - T_{\text{fus}}^*)$$

where $\Delta_{\text{fus}} \overline{H}(T_{\text{fus}}^*)$ is the molar enthalpy of fusion at the freezing point of the pure solvent (T_{fus}^*) and $\Delta \overline{C}_P^*$ is the difference in the molar heat capacities of liquid and solid solvent. Using Equation 25.18, show that

$$-\ln a_1 = \frac{\Delta_{\text{fus}} \overline{H}(T_{\text{fus}}^*)}{R(T_{\text{fus}}^*)^2}\theta + \frac{1}{R(T_{\text{fus}}^*)^2}\left(\frac{\Delta_{\text{fus}} \overline{H}(T_{\text{fus}}^*)}{T_{\text{fus}}^*} - \frac{\Delta \overline{C}_P^*}{2}\right)\theta^2 + \cdots$$

where $\theta = T_{\text{fus}}^* - T_{\text{fus}}$.

25-21. Take $\Delta_{\text{fus}} \overline{H}(T_{\text{fus}}^*) = 6.01 \text{ kJ·mol}^{-1}, \overline{C}_P^{\text{l}} = 75.2 \text{ J·K}^{-1}\text{·mol}^{-1}, \text{and } \overline{C}_P^{\text{s}} = 37.6 \text{ J·K}^{-1}\text{·mol}^{-1}$ to show that the equation for $-\ln a_1$ in the previous problem becomes

$$-\ln a_1 = (0.00968 \text{ K}^{-1})\theta + (5.2 \times 10^{-6} \text{ K}^{-2})\theta^2 + \cdots$$

for an aqueous solution. The freezing point depression of a 1.95-molal aqueous sucrose solution is 4.45°C. Calculate the value of a_1 at this concentration. Compare your result with the value in Table 25.2. The value you calculated in this problem is for 0°C, whereas the value in Table 25.2 is for 25°C, but the difference is fairly small because a_1 does not vary greatly with temperature (Problem 25–61).

25-22. The freezing point of a 5.0-molal aqueous glycerol (1,2,3-propanetriol) solution is −10.6°C. Calculate the activity of water at 0°C in this solution. (See Problems 25–20 and 25–21.)

25-23. Show that replacing T_{fus} by T_{fus}^* in the denominator of $(T_{\text{fus}} - T_{\text{fus}}^*)/T_{\text{fus}}^* T_{\text{fus}}$ (see Equation 25.20) gives $-\theta/(T_{\text{fus}}^*)^2 - \theta^2/(T_{\text{fus}}^*)^3 + \cdots$ where $\theta = T_{\text{fus}}^* - T_{\text{fus}}$.

25-24. Calculate the value of the freezing point depression constant for nitrobenzene, whose freezing point is 5.7°C and whose molar enthalpy of fusion is 11.59 kJ·mol⁻¹.

25-25. Use an argument similar to the one we used to derive Equations 25.22 and 25.23 to derive Equations 25.24 and 25.25.

25-26. Calculate the boiling point elevation constant for cyclohexane given that $T_{vap} = 354$ K and $\Lambda_{vap}\overline{H} = 29.97$ kJ·mol⁻¹.

25-27. A solution containing 1.470 g of dichlorobenzene in 50.00 g of benzene boils at 80.60°C at a pressure of 1.00 bar. The boiling point of pure benzene is 80.09°C, and the molar enthalpy of vaporization of pure benzene is 32.0 kJ·mol⁻¹. Determine the molecular mass of dichlorobenzene from these data.

25-28. Consider the following phase diagram for a typical pure substance. Label the region corresponding to each phase. Illustrate how this diagram changes for a dilute solution of a nonvolatile solute.

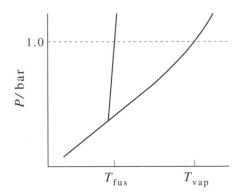

Now demonstrate that the boiling point increases and the freezing point decreases as a result of the dissolution of the solute.

25-29. A solution containing 0.80 g of a protein in 100 mL of a solution has an osmotic pressure of 2.06 torr at 25°C. What is the molecular mass of the protein?

25-30. Show that the osmotic pressure of an aqueous solution can be written as

$$\Pi = \frac{RT}{\overline{V}^*}\left(\frac{m}{55.506 \text{ mol·kg}^{-1}}\right)\phi$$

25-31. According to Table 25.2, the activity of the water in a 2.00-molal sucrose solution is 0.95807. What external pressure must be applied to the solution at 25.0°C to make the activity of the water in the solution the same as that in pure water at 25.0°C and 1 atm? Take the density of water to be 0.997 g·mL⁻¹.

25-32. Show that $a_2 = a_{\pm}^2 = m^2\gamma_{\pm}^2$ for a 2–2 salt such as $CuSO_4$ and that $a_2 = a_{\pm}^4 = 27m^4\gamma_{\pm}^4$ for a 1–3 salt such as $LaCl_3$.

25-33. Verify the following table:

Type of salt	Example	I_m
1–1	KCl	m
1–2	$CaCl_2$	$3m$
2–1	K_2SO_4	$3m$
2–2	$MgSO_4$	$4m$
1–3	$LaCl_3$	$6m$
3–1	Na_3PO_4	$6m$

Show that the general result for I_m is $|z_+z_-|(\nu_+ + \nu_-)m/2$.

25-34. Show that the inclusion of the factor ν in Equation 25.41 allows $\phi \to 1$ as $m \to 0$ for solutions of electrolytes as well as nonelectrolytes. [*Hint*: Realize that x_2 involves the total number of moles of solute particles (see Equation 25.44)].

25-35. Use Equation 25.41 and the Gibbs–Duhem equation to derive Equation 25.42.

25-36. The osmotic coefficient of $CaCl_2(aq)$ solutions can be expressed as

$$\phi = 1.0000 - (1.2083 \text{ kg}^{1/2} \cdot \text{mol}^{-1/2})m^{1/2} + (3.2215 \text{ kg} \cdot \text{mol}^{-1})m$$

$$- (3.6991 \text{ kg}^{3/2} \cdot \text{mol}^{-3/2})m^{3/2} + (2.3355 \text{ kg}^2 \cdot \text{mol}^{-2})m^2$$

$$- (0.67218 \text{ kg}^{5/2} \cdot \text{mol}^{-5/2})m^{5/2} + (0.069749 \text{ kg}^3 \cdot \text{mol}^{-3})m^3$$

$$0 \le m \le 5.00 \text{ mol} \cdot \text{kg}^{-1}$$

Use this expression to calculate and plot $\ln \gamma_\pm$ as a function of $m^{1/2}$.

25-37. Use Equation 25.43 to calculate $\ln \gamma_\pm$ for NaCl(aq) at 25°C as a function of molality and plot it versus $m^{1/2}$. Compare your results with those in Table 25.4.

25-38. In Problem 25–19, you determined $\ln \gamma_{2m}$ for sucrose by calculating the area under the curve of $\phi - 1$ versus m. When dealing with solutions of electrolytes, it is better numerically to plot $(\phi - 1)/m^{1/2}$ versus $m^{1/2}$ because of the natural dependence of ϕ on $m^{1/2}$. Show that

$$\ln \gamma_\pm = \phi - 1 + 2 \int_0^{m^{1/2}} \frac{\phi - 1}{m^{1/2}} dm^{1/2}$$

25-39. Use the data in Table 25.4 to calculate $\ln \gamma_\pm$ for NaCl(aq) at 25°C by plotting $(\phi - 1)/m^{1/2}$ against $m^{1/2}$ and determine the area under the curve by numerical integration (MathChapter G). Compare your values of $\ln \gamma_\pm$ with those you obtained in Problem 25-37 where you calculated $\ln \gamma_\pm$ from a curve-fit expression of ϕ as a polynomial in $m^{1/2}$.

25-40. Don Juan Pond in the Wright Valley of Antarctica freezes at $-57°C$. The major solute in the pond is $CaCl_2$. Estimate the concentration of $CaCl_2$ in the pond water.

25-41. A solution of mercury(II) chloride is a poor conductor of electricity. A 40.7-g sample of $HgCl_2$ is dissolved in 100.0 g of water, and the freezing point of the solution is found to be $-2.83°C$. Explain why $HgCl_2$ in solution is a poor conductor of electricity.

25-42. The freezing point of a 0.25-molal aqueous solution of Mayer's reagent, K_2HgI_4, is found to be $-1.41°C$. Suggest a possible dissociation reaction that takes place when K_2HgI_4 is dissolved in water.

25-43. Given the following freezing-point depression data, determine the number of ions produced per formula unit when the indicated substance is dissolved in water to produce a 1.00-molal solution.

Formula	$\Delta T/K$
$PtCl_2 \cdot 4NH_3$	5.58
$PtCl_2 \cdot 3NH_3$	3.72
$PtCl_2 \cdot 2NH_3$	1.86
$KPtCl_3 \cdot NH_3$	3.72
K_2PtCl_4	5.58

Interpret your results.

25-44. An aqueous solution of NaCl has an ionic strength of $0.315\ mol \cdot L^{-1}$. At what concentration will an aqueous solution of K_2SO_4 have the same ionic strength?

25-45. Derive the "practical" formula for κ^2 given by Equation 25.53.

25-46. Some authors define ionic strength in terms of molality rather than molarity, in which case

$$I_m = \frac{1}{2} \sum_{j=1}^{s} z_j^2 m_j$$

Show that this definition modifies Equation 25.53 for dilute solutions to be

$$\kappa^2 = \frac{2e^2 N_A (1000\ L \cdot m^{-3})\rho}{\varepsilon_0 \varepsilon_r kT} (I_m / mol \cdot kg^{-1})$$

where ρ is the density of the solvent (in $g \cdot mL^{-1}$).

25-47. Show that

$$\ln \gamma_\pm = -1.171 |z_+ z_-| (I_m / mol \cdot kg^{-1})^{1/2}$$

for an aqueous solution at 25°C, where I_m is the ionic strength expressed in terms of molality. Take ε_r to be 78.54 and the density of water to be $0.99707\ g \cdot mL^{-1}$.

25-48. Calculate the value of $\ln \gamma_\pm$ for a 0.010-molar NaCl(aq) solution at 25°C. The experimental value is -0.103. Take $\varepsilon_r = 78.54$ for $H_2O(l)$ at 25°C.

25-49. Derive the general equation

$$\phi = 1 + \frac{1}{m} \int_0^m m' d \ln \gamma_\pm$$

(*Hint*: See the derivation in Problem 25–35.) Use this result to show that

$$\phi = 1 + \frac{\ln \gamma_\pm}{3}$$

for the Debye–Hückel theory.

In the next nine problems we will develop the Debye–Hückel theory of ionic solutions and derive Equations 25.48 and 25.49.

25-50. In the Debye–Hückel theory, the ions are modeled as point ions, and the solvent is modeled as a continuous medium (no structure) with a relative permittivity ε_r. Consider an ion of type i (i = a cation or an anion) situated at the origin of a spherical coordinate system. The presence of this ion at the origin will attract ions of opposite charge and repel ions of the same charge. Let $N_{ij}(r)$ be the number of ions of type j (j = a cation or an anion) situated at a distance r from the central ion of type i (a cation or an anion). We can use a Boltzmann factor to say that

$$N_{ij}(r) = N_j e^{-w_{ij}(r)/k_B T}$$

where N_j/V is the bulk number density of j ions and $w_{ij}(r)$ is the interaction energy of an i ion with a j ion. This interaction energy will be electrostatic in origin, so let $w_{ij}(r) = q_j \psi_i(r)$, where q_j is the charge on the ion of type j and $\psi_i(r)$ is the electrostatic potential due to the central ion of type i.

A fundamental equation from physics that relates a spherically symmetric electrostatic potential $\psi_i(r)$ to a spherically symmetric charge density $\rho_i(r)$ is Poisson's equation

$$\frac{1}{r^2} \frac{d}{dr}\left(r^2 \frac{d\psi_i}{dr}\right) = -\frac{\rho_i(r)}{\varepsilon_0 \varepsilon_r} \tag{1}$$

where ε_r is the relative permittivity of the solvent. In our case, $\rho_i(r)$ is the charge density around the central ion. First, show that

$$\rho_i(r) = \frac{1}{V} \sum_j q_j N_{ij}(r) = \sum_j q_j C_j e^{-q_j \psi_i(r)/k_B T}$$

where C_j is the bulk number density of species j ($C_j = N_j/V$). Linearize the exponential term and use the condition of electroneutrality to show that

$$\rho_i(r) = -\psi_i(r) \sum_j \frac{q_j^2 C_j}{k_B T} \tag{2}$$

Now substitute $\rho_i(r)$ into Poisson's equation to get

$$\frac{1}{r^2} \frac{d}{dr}\left(r^2 \frac{d\psi_i}{dr}\right) = \kappa^2 \psi_i(r) \tag{3}$$

where

$$\kappa^2 = \sum_j \frac{q_j^2 C_j}{\varepsilon_0 \varepsilon_r k_B T} = \sum_j \frac{q_j^2}{\varepsilon_0 \varepsilon_r k_B T} \left(\frac{N_j}{V} \right) \tag{4}$$

Show that Equation 3 can be written as

$$\frac{d^2}{dr^2}[r\psi_i(r)] = \kappa^2[r\psi_i(r)]$$

Now show that the only solution for $\psi_i(r)$ that is finite for large values of r is

$$\psi_i(r) = \frac{Ae^{-\kappa r}}{r} \tag{5}$$

where A is a constant. Use the fact that if the concentration is very small, then $\psi_i(r)$ is just Coulomb's law and so $A = q_i/4\pi\varepsilon_0\varepsilon_r$ and

$$\psi_i(r) = \frac{q_i e^{-\kappa r}}{4\pi\varepsilon_0\varepsilon_r r} \tag{6}$$

Equation 6 is a central result of the Debye–Hückel theory. The factor of $e^{-\kappa r}$ modulates the resulting Coulombic potential, so Equation 6 is called a *screened Coulombic potential*.

25-51. Use Equations 2 and 6 of the previous problem to show that the net charge in a spherical shell of radius r surrounding a central ion of type i is

$$p_i(r)dr = \rho_i(r)4\pi r^2 dr = -q_i\kappa^2 r e^{-\kappa r}dr$$

as in Equation 25.54. Why is

$$\int_0^\infty p_i(r)dr = -q_i$$

25-52. Use the result of the previous problem to show that the most probable value of r is $1/\kappa$.

25-53. Show that

$$r_{mp} = \frac{1}{\kappa} = \frac{304 \text{ pm}}{(c/\text{mol} \cdot \text{L}^{-1})^{1/2}}$$

where c is the molarity of an aqueous solution of a 1–1 electrolyte at 25°C. Take $\varepsilon_r = 78.54$ for $H_2O(l)$ at 25°C.

25-54. Show that

$$r_{mp} = \frac{1}{\kappa} = 430 \text{ pm}$$

for a 0.50-molar aqueous solution of a 1–1 electrolyte at 25°C. Take $\varepsilon_r = 78.54$ for $H_2O(l)$ at 25°C.

25-55. How does the thickness of the ionic atmosphere compare for a 1–1 electrolyte and a 2–2 electrolyte?

25-56. In this problem, we will calculate the total electrostatic energy of an electrolyte solution in the Debye–Hückel theory. Use the equations in Problem 25–50 to show that the number of ions of type j in a spherical shell of radii r and $r + dr$ about a central ion of type i is

$$\left(\frac{N_{ij}(r)}{V}\right) 4\pi r^2 dr = C_j e^{-q_j \psi_i(r)/k_B T} 4\pi r^2 dr \approx C_j \left(1 - \frac{q_j \psi_i(r)}{k_B T}\right) 4\pi r^2 dr \quad (1)$$

The total Coulombic interaction between the central ion of type i and the ions of type j in the spherical shell is $N_{ij}(r) u_{ij}(r) 4\pi r^2 dr / V$ where $u_{ij}(r) = q_i q_j / 4\pi \varepsilon_0 \varepsilon_r r$. To determine the electrostatic interaction energy of all the ions in the solution with the central ion (of type i), U_i^{el}, sum $N_{ij}(r) u_{ij}(r)/V$ over all types of ions in a spherical shell and then integrate over all spherical shells to get

$$U_i^{el} = \int_0^\infty \left(\sum_j \frac{N_{ij}(r) u_{ij}(r)}{V}\right) 4\pi r^2 dr$$

$$= \sum_j \frac{C_j q_i q_j}{\varepsilon_0 \varepsilon_r} \int_0^\infty \left(1 - \frac{q_j \psi_i(r)}{k_B T}\right) r\, dr$$

Use electroneutrality to show that

$$U_i^{el} = -q_i \kappa^2 \int_0^\infty \psi_i(r) r\, dr$$

Now, using Equation 6 of Problem 25–50, show that the interaction of all ions with the central ion (of type i) is given by

$$U_i^{el} = -\frac{q_i^2 \kappa^2}{4\pi \varepsilon_0 \varepsilon_r} \int_0^\infty e^{-\kappa r} dr = -\frac{q_i^2 \kappa}{4\pi \varepsilon_0 \varepsilon_r}$$

Now argue that the total electrostatic energy is

$$U^{el} = \frac{1}{2} \sum_i N_i U_i^{el} = -\frac{V k_B T \kappa^3}{8\pi}$$

Why is there a factor of 1/2 in this equation? Wouldn't you be overcounting the energy otherwise?

25-57. We derived an expression for U^{el} in the previous problem. Use the Gibbs–Helmholtz equation for A (Problem 22–23) to show that

$$A^{el} = -\frac{V k_B T \kappa^3}{12\pi}$$

25-58. If we assume that the electrostatic interactions are the sole cause of the nonideality of an electrolyte solution, then we can say that

$$\mu_j^{el} = \left(\frac{\partial A^{el}}{\partial n_j}\right)_{T,V} = RT \ln \gamma_j^{el}$$

or that

$$\mu_j^{el} = \left(\frac{\partial A^{el}}{\partial N_j}\right)_{T,V} = k_B T \ln \gamma_j^{el}$$

Use the result you got for A^{el} in the previous problem to show that

$$k_B T \ln \gamma_j^{el} = -\frac{\kappa q_j^2}{8\pi \varepsilon_0 \varepsilon_r}$$

Use the formula

$$\ln \gamma_{\pm} = \frac{v_+ \ln \gamma_+ + v_- \ln \gamma_-}{v_+ + v_-}$$

to show that

$$\ln \gamma_{\pm} = -\left(\frac{v_+ q_+^2 + v_- q_-^2}{v_+ + v_-}\right)\frac{\kappa}{8\pi \varepsilon_0 \varepsilon_r k_B T}$$

Use the electroneutrality condition $v_+ q_+ + v_- q_- = 0$ to rewrite $\ln \gamma_{\pm}$ as

$$\ln \gamma_{\pm} = -|q_+ q_-|\frac{\kappa}{8\pi \varepsilon_0 \varepsilon_r k_B T}$$

in agreement with Equation 25.49.

25-59. Derive Equation 25.56 from Equation 25.49.

25-60. Show that Equation 25.59 reduces to Equation 25.49 for small concentrations.

25-61. In this problem, we will investigate the temperature dependence of activities. Starting with the equation $\mu_1 = \mu_1^* + RT \ln a_1$, show that

$$\left(\frac{\partial \ln a_1}{\partial T}\right)_{P,x_1} = \frac{\overline{H}_1^* - \overline{H}_1}{RT^2}$$

where $\overline{H}_1^*$ is the molar enthalpy of the pure solvent (at one bar) and $\overline{H}_1$ is its partial molar enthalpy in the solution. The difference between $\overline{H}_1^*$ and $\overline{H}_1$ is small for dilute solutions, so a_1 is fairly independent of temperature.

25-62. Henry's law says that the pressure of a gas in equilibrium with a nonelectrolyte solution of the gas in a liquid is proportional to the molality of the gas in the solution for sufficiently dilute solutions. What form do you think Henry's law takes on for a gas such as HCl(g) dissolved in water? Use the following data for HCl(g) at 25°C to test your prediction.

$P_{HCl}/10^{-11}$ bar	$m_{HCl}/10^{-3}$ mol·kg^{-1}
0.147	1.81
0.238	2.32
0.443	3.19
0.663	3.93
0.851	4.47
1.08	5.06
1.62	6.25
1.93	6.84
2.08	7.12

Gilbert Newton Lewis was born in West Newton, Massachusetts, on October 25, 1875, and died in 1946. In 1899, he received his Ph.D. from Harvard University, and after spending a year studying in Germany, he returned to Harvard as an instructor. Lewis left Harvard in 1904 to become Superintendent of Weights and Measures in the Philippines, and a year later he moved to The Massachusetts Institute of Technology. In 1912, he accepted the position of Dean of the College of Chemistry at the University of California at Berkeley, which he developed into one of the finest teaching and research departments in the world. He remained at Berkeley for the rest of his life, suffering a fatal heart attack in his laboratory. Lewis was one of America's outstanding chemists, certainly the finest not to receive a Nobel Prize. Lewis made many important contributions in chemistry. In the 1920s, he introduced Lewis formulas and described a covalent bond (which he named) as a shared pair of electrons. His work on the application of thermodynamics to physical chemistry culminated in his outstanding 1923 text, coauthored with Merle Randall, *Thermodynamics and the Free Energy of Chemical Substances*, from which a generation of chemists learned thermodynamics. Lewis was a dynamic individual who was responsible for development of many outstanding chemists, several of whom became members of the Berkeley faculty. His department produced a remarkable number of Nobel Prize winners.

CHAPTER 26

Chemical Equilibrium

One of the most important applications of thermodynamics is to chemical reactions at equilibrium. Thermodynamics enables us to predict with confidence the equilibrium pressures or concentrations of reaction mixtures. In this chapter we shall derive a relation between the standard Gibbs energy change and the equilibrium constant for a chemical reaction. We shall also learn how to predict the direction in which a chemical reaction will proceed if we start with arbitrary concentrations of reactants and products. We have developed all the necessary thermodynamic concepts in previous chapters. The underlying fundamental idea is that $\Delta G = 0$ for a system in equilibrium at constant temperature and pressure, and that the sign of ΔG determines whether or not a given process or chemical reaction will occur spontaneously at constant T and P.

26–1. Chemical Equilibrium Results When the Gibbs Energy Is a Minimum with Respect to the Extent of Reaction

For simplicity, we shall discuss gas-phase reactions first. Consider the general gas-phase reaction, which is described by the balanced equation

$$\nu_A A(g) + \nu_B B(g) \rightleftharpoons \nu_Y Y(g) + \nu_Z Z(g)$$

We define a quantity ξ, called the *extent of reaction*, such that the numbers of moles of the reactants and products are given by

$$\underbrace{\begin{aligned} n_A &= n_{A0} - \nu_A \xi \\ n_B &= n_{B0} - \nu_B \xi \end{aligned}}_{\text{reactants}} \qquad \underbrace{\begin{aligned} n_Y &= n_{Y0} + \nu_Y \xi \\ n_Z &= n_{Z0} + \nu_Z \xi \end{aligned}}_{\text{products}} \qquad (26.1)$$

where n_{j0} is the initial number of moles for each species. Recall from Chapter 19 that stoichiometric coefficients do not have units. Consequently, Equations 26.1 indicate that ξ has units of moles. As the reaction proceeds from reactants to products,

ξ varies from 0 to some maximum value dictated by the stoichiometry of the reaction. For example, if n_{A0} and n_{B0} in Equations 26.1 are equal to ν_A moles and ν_B moles, respectively, then ξ will vary from 0 to one mole. Differentiation of Equations 26.1 gives

$$\underbrace{\begin{aligned} dn_A &= -\nu_A d\xi \\ dn_B &= -\nu_B d\xi \end{aligned}}_{\text{reactants}} \qquad \underbrace{\begin{aligned} dn_Y &= \nu_Y d\xi \\ dn_Z &= \nu_Z d\xi \end{aligned}}_{\text{products}} \qquad (26.2)$$

The negative signs indicate that the reactants are disappearing and the positive signs indicate that the products are being formed as the reaction progresses from reactants to products.

Now let's consider a system containing reactants and products at constant T and P. The Gibbs energy for this multicomponent system is a function of T, P, n_A, n_B, n_Y, and n_Z, which we can express mathematically as $G = G(T, P, n_A, n_B, n_Y, n_Z)$. The total derivative of G is given by

$$dG = \left(\frac{\partial G}{\partial T}\right)_{P,n_j} dT + \left(\frac{\partial G}{\partial P}\right)_{T,n_j} dP + \left(\frac{\partial G}{\partial n_A}\right)_{T,P,n_{j\neq A}} dn_A$$

$$+ \left(\frac{\partial G}{\partial n_B}\right)_{T,P,n_{j\neq B}} dn_B + \left(\frac{\partial G}{\partial n_Y}\right)_{T,P,n_{j\neq Y}} dn_Y + \left(\frac{\partial G}{\partial n_Z}\right)_{T,P,n_{j\neq Z}} dn_Z$$

where the subscript n_j in the first two partial derivatives stands for n_A, n_B, n_Y, and n_Z. Using Equations 22.31 for $(\partial G/\partial T)_{P,n_j}$ and $(\partial G/\partial P)_{T,n_j}$, dG becomes

$$dG = -SdT + VdP + \mu_A dn_A + \mu_B dn_B + \mu_Y dn_Y + \mu_Z dn_Z$$

where

$$\mu_A = \left(\frac{\partial G}{\partial n_A}\right)_{T,P,n_B,n_Y,n_Z}$$

with similar expressions for μ_B, μ_Y, and μ_Z. For a reaction that takes place at constant T and P, dG becomes

$$dG = \sum_j \mu_j dn_j = \mu_A dn_A + \mu_B dn_B + \mu_Y dn_Y + \mu_Z dn_Z \quad \text{(constant } T \text{ and } P) \quad (26.3)$$

Substitute Equations 26.2 into Equation 26.3 to obtain

$$dG = -\nu_A \mu_A d\xi - \nu_B \mu_B d\xi + \nu_Y \mu_Y d\xi + \nu_Z \mu_Z d\xi$$
$$= (\nu_Y \mu_Y + \nu_Z \mu_Z - \nu_A \mu_A - \nu_B \mu_B) d\xi \quad \text{(constant } T \text{ and } P) \quad (26.4)$$

or

$$\left(\frac{\partial G}{\partial \xi}\right)_{T,P} = \nu_Y \mu_Y + \nu_Z \mu_Z - \nu_A \mu_A - \nu_B \mu_B \qquad (26.5)$$

We shall denote the right side of Equation 26.5 by $\Delta_r G$, so that

$$\left(\frac{\partial G}{\partial \xi}\right)_{T,P} = \Delta_r G = v_Y \mu_Y + v_Z \mu_Z - v_A \mu_A - v_B \mu_B \qquad (26.6)$$

The quantity $\Delta_r G$ is defined as the change in Gibbs energy when the extent of reaction changes by one mole. The units of $\Delta_r G$ are then $J \cdot mol^{-1}$. The quantity $\Delta_r G$ has meaning only if the balanced chemical equation is specified.

If we assume that all the partial pressures are low enough that we can consider each species to behave ideally, then we can use Equation 23.33 [$\mu_j(T, P) = \mu_j^\circ(T) + RT \ln(P_j/P^\circ)$] for the $\mu_j(T, P)$, in which case Equation 26.6 becomes

$$\Delta_r G = v_Y \mu_Y^\circ(T) + v_Z \mu_Z^\circ(T) - v_A \mu_A^\circ(T) - v_B \mu_B^\circ(T)$$

$$+ RT \left(v_Y \ln \frac{P_Y}{P^\circ} + v_Z \ln \frac{P_Z}{P^\circ} - v_A \ln \frac{P_A}{P^\circ} - v_B \ln \frac{P_B}{P^\circ} \right)$$

or

$$\Delta_r G = \Delta_r G^\circ + RT \ln Q \qquad (26.7)$$

where

$$\Delta_r G^\circ(T) = v_Y \mu_Y^\circ(T) + v_Z \mu_Z^\circ(T) - v_A \mu_A^\circ(T) - v_B \mu_B^\circ(T) \qquad (26.8)$$

and

$$Q = \frac{(P_Y/P^\circ)^{v_Y}(P_Z/P^\circ)^{v_Z}}{(P_A/P^\circ)^{v_A}(P_B/P^\circ)^{v_B}} \qquad (26.9)$$

The quantity $\Delta_r G^\circ(T)$ is the change in standard Gibbs energy for the reaction between unmixed reactants in their standard states at temperature T and a pressure of one bar to form unmixed products in their standard states at the same temperature T and a pressure of one bar. Because the standard pressure P° in Equation 26.9 is taken to be one bar, the P°'s are usually not displayed. It must be remembered, however, that all the pressures are referred to one bar, and that Q consequently is unitless.

When the reaction system is in equilibrium, the Gibbs energy must be a minimum with respect to any displacement of the reaction from its equilibrium position, and so Equation 26.5 becomes

$$\left(\frac{\partial G}{\partial \xi}\right)_{T,P} = \Delta_r G = 0 \quad \text{(equilibrium)} \qquad (26.10)$$

Setting $\Delta_r G = 0$ in Equation 26.7 gives

$$\Delta_r G^\circ(T) = -RT \ln \left(\frac{P_Y^{v_Y} P_Z^{v_Z}}{P_A^{v_A} P_B^{v_B}} \right)_{eq} = -RT \ln K_P(T) \qquad (26.11)$$

where

$$K_p(T) = \left(\frac{P_Y^{v_Y} P_Z^{v_Z}}{P_A^{v_A} P_B^{v_B}} \right)_{eq} \tag{26.12}$$

and where the subscript eq emphasizes that the pressures in Equations 26.11 and 26.12 are the pressures at *equilibrium*. The quantity $K_p(T)$ is called the *equilibrium constant* of the reaction. Although we have used an eq subscript for emphasis, this notation is not normally used and $K_p(T)$ is written without the subscript. Equilibrium-constant expressions imply that the pressures are their equilibrium values. The value of K_p cannot be evaluated unless the balanced chemical reaction to which it refers and the standard states of each of the reactants and products are given.

EXAMPLE 26–1
Write out the equilibrium-constant expression for the reaction that is represented by the equation

$$3\,H_2(g) + N_2(g) \rightleftharpoons 2\,NH_3(g)$$

SOLUTION: According to Equation 26.12,

$$K_p(T) = \frac{P_{NH_3}^2}{P_{H_2}^3 P_{N_2}}$$

where all the pressures are referred to the standard pressure of one bar. Note that if we had written the equation for the reaction as

$$\tfrac{3}{2}\,H_2(g) + \tfrac{1}{2}\,N_2(g) \rightleftharpoons NH_3(g)$$

then we would have obtained

$$K_p(T) = \frac{P_{NH_3}}{P_{H_2}^{3/2} P_{N_2}^{1/2}}$$

which is the square root of our previous expression. Thus, we see that the form of $K_p(T)$ and its subsequent numerical value depend upon how we write the chemical equation that describes the reaction.

26–2. An Equilibrium Constant Is a Function of Temperature Only

Equation 26.11 says that regardless of the initial pressures of the reactants and products, at equilibrium the ratio of their partial pressures raised to their respective stoichiometric coefficients will be a fixed value at a given temperature. Consider the reaction described by

$$PCl_5(g) \rightleftharpoons PCl_3(g) + Cl_2(g) \tag{26.13}$$

The equilibrium-constant expression for this reaction is

$$K_P(T) = \frac{P_{PCl_3} P_{Cl_2}}{P_{PCl_5}} \tag{26.14}$$

Suppose that initially we have one mole of $PCl_5(g)$ and no $PCl_3(g)$ or $Cl_2(g)$. When the reaction occurs to an extent ξ, there will be $(1 - \xi)$ moles of $PCl_5(g)$, ξ moles of $PCl_3(g)$, and ξ moles of $Cl_2(g)$ in the reaction mixture and the total number of moles will be $(1 + \xi)$. If we let ξ_{eq} be the extent of reaction at equilibrium, then the partial pressures of each species will be

$$P_{PCl_3} = P_{Cl_2} = \frac{\xi_{eq} P}{1 + \xi_{eq}}$$

$$P_{PCl_5} = \frac{(1 - \xi_{eq}) P}{1 + \xi_{eq}}$$

where P is the total pressure. The equilibrium-constant expression is

$$K_P(T) = \frac{\xi_{eq}^2}{1 - \xi_{eq}^2} P \tag{26.15}$$

It might appear from this result that $K_P(T)$ depends upon the total pressure, but this is not so. As Equation 26.11 shows, $K_P(T)$ is a function of only the temperature, and so is a constant value at a fixed temperature. Therefore, if P changes, then ξ_{eq} must change so that $K_P(T)$ in Equation 26.15 remains constant. Figure 26.1 shows ξ_{eq} plotted against P at 200°C, where $K_P = 5.4$. Note that ξ_{eq} decreases uniformly with increasing P, indicating that the equilibrium is shifted from the product side to the reactant side of Equation 26.13 or that less PCl_5 is dissociated. This effect of pressure on the position of equilibrium is an example of *Le Châtelier's principle*, which you learned in general chemistry. Le Châtelier's principle can be stated as follows: If a chemical reaction at equilibrium is subjected to a change in conditions that displaces it from equilibrium, then the reaction adjusts toward a new equilibrium state. The reaction proceeds in the direction that — at least partially — offsets the change in conditions. Thus, an increase in pressure shifts the equilibrium in Equation 26.13 such that the total number of moles decreases.

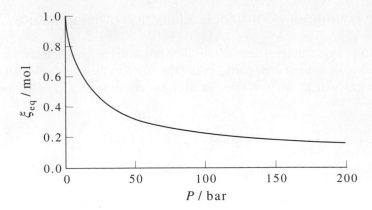

FIGURE 26.1
A plot of the fraction of $PCl_5(g)$ that is dissociated at equilibrium, ξ_{eq}, against total pressure P for the reaction given by Equation 26.13 at 200°C.

EXAMPLE 26–2
Consider the association of potassium atoms in the vapor phase to form dimers.

$$2\,K(g) \rightleftharpoons K_2(g)$$

Suppose we start with 2 moles of K(g) and no dimers. Derive an expression for $K_p(T)$ in terms of ξ_{eq}, the extent of reaction at equilibrium, and the pressure P.

SOLUTION: At equilibrium, there will be $2(1 - \xi_{eq})$ moles of K(g) and ξ_{eq} moles of $K_2(g)$. The total number of moles will be $(2 - \xi_{eq})$. The partial pressure of each species will be

$$P_K = \frac{2(1 - \xi_{eq})P}{2 - \xi_{eq}}$$

$$P_{K_2} = \frac{\xi_{eq}P}{2 - \xi_{eq}}$$

and

$$K_p(T) = \frac{P_{K_2}}{P_K^2} = \frac{\xi_{eq}(2 - \xi_{eq})}{4(1 - \xi_{eq})^2 P}$$

If P decreases, then $\xi_{eq}(2 - \xi_{eq})/4(1 - \xi_{eq})^2$ must decrease, which occurs by ξ_{eq} decreasing. If P increases, then $\xi_{eq}(2 - \xi_{eq})/4(1 - \xi_{eq})^2$ must increase, which occurs by ξ_{eq} increasing [$(1 - \xi_{eq})$ becoming smaller].

We subscripted the equilibrium constant defined by Equation 26.12 with a P to emphasize that it is expressed in terms of equilibrium pressures. We can also express the equilibrium constant in terms of densities or concentrations by using the ideal-gas relation $P = cRT$ where c is the concentration, n/V. Thus, we can rewrite K_P as

$$K_P = \frac{c_Y^{\nu_Y} c_Z^{\nu_Z}}{c_A^{\nu_A} c_B^{\nu_B}} \left(\frac{RT}{P^\circ} \right)^{\nu_Y + \nu_Z - \nu_A - \nu_B} \tag{26.16}$$

Just as we relate the pressures in the expression for K_P to some standard pressure P°, we must relate the concentrations in Equation 26.16 to some standard concentration c°, often taken to be 1 $mol \cdot L^{-1}$. If we multiply and divide each concentration in Equation 26.16 by c°, we can write

$$K_P = K_c \left(\frac{c^\circ RT}{P^\circ} \right)^{\nu_Y + \nu_Z - \nu_A - \nu_B} \tag{26.17}$$

where

$$K_c = \frac{(c_Y/c^\circ)^{\nu_Y} (c_Z/c^\circ)^{\nu_Z}}{(c_A/c^\circ)^{\nu_A} (c_B/c^\circ)^{\nu_B}} \tag{26.18}$$

Both K_P and K_c in Equation 26.17 are unitless, as is the factor $(c^\circ RT/P^\circ)^{\nu_Y + \nu_Z - \nu_A - \nu_B}$. The actual choices of P° and c° determine the units of R to use in Equation 26.17. If P° is taken to be one bar and c° to be one $mol \cdot L^{-1}$ (as is often the case), then the factor $c^\circ RT/P^\circ = RT/L \cdot bar \cdot mol^{-1}$ and R must be expressed as 0.083145 $L \cdot bar \cdot mol^{-1}$.

Equation 26.17 provides a relation between K_P and K_c for ideal gases. Just as we don't display the P°'s in Equation 26.9 because most often $P^\circ =$ one bar, we don't display the P°s and c°s in Equation 26.18 because most often $c^\circ =$ one $mol \cdot L^{-1}$. You must always be aware, however, of which reference states are being used in K_P and K_c when converting the numerical value of one to the other.

EXAMPLE 26–3
The value of $K_P(T)$ (based upon a standard state of one bar) for the reaction described by

$$NH_3(g) \rightleftharpoons \tfrac{3}{2}H_2(g) + \tfrac{1}{2}N_2(g)$$

is 1.36×10^{-3} at 298.15 K. Determine the corresponding value of $K_c(T)$ (based upon a standard state of one $mol \cdot L^{-1}$).

SOLUTION: In this case, $\nu_A = 1$, $\nu_Y = 3/2$, and $\nu_Z = 1/2$, so Equation 26.17 gives

$$K_P(T) = K_c(T) \left(\frac{c^\circ RT}{P^\circ} \right)^1$$

The conversion factor at 298.15 K is

$$\frac{c^\circ RT}{P^\circ} = \frac{(1 \ mol \cdot L^{-1})(0.083145 \ L \cdot bar \cdot mol^{-1} \cdot K^{-1})(298.15 \ K)}{1 \ bar}$$

$$= 24.79$$

and so $K_c = K_P/24.79 = 5.49 \times 10^{-5}$.

26–3. Standard Gibbs Energies of Formation Can Be Used to Calculate Equilibrium Constants

Notice that combining Equations 26.8 and 26.11 gives a relation between $\mu_j^\circ(T)$, the standard chemical potentials of the reactants and products, and the equilibrium constant, K_p. In particular, K_p is related to the difference between the standard chemical potentials of the products and reactants. Because a chemical potential is an energy (it is the molar Gibbs energy of a pure substance), its value must be referred to some (arbitrary) zero of energy. A convenient choice of a zero of energy is based on the procedure that we used to set up a table of standard molar enthalpies of formation (Table 19.2) in Section 19–11. Recall that we defined the standard molar enthalpy of formation of a substance as the energy as heat involved when one mole of the substance is formed directly from its constituent elements in their most stable form at one bar and the temperature of interest. For example, the value of $\Delta_r H$ for

$$H_2(g) + \tfrac{1}{2} O_2(g) \rightleftharpoons H_2O(l)$$

is -285.8 kJ when all the species are at 298.15 K and one bar, and so we write $\Delta_f H^\circ[H_2O(l)] = -285.8$ kJ·mol^{-1} at 298.15 K. By convention, we also have that $\Delta_f H^\circ[H_2(g)] = \Delta_f H^\circ[O_2(g)] = 0$ for $H_2(g)$ and $O_2(g)$ at 298.15 K and one bar. We also set up a table of practical absolute entropies of substances (Table 21.2) in Section 21–9, and so because

$$\Delta_r G^\circ = \Delta_r H^\circ - T \Delta_r S^\circ$$

we can also set up a table of values of $\Delta_f G^\circ$. Then for a reaction such as

$$\nu_A A + \nu_B B \longrightarrow \nu_Y Y + \nu_Z Z$$

we have

$$\Delta_r G^\circ = \nu_Y \Delta_f G^\circ[Y] + \nu_Z \Delta_f G^\circ[Z] - \nu_A \Delta_f G^\circ[A] - \nu_B \Delta_f G^\circ[B] \tag{26.19}$$

Table 26.1 lists values of $\Delta_f G^\circ$ at 298.15 K and one bar for a variety of substances, and much more extensive tables are available (see Section 26.9).

EXAMPLE 26–4
Using the data in Table 26.1, calculate $\Delta_r G^\circ(T)$ and K_p at 298.15 K for

$$NH_3(g) \rightleftharpoons \tfrac{3}{2} H_2(g) + \tfrac{1}{2} N_2(g)$$

SOLUTION: From Equation 26.19,

$$
\begin{aligned}
\Delta_r G^\circ &= \left(\tfrac{3}{2}\right) \Delta_f G^\circ[H_2(g)] + \left(\tfrac{1}{2}\right) \Delta_f G^\circ[N_2(g)] - (1)\Delta_f G^\circ[NH_3(g)] \\
&= \left(\tfrac{3}{2}\right)(0) + \left(\tfrac{1}{2}\right)(0) - (1)(-16.367 \text{ kJ·mol}^{-1}) \\
&= 16.367 \text{ kJ·mol}^{-1}
\end{aligned}
$$

TABLE 26.1

Standard molar Gibbs energies of formation, $\Delta_f G°$, for various substances at 298.15 K and one bar.

Substance	Formula	$\Delta_f G°/\text{kJ}\cdot\text{mol}^{-1}$
acetylene	$C_2H_2(g)$	209.20
ammonia	$NH_3(g)$	−16.367
benzene	$C_6H_6(l)$	124.35
bromine	$Br_2(g)$	3.126
butane	$C_4H_{10}(g)$	−17.15
carbon(diamond)	$C(s)$	2.900
carbon(graphite)	$C(s)$	0
carbon dioxide	$CO_2(g)$	−394.389
carbon monoxide	$CO(g)$	−137.163
ethane	$C_2H_6(g)$	−32.82
ethanol	$C_2H_5OH(l)$	−174.78
ethene	$C_2H_4(g)$	68.421
glucose	$C_6H_{12}O_6(s)$	−910.52
hydrogen bromide	$HBr(g)$	−53.513
hydrogen chloride	$HCl(g)$	−95.300
hydrogen fluoride	$HF(g)$	−274.646
hydrogen iodide	$HI(g)$	1.560
hydrogen peroxide	$H_2O_2(l)$	−105.445
iodine	$I_2(g)$	19.325
methane	$CH_4(g)$	−50.768
methanol	$CH_3OH(l)$	−166.27
	$CH_3OH(g)$	−161.96
nitrogen oxide	$NO(g)$	86.600
nitrogen dioxide	$NO_2(g)$	51.258
dinitrogen tetraoxide	$N_2O_4(g)$	97.787
	$N_2O_4(l)$	97.521
propane	$C_3H_8(g)$	−23.47
sucrose	$C_{12}H_{22}O_{11}(s)$	−1544.65
sulfur dioxide	$SO_2(g)$	−300.125
sulfur trioxide	$SO_3(g)$	−371.016
tetrachloromethane	$CCl_4(l)$	−65.21
	$CCl_4(g)$	−53.617
water	$H_2O(l)$	−237.141
	$H_2O(g)$	−228.582

and from Equation 26.11

$$\ln K_p(T) = -\frac{\Delta_r G^\circ}{RT} = -\frac{16.367 \times 10^3 \text{ J} \cdot \text{mol}^{-1}}{(8.3145 \text{ J} \cdot \text{K}^{-1} \cdot \text{mol}^{-1})(298.15 \text{ K})}$$

$$= -6.602$$

or $K_p = 1.36 \times 10^{-3}$ at 298.15 K.

26–4. A Plot of the Gibbs Energy of a Reaction Mixture Against the Extent of Reaction Is a Minimum at Equilibrium

In this section we shall treat a concrete example of the Gibbs energy of a reaction mixture as a function of the extent of reaction. Consider the thermal decomposition of $N_2O_4(g)$ to $NO_2(g)$ at 298.15 K, which we represent by the equation

$$N_2O_4(g) \rightleftharpoons 2NO_2(g)$$

Suppose we start with one mole of $N_2O_4(g)$ and no $NO_2(g)$. Then as the reaction proceeds, $n_{N_2O_4}$, the number of moles of $N_2O_4(g)$, will be given by $1 - \xi$ and n_{NO_2} will be given by 2ξ. Note that $n_{N_2O_4} = 1$ mol and $n_{NO_2} = 0$ when $\xi = 0$ and that $n_{N_2O_4} = 0$ and $n_{NO_2} = 2$ mol when $\xi = 1$ mol. The Gibbs energy of the reaction mixture is given by

$$G(\xi) = (1 - \xi)\overline{G}_{N_2O_4} + 2\xi\overline{G}_{NO_2}$$

$$= (1 - \xi)G^\circ_{N_2O_4} + 2\xi G^\circ_{NO_2} + (1 - \xi)RT \ln P_{N_2O_4} + 2\xi RT \ln P_{NO_2} \tag{26.20}$$

If the reaction is carried out at a constant total pressure of one bar, then

$$P_{N_2O_4} = x_{N_2O_4} P_{\text{total}} = x_{N_2O_4} \quad \text{and} \quad P_{NO_2} = x_{NO_2}$$

The total number of moles in the reaction mixture is $(1 - \xi) + 2\xi = 1 + \xi$, and so we have

$$P_{N_2O_4} = x_{N_2O_4} = \frac{1 - \xi}{1 + \xi} \quad \text{and} \quad P_{NO_2} = x_{NO_2} = \frac{2\xi}{1 + \xi}$$

Thus, Equation 26.20 becomes

$$G(\xi) = (1 - \xi)G^\circ_{N_2O_4} + 2\xi G^\circ_{NO_2} + (1 - \xi)RT \ln \frac{1 - \xi}{1 + \xi} + 2\xi RT \ln \frac{2\xi}{1 + \xi}$$

According to Section 26–3, we can choose our standard states such that $G^\circ_{N_2O_4} = \Delta_f G^\circ_{N_2O_4}$ and $G^\circ_{NO_2} = \Delta_f G^\circ_{NO_2}$, so $G(\xi)$ becomes

$$G(\xi) = (1 - \xi)\Delta_f G^\circ_{N_2O_4} + 2\xi\Delta_f G^\circ_{NO_2} + (1 - \xi)RT \ln \frac{1 - \xi}{1 + \xi} + 2\xi RT \ln \frac{2\xi}{1 + \xi} \tag{26.21}$$

Equation 26.21 gives the Gibbs energy of the reaction mixture, G, as a function of the extent of the reaction, ξ. Using the values of $\Delta_f G^\circ_{N_2O_4}$ and $\Delta_f G^\circ_{NO_2}$ given in Table 26.1, Equation 26.21 becomes

$$G(\xi) = (1 - \xi)(97.787 \text{ kJ} \cdot \text{mol}^{-1}) + 2\xi(51.258 \text{ kJ} \cdot \text{mol}^{-1})$$

$$+ (1 - \xi)RT \ln \frac{1 - \xi}{1 + \xi} + 2\xi RT \ln \frac{2\xi}{1 + \xi} \qquad (26.22)$$

where $RT = 2.4790 \text{ kJ} \cdot \text{mol}^{-1}$. Figure 26.2 shows $G(\xi)$ plotted against ξ. The minimum in the plot, or the equilibrium state, occurs at $\xi_{eq} = 0.1892$ mol. Thus, the reaction will proceed from $\xi = 0$ to $\xi = \xi_{eq} = 0.1892$ mol, where equilibrium is established.

The equilibrium constant is given by

$$K_P = \frac{P^2_{NO_2}}{P_{N_2O_4}} = \frac{[2\xi_{eq}/(1 + \xi_{eq})]^2}{(1 - \xi_{eq})/(1 + \xi_{eq})} = \frac{4\xi^2_{eq}}{1 - \xi^2_{eq}} = 0.148$$

We can compare this result to the one that we obtain from $\Delta_r G^\circ = -RT \ln K_P$, or

$$\ln K_P = -\frac{\Delta_r G^\circ}{RT}$$

$$= \frac{(2)(\Delta_f G^\circ[NO_2(g)]) - (1)(\Delta_f G^\circ[N_2O_4(g)])}{(8.3145 \text{ J} \cdot \text{K}^{-1} \cdot \text{mol}^{-1})(298.15 \text{ K})}$$

$$= -\frac{4.729 \times 10^3 \text{ J} \cdot \text{mol}^{-1}}{(8.3145 \text{ J} \cdot \text{K}^{-1} \cdot \text{mol}^{-1})(298.15 \text{ K})} = -1.9076$$

or $K_P = 0.148$.

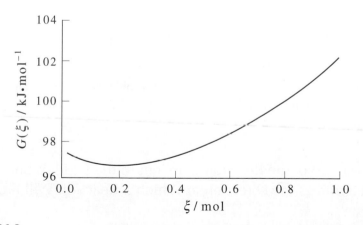

FIGURE 26.2
A plot of the Gibbs energy of the reaction mixture versus the extent of reaction for $N_2O_4(g) \rightleftharpoons 2 NO_2(g)$ at 298.15 K and one bar.

We can also differentiate Equation 26.22 with respect to ξ explicitly to obtain

$$
\left(\frac{\partial G}{\partial \xi}\right)_{T,P} = (2)(51.258\ \text{kJ·mol}^{-1}) - 97.787\ \text{kJ·mol}^{-1} - RT \ln \frac{1-\xi}{1+\xi}
$$

$$
+2RT \ln \frac{2\xi}{1+\xi} + (1-\xi)RT \left(\frac{1+\xi}{1-\xi}\right)\left[-\frac{1}{1+\xi} - \frac{1-\xi}{(1+\xi)^2}\right]
$$

$$
+2\xi\, RT \left(\frac{1+\xi}{2\xi}\right)\left[\frac{2}{1+\xi} - \frac{2\xi}{(1+\xi)^2}\right] \tag{26.23}
$$

We can replace $(1-\xi)/(1+\xi)$ in the first logarithm term by $P_{\text{N}_2\text{O}_4}$ and $2\xi/(1+\xi)$ in the second logarithm term by P_{NO_2}. Furthermore, a little algebra shows that the last two terms add up to zero, and so Equation 26.23 becomes

$$
\left(\frac{\partial G}{\partial \xi}\right)_{T,P} = \Delta_r G^\circ + RT \ln \frac{P_{\text{NO}_2}^2}{P_{\text{N}_2\text{O}_4}}
$$

At equilibrium, $\partial G/\partial \xi = 0$ and we get Equation 26.11.

We can also evaluate ξ_{eq} explicitly by setting Equation 26.23 equal to zero. Using the fact that the last two terms in Equation 26.23 add up to zero, we have

$$
\frac{(2)(51.258\ \text{kJ·mol}^{-1}) - 97.787\ \text{kJ·mol}^{-1}}{(8.3145\ \text{J·mol}^{-1}\text{·K}^{-1})(298.15\ \text{K})} = \ln\left(\frac{1-\xi_{eq}}{1+\xi_{eq}}\right) - \ln \frac{4\xi_{eq}^2}{(1+\xi_{eq})^2}
$$

or

$$
1.9076 = \ln\left(\frac{1-\xi_{eq}^2}{4\xi_{eq}^2}\right)
$$

or

$$
\frac{1-\xi_{eq}^2}{4\xi_{eq}^2} = e^{1.9076} = 6.7371
$$

or $\xi_{eq} = 0.1892$, in agreement with Figure 26.2. Problems 26–18 through 26–21 ask you to carry out a similar analysis for two other gas-phase reactions.

26–5. The Ratio of the Reaction Quotient to the Equilibrium Constant Determines the Direction in Which a Reaction Will Proceed

Consider the general reaction described by the equation

$$
\nu_A A(g) + \nu_B B(g) \rightleftharpoons \nu_Y Y(g) + \nu_Z Z(g)
$$

Equation 26.7 for this reaction scheme is

$$\Delta_r G(T) = \Delta_r G^\circ(T) + RT \ln \frac{P_Y^{v_Y} P_Z^{v_Z}}{P_A^{v_A} P_B^{v_B}} \tag{26.24}$$

Realize that the pressures in this equation are not necessarily equilibrium pressures, but are arbitrary. Equation 26.24 gives the value of $\Delta_r G$ when v_A moles of A(g) at pressure P_A react with v_B moles of B(g) at pressure P_B to produce v_Y moles of Y(g) at pressure P_Y and v_Z moles of Z(g) at pressure P_Z. If all the pressures happen to be one bar, then the logarithm term in Equation 26.24 will be zero and $\Delta_r G$ will be equal to $\Delta_r G^\circ$; in other words, the Gibbs energy change will be equal to the standard Gibbs energy change. If, on the other hand, the pressures are the equilibrium pressures, then $\Delta_r G$ will equal zero and we obtain Equation 26.11.

We can write Equation 26.24 in more concise form by introducing a quantity called the *reaction quotient* Q_P (see Equation 26.9)

$$Q_P = \frac{P_Y^{v_Y} P_Z^{v_Z}}{P_A^{v_A} P_B^{v_B}} \tag{26.25}$$

and using Equation 26.11 for $\Delta_r G^\circ$:

$$\begin{aligned} \Delta_r G &= -RT \ln K_P + RT \ln Q_P \\ &= RT \ln(Q_P/K_P) \end{aligned} \tag{26.26}$$

Realize that even though Q_P has the *form* of an equilibrium constant, the pressures are arbitrary.

At equilibrium, $\Delta_r G = 0$ and $Q_P = K_P$. If $Q_P < K_P$, then Q_P must increase as the system proceeds toward equilibrium, which means that the partial pressures of the products must increase and those of the reactants must decrease. In other words, the reaction proceeds from left to right as written. In terms of $\Delta_r G$, if $Q_P < K_P$, then $\Delta_r G < 0$, indicating that the reaction is spontaneous from left to right as written. Conversely, if $Q_P > K_P$, then Q_P must decrease as the reaction proceeds to equilibrium and so the pressures of the products must decrease and those of the reactants must increase. In terms of $\Delta_r G$, if $Q_P > K_P$, then $\Delta_r G > 0$, indicating that the reaction is spontaneous from right to left as written.

EXAMPLE 26–5

The equilibrium constant for the reaction described by

$$2\,SO_2(g) + O_2(g) \rightleftharpoons 2\,SO_3(g)$$

is $K_P = 10$ at 960 K. Calculate $\Delta_r G$ and indicate in which direction the reaction will proceed spontaneously for

$$2\,SO_2(1.0 \times 10^{-3}\ \text{bar}) + O_2(0.20\ \text{bar}) \rightleftharpoons 2\,SO_3(1.0 \times 10^{-4}\ \text{bar})$$

SOLUTION: We first calculate the reaction quotient under these conditions. According to Equation 26.25,

$$Q_P = \frac{P_{SO_3}^2}{P_{SO_2}^2 P_{O_2}} = \frac{(1.0 \times 10^{-4})^2}{(1.0 \times 10^{-3})^2 (0.20)} = 5.0 \times 10^{-2}$$

Note that these quantities are unitless because the pressures are taken relative to one bar. Using Equation 26.26, we have

$$\Delta_r G = RT \ln \frac{Q_P}{K_P}$$

$$= (8.314 \text{ J} \cdot \text{K}^{-1} \cdot \text{mol}^{-1})(960 \text{ K}) \ln \frac{5.0 \times 10^{-2}}{10}$$

$$= -42.3 \text{ kJ} \cdot \text{mol}^{-1}$$

The fact that $\Delta_r G < 0$ implies that the reaction will proceed from left to right as written. This may also be seen from the fact that $Q_P < K_P$.

26–6. The Sign of $\Delta_r G$ And Not That of $\Delta_r G°$ Determines the Direction of Reaction Spontaneity

It is important to appreciate the difference between $\Delta_r G$ and $\Delta_r G°$. The superscript °
on $\Delta_r G°$ emphasizes that this is the value of $\Delta_r G$ when all the reactants and products
are unmixed at partial pressures equal to one bar; $\Delta_r G°$ is the *standard* Gibbs energy
change. If $\Delta_r G° < 0$, then $K_P > 1$, meaning that the reaction will proceed from reactants to products if all the species are mixed at one bar partial pressures. If $\Delta_r G° > 0$,
then $K_P < 1$, meaning that the reaction will proceed from products to reactants if all
the species are mixed at one bar partial pressures. The fact that $\Delta_r G° > 0$ does *not*
mean that the reaction will not proceed from reactants to products if the species are
mixed under all conditions. For example, consider the reaction described by

$$N_2O_4(g) \rightleftharpoons 2 NO_2(g)$$

for which $\Delta_r G° = 4.729 \text{ kJ} \cdot \text{mol}^{-1}$ at 298.15 K. The corresponding value of $K_P(T)$
is 0.148. The fact that $\Delta_r G° = +4.729 \text{ kJ} \cdot \text{mol}^{-1}$ does *not* mean that no $N_2O_4(g)$ will
dissociate when we place some of it in a reaction vessel at 298.15 K. The value of $\Delta_r G$
for the dissociation of $N_2O_4(g)$ is given by

$$\Delta_r G = \Delta_r G° + RT \ln Q_P$$

$$= 4.729 \text{ kJ} \cdot \text{mol}^{-1} + (2.479 \text{ kJ} \cdot \text{mol}^{-1}) \ln \frac{P_{NO_2}^2}{P_{N_2O_4}} \qquad (26.27)$$

Let's say that we fill a container with $N_2O_4(g)$ and no $NO_2(g)$. Initially then, the logarithm term and $\Delta_r G$ in Equation 26.27 will be essentially negative infinity. Therefore, the dissociation of $N_2O_4(g)$ takes place spontaneously. The partial pressure of $N_2O_4(g)$ decreases and that of $NO_2(g)$ increases until equilibrium is reached. The equilibrium state is determined by the condition $\Delta_r G = 0$, at which point $Q_P = K_P$. Thus, initially $\Delta_r G$ has a large negative value and increases to zero as the reaction goes to equilibrium.

We should point out here that even though $\Delta_r G < 0$, the reaction may not occur at a detectable rate. For example, consider the reaction given by

$$2\,H_2(g) + O_2(g) \rightleftharpoons 2\,H_2O(l)$$

The value of $\Delta_r G°$ at 25°C for this reaction is -237 kJ per mole of $H_2O(l)$ formed. Consequently, $H_2O(l)$ at one bar and 25°C is much more stable than a mixture of $H_2(g)$ and $O_2(g)$ under those conditions. Yet, a mixture of $H_2(g)$ and $O_2(g)$ can be kept indefinitely. If a spark or a catalyst is introduced into this mixture, however, then the reaction occurs explosively. This observation serves to illustrate an important point: The "no" of thermodynamics is emphatic. If thermodynamics says that a certain process will not occur spontaneously, then it will not occur. The "yes" of thermodynamics, on the other hand, is actually a "maybe". The fact that a process will occur spontaneously does not imply that it will necessarily occur at a detectable rate. We shall study the rates of chemical reactions in Chapters 28 through 31.

26–7. The Variation of an Equilibrium Constant with Temperature Is Given by the Van't Hoff Equation

We can use the Gibbs-Helmoltz equation (Equation 22.61)

$$\left(\frac{\partial \Delta G°/T}{\partial T}\right)_P = -\frac{\Delta H°}{T^2} \tag{26.28}$$

to derive an equation for the temperature dependence of $K_P(T)$. Substitute $\Delta G°(T) = -RT \ln K_P(T)$ into Equation 26.28 to obtain

$$\left(\frac{\partial \ln K_P(T)}{\partial T}\right)_P = \frac{d \ln K_P(T)}{dT} = \frac{\Delta_r H°}{RT^2} \tag{26.29}$$

Note that if $\Delta_r H° > 0$ (endothermic reaction), then $K_P(T)$ increases with temperature, and if $\Delta_r H° < 0$ (exothermic reaction), then $K_P(T)$ decreases with increasing temperature. This is another example of Le Châtelier's principle.

Equation 26.29 can be integrated to give

$$\ln \frac{K_P(T_2)}{K_P(T_1)} = \int_{T_1}^{T_2} \frac{\Delta_r H°(T)dT}{RT^2} \tag{26.30}$$

If the temperature range is small enough that we can consider $\Delta_r H^\circ$ to be a constant, then we can write

$$\ln \frac{K_P(T_2)}{K_P(T_1)} = -\frac{\Delta_r H^\circ}{R}\left(\frac{1}{T_2} - \frac{1}{T_1}\right) \tag{26.31}$$

Equation 26.31 suggests that a plot of $\ln K_P(T)$ versus $1/T$ should be a straight line with a slope of $-\Delta_r H^\circ/R$ over a sufficiently small temperature range. Figure 26.3 shows such a plot for the reaction $H_2(g) + CO_2(g) \rightleftharpoons CO(g) + H_2O(g)$ over the temperature range 600°C to 900°C.

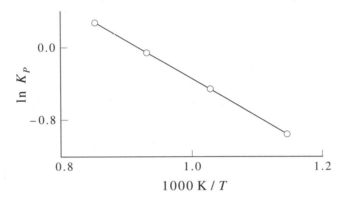

FIGURE 26.3
A plot of $\ln K_P(T)$ versus $1/T$ for the reaction $H_2(g) + CO_2(g) \rightleftharpoons CO(g) + H_2O(g)$ over the temperature range 600°C to 900°C. The circles represent experimental data

EXAMPLE 26–6
Given that $\Delta_r H^\circ$ has an average value of -69.8 kJ·mol^{-1} over the temperature range 500 K to 700 K for the reaction described by

$$PCl_3(g) + Cl_2(g) \rightleftharpoons PCl_5(g)$$

estimate K_P at 700 K given that $K_P = 0.0408$ at 500 K.

SOLUTION: We use Equation 26.31 with the above values

$$\ln \frac{K_P}{0.0408} = -\frac{-69.8 \times 10^3 \text{ J·mol}^{-1}}{8.3145 \text{ J·K}^{-1}\cdot\text{mol}^{-1}}\left(\frac{1}{700 \text{ K}} - \frac{1}{500 \text{ K}}\right)$$
$$= -4.80$$

or

$$K_P(T) = (0.0408)e^{-4.80} = 3.36 \times 10^{-4}$$

Note that the reaction is exothermic and so $K_P(T = 700 \text{ K})$ is less than $K_P(T = 500 \text{ K})$.

In Section 19–12 we discussed the temperature variation of $\Delta_r H°$. In particular, we derived the equation

$$\Delta_r H°(T_2) = \Delta_r H°(T_1) + \int_{T_1}^{T_2} \Delta C_P°(T)dT \qquad (26.32)$$

where $\Delta C_P°$ is the difference between the heat capacities of the products and reactants. Experimental heat capacity data over temperature ranges are often presented as polynomials in the temperature, and if this is the case, then $\Delta_r H°(T)$ can be expressed in the form (see Example 19–13)

$$\Delta_r H°(T) = \alpha + \beta T + \gamma T^2 + \delta T^3 + \cdots \qquad (26.33)$$

If this form for $\Delta_r H°(T)$ is substituted into Equation 26.29, and both sides integrated indefinitely, then we find that

$$\ln K_P(T) = -\frac{\alpha}{RT} + \frac{\beta}{R} \ln T + \frac{\gamma}{R}T + \frac{\delta T^2}{2R} + A \qquad (26.34)$$

The constants α through δ are known from Equation 26.33 and A is an integration constant that can be evaluated from a knowledge of $K_P(T)$ at some particular temperature. We could also have integrated Equation 26.29 from some temperature T_1 at which the value of $K_P(T)$ is known to an arbitrary temperature T to obtain

$$\ln K_P(T) = \ln K_P(T_1) + \int_{T_1}^{T} \frac{\Delta_r H°(T')dT'}{RT'^2} \qquad (26.35)$$

Equations 26.34 and 26.35 are generalizations of Equation 26.31 to the case where the temperature dependence of $\Delta_r H°$ is not ignored. Equation 26.34 shows that if $\ln K_P(T)$ is plotted against $1/T$, then the slope is not constant, but has a slight curvature. Figure 26.4 shows $\ln K_P(T)$ plotted versus $1/T$ for the ammonia synthesis reaction. Note that $\ln K_P(T)$ does not vary linearly with $1/T$, showing that $\Delta_r H°$ is temperature dependent.

EXAMPLE 26–7
Consider the reaction described by

$$\tfrac{1}{2}N_2(g) + \tfrac{3}{2}H_2(g) \rightleftharpoons NH_3(g)$$

The molar heat capacities of $N_2(g)$, $H_2(g)$, and $NH_3(g)$ can be expressed in the form

$$C_P°[N_2(g)]/J \cdot K^{-1} \cdot mol^{-1} = 24.98 + 5.912 \times 10^{-3}T - 0.3376 \times 10^{-6}T^2$$

$$C_P°[H_2(g)]/J \cdot K^{-1} \cdot mol^{-1} = 29.07 - 0.8368 \times 10^{-3}T + 2.012 \times 10^{-6}T^2$$

$$C_P°[NH_3(g)]/J \cdot K^{-1} \cdot mol^{-1} = 25.93 + 32.58 \times 10^{-3}T - 3.046 \times 10^{-6}T^2$$

over the temperature range 300 K to 1500 K. Given that $\Delta_f H°[NH_3(g)] = -46.11$ kJ·mol^{-1} at 300 K and that $K_P = 6.55 \times 10^{-3}$ at 725 K, derive a general expression for the variation of $K_P(T)$ with temperature in the form of Equation 26.34.

SOLUTION: We first use Equation 26.32

$$\Delta_r H°(T_2) = \Delta_r H°(T_1) + \int_{T_1}^{T_2} \Delta C_P°(T)dT$$

with $T_1 = 300$ K and $\Delta_r H°(T_1 = 300$ K$) = -46.11$ kJ·mol^{-1} and

$$\Delta C_P° = C_P°[NH_3(g)] - \frac{1}{2}C_P°[N_2(g)] - \frac{3}{2}C_P°[H_2(g)]$$

Integration gives

$$\Delta_r H°(T)/\text{J·mol}^{-1} = -46.11 \times 10^3 + \int_{300\ K}^{T} \Delta C_P°(T)dT$$

$$= -46.11 \times 10^3 - 31.17(T - 300)$$

$$+ \frac{30.88 \times 10^{-3}}{2}(T^2 - (300)^2) - \frac{5.895 \times 10^{-6}}{3}(T^3 - (300)^3)$$

or

$$\Delta_r H°(T)/\text{J·mol}^{-1} = -38.10 \times 10^3 - 31.17T + 15.44 \times 10^{-3}T^2 - 1.965 \times 10^{-6}T^3$$

Now we use Equation 26.35 with $T_1 = 725$ K and $K_P(T = 725$ K$) = 6.55 \times 10^{-3}$.

$$\ln K_P(T) = \ln K_P(T = 725\ \text{K}) + \int_{725}^{T} \frac{\Delta_r H°(T')}{RT'^2}dT'$$

$$= -5.028 + \frac{1}{R}\left[+38.10\left(\frac{1}{T} - \frac{1}{725}\right) - 31.17(\ln T - \ln 725) \right.$$

$$\left. + 15.44 \times 10^{-3}(T - 725) - \frac{1.965 \times 10^{-6}}{2}(T^2 - (725)^2) \right]$$

$$= 12.06 + \frac{4583}{T} - 3.749 \ln T + 1.857 \times 10^{-3}T - 0.118 \times 10^{-6}T^2$$

This equation was used to generate Figure 26.4. At 600 K, $\ln K_P = -3.21$, or $K_P = 0.040$, in excellent agreement with the experimental value of 0.041.

It is interesting to compare the results of this section to those of Section 23–4, where we derived the Clausius-Clapeyron equation, Equation 23.13. Note that Equations 26.31 and 23.13 are essentially the same because the vaporization of a liquid can be represented by the "chemical equation"

$$X(l) \rightleftharpoons X(g)$$

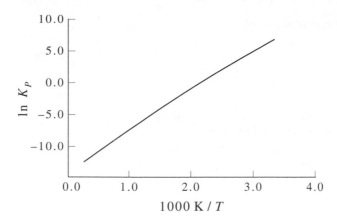

FIGURE 26.4
A plot of $\ln K_p(T)$ versus $1/T$ for the ammonia synthesis reaction, $\frac{3}{2}H_2(g) + \frac{1}{2}N_2(g) \rightleftharpoons NH_3(g)$.

26–8. We Can Calculate Equilibrium Constants in Terms of Partition Functions

An important chemical application of statistical thermodynamics is the calculation of equilibrium constants in terms of molecular parameters. Consider the general homogeneous gas-phase chemical reaction

$$\nu_A A(g) + \nu_B B(g) \rightleftharpoons \nu_Y Y(g) + \nu_Z Z(g)$$

in a reaction vessel at fixed volume and temperature. In this case we have (cf. Equation 23.26)

$$dA = \mu_A dn_A + \mu_B dn_B + \mu_Y dn_Y + \mu_Z dn_Z \qquad \text{(constant } T \text{ and } V)$$

instead of Equation 26.3. Introducing the extent of reaction through Equations 26.2, however, leads to the same condition for chemical equilbrium as in Section 26–1,

$$\nu_Y \mu_Y + \nu_Z \mu_Z - \nu_A \mu_A - \nu_B \mu_B = 0 \qquad (26.36)$$

We now introduce statistical thermodynamics through the relation between the chemical potential and a partition function. In a mixture of ideal gases, the species are independent, and so the partition function of the mixture is a product of the partition functions of the individual components. Thus

$$Q(N_A, N_B, N_Y, N_Z, V, T) = Q(N_A, V, T)Q(N_B, V, T)Q(N_Y, V, T)Q(N_Z, V, T)$$

$$= \frac{q_A(V, T)^{N_A}}{N_A!} \frac{q_B(V, T)^{N_B}}{N_B!} \frac{q_Y(V, T)^{N_Y}}{N_Y!} \frac{q_Z(V, T)^{N_Z}}{N_Z!}$$

The chemical potential of each species is given by an equation such as (Problem 26–33)

$$\mu_A = -RT \left(\frac{\partial \ln Q}{\partial N_A} \right)_{N_j, V, T} = -RT \ln \frac{q_A(V, T)}{N_A} \tag{26.37}$$

where Stirling's approximation has been used for $N_A!$. The N_j subscript on the partial derivative indicates that the numbers of particles of the other species are held fixed. Equation 26.37 simply says that the chemical potential of one species of an ideal gas mixture is calculated as if the other species were not present. This, of course, is the case for an ideal gas mixture.

If we substitute Equation 26.37 into Equation 26.36, then we get

$$\frac{N_Y^{\nu_Y} N_Z^{\nu_Z}}{N_A^{\nu_A} N_B^{\nu_B}} = \frac{q_Y^{\nu_Y} q_Z^{\nu_Z}}{q_A^{\nu_A} q_B^{\nu_B}} \tag{26.38}$$

For an ideal gas, the molecular partition function is of the form $f(T)V$ (Section 18–6) so that q/V is a function of temperature only. If we divide each factor on both sides of Equation 26.38 by V^{ν_j} and denote the number density N_j/V by ρ_j, then we have

$$K_c(T) = \frac{\rho_Y^{\nu_Y} \rho_Z^{\nu_Z}}{\rho_A^{\nu_A} \rho_B^{\nu_B}} = \frac{(q_Y/V)^{\nu_Y} (q_Z/V)^{\nu_Z}}{(q_A/V)^{\nu_A} (q_B/V)^{\nu_B}} \tag{26.39}$$

Note that K_c is a function of temperature only. Recall that $K_P(T)$ and $K_c(T)$ are related by (Equation 26.17)

$$K_P(T) = \frac{P_Y^{\nu_Y} P_Z^{\nu_Z}}{P_A^{\nu_A} P_B^{\nu_B}} = K_c(T) \left(\frac{c^\circ RT}{P^\circ} \right)^{\nu_Y + \nu_Z - \nu_A - \nu_B}$$

By means of Equation 26.17 and Equation 26.39, along with the results of Chapter 18, we can calculate equilibrium constants in terms of molecular parameters. This is best illustrated by means of examples.

A. A Chemical Reaction Involving Diatomic Molecules

We shall calculate the equilibrium constant for the reaction

$$H_2(g) + I_2(g) \rightleftharpoons 2\,HI(g)$$

from 500 K to 1000 K. The equilibrium constant is given by

$$K(T) = \frac{(q_{HI}/V)^2}{(q_{H_2}/V)(q_{I_2}/V)} = \frac{q_{HI}^2}{q_{H_2} q_{I_2}} \tag{26.40}$$

Using Equation 18.39 for the molecular partition functions gives

$$K(T) = \left(\frac{m_{HI}^2}{m_{H_2} m_{I_2}}\right)^{3/2} \left(\frac{4\Theta_{rot}^{H_2}\Theta_{rot}^{I_2}}{(\Theta_{rot}^{HI})^2}\right) \frac{(1 - e^{-\Theta_{vib}^{H_2}/T})(1 - e^{-\Theta_{vib}^{I_2}/T})}{(1 - e^{-\Theta_{vib}^{HI}/T})^2}$$
$$\times \exp\frac{2D_0^{HI} - D_0^{H_2} - D_0^{I_2}}{RT} \qquad (26.41)$$

where we have replaced D_e in Equation 18.39 by $D_0 + h\nu/2$ (Figure 18.2). All the necessary parameters are given in Table 18.2. Table 26.2 gives the numerical values of $K_P(T)$ and Figure 26.5 shows $\ln K$ plotted versus $1/T$. From the slope of the line in Figure 26.5 we get $\Delta_r \overline{H} = -12.9 \text{ kJ} \cdot \text{mol}^{-1}$ compared to the experimental value of $-13.4 \text{ kJ} \cdot \text{mol}^{-1}$. The discrepancy is due to the inadequacy of the rigid rotator-harmonic oscillator approximation at these temperatures.

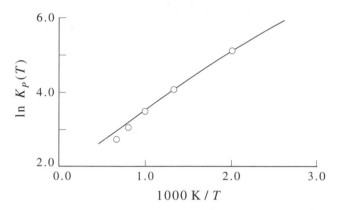

FIGURE 26.5
The logarithm of the equilibrium constant versus $1/T$ for the reaction $H_2(g) + I_2(g) \rightleftharpoons 2 \text{ HI}(g)$. The line is calculated from Equation 26.41 and the circles are the experimental values.

TABLE 26.2
The values of $K_P(T)$ for the reaction described by $H_2(g) + I_2(g) \rightleftharpoons 2 \text{HI}(g)$ calculated according to Equation 26.41.

T/K	$K_P(T)$	$\ln K_P(T)$
500	138	4.92
750	51.1	3.93
1000	28.5	3.35
1250	19.1	2.95
1500	14.2	2.65

B. A Reaction Involving Polyatomic Molecules

As an example of a reaction involving a polyatomic molecule, consider the reaction

$$H_2(g) + \tfrac{1}{2} O_2(g) \rightleftharpoons H_2O(g)$$

whose equilibrium constant is given by

$$K_c(T) = \frac{(q_{H_2O}/V)}{(q_{H_2}/V)(q_{O_2}/V)^{1/2}} \tag{26.42}$$

It is almost as convenient to calculate each partition function separately as to substitute them into K_c first. The necessary parameters are given in Tables 18.2 and 18.4. At 1500 K, the three partition functions are (Equations 18.39 and 18.60)

$$\frac{q_{H_2}(T, V)}{V} = \left(\frac{2\pi m_{H_2} k_B T}{h^2}\right)^{3/2} \left(\frac{T}{2\Theta_{rot}^{H_2}}\right) (1 - e^{-\Theta_{vib}^{H_2}/T})^{-1} e^{D_0^{H_2}/RT}$$

$$= 2.80 \times 10^{32} e^{D_0^{H_2}/RT} \ m^{-3} \tag{26.43}$$

$$\frac{q_{O_2}(T, V)}{V} = \left(\frac{2\pi m_{O_2} k_B T}{h^2}\right)^{3/2} \left(\frac{T}{2\Theta_{rot}^{O_2}}\right) (1 - e^{-\Theta_{vib}^{O_2}/T})^{-1} 3 e^{D_0^{O_2}/RT}$$

$$= 2.79 \times 10^{36} e^{D_0^{O_2}/RT} \ m^{-3} \tag{26.44}$$

and

$$\frac{q_{H_2O}(T, V)}{V} = \left(\frac{2\pi m_{H_2O} k_B T}{h^2}\right)^{3/2} \frac{\pi^{1/2}}{\sigma} \left(\frac{T^3}{\Theta_{rot,A}^{H_2O} \Theta_{rot,B}^{H_2O} \Theta_{rot,C}^{H_2O}}\right)^{1/2} \prod_{j=1}^{3} (1 - e^{-\Theta_{vib,j}^{H_2O}/T})^{-1} e^{D_0^{H_2O}/RT}$$

$$= 5.33 \times 10^{35} e^{D_0^{H_2O}/RT} \ m^{-3} \tag{26.45}$$

The factor of 3 occurs in q_{O_2}/V because the ground state of O_2 is $^3\Sigma_g^-$.

Notice that each of the above $q(T, V)/V$ has units of m^{-3}. This tells us that the reference state in this (molecular) case is a concentrations of one molecule per cubic meter, or that $c^\circ =$ one molecule$\cdot m^{-3}$. Using the values of D_0 from Table 18.2 and 18.4, the value of K_c at 1500 K is $K_c = 2.34 \times 10^{-7}$. To convert to K_P, we divide K_c by

$$\left(\frac{c^\circ RT}{N_A P^\circ}\right)^{1/2} = \left[\frac{(1 \ m^{-3})(8.3145 \ J\cdot mol^{-1}\cdot K^{-1})(1500 \ K)}{(6.022 \times 10^{23} \ mol^{-1})(10^5 \ Pa)}\right]^{1/2}$$

$$= 4.55 \times 10^{-13}$$

to obtain $K_P = 5.14 \times 10^5$, based upon a one bar standard state.

TABLE 26.3

The logarithm of the equilibrium constant for the reaction $H_2(g) + \frac{1}{2} O_2(g) \rightleftharpoons H_2O(g)$

T/K	$\ln K_p(\text{calc})$	$\ln K_p(\text{exp})$
1000	23.5	23.3
1500	13.1	13.2
2000	8.52	8.15

Table 26.3 compares the calculated values of $\ln K_p$ with experimental data. Although the agreement is fairly good, the agreement can be considerably improved by using more sophisticated spectroscopic models. At high temperatures, the rotational energies of the molecules are high enough to warrant centrifugal distortion effects and other extensions of the simple rigid rotator-harmonic oscillator approximation.

26–9. Molecular Partition Functions and Related Thermodynamic Data Are Extensively Tabulated

In the previous section we have seen that the rigid rotator-harmonic oscillator approximation can be used to calculate equilibrium constants in reasonably good agreement with experiment, and because of the simplicity of the model, the calculations involved are not extensive. If greater accuracy is desired, however, one must include corrections to the rigid rotator-harmonic oscillator model, and the calculations become increasingly more laborious. It is natural, then, that a number of numerical tables of partition functions has evolved, and in this section we shall discuss the use of such tables. These tables are actually much more extensive than a compilation of partition functions. They include many experimentally determined values of thermodynamic properties, often complemented by theoretical calculations. The thermodynamic tables that we are about to discuss in this section, then, represent a collection of the thermodynamic and/or statistical thermodynamic properties of many substances.

One of the most extensive tabulations of the thermochemical properties of substances is an American Chemical Society publication, *Journal of Physical Chemical Reference Data*, volume 14, supplement 1, 1985, usually referred to as the JANAF (*j*oint, *a*rmy, *n*avy, *air* *f*orce) tables. Each species listed has about a full page of thermodynamic/spectroscopic data, and Table 26.4 is a replica of the entry for ammonia. Note that the fourth and fifth columns of thermodynamic data are headed by $-\{G° - H°(T_r)\}/T$ and $H° - H°(T_r)$. Recall that the value of an energy must be referred to some fixed reference point (such as a zero of energy). The reference point used in the JANAF tables is the standard molar enthalpy at 298.15 K. Consequently, $G°(T)$ and $H°(T)$ are expressed relative to that value, as expressed

TABLE 26.4

A replica of the page of $NH_3(g)$ data in the JANAF tables

JANAF THERMOCHEMICAL TABLES

Ammonia (NH₃)

Enthalpy Reference Temperature = T_r = 298.15 K   Standard State Pressure = $p°$ = 0.1 MPa

T/K	$C°_p$	$S°$	$-(G°-H°_{T_r})/T$	$H°-H°_{T_r}$	$\Delta_f H°$	$\Delta_f G°$	Log K_f
	J K⁻¹ mol⁻¹			kJ mol⁻¹			
0	0.	0.	INFINITE	-10.045	-38.907	-38.907	INFINITE
100	33.284	155.840	223.211	-6.737	-41.550	-34.034	17.777
200	33.757	178.990	195.962	-3.394	-43.703	-25.879	6.707
298.15	35.652	192.774	192.774	0.	-45.898	-16.367	2.867
300	35.701	192.995	192.775	0.066	-45.939	-16.183	2.818
400	38.716	203.663	194.209	3.781	-48.041	-5.941	0.776
500	42.048	212.659	197.021	7.819	-49.857	4.800	-0.501
600	45.293	220.615	200.302	12.188	-51.374	15.879	-1.382
700	48.354	227.829	203.727	16.872	-52.618	27.190	-2.029
800	51.235	234.476	207.160	21.853	-53.616	38.584	-2.519
900	53.769	240.669	210.543	27.113	-54.411	50.047	-2.916
1000	56.491	246.486	213.849	32.637	-55.013	61.910	-3.234
1100	58.859	251.983	217.069	38.406	-55.746	73.625	-3.496
1200	61.048	257.199	220.197	44.402	-55.917	85.373	-3.716
1300	63.057	262.166	223.236	50.609	-55.982	97.141	-3.903
1400	64.893	266.907	226.187	57.008	-55.954	108.918	-4.064
1500	66.584	271.442	229.054	63.582	-55.833	120.696	-4.203
1600	68.079	275.788	231.840	70.315	-55.847	132.469	-4.325
1700	69.452	279.957	234.549	77.193	-55.672	144.234	-4.432
1800	70.695	283.962	237.184	84.201	-55.457	155.986	-4.527
1900	71.818	287.815	239.748	91.328	-55.201	167.725	-4.611
2000	72.833	291.525	242.244	98.561	-54.833	179.417	-4.687
2100	73.751	295.101	244.677	105.891	-54.473	191.152	-4.755
2200	74.581	298.552	247.048	113.309	-54.084	202.840	-4.818
2300	75.330	301.884	249.360	120.805	-53.671	214.509	-4.872
2400	76.009	305.104	251.616	128.372	-53.238	226.160	-4.922
2500	76.628	308.220	253.818	136.005	-52.789	237.792	-4.968
2600	77.174	311.236	255.969	143.695	-52.329	249.406	-5.011
2700	77.672	314.158	258.070	151.438	-51.860	261.003	-5.049
2800	78.132	316.991	260.124	159.228	-51.383	272.585	-5.088
2900	78.529	319.740	262.132	167.062	-50.909	284.143	-5.118
3000	78.902	322.409	264.097	174.933	-50.433	295.689	-5.148
3100	79.228	325.001	266.020	182.840	-49.959	307.218	-5.177
3200	79.521	327.521	267.903	190.778	-49.491	318.733	-5.203
3300	79.785	329.972	269.747	198.744	-49.030	330.233	-5.227
3400	80.011	332.358	271.554	206.734	-48.534	341.721	-5.250
3500	80.232	334.680	273.324	214.745	-48.139	353.191	-5.271
3600	80.400	336.942	275.060	222.776	-47.713	364.652	-5.291
3700	80.550	339.147	276.763	230.824	-47.302	376.101	-5.310
3800	80.684	341.297	278.433	238.886	-46.908	387.539	-5.327
3900	80.793	343.395	280.072	246.960	-46.531	398.967	-5.344
4000	80.881	345.441	281.680	255.043	-46.180	410.385	-5.359
4100	80.956	347.439	283.260	263.136	-45.847	421.795	-5.374
4200	81.006	349.391	284.811	271.234	-45.539	433.198	-5.388
4300	81.048	351.181	286.335	279.337	-45.254	444.593	-5.401
4400	81.065	353.161	287.833	287.442	-44.996	455.981	-5.413
4500	81.073	354.983	289.305	295.550	-44.764	467.364	-5.425
4600	81.057	356.765	290.752	303.656	-44.561	478.743	-5.436
4700	81.032	358.508	292.175	311.761	-44.387	490.117	-5.447
4800	80.990	360.213	293.575	319.862	-44.242	501.488	-5.457
4900	80.911	361.882	294.952	327.958	-44.129	512.856	-5.467
5000	80.856	363.517	296.307	336.048	-44.047	524.223	-5.477
5100	80.751	365.117	297.641	344.127	-43.999	535.587	-5.486
5200	80.751	366.685	298.954	352.202	-43.979	546.951	-5.494
5300	80.751	368.223	300.246	360.277	-43.982	558.315	-5.503
5400	80.751	369.732	301.519	368.352	-44.049	569.680	-5.511
5500	80.751	371.214	302.773	376.428	-44.112	581.044	-5.518
5600	80.751	372.669	304.008	384.503	-44.193	592.410	-5.526
5700	80.751	374.098	305.225	392.578	-44.291	603.778	-5.533
5800	80.751	375.503	306.425	400.653	-44.401	615.147	-5.540
5900	80.751	376.883	307.607	408.728	-44.516	626.516	-5.547
6000	80.751	378.240	308.773	416.803	-44.531	637.889	-5.553

PREVIOUS: June 1977 (1 atm)   CURRENT: June 1977 (1 bar)

Ammonia (NH₃)

AMMONIA (NH₃)   IDEAL GAS W_t = 17.03052 g mol⁻¹

$\Delta_a H°(0\ K)$ = -38.907 ± 0.4 kJ mol⁻¹
$\Delta_a H°(298.15\ K)$ = -45.898 ± 0.4 kJ mol⁻¹

$S°(298.15\ K)$ = 192.774 ± 0.025 J K⁻¹ mol⁻¹

Vibrational Frequencies and Degeneracies

ν_i, cm⁻¹	ν_i, cm⁻¹	ν_i, cm⁻¹
3506(1)	1022(1)	3577(2)
	1691(2)	

σ(internal) = 2
σ(external) = 3

Ground State Quantum Weight: 1
Point Group: C_{3v}
Bond Length: N-H = 1.0124 Å
Bond Angle: 106.67°
Product of the Moments of Inertia: $I_A I_B I_C$ = 0.0348 x 10⁻¹¹⁷ g³ cm⁶

Enthalpy of Formation

2nd and 3rd law analyses of equilibrium data for the reaction $1/2N_2(g)+3/2H_2(g) = NH_3(g)$ cited in the previous JANAF evaluation (1) plus more recent work of Schulz and Schaefer (6) were made using the revised thermal functions for $NH_3(g)$. All of the previously cited work in reaction calorimetry plus the early work of Berthelot (7, 8) and Thomsen (9) were reevaluated. No significant differences in the 3rd law calculations or in the corrections to the flow calorimetry data or in the equilibrium data were found. Thus, the 0.1 kcal discrepancy between the results of the Haber and Tamaru (12) and Wittig and Schmatz (13) were found. The previous JANAF selection (1) for $\Delta_f H°(298.15\ K)$ of $NH_3(g)$ was adopted. A recent evaluation (14) which includes new indirect calorimetry (unpublished) further confirms this selection.

Source	Method	$\Delta_f H°(298.15\ K)$ kcal mol⁻¹	$\Delta S°(298.15\ K)$ cal K⁻¹ mol⁻¹	$\Delta S°$(obs.–calc. 298.15 K)* cal K⁻¹ mol⁻¹
Larson, Dodge (2)	$K°_p$(a) from $K°_p$(10–1,000 atm, 600–800 K)	-10.88		+0.24±0.15ᵃ
Haber et al. (3)	$K°_p$(a) from $K°_p$(30 atm, 800–1200 K)	-10.88±0.15		-0.02±0.15ᶜ
Haber, Maschke (5)	$K°_p$(1 atm, 900–1400 K)	-10.85		0.20±0.19ᶜ
Schulz, Schaefer (6)	$K°_p$(1 atm, 567–673 K)	-10.87		0.14±0.3ᵈ
Berthelot (7)	Indirect: Reaction of Br₂(aq) and NH₃(aq)	-11.4		
Berthelot (8)	Indirect: Reaction of O₂(g) with NH₃(g)	-12.1		
Thomsen (9)	Indirect: Reaction of O₂(g) with NH₃(g)	-11.9		
Becker, Roth (10)	Indirect: Heat of combustion oxalates	-11.00±0.15		
Haber et al. (11)	Flow calorimetry at 298 K	-11.10±0.05		
Haber, Tamaru (12)	Flow calorimetry (793–932 K)	-11.00±0.008		
Wittig, Schmatz (13)	Flow calorimetry at 832 K	-10.99±0.05		

2nd law analysis assuming $\Delta C°_p$(cal K⁻¹ mol⁻¹) equals (a): -2.672+0.00591(T-700), (b): -1.336+0.00404(T-1000). (c): -0.855+0.00305(T-1100), (d): 3.287+0.00651(T-600).

Heat Capacity and Entropy

The thermodynamic functions differ from those of the 1965 JANAF table (1) in being taken directly from the later and more complete work of Haar (15). Haar treated in detail the contribution of the highly anharmonic out-of-plane vibrational mode, including its large coupling with rotation and its coupling with the other vibrational modes. Haar's values of $C°_p$ pass through a shallow maximum between 4000 and 5000 K; they were extrapolated from 5000 to 6000 K by assuming a constant value (19.300 cal K⁻¹ mol⁻¹). A summary of Haar's estimated uncertainties and of the differences between the 1965 table from the present table (in cal K⁻¹ mol⁻¹) is as follows:

	Uncertainties (Haar, 15)		1965 Table minus This Table	
T,K	$C°_p$	$S°$	$C°_p$	$S°$
1000	0.006	0.006	-0.034	-0.033
3000	0.10	0.10	+0.142	-0.122
5000	0.8	0.04	+1.775	+0.265

The National Bureau of Standards prepared this table (16) by critical analysis of data existing in 1973. Using the results of Haar (15) and $\Delta_f H°$ selected by NBS (16), we recalculate the table in terms of R=1.987192 cal K⁻¹ mol⁻¹ (17) and current JANAF reference states for the elements.

References

1. JANAF Thermochemical Tables, 2nd ed., NSRDS-NBS 37, 1971.
2. A. T. Larson and R. L. Dodge, J. Amer. Chem. Soc. 45, 2918 (1923).
3. A. T. Larson, J. Amer. Chem. Soc. 46, 367 (1924).
4. F. Haber and S. Tamaru, Z. Elektrochem. 21, 89 (1915).
5. C. Stephenson and H. O. McMahon, J. Amer. Chem. Soc. 61, 437 (1939).
6. F. Haber and A. Maschke, Z. Electrochem. 21, 128 (1915).
7. M. Berthelot, Compt. Rend. 92, ... ; Ann. Chim. Phys. (5) 20, 244 (1880).
8. M. Berthelot, Ann. Chim. Phys. (5) 20, 247 (1880).
9. J. Thomsen, "Thermochemical Investigations," Vol. II, p. 68, Johann A. Barth, Leipzig, 1883.
10. G. Becker and W. A. Roth, Z. Elektrochem. 40, 836 (1934).
11. F. Haber, S. Tamaru, and Ch. Ponnaz, Z. Elektrochem. 21, 206 (1915).
12. F. Haber and S. Tamaru, Z. Elektrochem. 21, 191 (1915).
13. E. Wittig and W. Schmatz, Z. Elektrochem. 63, 425 (1959); July 1972.
14. ...
15. L. Haar, J. Res. Nat. Bur. Stand. 78A, 207 (1968).
16. L. Haar, J. Res. Nat. Bur. Stand., Rept. 10904, 339, July, 1971.
17. S. Abramowitz et al., U. S. Nat. Bur. Stand. Special Publication; evaluation of ICSU-CODATA Task Group on Key Values for ... CODATA Task Group on Fundamental Constants, CODATA Bulletin 11, December, 1973.

H₃N₁(g)

by the headings $-\{G°(T) - H°(298.15 \text{ K})\}/T$ and $H°(T) - H°(298.15 \text{ K})$. Table 26.4 gives $-\{G°(T) - H°(298.15 \text{ K})\}/T$ for ammonia at a number of temperatures. The ratio $\{G°(T) - H°(298.15 \text{ K})\}/T$ rather than $\{G°(T) - H°(298.15 \text{ K})\}$ is given because $\{G°(T) - H°(298.15 \text{ K})\}/T$ varies more slowly with temperature, and hence the tables are easier to interpolate. It is not necessary to specify a reference point for the heat capacity or the entropy, as indicated by the headings to the second and third columns. The sixth and seventh columns give values of $\Delta_f H°$ and $\Delta_f G°$ at various temperatures. We learned in Section 26–3 that these data can be used to calculate values of $\Delta_r H°$, $\Delta_r G°$, and equilibrium constants of reactions.

Because $G°(T)$ and $H°(T)$ are expressed relative to $H°(298.15 \text{ K})$ in Table 26.4, we must express the molecular partition function $q(V, T)$ relative to a zero of energy. Recall that in Section 23–5 we wrote $q(V, T)$ as

$$q(V, T) = \sum_j e^{-\varepsilon_j/k_B T} = e^{-\varepsilon_0/k_B T} + e^{-\varepsilon_1/k_B T} + \cdots$$

$$= e^{-\varepsilon_0/k_B T}(1 + e^{-(\varepsilon_1 - \varepsilon_0)/k_B T} + \cdots)$$

$$= e^{-\varepsilon_0/k_B T} q^0(V, T) \tag{26.46}$$

where $q^0(V, T)$ is a molecular partition function in which the ground state energy is taken to be zero. If we substitute Equation 26.46 into Equation 17.41, then we obtain

$$U = \langle E \rangle = N k_B T^2 \left(\frac{\partial \ln q}{\partial T}\right)_V$$

$$= N\varepsilon_0 + N k_B T^2 \left(\frac{\partial \ln q^0}{\partial T}\right)_V \tag{26.47}$$

For one mole of an ideal gas, $\overline{H} = H°(T) = \overline{U} + P\overline{V} = \overline{U} + RT$, and so Equation 26.47 becomes

$$H°(T) = H_0° + RT^2 \left(\frac{\partial \ln q^0}{\partial T}\right)_V + RT \tag{26.48}$$

where $H_0° = N_A \varepsilon_0$. Because $q^0(V, T)$ is the molecular partition function in which the ground state energy is taken to be zero, $q^0(V, T)$ is given by either Equation 18.57 or 18.60, without the factors of $e^{-\Theta_{\text{vib},i}/2T}$ and $e^{D_e/k_B T}$, which represent the ground state of the molecule. Using either Equation 18.57 or 18.60, Equation 26.48 becomes

$$H°(T) - H_0° = \frac{3}{2}RT + \frac{2}{2}RT + \sum_j \frac{R\Theta_{\text{vib},j}}{e^{\Theta_{\text{vib},j}/T} - 1} + RT$$

$$= \frac{7}{2}RT + \sum_j \frac{R\Theta_{\text{vib},j}}{e^{\Theta_{\text{vib},j}/T} - 1} \quad \text{(linear molecule)} \tag{26.49a}$$

or

$$H°(T) - H_0° = \frac{3}{2}RT + \frac{3}{2}RT + \sum_j \frac{R\Theta_{\text{vib},j}}{e^{\Theta_{\text{vib},j}/T} - 1} + RT$$

$$= 4RT + \sum_j \frac{R\Theta_{\text{vib},j}}{e^{\Theta_{\text{vib},j}/T} - 1} \quad \text{(nonlinear molecule)} \quad (26.49b)$$

Note that there are no terms involving $\Theta_{\text{vib},j}/2T$ or D_e/k_BT in Equations 26.49 as there are in Equations 18.58 and 18.61 because we have taken the energy of the ground vibrational state to be zero.

We can use Equation 26.49b and the parameters in Table 18.4 to calculate $H°(298.15\text{ K}) - H_0°$ for ammonia

$$H°(298.15\text{ K}) - H_0° = 4(8.3145\text{ J·mol}^{-1}\text{·K}^{-1})(298.15\text{ K})$$

$$+(8.3145\text{ J·mol}^{-1}\text{·K}^{-1})\left[\frac{4800\text{ K}}{e^{4800/298.15} - 1} \right.$$

$$\left. + \frac{1360\text{ K}}{e^{1360/298.15} - 1} + \frac{(2)(4880\text{ K})}{e^{4880/298.15} - 1} + \frac{(2)(2330\text{ K})}{e^{2330/298.15} - 1} \right]$$

$$= 10.05\text{ kJ·mol}^{-1}$$

The very first entry in the fifth column in Table 26.4 is $-10.045\text{ kJ·mol}^{-1}$. This value represents $H°(0\text{ K}) - H°(298.15\text{ K})$, which is the negative of $H°(298.15\text{ K}) - H°(0\text{ K})$ that we just calculated because $H_0° = H°(0\text{ K})$. Thus, the value given by Equation 26.49b and the value given in Table 26.4 are in excellent agreement.

EXAMPLE 26–8
Use Equation 26.49b and the parameters in Table 18.4 to calculate $H°(T) - H_0°$ for $NH_3(g)$ at 1000 K and one bar. Compare your result to Table 26.4.

SOLUTION: Equation 26.49b gives

$$H°(1000\text{ K}) - H_0° = 42.290\text{ kJ·mol}^{-1}$$

Table 26.4 gives

$$H_0° - H°(298.15\text{ K}) = H°(0\text{ K}) - H°(298.15\text{ K}) = -10.045\text{ kJ·mol}^{-1} \quad (1)$$

and

$$H°(1000\text{ K}) - H°(298.15\text{ K}) = 32.637\text{ kJ·mol}^{-1} \quad (2)$$

If we subtract Equation 1 from Equation 2, then we obtain

$$H°(1000\text{ K}) - H_0° = 42.682\text{ kJ·mol}^{-1}$$

The value obtained from Table 26.4 is more accurate than the value calculated from Equation 26.49b. At 1000 K, the ammonia molecule is excited enough that the rigid rotator-harmonic oscillator approximation begins to become unsatisfactory.

We can also use the data in Table 26.4 to calculate the value of $q^0(V, T)$ for ammonia. Recall from Section 23–5 that we derived the equation (Equation 23.36)

$$\mu^\circ(T) - E_0^\circ = -RT \ln \left\{ \left(\frac{q^0}{V} \right) \frac{RT}{N_A P^\circ} \right\} \tag{26.50}$$

where $E_0^\circ = N_A \varepsilon_0 = H_0^\circ$ and $P^\circ = 1$ bar $= 10^5$ Pa. Equation 26.50 is valid only for an ideal gas, and recall that $q(V, T)/V$, or $q^0(V, T)/V$, is a function of temperature only for an ideal gas. Equation 26.50 clearly displays the fact that the chemical potential is calculated relative to some zero of energy.

Because $G^\circ = \mu^\circ$ for a pure substance, we can write Equation 26.50 as

$$G^\circ - H_0^\circ = -RT \ln \left\{ \left(\frac{q^0}{V} \right) \frac{RT}{N_A P^\circ} \right\} \tag{26.51}$$

It is easy to show that $G^\circ \to H_0^\circ$ as $T \to 0$ (because $T \ln T \to 0$ as $T \to 0$), and so H_0° is also the standard Gibbs energy at 0 K.

According to Equation 26.51

$$\frac{q^0}{V} \frac{RT}{N_A P^\circ} = e^{-(G^\circ - H_0^\circ)/RT}$$

or

$$\frac{q^0(V, T)}{V} = \frac{N_A P^\circ}{RT} e^{-(G^\circ - H_0^\circ)/RT} \tag{26.52a}$$

where $P^\circ = 10^5$ Pa. The fourth column in Table 26.4 gives $-\{G^\circ - H^\circ(298.15 \text{ K})\}/T$ instead of $-(G^\circ - H_0^\circ)/T$, but the first entry of the fifth column gives $H_0^\circ - H^\circ(298.15 \text{ K})$. Therefore, the exponential in Equation 26.52a can be obtained from

$$-\underbrace{\frac{(G^\circ - H_0^\circ)}{T}}_{\substack{\text{exponent in} \\ \text{Equation 26.52a}}} = -\underbrace{\frac{(G^\circ - H^\circ(298.15 \text{ K}))}{T}}_{\substack{\text{fourth column in} \\ \text{Table 26.4}}} + \underbrace{\frac{(H_0^\circ - H^\circ(298.15 \text{ K}))}{T}}_{\substack{\text{first entry of fifth} \\ \text{column in Table 26.4} \\ \text{divided by } T}} \tag{26.52b}$$

Let's use Equations 26.52 to calculate $q^0(V, T)$ for ammonia at 500 K. Substituting the data in Table 26.4 into Equation 26.52b gives

$$-\frac{(G° - H_0°)}{500 \text{ K}} = 197.021 \text{ J·K}^{-1}·\text{mol}^{-1} + \frac{-10.045 \text{ kJ·mol}^{-1}}{500 \text{ K}}$$

$$= 176.931 \text{ J·K}^{-1}·\text{mol}^{-1}$$

If we substitute this value into Equation 26.52a, then we obtain

$$\frac{q^0(V, T)}{V} = \frac{(6.022 \times 10^{23} \text{ mol}^{-1})(10^5 \text{ Pa})}{(8.314 \text{ J·mol}^{-1}·\text{K}^{-1})(500 \text{ K})} e^{(176.931 \text{ J·K}^{-1}·\text{mol}^{-1})/8.314 \text{ J·mol}^{-1}·\text{K}^{-1}}$$

$$= 2.53 \times 10^{34} \text{ m}^{-3}$$

Equation 18.60 gives (Problem 26–48)

$$\frac{q^0(V, T)}{V} = 2.59 \times 10^{34} \text{ m}^{-3}$$

The value given by Equations 26.52 is the more accurate because Equation 18.60 is based on the rigid rotator-harmonic oscillator approximation.

EXAMPLE 26–9

The JANAF tables give $-[G° - H°(298.15 \text{ K})]/T = 231.002 \text{ J·mol}^{-1}·\text{K}^{-1}$ and $H_0° - H°(298.15 \text{ K}) = -8.683 \text{ kJ·mol}^{-1}$ for $O_2(g)$ at 1500 K. Use these data and Equations 26.52 to calculate $q^0(V, T)/V$ for $O_2(g)$ at 1500 K.

SOLUTION: Equation 26.52b gives

$$-\frac{G° - H_0°}{T} = 231.002 \text{ J·mol}^{-1}·\text{K}^{-1} + \frac{-8.683 \text{ kJ·mol}^{-1}}{1500 \text{ K}}$$

$$= 225.093 \text{ J·mol}^{-1}·\text{K}^{-1}$$

and Equation 26.52a gives

$$\frac{q^0(V, T)}{V} = \frac{(6.022 \times 10^{23} \text{ mol}^{-1})(10^5 \text{ Pa})}{(8.314 \text{ J·mol}^{-1}·\text{K}^{-1})(1500 \text{ K})} e^{(225.093 \text{ J·K}^{-1}·\text{mol}^{-1})/8.314 \text{ J·mol}^{-1}·\text{K}^{-1}}$$

$$= 2.76 \times 10^{36} \text{ m}^{-3}$$

The value calculated in the previous section is $2.79 \times 10^{36} \text{ m}^{-3}$.

Lastly, the thermodynamic data in the JANAF tables can also be used to calculate values of D_0 for molecules. Table 26.4 gives $\Delta_f H°(0 \text{ K}) = -38.907 \text{ kJ·mol}^{-1}$ for $NH_3(g)$. The chemical equation that represents this process is

$$\frac{3}{2} H_2(g) + \frac{1}{2} N_2(g) \rightleftharpoons NH_3(g) \qquad \Delta_f H°(0 \text{ K}) = -38.907 \text{ kJ·mol}^{-1} \qquad (1)$$

The entries in the JANAF tables for H(g) and N(g) give $\Delta_f H^\circ(0\ K) = 216.035\ kJ \cdot mol^{-1}$ and $470.82\ kJ \cdot mol^{-1}$, respectively. These values correspond to the equations

$$\tfrac{1}{2} H_2(g) \rightleftharpoons H(g) \qquad \Delta_f H^\circ(0\ K) = 216.035\ kJ \cdot mol^{-1} \tag{2}$$

and

$$\tfrac{1}{2} N_2(g) \rightleftharpoons N(g) \qquad \Delta_f H^\circ(0\ K) = 470.82\ kJ \cdot mol^{-1} \tag{3}$$

If we subtract Equation 1 from the sum of Equation 3 and three times Equation 2, then we obtain

$$NH_3(g) \rightleftharpoons N(g) + 3\,H(g)$$

$$\Delta_f H^\circ(0\ K) = 38.907\ kJ \cdot mol^{-1} + (3)(216.035\ kJ \cdot mol^{-1}) + 470.82\ kJ \cdot mol^{-1}$$
$$= 1157.83\ kJ \cdot mol^{-1}$$

The value given in Table 18.4 is $1158\ kJ \cdot mol^{-1}$.

EXAMPLE 26–10
The JANAF tables give $\Delta_f H^\circ(0\ K)$ for HI(g), H(g), and I(g) to be $28.535\ kJ \cdot mol^{-1}$, $216.035\ kJ \cdot mol^{-1}$, and $107.16\ kJ \cdot mol^{-1}$, respectively. Calculate the value of D_0 for HI(g).

SOLUTION: The above data can be presented as

$$\tfrac{1}{2} H_2(g) + \tfrac{1}{2} I_2(s) \rightleftharpoons HI(g) \qquad \Delta_f H^\circ(0\ K) = 28.535\ kJ \cdot mol^{-1} \tag{1}$$

$$\tfrac{1}{2} H_2(g) \rightleftharpoons H(g) \qquad \Delta_f H^\circ(0\ K) = 216.035\ kJ \cdot mol^{-1} \tag{2}$$

$$\tfrac{1}{2} I_2(s) \rightleftharpoons I(g) \qquad \Delta_f H^\circ(0\ K) = 107.16\ kJ \cdot mol^{-1} \tag{3}$$

If we subtract Equation 1 from the sum of Equations 2 and 3, then we obtain

$$HI(g) \rightleftharpoons H(g) + I(g) \quad \Delta_f H^\circ(0\ K) = 294.66\ kJ \cdot mol^{-1}$$

The value given in Table 18.2 is $294.7\ kJ \cdot mol^{-1}$.

The thermodynamic tables contain a great deal of thermodynamic and/or statistical thermodynamic data. Their use requires a little practice, but it is well worth the effort. Problems 26–45 through 26–58 are meant to supply this practice.

26–10. Equilibrium Constants for Real Gases Are Expressed in Terms of Partial Fugacities

Up to this point in this chapter, we have discussed equilibria in systems of ideal gases only. In this section, we shall discuss equilibria in systems of nonideal gases. In Section 22–8 we introduced the idea of fugacity through the equation

$$\mu(T, P) = \mu^\circ(T) + RT \ln \frac{f}{f^\circ} \tag{26.53}$$

where $\mu^\circ(T)$ is the chemical potential of the corresponding ideal gas at one bar. Once again to simplify the notation we shall not display the f° in the rest of this chapter. Therefore, Equation 26.53 can be written in the form

$$\mu(T, P) = \mu^\circ(T) + RT \ln f \tag{26.54}$$

Consequently, we must keep in mind that f is taken relative to its standard state. In a mixture of gases, we would have

$$\mu_j(T, P) = \mu_j^\circ(T) + RT \ln f_j \tag{26.55}$$

Because the molecules in a mixture of gases in which the gases do not behave ideally are not independent of one another, the partial fugacity of each gas generally depends upon the concentrations of all the other gases in the mixture.

Now let's consider the general gas-phase reaction

$$\nu_A A(g) + \nu_B B(g) \rightleftharpoons \nu_Y Y(g) + \nu_Z Z(g)$$

The change in Gibbs energy upon converting the reactants at arbitrary partial pressures to products at arbitrary partial pressures is

$$\Delta_r G = \nu_Y \mu_Y + \nu_Z \mu_Z - \nu_A \mu_A - \nu_B \mu_B$$

If we substitute Equation 26.55 into this equation, then we get

$$\Delta_r G = \Delta_r G^\circ + RT \ln \frac{f_Y^{\nu_Y} f_Z^{\nu_Z}}{f_A^{\nu_A} f_B^{\nu_B}} \tag{26.56}$$

where

$$\Delta_r G^\circ = \nu_Y \mu_Y^\circ + \nu_Z \mu_Z^\circ - \nu_A \mu_A^\circ - \nu_B \mu_B^\circ$$

Note that Equation 26.56 is the generalization of Equation 26.24 to a system of nonideal gases. Realize that the values of the fugacities at this point are arbitrary, and not necessarily equilibrium values. If the reaction system is in equilibrium, then $\Delta_r G = 0$ and all the fugacities take on their equilibrium values. Equation 26.56 becomes

$$\Delta_r G^\circ(T) = -RT \ln K_f \tag{26.57}$$

where the equilibrium constant K_f is given by

$$K_f(T) = \left(\frac{f_Y^{\nu_Y} f_Z^{\nu_Z}}{f_A^{\nu_A} f_B^{\nu_B}} \right)_{eq} \tag{26.58}$$

Once again notice that the equilibrium constant is a function of temperature only, as dictated by Equation 26.57.

The equilibrium constant defined by Equation 26.57 is called a *thermodynamic equilibrium constant*. Equation 26.57, which relates K_f to $\Delta_r G°$ is exact, being valid for real gases as well as ideal gases. At low pressures we can replace the partial fugacities by partial pressures to obtain K_p, but we should expect this approximation to fail at high pressures. The formulas to calculate partial fugacities from equation-of-state data are extensions of the formulas in Section 22–8 where we calculated fugacities for pure gases. In order to obtain the partial fugacities to use in Equation 26.58 we need rather extensive pressure-volume data for the mixture of reacting gases. These data are available for the important industrial reaction

$$\tfrac{1}{2}N_2(g) + \tfrac{3}{2}H_2(g) \rightleftharpoons NH_3(g)$$

Table 26.5 shows both K_p and K_f as a function of the total pressure of the reaction mixture. Note that K_p is not a constant, but that K_f is fairly constant with increasing total pressure. The results shown in Table 26.5 emphasize that we must use fugacities and not pressures when dealing with systems at high pressures.

TABLE 26.5
Values of K_p and K_f as a function of total pressure for the ammonia synthesis equilibrium at 450°C.

total pressure/bar	$K_p/10^{-3}$	$K_f/10^{-3}$
10	6.59	6.55
30	6.76	6.59
50	6.90	6.50
100	7.25	6.36
300	8.84	6.08
600	12.94	6.42

EXAMPLE 26–11
The equilibrium constants K_p and K_f can be related by a quantity K_γ, such that $K_f = K_\gamma K_p$ and K_γ has the form of an equilibrium constant, but involving activity coefficients, γ_j. First derive an expression for K_γ and then evaluate it at the various pressures given in Table 26.5.

SOLUTION: The relation between pressure and fugacity is given by

$$f_j = \gamma_j P_j$$

If we substitute this expression into Equation 26.58, then we obtain

$$K_f = \frac{(\gamma_Y^{\nu_Y} P_Y^{\nu_Y})(\gamma_Z^{\nu_Z} P_Z^{\nu_Z})}{(\gamma_A^{\nu_A} P_A^{\nu_A})(\gamma_B^{\nu_B} P_B^{\nu_B})}$$

$$= \left(\frac{\gamma_Y^{\nu_Y} \gamma_Z^{\nu_Z}}{\gamma_A^{\nu_A} \gamma_B^{\nu_B}}\right) \cdot \left(\frac{P_Y^{\nu_Y} P_Z^{\nu_Z}}{P_A^{\nu_A} P_B^{\nu_B}}\right) = K_\gamma \cdot K_P$$

where we have used the standard state $f^\circ = P^\circ = 1$ bar. Using the data in Table 26.5, we see that

P/bar	10	30	50	100	300	600
K_γ	0.994	0.975	0.942	0.877	0.688	0.496

The deviation of K_γ from unity is a measure of the nonideality of the system.

26–11. Thermodynamic Equilibrium Constants Are Expressed in Terms of Activities

In the previous section we discussed the condition of equilibrium for a reaction system consisting of real gases. The central result was the introduction of K_f, in which the equilibrium constant is expressed in terms of partial fugacites. In this section we shall derive a similar expression for general equilibrium systems, consisting of gases, solids, liquids, and/or solutions. The starting point is Equation 24.35, which we write as

$$\mu_j = \mu_j^\circ(T) + RT \ln a_j \tag{26.59}$$

where a_j is the activity of species j and μ_j° is the chemical potential of the standard state. This equation essentially defines the activity, a_j. Recall that we discussed two different standard states in Chapters 24 and 25: a Raoult's law standard state, in which $a_j \to x_j$ as $x_j \to 1$, in which case $\mu_j^\circ = \mu_j^*$, the chemical potential of pure component j; and a Henry's law standard state, in which $a_j \to m_j$ or $a_j \to c_j$ as $m_j \to 0$ or $c_j \to 0$, in which case μ_j° is the chemical potential of the (hypothetical) corresponding ideal solution at unit molality or unit molarity. Although Equation 26.55 is restricted to gases, Equation 26.59 is general. In fact, we can include Equation 26.55 as a special case of Equation 26.59 by defining the activity of a gas by the relation $a_j = f_j/f_j^\circ$. In this case, $\mu_j^\circ(T)$ in Equation 26.59 is the corresponding (hypothetical) ideal gas at one bar and at the temperature of interest. Agreeing to set $a_j = f_j/f_j^\circ$ simply allows us to treat gases, liquids, solids, (and solutions) in the same notation.

Now let's consider the general reaction

$$\nu_A A + \nu_B B \rightleftharpoons \nu_Y Y + \nu_Z Z$$

The change in Gibbs energy for converting A and B in arbitrary states to Y and Z in arbitrary states is given by

$$\Delta_r G = \nu_Y \mu_Y + \nu_Z \mu_Z - \nu_A \mu_A - \nu_B \mu_B$$

If we substitute Equation 26.59 into this equation, then we obtain

$$\Delta_r G = \Delta_r G^\circ + RT \ln \frac{a_Y^{\nu_Y} a_Z^{\nu_Z}}{a_A^{\nu_A} a_B^{\nu_B}} \tag{26.60}$$

where

$$\Delta_r G^\circ = \nu_Y \mu_Y^\circ + \nu_Z \mu_Z^\circ - \nu_A \mu_A^\circ - \nu_B \mu_B^\circ$$

Equation 26.60 is called the *Lewis equation*, after the great thermodynamicist G. N. Lewis, who first introduced the concept of activity and pioneered the rigorous thermodynamic analysis of chemical equilibria. Note that Equation 26.60 is a generalization of Equation 26.56 to a non-ideal system, which may consist of condensed phases and solutions as well as gases. Realize that the activities at this point are arbitrary, and not necessarily the equilibrium activites. Just as we did in Section 26–5 for the case of a reaction system of ideal gases, we introduce a reaction quotient, or an *activity quotient*, in this case, by

$$Q_a = \frac{a_Y^{\nu_Y} a_Z^{\nu_Z}}{a_A^{\nu_A} a_B^{\nu_B}} \tag{26.61}$$

Using this notation, we can write Equation 26.60 as

$$\Delta_r G = \Delta_r G^\circ + RT \ln Q_a \tag{26.62}$$

According to Equation 26.59, $a_j = 1$ when a substance is in its standard state. Therefore, if all the reactants and products in a reaction mixture are in their standard states, then all the $a_j = 1$ in Equation 26.61 and so $Q_a = 1$, giving $\Delta_r G = \Delta_r G^\circ$. If the reaction system is at equilibrium at fixed T and P, then $\Delta_r G = 0$, and we have

$$\Delta_r G^\circ = -RT \ln Q_{a,eq} \tag{26.63}$$

where $Q_{a,eq}$ denotes Q_a in which all the activities have their equilibrium values. In analogy with Section 26–5, we denote $Q_{a,eq}$ by K_a

$$K_a(T) = \left(\frac{a_Y^{\nu_Y} a_Z^{\nu_Z}}{a_A^{\nu_A} a_B^{\nu_B}} \right)_{eq} \tag{26.64}$$

which we call a *thermodynamic equilibrium constant*. Equation 26.57 becomes

$$\Delta_r G^\circ = -RT \ln K_a \tag{26.65}$$

Equation 26.65 is completely general and rigorous, and applies to any system in equilibrium. Note that for a reaction involving only gases, $a_i = f_i$, and $K_a(T) = K_f(T)$, Equation 26.58, and Equation 26.65 is equivalent to Equation 26.57. Equations 26.64 and 26.65 are more general than Equations 26.57 and 26.58 because the reactants can be in any phase. The application of this equation is best done by example.

Let's consider a heterogeneous system such as the water-gas reaction

$$C(s) + H_2O(g) \overset{1000^\circ C}{\rightleftharpoons} CO(g) + H_2(g)$$

which is used in the industrial production of hydrogen. The (thermodynamic) equilibrium constant for this equation is

$$K_a = \frac{a_{CO(g)} a_{H_2(g)}}{a_{C(s)} a_{H_2O(g)}} = \frac{f_{CO(g)} f_{H_2(g)}}{a_{C(s)} f_{H_2O(g)}}$$

Although we have dealt with fugacities of gases earlier, we have not dealt with activities of pure solids and liquids. We must first choose a standard state for a pure condensed phase, which we choose to be the pure substance in its normal state at one bar and at the temperature of interest. To calculate the activity, we start with

$$\left(\frac{\partial \mu}{\partial P}\right)_T = \overline{V} \tag{26.66}$$

and the constant-temperature derivative of Equation 26.59

$$d\mu = RT \, d \ln a \qquad \text{(constant } T) \tag{26.67}$$

If we write Equation 26.66 as

$$d\mu = \overline{V} dP \qquad \text{(constant } T)$$

and introduce Equation 26.67, then we have

$$d \ln a = \frac{\overline{V}}{RT} dP \qquad \text{(constant } T)$$

We now integrate from the chosen standard state ($a = 1$, $P = 1$ bar) to an arbitrary state to obtain

$$\int_{a=1}^{a} d \ln a' = \int_{1}^{P} \frac{\overline{V}}{RT} dP' \qquad \text{(constant } T)$$

or

$$\ln a = \frac{1}{RT} \int_1^P \overline{V} dP' \qquad \text{(constant } T\text{)} \qquad (26.68)$$

For a condensed phase, $\overline{V}$ is essentially a constant over a moderate pressure range, and so Equation 26.68 becomes

$$\ln a = \frac{\overline{V}}{RT}(P - 1) \qquad (26.69)$$

EXAMPLE 26–12

Calculate the activity of $C(s)$ in the form of coke at 100 bar and 1000°C.

SOLUTION: The density of coke at 1000°C is about 1.5 g·cm^{-3}, and so its molar volume, $\overline{V}$, is 8.0 cm^3·mol^{-1}. From Equation 26.69

$$\ln a = \frac{(8.0 \text{ cm}^3 \cdot \text{mol}^{-1})(1 \text{ dm}^3/1000 \text{ cm}^3)(99 \text{ bar})}{(0.08206 \text{ dm}^3 \cdot \text{bar} \cdot \text{K}^{-1} \cdot \text{mol}^{-1})(1273 \text{ K})} = 0.0076$$

or $a = 1.01$. Note that the activity is essentially unity even at 100 bar.

According to Example 26–12, the activity of a pure condensed phase is unity at moderate pressures. Consequently, the activities of pure solids and liquids are normally not included in equilibrium constant expressions (as you may recall from general chemistry). For example, for the reaction

$$C(s) + H_2O(g) \rightleftharpoons CO(g) + H_2(g)$$

the equilibrium constant is given by

$$K = \frac{f_{CO(g)} f_{H_2(g)}}{f_{H_2O(g)}} \approx \frac{P_{CO(g)} P_{H_2(g)}}{P_{H_2O(g)}}$$

if the pressures are low enough. However, there are cases where the activities cannot be set to unity, as the following Example shows.

EXAMPLE 26–13

The change in the standard molar Gibbs energy for the conversion of graphite into diamond is 2.900 kJ·mol^{-1} at 298.15 K. The density of graphite is 2.27 g·cm^{-3} and that of diamond is 3.52 g·cm^{-3} at 298.15 K. At what pressure will these two forms of carbon be at equilibrium at 298.15 K?

SOLUTION: We can represent the process by the chemical equation

$$C(\text{graphite}) \rightleftharpoons C(\text{diamond})$$

for which

$$\Delta_r G^\circ = -RT \ln K_a = -RT \ln \frac{a_{\text{diamond}}}{a_{\text{graphite}}}$$

Using Equation 26.69, we have

$$\Delta_r G^\circ = -RT \left[\frac{\Delta \overline{V}}{RT} (P - 1) \right]$$

or

$$\frac{2900 \text{ J} \cdot \text{mol}^{-1}}{(8.3145 \text{ J} \cdot \text{mol}^{-1} \cdot \text{K}^{-1})(298.15 \text{ K})} =$$

$$-\frac{(3.41 \text{ cm}^3 \cdot \text{mol}^{-1} - 5.29 \text{ cm}^3 \cdot \text{mol}^{-1})(1 \text{ dm}^3 / 1000 \text{ cm}^3)(P - 1) \text{ bar}}{(0.083145 \text{ dm}^3 \cdot \text{bar} \cdot \text{mol}^{-1} \cdot \text{K}^{-1})(298.15 \text{ K})}$$

Solving the expression for P gives

$$P = 1.54 \times 10^4 \text{ bar} \approx 15\,000 \text{ bar}$$

26–12. The Use of Activities Makes a Significant Difference in Solubility Calculations Involving Ionic Species

Equation 26.65 can also be applied to reactions that take place in solution. For example, let's consider the dissociation of an aqueous solution that is 0.100 molar in acetic acid, $CH_3COOH(aq)$, for which $K = 1.74 \times 10^{-5}$ on a molarity scale. The equation for the reaction is

$$CH_3COOH(aq) + H_2O(l) \rightleftharpoons H_3O^+(aq) + CH_3COO^-(aq)$$

and the equilibirum-constant expression is

$$K_a = \frac{a_{H_3O^+} a_{CH_3COO^-}}{a_{CH_3COOH} a_{H_2O}} = \frac{a_{H_3O^+} a_{CH_3COO^-}}{a_{CH_3COOH}} = 1.74 \times 10^{-5} \qquad (26.70)$$

Being a neutral species at a concentration of around 0.100 molar, the undissociated acetic acid has an activity coefficient of essentially unity and so $a_{\text{HAc}} = c_{\text{HAc}}$. For the ions, we use the fact that (Table 25.3)

$$a_{H^+} a_{CH_3COO^-} = c_{H^+} c_{Ac^-} \gamma_\pm^2$$

and so Equation 26.70 becomes

$$\frac{c_{H_3O^+} c_{Ac^-}}{c_{\text{HAc}}} = \frac{1.74 \times 10^{-5}}{\gamma_\pm^2} \qquad (26.71)$$

As a first approximation, we shall set all the activity coefficients equal to unity and write

$$K_c = \frac{c_{H_3O^+} c_{Ac^-}}{c_{HAc}} = 1.74 \times 10^{-5} \text{ mol·L}^{-1}$$

From the following set-up

	$CH_3COOH(aq)$	$+$	$H_2O(l)$	$\rightleftharpoons$	$H_3O^+(aq)$	$+$	$CH_3COO^-(aq)$
initial	0.100 mol·L^{-1}		—		≈ 0		0
equilibrium	$0.100 \text{ mol·L}^{-1} - x$		—		x		x

we get

$$\frac{x^2}{0.100 \text{ mol·L}^{-1} - x} = 1.74 \times 10^{-5} \text{ mol·L}^{-1}$$

or $x = 1.31 \times 10^{-3}$ mol·L^{-1}, for a pH of 2.88. This is the type of calculation that is done in general chemistry.

Now let's not set $\gamma_\pm$ equal to unity. For $\gamma_\pm$, we shall use Equation 25.57

$$\ln \gamma_\pm = -\frac{1.173|z_+ z_-|(I_c/\text{mol·L}^{-1})^{1/2}}{1 + (I_c/\text{mol·L}^{-1})^{1/2}}$$

where the ionic strength I_c is given by

$$I_c = \frac{1}{2}(c_{H^+} + c_{Ac^-}) = c_{H^+} = c_{Ac^-}$$

In order to calculate I_c we must know c_{H^+} or c_{Ac^-}, but we cannot determine either of these from Equation 26.71 because it contains $\gamma_\pm^2$. We can solve this problem by iteration, however. We first calculate $\gamma_\pm$ using the values of c_{H^+} and c_{Ac^-} that we obtained above by letting $\gamma_\pm = 1$:

$$\ln \gamma_\pm = -\frac{1.173(1.31 \times 10^{-3})^{1/2}}{1 + (1.31 \times 10^{-3})^{1/2}} = -0.0410$$

or $\gamma_\pm^2 = 0.921$. We now use this value in the right-hand side of Equation 26.71, and write

$$\frac{x^2}{0.100 \text{ mol·L}^{-1} - x} = \frac{1.74 \times 10^{-5} \text{ mol·L}^{-1}}{0.921}$$

Solving for x, we find that $x = 1.365 \times 10^{-3}$ mol·L^{-1}. We now use this value to calculate a new value of $\gamma_\pm^2$ ($= 0.920$), and use this value in Equation 26.71 to calculate a new value of x ($= 1.366 \times 10^{-3}$ mol·L^{-1}). Cycling through once more gives $\gamma_\pm^2 = 0.920$ and $x = 1.366 \times 10^{-3}$ mol·L^{-1}, and so we find that $x = 1.37 \times 10^{-3}$ mol·L^{-1} (to three significant figures) and pH $= 2.86$. Thus we see that we calculate a pH of

2.86 using activities and a pH of 2.88 ignoring activities, not a significant difference. Fortunately the myriad of pH calculations that you did in general chemistry were sufficiently accurate. This is not necessarily the case for solubility calculations, as we shall now see.

The solubility product, K_{sp}, of $BaF_2(s)$ in water at 25°C is 1.7×10^{-6}, and the associated chemical equation is

$$BaF_2(s) \rightleftharpoons Ba^{2+}(aq) + 2\,F^-(aq)$$

The equilibrium constant expression is

$$a_{Ba^{2+}}a_{F^-}^2 = K_{sp} = 1.7 \times 10^{-6}$$

Using the formula (Table 25.3)

$$a_{Ba^{2+}}a_{F^-}^2 = c_{Ba^{2+}}c_{F^-}^2\gamma_{\pm}^3$$

we have

$$c_{Ba^{2+}}c_{F^-}^2 = \frac{1.7 \times 10^{-6}}{\gamma_{\pm}^3} \qquad (26.72)$$

If we set $\gamma_{\pm} = 1$, and let s be the solubility of $BaF_2(s)$, then $c_{Ba^{2+}} = s$ and $c_{F^-} = 2s$, and we have

$$(s)(2s)^2 = 1.7 \times 10^{-6}\ mol^3 \cdot L^{-3}$$

or $s = (1.7 \times 10^{-6}\ mol^3 \cdot L^{-3}/4)^{1/3} = 7.52 \times 10^{-3}\ mol \cdot L^{-1}$. We now calculate the ionic strength using this value of s to obtain

$$I_c = \frac{1}{2}(4s + 2s) = 3s = 0.0226\ mol \cdot L^{-1}$$

Using this value of I_c in Equation 25.57 gives $\gamma_{\pm} = 0.736$. Substitute this value into Equation 26.55 to get

$$4s^3 = \frac{1.7 \times 10^{-6}\ mol^3 \cdot L^{-3}}{0.399}$$

and so $s = 0.0102\ mol \cdot L^{-1}$. Cycling through again gives $\gamma_{\pm} = 0.705$ and $s = 0.0107$ $mol \cdot L^{-1}$. Once more gives $\gamma_{\pm} = 0.700$ and $s = 0.0107\ mol \cdot L^{-1}$ and one last iteration gives $\gamma_{\pm} = 0.700$ and $s = 0.011\ mol \cdot L^{-1}$ to two significant figures. Notice that in this case there is over a 30% difference between calculating s with and without the inclusion of activity coefficients.

EXAMPLE 26–14
Calculate the solubility of $TlBrO_3(s)$ in pure water and in an aqueous solution that is $0.500\ mol \cdot L^{-1}$ in $KNO_3(aq)$. $K_{sp} = 1.72 \times 10^{-4}$ for $TlBrO_3(s)$.

SOLUTION: The equation for the dissolution of $TlBrO_3(s)$ is

$$TlBrO_3(s) \rightleftharpoons Tl^+(aq) + BrO_3^-(aq)$$

with

$$a_{Tl^+}a_{BrO_3^-} = c_{Tl^+}c_{BrO_3^-}\gamma_\pm^2 = s^2\gamma_\pm^2 = 1.72 \times 10^{-4}$$

Letting $\gamma_\pm = 1$ at first, we find that $s = 0.0131 \text{ mol·L}^{-1}$. Using this value of s, we get $I_c = s$ and $\gamma_\pm = 0.887$ for $TlBrO_3(s)$ in pure water. Using this value of $\gamma_\pm$ in the K_{sp} expression gives $s = 0.0148 \text{ mol·L}^{-1}$. Subsequent iterations give $s = 0.0149 \text{ mol·L}^{-1}$.
For the case with 0.500 mol·L^{-1} $KNO_3(aq)$, we write

$$I_c = \frac{1}{2}(s + s + 0.500 \text{ mol·L}^{-1} + 0.500 \text{ mol·L}^{-1}) = s + 0.500 \text{ mol·L}^{-1}$$

Because s is much less than 0.500 mol·L^{-1}, we intially let $I_c = 0.500 \text{ mol·L}^{-1}$, which gives $\gamma_\pm = 0.616$. Using this value in the solubility product expression gives $s = 0.0213 \text{ mol·L}^{-1}$. Now $I_c = 0.5213 \text{ mol·L}^{-1}$ and $\gamma_\pm = 0.612$ and $s = 0.0214 \text{ mol·L}^{-1}$. Subsequent iterations give $s = 0.0214 \text{ mol·L}^{-1}$. Notice that the solubility of $TlBrO_3(s)$ is significantly enhanced in the 0.500 molar $KNO_3(aq)$ even though the $KNO_3(aq)$ does not participate in the dissolution reaction. If we had not included the activity coefficients, we would have gotten no effect at all.

Problems

26-1. Express the concentrations of each species in the following chemical equations in terms of the extent of reaction, ξ. The initial conditions are given under each equation.

a.
$$SO_2Cl_2(g) \rightleftharpoons SO_2(g) + Cl_2(g)$$
(1) n_0 0   0
(2) n_0 n_1 0

b.
$$2 SO_3(g) \rightleftharpoons 2 SO_2(g) + O_2(g)$$
(1) n_0 0   0
(2) n_0 0 n_1

c.
$$N_2(g) + 2 O_2(g) \rightleftharpoons N_2O_4(g)$$
(1) n_0 $2n_0$ 0
(2) n_0 n_0 0

26-2. Write out the equilibrium-constant expression for the reaction that is described by the equation

$$2 SO_2(g) + O_2(g) \rightleftharpoons 2 SO_3(g)$$

Compare your result to what you get if the reaction is represented by

$$SO_2(g) + \tfrac{1}{2} O_2(g) \rightleftharpoons SO_3(g)$$

26-3. Consider the dissociation of $N_2O_4(g)$ into $NO_2(g)$ described by

$$N_2O_4(g) \rightleftharpoons 2\,NO_2(g)$$

Assuming that we start with n_0 moles of $N_2O_4(g)$ and no $NO_2(g)$, show that the extent of reaction, ξ_{eq}, at equilibrium is given by

$$\frac{\xi_{eq}}{n_0} = \left(\frac{K_P}{K_P + 4P}\right)^{1/2}$$

Plot ξ_{eq}/n_0 against P given that $K_P = 6.1$ at $100°C$. Is your result in accord with Le Châtelier's principle?

26-4. In Problem 26–3 you plotted the extent of reaction at equilibrium against the total pressure for the dissociation of $N_2O_4(g)$ to $NO_2(g)$. You found that ξ_{eq} decreases as P increases, in accord with Le Châtelier's principle. Now let's introduce n_{inert} moles of an inert gas into the system. Assuming that we start with n_0 moles of $N_2O_4(g)$ and no $NO_2(g)$, derive an expression for ξ_{eq}/n_0 in terms of P and the ratio $r = n_{inert}/n_0$. As in Problem 26–3, let $K_P = 6.1$ and plot ξ_{eq}/n_0 versus P for $r = 0$, $r = 0.50$, $r = 1.0$, and $r = 2.0$. Show that introducing an inert gas into the reaction mixture at constant pressure has the same effect as lowering the pressure. What is the effect of introducing an inert gas into a reaction system at constant volume?

26-5. Re-do Problem 26–3 with n_0 moles of $N_2O_4(g)$ and n_1 moles of $NO_2(g)$ initially. Let $n_1/n_0 = 0.50$ and 2.0.

26-6. Consider the ammonia-synthesis reaction, which can be described by

$$N_2(g) + 3\,H_2(g) \rightleftharpoons 2\,NH_3(g)$$

Suppose initially there are n_0 moles of $N_2(g)$ and $3n_0$ moles of $H_2(g)$ and no $NH_3(g)$. Derive an expression for $K_P(T)$ in terms of the equilibrium value of the extent of reaction, ξ_{eq}, and the pressure, P. Use this expression to discuss how ξ_{eq}/n_0 varies with P and relate your conclusions to Le Châtelier's principle.

26-7. Nitrosyl chloride, NOCl, decomposes according to

$$2\,NOCl(g) \rightleftharpoons 2\,NO(g) + Cl_2(g)$$

Assuming that we start with n_0 moles of $NOCl(g)$ and no $NO(g)$ or $Cl_2(g)$, derive an expression for K_P in terms of the equilibrium value of the extent of reaction, ξ_{eq}, and the pressure, P. Given that $K_P = 2.00 \times 10^{-4}$, calculate ξ_{eq}/n_0 when $P = 0.080$ bar. What is the new value of ξ_{eq}/n_0 at equilibrium when $P = 0.160$ bar? Is this result in accord with Le Châtelier's principle?

26-8. The value of K_P at $1000°C$ for the decomposition of carbonyl dichloride (phosgene) according to

$$COCl_2(g) \rightleftharpoons CO(g) + Cl_2(g)$$

is 34.8 if the standard state is taken to be one bar. What would the value of K_p be if for some reason the standard state were taken to be 0.500 bar? What does this result say about the numerical values of equilibrium constants?

26-9. Most gas-phase equilibrium constants in the recent chemical literature were calculated assuming a standard state pressure of one atmosphere. Show that the corresponding equilibrium constant for a standard state pressure of one bar is given by

$$K_p(\text{bar}) = K_p(\text{atm})(1.01325)^{\Delta \nu}$$

where $\Delta \nu$ is the sum of the stoichiometric coefficients of the products minus that of the reactants.

26-10. Using the data in Table 26.1, calculate $\Delta_r G^\circ(T)$ and $K_p(T)$ at 25°C for

(a) $N_2O_4(g) \rightleftharpoons 2\,NO_2(g)$
(b) $H_2(g) + I_2(g) \rightleftharpoons 2\,HI(g)$
(c) $3\,H_2(g) + N_2(g) \rightleftharpoons 2\,NH_3(g)$

26-11. Calculate the value of $K_c(T)$ based upon a one $\text{mol} \cdot \text{L}^{-1}$ standard state for each of the equations in Problem 26–10.

26-12. Derive a relation between K_p and K_c for the following:

(a) $CO(g) + Cl_2(g) \rightleftharpoons COCl_2(g)$
(b) $CO(g) + 3\,H_2(g) \rightleftharpoons CH_4(g) + H_2O(g)$
(c) $2\,BrCl(g) \rightleftharpoons Br_2(g) + Cl_2(g)$

26-13. Consider the dissociation reaction of $I_2(g)$ described by

$$I_2(g) \rightleftharpoons 2\,I(g)$$

The total pressure and the partial pressure of $I_2(g)$ at 1400°C have been measured to be 36.0 torr and 28.1 torr, respectively. Use these data to calculate K_p (one bar standard state) and K_c (one $\text{mol} \cdot \text{L}^{-1}$ standard state) at 1400°C.

26-14. Show that

$$\frac{d \ln K_c}{dT} = \frac{\Delta_r U^\circ}{RT^2}$$

for a reaction involving ideal gases.

26-15. Consider the gas-phase reaction for the synthesis of methanol from $CO(g)$ and $H_2(g)$

$$CO(g) + 2\,H_2(g) \rightleftharpoons CH_3OH(g)$$

The value of the equilibrium constant K_p at 500 K is 6.23×10^{-3}. Initially equimolar amounts of $CO(g)$ and $H_2(g)$ are introduced into the reaction vessel. Determine the value of ξ_{eq}/n_0 at equilibrium at 500 K and 30 bar.

26-16. Consider the two equations

(1) $CO(g) + H_2O(g) \rightleftharpoons CO_2(g) + H_2(g)$ K_1

(2) $CH_4(g) + H_2O(g) \rightleftharpoons CO(g) + 3H_2(g)$ K_2

Show that $K_3 = K_1 K_2$ for the sum of these two equations

(3) $CH_4(g) + 2H_2O(g) \rightleftharpoons CO_2(g) + 4H_2(g)$ K_3

How do you explain the fact that you would add the values of $\Delta_r G^\circ$ but multiply the equilibrium constants when adding Equations 1 and 2 to get Equation 3.

26-17. Given:

$$2\,BrCl(g) \rightleftharpoons Cl_2(g) + Br_2(g) \qquad K_P = 0.169$$
$$2\,IBr(g) \rightleftharpoons Br_2(g) + I_2(g) \qquad K_P = 0.0149$$

Determine K_P for the reaction

$$BrCl(g) + \tfrac{1}{2}I_2(g) \rightleftharpoons IBr(g) + \tfrac{1}{2}Cl_2(g)$$

26-18. Consider the reaction described by

$$Cl_2(g) + Br_2(g) \rightleftharpoons 2\,BrCl(g)$$

at 500 K and a total pressure of one bar. Suppose that we start with one mole each of $Cl_2(g)$ and $Br_2(g)$ and no $BrCl(g)$. Show that

$$G(\xi) = (1 - \xi)G^\circ_{Cl_2} + (1 - \xi)G^\circ_{Br_2} + 2\xi G^\circ_{BrCl} + 2(1 - \xi)RT \ln \frac{1 - \xi}{2} + 2\xi RT \ln \xi$$

where ξ is the extent of reaction. Given that $G^\circ_{BrCl} = -3.694$ kJ·mol^{-1} at 500 K, plot $G(\xi)$ versus ξ. Differentiate $G(\xi)$ with respect to ξ and show that the minimum value of $G(\xi)$ occurs at $\xi_{eq} = 0.549$. Also show that

$$\left(\frac{\partial G}{\partial \xi}\right)_{T,P} = \Delta_r G^\circ + RT \ln \frac{P^2_{BrCl}}{P_{Cl_2} P_{Br_2}}$$

and that $K_P = 4\xi^2_{eq}/(1 - \xi_{eq})^2 = 5.9$.

26-19. Consider the reaction described by

$$2\,H_2O(g) \rightleftharpoons 2\,H_2(g) + O_2(g)$$

at 4000 K and a total pressure of one bar. Suppose that we start with two moles of $H_2O(g)$ and no $H_2(g)$ or $O_2(g)$. Show that

$$G(\xi) = 2(1 - \xi)G^\circ_{H_2O} + 2\xi G^\circ_{H_2} + \xi G^\circ_{O_2} + 2(1 - \xi)RT \ln \frac{2(1 - \xi)}{2 + \xi}$$

$$+ 2\xi RT \ln \frac{2\xi}{2 + \xi} + \xi RT \ln \frac{\xi}{2 + \xi}$$

where ξ is the extent of reaction. Given that $\Delta_f G°[H_2O(g)] = -18.334 \, kJ \cdot mol^{-1}$ at 4000 K, plot $G(\xi)$ against ξ. Differentiate $G(\xi)$ with respect to ξ and show that the minimum value of $G(\xi)$ occurs at $\xi_{eq} = 0.553$. Also show that

$$\left(\frac{\partial G}{\partial \xi}\right)_{T,P} = \Delta_r G° + RT \ln \frac{P_{H_2}^2 P_{O_2}}{P_{H_2O}^2}$$

and that $K_P = \xi_{eq}^3/(2 + \xi_{eq})(1 - \xi_{eq})^2 = 0.333$.

26-20. Consider the reaction described by

$$3 \, H_2(g) + N_2(g) \rightleftharpoons 2 \, NH_3(g)$$

at 500 K and a total pressure of one bar. Suppose that we start with three moles of $H_2(g)$, one mole of $N_2(g)$, and no $NH_3(g)$. Show that

$$G(\xi) = (3 - 3\xi)G_{H_2}° + (1 - \xi)G_{N_2}° + 2\xi G_{NH_3}°$$

$$+(3 - 3\xi)RT \ln \frac{3 - 3\xi}{4 - 2\xi} + (1 - \xi)RT \ln \frac{1 - \xi}{4 - 2\xi} + 2\xi RT \ln \frac{2\xi}{4 - 2\xi}$$

where ξ is the extent of reaction. Given that $G_{NH_3}° = 4.800 \, kJ \cdot mol^{-1}$ at 500 K (see Table 26.4), plot $G(\xi)$ versus ξ. Differentiate $G(\xi)$ with respect to ξ and show that the minimum value of $G(\xi)$ occurs at $\xi_{eq} = 0.158$. Also show that

$$\left(\frac{\partial G}{\partial \xi}\right)_{T,P} = \Delta_r G° + RT \ln \frac{P_{NH_3}^2}{P_{H_2}^3 P_{N_2}}$$

and that $K_P = 16\xi_{eq}^2(2 - \xi_{eq})^2/27(1 - \xi_{eq})^4 = 0.10$.

26-21. Suppose that we have a mixture of the gases $H_2(g)$, $CO_2(g)$, $CO(g)$, and $H_2O(g)$ at 1260 K, with $P_{H_2} = 0.55$ bar, $P_{CO_2} = 0.20$ bar, $P_{CO} = 1.25$ bar, and $P_{H_2O} = 0.10$ bar. Is the reaction described by the equation

$$H_2(g) + CO_2(g) \rightleftharpoons CO(g) + H_2O(g) \qquad K_P = 1.59$$

at equilibrium under these conditions? If not, in what direction will the reaction proceed to attain equilibrium?

26-22. Given that $K_P = 2.21 \times 10^4$ at 25°C for the equation

$$2 \, H_2(g) + CO(g) \rightleftharpoons CH_3OH(g)$$

predict the direction in which a reaction mixture for which $P_{CH_3OH} = 10.0$ bar, $P_{H_2} = 0.10$ bar, and $P_{CO} = 0.0050$ bar proceeds to attain equilibrium.

26-23. The value of K_P for a gas-phase reaction doubles when the temperature is increased from 300 K to 400 K at a fixed pressure. What is the value of $\Delta_r H°$ for this reaction?

26-24. The value of $\Delta_r H°$ is 34.78 kJ·mol^{-1} at 1000 K for the reaction described by

$$H_2(g) + CO_2(g) \rightleftharpoons CO(g) + H_2O(g)$$

Given that the value of K_p is 0.236 at 800 K, estimate the value of K_p at 1200 K, assuming that $\Delta_r H°$ is independent of temperature.

26-25. The value of $\Delta_r H°$ is -12.93 kJ·mol^{-1} at 800 K for

$$H_2(g) + I_2(g) \rightleftharpoons 2\,HI(g)$$

Assuming that $\Delta_r H°$ is independent of temperature, calculate K_p at 700 K given that $K_P = 29.1$ at 1000 K.

26-26. The equilibrium constant for the reaction described by

$$2\,HBr(g) \rightleftharpoons H_2(g) + Br_2(g)$$

can be expressed by the empirical formula

$$\ln K = -6.375 + 0.6415\ln(T/K) - \frac{11790\text{ K}}{T}$$

Use this formula to determine $\Delta_r H°$ as a function of temperature. Calculate $\Delta_r H°$ at 25°C and compare your result to the one you obtain from Table 19.2.

26-27. Use the following data for the reaction described by

$$2\,HI(g) \rightleftharpoons H_2(g) + I_2(g)$$

to obtain $\Delta_r H°$ at 400°C.

T/K	500	600	700	800
$K_P/10^{-2}$	0.78	1.24	1.76	2.31

26-28. Consider the reaction described by

$$CO_2(g) + H_2(g) \rightleftharpoons CO(g) + H_2O(g)$$

The molar heat capacitites of $CO_2(g)$, $H_2(g)$, $CO(g)$, and $H_2O(g)$ can be expressed by

$$\overline{C}_p[CO_2(g)]/R = 3.127 + (5.231 \times 10^{-3}\text{ K}^{-1})T - (1.784 \times 10^{-6}\text{ K}^{-2})T^2$$

$$\overline{C}_p[H_2(g)]/R = 3.496 - (1.006 \times 10^{-4}\text{ K}^{-1})T + (2.419 \times 10^{-7}\text{ K}^{-2})T^2$$

$$\overline{C}_p[CO(g)]/R = 3.191 + (9.239 \times 10^{-4}\text{ K}^{-1})T - (1.41 \times 10^{-7}\text{ K}^{-2})T^2$$

$$\overline{C}_p[H_2O(g)]/R = 3.651 + (1.156 \times 10^{-3}\text{ K}^{-1})T + (1.424 \times 10^{-7}\text{ K}^{-2})T^2$$

over the temperature range 300 K to 1500 K. Given that

substance	$CO_2(g)$	$H_2(g)$	$CO(g)$	$H_2O(g)$
$\Delta_f H°/$kJ·mol^{-1}	-393.523	0	-110.516	-241.844

at 300 K and that $K_p = 0.695$ at 1000 K, derive a general expression for the variation of $K_p(T)$ with temperature in the form of Equation 26.34.

26-29. The temperature dependence of the equilibrium constant K_p for the reaction described by

$$2\,C_3H_6(g) \rightleftharpoons C_2H_4(g) + C_4H_8(g)$$

is given by the equation

$$\ln K_p(T) = -2.395 - \frac{2505\text{ K}}{T} + \frac{3.477 \times 10^6\text{ K}^2}{T^2} \qquad 300\text{ K} < T < 600\text{ K}$$

Calculate the values of $\Delta_r G^\circ$, $\Delta_r H^\circ$, and $\Delta_r S^\circ$ for this reaction at 525 K.

26-30. At 2000 K and one bar, water vapor is 0.53% dissociated. At 2100 K and one bar, it is 0.88% dissociated. Calculate the value of $\Delta_r H^\circ$ for the dissociation of water at one bar, assuming that the enthalpy of reaction is constant over the range from 2000 K to 2100 K.

26-31. The following table gives the standard molar Gibbs energy of formation of Cl(g) at three different temperatures.

T/K	1000	2000	3000
$\Delta_f G^\circ/\text{kJ}\cdot\text{mol}^{-1}$	65.288	5.081	−56.297

Use these data to determine the value of K_p at each temperature for the reaction described by

$$\tfrac{1}{2}\,Cl_2(g) \rightleftharpoons Cl(g)$$

Assuming that $\Delta_r H^\circ$ is temperature independent, determine the value of $\Delta_r H^\circ$ from these data. Combine your results to determine $\Delta_r S^\circ$ at each temperature. Interpret your results.

26-32. The following experimental data were determined for the reaction described by

$$SO_3(g) \rightleftharpoons SO_2(g) + \tfrac{1}{2}\,O_2(g)$$

T/K	800	825	900	953	1000
$\ln K_p$	−3.263	−3.007	−1.899	−1.173	−0.591

Calculate $\Delta_r G^\circ$, $\Delta_r H^\circ$, and $\Delta_r S^\circ$ for this reaction at 900 K. State any assumptions that you make.

26-33. Show that

$$\mu = -RT \ln \frac{q(V, T)}{N}$$

if

$$Q(N, V, T) = \frac{[q(V, T)]^N}{N!}$$

26-34. Use Equation 26.40 to calculate $K(T)$ at 750 K for the reaction described by $H_2(g) + I_2(g) \rightleftharpoons 2\,HI(g)$. Use the molecular parameters given in Table 18.2. Compare your value to the one given in Table 26.2 and the experimental value shown in Figure 26.5.

26-35. Use the statistical thermodynamic formulas of Section 26–8 to calculate $K_p(T)$ at 900 K, 1000 K, 1100 K, and 1200 K for the association of Na(g) to form dimers, $Na_2(g)$ according to the equation

$$2\,Na(g) \rightleftharpoons Na_2(g)$$

Use your result at 1000 K to calculate the fraction of sodium atoms that form dimers at a total pressure of one bar. The experimental values of $K_p(T)$ are

T/K	900	1000	1100	1200
K_p	1.32	0.47	0.21	0.10

Plot $\ln K_p$ against $1/T$ to determine the value of $\Delta_r H°$.

26-36. Using the data in Table 18.2, calculate K_p at 2000 K for the reaction described by the equation

$$CO_2(g) \rightleftharpoons CO(g) + \tfrac{1}{2}O_2(g)$$

The experimental value is 1.3×10^{-3}.

26-37. Using the data in Tables 18.2 and 18.4, calculate the equilibrium constant for the water gas reaction

$$CO_2(g) + H_2(g) \rightleftharpoons CO(g) + H_2O(g)$$

at 900 K and 1200 K. The experimental values at these two temperatures are 0.43 and 1.37, respectively.

26-38. Using the data in Tables 18.2 and 18.4, calculate the equilibrium constant for the reaction

$$3\,H_2(g) + N_2(g) \rightleftharpoons 2\,NH_3(g)$$

at 700 K. The accepted value is 8.75×10^{-5} (see Table 26.4).

26-39. Calculate the equilibrium constant K_p for the reaction

$$I_2(g) \rightleftharpoons 2\,I(g)$$

using the data in Table 18.2 and the fact that the ground electronic state of the iodine atom is $^2P_{3/2}$ and that the first excited electronic state $(^2P_{1/2})$ lies 7580 cm^{-1} higher. The experimental values of K_p are

T/K	800	900	1000	1100	1200
K_p	3.05×10^{-5}	3.94×10^{-4}	3.08×10^{-3}	1.66×10^{-2}	6.79×10^{-2}

Plot $\ln K_p$ against $1/T$ to determine the value of $\Delta_r H°$. The experimental value is 153.8 kJ·mol^{-1}.

26-40. Consider the reaction given by

$$H_2(g) + D_2(g) \rightleftharpoons 2\,HD(g)$$

Using the Born-Oppenheimer approximation and the molecular parameters in Table 18.2, show that

$$K(T) = 4.24e^{-77.7 \text{ K}/T}$$

Compare your predictions using this equation to the data in the JANAF tables.

26-41. Using the harmonic oscillator-rigid rotator approximation, show that

$$K(T) = \left(\frac{m_{H_2} m_{Br_2}}{m_{HBr}^2}\right)^{3/2} \left(\frac{\sigma_{HBr}^2}{\sigma_{H_2} \sigma_{Br_2}}\right) \left(\frac{(\Theta_{rot}^{HBr})^2}{\Theta_{rot}^{H_2} \Theta_{rot}^{Br_2}}\right)$$

$$\times \frac{(1 - e^{-\Theta_{vib}^{HBr}/T})^2}{(1 - e^{-\Theta_{vib}^{H_2}/T})(1 - e^{-\Theta_{vib}^{Br_2}/T})} e^{(D_0^{H_2}+D_0^{Br_2}-2D_0^{HBr})/RT}.$$

for the reaction described by

$$2\,HBr(g) \rightleftharpoons H_2(g) + Br_2(g)$$

Using the values of Θ_{rot}, Θ_{vib}, and D_0 given in Table 18.2, calculate K at 500 K, 1000 K, 1500 K, and 2000 K. Plot $\ln K$ against $1/T$ and determine the value of $\Delta_r H°$.

26-42. Use Equation 26.49b to calculate $H°(T) - H_0°$ for $NH_3(g)$ from 300 K to 6000 K and compare your values to those given in Table 26.4 by plotting them on the same graph.

26-43. Use the JANAF tables to calculate K_P at 1000 K for the reaction described by

$$H_2(g) + I_2(g) \rightleftharpoons 2\,HI(g)$$

Compare your results to the value given in Table 26.2.

26-44. Use the JANAF tables to plot $\ln K_P$ versus $1/T$ from 900 K to 1200 K for the reaction described by

$$2\,Na(g) \rightleftharpoons Na_2(g)$$

and compare your results to those obtained in Problem 26–35.

26-45. In Problem 26–36 we calculated K_P for the decomposition of $CO_2(g)$ to $CO(g)$ and $O_2(g)$ at 2000 K. Use the JANAF tables to calculate K_P and compare your result to the one that you obtained in Problem 26–36.

26-46. You calculated K_P at 700 K for the ammonia synthesis reaction in Problem 26–38. Use the data in Table 26.4 to calculate K_P and compare your result to the one that you obtained in Problem 26–38.

26-47. The JANAF tables give the following data for I(g) at one bar:

T/K	800	900	1000	1100	1200
$\Delta_f G°/\text{kJ·mol}^{-1}$	34.580	29.039	24.039	18.741	13.428

Calculate K_p for the reaction described by

$$I_2(g) \rightleftharpoons 2\,I(g)$$

and compare your results to the values given in Problem 26–39.

26-48. Use Equation 18.60 to calculate the value of $q^0(V, T)/V$ given in the text (page 1076) for $NH_3(g)$ at 500 K.

26-49. The JANAF tables give the following data for Ar(g) at 298.15 K and one bar:

$$-\frac{G^\circ - H^\circ(298.15\ \text{K})}{T} = 154.845\ \text{J}\cdot\text{mol}^{-1}\cdot\text{K}^{-1}$$

and

$$H^\circ(0\ \text{K}) - H^\circ(298.15\ \text{K}) = -6.197\ \text{kJ}\cdot\text{mol}^{-1}$$

Use these data to calculate $q^0(V, T)/V$ and compare your result to what you obtain using Equation 18.13.

26-50. Use the JANAF tables to calculate $q^0(V, T)/V$ for $CO_2(g)$ at 500 K and one bar and compare your result to what you obtain using Equation 18.57 (with the ground state energy taken to be zero).

26-51. Use the JANAF tables to calculate $q^0(V, T)/V$ for $CH_4(g)$ at 1000 K and one bar and compare your result to what you obtain using Equation 18.60 (with the ground state energy taken to be zero).

26-52. Use the JANAF tables to calculate $q^0(V, T)/V$ for $H_2O(g)$ at 1500 K and one bar and compare your result to what you obtain using Equation 26.45. Why do you think there is some discrepancy?

26-53. The JANAF tables give the following data:

	H(g)	Cl(g)	HCl(g)
$\Delta_f H^\circ(0\ \text{K})/\text{kJ}\cdot\text{mol}^{-1}$	216.035	119.621	−92.127

Use these data to calculate D_0 for HCl(g) and compare your value to the one in Table 18.2.

26-54. The JANAF tables give the following data:

	C(g)	H(g)	CH$_4$(g)
$\Delta_f H^\circ(0\ \text{K})/\text{kJ}\cdot\text{mol}^{-1}$	711.19	216.035	−66.911

Use these data to calculate D_0 for $CH_4(g)$ and compare your value to the one in Table 18.4.

26-55. Use the JANAF tables to calculate D_0 for $CO_2(g)$ and compare your result to the one given in Table 18.4.

26-56. A determination of K_γ (see Example 26–11) requires a knowledge of the fugacity of each gas in the equilibrium mixture. These data are not usually available, but a useful approximation is to take the fugacity coefficient of a gaseous constituent of a mixture to be equal to the value for the pure gas at the *total pressure of the mixture*. Using this approximation, we can use Figure 22.11 to determine γ for each gas and then calculate K_γ. In this problem we shall apply this approximation to the data in Table 26.5. First use Figure 22.11 to estimate that $\gamma_{H_2} = 1.05$, $\gamma_{N_2} = 1.05$, and that $\gamma_{NH_3} = 0.95$ at a total pressure of 100 bar and a temperature of 450°C. In this case $K_\gamma = 0.86$, in fairly good agreement with the value given in Example 26–11. Now calculate K_γ at 600 bar and compare your result with the value given in Example 26–11.

26-57. Recall from general chemistry that Le Châtelier's principle says that pressure has no effect on a gaseous equilibrium system such as

$$CO(g) + H_2O(g) \rightleftharpoons H_2(g) + CO_2(g)$$

in which the total number of moles of reactants is equal to the total number of moles of product in the chemical equation. The thermodynamic equilibrium constant in this case is

$$K_f = \frac{f_{CO_2} f_{H_2}}{f_{CO} f_{H_2O}} = \frac{\gamma_{CO_2} \gamma_{H_2}}{\gamma_{CO} \gamma_{H_2O}} \frac{P_{CO_2} P_{H_2}}{P_{CO} P_{H_2O}} = K_\gamma K_P$$

If the four gases behaved ideally, then pressure would have no effect on the position of equilibrium. However, because of deviations from ideal behavior, a shift in the equilibrium composition will occur when the pressure is changed. To see this, use the approximation introduced in Problem 26–56 to estimate K_γ at 900 K and 500 bar. Note that K_γ under these conditions is greater than K_γ at one bar, where $K_\gamma \approx 1$ (ideal behavior). Consequently, argue that an increase in pressure causes the equilibrium to shift to the left in this case.

26-58. Calculate the activity of $H_2O(l)$ as a function of pressure from one bar to 100 bar at 20.0°C. Take the density of $H_2O(l)$ to be 0.9982 g·mL^{-1} and assume that it is incompressible.

26-59. Consider the dissociation of HgO(s,red) to Hg(g) and $O_2(g)$ according to

$$HgO(s, red) \rightleftharpoons Hg(g) + \tfrac{1}{2} O_2(g)$$

If we start with only HgO(s,red), then assuming ideal behavior, show that

$$K_P = \frac{2}{3^{3/2}} P^{3/2}$$

where P is the total pressure. Given the following "dissociation pressure" of HgO(s,red) at various temperatures, plot $\ln K_P$ versus $1/T$.

$t/°C$	P/atm	$t/°C$	P/atm
360	0.1185	430	0.6550
370	0.1422	440	0.8450
380	0.1858	450	1.067
390	0.2370	460	1.339
400	0.3040	470	1.674
410	0.3990	480	2.081
420	0.5095		

An excellent curve fit to the plot of $\ln K_P$ against $1/T$ is given by

$$\ln K_P = -172.94 + \frac{4.0222 \times 10^5 \text{ K}}{T} - \frac{2.9839 \times 10^8 \text{ K}^2}{T^2} + \frac{7.0527 \times 10^{10} \text{ K}^3}{T^3}$$

$$630 \text{ K} < T < 750 \text{ K}$$

Use this expression to determine $\Delta_r H°$ as a function of temperature in the interval $630 \text{ K} < T < 750 \text{ K}$. Given that

$$C_P°[O_2(g)]/R = 4.8919 - \frac{829.931 \text{ K}}{T} - \frac{127962 \text{ K}^2}{T^2}$$

$$C_P°[Hg(g)]/R = 2.500$$

$$C_P°[HgO(s, \text{red})]/R = 5.2995$$

in the interval $298 \text{ K} < T < 750 \text{ K}$, calculate $\Delta_r H°$, $\Delta_r S°$, and $\Delta_r G°$ at 298 K.

26-60. Consider the dissociation of $Ag_2O(s)$ to $Ag(s)$ and $O_2(g)$ according to

$$Ag_2O(s) \rightleftharpoons 2 \, Ag(s) + \tfrac{1}{2} O_2(g)$$

Given the following "dissociation pressure" data:

$t/°C$	173	178	183	188
P/torr	422	509	605	717

Express K_P in terms of P (in torr) and plot $\ln K_P$ versus $1/T$. An excelllent curve fit to these data is given by

$$\ln K_P = 0.9692 + \frac{5612.7 \text{ K}}{T} - \frac{2.0953 \times 10^6 \text{ K}^2}{T^2}$$

Use this expression to derive an equation for $\Delta_r H°$ from $445 \text{ K} < T < 460 \text{ K}$. Now use the following heat capacity data:

$$C_P°[O_2(g)]/R = 3.27 + (5.03 \times 10^{-4} \text{ K}^{-1})T$$

$$C_P°[Ag(s)]/R = 2.82 + (7.55 \times 10^{-4} \text{ K}^{-1})T$$

$$C_P°[Ag_2O(s)]/R = 6.98 + (4.48 \times 10^{-3} \text{ K}^{-1})T$$

to calculate $\Delta_r H°$, $\Delta_r S°$, and $\Delta_r G°$ at 298 K.

26-61. Calcium carbonate occurs as two crystalline forms, calcite and aragonite. The value of $\Delta_r G°$ for the transition

$$CaCO_3(\text{calcite}) \rightleftharpoons CaCO_3(\text{aragonite})$$

is $+1.04 \text{ kJ·mol}^{-1}$ at 25°C. The density of calcite at 25°C is 2.710 g·cm^{-3} and that of aragonite is 2.930 g·cm^{-3}. At what pressure will these two forms of $CaCO_3$ be at equilbrium at 25°C.

26-62. The decomposition of ammonium carbamate, NH_2COONH_4 takes place according to

$$NH_2COONH_4(s) \rightleftharpoons 2\,NH_3(g) + CO_2(g)$$

Show that if all the $NH_3(g)$ and $CO_2(g)$ result from the decomposition of ammonium carbamate, then $K_P = (4/27)P^3$, where P is the total pressure at equilibrium.

26-63. Calculate the solubility of LiF(s) in water at 25°C. Compare your result to the one you obtain by using concentrations instead of activities. Take $K_{sp} = 1.7 \times 10^{-3}$.

26-64. Calculate the solubility of $CaF_2(s)$ in a solution that is 0.0150 molar in $MgSO_4(aq)$. Take $K_{sp} = 3.9 \times 10^{-11}$ for $CaF_2(s)$.

26-65. Calculate the solubility of $CaF_2(s)$ in a solution that is 0.050-molar in NaF(aq). Compare your result to the one you obtain by using concentrations instead of activities. Take $K_{sp} = 3.9 \times 10^{-11}$ for $CaF_2(s)$.

James Clerk Maxwell was born in Edinburgh, Scotland, on November 13, 1831, but was raised at the family estate in Glenlair, 30 miles south of Glasgow, and died there in 1879. Maxwell is one of the foremost scientists of modern times, making contributions to many fields of science. In 1873, in his *Treatise on Electricity and Magnetism*, he presented his theory of electricity and magnetism in mathematical form, which is succinctly summarized by the famous Maxwell's equations. Maxwell calculated that the speed of propagation of an electromagnetic field is the same as that of the speed of light, which lead him to postulate that light is an electromagnetic phenomenon. Maxwell also performed experiments on color vision and demonstrated the first color photograph to the Royal Society in 1861. He also applied the methods of probability to describe the properties of gases, and he was able to show that the velocities of the molecules of a gas follow what we now call a Maxwell-Boltzmann distribution. Maxwell also investigated, both theoretically and experimentally with his wife, the effect of temperature and pressure on the viscosity, thermal conductivity, and diffusion of gases. These experiments gave a means to estimate the values of the Avogadro constant and atomic properties such as size and mass. In 1871, Maxwell became the first Cavendish Professor of Physics at Cambridge University. He died of abdominal cancer, the same disease that caused the death of his mother at the same age.

CHAPTER 27

The Kinetic Theory of Gases

The fact that *all* gases obey the ideal gas equation when the pressure is sufficiently low implies that the form of the equation is independent of the nature of the gas itself. In this chapter we shall introduce a simple model of gases in which we picture the molecules of a gas to be in constant, incessant motion, colliding with each other and with the walls of the container. Because this model focuses on the motion of the molecules, it is called the *kinetic theory of gases*. For simplicity, we shall assume that the molecules behave as hard spheres, so that there are no interactions between the particles except for the very short durations of time when they collide with each other. In the first section we shall present a simplified treatment of the collisions of the individual molecules with the walls of their container and show how this leads to the ideal gas equation. Then we shall derive an expression for the distribution of the speeds of the molecules in a gas, the so-called Maxwell-Boltzmann distribution. Then we shall consider the collisions of the molecules with the walls in a somewhat more detailed treatment than in the first section, and derive an expression for the collision frequency of the molecules with the walls. Finally, we shall introduce the concept of the mean-free-path and derive expressions for the frequency of collisions of a single molecule and the total collision frequency (per unit volume) of all the molecules.

27–1. The Average Translational Kinetic Energy of the Molecules in a Gas Is Directly Proportional to the Kelvin Temperature

The pressure that a gas exerts on the walls of its container is due to the collisions that the particles of the gas make with the walls. Let's consider one of the molecules of the gas (call it molecule 1) as it moves throughout its container, as illustrated in Figure 27.1. We have assumed that the container is a rectangular parallelepiped of sides a, b, and c for simplicity, but it is not necessary to do so. The velocity of the molecule has components u_{1x}, u_{1y}, and u_{1z}. We can treat the motion along the x-direction first and

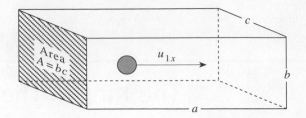

FIGURE 27.1
A molecule with its x-component of velocity equal to u_{1x} moving perpendicular to one face of a rectangular parallelepiped of lengths a, b, and c.

then extend the result to an arbitrary direction afterwards. Suppose that the molecule is moving from left to right in Figure 27.1 so that u_{1x} is positive. The x-component of the momentum of the particle is mu_{1x}. We assume that when the particle collides with the right-hand wall shown in Figure 27.1, the motion of the particle is reversed so that its momentum now is $-mu_{1x}$. In other words, we are assuming here that the collisions of the particle with the walls are perfectly elastic. The change in momentum, $\Delta(mu_{1x})$, then is $\Delta(mu_{1x}) = mu_{1x} - (-mu_{1x}) = 2mu_{1x}$. If the distance between the two walls perpendicular to the x-direction is a, then the time elapsed between collisions with the right-hand wall is $\Delta t = 2a/u_{1x}$ because the molecule travels a distance $2a$ to arrive back at the right-hand wall. Recall that Newton's second law of motion says that the rate of change of momentum is equal to a force. The rate of change of momentum due to collisions with the right-hand wall is

$$\frac{\Delta(mu_{1x})}{\Delta t} = \frac{2mu_{1x}}{2a/u_{1x}} = \frac{mu_{1x}^2}{a} \tag{27.1}$$

and so the force that molecule 1 exerts on the right-hand wall is

$$F_1 = \frac{mu_{1x}^2}{a}$$

The area of the wall is bc (see Figure 27.1) and so the pressure exerted on the wall is

$$P_1 = \frac{F_1}{bc} = \frac{mu_{1x}^2}{abc} = \frac{mu_{1x}^2}{V} \tag{27.2}$$

where $V = abc$ is the volume of the container.

Each of the other molecules exerts a similar pressure, and so the total pressure on the right-hand wall is

$$P = \sum_{j=1}^{N} P_j = \sum_{j=1}^{N} \frac{mu_{jx}^2}{V} = \frac{m}{V} \sum_{j=1}^{N} u_{jx}^2 \tag{27.3}$$

where N is the total number of molecules. The sum of the u_{jx}^2 divided by N is the average value of u_x^2, and if we denote the average by $\langle u_x^2 \rangle$, then we can write

$$\langle u_x^2 \rangle = \frac{1}{N} \sum_{j=1}^{N} u_{jx}^2 \tag{27.4}$$

If we introduce Equation 27.4 into Equation 27.3, then we obtain

$$PV = Nm \langle u_x^2 \rangle \tag{27.5}$$

We arbitrarily chose to work with the x-direction, but we could just as well have chosen the y- or z-direction. Because the x-, y-, and z-directions are equivalent, it must be that

$$\langle u_x^2 \rangle = \langle u_y^2 \rangle = \langle u_z^2 \rangle \tag{27.6}$$

Equation 27.6 is a statement of the fact that a homogeneous gas is *isotropic*; it has the same properties in any direction. Furthermore, the total speed u of any molecule satisfies

$$u^2 = u_x^2 + u_y^2 + u_z^2$$

and so

$$\langle u^2 \rangle = \langle u_x^2 \rangle + \langle u_y^2 \rangle + \langle u_z^2 \rangle \tag{27.7}$$

Equation 27.7 along with Equation 27.6 says that

$$\langle u_x^2 \rangle = \langle u_y^2 \rangle = \langle u_z^2 \rangle = \tfrac{1}{3} \langle u^2 \rangle \tag{27.8}$$

We substitute this result into Equation 27.5 to obtain

$$PV = \tfrac{1}{3} Nm \langle u^2 \rangle \tag{27.9}$$

Equation 27.9 is a fundamental equation of the kinetic theory of gases, relating a macroscopic property, PV, on the left-hand side with a molecular property, $m \langle u^2 \rangle$, on the right-hand side. We learned in Chapter 18 that the average translational (kinetic) energy of an ideal gas is $\tfrac{3}{2}RT$ per mole, or $\tfrac{3}{2}k_B T$ per molecule. In an equation, we have

$$\tfrac{1}{2} m \langle u^2 \rangle = \tfrac{3}{2} k_B T$$

or, if we multiply both sides by the Avogadro constant

$$\tfrac{1}{2} N_A m \langle u^2 \rangle = \tfrac{3}{2} RT \tag{27.10}$$

The product $N_A m = M$, the molar mass of the gas. Consequently, we can write

$$\tfrac{1}{3} M \langle u^2 \rangle = RT \tag{27.11}$$

If we substitute Equation 27.11 into Equation 27.9, then we obtain the ideal gas equation.

EXAMPLE 27–1
Use Equation 27.10 to calculate the average translational energy of one mole of an ideal gas at 25°C.

SOLUTION: We use $R = 8.314 \text{ J} \cdot \text{mol}^{-1} \cdot \text{K}^{-1}$ and obtain

$$\langle \text{KE} \rangle = \tfrac{3}{2}(8.314 \text{ J} \cdot \text{mol}^{-1} \cdot \text{K}^{-1})(298 \text{ K}) - 3.72 \text{ kJ} \cdot \text{mol}^{-1}$$

We can use Equation 27.11 to estimate the average speed of a gas molecule at a temperature T. We first solve Equation 27.11 for $\langle u^2 \rangle$ to obtain

$$\langle u^2 \rangle = \frac{3RT}{M} \tag{27.12}$$

The units of $\langle u^2 \rangle$ are $\text{m}^2 \cdot \text{s}^{-2}$. To obtain a quantity that has units of $\text{m} \cdot \text{s}^{-1}$, we take the square root of $\langle u^2 \rangle$:

$$\langle u^2 \rangle^{1/2} = \left(\frac{3RT}{M} \right)^{1/2} \tag{27.13}$$

The quantity $\langle u^2 \rangle^{1/2}$ is the square root of the mean value of u^2 and is called the *root-mean-square* speed. If we denote the root-mean-square speed by u_{rms}, then Equation 27.13 becomes

$$u_{\text{rms}} = \left(\frac{3RT}{M} \right)^{1/2} \tag{27.14}$$

EXAMPLE 27–2
Calculate the root-mean-square speed of a nitrogen molecule at 25°C.

SOLUTION: We must use a value of R having the units such that u_{rms} will have units of $\text{m} \cdot \text{s}^{-1}$. If we use the value $R = 8.314 \text{ J} \cdot \text{mol}^{-1} \cdot \text{K}^{-1}$ and be sure to express the molar mass in units of $\text{kg} \cdot \text{mol}^{-1}$, then u_{rms} will have units of $\text{m} \cdot \text{s}^{-1}$. Therefore

$$u_{\text{rms}} = \left(\frac{3 \times 8.314 \text{ J} \cdot \text{mol}^{-1} \cdot \text{K}^{-1} \times 298 \text{ K}}{0.02802 \text{ kg} \cdot \text{mol}^{-1}} \right)^{1/2}$$

$$= \left(2.65 \times 10^5 \, \frac{\text{J}}{\text{kg}} \right)^{1/2} = \left(2.65 \times 10^5 \, \frac{\text{kg} \cdot \text{m}^2 \cdot \text{s}^{-2}}{\text{kg}} \right)^{1/2}$$

$$= 515 \text{ m} \cdot \text{s}^{-1}$$

Notice that we have used the fact that $1 \text{ J} = 1 \text{ kg} \cdot \text{m}^2 \cdot \text{s}^{-2}$.

We have called u_{rms} an estimate of the average speed because generally $\langle u^2 \rangle \neq \langle u \rangle^2$, and so $u_{rms} \neq \langle u \rangle$. We shall see in Section 27–3, however, that u_{rms} and $\langle u \rangle$ differ by less than 10%. Typical average molecular speeds at room temperature are of the order of hundreds of meters per second, as shown in Table 27.1. Although we shall not prove it here, the speed of sound, u_{sound}, in a monatomic ideal gas is given by

$$u_{sound} = \left(\frac{5RT}{3M} \right)^{1/2} \tag{27.15}$$

which differs from u_{rms} by about 30%. The speed of sound in argon at 25°C is $346 \text{ m} \cdot \text{s}^{-1}$, or 770 miles per hour.

Before leaving this section, we should consider the assumptions that we made in deriving Equation 27.9. We assumed that the collisions with the wall are perfectly elastic. This cannot really be the case because the wall is made up of molecules, which are in thermal motion, and so some collisions will be more or less energetic than others, depending upon the direction of the motion of the molecules of the wall with respect to the colliding molecule. On the average, however, the gas molecules will bounce off the wall with the same speed they had beforehand because the molecules of the wall must be at the same temperature and so have the same average translational energy as the gas molecules if the system is in thermal equilibrium. We also tacitly assumed that the molecules of the gas do not collide with each other as they travel from one wall to the other in Figure 27.1. But if the gas is in equilibrium, on the average, any collision that deflects the path of a molecule from that shown in Figure 27.1 will be balanced by a collision that replaces the molecule.

TABLE 27.1
Average speeds (Equation 27.42) and root-mean-square speeds (Equation 27.14) of gas molecules at 25°C. Notice that the ratio of $\langle u \rangle$ to u_{rms} is about 0.92.

Gas	$\langle u \rangle / \text{m} \cdot \text{s}^{-1}$	$u_{rms} / \text{m} \cdot \text{s}^{-1}$
NH_3	609	661
CO_2	379	411
He	1260	1360
H_2	1770	1920
CH_4	627	681
N_2	475	515
O_2	444	482
SF_6	208	226

It so happens that many of the quantities of the kinetic theory of gases can be derived at a number of levels of rigor, varying from very elementary treatments, in which all the molecules are assumed to have the same (average) speed and move only along the x-, y-, and z-directions, to very sophisticated treatments that make no unnecessary assumptions. An interesting thing is that the results of these various derivations differ only by constant factors of the order of unity. One can generate pages of algebra to get a more exact equation, which may have the very same dependence on temperature and pressure as the simple equation but differs by a factor like $2^{1/2}$ or 3/8. In order to introduce the basic ideas of the kinetic theory of gases in this chapter, we shall usually present the more elementary derivations, but we shall present a slightly fancier derivation of Equation 27.9 in Section 27–4.

27–2. The Distribution of the Components of Molecular Speeds Are Described by a Gaussian Distribution

As we implied in the previous section, all molecules in a gas do not have the same speed. Experimentally, the molecular speeds in a gas are described by the curves in Figure 27.2, where the distribution of molecular speeds is plotted against u. Notice that a greater fraction of molecules has higher speeds as the temperature increases. In this section we derive a theoretical equation for the distribution of the components of molecular velocities and in the next section we shall derive an equation for the distribution of molecular speeds. These distributions were first derived somewhat heuristically by the Scottish physicist James Clerk Maxwell in 1860 and later more rigorously by the Austrian physicist Ludwig Boltzmann. They are collectively now called the *Maxwell-Boltzmann distribution*. It is interesting to note that Maxwell derived the distribution law long before it was verified experimentally.

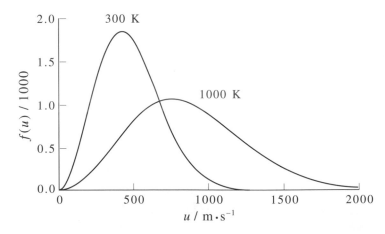

FIGURE 27.2
The distribution of molecular speeds in nitrogen at 300 K and 1000 K.

   Let $h(u_x, u_y, u_z)du_x du_y du_z$ be the fraction of molecules that have velocity components between u_x and $u_x + du_x$, u_y and $u_y + du_y$, and u_z and $u_z + du_z$, or the probability that any one molecule has such velocity components. A key step in Maxwell's derivation is to assume that the probability that the x-component of the velocity of a molecule has a given value is completely independent of the values of the y-component or the z-component. In other words, he assumed that the probability distributions in each of the three directions are independent of each other. This assumption, which is perhaps less than obvious, can be avoided at the expense of a much more lengthy derivation, but it turns out to be correct. In terms of an equation, the assumption that the three components of the velocity are statistically independent becomes

$$h(u_x, u_y, u_z) = f(u_x)f(u_y)f(u_z) \tag{27.16}$$

where $f(u_x)$, $f(u_y)$, and $f(u_z)$ are the probability distributions of the individual components. The probability distribution in each of the three directions is the same because the gas is isotropic. Furthermore, because the gas is isotropic, the function $h(u_x, u_y, u_z)$ must depend only upon the speed or the magnitude of the velocity $\mathbf{u}$, whose square is given by (MathChapter C)

$$\mathbf{u} \cdot \mathbf{u} = u^2 = u_x^2 + u_y^2 + u_z^2 \tag{27.17}$$

Therefore, we can write Equation 27.16 as

$$h(u) = h(u_x, u_y, u_z) = f(u_x)f(u_y)f(u_z) \tag{27.18}$$

Taking the logarithm of Equation 27.18 gives

$$\ln h(u) = \ln f(u_x) + \ln f(u_y) + \ln f(u_z) \tag{27.19}$$

Differentiating Equation 27.19 with respect to u_x gives

$$\left(\frac{\partial \ln h(u)}{\partial u_x} \right)_{u_y, u_z} = \frac{d \ln f(u_x)}{du_x} \tag{27.20}$$

Because the function h depends upon u, we would like to rewrite it as a derivative with respect to u rather than u_x. To do this, we write

$$\left(\frac{\partial \ln h}{\partial u_x} \right)_{u_y, u_z} = \frac{d \ln h}{du} \left(\frac{\partial u}{\partial u_x} \right)_{u_y, u_z} = \frac{u_x}{u} \frac{d \ln h}{du} \tag{27.21}$$

where we have used Equation 27.17 to replace $\partial u / \partial u_x$ by u_x/u (Problem 27–10). Substituting Equation 27.21 into the left side of Equation 27.20 gives

$$\frac{d \ln h(u)}{u \, du} = \frac{d \ln f(u_x)}{u_x \, du_x}$$

The three probability distributions $f(u_x)$, $f(u_y)$, and $f(u_z)$ are all the same, so

$$\frac{d \ln h(u)}{u \, du} = \frac{d \ln f(u_x)}{u_x \, du_x} = \frac{d \ln f(u_y)}{u_y \, du_y} = \frac{d \ln f(u_z)}{u_z \, du_z} \qquad (27.22)$$

Because u_x, u_y, and u_z are independent of one another, Equation 27.22 must be equal to a constant. Defining this constant to be -2γ, we find that

$$\frac{d \ln f(u_j)}{u_j \, du_j} = -2\gamma \qquad j = x, y, z \qquad (27.23)$$

or upon integration

$$f(u_j) = A e^{-\gamma u_j^2} \qquad j = x, y, z \qquad (27.24)$$

We have written $-\gamma$ instead of γ in Equation 27.23 in anticipation that γ must be a positive quantity. (See Problem 27–11.)

We shall now use $f(u_x)$ as a specific example to determine the two constants A and γ. We can determine A in terms of γ by realizing that

$$\int_{-\infty}^{\infty} f(u_x) du_x = 1 \qquad (27.25)$$

because $f(u_x)$ is a probability distribution. Substituting Equation 27.24 into Equation 27.25 gives

$$A \int_{-\infty}^{\infty} e^{-\gamma u_x^2} du_x = 1 \qquad (27.26)$$

The integrand $f(u_x) = e^{-\gamma u_x^2}$ is an even function of u_x (MathChapter B) and so

$$A \int_{-\infty}^{\infty} e^{-\gamma u_x^2} du_x = 2A \int_{0}^{\infty} e^{-\gamma u_x^2} du_x \qquad (27.27)$$

We have encountered this integral many times (e.g., MathChapter B), and using Equation B.16, we find that

$$A \int_{-\infty}^{\infty} e^{-\gamma u_x^2} du_x = 2A \int_{0}^{\infty} e^{-\gamma u_x^2} du_x = 2A \left(\frac{\pi}{4\gamma} \right)^{1/2} = 1 \qquad (27.28)$$

or $A = (\gamma/\pi)^{1/2}$. Therefore, $f(u_x)$ is given by

$$f(u_x) = \left(\frac{\gamma}{\pi} \right)^{1/2} e^{-\gamma u_x^2} \qquad (27.29)$$

with a similar result for $f(u_y)$ and $f(u_z)$.

We can now determine γ by using Equations 27.8 and 27.12, which together say that $\langle u_x^2 \rangle = RT/M$. In terms of $f(u_x)$, the average of u_x^2 is given by (MathChapter B)

$$\langle u_x^2 \rangle = \frac{RT}{M} = \int_{-\infty}^{\infty} u_x^2 f(u_x) du_x = \left(\frac{\gamma}{\pi} \right)^{1/2} \int_{-\infty}^{\infty} u_x^2 e^{-\gamma u_x^2} du_x \tag{27.30}$$

Notice once again that the integrand in Equation 27.30 is an even function of u_x, so that

$$\langle u_x^2 \rangle = \frac{RT}{M} = 2 \int_0^{\infty} u_x^2 f(u_x) du_x = 2 \left(\frac{\gamma}{\pi} \right)^{1/2} \int_0^{\infty} u_x^2 e^{-\gamma u_x^2} du_x \tag{27.31}$$

The integral here is discussed in MathChapter B, and using Equation B.20 gives

$$\frac{RT}{M} = 2 \left(\frac{\gamma}{\pi} \right)^{1/2} \cdot \frac{1}{4\gamma} \left(\frac{\pi}{\gamma} \right)^{1/2} = \frac{1}{2\gamma}$$

or that $\gamma = M/2RT$. Therefore Equation 27.29 becomes

$$f(u_x) = \left(\frac{M}{2\pi RT} \right)^{1/2} e^{-Mu_x^2/2RT} \tag{27.32}$$

Equation 27.32 is plotted in Figure 27.3. The areas under the curves in Figure 27.3 are unity because the probability distribution is normalized. Figure 27.3 shows that as the temperature increases, more molecules are likely to be found with higher values of u_x. Note that $f(u_x)$ plotted in Figure 27.3 does not look like the experimental curves shown in Figure 27.2. This is because $f(u_x)$ is the distribution function for one component of the molecular velocity, whereas the curves in Figure 27.2 represent the distribution in the overall molecular speed, which is given by $u = (u_x^2 + u_y^2 + u_z^2)^{1/2}$. The range of a component of velocity is $-\infty$ to ∞ as shown in Figure 27.3 because

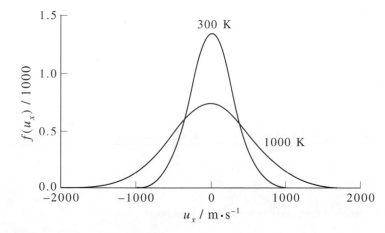

FIGURE 27.3
The distribution of a component of the velocity of a nitrogen molecule at 300 K and 1000 K.

the molecule may be moving in a positive or a negative direction. The range of the overall *speed* is 0 to ∞ as shown in Figure 27.2 because the *length* of the velocity vector, $u = (u_x^2 + u_y^2 + u_z^2)^{1/2}$, is an intrinsically positive quantity. We shall derive an expression for the distribution of molecular speeds in the next section.

We have written $f(u_x)$ in Equation 27.32 in terms of the molar mass M and the molar gas constant R. Because $f(u_x)$ describes the probability distribution of the components of *molecular* velocities, it is common to re-write Equation 27.32 in the form

$$f(u_x) = \left(\frac{m}{2\pi k_B T}\right)^{1/2} e^{-mu_x^2/2k_B T} \tag{27.33}$$

where m is the mass (in kilograms) of one molecule and k_B is the Boltzmann constant. As we have seen, the Boltzmann constant appears in a great many of the equations of physical chemistry, often in the combination $k_B T$, which has units of energy. In fact, notice that the argument of the exponential function in Equation 27.33 is the x-component of the kinetic energy divided by $k_B T$, which is unitless as it must be. Notice also that we have simply replaced M/R in Equation 27.32 with m/k_B by dividing both M and R by the Avogadro constant.

We can use Equation 27.33 to calculate the average value of u_x, which is given by

$$\langle u_x \rangle = \int_{-\infty}^{\infty} u_x f(u_x) du_x = \left(\frac{m}{2\pi k_B T}\right)^{1/2} \int_{-\infty}^{\infty} u_x e^{-mu_x^2/2k_B T} du_x \tag{27.34}$$

The integrand is an odd function of u_x, and so $\langle u_x \rangle = 0$. Physically, this result is due to the fact that a molecule is equally likely to be moving in a positive x-direction as in a negative x-direction.

EXAMPLE 27–3
Determine the average value of u_x^2 and of $\frac{1}{2}mu_x^2$, the x-component of the kinetic energy.

SOLUTION: The average value of u_x^2 is given by

$$\langle u_x^2 \rangle = \left(\frac{m}{2\pi k_B T}\right)^{1/2} \int_{-\infty}^{\infty} u_x^2 e^{-mu_x^2/2k_B T} du_x$$

Because the integrand is an even function of u_x, we can write

$$\langle u_x^2 \rangle = 2\left(\frac{m}{2\pi k_B T}\right)^{1/2} \int_{0}^{\infty} u_x^2 e^{-mu_x^2/2k_B T} du_x$$

Using Equation B.20 with $\alpha = m/2k_B T$, we find that

$$\langle u_x^2 \rangle = \frac{k_B T}{m} = \frac{RT}{M}$$

The average x-component of the kinetic energy of a molecule is

$$\tfrac{1}{2}m\langle u_x^2\rangle = \tfrac{1}{2}k_BT \tag{27.35}$$

with a similar result for the y- and z-components.

Equation 27.35 implies that

$$\tfrac{1}{2}m\langle u_x^2\rangle = \tfrac{1}{2}m\langle u_y^2\rangle = \tfrac{1}{2}m\langle u_z^2\rangle = \tfrac{1}{2}k_BT$$

The total kinetic energy is given by

$$\tfrac{1}{2}m\langle u^2\rangle = \tfrac{3}{2}k_BT$$

These two equations show that the total kinetic energy of $3k_BT/2$ is divided equally into the x-, y-, and z-components, as you might expect because the gas is isotropic.

Most experimental observations depend upon averages of molecular speeds, but there are a few that depend upon the entire distribution itself. One of these involves the shapes of the spectral lines in the emission spectra of atoms and molecules. Ideally, spectral lines are very narrow, being broadened by the finite lifetime of the excited states. However, lifetime broadening is often not the major source of the observed width of a spectral line. The lines are also broadened because of the motion of the molecules emitting the radiation. If an atom or molecule at rest emits radiation of frequency v_0, then due to the Doppler effect, the frequency measured by a stationary observer will be

$$v \approx v_0\left(1 + \frac{u_x}{c}\right) \tag{27.36}$$

if the atom or molecule is moving away or toward the observer with speed u_x, where c is the speed of light. If one observes the radiation emitted from a gas at a temperature T, then it is found that the spectral line at v_0 will be spread out by the Maxwell distribution of u_x of the molecule emitting the radiation. Using Equation 27.36, the distribution in u_x can be converted to a distribution in v. Substituting the relation $u_x = c(v - v_0)/v_0$ from Equation 27.36 into Equation 27.33 gives

$$I(v) \propto e^{-mc^2(v-v_0)^2/2v_0^2k_BT} \tag{27.37}$$

for the observed shape of the spectral line. The form of $I(v)$ is that of a Gaussian curve centered at v_0 with a variance given by (see MathChapter B)

$$\sigma^2 = \frac{v_0^2 k_B T}{mc^2} = \frac{v_0^2 RT}{Mc^2}$$

where M is the molar mass. Sodium emits light of frequency 5×10^{14} Hz corresponding to the transition from the $3p\,^2P_{3/2}$ excited state to the $3s\,^2S_{1/2}$ ground state. The emission from a cell containing a low pressure of sodium vapor at 500 K, σ, which is a measure of the width of this spectral emission, is about 7×10^8 Hz. If the sodium atoms were stationary, then the measured value of σ would be about 1.0×10^6 Hz. The broadening of spectral lines due to the distribution of molecular velocities is called *Doppler broadening*.

27–3. The Distribution of Molecular Speeds Is Given by the Maxwell–Boltzmann Distribution

So far we have derived the probability distribution for a given component of the molecular velocity. Because a homogeneous gas is isotropic, the direction in which a molecule moves has no physical consequence on the properties of the gas; only the magnitude of **u**, the speed, is relevant. Therefore, in this section we shall derive the distribution of molecular *speeds*. We define a function $F(u)$ by

$$F(u)du = f(u_x)du_x f(u_y)du_y f(u_z)du_z \tag{27.38}$$

If we substitute Equation 27.33 and its analogs for u_y and u_z into Equation 27.38, then we obtain

$$F(u)du = \left(\frac{m}{2\pi k_\text{B}T}\right)^{3/2} e^{-m(u_x^2+u_y^2+u_z^2)/2k_\text{B}T} du_x du_y du_z \tag{27.39}$$

We need to convert the right-hand side of Equation 27.39 to the form $F(u)du$, which is the probability that a molecule has a speed between u and $u + du$. In order to do this, consider a rectangular coordinate system in which the distances along the axes are u_x,

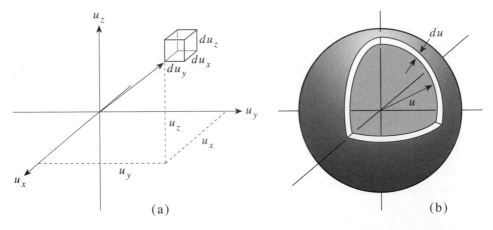

(a)   (b)

FIGURE 27.4
An illustration of velocity space. (a) A cartesian representation in which a point is specified by the values of u_x, u_y, and u_z and the differential "volume" element is $du_x du_y du_z$. The molecular velocity is a vector of length $(u_x^2 + u_y^2 + u_z^2)^{1/2}$. (b) A spherical representation in which the "volume" element is a spherical shell of radius u and thickness, du, with a volume $4\pi u^2 du$.

u_y, and u_z, the three components of the velocity, as shown in Figure 27.4a. The molecular velocity, $\mathbf{u}$, which is a vector quantity with components u_x, u_y, and u_z, is shown in the figure, and the length of $\mathbf{u}$ is $u = (u_x^2 + u_y^2 + u_z^2)^{1/2}$. The space described by this coordinate system is called a *velocity space*, and is simply the analog of the three-dimensional space described by the x, y, z coordinate system. Just as $dxdydz$ is an infinitesimal volume element in ordinary space, $du_x du_y du_z$ is an infinitesimal "volume" element in velocity space. Because a gas is isotropic, it is more convenient to use spherical coordinates rather than cartesian coordinates (see Figure 27.4) to describe the distribution of molecular speeds. In ordinary space, the infinitesimal volume element is $4\pi r^2 dr$, which is the volume of a spherical shell of radius r and thickness dr. In our velocity space, the analogous infinitesimal volume element is $4\pi u^2 du$ (Figure 27.4b). Thus, in Equation 27.39 we replace $u_x^2 + u_y^2 + u_z^2$ by u^2 and $du_x du_y du_z$ by $4\pi u^2 du$ to obtain

$$F(u)du = 4\pi \left(\frac{m}{2\pi k_{\mathrm{B}}T} \right)^{3/2} u^2 e^{-mu^2/2k_{\mathrm{B}}T} du \tag{27.40}$$

Equation 27.40 gives the probability distribution of a molecule having a speed between u and $u + du$. Note that unlike Equation 27.39 for the probability distribution of a component of the velocity, Equation 27.40 has a factor of u^2. In addition, realize that although the range of a component of the speed is $-\infty$ to ∞, the range of u, which is an intrinsically positive quantity, is 0 to ∞.

EXAMPLE 27–4
Show that Equation 27.40 is normalized.

SOLUTION: We use Equation B.20 with $\alpha = m/2k_{\mathrm{B}}T$:

$$\int_0^\infty F(u)du = 4\pi \left(\frac{m}{2\pi k_{\mathrm{B}}T} \right)^{3/2} \int_0^\infty u^2 e^{-mu^2/2k_{\mathrm{B}}T} du$$

$$= 4\pi \left(\frac{m}{2\pi k_{\mathrm{B}}T} \right)^{3/2} \cdot \frac{k_{\mathrm{B}}T}{2m} \cdot \left(\frac{2\pi k_{\mathrm{B}}T}{m} \right)^{1/2}$$

$$= 1$$

We can also calculate averages of u. For example, the average speed is given by (MathChapter B)

$$\langle u \rangle = \int_0^\infty u F(u)du = 4\pi \left(\frac{m}{2\pi k_{\mathrm{B}}T} \right)^{3/2} \int_0^\infty u^3 e^{-mu^2/2k_{\mathrm{B}}T} du \tag{27.41}$$

The appropriate standard integral is (See Table 27.2 for a collection of the integrals that we have used.)

$$\int_0^\infty x^{2n+1} e^{-\alpha x^2} dx = \frac{n!}{2\alpha^{n+1}}$$

and so Equation 27.41 becomes

$$\langle u \rangle = 4\pi \left(\frac{m}{2\pi k_B T} \right)^{3/2} \cdot \frac{1!}{2} \left(\frac{2k_B T}{m} \right)^2 = \left(\frac{8k_B T}{\pi m} \right)^{1/2} = \left(\frac{8RT}{\pi M} \right)^{1/2} \qquad (27.42)$$

Notice that this value differs slightly from $u_{rms} = (3k_B T/m)^{1/2}$; in fact, the ratio of $\langle u \rangle$ to u_{rms} is $(8/3\pi)^{1/2} = 0.92$.

We can derive the value of u_{rms} directly from Equation 27.40:

$$\langle u^2 \rangle = \int_0^\infty u^2 F(u) du = 4\pi \left(\frac{m}{2\pi k_B T} \right)^{3/2} \int_0^\infty u^4 e^{-mu^2/2k_B T} du$$

By referring to Table 27.2, we see that

$$\langle u^2 \rangle = 4\pi \left(\frac{m}{2\pi k_B T} \right)^{3/2} \cdot \frac{1 \cdot 3}{8} \left(\frac{2k_B T}{m} \right)^2 \left(\frac{2\pi k_B T}{m} \right)^{1/2}$$

$$= \frac{3k_B T}{m}$$

By definition, $u_{rms} = \langle u^2 \rangle^{1/2} = (3k_B T/m)^{1/2} = (3RT/M)^{1/2}$, which is what we obtained earlier.

One other characteristic speed is the most probable speed. The most probable speed, u_{mp}, is given by the maximum value of $F(u)$, which is found by setting the derivative of $F(u)$ equal to zero.

$$\frac{dF(u)}{du} = 4\pi \left(\frac{m}{2\pi k_B T} \right)^{3/2} \left[2u - \frac{mu^3}{k_B T} \right] e^{-mu^2/2k_B T} = 0$$

TABLE 27.2
Some integrals that occur frequently in the kinetic theory of gases.

$$\int_0^\infty x^{2n} e^{-\alpha x^2} dx = \frac{1 \cdot 3 \cdot 5 \cdots (2n-1)}{2^{n+1} \alpha^n} \left(\frac{\pi}{\alpha} \right)^{1/2} \qquad n \geq 1$$

$$\int_0^\infty x^{2n+1} e^{-\alpha x^2} dx = \frac{n!}{2\alpha^{n+1}} \qquad n \geq 0$$

$$\int_0^\infty x^{n/2} e^{-\alpha x} dx = \frac{n(n-2)(n-4) \cdots (1)}{(2\alpha)^{(n+1)/2}} \left(\frac{\pi}{\alpha} \right)^{1/2} \qquad n \text{ odd}$$

$$= \frac{(n/2)!}{\alpha^{(n+2)/2}} \qquad n \text{ even}$$

For $dF(u)/du$ to be equal to zero, the factor in brackets must equal zero, and so we have

$$u_{mp} = \left(\frac{2k_B T}{m}\right)^{1/2} = \left(\frac{2RT}{M}\right)^{1/2} \tag{27.43}$$

Notice that all the characteristic speeds that we have encountered, u_{rms}, $\langle u \rangle$, and u_{mp}, are of the form $(\text{constant} \cdot k_B T/m)^{1/2}$ or $(\text{constant} \cdot RT/M)^{1/2}$.

We can express the Maxwell-Boltzmann distribution in terms of kinetic energy, $\varepsilon = mu^2/2$, rather than speed by setting $u = (2\varepsilon/m)^{1/2}$. In this case, $du = d\varepsilon/(2m\varepsilon)^{1/2}$ and Equation 27.40 becomes

$$F(\varepsilon)d\varepsilon = 4\pi \left(\frac{m}{2\pi k_B T}\right)^{3/2} \cdot \frac{2\varepsilon}{m} \cdot e^{-\varepsilon/k_B T} \frac{d\varepsilon}{(2m\varepsilon)^{1/2}}$$

$$= \frac{2\pi}{(\pi k_B T)^{3/2}} \varepsilon^{1/2} e^{-\varepsilon/k_B T} d\varepsilon \tag{27.44}$$

EXAMPLE 27–5

Show that the distribution given by Equation 27.44 is normalized.

SOLUTION: We need to show that

$$\int_0^\infty F(\varepsilon)d\varepsilon = \frac{2\pi}{(\pi k_B T)^{3/2}} \int_0^\infty \varepsilon^{1/2} e^{-\varepsilon/k_B T} d\varepsilon = 1$$

The necessary integral here is the third entry in Table 27.2 with $n = 1$.

$$\int_0^\infty x^{1/2} e^{-\alpha x} dx = \frac{1}{2\alpha}\left(\frac{\pi}{\alpha}\right)^{1/2}$$

and so

$$\int_0^\infty F(\varepsilon)d\varepsilon = \frac{2\pi}{(\pi k_B T)^{3/2}} \int_0^\infty \varepsilon^{1/2} e^{-\varepsilon/k_B T} d\varepsilon$$

$$= \frac{2\pi}{(\pi k_B T)^{3/2}} \cdot \frac{k_B T}{2} \cdot (\pi k_B T)^{1/2} = 1$$

In addition it is straightforward to show that (we use the third entry in Table 27.2 with $n = 3$)

$$\langle \varepsilon \rangle = \int_0^\infty \varepsilon f(\varepsilon)d\varepsilon = \frac{2\pi}{(\pi k_B T)^{3/2}} \int_0^\infty \varepsilon^{3/2} e^{-\varepsilon/k_B T} d\varepsilon$$

$$= \frac{2\pi}{(\pi k_B T)^{3/2}} \cdot 3\left(\frac{k_B T}{2}\right)^2 \cdot (\pi k_B T)^{1/2} = \frac{3}{2}k_B T$$

in agreement with Equation 27.10.

27–4. The Frequency of Collisions that a Gas Makes with a Wall Is Proportional to its Number Density and to the Average Molecular Speed

In this section we shall derive an expression for the frequency of collisions that the molecules of a gas make with the walls of its container. Such a quantity is central to the theory of the rates of surface reactions. The geometry that we shall use to derive the desired equation is shown in Figure 27.5. Figure 27.5 shows an oblique cylinder of slant height udt with a base area A making an angle θ with the normal to the wall. This cylinder has been constructed to enclose all those molecules that will strike the area at an angle θ with a speed u in the time interval dt. The volume of such a cylinder is its base area (A) times its vertical height $(u\cos\theta dt)$ or $(Audt)\cos\theta$. The number of molecules in this cylinder is $\rho(Audt)\cos\theta$, where ρ is the number density, N/V. The fraction of molecules that have a speed between u and $u+du$ is $F(u)du$ and the fraction travelling within a solid angle bounded by θ and $\theta+d\theta$ and ϕ and $\phi+d\phi$ is $\sin\theta d\theta d\phi/4\pi$, where the factor 4π represents a complete solid angle (see MathChapter D). The product of the above factors gives the number of molecules, dN_{coll}, colliding with the area A from the specified direction in the time interval dt:

$$dN_{\text{coll}} = \rho(Audt)\cos\theta \cdot F(u)du \cdot \frac{\sin\theta d\theta d\phi}{4\pi} \qquad (27.45)$$

If we divide Equation 27.45 by Adt, then we have

$$dz_{\text{coll}} = \frac{1}{A}\frac{dN_{\text{coll}}}{dt} = \frac{\rho}{4\pi} uF(u)du \cdot \cos\theta \sin\theta d\theta d\phi \qquad (27.46)$$

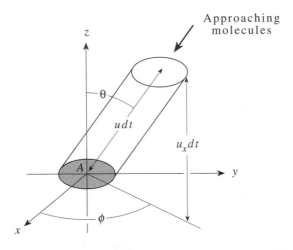

FIGURE 27.5
The geometry used to calculate the rate of collisions of the molecules of a gas with the walls of the container. Note that θ varies from 0 to $\pi/2$ because molecules strike the wall from only one side.

where dz_{coll} is the number of the collisions per unit time per unit area with the wall by molecules whose speeds are in the range u and $u + du$ and whose direction lies within the solid angle $\sin\theta d\theta d\phi$. Notice that Equation 27.46 has a factor of u^3 ($F(u)$ has a factor of u^2), as compared to a factor of u^2 in Equation 27.40. Figure 27.6 shows the two (unnormalized) functions $u^2 e^{-mu^2/2k_BT}$ and $u^3 e^{-mu^2/2k_BT}$ plotted against the speed u. Notice that the function $u^3 e^{-mu^2/2k_BT}$ peaks at higher speeds than does $u^2 e^{-mu^2/2k_BT}$. (Problem 27–28 has you show that $u^3 e^{-mu^2/2k_BT}$ peaks at $u_{mp} = (3k_BT/m)^{1/2}$ and that $u^2 e^{-mu^2/2k_BT}$ peaks at $u_{mp} = (2k_BT/m)^{1/2}$.) From a physical point of view, this means that the molecules that strike a plane of area A are travelling at higher speeds than the molecules in a gas in general. The reason for this is that the molecules travelling at higher speeds are more likely to strike the area A in a given time.

If we integrate Equation 27.46 over all possible speeds and directions, then we obtain

$$z_{coll} = \frac{\rho}{4\pi} \int_0^\infty u F(u) du \int_0^{\pi/2} \cos\theta \sin\theta d\theta \int_0^{2\pi} d\phi \qquad (27.47)$$

Notice that we integrate θ from 0 to $\pi/2$ because the molecules strike the wall from one side only. The integral involving u is equal to $\langle u \rangle$, the integral over θ is equal to $1/2$, and that over ϕ is 2π, so that we we have for the collision frequency per unit area, z_{coll}

$$z_{coll} = \frac{1}{A}\frac{dN_{coll}}{dt} = \rho\frac{\langle u \rangle}{4} \qquad (27.48)$$

Problems 27–49 through 27–52 discuss several applications of Equation 27.48.

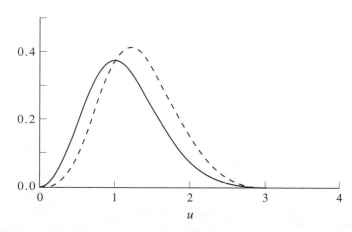

FIGURE 27.6
A plot of $u^2 e^{-mu^2/2k_BT}$ (solid line) and $u^3 e^{-mu^2/2k_BT}$ (dashed line) against the speed u, in units of $(k_BT/m)^{1/2}$. Notice that $u^3 e^{-mu^2/2k_BT}$ peaks at higher values of u than does $u^2 e^{-mu^2/2k_BT}$.

EXAMPLE 27–6
Use Equation 27.48 to calculate the collision frequency per unit area for nitrogen at 25°C and one bar.

SOLUTION: The number density is given by

$$\rho = \frac{N_A n}{V} = \frac{N_A P}{RT} = \frac{(6.022 \times 10^{23} \text{ mol}^{-1})(1 \text{ bar})}{(0.08314 \text{ L·bar·mol}^{-1}\text{·K}^{-1})(298 \text{ K})}$$

$$= 2.43 \times 10^{22} \text{ L}^{-1} = 2.43 \times 10^{25} \text{ m}^{-3}$$

and

$$\langle u \rangle = \left(\frac{8RT}{\pi M}\right)^{1/2} = \left(\frac{8(8.314 \text{ J·K}^{-1}\text{·mol}^{-1})(298 \text{ K})}{\pi(0.02802 \text{ kg})}\right)^{1/2}$$

$$= 475 \text{ m·s}^{-1}$$

Therefore

$$z_{coll} = 2.88 \times 10^{27} \text{ s}^{-1}\text{·m}^{-2} = 2.88 \times 10^{23} \text{ s}^{-1}\text{·cm}^{-2}$$

We can use Equation 27.48 to rederive Equation 27.9. The component of momentum perpendicular to the wall is $mu \cos\theta$, and if we assume that the collisions with the wall are elastic, then the change of momentum upon each collision is $2mu \cos\theta$ (Figure 27.7). The pressure exerted on the wall by those molecules whose speed is between u and $u + du$ and whose direction lies in the solid angle $\sin\theta d\theta d\phi$ is equal to the product of the momentum change per collision and the frequency of collisions per unit area (Equation 27.46)

$$dP = (2mu \cos\theta)dz_{coll}$$

$$= (2mu \cos\theta)\frac{\rho}{4\pi}uF(u)du \cos\theta \sin\theta d\theta d\phi$$

$$= \rho\left(\frac{m}{2\pi k_B T}\right)^{3/2}(2mu \cos\theta)u^3 e^{-mu^2/2k_B T}du \cos\theta \sin\theta d\phi$$

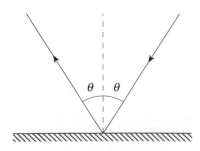

FIGURE 27.7
An elastic collision of a molecule with a wall. The component of velocity perpendicular to the wall is reversed in the collision. Thus, the total change in momentum is $2mu \cos\theta$.

We integrate this expression over all values of θ and ϕ (remember that $0 \leq \theta \leq \pi/2$)

$$\int_0^{\pi/2} \cos^2 \theta \sin \theta d\theta \int_0^{2\pi} d\phi = \frac{2\pi}{3}$$

and use the fact that

$$4\pi \left(\frac{m}{2\pi k_B T} \right)^{3/2} \int_0^\infty u^4 e^{-mu^2/2k_B T} du = \langle u^2 \rangle$$

to get

$$P = \frac{1}{3}\rho m \langle u^2 \rangle = \frac{1}{3V} Nm \langle u^2 \rangle$$

in agreement with Equation 27.9.

27–5. The Maxwell–Boltzmann Distribution Has Been Verified Experimentally

The Maxwell-Boltzmann distribution has been verified experimentally in a number of different experiments, but one of the most straightforward is due to Kusch and his coworkers at Columbia University in the 1950s. Their experimental set-up, sketched in Figure 27.8, consists of a furnace with a very small hole that allowed a beam of atoms (such as potassium) to emerge into an evacuated chamber. The beam passed through a pair of collimating slits and then through a velocity-selector, which allowed only those atoms with a given speed to reach a detector. The velocity selector (Figure 27.9) consisted of a set of rotating discs with slits cut in them in such a way that only those atoms with the right speed could pass through. Atoms of a given speed could be

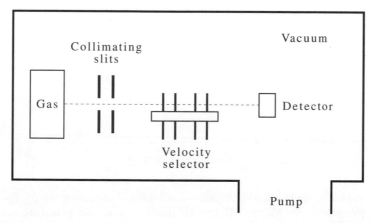

FIGURE 27.8
A schematic illustration of an apparatus used for an experimental test of the Maxwell-Boltzmann distribution.

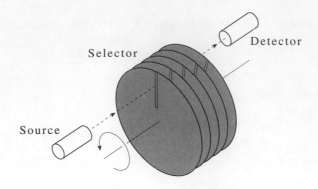

FIGURE 27.9
A sketch of a velocity selector. Only those atoms travelling at the right speed can pass through the set of rotating disks.

selected by rotating the disks at the appropriate frequency. The intensity measured at the detector gives the relative fraction of atoms with a given speed.

A comparison of the experimental results for gaseous potassium atoms and the prediction of the Maxwell-Boltzmann distribution is shown in Figure 27.10. The circles are the experimental data and the solid line is the predicted measured flux of potassium atoms as a function of the speed based upon the Maxwell-Boltzmann distribution of the emerging beam. The agreement between the two is seen to be excellent. Kusch was awarded the Nobel Prize in physics in 1955 for his work involving atomic and molecular beams.

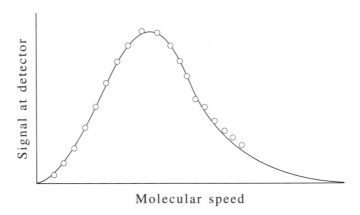

FIGURE 27.10
An experimental test of the Maxwell-Boltzmann distribution of molecular speeds. The solid line is computed according to the Maxwell-Boltzmann distribution and the points are experimental data of Miller and Kusch.

27–6. The Mean Free Path Is the Average Distance a Molecule Travels Between Collisions

When we discuss the theory of the rates of gas-phase chemical reactions in Chapter 30, we shall need to know about the frequency of collisions between the molecules in a gas. First let's consider the frequency of collisions of a single gas-phase molecule. As usual, we shall treat the molecules as hard spheres of diameter d. Furthermore, we shall assume that all the other molecules are stationary, and then take into account that all the molecules are moving relative to each other at the end of the derivation. As our molecule travels along, it sweeps out a cylinder of diameter $2d$ such that it will collide with any molecule whose center lies within this cylinder. This so-called *collision cylinder* is shown in Figure 27.11. Because there will be a collison if the center of our molecule comes within a distance d of the center of one of the other molecules, each of these molecules presents a target of effective radius d, and hence an area or a *collision cross section* equal to πd^2. Figure 27.11 illustrates that the radius of the collision cylinder is d, which is the diameter of the molecules. We shall denote the hard-sphere collision cross section πd^2 by the Greek letter σ. The volume of the collision cylinder is equal to its cross section (σ) times its length ($\langle u \rangle dt$), or $\sigma \langle u \rangle dt$. Because a collision will occur whenever the center of another molecule lies within this cylinder, the number of collisions that this molecule makes is equal to the number of molecules within the collision cylinder. If the number density of molecules is ρ, then the number of collisions in the time interval dt is

$$dN_{\text{coll}} = \rho \sigma \langle u \rangle dt$$

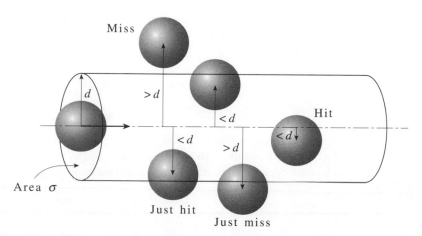

FIGURE 27.11
The collision cylinder swept out by a gas molecule as it travels through the gas. A collision will occur whenever the center of another molecule lies within the cylinder.

or the collision frequency, z_A, is

$$z_A = \frac{dN_{coll}}{dt} = \rho\sigma\langle u \rangle = \rho\sigma\left(\frac{8k_B T}{\pi m}\right)^{1/2} \tag{27.49}$$

Equation 27.49 is not quite correct because we assumed that all the molecules except the one that we were considering were stationary. We learned in Section 5–2 that we can treat the motion of two bodies of masses m_1 and m_2 moving with respect to each other by the motion of one body with a reduced mass $\mu = m_1 m_2/(m_1 + m_2)$ moving with respect to the other one being fixed. Thus, we can take into account that all the molecules are moving relative to each other by replacing m by μ in Equation 27.49. If the masses of the two colliding molecules are the same, then $\mu = m/2$ and the average relative speed, $\langle u_r \rangle$, is given by

$$\langle u_r \rangle = 2^{1/2}\langle u \rangle$$

Thus, the correct expression for z_A is

$$z_A = \rho\sigma\langle u_r \rangle = 2^{1/2}\rho\sigma\langle u \rangle \tag{27.50}$$

EXAMPLE 27–7
Use Equation 27.50 to calculate the collision frequency of a single nitrogen molecule in nitrogen at 25°C and one bar.

SOLUTION: According to Table 27.3, $\sigma = 0.450 \times 10^{-18}$ m² for nitrogen. The number density of nitrogen at 25°C and one bar was calculated in Example 27–6 to be $\rho = 2.43 \times 10^{25}$ m^{-3} and the average speed was calculated to be $\langle u \rangle = 475$ m·s^{-1}. The collision frequency then is

$$z_A = 2^{1/2}(2.43 \times 10^{25} \text{ m}^{-3})(0.450 \times 10^{-18} \text{ m}^2)(475 \text{ m·s}^{-1})$$
$$= 7.3 \times 10^9 \text{ s}^{-1}$$

To put this result into physical perspective, recall from Chapter 5 that a typical vibrational frequency of a diatomic molecule is about 10^{13}–10^{14} s^{-1}, and so we see that a typical diatomic molecule vibrates thousands of times between collisions (one bar and 25°C).

We should point out that the reciprocal of z_A is a measure of the average time between collisions. Thus, at one bar (Example 27–7), on average a nitrogen molecule has a collision every 1.4×10^{-10} s at 25°C.

We can determine the average distance that a molecule travels between collisions, its *mean free path*, l, by realizing that if a molecule travels at an average speed of $\langle u \rangle$

TABLE 27.3
Collision diameters, d(pm) and collision cross sections σ (nm^2) for various molecules.

Gas	d/pm	σ/nm^2
He	210	0.140
Ar	370	0.430
Xc	490	0.750
H_2	270	0.230
N_2	380	0.450
O_2	360	0.410
Cl_2	540	0.920
CH_4	410	0.530
C_2H_4	430	0.580

meters per second and makes z_A collisions in one second, then the average distance travelled between collisions is given by

$$l = \frac{\langle u \rangle}{z_A} = \frac{\langle u \rangle}{2^{1/2} \rho \sigma \langle u \rangle} = \frac{1}{2^{1/2} \rho \sigma}$$

If we replace ρ by its ideal gas value ($\rho = PN_A/RT$), then we have

$$l = \frac{RT}{2^{1/2} N_A \sigma P} \tag{27.51}$$

Equation 27.51 shows that at a given temperature, the mean free path is inversely proportional to the pressure. For nitrogen at $25°$C and one bar, l is equal to 6.5×10^{-8} m, which is about 200 times the effective diameter of a nitrogen molecule.

EXAMPLE 27–8
Calculate the mean free path of a hydrogen molecule at 298 K at the low pressure of 10^{-5} torr.

SOLUTION: According to Table 27.3, $\sigma = 0.230 \times 10^{-18}$ m^2 for H_2. Using Equation 27.51 gives

$$l = \frac{(0.08206 \text{ L·atm·mol}^{-1}\text{·K}^{-1})(298 \text{ K})}{2^{1/2}(6.022 \times 10^{23} \text{ mol}^{-1})(0.230 \times 10^{-18} \text{ m}^2)(1 \times 10^{-5}\text{torr})(1\text{atm}/760 \text{ torr})}$$
$$= 9500 \text{ L·m}^{-2} = 9.5 \text{ m}$$

where we have used the fact that $1 \text{ L} = 10^{-3}$ m^3.

We can get another physical interpretation of mean free path from the following argument. Once again, consider the cylinder that a molecule sweeps out as it moves along and let the direction of motion be along the x-axis. Furthermore, consider each molecule whose center lies within the cylinder to be a target. The number of such targets in a plane of unit area perpendicular to the x-direction and of thickness dx is ρdx, where ρ is the number density of molecules in the gas. Neglecting overlap, the total target area presented by these molecules is the collision cross section of each target (σ) times the total number of targets (ρdx), or $\sigma \rho dx$. The probability that our one molecule will suffer a collision then is the ratio of this area to the total area (unit area):

$$\text{probability of a collision} = \sigma \rho dx \tag{27.52}$$

Now consider a beam of n_0 molecules travelling with equal velocities in the positive direction and let them all start at $x = 0$. Furthermore, let $n(x)$ be the number of molecules that travel a distance x without a collision. The number of molecules that undergo a collision between x and $x + dx$ is the number of molecules reaching x, $n(x)$, multiplied by the probability of a collision in dx (Equation 27.52), and so

$$\begin{pmatrix} \text{number of molecules that} \\ \text{undergo a collision between} \\ x \text{ and } x + dx \end{pmatrix} = n(x)\sigma \rho dx$$

But a collision removes molecules from the beam. So this quantity is also equal to $n(x) - n(x + dx)$, the number that reach x minus the number that reach $x + dx$. Therefore, we can write

$$n(x) - n(x + dx) = \sigma \rho n(x) dx$$

We divide both sides by dx and use the definition

$$\frac{n(x + dx) - n(x)}{dx} = \frac{dn}{dx}$$

to get

$$\frac{dn}{dx} = -\sigma \rho n(x)$$

The solution to this equation is

$$n(x) = n_0 e^{-\sigma \rho x} \tag{27.53}$$

But $\sigma \rho$ is just the reciprocal of the mean free path (without the factor of $2^{1/2}$ that comes about when we allow all the molecules to be moving), and so we can write Equation 27.53 as

$$n(x) = n_0 e^{-x/l} \tag{27.54}$$

The number of molecules that collide in the interval x and $x + dx$ is $n(x) - n(x + dx)$, and so the probability $p(x)dx$ that one of the initial n_0 molecules will collide in this interval is

$$p(x)dx = \frac{n(x) - n(x + dx)}{n_0} = -\frac{1}{n_0}\frac{dn}{dx}dx$$

$$= \frac{1}{l}e^{-x/l}dx \tag{27.55}$$

It is easy to show that Equation 27.55 is normalized and also to show that $\langle x \rangle = l$, as you might expect.

EXAMPLE 27–9
Determine the distance at which one half of the molecules will have been scattered from a beam consisting initially of n_0 molecules.

SOLUTION: We shall use Equation 27.55. Let the distance be d and write

$$\frac{1}{l}\int_0^d e^{-x/l}dx = \tfrac{1}{2} = 1 - e^{-d/l}$$

or $d = l(\ln 2) = 0.693l$. Thus, one half of the molecules will be scattered before they travel 70% of the mean-free-path.

Figure 27.12 shows the probability that a molecule collides before it travels a distance x plotted against x/l.

One other quantity that we shall introduce in this section is the total collision frequency per unit volume, Z_{AA}, among all the molecules in a gas. This is another quantity that is involved in the theory of the rates of gas-phase reactions. If z_A is the

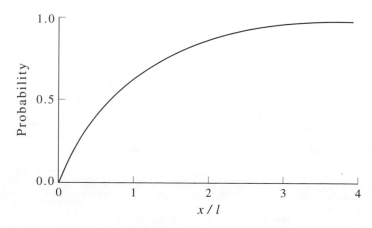

FIGURE 27.12
The probability that a molecule will collide before it travels a distance x versus x/l.

collision frequency of one particular molecule, then the total collision frequency per unit volume is obtained by multiplying z_A by the number density of molecules, ρ, and then dividing by 2 in order to avoid counting a collision between a pair of similar molecules as two distinct collisions. Thus we have from Equation 27.50

$$Z_{AA} = \tfrac{1}{2}\rho z_A = \tfrac{1}{2}\sigma \langle u_r\rangle \rho^2 = \frac{\sigma \langle u\rangle \rho^2}{2^{1/2}} \tag{27.56}$$

For nitrogen at 25°C and one bar, $Z_{AA} = 8.9 \times 10^{34}$ s$^{-1}\cdot$m^{-3}. In a gas consisting of two types of molecules, say A and B, then the collision frequency per unit volume is given by

$$Z_{AB} = \sigma_{AB}\langle u_r\rangle \rho_A \rho_B \tag{27.57}$$

where

$$\sigma_{AB} = \pi\left(\frac{d_A + d_B}{2}\right)^2 \qquad \text{and} \qquad \langle u_r\rangle = (8k_B T/\pi\mu)^{1/2} \tag{27.58}$$

where μ, the reduced mass, is equal to $m_A m_B/(m_A + m_B)$.

EXAMPLE 27–10
Calculate the frequency of nitrogen–nitrogen collisions in one cubic centimeter of air at one bar and 20°C. Assume that 80% of the molecules are nitrogen molecules.

SOLUTION: The partial pressure of nitrogen is 0.80 bar and the number density is

$$\rho = \frac{N_A P_{N_2}}{RT} = \frac{(6.022 \times 10^{23} \text{ mol}^{-1})(0.80 \text{ bar})}{(0.08314 \text{ L}\cdot\text{bar}\cdot\text{mol}^{-1}\cdot\text{K}^{-1})(293 \text{ K})}$$
$$= 2.0 \times 10^{22} \text{ L}^{-1} = 2.0 \times 10^{25} \text{ m}^{-3}$$

The average speed is

$$\langle u\rangle = \left(\frac{8RT}{\pi M}\right)^{1/2} = \left[\frac{8(8.314 \text{ J}\cdot\text{K}^{-1})(293 \text{ K})}{\pi(0.02802 \text{ kg})}\right]^{1/2}$$
$$= 470 \text{ m}\cdot\text{s}^{-1}$$

We use $\sigma_{N_2} = 4.50 \times 10^{-19}$ m^2 from Table 27.3 and so

$$Z_{N_2,N_2} = \frac{(4.50 \times 10^{-19} \text{ m}^2)(470 \text{ m}\cdot\text{s}^{-1})(2.0 \times 10^{25} \text{ m}^{-3})^2}{2^{1/2}}$$
$$= 6.0 \times 10^{34} \text{ s}^{-1}\cdot\text{m}^{-3} = 6.0 \times 10^{28} \text{ s}^{-1}\cdot\text{cm}^{-3}$$

27–7. The Rate of a Gas–Phase Chemical Reaction Depends Upon the Rate of Collisions in Which the Relative Kinetic Energy Exceeds Some Critical Value.

In Example 27–10, we calculated that the number of collisions between molecules at one bar and 20°C is about 6×10^{28} $\text{s}^{-1} \cdot \text{cm}^{-3}$, or about 10^8 $\text{mol} \cdot \text{dm}^{-3} \cdot \text{s}^{-1}$. Consider now a gas-phase chemical reaction $A + B \rightarrow$ products. If every collision led to a reaction, then the reaction would proceed at a rate of 10^8 $\text{mol} \cdot \text{dm}^{-3} \cdot \text{s}^{-1}$ and one mole per liter would be consumed in 10^{-8} seconds, which is much faster than most rates of chemical reactions. When we study the theory of the rates of chemical reactions in the gas phase in Chapter 30, one of the assumptions that we shall make is that the *relative* energy of two colliding molecules must exceed a certain critical value in order for a reaction to occur. Thus, we shall need to know not just the total frequency of collisions, Equation 27.57, but those in which the relative energy of the two colliding molecules exceeds a certain critical value.

To derive this result, we shall start with Equation 27.46 for the collision frequency of the molecules of a gas with a wall. Although the collisions of the molecules of a gas with a wall is certainly not the same as the collisions of the molecules of a gas with each other, the physical result that molecules travelling at higher speeds are more likely to strike the wall in a given time carries over to the case of collisions between molecules. Mathematically, the importance of molecules travelling at higher speeds is seen through the factor $u^3 e^{-mu^2/2k_B T}$, as shown in Figure 27.6. We can take into account that the molecules collide with each other rather than with a stationary wall by replacing m by the reduced mass $\mu = m_A m_B/(m_A + m_B)$. Thus, we shall write that the collision frequency per unit volume between molecules A and B in which they collide with a relative speed between u_r and $u_r + du_r$ is proportional to $u_r^3 e^{-\mu u_r^2/2k_B T}$, or

$$
\begin{aligned}
dZ_{AB} &\propto u_r^3 e^{-\mu u_r^2/2k_B T} du_r \\
&= A u_r^3 e^{-\mu u_r^2/2k_B T} du_r
\end{aligned}
\tag{27.59}
$$

where A is a proportionality constant. Equation 27.59 is simply the differential form of Equation 27.57. We can determine A by requiring that the integral of Equation 27.59 over all relative speeds be equal to Z_{AB} given by Equation 27.57. Therefore, we write

$$
\begin{aligned}
\sigma_{AB} \rho_A \rho_B \left(\frac{8 k_B T}{\pi \mu} \right)^{1/2} &= A \int_0^\infty u_r^3 e^{-\mu u_r^2/2k_B T} du_r \\
&= 2A \left(\frac{k_B T}{\mu} \right)^2
\end{aligned}
\tag{27.60}
$$

where the integral is given in Table 27.2. Solving Equation 27.60 for A gives

$$
A = \sigma_{AB} \rho_A \rho_B \left(\frac{\mu}{k_B T} \right)^{3/2} \left(\frac{2}{\pi} \right)^{1/2}
$$

and so Equation 27.59 becomes

$$dZ_{AB} = \sigma_{AB}\rho_A\rho_B \left(\frac{\mu}{k_BT}\right)^{3/2} \left(\frac{2}{\pi}\right)^{1/2} e^{-\mu u_r^2/2k_BT} u_r^3 du_r \tag{27.61}$$

This expression represents the collision frequency per unit volume between molecules of types A and B with relative speeds in the range u_r and $u_r + du_r$. Notice that this distribution has a factor of u_r^3, which reflects the fact that molecules with higher relative speeds collide more frequently. The factor

$$\left(\frac{\mu}{k_BT}\right)^{3/2} \left(\frac{2}{\pi}\right)^{1/2} e^{-\mu u_r^2/2k_BT} u_r^3 du_r$$

in Equation 27.61 is proportional to the probability that the molecules have a relative speed between u_r and $u_r + du_r$.

EXAMPLE 27–11

Derive an expression for the collision frequency per unit volume in which the relative kinetic energy exceeds some critical value ε_c.

SOLUTION: We start with Equation 27.61 and convert to relative kinetic energy, $\varepsilon_r = \mu u_r^2/2$. Solving for u_r,

$$u_r = (2\varepsilon_r/\mu)^{1/2} \qquad du_r = (1/2\mu\varepsilon_r)^{1/2} d\varepsilon_r$$

and substituting into Equation 27.61 gives

$$dZ_{AB} = \sigma_{AB}\rho_A\rho_B \left(\frac{8}{\pi\mu}\right)^{1/2} \left(\frac{1}{k_BT}\right)^{3/2} \varepsilon_r e^{-\varepsilon_r/k_BT} d\varepsilon_r \tag{27.62}$$

This expression represents the collision frequency per unit volume in which the relative kinetic energies of the colliding particles are between ε_r and $\varepsilon_r + d\varepsilon_r$.

To find the collision frequency per unit volume in which the realtive kinetic energy exceeds ε_c, we integrate Equation 27.62 from ε_c to ∞, using

$$\int_{\varepsilon_c}^{\infty} \varepsilon_r e^{-\varepsilon_r/k_BT} d\varepsilon_r = (k_BT)^2 \left(1 + \frac{\varepsilon_c}{k_BT}\right) e^{-\varepsilon_c/k_BT}$$

to get

$$Z_{AB}(\varepsilon_r > \varepsilon_c) = \sigma_{AB}\rho_A\rho_B \left(\frac{8k_BT}{\pi\mu}\right)^{1/2} \left(1 + \frac{\varepsilon_c}{k_BT}\right) e^{-\varepsilon_c/k_BT} \tag{27.63}$$

Note that this quantity varies essentially as $e^{-\varepsilon_c/k_BT}$

Problems

27-1. Calculate the average translational energy of one mole of ethane at 400 K, assuming ideal behavior. Compare your result to $\overline{U}^{\text{id}}$ for ethane at 400 K given in Figure 22.3.

27-2. Calculate the root-mean-square speed of a nitrogen molecule at 200 K, 300 K, 500 K, and 1000 K.

27-3. If the temperature of a gas is doubled, by how much is the root-mean-square speed of the molecules increased?

27-4. The speed of sound in air at sea level at 20°C is about 770 mph. Compare this value with the root-mean-square speed of nitrogen and oxygen molecules at 20°C.

27-5. Arrange the following gases in order of increasing root-mean-square speed at the same temperature: O_2, N_2, H_2O, CO_2, NO_2, $^{235}UF_6$, and $^{238}UF_6$.

27-6. Consider a mixture of $H_2(g)$ and $I_2(g)$. Calculate the ratio of the root-mean-square speed of $H_2(g)$ and $I_2(g)$ molecules in the reaction mixture.

27-7. The speed of sound in an ideal monatomic gas is given by

$$u_{\text{sound}} = \left(\frac{5RT}{3M} \right)^{1/2}$$

Derive an equation for the ratio $u_{\text{rms}}/u_{\text{sound}}$. Calculate the root-mean-square speed for an argon atom at 20°C and compare your answer to the speed of sound in argon.

27-8. Calculate the speed of sound in argon at 25°C.

27-9. The speed of sound in an ideal polyatomic gas is given by

$$u_{\text{sound}} = \left(\frac{\gamma RT}{M} \right)^{1/2}$$

where $\gamma = C_P/C_V$. Calculate the speed of sound in nitrogen at 25°C.

27-10. Use Equation 27.17 to prove that $\partial u/\partial u_x = u_x/u$.

27-11. Give a physical argument why γ in Equation 27.24 must be a positive quantity.

27-12. We can use Equation 27.33 to calculate the probability that the x-component of the velocity of a molecule lies within some range. For example, show that the probability that $-u_{x0} \leq u_x \leq u_{x0}$ is given by

$$\text{Prob}\{-u_{x0} \leq u_x \leq u_{x0}\} = \left(\frac{m}{2\pi k_B T} \right)^{1/2} \int_{-u_{x0}}^{u_{x0}} e^{-mu_x^2/2k_B T} du_x$$

$$= 2 \left(\frac{m}{2\pi k_B T} \right)^{1/2} \int_0^{u_{x0}} e^{-mu_x^2/2k_B T} du_x$$

Now let $mu_x^2/2k_B T = w^2$ to get the cleaner-looking expression

$$\text{Prob}\{-u_{x0} \leq u_x \leq u_{x0}\} = \frac{2}{\pi^{1/2}} \int_0^{w_0} e^{-w^2} dw$$

where $w_0 = (m/2k_B T)^{1/2} u_{x0}$.

It so happens that the above integral cannot be evaluated in terms of any function that we have encountered up to now. It is customary to express the integral in terms of a new function called the *error function*, which is defined by

$$\text{erf}(z) = \frac{2}{\pi^{1/2}} \int_0^z e^{-x^2} dx \tag{1}$$

The error function can be evaluated as a function of z by evaluating its defining integral numerically. Some values of erf(z) are

z	erf(z)	z	erf(z)
0.20	0.22270	1.20	0.91031
0.40	0.42839	1.40	0.95229
0.60	0.60386	1.60	0.97635
0.80	0.74210	1.80	0.98909
1.00	0.84270	2.00	0.99532

Now show that

$$\text{Prob}\{-u_{x0} \leq u_x \leq u_{x0}\} = \text{erf}(w_0)$$

Calculate the probability that $-(2k_B T/m)^{1/2} \leq u_x \leq (2k_B T/m)^{1/2}$?

27-13. Use the result of Problem 27–12 to show that

$$\text{Prob}\{|u_x| \geq u_{x0}\} = 1 - \text{erf}(w_0)$$

27-14. Use the result of Problem 27–12 to calculate $\text{Prob}\{u_x \geq +(k_B T/m)^{1/2}\}$ and $\text{Prob}\{u_x \geq +(2k_B T/m)^{1/2}\}$.

27-15. Use the result of Problem 27–12 to plot the probability that $-u_{x0} \leq u_x \leq u_{x0}$ against $u_{x0}/(2k_B T/m)^{1/2}$.

27-16. Use Simpson's rule or any other numerical integration routine to verify the values of erf(z) given in Problem 27–12. Plot erf(z) against z.

27-17. Derive an expression for the average value of the positive values of u_x.

27-18. This problem deals with the idea of the *escape velocity* of a particle from a body such as the Earth's surface. Recall from your course in physics that the potential energy of two masses, m_1 and m_2, separated by a distance r is given by

$$V(r) = -\frac{Gm_1 m_2}{r}$$

(note the similarity with Coulomb's law) where $G = 6.67 \times 10^{-11}$ J·m·kg^{-1} is called the gravitional constant. Suppose a particle of mass m has a velocity u perpendicular to the Earth's surface. Show that the minimum velocity that the particle must have in order to escape the Earth's surface (its *escape velocity*) is given by

$$u = \left(\frac{2GM_{earth}}{R_{earth}} \right)^{1/2}$$

Given that $M_{earth} = 5.98 \times 10^{24}$ kg is the mass of the Earth and $R_{earth} = 6.36 \times 10^6$ m is its mean radius, calculate the escape velocity of a hydrogen molecule and a nitrogen molecule. What temperature would each of these molecules have to have so that their average speed exceeds their escape velocity?

27-19. Repeat the calculation in the previous problem for the moon's surface. Take the mass of the moon to be 7.35×10^{22} kg and its radius to be 1.74×10^6 m.

27-20. Show that the variance of Equation 27.37 is given by $\sigma^2 = v_0^2 k_B T/mc^2$. Calcuate σ for the $3p\,^2P_{3/2}$ to $3s\,^2S_{1/2}$ transition in atomic sodium vapor (see Figure 8.4) at 500 K.

27-21. Show that the distribution of speeds for a two-dimensional gas is given by

$$F(u)du = \frac{m}{k_B T} u e^{-mu^2/2k_B T} du$$

(Recall that the area element in plane polar coordinates is $r\,dr\,d\theta$.)

27-22. Use the formula in the previous problem to derive formulas for $\langle u \rangle$ and $\langle u^2 \rangle$ for a two-dimensional gas. Compare your result for $\langle u^2 \rangle$ to $\langle u_x^2 \rangle + \langle u_y^2 \rangle$.

27-23. Use the formula in Problem 27–21 to calculate the probability that $u \geq u_0$ for a two-dimensional gas.

27-24. Show that the probability that a molecule has a speed less than or equal to u_0 is given by

$$\text{Prob}\{u \leq u_0\} = \frac{4}{\pi^{1/2}} \int_0^{x_0} x^2 e^{-x^2} dx$$

where $x_0 = (m/2k_B T)^{1/2} u_0$. This integral cannot be expressed in terms of any simple function and must be integrated numerically. Use Simpson's rule or any other integration routine to evaluate $\text{Prob}\{u \leq (2k_B T/m)^{1/2}\}$.

27-25. Using Simpson's rule or any other integration routine, plot $\text{Prob}\{u \leq u_0\}$ against $u_0/(m/2k_B T)^{1/2}$. (see Problem 27–24.)

27-26. What is the most probable kinetic energy for a molecule in the gas phase?

27-27. Derive an expression for $\sigma_\varepsilon^2 = \langle \varepsilon^2 \rangle - \langle \varepsilon \rangle^2$ from Equation 27.44. Now form the ratio $\sigma_\varepsilon/\langle \varepsilon \rangle$. What does this say about the fluctuations in ε?

27-28. Compare the most probable speed of a molecule that collides with a small surface area with the most probable speed of a molecule in the bulk of the gas phase.

27-29. Use Equation 27.48 to calculate the collision frequency per unit area for helium at 100 K and 10^{-6} torr.

27-30. Calculate the average speed of a molecule that strikes a small surface area. How does this value compare to the average speed of all the molecules?

27-31. How long will it take for an initially clean surface to become 1.0% covered if it is bathed by an atmosphere of nitrogen at 77 K and one bar? Assume that every nitrogen molecule that strikes the surface sticks and that a nitrogen molecule covers an area of 1.1×10^5 pm^2.

27-32. Calculate the number of methane molecules at 25°C and one torr that strike a 1.0 cm^2 surface in one millisecond.

27-33. Consider the velocity selector shown in Figure 27.9. Let the distance between successive disks be h, the rotational frequency be v (in units of Hz), and the angle between the slits of successive disks be θ (in degrees). Derive the following condition for a molecule traveling with speed u to pass through successive slits:

$$u = \frac{360vh}{\theta}$$

Typical values of h and θ are 2 cm and 2°, respectively, so $u = 3.6v$. By varying v from 0 to about 500 Hz, you can select speeds from 0 to over 1500 m·s^{-1}.

27-34. The figure below illustrates another method that has been used to determine the distribution of molecular speeds. A pulse of molecules collimated from a hot oven enters a rotating hollow drum. Let R be the radius of the drum, v be its rotational frequency, and s be the distance through which the drum rotates during the time it takes for a molecule to travel from the entrance slit to the inner surface of the drum. Show that

$$s = \frac{4\pi R^2 v}{u}$$

where u is the speed of the molecule.

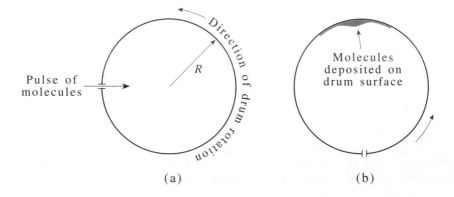

(a)   (b)

Use Equation 27.46 to show that the distribution of molecular speeds emerging from the oven is proportional to $u^3 e^{-mu^2/2k_B T} du$. Now show that the distribution of molecules striking the inner surface of the cylinder is given by

$$I(s)ds = \frac{A}{s^5} e^{-m(4\pi R^2 v)^2/2k_B Ts^2} ds$$

where A is simply a proportionality constant. Plot I versus s for various values of $4\pi R^2 v/(2k_B T/m)^{1/2}$, say 0.1, 1, and 3. Experimental data are quantitatively described by the above equation.

27-35. Use Equation 27.49 to calculate the collision frequency of a single hydrogen molecule at 25°C and (a) one torr and (b) one bar.

27-36. On the average, what is the time between collisions of a xenon atom at 300 K and (a) one torr and (b) one bar.

27-37. What is the probability that an oxygen molecule at 25°C and one bar will travel (a) 1.00×10^{-5} mm, (b) 1.00×10^{-3} mm, and (c) 1.00 mm without undergoing a collision?

27-38. Repeat the calculation in the previous problem for a pressure of one torr.

27-39. At an altitude of 150 km, the pressure is about 2×10^{-6} torr and the temperature is about 500 K. Assuming for simplicity that the air consists entirely of nitrogen, calculate the mean free path under these conditions. What is the average collision frequency?

27-40. The following table gives the pressure and temperature of the Earth's upper atmosphere as a function of altitude:

altitude/km	P/mbar	T/K
20.0	56	220
40.0	3.2	260
60.0	0.28	260
80.0	0.013	180

Assuming for simplicity that air consists entirely of nitrogen, calculate the mean free path at each of these conditions.

27-41. Interstellar space has an average temperature of about 10 K and an average density of hydrogen atoms of about one hydrogen atom per cubic meter. Compute the mean free path of a hydrogen atom in interstellar space. Take the diameter of a hydrogen atom to be 100 pm.

27-42. Calculate the pressures at which the mean free path of a hydrogen molecule will be 100 μm, 1.00 mm, and 1.00 m at 20°C.

27-43. Derive an expression for the distance, d, at which a fraction f of the molecules will have been scattered from a beam consisting initially of n_0 molecules. Plot d against f.

27-44. Calculate the frequency of nitrogen–oxygen collisions per dm^3 in air at the conditions given in Problem 27–40. Assume in this case that 80% of the molecules are nitrogen molecules.

27-45. Use Equation 27.58 to show that

$$\langle u_r \rangle = (\langle u_A \rangle^2 + \langle u_B \rangle^2)^{1/2}$$

27-46. Modify the derivation of Equation 27.49 to consider the collision frequency of a molecule of type A with B molecules in a mixture of A and B. Derive Equation 27.57 directly from your answer.

27-47. Consider a mixture of methane and nitrogen in a $10.0\ dm^3$ container at 300 K with partial pressures $P_{CH_4} = 65.0$ mbar and $P_{N_2} = 30.0$ mbar. Use the equation that you derived in the previous problem to calculate the collision frequency of a methane molecule with nitrogen molecules. Also calculate the frequency of methane–nitrogen collisions per dm^3.

27-48. Calculate the average relative kinetic energy with which the molecules in a gas collide.

The following four problems deal with molecular effusion.

27-49. Equation 27.48 gives us the frequency of collisions that the molecules of a gas make with a surface area of the walls of the container. Suppose now we make a very small hole in the wall. If the mean free path of the gas is much larger than the width of the hole, any molecule that strikes the hole will leave the container without undergoing any collisions along the way. In this case, the molecules leave the container individually, independently of the others. The rate of flow through the hole will be small enough that the remaining gas is unaffected, and remains essentially in equilibrium. This process is called *molecular effusion*. Equation 27.48 can be applied to calculate the rate of molecular effusion. Show that Equation 27.48 can be expressed as

$$\text{effusion flux} = \frac{P}{(2\pi m k_B T)^{1/2}} = \frac{N_A P}{(2\pi M R T)^{1/2}} \tag{1}$$

where P is the pressure of the gas. Calculate the number of nitrogen molecules that effuse per second through a round hole of 0.010 mm diameter if the gas is at 25°C and one bar.

27-50. Equation 1 of the previous problem can be used to determine vapor pressures of substances with very low vapor pressures. This was done by Irving Langmuir to measure the vapor pressure of tungsten at various temperatures in his investigation of tungsten filaments in light bulbs and vacuum tubes. (Langmuir, who was awarded the Nobel Prize in chemistry in 1932, worked for General Electric.) He estimated the rate of effusion by weighing the tungsten filament at the beginning and the end of each experimental run. Langmuir did these experiments around 1913, but his data appear in the *CRC Handbook of Chemistry and Physics* to this day. Use the following data to determine the vapor pressure of tungsten at each temperature and then determine the molar enthalpy of vaporization of tungsten.

T/K	effusion flux/g$\cdot$m$^{-2}\cdot$s^{-1}
1200	3.21×10^{-23}
1600	1.25×10^{-14}
2000	1.76×10^{-9}
2400	4.26×10^{-6}
2800	1.10×10^{-3}
3200	6.38×10^{-3}

27-51. The vapor pressure of mercury can be determined by the effusion technique described in the previous problem. Given that 0.126 mg of mercury passed through a small hole of area 1.65 mm^2 in 2.25 hours at 0°C, calculate the vapor pressure of mercury in torr.

27-52. We can use Equation 1 of Problem 27–49 to derive an expression for the pressure as a function of time for an ideal gas that is effusing from its container. First show that

$$\text{rate of effusion} = -\frac{dN}{dt} = \frac{PA}{(2\pi m k_\text{B} T)^{1/2}}$$

where N is the number of molecules effusing and A is the area of the hole. At constant T and V,

$$\frac{dN}{dt} = \frac{d}{dt}\left(\frac{PV}{k_\text{B}T}\right) = \frac{V}{k_\text{B}T}\frac{dP}{dt}$$

Now show that

$$P(t) = P(0)e^{-\alpha t}$$

where $\alpha = (k_\text{B}T/2\pi m)^{1/2}A/V$. Note that the pressure of the gas decreases exponentially with time.

27-53. How would you interpret the velocity distribution

$$h(v_x, v_y, v_z) = \left(\frac{m}{2\pi k_\text{B}T}\right)^{3/2} \exp\left[-\frac{m}{2k_\text{B}T}\left\{(v_x - a)^2 + (v_y - b)^2 + (v_z - c)^2\right\}\right]$$

Svante Arrhenius was born in Wijk near Uppsala, Sweden, on February 19, 1859, and died in 1927. Arrhenius' name appears in all physical chemistry texts because of his equation that describes the temperature dependence of a reaction rate constant in terms of activation energy. Even more famous, however, is his work dealing with the properties of solutions of weak electrolytes. He received his doctorate in 1884 from the University of Uppsala for his dissertation on the theory of electrolytic solutions. His thesis work was quite controversial and was not immediately accepted; in fact, he barely obtained his doctorate. He then received a traveling scholarship, which allowed him to spend five years studying in Europe with Ostwald, Boltzmann, and van't Hoff. Upon his return, he could not obtain a university position because of the controversy surrounding his thesis, and he became a teacher at the Technical High School in Stockholm. Two years later, his position was elevated to professor at the University of Stockholm after he underwent an oral examination by a hostile committee. Ostwald and van't Hoff were instrumental in gaining acceptance of his work and they published his paper "On the Dissociation of Substances in Aqueous Solutions" in the first issue of their journal, *Zeitschrift für Physikalische Chemie*. In 1904, Arrhenius became the first director of the newly created Nobel Institute for Physical Research in Stockholm. In 1903, he was awarded the Nobel Prize for chemistry "for his electrolytic theory of dissociations."

CHAPTER 28

Chemical Kinetics I: Rate Laws

This chapter begins our study of the area of physical chemistry known as chemical kinetics. Our development of chemical kinetics will differ from our presentation of quantum mechanics and thermodynamics. In developing quantum mechanics, we started with a small set of postulates, and classical thermodynamics is built upon just three laws. If we had our choice, we would certainly develop chemical kinetics starting with a few very simple principles. Unfortunately, this is not yet possible. The field of chemical kinetics has not yet matured to a point where a set of unifying principles has been identified, but the current search for such a set contributes to the excitement of modern research in the field.

Presently, there are many different theoretical models for describing how chemical reactions occur. None is perfect, but each has its merits. Several provide a microscopic picture of how chemical reactions take place. Thus, in chemical kinetics, you must become familiar with different ideas and sometimes concepts that seem unrelated. Bear in mind that this situation is common in scientific disciplines in which further research is needed in order to provide a more fundamental understanding of the subject.

This chapter presents some of the phenomenological concepts of chemical kinetics. You will learn that the time-dependence of the reactant and product concentrations during a chemical reaction can be described by differential equations known as rate laws. A rate law serves to define a rate constant, which is one of the most important parameters used to describe the dynamics of chemical reactions. You will also learn that rate laws are determined from experimental data, and we will discuss some of the experimental techniques used to deduce rate laws. We will examine several rate laws and show how they can be integrated to give mathematical expressions for the time-dependent concentrations. Finally, you will learn that rate constants are temperature dependent and how to describe this behavior mathematically.

28–1. The Time Dependence of a Chemical Reaction Is Described by a Rate Law

Consider the general chemical reaction described by

$$v_A A + v_B B \longrightarrow v_Y Y + v_Z Z \tag{28.1}$$

Recall that we defined the extent of reaction, ξ, in Chapter 26 such that

$$n_A(t) = n_A(0) - v_A \xi(t) \qquad n_B(t) = n_B(0) - v_B \xi(t)$$

$$n_Y(t) = n_Y(0) + v_Y \xi(t) \qquad n_Z(t) = n_Z(0) + v_Z \xi(t) \tag{28.2}$$

where $n_j(0)$ denotes the initial values of n_j. The extent of reaction, ξ, has units of moles and connects the amount of reaction that has occurred to the stoichiometry dictated by the balanced chemical equation. The change in $n_j(t)$ with time is then given by

$$\frac{dn_A(t)}{dt} = -v_A \frac{d\xi(t)}{dt} \qquad \frac{dn_B(t)}{dt} = -v_B \frac{d\xi(t)}{dt}$$

$$\frac{dn_Y(t)}{dt} = v_Y \frac{d\xi(t)}{dt} \qquad \frac{dn_Z(t)}{dt} = v_Z \frac{d\xi(t)}{dt} \tag{28.3}$$

Most experimental techniques measure concentration as a function of time. If the volume, V, of the system is constant, then dividing Equations 28.3 by V gives the corresponding expressions for the time-dependent concentrations,

$$\frac{1}{V}\frac{dn_A(t)}{dt} = \frac{d[A]}{dt} = -\frac{v_A}{V}\frac{d\xi(t)}{dt} \qquad \frac{1}{V}\frac{dn_B(t)}{dt} = \frac{d[B]}{dt} = -\frac{v_B}{V}\frac{d\xi(t)}{dt}$$

$$\frac{1}{V}\frac{dn_Y(t)}{dt} = \frac{d[Y]}{dt} = \frac{v_Y}{V}\frac{d\xi(t)}{dt} \qquad \frac{1}{V}\frac{dn_Z(t)}{dt} = \frac{d[Z]}{dt} = \frac{v_Z}{V}\frac{d\xi(t)}{dt} \tag{28.4}$$

where [A], for example, is equal to $n_A(t)/V$. The above expressions are used to define the *rate of reaction*, $v(t)$:

$$v(t) = -\frac{1}{v_A}\frac{d[A]}{dt} = -\frac{1}{v_B}\frac{d[B]}{dt} = \frac{1}{v_Y}\frac{d[Y]}{dt} = \frac{1}{v_Z}\frac{d[Z]}{dt} = \frac{1}{V}\frac{d\xi}{dt} \tag{28.5}$$

Note that all the quantities in Equation 28.5 are positive. For example, the rate of reaction for

$$2\,NO(g) + O_2(g) \longrightarrow 2\,NO_2(g) \tag{28.6}$$

is given by

$$v(t) = -\frac{1}{2}\frac{d[NO]}{dt} = -\frac{d[O_2]}{dt} = \frac{1}{2}\frac{d[NO_2]}{dt} \tag{28.7}$$

For most chemical reactions, $v(t)$ is related to the concentrations of the various chemical species present at time t. The relationship between $v(t)$ and the concentrations is called the *rate law*. Rate laws must be determined from experimental measurements. Rate laws cannot, in general, be deduced from the balanced chemical reaction. For example, experimental studies reveal that the reaction between nitrogen monoxide and oxygen to form nitrogen dioxide, Equation 28.6, obeys the rate law

$$v(t) = k[NO]^2[O_2] \tag{28.8}$$

where k is a constant. Equation 28.8 shows that the rate is proportional to $[NO]^2[O_2]$. The proportionality constant, k, is called the *rate constant* for the reaction. For this particular rate law, the rate depends differently on the concentrations of the two reactants. A doubling of the oxygen concentration results in a doubling of the reaction rate, whereas a doubling of the nitrogen monoxide concentration causes a quadrupling of the reaction rate.

The rate laws often have the form

$$v(t) = k[A]^{m_A}[B]^{m_B} \cdots \tag{28.9}$$

where $[A]$, $[B]$, $\cdots$ are the concentrations of the various reactants and the exponents, or *orders*, m_A, m_B, $\cdots$, are constants (see Table 28.1). We say that the rate law given by Equation 28.9 is m_Ath order in A, m_Bth order in B, etc. For example, the rate law given by Equation 28.8 is second order in NO and first order in O_2, and the rate law given by the third entry in Table 28.1 is 3/2 order in CH_3CHO. For many of the reactions listed in Table 28.1, the order of the reactant differs from its stoichiometric coefficient in the balanced chemical reaction; this is often the case. The examples listed in Table 28.1 are all gas-phase chemical reactions; however, the rate law concept applies to all reactions regardless of the phases of the reactants, products, and surrounding medium. When the rate law can be written as in Equation 28.9, the sum of the exponents is commonly referred to as the overall order of the chemical reaction. For example, the rate law given in Equation 28.8 has an overall reaction order of three.

TABLE 28.1

Examples of gas-phase chemical reactions and their corresponding rate laws

Chemical reaction	Rate law
$H_2(g) + I_2(g) \rightarrow 2\,HI(g)$	$v = k[H_2][I_2]$
$2\,NO(g) + O_2(g) \rightarrow 2\,NO_2(g)$	$v = k[NO]^2[O_2]$
$CH_3CHO(g) \rightarrow CH_4(g) + CO(g)$	$v = k[CH_3CHO]^{3/2}$
$NO_2(g) + CO(g) \rightarrow CO_2(g) + NO(g)$	$v = k[NO_2]^2$
$Cl_2(g) + CO(g) \rightarrow Cl_2CO(g)$	$v = k[Cl_2]^{3/2}[CO]$
$2\,NO(g) + 2\,H_2(g) \rightarrow N_2(g) + 2\,H_2O(g)$	$v = k[NO]^2[H_2]$

TABLE 28.2
The orders and units of reaction rate constants, k, for different rate laws

Rate law	Order	Units of k
$v = k$	0	$\text{dm}^{-3} \cdot \text{mol} \cdot \text{s}^{-1}$
$v = k[A]$	1	s^{-1}
$v = k[A]^2$	2	$\text{dm}^3 \cdot \text{mol}^{-1} \cdot \text{s}^{-1}$
$v = k[A][B]$	1 in [A]	
	1 in [B]	
	overall: 2	$\text{dm}^3 \cdot \text{mol}^{-1} \cdot \text{s}^{-1}$
$v = k[A]^{1/2}$	1/2	$\text{dm}^{-3/2} \cdot \text{mol}^{1/2} \cdot \text{s}^{-1}$
$v = k[A][B]^{1/2}$	1 in [A]	
	1/2 in [B]	
	overall: 3/2	$\text{dm}^{3/2} \cdot \text{mol}^{-1/2} \cdot \text{s}^{-1}$

The units of a rate constant depend on the form of the rate law. Table 28.2 gives some examples of rate laws, their overall orders, and the corresponding units of their reaction rate constants.

EXAMPLE 28–1
The standard SI unit of concentration is $\text{mol} \cdot \text{dm}^{-3}$. In the scientific literature, we frequently encounter units of $\text{mol} \cdot \text{L}^{-1}$ for reactions in solution and $\text{molecule} \cdot \text{cm}^{-3}$ for gas-phase reactions. A liter is equivalent to a dm^3, so $\text{mol} \cdot \text{L}^{-1}$ is equivalent to $\text{mol} \cdot \text{dm}^{-3}$, but how do we convert the older unit of $\text{molecule} \cdot \text{cm}^{-3}$ to SI units?

SOLUTION: One $\text{molecule} \cdot \text{cm}^{-3}$ corresponds to

$$(1 \text{ molecule} \cdot \text{cm}^{-3}) \left(\frac{1}{6.022 \times 10^{23} \text{ molecule} \cdot \text{mol}^{-1}} \right) \left(\frac{10 \text{ cm}}{\text{dm}} \right)^3$$
$$= 1.661 \times 10^{-21} \text{ mol} \cdot \text{dm}^{-3}$$

Thus, for example, a concentration of $2.00 \times 10^{20} \text{ molecule} \cdot \text{cm}^{-3}$ corresponds to

$$(2.00 \times 10^{20} \text{ molecule} \cdot \text{cm}^{-3}) \left(\frac{1.661 \times 10^{-21} \text{ mol} \cdot \text{dm}^{-3}}{1 \text{ molecule} \cdot \text{cm}^{-3}} \right) = 0.332 \text{ mol} \cdot \text{dm}^{-3}$$

Many rate laws cannot be written in the form of Equation 28.9. For example, the rate law for the reaction

$$H_2(g) + Br_2(g) \longrightarrow 2\,HBr(g) \tag{28.10}$$

is given by

$$v(t) = \frac{k'[H_2][Br_2]^{1/2}}{1 + k''[HBr][Br_2]^{-1}} \tag{28.11}$$

where k' and k'' are constants. In this case, the concept of reaction order has no meaning. We will show in Chapter 29 that such a complicated rate law tells us that the chemical reaction occurs by a multistep process.

28–2. Rate Laws Must Be Determined Experimentally

In this section, we will examine two experimental techniques that chemists use to determine rate laws. For discussion purposes, we consider the general chemical equation

$$\nu_A A + \nu_B B \longrightarrow \nu_Y Y + \nu_Z Z \tag{28.12}$$

and assume that the rate law has the form

$$v = k[A]^{m_A}[B]^{m_B} \tag{28.13}$$

If we knew the reaction orders m_A and m_B, then a measurement of the rate as a function of concentrations would enable us to determine the rate constant k. Our problem, then, is how to determine the values of m_A and m_B.

Suppose the initial reaction mixture has a large excess concentration of A. In this case, the concentration of A remains essentially constant as the reaction takes place. Equation 28.13 therefore simplifies to

$$v = k'[B]^{m_B} \tag{28.14}$$

where $k' = k[A]^{m_A}$ is a constant. The order of B can then be determined by measuring the rate as a function of [B]. The only requirement is that A always be in large excess so that k' remains constant. Likewise, if B is initially present in large excess, then Equation 28.13 simplifies to

$$v = k''[A]^{m_A} \tag{28.15}$$

where $k'' = k[B]^{m_B}$ is a constant. The order of A can then be determined by measuring the rate as a function of [A]. This technique is called the *method of isolation* and can be extended to reactions that involve more than two reactants.

Sometimes it is not possible to have a reactant in excess. We still need to be able to determine the orders of the various reactants but cannot use the method of isolation. Ideally, if we have many measurements of the rate, $d[A]/dt$, at various concentrations of [A] and [B], the orders of the reactants and the rate constant could be determined directly by fitting the data to Equation 28.13. Unfortunately, it is not possible to measure the differential, $d[A]/dt$. We can, however, measure the concentration change for a finite time period, Δt; in other words, we can measure $\Delta[A]/\Delta t$. If we equate such a measurement to the reaction rate, then

$$v = -\frac{d[A]}{v_A dt} \approx -\frac{\Delta[A]}{v_A \Delta t} = k[A]^{m_A}[B]^{m_B} \tag{28.16}$$

The approximate equality between $d[A]/dt$ and $\Delta[A]/\Delta t$ is more accurate the shorter the time period of the measurement and is exact as $\Delta t \to 0$ (this is just the definition of a derivative). Consider what happens if two different measurements of the initial rate (from $t = 0$ to $t = t$) are made in which the initial concentration of A, $[A]_0$, is the same and the initial concentration of B is varied. The rates of reaction for these two sets of initial conditions are given by

$$v_1 = -\frac{1}{v_A}\left(\frac{\Delta[A]}{\Delta t}\right)_1 = k[A]_0^{m_A}[B]_1^{m_B} \tag{28.17}$$

and

$$v_2 = -\frac{1}{v_A}\left(\frac{\Delta[A]}{\Delta t}\right)_2 = k[A]_0^{m_A}[B]_2^{m_B} \tag{28.18}$$

where the subscripts 1 and 2 are used to distinguish between the two different experiments with different initial concentrations of [B]. If we divide Equation 28.17 by Equation 28.18, take the logarithm of both sides, and then solve for m_B, we obtain

$$m_B = \frac{\ln \dfrac{v_1}{v_2}}{\ln \dfrac{[B]_1}{[B]_2}} \tag{28.19}$$

Clearly, if [B] is held fixed and the initial concentration of A is varied, the order m_A can be determined in a similar manner. This procedure for determining the reaction orders is called the *method of initial rates*.

EXAMPLE 28–2
Consider the following initial rate data for the reaction

$$2\,NO_2(g) + F_2(g) \longrightarrow 2\,NO_2F(g)$$

Run	$[NO_2]_0/mol \cdot dm^{-3}$	$[F_2]_0/mol \cdot dm^{-3}$	$v_0/mol \cdot dm^{-3} \cdot s^{-1}$
1	1.15	1.15	6.12×10^{-4}
2	1.72	1.15	1.36×10^{-3}
3	1.15	2.30	1.22×10^{-3}

where $[NO_2]_0$ and $[F_2]_0$ are the intial concentrations of $NO_2(g)$ and $F_2(g)$ and v_0 is the initial rate. Determine the reaction rate law and the value of the rate constant.

SOLUTION: We assume the rate law has the form

$$v = k[NO_2]^{m_{NO_2}}[F_2]^{m_{F_2}}$$

To use the method of initial rates, we must also assume that the measured initial rate is given by the rate law in which the concentrations are the initial concentrations, or

$$v_0 = k[NO_2]_0^{m_{NO_2}}[F_2]_0^{m_{F_2}} \tag{1}$$

To determine m_{F_2} we measure the initial rate at a constant initial concentration of NO_2, $[NO_2]_0$, and vary the initial concentration of F_2, $[F_2]_0$. Runs 1 and 3 in the above table provide the results of such an experiment. Using Equation 28.19, we obtain

$$m_{F_2} = \frac{\ln \dfrac{6.12 \times 10^{-4}\ \text{mol} \cdot \text{dm}^{-3} \cdot \text{s}^{-1}}{1.22 \times 10^{-3}\ \text{mol} \cdot \text{dm}^{-3} \cdot \text{s}^{-1}}}{\ln \dfrac{1.15\ \text{mol} \cdot \text{dm}^{-3}}{2.30\ \text{mol} \cdot \text{dm}^{-3}}}$$

$$= \frac{-0.690}{-0.693} = 0.996$$

To determine m_{NO_2}, we carry out two experiments at a constant value of $[F_2]_0$ and vary the value of $[NO_2]_0$. Runs 1 and 2 are two such experiments. Using an expression analogous to Equation 28.19 for m_{F_2}, we find

$$m_{NO_2} = \frac{\ln \dfrac{1.36 \times 10^{-3}\ \text{mol} \cdot \text{dm}^{-3} \cdot \text{s}^{-1}}{6.12 \times 10^{-4}\ \text{mol} \cdot \text{dm}^{-3} \cdot \text{s}^{-1}}}{\ln \dfrac{1.72\ \text{mol} \cdot \text{dm}^{-3}}{1.15\ \text{mol} \cdot \text{dm}^{-3}}}$$

$$= \frac{0.799}{0.403} = 1.98$$

Assuming the orders are integer valued, the rate law is

$$v = k[NO_2]^2[F_2]^1$$

Solving Equations 1 for the rate constant gives

$$k = \frac{v_0}{[NO_2]_0^2[F_2]_0^1}$$

Using the first set of data in the table, we obtain

$$k = \frac{6.12 \times 10^{-4} \text{ mol} \cdot \text{dm}^{-3} \cdot \text{s}^{-1}}{(1.15 \text{ mol} \cdot \text{dm}^{-3})^2 (1.15 \text{ mol} \cdot \text{dm}^{-3})}$$

$$= 4.02 \times 10^{-4} \text{ dm}^6 \cdot \text{mol}^{-2} \cdot \text{s}^{-1}$$

The other two sets of data give 4.00×10^{-4} $\text{dm}^6 \cdot \text{mol}^{-2} \cdot \text{s}^{-1}$ and 4.01×10^{-4} $\text{dm}^6 \cdot \text{mol}^{-2} \cdot \text{s}^{-1}$. The average rate constant for the three sets of data is 4.01×10^{-4} $\text{dm}^6 \cdot \text{mol}^{-2} \cdot \text{s}^{-1}$.

In using either the method of isolation or the method of initial rates, we have tacitly assumed that the reactants can be mixed in any desired proportions and the reaction rate can then be measured. In the laboratory, two solutions can be thoroughly mixed in approximately a millisecond. For many reactions, the time required to mix the reactants is long compared with the reaction process itself, and thus the rate law and rate constant cannot be determined using either of the techniques discussed in this section. To study fast reactions, different experimental approaches must be used. We will discuss some techniques that are used to study faster reactions, called *relaxation methods*, in Section 28–7. But first, we need to learn more about the properties of rate laws.

28–3. First-Order Reactions Show an Exponential Decay of Reactant Concentration with Time

Consider the reactions given by Equation 28.20, where A and B denote the reactants:

$$A + B \longrightarrow \text{products} \tag{28.20}$$

This chemical equation does not tell us anything about its rate law. Suppose the rate law is first order in [A]. Then

$$v(t) = -\frac{d[A]}{dt} = k[A] \tag{28.21}$$

If the concentration of A is $[A]_0$ at time $t = 0$ and [A] at time t, this equation can be integrated to give

$$\ln \frac{[A]}{[A]_0} = -kt \tag{28.22}$$

or

$$[A] = [A]_0 e^{-kt} \tag{28.23}$$

Equation 28.23 shows that [A] decays exponentially with time from its initial value of $[A]_0$ to zero (see Figure 28.1a). Rearranging Equation 28.22 gives

$$\ln[A] = \ln[A]_0 - kt \tag{28.24}$$

which shows that a plot of ln [A] versus t will yield a straight line of slope $-k$ and intercept $\ln[A]_0$ (see Figure 28.1b).

The chemical reaction given by

$$N_2O_5(g) \longrightarrow 2\,NO_2(g) + \tfrac{1}{2}O_2(g)$$

obeys the first-order rate law

$$v(t) = -\frac{d[N_2O_5]}{dt} = k[N_2O_5]$$

Table 28.3 gives the measured concentration of $N_2O_5(g)$ as a function of time for this reaction at 318 K. Figure 28.2 shows a plot of $\ln[N_2O_5]$ versus time. The plot is linear, which is expected for a first-order reaction (Equation 28.24). From the slope of the line in Figure 28.2, we obtain a rate constant of $k = 3.04 \times 10^{-2}\ \text{min}^{-1}$.

The length of time required for half of the reactant to disappear is called the *half-life* of the reaction and is written as $t_{1/2}$. For the first-order reaction considered above, Equation 28.22 can be used to derive a relationship between the rate constant k and the

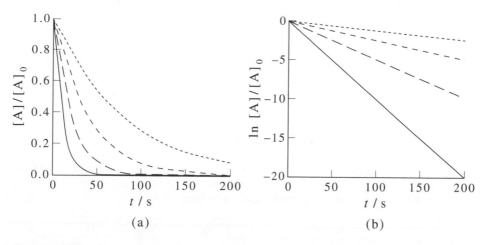

FIGURE 28.1
Kinetic plots for a first-order chemical reaction. (a) [A] is plotted as a function of t for values of the rate constant k of 0.0125 s^{-1} (dotted line), 0.0250 s^{-1} (dashed line), 0.0500 s^{-1} (long dashed line), and 0.100 s^{-1} (solid line). (b) The curves in part (a) are plotted as ln [A] versus t. The slope of the line is equal to $-k$ (see Equation 28.24).

TABLE 28.3
The concentration $[N_2O_5]$ and $\ln[N_2O_5]$ as a function of time for the reaction $N_2O_5(g) \rightarrow 2\,NO_2(g) + \frac{1}{2}\,O_2(g)$ at 318 K.

t/min	$[N_2O_5]/10^{-2}$ mol·dm^{-3}	$\ln([N_2O_5]/\text{mol·dm}^{-3})$
0	1.24	−4.39
10	0.92	−4.69
20	0.68	−4.99
30	0.50	−5.30
40	0.37	−5.60
50	0.28	−5.88
60	0.20	−6.21
70	0.15	−6.50
80	0.11	−6.81
90	0.08	−7.13
100	0.06	−7.42

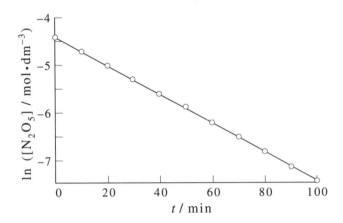

FIGURE 28.2
A plot of $\ln[N_2O_5]$ versus time for the reaction $N_2O_5(g) \rightarrow 2\,NO_2(g) + \frac{1}{2}\,O_2(g)$ at 318 K. The plot gives a straight line, consistent with a first-order rate law. The slope of the plot gives $k = 3.04 \times 10^{-2}$ min^{-1}. (The data plotted are given in Table 28.3.)

half-life $t_{1/2}$. At time $t = t_{1/2}$, the concentration of A equals $[A]_0/2$. Substituting these values into Equation 28.22 gives us

$$\ln \frac{1}{2} = -kt_{1/2}$$

or

$$t_{1/2} = \frac{\ln 2}{k} = \frac{0.693}{k} \tag{28.25}$$

Also note that the half-life of a first-order reaction is independent of the initial amount of the reactant, $[A]_0$. Figure 28.3 shows a plot of $[N_2O_5]$ as a function of time, where the values of $[N_2O_5]$ given in Table 28.3 are indicated in units of reaction half-lives.

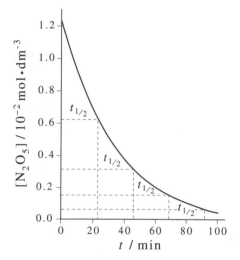

FIGURE 28.3
A plot of $[N_2O_5]$ versus time for the reaction $N_2O_5(g) \rightarrow 2\,NO_2(g) + \frac{1}{2}O_2(g)$ at 318 K. The solid line is the best fit of Equation 28.23 to the data given in Table 28.3. The concentration of N_2O_5 in increments of the reaction half-life (23 min) is also indicated.

Examples of gas-phase chemical reactions that exhibit first-order rate laws are given in Table 28.4 along with their measured rate constants. Notice that even though the time-dependence of the reactant concentrations for all these reactions can be described by Equation 28.23, the values of the rate constants vary over many orders of magnitude. Hence, a particular rate law does not provide any information on the magnitude of the rate constant.

EXAMPLE 28–3
The rate law for the reaction described by

$$N_2O_2(g) \longrightarrow 2\,NO(g)$$

is first order in the concentration of $N_2O_2(g)$. Derive an expression for the time-dependent behavior of $[NO]$, the product concentration.

SOLUTION: The rate of formation of NO is given by the rate law

$$v = \frac{1}{2}\frac{d[NO]}{dt} = k[N_2O_2] \tag{1}$$

The rate law for the disappearance of N_2O_2 is first order, so we can use Equation 28.23 to describe $[N_2O_2]$, which allows us to rewrite Equation 1 as

$$\frac{d[NO]}{dt} = 2k[N_2O_2]_0 e^{-kt}$$

Separating the time and concentration variables gives us

$$d[NO] = 2k[N_2O_2]_0 e^{-kt} dt$$

Integrating $[NO]$ from $[NO]_0 = 0$ to $[NO]$ and the time variable from 0 to t gives us (remember that $[N_2O_2]_0$ is a constant)

$$[NO] = 2[N_2O_2]_0(1 - e^{-kt})$$

T A B L E 28.4
The reaction rate constants, k, for a variety of first-order gas-phase chemical reactions at 500 K and 700 K.

Reaction	k/s^{-1} at 500 K	k/s^{-1} at 700 K
Isomerizations		
cyclopropane $\rightarrow$ propene	7.85×10^{-14}	1.13×10^{-5}
cyclopropene $\rightarrow$ propyne	5.67×10^{-4}	13.5
cis-but-2-ene $\rightarrow$ *trans*-but-2-ene	2.20×10^{-14}	1.50×10^{-6}
$CH_3NC \rightarrow CH_3CN$	6.19×10^{-4}	38.5
vinyl allyl ether $\rightarrow$ pent-4-enal	2.17×10^{-2}	141
Decompositions		
cyclobutane $\rightarrow$ 2 ethene	1.77×10^{-12}	1.12×10^{-4}
ethyleneoxide $\rightarrow CH_3CHO, CH_2O, CH_2CO$	1.79×10^{-11}	2.19×10^{-4}
ethyl fluoride $\rightarrow HF + $ ethene	1.57×10^{-13}	4.68×10^{-6}
ethyl chloride $\rightarrow HCl + $ ethene	3.36×10^{-12}	6.20×10^{-5}
ethyl bromide $\rightarrow HBr + $ ethene	8.06×10^{-11}	4.32×10^{-4}
ethyl iodide $\rightarrow HI + $ ethene	1.07×10^{-9}	4.06×10^{-3}
isopropyl ether $\rightarrow$ propene + isopropylalcohol	6.76×10^{-14}	5.44×10^{-3}

28–4. The Rate Laws for Different Reaction Orders Predict Different Behaviors for Time-Dependent Reactant Concentrations

What are the time dependencies of concentrations for rate laws that are not first order? Do the concentrations of the reactants still decay exponentially with time? If the reactant concentrations were to decay exponentially regardless of order, experimental measurements of concentration as a function of time would not provide any insight into the reaction order. If, however, different reaction orders exhibit different functional forms for the time dependence of reactant concentrations, experimental measurements of the rate as a function of initial concentrations can in principle be used to deduce information about reaction order.

Consider the equation

$$A + B \longrightarrow \text{products} \tag{28.26}$$

where experimental data reveal that the rate law is

$$-\frac{d[A]}{dt} = k[A]^2 \tag{28.27}$$

We now want to derive an expression for [A] from Equation 28.27. By separating the concentration and time variables and then integrating the resulting expression assuming that the initial concentration of A at $t = 0$ is $[A]_0$ and at time t later is [A], we get

$$\frac{1}{[A]} = \frac{1}{[A]_0} + kt \tag{28.28}$$

This result predicts that for a second-order reaction, a plot of $1/[A]$ versus t will yield a straight line of slope k and intercept $1/[A]_0$.

The reaction

$$NOBr(g) \longrightarrow NO(g) + \tfrac{1}{2} Br_2(g)$$

is found to obey the rate law

$$v = k[NOBr]^2$$

Table 28.5 gives the time-dependent concentration of NOBr(g), and Figure 28.4 shows a plot of $1/[NOBr]$ versus time. The graph shows that $1/[NOBr]$ varies linearly with time, consistent with Equation 28.28. The value of the rate constant, given by the slope of the line, is equal to $2.01 \text{ dm}^3 \cdot \text{mol}^{-1} \cdot \text{s}^{-1}$.

TABLE 28.5
Kinetic data for the reaction $NOBr(g) \rightarrow NO(g) + \tfrac{1}{2} Br_2(g)$.

t/s	$[NOBr]/\text{mol} \cdot \text{dm}^{-3}$	$[NOBr]^{-1}/\text{mol}^{-1} \cdot \text{dm}^3$
0	0.0250	40.0
6.2	0.0191	52.3
10.8	0.0162	61.7
14.7	0.0144	69.9
20.0	0.0125	80.0
24.6	0.0112	89.3

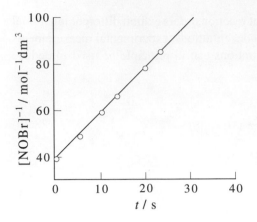

FIGURE 28.4
A plot of 1/[NOBr] versus time for the reaction $NOBr(g) \rightarrow NO(g) + \frac{1}{2} Br_2(g)$. The experimental data are given in Table 28.5. The linear dependence of 1/[NOBr] on time is consistent with a second-order rate law, Equation 28.28. The value of the rate constant, given by the slope of the line, is equal to $2.01 \, dm^3 \cdot mol^{-1} \cdot s^{-1}$.

The following example shows how the reaction rate law can be found by combining the method of isolation and the predictions of integrated rate laws.

EXAMPLE 28–4
The reaction between carbon disulfide and ozone

$$CS_2(g) + 2 O_3(g) \longrightarrow CO_2(g) + 2 SO_2(g)$$

was studied using a large excess of CS_2. The pressure of ozone as a function of time is given in the following table. Is the reaction first order or second order with respect to ozone?

Time/s	Ozone pressure/torr
0	1.76
30	1.04
60	0.79
120	0.52
180	0.37
240	0.29

SOLUTION: We first assume that the rate law has the general form

$$v = k[CS_2]^{m_{CS_2}}[O_3]^{m_{O_3}}$$

Because CS_2 is present in excess, $[CS_2]$ is essentially constant and we can write

$$v = k'[O_3]^{m_{O_3}} \propto P_{O_3}^{m_{O_3}}$$

(This is the method of isolation.) From Section 28–3, if $m_{O_3} = 1$ (first order), then $\ln P_{O_3}$ versus time is linear. If $m_{O_3} = 2$ (second order), then $1/P_{O_3}$ versus time is linear. These two plots are presented below. The plot of $\ln P_{O_3}$ versus time is not linear, whereas the plot of $1/P_{O_3}$ versus time is linear. Therefore, the reaction is second order in ozone concentration.

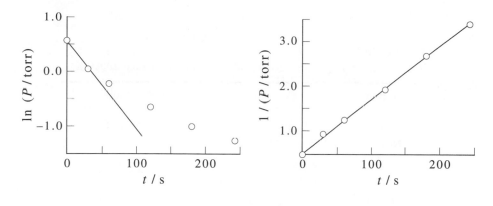

The half-life of a second-order reaction can be determined from Equation 28.28. Setting $t = t_{1/2}$ and $[A]_{t_{1/2}} = [A]_0/2$ gives us

$$t_{1/2} = \frac{1}{k[A]_0} \tag{28.29}$$

Notice that the half-life depends on the initial concentration of the reactant for a second-order reaction. This relation is different from that found for a first-order reaction, for which the half-life is independent of concentration (Equation 28.25).

Finally, consider the reaction

$$A + B \longrightarrow products \tag{28.30}$$

for the case in which experimental data reveal that the rate law is

$$-\frac{d[A]}{dt} = -\frac{d[B]}{dt} = k[A][B] \tag{28.31}$$

This rate law is first order in each reactant and second order overall. The rate law given by Equation 28.31 is more difficult to integrate, and the details are addressed in Problem 28–24. The resulting integrated rate law is

$$kt = \frac{1}{[A]_0 - [B]_0} \ln \frac{[A][B]_0}{[B][A]_0} \tag{28.32}$$

If $[A]_0 = [B]_0$, Equation 28.32 is indeterminate. Problem 28–25 shows that the integrated rate law for the case when $[A]_0 = [B]_0$ is

$$\frac{1}{[A]} = \frac{1}{[A]_0} + kt \quad \text{or} \quad \frac{1}{[B]} = \frac{1}{[B]_0} + kt \tag{28.33}$$

consistent with Equation 28.28. Table 28.6 lists the rate constants for a variety of reactions that follow a second-order rate law.

TABLE 28.6
Reaction rate constants for second-order gas-phase reactions at 500 K.

Reaction	$k/\mathrm{dm^3 \cdot mol^{-1} \cdot s^{-1}}$
$2\,\mathrm{HI}(g) \rightarrow \mathrm{H_2}(g) + \mathrm{I_2}(g)$	4.91×10^{-9}
$2\,\mathrm{NOCl}(g) \rightarrow 2\,\mathrm{NO}(g) + \mathrm{Cl_2}(g)$	0.363
$\mathrm{NO_2}(g) + \mathrm{O_3}(g) \rightarrow \mathrm{NO_3}(g) + \mathrm{O_2}(g)$	5.92×10^6
$\mathrm{NO}(g) + \mathrm{Cl_2}(g) \rightarrow \mathrm{NOCl}(g) + \mathrm{Cl}(g)$	5.32
$\mathrm{NO}(g) + \mathrm{O_3}(g) \rightarrow \mathrm{NO_2}(g) + \mathrm{O_2}(g)$	5.70×10^7
$\mathrm{O_3}(g) + \mathrm{C_3H_8}(g) \rightarrow \mathrm{C_3H_7O}(g) + \mathrm{HO_2}(g)$	14.98

28–5. Reactions Can Also Be Reversible

Consider the isomerization reaction of *cis*-1,2-dichloroethene to form *trans*-1,2-di-chloroethene, If we start with a sample of pure *cis*-1,2-dichloroethene, we find that the reaction does not go to completion but generates an equilibrium mixture between the two geometrical isomers. Similarly, if we start with pure *trans*-1,2-dichloroethene, we obtain the same equilibrium mixture. In both experiments, the final concentrations of the two isomers are dictated by the equilibrium constant for the reaction. When a reaction occurs in both directions, we say that the reaction is *reversible*. (Be careful not to confuse this definition of the word "reversible" with that for a thermodynamic process.)

To indicate that a kinetic process is reversible, we explicitly draw two arrows, one representing the forward reaction and one representing the reverse reaction. The rate constants for the two reactions, k_1 and k_{-1}, are written next to the arrow they are associated with. We will use a positive subscript for the rate constant of a forward reaction and a negative subscript for a rate constant of a reverse reaction. The reaction discussed above is a specific example of the general reaction scheme

$$\mathrm{A} \underset{k_{-1}}{\overset{k_1}{\rightleftharpoons}} \mathrm{B} \tag{28.34}$$

For any initial concentrations, $[\mathrm{A}]_0$ and $[\mathrm{B}]_0$, the chemical system must go to equilibrium. At equilibrium, the ratio of the concentrations of A and B is given by the equilibrium-constant expression,

$$K_c = \frac{[\mathrm{B}]_{eq}}{[\mathrm{A}]_{eq}} \tag{28.35}$$

For the concentrations of A and B to remain constant at their equilibrium values, both $d[A]/dt$ and $d[B]/dt$ must equal zero. Thus, the kinetic conditions for Equation 28.34 to be at equilibrium is that

$$-\frac{d[A]}{dt} = \frac{d[B]}{dt} = 0 \tag{28.36}$$

This result is important, and we will use it extensively in the next chapter when we discuss reaction mechanisms. Even though the concentrations of A and B are constant at equilibrium, A is converted to B and B is converted to A but in such a way that there is no *net* change in the concentrations of either A or B. The equilibrium state is a *dynamic equilibrium*.

We now examine the specific case in which the rate law for Equation 28.34 is first order in both [A] and [B]. The rate is then given by

$$-\frac{d[A]}{dt} = k_1[A] - k_{-1}[B] \tag{28.37}$$

Unlike the rate laws considered so far, Equation 28.37 expresses the rate as a sum of two terms. The first term is the rate at which A reacts to form B. The second term is the rate at which B reacts to form A. The difference in sign of these two terms reflects that the forward reaction depletes the concentration of A and the back reaction increases the concentration of A with time.

If $[A] = [A]_0$ and $[B] = 0$ at time $t = 0$, the stoichiometry of Equation 28.34 requires that $[B] = [A]_0 - [A]$, and Equation 28.37 becomes

$$-\frac{d[A]}{dt} = (k_1 + k_{-1})[A] - k_{-1}[A]_0 \tag{28.38}$$

Integrating this rate law subject to the above initial conditions gives us (Problem 28–32)

$$[A] = ([A]_0 - [A]_{eq})e^{-(k_1+k_{-1})t} + [A]_{eq} \tag{28.39}$$

where $[A]_{eq}$ is the concentration of A at equilibrium. By bringing $[A]_{eq}$ to the left side and then taking logarithms, we can rewrite Equation 28.39 as

$$\ln([A] - [A]_{eq}) = \ln([A]_0 - [A]_{eq}) - (k_1 + k_{-1})t \tag{28.40}$$

which tells us that a plot of $\ln([A] - [A]_{eq})$ as a function of time has a slope of $-(k_1 + k_{-1})$ and an intercept of $\ln([A]_0 - [A]_{eq})$. From such an analysis of the kinetic data, the sum of the rate constants $k_1 + k_{-1}$ is determined. Generally, however, we want to determine k_1 and k_{-1} individually. We can do this by taking advantage of the connection between the rate law and the equilibrium constant. At equilibrium, $-d[A]/dt = 0$ and the rate law, Equation 28.37, becomes

$$k_1[A]_{eq} = k_{-1}[B]_{eq} \tag{28.41}$$

or

$$\frac{k_1}{k_{-1}} = \frac{[B]_{eq}}{[A]_{eq}} = K_c \tag{28.42}$$

If we know the sum $k_1 + k_1^{-1}$ and the value of K_c, both rate constants can be uniquely determined. Figure 28.5 shows a plot of $[A]/[A]_0$ and $[B]/[A]_0$ for the reversible reaction given by Equation 28.34 subject to the inital conditions $[A] = [A]_0$ and $[B] = 0$ at time $t = 0$. Because $[A] + [B] = [A]_0$, the concentration of B is given by $[A]_0 - [A]$. The rate constants for the forward and reverse reactions are $k_1 = 2.25 \times 10^{-2}$ s^{-1} and $k_{-1} = 1.50 \times 10^{-2}$ s^{-1}. The value of $[A]$ decreases from $[A]/[A]_0 = 1.000$ to $[A]/[A]_0 = [A]_{eq}/[A]_0 = 0.400$, and the value of $[B]$ increases from $[B]_0/[A]_0 = 0$ to $[B]/[A]_0 = [B]_{eq}/[A]_0 = 0.600$. At equilibrium, the values of the concentrations satisfy the relationship $K_c = [B]_{eq}/[A]_{eq} = k_1/k_{-1} = 1.50$.

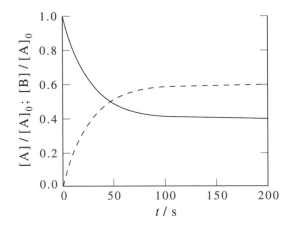

FIGURE 28.5
The time dependence of $[A]/[A]_0$ (solid line) and $[B]/[A]_0$ (dashed line) for the reversible reaction given by Equation 28.34 subject to the inital conditions $[A] = [A]_0$ and $[B] = 0$ at time $t = 0$. The rate constants for the forward and reverse reaction are $k_1 = 2.25 \times 10^{-2}$ s^{-1} and $k_{-1} = 1.50 \times 10^{-2}$ s^{-1}.

EXAMPLE 28–5
The reaction *cis*-2-butene to *trans*-2-butene is first order in both directions. At 25°C, the equilibrium constant is 0.406 and the forward rate constant is 4.21×10^{-4} s^{-1}. Starting with a sample of the pure *cis* isomer with $[cis]_0 = 0.115$ mol·dm^{-3}, how long would it take for half the equilibrium amount of the *trans* isomer to form?

SOLUTION: We represent the reaction by

$$cis \underset{k_{-1}}{\overset{k_1}{\rightleftharpoons}} trans$$

Because the reaction is first order in both directions,

$$K_c = \frac{[trans]_{eq}}{[cis]_{eq}} = \frac{k_1}{k_{-1}} = 0.406$$

Solving for the reverse rate constant gives

$$k_{-1} = \frac{4.21 \times 10^{-4} \text{ s}^{-1}}{0.406} = 1.04 \times 10^{-3} \text{ s}^{-1}$$

By mass balance, $[cis]_0 = [cis]_{eq} + [trans]_{eq}$. Therefore,

$$\frac{[trans]_{eq}}{[cis]_{eq}} = \frac{[trans]_{eq}}{[cis]_0 - [trans]_{eq}} = \frac{[trans]_{eq}}{0.115 \text{ mol·dm}^{-3} - [trans]_{eq}} = 0.406$$

or $[trans]_{eq} = 0.0332$ mol·dm^{-3}. The concentration of *cis* at equilibrium is then 0.115 mol·dm^{-3} − 0.0332 mol·dm^{-3} = 0.082 mol·dm^{-3}. Half the equilibrium amount of the *trans* isomer is 0.0166 mol·dm^{-3}, which corresponds to a *cis* concentration of 0.115 mol·dm^{-3} − 0.0166 mol·dm^{-3} = 0.098 mol·dm^{-3}. Equation 28.40 can now be used to find the time required for half the equilibrium amount of the *trans* isomer to form. Solving Equation 28.40 for time, t, gives

$$t = \frac{1}{k_1 + k_{-1}} \ln \frac{[cis]_0 - [cis]_{eq}}{[cis] - [cis]_{eq}}$$

From the above calculations, $[cis]_0 = 0.115$ mol·dm^{-3}, $[cis]_{eq} = 0.082$ mol·dm^{-3}, and $[cis] = 0.098$ mol·dm^{-3} when the concentration of the *trans* isomer is at half its equilibrium value. Substituting these values into the above equation gives

$$t = \left(\frac{1}{4.21 \times 10^{-4} \text{ s}^{-1} + 1.04 \times 10^{-3} \text{ s}^{-1}} \right) \ln \frac{0.115 \text{ mol·dm}^{-3} - 0.082 \text{ mol·dm}^{-3}}{0.098 \text{ mol·dm}^{-3} - 0.082 \text{ mol·dm}^{-3}}$$

$$= 490 \text{ s}$$

28–6. The Rate Constants of a Reversible Reaction Can Be Determined Using Relaxation Methods

In Section 28–2, we discussed two experimental techniques used to determine the rate law for a chemical reaction provided that the half-life is long compared with the time needed to mix the two reactants. The same constraint applies to the study of reversible reactions. If equilibrium is achieved faster than the reactants can be thoroughly mixed, the method of isolation or the method of initial rates cannot be used to determine the rate law. For example, suppose we wanted to study the reaction

$$H^+(aq) + OH^-(aq) \underset{k_{-1}}{\overset{k_1}{\rightleftharpoons}} H_2O(l) \tag{28.43}$$

We could consider mixing a strong acid with a strong base and then monitoring the time dependence of the pH of the solution as the neutralization reaction occurs. Unfortunately, it requires about a millisecond to thoroughly mix two solutions, and this amount of time is orders of magnitude longer than the time needed for Equation 28.43 to reach equilibrium (Example 28–6).

EXAMPLE 28–6
The rate constant k_1 for the reaction described by

$$H^+(aq) + OH^-(aq) \underset{k_{-1}}{\overset{k_1}{\rightleftharpoons}} H_2O(l)$$

is 1.4×10^{11} $dm^3 \cdot mol^{-1} \cdot s^{-1}$. Calculate the half life of this reaction if the initial conditions are
(a) $[H^+]_0 = [OH^-]_0 = 0.10$ $mol \cdot dm^{-3}$
(b) $[H^+]_0 = [OH^-]_0 = 1.0 \times 10^{-7}$ $mol \cdot dm^{-3}$

SOLUTION: This reaction goes essentially to completion, so it is of the form of Equation 28.30 with $[A]_0 = [B]_0$. The integrated rate law is given by Equation 28.33, and if we set $[A] = [A]_0/2$ or $[B] = [B]_0/2$ in Equation 28.33, then we have that $t_{1/2} = 1/(k_1[A]_0) = 1/(k_1[B]_0)$ as in Equation 28.29, Therefore, for the initial conditions in (a), we have

$$t_{1/2} = \frac{1}{k_1[A]_0} = \frac{1}{(1.4 \times 10^{11}\ dm^3 \cdot mol^{-1} \cdot s^{-1})(0.10\ mol \cdot dm^{-3})}$$
$$= 7.1 \times 10^{-11}\ s$$

For the initial conditions in (b), we have

$$t_{1/2} = \frac{1}{k_1[A]_0} = \frac{1}{(1.4 \times 10^{11}\ dm^3 \cdot mol^{-1} \cdot s^{-1})(1.0 \times 10^{-7}\ mol \cdot dm^{-3})}$$
$$= 7.1 \times 10^{-5}\ s$$

In both cases [note that (b) corresponds to pure water at 298 K], the half-life is much shorter than the time required to mix the reactants (10^{-3} s). Thus, we cannot use mixing techniques to study this reaction.

The limitations revealed in Example 28–6 can be overcome using experimental techniques known as *relaxation methods*. Relaxation methods are fundamentally different from the experimental approaches discussed so far in this chapter. The general idea is to start with a chemical system initially in equilibrium at some specified temperature and pressure. The conditions are then suddenly changed so that the system is no longer at equilibrium. There are many different ways to shift the equilibrium. Temperature, pressure, pH, and pOH jump methods have been developed and used effectively to study kinetic processes. Here, we examine the most common relaxation method used to study reaction kinetics in solution, the *temperature-jump relaxation technique*. In a temperature-jump experiment, the temperature of the equilibrium reaction mixture is suddenly changed at constant pressure. Following the sudden change in temperature, the chemical system responds by relaxing to a new equilibrium state that corresponds to the new temperature. We will see that the rate constants for the forward and reverse reactions are related to the time required for the system to relax to its new equilibrium state.

Experimentally, the temperature of a solution can be increased by about 5 K in one microsecond by discharging a high-voltage capacitor through the reaction solution. Given that the equilibrium constant depends exponentially on the inverse of the temperature (recall from Section 26–1 that $\ln K_P = -\Delta_r G^\circ / RT$), such a perturbation can cause a significant change in equilibrium concentrations.

Before discussing the acid-base reaction of water (Equation 28.43), we first consider the simple general equilibrium reaction described by

$$A \underset{k_{-1}}{\overset{k_1}{\rightleftharpoons}} B \qquad (28.44)$$

where the rates for the forward and reverse reactions are assumed to be first order in the reactants. Initially, this chemical system is in equilibrium at temperature T_1, where the concentrations of A and B are $[A]_{1,eq}$ and $[B]_{1,eq}$, respectively. Now consider what happens if we rapidly jump the temperature from T_1 to T_2. From Equation 26.31, we see that the equilibrium concentration of B increases following the temperature jump if $\Delta_r H^\circ$ for the reaction is positive and decreases if $\Delta_r H^\circ$ is negative. (If $\Delta_r H^\circ = 0$, the equilibrium constant is temperature independent and we cannot learn anything by performing a temperature-jump relaxation experiment.) For the purpose of plotting the time evolution of the reaction following the temperature jump, we will assume that $\Delta_r H^\circ < 0$, and we denote the equilibrium concentrations at T_2 by $[A]_{2,eq}$ and $[B]_{2,eq}$.

We have assumed that the rates of the forward and reverse reactions are first order in the reactants, so the rate law for Equation 28.44 is

$$\frac{d[B]}{dt} = k_1[A] - k_{-1}[B] \qquad (28.45)$$

We now want to write Equation 28.45 so that it applies to a system that has just been subjected to a perturbation to a new equilibrium state. To do this, let

$$[A] = [A]_{2,\text{eq}} + \Delta[A]$$

$$[B] = [B]_{2,\text{eq}} + \Delta[B] \tag{28.46}$$

Substitute these equations into Equation 28.45 to obtain (recall that $[A]_{2,\text{eq}}$ and $[B]_{2,\text{eq}}$ are constants)

$$\frac{d\Delta[B]}{dt} = k_1[A]_{2,\text{eq}} + k_1\Delta[A] - k_{-1}[B]_{2,\text{eq}} - k_{-1}\Delta[B] \tag{28.47}$$

According to Equation 28.44, the sum of the concentrations of A and B remains constant during the experiment, so $\Delta([A] + [B]) = \Delta[A] + \Delta[B] = 0$. Furthermore, $[A]_{2,\text{eq}}$ and $[B]_{2,\text{eq}}$ satisfy Equation 28.42 $(k_1[A]_{2,\text{eq}} = k_{-1}[B]_{2,\text{eq}})$, so Equation 28.47 becomes

$$\frac{d\Delta[B]}{dt} = -(k_1 + k_{-1})\Delta[B] \tag{28.48}$$

Integrating Equation 28.48 subject to the condition that $[B] = [B]_{1,\text{eq}}$ at $t = 0$, or that $\Delta[B]$ at time $t = 0$ is equal to $\Delta[B]_0 = [B]_{1,\text{eq}} - [B]_{2,\text{eq}}$ gives us

$$\Delta[B] = \Delta[B]_0 e^{-(k_1+k_{-1})t} = \Delta[B]_0 e^{-t/\tau} \tag{28.49}$$

where

$$\tau = \frac{1}{k_1 + k_{-1}} \tag{28.50}$$

is called the *relaxation time*. Note that τ has units of time and is a measure of how long it takes for $\Delta[B]$ to decay to $1/e$ of its initial value.

Figure 28.6 shows a plot of $\Delta[B]$ versus time for a typical temperature-jump experiment. From Equation 28.49, we see that a plot of $\ln(\Delta[B]/\Delta[B]_0)$ versus t is linear and has a slope of $-(k_1 + k_{-1})$, which is the negative of the sum of the forward and reverse rate constants for the reaction at T_2. If we know the equilibrium constant at T_2 and the rate laws for the forward and reverse reactions, then the rate constants k_1 and k_{-1} can be independently determined.

We now return to consider the chemical reaction given by Equation 28.43. The general form of this specific reaction is

$$A + B \underset{k_{-1}}{\overset{k_1}{\rightleftharpoons}} P \tag{28.51}$$

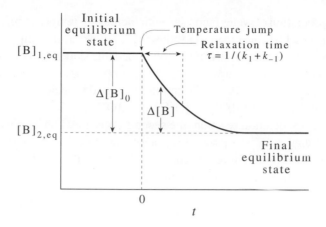

FIGURE 28.6
The time-dependent change in [B] following a temperature-jump experiment for the chemical system given by Equation 28.44, where the rate laws for the forward and reverse reaction are first order in the reactants. The plot assumes that $\Delta_r H° < 0$, whereby $[B]_{2,eq} < [B]_{1,eq}$. Following the temperature jump, the value of [B] decays exponentially from $[B]_{1,eq}$ to $[B]_{2,eq}$. The time constant for this exponential decay is given by $1/(k_1 + k_{-1})$

If we assume that both the forward and reverse reactions are first order in their respective reactants, the rate law is

$$\frac{d[P]}{dt} = k_1[A][B] - k_{-1}[P] \tag{28.52}$$

If we take $\Delta[P]$ to be equal to $[P] - [P]_{2,eq}$, then (Problem 28–33)

$$\Delta[P] = \Delta[P]_0 e^{-t/\tau} \tag{28.53}$$

where τ, the relaxation time, is given by

$$\tau = \frac{1}{k_1([A]_{2,eq} + [B]_{2,eq}) + k_{-1}} \tag{28.54}$$

Equation 28.53 predicts that a plot of $\ln(\Delta[P]/\Delta[P]_0)$ against t is linear and has a slope of $-k_1([A]_{2,eq} + [B]_{2,eq}) - k_{-1}$. For chemical reactions that can be described by Equation 28.51 and obey the rate law given by Equation 28.52, the values of k_1 and k_{-1} can be uniquely determined by plotting $\Delta[P]$ against t for samples containing different total concentrations, $[A]_{2,eq} + [B]_{2,eq}$. The dissociation reaction of water, Equation 28.43, satisfies these conditions. It is important to realize that the dynamics of the dissociation of water could not be studied until relaxation techniques were developed.

Because the dissociation of water increases with increasing temperature, the concentrations of $H^+(aq)$ and $OH^-(aq)$ increase following a temperature jump, causing a measurable increase in the conductivity of the solution. Time-dependent conductivity measurements following a temperature jump to a final temperature of $T_2 = 298$ K revealed a relaxation time of $\tau = 3.7 \times 10^{-5}$ s. From this measured relaxation time and the equilibrium constant ($K_c = [H_2O]/K_w = [H_2O]/[H^+][OH^-] = 5.49 \times 10^{15}$ $\text{mol}^{-1} \cdot \text{dm}^{-3}$ at 298 K) for the dissociation reaction, the second-order rate constant was found to be $k_1 = 1.4 \times 10^{11}$ $\text{dm}^3 \cdot \text{mol}^{-1} \cdot \text{s}^{-1}$, which is one of the largest rate constants ever measured. Table 28.7 presents a list of rate constants for reversible acid-base reactions that were determined from relaxation measurements.

TABLE 28.7

Rate constants for reversible acid-base reactions in water at 298 K.

Reaction	$k_1/\text{dm}^3 \cdot \text{mol}^{-1} \cdot \text{s}^{-1}$	k_{-1}/s^{-1}
$H^+(aq) + OH^-(aq) \rightleftharpoons H_2O(l)$	1.4×10^{11}	2.5×10^{-5}
$H^+(aq) + HCO_3^-(aq) \rightleftharpoons H_2CO_3(aq)$	4.7×10^{10}	8×10^6
$H^+(aq) + CH_3COO^-(aq) \rightleftharpoons CH_3COOH(aq)$	4.5×10^{10}	7.8×10^5
$H^+(aq) + C_6H_5COO^-(aq) \rightleftharpoons C_6H_5COOH(aq)$	3.5×10^{10}	2.2×10^6
$H^+(aq) + NH_3(aq) \rightleftharpoons NH_4^+(aq)$	4.3×10^{10}	2.5×10^1
$H^+(aq) + Me_3N(aq) \rightleftharpoons Me_3NH^+(aq)$	2.5×10^{10}	4
$H^+(aq) + HCO_3^-(aq) \rightleftharpoons CO_2(aq) + H_2O(l)$	5.6×10^4	4.3×10^{-2}

EXAMPLE 28–7

Use the data in Table 28.7 to calculate the relaxation time for the reaction

$$H^+(aq) + C_6H_5COO^-(aq) \rightleftharpoons C_6H_5COOH(aq)$$

for a temperature-jump experiment to a final temperature of 298 K. The solution was initally prepared by adding 0.015 moles of benzoic acid to water such that a liter of total solution was made. Assume that both the forward and reverse reactions are first order in each of the reactants.

SOLUTION: If we assume that both the forward and reverse reactions are first order in their respective reactants, the relaxation time is given by Equation 28.54 or

$$\tau = \frac{1}{k_1([H^+]_{2,eq} + [C_6H_5COO^-]_{2,eq}) + k_{-1}} \tag{1}$$

From Table 28.7, $k_1 = 3.5 \times 10^{10}$ dm$^3\cdot$mol$^{-1}\cdot$s^{-1} and $k_{-1} = 2.2 \times 10^6$ s^{-1} at 298 K. The equilibrium constant is $K_c = k_1/k_{-1} = 1.6 \times 10^4$ dm$^3\cdot$mol^{-1}. The initial concentration of benzoic acid is 0.015 mol$\cdot$dm^{-3}, and therefore at equilibrium (at 298 K)

$$K_c = 1.6 \times 10^4 \text{ dm}^3\cdot\text{mol}^{-1} = \frac{0.015 \text{ mol}\cdot\text{dm}^{-3} - x}{x^2}$$

where x is the concentration of dissociated acid. Solving the above expression for x gives

$$x = [\text{H}^+]_{2,\text{eq}} = [\text{C}_6\text{H}_5\text{COO}^-]_{2,\text{eq}} = 9.4 \times 10^{-4} \text{ mol}\cdot\text{dm}^{-3}$$

Substituting these values into Equation 1 gives a relaxation time of

$$\tau = \frac{1}{(3.5 \times 10^{10} \text{ dm}^3\cdot\text{mol}^{-1}\cdot\text{s}^{-1})((2)(9.4 \times 10^{-4} \text{ mol}\cdot\text{dm}^{-3})) + 2.2 \times 10^6 \text{ s}^{-1}}$$

$$= 1.5 \times 10^{-8} \text{ s}$$

28–7. Rate Constants Are Usually Strongly Temperature Dependent

The rates of chemical reactions almost always depend strongly upon the temperature. Figure 28.7 shows the temperature dependence of the rates of several types of reactions. The temperature dependence shown in (a) is the most common, and is the one we will discuss in detail. The other two curves in the figure illustrate (b) a reaction that becomes explosive at some threshold temperature and (c) an enzyme-controlled reaction, where the enzyme becomes deactivated at higher temperatures.

The temperature dependence of the rate of a reaction is due to the temperature dependence of the rate constant of the reaction. For the commonly occuring case

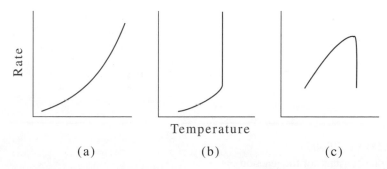

FIGURE 28.7
Some examples of the temperature dependence of reaction rates. (a) The most commonly occuring type, where the rate increases essentially exponentially with the reciprocal of the temperature. (b) A reaction that becomes explosive at some threshold temperature. (c) An enzyme-controlled reaction, where the enzyme becomes deactivated at higher temperatures.

illustrated in Figure 28.7a, the temperature dependence of the rate constant is described approximately by the empirical equation

$$\frac{d \ln k}{dT} = \frac{E_a}{RT^2} \tag{28.55}$$

where E_a has units of energy. If E_a is independent of temperature, Equation 28.55 can be integrated to give

$$\ln k = \ln A - \frac{E_a}{RT} \tag{28.56}$$

or

$$k = A e^{-E_a/RT} \tag{28.57}$$

where A is a constant. The constant A is commonly called the *pre-exponential factor*, and E_a is called the *activation energy*. Equation 28.56 predicts that a plot of $\ln k$ as a function of $1/T$ is linear with an intercept of $\ln A$ and a slope of $-E_a/R$. Figure 28.8 shows a plot of $\ln k$ versus $1/T$ for the reaction $2\,HI(g) \rightarrow H_2(g) + I_2(g)$. The solid line is the best fit of a straight line to the experimental data (circles). The slope of this best-fit line gives an activation energy of $184\ kJ \cdot mol^{-1}$. From the intercept, A is found to be $7.94 \times 10^{10}\ dm^3 \cdot mol^{-1} \cdot s^{-1}$.

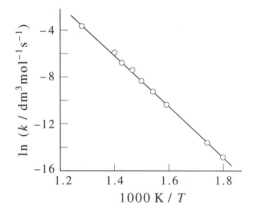

FIGURE 28.8
A plot of $\ln k$ against $1/T$ for the reaction $2\,HI(g) \rightarrow H_2(g) + I_2(g)$. The best fit of the experimental data to a straight line gives $A = 7.94 \times 10^{10}\ dm^3 \cdot mol^{-1} \cdot s^{-1}$ and $E_a = 184\ kJ \cdot mol^{-1}$.

EXAMPLE 28–8
The rate constant for the reaction

$$2\,HI(g) \longrightarrow H_2(g) + I_2(g)$$

is $1.22 \times 10^{-6}\ dm^3 \cdot mol^{-1} \cdot s^{-1}$ at 575 K and $2.50 \times 10^{-3}\ dm^3 \cdot mol^{-1} \cdot s^{-1}$ at 716 K. Estimate the value of E_a from these data.

SOLUTION: Assuming the activation energy and the pre-exponential factor are temperature independent, the rate constants $k(T_1)$ and $k(T_2)$ for the same reaction at temperatures T_1 and T_2, respectively, are given by

$$k(T_1) = Ae^{-E_a/RT_1} \quad \text{and} \quad k(T_2) = Ae^{-E_a/RT_2}$$

Dividing the first equation by the second equation and taking the logarithm of the results gives us

$$\ln \frac{k(T_1)}{k(T_2)} = \frac{E_a}{R}\left(\frac{1}{T_2} - \frac{1}{T_1}\right)$$

Solving this expression for E_a and substituting in the above data gives us

$$E_a = R\left(\frac{T_1 T_2}{T_1 - T_2}\right)\ln\frac{k(T_1)}{k(T_2)}$$

$$= (8.315 \text{ J·K}^{-1}\text{·mol}^{-1})\left(\frac{(716 \text{ K})(575 \text{ K})}{575 \text{ K} \quad 716 \text{ K}}\right)\ln\frac{1.22 \times 10^{-6} \text{ dm}^3\text{·mol}^{-1}\text{·s}^{-1}}{2.50 \times 10^{-3} \text{ dm}^3\text{·mol}^{-1}\text{·s}^{-1}}$$

$$= 185 \text{ kJ·mol}^{-1}$$

In the 1880s, the Swedish chemist Svante Arrhenius found that Equation 28.57 describes the temperature dependence of the rate constants for many reactions, and he used it to develop a general model for how reactions occur. Arrhenius noticed that the magnitude of the temperature effect on reaction rates was much too large to be explained in terms of only a change in the translational energy of the reactants. Thus, for a reaction to occur, it requires more than just a collision between reactants. Because of his contribution to the field of chemical kinetics, Equation 28.57 is now called the *Arrhenius equation*.

If we think of the activation energy as the energy that must be provided to enable the reactant(s) to react, we can describe a chemical reaction in terms of the simple energy diagram shown in Figure 28.9. We say that the chemical reaction proceeds from reactant to products along a *reaction coordinate*. The reaction coordinate is generally multidimensional, representing the bond lengths and bond angles associated with the chemical process. In some instances, the reaction coordinate is obvious. For example, for the thermal dissociation of $I_2(g)$, the reaction coordinate is the I–I bond length. For most chemical reactions, however, the reaction coordinate is difficult to visualize.

Although the Arrhenius equation is used extensively to determine the activation energies of chemical reactions, the plot of $\ln k$ versus $1/T$ for some reactions is not linear. Such nonlinear behavior can now be justified theoretically, and many modern theories of reaction rates predict that rate constants behave like

$$k = aT^m e^{-E'/RT} \tag{28.58}$$

where a, E', and m are temperature-independent constants. Depending on the assumptions of the rate theory, the constant m takes on different values, e.g., 1, 1/2, $-1/2$. If m is known, the constant E' can be found from the slope of a plot of $\ln(k/T^m)$ versus

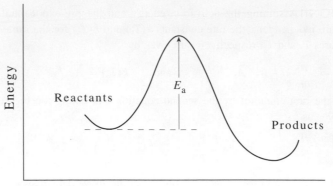

FIGURE 28.9
A schematic drawing of the energy profile of a chemical reaction. To transform into products, the reactants must acquire energy in excess the activation barrier. The reaction coordinate represents the changes in bond lengths and bond angles that occur as the chemical reaction proceeds from reactants to products.

$1/T$. If m is not known, the value of m is very difficult to determine from experimental data because the exponential dependence of $k(T)$ on $1/T$ usually dominates the power dependence on T. In the next section, we will explore one commonly used model, transition-state theory, that predicts an equation of the form given by Equation 28.58.

EXAMPLE 28–9
What is the relationship between the Arrhenius activation energy, E_a, and pre-exponential factor, A, and the constants m, a, and E' in Equation 28.58?

SOLUTION: Using Equation 28.55, we can define the activation energy to be

$$E_a = RT^2 \frac{d \ln k}{dT} \tag{28.59}$$

Substituting Equation 28.58 into this expression gives

$$E_a = E' + mRT$$

Solving for E' and substituting the result into Equation 28.58 and then comparing the result with Equation 28.57 shows that

$$A = aT^m e^m$$

28–8. Transition-State Theory Can Be Used to Estimate Reaction Rate Constants

In this section, we will briefly discuss a theory of reaction rates called *activated-complex theory* or *transition-state theory*. This theory was developed in the 1930s, principally by Henry Eyring, and focuses on the transient species in the vicinity of the top of the activation barrier to reaction. This species is called the *activated complex* or *transition state*, from which the theory derives its names.

Consider the reaction

$$A + B \longrightarrow P$$

where the rate law is given by

$$\frac{d[P]}{dt} = k[A][B] \tag{28.60}$$

Activated complex theory proposes that the reactants and the activated complex are in equilibrium with each other and that we model the reaction by the two-step process

$$A + B \rightleftharpoons AB^{\ddagger} \longrightarrow P \tag{28.61}$$

The species $AB^{\ddagger}$ is the activated complex. The equilibrium-constant expression between the reactants and the activated complex is (Section 26–2)

$$K_c^{\ddagger} = \frac{[AB^{\ddagger}]/c^{\circ}}{[A]/c^{\circ}[B]/c^{\circ}} = \frac{[AB^{\ddagger}]c^{\circ}}{[A][B]} \tag{28.62}$$

where c° is the standard-state concentration (often taken to be 1.00 mol·dm^{-3}). Using the results of Section 26–8, $K_c^{\ddagger}$ may be written in terms of partition functions

$$K_c^{\ddagger} = \frac{(q^{\ddagger}/V)c^{\circ}}{(q_A/V)(q_B/V)} \tag{28.63}$$

where the q_A, q_B, and $q^{\ddagger}$ are the partition functions of A, B, and $AB^{\ddagger}$, respectively.

The activated complex is assumed to be stable throughout a small region of width δ centered at the barrier top (Figure 28.10). The two-step process given by Equation 28.61 predicts that the rate of the reaction will be the product of the concentration of the activated complex $[AB^{\ddagger}]$ and v_c, the frequency with which these complexes cross over the barrier top, or

$$\frac{d[P]}{dt} = v_c[AB^{\ddagger}] \tag{28.64}$$

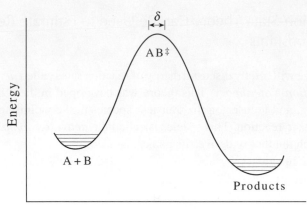

FIGURE 28.10
A one-dimensional energy diagram for the reaction given by Equation 28.61. The activated complex, $AB^{\ddagger}$, is defined to exist in the small region δ, centered around the barrier top.

Equations 28.64 and 28.60 give two different, yet equivalent, expressions for the reaction rate. Solving Equation 28.62 for $[AB^{\ddagger}]$ and substituting the resulting expression into Equation 28.64, and then equating the result to Equation 28.60 gives us

$$\frac{d[P]}{dt} = k[A][B] = v_c[AB^{\ddagger}] = v_c \frac{[A][B]K_c^{\ddagger}}{c^{\circ}}$$

or

$$k = \frac{v_c K_c^{\ddagger}}{c^{\circ}} \tag{28.65}$$

Note that k has units of $(\text{concentration})^{-1} \cdot \text{s}^{-1}$.

In writing Equation 28.64, we have implied that the motion of the reacting system over the barrier top is a one-dimensional translational motion. The translational partition function, q_{trans}, corresponding to one-dimensional translational motion is (Problem 18–3)

$$q_{\text{trans}} = \frac{(2\pi m^{\ddagger} k_B T)^{1/2}}{h} \delta \tag{28.66}$$

where $m^{\ddagger}$ is the mass of the activated complex. We can write the partition function of the activated complex as $q^{\ddagger} = q_{\text{trans}} q_{\text{int}}^{\ddagger}$, where $q_{\text{int}}^{\ddagger}$ accounts for all the remaining degrees of freedom of the activated complex. We can now rewrite Equation 28.63 as

$$K_c^{\ddagger} = \frac{(2\pi m^{\ddagger} k_B T)^{1/2}}{h} \delta \frac{(q_{\text{int}}^{\ddagger}/V)c^{\circ}}{(q_A/V)(q_B/V)} \tag{28.67}$$

Substituting Equation 28.67 into Equation 28.65 gives us the following expression for the reaction rate constant

$$k = v_c \frac{(2\pi m^{\ddagger} k_B T)^{1/2}}{hc^{\circ}} \delta \frac{(q_{int}^{\ddagger}/V)c^{\circ}}{(q_A/V)(q_B/V)} \tag{28.68}$$

Equation 28.68 contains two quantities, v_c and δ, that are not well defined and are difficult to determine. Their product, however, can be equated to the average speed with which the activated complex crosses the barrier, $\langle u_{ac} \rangle$, where $\langle u_{ac} \rangle = v_c \delta$. Because we have assumed that the reactants and activated complex are in equilibrium, we can use a (one-dimensional) Maxwell-Boltzmann distribution (Equation 27.33) to calculate $\langle u_{ac} \rangle$, or

$$\langle u_{ac} \rangle = \int_0^{\infty} u f(u) du = \left(\frac{m^{\ddagger}}{2\pi k_B T} \right)^{1/2} \int_0^{\infty} u e^{-m^{\ddagger} u^2 / 2 k_B T} du = \left(\frac{k_B T}{2\pi m^{\ddagger}} \right)^{1/2} \tag{28.69}$$

Note that we have integrated over only positive values of u because we are considering only those activated complexes that pass over the barrier in the direction of reactants to products. Substituting Equation 28.69 into Equation 28.68 for $v_c \delta$ gives us the transition-state theory expression for the rate constant

$$k = \frac{k_B T}{hc^{\circ}} \frac{(q_{int}^{\ddagger}/V)c^{\circ}}{(q_A/V)(q_B/V)} = \frac{k_B T}{hc^{\circ}} K^{\ddagger} \tag{28.70}$$

where $K^{\ddagger}$ is the "equilibrium constant" for the formation of the transition state from the reactants, but the motion along the reaction coordinate excluded in $q_{int}^{\ddagger}$.

We define the standard Gibbs energy of activation, $\Delta^{\ddagger} G^{\circ}$, to be the change in Gibbs energy in going from the reactants at a concentration c° to the transition state at a concentration c°. The relation between $\Delta^{\ddagger} G^{\circ}$ and $K^{\ddagger}$ is

$$\Delta^{\ddagger} G^{\circ} = -RT \ln K^{\ddagger} \tag{28.71}$$

We can use Equation 28.71 to express the rate constant k in terms of $\Delta^{\ddagger} G^{\circ}$. Solving Equation 28.71 for $K^{\ddagger}$ and substituting the result into Equation 28.70 gives us

$$k(T) = \frac{k_B T}{hc^{\circ}} e^{-\Delta^{\ddagger} G^{\circ} / RT} \tag{28.72}$$

We can express $\Delta^{\ddagger} G^{\circ}$ in terms of $\Delta^{\ddagger} H^{\circ}$, the standard enthalpy of activation, and $\Delta^{\ddagger} S^{\circ}$, the standard entropy of activation, by introducing

$$\Delta^{\ddagger} G^{\circ} = \Delta^{\ddagger} H^{\circ} - T \Delta^{\ddagger} S^{\circ} \tag{28.73}$$

which, upon substitution into Equation 28.72, gives us

$$k(T) = \frac{k_B T}{hc^{\circ}} e^{\Delta^{\ddagger} S^{\circ} / R} e^{-\Delta^{\ddagger} H^{\circ} / RT} \tag{28.74}$$

We can express the Arrhenius activation energy E_a in terms of $\Delta^\ddagger H^\circ$ and the Arrhenius A factor in terms of $\Delta^\ddagger S^\circ$. Differentiating the logarithm of Equation 28.70 with respect to temperature gives

$$\frac{d\ln k}{dT} = \frac{1}{T} + \frac{d\ln K^\ddagger}{dT} \tag{28.75}$$

Using the fact that $d\ln K_c/dT = \Delta U^\circ/RT^2$ for an ideal-gas system (see Problem 26–14), we can rewrite Equation 28.75 as

$$\frac{d\ln k}{dT} = \frac{1}{T} + \frac{\Delta^\ddagger U^\circ}{RT^2} \tag{28.76}$$

Furthermore, $\Delta^\ddagger H^\circ = \Delta^\ddagger U^\circ + \Delta^\ddagger PV = \Delta^\dagger U^\circ + RT\Delta^\ddagger n = \Delta^\ddagger U^\circ - RT$, for the reaction given by Equation 28.61, and so we can rewrite Equation 28.76 as

$$\frac{d\ln k}{dT} = \frac{\Delta^\ddagger H^\circ + 2RT}{RT^2} \tag{28.77}$$

Comparing Equation 28.77 with 28.55 gives us

$$E_a = \Delta^\ddagger H^\circ + 2RT \tag{28.78}$$

Solving this expression for $\Delta^\ddagger H^\circ$ and substituting the result into Equation 28.74 gives us

$$k(T) = \frac{e^2 k_B T}{hc^\circ} e^{\Delta^\ddagger S^\circ/R} e^{-E_a/RT} \tag{28.79}$$

Thus, in terms of the thermodynamic interpretation of transition-state theory, the Arrhenius A factor is given by

$$A = \frac{e^2 k_B T}{hc^\circ} e^{\Delta^\ddagger S^\circ/R} \tag{28.80}$$

EXAMPLE 28–10

The Arrhenius activation energy and pre-exponential factor for the reaction

$$H(g) + Br_2(g) \longrightarrow HBr(g) + Br(g)$$

are 15.5 kJ·mol^{-1} and 1.09×10^{11} dm^3·mol^{-1}·s^{-1}, respectively. What are the values $\Delta^\ddagger H^\circ$ and $\Delta^\ddagger S^\circ$ at 1000 K based on a standard state of 1.00 mol·dm^{-3}? Assume ideal-gas behavior.

SOLUTION: Equations 28.78 and 28.80 give

$$\Delta^\ddagger H^\circ = E_a - 2RT$$

$$= 15.5 \text{ kJ·mol}^{-1} - (2)(8.314 \text{ J·mol}^{-1}\cdot\text{K}^{-1})(1000 \text{ K})$$

$$= -1.13 \text{ kJ·mol}^{-1}$$

and

$$\Delta^{\ddagger}S^{\circ} = R \ln \frac{h A c^{\circ}}{e^2 k_{\mathrm{B}} T}$$

$$= (8.314 \text{ J·mol}^{-1}\text{·K}^{-1})$$

$$\times \ln \left\{ \frac{(6.626 \times 10^{-34} \text{ J·s})(1.09 \times 10^{11} \text{ dm}^3\text{·mol}^{-1}\text{·s}^{-1})(1.00 \text{ mol·dm}^{-3})}{e^2(1.381 \times 10^{-23} \text{ J·K}^{-1})(1000 \text{ K})} \right\}$$

$$= -60.3 \text{ J·K}^{-1}\text{·mol}^{-1}$$

Note that the value of $\Delta^{\ddagger}S^{\circ}$, like that of an equilibrium constant, depends upon the choice of standard state.

Values of $\Delta^{\ddagger}S^{\circ}$ give information about the relative structures of the activated complex and the reactants. A positive value indicates that the structure of the activated complex is less ordered than the reactants, whereas a negative value indicates that its structure is more ordered than the reactants.

Problems

28-1. For each of the following chemical reactions, calculate the equilibrium extent of reaction at 298.15 K and one bar. (See Section 26–4.)

    (a) $H_2(g) + Cl_2(g) \rightleftharpoons 2 HCl(g)$ $\Delta_r G^{\circ} = -190.54 \text{ kJ·mol}^{-1}$

        Initial amounts: one mole of $H_2(g)$ and $Cl_2(g)$ and no HCl(g).

    (b) $N_2(g) + O_2(g) \rightleftharpoons 2 NO(g)$ $\Delta_r G^{\circ} = 173.22 \text{ kJ·mol}^{-1}$

        Initial amounts: one mole of $N_2(g)$ and $O_2(g)$ and no NO(g).

28-2. Dinitrogen oxide, N_2O, decomposes according to the equation

$$2 N_2O(g) \longrightarrow 2 N_2(g) + O_2(g)$$

Under certain conditions at 900 K, the rate of reaction is $6.16 \times 10^{-6} \text{ mol·dm}^{-3}\text{·s}^{-1}$. Calculate the values of $d[N_2O]/dt$, $d[N_2]/dt$, and $d[O_2]/dt$.

28-3. Suppose the reaction in Problem 28–2 is carried out in a 2.67 dm^3 container. Calculate the value of $d\xi/dt$ corresponding to the rate of reaction of $6.16 \times 10^{-6} \text{ mol·dm}^{-3}\text{·s}^{-1}$.

28-4. The oxidation of hydrogen peroxide by permanganate occurs according to the equation

$$2 KMnO_4(aq) + 3 H_2SO_4(aq) + 5 H_2O_2(aq) \longrightarrow 2 MnSO_4(aq)$$

$$+ 8 H_2O(l) + 5 O_2(g) + K_2SO_4(aq)$$

Define v, the rate of reaction, in terms of each of the reactants and products.

28-5. The second-order rate constant for the reaction

$$O(g) + O_3(g) \longrightarrow 2 O_2(g)$$

is 1.26×10^{-15} cm$^3 \cdot$molecule$^{-1} \cdot$s^{-1}. Determine the value of the rate constant in units of dm$^3 \cdot$mol$^{-1} \cdot$s^{-1}.

28-6. The definition of the rate of reaction in terms of molar concentration (Equation 28.5) assumes that the volume remains constant during the course of the reaction. Derive an expression for the rate of reaction in terms of the molar concentration of a reactant A for the case in which the volume changes during the course of the reaction.

28-7. Derive the integrated rate law for a reaction that is zero order in reactant concentration.

28-8. Determine the rate law for the reaction described by

$$NO(g) + H_2(g) \longrightarrow products$$

from the initial rate data tabulated below.

$P_0(H_2)$/torr	$P_0(NO)$/torr	v_0/torr$\cdot$s^{-1}
400	159	34
400	300	125
289	400	160
205	400	110
147	400	79

Calculate the rate constant for this reaction.

28-9. Sulfuryl chloride decomposes according to the equation

$$SO_2Cl_2(g) \longrightarrow SO_2(g) + Cl_2(g)$$

Determine the order of the reaction with respect to $SO_2Cl_2(g)$ from the following initial-rate data collected at 298.15 K

$[SO_2Cl_2]_0$/mol$\cdot$dm^{-3}	0.10	0.37	0.76	1.22
v_0/mol$\cdot$dm$^{-3} \cdot$s^{-1}	2.24×10^{-6}	8.29×10^{-6}	1.71×10^{-5}	2.75×10^{-5}

Calculate the rate constant for this reaction at 298.15 K.

28-10. Consider the reaction described by

$$Cr(H_2O)_6^{3+}(aq) + SCN^-(aq) \longrightarrow Cr(H_2O)_5(SCN)^{2+}(aq) + H_2O(l)$$

for which the following initial rate data were obtained at 298.15 K.

$[Cr(H_2O)_6^{3+}]_0$/mol$\cdot$dm^{-3}	$[SCN^-]_0$/mol$\cdot$dm^{-3}	v_0/mol$\cdot$dm$^{-3} \cdot$s^{-1}
1.21×10^{-4}	1.05×10^{-5}	2.11×10^{-11}
1.46×10^{-4}	2.28×10^{-5}	5.53×10^{-11}
1.66×10^{-4}	1.02×10^{-5}	2.82×10^{-11}
1.83×10^{-4}	3.11×10^{-5}	9.44×10^{-11}

Determine the rate law for the reaction and the rate constant at 298.15 K. Assume the orders are integers.

28-11. Consider the base-catalyzed reaction

$$OCl^-(aq) + I^-(aq) \longrightarrow OI^-(aq) + Cl^-(aq)$$

Use the following initial-rate data to determine the rate law and the corresponding rate constant for the reaction.

$[OCl^-]/mol \cdot dm^{-3}$	$[I^-]/mol \cdot dm^{-3}$	$[OH^-]/mol \cdot dm^{-3}$	$v_0/mol \cdot dm^{-3} \cdot s^{-1}$
1.62×10^{-3}	1.62×10^{-3}	0.52	3.06×10^{-4}
1.62×10^{-3}	2.88×10^{-3}	0.52	5.44×10^{-4}
2.71×10^{-3}	1.62×10^{-3}	0.84	3.16×10^{-4}
1.62×10^{-3}	2.88×10^{-3}	0.91	3.11×10^{-4}

28-12. The reaction

$$SO_2Cl_2(g) \longrightarrow SO_2(g) + Cl_2(g)$$

is first order and has a rate constant of 2.24×10^{-5} s^{-1} at 320°C. Calculate the half-life of the reaction. What fraction of a sample of $SO_2Cl_2(g)$ remains after being heated for 5.00 hours at 320°C? How long will a sample of $SO_2Cl_2(g)$ need to be maintained at 320°C to decompose 92.0% of the initial amount present?

28-13. The half-life for the following gas-phase decomposition reaction

is found to be independent of the initial concentration of the reactant. Determine the rate law and integrated rate law for this reaction.

28-14. Hydrogen peroxide, H_2O_2, decomposes in water by a first-order kinetic process. A 0.156-mol·dm^{-3} solution of H_2O_2 in water has an initial rate of 1.14×10^{-5} mol·dm^{-3}·s^{-1}. Calculate the rate constant for the decomposition reaction and the half-life of the decomposition reaction.

28-15. A first-order reaction is 24.0% complete in 19.7 minutes. How long will the reaction take to be 85.5% complete? Calculate the rate constant for the reaction.

28-16. The nucleophilic substitution reaction

$$PhSO_2SO_2Ph(sln) + N_2H_4(sln) \longrightarrow PhSO_2NHNH_2(sln) + PhSO_2H(sln)$$

was studied in cyclohexane solution at 300 K. The rate law was found to be first order in $PhSO_2SO_2Ph$. For an initial concentration of $[PhSO_2SO_2Ph]_0 = 3.15 \times 10^{-5}$ mol·dm^{-3}, the following rate data were observed. Determine the rate law and the rate constant for this reaction.

$[N_2H_4]_0/10^{-2}$ mol·dm^{-3}	0.5	1.0	2.4	5.6
$v/mol \cdot dm^{-3} \cdot s^{-1}$	0.085	0.17	0.41	0.95

28-17. Show that if A reacts to form either B or C according to

$$A \xrightarrow{k_1} B \quad \text{or} \quad A \xrightarrow{k_2} C$$

then

$$[A] = [A]_0 e^{-(k_1 + k_2)t}$$

Now show that $t_{1/2}$, the half-life of A, is given by

$$t_{1/2} = \frac{0.693}{k_1 + k_2}$$

Show that $[B]/[C] = k_1/k_2$ for all times t. For the set of initial conditions $[A] = [A]_0$, $[B]_0 = [C]_0 = 0$, and $k_2 = 4k_1$, plot $[A]$, $[B]$, and $[C]$ as a function of time on the same graph.

The following six problems deal with the decay of radioactive isotopes, which is a first-order process. Therefore, if $N(t)$ is the number of a radioactive isotope at time t, then $N(t) = N(0)e^{-kt}$, where $N(0)$ is the corresponding number at $t = 0$. In dealing with radioactive decay, we use the half-life, $t_{1/2} = 0.693/k$, almost exclusively to describe the rate of decay (the kinetics of decay).

28-18. You order a sample of Na_3PO_4 containing the radioisotope phosphorus–32 ($t_{1/2} = 14.3$ days). If the shipment is delayed in transit for two weeks, how much of the original activity will remain when you receive the sample?

28-19. Copper–64 ($t_{1/2} = 12.8$ h) is used in brain scans for tumors and in studies of Wilson's disease (a genetic disorder characterized by the inability to metabolize copper). Calculate the number of days required for an administered dose of copper–64 to drop to 0.10% of the initial value injected. Assume no loss of copper–64 except by radioactive decay.

28-20. Sulfur–38 can be incorporated into proteins to follow certain aspects of protein metabolism. If a protein sample initially has an activity of $10\,000$ disintegrations $\cdot$ min^{-1}, calculate the activity 6.00 h later. The half-life of sulfur–38 is 2.84 h. *Hint*: Use the fact that the rate of decay is proportional to $N(t)$ for a first-order process.

28-21. The radioisotope phosphorus–32 can be incorporated into nucleic acids to follow certain aspects of their metabolism. If a nucleic acid sample initially has an activity of $40\,000$ disintegrations $\cdot$ min^{-1}, calculate the activity 220 h later. The half-life of phosphorus–32 is 14.28 d. *Hint*: Use the fact that the rate of decay is proportional to $N(t)$ for a first-order process.

28-22. Uranium–238 decays to lead–206 with a half-life of 4.51×10^9 y. A sample of ocean sediment is found to contain 1.50 mg of uranium–238 and 0.460 mg of lead-206. Estimate the age of the sediment assuming that lead–206 is formed only by the decay of uranium and that lead–206 does not itself decay.

28-23. Potassium-argon dating is used in geology and archeology to date sedimentary rocks. Potassium–40 decays by two different paths

$$^{40}_{19}K \longrightarrow {}^{40}_{20}Ca + {}^{0}_{-1}e \quad (89.3\%)$$

$$^{40}_{19}K \longrightarrow {}^{40}_{18}Ar + {}^{0}_{1}e \quad (10.7\%)$$

The overall half-life for the decay of potassium–40 is 1.3×10^9 y. Estimate the age of sedimentary rocks with an argon–40 to potassium–40 ratio of 0.0102. (See Problem 28–17.)

28-24. In this problem, we will derive Equation 28.32 from the rate law (Equation 28.31)

$$-\frac{d[A]}{dt} = k[A][B] \tag{1}$$

Use the reaction stoichiometry of Equation 28.30 to show that $[B] = [B]_0 - [A]_0 + [A]$. Use this result to show that Equation 1 can be written as

$$-\frac{d[A]}{dt} = k[A]\{[B]_0 - [A]_0 + [A]\}$$

Now separate the variables and then integrate the resulting equation subject to its initial conditions to obtain the desired result, Equation 28.32:

$$kt = \frac{1}{[A]_0 - [B]_0} \ln \frac{[A][B]_0}{[B][A]_0}$$

28-25. Equation 28.32 is indeterminate if $[A]_0 = [B]_0$. Use L'Hopital's rule to show that Equation 28.32 reduces to Equation 28.33 when $[A]_0 = [B]_0$. (*Hint*: Let $[A] = [B] + x$ and $[A]_0 = [B]_0 + x$.).

28-26. Uranyl nitrate decomposes according to

$$UO_2(NO_3)_2(aq) \longrightarrow UO_3(s) + 2NO_2(g) + \tfrac{1}{2}O_2(g)$$

The rate law for this reaction is first order in the concentration of uranyl nitrate. The following data were recorded for the reaction at 25.0°C.

t/min	0	20.0	60.0	180.0	360.0
$[UO_2(NO_3)_2]/\text{mol}\cdot\text{dm}^{-3}$	0.01413	0.01096	0.00758	0.00302	0.00055

Calculate the rate constant for this reaction at 25.0°C.

28-27. The data for the decomposition of uranyl nitrate (Problem 28–26) at 350°C are tabulated below

t/min	0	6.0	10.0	17.0	30.0	60.0
$[UO_2(NO_3)_2]/\text{mol}\cdot\text{dm}^{-3}$	0.03802	0.02951	0.02089	0.01259	0.00631	0.00191

Calculate the rate constant for this reaction at 350°C.

28-28. The following data are obtained for the reaction

$$N_2O(g) \longrightarrow N_2(g) + \tfrac{1}{2}O_2(g)$$

at 900 K.

t/s	0	3146	6494	13933
$[N_2O]/mol \cdot dm^{-3}$	0.521	0.416	0.343	0.246

The rate law for this reaction is second order in N_2O concentration. Calculate the rate constant for this decomposition reaction.

28-29. Consider a chemical reaction

$$A \longrightarrow products$$

that obeys the rate law

$$-\frac{d[A]}{dt} = k[A]^n$$

where n, the reaction order, can be any number except $n = 1$. Separate the concentration and time variables and then integrate the resulting expression assuming the concentration of A is $[A]_0$ at time $t = 0$ and is $[A]$ at time t to show that

$$kt = \frac{1}{n-1}\left(\frac{1}{[A]^{n-1}} - \frac{1}{[A]_0^{n-1}}\right) \qquad n \neq 1 \qquad (1)$$

Use Equation 1 to show that the half-life of a reaction of order n is

$$kt_{1/2} = \frac{1}{n-1}\frac{2^{n-1}-1}{[A]_0^{n-1}} \qquad n \neq 1 \qquad (2)$$

Show that this result reduces to Equation 28.29 when $n = 2$.

28-30. Show that Equation 1 of Problem 28–29 can be written in the form

$$\frac{\left(\dfrac{[A]_0}{[A]}\right)^x - 1}{x} = k[A]_0^x t$$

where $x = n - 1$. Now use L'Hopital's rule to show that

$$\ln\frac{[A]}{[A]_0} = -kt$$

for $n = 1$. (Remember that $da^x/dx = a^x \ln a$.)

28-31. The following data were obtained for the reaction

$$N_2O(g) \longrightarrow N_2(g) + \tfrac{1}{2}O_2(g)$$

$[N_2O]_0/mol \cdot dm^{-3}$	1.674×10^{-3}	4.458×10^{-3}	9.300×10^{-3}	1.155×10^{-2}
$t_{1/2}/s$	1200	470	230	190

Assume the rate law for this reaction is

$$-\frac{d[\text{N}_2\text{O}]}{dt} = k[\text{N}_2\text{O}]^n$$

and use Equation 2 of Problem 28–29 to determine the reaction order of N_2O by plotting $\ln t_{1/2}$ against $\ln[\text{A}]_0$. Calculate the rate constant for this decomposition reaction.

28-32. We will derive Equation 28.39 from Equation 28.38 in this problem. Rearrange Equation 28.38 to become

$$\frac{d[\text{A}]}{(k_1 + k_{-1})[\text{A}] - k_{-1}[\text{A}]_0} = -dt$$

and integrate to obtain

$$\ln\{(k_1 + k_{-1})[\text{A}] - k_{-1}[\text{A}]_0\} = -(k_1 + k_{-1})t + \text{constant}$$

or

$$(k_1 + k_{-1})[\text{A}] - k_{-1}[\text{A}]_0 = ce^{-(k_1 + k_{-1})t}$$

where c is a constant. Show that $c = k_1[\text{A}]_0$ and that

$$(k_1 + k_{-1})[\text{A}] - k_{-1}[\text{A}]_0 = k_1[\text{A}]_0 e^{-(k_1 + k_{-1})t} \tag{1}$$

Now let $t \to \infty$ and show that

$$[\text{A}]_0 = \frac{(k_1 + k_{-1})[\text{A}]_{\text{eq}}}{k_{-1}}$$

and

$$[\text{A}]_0 - [\text{A}]_{\text{eq}} = \frac{k_1[\text{A}]_{\text{eq}}}{k_{-1}} = \frac{k_1[\text{A}]_0}{k_1 + k_{-1}}$$

Use these results in Equation 1 to obtain Equation 28.39.

28-33. Consider the general chemical reaction

$$\text{A} + \text{B} \underset{k_{-1}}{\overset{k_1}{\rightleftharpoons}} \text{P}$$

If we assume that both the forward and reverse reactions are first order in their respective reactants, the rate law is given by (Equation 28.52)

$$\frac{d[\text{P}]}{dt} = k_1[\text{A}][\text{B}] - k_{-1}[\text{P}] \tag{1}$$

Now consider the response of this chemical reaction to a temperature jump. Let $[\text{A}] = [\text{A}]_{2,\text{eq}} + \Delta[\text{A}]$, $[\text{B}] = [\text{B}]_{2,\text{eq}} + \Delta[\text{B}]$, and $[\text{P}] = [\text{P}]_{2,\text{eq}} + \Delta[\text{P}]$, where the subscript "2,eq" refers to the new equilibrium state. Now use the fact that $\Delta[\text{A}] = \Delta[\text{B}] = -\Delta[\text{P}]$ to show that Equation 1 becomes

$$\frac{d\Delta[\text{P}]}{dt} = k_1[\text{A}]_{2,\text{eq}}[\text{B}]_{2,\text{eq}} - k_{-1}[\text{P}]_{2,\text{eq}}$$

$$- \{k_1([\text{A}]_{2,\text{eq}} + [\text{B}]_{2,\text{eq}}) + k_{-1}\}\Delta[\text{P}] + O(\Delta[\text{P}]^2)$$

Show that the first terms on the right side of this equation cancel and that Equations 28.53 and 28.54 result.

28-34. The equilibrium constant for the reaction

$$H^+(aq) + OH^-(aq) \underset{k_{-1}}{\overset{k_1}{\rightleftharpoons}} H_2O(l)$$

at 25°C is $K_c = [H_2O]/[H^+][OH^-] = 5.49 \times 10^{15}$ mol$^{-1} \cdot$dm^3. The time-dependent conductivity of the solution following a temperature jump to a final temperature of 25°C shows a relaxation time of $\tau = 3.7 \times 10^{-5}$ s. Determine the values of the rate constants k_1 and k_{-1}. At 25°C, the density of water is $\rho = 0.997$ g$\cdot$cm^{-3}.

28-35. The equilibrium constant for the reaction

$$D^+(aq) + OD^-(aq) \underset{k_{-1}}{\overset{k_1}{\rightleftharpoons}} D_2O(l)$$

at 25°C is $K_c = 4.08 \times 10^{16}$ mol$^{-1} \cdot$dm^3. The rate constant k_{-1} is independently found to be 2.52×10^{-6} s^{-1}. What do you predict for the observed relaxation time for a temperature-jump experiment to a final temperature of 25°C? The density of D_2O is $\rho = 1.104$ g$\cdot$cm^{-3} at 25°C.

28-36. Consider the chemcial reaction described by

$$2\,A(aq) \underset{k_{-1}}{\overset{k_1}{\rightleftharpoons}} D(aq)$$

If we assume the forward reaction is second order and the reverse reaction is first order, the rate law is given by

$$\frac{d[D]}{dt} = k_1[A]^2 - k_{-1}[D] \tag{1}$$

Now consider the response of this chemical reaction to a temperature jump. Let $[A] = [A]_{2,eq} + \Delta[A]$ and $[D] = [D]_{2,eq} + \Delta[D]$, where the subscript "2,eq" refers to the new equilibrium state. Now use the fact that $\Delta[A] = -2\Delta[D]$ to show that Equation 1 becomes

$$\frac{d\Delta[D]}{dt} = -(4k_1[A]_{2,eq} + k_{-1})\Delta[D] + O(\Delta[D]^2)$$

Show that if we ignore the $O(\Delta[D]^2)$ term, then

$$\Delta[D] = \Delta[D]_0 e^{-t/\tau}$$

where $\tau = 1/(4k_1[A]_{2,eq} + k_{-1})$.

28-37. In Problem 28–36, you showed that the relaxation time for the dimerization reaction

$$2\,A(aq) \underset{k_{-1}}{\overset{k_1}{\rightleftharpoons}} D(aq)$$

is given by $\tau = 1/(4k_1[A]_{2,eq} + k_{-1})$. Show that this equation can be rewritten as

$$\frac{1}{\tau^2} = k^2_{-1} + 8k_1 k_{-1}[S]_0$$

where $[S]_0 = 2[D] + [A] = 2[D]_{2,eq} + [A]_{2,eq}$.

28-38. The first step in the assembly of the protein yeast phosphoglycerate mutase is a reversible dimerization of a polypeptide,

$$2\,A(aq) \underset{k_{-1}}{\overset{k_1}{\rightleftharpoons}} D(aq)$$

where A is the polypeptide and D is the dimer. Suppose that a 1.43×10^{-5} mol·dm^{-3} solution of A is prepared and allowed to come to equilibrium at 280 K. Once equilibrium is achieved, the temperature of the solution is jumped to 293 K. The rate constants k_1 and k_{-1} for the dimerization reaction at 293 K are 6.25×10^3 dm^3·mol^{-1}·s^{-1} and 6.00×10^{-3} s^{-1}, respectively. Calculate the value of the relaxation time observed in the experiment. (*Hint*: See Problem 28–37.)

28-39. Does the Arrhenius A factor always have the same units as the reaction rate constant?

28-40. Use the results of Problems 28–26 and 28–27 to calculate the values of E_a and A for the decomposition of uranyl nitrate.

28-41. The experimental rate constants for the reaction described by

$$OH(g) + ClCH_2CH_2Cl(g) \longrightarrow H_2O(g) + ClCHCH_2Cl(g)$$

at various temperatures are tabulated below.

T/K	292	296	321	333	343	363
$k/10^8$ dm^3·mol^{-1}·s^{-1}	1.24	1.32	1.81	2.08	2.29	2.75

Determine the values of the Arrhenius parameters A and E_a for this reaction.

28-42. The Arrhenius parameters for the reaction described by

$$HO_2(g) + OH(g) \longrightarrow H_2O(g) + O_2(g)$$

are $A = 5.01 \times 10^{10}$ dm^3·mol^{-1}·s^{-1} and $E_a = 4.18$ kJ·mol^{-1}. Determine the value of the rate constant for this reaction at 298 K.

28-43. At what temperature will the reaction described in Problem 28–42 have a rate constant that is twice that at 298 K?

28-44. The rate constants for the reaction

$$CHCl_2(g) + Cl_2(g) \longrightarrow CHCl_3(g) + Cl(g)$$

at different temperatures are tabulated below

T/K	357	400	458	524	533	615
$k/10^7 \text{ dm}^3 \cdot \text{mol}^{-1} \cdot \text{s}^{-1}$	1.72	2.53	3.82	5.20	5.61	7.65

Calculate the values of the Arrhenius parameters A and E_a for this reaction.

28-45. The rate constant for the chemical reaction

$$2\,N_2O_5(g) \longrightarrow 4\,NO_2(g) + O_2(g)$$

doubles from 22.50°C to 27.47°C. Determine the activation energy of the reaction. Assume the pre-exponential factor is independent of temperature.

28-46. Show that if A reacts to form either B or C according to

$$A \xrightarrow{k_1} B \qquad \text{or} \qquad A \xrightarrow{k_2} C$$

then E_a, the observed activation energy for the disappearance of A, is given by

$$E_a = \frac{k_1 E_1 + k_2 E_2}{k_1 + k_2}$$

where E_1 is the activation energy for the first reaction and E_2 is the activation energy for the second reaction.

28-47. Cyclohexane interconverts between a "chair" and a "boat" structure. The activation parameters for the reaction from the chair to the boat form of the molecule are $\Delta^{\ddagger} H^{\circ} = 31.38 \text{ kJ} \cdot \text{mol}^{-1}$ and $\Delta^{\ddagger} S^{\circ} = 16.74 \text{ J} \cdot \text{K}^{-1}$. Calculate the standard Gibbs energy of activation and the rate constant for this reaction at 325 K.

28-48. The gas-phase rearrangement reaction

$$\text{vinyl allyl ether} \longrightarrow \text{allyl acetone}$$

has a rate constant of $6.015 \times 10^{-5} \text{ s}^{-1}$ at 420 K and a rate constant of $2.971 \times 10^{-3} \text{ s}^{-1}$ at 470 K. Calculate the values of the Arrhenius parameters A and E_a. Calculate the values of $\Delta^{\ddagger} H^{\circ}$ and $\Delta^{\ddagger} S^{\circ}$ at 420 K. (Assume ideal-gas behavior.)

28-49. The kinetics of a chemical reaction can be followed by a variety of experimental techniques, including optical spectroscopy, NMR spectroscopy, conductivity, resistivity, pressure changes, and volume changes. When using these techniques, we do not measure the concentration itself but we know that the observed signal is proportional to the concentration; the exact proportionality constant depends on the experimental technique and the species present in the chemical system. Consider the general reaction given by

$$\nu_A A + \nu_B B \longrightarrow \nu_Y Y + \nu_Z Z$$

where we assume that A is the limiting reagent so that $[A] \to 0$ as $t \to \infty$. Let p_i be the proportionality constant for the contribution of species i to S, the measured signal from the instrument. Explain why at any time t during the reaction, S is given by

$$S(t) = p_A[A] + p_B[B] + p_Y[Y] + p_Z[Z] \tag{1}$$

Show that the initial and final readings from the instrument are given by

$$S(0) = p_A[A]_0 + p_B[B]_0 + p_Y[Y]_0 + p_Z[Z]_0 \tag{2}$$

and

$$S(\infty) = p_B\left([B]_0 - \frac{v_B}{v_A}[A]_0\right) + p_Y\left([Y]_0 + \frac{v_Y}{v_A}[A]_0\right) + p_Z\left([Z]_0 + \frac{v_Z}{v_A}[A]_0\right) \tag{3}$$

Combine Equations 1 through 3 to show that

$$[A] = [A]_0 \frac{S(t) - S(\infty)}{S(0) - S(\infty)}$$

(*Hint*: Be sure to express [B], [Y], and [Z] in terms of their initial values, [A] and $[A]_0$.)

28-50. Use the result of Problem 28–49 to show that for the first-order rate law, $v = k[A]$, the time-dependent signal is given by

$$S(t) = S(\infty) + [S(0) - S(\infty)]e^{-kt}$$

28-51. Use the result of Problem 28–49 to show that for the second-order rate law, $v = k[A]^2$, the time-dependent signal is given by

$$S(t) = S(\infty) + \frac{S(0) - S(\infty)}{1 + [A]_0 kt}$$

28-52. Because there is a substantial increase in the volume of the solution as the reaction proceeds, the decomposition of diacetone alcohol can be followed by a dilatometer, a device that measures the volume of a sample as a function of time. The instrument readings at various times are tabulated below.

Time/s	0	24.4	35.0	48.0	64.8	75.8	133.4	∞
S/arbitrary units	8.0	20.0	24.0	28.0	32.0	34.0	40.0	43.3

Use the expressions derived in Problems 28–50 and 28–51 to determine if the decomposition reaction is a first- or second-order process.

28-53. In Problem 28–49, we assumed that A reacted completely so that $[A] \to 0$ as $t \to \infty$. Show that if the reaction does not go to completion but establishes an equilibrium instead, then

$$[A] = [A]_{eq} + \{[A]_0 - [A]_{eq}\}\frac{S(t) - S(\infty)}{S(0) - S(\infty)}$$

where $[A]_{eq}$ is the equilibrium concentration of A.

Sherwood Rowland (top left), Mario Molina (top right), and Paul Crutzen (bottom) received the Nobel Prize for chemistry in 1995 "for their work in atmospheric chemistry, particularly concerning the formation and decomposition of ozone." **F. Sherwood Rowland** was born in Delaware, Ohio, on June 28, 1927, and received his Ph.D. from the University of Chicago in 1951. After serving as an instructor at Princeton University, he joined the faculty at the University of Kansas in 1956. In 1964, he went to the University of California at Irvine, where he was the founding chairman and where he remains today. His current research interests are atmospheric and environmental chemistry. **Mario J. Molina** was born in Mexico City, Mexico, on March 19, 1943. After receiving his Ph.D. from the University of California at Berkeley in 1972, he joined Rowland as a postdoctoral fellow at UC Irvine and worked with him on the studies of chlorine and fluorochloromethane in the atmosphere. In 1989, he joined the faculty of The Massachusetts Institute of Technology, where he currently remains. Molina donated a major part of his share of the Nobel Prize to help scholars from developing nations conduct environmental research at MIT. **Paul J. Crutzen** was born in Amsterdam, the Netherlands, on December 3, 1933. He received his Ph.D. in meteorology in 1968 and his D.Sc. in 1973 from the University of Stockholm. After spending 1974 to 1980 at the National Center for Atmospheric Research at Boulder, Colorado, he returned to his current position at the Max Planck Institute for Chemistry at Mainz.

CHAPTER 29

Chemical Kinetics II: Reaction Mechanisms

In this chapter, we consider how reactants are converted to products. We begin by discussing elementary reactions, which are defined as chemical reactions that occur in a single step. We will show that the rate law for an elementary reaction can be deduced from the reaction stoichiometry. We will then discuss complex reactions, or reactions that do not occur by a single step. One of the major goals of chemical kinetics is to determine the *mechanism*, or sequence of elementary reactions, by which a complex reaction occurs. We will discuss several commonly encountered reaction mechanisms and learn some of the approximations used to derive the rate law for a complex reaction from a proposed reaction mechanism. We will then examine the mechanisms of "unimolecular" reactions and chain reactions. Finally, we will discuss chemical catalysis, with an emphasis on the biochemical reactions catalyzed by enzymes.

29–1. A Mechanism Is a Sequence of Single-Step Chemical Reactions Called Elementary Reactions

Many chemical reactions involve reaction intermediates, and the overall kinetic process can be written as

$$\text{reactants} \longrightarrow \text{intermediates} \longrightarrow \text{products}$$

As an example of this type of reaction, consider the chemical reaction given by

$$NO_2(g) + CO(g) \xrightarrow{k_{obs}} NO(g) + CO_2(g) \tag{29.1}$$

This reaction does not occur in a single step but proceeds by the following two-step process:

$$NO_2(g) + NO_2(g) \xrightarrow{k_1} NO_3(g) + NO(g) \tag{29.2}$$

$$NO_3(g) + CO(g) \xrightarrow{k_2} NO_2(g) + CO_2(g) \tag{29.3}$$

1181

Neither of the two steps, Equations 29.2 and 29.3, is thought to involve any reaction intermediates. A reaction that does not involve any intermediates occurs in a single step and is called an *elementary reaction*. We say that the chemical reaction given by Equation 29.1 is a *complex reaction* whose *mechanism* is given by the sequence of elementary chemical reactions given by Equations 29.2 and 29.3.

We want to distinguish complex reactions from elementary reactions. From this point on, the arrows $\Rightarrow$ and $\Leftarrow$ will be used to indicate reactions currently thought to be elementary. The arrows $\rightarrow$ and $\leftarrow$ will be used to indicate complex reactions. Only a few complex reactions have been studied in great enough detail that the elementary steps are known with certainty.

The *molecularity* of an elementary reaction is defined to be the number of reactant molecules involved in the chemical reaction. Elementary reactions that involve one, two, and three molecules are termed *unimolecular*, *bimolecular*, and *termolecular* reactions, respectively. These terms should be used to describe only elementary reactions. In Chapter 28, we learned that rate laws must be determined experimentally; however, we will now see that the rate law for an elementary reaction can be deduced from the balanced chemical equation itself.

Because an elementary reaction does not involve the formation of a reaction intermediate, the products must be formed directly from the reactants. Thus, for a unimolecular reaction,

$$A \Longrightarrow \text{products}$$

the rate can depend only on the concentration of A molecules that are available to react. Therefore, the rate law for a unimolecular reaction is first order in the reactant, or

$$v = k[A]$$

For both bimolecular and termolecular reactions, the reactants must collide for the reaction to occur. To have a collision and also not form any reaction intermediates, an elementary reaction must be one in which all the reactants collide simultaneously, with the reaction occuring immediately upon this collision. The rate of the reaction will therefore depend on the collision frequency between the required reactants. In our study of the kinetic theory of gases in Chapter 27, we learned that the collision frequency is proportional to the number densities, or the concentrations, of the colliding molecules (Equation 27.57). Therefore, for the bimolecular reaction

$$A + B \Longrightarrow \text{products}$$

the rate of reaction must be given by

$$v = k[A][B]$$

The rate law for a bimolecular reaction is first order in each of the two reactants and second order overall.

Similarly, the termolecular reaction

$$A + B + C \Longrightarrow \text{products}$$

must have a rate law that is first order in each of the three reactants and third order overall, or

$$v = k[A][B][C]$$

The probability of having a simultaneous collision between all the reactants decreases with increasing molecularity of the reaction. No elementary reaction with a molecularity greater than three is known, and the overwhelming majority of elementary reactions are bimolecular.

EXAMPLE 29–1

Deduce the rate laws for the following reactions.

a) $2\,NO(g) + O_2(g) \xrightarrow{k} N_2O_4(g)$

b) $O_3(g) + Cl(g) \xrightarrow{k} ClO(g) + O_2(g)$

c) $NO_2(g) + F_2(g) \xrightarrow{k} NO_2F(g) + F(g)$

SOLUTION: a) The type of arrow used in this chemical equation indicates that this reaction is not an elementary reaction. Therefore, we need experimental data to deduce the rate law.

b) The type of arrow used in the chemical equation indicates that this reaction is an elementary reaction. Therefore, the rate law is

$$v = k[O_3][Cl]$$

c) This reaction is also a bimolecular elementary reaction and the rate law is

$$v = k[NO_2][F_2]$$

29–2. The Principle of Detailed Balance States that when a Complex Reaction Is at Equilibrium, the Rate of the Forward Process Is Equal to the Rate of the Reverse Process for Each and Every Step of the Reaction Mechanism

We now show that the equilibrium constant for an elementary reaction is equal to the ratio of the forward and reverse rate constants. Consider the general reversible reaction for which the forward process and the reverse process are each bimolecular

$$A + B \underset{k_{-1}}{\overset{k_1}{\rightleftharpoons}} C + D \tag{29.4}$$

We will encounter such reactions many times in our study of chemical kinetics. We refer to this type of reaction as a *reversible elementary reaction*, which signifies that the reaction occurs in both the forward and reverse directions to a significant extent and that the reaction in each direction is an elementary reaction. Because both the forward and reverse reactions of the chemical reaction in Equations 29.4 are elementary bimolecular reactions, the rates of the forward and reverse reactions, v_1 and v_{-1}, are

$$v_1 = k_1[A][B]$$

$$v_{-1} = k_{-1}[C][D]$$

At equilibrium, $v_1 = v_{-1}$, so

$$k_1[A]_{eq}[B]_{eq} = k_{-1}[C]_{eq}[D]_{eq} \tag{29.5}$$

The subscript "eq" emphasizes that the concentrations of A, B, C, and D are those at equilibrium. The equilibrium constant K_c is given by

$$K_c = \frac{[C]_{eq}[D]_{eq}}{[A]_{eq}[B]_{eq}}$$

so Equation 29.5 becomes

$$\frac{k_1}{k_{-1}} = \frac{[C]_{eq}[D]_{eq}}{[A]_{eq}[B]_{eq}} = K_c \tag{29.6}$$

The relationship $K_c = k_1/k_{-1}$ holds for all reversible elementary reactions and is commonly called the *principle of detailed balance*. This principle applies only to elementary reactions at equilibrium. If the reaction is not an elementary reaction, K_c need not be equal to k_1/k_{-1}.

Although the principle of detailed balance does not apply to a complex reaction, it must apply to each step of the mechanism of a complex reaction because each step is, by definition, an elementary reaction. This is an important point and must be kept in mind when deriving an expression for the equilibrium constant from rate laws. As an example, consider the reversible equilibrium reaction

$$A \rightleftharpoons B \tag{29.7}$$

Assume that the mechanism for this reaction consists of the following two competing steps:

$$A + C \underset{k_{-1}}{\overset{k_1}{\rightleftharpoons}} B + C \tag{29.8}$$

and

$$A \underset{k_{-2}}{\overset{k_2}{\rightleftharpoons}} B \qquad (29.9)$$

We will consider several real examples of such a reaction mechanism when we discuss enzyme catalysis in Section 29–9 and surface catalysis in Chapter 31. Note that one of the elementary steps of the reaction mechanism is written in a form that is identical to the overall complex reaction. The difference is that Equations 29.7 accounts for all the possible chemical pathways for the reaction $A \rightleftharpoons B$. Because there are two pathways by which the reaction can occur, Equations 29.7 cannot be an elementary reaction. However, an elementary reaction equivalent to Equations 29.7 can be one of the possible reaction pathways. This is why the elementary step, Equations 29.9, has the same form as the overall complex reaction.

According to the principle of detailed balance, when the overall reaction, Equations 29.7, is at equilibrium, each step of the reaction mechanism must also be at equilibrium. Therefore, at equilibrium,

$$v_1 = k_1[A]_{eq}[C]_{eq} = v_{-1} = k_{-1}[B]_{eq}[C]_{eq} \qquad (29.10)$$

and

$$v_2 = k_2[A]_{eq} = v_{-2} = k_{-2}[B]_{eq} \qquad (29.11)$$

The equilibrium conditions given by Equations 29.10 and 29.11 become

$$\frac{[B]_{eq}}{[A]_{eq}} = K_c = \frac{k_1}{k_{-1}} \qquad (29.12)$$

and

$$\frac{[B]_{eq}}{[A]_{eq}} = K_c = \frac{k_2}{k_{-2}} \qquad (29.13)$$

Equating Equations 29.12 and 29.13 gives us

$$\frac{k_1}{k_{-1}} = \frac{k_2}{k_{-2}} \qquad (29.14)$$

Because of the principle of detailed balance, the four rate constants k_1, k_{-1}, k_2, and k_{-2} are not independent of one another. The equation for the overall reaction is given by the sum of two steps of the mechanism (Equations 29.8 and 29.9), so we also have

$$v_1 + v_2 = v_{-1} + v_{-2} \qquad (29.15)$$

at equilibrium. Example 29–2 uses Equation 29.15 to derive the equilibrium constant for the overall reaction. This derivation demonstrates the importance of detailed balance in treating the kinetics of equilibrium reactions.

EXAMPLE 29–2

Show that the equilibrium condition given by Equation 29.15 also gives

$$\frac{[B]_{eq}}{[A]_{eq}} = \frac{k_1}{k_{-1}}$$

for the reaction described by Equations 29.7 through 29.9.

SOLUTION: Equation 29.15 states that at equilibrium

$$v_1 + v_2 = v_{-1} + v_{-2}$$

The rate laws for the elementary steps (Equations 29.8 and 29.9) are as follows:

$$v_1 = k_1[A][C]$$
$$v_{-1} = k_{-1}[B][C]$$
$$v_2 = k_2[A]$$
$$v_{-2} = k_{-2}[B]$$

Substituting these expressions into Equation 29.15 gives us at equilibrium

$$k_1[A]_{eq}[C]_{eq} + k_2[A]_{eq} = k_{-1}[B]_{eq}[C]_{eq} + k_{-2}[B]_{eq}$$

which can be rearranged to give

$$K_c = \frac{[B]_{eq}}{[A]_{eq}} = \frac{k_1[C]_{eq} + k_2}{k_{-1}[C]_{eq} + k_{-2}} \tag{1}$$

Note that this expression depends on $[C]_{eq}$. To eliminate $[C]_{eq}$ from this equation, we use the relationship between the rate constants, Equation 29.14, that results from the application of the principle of detailed balance to this kinetic mechanism. Specifically, factoring k_1 out of the numerator and k_{-1} out of the denominator of Equation (1) gives us

$$K_c = \frac{[B]_{eq}}{[A]_{eq}} = \frac{k_1([C]_{eq} + k_2/k_1)}{k_{-1}([C]_{eq} + k_{-2}/k_{-1})} \tag{2}$$

Equation 29.14 shows that

$$\frac{k_2}{k_1} = \frac{k_{-2}}{k_{-1}}$$

which upon substitution into Equation 2, gives us

$$K_c = \frac{[B]_{eq}}{[A]_{eq}} = \frac{k_1}{k_{-1}}$$

as expected.

EXAMPLE 29–3

The chemical reaction described by

$$H_2(g) + 2\,ICl(g) \rightleftharpoons 2\,HCl(g) + I_2(g) \tag{1}$$

occurs by the two-step mechanism

$$H_2(g) + ICl(g) \underset{k_{-1}}{\overset{k_1}{\rightleftharpoons}} HI(g) + IICl(g) \tag{2}$$

$$HI(g) + ICl(g) \underset{k_{-2}}{\overset{k_2}{\rightleftharpoons}} HCl(g) + I_2(g) \tag{3}$$

Use the principle of detailed balance to show that the equilibrium constant for Equation 1 is the product of the equilibrium constants of Equations 2 and 3.

SOLUTION: The equilibrium constant for Equation 1, $K_{c,1}$, is given by

$$K_{c,1} = \frac{[HCl]_{eq}^2[I_2]_{eq}}{[H_2]_{eq}[ICl]_{eq}^2}$$

When the reaction is at equilibrium, the principle of detailed balance requires that Equations 2 and 3 also be at equilibrium. The equilibrium-constant expressions for Equations 2 and 3 are

$$K_{c,2} = \frac{[HI]_{eq}[HCl]_{eq}}{[H_2]_{eq}[ICl]_{eq}}$$

and

$$K_{c,3} = \frac{[HCl]_{eq}[I_2]_{eq}}{[HI]_{eq}[ICl]_{eq}}$$

The product $K_{c,2}K_{c,3}$ is

$$K_{c,2}K_{c,3} = \left(\frac{[HI]_{eq}[HCl]_{eq}}{[H_2]_{eq}[ICl]_{eq}}\right)\left(\frac{[HCl]_{eq}[I_2]_{eq}}{[HI]_{eq}[ICl]_{eq}}\right)$$

$$= \frac{[HCl]_{eq}^2[I_2]_{eq}}{[H_2]_{eq}[ICl]_{eq}^2}$$

$$= K_{c,1}$$

Note that the overall reaction, Equation 1, is the sum of the reactions given by Equations 2 and 3. The equilibrium constant for Equation 1, on the other hand, is the product of the equilibrium constants for Equations 2 and 3.

29–3. When Are Consecutive and Single-Step Reactions Distinguishable?

Consider the thermal decomposition of gaseous OClO to form chlorine atoms and oxygen molecules

$$OClO(g) \rightleftharpoons Cl(g) + O_2(g) \tag{29.16}$$

This reaction occurs by the following two-step mechanism

$$OClO(g) \underset{k_{-1}}{\overset{k_1}{\rightleftharpoons}} ClOO(g)$$

$$ClOO(g) \underset{k_{-2}}{\overset{k_2}{\rightleftharpoons}} Cl(g) + O_2(g) \tag{29.17}$$

Experimental studies of these reactions show that $v_1 \gg v_{-1}$ and that $v_2 \gg v_{-2}$. Because of the relative magnitudes of these reaction rates, the overall reaction goes essentially to completion, and an excellent approximation is to ignore the back reactions and model the mechanism of this reaction by a sequence of two irreversible elementary reactions, or

$$OClO(g) \overset{k_1}{\Longrightarrow} ClOO(g) \overset{k_2}{\Longrightarrow} Cl(g) + O_2(g)$$

Many complex reactions occur by such a sequence of elementary reactions.

Consider a general complex reaction described by

$$A \overset{k_{obs}}{\longrightarrow} P \tag{29.18}$$

where k_{obs} is the experimentally observed rate constant for the reaction. Of course, we cannot determine the rate law from this chemical equation, but suppose that the reaction occurs by the two-step mechanism

$$A \overset{k_1}{\Longrightarrow} I \tag{29.19}$$

$$I \overset{k_2}{\Longrightarrow} P \tag{29.20}$$

(This mechanism is often written on a single line as $A \overset{k_1}{\Longrightarrow} I \overset{k_2}{\Longrightarrow} P$.) Because each step of this mechanism is an elementary reaction, the rate laws for each species, A, I, and P are

$$\frac{d[A]}{dt} = -k_1[A] \tag{29.21}$$

$$\frac{d[I]}{dt} = k_1[A] - k_2[I] \tag{29.22}$$

$$\frac{d[P]}{dt} = k_2[I] \tag{29.23}$$

These coupled differential equations (the solution to Equation 29.22 depends on the solution to Equation 29.21, and the solution to Equation 29.23 depends on the solution to Equation 29.22) can be solved analytically (Problem 29–5). Equations 29.24 through 29.26 give the solutions assuming that the initial concentrations at time $t = 0$ are $[A] = [A]_0$ and $[I]_0 = [P]_0 = 0$.

$$[A] = [A]_0 e^{-k_1 t} \tag{29.24}$$

$$[I] = \frac{k_1 [A]_0}{k_2 - k_1} (e^{-k_1 t} - e^{-k_2 t}) \tag{29.25}$$

$$[P] = [A]_0 - [A] - [I] = [A]_0 \left\{ 1 + \frac{1}{k_1 - k_2} (k_2 e^{-k_1 t} - k_1 e^{-k_2 t}) \right\} \tag{29.26}$$

One question that should be considered is whether or not it is always possible to distinguish the individual steps of a sequential reaction. In other words, when can this two-step consecutive reaction mechanism be distinguished unambiguously from the one-step reaction,

$$A \xrightarrow{k_1} P$$

Note that $[A]$ decays exponentially with time for both the single-step and two-step reaction schemes. Thus, measurement of the decay kinetics of $[A]$ will not provide data that can distinguish between a single-step and a multistep process. However, the number of steps involved affects the appearance of the product. For a single-step reaction, $[P]$ is given by (see Example 28–3)

$$[P] = [A]_0 (1 - e^{-k_1 t}) \tag{29.27}$$

Equation 29.27 appears to be different from Equation 29.26, but consider what happens if k_2 is much greater than k_1. When $k_2 \gg k_1$, we can neglect k_1 compared with k_2 in the denominator of Equation 29.26, and the term involving $e^{-k_2 t}$ will decay much faster than the term involving $e^{-k_1 t}$, so $[P]$ given by Equation 29.26 becomes

$$[P] = [A]_0 \left\{ 1 + \frac{1}{k_1 - k_2} (k_2 e^{-k_1 t} - k_1 e^{-k_2 t}) \right\}$$

$$\approx [A]_0 \left\{ 1 + \frac{1}{-k_2} k_2 e^{-k_1 t} \right\}$$

$$= [A]_0 (1 - e^{-k_1 t})$$

This result is identical to Equation 29.27. The single-step and a two-step reaction mechanism are therefore indistinguishable when $k_2 \gg k_1$. Thus, the observation of identical rate constants for the decay of the reactant and the growth of the product does not necessarily mean that no chemical intermediates arise along the reaction path. This exemplifies one of the difficulties in establishing that a chemical reaction is really an elementary reaction.

If one step in a reaction mechanism is much slower than any of the other steps, that step effectively controls the overall reaction rate and is called the *rate-determining step*. Not all reaction mechanisms have a rate-determining step, but when one does occur, the overall reaction rate is limited by the rate-determining step. For example, reconsider the reaction between $NO_2(g)$ and $CO(g)$ to form $NO(g)$ and $CO_2(g)$ (Equation 29.1):

$$NO_2(g) + CO(g) \xrightarrow{k_{obs}} NO(g) + CO_2(g)$$

Recall from Section 29–1 that this reaction occurs by the two-step mechanism

$$NO_2(g) + NO_2(g) \xRightarrow{k_1} NO_3(g) + NO(g)$$

$$NO_3(g) + CO(g) \xRightarrow{k_2} NO_2(g) + CO_2(g)$$

The first step turns out to be much slower than the second, or $v_1 \ll v_2$. Because the reaction proceeds through both steps sequentially, the first step acts as a bottleneck and therefore is rate determining. In this particular case, the rate of the overall reaction will be given by the rate of the rate-determining step, or

$$v = k_1[NO_2]^2$$

which is the experimentally observed rate law. In effect, the $CO(g)$ molecules have to wait around for $NO_3(g)$ molecules to be produced. Once formed, the $NO_3(g)$ molecules are consumed rapidly by reaction with $CO(g)$.

EXAMPLE 29–4

Can a single-step and two-step mechanism be distinguished when the second step of the two-step reaction scheme is rate determining?

SOLUTION: Before we examine the rate laws, let us use our intuition to answer this question. Consider what happens if the rate of the second step of a two-step mechanism is rate determining. In this case, the reactant will disappear before any appreciable amount of product is formed. On the other hand, for a single-step process, the rate at which the reactants disappear and the rate at which the product is formed must be the same. Therefore, we expect that there are conditions under which these two processes can be distinguished if we monitor both the decay of A and the formation of P.

Let us first examine the exact solution for [P] (Equation 29.26)

$$[P] = [A]_0 \left\{ 1 + \frac{1}{k_1 - k_2}(k_2 e^{-k_1 t} - k_1 e^{-k_2 t}) \right\}$$

Consider what happens to this expression when $k_2 \ll k_1$. First,

$$\frac{1}{k_1 - k_2} \approx \frac{1}{k_1}$$

Second,

$$k_2 e^{-k_1 t} - k_1 e^{-k_2 t} \approx -k_1 e^{-k_2 t}$$

because the factor containing $e^{-k_1 t}$ will decay much more rapidly than the factor containing $e^{-k_2 t}$. Thus, when $k_2 \ll k_1$, Equation 29.26 can be simplified to

$$[\text{P}] = [\text{A}]_0 \left\{ 1 + \frac{1}{k_1}(-k_1 e^{-k_2 t}) \right\} = [\text{A}]_0 (1 - e^{-k_2 t})$$

This equation has the same functional form as [P] for a single-step reaction (Equation 29.27), except that it depends upon k_2, the rate constant for the second step of the mechanism. For a single-step reaction, the kinetics of A and P depend upon the same rate constant. For the two-step mechanism in which the second step is rate determining, the kinetics of A depend upon k_1 and the kinetics of P depend upon k_2. Therefore, if we measure both the decay kinetics of A and the formation kinetics of P, we can distinguish between a single-step and two-step mechanism when the second step of the two-step reaction scheme is rate determining.

29–4. The Steady-State Approximation Simplifies Rate Expressions by Assuming that d[I]/dt = 0, where I Is a Reaction Intermediate

Reconsider the reaction mechanism

$$\text{A} \overset{k_1}{\Longrightarrow} \text{I} \overset{k_2}{\Longrightarrow} \text{P} \tag{29.28}$$

for the initial conditions $[\text{A}] = [\text{A}]_0$, and $[\text{I}]_0 = [\text{P}]_0 = 0$. In Section 29–3, we discussed the time-dependent behavior of the concentration of the reactant, [A], and product, [P]. We now discuss the time-dependent behavior of the concentration of the intermediate, [I]. The concentration of I varies with the relative magnitudes of the rate constants k_1 and k_2. Equation 29.25 gives the dependence of [I] on the reaction rate constants, k_1 and k_2, and Figure 29.1 shows plots of [I] versus time for two different relationships between k_1 and k_2. The data plotted in Figure 29.1a show that if $k_1 = 10 k_2$, then [I] builds up and then decays. In other words, the value of [I] changes significantly over the course of the reaction. In contrast, if the second step is much faster than the first step, very little of the intermediate can build up. This type of behavior is shown in Figure 29.1b, where we have taken $k_2 = 10 k_1$; here, we see that [I] builds up quickly to a very small concentration that remains relatively constant during the course of the reaction. In this latter case, we can reasonably make the approximation that $d[\text{I}]/dt = 0$, which means that we can equate the rate equation associated with this intermediate to

1192

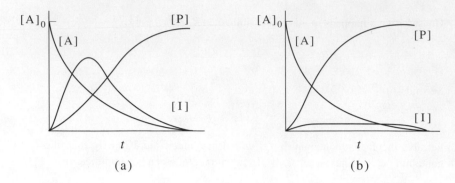

(a)   (b)

FIGURE 29.1
Concentration profiles for the consecutive reaction scheme $A \overset{k_1}{\Rightarrow} I \overset{k_2}{\Rightarrow} P$ with initial concentrations $[A] = [A]_0$, and $[I]_0 = [P]_0 = 0$. (a) $k_1 = 10\,k_2$: The concentration of I rises and then decays, changing significantly during the course of the reaction; (b) $k_2 = 10\,k_1$: The concentration of I rapidly builds up to a constant, but negligible, concentration that persists for a large extent of the reaction. In this case, the steady-state approximation can be applied to [I].

zero. This procedure is called the *steady-state approximation* and can greatly simplify the mathematics associated with a particular kinetic model.

For the above two-step mechanism, the rate laws for A, I, and P are given by Equations 29.21 through 29.23. If we invoke the steady-state approximation, then $d[I]/dt = 0$ and Equation 29.22 becomes

$$[I]_{ss} = \frac{k_1[A]}{k_2} \tag{29.29}$$

where the subscript "ss" is used to emphasize that this is the concentration of I obtained by assuming the steady-state approximation. The time-dependent concentration of A is given by Equation 29.24,

$$[A] = [A]_0 e^{-k_1 t}$$

which upon substitution into Equation 29.29 gives

$$[I]_{ss} = \frac{k_1}{k_2}[A]_0 e^{-k_1 t} \tag{29.30}$$

Note that the steady-state approximation assumes that $d[I]/dt = 0$. But the result of using the steady-state approximation, Equation 29.30, is a time-dependent expression for [I]. We therefore need to consider when the expression in Equation 29.30 satisfies the assumption that $d[I]/dt = 0$. Evaluating $d[I]/dt$ from Equation 29.30 gives us

$$\frac{d[I]_{ss}}{dt} = -\frac{k_1^2}{k_2}[A]_0 e^{-k_1 t} \tag{29.31}$$

We see that the differential $d[I]/dt$ approaches zero as $k_1^2[A]_0/k_2$ goes to zero. Therefore, the steady-state approximation is a reasonable assumption for treating the kinetics of the reaction scheme given by Equation 29.28 if $k_2 \gg k_1^2[A]_0$.

The concentration of P is given by $[A]_0 - [A] - [I]$ or it can also be found by substituting Equation 29.30 into Equation 29.23 and integrating (Problem 29–6). Both approaches give

$$[P] = [A]_0(1 - e^{-k_1 t}) \tag{29.32}$$

If we compare Equation 29.32 with the exact solution for [P] (Equation 29.26), we see that the exact solution reduces to Equation 29.32 only if $k_2 \gg k_1$. In other words, we have found that for this two-step mechanism, the steady-state assumption corresponds to the case in which the intermediate is so reactive that $[I] \approx 0$.

Figure 29.2 shows a plot of the calculated time-dependent concentrations of A, I, and P for $k_2 = 10k_1$ using the exact expressions and those expressions obtained from applying the steady-state approximation. The plot shows that the approximate solutions are in excellent agreement with the exact solutions. Problem 29–7 asks you to compute the exact and approximate solutions for $k_2 = 2k_1$. As you might expect from the above discussion, you will learn by doing this problem that the steady-state approximation is a poor approximation when the rate constants of the two steps are comparable.

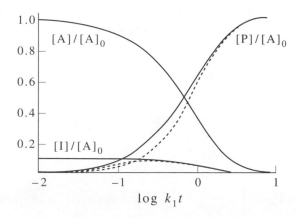

FIGURE 29.2
The quantities $[A]/[A]_0$, $[I]/[A]_0$, and $[P]/[A]_0$ are plotted as a function of $\log(k_1 t)$ for the case in which $k_2 = 10k_1$ for the reaction scheme $A \overset{k_1}{\Rightarrow} I \overset{k_2}{\Rightarrow} P$. The solid lines are the concentrations obtained using the steady-state approximation. The dashed lines are the exact solutions to the rate equations. The logarithmic time scale exaggerates the differences between the approximate and exact solutions. The agreement between the exact solutions and those obtained using the steady-state approximation is nearly quantitative in this case.

EXAMPLE 29–5
The decomposition of ozone

$$2\,O_3(g) \longrightarrow 3\,O_2(g)$$

occurs by the reaction mechanism

$$M(g) + O_3(g) \underset{k_{-1}}{\overset{k_1}{\rightleftharpoons}} O_2(g) + O(g) + M(g)$$

$$O(g) + O_3(g) \overset{k_2}{\Longrightarrow} 2\,O_2(g)$$

where M is a molecule that can exchange energy with the reacting ozone molecule through a collision, but M itself does not react. Use this mechanism to derive the rate law for $d[O_3]/dt$ assuming that the intermediate $O(g)$ concentration can be treated by the steady-state approximation.

SOLUTION: The rate equations for $O_3(g)$ and $O(g)$ are

$$\frac{d[O_3]}{dt} = -k_1[O_3][M] + k_{-1}[O_2][O][M] - k_2[O][O_3]$$

and

$$\frac{d[O]}{dt} = k_1[O_3][M] - k_{-1}[O_2][O][M] - k_2[O][O_3]$$

Invoking the steady-state approximation for the intermediate $O(g)$ means that we set $d[O]/dt = 0$. Setting $d[O]/dt = 0$ and solving the resulting expression for $[O]$ gives us

$$[O] = \frac{k_1[O_3][M]}{k_{-1}[O_2][M] + k_2[O_3]}$$

Substituting this result into the rate equation for O_3 then gives us

$$\frac{d[O_3]}{dt} = -\frac{2k_1 k_2[O_3]^2[M]}{k_{-1}[O_2][M] + k_2[O_3]}$$

Note the complexity of this rate law compared with what we have encountered so far.

Because the steady-state approximation simplifies the mathematics, you might be tempted to use it solely for that reason. As we have pointed out, however, this approximation makes assumptions about the relative magnitudes of the rate constants for different steps of the reaction mechanism. The validity of such assumptions must be experimentally confirmed before the steady-state approximation should be used.

29–5. The Rate Law for a Complex Reaction Does Not Imply a Unique Mechanism

Recall that the rate law of a complex reaction provides information on how the rate of reaction depends on concentration, but it does not tell us how the reaction occurs. In general chemistry, you may have learned that the rate law for a complex reaction can be derived by combining the rate laws for the individual elementary steps. Problems 29–9 through 29–17 involve deriving the rate laws for a number of complex reactions from the reaction mechanisms. Here we explore the question of whether or not the empirically determined rate law implies a unique reaction mechanism. Consider the oxidation of nitrogen monoxide to form nitrogen dioxide according to

$$2\,NO(g) + O_2(g) \xrightarrow{k_{obs}} 2\,NO_2(g) \tag{29.33}$$

Measurements of the rate of appearance of the $NO_2(g)$ product reveal a rate law of the form

$$\frac{1}{2}\frac{d[NO_2]}{dt} = k_{obs}[NO]^2[O_2] \tag{29.34}$$

Note that this rate law is consistent with the conclusion that the reaction is an elementary termolecular reaction. Experimental studies confirm, however, that the reaction given by Equation 29.33 is not an elementary reaction, which we have indicated by using the appropriate arrow in writing the chemical equation.

We will now consider two possible mechanism for this reaction. We will derive the rate laws associated with these two reaction mechanisms, and then compare these rate laws to the experimentally observed rate law.

Mechanism 1:

$$NO(g) + O_2(g) \underset{k_{-1}}{\overset{k_1}{\rightleftharpoons}} NO_3(g) \qquad \text{(fast equilibrium)} \tag{29.35}$$

$$NO_3(g) + NO(g) \xrightarrow{k_2} 2\,NO_2(g) \qquad \text{(rate determining)} \tag{29.36}$$

Mechanism 1 proposes that the first step of the reaction is a rapidly established equilibrium between the reactants and the nitrogen trioxide radical (NO_3). The second step is a slow reaction between nitrogen trioxide and nitrogen monoxide, and this step is rate determining.

The first step of this mechanism (Equation 29.35) is assumed to establish equilibrium and to remain there as the subsequent step of the mechanism takes place, so we can write

$$K_{c,1} = \frac{k_1}{k_{-1}} = \frac{[NO_3]}{[NO][O_2]} \tag{29.37}$$

The rate law for the second step of the mechanism (Equation 29.36) is

$$\frac{1}{2}\frac{d[NO_2]}{dt} = k_2[NO_3][NO] \tag{29.38}$$

Because the second step of the mechanism is rate determining, Equation 29.38 gives the rate law of the overall reaction. We now need to relate the concentration of the intermediate species, NO_3, to the concentrations of the reactants. This relationship can be determined using the equilibrium condition given by Equation 29.37. Solving Equation 29.37 for $[NO_3]$ and substituting the result into Equation 29.38 gives us the following rate law for the chemical reaction.

$$\frac{1}{2}\frac{d[NO_2]}{dt} = k_2K_{c,1}[NO]^2[O_2] \tag{29.39}$$

This rate law is in agreement with the experimental rate law, Equation 29.34, if $k_{obs} = k_2K_{c,1}$. The experimentally determined rate constant is, therefore, not a rate constant for any one particular step of the reaction mechanism but is a product of the rate constant for the second step of the mechanism and the equilibrium constant of the first step of the mechanism.

We now consider a different proposed mechanism for how the reaction in Equation 29.33 occurs.

Mechanism 2:

$$NO(g) + NO(g) \underset{k_{-1}}{\overset{k_1}{\rightleftharpoons}} N_2O_2(g) \qquad (N_2O_2(g) \text{ is in steady state}) \tag{29.40}$$

$$N_2O_2(g) + O_2(g) \overset{k_2}{\Longrightarrow} 2NO_2(g) \tag{29.41}$$

Mechanism 2 involves the formation of the intermediate chemical species $N_2O_2(g)$. It is assumed that the steady-state approximation is valid, or in other words, the concentration of $N_2O_2(g)$ is time independent, so that $d[N_2O_2]/dt = 0$. Using this mechanism, the rate laws for $[NO]$ and $[N_2O_2]$ are

$$\frac{1}{2}\frac{d[NO]}{dt} = -k_1[NO]^2 + k_{-1}[N_2O_2] \tag{29.42}$$

and

$$\frac{d[N_2O_2]}{dt} = -k_{-1}[N_2O_2] - k_2[N_2O_2][O_2] + k_1[NO]^2 \tag{29.43}$$

and the rate of reaction is given by

$$\frac{1}{2}\frac{d[NO_2]}{dt} = k_2[N_2O_2][O_2] \tag{29.44}$$

The rate of appearance of $NO_2(g)$, Equation 29.44, depends upon the concentration of the intermediate species, $N_2O_2(g)$. Once again, we need to express the rate in terms of the concentrations of the reactants, [NO] and $[O_2]$, to compare the predicted and experimental rate laws. Using the steady-state approximation for the intermediate species N_2O_2 means that we can set Equation 29.43 equal to zero. Setting Equation 29.43 equal to zero and solving for $[N_2O_2]$ gives

$$[N_2O_2] = \frac{k_1[NO]^2}{k_{-1} + k_2[O_2]} \tag{29.45}$$

Recall that the use of the steady-state approximation requires that $[N_2O_2]$ be constant in time. One way this condition can be met is if v_{-1}, the rate of the back reaction in Equation 29.40, is much larger than both v_1, the rate of the forward reaction in Equation 29.40, and v_2, the rate of the reaction in Equation 29.41. Then, only a negligible and essentially constant amount of N_2O_2 is ever present, satisfying the steady-state assumption. Under these conditions, $k_{-1}[N_2O_2] \gg k_2[N_2O_2][O_2]$ or $k_{-1} \gg k_2[O_2]$, and Equation 29.45 simplifies to

$$[N_2O_2] = \frac{k_1}{k_{-1}}[NO]^2$$

Substituting this result into Equation 29.44 gives

$$\frac{1}{2}\frac{d[NO_2]}{dt} = \frac{k_2k_1}{k_{-1}}[NO]^2[O_2] = k_2K_{c,1}[NO]^2[O_2] \tag{29.46}$$

This rate law also agrees with the experimental rate law, Equation 29.34, if $k_{obs} = k_2K_{c,1}$. We have found that both mechanisms are in agreement with the observed rate law. To distinguish between these two reaction mechanisms would then require additional information. For example, if you showed that $NO_3(g)$ existed in the reaction flask, then you could discount Mechanism 2. Another approach could involve introducing reagents to the reaction mixture that would react with the reactive $NO_3(g)$ radical to form stable products that could be isolated and characterized, thereby proving that $NO_3(g)$ was generated in the reaction flask. Currently, the experimental data favor Mechanism 2.

EXAMPLE 29–6
In the above discussion, the concentration of $N_2O_2(g)$ satisfies the steady-state approximation if v_{-1} is much larger than both v_1 and v_2. The steady-state approximation would also apply if v_2 is much larger than both v_1 and v_{-1}. What is the predicted rate law for this reaction mechanism under these latter conditions?

SOLUTION: Equation 29.45 gives an expression for $[N_2O_2]$ when its concentration is treated by the steady-state approximation. If v_2 is much larger than both v_1 and v_{-1}, then $k_2[O_2] \gg k_{-1}$, and Equation 29.45 simplifies to

$$[N_2O_2] = \frac{k_1[NO]^2}{k_2[O_2]}$$

Substituting this result into Equation 29.44 gives us the rate law

$$\frac{1}{2}\frac{d[NO_2]}{dt} = \frac{k_2 k_1 [NO]^2 [O_2]}{k_2[O_2]} = k_1[NO]^2$$

This rate law is different from the experimental rate law. This result tells us that even though there are two possible relationships between the rates of the steps of the reaction mechanism such that the steady-state approximation applies to the reaction intermediate species $N_2O_2(g)$, only one of these sets of conditions is consistent with the observed rate law.

Our study of the oxidation of nitrogen monoxide points out some of the difficulties in deriving mechanisms to explain experimental rate laws. First, although the experimentally determined rate law has the mathematical form of an elementary reaction (Equation 29.34), it is not an elementary reaction. This confirms once again that the observed rate law is not by itself sufficient to prove that a reaction is elementary. Second, the experimental rate law can be accounted for by two different mechanisms, so we see that a rate law does not imply a unique mechanism. A mechanism is only a hypothesis of how the reaction proceeds. The ability of a mechanism to account for an experimental rate law is only the first step in establishing that the mechanism is correct. Ultimately, verifying a reaction mechanism requires intensive experimental verification of each elementary step.

29–6. The Lindemann Mechanism Explains How Unimolecular Reactions Occur

If the reaction described by

$$CH_3NC(g) \xrightarrow{k_{obs}} CH_3CN(g) \tag{29.47}$$

is an elementary reaction, then it must obey the rate law

$$\frac{d[CH_3NC]}{dt} = -k_{obs}[CH_3NC] \tag{29.48}$$

A close examination of this and many other supposed unimolecular reactions reveals that the rate law given by Equation 29.48 is valid only at high concentration. At low concentration, the experimental data for this reaction are consistent with the second-order rate law

$$\frac{d[CH_3NC]}{dt} = -k_{obs}[CH_3NC]^2 \qquad (29.49)$$

Equation 29.49 is not the rate law for a unimolecular reaction, and we are forced to reexamine the initial statement that reactions such as that given by Equation 29.47 are elementary.

The data given in Table 29.1 show that the activation energies of "unimolecular" reactions can be quite large compared with $k_B T$. To understand how these reactions occur, we need to identify the source of energy that enables the reacting molecule to overcome the energy barrier to reaction. A mechanism that predicts the rate law given by Equation 29.48 at high gas concentration and the rate law given by Equation 29.49 at low gas concentration was proposed independently by the British chemists J. A. Christiansen in 1921 and F. A. Lindemann in 1922. Their work underlies the current theory of unimolecular reaction rates. The mechanism is generally referred to as the *Lindemann mechanism.*

Lindemann proposed that the energy source for a unimolecular reaction such as that described by Equation 29.47 results from bimolecular collisions. He further postulated that there must be a time lag between the collision (or energizing step) and the ensuing reaction. Depending on the collision rate in the gas and the time lag before reaction,

TABLE 29.1

Arrhenius parameters for unimolecular reactions. The rate constants for these reactions at 500 K and 700 K are given in Table 28.4.

Reaction	$\ln(A/s^{-1})$	$E_a/kJ \cdot mol^{-1}$
Isomerizations		
Cyclopropane $\Rightarrow$ propene	35.7	274
Cyclopropene $\Rightarrow$ propyne	29.9	147
cis-But-2-ene $\Rightarrow$ *trans*-but-2-ene	31.8	263
$CH_3NC \rightarrow CH_3CN$	31.3	131
Vinyl allyl ether $\Rightarrow$ pent-4-enal	26.9	128
Decompositions		
Cyclobutane $\Rightarrow$ 2 ethene	35.9	262
Ethylene oxide $\Rightarrow$ CH_3CHO, CH_2O, CH_2CO	32.5	238
Ethyl fluoride $\Rightarrow$ HF + ethene	30.9	251
Ethyl chloride $\Rightarrow$ HCl + ethene	32.2	244
Ethyl bromide $\Rightarrow$ HBr + ethene	31.1	226
Ethyl iodide $\Rightarrow$ HI + ethene	32.5	221
Isopropyl ether $\Rightarrow$ propene + isopropanol	33.6	266

the molecule could possibly undergo a deactivating bimolecular collision before it has a chance to react. In terms of chemical equations, the Lindemann mechanism for unimolecular reactions of the form $A(g) \rightarrow B(g)$ is

$$A(g) + M(g) \underset{k_{-1}}{\overset{k_1}{\rightleftharpoons}} A(g)^* + M(g) \tag{29.50}$$

$$A(g)^* \overset{k_2}{\Longrightarrow} B(g) \tag{29.51}$$

The symbol $A(g)^*$ in Equations 29.50 and 29.51 represents an energized reactant molecule, and $M(g)$ is the collision partner. The molecule $M(g)$ can be a second reactant molecule, a product molecule, or a nonreactive buffer gas such as $N_2(g)$ or $Ar(g)$.

Based on the Lindemann mechanism, the rate of product formation is given by

$$\frac{d[B]}{dt} = k_2[A^*] \tag{29.52}$$

Because collisions both energize $A(g)$ and de-energize $A(g)^*$, the concentration of $A(g)^*$ at any given time will be very small, and we can reasonably invoke the steady-state approximation. In that case,

$$\frac{d[A^*]}{dt} = 0 = k_1[A][M] - k_{-1}[A^*][M] - k_2[A^*] \tag{29.53}$$

Equation 29.53 can be solved for $[A^*]$,

$$[A^*] = \frac{k_1[M][A]}{k_2 + k_{-1}[M]} \tag{29.54}$$

Substituting Equation 29.54 into Equation 29.52 gives us the following rate law for the overall reaction

$$\frac{d[B]}{dt} = -\frac{d[A]}{dt} = \frac{k_1 k_2[M][A]}{k_2 + k_{-1}[M]} = k_{obs}[A] \tag{29.55}$$

where

$$k_{obs} = \frac{k_1 k_2[M]}{k_2 + k_{-1}[M]} \tag{29.56}$$

We see that k_{obs} depends on $[M]$ and is therefore concentration dependent. At sufficiently high concentrations, we expect that v_{-1}, the rate of collisional deactivation, will be greater than v_2, the rate of reaction. In this case, we have $k_{-1}[M][A^*] \gg k_2[A^*]$ or $k_{-1}[M] \gg k_2$ and k_{obs} simplifies to

$$k_{obs} = \frac{k_1 k_2}{k_{-1}} \tag{29.57}$$

The rate law for the overall reaction (Equation 29.55) then becomes $d[B]/dt = k_1 k_2 [A]/k_{-1}$. In this high-concentration limit, the reaction rate is first order in A. At sufficiently low concentrations, we expect that v_2, the rate of reaction, will be greater than v_{-1}, the rate of collisional deactivation. This means that $k_2 \gg k_{-1}[M]$, and therefore at low concentration, k_{obs} simplifies to

$$k_{obs} = k_1 [M] \tag{29.58}$$

The rate law for the overall reaction in this case becomes $d[B]/dt = k_1 [M][A]$. In this low-concentration limit, the rate law is first order in both A and M and has an overall reaction order of two. One of the great successes of the Lindemann mechanism was its ability to predict the experimentally observed change from first-order kinetics to second-order kinetics with decreasing concentration. Figure 29.3 shows a plot of the observed rate constant, k_{obs}, for the isomerization reaction $CH_3NC(g) \rightarrow CH_3CN(g)$ at 472.5 K as a function of $[CH_3NC]$. The low-concentration data show that k_{obs} depends linearly on concentration (Equation 29.58), and data at high concentration show that k_{obs} is independent of concentration (Equation 29.57). In the region between these two limiting behaviors, k_2 is comparable with $k_{-1}[M]$, and neither of the limiting expressions discussed above describes the kinetics.

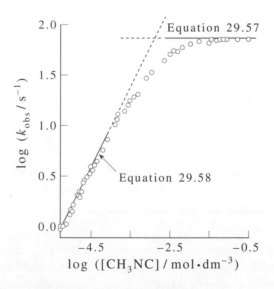

FIGURE 29.3
A plot of the concentration dependence of the unimolecular rate constant for the isomerization reaction of methylisocyanide at 472.5 K. At low concentration, the rate constant varies linearly with concentration, as predicted by Equation 29.58. At high concentration, the rate constant is independent of concentration, in agreement with Equation 29.57.

EXAMPLE 29–7
This example explores the connection between the observed activation parameters for a chemical reaction and the activation parameters of the individual steps of the reaction mechanism. Specifically, suppose that the rate constant for each step of the Lindemann mechanism shows Arrhenius behavior. How are the measured values of A and E_a related to the pre-exponential factors and activation energies for the individual steps of the mechanism for the high-concentration reaction?

SOLUTION: At high concentration, $k_{obs} = k_1 k_2 / k_{-1}$. Denoting the experimentally measured values of A and E_a by A_{obs} and $E_{a,obs}$, we have (Equation 28.57)

$$k_{obs} = A_{obs} e^{-E_{a,obs}/RT}$$

If each step of the reaction mechanism shows Arrhenius behavior, then each of the rate constants, k_1, k_{-1}, and k_2 can be written in terms of an Arrhenius equation,

$$k_1 = A_1 e^{-E_{a,1}/RT}$$

$$k_{-1} = A_{-1} e^{-E_{a,-1}/RT}$$

$$k_2 = A_2 e^{-E_{a,2}/RT}$$

Substituting these equations into $k_{obs} = k_1 k_2 / k_{-1}$ shows that

$$E_{a,obs} = E_{a,1} + E_{a,2} - E_{a,-1}$$

and

$$A_{obs} = \frac{A_1 A_2}{A_{-1}}$$

We now see that measured values for A_{obs} and $E_{a,obs}$ do not correspond to a single step of the reaction but are influenced by each step of the reaction mechanism.

Although the Lindemann mechanism predicts the correct qualitative behavior of the reaction rate with concentration, it fails to provide quantitative agreement with experimental data over a range of concentrations. This failure results from the fact that this mechanism does not address the details of how the energy transfer process takes place, only that it occurs. Modern theories of both intramolecular and intermolecular energy transfers are able to bring these general ideas into quantitative agreement with the observed rates of chemical reactions.

29–7. Some Reaction Mechanisms Involve Chain Reactions

Consider the reaction between hydrogen and bromine to produce hydrogen bromide. The balanced chemical equation that describes this reaction is

$$H_2(g) + Br_2(g) \rightleftharpoons 2\,HBr(g) \tag{29.59}$$

The experimentally determined rate law is

$$\frac{1}{2}\frac{d[HBr]}{dt} = \frac{k[H_2][Br_2]^{1/2}}{1 + k'[HBr][Br_2]^{-1}} \tag{29.60}$$

where k and k' are constants. The rate law given by Equation 29.60 depends on both the reactant and product concentrations. Because the product appears in the denominator of the rate expression, its accumulation decreases the reaction rate.

Detailed kinetic studies of this reaction have resulted in the following proposed mechanism:

Initiation: $\quad\quad\quad Br_2(g) + M(g) \xrightarrow{k_1} 2\,Br(g) + M(g) \tag{1}$

Propagation: $\quad\quad Br(g) + H_2(g) \xrightarrow{k_2} HBr(g) + H(g) \tag{2}$

$\quad\quad\quad\quad\quad\quad H(g) + Br_2(g) \xrightarrow{k_3} HBr(g) + Br(g) \tag{3}$

Inhibition: $\quad\quad\quad HBr(g) + H(g) \xrightarrow{k_{-2}} Br(g) + H_2(g) \tag{4}$

$\quad\quad\quad\quad\quad\quad HBr(g) + Br(g) \xrightarrow{k_{-3}} H(g) + Br_2(g) \tag{5}$

Termination: $\quad\quad 2\,Br(g) + M(g) \xrightarrow{k_{-1}} Br_2(g) + M(g) \tag{6}$

The first step, Equation 1, is a bimolecular reaction in which $M(g)$ is a molecule that collides with the $Br_2(g)$ molecule, thereby imparting the energy necessary to break the chemical bond. Equations 2 through 5 reveal how $HBr(g)$ is both formed and destroyed. Note that one of the products of Equation 2 is a reactant in Equation 3. Both formation reactions of $HBr(g)$ generate a chemical species that can go on to react to form $HBr(g)$. These reactions therefore serve to propagate the further formation of $HBr(g)$. Reactions of this type are called *chain reactions*. Now consider the back reactions of Equations 2 and 3, which are given in Equations 4 and 5. These two reactions destroy HBr and thereby inhibit product formation. Note that one of the products of Equation 4 is a reactant in Equation 5 and that one of the products in Equation 5 is a reactant in Equation 4. The inhibition reactions are also chain reactions. These inhibition reactions have been studied in great detail. The reaction $HBr(g) + Br(g)$ (Equation 5) is endothermic by nearly 170 $kJ\cdot mol^{-1}$, whereas the reaction $HBr(g) + H(g)$ (Equation 4) is exothermic by approximately 70 $kJ\cdot mol^{-1}$. Because the reaction given by Equation 5 requires such a relatively large input of energy compared with the reaction given by Equation 4, the contribution of the reaction in Equation 5 to the overall chemistry is negligible and assuming that $k_{-3} \approx 0$ is a good approximation.

We now derive the rate law that corresponds to this mechanism. We will want to compare our result with the experimentally determined rate law, Equation 29.60, so we must derive an expression for $d[HBr]/dt$ in terms of the reactants $[H_2]$ and $[Br_2]$ and the product $[HBr]$ from the above mechanism. Because each step of the above mechanism is an elementary reaction, we can write the rate laws for $[HBr]$, $[H]$, and $[Br]$. Using the mechanism given by Equations 1 through 4, and 6, the rate laws for $[HBr]$, $[H]$, and $[Br]$ ignoring Equation 5 are

$$\frac{d[HBr]}{dt} = k_2[Br][H_2] - k_{-2}[HBr][H] + k_3[H][Br_2] \tag{29.61}$$

$$\frac{d[H]}{dt} = k_2[Br][H_2] - k_{-2}[HBr][H] - k_3[H][Br_2] \tag{29.62}$$

and

$$\frac{d[Br]}{dt} = 2k_1[Br_2][M] - 2k_{-1}[Br]^2[M] - k_2[Br][H_2]$$
$$+ k_{-2}[HBr][H] + k_3[H][Br_2] \tag{29.63}$$

The factor of 2 in the first two terms arises because of the stoichiometry of Equations 1 and 6. {For Equation 1, $(1/2)d[Br]/dt = k_1[Br_2][M]$, or $d[Br]/dt = 2k_1[Br_2][M]$.} To simplify this problem, we apply the steady-state approximation to the two reactive intermediates Br(g) and H(g) so that $d[Br]/dt = d[H]/dt = 0$. Recall from the discussion in Section 29–4 that this approximation needs to be justified by independent experimental measurements. Deriving a rate law consistent with the experimental rate law does not in itself justify the use of such an approximation. Applying the steady-state approximation to $[H]$ gives us

$$\frac{d[H]}{dt} = 0 = k_2[Br][H_2] - k_{-2}[HBr][H] - k_3[H][Br_2] \tag{29.64}$$

and likewise for $[Br]$ we obtain

$$\frac{d[Br]}{dt} = 0 = 2k_1[Br_2][M] - 2k_{-1}[Br]^2[M] - k_2[Br][H_2]$$
$$+ k_{-2}[HBr][H] + k_3[H][Br_2] \tag{29.65}$$

The goal now is to use Equations 29.64 and 29.65 to find expressions for $[H]$ and $[Br]$ in terms of the reactants and products of the reaction. Then we can substitute these expressions into Equation 29.61 and thereby obtain the predicted rate law of the overall reaction in terms of the concentrations of the reactants and products.

Note that the three terms on the right side of Equation 29.64 are the negatives of the last three terms on the right side of Equation 29.65. By adding Equations 29.64 to 29.65, we obtain

$$0 = 2k_1[Br_2][M] - 2k_{-1}[Br]^2[M]$$

and solving this expression for [Br] gives us

$$[\text{Br}] = \left(\frac{k_1}{k_{-1}}\right)^{1/2} [\text{Br}_2]^{1/2} = (K_{c,1})^{1/2}[\text{Br}_2]^{1/2} \tag{29.66}$$

We can obtain an expression for [H] in terms of the concentrations of the reactants and products by substituting Equation 29.66 into Equation 29.64. This procedure gives us

$$[\text{H}] = \frac{k_2 K_{c,1}^{1/2}[\text{H}_2][\text{Br}_2]^{1/2}}{k_{-2}[\text{HBr}] + k_3[\text{Br}_2]} \tag{29.67}$$

Combining Equations 29.61, 29.66, and 29.67 gives us the rate law

$$\frac{1}{2}\frac{d[\text{HBr}]}{dt} = \frac{k_2 K_{c,1}^{1/2}[\text{H}_2][\text{Br}_2]^{1/2}}{1 + (k_{-2}/k_3)[\text{HBr}][\text{Br}_2]^{-1}} \tag{29.68}$$

This rate law has the same functional form as the experimentally derived rate law. Comparing Equations 29.68 and 29.60 shows us that the measured constants k and k' are related to the rate constants of the reaction mechanism by $k = k_2 K_{c,1}^{1/2}$ and $k' = k_{-2}/k_3$.

EXAMPLE 29–8

In the initial stages of the reaction described by

$$\text{H}_2(g) + \text{Br}_2(g) \rightleftharpoons 2\,\text{HBr}(g)$$

the observed rate law is

$$\frac{1}{2}\frac{d[\text{HBr}]}{dt} = k_{\text{obs}}[\text{H}_2][\text{Br}_2]^{1/2}$$

Show that this result is consistent with the rate law given by Equation 29.68, and determine k_{obs} in terms of the rate constants of the mechanism.

SOLUTION: In the initial stages of the reaction $[\text{HBr}] \ll [\text{Br}_2]$, and as a result

$$\frac{k_{-2}}{k_3}[\text{HBr}][\text{Br}_2]^{-1} \ll 1$$

This result allows us to simplify the denominator of Equation 29.68, obtaining the result

$$\frac{1}{2}\frac{d[\text{HBr}]}{dt} = k_2 K_{c,1}^{1/2}[\text{H}_2][\text{Br}_2]^{1/2}$$

The measured rate constant, k_{obs}, is equal to $k_2 K_{c,1}^{1/2}$.

Problems 29–24 through 29–32 consider several different types of chemical reactions that involve chain reactions.

29–8. A Catalyst Affects the Mechanism and Activation Energy of a Chemical Reaction

We know that the rate of a reaction can usually be increased by increasing the temperature and that there are practical limitations to the effects of temperature (Section 28–7). For example, reactions in solution are constrained to the temperature range between the melting and boiling point of the solvent. An entirely different approach to making reactions go faster would be to enable the reaction to proceed by a different mechanism that has a lower activation energy. This is the general idea behind a chemical catalyst. *A catalyst* is a substance that participates in the chemical reaction but is not consumed in the process. By participation in the reaction, a catalyst provides a new mechanism by which the reaction can occur. The trick is to develop a catalyst that gives rise to a reaction path with a negligible activation barrier. If the catalyst is in the same phase as the reactants and products, the reaction is an example of *homogeneous catalysis*. If the catalyst is in a different phase from the reactants and products, the reaction is an example of *heterogeneous catalysis*.

Because a catalyst is not consumed by the chemical reaction, the exothermicity or endothermicity of the chemical reaction is not altered by the presence of a catalyst. Figure 29.4 illustrates how a changes in mechanism can influence the reaction rate. Here we see that the mechanism made possible by the catalyst has a lower activation energy than the reaction mechanism in the absence of the catalyst. Because the reaction rate depends exponentially on the activation energy (see Section 28–6), small changes in the height of the activation barrier result in substantial changes in the reaction rate. Because the mechanisms of the catalyzed and uncatalyzed reactions are different, they

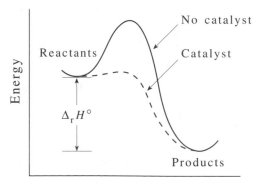

FIGURE 29.4
A schematic illustration of the energy curves for an exothermic reaction with and without a catalyst. The role of the catalyst is to lower the activation energy of a chemical reaction by making a new mechanism accessible by which the reaction can occur. Because the mechanisms of the catalyzed and uncatalyzed reactions are different, they will occur along different reaction coordinates.

correspond to different reaction coordinates. Hence, we use the plural "coordinates" for the horizontal axis label of Figure 29.4. Figure 29.4 shows that the catalyst lowers the activation energy for both the forward and reverse reactions so that the rates of both forward and reverse reaction rates are increased.

Consider the reaction,

$$A \longrightarrow products$$

Addition of a catalyst creates a new reaction pathway that competes with that of the uncatalyzed mechanism. The overall reaction mechanism then involves two competing reactions

$$A \xrightarrow{k} products$$

$$A + catalyst \xrightarrow{k_{cat}} products + catalyst$$

If each of these competing reactions is an elementary process, the rate law for the overall reaction is given by the sum of the two terms.

$$-\frac{d[A]}{dt} = k[A] + k_{cat}[A][catalyst]$$

The first term on the right side of this equation in the rate expression is the rate law for the uncatalyzed reaction, and the second term is the rate law for the mechanism involving the catalyst. In most cases, catalysts enhance reaction rates by many orders of magnitude, and therefore only the rate law for the catalyzed reaction need be considered in analyzing experimental data.

As an example of homogeneous catalysis, consider the oxidation-reduction reaction between aqueous cerium(IV) ions, $Ce^{4+}(aq)$, and aqueous thallium(I) ions, $Tl^+(aq)$:

$$2\,Ce^{4+}(aq) + Tl^+(aq) \longrightarrow 2\,Ce^{3+}(aq) + Tl^{3+}(aq)$$

In the absence of a catalyst, this reaction is a termolecular elementary reaction; it occurs very slowly and its rate law is

$$v = k[Tl^+][Ce^{4+}]^2$$

The slow rate is a consequence of the requirement that the reactive event involves a simultaneous collision of one thallium and two cerium ions, which has a low probability of occurring. Addition of $Mn^{2+}(aq)$ to the solution catalyzes the above reaction. The facile oxidation and reduction reactions of the manganese ion open up a new reaction pathway by which cerium(IV) can oxidize thallium(I). The new reaction pathway involves only bimolecular reactions and occurs by the following mechanism

$$Ce^{4+}(aq) + Mn^{2+}(aq) \xrightarrow{k_{cat}} Mn^{3+}(aq) + Ce^{3+}(aq) \quad \text{(rate determining)}$$

$$Ce^{4+}(aq) + Mn^{3+}(aq) \Longrightarrow Mn^{4+}(aq) + Ce^{3+}(aq)$$

$$Mn^{4+}(aq) + Tl^{+}(aq) \Longrightarrow Mn^{2+}(aq) + Tl^{3+}(aq)$$

Because the first step of this reaction mechanism is rate determining, the rate law for the catalyzed reaction is

$$v = k_{cat}[Ce^{4+}][Mn^{2+}]$$

The overall rate law for the reaction in the presence of the manganese catalyst is then given by

$$v = k[Tl^{+}][Ce^{4+}]^{2} + k_{cat}[Ce^{4+}][Mn^{2+}]$$

The first term in the rate expression is the rate law for the uncatalyzed reaction, and the second term is the rate law for the mechanism involving the catalyst.

As an example of heterogeneous catalysis, we consider the synthesis of ammonia from $H_2(g)$ and $N_2(g)$

$$3\,H_2(g) + N_2(g) \longrightarrow 2\,NH_3(g)$$

The activation barrier for this reaction in the gas phase is roughly given by the dissociation energy of the $N_2(g)$ bond, ≈ 940 kJ·mol^{-1}. Even though $\Delta_r G^{\circ}$ at 300 K is -32.4 kJ·mol^{-1} for this reaction, the barrier to reaction is so large that a mixture of $H_2(g)$ and $N_2(g)$ can be stored indefinitely without producing any appreciable amount of ammonia. In the presence of an iron surface, however, the net activation energy for the synthesis of ammonia from $H_2(g)$ and $N_2(g)$ is ≈ 80 kJ·mol^{-1}, more than an order of magnitude less than that of the gas-phase reaction. The mechanism of the surface-catalyzed synthesis of ammonia is fairly complicated and will be discussed in detail along with other heterogeneous surface-catalyzed gas-phase reactions in the second half of Chapter 31.

As a final example, we consider the destruction of stratospheric ozone by chlorine atoms. A naturally occurring reaction for the destruction of ozone in the stratosphere is given by

$$O_3(g) + O(g) \Longrightarrow 2\,O_2(g)$$

In the presence of chlorine atoms, the following two reactions readily occur

$$O_3(g) + Cl(g) \Longrightarrow ClO(g) + O_2(g)$$

$$ClO(g) + O(g) \Longrightarrow O_2(g) + Cl(g)$$

The net result of this two-step cycle is the destruction of an ozone molecule without consuming the chlorine atom. Therefore, the chlorine atom is a catalyst for ozone destruction. Because the reactants are all in the gas-phase, this is an example of a homogeneous catalysis reaction. Eventually, the chlorine atoms react with other molecules in the stratosphere. In fact, at any given time, most of the chlorine in the stratosphere is bound up in the reservoir molecules $HCl(g)$ and $ClONO_2(g)$, which are formed by the reactions

$$Cl(g) + CH_4(g) \Longrightarrow HCl(g) + CH_3(g)$$

$$ClO(g) + NO_2(g) \Longrightarrow ClONO_2$$

In the gas phase, these reservoir molecules are fairly unreactive toward one another. The surface of the polar stratospheric clouds catalyzes the reaction between $HCl(g)$ and $ClONO_2(g)$, however, forming molecular chlorine by the reaction

$$HCl(g) + ClONO_2(g) \longrightarrow Cl_2(g) + HNO_3(g) \quad \left(\begin{array}{c} \text{surface of polar} \\ \text{stratospheric clouds} \end{array} \right)$$

Because the reactants and cloud particles are in different phases, this reaction is a heterogeneous catalyzed reaction. The $Cl_2(g)$ formed by this reaction is photodissociated by the sunlight, thereby regenerating destructive chlorine atoms.

29–9. The Michaelis–Menten Mechanism Is a Reaction Mechanism for Enzyme Catalysis

One of the most important classes of catalyzed reactions consists of the biological processes that involve enzymes. *Enzymes* are protein molecules that catalyze specific biochemical reactions. Without enzymes, many of the reactions necessary to sustain life would occur at negligible rates and life as we know it would cease to exist. The reactant molecule acted upon by an enzyme is called the *substrate*. The region of the enzyme where the substrate reacts is called the *active site*. The active site is only a small part of the enzyme molecule. For example, consider the enzyme hexokinase, which catalyzes the reaction of glucose to glucose-6-phosphate. The overall chemical reaction that occurs is given by

Glucose   Glucose 6-phosphate

where ATP and ADP are abbreviations for the molecules adenosine triphosphate and adenosine diphosphate, respectively. Figure 29.5a shows a space-filling model of hexokinase. We see that the enzyme has a cleft, which is the location of the active site of the enzyme. The glucose molecule enters this cleft, and the enzyme closes around the active site. Figure 29.5b shows the corresponding space-filling model when the active site is occupied by a glucose molecule and the enzyme has closed around the substrate. The specificity of an enzyme depends in part on the geometry of the active site and the spatial constraints imposed on this region by the overall structure of the enzyme molecule.

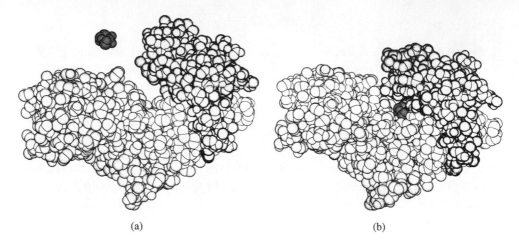

(a)   (b)

FIGURE 29.5
Space-filling model of the two conformations of hexokinase. (a) The active site is not occupied. There is a cleft in the enzyme structure that allows the substrate molecule (glucose) to access the active site. (b) The active site is occupied. The enzyme has closed around the substrate.

Experimental studies reveal that the rate law for many enzyme-catalyzed reactions has the form

$$-\frac{d[S]}{dt} = \frac{k[S]}{K + [S]} \tag{29.69}$$

where [S] is the substrate concentration and k and K are constants. A simple mechanism that accounts for this rate law was proposed by Leonor Michaelis and Maude Menten in 1913. Their mechanism, given by Equation 29.70, is a two-step process that involves formation of an intermediate complex between the enzyme and the substrate, denoted by ES (see, for example, Figure 29.5b)

$$E + S \underset{k_{-1}}{\overset{k_1}{\rightleftharpoons}} ES \underset{k_{-2}}{\overset{k_2}{\rightleftharpoons}} E + P \tag{29.70}$$

The Michaelis-Menten mechanism gives rise to the following rate expressions for [S], [ES], and [P]

$$-\frac{d[\text{S}]}{dt} = k_1[\text{E}][\text{S}] - k_{-1}[\text{ES}] \tag{29.71}$$

$$-\frac{d[\text{ES}]}{dt} = (k_2 + k_{-1})[\text{ES}] - k_1[\text{E}][\text{S}] - k_{-2}[\text{E}][\text{P}] \tag{29.72}$$

$$\frac{d[\text{P}]}{dt} = k_2[\text{ES}] - k_{-2}[\text{E}][\text{P}] \tag{29.73}$$

For this reaction scheme, the enzyme exists either as free enzyme, [E], or as part of an enzyme-substrate complex, [ES]. Because the enzyme is a catalyst and is not consumed by the reaction process, the sum of these two concentrations is constant and equal to the initial enzyme concentration, $[\text{E}]_0$; in an equation

$$[\text{E}]_0 = [\text{ES}] + [\text{E}] \tag{29.74}$$

We can use Equation 29.74 to rewrite Equation 29.72 as

$$-\frac{d[\text{ES}]}{dt} = [\text{ES}](k_1[\text{S}] + k_{-1} + k_2 + k_{-2}[\text{P}]) - k_1[\text{S}][\text{E}]_0 - k_{-2}[\text{P}][\text{E}]_0 \tag{29.75}$$

When the enzyme is mixed with a large excess of substrate, there is an initial period during which the concentration of the enzyme-substrate complex, [ES], builds up. Michaelis and Menten postulated that the equilibrium concentration of this complex is rapidly achieved, after which [ES] remains essentially constant during the course of the reaction, satisfying the requirement of the steady-state approximation for this interme-diate species. Assuming the steady-state approximation enables us to set $d[\text{ES}]/dt = 0$, after which Equation 29.75 can be solved to give the following expression for [ES] in terms of the reaction rate constants and $[\text{E}]_0$, [S], and [P]

$$[\text{ES}] = \frac{k_1[\text{S}] + k_{-2}[\text{P}]}{k_1[\text{S}] + k_{-2}[\text{P}] + k_{-1} + k_2}[\text{E}]_0 \tag{29.76}$$

Substituting this result into Equation 29.71 and using Equation 29.74 gives us

$$v = -\frac{d[\text{S}]}{dt} = \frac{k_1 k_2[\text{S}] - k_{-1}k_{-2}[\text{P}]}{k_1[\text{S}] + k_{-2}[\text{P}] + k_{-1} + k_2}[\text{E}]_0 \tag{29.77}$$

If the experimental measurements of the reaction rate are taken during the time period when only a small percentage (1–3%) of the substrate is converted to product, then $[\text{S}] \approx [\text{S}]_0$ and $[\text{P}] \approx 0$, and Equation 29.77 simplifies to

$$v = -\frac{d[\text{S}]}{dt} = \frac{k_1 k_2[\text{S}]_0[\text{E}]_0}{k_1[\text{S}]_0 + k_{-1} + k_2} = \frac{k_2[\text{S}]_0[\text{E}]_0}{K_m + [\text{S}]_0} \tag{29.78}$$

where $K_m = (k_{-1} + k_2)/k_1$. The quantity K_m is called the *Michaelis constant*. Enzyme kinetics are generally studied by measuring the initial rate as a function of substrate concentration for a fixed enzyme concentration, the conditions necessary to validate that Equation 29.78 is applicable.

Equation 29.78 shows that the initial rate for an enzyme-catalyzed reaction is first order in substrate at low substrate concentrations ($K_m \gg [S]_0$) and then becomes zero order in substrate at high substrate concentrations ($K_m \ll [S]_0$). The zero-order rate law occurs because there is so much substrate relative to enzyme that essentially all the enzyme molecules are tied up with substrate at any instant, so the rate is independent of substrate concentration. At high values of $[S]_0$, Equation 29.78 becomes

$$-\frac{d[S]}{dt} = k_2[E]_0 \tag{29.79}$$

which is the maximum rate the reaction can achieve. Thus, v_{max}, the maximum rate for the Michaelis-Menton mechanism, is given by $v_{max} = k_2[E]_0$.

The *turnover number* is defined as the maximum rate divided by the concentration of enzyme active sites. The turnover number is therefore the maximum number of substrate molecules that can be converted into product molecules per unit time by an enzyme molecule. The concentration of enzyme active sites is not necessarily equal to the concentration of enzyme present because some enzymes have more than one active site. If the enzyme has a single active site, the turnover number is given by $v_{max}/[E]_0 = k_2$. Table 29.2 lists the turnover numbers of a few enzymes.

TABLE 29.2
Turnover numbers of some enzymes.

Enzyme	Substrate	Turnover number/s^{-1}
Catalase	H_2O_2	4.0×10^7
Acetylcholinesterase	Acetylcholine	1.4×10^5
β-Lactamase	Penicillin	2000
Fumarase	Fumarate	800
Rec A protein	ATP	0.4

EXAMPLE 29–9

The enzyme carbonic anhydrase catalyzes both the forward and the reverse reactions for the hydration of CO_2 according to

$$H_2O(l) + CO_2(aq) \rightleftharpoons HCO_3^-(aq) + H^+(aq)$$

Carbon dioxide is produced in tissue as one of the final products of respiration. It then diffuses into the blood system, where it is converted to the bicarbonate ion by carbonic anhydrase. The reverse reactions occur in the lungs, where $CO_2(g)$ is expelled. Carbonic anhydrase has a single active site, and its molecular mass is $30\,000 \text{ g·mol}^{-1}$. If $8.0\ \mu\text{g}$ of carbonic anhydrase catalyze the hydration of 0.146 g of CO_2 in 30 seconds at $37°C$, what is the turnover number of the enzyme (in units of s^{-1})?

SOLUTION: To calculate the turnover number, we need to determine the ratio of the number of moles of CO_2 that react per second to the number of moles of enzyme present. The number of moles of enzyme present is

$$\text{number of moles of enzyme} = \frac{8.0 \times 10^{-6} \text{ g}}{30\,000 \text{ g·mol}^{-1}} = 2.7 \times 10^{-10} \text{ mol}$$

The number of moles of CO_2 reacted in 30 seconds is given by

$$\frac{0.146 \text{ g}}{44 \text{ g·mol}^{-1}} = 3.3 \times 10^{-3} \text{ mol}$$

or a rate of $1.1 \times 10^{-4} \text{ mol·s}^{-1}$. The turnover number is then

$$\text{turnover number} = \frac{1.1 \times 10^{-4} \text{ mol·s}^{-1}}{2.7 \times 10^{-10} \text{ mol}} = 4.1 \times 10^{5} \text{ s}^{-1}$$

Thus, we see that each carbonic anhydrase molecule converts $410\,000$ CO_2 molecules to $HCO_3^-(aq)$ per second! It is one of the fastest working enzymes known. (See Problem 29–40, however.)

Problems

29-1. Give the units of the rate constant for a unimolecular, bimolecular, and termolecular reaction.

29-2. Determine the rate law for the following reaction

$$F(g) + D_2(g) \overset{k}{\Longrightarrow} FD(g) + D(g)$$

Give the units of k. Determine the molecularity of this reaction.

29-3. Determine the rate law for the reaction

$$I(g) + I(g) + M(g) \overset{k}{\Longrightarrow} I_2(g) + M(g)$$

where M is any molecule present in the reaction container. Give the units of k. Determine the molecularity of this reaction. Is this reaction identical to

$$I(g) + I(g) \overset{k}{\Longrightarrow} I_2(g)$$

Explain.

29-4. For $T < 500$ K, the reaction

$$NO_2(g) + CO(g) \overset{k_{obs}}{\longrightarrow} CO_2(g) + NO(g)$$

has the rate law

$$\frac{d[CO_2]}{dt} = k_{obs}[NO_2]^2$$

Show that the following mechanism is consistent with the observed rate law

$$NO_2(g) + NO_2(g) \overset{k_1}{\Longrightarrow} NO_3(g) + NO(g) \qquad \text{(rate determining)}$$

$$NO_3(g) + CO(g) \overset{k_2}{\Longrightarrow} CO_2(g) + NO_2(g)$$

Express k_{obs} in terms of k_1 and k_2.

29-5. Solve Equation 29.21 to obtain $[A] = [A]_0 e^{-k_1 t}$, and substitute this result into Equation 29.22 to obtain

$$\frac{d[I]}{dt} + k_2[I] = k_1[A]_0 e^{-k_1 t}$$

This equation is of the form (see the *CRC Handbook of Standard Mathematical Tables*, for example)

$$\frac{dy(x)}{dx} + p(x)y(x) = q(x)$$

a linear, first-order differential equation whose general solution is

$$y(x)e^{h(x)} = \int q(x)e^{h(x)}dx + c$$

where $h(x) = \int p(x)dx$ and c is a constant. Show that this solution leads to Equation 29.25.

29-6. Verify that Equation 29.32 is obtained if Equation 29.30 is substituted into Equation 29.23 and the resulting expression is integrated.

29-7. Consider the reaction mechanism

$$A \overset{k_1}{\Longrightarrow} I \overset{k_2}{\Longrightarrow} P$$

where $[A] = [A]_0$ and $[I]_0 = [P]_0 = 0$ at time $t = 0$. Use the exact solution to this kinetic scheme (Equations 29.24 through 29.26) to plot the time dependence of $[A]/[A]_0$, $[I]/[A]_0$, and $[P]/[P]_0$ versus $\log k_1 t$ for the case $k_2 = 2k_1$. On the same graph, plot the time dependence of $[A]/[A]_0$, $[I]/[I]_0$, and $[P]/[P]_0$ using the expressions for $[A]$, $[I]$, and $[P]$ obtained assuming the steady-state approximation for $[I]$. Based on your results, can you use the steady-state approximation to model the kinetics of this reaction mechanism when $k_2 = 2k_1$?

29-8. Consider the mechanism for the decomposition of ozone presented in Example 29–5. Explain why either (a) $v_{-1} \gg v_2$ and $v_{-1} \gg v_1$ or (b) $v_2 \gg v_{-1}$ and $v_2 \gg v_1$ must be true for the steady-state approximation to apply. The rate law for the decomposition reaction is found to be

$$\frac{d[O_3]}{dt} = -k_{obs}[O_3][M]$$

Is this rate law consistent with the conditions given by either (a) or (b) or both?

29-9. Consider the reaction mechanism

$$A + B \underset{k_{-1}}{\overset{k_1}{\rightleftharpoons}} C \tag{1}$$

$$C \overset{k_2}{\Longrightarrow} P \tag{2}$$

Write the expression for $d[P]/dt$, the rate of product formation. Assume equilibrium is established in the first reaction before any appreciable amount of product is formed, and thereby show that

$$\frac{d[P]}{dt} = k_2 K_c [A][B]$$

where K_c is the equilibrium constant for step (1) of the reaction mechanism. This assumption is called the *fast-equilibrium approximation*.

29-10. The rate law for the reaction of *para*-hydrogen to *ortho*-hydrogen

$$para\text{-}H_2(g) \overset{k_{obs}}{\longrightarrow} ortho\text{-}H_2(g)$$

is

$$\frac{d[ortho\text{-}H_2]}{dt} = k_{obs}[para\text{-}H_2]^{3/2}$$

Show that the following mechanism is consistent with this rate law.

$$para\text{-}H_2(g) \underset{k_{-1}}{\overset{k_1}{\rightleftharpoons}} 2\,H(g) \qquad \text{(fast equilibrium)} \tag{1}$$

$$H(g) + para\text{-}H_2(g) \overset{k_2}{\Longrightarrow} ortho\text{-}H_2(g) + H(g) \tag{2}$$

Express k_{obs} in terms of the rate constants for the individual steps of the reaction mechanism.

29-11. Consider the decomposition reaction of $N_2O_5(g)$

$$2\,N_2O_5(g) \overset{k_{obs}}{\longrightarrow} 4\,NO_2(g) + O_2(g)$$

A proposed mechanism for this reaction is

$$N_2O_5(g) \underset{k_{-1}}{\overset{k_1}{\rightleftharpoons}} NO_2(g) + NO_3(g)$$

$$NO_2(g) + NO_3(g) \overset{k_2}{\Longrightarrow} NO(g) + NO_2(g) + O_2(g)$$

$$NO_3(g) + NO(g) \overset{k_3}{\Longrightarrow} 2\,NO_2(g)$$

Assume that the steady-state approximation applies to both the $NO(g)$ and $NO_3(g)$ reaction intermediates to show that this mechanism is consistent with the experimentally observed rate law

$$\frac{d[O_2]}{dt} = k_{obs}[N_2O_5]$$

Express k_{obs} in terms of the rate constants for the individual steps of the reaction mechanism.

29-12. The rate law for the reaction between $CO(g)$ and $Cl_2(g)$ to form phosgene (Cl_2CO)

$$Cl_2(g) + CO(g) \xrightarrow{k_{obs}} Cl_2CO(g)$$

is

$$\frac{d[Cl_2CO]}{dt} = k_{obs}[Cl_2]^{3/2}[CO]$$

Show that the following mechanism is consistent with this rate law.

$$Cl_2(g) + M(g) \underset{k_{-1}}{\overset{k_1}{\rightleftharpoons}} 2\,Cl(g) + M(g) \qquad \text{(fast equilibrium)}$$

$$Cl(g) + CO(g) + M(g) \underset{k_{-2}}{\overset{k_2}{\rightleftharpoons}} ClCO(g) + M(g) \qquad \text{(fast equilibrium)}$$

$$ClCO(g) + Cl_2(g) \xrightarrow{k_3} Cl_2CO(g) + Cl(g) \qquad \text{(slow)}$$

where M is any gas molecule present in the reaction container. Express k_{obs} in terms of the rate constants for the individual steps of the reaction mechanism.

29-13. Nitramide (O_2NNH_2) decomposes in water according to the chemical equation

$$O_2NNH_2(aq) \xrightarrow{k_{obs}} N_2O(g) + H_2O(l)$$

The experimentally determined rate law for this reaction is

$$\frac{d[N_2O]}{dt} = k_{obs} \frac{[O_2NNH_2]}{[H^+]}$$

A proposed mechanism for this reaction is

$$O_2NNH_2(aq) \underset{k_{-1}}{\overset{k_1}{\rightleftharpoons}} O_2NNH^-(aq) + H^+(aq) \qquad \text{(fast equilibrium)}$$

$$O_2NNH^-(aq) \xrightarrow{k_2} N_2O(g) + OH^-(aq) \qquad \text{(slow)}$$

$$H^+(aq) + OH^-(aq) \xrightarrow{k_3} H_2O(l) \qquad \text{(fast)}$$

Is this mechanism consistent with the observed rate law? If so, what is the relationship between k_{obs} and the rate constants for the individual steps of the mechanism?

29-14. What would you predict for the rate law for the reaction mechanism in Problem 29–13 if, instead of a fast equilibrium followed by a slow step, you assumed that the concentration of $O_2NNH^-(aq)$ was such that the steady-state approximation could be applied to this reaction intermediate?

29-15. The rate law for the hydrolysis of ethyl acetate by aqueous sodium hydroxide at 298 K

$$CH_3COOCH_2CH_3(aq) + OH^-(aq) \xrightarrow{k_{obs}} CH_3CO_2^-(aq) + CH_3CH_2OH(aq)$$

is

$$\frac{d[CH_3CH_2OH]}{dt} = k_{obs}[OH^-][CH_3COOCH_2CH_3]$$

Despite the form of this rate law, this reaction is not an elementary reaction but is believed to occur by the following mechanism

$$CH_3COOCH_2CH_3(aq) + OH^-(aq) \underset{k_{-1}}{\overset{k_1}{\rightleftharpoons}} CH_3CO^-(OH)OCH_2CH_3(aq)$$

$$CH_3CO^-(OH)OCH_2CH_3(aq) \xrightarrow{k_2} CH_3CO_2H(aq) + CH_3CH_2O^-(aq)$$

$$CH_3CO_2H(aq) + CH_3CH_2O^-(aq) \xrightarrow{k_3} CH_3CO_2^-(aq) + CH_3CH_2OH(aq)$$

Under what conditions does this mechanism give the observed rate law? For those conditions, express k_{obs} in terms of the rate constants for the individual steps of the reaction mechanism.

29-16. The decomposition of perbenzoic acid in water

$$2\,C_6H_5CO_3H(aq) \rightleftharpoons 2\,C_6H_5CO_2H(aq) + O_2(g)$$

is proposed to occur by the following mechanism

$$C_6H_5CO_3H(aq) \underset{k_{-1}}{\overset{k_1}{\rightleftharpoons}} C_6H_5CO_3^-(aq) + H^+(aq)$$

$$C_6H_5CO_3H(aq) + C_6H_5CO_3^-(aq) \xrightarrow{k_2} C_6H_5CO_2H(aq)$$
$$+ C_6H_5CO_2^-(aq) + O_2(g)$$

$$C_6H_5CO_2^-(aq) + H^+(aq) \xrightarrow{k_3} C_6H_5CO_2H(aq)$$

Derive an expression for the rate of formation of O_2 in terms of the reactant concentration and $[H^+]$.

29-17. The rate law for the reaction described by

$$2\,H_2(g) + 2\,NO(g) \xrightarrow{k_{obs}} N_2(g) + 2\,H_2O(g)$$

is

$$\frac{d[N_2]}{dt} = k_{obs}[H_2][NO]^2$$

Below is a proposed mechanism for this reaction

$$H_2(g) + NO(g) + NO(g) \xrightarrow{k_1} N_2O + H_2O(g)$$

$$H_2(g) + N_2O(g) \xrightarrow{k_2} N_2(g) + H_2O(g)$$

Under what conditions does this mechanism give the observed rate law? Express k_{obs} in terms of the rate constants for the individual steps of the mechanism.

29-18. A second proposed mechanism for the reaction discussed in Problem 29–17 is

$$NO(g) + NO(g) \underset{k_{-1}}{\overset{k_1}{\rightleftharpoons}} N_2O_2(g)$$

$$H_2(g) + N_2O_2(g) \overset{k_2}{\Longrightarrow} N_2O(g) + H_2O(g)$$

$$H_2(g) + N_2O(g) \overset{k_3}{\Longrightarrow} N_2(g) + H_2O(g)$$

Under what conditions does this mechanism give the observed rate law? Express k_{obs} in terms of the rate constants for the individual steps of the mechanism. Do you favor this mechanism or that given in Problem 29–17? Explain your reasoning.

29-19. An alternative mechanism for the chemical reaction

$$Cl_2(g) + CO(g) \overset{k_{obs}}{\longrightarrow} Cl_2CO(g)$$

(see Problem 29–12) is

$$Cl_2(g) + M(g) \underset{k_{-1}}{\overset{k_1}{\rightleftharpoons}} 2\,Cl(g) + M(g) \qquad \text{(fast equilibrium)}$$

$$Cl(g) + Cl_2(g) \underset{k_{-2}}{\overset{k_2}{\rightleftharpoons}} Cl_3(g) \qquad \text{(fast equilibrium)}$$

$$Cl_3(g) + CO(g) \overset{k_3}{\Longrightarrow} Cl_2CO(g) + Cl(g)$$

where M is any molecule present in the reaction chamber. Show that this mechanism also gives the observed rate law. How would you go about determining whether this mechanism or the one given in Problem 29–12 is correct?

29-20. The Lindemann reaction mechanism for the isomerization reaction

$$CH_3NC(g) \longrightarrow CH_3CN(g)$$

is

$$CH_3NC(g) + M(g) \underset{k_{-1}}{\overset{k_1}{\rightleftharpoons}} CH_3NC^*(g) + M(g)$$

$$CH_3NC^*(g) \overset{k_2}{\Longrightarrow} CH_3CN(g)$$

Under what conditions does the steady-state approximation apply to CH_3NC^*?

29-21. In Section 29–6 we examined the unimolecular reaction

$$CH_3NC(g) \Longrightarrow CH_3CN(g)$$

Consider this reaction carried out in the presence of a helium buffer gas. The collision of a CH_3NC molecule with either another CH_3NC molecule or a helium atom can energize the molecule, thereby leading to reaction. If the energizing reactions involving a CH_3NC

molecule and a He atom occur with different rates, the reaction mechanism would be given by

$$CH_3NC(g) + CH_3NC(g) \underset{k_{-1}}{\overset{k_1}{\rightleftharpoons}} CH_3NC^*(g) + CH_3NC(g)$$

$$CH_3NC(g) + He(g) \underset{k_{-2}}{\overset{k_2}{\rightleftharpoons}} CH_3NC^*(g) + He(g)$$

$$CH_3NC^*(g) \overset{k_3}{\Longrightarrow} CH_3CN$$

Apply the steady-state approximation to the intermediate species, $CH_3NC^*(g)$, to show that

$$\frac{d[CH_3CN]}{dt} = \frac{k_3(k_1[CH_3NC]^2 + k_2[CH_3NC][He])}{k_{-1}[CH_3NC] + k_{-2}[He] + k_3}$$

Show that this equation is equivalent to Equation 29.55 when $[He] = 0$.

29-22. Consider the reaction and mechanism given in Problem 29–10. The activation energy for the dissociation of $H_2(g)$ [step (1)] is given by D_0, the dissociation energy. If the activation energy of step (2) of the mechanism is E_2, show that $E_{a,obs}$, the experimentally determined activation energy, is given by

$$E_{a,obs} = E_2 + \frac{D_0}{2}$$

Also show that A_{obs}, the experimentally determined Arrhenius pre-exponential factor, is given by

$$A_{obs} = A_2 \left(\frac{A_1}{A_{-1}} \right)^{1/2}$$

where A_i is the Arrhenius pre-exponential factor corresponding to the rate constant k_i.

29-23. The thermal decomposition of ethylene oxide occurs by the mechanism

$$H_2COCH_2(g) \overset{k_1}{\Longrightarrow} H_2COCH(g) + H(g)$$

$$H_2COCH(g) \overset{k_2}{\Longrightarrow} CH_3(g) + CO(g)$$

$$CH_3(g) + H_2COCH_2(g) \overset{k_3}{\Longrightarrow} H_2COCH(g) + CH_4(g)$$

$$CH_3(g) + H_2COCH(g) \overset{k_4}{\Longrightarrow} products$$

Which of these reaction(s) are the initiation, propagation, and termination step(s) of the reaction mechanism? Show that if the intermediates CH_3 and H_2COCH are treated by the steady-state approximation, the rate law, $d[products]/dt$, is first order in ethylene oxide concentration.

The next six problems examine the kinetics of the thermal decomposition of acetaldehyde.

29-24. A proposed mechanism for the thermal decomposition of acetaldehyde

$$CH_3CHO(g) \overset{k_{obs}}{\longrightarrow} CH_4(g) + CO(g)$$

is

$$CH_3CHO(g) \xrightarrow{k_1} CH_3(g) + CHO(g) \tag{1}$$

$$CH_3(g) + CH_3CHO(g) \xrightarrow{k_2} CH_4(g) + CH_3CO(g) \tag{2}$$

$$CH_3CO(g) \xrightarrow{k_3} CH_3(g) + CO(g) \tag{3}$$

$$2CH_3(g) \xrightarrow{k_4} C_2H_6 \tag{4}$$

Is this reaction a chain reaction? If so, identify the initiation, propagation, inhibition, and termination step(s). Determine the rate laws for $CH_4(g)$, $CH_3(g)$, and $CH_3CO(g)$. Show that if you assume the steady-state approximation for the intermediate species, $CH_3(g)$ and $CH_3CO(g)$, the rate law for methane formation is given by

$$\frac{d[CH_4]}{dt} = \left(\frac{k_1}{2k_4}\right)^{1/2} k_2[CH_3CHO]^{3/2}$$

29-25. Suppose that we replace the termination step (Equation 4) of the mechanism in Problem 29–24 with the termination reaction

$$2CH_3CO(g) \xrightarrow{k_4} CH_3COCOCH_3$$

Determine the rate laws for $CO(g)$, $CH_3(g)$, and $CH_3CO(g)$. Once again, assume that the steady-state approximation can be applied to the intermediates $CH_3(g)$ and $CH_3CO(g)$, and show that in this case the rate of formation of CO is given by

$$\frac{d[CO]}{dt} = \left(\frac{k_1}{k_4}\right)^{1/2} k_3[CH_3CHO]^{1/2}$$

29-26. The chain length γ of a chain reaction is defined as the rate of the overall reaction divided by the rate of the initiation step. Give a physical interpretation of the chain length. Show that γ for the reaction mechanism and rate law given in Problem 29–25 is

$$\gamma = k_3 \left(\frac{1}{k_1 k_4}\right)^{1/2} [CH_3CHO]^{-1/2}$$

29-27. Show that the chain length γ (see Problem 29–26) for the reaction mechanism and the rate law given in Problem 29–24 is

$$\gamma = k_2 \left(\frac{1}{k_1 k_4}\right)^{1/2} [CH_3CHO]^{1/2}$$

29-28. Consider the mechanism for the thermal decomposition of acetaldehyde given in Problem 29–24. Show that E_{obs}, the measured Arrhenius activation energy for the overall reaction, is given by

$$E_{obs} = E_2 + \tfrac{1}{2}(E_1 - E_4)$$

where E_i is the activation energy of the ith step of the reaction mechanism. How is A_{obs}, the measured Arrhenius pre-exponential factor for the overall reaction, related to the Arrhenius pre-exponential factors for the individual steps of the reaction mechanism?

29-29. Consider the mechanism for the thermal decomposition of acetaldehyde given in Problem 29–25. Show that E_{obs}, the measured Arrhenius activation energy for the overall reaction, is given by

$$E_{obs} = E_3 + \tfrac{1}{2}(E_1 - E_4)$$

where E_i is the activation energy of the ith step of the reaction mechanism. How is A_{obs}, the measured Arrhenius pre-exponential factor for the overall reaction, related to the Arrhenius pre-exponential factors for the individual steps of the reaction mechanism?

29-30. Consider the reaction between $H_2(g)$ and $Br_2(g)$ discussed in Section 29–7. Justify why we ignored the $H_2(g)$ dissociation reaction in favor of the $Br_2(g)$ dissociation reaction as being the initiating step of the reaction mechanism.

29-31. In Section 29–7, we considered the chain reaction between $H_2(g)$ and $Br_2(g)$. Consider the related chain reaction between $H_2(g)$ and $Cl_2(g)$.

$$Cl_2(g) + H_2(g) \longrightarrow 2\,HCl(g)$$

The mechanism for this reaction is

$$Cl_2(g) + M(g) \overset{k_1}{\Longrightarrow} 2\,Cl(g) + M(g) \tag{1}$$

$$Cl(g) + H_2(g) \overset{k_2}{\Longrightarrow} HCl(g) + H(g) \tag{2}$$

$$H(g) + Cl_2(g) \overset{k_3}{\Longrightarrow} HCl(g) + Cl(g) \tag{3}$$

$$2\,Cl(g) + M(g) \overset{k_4}{\Longrightarrow} Cl_2(g) + M(g) \tag{4}$$

Label the initiation, propagation, and termination step(s). Use the following bond dissociation data to explain why it is reasonable not to include the analogous inhibition steps in this mechanism that are included in the mechanism for the chain reaction involving $Br_2(g)$

Molecule	$D_0/kJ \cdot mol^{-1}$
H_2	432
HBr	363
HCl	428
Br_2	190
Cl_2	239

29-32. Derive the rate law for $v = (1/2)(d[HCl]/dt)$ for the mechanism of the

$$Cl_2(g) + H_2(g) \longrightarrow 2\,HCl(g)$$

reaction given in Problem 29–31.

29-33. It is possible to initiate chain reactions using photochemical reactions. For example, in place of the thermal initiation reaction for the $Br_2(g) + H_2(g)$ chain reaction

$$Br_2(g) + M \overset{k_1}{\Longrightarrow} 2\,Br(g) + M$$

we could have the photochemical initiation reaction

$$Br_2(g) + h\nu \Longrightarrow 2\,Br(g)$$

If we assume that all the incident light is absorbed by the Br_2 molecules and that the quantum yield for photodissociation is 1.00, then how does the photochemical rate of dissociation of Br_2 depend on I_{abs}, the number of photons per unit time per unit volume? How does $d[Br]/dt$, the rate of formation of Br, depend on I_{abs}? If you assume that the chain reaction is initiated only by the photochemical generation of Br, then how does $d[HBr]/dt$ depend on I_{abs}?

29-34. In Section 29–9, we derived the Michaelis-Menton rate law for enzyme catalysis. The derivation presented there is limited to the case in which only the rate of the initial reaction is measured so that $[S] = [S]_0$ and $[P] = 0$. We will now determine the Michaelis-Menton rate law by a different approach. Recall that the Michaelis-Menton mechanism is

$$E + S \underset{k_{-1}}{\overset{k_1}{\rightleftharpoons}} ES$$

$$ES \overset{k_2}{\Longrightarrow} E + P$$

The rate law for this reaction is $v = k_2[ES]$. Write the rate expression for $[ES]$. Show that if you apply the steady-state approximation to this intermediate, then

$$[ES] = \frac{[E][S]}{K_m} \tag{1}$$

where K_m is the Michaelis constant. Now show that

$$[E]_0 = [E] + \frac{[E][S]}{K_m} \tag{2}$$

(*Hint*: The enzyme is not consumed.) Solve Equation 2 for $[E]$ and substitute the result into Equation 1 and thereby show that

$$v = \frac{k_2[E]_0[S]}{K_m + [S]} \tag{3}$$

If the rate is measured during a time period when only a small amount of substrate is consumed, then $[S] = [S]_0$ and Equation 3 reduces to the Michaelis-Menton rate law given by Equation 29.78.

29-35. The ability of enzymes to catalyze reactions can be hindered by *inhibitor molecules*. One of the mechanisms by which an inhibitor molecule works is by competing with the substrate molecule for binding to the active site of the enzyme. We can include this inhibition reaction in a modified Michaelis-Menton mechanism for enzyme catalysis.

$$E + S \underset{k_{-1}}{\overset{k_1}{\rightleftharpoons}} ES \tag{1}$$

$$E + I \underset{k_{-2}}{\overset{k_2}{\rightleftharpoons}} EI \tag{2}$$

$$ES \overset{k_3}{\Longrightarrow} E + P \tag{3}$$

In Equation 2, I is the inhibitor molecule and EI is the enzyme-inhibitor complex. We will consider the case where reaction (2) is always in equilibrium. Determine the rate laws for [S], [ES], [EI], and [P]. Show that if the steady-state assumption is applied to ES, then

$$[ES] = \frac{[E][S]}{K_m}$$

where K_m is the Michaelis constant, $K_m = (k_{-1} + k_3)/k_1$. Now show that material balance for the enzyme gives

$$[E]_0 = [E] + \frac{[E][S]}{K_m} + [E][I]K_I$$

where $K_I = [EI]/[E][I]$ is the equilibrium constant for step (2) of the above reaction mechanism. Use this result to show that the initial reaction rate is given by

$$v = \frac{d[P]}{dt} = \frac{k_3[E]_0[S]}{K_m + [S] + K_m K_I[I]} \approx \frac{k_3[E]_0[S]_0}{K'_m + [S]_0} \tag{4}$$

where $K'_m = K_m(1 + K_I[I])$. Note that the second expression in Equation 4 has the same functional form as the Michaelis-Menton equation. Does Equation 4 reduce to the expected result when $[I] \to 0$?

29-36. Antibiotic-resistant bacteria have an enzyme, penicillinase, that catalyzes the decomposition of the antibiotic. The molecular mass of penicillinase is $30\,000$ g·mol^{-1}. The turnover number of the enzyme at 28°C is 2000 s^{-1}. If 6.4 μg of penicillinase catalyzes the destruction of 3.11 mg of amoxicillin, an antibiotic with a molecular mass of 364 g·mol^{-1}, in 20 seconds at 28°C, how many active sites does the enzyme have?

29-37. Show that the inverse of Equation 29.78 is

$$\frac{1}{v} = \frac{1}{v_{max}} + \frac{K_m}{v_{max}} \frac{1}{[S]_0} \tag{1}$$

This equation is called the *Lineweaver-Burk equation*. In Example 29–9, we examined the reaction for the hydration of CO_2 that is catalyzed by the enzyme carbonic anhydrase. For a total enzyme concentration of 2.32×10^{-9} mol·dm^{-3}, the following data were obtained.

v/mol·dm^{-3}·s^{-1}	$[CO_2]_0/10^{-3}$ mol·dm^{-3}
2.78×10^{-5}	1.25
5.00×10^{-5}	2.50
8.33×10^{-5}	5.00
1.66×10^{-4}	20.00

Plot these data according to Equation 1, and determine the values of K_m, the Michaelis constant, and k_2, the rate constant for product formation from the enzyme-substrate complex from the slope and intercept of the best-fit line to the plotted data.

29-38. Carbonic anhydrase catalyzes the reaction

$$H_2O(l) + CO_2(g) \rightleftharpoons H_2CO_3(aq)$$

Data for the reverse dehydration reaction using a total enzyme concentration of 2.32×10^{-9} mol·dm^{-3} are given below

v/mol·dm^{-3}·s^{-1}	$[H_2CO_3]_0/10^{-3}$ mol·dm^{-3}
1.05×10^{-5}	2.00
2.22×10^{-5}	5.00
3.45×10^{-5}	10.00
4.17×10^{-5}	15.00

Use the approach discussed in Problem 29–37 to determine the values of K_m, the Michaelis constant, and k_2, the rate of product formation from the enzyme substrate complex.

29-39. Show that the Michaelis-Menton mechanism for enzyme catalysis gives $v = (1/2)v_{max}$ when $[S]_0 = K_m$.

29-40. The protein catalase catalyzes the reaction

$$2\,H_2O_2(aq) \longrightarrow 2\,H_2O(l) + O_2(g)$$

and has a Michaelis constant of $K_m = 25 \times 10^{-3}$ mol·dm^{-3} and a turnover number of 4.0×10^7 s^{-1}. Calculate the initial rate of this reaction if the total enzyme concentration is 0.016×10^{-6} mol·dm^{-3} and the initial substrate concentration is 4.32×10^{-6} mol·dm^{-3}. Calculate v_{max} for this enzyme. Catalase has a single active site.

29-41. The presence of 4.8×10^{-6} mol·dm^{-3} of a competitive inhibitor decreases the initial rate calculated in Problem 29–40 by a factor of 3.6. Calculate K_I, the equilibrium constant for the binding reaction between the enzyme and the inhibitor. (*Hint*: See Problem 29–35.)

29-42. The turnover number for acetylcholinesterase, an enzyme with a single active site that metabolizes acetylcholine, is 1.4×10^4 s^{-1}. How many grams of acetylcholine can 2.16×10^{-6} g of acetylcholinesterase metabolize in one hour? (Take the molecular mass of the enzyme to be 4.2×10^4 g·mol^{-1}; acetylcholine has the molecular formula $C_7NO_2H_{16}^+$.)

29-43. Consider the following mechanism for the recombination of bromine atoms to form molecular bromine

$$2\,Br(g) \underset{k_{-1}}{\overset{k_1}{\rightleftharpoons}} Br_2^*(g)$$

$$Br_2^*(g) + M(g) \overset{k_2}{\Longrightarrow} Br_2(g) + M(g)$$

The first step results in formation of an energized bromine molecule. This excess energy is then removed by a collision with a molecule M in the sample. Show that if the steady-state approximation is applied to $Br_2^*(g)$, then

$$\frac{d[Br]}{dt} = -\frac{2k_1k_2[Br]^2[M]}{k_{-1} + k_2[M]}$$

Determine the limiting expression for $d[Br]/dt$ when $v_2 \gg v_{-1}$. Determine the limiting expression for $d[Br]/dt$ when $v_2 \ll v_{-1}$.

29-44. A mechanism for the recombination of bromine atoms to form molecular bromine is given in Problem 29–43. When this reaction occurs in the presence of a large excess of buffer gas, a negative activation energy is measured. Because M(g), the buffer gas molecule, is responsible for the deactivation of $Br_2^*(g)$ but is not consumed itself by the reaction, we can consider it to be a catalyst. Below are the measured rate constants for this reaction in the presence of the same concentration of excess $Ne(g)$ and $CCl_4(g)$ buffer gases at several temperatures. Which gas is the better catalyst for this reaction?

T/K	Ne $k_{obs}/mol^{-2} \cdot dm^6 \cdot s^{-1}$	CCl_4 $k_{obs}/mol^{-2} \cdot dm^6 \cdot s^{-1}$
367	1.07×10^9	1.01×10^{10}
349	1.15×10^9	1.21×10^{10}
322	1.31×10^9	1.64×10^{10}
297	1.50×10^9	2.28×10^{10}

Why do you think there is a difference in the "catalytic" behavior of these two buffer gases?

29-45. The standard Gibbs energy change of reaction for

$$2H_2(g) + O_2(g) \longrightarrow 2H_2O(g)$$

is -457.2 kJ at 298 K. At room temperature, however, this reaction does not occur and mixtures of gaseous hydrogen and oxygen are stable. Explain why this is so. Is such a mixture indefinitely stable?

29-46. The HF(g) chemical laser is based on the reaction

$$H_2(g) + F_2(g) \longrightarrow 2HF(g)$$

The mechanism for this reaction involves the elementary steps

$$\Delta_r H°/kJ \cdot mol^{-1} \text{ at } 298K$$

(1) $F_2(g) + M(g) \underset{k_{-1}}{\overset{k_1}{\rightleftharpoons}} 2F(g) + M(g)$ $+159$

(2) $F(g) + H_2(g) \underset{k_{-2}}{\overset{k_2}{\rightleftharpoons}} HF(g) + H(g)$ -134

(3) $H(g) + F_2(g) \overset{k_3}{\Longrightarrow} HF(g) + F(g)$ -411

Comment on why the reaction $H_2(g) + M(g) \rightarrow 2H(g) + M(g)$ is not included in the mechanism of the HF(g) laser even though it produces a reactant that could participate in step (3) of the reaction mechanism. Derive the rate law for $d[HF]/dt$ for the above mechanism assuming that the steady-state approximation can be applied to both intermediate species, F(g) and H(g).

29-47. A mechanism for ozone creation and destruction in the stratosphere is

$$O_2(g) + h\nu \overset{j_1}{\Longrightarrow} O(g) + O(g)$$

$$O(g) + O_2(g) + M(g) \overset{k_2}{\Longrightarrow} O_3(g) + M(g)$$

$$O_3(g) + h\nu \overset{j_3}{\Longrightarrow} O_2(g) + O(g)$$

$$O(g) + O_3(g) \overset{k_4}{\Longrightarrow} O_2(g) + O_2(g)$$

where we have used the symbol j to indicate that the rate constant is for a photochemical reaction. Determine the rate expressions for $d[O]/dt$ and $d[O_3]/dt$. Assume that both intermediate species, $O(g)$ and $O_3(g)$, can be treated by the steady-state approximation and thereby show that

$$[O] = \frac{2j_1[O_2] + j_3[O_3]}{k_2[O_2][M] + k_4[O_3]} \tag{1}$$

and

$$[O_3] = \frac{k_2[O][O_2][M]}{j_3 + k_4[O]} \tag{2}$$

Now substitute Equation 1 into Equation 2 and solve the resulting quadratic formula for $[O_3]$ to obtain

$$[O_3] = [O_2]\frac{j_1}{2j_3}\left\{\left(1 + 4\frac{j_3}{j_1}\frac{k_2}{k_4}[M]\right)^{1/2} - 1\right\}$$

Typical values for these parameters at an altitude of 30 km are $j_1 = 2.51 \times 10^{-12}$ s^{-1}, $j_3 = 3.16 \times 10^{-4}$ s^{-1}, $k_2 = 1.99 \times 10^{-33}$ $cm^6 \cdot molecule^{-2} \cdot s^{-1}$, $k_4 = 1.26 \times 10^{-15}$ $cm^3 \cdot molecule^{-1} \cdot s^{-1}$, $[O_2] = 3.16 \times 10^{17}$ $molecule \cdot cm^{-3}$, and $[M] = 3.98 \times 10^{17}$ $molecule \cdot cm^{-3}$. Find $[O_3]$ and $[O]$ at an altitude of 30 km using Equations 1 and 2. Was the use of the steady-state assumption justified?

In the next four problems, we will examine the explosive reaction

$$2\,H_2(g) + O_2(g) \rightleftharpoons 2\,H_2O(g)$$

29-48. A simplified mechanism for this reaction is

$$\text{electric spark} + H_2(g) \Longrightarrow 2\,H(g) \tag{1}$$

$$H(g) + O_2(g) \overset{k_1}{\Longrightarrow} OH(g) + O(g) \tag{2}$$

$$O(g) + H_2(g) \overset{k_2}{\Longrightarrow} OH(g) + H(g) \tag{3}$$

$$H_2(g) + OH(g) \overset{k_3}{\Longrightarrow} H_2O(g) + H(g) \tag{4}$$

$$H(g) + O_2(g) + M(g) \overset{k_4}{\Longrightarrow} HO_2(g) + M(g) \tag{5}$$

A reaction that produces more molecules that can participate in chain-propagation steps than it consumes is called a branching chain reaction. Label the branching chain reaction(s), inititation reaction(s), propagation reaction(s), and termination reaction(s) for this mechanism. Use the following bond dissociation energies to evaluate the energy change for steps (2) and (3)

Molecule	$D_0/kJ \cdot mol^{-1}$
H_2	432
O_2	493
OH	424

29-49. Using the mechanism given in Problem 29–48, determine the rate expression for [H] when the initiation step involves an electric spark that gives rise to a rate I_0 of the hydrogen atom production. Determine the rate expresions for [OH] and [O]. Assume that $[O] \approx [OH] \ll [H]$, so now we can apply the steady-state approximation to the intermediate species, O(g) and OH(g). Show that this use of the steady-state approximation gives

$$[O] = \frac{k_1[H][O_2]}{k_2[H_2]} \quad \text{and} \quad [OH] = \frac{2k_1[H][O_2]}{k_3[H_2]}$$

Use these results and your rate expression for [H] to show that

$$\frac{d[H]}{dt} = I_0 + (2k_1[O_2] - k_4[O_2][M])[H]$$

29-50. Consider the result of Problem 29–49. The rate of hydrogen atom production has a functional dependence of

$$\frac{d[H]}{dt} = I_0 + (\alpha - \beta)[H] \tag{1}$$

Which step(s) of the chemical reaction are responsible for the magnitudes of α and β? We can envision two solutions to this rate law, one for $\alpha > \beta$ and one for $\alpha < \beta$. For $\alpha < \beta$ show that the solution to Equation 1 becomes

$$[H] = \frac{I_0}{\beta - \alpha}(1 - e^{-(\beta - \alpha)t})$$

Plot [H] as a function of time. Determine the slope of the plot at short times. Determine the final steady-state value of [H].

29-51. We now consider the solution to the equation (Problem 29–50)

$$\frac{d[H]}{dt} = I_0 + (\alpha - \beta)[H]$$

when $\alpha > \beta$. Show that the solution to this differential equation is given by

$$[H] = \frac{I_0}{\alpha - \beta}(e^{(\alpha - \beta)t} - 1)$$

Plot [H] as a function of time. Describe the differences observed between this plot and that obtained in Problem 29–50. Which case do you think is characteristic of a chemical explosion?

Yuan Lee (top left), Dudley Herschbach (right), and John Polanyi (bottom left) were awarded the Nobel Prize for chemistry in 1986 "for their contributions to the understanding of the dynamics of chemical elementary processes." **Yuan T. Lee** was born in Hsinchu, Taiwan, on November 29, 1936. He received his Ph.D. from the University of California at Berkeley under Dudley Herschbach in 1965. After spending six years at the University of Chicago, he returned to Berkeley as a professor of chemistry. He returned to Taiwan in 1994 as President of Academia Sinica, which is similar to being president of the National Academy of Sciences. While Lee continues to study the dynamics of simple elementary reactions, he has also turned his attention to the exploration of the reactions of large molecules in molecular beams. **Dudley Herschbach** was born in San Jose, California, on June 18, 1932. He received his Ph.D. in chemical physics from Harvard University under E. Bright Wilson, Jr. in 1958. After teaching several years at the University of California at Berkeley, he returned to Harvard in 1963, where he remains today. He pioneered the use of molecular beams to study chemical kinetics, especially the reaction between potassium and iodomethane. Herschbach has demonstrated a special concern for undergraduates by serving as comaster with his wife at one of the residence halls at Harvard, a job that involves 40 hours a week outside of teaching and research. **John C. Polanyi** was born on January 23, 1929, in Berlin, Germany, but grew up in Manchester, England. He received his Ph.D. from the University of Manchester in 1952. In 1956, he joined the faculty of the University of Toronto, Canada, where he is today. Polanyi developed the technique of infrared chemiluminescence to study the products of reactions. In addition to his scientific papers, he has published almost one hundred articles on science policy, on the control of armaments, and on the impact of science on society.

CHAPTER 30

Gas-Phase Reaction Dynamics

Bimolecular gas-phase reactions are among the simplest elementary kinetic processes that occur in nature. In this chapter, we will examine some of the current models that are used to describe the molecular aspects of bimolecular gas-phase reactions. First, we will modify the collision theory presented in Chapter 27 and define the rate constant in terms of a reaction cross section. We will then examine experimentally measured reaction cross sections for several gas-phase reactions. The simplest gas-phase reaction is the hydrogen exchange reaction $H_A + H_B–H_C \Rightarrow H_A–H_B + H_C$. This reaction has been studied in great detail and the experimental data for it are often used to test theories of gas-phase chemical reactions.

In this chapter, however, we have chosen to focus our discussion on the reaction $F(g) + D_2(g) \Rightarrow DF(g) + D(g)$. From a study of this reaction, we will not only learn the same concepts that underlie the $H(g) + H_2(g)$ exchange reaction but will also learn about molecular processes that can occur in reactions in which $\Delta_r U° < 0$. The reaction $F(g) + D_2(g)$ therefore serves as an excellent system for studying the molecular details of gas-phase reactions. We will examine data obtained from crossed molecular beam spectroscopy experiments and learn how such measurements reveal the chemical dynamics of reactive collisions. We will then see that contemporary quantum-mechanical calculations can provide a detailed description of the reaction path by which the $F(g) + D_2(g)$ reactants become the $DF(g) + D(g)$ products.

30–1. The Rate of a Bimolecular Gas-Phase Reaction Can Be Calculated Using Hard-Sphere Collision Theory and an Energy-Dependent Reaction Cross Section

The rate of the general bimolecular elementary gas-phase reaction

$$A(g) + B(g) \xrightarrow{k} \text{products} \qquad\qquad (30.1)$$

is given by

$$v = -\frac{d[A]}{dt} = k[A][B] \tag{30.2}$$

Hard-sphere collision theory can be used to estimate the rate constant k. Using the naive assumption that every collision between the hard spheres A and B yields products, the rate of reaction is given by the collision frequency per unit volume (Equation 27.57)

$$v = Z_{AB} = \sigma_{AB} \langle u_r \rangle \rho_A \rho_B \tag{30.3}$$

In Equation 30.3, σ_{AB} is the hard-sphere collision cross section of A and B molecules, $\langle u_r \rangle$ is the average relative speed of a colliding pair of A and B molecules, and ρ_A and ρ_B are the number densities of A and B molecules in the sample, respectively. Recall from Section 27–6 that the hard-sphere collision cross section σ_{AB} is given by $\sigma_{AB} = \pi d_{AB}^2$, where d_{AB} is the sum of the radii of the two colliding spheres. Being a collision frequency per unit volume, Z_{AB} has the units of collisions·m^{-3}·s^{-1}, where the "units" collisions is not usually included. Because we are assuming that every collision leads to a reaction, Z_{AB} also gives us the number of product molecules formed per unit volume per unit time. Comparison of Equations 30.2 and 30.3 shows that we can define the rate constant as

$$k = \sigma_{AB} \langle u_r \rangle \tag{30.4}$$

The units of k are given by the units of $Z_{AB}/\rho_A \rho_B$, or molecules·m^{-3}·s^{-1}/(molecules·m^{-3})^2 = molecules^{-1}·m^3·s^{-1}. To obtain k in the more commonly used units of dm^3·mol^{-1}·s^{-1}, we need to multiply the right side of Equation 30.4 by N_A and by $(10\ dm·m^{-1})^3$, giving

$$k = (1000\ dm^3 · m^{-3}) N_A \sigma_{AB} \langle u_r \rangle \tag{30.5}$$

where σ_{AB} has units of m^2 and $\langle u_r \rangle$ has units of $m·s^{-1}$.

EXAMPLE 30–1

Use hard-sphere collision theory to calculate the rate constant for the reaction

$$H_2(g) + C_2H_4(g) \Longrightarrow C_2H_6(g)$$

at 298 K. Express the rate constant in units of $dm^3·mol^{-1}·s^{-1}$.

SOLUTION: The hard-sphere collision theory rate constant in units of $dm^3·mol^{-1}·s^{-1}$ is given by Equation 30.5. Using the first of Equations 27.58 and the data in Table 27.3 gives

$$\sigma_{AB} = \pi d_{AB}^2 = \pi \left(\frac{270\ pm + 430\ pm}{2} \right)^2$$

$$= 3.85 \times 10^{-19}\ m^2$$

The average relative speed of the reactants is given by the second of Equations 27.58

$$\langle u_r \rangle = \left(\frac{8 k_B T}{\pi \mu} \right)^{1/2}$$

The reduced mass is

$$\mu = \frac{m_{H_2} m_{C_2H_4}}{m_{H_2} + m_{C_2H_4}} = 3.12 \times 10^{-27} \text{ kg}$$

and therefore,

$$\langle u_r \rangle = \left[\frac{(8)(1.381 \times 10^{-23} \text{ J·K}^{-1})(298 \text{ K})}{(\pi)(3.12 \times 10^{-27} \text{ kg})} \right]^{1/2}$$
$$= 1.83 \times 10^3 \text{ m·s}^{-1}$$

Substituting our calculated values of σ_{AB} and $\langle u_r \rangle$ into Equation 30.5 gives

$$k = (1000 \text{ dm}^3 \cdot \text{m}^{-3})(6.022 \times 10^{23} \text{ mol}^{-1})(3.85 \times 10^{-19} \text{ m}^2)(1.83 \times 10^3 \text{ m·s}^{-1})$$
$$= 4.24 \times 10^{11} \text{ dm}^3 \cdot \text{mol}^{-1} \cdot \text{s}^{-1}$$

The experimental rate constant for this reaction at 298 K is $3.49 \times 10^{-26} \text{ dm}^3 \cdot \text{mol}^{-1} \cdot \text{s}^{-1}$, more than 30 orders of magnitude smaller than the hard-sphere collision theory prediction!

As we noted in Section 27–7, and demonstrated in Example 30–1, the calculated rate constants using naive hard-sphere collision theory are often significantly larger than the experimental rate constants. In addition, because $\langle u_r \rangle \propto T^{1/2}$, Equation 30.4 predicts that k should show a temperature dependence of $T^{1/2}$, whereas the Arrhenius equation predicts and experimental measurements generally show that k is exponentially dependent on $1/T$.

In deriving the naive hard-sphere collision theory, we assumed that each pair of reactants approaches one another with a relative speed of $\langle u_r \rangle$. In a mixture of reactive gases, pairs of reactant molecules approach one another with a distribution of speeds. As two molecules collide, the valence electrons of the two molecules repel one another, so no reaction can occur unless the relative speed is sufficient to overcome this repulsive force. We make our first improvement to collision theory by taking into account the dependence of the reaction rate on the relative speed, or the energy, of the collision. We take this dependence into account by arbitrarily introducing in place of the collision cross section σ_{AB} in Equation 30.4 a reaction cross section, denoted by $\sigma_r(u_r)$, that depends on the relative speed of the reactants. We will thus write the rate constant for molecules that collide with a relative speed of u_r by an expression similar to Equation 30.4,

$$k(u_r) = u_r \sigma_r(u_r) \tag{30.6}$$

To calculate the observed rate constant, we must average over all possible collision speeds, so we write the observed rate constant as

$$k = \int_0^\infty du_r f(u_r) k(u_r) = \int_0^\infty du_r u_r f(u_r) \sigma_r(u_r) \tag{30.7}$$

where $f(u_r)$ is the distribution of relative speeds in the gas sample. From the kinetic theory of gases (Section 27–7), $u_r f(u_r) du_r$ is given by

$$u_r f(u_r) du_r = \left(\frac{\mu}{k_B T}\right)^{3/2} \left(\frac{2}{\pi}\right)^{1/2} u_r^3 e^{-\mu u_r^2 / 2 k_B T} du_r \tag{30.8}$$

To compare Equation 30.7 with the traditional Arrhenius form of k, we need to change the dependent variable from u_r to E_r, the relative kinetic energy. The relative speed, u_r, is related to the relative kinetic energy, E_r, by

$$E_r = \tfrac{1}{2}\mu u_r^2$$

so

$$u_r = \left(\frac{2E_r}{\mu}\right)^{1/2} \quad \text{and} \quad du_r = \left(\frac{1}{2\mu E_r}\right)^{1/2} dE_r \tag{30.9}$$

Using the relationships given in Equations 30.9, we can use Equation 30.8 to write

$$u_r f(u_r) du_r = \left(\frac{2}{k_B T}\right)^{3/2} \left(\frac{1}{\mu\pi}\right)^{1/2} E_r e^{-E_r / k_B T} dE_r \tag{30.10}$$

Substituting Equation 30.10 into Equation 30.7 gives us

$$k = \left(\frac{2}{k_B T}\right)^{3/2} \left(\frac{1}{\mu\pi}\right)^{1/2} \int_0^\infty dE_r E_r e^{-E_r / k_B T} \sigma_r(E_r) \tag{30.11}$$

To evaluate k, we need a model for $\sigma_r(E_r)$, the energy dependence of the reaction cross section. The simplest model is to assume that only those collisions for which the relative kinetic energy exceeds a threshold energy, E_0, are reactive. In this case,

$$\sigma_r(E_r) = \begin{cases} 0 & E_r < E_0 \\ \pi d_{AB}^2 & E_r \geq E_0 \end{cases} \tag{30.12}$$

and

$$k = \left(\frac{2}{k_B T}\right)^{3/2} \left(\frac{1}{\mu\pi}\right)^{1/2} \int_{E_0}^{\infty} dE_r E_r e^{-E_r/k_B T} \pi d_{AB}^2$$

$$= \left(\frac{8k_B T}{\mu\pi}\right)^{1/2} \pi d_{AB}^2 e^{-E_0/k_B T} \left(1 + \frac{E_0}{k_B T}\right)$$

$$= \langle u_r \rangle \sigma_{AB} e^{-E_0/k_B T} \left(1 + \frac{E_0}{k_B T}\right) \tag{30.13}$$

where $\sigma_{AB} = \pi d_{AB}^2$ is the hard-sphere collision cross section. Equation 30.13 is identical to that obtained for the collision rate between pairs of hard spheres whose relative energies exceed a certain threshold energy, E_0 (Equation 27.63). This result is expected because we have assumed that all collisions for which $E_r \geq E_0$ are reactive. The important point to recognize here is that the present treatment accounts for the energy requirements of the reaction through $\sigma_r(E_r)$. We can therefore explore different models for $\sigma_r(E_r)$, which will then give different expressions for the rate constant. The validity of any given model must, of course, be tested experimentally.

EXAMPLE 30–2

We saw in Example 30–1 that Equation 30.5 gives a value of $4.24 \times 10^{11} \, \text{dm}^3 \cdot \text{mol}^{-1} \cdot \text{s}^{-1}$ at 298K compared with the experimental value of $3.49 \times 10^{-26} \, \text{dm}^3 \cdot \text{mol}^{-1} \cdot \text{s}^{-1}$ for the reaction $H_2(g) + C_2H_4(g) \Longrightarrow C_2H_6(g)$. What value of E_0 in Equation 30.13 gives the experimental value of k?

SOLUTION: Letting $k = 3.49 \times 10^{-26} \, \text{dm}^3 \cdot \text{mol}^{-1} \cdot \text{s}^{-1}$ in Equation 30.13 gives us

$$\frac{3.49 \times 10^{-26} \, \text{dm}^3 \cdot \text{mol}^{-1} \cdot \text{s}^{-1}}{4.24 \times 10^{11} \, \text{dm}^3 \cdot \text{mol}^{-1} \cdot \text{s}^{-1}} = e^{-E_0/k_B T} \left(1 + \frac{E_0}{k_B T}\right)$$

Letting $x = E_0/k_B T$, we have

$$8.23 \times 10^{-38} = e^{-x}(1 + x)$$

which is satisfied by $x = 89.9$. At 298 K,

$$E_0 = x k_B T$$
$$= (89.9)(1.381 \times 10^{-23} \, \text{J} \cdot \text{K}^{-1})(298 \, \text{K})$$
$$= 3.70 \times 10^{-19} \, \text{J} = 223 \, \text{kJ} \cdot \text{mol}^{-1}$$

The experimental value of the activation energy is $180 \, \text{kJ} \cdot \text{mol}^{-1}$.

30–2. A Reaction Cross Section Depends Upon the Impact Parameter

The simple energy-dependent reaction cross section given by Equation 30.12 is not realistic. To see why this is so, consider the following two collision geometries

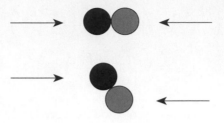

where the arrows indicate the direction from which the molecules approach the point of collision. When the relative collision energy is the same for these two cases, Equation 30.12 predicts that both collision geometries will have the same reaction cross section. But in the bottom case, the particles graze one another, and in the top case, the two particles collide head-on. The grazing collision provides almost no energy for reaction because most of the energy remains in the forward translational motion of each reactant. In contrast, for the head-on collision, the molecules come to a stop and, in principle, all the relative kinetic energy becomes available for reaction. These two collision geometries suggest that a more reasonable model for $\sigma_r(E_r)$ is one in which the cross section depends on the component of the relative kinetic energy that lies along the line that joins the centers of the colliding molecules, as illustrated in Figure 30.1. This is called the *line-of-centers model* for $\sigma_r(E_r)$. If we denote the relative kinetic energy along the line of centers by E_{loc}, then we are assuming that a reaction occurs when $E_{loc} > E_0$.

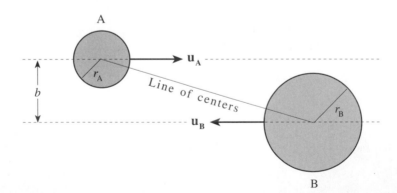

FIGURE 30.1
The collision geometry between two hard spheres. Molecules A and B, of radii r_A and r_B, respectively, approach one another with a relative velocity of $\mathbf{u}_r = \mathbf{u}_A - \mathbf{u}_B$. The distance between the lines drawn through the centers of the two molecules that lie along their respective velocity vectors (dashed lines) is given by b and is called the impact parameter. The relative kinetic energy in the direction of the line that joins the centers of the two spheres is E_{loc}.

To determine $\sigma_r(E_r)$ for the line-of-centers model, consider the geometry shown in Figure 30.1. The A and B molecules travel toward one another with a relative velocity $\mathbf{u}_r = \mathbf{u}_A - \mathbf{u}_B$ and hence a relative kinetic energy of $E_r = (1/2)\mu u_r^2$. We now draw a line through the center of each molecule that lies along the velocity vector of that molecule (the dashed lines in Figure 30.1). The *impact parameter b* is defined as the perpendicular distance between these two dashed lines. We see that the two molecules will collide only if the impact parameter is less than the sum of the radii of the colliding molecules, or less than the collision diameter d_{AB}. In an equation, a collision will occur if $b < r_A + r_B = d_{AB}$. If the impact parameter exceeds the collision diameter, $b > d_{AB}$, the molecules will miss one another as they pass. For a fixed relative kinetic energy between A and B, the kinetic energy along the line of centers when a collision occurs depends on the impact parameter. For example if $b = 0$, the two molecules hit head-on, and all the relative kinetic energy lies along the line of centers, $E_{loc} = E_r$. At the other extreme, when $b \geq d_{AB}$, none of the relative kinetic energy lies along the line of centers as the two reactants pass one another without colliding and the collision cross section must then be equal to zero.

The derivation of the energy dependence of the reaction cross section for the line-of-centers model is a bit involved geometrically. The final result is

$$\sigma_r(E_r) = \begin{cases} 0 & E_r < E_0 \\ \pi d_{AB}^2 \left(1 - \dfrac{E_0}{E_r}\right) & E_r \geq E_0 \end{cases} \tag{30.14}$$

Note that Equation 30.14 for $\sigma_r(E_r)$ differs from Equation 30.12 by a multiplicative factor of $(1 - E_0/E_r)$.

The measured energy dependence of the reaction cross section for the chemical reaction,

$$Ne^+(g) + CO(g) \Longrightarrow Ne(g) + C^+(g) + O(g)$$

is shown in Figure 30.2. The cross section for this reaction exhibits a threshold energy of about 8 kJ·mol^{-1}. Below a collision energy of about 8 kJ·mol^{-1}, no reaction occurs. Above this energy, the reaction cross section increases with increasing collision energy and then levels off when the collision energy is greater than about 60 kJ·mol^{-1}. This type of behavior is consistent with that predicted by the lines-of-center model for the reaction cross section (Equation 30.14).

Substituting Equation 30.14 for the cross section into Equation 30.11 gives the following expression for the rate constant (Problem 30–3).

$$k = \left(\frac{8k_B T}{\mu\pi}\right)^{1/2} \pi d_{AB}^2 e^{-E_0/k_B T} = \langle u_r\rangle \sigma_{AB} e^{-E_0/k_B T} \tag{30.15}$$

Note that this expression for k differs from Equation 30.13 by a factor of $(1 + E_0/k_B T)$.

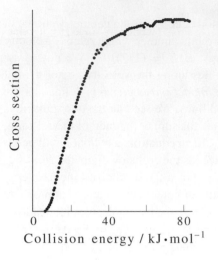

FIGURE 30.2
The experimentally determined cross section as a function of collision energy for the reaction $Ne^+(g) + CO(g) \Rightarrow Ne(g) + C^+(g) + O(g)$. The reaction cross section shows a threshold energy of ≈ 8 kJ·mol^{-1}. The dependence of the reaction cross section on the relative kinetic energy of the collision is consistent with the line-of-centers model.

EXAMPLE 30–3
How is the threshold energy E_0 for the line-of-centers collision theory rate constant related to the Arrhenius activation energy, E_a?

SOLUTION: The Arrhenius activation energy, E_a, is given by (see Equation 28.55)

$$E_a = k_B T^2 \frac{d \ln k}{dT}$$

Using the rate constant from Equation 30.15 in Equation 28.55 gives us

$$E_a = k_B T^2 \frac{d}{dT} \left\{ \ln \left[\left(\frac{8k_B T}{\pi \mu} \right)^{1/2} \pi d_{AB}^2 \right] - \frac{E_0}{k_B T} \right\}$$

$$= k_B T^2 \frac{d}{dT} \left\{ \ln T^{1/2} - \frac{E_0}{k_B T} + \text{ terms not involving } T \right\}$$

$$= E_0 + \tfrac{1}{2} k_B T$$

Combining this result with the collision-theory rate constant given by Equation 30.15 also shows that the Arrhenius A factor is given by

$$A = \langle u_r \rangle \sigma_{AB} e^{1/2}$$

Table 30.1 lists observed and calculated pre-exponential factors for several bi-molecular reactions. The calculated values using Equation 30.15 exceed the experimentally determined values of the pre-exponential factors, often by several orders of magnitude. In recent years, the function $\sigma_r(E_r)$ has been determined experimentally for many chemical reactions over a large range of collision energies. Although most reactions exhibit a threshold energy, the general shape of the energy dependence of

TABLE 30.1

Experimental Arrhenius pre-exponential factors and activation energies for some bimolecular gas-phase reactions. The experimental pre-exponential factors are compared with those calculated using hard-sphere collision theory.

Reaction	$A/dm^3 \cdot mol^{-1} \cdot s^{-1}$		$E_a/kJ \cdot mol^{-1}$
	Observed	Calculated	
$NO(g) + O_3(g) \rightarrow NO_2(g) + O_2(g)$	7.94×10^8	5.01×10^{10}	10.5
$NO(g) + O_3(g) \rightarrow NO_3(g) + O(g)$	6.31×10^9	6.31×10^{10}	29.3
$F_2(g) + ClO_2(g) \rightarrow FClO_2(g) + F(g)$	3.16×10^7	5.01×10^{10}	35.6
$2\,ClO(g) \rightarrow Cl_2(g) + O_2(g)$	6.31×10^7	2.50×10^{10}	0
$H_2(g) + C_2H_4(g) \rightarrow C_2H_6(g)$	1.24×10^6	7.30×10^{11}	180

the reaction cross section is not well approximated by Equation 30.14. The conclusion from these studies is that the molecular details of gas-phase reactions cannot be described accurately by the simple hard-sphere collision theories we have discussed so far.

30–3. The Rate Constant for a Gas-Phase Chemical Reaction May Depend on the Orientations of the Colliding Molecules

The data in Table 30.1 show that hard-sphere collision theory does not accurately account for the magnitude of the Arrhenius A factor. One of the fundamental flaws of this model is the assumption that every collision of sufficient energy is reactive. In addition to an energy requirement, the reacting molecules may need to collide with a specific orientation for the chemical reaction to occur. Several experimental studies have verified the importance of molecular orientation in determining whether or not a collision is reactive. For example, consider the reaction

$$Rb(g) + CH_3I(g) \Longrightarrow RbI(g) + CH_3(g)$$

Experimental studies reveal that this reaction occurs only when the rubidium atom collides with the iodomethane molecule in the vicinity of the iodine atom (Figure 30.3). Collisions between the rubidium atom and the methyl end of the molecule do not lead to reaction. This set of collision geometries is indicated by the cone of nonreactivity in Figure 30.3. Because hard-sphere collision theory does not take the collision geometry into account, the theory must overestimate the rate constant for reactions that are orientation dependent. Such a steric requirement is physically important for many chemical reactions; however, steric factors alone cannot account for the significant differences observed between the experimental and calculated Arrhenius A factors in Table 30.1.

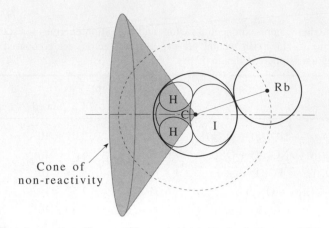

FIGURE 30.3
The elementary reaction $Rb(g) + CH_3I(g) \Rightarrow RbI(g) + CH_3(g)$ occurs only for a subset of the possible collision geometries. The rubidium atom must collide with the iodomethane molecule in the vicinity of the iodine atom for a reaction to occur. For those reactants that collide within the cone of nonreactivity, no reaction occurs.

30–4. The Internal Energy of the Reactants Can Affect the Cross Section of a Reaction

The reaction cross section for many gas-phase reactions is dependent on the internal energy of the reacting molecules. Consider the data plotted in Figure 30.4. In this figure, the reaction cross section for the reaction of the hydrogen molecular ion with atomic helium

$$H_2^+(g) + He(g) \Longrightarrow HeH^+(g) + H(g)$$

is plotted as a function of total energy. The total energy available for this reaction is the combined kinetic and vibrational energy of the reactants. Each curve plotted in Figure 30.4 corresponds to the reactant $H_2^+(g)$ in a specific vibrational state. Several interesting features appear in these data. For the vibrational states $v = 0$ to $v = 3$, there is a threshold energy of about 70 kJ·mol^{-1}. For $H_2^+(g)$ molecules that have a vibrational quantum number of $v = 0$ to $v = 3$, the total vibrational energy is less than E_0. Additional translational energy is required for the reaction to occur, and the data reveal an energy threshold for the reaction. However, $H_2^+(g)$ molecules that have a vibrational quantum number of $v = 4$ or $v = 5$ have sufficient internal energy to react because $E_{vib} > E_0$, and so additional translational energy is not needed for these molecules to react. This is why a threshold energy is not observed when the $H_2^+(g)$ molecule has a vibrational quantum number of $v \geq 4$. At a constant total energy, the reaction cross sections for $H_2^+(g)$ in the $v = 4$ or $v = 5$ level are much larger than that observed for this reactant in the $v = 0$ to $v = 3$ levels. We see that for a constant total energy, the value of σ_r depends strongly on the vibrational state of the reactant.

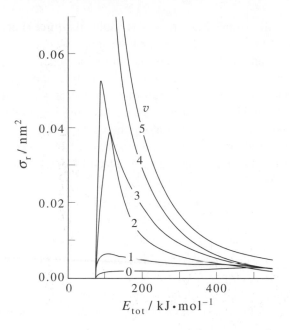

FIGURE 30.4
The reaction cross section for $H_2^+(g) + He(g) \Rightarrow HeH^+(g) + H(g)$ is plotted as a function of total energy. The different curves plotted correspond to the reactant molecule $H_2^+(g)$ in different vibrational states; v is the vibrational quantum number. At a fixed total energy, the reaction cross section depends on the vibrational state of $H_2^+(g)$, demonstrating the importance of the internal modes on the cross section for the reaction.

We know from our study of quantum mechanics that the internal energy of a molecule is distributed among the discrete rotational, vibrational, and electronic states. Data such as that shown in Figure 30.4 tell us that chemical reactivity depends not only on the total energy of the reacting molecules but also on how that energy is distributed among these internal energy levels. Simple hard-sphere collision theory considers only the translational energy of the reacting molecules. Energy can also be exchanged between the different degrees of freedom during the reactive collision; for example, vibrational energy can be converted into translational energy and vice versa. To understand gas-phase reaction dynamics we must consider how all the degrees of freedom of the reacting systems evolve during a reactive collision.

30–5. A Reactive Collision Can Be Described in a Center-of-Mass Coordinate System

Consider the collision and subsequent scattering process for the bimolecular reaction

$$A(g) + B(g) \Longrightarrow C(g) + D(g)$$

For simplicity, we will assume there are no intermolecular forces between the separated reactants and separated products. Before the collision, molecules A and B are traveling with a velocity of $\mathbf{u}_A$ and $\mathbf{u}_B$, respectively. The collision generates molecules C and D, which then move away from each other with velocities $\mathbf{u}_C$ and $\mathbf{u}_D$, respectively. We will describe the collision process in the center-of-mass coordinate system. The idea is to view the collision from the center of mass of the two colliding molecules. Recall from Section 5–2 that the center of mass lies along the vector $\mathbf{r} = \mathbf{r}_A - \mathbf{r}_B$ that connects the centers of the two colliding molecules. The location of $\mathbf{R}$, the center of mass, along this vector depends on the masses of the two molecules, and is defined by

$$\mathbf{R} = \frac{m_A \mathbf{r}_A + m_B \mathbf{r}_B}{M} \tag{30.16}$$

where M is the total mass, $M = m_A + m_B$. If the two masses are equal, then $m_A = m_B$, and the center of mass sits halfway between A and B on the vector $\mathbf{r}$. If $m_A > m_B$ then the center of mass sits closer to A than to B.

The velocity is the time derivative of the position vector and therefore, $\mathbf{u}_{cm}$, the velocity of the center of mass, is defined by the time derivative of Equation 30.16, or

$$\mathbf{u}_{cm} = \frac{m_A \mathbf{u}_A + m_B \mathbf{u}_B}{M} \tag{30.17}$$

The total kinetic energy is given by the sum of the kinetic energies of the reactants.

$$KE_{react} = \tfrac{1}{2} m_A u_A^2 + \tfrac{1}{2} m_B u_B^2 \tag{30.18}$$

Example 30–4 shows that Equation 30.18 can be rewritten as

$$KE_{react} = \tfrac{1}{2} M u_{cm}^2 + \tfrac{1}{2} \mu u_r^2 \tag{30.19}$$

where μ is the reduced mass and $u_r = |\mathbf{u}_r| = |\mathbf{u}_A - \mathbf{u}_B|$ is the relative speed of the two molecules. If there are no external forces acting on the reactant molecules, the kinetic energy of the center of mass is constant (Section 5.2).

EXAMPLE 30–4
Show that Equation 30.19 follows from Equation 30.18.

SOLUTION: We start with Equation 30.18

$$KE_{react} = \tfrac{1}{2} m_A u_A^2 + \tfrac{1}{2} m_B u_B^2 \tag{1}$$

and want to rewrite this equation in terms of u_{cm} and u_r. The equations for $\mathbf{u}_{cm}$ and $\mathbf{u}_r$ are

$$\mathbf{u}_{cm} = \frac{m_A}{M} \mathbf{u}_A + \frac{m_B}{M} \mathbf{u}_B$$

and

$$\mathbf{u}_r = \mathbf{u}_A - \mathbf{u}_B$$

If we multiply $\mathbf{u}_{cm}$ by M/m_B and then add the result to $\mathbf{u}_r$, we obtain

$$\mathbf{u}_A = \mathbf{u}_{cm} + \frac{m_B}{M}\mathbf{u}_r \tag{2}$$

Similarly, multiplying $\mathbf{u}_{cm}$ by M/m_A and subtracting from $\mathbf{u}_r$ gives us

$$\mathbf{u}_B = \mathbf{u}_{cm} - \frac{m_A}{M}\mathbf{u}_r \tag{3}$$

Substituting Equations 2 and 3 into Equation 1 gives

$$
\begin{aligned}
KE_{react} &= \frac{m_A}{2}\left(\mathbf{u}_{cm} + \frac{m_B}{M}\mathbf{u}_r\right)^2 + \frac{m_B}{2}\left(\mathbf{u}_{cm} - \frac{m_A}{M}\mathbf{u}_r\right)^2 \\
&= \tfrac{1}{2}Mu_{cm}^2 + \tfrac{1}{2}\mu u_r^2
\end{aligned}
$$

where μ is the reduced mass, $\mu = m_A m_B/M$.

Figure 30.5 shows a series of snapshots of a bimolecular collision as viewed along the motion of the center of mass. Figure 30.5 implies that the center-of-mass motion is constant during the entire collision, a fact we will soon prove. The relative velocity,

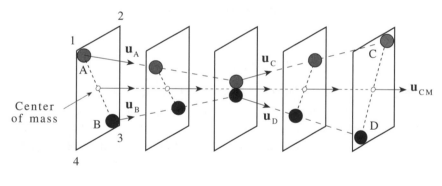

—————— Increasing Time ——————▶

FIGURE 30.5
The details of a bimolecular collision viewed at various times along the center-of-mass motion. The velocities of the reactants A and B and products C and D can be divided into a component that lies along the center of mass and a relative velocity component that lies in the plane defined by 1234. The center-of-mass velocity remains constant before, during, and after the collision, and therefore the molecules remain in a plane that travels at the speed of the center of mass. Only the relative component of the velocity is important in determining the energy available for the reaction. In the left two snapshots, molecules A and B are approaching one another in the 1234 plane. The collision occurs in the middle snapshot. The right two snapshots show the products moving apart in the 1234 plane. The direction of the relative velocity of the reactants and products can be different.

on the other hand, changes during the collision. The colliding molecules move in the plane defined by 1234, which itself is moving at the velocity of the center of mass. Equation 30.19 tells us that the kinetic energy is composed of two contributions; one due to the motion of the center of mass and one due to the relative motion of the two colliding molecules. Only the component of the kinetic energy that lies along the collision direction, or $(1/2)\mu u_r^2$, is available for the reaction. The center-of-mass velocity does not affect the distance between the two reacting molecules and therefore has no effect on the chemical reaction. After the collision, the center of mass is given by

$$\mathbf{R} = \frac{m_C \mathbf{r}_C + m_D \mathbf{r}_D}{M} \tag{30.20}$$

and the center-of-mass velocity is given by

$$\mathbf{u}_{cm} = \frac{m_C \mathbf{u}_C + m_D \mathbf{u}_D}{M} \tag{30.21}$$

As shown in Figure 30.5, the product molecules of a collision can move away from the center of mass in a direction different from that by which the reactants approached the center of mass. The kinetic energy of the products is (Problem 30–12)

$$KE_{prod} = \tfrac{1}{2} M u_{cm}^2 + \tfrac{1}{2} \mu' u_r'^2 \tag{30.22}$$

where μ' and u_r' are the reduced mass and relative speed of the product molecules. There are no primes on M and u_{cm} because the total mass is conserved and the center-of-mass velocity does not change during the collision. Linear momentum must be conserved in the collision, so

$$m_A \mathbf{u}_A + m_B \mathbf{u}_B = m_C \mathbf{u}_C + m_D \mathbf{u}_D \tag{30.23}$$

Using Equation 30.23, we see that Equations 30.21 and 30.17 are the same, confirming that the velocity of the center of mass is unaffected by the reactive collision. The energy associated with the motion of the center of mass is therefore constant, and from this point on, we will ignore this constant contribution to the total kinetic energy.

Because energy must be conserved,

$$E_{react,int} + \tfrac{1}{2} \mu u_r^2 = E_{prod,int} + \tfrac{1}{2} \mu' u_r'^2 \tag{30.24}$$

where $E_{react,int}$ and $E_{prod,int}$ are the total internal energies of the reactants and products, respectively. This internal energy takes into account all the degrees of freedom other than translation.

EXAMPLE 30–5
Consider the reaction

$$F(g) + D_2(g) \Longrightarrow DF(g) + D(g)$$

where the relative kinetic energy of the reactants is $KE_{react} = 7.62 \text{ kJ·mol}^{-1}$. Treating the reactants and products as hard spheres gives $E_{prod,int} - E_{react,int} = D_e(D_2) - D_e(DF) = -140 \text{ kJ·mol}^{-1}$. Calculate the relative speed of the products. Then use Equations 1 and 2 of Problem 30–11 to determine the values of $|\mathbf{u}_{DF} - \mathbf{u}_{cm}|$ and $|\mathbf{u}_D - \mathbf{u}_{cm}|$, the speeds of each product relative to the center of mass. [Recall that D_e is the energy difference between the minimum of the potential-energy curve and the dissociated atoms in their ground states (see Section 13–6).]

SOLUTION: The relative kinetic energy of the reactants corresponds to a relative speed of

$$u_r = \left(\frac{2 \, KE_{react}}{\mu} \right)^{1/2}$$

The reduced mass of the reactants is

$$\mu = \frac{m_{D_2} m_F}{m_{D_2} + m_F} = 5.52 \times 10^{-27} \text{ kg}$$

so

$$u_r = \left[\frac{(2)(7.62 \times 10^3 \text{ J·mol}^{-1})}{(5.52 \times 10^{-27} \text{ kg})(6.022 \times 10^{23} \text{ mol}^{-1})} \right]^{1/2}$$

$$= 2.14 \times 10^3 \text{ m·s}^{-1}$$

The relative speed of the products can now be found using Equation 30.24. Solving this equation for u_r' gives us

$$u_r' = \left(\frac{\mu}{\mu'} u_r^2 - \frac{2(E_{prod,int} - E_{react,int})}{\mu'} \right)^{1/2} \tag{1}$$

where μ' is the reduced mass of the products

$$\mu' = \frac{m_{DF} m_D}{m_{DF} + m_D} = 3.05 \times 10^{-27} \text{ kg}$$

Thus

$$u_r' = \left[\frac{5.52 \times 10^{-27} \text{ kg}}{3.05 \times 10^{-27} \text{ kg}} (2.14 \times 10^3 \text{ m·s}^{-1})^2 \right.$$

$$\left. - \frac{(2)(-1.40 \times 10^5 \text{ J·mol}^{-1})}{(3.05 \times 10^{-27} \text{ kg})(6.022 \times 10^{23} \text{ mol}^{-1})} \right]^{1/2}$$

$$= 1.27 \times 10^4 \text{ m·s}^{-1}$$

The speeds of the products relative to the center of mass, $|\mathbf{u}_{DF} - \mathbf{u}_{cm}|$ and $|\mathbf{u}_D - \mathbf{u}_{cm}|$, are given by Equations 1 and 2 of Problem 30.11.

$$|\mathbf{u}_{DF} - \mathbf{u}_{cm}| = \frac{m_D}{M} |\mathbf{u}_r'| = \frac{m_D}{M} u_r'$$

$$= \frac{2.014 \text{ amu}}{23.03 \text{ amu}} (1.27 \times 10^4 \text{ m·s}^{-1})$$

$$= 1.11 \times 10^3 \text{ m·s}^{-1}$$

and

$$|\mathbf{u}_D - \mathbf{u}_{cm}| = \frac{m_{DF}}{M}|\mathbf{u}'_r| = \frac{m_{DF}}{M}u'_r$$

$$= \frac{21.01 \text{ amu}}{23.03 \text{ amu}}(1.27 \times 10^4 \text{ m·s}^{-1})$$

$$= 1.16 \times 10^4 \text{ m·s}^{-1}$$

The energy and momentum conservation laws enable us to define the velocity of the products but not the angle between the vectors $\mathbf{u}_r$ and $\mathbf{u}'_r$. In principle, the product molecules can scatter in any direction from the point of collision. We will learn, however, that many reactions exhibit highly anisotropic scattering angles. Such data provide unique insight into the molecular details of the reactive collision. Before we examine how we theoretically describe the angular distribution of the products, we will discuss some of the experimental approaches used to provide data on reactive collisions.

30–6. Reactive Collisions Can Be Studied Using Crossed Molecular Beam Machines

One of the most important experimental techniques used to study the molecular dynamics of bimolecular gas-phase reactions is the *crossed molecular beam method*. The basic design of a crossed molecular beam apparatus is shown in Figure 30.6a. The experimental device is designed to cross a beam of A molecules with a beam of B molecules at a specific location inside a large vacuum chamber. The product molecules are then detected using a mass spectrometer. In some crossed molecular beam machines, the detector can be rotated in the plane defined by the two molecular beams, thereby allowing the measurement of the angular distribution of the scattered products. The mass spectrometer can also be set to measure a specific molecular mass so that individual product molecules are detected.

Supersonic molecular beams are used to produce the velocities of the molecules in the reactant beams. A schematic diagram of a supersonic molecular beam source is shown in Figure 30.6b. A supersonic molecular beam can be generated by taking a high-pressure, dilute mixture of the reactant molecule of interest in an inert carrier gas (He and Ne are commonly used) and pulsing the mixture through a small nozzle into the vacuum chamber. A small pinhole, known as a skimmer, is located a few centimeters away from where the molecules enter the vacuum chamber through the nozzle. Only those molecules that pass through the small hole in the skimmer enter the remainder of the vacuum chamber. This procedure creates a collimated beam of molecules. The beam is supersonic because the pressure conditions inside the vacuum chamber are

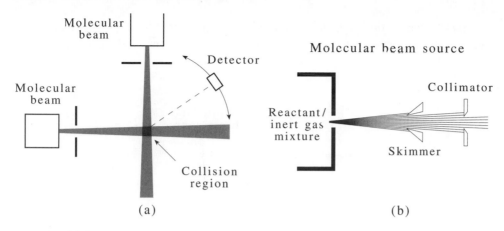

FIGURE 30.6
(a) A schematic diagram of a crossed molecular beam machine. Each reactant is introduced into the vacuum chamber by a molecular beam source. The two molecular beams collide at the collision region. The product molecules then travel away from the collision region. A mass spectrometer detector is located at a fixed distance away from the collision region and is used to detect product molecules. The detector can be moved so that the number of molecules leaving the collision region at different angles can be determined. (b) A schematic drawing of a supersonic molecular beam source. The reactant is expanded along with an inert gas through a small orifice into the vacuum chamber. A skimmer is used so that a collimated beam of molecules is directed toward the collision region.

such that the molecules in the beam move at a speed greater than the speed of sound (Problems 30–13 and 30–14).

A supersonic molecular beam has several important advantages that make it ideal for crossed-beam studies. Figure 30.7 shows a plot of the Maxwell-Boltzmann distribution of molecular speeds for $N_2(g)$ at 300 K and the speed distribution observed from a supersonic molecular beam of $N_2(g)$ in helium at 300 K. The supersonic expansion generates a collection of molecules with a high translational energy but a very small spread in molecular speeds. In addition, molecules can be prepared with low rotational and vibrational energies.

Thus, in crossed molecular-beam experiments, u_r is specified by the velocities of the reactants. By varying the conditions under which the molecular beams are generated, experimentalists can change the relative velocities of the reactants and thereby change the collision energy. By measuring the product yield as a function of collision energy, the energy-dependent reaction cross section, $\sigma_r(E_r)$, can be determined.

The product molecules formed from reactive collisions travel away from the collision region. Their motion is determined by the conservation laws for mass, linear momentum, and energy. If we measure the number of molecules of a particular reaction product that arrive at the detector as a function of time after the collision, we can resolve the velocity distribution of the product molecules. If we measure the total number of product molecules as a function of the scattering angle, we can determine the angular distribution of the product molecules. From these two types of experiments, many of the molecular details of gas-phase reactive collisions can be determined.

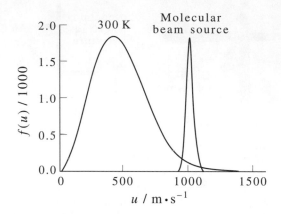

FIGURE 30.7
The Maxwell-Boltzmann velocity distribution of $N_2(g)$ molecules at 300 K is compared with the velocity distribution generated by a supersonic expansion of a gaseous mixture of $N_2(g)$ in He(g) at 300 K. The molecular beam produces a narrow, nonequilibrium velocity distribution.

30–7. The Reaction $F(g) + D_2(g) \Rightarrow DF(g) + D(g)$ Can Produce Vibrationally Excited DF(g) Molecules

In this and several of the following sections, we will be concerned with the reaction

$$F(g) + D_2(g) \Longrightarrow DF(g) + D(g) \tag{30.25}$$

Figure 30.8 shows a one-dimensional energy diagram for this reaction. The energy diagram reflects only the changes in potential energy. Diagrams that indicate how the potential energy changes as the reaction proceeds along the reaction coordinate are called *potential energy diagrams*. The energy of the lowest vibrational state of $D_2(g)$ and the energies of the first six vibrational states of DF(g) are also shown. In drawing these energy states, we have assumed that the vibrational motion of both $D_2(g)$ and DF(g) is harmonic.

We consider here the reaction energetics when the reactant $D_2(g)$ is in its ground vibrational state, with an internal energy of $(1/2)h\nu_{D_2}$. Figure 30.8 shows that the reaction can produce DF(g) in several of its low vibrational states. We will write the overall reaction as

$$F(g) + D_2(v = 0) \Longrightarrow DF(v) + D(g) \tag{30.26}$$

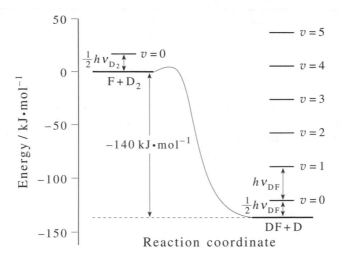

FIGURE 30.8
A potential-energy diagram for the reaction $F(g) + D_2(v = 0) \Rightarrow DF(v) + D(g)$. The vibrational states of the $D_2(g)$ reactant and $DF(g)$ product are indicated and labeled by their vibrational quantum numbers. The potential-energy diagram shows that the difference between the ground electronic state energies of $D_2(g)$ and $DF(g)$ is $D_e(D_2) - D_e(DF) = -140 \text{ kJ} \cdot \text{mol}^{-1}$. The reaction has an activation energy barrier of about $7 \text{ kJ} \cdot \text{mol}^{-1}$.

where the vibrational state of the reactant is specified but the vibrational state(s) of the product is left unspecified and must be determined experimentally. The total energy available for the reaction, E_{tot}, is the sum of the internal energy of the reactants, E_{int}, and the relative translational energy of the reactants, E_{trans}. Because energy must be conserved,

$$E_{tot} = E_{trans} + E_{int} = E'_{trans} + E'_{int} \tag{30.27}$$

where E'_{int} and E'_{trans} are the internal energy (rotational, vibrational, and electronic) and relative translational energy of the product molecules, respectively. For a given total energy, a change in the internal energy of the products, E'_{int}, must be balanced by a corresponding change in their relative translational energy, E'_{trans}. Thus for a fixed total collision energy, $DF(g)$ molecules generated in different vibrational states move away from the collision region with different velocities. We will find it useful to consider separately the rotational, vibrational, and electronic contributions to E_{int} and E'_{int}. We can write Equation 30.27 as

$$E_{tot} = E_{trans} + E_{rot} + E_{vib} + E_{elec} = E'_{trans} + E'_{rot} + E'_{vib} + E'_{elec} \tag{30.28}$$

For the reaction given by Equation 30.26, where the reactants and products are in their ground electronic states, $E_{elec} = -D_e(D_2)$ and $E'_{elec} = -D_e(DF)$.

EXAMPLE 30–6

Consider the reaction

$$F(g) + D_2(v = 0) \Longrightarrow DF(v) + D(g)$$

where the relative translational energy of the reactants is 7.62 kJ·mol^{-1}. Assuming the reactants and products are in their ground electronic states and ground rotational states, determine the range of possible vibrational states of the product, DF(g). Treat the vibrational motion of both $D_2(g)$ and DF(g) as harmonic with $\tilde{v}_{D_2} = 2990$ cm^{-1} and $\tilde{v}_{DF} = 2907$ cm^{-1}. [$D_e(D_2) - D_e(DF) = -140$ kJ·mol^{-1}.]

SOLUTION: Energy must be conserved by this reaction. Using Equation 30.28 and assuming that the reactants and products are in their ground electronic and rotational states gives us

$$E_{trans} + E_{vib} - D_e(D_2) = E'_{trans} + E'_{vib} - D_e(DF)$$

Solving for E'_{trans} gives us

$$E'_{trans} = E_{trans} + E_{vib} - E'_{vib} - [D_e(D_2) - D_e(DF)] \tag{1}$$

The reactant $D_2(g)$ is in its ground vibrational state, so $E_{vib} = \frac{1}{2}hv_{D_2} = 17.9$ kJ·mol^{-1}. Thus, Equation 1 gives

$$E'_{trans} = 7.62 \text{ kJ·mol}^{-1} + 17.9 \text{ kJ·mol}^{-1} - E'_{vib} + 140 \text{ kJ·mol}^{-1}$$

$$= 166 \text{ kJ·mol}^{-1} - E'_{vib}$$

Translational energy is an intrinsically positive quantity, so the reaction occurs only if $E'_{vib} < 166$ kJ·mol^{-1}. Assuming that the vibrational motion of DF(g) is harmonic gives us

$$E'_{vib} = \left(v + \tfrac{1}{2}\right)hv_{DF} = \left(v + \tfrac{1}{2}\right)(34.8 \text{ kJ·mol}^{-1}) < 166 \text{ kJ·mol}^{-1}$$

from which we find that $v \leq 4$. We will see shortly that this result is in agreement with experimental data.

30–8. The Velocity and Angular Distribution of the Products of a Reactive Collision Provide a Molecular Picture of the Chemical Reaction

We will now examine the crossed molecular beam data for the reaction described by $F(g) + D_2(v = 0) \Longrightarrow DF(v) + D(g)$ for the case in which the relative translational energy of the reactants is 7.62 kJ·mol^{-1}. In Example 30–6, we found that for this value of the relative translational energy of the reactants, the product DF(g) could be produced in the vibrational states from $v = 0$ through $v = 4$. We will now describe this reaction using the center-of-mass coordinate system presented in Section 30–5.

After a reactive collision between F(g) and D$_2$(g), the velocities of both the DF(g) molecule and the D(g) atom are determined by the dynamics of the reactant collision and point away from the center of mass. We know from Section 30–5 that the product velocities, $\mathbf{u}'_{DF}$ and $\mathbf{u}'_D$, are not independent and are related by conservation laws. In principle, the products can separate in any direction consistent with the conservation laws for mass, momentum, and energy. In practice, only a small subset of these allowed directions are observed for this reaction. We need to find a way to describe the angular dependence of how the product molecules leave the site of the reactive collision.

Figure 30.9 examines the collision between molecules A and B viewed along the relative velocity vector for a fixed value of the impact parameter b. For simplicity, we take molecule B to be fixed in space and then look at how it collides with an approaching molecule A that is moving at the relative velocity $\mathbf{u}_r$. Because molecule B is spherical, the scattering center is cylindrically symmetric to molecule A. This fact means that the angle ϕ for the collision between molecules A and B takes on all possible values with equal probability. Unlike the angle ϕ, the angle θ in Figure 30.9 depends on the details of the chemical reaction process. We will see that this angle can be determined experimentally from crossed molecular beam data.

Before we examine the experimental crossed molecular beam data for the F(g) + D$_2$(g) reaction, we need to consider how the experimental data depend on the internal vibrational energy of the DF(g) molecule. Only a fixed amount of energy is available to the products, and this energy must be partitioned between the internal states and the translational kinetic energy of the product molecules. Therefore, the translational kinetic energy, and hence the velocity, of the DF(g) product must decrease when excited vibrational states become populated.

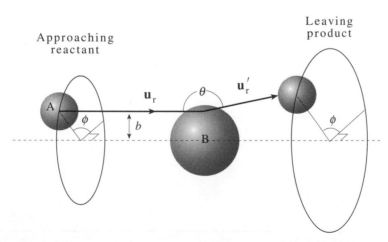

FIGURE 30.9
The angular distribution resulting from the bimolecular collision between molecules A and B as seen from the B molecule before and after the collision. For a fixed value of b, the impact parameter, the reactants and products take on all possible angles ϕ with equal probability, thereby forming a cone around the relative velocity vector $\mathbf{u}_r$. The angle θ, however, depends on the dynamics of the reaction and must be determined experimentally.

EXAMPLE 30–7

Re-examine the $F(g) + D_2(v = 0) \Longrightarrow DF(v) + D(g)$ reaction, where the relative translational energy of the reactants is 7.62 kJ·mol^{-1} and $D_e(D_2) - D_e(DF) = -140$ kJ·mol^{-1}. Determine the value of $|\mathbf{u}_{DF} - \mathbf{u}_{cm}|$ for DF(g) molecules produced in the vibrational levels from $v = 0$ through $v = 4$. (Assume that the reactants and products are in their ground electronic and rotational states and that the vibrational motion of both D_2 and DF is harmonic with $\tilde{\nu}_{D_2} = 2990$ cm^{-1} and $\tilde{\nu}_{DF} = 2907$ cm^{-1}.)

SOLUTION: Assuming that the reactants are in their ground electronic and rotational states, Equation 30.28 gives us

$$E'_{trans} + E'_{vib} = E_{trans} + E_{vib} - [D_e(D_2) - D_e(DF)]$$
$$= 7.62 \text{ kJ·mol}^{-1} + 17.9 \text{ kJ·mol}^{-1} + 140 \text{ kJ·mol}^{-1}$$
$$= 166 \text{ kJ·mol}^{-1}$$

If we assume that the vibrational motion of the DF(g) molecule is harmonic, then

$$E'_{trans} + E'_{vib} = \tfrac{1}{2}\mu' u'^2_r + \left(v + \tfrac{1}{2}\right)(34.8 \text{ kJ·mol}^{-1}) = 166 \text{ kJ·mol}^{-1} \qquad (1)$$

The reduced mass of the products is $\mu' = 1.84 \times 10^{-3}$ kg·mol^{-1} (Example 30–5). Solving Equation 1 for u'_r gives

$$u'_r = \left\{ \left(\frac{2}{1.84 \times 10^{-3} \text{ kg·mol}^{-1}} \right) \left(1.66 \times 10^5 \right. \right.$$
$$\left. \left. - \left[v + \tfrac{1}{2}\right][3.48 \times 10^4] \right) \text{ J·mol}^{-1} \right\}^{1/2}$$

Problem 30–11 shows that the relative velocity of the DF(g) molecule and the center-of-mass is given by

$$|\mathbf{u}_{DF} - \mathbf{u}_{cm}| = \frac{m_D}{M}|\mathbf{u}'_r| = \frac{m_D}{M}u'_r$$

The values of u'_r and $|\mathbf{u}_{DF} - \mathbf{u}_{cm}|$ for DF(g) produced in the vibrational states $v = 0$ through $v = 4$ are tabulated below.

v	$u'_r/10^4$ m·s^{-1}	$|\mathbf{u}_{DF} - \mathbf{u}_{cm}|/10^2$ m·s^{-1}
0	1.27	11.1
1	1.11	9.71
2	0.927	8.11
3	0.693	6.06
4	0.320	2.80

Example 30–7 shows that the velocity of the DF(g) molecule depends on the vibrational state of the product. This means that in a crossed molecular beam apparatus, the time required for the DF(g) molecule to travel from the collision region to the mass spectrometer will depend on its vibrational state. Figure 30.10 illustrates the type of

data observed if the mass spectrometer signal is plotted as a function of time. The graph reveals four distinguishable peaks. These four peaks correspond to product molecules that traveled away from the reaction site in the same direction but with different speeds, and therefore arrived at the mass spectrometer at different times. The first peak corresponds to those molecules that left the reaction site with the highest speed. These molecules have the largest translational energy and therefore the least amount of internal vibrational energy. Subsequent peaks correspond to molecules with a smaller translational energy and greater internal vibrational energy. The area under a peak is proportional to the total number of product molecules in that vibrational state. If we compare the areas of the different peaks, the relative populations of the different vibrational states can be determined.

The dependence of reaction product formation on the angle θ (shown in Figure 30.9) can be determined by moving the detector in the plane defined by the two molecular beams (see Figure 30.6a). Thus, we can determine the relative populations of each vibrational state for all possible scattering angles. Rather than in a three-dimensional picture that depicts all the reaction trajectories, the data are commonly represented in a two-dimensional polar contour plot. Figure 30.11 shows the contour plot for the reaction between F(g) and $D_2(v = 0)$, for which the relative translational energy of the reactants is 7.62 kJ·mol^{-1}. The center of mass sits at the center of the contour plot. The distance from the origin to any point in the polar plot is the speed of the DF(g) molecule relative to the center of mass, $|\mathbf{u}_{DF} - \mathbf{u}_{cm}|$. Below the plot, the arrows indicate the directions with which the reactants approach each other. The horizontal axis of the contour plot lies along the relative velocity vector of the reactants. The

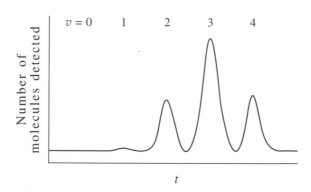

FIGURE 30.10
The number of DF(g) molecules detected by a mass spectrometer is plotted as a function of time after the reaction between F(g) and $D_2(g)$ in a crossed molecular beam study. The initial relative kinetic energy of the reactants is 7.62 kJ·mol^{-1}. The DF(g) molecules with the highest translational energy, and therefore the least amount of vibrational energy, arrive at the detector first. Because the total energy is constant, DF(g) molecules produced in an excited vibrational state must have a lower translational energy. Therefore, the different peaks observed in the plot correspond to DF(g) molecules in different vibrational states. There is no peak at $v = 0$ because no DF(g) molecules are produced in the $v = 0$ vibrational state under these conditions.

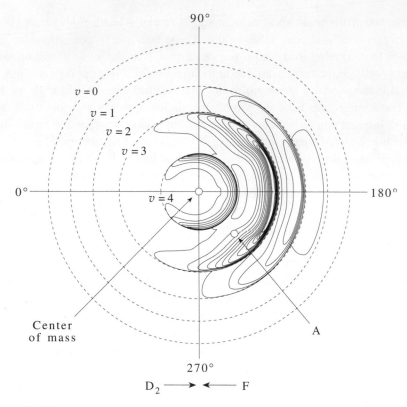

FIGURE 30.11

A contour map of the angular and speed distributions for the product molecule DF for the reaction $F(g) + D_2(v = 0)$, for which the relative translational energy of the reactants is $7.62 \, \text{kJ} \cdot \text{mol}^{-1}$. The center of mass is fixed at the origin. The dashed circles correspond to the maximum relative speeds a DF(g) molecule can have for the indicated vibrational state. The data reveal that the product molecules preferentially scatter back in the direction of the incident fluorine atom, a scattering angle of $\theta = 180°$. The arrows at the bottom of the figure show the direction with which each reactant molecule approaches each other.

angles indicated in Figure 30.11 are the scattering angles θ. In an atom-molecule reaction, we take $\theta = 0°$ to lie along the direction defined by the trajectory of the incident atom. An angle of $\theta = 0°$ corresponds to a collision in which the F(g) atom collides with the $D_2(g)$ molecule and the DF(g) product molecule travels in the same direction as the incident F(g) atom. An angle of $\theta = 180°$ corresponds to a collision in which the F(g) atom collides with the $D_2(g)$ molecule, reacts, and then the DF(g) molecule bounces back opposite to the incident direction of the F(g) atom (Figure 30.12).

The contours in Figure 30.11 represent a constant number of DF(g) product molecules. The dashed circles in Figure 30.11 correspond to the maximum relative speed allowed for a product molecule in a given vibrational state. An increase in the diameter of this circle corresponds to an increase in the relative speed of the product molecule. Recall that the total energy is fixed, so the speed of the DF(g) molecule decreases with increasing vibrational quantum number. Thus, the diameters of the circles

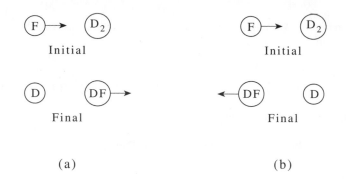

FIGURE 30.12
An illustration of the atom-molecule reaction $F(g) + D_2(g)$ in which (a) $\theta = 0°$ and (b) $\theta = 180°$.

decrease with increasing vibrational quantum number. Notice that the data show that a large number of product molecules have speeds between the dashed circles. The dashed circles shown in Figure 30.11 correspond to the case where there is internal energy only in the vibrational states of the molecule, in which case the rotational energy corresponding to these circles is $E_{rot} = 0$, with $J = 0$. If DF(g) is produced in an excited rotational state, we would expect to observe a speed that has an value intermediate between two of the dashed circles. For example, the region between the dashed circles labeled $v = 3$ and $v = 4$ in Figure 30.11 (see point A in the figure) corresponds to a DF(g) molecule that has a vibrational quantum number $v = 3$ but is also rotationally excited. If we know the energy spacing of the rotational states, the rotational energy distribution can also be determined from the contour map, see Example 30–8.

EXAMPLE 30–8
From the analysis of the speed contour plot shown in Figure 30.11, Point A is found to correspond to a DF(g) molecule with a total rotational and vibrational energy of 11 493.6 cm^{-1}. Using the following data for DF(g), determine the rotational level of the molecule assuming the vibrational quantum number is $v = 3$.

$\tilde{v}_e/\text{cm}^{-1}$	$\tilde{v}_e\tilde{x}_e/\text{cm}^{-1}$	$\tilde{B}_e/\text{cm}^{-1}$	$\tilde{\alpha}_e/\text{cm}^{-1}$
2998.3	45.71	11.007	0.293

SOLUTION: The vibrational and rotational energies of a diatomic molecule are given by Equations 13.21 and 13.17

$$E_{vib}(v) = \tilde{v}_e\left(v + \tfrac{1}{2}\right) - \tilde{v}_e\tilde{x}_e\left(v + \tfrac{1}{2}\right)^2$$
$$E_{rot}(J, v) = \left[\tilde{B}_e - \tilde{\alpha}_e\left(v + \tfrac{1}{2}\right)\right]J(J + 1)$$

The total rotational-vibrational energy of the DF(g) molecule is then the sum of the vibrational and rotational energies:

$$
\begin{aligned}
E_{\text{vib}}(v) &+ E_{\text{rot}}(J, v) \\
&= \tilde{\nu}_e \left(v + \tfrac{1}{2}\right) - \tilde{\nu}_e \tilde{x}_e \left(v + \tfrac{1}{2}\right)^2 + \left[\tilde{B}_e - \tilde{\alpha}_e \left(v + \tfrac{1}{2}\right)\right] J(J+1) \quad (1)
\end{aligned}
$$

Substituting the above spectroscopic data into Equation 1 and setting $v = 3$ gives us

$$11\,493.6 \text{ cm}^{-1} = 9934.1 \text{ cm}^{-1} + (9.982 \text{ cm}^{-1})J(J+1)$$

which simplifies to

$$J(J+1) = 156$$

from which we find that $J = 12$. Point A on the contour plot given in Figure 30.11 corresponds to a population of DF(g) molecules with the quantum numbers $v = 3$ and $J = 12$.

The experimental data in Figure 30.11 reveal three important features of the reaction. First, we see that the product preferentially scatters backward, toward the direction of the incident fluorine atom, a scattering angle of $\theta = 180°$. These data suggest that the fluorine atom undergoes a nearly head-on collision with the $D_2(g)$ molecule and then bounces backward after abstracting one of the deuterium atoms. This type of reaction is called a *rebound reaction*. Second, an analysis of this contour map reveals that the most probable product of the reaction is DF($v = 3$). Third, there is considerable population between the dashed circles, indicating that a variety of rotational levels of the DF(g) molecule are populated by the reaction.

The relative populations of the first five vibrational states deserve a bit more attention. Note the lack of contour lines between the dashed circles for $v = 0$ and $v = 1$. This result means that no product molecules are formed in the ground vibrational state. The populations determined from the contour diagram are given in Table 30.2 and corresponds to those shown in Figure 30.11. This product distribution cannot be

TABLE 30.2

The observed relative populations of the vth vibrational state to the $v = 3$ state of DF(v) for the reaction $F(g) + D_2(g) \Rightarrow DF(g) + D(g)$, in which the relative translational energy of the reactants is $7.62 \text{ kJ} \cdot \text{mol}^{-1}$.

Vibrational quantum number	Relative population
0	0.00
1	0.02
2	0.44
3	1.00
4	0.49

described by a Boltzmann distribution (see Example 30–9), and we say that the reaction generates a nonequilibrium product distribution.

EXAMPLE 30–9

Determine the populations of DF(v) relative to DF($v = 3$) for $v = 0$ through $v = 4$, assuming that the overall distribution is in thermal equilibrium at 300 K. (Assume the vibrational motion of DF(g) is harmonic with $\tilde{\nu}_{DF} = 2907$ cm^{-1}.)

SOLUTION: If the DF(g) molecules are at thermal equilibrium, the ratio of the populations of DF(g) molecules in two vibrational levels is given by the Boltzmann distribution. Thus,

$$\frac{N(v)}{N(v = 3)} = \frac{e^{-(v+1/2)h\nu_{DF}/k_BT}}{e^{-(3+1/2)h\nu_{DF}/k_BT}} = e^{-(v-3)h\nu_{DF}/k_BT}$$

The calculated relative populations are tabulated below

Vibrational quantum number	$N(v)/N(v = 3)$
0	1.44×10^{18}
1	1.28×10^{12}
2	1.75×10^{6}
3	1.00
4	8.84×10^{-7}

The values of $N(v)/N(v = 3)$ for a sample at thermal equilibrium are significantly different from the relative populations determined for the chemical reaction F(g) + D$_2$(g) $\Rightarrow$ DF(g) + D(g) (see Table 30.2). Note in particular that the $v = 0$ state is the most populous in the thermal distribution, whereas the reaction generates no population in the $v = 0$ state. The vibrational distribution of the reaction product cannot be described by a Boltzmann distribution.

30–9. Not All Gas-Phase Chemical Reactions Are Rebound Reactions

Figure 30.13 shows the velocity contour map for the reaction

$$K(g) + I_2(g) \longrightarrow KI(g) + I(g)$$

in which the initial relative translational energy between the reactants is 15.13 kJ·mol^{-1}. Unlike in the F(g) + D$_2$(g) reaction, we can see that the product diatomic molecule in this case, KI(g), is preferentially scattered in the forward direction, along the direction of the incident K(g) atom. This type of reaction, in which the incident atom abstracts part of a molecule and keeps going in the forward direction, is called a *stripping reaction*.

The mechanism of stripping reactions is interesting. The reaction cross section for the K(g) + I$_2$(g) reaction is 1.25×10^{6} pm^2. Assuming the radius of K(g) and I$_2$(g) are 205 pm and 250 pm, respectively, the hard-sphere collision cross section

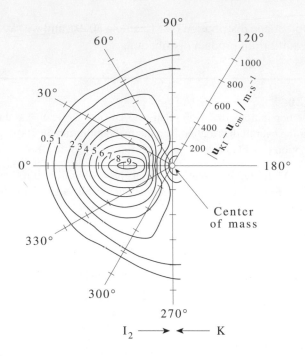

FIGURE 30.13
A contour map of the angular and velocity distributions for the product molecule KI(g) for the reaction $K(g) + I_2(g) \rightarrow KI(g) + I(g)$, in which the relative translational energy of the reactants is 15.13 kJ·mol^{-1}. In this stripping reaction, the product molecules continue going in the direction of the incident potassium atom, a scattering angle near $\theta = 0°$. The numbers that label the contours are a measure of the relative number of KI(g) molecules.

is $\pi d_{AB}^2 = 6.5 \times 10^5$ pm^2. The measured reaction cross section is twice as large as the hard-sphere estimate. If the approaching potassium atom and iodine molecule were to travel in straight lines at the maximum impact parameter corresponding to this experimentally determined reaction cross-section, these reactants would miss one another. The fact that a reaction occurs indicates that the trajectories of the reacting molecules are affected by a long-range potential that draws them together. We would not expect the van der Waals interactions between the potassium atom and the iodine molecule to be strong enough to cause such a large effect. Research shows that this reaction involves the transfer of an electron between the two reactants, which takes place before the reactants collide. Thus, the first step of the reaction occurs when the reactants are still separated and produces a pair of ions,

$$K(g) + I_2(g) \longrightarrow K^+(g) + I_2^-(g)$$

The ions are then attracted to one another through a Coulomb potential. The more energetically stable products KI(g) + I(g) are formed when the two ions collide. The KI(g) moves off in the same direction as the incident potassium ion. This mechanism has been coined the *harpoon mechanism* because the potassium atom uses its electron like a harpoon to draw in the $I_2(g)$ molecule.

Figure 30.14 shows the results for the reactive scattering reaction

$$O(g) + Br_2(g) \Longrightarrow BrO(g) + Br(g)$$

carried out with a relative translational kinetic energy of 12.55 kJ·mol^{-1}. The data show that the product molecule BrO(g) is forward and back scattered with equal intensity. Thus, neither mechanism discussed so far can account for this observed behavior. In fact, no simple hit-and-run collision picture can explain this result. To display such a symmetric forward- and back-scattered product distribution, the reacting molecules need to "forget" the original collision geometry. This is possible only if the collision results in the formation of an atom-molecule complex, whose lifetime is long compared with its rotational period. This long lifetime allows the complex to rotate many times before generating products. In this case, the angular distribution of the product molecules becomes independent of their initial collision geometry.

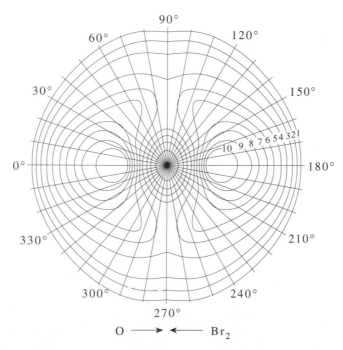

FIGURE 30.14
A contour map of the angular and velocity distributions for the product molecule OBr(g) from the reaction of $O(g) + Br_2(g)$, in which the relative translational energy of the reactants is 12.55 kJ·mol^{-1}. The numbers that label the contours indicate the relative number of BrO(g) molecules observed.

30–10. The Potential-Energy Surface for the Reaction $F(g) + D_2(g) \Rightarrow DF(g) + D(g)$ Can Be Calculated Using Quantum Mechanics

In Chapter 9, we learned that the potential energy of a diatomic molecule depends on only the distance between the two bonded atoms. Thus, the potential-energy surface for a diatomic molecule such as $D_2(g)$ or $DF(g)$ can be plotted in two dimensions by plotting the potential energy as a function of the bond length. The word "surface" is a misnomer in this case. A diatomic molecule has only one geometric parameter, the bond length. Using the term "potential-energy curve" when the potential energy depends on a single parameter and the word "surface" when the potential energy depends on more than one geometric parameter is more appropriate. Figure 30.15 shows the potential-energy curve for $D_2(g)$.

The potential energy of a polyatomic molecule depends on more than one variable because there is more than one bond length that can be varied. We will also need to specify the bond angles(s). For example, consider a water molecule. The geometry of a water molecule is completely specified by three geometric parameters, r_{O-H_A}, r_{O-H_B}, and the angle α between the two O–H bonds.

$$r_{O-H_A} \quad O \quad r_{O-H_B}$$
$$H_A \quad \alpha \quad H_B$$

The potential energy of a water molecule is a function of these three parameters, or $V = V(r_{O-H_A}, r_{O-H_B}, \alpha)$. A plot of the complete potential-energy surface of a water molecule therefore requires four axes, one axis for the value of the potential energy

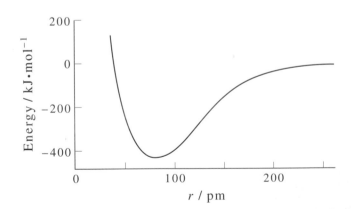

FIGURE 30.15
The potential-energy curve of $D_2(g)$. The zero of energy is defined to be that of the two separated atoms. The minimum of the potential-energy curve corresponds to the equilibrium bond length of the $D_2(g)$ molecule.

and one axis for each of the three geometric parameters. The potential-energy surface is four-dimensional. Because we are limited to three dimensions for plotting functions, we cannot draw the entire potential-energy surface of a water molecule in a single plot. We can, however, draw parts of the potential-energy surface. We can fix one of the geometric parameters, for example, the angle α, and then draw a three-dimensional plot of $V(r_{O-H_A}, r_{O-H_B}, \alpha = \text{constant})$. Such a plot is a cross-sectional cut of the full potential-energy surface. A cross-sectional plot teaches us how the potential energy of the molecule changes when we vary some of the geometric variables while holding others constant. For example, a three-dimensional plot of $V(r_{O-H_A}, r_{O-H_B}, \alpha = \text{constant})$ as a function of r_{O-H_A} and r_{O-H_B} tells us how the potential energy of a water molecule changes when the bond lengths r_{O-H_A} and r_{O-H_B} are varied at a constant bond angle of α. If we made a series of cross-sectional plots for different values of α, we could see how the potential energy depends upon the bond angle.

We encounter a similar limitation in viewing the potential-energy surfaces for simple chemical reactions as we did for a water molecule. Let us return to a discussion of the chemical reaction

$$F(g) + D_A D_B(g) \Longrightarrow D_A F(g) + D_B(g)$$

where the subscripts A and B are used so that we can differentiate between the two deuterium atoms. When the reactants are at infinite separation, there are no attractive or repulsive forces between the fluorine atom and the D$_2$(g) molecule, so the potential-energy surface for the reaction is the same as that for an isolated D$_2$(g) molecule. Likewise, when the products are at infinite separation, the potential-energy surface for the reaction is the same as that for the isolated DF(g) molecule. As the reaction occurs, however, r_{DF}, the distance between the fluorine atom and D$_A$, decreases and r_{D_2}, the distance between D$_A$ and D$_B$, increases, and the potential energy depends on both distances. The potential energy also depends on the angle at which the fluorine atom approaches the D$_2$(g) molecule. We define the collision angle β between the fluorine atom and D$_2$(g) molecule to be that between the lines that lie along the F–D$_A$ and D$_A$–D$_B$ bonds. In Figure 30.16, we show three different ways the fluorine atom can approach the D$_2$(g) molecule: linear ($\beta = 180°$), bent ($\beta = 135°$), and perpendicular ($\beta = 90°$).

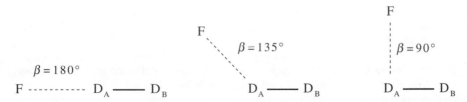

FIGURE 30.16
Three different collision angles, β, for the reactants F(g) + D$_2$(g).

Because the potential-energy surface for this reaction depends on two distances (r_{DF} and r_{D_2}) and one collision angle (β), a four-dimensional coordinate system is needed to plot the complete surface. To view the potential-energy surface, we have to fix the value of one of the geometric parameters and then plot the dependence of the potential energy on the two remaining variables. We could make a series of such plots for different values of the fixed variable to see how the potential surface depends on all three geometric parameters.

The potential-energy surface for a chemical reaction can be calculated using the electronic structure techniques discussed in Chapter 11 for polyatomic molecules. By performing such a calculation for a number of different nuclear configurations, we can obtain the potential energy as a function of the nuclear coordinates. Figure 30.17 presents a contour diagram of the calculated potential-energy surface for the reaction

$$F(g) + D_2(g) \Longrightarrow DF(g) + D(g)$$

where the collision angle β is set to 180°, the experimental value determined from the crossed-molecular beam data (see Section 30–8). Such a geometry is said to be collinear. Each line in the contour map corresponds to a constant value of the energy. The zero of energy has been arbitrarily assigned to the reactants at infinite separation.

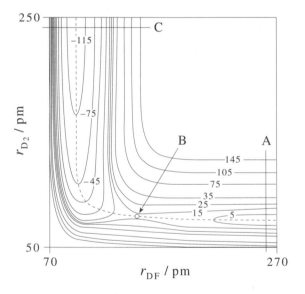

FIGURE 30.17
An energy contour map for the reaction $F(g) + D_2(g) \Rightarrow DF(g) + D(g)$ at a collision geometry of $\beta = 180°$, using the Born-Oppenheimer approximation to calculate the potential-energy surface. The numbers for the energy contours are in units of $kJ \cdot mol^{-1}$. The zero of energy is defined as the infinitely separated reactants. Point B is the location of the transition state of the reaction. The cross-sectional cuts through this surface indicated by the lines at A and C correspond to potential-energy curves for the isolated $D_2(g)$ and $DF(g)$ molecules, respectively. The dashed line is the mimimum energy pathway for the reaction.

At r_{DF} = A in Figure 30.17, the reactants are at a large separation, and the potential-energy surface is identical to the potential-energy curve for an isolated $D_2(g)$ molecule and F(g) atom. In other words, if we plotted the cross-sectional plot of this surface $V(r_{D_2}, r_{DF}, \beta = 180°)$ as a function of r_{D_2}, we would obtain Figure 30.15. Likewise, at r_{D_2} = C in Figure 30.17, the products are at a large separation and a cross-sectional plot of $V(r_{D_2}, r_{DF}, \beta = 180°)$ as a function of r_{DF} is identical to the potential-energy curve for an isolated DF(g) molecule.

Let's now follow the minimum energy path from the reactants to the products, given by the dashed line in Figure 30.17. We see that as the reactants approach one another, the distance r_{D_2} remains fairly constant, the distance r_{DF} decreases, and the potential energy increases, reaching a maximum at point B. After passing through point B, the products have been formed, the distance r_{DF} decreases slightly and then remains constant, the distance r_{D_2} increases, and the potential energy decreases. The calculated potential-energy surface has an energy barrier between the reactants and products. The minimum height of this energy barrier (about 7 kJ·mol^{-1}) occurs at point B, which is called the *transition state*. The transition state separates the reactants from the products. It sits at a unique point on the potential-energy surface. If we follow the minimum energy path from the transition state to either the separated reactants or the separated products, the energy decreases. If we move away from the transition state in the direction perpendicular to this minimum energy path, the energy increases. Thus, in one direction, the transition state is an energy maximum; in the perpendicular direction, the transition state is an energy minimum. Such points are called *saddle points* because the surface in the vicinity of the point has the shape of a saddle. The transition state for a chemical reaction usually sits at a saddle point on the potential-energy surface.

Problems

30-1. Calculate the hard-sphere collision theory rate constant for the reaction

$$NO(g) + Cl_2(g) \Longrightarrow NOCl(g) + Cl(g)$$

at 300 K. The collision diameters of NO and Cl_2 are 370 pm and 540 pm, respectively. The Arrhenius parameters for the reaction are $A = 3.981 \times 10^9$ dm^3·mol^{-1}·s^{-1} and $E_a = 84.9$ kJ·mol^{-1}. Calculate the ratio of the hard-sphere collision theory rate constant to the experimental rate constant at 300 K.

30-2. Compare a plot of $\sigma_r(E_r)/\pi d_{AB}^2$ given by Equation 30.14 to the data shown in Figure 30.2.

30-3. Show that Equation 30.15, the rate constant for the line-of-centers model, is obtained by substituting Equation 30.14, the reaction cross section for the line-of-centers model, into Equation 30.11 and then integrating the resulting expression.

30-4. The Arrhenius parameters for the reaction

$$NO(g) + O_3(g) \Longrightarrow NO_2(g) + O_2(g)$$

are $A = 7.94 \times 10^9 \, dm^3 \cdot mol^{-1} \cdot s^{-1}$ and $E_a = 10.5 \, kJ \cdot mol^{-1}$. Assuming the line-of-centers model, calculate the values of E_0, the threshold energy, and σ_{AB}, the hard-sphere reaction cross section, for this reaction at 1000 K.

30-5. Consider the following bimolecular reaction at 3000 K:

$$CO(g) + O_2(g) \Longrightarrow CO_2(g) + O(g)$$

The experimentally determined Arrhenius pre-exponential factor is $A = 3.5 \times 10^9 \, dm^3 \cdot mol^{-1} \cdot s^{-1}$, and the activation energy is $E_a = 213.4 \, kJ \cdot mol^{-1}$. The hard-sphere collision diameter of O_2 is 360 pm and that for CO is 370 pm. Calculate the value of the hard sphere line-of-centers model rate constant at 3000 K and compare it with the experimental rate constant. Also compare the calculated and experimental A values.

30-6. The threshold energy, E_0, for the reaction

$$H_2^+(g) + He(g) \Longrightarrow HeH^+(g) + H(g)$$

is 70.0 $kJ \cdot mol^{-1}$. Determine the lowest vibrational level of $H_2^+(g)$ such that the vibrational energy of the reactants exceeds E_0. The spectroscopic constants for H_2^+ are $\tilde{\nu}_e = 2321.7 \, cm^{-1}$ and $\tilde{\nu}_e \tilde{x}_e = 66.2 \, cm^{-1}$.

30-7. Calculate the total kinetic energy of an F(g) atom moving at a speed of 2500 $m \cdot s^{-1}$ toward a head-on collision with a stationary $D_2(g)$ molecule. (Assume the reactants are hard spheres.)

30-8. A F(g) atom and a $D_2(g)$ molecule are moving toward a head-on collision with one another. The F(g) atom has a speed of 1540 $m \cdot s^{-1}$. Calculate the speed of the $D_2(g)$ molecule so that the total kinetic energy is the same as that in Problem 30–7. (Assume the reactants are hard spheres.)

30-9. In Problem 30–7, you calculated the total kinetic energy for a F(g) atom moving at a speed of 2500 $m \cdot s^{-1}$ toward a head-on collision with a stationary $D_2(v = 0)$ molecule. Determine the ratio of the total kinetic energy to the zero-point vibrational energy of the $D_2(g)$ molecule given that $\tilde{\nu}_{D_2} = 2990 \, cm^{-1}$.

30-10. Consider the head-on collision between a F(g) atom and a stationary $D_2(g)$ molecule. Estimate the minimum speed of the F(g) atom so that its kinetic energy exceeds the bond dissociation energy of $D_2(g)$. (The value of D_0 for D_2 is 435.6 $kJ \cdot mol^{-1}$.)

30-11. Following Example 30–4, show that the equations

$$\mathbf{u}_{cm} = \frac{m_C}{M} \mathbf{u}_C + \frac{m_D}{M} \mathbf{u}_D$$

and

$$\mathbf{u}_r = \mathbf{u}_C - \mathbf{u}_D$$

lead to

$$\mathbf{u}_C = \mathbf{u}_{cm} + \frac{m_D}{M} \mathbf{u}_r \tag{1}$$

and

$$\mathbf{u}_D = \mathbf{u}_{cm} - \frac{m_C}{M} \mathbf{u}_r \tag{2}$$

30-12. Derive Equation 30.22.

30-13. The speed of sound, u_s, in a fluid is given by

$$u_s^2 = \frac{\gamma \overline{V}}{M \kappa_T} \tag{1}$$

where $\gamma = C_P / C_V$, M is the molar mass, and $\kappa_T = -(1/V)(\partial V / \partial P)_T$ is the isothermal compressibility of the fluid. Assuming ideal behavior, calculate the speed of sound in $N_2(g)$ at 25°C. Take $\overline{C}_P = 7R/2$. The measured value is 348 m·s^{-1}.

30-14. The speed of sound, u_s, in a fluid is given by Equation 1 of Problem 30–13. In addition, $\overline{C}_P$ and $\overline{C}_V$ are related by (Equation 22.27)

$$\overline{C}_P - \overline{C}_V = \frac{\alpha^2 T \overline{V}}{\kappa_T}$$

where $\alpha = (1/V)(\partial V / \partial T)_P$ is the coefficient of thermal expansion. Given that $\overline{C}_P = 135.6$ J·K^{-1}·mol^{-1}, $\kappa_T = 9.44 \times 10^{-10}$ Pa^{-1}, $\alpha = 1.237 \times 10^{-3}$ K^{-1}, and the density $\rho = 0.8765$ g·mL^{-1} for benzene at one atm and 20°C, calculate the speed of sound in benzene. The measured value is 1320 m·s^{-1}.

30-15. The peak speed of the molecules in a supersonic molecular beam of a carrier gas is well approximated by

$$u_{\text{peak}} = \left(\frac{2RT}{M} \right)^{1/2} \left(\frac{\gamma}{\gamma - 1} \right)^{1/2}$$

where T is the temperature of the source chamber of the gas mixture, M is the molar mass of the carrier gas, and γ is the ratio of the heat capacities, $\gamma = C_P / C_V$, of the carrier gas. Determine the peak velocity for a benzene molecule in a supersonic neon beam in which the source chamber of the gas is maintained at 300 K. Repeat the calculation for a helium beam under the same conditions. Assume that He(g) and Ne(g) can be treated as ideal gases.

30-16. Estimate the temperature required so that the average speed of a benzene molecule in a gas cell is the same as that for a benzene molecule in a helium supersonic molecular beam generated under the conditions stated in Problem 30–15.

30-17. Show that for the general reaction

$$A(g) + BC(g) \Longrightarrow AB(g) + C(g)$$

Equation 30.28 can be written as

$$E_{\text{tot}} = \tfrac{1}{2} \mu u_r^2 + F(J) + G(v) + T_e$$
$$= \tfrac{1}{2} \mu u_r'^2 + F'(J) + G'(v) + T_e'$$

within the harmonic oscillator-rigid rotator approximation where T_e, $G(v)$, and $F(J)$ are the electronic, vibrational, and rotational terms of the diatomic reactant, BC(g), and T_e', $G'(v)$, and $F'(J)$ are the corresponding terms for the diatomic product, AB(g).

30-18. Consider the reaction

$$Cl(g) + H_2(v = 0) \Longrightarrow HCl(v) + H(g)$$

where $D_e(H_2) - D_e(HCl) = 12.4 \text{ kJ·mol}^{-1}$. Assume there is no activation barrier to the reaction. Model the reactants as hard spheres (no vibrational motion) and calculate the minimum value of the relative speed required for reaction to occur. If we model $H_2(g)$ and $HCl(g)$ as hard-sphere harmonic oscillators with $\tilde{\nu}_{H_2} = 4159 \text{ cm}^{-1}$ and $\tilde{\nu}_{HCl} = 2886 \text{ cm}^{-1}$, calculate the minimum value of the relative speed required for reaction to occur.

30-19. The reaction $H(g) + F_2(v = 0) \Rightarrow HF(g) + F(g)$ produces vibrationally excited HF molecules. Determine the minimum value of the relative kinetic energy such that HF(g) molecules in the $v = 12$ vibrational state are produced. The following are the vibrational spectroscopic constants for HF and F_2: $\tilde{\nu}_e(HF) = 4138.32 \text{ cm}^{-1}$, $\tilde{\nu}_e(F_2) = 916.64 \text{ cm}^{-1}$, $\tilde{\nu}_e\tilde{x}_e(HF) = 89.88 \text{ cm}^{-1}$, $\tilde{\nu}_e\tilde{x}_e(F_2) = 11.24 \text{ cm}^{-1}$, $D_0(HF) = 566.2 \text{ kJ·mol}^{-1}$, and $D_0(F_2) = 154.6 \text{ kJ·mol}^{-1}$.

30-20. Consider the energetics of the reaction

$$F(g) + H_2(v = 0) \Longrightarrow HF(v) + H(g)$$

where the relative translational energy of the reactants is 7.62 kJ·mol^{-1}, and $D_e(H_2) - D_e(HF) = -140 \text{ kJ·mol}^{-1}$. Determine the range of possible vibrational states of the product HF(g) molecule. Assume the vibrational motion of both $H_2(g)$ and HF(g) is harmonic with $\tilde{\nu}_{H_2} = 4159 \text{ cm}^{-1}$ and $\tilde{\nu}_{HF} = 3959 \text{ cm}^{-1}$.

30-21. In Example 30–5 we calculated the speeds of the products relative to the center of mass, $|\mathbf{u}_{DF} - \mathbf{u}_{cm}|$ and $|\mathbf{u}_D - \mathbf{u}_{cm}|$, for the reaction $F(g) + D_2(g) \Longrightarrow DF(g) + D(g)$ assuming that the reactants and products could be treated as hard spheres. Now calculate these quantities taking into account the zero-point vibrational energies of $D_2(g)$ and DF(g). Assume that the vibrational motion of $D_2(g)$ and DF(g) is harmonic with $\tilde{\nu}_{D_2} = 2990 \text{ cm}^{-1}$ and $\tilde{\nu}_{DF} = 2907 \text{ cm}^{-1}$, respectively. How different are your results from the hard-sphere calculations presented in Example 30–5?

The following four problems consider the reaction

$$Cl(g) + HBr(v = 0) \Longrightarrow HCl(v) + Br(g)$$

where the relative translational energy of the reactants is 9.21 kJ·mol^{-1}, the difference $D_e(HBr) - D_e(HCl) = -67.2 \text{ kJ·mol}^{-1}$, and the activation energy for this reaction is $\approx 6 \text{ kJ·mol}^{-1}$.

30-22. Determine the range of possible vibrational states of the product molecule, HCl(g). The spectroscopic constants for HBr(g) and HCl(g) are

	$\tilde{\nu}_e/\text{cm}^{-1}$	$\tilde{\nu}_e\tilde{x}_e/\text{cm}^{-1}$
HBr	2648.98	45.22
HCl	2990.95	52.82

Draw a diagram for this reaction that is similar to that shown in Figure 30.8 for the $F(g) + D_2(g)$ reaction.

30-23. Calculate the value of $|\mathbf{u}_{HCl} - \mathbf{u}_{cm}|$, the speed of the HCl(g) molecule relative to the center of mass, for each of the possible vibrational states of HCl(g) in Problem 30–22.

30-24. Determine the speeds for a HCl(g) molecule relative to the center of mass $|\mathbf{u}_{HCl} - \mathbf{u}_{cm}|$ in the $v = 0$, $J = 0$ and $v = 0$, $J = 1$ states. The rotational constants for HCl(g) are $\tilde{B}_e = 10.59 \text{ cm}^{-1}$ and $\tilde{\alpha}_e = 0.307 \text{ cm}^{-1}$.

30-25. Using the data in Problem 30–24, determine the value of J_{min}, the minimum value of J, such that the kinetic energy of a HCl($v = 0$, $J = J_{min}$) molecule is greater than the kinetic energy of an HCl($v = 1$, $J = 0$) molecule. [Note that if this reaction produces HCl($v = 0$, $J \geq J_{min}$), then these molecules have relative speeds characteristic of an HCl($v = 1$) molecule, affecting the analysis of the product velocity contour plots.]

30-26. Using the data given in Table 13.2, estimate the minimum value of the relative speed of the reactants so that the following reactions occur:

$$HCl(v = 0) + Br(g) \Longrightarrow HBr(v = 0) + Cl(g)$$

and

$$HCl(v = 1) + Br(g) \Longrightarrow HBr(v = 0) + Cl(g)$$

30-27. Do the values of the radii of the dashed circles in Figure 30.11 increase, decrease, or remain the same as the relative translational energy of the reactants is increased from $7.62 \text{ kJ} \cdot \text{mol}^{-1}$? Determine the percentage change, if any, in the radius of the dashed circle for $v = 0$ if the relative translational energy is doubled from $7.62 \text{ kJ} \cdot \text{mol}^{-1}$ to $15.24 \text{ kJ} \cdot \text{mol}^{-1}$.

30-28. Figure 30.11 presents the contour map for the product molecule DF(g) for the reaction between F(g) and $D_2(v = 0)$. The dashed lines correspond to the expected speeds for DF(g) molecules in those vibrational states when J, the rotational quantum number, equals 0. The regions between two circles then correspond to molecules that are rotationally excited. Determine the minimum value for J such that a DF($v = 2$) molecule has a relative speed expected for a DF($v = 3$) molecule. The spectroscopic constants for DF(g) are given in Example 30–8. Does your result suggest that there could be a problem encountered in the analysis of the scattering data for this reaction?

30-29. For the reaction $Cl(g) + H_2(g) \Rightarrow HCl(g) + H(g)$, $D_e(H_2) - D_e(HCl) = 12.4$ $\text{kJ} \cdot \text{mol}^{-1}$. Assuming that the relative kinetic energy is $8.52 \text{ kJ} \cdot \text{mol}^{-1}$ and that the $H_2(g)$ reactant is prepared in a $v = 3$, $J = 0$ state, what are the possible vibrational states of HCl(g)? The vibrational spectroscopic constants for $H_2(g)$ and HCl(g) are $\tilde{\nu}_e(H_2) = 4401.21 \text{ cm}^{-1}$, $\tilde{\nu}_e(HCl) = 2990.95 \text{ cm}^{-1}$, $\tilde{\nu}_e\tilde{x}_e(H_2) = 121.34 \text{ cm}^{-1}$, and $\tilde{\nu}_e\tilde{x}_e(HCl) = 52.82 \text{ cm}^{-1}$.

30-30. Suppose that the reaction given in Problem 30–29 produces HCl(g) in $v = v_{max}$, the highest possible vibrational state under the given conditions. Determine the largest possible value of J for the HCl($v = v_{max}$, J) molecule. The rotational constants of HCl(g) are $\tilde{B}_e = 10.59 \text{ cm}^{-1}$ and $\tilde{\alpha}_e = 0.307 \text{ cm}^{-1}$.

30-31. Consider the product velocity distribution for the reaction between K(g) and $I_2(v = 0)$ at a relative translational energy of $15.13 \text{ kJ} \cdot \text{mol}^{-1}$ shown in Figure 30.13. Assume that the vibrational motion of $I_2(g)$ and KI(g) is harmonic with $\tilde{\nu}_{I_2} = 213 \text{ cm}^{-1}$

and $\tilde{v}_{KI} = 185$ cm^{-1}. Given that $D_e(I_2) - D_e(KI) = -171$ kJ·mol^{-1}, determine the maximum vibrational quantum number for the product KI(g). Now determine the speed of a KI($v = 0$) molecule relative to the center of mass. Repeat the calculation for the KI($v = 1$) molecule. Do the data in the contour map support a conclusion that KI(g) is produced in a distribution of vibrational levels?

30-32. Below is a plot of the LiCl(g) product velocity distribution for the reaction

$$\text{Li(g)} + \text{HCl}(v = 0) \Longrightarrow \text{LiCl}(v) + \text{H(g)}$$

recorded at a relative translational energy of 38.49 kJ·mol^{-1}. Is this reaction an example of a rebound reaction, a stripping reaction, or a reaction in which a long-lived complex (relative to the rotational period of the complex) is formed between the reactants before any product molecules are produced? Explain your reasoning.

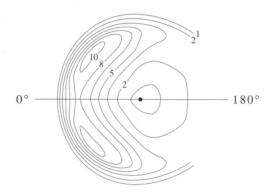

30-33. Shown below are the velocity contour plots of the N_2D^+(g) product recorded at two different relative translational energies for the reaction

$$N_2^+(g) + D_2(v = 0) \Longrightarrow N_2D^+(v) + D(g)$$

The scale between the two plots indicating 1000 m·s^{-1} applies to both plots.

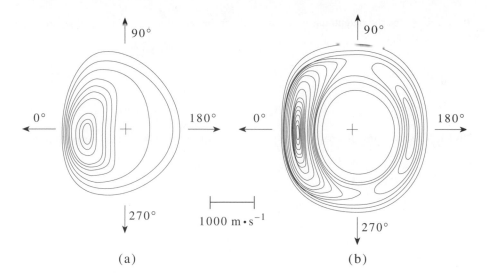

(a)   (b)

The value of $D_e(N_2D^+) - D_e(N_2^+) - D_e(D_2)$ is equal to 96 kJ·mol^{-1}. The relative translational energy of the reactants is 301.02 kJ·mol^{-1} and 781.49 kJ·mol^{-1} for the left and right contour plots, respectively. Propose an explanation for why N_2D^+(g) product molecules observed with low relative velocities are present in (a) but absent in (b).

30-34. The reaction between Ca(g) and F$_2$(g) generates an electronically excited product according to the equation

$$Ca(^1S_0) + F_2(g) \Longrightarrow CaF^*(B^2\Sigma^+) + F(g)$$

The radii of Ca($^1S_0$) and F$_2$(g) are 100 pm and 370 pm, respectively. Determine the hard-sphere collision cross section. The cross section for this reaction is $> 10^6$ pm^2. Propose a mechanism for this reaction.

30-35. Consider the reaction described in Problem 30–34. The product $CaF^*(B^2\Sigma^+)$ relaxes to its electronic ground state by fluorescence. Explain how you could determine the vibrational states of the product from a measurement of the fluorescence spectrum.

30-36. For the reaction

$$Ca(^1S_0) + F_2(g) \Longrightarrow CaF^*(B^2\Sigma^+) + F(g)$$

the peak of the fluorescence spectrum corresponds to emission from the $v' = 10$ level of the $B^2\Sigma^+$ state to the $v'' = 10$ level of the ground electronic state of CaF*. Calculate the wavelength of this emission line. The spectroscopic constants for the $B^2\Sigma^+$ state are $T_e = 18\,844.5$ cm^{-1}, $\tilde{v}'_e = 566.1$ cm^{-1}, and $\tilde{v}'_e\tilde{x}'_e = 2.80$ cm^{-1} and those for the ground state are $\tilde{v}''_e = 581.1$ cm^{-1} and $\tilde{v}''_e\tilde{x}''_e = 2.74$ cm^{-1}. In what part of the electromagnetic spectrum will you observe this emission?

30-37. Describe the potential-energy surfaces for the reactions

$$I(g) + H_2(v = 0) \Longrightarrow HI(v) + H(g)$$

and

$$I(g) + CH_4(v = 0) \Longrightarrow HI(v) + CH_3(g)$$

30-38. The following plot depicts the potential-energy surface for the isomerization reaction

$$OClO(g) \Longrightarrow ClOO(g)$$

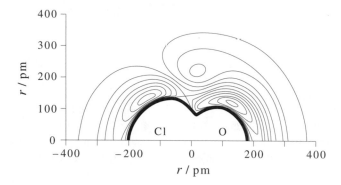

The contour map is a plot of the potential energy as a function of the location of an oxygen atom around a diatomic ClO of fixed bond length. The energy spacing betwen contour lines is 38.6 kJ·mol^{-1}. Label the location of the oxygen atom in the reactant (OClO) and product (ClOO) molecules. Draw the minimum energy path for the isomerization reaction. Which isomer is more stable? Estimate the range for the height of the activation barrier for this isomerization reaction from the potential-energy surface. Is the energy barrier to isomerization less than, greater than, or equal to the barrier for dissociation into $O(g) + ClO(g)$?

30-39. The *opacity function* $P(b)$ is defined as the fraction of collisions with impact parameter b that lead to reaction. The reaction cross section is related to the opacity function by

$$\sigma_r = \int_0^\infty 2\pi b P(b)db$$

Justify this expression. Assume that the opacity function is given by

$$P(b) = \begin{cases} 1 & b \leq d_{AB} \\ 0 & b > d_{AB} \end{cases}$$

Show that this opacity function gives the hard-sphere collision theory model for σ_r.

30-40. The opacity function is defined in Problem 30–39. Determine an expression for b_{max} in terms of d_{AB}, E_0, and E_r so that an opacity function given by

$$P(b) = \begin{cases} 1 & b \leq b_{max} \\ 0 & b > b_{max} \end{cases}$$

yields the reaction cross section $\sigma_r(E_r)$ for the line-of-centers model (Equation 30.14).

30-41. For the reaction $H(g) + H_2(g)$, the opacity function (defined in Problem 30–39) is

$$
P(b) = \begin{cases} A \cos \dfrac{\pi b}{2b_{max}} & b \leq b_{max} \\ 0 & b > b_{max} \end{cases}
$$

where A is a constant. Derive an expression for the reaction cross section in terms of b_{max}.

30-42. Explain how the $F(g) + D_2(g)$ reaction can be exploited to make a chemical laser. (*Hint:* See Table 30.2 and Section 15–4.)

30-43. A quantum-mechanical calculation of the potential-energy surface for the collinear hydrogen atom exchange reaction described by $H_A(g) + H_B H_C(g) \Longrightarrow H_A H_B(g) + H_C(g)$ gives a reaction barrier that lies 58.75 $kJ \cdot mol^{-1}$ above the bottom of the potential well of the reactants. Calculate the minimum relative speed for the collision between $H(g)$ and $H_2(v = 0)$ so that the hydrogen-atom exchange reaction occurs. Assume that the vibrational motion of $H_2(g)$ is harmonic.

30-44. Below is a drawing of the contour plot of the potential-energy surface of the collinear $H(g) + H_2(g)$ reaction in the vicinity of the transition state. We take r_{12} and r_{23} to be the bond length of H_2 reactant and product, respectively. Label the location of the transition state. Draw a dashed line that indicates the lowest energy path for the reaction. Draw a two-dimensional representation of the reaction path in which you plot $V(r_{12}, r_{23})$ as a function of $r_{12} - r_{23}$.

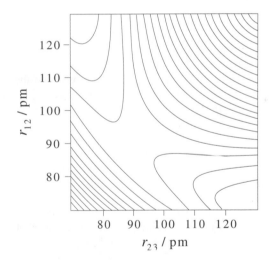

30-45. Repeat the calculation in Problem 30–43 for the reaction

$$
H(g) + D_2(v = 0) \Longrightarrow HD(v = 0) + D(g)
$$

Assume that the vibrational motion of $D_2(g)$ is harmonic.

Dorothy Crowfoot Hodgkin was born in Cairo, Egypt, on May 12, 1910, and died in 1994. She received her doctorate in 1934 from Cambridge University for her dissertation on X-ray crystallography of large molecules. While a graduate student, she recorded the first X-ray diffraction pattern from a protein (pepsin) crystal. She later determined the three-dimensional structures of a number of biologically important molecules such as cholesterol, penicillin, vitamin B_{12}, and zinc insulin (which contains almost 800 atoms). In 1965, Hodgkin was awarded the Order of Merit, the highest civilian honor in the United Kingdom, the second woman after Florence Nightingale to be thus honored. In 1970, she became the chancellor of Bristol University, where she established the Hodgkin Scholarship and Hodgkin House, in honor of her husband, for students from third world countries. She was actively involved in the campaign for peace and disarmament and served as the president of the Pugwash Conference on Science and World Affairs in the 1970s. In 1964, Hodgkin was awarded the Nobel Prize for chemistry "for her determinations by X-ray techniques of the structure of important biological substances."

CHAPTER 31

Solids and Surface Chemistry

In this chapter, we examine some of the modern topics in solid-state chemistry. In the first half, we discuss the structure of crystals. We will learn that X-ray diffraction can be used to determine the structure of atomic and molecular crystals. We will show that the X-ray diffraction pattern of a crystal reflects the periodic distribution of the electron density in the crystal. In the case of molecular crystals, the X-ray data allow for the determination of molecular bond lengths and bond angles.

In the second half of this chapter, we give an introduction to *surface chemistry*, which is the study of how the surfaces of solids catalyze chemical reactions. For example, the cracking of large molecules in crude oil is carried out in the presence of alumina silicate catalysts, commonly known as zeolites. Zeolites are particularly effective at converting olefins and cycloparaffins to the paraffins and aromatics used in gasoline and jet fuel. Improving the efficacy of catalytic reactions is an important area of both academic and industrial research. A mere 1% increase in the conversion efficiency of catalytic cracking would reduce U.S. oil imports by 22 million barrels per year.

The reaction between H_2 and N_2 to form NH_3 does not occur to any appreciable extent in the gas phase, but it readily occurs in the presence of an iron catalyst that is doped with K_2O and interspersed with Al_2O_3. This reaction is of great importance to society because ammonia is the starting point for the synthesis of all common bulk fertilizers. Understanding the details of these types of reactions requires a knowledge of how molecules chemically interact with surfaces.

31–1. The Unit Cell Is the Fundamental Building Block of a Crystal

Figure 31.1 shows the arrangement of the atoms in a crystal of copper. From this arrangement, we can see that the crystal possesses a periodic structure, and we should take advantage of this periodicity to describe its structure. We define a *unit cell* to be the smallest collection of atoms (or molecules) in the crystal such that the replication

1271

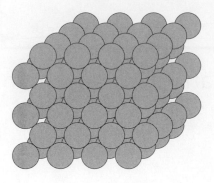

FIGURE 31.1
A schematic illustration of the location of the copper atoms in a copper crystal. Note the periodic arrangement of the atoms.

of the unit cell in three dimensions generates the entire crystal. In other words, we will describe the crystal by a repeating pattern of unit cells. Figure 31.2 illustrates in two dimensions how a unit cell can generate a crystal lattice. Clearly, a unit cell cannot have any arbitrary shape. For example, we cannot have a spherical unit cell because there would be gaps between the spheres when this unit cell is replicated in three dimensions. It is also impossible to generate a crystal lattice by a unit cell that has a five-fold symmetry axis (Problem 31–43). The unit cell must be a geometric structure that fills all space when replicated. Figure 31.3 shows the structure of the

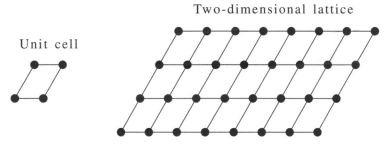

FIGURE 31.2
A two-dimensional illustration of the generation of a crystal lattice by a unit cell.

unit cell of crystalline copper. The unit cell for this crystalline arrangement is a cube. Copper atoms are centered on both the corners of the cube and at the faces of the cube. If we were to replicate this unit cell in three dimensions, we would obtain the structure shown in Figure 31.1. Note that each of the eight corner copper atoms is shared by eight neighboring unit cells, and each of the atoms centered at the six faces of the cube is shared by two neighboring unit cells (see Figure 31.3). Thus, there are $(1/8)8 + (1/2)6 = 4$ copper atoms per unit cell.

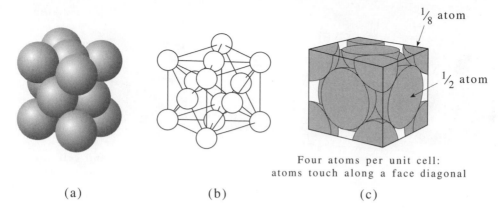

$\frac{1}{8}$ atom

$\frac{1}{2}$ atom

Four atoms per unit cell:
atoms touch along a face diagonal

(a)   (b)   (c)

FIGURE 31.3
The packing of copper atoms in a copper crystal. (a) The set of atoms that contribute to a unit cell of the crystal. The unit cell is a cube. (b,c) Copper atoms are centered on the corners and at the faces of the cube. Each copper atom is therefore shared between neighboring unit cells. (b) The unit cell for a three-dimensional lattice model of copper, where each atom of the crystal is associated with a lattice point. (c) The fractions of each copper atom shown in (a) that contribute to the unit cell of the crystal.

EXAMPLE 31–1
Figure 31.4 shows the unit cell of a potassium crystal. How many atoms are there in such a unit cell?

SOLUTION: As Figure 31.4 shows, there are atoms at each corner and one in the center of the (cubic) unit cell. The atoms at the corners are shared by eight unit cells, and the one in the center lies entirely within the unit cell. Therefore, there are $(1/8)8 + 1 = 2$ atoms per unit cell of potassium.

Figures 31.3 and 31.4 are two examples of cubic unit cells. The unit cell in Figure 31.3 is called a *face-centered cubic* unit cell because there are atoms centered at the faces of the cube in addition to the atoms at the corners. The unit cell shown in Figure 31.4 is called a *body-centered cubic* unit cell because there is an atom at the center of the cube in addition to the atoms at the corners. Only one other type of cubic unit cell is possible (Figure 31.5), one called a *primitive cubic* unit cell. Polonium is the only element whose crystals have a primitive cubic unit cell. Note that there is only one atom per unit cell in a primitive cubic unit cell.

So far we have discussed only cubic unit cells. The most general unit cell is a three-dimensional parallelepiped (Figure 31.6a). We will take the lower left corner of the unit cell as the origin of a coordinate system, with the positive **a**, **b**, and **c** axes pointing from this origin along the sides of the unit cell. We can describe the geometry of the unit cell by specifying a, b, and c, its lengths along the **a**, **b**, and **c** axes, respectively,

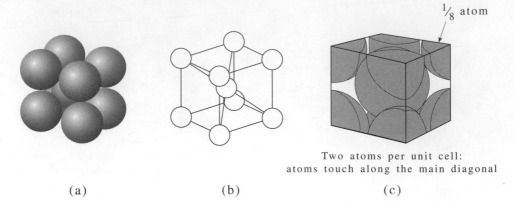

Two atoms per unit cell:
atoms touch along the main diagonal

(a)   (b)   (c)

FIGURE 31.4
The packing of potassium atoms in a potassium crystal. (a) The set of atoms that contribute to a unit cell of the crystal. The unit cell is a cube. (b,c) Potassium atoms are centered on the corners and at the center of the cube. (b) The unit cell for a three-dimensional lattice model of potassium, where each atom of the crystal is associated with a lattice point. (c) The fractions of each potassium atom shown in (a) that contribute to the unit cell of the crystal.

and its angles α, β, and γ between pairs of axes. Figure 31.6b shows that this unit cell generates a three-dimensional solid when replicated in three dimensions.

    You might think there is an infinite number of unit cells that can be used to generate crystal lattices, but in 1848, the French physicist August Bravais proved that only 14

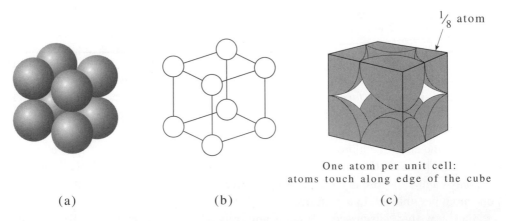

One atom per unit cell:
atoms touch along edge of the cube

(a)   (b)   (c)

FIGURE 31.5
The packing of polonium atoms in a polonium crystal. (a) The set of atoms that contribute to a unit cell of the crystal. The unit cell is a cube. (b,c) Polonium atoms are centered on the corners of the cube. (b) The unit cell for a three-dimensional lattice model of polonium, where each atom of the crystal is associated with a lattice point. (c) The fractions of each polonium atom shown in (a) that contribute to the unit cell of the crystal.

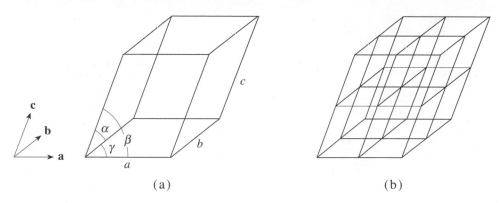

FIGURE 31.6
(a) The general shape of a unit cell. We take the bottom left corner of the unit cell to be the origin of the **a**, **b**, **c** coordinate system. The unit cell is defined by a, b, and c, its length along the **a**, **b**, and **c** axes, respectively, and the angles α, β, and γ between pairs of axes. (b) By replicating the unit cell in three dimensions, a crystal lattice is generated.

distinct unit cells are necessary to generate all possible crystal lattices. These 14 so-called *Bravais lattices* are shown in Figure 31.7. In this chapter, we will focus our discussion on the lattices that have orthogonal axes, $\alpha = \beta = \gamma = 90°$. Note that the three cubic Bravais lattices are the primitive cubic (Figure 31.5), the body-centered cubic (Figure 31.4), and the face-centered cubic (Figure 31.3) lattices.

EXAMPLE 31–2
Copper, which crystallizes as a face-centered cubic lattice, has a density of $8.930 \, \text{g} \cdot \text{cm}^{-3}$ at 20°C. Calculate the radius of a copper atom, assuming that the atoms touch along a face diagonal, as shown in Figure 31.3c. Such a radius is called the *crystallographic radius*.

SOLUTION: There are four atoms per unit cell, so the mass of a unit cell is

$$\text{mass unit cell} = \frac{(4)(63.55 \, \text{g} \cdot \text{mol}^{-1})}{6.022 \times 10^{23} \, \text{mol}^{-1}} = 4.221 \times 10^{-22} \, \text{g}$$

and so its volume is

$$V_{\text{unit cell}} = \frac{4.221 \times 10^{-22} \, \text{g}}{8.930 \, \text{g} \cdot \text{cm}^{-3}} = 4.727 \times 10^{-23} \, \text{cm}^3$$

Because the unit cell is cubic, a, the length of an edge, is given by the cube root of $V_{\text{unit cell}}$, or

$$a = (V_{\text{unit cell}})^{1/3} = 3.616 \times 10^{-8} \, \text{cm} = 361.6 \, \text{pm}$$

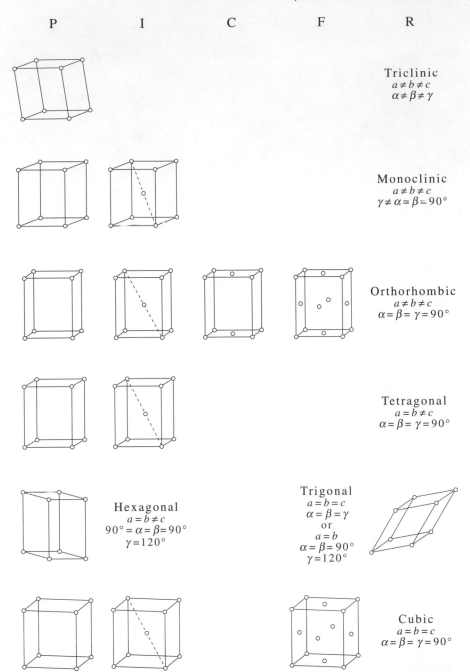

FIGURE 31.7

The 14 Bravais lattices. These 14 unit cells generate all the possible three-dimensional crystal lattices. The lattices are organized into columns, where P refers to a primitive unit cell (one lattice point per unit cell), I refers to a body-centered unit cell, C refers to an end-centered unit cell, F refers to a face-centered unit cell, and R refers to a rhombohedral unit cell. The 14 Bravais lattices are organized into seven classes (triclinic, monoclinic, orthorhombic, tetragonal, hexagonal, trigonal, and cubic) by the general geometric features of the lengths of the three sides of the parallelepiped and the angles between the **a**, **b**, and **c** axes of the unit cells.

Figure 31.3c shows that the effective radius of an atom in a face-centered cubic lattice is given by one-fourth of the length of the diagonal of a face. The length of the diagonal is given by

$$d = (2)^{1/2}a = 511.4 \text{ pm}$$

so the crystallographic radius of a copper atom is $(511.4 \text{ pm})/4 = 127.8$ pm.

EXAMPLE 31–3

What fraction of the volume of the unit cell is occupied by copper atoms? Assume each atom is a hard sphere in contact with its nearest neighbor.

SOLUTION: Recall that copper crystallizes in a face-centered cubic structure. Let a be the length of the sides of the cubic unit cell. The total volume of the unit cell is a^3. Consider one of the six identical faces of the unit cell shown below.

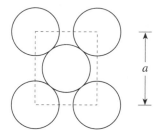

If r is the radius of a copper atom, then by the Pythagorean theorem, we have

$$(4r)^2 = a^2 + a^2$$

or

$$r = \left(\frac{1}{8}\right)^{1/2} a$$

The volume of a copper atom in terms of a, the length of the unit cell, is then

$$V = \frac{4}{3}\pi r^3 = \frac{\pi a^3}{6(8)^{1/2}}$$

There are a total of four copper atoms per unit cell, so the fraction of occupied volume is

$$\text{fraction occupied} = \frac{4V}{a^3} = 0.740$$

A crystal lattice is a network of points that reflects the symmetry of the representative crystal. The points are mathematical constructs and do not necessarily depict atoms. Generally, the lattice points may represent a single atom, a molecule, or even a collection of atoms or molecules in the crystal. The unit cell is then formed by connecting lattice points and is usually the smallest parallelepiped of lattice points such that the replication of the unit cell in three dimensions generates the entire lattice. For example, each atom in the copper crystal could be represented by a lattice point in the unit cell given by Figure 31.3b. In this case, we have simply replaced each copper atom in the crystal by a single lattice point. Now consider the face-centered cubic unit cell of crystalline C_{60} molecules, Figure 31.8. Rather than describing the location of each of the atoms of each C_{60} molecule, we can associate a lattice point with a single C_{60} molecule and then represent the unit cell by the simple structure shown in Figure 31.8b. In this case, the lattice point represents the location of a molecule in the crystal.

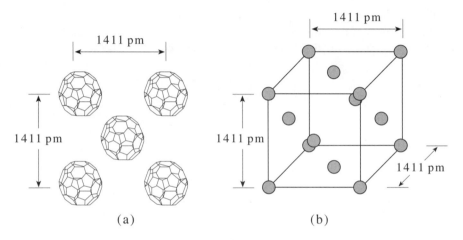

(a)   (b)

FIGURE 31.8
(a) A face of the face-centered cubic unit cell of crystalline C_{60}. A C_{60} molecule is centered on each corner and at the face. The center-to-center distance between C_{60} molecules on one of the sides of the cube is 1411 pm. If we associate each C_{60} molecule with a point on a three-dimensional lattice, the unit cell in terms of these lattice points is given by the structure shown in (b). Because each lattice point represents one C_{60} molecule, the distance between lattice points along the side of the unit cell is 1411 pm.

31–2. The Orientation of a Lattice Plane Is Described by its Miller Indices

The coordinates of the atoms contained in a unit cell are expressed in units of a, b, and c, the lengths of three edges of the unit cell. For example, consider the primitive cube unit cell (Figure 31.5). If we take the lattice point at the bottom left corner to be the origin of the crystal coordinate system, the coordinates of this point are $0a$, $0b$, $0c$, which we will write as (0,0,0). We see that moving a distance a along the **a** axis from the origin takes us to the lattice point $1a$, $0b$, $0c$, or (1,0,0). The remaining lattice points of the primitive cubic unit cell are (0,1,0), (0,0,1), (1,1,0), (1,0,1), (0,1,1), and (1,1,1).

EXAMPLE 31–4

What are the coordinates of the lattice points in the body-centered cubic unit cell?

SOLUTION: The body-centered cubic unit cell has lattice points at each of the eight corners of the cube and a lattice point at the center of the cube. The lattice points at the corners are separated by the length of the edges of the unit cell, so these lattice points have the same coordinates as those of the primitive cubic unit cell, or (0,0,0), (1,0,0), (0,1,0), (0,0,1), (1,1,0), (1,0,1), (0,1,1), and (1,1,1). The lattice point at the center of the cube is located at a distance of 1/2 the length of all three edges of the unit cell, or (1/2,1/2,1/2).

Because of the periodicity of the crystal lattice, we can view the lattice as being comprised of sets of equally spaced parallel planes containing lattice points (Figure 31.9). Although this particular description of the crystal lattice may seem to be just another arbitrary way of looking at the crystal structure, it is important for understanding X-ray diffraction patterns and relating these patterns to the distances and angles between atoms and molecules in the crystal. As for the coordinates of the lattice points, we would like to describe a set of parallel crystallographic planes in terms of the lengths of the three sides of the unit cell. Consider a plane that intersects the **a**, **b**, and **c** axes of a unit cell at points a', b', and c'. For example, the planes in Figure 31.9b intersect the **a** axis at a, intersect the **b** axis at b, and are parallel to the **c** axis (that is, the planes intersect the **c** axis at infinity). Thus, a', b', and c' in this case are a, b, and ∞. We designate the plane by the three indices

$$h = \frac{a}{a'} \qquad k = \frac{b}{b'} \qquad l = \frac{c}{c'} \qquad (31.1)$$

which in the case of Figure 31.9b are 1, 1, and 0, which we write as 110. Thus, we refer to the planes shown in Figure 31.9b as the 110 planes. Similarly, the planes in Figure 31.9c intersect the **a**, **b**, and **c** axes at $a' = a$, $b' = b$, and $c' = c$, so in this case $h = 1$, $k = 1$, and $l = 1$ and the planes are called the 111 planes.

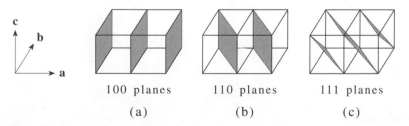

100 planes   110 planes   111 planes

(a)   (b)   (c)

FIGURE 31.9

Various sets of equally spaced parallel planes for a primitive cubic lattice. The indices *hkl* are associated with a family of parallel planes separated by a distance a/h along the **a** axis, b/k along the **b** axis, and c/l along the **c** axis, where a, b, and c are the lengths of the sides of the unit cell. (a) 100 planes (b) 110 planes (c) 111 planes.

The three indices h, k, and l that we use to specify parallel planes through a crystal lattice are called *Miller indices*. These indices uniquely specify a set of parallel planes within a crystal. The Miller indices are associated with a family of parallel planes separated by a distance a/h along the **a** axis, b/k along the **b** axis, and c/l along the **c** axis. Figure 31.10a illustrates the set of 220 planes of a cubic lattice. The darkened plane intersects the **a**, **b**, and **c** axes at $a' = a/2$, $b' = b/2$, and $c' = \infty$, so $h = 2$, $l = 2$, and $l = 0$ and the planes are called the 220 planes. The set of 220 planes are separated by a distance of $a/2$ and $b/2$ along the **a** and **b** axes of the crystal, respectively. Now consider the set of planes shown in Figure 31.10b. Taking the origin of the coordinate system of the unit cell to be at the lower left corner of the cube, the darkened plane intersects the crystal axes of the unit cell at $a' = a$, $b' = b$, and $c' = -c$, so Equations 31.1 give $h = k = 1$ and $l = -1$. By convention, we designate negative indices by the corresponding number with a bar over it, hence the designation $11\bar{1}$. Figure 31.10b illustrates the set of $11\bar{1}$ planes of a cubic lattice.

We give without proof that d, the perpendicular distance between adjacent hkl planes for a orthorhombic unit cell (see Figure 31.7), is given by

$$\frac{1}{d^2} = \frac{h^2}{a^2} + \frac{k^2}{b^2} + \frac{l^2}{c^2} \tag{31.2}$$

For a cubic unit cell ($a = b = c$), Equation 31.2 simplifies to

$$\frac{1}{d^2} = \frac{h^2 + k^2 + l^2}{a^2} \tag{31.3}$$

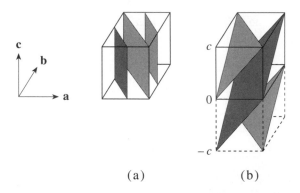

(a)   (b)

FIGURE 31.10
(a) An illustration of the set of parallel 220 planes of a cubic lattice. The highlighted plane intersects the crystal axes at $a' = a/2$, $b' = b/2$, and $c' = \infty$, hence the designation 220. (b) An illustration of the set of parallel $11\bar{1}$ planes of a cubic lattice. The highlighted plane intersects the crystal axes of the unit cell at $a' = a$, $b' = b$, and $c' = -c$, hence the designation $11\bar{1}$.

EXAMPLE 31–5

Consider an orthorhombic unit cell with dimensions $a = 487$ pm, $b = 646$ pm, and $c = 415$ pm. Calculate the perpendicular distance between (a) the 110 planes and (b) the 222 planes of this crystal.

SOLUTION: The sets of parallel 110 and 222 planes that cross an orthorhombic unit cell are shown below.

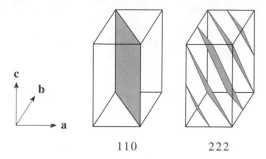

110   222

We can find the perpendicular distances between adjacent planes using Equation 31.2. For the 110 planes, we have

$$\frac{1}{d^2} = \frac{h^2}{a^2} + \frac{k^2}{b^2} + \frac{l^2}{c^2}$$

$$= \frac{1}{(487 \text{ pm})^2} + \frac{1}{(646 \text{ pm})^2} + \frac{0}{(415 \text{ pm})^2} = 6.61 \times 10^{-6} \text{ pm}^{-2}$$

or $d = 389$ pm. For the 222 planes, a similar calculation gives $d = 142$ pm.

31–3. The Spacing Between Lattice Planes Can Be Determined from X-Ray Diffraction Measurements

The structure of a crystal can be determined using the technique of X-ray diffraction. X-rays are generated by bombarding a metal target (often copper) with high-energy electrons inside a vacuum tube. Collisions between the high-energy electrons and copper atoms generate electronically excited copper cations, which then relax back to their ground state by emitting a photon. The radiation emitted consists of two closely spaced lines at 154.433 pm and 154.051 pm. One of these is then directed at a single crystal. The crystal mount can be rotated, enabling the experimenter to orient the incident X-rays with respect to the three crystallographic axes. Most of the X-rays pass straight through the crystal. A small amount of the radiation, however, is diffracted by the crystal, and the pattern of this diffracted light is recorded by a two-dimensional array detector. The image recorded on the detector is called the *diffraction pattern*.

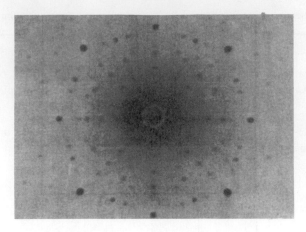

FIGURE 31.11
The X-ray diffraction pattern observed from a single crystalline sample of tungsten. The unit cell of tungsten is a body-centered cube.

Figure 31.11 shows the X-ray diffraction pattern from a single crystal of tungsten. We see that the pattern is a collection of spots of varying intensity. We will now learn that the positions and intensities of the diffraction spots are determined by the spacing between the different sets of parallel hkl planes of the crystal lattice.

Consider two lattice points A_1 and A_2 (Figure 31.12) that lie in neighboring hkl planes and lie along the **a** axis of the crystal and are separated by a'. (For example, the planes might be perpendicular to the plane of Figure 31.12.) Let α_0 be the angle of incidence of the X-ray beam and α be the angle of diffraction. Now consider an observer situated at the angle α. If $\alpha \neq \alpha_0$, the X-rays diffracted from lattice point A_1

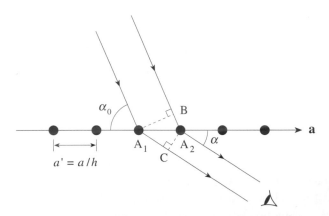

FIGURE 31.12
An illustration of the scattering from lattice points along the **a** axis contained in adjacent hkl planes. (For example, the hkl planes might be perpendicular to the plane of the figure.) The hkl planes are parallel, so α_0, the angle of incidence of the X-ray radiation, is the same for each lattice point. The X-rays then scatter from these lattice points at an angle α.

travel a different total path length in reaching the observer than those diffracted from lattice point A_2. This difference in the path length, Δ, is given by

$$\Delta = \overline{A_1C} - \overline{A_2B} \tag{31.4}$$

in Figure 31.12. If the distance Δ is equal to an integral multiple of the wavelength of the X-ray radiation, the two diffracted beams will interfere constructively. If Δ is not equal to an integral multiple of the wavelength of the X-ray radiation, the two beams interfere destructively. If we extend this argument to include diffraction from all the atoms in the row shown in Figure 31.12, then to observe a diffraction signal, the light diffracted from each atom in the row must interfere constructively. This means that the crystal plane must be oriented with respect to the incident X-rays so that Δ is equal to an integral multiple of the wavelength of the X-ray radiation, or that $\Delta = n\lambda$, where n is an integer. Now suppose that this condition is satisfied for a particular set of hkl planes in the crystal. From the geometry shown in Figure 31.12, we have $\overline{A_2B} = a' \cos \alpha_0$ and $\overline{A_1C} = a' \cos \alpha$, so we can write Equation 31.4 as

$$\Delta = a'(\cos \alpha - \cos \alpha_0) = n\lambda \tag{31.5}$$

The distance between lattice points along the **a** axis in neighboring hkl planes is given by $a' = a/h$, where a is the length of the unit cell along the **a** axis. We can therefore rewrite Equation 31.5 in terms of the Miller index and the unit cell length,

$$a(\cos \alpha - \cos \alpha_0) = nh\lambda \tag{31.6}$$

The diffraction spots that correspond to $n = 1$ are called *first-order reflections*; the diffraction spots that correspond to $n = 2$ are called *second-order reflections*, and so on.

There is an equation similar to Equation 31.6 for each of the other two axes of the crystal. If we take β_0 and γ_0 to be the angles of incidence of the X-ray radiation with respect to the **b** and **c** axes of the crystal and β and γ as the corresponding diffraction angles, the equations for first-order diffraction from the lattice points along the **b** and **c** axes in a set of parallel hkl planes are given by

$$b(\cos \beta - \cos \beta_0) = k\lambda \tag{31.7}$$

and

$$c(\cos \gamma - \cos \gamma_0) = l\lambda \tag{31.8}$$

Equations 31.6 through 31.8 were originally derived by the German physicist Max von Laue and are collectively called the *von Laue equations*.

As an example of how to use the von Laue equations, we will consider the diffraction pattern obtained when an X-ray beam is directed at a crystal whose unit cell is a primitive cubic. We will orient the crystal such that the incident X-rays are perpendicular to the

a axis of the crystal. In this case, α_0, the angle between the X-rays and the **a** axis of the crystal is 90°, and the von Laue equations for first-order diffraction become

$$a \cos \alpha = h\lambda \tag{31.9}$$

$$a(\cos \beta - \cos \beta_0) = k\lambda \tag{31.10}$$

$$a(\cos \gamma - \cos \gamma_0) = l\lambda \tag{31.11}$$

Consider the set of parallel planes $h00$. Equation 31.9 tells us that each value of h corresponds to a specific value of the scattering angle α. For $h = 0$, $\cos \alpha = 0$ and $\alpha = 90°$; for $h = 1$, $\cos \alpha = \lambda/a$; for $h = 2$, $\cos \alpha = 2\lambda/a$; etc. Furthermore, $\beta = \beta_0$ and $\gamma = \gamma_0$ because $k = 0$ and $l = 0$. Thus, the von Laue equations show that when the crystal is oriented with its **a** axis perpendicular to the incident X-rays, the $h00$ planes will give rise to a set of diffraction spots that lie along a line that is perpendicular to the direction of the incident X-rays and parallel to the **a** axis of the crystal (see Figure 31.13). For $h = 0$, $\cos \alpha = 0$, so $\alpha = 90°$, meaning the X-ray beam passes straight through the crystal (see Figure 31.13). Positive values of h give a series of diffraction spots through positive values of α, where α is given by $\cos \alpha = h\lambda/a$ with $h = 1$, 2, ... Negative values of h give a series of diffraction spots through negative values of α. Thus, we obtain the diffraction pattern shown in Figure 31.13.

Example 31–6 shows how we can use the spacing between the 000 and 100 diffraction spots to determine the spacing between lattice points along the **a** axis of the unit cell. If we were to collect diffraction information for the crystal oriented with the incident X-rays perpendicular to both the **b** and **c** axes, we could determine the lattice spacing along these axes in a similar manner.

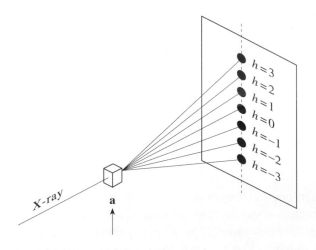

FIGURE 31.13
An illustration of the X-ray diffraction pattern from the $h00$ planes of a crystal in which the incident X-rays are perpendicular to the **a** axis of the crystal.

EXAMPLE 31–6

The detector in an X-ray diffractometer is located 5.00 cm from the crystal. A crystal whose unit cell is a primitive cube is oriented such that its **a** axis is perpendicular to the incident X-rays. The distance between the detected spots corresponding to diffraction from the origin and 100 planes of the crystal is 2.25 cm. The $\lambda = 154.433$ pm line of copper is used as the X-ray source. What is the length of the unit cell along the **a** axis?

SOLUTION: The geometry of this experiment is shown below.

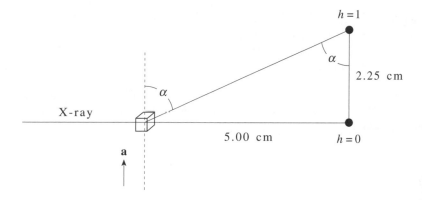

The incident X-rays are perpendicular to the **a** axis of the crystal, so the scattering pattern from the $h00$ planes will look like that shown in Figure 31.13. The scattering from the origin planes passes straight through the crystal. The angle α corresponding to the diffraction from the 100 planes is given by Equation 31.9

$$a \cos \alpha = \lambda$$

We can solve this equation for the cell constant a,

$$a = \frac{\lambda}{\cos \alpha} \tag{1}$$

The above drawing shows that $\tan \alpha = 5.00/2.25$ or $\alpha = \tan^{-1}(5.00/2.25) = 65.77°$. Substituting this into Equation 1 gives us

$$a = \frac{154.433 \text{ pm}}{\cos 65.77°} = 376.37 \text{ pm}$$

For an arbitrary hkl plane, the direction of diffraction with respect to the **a** axis is the same as that for the $h00$ planes. But there is also diffraction with respect to the **b** and **c** axes. Thus, the diffraction spots from an hkl plane will lie along the surface of a cone that makes an angle α with respect to the plane defined by the direction of the incident X-rays and **a** axis of the crystal (Figure 31.14). The exact location of the diffraction spot along this cone depends on the scattering angles β and γ, which depend on the values of k and l and are determined by Equations 31.7 and 31.8. Figure 31.15 illustrates the

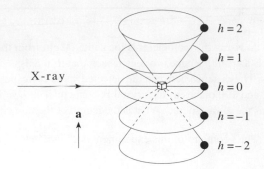

FIGURE 31.14
The scattering from the *hkl* planes in the crystal, where the incident X-rays are perpendicular to the **a** axis of the crystal. The filled dots represent the scattering from the *h*00 planes (see Figure 31.13). For *hkl* planes where *k* and/or *l* are not equal to zero, the scattering angle with respect to the crystal **a** axis is the same as that for the *h*00 plane. Thus, diffraction spots from the *hkl* planes lie along the surface of a cone of constant scattering angle α with respect to the crystal **a** axis. The exact location of the spot from a particular *hkl* plane is determined by the van Laue equations.

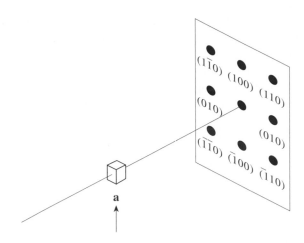

FIGURE 31.15
The diffraction spots from some of the *hkl* planes of a primitive cubic crystal whose **a** axis is oriented perpendicular to the incident X-rays. Each spot corresponds to unique values of the diffraction angles α, β, and γ. The diffraction angles can be determined by the van Laue equations.

location of the diffraction spots from some of the *hkl* planes of the primitive cubic crystal we have been discussing.

You might have learned in general chemistry that the English chemist William Bragg developed an alternative way of looking at X-ray diffraction. Bragg modeled the diffraction of X-rays from crystals as originating from the reflection of X-rays from the various sets of parallel *hkl* lattice planes. The derivation of his equation is given in Problem 31–29, and the result is

$$\lambda = 2 \left(\frac{d}{n} \right) \sin \theta \tag{31.12}$$

where θ is the angle of incidence (and reflection) of the X-rays with respect to the lattice plane, λ is the wavelength of the X-ray radiation, and $n = 1, 2, \ldots$ is the order of the reflection. Equation 31.3 gives d in terms of the Miller indices for a cubic unit cell, and so we can write Equation 31.12 as

$$\sin^2 \theta = \frac{n^2 \lambda^2}{4a^2} (h^2 + k^2 + l^2) \tag{31.13}$$

EXAMPLE 31–7
Silver crystallizes in a face-centered-cubic structure with a unit cell length of 408.6 pm. Use the Bragg equation to calculate the first few observed diffraction angles from the 111 planes using X-radiation with a wavelength of 154.433 pm.

SOLUTION: The smallest diffraction angle occurs for $n = 1$ (first-order diffraction), so Equation 31.13 gives

$$\sin^2 \theta = \frac{\lambda^2}{4a^2} (h^2 + k^2 + l^2) = \frac{(154.433 \text{ pm})^2}{4(408.6 \text{ pm})^2} \text{ (3)}$$
$$= 0.1071$$

or $\theta = 19.11°$. The next smallest diffraction angle occurs for $n = 2$ (second-order diffraction), so

$$\sin^2 \theta = (4)(0.1071) = 0.4284$$

or $\theta = 40.88°$.

Equation 31.12 can be derived from the von Laue equations, so the two approaches are equivalent ways of viewing the origin of the observed diffraction patterns (Problems 31-44 and 31–45).

We see from the von Laue equations that the angle of diffraction depends on the incident angle, the dimensions of the unit cell, the wavelength of the X-ray radiation, and the Miller indices. In practice, diffraction spots are not observed from all *hkl* planes of the crystal lattice. For example, an atomic crystal whose unit cell is body-centered cubic shows no diffraction from the *hkl* planes in which $h + k + l$ is an odd number. In addition, the intensity of the diffraction spots corresponding to different *hkl* planes can vary significantly. To understand which lattice planes give rise to diffraction spots and what determines the intensity of these spots, we need to examine the details of how atoms diffract X-rays.

31–4. The Total Scattering Intensity Is Related to the Periodic Structure of the Electron Density in the Crystal

X-rays are scattered by the electrons in crystals. Because both the number of electrons and the size of the atomic orbitals vary from atom to atom, different atoms have different scattering efficiencies. The *scattering factor* of an atom, f, is defined by

$$f = 4\pi \int_0^\infty \rho(r)\frac{\sin kr}{kr}r^2dr \qquad (31.14)$$

where $\rho(r)$ is the spherically symmetric electron density (number of electrons per unit volume) of the atom and $k = (4\pi/\lambda)\sin\theta$, where θ is the scattering angle and λ is the wavelength of the X-radiation. The wavelength of the X-radiation used to record diffraction patterns is comparable with the size of an atom, so the scattering from different regions of an atom interfere with each other. The integrand in Equation 31.14 takes this interference into account through the factor $\sin(kr)/kr$. Figure 31.16 shows a plot of f as a function of $\sin\theta/\lambda$ for different atoms.

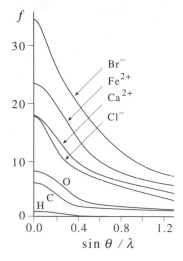

FIGURE 31.16
The dependence of the scattering factors on the number of electrons and the angle of diffraction. The scattering factor for $\theta = 0$ is equal to the total number of electrons on the atom or ion.

EXAMPLE 31–8
Show that the scattering factor of an atom in the direction $\theta = 0$ is equal to the total number of electrons on the atom.

SOLUTION: A scattering angle of $\theta = 0$ means that the X-rays pass straight through the atom. If $\theta = 0$, then $k = (4\pi/\lambda)\sin\theta = 0$, and the term $\sin(kr)/kr$ in Equation 31.14 is indeterminate. To evaluate the integrand, we need to evaluate $\lim_{kr \to 0}(\sin kr/kr)$. If we express $\sin kr$ in terms of a power series in kr (see Math-Chapter I), then

$$\lim_{kr \to 0}\frac{\sin kr}{kr} = \lim_{kr \to 0}\frac{kr - \dfrac{(kr)^3}{3!} + \cdots}{kr} = 1 + O[(kr)^2]$$

Thus, Equation 31.14 becomes

$$f = 4\pi \int_0^\infty \rho(r) r^2 dr$$

The integrand is the product of the electron density and the spherical volume element $4\pi r^2 dr$, which upon integration gives the total number of electrons in the atom.

Now let's consider the one-dimensional lattice shown in Figure 31.17. This lattice consists of two different types of atoms, 1 and 2, with scattering factors f_1 and f_2, respectively. The distance betweeen successive 1 atoms or successive 2 atoms is a/h, where a is the length of the unit cell along the **a** axis, and the distance between successive 1 and 2 atoms is x. If the crystal is oriented such that Equation 31.6, the von Laue equation governing scattering from atoms along the **a** axis, is satisfied, then the difference in path length traveled by X-rays diffracted by successive 1 atoms, Δ_{11}, (or successive 2 atoms, Δ_{22}) is given by (see Equation 31.5),

$$\Delta_{11} = \Delta_{22} = \frac{a}{h}(\cos\alpha - \cos\alpha_0) = \lambda \tag{31.15}$$

where we have taken $n = 1$. However, the difference in path length traveled by X-rays diffracted by successive 1 and 2 atoms is given by

$$\Delta_{12} = x(\cos\alpha - \cos\alpha_0) \tag{31.16}$$

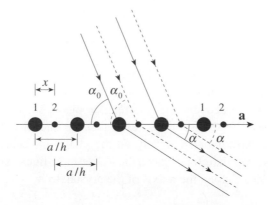

FIGURE 31.17
An illustration of the scattering from a lattice consisting of two different types of atoms. The distance between the successive 1 atoms and successive 2 atoms is a/h and the distance between successive 1 atoms and 2 atoms is x.

and is not equal to an integral number of wavelengths. The difference in path lengths traveled by the X-rays scattered from the successive 1 and 2 atoms can be found by rearranging Equation 31.15 to give

$$\cos\alpha - \cos\alpha_0 = \frac{\lambda h}{a}$$

and then substituting this result into Equation 31.16

$$\Delta_{12} = \frac{\lambda h x}{a} \tag{31.17}$$

This difference in path length corresponds to a phase difference between the diffracted beams from successive 1 and 2 atoms of

$$\phi = 2\pi\frac{\Delta_{12}}{\lambda} = 2\pi\frac{\lambda h x/a}{\lambda} = \frac{2\pi h x}{a} \tag{31.18}$$

The amplitude of the light scattered from successive 1 and 2 atoms is then

$$A = f_1\cos\omega t + f_2\cos(\omega t + \phi) \tag{31.19}$$

where f_1 and f_2 are the scattering factors of atoms 1 and 2, respectively, and ω is the angular frequency of the X-ray radiation. For convenience, we will use exponential functions instead of cosine functions (see MathChapter A) to describe the time-dependent behavior of the electric field, whereby we can write Equation 31.19 as

$$A = f_1 e^{i\omega t} + f_2 e^{i(\omega t + \phi)} \tag{31.20}$$

Recall that the detected intensity is proportional to the square of the magnitude of the amplitude (Problem 3-31), so

$$\begin{aligned} I \propto |A|^2 &= [f_1 e^{i\omega t} + f_2 e^{i(\omega t + \phi)}][f_1 e^{-i\omega t} + f_2 e^{-i(\omega t + \phi)}] \\ &= f_1^2 + f_1 f_2 e^{i\phi} + f_1 f_2 e^{-i\phi} + f_2^2 \\ &= f_1^2 + f_2^2 + 2f_1 f_2\cos\phi \end{aligned} \tag{31.21}$$

The first two terms of Equation 31.21 reflect the constructive interference of the X-rays scattered from the set of parallel planes through the 1 atoms and 2 atoms, respectively. The third term takes into account the interference of the scattering from these two sets of parallel planes. We see from this result that the intensity does not depend on the frequency of the X-radiation but only on the phase difference between the two diffracted beams. We can therefore ignore the $e^{i\omega t}$ term in Equation 31.20, and define $F(h)$, the structure factor along the **a** axis of the crystal, to be

$$F(h) = f_1 + f_2 e^{i\phi} = f_1 + f_2 e^{2\pi i h x/a} \tag{31.22}$$

where ϕ is given by Equation 31.18. The intensity (Equation 31.21) is then proportional to $|F(h)|^2$.

Generalizing Equation 31.22 to three dimensions for a unit cell that contains atoms of type j located at points x_j, y_j, and z_j, gives us

$$F(hkl) = \sum_j f_j e^{2\pi i(hx_j/a + ky_j/b + lz_j/c)} \tag{31.23}$$

where a, b, and c are the lengths of the three sides of the unit cell, f_j is the scattering factor of an atom of type j, and hkl are the Miller indices of the diffracting planes. The coordinates x_j, y_j, and z_j are commonly expressed in units of a, b, and c, the lengths of the unit cells, in which case Equation 31.23 can be written as

$$F(hkl) = \sum_j f_j e^{2\pi i(hx'_j + ky'_j + lz'_j)} \tag{31.24}$$

where $x'_j = x_j/a$, $y'_j = y_j/b$, and $z'_j = z_j/c$. The quantity $F(hkl)$ is called the *structure factor* of the crystal. Generalizing Equation 31.21 to three dimensions shows that the intensity of a diffraction spot from a crystal is proportional to the square the magnitude of the structure factor, so that $I \propto |F(hkl)|^2$. Thus, if $F(hkl) = 0$ for any set of Miller indices h, k, and l, then those planes will not give rise to an observable diffraction spot. The following Example illustrates such a result.

EXAMPLE 31–9

Derive an expression for the structure factor for a body-centered cubic unit cell of identical atoms. Do all the hkl planes of the crystal lattice give rise to diffraction spots?

SOLUTION: In Example 31–4, we showed that the coordinates of the lattice points in a body-centered cubic unit cell are (0,0,0), (1,0,0), (0,1,0), (0,0,1), (1,1,0), (1,0,1), (0,1,1), (1,1,1), and (1/2,1/2,1/2). The unit of distance is a, the length of the side of the cubic unit cell, so these coordinates correspond to those given in Equation 31.24. Each corner lattice point is shared by 8 unit cells, so we must multiply the scattering efficiency of each corner lattice point by 1/8. Using Equation 31.24 and setting $a = b = c$ because the unit cell is cubic gives us

$$F(hkl) = \tfrac{1}{8}f[e^{2\pi i(0+0+0)} + e^{2\pi i(h+0+0)} + e^{2\pi i(0+k+0)}$$
$$+ e^{2\pi i(0+0+l)} + e^{2\pi i(h+k+0)} + e^{2\pi i(h+0+l)} + e^{2\pi i(0+k+l)}$$
$$+ e^{2\pi i(h+k+l)}] + f[e^{2\pi i(h/2+k/2+l/2)}]$$

Now $e^{2\pi i} = \cos 2\pi + i \sin 2\pi = 1$ and $e^{\pi i} = -1$, so the above expression simplifies to

$$F(hkl) = \tfrac{1}{8}f[1^0 + 1^h + 1^k + 1^l + 1^{h+k} + 1^{h+l} + 1^{k+l} + 1^{h+k+l}] + f(-1)^{h+k+l}$$

But $1^n = 1$ for all n, so

$$F(hkl) = \tfrac{1}{8}f[8] + f(-1)^{h+k+l} = f[1 + (-1)^{h+k+l}]$$

If $h + k + l$ is an even number, then $F(hkl) = 2f$. If $h + k + l$ is an odd number, then $F(hkl) = 0$. Thus, only the lattice planes such that $h + k + l$ is even give rise to a diffraction spot. Problem 31–37 shows that there will be reflections for all integer values of h, k, and l for a primitive cubic unit cell and that there will be reflections if h, k, and l are either all even or all odd for a face-centered cubic unit cell.

Sodium chloride and potassium chloride crystallize in two interpenetrating face-centered cubic lattices (see Figure 31.18a for NaCl). There are 27 ions that contribute to the unit cell. Each cation centered on the corners of the unit cell is shared by eight unit cells. The cations centered at the faces are shared between two unit cells. Thus, there are $(1/8)8 + 1/2(6) = 4$ cations per unit cell. The anion in the center of the unit cell is completely contained within the unit cell. The remaining anions are centered on the edges of the unit cell and are therefore shared between four unit cells. Thus, there are $1 + (1/4)12 = 4$ chloride ions per unit cell, or four NaCl or KCl units per unit cell.

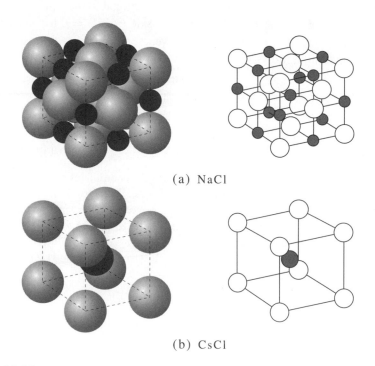

(a) NaCl

(b) CsCl

FIGURE 31.18
Space-filling and ball-and-stick representations of the unit cells of (a) NaCl and (b) CsCl. The different crystalline structures in the two cases are a direct consequence of the relative sizes of the cations and anions.

We can determine the structure factors for sodium chloride or potassium chloride using Equation 31.24. Let f_+ and f_- be the scattering factors for the cations and anions, respectively. By substituting the locations of the various ions contained in the unit cell into Equation 31.24 (Problem 31–41), the structure factor for sodium chloride or potassium chloride is found to be

$$F(hkl) = f_+[1 + (-1)^{h+k} + (-1)^{h+l} + (-1)^{k+l}]$$
$$+ f_-[(-1)^{h+k+l} + (-1)^h + (-1)^k + (-1)^l] \tag{31.25}$$

Equation 31.25 shows that

$$F(hkl) = 4(f_+ + f_-) \quad h, k, \text{ and } l \text{ are all even}$$
$$F(hkl) = 4(f_+ - f_-) \quad h, k, \text{ and } l \text{ are all odd} \tag{31.26}$$

If two of the indices are even and the third index is odd (or two of the indices are odd and the third index is even), then $F(hkl) = 0$. Because the intensity is proportional to the square of the magnitude of the structure factor, Equations 31.26 show that the intensity of the diffraction spots from the all-even hkl planes will be greater than the intensity of the diffraction spots from the all-odd hkl planes. This result is what is observed.

Equations 31.26 also reveal that if the scattering factors of the two ions are nearly the same, the scattering from the all-odd hkl planes will be very weak. As mentioned above, potassium chloride also crystallizes in a face-centered cubic lattice. But unlike NaCl(s), KCl(s) shows no diffraction spots corresponding to scattering from the hkl planes where $h, k,$ and l are all odd. Because K^+ and Cl^- are isoelectronic, the scattering factors of these two ions are essentially identical. As a result, the structure factor for the scattering from the all-odd hkl planes is essentially zero (see Equations 31.26).

Figure 31.18b shows the structure of CsCl(s), which is the same as for CsBr(s) and CsI(s). The scattering factor for this type of unit cell is (Problem 31–42)

$$F(hkl) = (f_+ + f_-) \quad \begin{array}{l} h, k, \text{ and } l \text{ are all even or} \\ \text{just one of them is even} \end{array}$$

$$F(hkl) = (f_+ - f_-) \quad \begin{array}{l} h, k, \text{ and } l \text{ are all odd or} \\ \text{just one of them is odd} \end{array}$$

31–5. The Structure Factor and the Electron Density Are Related by a Fourier Transform

In the previous section, we modeled the crystal as a set of atoms located at points (x_j, y_j, z_j) of the unit cell. We then defined the structure factor in terms of the scattering of X-rays from atoms located at each of the positions in the unit cell. In both atomic and molecular crystals, the electron density is not localized at individual points within the unit cell. Thus, our model for X-ray diffraction in terms of point scatterers is somewhat simplistic. Instead, we should consider the unit cell of the crystal to have a continuous electron density distribution $\rho(x, y, z)$. The structure factor (Equation 31.23) is no longer simply a sum over discrete atoms but now becomes an integral over the continuous electron density distribution in the unit cell:

$$F(hkl) = \int_0^a \int_0^b \int_0^c \rho(x, y, z)e^{2\pi i(hx/a+ky/b+lz/c)}dxdydz \qquad (31.27)$$

The entire crystal is built by replicating the unit cell in three dimensions. Each replicated unit cell has an identical structure factor, and so for a crystal of dimensions A, B, and C along the **a**, **b**, and **c** axes, respectively,

$$F(hkl) \propto \int_0^A \int_0^B \int_0^C \rho(x, y, z)e^{2\pi i(hx/a+ky/b+lz/c)}dxdydz$$

The electron density $\rho(x, y, z)$ is zero outside the crystal. Thus, the limits of these integrals can be changed to run from $-\infty$ to ∞ without affecting the value of the integral

$$F(hkl) \propto \int_{-\infty}^{\infty} \int_{-\infty}^{\infty} \int_{-\infty}^{\infty} \rho(x, y, z)e^{2\pi i(hx/a+ky/b+lz/c)}dxdydz \qquad (31.28)$$

Equation 31.28 shows that $F(hkl)$ is related to $\rho(x, y, z)$ by what is called a Fourier transform. One of the consequences of this Fourier transform relation is that $\rho(x, y, z)$ is given by

$$\rho(x, y, z) = \sum_{h=-\infty}^{\infty} \sum_{k=-\infty}^{\infty} \sum_{l=-\infty}^{\infty} F(hkl)e^{-2\pi i(hx/a+ky/b+lz/c)} \qquad (31.29)$$

As we learned in Section 31–4, $I(hkl)$, the intensity of scattered X-radiation from the hkl plane of the crystal, is proportional to the square of the magnitude of the structure factor, $I(hkl) \propto |F(hkl)|^2$. Experimental diffraction patterns give $|F(hkl)|^2$. To calculate $\rho(x, y, z)$ using Equation 31.29 requires that we determine $F(hkl)$. Because $F(hkl)$ is a complex quantity, we can write $F(hkl)$ in terms of the sum

$$F(hkl) = A(hkl) + iB(hkl) \qquad (31.30)$$

The intensity is then

$$I(hkl) \propto |F(hkl)|^2 = [A(hkl) + iB(hkl)][A(hkl) - iB(hkl)]$$
$$= [A(hkl)]^2 + [B(hkl)]^2 \qquad (31.31)$$

Unfortunately, $A(hkl)$ and $B(hkl)$ are not determined individually by a diffraction experiment. Only the sum of their squares is measured. The problem of determining $A(hkl)$ and $B(hkl)$ from the measurement of $I(hkl)$ is called the *phase problem*. Crystallographers have developed several methods for circumventing the phase problem. Figure 31.19 shows an electron-density map of benzoic acid that was determined from the X-ray diffraction pattern of a single crystal of benzoic acid. Each contour line in

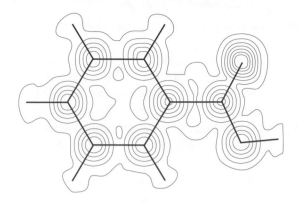

FIGURE 31.19
An electron-density map of a benzoic acid molecule determined from the X-ray diffraction pattern of a benzoic acid crystal. Each contour line corresponds to a constant value of the electron density. The location of the nuclei are readily deduced from this electron-density map and are represented by the vertices of the solid lines.

the figure corresponds to a constant value of the electron density. The positions of the nuclei are readily deduced from the electron-density map, from which bond length and bond angle information can be determined. Today, crystallographers can obtain and interpret electron-density maps of large chemical systems including strands of DNA and proteins.

31–6. A Gas Molecule Can Physisorb or Chemisorb to a Solid Surface

In 1834, the English chemist Michael Faraday suggested that the first step of a surface-catalyzed reaction was the sticking of the reactant molecule to the solid surface. Originally, it was believed that the main effect of the surface is to produce a local reactant concentration that is much higher than in the gas phase. Because the rate law depends on reactant concentration, this effect would then lead to an increased rate of reaction. Today, researchers have confirmed that the sticking of molecules to a surface is indeed the first step of surface-catalyzed reactions. As we will learn in the remainder of this chapter, however, the solid surface plays a much more important role than simply increasing the apparent concentration of reactant molecules.

A molecule approaching a surface experiences an attractive potential. The process of trapping molecules or atoms that are incident on a surface is called *adsorption*. The adsorbed molecule or atom is called the *adsorbate*, and the surface is referred to as the *substrate*. Adsorption is always an exothermic process, and so $\Delta_{ads} H < 0$.

There are two types of adsorption processes that need to be distinguished. The first is called *physisorption* (physical adsorption). In physisorption, the attractive forces between the substrate and the adsorbate arise from van der Waals interactions. This process leads to a weak interaction between the adsorbate and the substrate, and the strength of the substrate-adsorbate bond is typically less than 20 kJ·mol^{-1}. The adsorbate-substrate bond is long compared with the length of the bonds in the bulk solid.

The second type of adsorption is called *chemisorption* (chemical adsorption) and was first proposed by the American chemist Irving Langmuir in 1916. In chemisorption, the adsorbate is bound to the substrate by covalent or ionic forces, much like those that occur between the bonded atoms of a molecule. In chemisorption, a bond of the molecule is broken and new chemical bonds are formed between the molecular fragments and the substrate. Unlike in physisorption, the strength of the substrate-adsorbate bond for a chemisorbed substrate is large; values between 250 and 500 kJ·mol^{-1} are typical. In addition, the length of the substrate-adsorbate bond is shorter for a chemisorbed molecule than for a physisorbed molecule. Because chemisorption involves the formation of chemical bonds to the surface, only a single layer of molecules, or a *monolayer*, can chemisorb to the surface.

Lennard-Jones originally modeled the physisorbed and chemisorbed states in terms of one-dimensional potential-energy curves. Such a model assumes that the substrate has only one type of binding site and that neither the angle at which the adsorbate approaches the substrate nor the orientation of the adsorbate with respect to the substrate is important. If so, the potential energy depends only on z, the distance between the substrate and the adsorbate. Figure 31.20 shows a plot of one-dimensional potential-energy curves for the adsorption of a diatomic molecule, AB, on a surface. We define $V(z) = 0$ to correspond to the infinite separation of the substrate and the diatomic

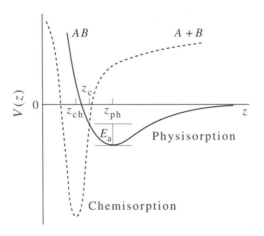

FIGURE 31.20
One-dimensional potential-energy curves for the physisorption of molecule AB (solid line) and the dissociative chemisorption of AB (dashed line). The quantity z is the distance from the surface. In the physisorbed state, the molecule AB is bound to the surface by van der Waals forces. In the chemisorbed state, the AB bond is broken, and the individual atoms are bound covalently to metal atoms on the surface. The points z_{ch} and z_{ph} are the surface-molecule bond lengths for a chemisorbed and physisorbed molecule, respectively. The two potential curves cross at z_c. The activation energy for the conversion from physisorption to chemisorption is measured from the bottom of the physisorbed potential and is E_a.

molecule. Consider first the physisorbed potential-energy curve. As the distance between the adsorbate and substrate decreases, the molecule experiences an attractive force, so the potential energy becomes negative. The potential energy reaches a minimum at z_{ph} (ph for physisorption), and for distances less than z_{ph}, the potential is repulsive. The distance z_{ph} is the equilibrium substrate-adsorbate bond length for the physisorbed molecule.

Now consider the chemisorbed potential-energy curve. The chemisorption of a diatomic molecule involves breaking the molecular bond between the two atoms and then forming bonds between the atomic fragments and the substrate. This process is commonly referred to as *dissociative chemisorption*. Compared with the physisorbed potential, the chemisorbed potential will have a deeper well depth and a shorter substrate-adsorbate bond length, z_{ch} (ch for chemisorption), (Figure 31.20). Direct desorption of the atoms from the substrate generates free atoms in the gas phase. Thus, for large values of z, the chemisorbed potential is positive and at infinite separation, the energy difference between the chemisorbed and physisorbed potential is simply the diatomic bond strength.

Because the substrate-adsorbate bond length for the physisorbed molecule is larger than the chemisorbed molecule, molecules that chemisorb to a surface can be initially trapped in a physisorbed state. In this case, the physisorbed molecule is referred to as a *precursor* to the chemisorbed molecule. We see that the two potential-energy curves in Figure 31.20 cross at a distance z_c from the surface. If the molecule can hop from one potential-energy surface to the other at the point z_c, we can think of the molecule moving from the physisorbed state to the chemisorbed state as a chemical reaction with an activation energy of E_a. For the curve crossing shown in Figure 31.20, the barrier to chemisorption is less than the strength of the substrate –AB bond. There are cases known, for example H_2 on the 110 surface of copper, for which the energy at the curve crossing is greater than the substrate –AB bond strength (Problem 31–46).

31–7. Isotherms Are Plots of Surface Coverage as a Function of Gas Pressure at Constant Temperature

A plot of surface coverage as a function of gas pressure at constant temperature is called an *adsorption isotherm*. In this section, we will learn that adsorption isotherms can be used to determine the equilibrium constant for the adsorption-desorption reaction, the concentration of surface sites available for adsorption, and the enthalpy of adsorption.

The simplest expression for an adsorption isotherm was first derived by Langmuir in 1918. Langmuir assumed that the adsorbed molecules do not interact with one another, that the enthalpy of adsorption was independent of surface coverage, and that there are a finite number of surface sites where a molecule can adsorb. The process of adsorption and desorption is depicted by the reversible elementary process

$$\text{A(g)} + \text{S(s)} \underset{k_d}{\overset{k_a}{\rightleftharpoons}} \text{A–S(s)} \qquad K_c = \frac{k_a}{k_d} = \frac{[\text{A–S}]}{[\text{A}][\text{S}]} \qquad (31.32)$$

where k_a and k_d are the rate constants for adsorption and desorption, respectively. The fact that k_a and k_d are constants independent of the extent of surface coverage implies that the adsorbed molecules do not interact with one another. Let σ_0 be the concentration of surface sites in units of m^{-2}. If the fraction of surface sites occupied by an adsorbate is θ, then σ, the adsorbate concentration on the surface, is $\theta\sigma_0$, and the concentration of empty surface sites is given by $\sigma_0 - \theta\sigma_0 = (1 - \theta)\sigma_0$. We now assume that the rate of desorption is proportional to the number of occupied surface sites and that the rate of adsorption from the gas phase is proportional to both the number of available (unoccupied) surface sites and the number density of molecules in the gas phase. Mathematically, the rates of desorption and adsorption are given by

$$\text{rate of desorption} = v_d = k_d\theta\sigma_0 \tag{31.33}$$

and

$$\text{rate of absorption} = v_a = k_a(1 - \theta)\sigma_0[A] \tag{31.34}$$

where [A] is the number density or the concentration of A(g). At equilibrium, these rates must be equal, so

$$k_d\theta = k_a(1 - \theta)[A]$$

or

$$\frac{1}{\theta} = 1 + \frac{1}{K_c[A]} \tag{31.35}$$

where $K_c = k_a/k_d$ is the concentration equilibrium constant for Equation 31.32. Generally the pressure of A(g) and not the concentration of A(g) is measured. If the pressure

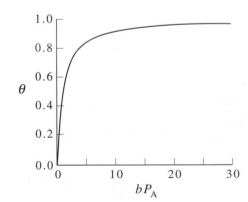

FIGURE 31.21
A plot of Equation 31.36, showing that the fraction of the surface covered, θ, is a nonlinear function of the gas pressure.

of A(g) is low enough that the ideal-gas law can be used, then $[A] = P_A/k_B T$. If we define $b = K_c/k_B T$, Equation 31.35 becomes

$$\frac{1}{\theta} = 1 + \frac{1}{bP_A} \tag{31.36}$$

Equation 31.36 is called the Langmuir adsorption isotherm. Figure 31.21 shows a plot of θ versus bP_A. Note that θ approaches unity, corresponding to the adsorption of a monolayer on the surface, as the pressure becomes large. Example 31–10 shows how b and the total number of surface sites available can be determined from the Langmuir adsorption isotherm.

EXAMPLE 31–10
Experimental adsorption data are often tabulated as the equivalent volume of gas, V, that will adsorb onto the surface at a particular temperature and pressure. Typically, this volume of adsorbed gas is tabulated as the volume that the gas would occupy under a pressure of one atmosphere at 273.15 K (0°C). Langmuir studied the adsorption of $N_2(g)$ onto a mica surface at 273.15 K. From the data presented below, determine the values of b and V_m, the volume of gas that corresponds to a monolayer coverage. Use this value of V_m to determine the total number of surface sites.

$P/10^{-12}$ torr	$V/10^{-8}$ m^3
2.55	3.39
1.79	3.17
1.30	2.89
0.98	2.62
0.71	2.45
0.46	1.95
0.30	1.55
0.21	1.23

SOLUTION: A monolayer of coverage corresponds to $\theta = 1$. When $\theta = 1$, the volume V_m is adsorbed onto the surface. The value of θ is then related to V_m by

$$\theta = \frac{V}{V_m}$$

Substituting this expression for θ into Equation 31.36 and rearranging the result gives us

$$\frac{1}{V} = \frac{1}{PbV_m} + \frac{1}{V_m}$$

From this equation, we see that a plot of $1/V$ versus $1/P$ will have a slope of $1/bV_m$ and an intercept of $1/V_m$. The figure below shows such a plot.

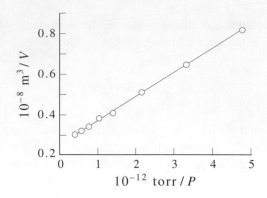

The intercept of the fitted line is 0.252, giving $V_m = 3.96 \times 10^{-8}$ m^3. The slope of the fitted line is 1.18×10^{-5} torr·m^{-3}, from which we find $b = 2.14 \times 10^{12}$ torr^{-1}.

At 0.00°C and 1.00 atm, 1 mol of gas occupies 2.24×10^{-2} m^3. Thus, the number of moles of gas in the volume V_m is

$$\frac{3.96 \times 10^{-8} \text{ m}^3}{2.24 \times 10^{-2} \text{ m}^3 \cdot \text{mol}^{-1}} = 1.77 \times 10^{-6} \text{ mol}$$

which corresponds to

$$(6.022 \times 10^{23} \text{ mol}^{-1})(1.77 \times 10^{-6} \text{ mol}) = 1.06 \times 10^{18} \text{ molecules}$$

Because each molecule occupies a single surface site, there are 1.06×10^{18} sites on the surface. If the mica substrate were a 0.010-m square, the concentration of surface sites would be

$$\sigma_0 = \frac{1.06 \times 10^{18} \text{ molecules}}{(0.010 \text{ m})^2} = 1.06 \times 10^{22} \text{ m}^{-2}$$

Figure 31.22 shows that experimental data for the adsorption of oxygen and carbon monoxide on a silica surface are well described by the Langmuir adsorption isotherm. Example 31–11 derives the adsorption isotherm for the case in which a diatomic

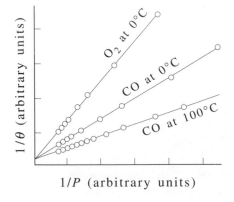

FIGURE 31.22
A plot of $1/\theta$, the inverse of the fraction of surface sites occupied, as a function of $1/P$ for $O_2(g)$ and CO adsorbed on silica. The data are well described by the Langmuir adsorption isotherm (Equation 31.36). The solid lines are the best fit of the Langmuir adsorption isotherm to the experimental data.

molecule dissociates upon adsorption to the surface. Langmuir adsorption isotherms can be derived for many different kinetic models of adsorption.

EXAMPLE 31–11

Derive the Langmuir adsorption isotherm for the case in which a diatomic molecule dissociates upon adsorption to the surface.

SOLUTION: This reaction can be written as

$$A_2(g) + 2S(s) \underset{k_d}{\overset{k_a}{\rightleftharpoons}} 2A\text{–}S(s) \qquad K_c = \frac{k_a}{k_d} = \frac{[A\text{–}S]^2}{[A_2][S]^2}$$

Because two surface sites are involved in the adsorption and desorption process, the rates of adsorption, v_a, and desorption, v_d, are

$$v_a = k_a[A_2](1-\theta)^2 \sigma_0^2$$

$$v_d = k_d \theta^2 \sigma_0^2$$

At equilibrium, these rates are equal, and so

$$k_a[A_2](1-\theta)^2 = k_d \theta^2$$

from which we find

$$\theta = \frac{K_c^{1/2}[A_2]^{1/2}}{1 + K_c^{1/2}[A_2]^{1/2}}$$

In terms of P_{A_2}, the pressure of A_2,

$$\theta = \frac{b_{A_2}^{1/2} P_{A_2}^{1/2}}{1 + b_{A_2}^{1/2} P_{A_2}^{1/2}} \qquad (1)$$

where $b_{A_2} = K_c/k_B T$. We can rewrite Equation 1 as

$$\frac{1}{\theta} = 1 + \frac{1}{b_{A_2}^{1/2} P_{A_2}^{1/2}}$$

from which we see that a plot of $1/\theta$ versus $1/P_{A_2}^{1/2}$ will yield a straight line of slope $1/b_{A_2}^{1/2}$ and intercept 1.

The reciprocal of the rate constant k_d in the Langmuir adsorption isotherm has an interesting physical interpretation. Consider the one-dimensional potential-energy curve shown in Figure 31.23. We see that an energy equal to $E_{ads} = -\Delta_{ads}H$ must be added to the system to break the adsorbate-substrate bond. Experimentally, k_d, the

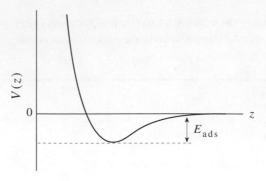

FIGURE 31.23
A one-dimensional potential-energy curve for molecular adsorption. The well depth, E_{ads}, is the negative of the heat of adsorption, $\Delta_{ads}H$.

rate constant for the desorption of a molecule from a surface, obeys an Arrhenius-like expression

$$k_d = \tau_0^{-1} e^{-E_{ads}/RT} \tag{31.37}$$

where $E_{ads} = -\Delta_{ads}H$, the enthalpy of adsorption, and τ_0 is a constant whose value is typically $\approx 10^{-12}$ s. The reciprocal of k_d, which has units of time, is called the *residence time*, τ, of a molecule on the surface. Equation 31.37 can be expressed in terms of τ by taking its reciprocal:

$$\tau = \tau_0 e^{E_{ads}/RT} \tag{31.38}$$

EXAMPLE 31–12
The enthalpy of adsorption of CO on palladium is -146 kJ·mol^{-1}. Estimate the residence time of a CO molecule on a palladium surface at 300 K and 500 K. (Assume that $\tau_0 = 1.0 \times 10^{-12}$ s.)

SOLUTION: The residence time is given by Equation 31.38

$$\tau = \tau_0 e^{E_{ads}/RT}$$

At $T = 300$ K

$$\tau = (1.0 \times 10^{-12} \text{ s}) \exp\left\{ \frac{146 \times 10^3 \text{ J·mol}^{-1}}{(8.314 \text{ J·mol}^{-1}\cdot\text{K}^{-1})(300 \text{ K})} \right\}$$
$$= 2.6 \times 10^{13} \text{ s}$$

and at $T = 500$ K

$$\tau = (1.0 \times 10^{-12} \text{ s}) \exp\left\{ \frac{146 \times 10^3 \text{ J·mol}^{-1}}{(8.314 \text{ J·mol}^{-1}\cdot\text{K}^{-1})(500 \text{ K})} \right\}$$
$$= 1800 \text{ s}$$

Notice that the residence time is very temperature sensitive.

Recall that the Langmuir adsorption isotherm applies only to a monolayer. In many cases, molecules can adsorb on top of other adsorbed molecules. There are models that can account for this multilayer adsorption; one is presented in Problem 31–68.

31–8. The Langmuir Adsorption Isotherm Can Be Used to Derive Rate Laws for Surface–Catalyzed Gas–Phase Reactions

Consider the surface catalysis of the first-order gas-phase reaction

$$A(g) \xrightarrow{k_{obs}} B(g)$$

such that the observed rate law is given by

$$\frac{d[B]}{dt} = k_{obs} P_A \tag{31.39}$$

We will propose that this reaction occurs by the following two-step mechanism

$$A(g) \xrightarrow{k_a} A(ads) \xrightarrow{k_1} B(g)$$

The first step is the adsorption of $A(g)$ onto the surface. Once adsorbed, the molecule reacts to form products, which then immediately desorb into the gas phase. The rate law for the second step of the reaction mechanism can be written as

$$\frac{d[B]}{dt} = k_1[A(ads)] = k_1 \sigma_A \tag{31.40}$$

where σ_A is the surface concentration of A. If there is a total of σ_0 surface sites, then $\sigma_A = \sigma_0 \theta$. Using a Langmuir adsorption isotherm to describe θ (Equations 31.35 and 31.36), the rate law becomes

$$\frac{d[B]}{dt} = k_1 \frac{\sigma_0 K_c[A]}{1 + K_c[A]} = k_1 \frac{\sigma_0 b P_A}{1 + b P_A} \tag{31.41}$$

At low gas pressures, $b P_A \ll 1$, and the rate law becomes first order in reactant pressure,

$$\frac{d[B]}{dt} = k_1 \sigma_0 b P_A = k_{obs} P_A \tag{31.42}$$

and we have rationalized the observed rate law.

For high gas pressures, $b P_A \gg 1$ and Equation 31.41 becomes zero order in the reactant pressure,

$$\frac{d[B]}{dt} = k_1 \sigma_0 = k_{obs} \tag{31.43}$$

Thus, the proposed mechanism makes the experimentally verifiable prediction that the rate should approach an upper limit as the pressure increases. Most reactions are studied at low pressure and according to Equation 31.42 the observed rate constant is equal to the product $k_1 \sigma_0 b$. To determine the rate constant, k_1, both σ_0 and b must be independently determined. We know from Example 31–10 that these values can be obtained from adsorption isotherm data.

The same general approach can be used to derive expressions for the rate laws for the surface catalysis of bimolecular gas-phase reactions. Two general mechanisms, the *Langmuir-Hinshelwood mechanism* and the *Eley-Rideal mechanism*, are commonly used to describe how surfaces catalyze bimolecular gas-phase reactions. Here we illustrate these models by applying them to the oxidation reaction of $CO(g)$ by $O_2(g)$ on a surface of platinum. The balanced equation for the oxidation reaction between $CO(g)$ and $O_2(g)$ is

$$2\,CO(g) + O_2(g) \longrightarrow 2\,CO_2(g) \tag{31.44}$$

The Langmuir-Hinshelwood mechanism for the above reaction is as follows:

$$CO(g) \rightleftharpoons CO(ads)$$

$$O_2(g) \rightleftharpoons 2\,O(ads)$$

$$CO(ads) + O(ads) \overset{k_3}{\Longrightarrow} CO_2(g)$$

In this mechanism, both reactants compete for surface sites. The $CO(g)$ molecule adsorbs molecularly and $O_2(g)$ dissociatively chemisorbs. A reaction then occurs between an adsorbed CO molecule and an adsorbed O atom, producing molecular $CO_2(g)$, which immediately desorbs from the platinum surface. If we assume ideal-gas behavior, that the first two steps of the mechanism are in instantaneous equilibrium during the course of the reaction, and that the third step of the above mechanism is rate determining, then the rate law for the Langmuir-Hinshelwood mechanism is (Problem 31–57)

$$v = \frac{k_3 b_{CO} b_{O_2}^{1/2} P_{CO} P_{O_2}^{1/2}}{(1 + b_{O_2}^{1/2} P_{O_2}^{1/2} + b_{CO} P_{CO})^2} \tag{31.45}$$

where k_3 is the rate constant for the third step of the mechanism, $b_{CO} = K_{CO}/k_B T$, $b_{O_2} = K_{O_2}/k_B T$, and K_{CO} and K_{O_2} are the equilibrium constants for the first two steps of the Langmuir-Hinshelwood mechanism.

EXAMPLE 31–13
Consider the rate law given by Equation 31.45. Determine the form of rate law if (a) the surface is sparsely covered with reactants, and (b) the adsorption of $CO(g)$ is much more extensive than the adsorption of $O_2(g)$ to the surface.

SOLUTION: This example asks that we consider limiting cases for the general rate expression given by Equation 31.45.

(a) If the surface is sparsely covered, then $b_{O_2}^{1/2} P_{O_2}^{1/2} + b_{CO} P_{CO} \ll 1$, so the denominator of Equation 31.45 is approximately 1 and the rate is

$$v = k_3 b_{CO} b_{O_2}^{1/2} P_{CO} P_{O_2}^{1/2}$$

(b) If the adsorption of CO(g) is much more extensive than the adsorption of O_2(g) to the surface, the denominator of the rate law in Equation 31.45 is dominated by the $b_{CO} P_{CO}$ term, and the rate law becomes

$$v = \frac{k_3 b_{CO} b_{O_2}^{1/2} P_{CO} P_{O_2}^{1/2}}{(b_{CO} P_{CO})^2} = \frac{k_3 b_{O_2}^{1/2} P_{O_2}^{1/2}}{b_{CO} P_{CO}}$$

The Eley-Rideal mechanism proposes that the oxidation occurs by the following three-step mechanism:

$$O_2(g) \rightleftharpoons 2\,O(ads)$$

$$CO(g) \rightleftharpoons CO(ads)$$

$$CO(g) + O(ads) \xrightarrow{k_3} CO_2(g)$$

Even though both CO(g) and O_2(g) can adsorb onto the surface, the reaction does not occur between adsorbed reactants. In the Eley-Rideal mechanism, the O_2(g) dissociatively chemisorbs to the surface. A subsequent collision between a gas-phase CO molecule and an adsorbed O atom then generates gaseous CO_2. In other words, the CO(g) molecule abstracts an O atom from the surface. If we assume ideal-gas behavior, that the first two steps of this mechanism are in instantaneous equilibrium during the course of the reaction, and that the third step of the above mechanism is rate determining, then the rate law for this mechanism is (Problem 31–58)

$$v = \frac{k_3 b_{CO} b_{O_2}^{1/2} P_{CO} P_{O_2}^{1/2}}{1 + b_{O_2}^{1/2} P_{O_2}^{1/2} + b_{CO} P_{CO}} \tag{31.46}$$

EXAMPLE 31–14
Consider the reaction given by Equation 31.44. Plot the predictions of the two models, Equations 31.45 and 31.46 as a function of the partial pressure of CO(g) at a fixed pressure of O_2(g).

SOLUTION: Let us first consider the limiting expressions for the rate laws when $P_{CO} \ll P_{O_2}$ and when $P_{CO} \gg P_{O_2}$.

(a) The Langmuir-Hinshelwood rate law: If $P_{CO} \ll P_{O_2}$, we can neglect $b_{CO}P_{CO}$ in the denominator of Equation 31.45 to obtain

$$v \approx \frac{k_3 b_{CO} b_{O_2}^{1/2} P_{CO} P_{O_2}^{1/2}}{(1 + b_{O_2}^{1/2} P_{O_2}^{1/2})^2} \propto P_{CO}$$

for a fixed pressure of $O_2(g)$. If $P_{CO} \gg P_{O_2}$, then

$$v \approx \frac{k_3 b_{CO} b_{O_2}^{1/2} P_{CO} P_{O_2}^{1/2}}{(1 + b_{CO}P_{CO})^2} \propto \frac{1}{P_{CO}}$$

for a fixed pressure of $O_2(g)$ and large values of P_{CO}.

(b) The Eley-Rideal rate law: If $P_{CO} \ll P_{O_2}$, then we can neglect $b_{CO}P_{CO}$ in the denominator of Equation 31.46 to obtain

$$v \approx \frac{k_3 b_{CO} b_{O_2}^{1/2} P_{CO} P_{O_2}^{1/2}}{1 + b_{O_2}^{1/2} P_{O_2}^{1/2}} \propto P_{CO}$$

for a fixed pressure of $O_2(g)$. If $P_{CO} \gg P_{O_2}$, then

$$v \approx \frac{k_3 b_{CO} b_{O_2}^{1/2} P_{CO} P_{O_2}^{1/2}}{1 + b_{CO}P_{CO}} \approx \text{constant}$$

for a fixed pressure of $O_2(g)$ and large values of P_{CO}. Thus, we see that both rate laws predict the same behavior when $P_{CO} \ll P_{O_2}$ but different behavior for $P_{CO} \gg P_{O_2}$. The variation of the rate with CO(g) concentration at fixed $O_2(g)$ concentrations for the two rate laws are shown below.

Langmuir-Hinshelwood mechanism

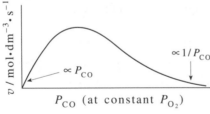

Eley-Rideal mechanism

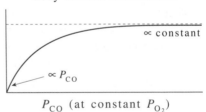

These two mechanisms can then be distinguished by measuring the rate of the reaction, dP_{CO_2}/dt, as a function of P_{CO} at a constant pressure of $O_2(g)$.

Detailed studies of the reaction given by Equation 31.44 show that it occurs by the Langmuir-Hinshelwood mechanism. To date, most of the surface-catalyzed bimolecular gas-phase reactions that have been studied in detail are believed to occur by the Langmuir-Hinshelwood mechanism, although some do occur by the Eley-Rideal mechanism.

31–9. The Structure of a Surface Is Different from that of a Bulk Solid

Up to this point, we have ignored the microscopic structure of the surface. The simplest model of a surface is to assume that the surface is perfectly flat and that the distance between atoms is the same as in the bulk solid. But are surfaces really flat, and are the distances between atoms unaffected by their location? These questions are central to a molecular understanding of surface chemistry. For example, if the surface is not flat and there are many different types of surface sites, the enthalpy of adsorption may vary among these sites. Different adsorption sites may have different barriers to desorption and may also exhibit different reactivity. Understanding chemical reactions on surfaces requires that we have a detailed understanding of the atomic structure of the surface.

As a result of the invention of various forms of surface-sensitive spectroscopies, we now know that most surfaces are not flat. The atomic structure of a surface is characterized by numerous irregularities that can give the surface a significant roughness. For example, Figure 31.24 shows a picture of a zinc surface obtained using a technique called *scanning electron microscopy*. Although this picture does not afford atomic resolution, it shows that the surface is not flat. Zinc atoms located along the ledges have a different number of neighboring atoms from those that sit in the middle of a terrace. In addition, the ledges are not straight but have kinks, so atoms at different points along a ledge can also have a different number of neighboring atoms. Figure 31.25 illustrates some of the structural imperfections found on surfaces. Terraces, steps, and adatoms (single atoms) create a number of distinguishable sites where molecules can adsorb.

Many surface-sensitive spectroscopies probe the surface with low-energy electrons. One of the most important of these techniques is known as *low-energy electron*

FIGURE 31.24
A scanning electron micrograph of a zinc surface. The surface is not flat but consists of a series of hexagonal-shaped terraces. The edge of each terrace is rugged, indicating that the zinc atoms do not sit in perfectly aligned rows along any given terrace.

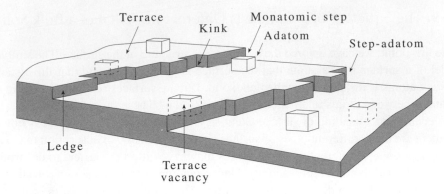

FIGURE 31.25

An illustration of some of the possible structural defects that occur on a surface. The surface is characterized by ledges, steps, and terraces. The steps can be one or many rows of atoms. The steps also need not be straight, which gives rise to kinks. Single atoms, or adatoms, may sit anywhere on a terrace. There can also be vacancies in the terrace, leaving small holes in the surface. These holes are indicated by the dotted cubes.

diffraction, or LEED, *spectroscopy*. Electrons with kinetic energies in the range of 5000 to 10 000 kJ·mol^{-1} are commonly called low-energy electrons and penetrate the surface of a metal to only about 500 pm. This penetration corresponds to only a few atomic layers. When low-energy electrons strike a surface, they scatter. Some of the electrons scatter elastically, that is, without any loss of energy; others scatter inelastically, exchanging their kinetic energy with the vibrational modes of the metal lattice. If the de Broglie wavelength of the electrons is comparable with the distances between atomic planes in the metal, the elastically scattered electrons can be diffracted. Because the electrons penetrate only a few atomic layers from the surface, the diffraction pattern is determined by the atomic structure near and at the surface.

The structure of a crystal surface depends upon how the crystal is cut. We specify the structure of the surface by specifying the three Miller indices *hkl* for the plane of the surface that corresponds to the crystal plane in the bulk metal. Thus a 111 surface means that the atoms on the face of the crystal have the same structure that they have in the 111 crystallographic plane (see Section 31–2). Figure 31.26 shows a LEED diffraction pattern for the 111 surface of platinum. The sharp diffraction spots in the LEED pattern can be used to determine the distance between atoms on and near the surface. The analysis of the LEED diffraction pattern is similar to that used to analyze X-ray diffraction patterns. LEED studies of many surfaces find that surface atoms generally occupy sites that are shifted from the atomic positions in the bulk. Most atomic metals exhibit a significant contraction, up to 40%, of the interlayer distance between the first and second layers of atoms. There is often a compensating expansion between the second and third layer of about 1%, and a smaller but measurable expansion between the third and fourth layer.

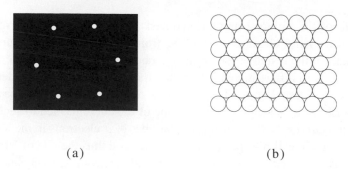

(a)   (b)

FIGURE 31.26
(a) A LEED pattern for the 111 surface of platinum. (b) A schematic diagram of the 111 surface of platinum.

31–10. The Reaction Between $H_2(g)$ and $N_2(g)$ to Produce $NH_3(g)$ Can Be Surface Catalyzed

One of the most thoroughly studied surface-catalyzed reactions is the synthesis of $NH_3(g)$ from $H_2(g)$ and $N_2(g)$,

$$3\,H_2(g) + N_2(g) \longrightarrow 2\,NH_3(g)$$

For this reaction to take place, the N_2 bond must be broken; thus, the activation barrier is on the order of the dissociation energy of N_2, which is $941.6\ kJ\cdot mol^{-1}$. Even though $\Delta_r G^\circ$ for this reaction is $-32.37\ kJ\cdot mol^{-1}$ at 300 K, the barrier to reaction is so large that a mixture of $H_2(g)$ and $N_2(g)$ can be stored indefinitely without producing any appreciable amount of ammonia. But on an iron surface, $N_2(g)$ dissociates with an activation energy of $\approx 10\ kJ\cdot mol^{-1}$. Iron also readily dissociatively chemisorbs $H_2(g)$. The adsorbed N atoms and H atoms then diffuse and can react to form NH(ads), NH_2(ads), and finally NH_3(ads), which desorbs into the gas phase. Experimental studies provide convincing support that the following steps contribute to the reaction mechanism:

$$H_2(g) + 2\,S(s) \rightleftharpoons 2\,H(ads)$$

$$N_2(g) \rightleftharpoons N_2(ads) \qquad\qquad \text{(physisorption)}$$

$$N_2(ads) + 2\,S(s) \rightleftharpoons 2\,N(ads) \qquad \text{(dissociative chemisorption)}$$

$$N(ads) + H(ads) \rightleftharpoons NH(ads)$$

$$NH(ads) + H(ads) \rightleftharpoons NH_2(ads)$$

$$NH_2(ads) + H(ads) \rightleftharpoons NH_3(ads)$$

$$NH_3(ads) \rightleftharpoons NH_3(g)$$

The rate of the surface-catalyzed synthesis of ammonia is sensitive to both the barrier for dissociative chemisorption of N_2 from the physisorbed precursor state and to the strength of the resulting metal-nitrogen bond. This activation energy and bond strength varies with different metals and therefore the rate of reaction depends on the particular catalyst used. Figure 31.27 shows the relative rates of ammonia synthesis for different transition-metal catalysts. The shape of the curve has led to such plots being called *volcano curves*. With an increasing number of *d* electrons in the metal catalyst, the strength of the metal-nitrogen bond decreases, and thus the rate of $NH_3(g)$ product should increase. But with increasing number of *d* electrons, the activation energy for the dissociative chemisorption of $N_2(g)$ increases, causing a decrease in the rate of $NH_3(g)$ production. These two opposing effects lead to the observed volcano curve.

The rate of a surface-catalyzed reaction is also sensitive to the particular surface, as specified by its Miller indices. Figure 31.28 shows the relative rates of $NH_3(g)$ production for five different surfaces of iron. A negligible yield of $NH_3(g)$ is observed on the smooth 110 surface, and the rough 111 surface has the highest yield. Thus,

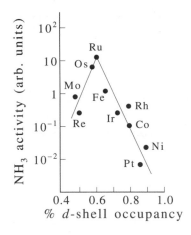

FIGURE 31.27
The relative rates of ammonia synthesis for different transition-metal catalysts. The shape of the plotted data is influenced by the opposing effects of the strength of the surface nitrogen bond and the activation energy of the dissociative chemisorption of $N_2(g)$ as the number of *d* electron in the metal catalyst is increased.

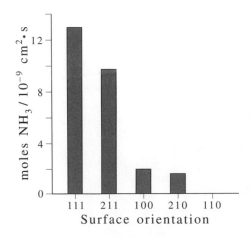

FIGURE 31.28
The relative rates of ammonia synthesis on different iron surfaces.

in addition to being sensitive to different metal catalysts, the synthesis of $NH_3(g)$ on an iron surface is sensitive to the microscopic structure of the surface. Extensive experimental studies show that this result stems from changes in the activation barriers for the dissociative chemisorption of $N_2(g)$ to the surface.

Problems

31-1. Polonium is the only metal that exists as a simple cubic lattice. Given that the length of a side of the unit cell of polonium is 334.7 pm at 25°C, calculate the density of polonium.

31-2. Consider the packing of hard spheres of radius R in a primitive cubic lattice, a face-centered cubic lattice, and a body-centered cubic lattice. Show that a, the length of the unit cell, and f, the fraction of the volume of the unit cell occupied by the spheres, are given as listed.

Unit cell	a	f
Primitive cubic	$2R$	$\pi/6$
Face-centered cubic	$4R/\sqrt{2}$	$\pi\sqrt{2}/6$
Body-centered cubic	$4R/\sqrt{3}$	$\pi\sqrt{3}/8$

31-3. Tantalum forms a body-centered cubic unit cell with $a = 330.2$ pm. Calculate the crystallographic radius of a tantalum atom.

31-4. Nickel forms a face-centered cubic unit cell with $a = 351.8$ pm. Calculate the crystallographic radius of a nickel atom.

31-5. Copper, which crystallizes as a face-centered cubic lattice, has a crystallographic radius of 127.8 pm. Calculate the density of copper.

31-6. Europium, which crystallizes as a body-centered cubic lattice, has a density of 5.243 $g \cdot cm^{-3}$ at 20°C. Calculate the crystallographic radius of a europium atom at 20°C.

31-7. Potassium crystallizes as a body-centered cubic lattice, and the length of a unit cell is 533.3 pm. Given that the density of potassium is 0.8560 $g \cdot cm^{-3}$, calculate the Avogadro constant.

31-8. Cerium crystallizes as a face-centered cubic lattice, and the length of a unit cell is 516.0 pm. Given that the density of cerium is 6.773 $g \cdot cm^{-3}$, calculate the Avogadro constant.

31-9. Given that the density of KBr is 2.75 $g \cdot cm^{-3}$ and that the length of an edge of a cubic unit cell is 654 pm, determine how many formula units of KBr there are in a unit cell. Does the unit cell have a NaCl or a CsCl structure? (See Figure 31.18.)

31-10. Crystalline potassium fluoride has the NaCl type of structure shown in Figure 31.18a. Given that the density of KF(s) is 2.481 $g \cdot cm^{-3}$ at 20°C, calculate the unit cell length and the nearest-neighbor distance in KF(s). (The nearest-neighbor distance is the shortest distance between the centers of any two adjacent ions in the lattice.)

31-11. The crystalline structure of sodium chloride can be described by two interpenetrating face-centered cubic structures (see Figure 31.18a) with four formula units per unit cell. Given that the length of a unit cell is 564.1 pm at 20°C, calculate the density of NaCl(s). The literature value is 2.163 g·cm^{-3}.

31-12. Determine the Miller indices of each set of lines shown in the figure below.

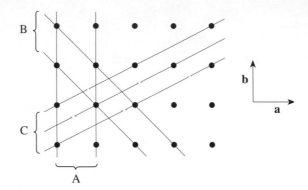

31-13. Determine the Miller indices of each set of lines shown in the figure below.

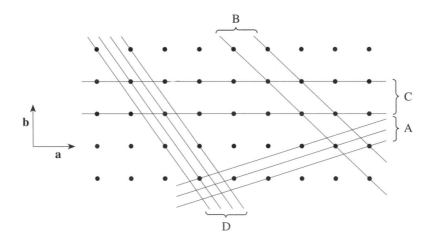

31-14. Sketch the following planes in a two-dimensional square lattice: (a) 01, (b) 21, (c) 1$\bar{1}$, (d) 32.

31-15. What is the relation between the 11 planes and the 1$\bar{1}$ planes of a two-dimensional square lattice?

31-16. What is the relation between the 1$\bar{1}$ planes and the $\bar{1}$1 planes of a two-dimensional square lattice?

31-17. In this problem, we will derive a two-dimensional version of Equation 31.2. Using the figure below, show that

$$\tan\alpha = \frac{b/k}{a/h} \quad \text{and} \quad \sin\alpha = \frac{d}{a/h}$$

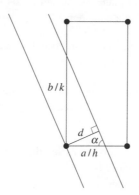

Now show that

$$\sin^2\alpha = \frac{\tan^2\alpha}{1+\tan^2\alpha}$$

and that

$$\frac{1}{d^2} = \frac{h^2}{a^2} + \frac{k^2}{b^2}$$

Equation 31.2 is the extension of this result to three dimensions.

31-18. Determine the Miller indices of the four planes shown in the figure below.

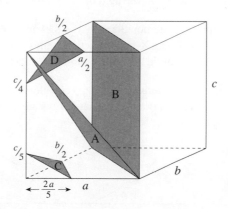

31-19. Sketch the following planes in a three-dimensional cubic lattice: (a) 011, (b) $1\bar{1}0$, (c) 211, (d) 222.

31-20. Determine the Miller indices of the plane that intersects the crystal axes at (a) $(a, 2b, 3c)$, (b) $(a, b, -c)$, and (c) $(2a, b, c)$.

31-21. Calculate the separation between the (a) 100 planes, (b) 111 planes, and (c) $12\overline{1}$ planes in a cubic lattice whose unit cell length is 529.8 pm.

31-22. The distance between the 211 planes in barium is 204.9 pm. Given that barium forms a body-centered cubic lattice, calculate the density of barium.

31-23. Gold crystallizes as a face-centered cubic crystal. Calculate the surface number density of gold atoms in the 100 planes. Take the length of the unit cell (Figure 31.3) to be 407.9 pm.

31-24. Chromium crystallizes as a body-centered cubic structure with a density of 7.20 g·cm^{-3} at 20°C. Calculate the length of a unit cell and the distance between successive 110, 200, and 111 planes.

31-25. A single crystal of NaCl is oriented such that the incident X-rays are perpendicular to the **a** axis of the crystal. The distance between the spots corresponding to diffraction from the origin and 100 planes is 14.8 mm, and the detector is located 52.0 mm from the crystal. Calculate the value of a, the length of the unit cell along the **a** axis. Take the wavelength of the X-radiation to be $\lambda = 154.433$ pm.

31-26. Silver crystallizes as a face-centered cubic structure with a unit cell length of 408.6 pm. The single crystal of silver is oriented such that the incident X-rays are perpendicular to the **c** axis of the crystal. The detector is located 29.5 mm from the crystal. What is the distance between the diffraction spots from the 001 and 002 planes on the face of the detector for (a) the $\lambda = 154.433$-pm line of copper, and (b) the $\lambda = 70.926$-pm line of a molybdenum X-ray source? Which X-ray source gives you the better spatial resolution between the diffraction spots?

31-27. The X-ray diffraction angles for the first-order diffraction spot from the 111 planes of a cubic crystal with $a = 380.5$ pm are observed to be $\alpha = 18.79°$, $\beta = 0°$, and $\gamma = 0°$. How is the crystal oriented? Take the wavelength of the X-radiation to be $\lambda = 154.433$ pm.

31-28. The unit cell of topaz is orthorhombic with $a = 839$ pm, $b = 879$ pm, and $c = 465$ pm. Calculate the values of the Bragg X-ray diffraction angles from the 110, 101, 111, and 222 planes. Take the wavelength of the X-radiation to be $\lambda = 154.433$ pm.

31-29. In this problem, we will derive the Bragg equation, Equation 31.12. William and Lawrence Bragg (father and son) assumed that X-rays are scattered by successive planes of atoms within a crystal (see the following figure).

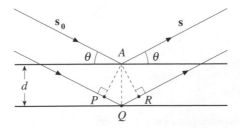

Each set of planes reflects the X-rays specularly; that is, the angle of incidence is equal to the angle of reflection, as shown in the figure. The X-radiation reflected from the lower plane in the figure travels a distance PQR longer than the X-radiation reflected by the upper layer. Show that $PQR = 2d \sin \theta$, and argue that $2d \sin \theta$ must be an integral number of wavelengths for constructive interference and hence a diffraction pattern to be observed.

31-30. The Bragg diffraction angle of the second-order reflection from the 222 planes of a potassium crystal is $\theta = 27.43°$ when X-radiation of wavelength $\lambda = 70.926$ pm is used. Given that potassium exists as a body-centered cubic lattice, determine the length of the unit cell and the density of the crystal.

31-31. The crystalline structure of $CuSO_4(s)$ is orthorhombic with unit cell dimensions of $a = 488.2$ pm, $b = 665.7$ pm, and $c = 831.6$ pm. Calculate the value of θ, the first-order Bragg diffraction angle, from the 100 planes, the 110 planes, and the 111 planes if $CuSO_4(s)$ is irradiated with X-rays with $\lambda = 154.433$ pm.

31-32. One experimental method of collecting X-ray diffraction data, (called the *powder method*) involves irradiating a crystalline powder rather than a single crystal. The various sets of reflecting planes in a powder will be essentially randomly oriented so that there will always be planes oriented such that they will reflect the monochromatic X-radiation. The crystallites whose particular hkl planes are oriented at the Bragg diffraction angle, θ, to the incident beam will reflect the beam constructively. In this problem, we will illustrate the procedure that can be used for indexing the planes that give rise to observed reflections and consequently leads to the determination of the type of unit cell. This method is limited to cubic, tetragonal, and orthorhombic crystals (all unit cell angles are $90°$). We will illustrate the method for a cubic unit cell.

First, show that the Bragg equation can be written as

$$\sin^2 \theta = \frac{\lambda^2}{4a^2}(h^2 + k^2 + l^2)$$

for a cubic unit cell. Then we tabulate the diffraction-angle data in order of increasing values of $\sin^2 \theta$. We then search for the smallest sets of h, k, and l that are in the same ratios as the values of $\sin^2 \theta$. We then compare these values of h, k, and l with the allowed values given in Problem 31–38 to determine the type of unit cell.

Lead is known to crystallize in one of the cubic structures. Suppose that a powder sample of lead gives Bragg reflections at the following angles: $15.66°$, $18.17°$, $26.13°$, $31.11°$, $32.71°$, and $38.59°$, using X-radiation with $\lambda = 154.433$ pm. Now form a table of increasing values of $\sin^2 \theta$, divide by the smallest value, convert the resulting values to integer values by multiplying by a common integer factor, and then determine the possible values of h, k, and l. For example, the first two entries in such a table are listed below.

$\sin^2 \theta$	Division by 0.0729	Conversion to integer value	Possible value of hkl
0.0729	1	3	111
0.0972	1.33	4	200

Complete this table, determine the type of cubic unit cell for lead, and determine its length.

31-33. The X-ray powder diffraction patterns of NaCl(s) and KCl(s), both of which have the structures given in Figure 31.18a, are shown below.

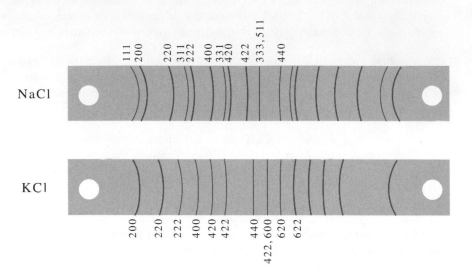

Given that NaCl and KCl have the same crystal structure, explain the differences between the two sets of data. Realize that the value of f_{K^+} is almost equal to f_{Cl^-} because K^+ and Cl^- are isoelectronic.

31-34. Iridium crystals have a cubic unit cell. The first six observed Bragg diffraction angles from a powered sample using X-rays with $\lambda = 165.8$ pm are $21.96°$, $25.59°$, $37.65°$, $45.74°$, $48.42°$, and $59.74°$. Use the method outlined in Problem 31–32 to determine the type of cubic unit cell and its length.

31-35. The density of tantallum at $20°C$ is 16.69 g·cm^{-3}, and its unit cell is cubic. Given that the first five observed Bragg diffraction angles are $\theta = 19.31°$, $27.88°$, $34.95°$, $41.41°$, and $47.69°$, find the type of unit cell and its length. Take the wavelength of the X-radiation to be $\lambda = 154.433$ pm.

31-36. The density of silver at $20°C$ is 10.50 g·cm^{-3}, and its unit cell is cubic. Given that the first five observed Bragg diffraction angles are $\theta = 19.10°$, $22.17°$, $32.33°$, $38.82°$, and $40.88°$, find the type of unit cell and its length. Take the wavelength of the X-radiation to be $\lambda = 154.433$ pm.

31-37. Derive an expression for the structure factor of a primitive cubic unit cell and a face-centered cubic unit cell. Show that there will be observed reflections for a primitive unit cell for all integer values of h, k, and l and reflections for a face-centered-cubic unit cell only if h, k, and l are either all even or all odd.

31-38. Use the results of the previous problem and Example 31–9 to verify the entries in the following table.

Miller indices (hkl)	Cubic lattice type for which a reflection is observed		
100	pc		
110	pc		bcc
111	pc	fcc	
200	pc	fcc	bcc
210	pc		
211	pc		bcc
220	pc	fcc	bcc
300	pc		
221	pc		
310	pc		bcc
311	pc	fcc	
222	pc	fcc	bcc
320	pc		
321	pc		bcc
400	pc	fcc	bcc

31-39. The X-ray diffraction pattern of a cubic crystalline substance shows data that correspond to reflections from the 110, 200, 220, 310, 222, and 400 planes. What type of cubic unit cell does the substance have? (*Hint*: See the table in Problem 31–38.)

31-40. Chromium is either a face-centered cubic or a body-centered cubic crystalline solid. Given that it has the following observed successive values of d: 203.8 pm, 144.2 pm, 117.7 pm, 102.0 pm, 91.20 pm, and 83.25 pm, determine the type of cubic unit cell, the length of the unit cell, and the density. (*Hint*: See the table in Problem 31–38.)

31-41. In this problem, we will derive the structure factor for a sodium chloride-type unit cell. First, show that the coordinates of the cations at the eight corners are $(0,0,0)$, $(1,0,0)$, $(0,1,0)$, $(0,0,1)$, $(1,1,0)$, $(1,0,1)$, $(0,1,1,)$, and $(1,1,1)$ and those at the six faces are $(\frac{1}{2},\frac{1}{2},0)$, $(\frac{1}{2},0,\frac{1}{2})$, $(0,\frac{1}{2},\frac{1}{2})$, $(\frac{1}{2},\frac{1}{2},1)$, $(\frac{1}{2},1,\frac{1}{2})$, and $(1,\frac{1}{2},\frac{1}{2})$. Similarly, show that the coordinates of the anions along the 12 edges are $(\frac{1}{2},0,0)$, $(0,\frac{1}{2},0)$, $(0,0,\frac{1}{2})$, $(\frac{1}{2},1,0)$, $(1,\frac{1}{2},0)$, $(0,\frac{1}{2},1)$, $(\frac{1}{2},0,1)$, $(1,0,\frac{1}{2})$, $(0,1,\frac{1}{2})$, $(\frac{1}{2},1,1)$, $(1,\frac{1}{2},1)$, and $(1,1,\frac{1}{2})$ and those of the anion at the center of the unit cell are $(\frac{1}{2},\frac{1}{2},\frac{1}{2})$. Now show that

$$F(hkl) = \frac{f_+}{8}[1 + e^{2\pi ih} + e^{2\pi ik} + e^{2\pi il} + e^{2\pi i(h+k)} + e^{2\pi i(h+l)} + e^{2\pi i(k+l)} + e^{2\pi i(h+k+l)}]$$

$$+ \frac{f_+}{2}[e^{\pi i(h+k)} + e^{\pi i(h+l)} + e^{\pi i(k+l)} + e^{\pi i(h+k+2l)} + e^{\pi i(h+2k+l)} + e^{\pi i(2h+k+l)}]$$

$$+ \frac{f_-}{4}[e^{\pi ih} + e^{\pi ik} + e^{\pi il} + e^{\pi i(h+2k)} + e^{\pi i(2h+k)} + e^{\pi i(k+2l)} + e^{\pi i(h+2l)} + e^{\pi i(2h+l)} + e^{\pi i(2k+l)}$$

$$+ e^{\pi i(h+2k+2l)} + e^{\pi i(2h+k+2l)} + e^{\pi i(2h+2k+l)}] + f_- e^{\pi i(h+k+l)}$$

$$= f_+[1 + (-1)^{h+k} + (-1)^{h+l} + (-1)^{k+l}]$$

$$+ f_-[(-1)^h + (-1)^k + (-1)^l + (-1)^{h+k+l}]$$

Finally, show that

$$F(hkl) = 4(f_+ + f_-)$$

if $h, k,$ and l are all even; that

$$F(hkl) = 4(f_+ - f_-)$$

if $h, k,$ and l are all odd, and that $F(hkl) = 0$ otherwise.

31-42. Show that

$$F(hkl) = f_+ + f_- \qquad \text{if } h, k, \text{ and } l \text{ are all even}$$
$$\text{or just one of them is even}$$

$$= f_+ - f_- \qquad \text{if all are odd}$$
$$\text{or just one is odd}$$

for the CsCl(s) crystal structure shown in Figure 31.18b. Cesium bromide and cesium iodide have the same crystal structure as cesium chloride. Compare the expected diffraction patterns of cesium chloride and cesium iodide. Recall that Cs^+ and I^- are isoelectronic.

31-43. In this problem, we will prove that a crystal lattice can have only one-, two-, three-, four-, and six-fold axes of symmetry. Consider the following figure, where P_1, P_2, and P_3 are three lattice points, each separated by the lattice vector **a**.

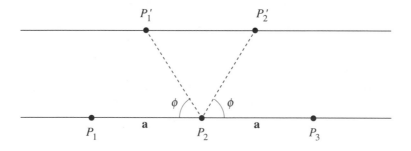

If the lattice has n-fold symmetry, then both a clockwise and a counter-clockwise rotation by $\phi = 360°/n$ about the point P_2 will lead to the points P_1' and P_2', which must be lattice points (because of the fact that the lattice has an n-fold axis of symmetry). Show that the vector distance $P_1'P_2'$ must satisfy the relation

$$2\mathbf{a}\cos\phi = N\mathbf{a}$$

where N is a positive or negative integer. Now show that the only values of ϕ that satisfy the above relation are $360°$ $(n = 1)$, $180°$ $(n = 2)$, $120°$ $(n = 3)$, $90°$ $(n = 4)$, and $60°$ $(n = 6)$, corresponding to $N = 2, -2, -1, 0,$ and 1, respectively.

31-44. The von Laue equations are often expressed in vector notation. The following figure illustrates the X-ray scattering from two lattice points P_1 and P_2.

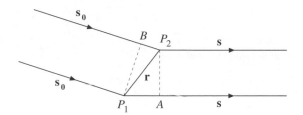

Let s_0 be a unit vector in the direction of the incident radiation and s be a unit vector in the direction of the scattered X-radiation. Show that the difference in the path lengths of the waves scattered from P_1 and P_2 is given by

$$\delta = P_1 A - P_2 B = \mathbf{r} \cdot \mathbf{s} - \mathbf{r} \cdot \mathbf{s}_0 = \mathbf{r} \cdot \mathbf{S}$$

where $\mathbf{S} = \mathbf{s} - \mathbf{s}_0$. Because P_1 and P_2 are lattice points, $\mathbf{r}$ must be expressible as $m\mathbf{a} + n\mathbf{b} + p\mathbf{c}$, where m, n, and p are integers, and $\mathbf{a}$, $\mathbf{b}$, and $\mathbf{c}$ are the unit cell axes. Show that the fact that δ must be an integral multiple of the wavelength λ leads to the equations

$$\mathbf{a} \cdot \mathbf{S} = h\lambda$$

$$\mathbf{b} \cdot \mathbf{S} = k\lambda$$

$$\mathbf{c} \cdot \mathbf{S} = l\lambda$$

where h, k, and l are integers. These equations are the von Laue equations in vector notation.

31-45. We can derive the Bragg equation from the von Laue equations derived in the previous problem. First show that $\mathbf{S} = \mathbf{s} - \mathbf{s}_0$ bisects the angle between $\mathbf{s}_0$ and $\mathbf{s}$ and is normal to the plane from which the X-radiation would be specularly reflected (the angle of incidence equals the angle of reflection). Now show that the distance from the origin of the $\mathbf{a}$, $\mathbf{b}$, and $\mathbf{c}$ axes to the hkl plane is given by

$$d = \frac{\mathbf{a}}{h} \cdot \frac{\mathbf{S}}{|\mathbf{S}|} = \frac{\mathbf{b}}{k} \cdot \frac{\mathbf{S}}{|\mathbf{S}|} = \frac{\mathbf{c}}{l} \cdot \frac{\mathbf{S}}{|\mathbf{S}|} = \frac{\lambda}{|\mathbf{S}|}$$

Last, show that $|\mathbf{S}| = [(\mathbf{s} - \mathbf{s}_0) \cdot (\mathbf{s} - \mathbf{s}_0)]^{1/2} = [2 - 2\cos 2\theta]^{1/2} = 2\sin\theta$, which leads to the Bragg equation $d = \lambda/2\sin\theta$.

31-46. The enthalpy of adsorption for H_2 adsorbed on a surface of copper is -54.4 kJ·mol^{-1}. The activation energy for going from the physisorbed state to the chemisorbed state is 29.3 kJ·mol^{-1}, and the curve crossing between these two potentials occurs at $V(z) = 21$ kJ·mol^{-1}. Draw a schematic representation similar to that in Figure 31.20 for the case of H_2 interacting with copper.

31-47. In Section 27–4, we showed that the collision frequency per unit area is (Equation 27.48)

$$z_{coll} = \frac{\rho \langle u \rangle}{4} \tag{1}$$

Use Equation 1 and the ideal-gas law to show that J_N, the number of molecules striking a surface of unit area (1 m^2) in one second, is

$$J_N = \frac{P N_A}{(2 \pi M R T)^{1/2}}$$

where M is the molar mass of the molecule, P is the pressure of the gas, and T is the temperature. How many nitrogen molecules will strike a 1.00-cm^2 surface in 1.00 s at 298.1 K and a gas pressure of 1.05×10^{-6} Pa?

31-48. One *langmuir* corresponds to an exposure of a surface to a gas at a pressure of 1.00×10^{-6} torr for 1 second at 298.15 K. Define one langmuir in units of pascals instead of torr. How many nitrogen molecules will strike a surface of area 1.00 cm^2 when exposed to 1.00 langmuir? (See Problem 31–47.)

31-49. If the density of surface sites is 2.40×10^{14} cm^{-2} and every molecule that strikes the surface adsorbs to one of these sites, determine the fraction of a monolayer created by the exposure of a 1.00-cm^2 surface to 1.00×10^{-4} langmuir of $N_2(g)$ at 298.15 K.

31-50. For conducting surface experiments it is important to maintain a clean surface. Suppose that a 1.50-cm^2 surface is placed inside a high-vacuum chamber at 298.15 K and the pressure inside the chamber is 1.00×10^{-12} torr. If the density of the surface sites is 1.30×10^{16} cm^{-2} and we assume that the only gas in the chamber is H_2O and that each of the H_2O molecules that strike the surface adsorbs to one surface site, how long will it be until 1.00% of the surface sites are occupied by water?

31-51. Use the results of Example 31–12 to determine the rate of desorption of CO from palladium at 300 K and 500 K.

31-52. The following data were obtained for the adsorption of $N_2(g)$ to a piece of solid graphite at 197 K. The tabulated volumes are the volumes that the adsorbed gas would occupy at 0.00°C and one bar

P/bar	3.54	10.13	16.92	26.04	29.94
$V/10^{-4}$ m^3	328	456	497	527	536

Determine the values of V_m and b using the Langmuir adsorption isotherm. The total mass of the carbon solid is 1325 g. Determine the fraction of the carbon atoms that are accessible as binding sites if you assume that each surface atom can adsorb one N_2 molecule.

31-53. The first-order surface reaction

$$A(g) \Longrightarrow A(ads) \Longrightarrow B(g)$$

has a rate of 1.8×10^{-4} mol·dm^{-3}·s^{-1}. The surface has a dimension of 1.00 cm by 3.50 cm. Calculate the rate of reaction if the dimensions of the two sides of the surface were each doubled. [Assume that A(g) is in excess.]

31-54. Consider the reaction scheme

$$A(g) + S \xrightarrow{k_1} A\text{–}S \xrightarrow{k_2} P(g)$$

for which the rate law is

$$v = k_2 \theta_A$$

where θ_A is the fraction of surface sites occupied by A molecules. Use the Langmuir adsorption isotherm (Equation 31.35) to obtain an expression for the reaction rate in terms of K_c and [A]. Under what conditions will the reaction be first order in the concentration of A?

31-55. Consider a surface-catalyzed bimolecular reaction between molecules A and B that has a rate law of the form

$$v = k_3 \theta_A \theta_B$$

where θ_A is the fraction of surface sites occupied by reactant A and θ_B is the fraction of surface sites occupied by reactant B. A mechanism consistent with this reaction is as follows:

$$A(g) + S(s) \underset{k_d^A}{\overset{k_a^A}{\rightleftharpoons}} A\text{–}S(s) \qquad \text{(fast equilibrium)} \qquad (1)$$

$$B(g) + S(s) \underset{k_d^B}{\overset{k_a^B}{\rightleftharpoons}} B\text{–}S(s) \qquad \text{(fast equilibrium)} \qquad (2)$$

$$A\text{–}S(s) + B\text{–}S(s) \xrightarrow{k_3} \text{products}$$

Take K_A and K_B to be the equilibrium constants for Equations 1 and 2, respectively. Derive expressions for θ_A and θ_B in terms of [A], [B], K_A, and K_B. Use your results to show that the rate law can be written as

$$v = \frac{k_3 K_A K_B [A][B]}{(1 + K_A[A] + K_B[B])^2}$$

31-56. Reconsider the surface-catalyzed bimolecular reaction in Problem 31–55. If A(g) and B(g) do not compete for surface sites, but instead each molecule uniquely binds to a different type of surface site, show that the rate law is given by

$$v = \frac{k_3 K_A K_B [A][B]}{(1 + K_A[A])(1 + K_B[B])}$$

31-57. In this problem we derive Equation 31.45, the rate law for the oxidation reaction $2\,CO(g) + O_2(g) \longrightarrow 2\,CO_2(g)$ assuming that the reaction occurs by the Langmuir-Hinshelwood mechanism. The overall rate law for this mechanism is

$$v = k_3 \theta_{CO} \theta_{O_2}$$

Show that

$$\theta_{O_2} = \frac{(K_{O_2}[O_2])^{1/2}}{1 + (K_{O_2}[O_2])^{1/2} + K_{CO}[CO]}$$

and

$$\theta_{CO} = \frac{K_{CO}[CO]}{1 + (K_{O_2}[O_2])^{1/2} + K_{CO}[CO]}$$

Use these expressions and the relationship $b = K_c/k_B T$ to obtain the rate law given by Equation 31.45. (Assume ideal-gas behavior.)

31-58. In this problem we derive Equation 31.46, the rate law for the oxidation reaction $2\,CO(g) + O_2(g) \longrightarrow 2\,CO_2(g)$ assuming that the reaction occurs by the Eley-Rideal mechanism. The overall rate law for this mechanism is

$$v = k_3 \theta_{O_2}[CO]$$

Assuming that both $CO(g)$ and $O_2(g)$ compete for adsorption sites, show that

$$v = \frac{k_3 K_{O_2}^{1/2}[O_2]^{1/2}[CO]}{1 + K_{O_2}^{1/2}[O_2]^{1/2} + K_{CO}[CO]}$$

Use the relationship between K_c and b and the ideal-gas law to show that this equation is equivalent to Equation 31.46.

31-59. The hydrogenation of ethene on copper obeys the rate law

$$v = \frac{k[H_2]^{1/2}[C_2H_4]}{(1 + K[C_2H_4])^2}$$

where k and K are constants. Mechanistic studies show that the reaction occurs by the Langmuir-Hinshelwood mechanism. How are k and K related to the rate constants for the individual steps of the reaction mechanism? What can you conclude about the relative adsorption of $H_2(g)$ and $C_2H_4(g)$ to the copper surface from the form of the observed rate law?

31-60. The iron-catalyzed exchange reaction

$$NH_3(g) + D_2(g) \longrightarrow NH_2D(g) + HD(g)$$

obeys the rate law

$$v = \frac{k[D_2]^{1/2}[NH_3]}{(1 + K[NH_3])^2}$$

Is this rate law consistent with either the Eley-Rideal or Langmuir-Hinshelwood mechanisms? How are k and K related to the rate constants of the individual steps of the mechanism you chose? What does the rate law tell you about the relative adsorption of $D_2(g)$ and $NH_3(g)$ to the iron surface?

31-61. Consider the surface-catalyzed exchange reaction

$$H_2(g) + D_2(g) \longrightarrow 2\,HD(g)$$

Experimental studies show that this reaction occurs by the Langmuir-Hinshelwood mechanism by which both $H_2(g)$ and $D_2(g)$ first dissociatively chemisorb to the surface. The rate-determining step is the reaction between the adsorbed H and D atoms. Derive an expression for the rate law for this reaction in terms of the gas-phase pressures of $H_2(g)$ and $D_2(g)$. (Assume ideal-gas behavior.)

31-62. LEED spectroscopy records the intensities and locations of electrons that are diffracted from a surface. For an electron to diffract, its de Broglie wavelength must be less than twice the distance between the atomic planes in the solid (see Section 31–9). Show that the de Broglie wavelength of an electron accelerated through a potential difference of ϕ volts is given by

$$\lambda/\text{pm} = \left(\frac{1.504 \times 10^6 \text{ V}}{\phi} \right)^{1/2}$$

31-63. The distance between the 100 planes of a nickel substrate, whose surface is a 100 plane, is 351.8 pm. Calculate the minimum accelerating potential so that electrons can diffract from the crystal. Calculate the kinetic energy of these electrons. (*Hint*: See Problem 31–62.)

31-64. The distance between the 111 surface of silver and the second layer of atoms is 235 pm, the same as in the bulk. If electrons with a kinetic energy of 8.77 eV strike the surface, will an electron diffraction pattern be observed? (*Hint*: See Problem 31–62.)

31-65. Figure 31.28 shows the relative rates of ammonia synthesis for five different surfaces of iron. Iron crystallizes as a body-centered cubic structure. Draw a schematic representation of the atomic arrangement of the 100, 110, and 111 surfaces. (*Hint*: See Figure 31.9.) Determine the center-to-center distance between nearest neighbor atoms on the surface in units of a, the dimension of the unit cell.

31-66. The *Freundlich adsorption isotherm* is given by

$$V = kP^a$$

where k and a are constants. Can the data in Problem 31–52 be described by the Freundlich adsorption isotherm? Determine the best-fit values of k and a to the data.

31-67. Show that if $\theta \ll 1$, the Langmuir adsorption isotherm reduces to the Freundlich adsorption isotherm (Problem 31–66) with $k = bV_m$ and $a = 1$.

31-68. Multilayer physisorption is often described by the *BET adsorption isotherm*

$$\frac{P}{V(P^* - P)} = \frac{1}{cV_m} + \frac{(c-1)P}{V_m cP^*}$$

where P^* is the vapor pressure of the adsorbate at the temperature of the experiment, V_m is the volume corresponding to a monolayer of coverage on the surface, V is the total volume adsorbed at pressure P, and c is a constant. Rewrite the equation for the BET adsorption isotherm in the form

$$\frac{V}{V_m} = f(P/P^*)$$

Plot V/V_m versus P/P^* for $c = 0.1,\ 1.0,\ 10,$ and 100. Discuss the shapes of the curves.

31-69. The energy of adsorption, E_{ads}, can be measured by the technique of *temperature programmed desorption* (TPD). In a TPD experiment, the temperature of a surface with bound adsorbate is changed according to the equation

$$T = T_0 + \alpha t \tag{1}$$

where T_0 is the initial temperature, α is a constant that determines the rate at which the temperature is changed, and t is the time. A mass spectrometer is used to measure the concentration of molecules that desorb from the surface. The analysis of TPD data depends on the kinetic model for desorption. Consider a first-order desorption process

$$M\text{–}S(s) \overset{k_d}{\Longrightarrow} M(g) + S(s)$$

Write an expression for the rate law for desorption. Use Equation 1, Equation 31.37, and your rate law to show that your rate law can be written as

$$\frac{d[M\text{–}S]}{dT} = -\frac{[M\text{–}S]}{\alpha}(\tau_0^{-1}e^{-E_{ads}/RT}) \tag{2}$$

With increasing temperature, $d[M\text{–}S]/dt$ initially increases, then reaches a maximum, after which it decreases. Let $T = T_{max}$ be the temperature corresponding to the maximum rate of desorption. Use Equation 2 to show that at T_{max}

$$\frac{E_{ads}}{RT_{max}^2} = \frac{\tau_0^{-1}}{\alpha}e^{-E_{ads}/RT_{max}} \tag{3}$$

(*Hint*: Remember that [M–S] is a function of T.)

31-70. Show that Equation 3 of Problem 31–69 can be written as

$$2\ln T_{max} - \ln \alpha = \frac{E_{ads}}{RT_{max}} + \ln \frac{E_{ads}}{R\tau_0^{-1}}$$

What are the slope and intercept of a plot of $(2\ln T_{max} - \ln \alpha)$ versus $1/T_{max}$? The maximum desorption rates of CO from the 111 surface of palladium as a function of the rate of heating of the palladium surface are given below. Determine the values of E_{ads}

and τ_0^{-1} from these data. Use the results to determine k_d, the desorption rate constant, at 600 K.

$\alpha/\text{K}\cdot\text{s}^{-1}$	T_{max}/K
26.0	500
20.1	496
16.5	493
11.0	487

31-71. At a heating rate of 10 $\text{K}\cdot\text{s}^{-1}$, the maximum rate of desorption of CO from a Pd(s) surface occurs at 625 K. Calculate the value of E_{ads}, assuming that the desorption is a first-order process and that $\tau_0 = 1.40 \times 10^{-12}$ s. (See Problems 31–69 and 31–70).

Answers to the Numerical Problems

Chapter 1

1-1. $\nu = 1.50 \times 10^{15}$ Hz,
$\tilde{\nu} = 5.00 \times 10^4$ cm^{-1}, $E = 9.93 \times 10^{-19}$ J

1-2. $\nu = 3 \times 10^{13}$ Hz, $\lambda = 1 \times 10^{-5}$ m, $E = 2 \times 10^{-20}$ J

1-3. $\nu = 2.0 \times 10^{10}$ Hz, $\lambda = 1.5 \times 10^{-2}$ m, $E = 1.3 \times 10^{-23}$ J

1-6. (a) $\lambda_{max} = 9.67 \times 10^{-6}$ m
(b) $\lambda_{max} = 9.67 \times 10^{-7}$ m
(c) $\lambda_{max} = 2.90 \times 10^{-7}$ m

1-7. $T = 1.12 \times 10^4$ K

1-8. $\lambda_{max} = 3 \times 10^{-10}$ m

1-9. $E = 2 \times 10^{-15}$ J

1-11. (a) 1.07×10^{16} photons
(b) 5.40×10^{15} photons
(c) 2.68×10^{15} photons

1-12. $\lambda_{max} = 1.0 \times 10^{-5}$ m

1-13. 4.738×10^{14} Hz; 3.139×10^{-19} J

1-14. 1.70×10^{15} photon$\cdot$s^{-1}

1-15. 5300 K

1-16. $\phi = 3.52 \times 10^{-19}$ J,
K.E. $= 1.33 \times 10^{-19}$ J

1-17. K.E. $= 2.89 \times 10^{-19}$ J

1-18. $\phi = 7.35 \times 10^{-19}$ J,
$\nu_0 = 1.11 \times 10^{15}$ Hz

1-19. $h = 6.60 \times 10^{-34}$ J$\cdot$s,
$\phi = 3.59 \times 10^{-19}$ J

1-20. 121.56 nm, 102.571 nm, 97.2526 nm

1-21. $n = 3$

1-22. $n = 2$

1-24. 91.2 nm, 2.18×10^{-18} J

1-25. (a) 0.123 nm (b) 2.86×10^{-3} nm
(c) 0.332 nm

1-26. (a) 1.602×10^{-17} J$\cdot$electron^{-1}, 1.23×10^{-10} m (b) 6.02×10^{-18} J

1-27. 0.082 V

1-28. 1.28×10^{-18} J/α particle, 5.08 pm

1-29. 2500 K

1-33. 54.4 eV, 5 250 kJ$\cdot$mol^{-1}

1-34. $v_1 = 2.187 \times 10^6$ m$\cdot$s^{-1}, $v_2 = 1.093 \times 10^6$ m$\cdot$s^{-1}, $v_3 = 7.290 \times 10^5$ m$\cdot$s^{-1}

1-35. 3.6×10^7 m$\cdot$s^{-1}

1-36. 6.6×10^{-23} kg$\cdot$m$\cdot$s^{-1} compared to 1.992×10^{-24} kg$\cdot$m$\cdot$s^{-1}

1-38. 2.9×10^{-23} s

1-39. 7×10^{-25} J

1-40. 7×10^{-22} J

1-44. H(g), 3.18×10^4 K

MathChapter A

A-1. (a) $2 - 11i$ (b) i (c) ie^{-2} (d) $2 - i\sqrt{2}$

A-2. (a) x (b) $x^2 - 4y^2$ (c) $4xy$ (d) $x^2 + 4y^2$
(e) 0

A-3. (a) $6e^{i\pi/2}$ (b) $3\sqrt{2}\, e^{-0.340i}$ (c) $\sqrt{5}\, e^{1.11i}$
(d) $\sqrt{\pi^2 + e^2}\, e^{0.713i}$

A-4. (a) $\dfrac{1}{\sqrt{2}} + i\dfrac{1}{\sqrt{2}}$ (b) $-3 + 3\sqrt{3}i$
(c) $\sqrt{2}(1 - i)$ (d) 2

A-12. $e^{-\pi/2}$

A-14. $x = 2,\ -1 \pm i\sqrt{3}$

Chapter 2

2-1. (a) $y(x) = c_1 e^{3x} + c_2 e^x$
(b) $y(x) = c_1 + c_2 e^{-6x}$ (c) $y(x) = c_1 e^{-3x}$
(d) $y(x) = c_1 e^{(-1+\sqrt{2})x} + c_2 e^{(-1-\sqrt{2})x}$
(e) $y(x) = c_1 e^{2x} + c_2 e^x$

2-2. (a) $y(x) = 2e^{2x}$
(b) $y(x) = -3e^{2x} + 2e^{3x}$ (c) $y(x) = 2e^{2x}$

2-4. (a) $x(t) = \dfrac{v_0}{\omega} \sin \omega t$
(b) $x(t) = A \cos \omega t + \dfrac{v_0}{\omega} \sin \omega t$

2-5. $c_1 = A \sin \phi = B \cos \psi$;
$c_2 = A \cos \phi = -B \sin \psi$

2-6. (a) $y(x) = e^{-x}(c_3 \cos x + c_4 \sin x)$
(b) $y(x) = e^{3x}(c_3 \cos 4x + c_4 \sin 4x)$
(c) $y(x) = e^{-\beta x}(c_3 \cos \omega x + c_4 \sin \omega x)$

1327

(d) $y(x) = e^{-2x}(\cos x - \sin x)$

2-7. The motion is oscillatory with frequency $(1/2\pi)(k/m)^{1/2}$ and amplitude $v_0(m/k)^{1/2}$.

2-9. $\psi(x) = A \sin \dfrac{n\pi x}{a}$ $n = 1, 2, \ldots$

2-13. $\psi(x, y) =$

$A \sin \dfrac{n_x \pi x}{a} \sin \dfrac{n_y \pi y}{b}$ $\begin{array}{l} n_x = 1, 2, \ldots \\ n_y = 1, 2, \ldots \end{array}$

2-14. $\psi(x, y, z) =$

$A \sin \dfrac{n_x \pi x}{a} \sin \dfrac{n_y \pi y}{b} \sin \dfrac{n_z \pi z}{c}$ $\begin{array}{l} n_x = 1, 2, \ldots \\ n_y = 1, 2, \ldots \\ n_z = 1, 2, \ldots \end{array}$

$E = \dfrac{h^2}{8m}\left(\dfrac{n_x^2}{a^2} + \dfrac{n_y^2}{b^2} + \dfrac{n_z^2}{c^2}\right)$

2-17. height $= v_0^2/2g$, time to return $= 2v_0/g$

2-19. $\theta(t) = A_0 e^{-\lambda t/2m} \cos(\omega t + \phi)$ where $\omega = \sqrt{(g/l) - (\lambda/2m)^2}$. If $\lambda^2 > 4m^2 g/l$, then the solution is
$\theta(t) = e^{-\lambda t/2m}(c_1 e^{\alpha t} + c_2 e^{-\alpha t})$, where $\alpha = \sqrt{(\lambda/2m)^2 - (g/l)}$ is real. Therefore, there is no oscillatory motion if $\lambda^2 > 4m^2 g/l$.

MathChapter B

B-1. $a/2$

B-2. $\dfrac{a^2}{12} - \dfrac{a^2}{2n^2\pi^2}$

B-3. $1/2$

B-6. $\left(\dfrac{8k_B T}{\pi m}\right)^{1/2}$

B-7. $\frac{3}{2}k_B T$

Chapter 3

3-1. (a) $\pm x^2$ (b) $e^{-ax}\left(x^3 - a^3\right)$ (c) $9/4$
(d) $6xy^2z^4 + 2x^3z^4 + 12x^3y^2z^2$

3-2. (a) Nonlinear (b) Nonlinear (c) Linear (d) Nonlinear (e) Linear (f) Nonlinear

3-3. (a) $-\omega^2$ (b) $i\omega$ (c) $\alpha^2 + 2\alpha + 3$ (d) 6

3-5. (a) $\hat{A}^2 = \dfrac{d^4}{dx^4}$

(b) $\hat{A}^2 = \dfrac{d^2}{dx^2} + 2x\dfrac{d}{dx} + 1 + x^2$

(c) $\hat{A}^2 = \dfrac{d^4}{dx^4} - 4x\dfrac{d^3 f(x)}{dx^3} + (4x^2 - 2)\dfrac{d^2}{dx^2} + 1$

3-9. $m^{-1/2}$

3-13. No

3-19. No

3-20. $\langle x \rangle = \dfrac{a}{2}$, $\langle x^2 \rangle = \dfrac{a^2}{3} - \dfrac{a^2}{8\pi^2}$

3-21. $\langle p \rangle = 0$, $\langle p^2 \rangle = \dfrac{h^2}{a^2}$

3-26. $1,2,1,1$; $(1,1,1)$; $(2,1,1)(1,2,1)$; $(2,2,1)$; $(1,1,2)$

3-27. 1.52×10^4 cm^{-1}, $[(25 - 20)h^2/8ma^2]$

3-32. $\sigma_p = 0, \sigma_x = \infty$

MathChapter C

C-1. $\sqrt{14}$

C-2. $\left(x^2 + y^2\right)^{1/2}$, $\left(x^2 + y^2 + z^2\right)^{1/2}$

C-3. $\cos\frac{\pi}{2} = 0$

C-6. $109°$

C-7. $5\mathbf{i} + 5\mathbf{j} - 5\mathbf{k}$; $-5\mathbf{i} - 5\mathbf{j} + 5\mathbf{k}$

Chapter 4

4-1. (a) $A = (1/\pi)^{1/4}$ (b) not normalizable (c) $A = (1/2\pi)^{1/2}$ (d) not normalizable (e) $A = 2$

4-2. (a) $A = 2/\sqrt{\pi}$ (b) normalized (c) normalized

4-5. (a) unacceptable (not normalizable) (b) acceptable (c) acceptable (d) unacceptable (not normalizable)

4-6. $\langle E \rangle = 6\hbar^2/ma^2$, $\langle E^2 \rangle = 126\hbar^4/m^2a^4$, and $\sigma_E^2 = \dfrac{90\hbar^4}{m^2a^4}$

4-8. $\langle p \rangle = 0$, $\sigma_p^2 = \dfrac{h^2}{4}\left(\dfrac{n_x^2}{a^2} + \dfrac{n_y^2}{b^2}\right)$

4-9. $\langle E \rangle = \dfrac{5\hbar^2}{m}\left(\dfrac{1}{a^2} + \dfrac{1}{b^2}\right)$

4-14. (a) commutes (b) does not commute (c) does not commute (d) does not commute

4-15. $\hat{P}$ and $\hat{Q}$ must commute

4-16. (a) $[\hat{A}, \hat{B}] = \hat{A}\hat{B}f - \hat{B}\hat{A}f = 2\dfrac{d}{dx}$ (b) $[\hat{A}, \hat{B}] = 2$ (c) $[\hat{A}, \hat{B}]f = -f(0)$

(d) $[\hat{A}, \hat{B}] = 4x\dfrac{d}{dx} + 3$

4-17. The subscripts occur as x, y, z and as cyclic permutations of x, y, z.

4-20. $\left[\hat{X}, \hat{P}_y\right] = 0$, $\left[\hat{X}, \hat{P}_x\right] = -i\hbar$,

$\left[\hat{Y}, \hat{P}_y\right] = -i\hbar$, $\left[\hat{Y}, \hat{P}_x\right] = 0$

4-21. Yes

4-22. Yes

4-23. $0, 0$

4-24. $[\hat{A}, \hat{B}] = 0$

4-26. The results gives Newton's equation, in an average sense.

4-29. No

4-32. id/dx is Hermitian, d^2/dx^2 is Hermitian, id^2/dx^2 is not Hermitian

4-36. 0.52 ($v_0 = 1.966$)

4-37. $4/(4 + v_0)$

MathChapter D

D-2. $(1, \frac{\pi}{2}, 0)$, $(1, \frac{\pi}{2}, \frac{\pi}{2})$, $(1, 0, \phi)$, $(1, \pi, \phi)$

D-3. (a) a sphere of radius 5 centered at the origin (b) a cone about the z axis (c) the y-z plane

D-4. $2\pi a^3/3$

D-5. $2\pi a^2$

D-6. $4/15$

D-10. $0, 1/3$

D-11. $8\pi/3$

Chapter 5

5-3. the period, which is the time it takes to undergo one cycle, $= 2\pi/\omega = 1/\nu$

5-7. $9.104\,432 \times 10^{-31}$ kg; 0.05%

5-9. 479 N·m^{-1}

5-10. 1.81×10^{10} m^{-1}

5-11. $V(x) = D[\beta^2 x^2 - \beta^3 x^3 + \frac{7}{12}\beta^4 x^4 + O(x^5)]$; $\gamma = -6D\beta^3$

5-12. $\tilde{x} = 0.01962$; $\tilde{\nu}\tilde{x} = 56.59$ cm^{-1}

5-13. $k = 385$ N·m^{-1}, $\tau = 1/\nu = 1.30 \times 10^{-14}$ s

5-14. $\tilde{\nu}_{\mathrm{obs}} = 321$ cm^{-1}, $\varepsilon_0 = 3.19 \times 10^{-21}$ J

5-22. $A_{\mathrm{rms}} = (\hbar/4\pi c\tilde{\nu}\mu)^{1/2}$; H$_2$: 8.719 pm; ^{35}Cl^{35}Cl: 4.170 pm; ^{14}N^{14}N: 3.215 pm

5-33. $\mu \approx 10^{-25}$ kg, $r = 10^{-10}$ m, and so $I \approx 10^{-45}$ kg·m^2, $B \approx 10^{10}$ Hz

5-34. 3.35×10^{-47} kg·m^2, 142 pm

5-35. 113 pm

5-37. (a) 0; (b) $2\hbar^2\left(\dfrac{3}{4\pi}\right)^{1/2}\cos\theta$; (c) and (d) $2\hbar^2\left(\dfrac{3}{8\pi}\right)^{1/2}\sin\theta e^{\pm i\phi}$; all four functions are eigenfunctions of $\hat{L}^2$

5-44. The mass of an electron in Equation 1.22 is replaced by the reduced mass of a hydrogen atom

5-45. $109\,677.6$ cm^{-1}

5-46. $\mu = 9.106\,909 \times 10^{-31}$ kg; $109\,707.3$ cm^{-1}

5-47. $1.000\,270$

Chapter 6

6-10. The charge distribution of a completely filled subshell is spherically symmetric.

6-12. See Equations 6.37.

6-15. otherwise L_x and L_y would be known exactly; no

6-20. 0.762

6-21. $1.3a_0$; $2.7a_0$

6-28. $\langle r\rangle_{20} = 6a_0$; $\langle r\rangle_{21} = 5a_0$

6-29. E_2; they are not unique

6-32. Yes (see Problem 6-10)

6-33. n^2

6-34. 1312 kJ·mol^{-1}; 5248 kJ·mol^{-1}

6-35. the two values of E_n differ by m_e/μ, where μ is the reduced mass

6-36. $0.999\,728$

6-44. C·m^2·s^{-1}, $9.274\,007 \times 10^{-24}$ C·m^2·s^{-1}

6-46. 1.391×10^{-22} J versus 1.635×10^{-18} J

6-47. 35 possible transitions; 5 transitions for $\Delta m = 0$; 10 transitions for $\Delta m = \pm 1$

MathChapter E

E-1. $5, 5, 5$

E-2. $-5, -5$

E-3. $0, 0$

E-4. $x^4 - 3x^2 = 0$; $x = 0, 0, \pm\sqrt{3}$

E-5. $x^4 - 4x^2 = 0$; $x = 0, 0, \pm 2$

E-6. $\cos^4 \theta + \sin^2 \theta = 1$

E-7. $(\frac{9}{5}, \frac{1}{5})$

E-8. $(1, 3, -4)$

Chapter 7

7-3. $\beta = (k\mu/7\hbar^2)^{1/2}$,
  $E = (7^{1/2}/5)\hbar(k/\mu)^{1/2} = 0.529\,\hbar(k/\mu)^{1/2}$

7-4. 0

7-5. $\alpha_{min} = 3m_e e^2/8\varepsilon_0 \pi \hbar^2$;
  $E_{min} = -(3/8)(e^2/4\pi\varepsilon_0 a_0)$

7-6. $c_2 = 0$, $\alpha = 1/a_0$, $E_{min} = E_{exact}$

7-7. $\frac{1}{2}\hbar(k/\mu)^{1/2}$, the exact result

7-8. $\alpha_{min} = (\mu k)^{1/2}/2\hbar$, $E_{min} = \frac{3}{2}\hbar(k/\mu)^{1/2}$;
  $\alpha_{min} = (3\mu k/\hbar^2)^{1/4}$, $E_{min} = 3^{1/2}\hbar(k/\mu)^{1/2}$;
  $e^{-\alpha r^2} = e^{-\alpha x^2}e^{-\alpha y^2}e^{-\alpha z^2}$ is the exact
  function

7-9. $\alpha_{min} = (6\mu c/\hbar^2)^{1/3}$, $E_{min} = \left(\frac{81}{256}\right)^{1/3}\frac{c^{1/3}\hbar^{4/3}}{\mu^{2/3}}$

7-10. $\lambda_{min} = \left[\left(\frac{\pi^2}{6} - 1\right)\frac{k\mu}{2\hbar^2}\right]^{1/4}$;

  $E_{min} = \frac{1}{2^{1/2}}\left(\frac{\pi^2}{6} - 1\right)^{1/2}\hbar\omega = 0.568\,\hbar\omega$

7-11. $E = \frac{h^2}{8ma^2} + V_0 a\left(\frac{1}{4} + \frac{1}{\pi^2}\right)$ and
  $\frac{h^2}{2ma^2} + \frac{V_0 a}{4}$. The relative magnitude of
  these two roots depends upon the relative
  magnitudes of h^2/ma^2 and $V_0 a$.

7-12. $l/a = 1.6546$, $E_{min} = 0.6816\hbar^2/ma^2$;
  $l/a = 0.68353$, $E_{min} = 0.6219\hbar^2/ma^2$

7-13. $\lambda a = 0.92423$, $E_{min} = 0.6381\hbar^2/ma^2$;
  $\lambda a = 1.1689$, $E_{min} = 0.8432\hbar^2/ma^2$

7-14. $5\hbar^2/ma^2$

7-15. $7\hbar^2/ma^2$

7-16.

$$E = \frac{3}{2}\hbar\omega + \frac{7\delta}{32\alpha^2}$$

$$-\frac{1}{2}\left[(2\hbar\omega)^2 + \frac{3}{2}\frac{\hbar\omega\delta}{\alpha^2} + \frac{11}{64}\frac{\delta^2}{\alpha^4}\right]^{1/2}$$

$$= \frac{1}{2}\hbar\omega + \frac{\delta}{32\alpha^2} + O(\delta^2)$$

7-21. $b/32\alpha^2$

7-22. $b/2$

7-23. $\Delta E = 0$,

7-24. $\Delta E = \frac{V_0 a}{2}\left(\frac{1}{2} + \frac{1 - \cos n\pi}{n^2\pi^2}\right)$

7-25. $\Delta E = \frac{3c}{4\alpha^2} - \frac{k}{4\alpha}$

7-27. $a = D\beta^2$, $b = -D\beta^3$, $c = 7D\beta^4/12$;
  $\Delta E(v=0) = \frac{3c}{4\alpha^2}$; $\Delta E(v=1) = \frac{15c}{4\alpha^2}$;
  $\Delta E(v=2) = \frac{39c}{4\alpha^2}$

Chapter 8

8-9. $S_{100} = 2\zeta^{3/2}e^{-\zeta r}/(4\pi)^{1/2}$
  $S_{200} = (2\zeta)^{5/2}re^{-\zeta r}/(96\pi)^{1/2}$
  $S_{210} = (3)^{1/2}(2\zeta)^{5/2}re^{-\zeta r}\cos\theta/(96\pi)^{1/2}$

8-17. because the effective Hamiltonian
  operator depends only upon r

8-18. because the radial dependence of the
  effective Hamiltonian operator differs
  from the Hamiltonian operator of a
  hydrogen atom

8-22. 0 and 1/2

8-23. The eigenvalues are zero.

8-25. $E(\text{triplet}) = -2.12414E_h = -466\,195\,\text{cm}^{-1}$
  $E(\text{singlet}) = -2.03635E_h = -446\,927\,\text{cm}^{-1}$
  $E(\text{ground state}) = -2.7500E_h = -603\,555\,\text{cm}^{-1}$
  $E(\text{triplet} \rightarrow \text{ground state}) = 137\,370\,\text{cm}^{-1}$
  $E(\text{singlet} \rightarrow \text{ground state}) = 156\,630\,\text{cm}^{-1}$

8-26. $^2P_{3/2}, ^2P_{1/2}$

8-28. $(1 \times 1)(^1S) + (3 \times 3)(^3P) + (1 \times 5)(^1D) = 15$

8-29. 45; $(1 \times 1)(^1S) + (1 \times 5)(^1D) + (3 \times 3)(^3P) + (3 \times 7)(^3F) + (1 \times 9)(^1G) = 45$

8-30. $^1P_1, ^3P_2, ^3P_1, ^3P_0; ^3P_0$

8-31. 20; $^1D_2, ^3D_3, ^3D_2, ^3D_1; ^3D_1$

8-32. $^1S_0, ^1D_2, ^1G_4, ^3P_2, ^3P_1, ^3P_0, ^3F_4, ^3F_3, ^3F_2; ^3F_2$

8-33. $^2P_{3/2}, ^2P_{1/2}, ^2D_{5/2}, ^2D_{3/2}, ^4S_{3/2}; ^4S_{3/2}$

8-34. $[\text{Ne}]3s^2; ^1S_0$

8-35. $^3F_2$ (see Problem 8-32)

8-36. $^1S_0$

8-37. $^3P_2$: fivefold degenerate, $^3P_1$: threefold
  degenerate, $^3P_0$: singly degenerate, $^1P_1$:
  threefold degenerate

8-38. $2p \to 1s$, 0.355 cm^{-1}; $3p \to 1s$, 0.108 cm^{-1}; $4p \to 1s$, 0.046 cm^{-1}

8-39. 18 459 Å

sharp series	principal series
11 404 Å	5 595.9 Å
11 381 Å	5 889.9 Å
6 160.7 Å	3 303.0 Å
6 154.1 Å	3 302.4 Å
5 153.4 Å	2 853.0 Å
5 148.8 Å	2 852.8 Å

diffuse series	fundamental series
8 194.7 Å	18 459 Å
8 183.2 Å	
5 688.1 Å	12 679 Å
5 682.6 Å	
4 982.8 Å	
4 978.5 Å	

8-46. $2.9/n^3$ cm^{-1}

8-47. $^2P_{3/2}$. Spin orbit coupling increases with increasing atomic number.

Chapter 9

9-4. $(1s_A - 1s_B)/\sqrt{2(1-S)}$

9-12. N_2 has a bond order of 3; N_2^+ has a bond order of 5/2.

9-13. F_2 has a bond order of 1; F_2^+ has a bond order of 3/2.

9-14. The relative bond orders are 3, 5/2, and 5/2.

9-15. The bond order of C_2 is 2; that of C_2^- is 5/2.

9-17. The bond order of NO^+ is 3; that of NO is 5/2.

9-18. 3

9-21. 6, Cr_2

9-22. $2p_{x,O}$

9-23. 3.40×10^{-18} J

9-25. bond order: 3; diamagnetic

9-26. bond order: 1; paramagnetic

9-30. See Table 9.6.

9-31. See Table 9.6.

9-32. $^3\Pi$; $^1\Pi$

9-33. $-\frac{1}{2} E_h$ per atom, 0.625

9-34. 19.4×10^{-30} C·m

9-35. 19.4×10^{-30} C·m

9-36. 2.55×10^{-29} C·m, 76.0%

9-38. $0.181e$

9-39. $0.43e$ on H, $-0.43e$ on F; $0.17e$ on H, $-0.17e$ on Cl; $0.12e$ on H, $-0.12e$ on Br; $0.054e$ on H, $-0.054e$ on I

9-41. $\nu_{D_2} = 9.332 \times 10^{13}$ Hz, $D_0^{D_2} = 439.8$ kJ·mol^{-1}

Chapter 10

10-4. $120°$

10-6. $\xi_2 = (2s - 2p_z)/\sqrt{2}$

10-9. $109.47°$

10-11. $104.5°$

10-12. $\psi_1 = 0.8945(0.5004 \cdot 2s + 0.6122 \cdot 2p_z + 0.7907 \cdot 2p_y)$; $\psi_2 = 0.8945(0.5004 \cdot 2s + 0.6122 \cdot 2p_z - 0.7907 \cdot 2p_y)$

10-13. Both descriptions are correct.

10-15. Small changes in the bond angle will not result in net overlap between the lobes of the orbitals, so the energies will remain unchanged.

10-18. (a) linear (b) linear (c) bent

10-19. (a) bent (b) bent (c) linear

10-20. $1s_X$, a nonbonding core electron orbital

10-22. Take the molecule to sit in the x-y plane, then the orbitals involved are $2s$, $2p_x$, and $2p_y$ on the central X atom and the $1s$ orbitals on each hydrogen atom.

10-23. (a) planar (b) pyramidal (c) pyramidal (d) pyramidal

10-25. $E_\pm = (\alpha \pm \beta)/(1 \pm S)$; $\psi_\pm = (2p_{zA} \pm 2p_{zB})/[\sqrt{2(1 \pm S)}]$

10-30. $x^4 - 4x^2 = 0$; $x = 2, 0, 0, -2$; $E = \alpha + 2\beta, \alpha, \alpha, \alpha - 2\beta$ The ground state is predicted to be a triplet state; the two molecules have the same stability. Cyclobutadiene has no delocalization energy.

10-31. $x^4 - 3x^2 = 0$; $x = \sqrt{3}, 0, 0, -\sqrt{3}$;
$E = \alpha + \sqrt{3}\beta, \alpha, \alpha, \alpha - \sqrt{3}\beta$;
$E_\pi = 2(\alpha + \sqrt{3}\beta) + 2\alpha = 4\alpha + 2\sqrt{3}\beta$;
$E_{\text{deloc}} = -0.5359\beta$

10-32. $x^4 - 5x^2 + 4 = 0$;
$x = 1, 0, -\frac{1}{2} \pm \frac{1}{2}\sqrt{17}$;
$E_\pi = 2(\alpha + 2.562\beta) + 2\alpha = 4\alpha + 5.124\beta$; $E_{\text{deloc}} = 1.124\beta$

10-35. $10\alpha + 13.68\beta$

10-36. 3.68β

10-37. For the triangular geometry,
$x^3 - 3x + 2 = 0$. $E_{\text{H}_3^+} = 2\alpha + 4\beta$,
$E_{\text{H}_3} = 3\alpha + 3\beta$, $E_{\text{H}_3^-} = 4\alpha + 2\beta$; for the
linear geometry, $x^3 - 2x = 0$.
$E_{\text{H}_3^+} = 2\alpha + 2\sqrt{2}\beta$, $E_{\text{H}_3} = 3\alpha + 2\sqrt{2}\beta$,
$E_{\text{H}_3^-} = 4\alpha + 2\sqrt{2}\beta$; therefore, H_3^+ is
triangular, H_3^- is linear, and H_3 is
triangular.

10-41. radical: $E_{\text{deloc}} = 0.828\beta$; $q_1 = q_2 = q_3 = 1$; $P_{12}^\pi = P_{23}^\pi = 0.707$
cation:
$E_{\text{deloc}} = 0.828\beta$; $q_1 = q_3 = 1/2$; $q_2 = 1$;
$P_{12}^\pi = P_{23}^\pi = 0.707$
anion:
$E_{\text{deloc}} = 0.828\beta$; $q_1 = q_3 = 3/2$; $q_2 = 1$;
$P_{12}^\pi = P_{23}^\pi = 0.707$

10-42. $q_n = 1$; $P_{rs}^\pi = 2/3$. Benzene is a
symmetric molecule

10-43. (b) for hexatriene
$E_1 = \alpha + 1.802\beta$ $E_2 = \alpha + 1.247\beta$
$E_3 = \alpha + 0.4450\beta$ $E_4 = \alpha - 0.4450\beta$ $E_5 = \alpha - 1.247\beta$
$E_6 = \alpha - 1.802\beta$; $E_{\text{deloc}} = 0.9880\beta$;
E_{deloc} per carbon atom = 0.1647 for
octatetraene
$E_1 = \alpha + 1.879\beta$ $E_2 = \alpha + 1.532\beta$
$E_3 = \alpha + \beta$ $E_4 = \alpha + 0.3473\beta$ $E_5 = \alpha - 0.3473\beta$ $E_6 = \alpha - \beta$ $E_7 = \alpha - 1.532\beta$ $E_8 = \alpha - 1.879\beta$;
$E_{\text{deloc}} = 1.517\beta$; E_{deloc} per carbon atom = 0.1896 (c) benzene

10-45.
$E_1 - E_N = 2\beta(\cos\frac{\pi}{N+1} - \cos\frac{\pi N}{N+1}) \to 2\beta(\cos 0 - \cos\pi) = 4\beta$

10-46. a conductor

Chapter 11

11-14. $\alpha = \alpha(\zeta = 1.00)\zeta^2 = (0.2709)(1.24)^2 = 0.4166$

11-15. $\alpha_{1s1} = (0.10982)(1.24)^2 = 0.1688$;
$\alpha_{1s2} = (0.40578)(1.24)^2 = 0.6239$;
$\alpha_{1s3} = (2.2777)(1.24)^2 = 3.425$

11-16. $\phi_{3s} = -2.51831\phi_{3s}^{\text{GF}}(r, 3.18649) + 6.15890\phi_{3s}^{\text{GF}}(r, 1.19427) + 1.06018\phi_{3s}^{\text{GF}}(r, 4.2037) + d'_{3p}\phi_{3s}^{\text{GF}}(r, 1.42657)$;
$\phi_{3p} = -1.42993\phi_{3p}^{\text{GF}}(r, 3.18649) + 3.23572\phi_{3p}^{\text{GF}}(r, 1.19427) + 7.43507\phi_{3p}^{\text{GF}}(r, 4.20377) + d'_{3p}\phi_{3p}^{\text{GF}}(r, 1.42657)$

11-18. C (0, 0, 0); Cl (0, 0, 178.1); H_a (−103.66, 0, −35.59); H_b (51.84, −89.78, −35.59); H_c (51.84, 89.78, −35.59)

11-19. $k_{\text{H}_2} = 641.4$ N·m^{-1}, $k_{\text{CO}} = 2403$ N·m^{-1}, $k_{\text{N}_2} = 3151$ N·m^{-1}

11-20. (a) $\phi(r) = (128\alpha^5/\pi^3)^{1/4}xe^{-\alpha r^2}$;
(b) $\phi(r) = (2048\alpha^7/9\pi^3)^{1/4}x^2 e^{-\alpha r^2}$

11-21. $2p_y$, no radial nodes

11-23. $3d_{x^2-y^2}$

11-24. Using a sum of three functions with different values of zeta to fit one of the functions in a basis set.

11-25. $0, -2, +1$

11-26. 4.70×10^{-30} C·m;
4.702×10^{-30} C·m

11-28. $(1a_1)^2(2a_1)^2(3a_1)^2(4a_1)^2(1b_2)^2(5a_1)^2$
$(1b_1)^2(2b_2)^2$; $3a_1$; inner valence electrons

11-29. 1 D $= 1 \times 10^{-18}$ esu·cm $=$
$(1 \times 10^{-18}$ esu·cm$)(1.6022 \times 10^{-19}$ C$/4.803 \times 10^{-10}$ esu$)($m$/100$ cm$) = 3.336 \times 10^{-30}$ C·m

MathChapter F

F-1. $\mathbf{C} = \begin{pmatrix} 5 & -3 & -2 \\ -11 & 4 & -6 \\ -3 & -1 & -1 \end{pmatrix}$;

$$D = \begin{pmatrix} -7 & 6 & 1 \\ 19 & -2 & 12 \\ 6 & 5 & 5 \end{pmatrix}$$

F-4. If A, B, and C corrrespond to $\hat{L}_x$, $\hat{L}_y$, and $\hat{L}_z$, respectively, then the results are similar to the commutation relations of $\hat{L}_x$, $\hat{L}_y$, and $\hat{L}_z$.

F-7. $\det C_3 = 1$; $\operatorname{Tr} C_3 = -1$; $\det \sigma_v = -1$; $\operatorname{Tr} \sigma_v = 0$; $\det \sigma_v' = -1$; $\operatorname{Tr} \sigma_v' = 0$; $\det \sigma_v'' = -1$; $\operatorname{Tr} \sigma_v'' = 0$

F-8. $C_3, \sigma_v, \sigma_v', \sigma_v''$

F-9. $A^{-1} = \begin{pmatrix} 0 & \sqrt{2} \\ \sqrt{2} & -1 \end{pmatrix}$;

$$A^{-1} = -\frac{1}{4} \begin{pmatrix} 1 & -2 & -1 \\ 1 & -6 & 3 \\ -2 & 4 & -2 \end{pmatrix}$$

F-12. $x = 24/13$, $y = -19/13$, and $z = -8/13$

Chapter 12

12-3. $E, C_3, 3C_2, \sigma_h, 3\sigma_v, S_3$

12-7. $\hat{C}_4^3$ is a counter-clockwise rotation by $270°$ and $\hat{C}_4^{-1}$ is a clockwise rotation by $90°$.

12-9. 16

12-10. 24

12-11. $\hat{C}_2, \hat{\sigma}_v', \hat{\sigma}_v$

12-12. $\hat{\sigma}_v', \hat{\sigma}_v$

12-15. See Equation F.2.

12-16. $\hat{E}u_x = u_x$, $\hat{C}_2 u_x = -u_x$, $\hat{\sigma}_v u_x = u_x$, $\hat{\sigma}_v' u_x = -u_x$

12-17. $\hat{E} R_x = R_x$, $\hat{C}_2 R_x = -R_x$, $\hat{\sigma}_v R_x = -R_x$, $\hat{\sigma}_v' R_x = R_x$

12-21. $\Gamma = 4A_2 + 2E + 2T_1 + T_2$

12-22. $\Gamma = 8A_1 + 5A_2 + 6B_1 + 8B_2$

12-23. $\Gamma = A_1' + A_2' + 3E' + 2A_2'' + E''$

12-24. no

12-25. $\phi_2 = \psi_1 - \psi_2 + \psi_3 - \psi_4 + \psi_5 - \psi_6$
$\hat{E}\phi_2 = \phi_2$; $\hat{C}_6\phi_2 =$
$\psi_6 - \psi_1 + \psi_2 - \psi_3 + \psi_4 - \psi_5 = -\phi_2$;
$\hat{C}_3\phi_2 = \psi_5 - \psi_6 + \psi_1 - \psi_2 + \psi_3 - \psi_4 = \phi_2$; $\hat{C}_2\phi_2 =$
$\psi_4 - \psi_5 + \psi_6 - \psi_1 + \psi_2 - \psi_3 = -\phi_2$;
$\hat{C}_2' = -\psi_1 + \psi_6 - \psi_5 + \psi_4 - \psi_3 + \psi_2 =$

$-\phi_2$; $\hat{C}_2''\phi_2 =$
$-\psi_2 + \psi_1 - \psi_6 + \psi_5 - \psi_4 + \psi_3 = \phi_2$:
$\hat{i}\phi_2 = -\psi_4 + \psi_5 - \psi_6 + \psi_1 - \psi_2 + \psi_3 = \phi_2$; $\hat{\sigma}\phi_2 =$
$\psi_2 - \psi_1 + \psi_6 - \psi_5 + \psi_4 - \psi_3 = -\phi_2$

12-26. $H_{33} = \alpha + \beta$, $H_{34} = 0$, $H_{44} = \alpha + \beta$, $S_{33} = S_{44} = 1$, $S_{34} = 0$; $(x+1)^2 = 0$

12-27. Number of nodes of
$\phi_2 > \phi_4 = \phi_3 > \phi_6 = \phi_5 > \phi_1$; $\phi_1(A_{2u})$, $\phi_2(B_{2g})$, $\phi_3(E_{1g})$, $\phi_4(E_{1g})$, $\phi_5(E_{2u})$, $\phi_6(E_{2u})$

12-29. $\Gamma = B_{2g} + E_{1g} + A_{2u} + E_{2u}$; breaks down into two 1×1 and two 2×2 determinants

12-30. $\Gamma = A_{2u} + B_{1u} + E_g$ breaks down into two 1×1 determinants and one 2×2 determinant

12-31. $\phi_1 = (\psi_1 + \psi_3)/\sqrt{2}$, $\phi_2 = \psi_2$, $\phi_3 = (\psi_1 - \psi_3)/\sqrt{2}$.
$H_{11} = H_{22} = H_{33} = \alpha$, $H_{12} = \sqrt{2}\beta$, $H_{13} = H_{23} = 0$; $(x^2 - 2)(x) = 0$;
$E = \alpha \pm \sqrt{2}\beta, \alpha$

12-35. See Problem 12-36.

12-36. $\hat{E}f_{\text{even}}(x) = f_{\text{even}}$, $\hat{\sigma}f_{\text{even}}(x) = f_{\text{even}}$;
$\hat{E}f_{\text{odd}}(x) = f_{\text{odd}}(x)$, $\hat{\sigma}f_{\text{odd}}(x) = -f_{\text{odd}}(x)$

12-37. $\Gamma = 4 \ 1 - 2 - 4 - 1 \ 2$,
$\Gamma = 2A_2'' + E''$;
$\phi_1 = (\psi_2 + \psi_3 + \psi_4)/\sqrt{3}$, ψ_1,
$(2\psi_2 - \psi_3 - \psi_4)/\sqrt{6}$,
$(2\psi_3 - \psi_2 - \psi_4)/\sqrt{6}$; $H_{11} = H_{22} = \alpha$,
$H_{12} = \sqrt{3}\beta$, $H_{33} = H_{44} = \alpha$,
$H_{34} = -\alpha/2$, $S_{34} = -1/2$.
$E_\pi = 2(\alpha + \sqrt{3}\beta) + 2\alpha = 4\alpha + 2\sqrt{3}\beta$

12-38. $\Gamma = 4 \ 0 \ 0 - 4$, $\Gamma = 2B_g + 2A_u$;
$\phi_1 = (\psi_1 - \psi_4)/\sqrt{2}$,
$\phi_2 = (\psi_2 - \psi_3)/\sqrt{2}$,
$\phi_3 = (\psi_1 + \psi_4)/\sqrt{2}$, $\phi_4 = (\psi_2 + \psi_3)/\sqrt{2}$,
$H_{11} = \alpha - \beta$, $H_{22} = \alpha$, $H_{12} = 0$,
$H_{33} = \alpha + \beta$, $H_{44} = \alpha$, $H_{34} = 2\beta$,
$(x - 1)(x)(x^2 + x - 4) = 0$; $E_\pi = 2(\alpha + 2.562\beta) + 2\alpha = 4\alpha + 5.124\beta$

12-39. See Problem 12-31.

Chapter 13

13-1. 127 pm

13-2. 305 pm

13-3. $1.96 \times 10^{11} \text{ s}^{-1}$, 1.96×10^5 MHz, 6.54 cm^{-1}

13-4. $1.36 \times 10^{11} \text{ revolution} \cdot \text{s}^{-1}$

13-5. $2\tilde{B} = 2.96 \text{ cm}^{-1}$

13-6. $1.896 \times 10^{-46} \text{ kg} \cdot \text{m}^2$

13-7. $84.0 \text{ N} \cdot \text{m}^{-1}$; 1.20×10^{-13} s

13-8. 321 cm^{-1}, 3.19×10^{-21} J

13-9. 3.21 pm

13-11.
$\tilde{\nu}_R = 2160.0 \text{ cm}^{-1} + (3.87 \text{ cm}^{-1})(J+1)$;
$\tilde{\nu}_R = 2160.0 \text{ cm}^{-1} - (3.87 \text{ cm}^{-1})J$

13-12.
$\tilde{\nu}_R = 936.7 \text{ cm}^{-1} + (1.52 \text{ cm}^{-1})(J+1)$
$J = 0, 1, 2, \ldots$; $\tilde{\nu}_P = 936.7 \text{ cm}^{-1} - (1.52 \text{ cm}^{-1})(J+1)$ $J = 0, 1, 2, \ldots$

13-13. $\tilde{\nu}_R(0 \to 1) = 2905.57 \text{ cm}^{-1}$;
$\tilde{\nu}_R(1 \to 2) = 2925.22 \text{ cm}^{-1}$;
$\tilde{\nu}_P(1 \to 0) = 2864.43 \text{ cm}^{-1}$;
$\tilde{\nu}_P(2 \to 1) = 2842.93 \text{ cm}^{-1}$

13-14. $\tilde{B}_0 = 8.35 \text{ cm}^{-1}$; $\tilde{B}_1 = 8.12 \text{ cm}^{-1}$;
$\tilde{B}_e = 8.47 \text{ cm}^{-1}$; $\tilde{\alpha}_e = 0.23 \text{ cm}^{-1}$

13-15. $\tilde{B}_{\text{HI}} = 6.428 \text{ cm}^{-1}$;
$I_{\text{HI}} = 4.355 \times 10^{-47} \text{ kg} \cdot \text{m}^2$;
$R_{e,\text{HI}} = 161.9$ pm; $\tilde{B}_{\text{DI}} = 3.254 \text{ cm}^{-1}$;
$I_{\text{DI}} = 8.602 \times 10^{-47} \text{ kg} \cdot \text{m}^2$;
$R_{e,\text{DI}} = 161.7$ pm

13-16. yes

Molecule	$^{74}\text{Ge}^{32}\text{S}$	$^{72}\text{Ge}^{32}\text{S}$
$(\tilde{\nu}, J=0)/\text{cm}^{-1}$	0.372 372	0.375 496
$(\tilde{\nu}, J=1)/\text{cm}^{-1}$	0.744 742	0.750 990
$\Delta\tilde{\nu}/\text{cm}^{-1}$	0.372 370	0.375 494

13-17.
$\tilde{B} = 10.40 \text{ cm}^{-1}$; $\tilde{D} = 4.55 \times 10^{-4} \text{ cm}^{-1}$

13-18.
$\tilde{B} = 1.9227 \text{ cm}^{-1}$; $\tilde{D} = 6.53 \times 10^{-6} \text{ cm}^{-1}$

13-19. $0 \to 1 : 21.1847 \text{ cm}^{-1}$; $1 \to 2 :$
42.3566 cm^{-1}; $2 \to 3 :$
63.5030 cm^{-1}; $3 \to 4 : 84.6110 \text{ cm}^{-1}$

13-20. $\tilde{D}_e = 37\,200 \text{ cm}^{-1}$

13-21. 69 or 70

13-22. $490 \text{ N} \cdot \text{m}^{-1}$

13-24. $\tilde{\nu}_e = 2169.0 \text{ cm}^{-1}$; $\tilde{x}_e\tilde{\nu}_e = 13.0 \text{ cm}^{-1}$

13-25. 2558.539 cm^{-1}; 5026.642 cm^{-1};
7404.309 cm^{-1}; 9691.54 cm^{-1}

13-26. $\tilde{\nu}_e = 2989 \text{ cm}^{-1}$; $\tilde{\nu}_e\tilde{x}_e = 51.6 \text{ cm}^{-1}$

13-27. $\tilde{\nu}_e = 384.1 \text{ cm}^{-1}$; $\tilde{\nu}_e\tilde{x}_e = 1.45 \text{ cm}^{-1}$

13-28. $\tilde{T}_e = 65\,080.3 \text{ cm}^{-1}$

13-29. $E'_0 = 65\,833.13 \text{ cm}^{-1}$,
$E''_0 = 1081.58 \text{ cm}^{-1}$,
$\tilde{\nu}_{\text{obs}}(0 \to 0) = 64\,751.55 \text{ cm}^{-1}$,
$\tilde{\nu}_{\text{obs}}(0 \to 1) = 66\,230.85 \text{ cm}^{-1}$,
$\tilde{\nu}_{\text{obs}}(0 \to 2) = 67\,675.35 \text{ cm}^{-1}$,
$\tilde{\nu}_{\text{obs}}(0 \to 3) = 69\,085.05 \text{ cm}^{-1}$

13-30. $\tilde{\alpha}_e = 0.00592 \text{ cm}^{-1}$; 0.82004 cm^{-1}

13-31. $\tilde{\nu}'_e = 1126.2 \text{ cm}^{-1}$; $\tilde{x}'_e\tilde{\nu}'_e = 8.0 \text{ cm}^{-1}$

13-32. $\tilde{\nu}'_e = 267.76 \text{ cm}^{-1}$; $\tilde{x}'_e\tilde{\nu}'_e = 0.04 \text{ cm}^{-1}$

13-33. (a) 3, 3, 9 (b) 3, 2, 4 (c) 3, 3, 30 (d) 3, 3, 6

13-34. HCl, CH_3I, and H_2O

13-35. symmetric top, spherical top, asymmetric top, spherical top

13-36. prolate, oblate, oblate, prolate

13-38.
$\lambda^2 - \lambda(2\cos^2\theta + 8\sin^2\theta + 8\cos^2\theta + 2\sin^2\theta) + 16\cos^4\theta + 68\cos^2\theta\sin^2\theta + 16\sin^4\theta - 36\sin^2\theta\cos^2\theta =$
$\lambda^2 - 10\lambda + 16(\cos^2\theta + \sin^2\theta)^2 =$
$\lambda^2 - 10\lambda + 16 = 0$; $\lambda = 2, 8$

13-42. $\sqrt{5}/2$

13-43. $1/\sqrt{2}$

13-44.
$\Gamma_{3N} = 12\ 0\ 2\ 2\ 2 = 3A_1 + A_2 + 4E$;
$\Gamma_{\text{vib}} = 2A_1 + 2E$; all modes infrared active

13-45. $\Gamma_{3N} = 15\ -1\ 3\ 3 =$
$5A_1 + 2A_2 + 4B_1 + 4B_2$;
$\Gamma_{\text{vib}} = 4A_1 + A_2 + 2B_1 + 2B_2$; the A_2 mode is infrared inactive.

13-46. $\Gamma_{3N} = 18\ 0\ 0\ 6 =$
$6A_g + 3B_g + 3A_u + 6B_u$;
$\Gamma_{\text{vib}} = 5A_g + B_g + 2A_u + 4B_u$; A_g and B_g modes infrared inactive

13-47. $\Gamma_{3N} = 15\ 1\ -1\ -3\ -1\ -3\ -1\ 5\ 3\ 1 = A_{1g} + A_{2g} + B_{1g} + B_{2g} + E_g + 2A_u + B_{2u} + 3E_u$; $\Gamma_{\text{vib}} = A_{1g} + B_{1g} + B_{2g} + A_{2u} + B_{2u} + 2E_u$; A_{1g}, B_{1g}, B_{2g}, and B_{2u} modes infrared inactive

13-48. $\Gamma_{3N} = 15 \ 0 \ -1 \ -1 \ 3 =$
$A_1 + E + T_1 + 3T_2$;
$\Gamma_{\text{vib}} = A_1 + E + 2T_2$; A_1 and E modes infrared inactive

13-50. $\tilde{B}_0 = 1.9163 \text{ cm}^{-1}$; $\tilde{B}_1 = $
1.8986 cm^{-1}; $\tilde{B}_e = 1.92515 \text{ cm}^{-1}$;
$\tilde{I}_e = 1.454 \times 10^{-46} \text{ kg·m}^2$;
$R_e = 113.0 \text{ pm}$

Chapter 14

14-2. 8.41 T

14-3. 6.341 T

14-4. ^{1}H 7.05 T, ^{2}H 45.9 T, ^{13}C 28.0 T, ^{14}N
97.5 T, ^{31}P 17.4 T

14-5. 500 MHz or 12 T

14-6. 2200 Hz

14-7. Use the equation
$\Delta \nu = (60.0 \text{ MHz})\Delta \delta \times 10^{-6}$.

14-15. γ has units of $\text{T}^{-1}\cdot\text{s}^{-1}$, B_0 has units
of T, I has units of J·s, and J has units of
s^{-1}.

14-26. The centers of the doublets occur at
$\nu_0(1 - \sigma_1)$ and $\nu_0(1 - \sigma_2)$, so they are
separated by $\nu_0|\sigma_1 - \sigma_2|$.

14-33. Let $J = 0$ in Table 14.6.
$\nu_{1\to2} = \nu_{3\to4} = \nu_0(1 - \sigma_1)$;
$\nu_{1\to3} = \nu_{2\to4} = \nu_0(1 - \sigma_2)$

Chapter 15

15-1. fluorescence

15-3. $\text{J}^{-1}\cdot\text{m}^2\cdot\text{s}^{-1}$, $\text{J}^{-1}\cdot\text{m}^4\cdot\text{s}^{-1}$

15-9. $1.1 \times 10^{-29} \text{ C·m}$

15-10. $9.98 \times 10^{19} \text{ J}^{-1}\cdot\text{m}^2\cdot\text{s}^{-2}$;
$7.68 \times 10^{-30} \text{ C·m}$

15-12. $(A_{32} + A_{31})^{-1}$

15-13. 81.9 ns

15-14. $^3\text{S}_1$

15-15. $0.904 \ E_h = 198\,000 \text{ cm}^{-1}$

15-16. 3392.242 nm (in vacuo); 3391.3 in
air

15-18. 36; $(3 \times 5) + (1 \times 5) + (3 \times 3) +$
$(1 \times 3) + (3 \times 1) + (1 \times 1) = 36$

15-19. 560 kW; 5.49×10^{18} photons

15-20. 3.71×10^{20} atoms; $E = 106$ J;
1.06×10^{12} W

15-21. 760-nm pulse

15-22. 5.36×10^{19} photons, 1.3%

15-23. forbidden in the harmonic oscillator
approximation.

15-24. 6.52×10^{21} photons

15-25. 961.57 cm^{-1}, 962.34 cm^{-1}; forbidden
in the rigid rotator approximation.

15-26. 160 nm;
1.0×10^{-4} J; 8.10×10^{13} photons

15-27. $17\,300 \text{ cm}^{-1}$; $J = 0$

15-28. 0.30

15-29. 79.7 s

15-30. 1.10×10^{15} molecules destroyed·s^{-1};
9.12 W

15-31. 1.965×10^5 J

15-32.
$\Delta\nu = 1.59 \times 10^{13} \text{ s}^{-1}$; $\Delta\nu = 159 \text{ s}^{-1}$; no

15-33. 610 m·s^{-1}

15-34. 388 nm; 15.12 cm^{-1}; no

15-35. 4.06×10^{13} photons

15-36. 532.05 nm; 8.035×10^{17} photons;
4.018×10^{17} photons

15-37. 358.4 nm

15-38. A: unitless; ε: $\text{m}^2\cdot\text{mol}^{-1}$; $A = 0.602$;
$\varepsilon = 629 \text{ m}^2\cdot\text{mol}^{-1}$; 8.3%

15-39. m^2; $\sigma = 2.303\varepsilon/N_A$;
$2.41 \times 10^{-21} \text{ m}^2$

15-40. $\text{m}^2\cdot\text{mol}^{-1}$; $\kappa = 2.303\varepsilon$;
$\kappa = 1450 \text{ m}^2\cdot\text{mol}^{-1}$

15-41. m·mol^{-1}; $\alpha^{1/2} = 1.66/\Delta\tilde{\nu}_{1/2}$

MathChapter G

G-1. 0.8596

G-2. 1.4142

G-3. 4.965

G-4. 0.615 atm

G-5. 0.077 780

G-6. 0.3473, 1.532, -1.879

G-7. 0.0750

G-11. $\ln 2 = 0.693\,147$; $n = 10$

G-12. 0.886 2269

G-13. 6.493 94

Chapter 16

16-1. 2.98×10^6 atm, 3.02×10^6 bar

16-2. 7.39×10^2 torr, 0.972 atm

16-3. 1.00 atm

16-4. $-40°$

16-6. 3.24×10^4 molecules, 1.85×10^{19} cm$^3 \cdot$mol^{-1}

16-7. 44.10

16-9. $y_{H_2} = 0.77$, $y_{N_2} = 0.23$

16-10. 2.2 bar, 2.2 bar

16-11. Cl$_2$

16-12. 62.3639 dm$^3 \cdot$torr$\cdot$K$^{-1} \cdot$mol^{-1}

16-15. 0.04998 dm$^3 \cdot$mol^{-1}; 0.03865 dm$^3 \cdot$mol^{-1}; 0.01663 dm$^3 \cdot$mol^{-1}

16-16. 353 bar; 8008 bar; 438 bar; 284 bar

16-17. 1031 bar; 411 bar

16-18. 10.00 mol$\cdot$L^{-1}; 10.28 mol$\cdot$L^{-1}

16-19. 1570 bar; -4250 bar

16-21. R-K: 345 bar; P-R: 129 bar

16-22. 0.07073 L$\cdot$mol^{-1}, 0.07897 L$\cdot$mol^{-1}, 0.2167 L$\cdot$mol^{-1}; 14.14 mol$\cdot$L^{-1}; 4.615 mol$\cdot$L^{-1}.

16-23. RK: 20.13 mol$\cdot$L^{-1}, 5.148 mol$\cdot$L^{-1}; PR: 23.61 mol$\cdot$L^{-1}, 5.564 mol$\cdot$L^{-1}

16-24. vdW: 4.786 mol$\cdot$L^{-1}, 0.5741 mol$\cdot$L^{-1}; RK: 6.823 mol$\cdot$L^{-1}, 0.6078 mol$\cdot$L^{-1}; PR: 8.116 mol$\cdot$L^{-1}, 0.6321 mol$\cdot$L^{-1}

16-35. $\overline{V} \approx 78.5$ cm$^3 \cdot$mol^{-1}

16-37. 0.00150 bar^{-1}

16-38. -5.33×10^{-3} dm$^3 \cdot$mol^{-1}

16-40. 1 kJ vs. 100 kJ

16-43. Yes

16-44. -15.15 cm$^3 \cdot$mol^{-1}

16-45. -60 cm$^3 \cdot$mol^{-1}

16-52.
$(\text{C}\cdot\text{m})^2(\text{m}^3)/(\text{C}^2\cdot\text{s}^2\cdot\text{kg}^{-1}\cdot\text{m}^{-3})(\text{m}^6) = $
kg$\cdot$m$^2\cdot$s^{-1} = J

16-54.
5.86×10^{-78} J$\cdot$m^6; 1.35×10^{-77} J$\cdot$m^6

MathChapter H

H-1. $\kappa = 1/P$

H-2. $\alpha = 1/T$

H-4. $\frac{3}{2}Nk_B T = \frac{3}{2}nRT$

H-7. 0; a/V^2; $3A/2T^{1/2}V(V+B)$

H-9. 0; 0; $-3A/4T^{3/2}V(V+B)$

H-10. exact

H-11. inexact; exact

Chapter 17

17-7. $-\hbar\gamma B_z$, 0

17-8. $\exp(-0.010\ \text{K}/T)$

17-10. $\langle E \rangle = Nk_B T$

17-11. $\langle E \rangle = \frac{3}{2}Nk_B T - \frac{aN^2}{V}$

17-12. $\frac{3}{2}Nk_B T$, $P = \dfrac{Nk_B T}{V - b}$

17-13. Nk_B

17-14. $\frac{3}{2}Nk_B$

17-18. $\overline{C}_V$ is a function of $T^* = T/\Theta_E$, where $\Theta_E = h\nu/k_B$

17-23. (a) 6 (b) 9 (c) 12

17-24. 9 total, 3 allowed

17-25. 6 allowed terms

17-26. 27 total, 1 allowed

17-27. 10 allowed terms

17-28. 1.94×10^{-6}

17-29. 0.0928

17-30. 1420

17-31. 0.286

17-35. $f_v = e^{-h\nu v/k_B T}(1 - e^{-h\nu/k_B T})$

17-36. 1.000, 0.9962, 0.9650

17-41.

Term symbol	Fraction of atoms, 1000 K	Fraction of atoms, 2500 K
$^2S_{1/2}$	1.00	1.00
$^2P_{1/2}$	2.55×10^{-11}	5.79×10^{-5}
$^2P_{3/2}$	4.97×10^{-11}	1.15×10^{-4}
$^2S_{1/2}$	8.27×10^{-17}	3.69×10^{-7}

17-42. 29.06 J$\cdot$mol$^{-1} \cdot$K^{-1}

17-43. 3420 K

MathChapter I

I-1. $1.25 \times 10^{-3}\%$, $4.97 \times 10^{-3}\%$, $\ldots$, 0.468%

I-2. 0.2498%, 0.4992%, $\ldots$, 4.921%

I-3. $1 + \dfrac{x}{2} - \dfrac{x^2}{8} + O(x^3)$

I-4. $\dfrac{e^{-\frac{1}{2}\beta h\nu}}{1 - e^{-\beta h\nu}}$

I-6. 1

I-10. 1

I-11. $\dfrac{a^3}{3} - \dfrac{a^4}{4} + \cdots$

I-12.
$[x^{2n+3}/(2n+3)!]/[x^{2n+1}/(2n+1)!] \longrightarrow$
x^2/n^2

Chapter 18

18-4. $f_2(T = 300\text{K}) = 4.8 \times 10^{-36}$;
$f_2(T = 1000\text{K}) = 2.5 \times 10^{-11}$;
$f_2(T = 2000\text{K}) = 5.0 \times 10^{-6}$

18-5. $f_2(T = 300\text{K}) = 9.0 \times 10^{-32}$;
$f_2(T = 1000\text{K}) = 4.9 \times 10^{-10}$;
$f_2(T = 2000\text{K}) = 2.2 \times 10^{-5}$

18-7. $D_e = D_0 + \frac{1}{2}R\Theta_{\text{vib}}$ CO:
1083 kJ·mol^{-1}; NO: $638.1 \text{ kJ·mol}^{-1}$;
K_2: 54.1 kJ·mol^{-1}

18-8. 6332 K, 4478 K

18-10. $f_{\nu>0} = 7.6 \times 10^{-7}$ at 300 K;
$f_{\nu>0} = 1.46 \times 10^{-2}$ at 1000 K

18-11. $f_{\nu>0} = e^{-\Theta_{\text{vib}}/T} = 1.01 \times 10^{-9}$ for
H_2; 0.0683 for Cl_2; 0.358 for I_2; etc.

18-12. 87.6 K, 43.8 K

18-13. 9 or 10

18-14. N_2: 0.32%; H_2: 9.45%

18-16. $\approx 20\%$

18-18. NO(g) at 300 K, $J_{\text{max}} = 7$; at 1000 K
$J_{\text{max}} = 14$

18-21. $\Theta_{\text{vib},j} = 5360$ K:
$(\overline{C}_{Vj}/R) = 1.05 \times 10^{-2}$;
$\Theta_{\text{vib},j} = 5160$ K: $(\overline{C}_{Vj}/R) = 1.36 \times 10^{-2}$;
$\Theta_{\text{vib},j} = 2290$ K: $(\overline{C}_{Vj}/R) = 3.35 \times 10^{-1}$

18-22. $\Theta_{\text{elec},1} = 227.6$ K; $\Theta_{\text{elec},2} = 325.9$ K;
$q_{\text{elec}} = 5 + 3e^{-227.6 \text{ K}/T} + e^{-325.9 \text{ K}/T} = $
8.803 at 5000 K

18-23. 2, 1, 12, 24, 2, 4

18-24. $\Theta_{\text{rot}} = 2.141$ K; $\Theta_{\text{vib},1} = 3016$ K;
$\Theta_{\text{vib},2} = 1026$ K; $\Theta_{\text{vib},3} = 4765$ K;
$\overline{C}_V/R = 6.21$

18-25. 2; $2.368 \times 10^{-46} \text{ kg·m}^2$; 1.702 K;
2842 K; 4849 K; 4715 K; 1049 K;
863.3 K; $4.34R$

18-28. $I = 6.746 \times 10^{-46} \text{ kg·m}^2$,
$\Theta_{\text{rot}} = 0.597$ K

18-29. see Table 18.4; $5.304R$

18-34. $\Theta_{\text{vib,D}_2} = 4480$ K; $\Theta_{\text{rot,D}_2} = 42.7$ K;
$\Theta_{\text{vib,HD}} = 5484$K; $\Theta_{\text{rot,HD}} = 64.0$ K

18-35. $\ln q_{\text{rot}}(T) = \ln \dfrac{T}{\Theta_{\text{rot}}} + \dfrac{1}{3}\left(\dfrac{\Theta_{\text{rot}}}{T}\right)$
$+ \dfrac{1}{90}\left(\dfrac{\Theta_{\text{rot}}}{T}\right)^2 + \cdots$

18-37. yes; 2140 K; no

18-38. 4 degrees of freedom;
$U = \dfrac{3}{2}RT + \dfrac{R\Theta_{\text{vib}}}{2} + \dfrac{R\Theta_{\text{vib}}e^{-\Theta_{\text{vib}}/T}}{1 - e^{-\Theta_{\text{vib}}/T}}$

18-39. (a) $3R/2$ (b) $7R/2$ (c) $6R$ (d) $13R/2$
(e) $12R$

18-40. 0.52%

Chapter 19

19-1. KE = 9.80 kJ; $u = 44.3 \text{ m·s}^{-1}$;
22.2°C

19-2. 15.0 bar; 3000 J

19-3. 28.8 bar; 3.60 J

19-4. 4.01 kJ

19-5. -1.73 kJ

19-6. 11.4 kJ

19-7. $+413$ J; $+309$ J; they differ because w
is a path function

19-9. $-3.93 \text{ kJ·mol}^{-1}$

19-10. $-3.92 \text{ kJ·mol}^{-1}$

19-12. $V_1 = 11.35$ L; $V_2 = 22.70$ L; $T_2 = $
1090 K; $\Delta U = 10.2 \text{ kJ·mol}^{-1}$; $\Delta H = $
17.0 kJ·mol^{-1}; $q = 13.6$ kJ; $w = -3.40$ kJ

19-13. 418 J

19-19. $T_2 = 226$ K, $w = -898$ J

19-20. 519 K

19-21. 421 K

19-22. $q_P = 122.9 \text{ kJ·mol}^{-1}$,
$\Delta H = 122.9 \text{ kJ·mol}^{-1}$,
$\Delta U = 113.1 \text{ kJ·mol}^{-1}$,
$w = -9.8 \text{ kJ·mol}^{-1}$;
$q_V = 113.1 \text{ kJ·mol}^{-1}$,
$\Delta H = 122.9 \text{ kJ·mol}^{-1}$,
$\Delta U = 113.1 \text{ kJ·mol}^{-1}$, $w = 0$

19-23. $\Delta_r U^\circ = 288.3 \text{ kJ·mol}^{-1}$

19-24. 74.6 kg

19-25. 295 K

19-26. 3340 kJ

19-35. $\Delta_r H = 416$ kJ

19-36. $\Delta_r H = -521.6$ kJ

19-37. $\Delta_r H = +2.9$ kJ

19-38.
$\Delta_r H°[\text{fructose}] = +1249.3$ kJ·mol^{-1}

19-39. methanol: -22.7 kJ·g^{-1};
$N_2H_4(l) = -19.4$ kJ·g^{-1}

19-40. 32.5 kJ

19-41. (a) -44.14 kJ, exothermic
(b) -429.87 kJ, exothermic

19-42. 43.8 kJ·mol^{-1}; 44.0 kJ·mol^{-1} from Table 19.2

19-43. 136.964 kJ·mol^{-1}

19-44. -394.378 kJ·mol^{-1}

19-46. 4040 K

19-48. 64.795 kJ·mol^{-1}

19-49. $1.50R$

19-50. -13.3 kJ; -15.7 kJ

19-53. Drops by 30 K

MathChapter J

J-1. $x^5 + 5x^4 + 10x^3 + 10x^2 + 5x + 1$

J-2. $x^2 + 2xy + 2xz + y^2 + 2yz + z^2$

J-3. $x^4 + 4x^3y + 4x^3z + 6x^2y^2 + 12x^2yz + 6x^2z^2 + 4xy^3 + 12xy^2z + 12xyz^2 + 4yz^3 + 6y^2z^2 + 4y^3z + y^4 + 4xz^3 + z^4$

J-4. 6

J-5. Each number in a row is the sum of the two numbers above it.

J-6. 84

J-7. 1.12×10^{-5} vs. 0.0194 in Table J.1

Chapter 20

20-2. dz/y

20-6.
$$q_{rev} = \int_{T_1}^{T_4} \overline{C}_V(T)dT + \int_{T_4}^{T_1} \overline{C}_V(T)dT$$
$$-\int_{\overline{V}_1}^{\overline{V}_2} P_2 d\overline{V}; \Delta\overline{S} = R \ln \frac{V_2}{V_1}$$

20-8. 5.76 J·K^{-1}; positive because the gas is expanding

20-9. 19.1 J·K^{-1}; positive because the gas is expanding

20-10. $q_{rev} = -P_1(\overline{V}_2 - \overline{V}_1)$;
$$\Delta S = R \ln \left(\frac{\overline{V}_2 - b}{\overline{V}_1 - b} \right)$$

20-12. $q_{rev} = -P_2(\overline{V}_2 - \overline{V}_1)$;
$$\Delta S = R \ln \left(\frac{\overline{V}_2 - b}{\overline{V}_1 - b} \right)$$

20-13. $\Delta S = 37.4$ J·K^{-1}

20-14. $\Delta\overline{S} = 30.6$ J·K^{-1}·mol^{-1}

20-17. ΔS can be positive or negative for an isothermal process; $\Delta S = -5.76$ J·K^{-1}

20-18. $\Delta S = 217.9$ J·K^{-1}

20-19. $\Delta S = 44.0$ J·K^{-1}

20-25. $\Delta S_{sys} = 13.4$ J·K^{-1};
$\Delta S_{surr} = -13.4$ J·K^{-1}; $\Delta S_{tot} = 0$

20-26. $\Delta S_{surr} = 0$; $\Delta S_{sys} = 13.4$ J·K^{-1};
$\Delta S_{tot} = 13.4$ J·K^{-1}

20-27. $\Delta\overline{S} = 192.78$ J·K^{-1}·mol^{-1}

20-28. $y_1 = 0.5$

20-29. $\Delta_{mix}\overline{S} = 5.29$ J·K^{-1}

20-33. $\Delta S = 95.6$ J·K^{-1}·mol^{-1}

20-37. $\exp(-1.5 \times 10^{17})$

20-38. $(1/2)^{N_A}$

20-40. 164.1 J·K^{-1}·mol^{-1}

20-41. 191.6 J·K^{-1}·mol^{-1}

20-42. 213.8 J·K^{-1}·mol^{-1}

20-43. 193.1 J·K^{-1}·mol^{-1}

20-45. 21% at 1 atm; 41% at 25 atm

Chapter 21

21-2. 37.5 J·K^{-1}

21-3. 192.6 J·K^{-1}

21-4. 38.75 J·K^{-1}

21-5. 44.51 J·K^{-1}

21-10.

Substance	$\Delta_{vap}\overline{S}/$ J·K^{-1}·mol^{-1}
Pentane	83.41
Hexane	84.39
Heptane	85.5
Ethylene oxide	89.9
Benzene	86.97
Diethyl ether	86.2
Tetrachloromethane	85.2
Mercury	93.85
Bromine	90.3

21-11.

Substance	$\Delta_{fus}\overline{S}/$ J·K^{-1}·mol^{-1}
Pentane	58.7
Hexane	73.5
Heptane	77.6
Ethylene oxide	32.0
Benzene	35.7
Diethyl ether	46.3
Tetrachloromethane	13
Mercury	9.77
Bromine	40

21-14. 192.05 J·K^{-1}·mol^{-1}

21-16. 223.2 J·K^{-1}·mol^{-1} compared to 223.1 J·K^{-1}·mol^{-1}

21-18. 237.8 J·K^{-1}·mol^{-1}

21-20. 196.7 J·K^{-1}·mol^{-1}

21-21. 139.3 J·K^{-1}·mol^{-1}

21-22. 272.6 J·K^{-1}·mol^{-1}

21-23. 274.3 J·K^{-1}·mol^{-1}

21-24. 154.7 J·K^{-1}·mol^{-1}; residual entropy

21-25. 185.6 J·K^{-1}·mol^{-1}

21-30. 222.8 J·K^{-1}·mol^{-1}

21-31. 159.9 J·K^{-1}·mol^{-1}; residual entropy

21-32. 193.1 J·K^{-1}·mol^{-1}

21-33. 245.4 J·K^{-1}·mol^{-1}

21-34. 173.7 J·K^{-1}·mol^{-1}

21-35.

$S°(H_2, 298.15K) = 130.3$ J·K^{-1}·mol^{-1};
$S°(D_2, 298.15K) = 144.7$ J·K^{-1}·mol^{-1};
$S°(HD, 298K) = 143.5$ J·K^{-1}·mol^{-1}

21-36. 253.6 J·K^{-1}·mol^{-1}. The experimental value is 253.7 J·K^{-1}·mol^{-1}.

21-37. 234.3 J·K^{-1}·mol^{-1}. The experimental value is 240.1 J·K^{-1}·mol^{-1}. The difference is due to residual entropy.

21-38. -172.7 J·K^{-1}·mol^{-1}

21-39. -49.6 J·K^{-1}·mol^{-1}

21-40. (a) CO_2 (b) $CH_3CH_2CH_3$ (c) $CH_3CH_2CH_2CH_2CH_3$

21-41. (a) D_2O (b) CH_3CH_2OH (c) $CH_3CH_2CH_2CH_2NH_2$

21-42. (d) > (a) > (b) > (c)

21-43. (c) > (b) ≈ (d) > (a)

21-44. translational for both

21-45. 239.5 J·K^{-1}·mol^{-1}

21-46. 188.2 J·K^{-1}·mol^{-1}

21-47. (a) 2.86 J·K^{-1}·mol^{-1}

(b) -242.9 J·K^{-1}·mol^{-1}

(c) -112.0 J·K^{-1}·mol^{-1}

21-48. (a) -332.3 J·K^{-1}·mol^{-1}

(b) 252.66 J·K^{-1}·mol^{-1}

(c) -173 J·K^{-1}·mol^{-1}

Chapter 22

22-1. $\Delta_{vap}\overline{G}(80.09°C) = 0$;

$\Delta_{vap}\overline{G}(75.0°C) = 0.441$ kJ·mol^{-1};

$\Delta_{vap}\overline{G}(85.0°C) = -0.428$ kJ·mol^{-1}

22-2. $\Delta_{vap}\overline{G}(80.09°C) = 0$;

$\Delta_{vap}\overline{G}(75.0°C) = +444$ J·mol^{-1};

$\Delta_{vap}\overline{G}(85.0°C) = -425$ J·mol^{-1}; no

22-5. $P\overline{V} = RT$ and $P(\overline{V} - b) = RT$

22-7. -0.0513 kJ·mol^{-1}

22-8. R

22-9. 7.87×10^{-3} dm^3·bar^{-1}·K^{-1} = 0.787 J·K^{-1}·mol^{-1}

22-13. -0.0552 kJ·mol^{-1}

22-16. $(\partial\overline{C}_P/\partial P)_T = 4.47 \times 10^{-4}$ dm^3·mol^{-1}·K^{-1}; $\overline{C}_P = 25.21$ J·K^{-1}·mol^{-1}

22-17. 138.1 J·mol^{-1}

22-19. V and U

22-20. 0.0156 J·K^{-1}·mol^{-1}

22-21. 0.866 J·K^{-1}·mol^{-1}

22-22. 0.466 J·K^{-1}·mol^{-1}

22-30. $\gamma \approx 0.63$

22-51.

Gas	Ar	N$_2$	CO$_2$
μ_{JT}(theor.)/K·atm^{-1}	0.44	0.24	1.38
μ_{JT}(exp.)/K·atm^{-1}	0.43	0.26	1.3
Percent Difference	3.4	6.6	6.6

22-52.

Gas	Ar	N_2	CO_2
T_i(theor.)/K	791	634	1310
T_i(exp.)/K	794	621	1500
Percent Difference	0.378	2.09	12.7

22-53. Ar: 42.6 K, N_2: 25.7 K, CO_2: 129 K.

22-55. $-19.7 \text{ J·K}^{-1}\text{·mol}^{-1}$ versus $-19.1 \text{ J·K}^{-1}\text{·mol}^{-1}$ for an ideal gas.

22-56. $-20.0 \text{ J·K}^{-1}\text{·mol}^{-1}$ versus $-19.1 \text{ J·K}^{-1}\text{·mol}^{-1}$ for an ideal gas.

Chapter 23

23-1. No. Its normal melting point is higher than the triple point temperature.

23-4. 11.1 torr, 172.4 K

23-6. 1556 bar

23-9. 352.8 K

23-10. $T_c = 305.4$ K

23-16. $T_c = 152$ K

23-17. $\Delta_{\text{vap}}\overline{H} = 35.26 \text{ kJ·mol}^{-1}$

23-20. $dT/dP = 27.9 \text{ K·atm}^{-1}$; at 2 atm, the boiling point is about $127.9°C$

23-21. 29.5 kJ·mol^{-1}

23-22. $59.62 \text{ kJ·mol}^{-1}$

23-23. $383 \text{ cm}^3\text{·mol}^{-1}$

23-27. 1070 torr

23-28. 41.2 kJ·mol^{-1}

23-29. $T_{\text{vap}} = 2010$ K; $\Delta_{\text{vap}}\overline{H} = 179.6 \text{ kJ·mol}^{-1}$

23-30. $T_{\text{sub}} = 386$ K; $\Delta_{\text{sub}}\overline{H} = 62.3 \text{ kJ·mol}^{-1}$

23-31. $50.96 \text{ kJ·mol}^{-1}$

23-32. $410.8 \text{ kJ·mol}^{-1}$

23-33. $\Delta_{\text{sub}}\overline{H} = 27.6 \text{ kJ·mol}^{-1}$

23-34. 1.12

23-36. $\Delta_r H° = 1895 \text{ J·mol}^{-1}$; $\Delta_r S° = -3.363 \text{ J·mol}^{-1}$; $P = 15\,000$ atm

23-37. $-42.72 \text{ kJ·mol}^{-1}$

23-39. $-48.43 \text{ kJ·mol}^{-1}$

23-40. $-50.25 \text{ kJ·mol}^{-1}$

23-41. $-45.53 \text{ kJ·mol}^{-1}$

23-42. 0.0315 atm

23-43. 0.0315 atm

23-45. 0.0313 atm

Chapter 24

24-4. $G = \mu n = U - TS + PV$

24-5. $G = \mu n = A + PV$

24-16. $n^1/n^{\text{vap}} = 0.58$

24-18. The vapor phase is richer in the more volatile component.

24-20. $x_1 = 0.463$; $y_1 = 0.542$

24-26. $P_{\text{total}} = 140$ torr; $y_1 = 0.26$

24-27. $y_1 < x_1$ because $P_1^* < P_2^*$

24-29. $P_1^* = 120$ torr; $P_2^* = 140$ torr; $k_{\text{H},1} = 162$ torr; $k_{\text{H},2} = 180$ torr

24-44. Yes

24-45. No

24-47. $\overline{G}^{\text{E}}/R = 0.8149 x_1 x_2 (1 + 0.4183 x_1)$ is not symmetric about $x_1 = x_2 = 1/2$.

24-48. $a_{\text{tri}}^{(\text{R})} = 0.181$; $\gamma_{\text{tri}}^{(\text{R})} = 0.631$

24-49. $P_1^* = 78.8$ torr; $P_1 = 30.6$ torr; $a_1^{(\text{R})} = 0.39$; $\gamma_1^{(\text{R})} = 1.6$; $k_{\text{H},1} = 180.7$ torr; $a_1^{(\text{H})} = 0.17$; $\gamma_1^{(\text{H})} = 0.68$

24-52. No

24-57. $\overline{G}^{\text{E}}/RT = x_1 x_2 [\alpha + \beta(1 - x_1/2)]$

Chapter 25

25-1. 4.78 mol·L^{-1}; 7.24 mol·kg^{-1}; molality is independent of temperature

25-2. 18.4 mol·L^{-1}

25-3. 1.7 g·mL^{-1}

25-4. 0.00893

25-6. $0.060 \text{ mol·kg}^{-1}$; $0.313 \text{ mol·kg}^{-1}$; $0.660 \text{ mol·kg}^{-1}$; $1.484 \text{ mol·kg}^{-1}$; $3.960 \text{ mol·kg}^{-1}$

25-7. 0.73 mol·kg^{-1}

25-9. 2.83 mol·L^{-1}

25-15. $x_1 = 0.9487$; $\gamma_1 = 0.983$

25-18. $\gamma_{2m} = 1.186$

25-19. $\phi - 1 = 0.2879$; the integral $= 0.272$; $\ln \gamma_{2m} = 0.560$; $\gamma_{2m} = 1.75$

25-21. 0.958

25-22. 0.902

25-24. $6.87 \text{ K·kg·mol}^{-1}$

25-26. $2.93 \text{ K·kg·mol}^{-1}$

25-27. $K_b = 2.53 \text{ K·kg·mol}^{-1}$; 147

25-29. 72 000

25-31. 58.0 atm

25-40. 10 mol·kg^{-1}

25-41. The value of ν comes out to be 1.02, which means that $HgCl_2(aq)$ is undissociated at the given conditions.

25-42. $\nu = 3$;
$$K_2HgI_4(aq) \longrightarrow 2\,K^+(aq) + HgI_4^{2-}(aq)$$

25-43. $Pt(NH_3)_4^{2+}(aq)$, $2\,Cl^-(aq)$
$Pt(NH_3)_3Cl^+(aq)$, $Cl^-(aq)$
$Pt(NH_3)_2Cl_2(aq)$
$K^+(aq)$, $Pt(NH_3)Cl_3^-(aq)$
$2\,K^+(aq)$, $PtCl_4^{2-}(aq)$

25-44. One third of $0.315\ mol \cdot L^{-1}$, or $0.105\ mol \cdot L^{-1}$

25-48. 0.889

25-51. electroneutrality

25-55. The thickness of the ionic atmosphere of a 1–1 electrolyte is twice as large as that of a 2–2 electrolyte.

25-62. The pressure is proportional to the molality squared because $HCl(aq)$ dissociates into $H^+(aq)$ and $Cl^-(aq)$.

Chapter 26

26-1. (a) (1); $n_0 - \xi, \xi, \xi$; (2); $n_0 - \xi$, $n_1 + \xi, \xi$
(b) (1); $n_0 - 2\xi, 2\xi, \xi$ (2); $n_0 - 2\xi, 2\xi$, $n_1 + \xi$
(c) (1); $n_0 - \xi, 2n_0 - 2\xi, \xi$ (2); $n_0 - \xi$, $n_0 - 2\xi, \xi$

26-2. The equilibrium-constant expression for the second reaction is the square root of that of the first reaction.

26-3. Yes, ξ_{eq} decreases with increasing P.

26-6. ξ_{eq} increases as P increases, in accord with Le Châtelier's principle

26-7. $\xi_{eq}/n_0 = 0.0783$ at $P = 0.080$ bar; $\xi_{eq}/n_0 = 0.0633$ at $P = 0.160$ bar; ξ_{eq} decreases as P increases, in accord with Le Châtelier's principle

26-8. $K_P = 17.4$; the values of equilibrium constants depend upon the reference state

26-10. (a) $\Delta_r G° = 4.729\ kJ \cdot mol^{-1}$; $K_P = 0.148$
(b) $\Delta_r G° = -16.205\ kJ \cdot mol^{-1}$; $K_P = 690$
(c) $\Delta_r G° = -32.734\ kJ \cdot mol^{-1}$; $K_P = 6.80 \times 10^5$

26-11. (a) $K_c = 5.97 \times 10^{-3}$; (b) $K_c = 690$; (c) $K_c = 4.17 \times 10^8$

26-13. $K_P = 2.94 \times 10^{-3}$; $K_c = 2.11 \times 10^{-5}$

26-15. $\xi_{eq} = 0.31$

26-17. $K_3 = 3.37$

26-21. The reaction as written will proceed to the right.

26-22. The reaction as written will proceed to the left.

26-23. $\Delta_r H° = 6.91\ kJ \cdot mol^{-1}$

26-24. $K_P = 1.35$

26-25. $K_P = 14.9$

26-26. $\Delta_r H° = 99.6\ kJ \cdot mol^{-1}$

26-27. $\Delta_r H° = 12.02\ kJ \cdot mol^{-1}$

26-29. $\Delta_r G° = -23.78\ kJ \cdot mol^{-1}$; $\Delta_r H° = -89.30\ kJ \cdot mol^{-1}$; $\Delta_r S° = -124.8\ J \cdot mol^{-1} \cdot K^{-1}$

26-30. $\Delta_r H° = 266.5\ kJ \cdot mol^{-1}$

26-31. $K_P = 3.889 \times 10^{-4}, 0.7367, 9.554$; $\Delta_r H° = 125.9\ kJ \cdot mol^{-1}$; $\Delta_r S° = 60.61\ J \cdot K^{-1} \cdot mol^{-1}, 60.41\ J \cdot K^{-1} \cdot mol^{-1}$, $60.73\ J \cdot K^{-1} \cdot mol^{-1}$

26-32. $\Delta_r G° = 14.21\ kJ \cdot mol^{-1}$, $\Delta_r H° = 90.2\ kJ \cdot mol^{-1}$, $\Delta_r S° = 84.5\ J \cdot K^{-1} \cdot mol^{-1}$

26-34. $K = 52.29$

26-35. $K_P(900\ K) = 1.47$, $K_P(1000\ K) = 0.52$, $K_P(1100\ K) = 0.22$, $K_P(1200\ K) = 0.11$,; $\Delta_r H° = -76.8\ kJ \cdot mol^{-1}$

26-36. 1.46×10^{-3}

26-37. $K_P(900\ K) = 0.56$; $K_P(1200\ K) = 1.66$

26-38. $K_P = 12.3 \times 10^{-5}$

26-39. $\Delta_r H° = 153.8\ kJ \cdot mol^{-1}$

26-41. $\Delta_r H° = 98.8\ kJ \cdot mol^{-1}$

26-43. $K = 3.37$

26-44. At 900 K, $K_P(JANAF) = 1.28$, $K_P(Problem\ 24\text{-}35) = 1.47$
At 1000 K, $K_P(JANAF) = 0.472$, $K_P(Problem\ 24\text{-}35) = 0.52$
At 1100 K, $K_P(JANAF) = 0.208$, $K_P(Problem\ 24\text{-}35) = 0.22$

26-45. K_P(JANAF) $= 1.32 \times 10^{-3}$,
K_P(Problem 24-36) $= 1.46 \times 10^{-3}$

26-46. K_P(JANAF) $= 8.75 \times 10^{-5}$,
K_P(Problem 24-38) $= 12.3 \times 10^{-5}$

26-47.

T/K	K_P(JANAF)	K_P(Problem 24-39)
800	3.05×10^{-5}	3.14×10^{-5}
900	4.26×10^{-4}	4.08×10^{-4}
1000	3.08×10^{-3}	3.19×10^{-3}
1100	1.66×10^{-2}	1.72×10^{-2}
1200	6.78×10^{-2}	7.07×10^{-2}

26-49. 2.443×10^{32} m^{-3}

26-50. 3.84×10^{35} m^{-3} versus
3.86×10^{35} m^{-3}

26-51. 1.87×10^{35} m^{-3} versus
1.91×10^{35} m^{-3}

26-52. 5.66×10^{35} m^{-3} versus
5.51×10^{35} m^{-3}

26-53. $D_0 = 427.8$ kJ$\cdot$mol^{-1}

26-54. $D_0 = 1642$ kJ$\cdot$mol^{-1}

26-55. $D_0 = 1598$ kJ$\cdot$mol^{-1}

26-56. $K_\gamma \approx 0.53$

26-57. $K_\gamma \approx 1.1$; therefore K_P at 500 bar
must be smaller than K_P at one bar

26-58. $\ln a = 1.08$ at 100 bar

26-59. $\Delta_r H^\circ = 159.2$ kJ$\cdot$mol^{-1};
$\Delta_r S^\circ = 217.7$ J$\cdot$mol$^{-1}\cdot$K^{-1};
$\Delta_r G^\circ = 94.3$ kJ$\cdot$mol^{-1}

26-60. $\Delta_r H^\circ = 31.67$ kJ$\cdot$mol^{-1};
$\Delta_r S^\circ = 96.51$ J$\cdot$mol$^{-1}\cdot$K^{-1};
$\Delta_r G^\circ = 2.910$ kJ$\cdot$mol^{-1}

26-61. 3800 bar

26-63. 0.051 mol$\cdot$L^{-1}

26-64. 3.4×10^{-4} mol$\cdot$L^{-1}

26-65. 3.3×10^{-4} mol$\cdot$L^{-1}

Chapter 27

27-1. 4.99 kJ$\cdot$mol^{-1} compared to
14.55 kJ$\cdot$mol^{-1}

27-2. 421.9 m$\cdot$s^{-1}, 516.8 m$\cdot$s^{-1},
667.2 m$\cdot$s^{-1}, 943.5 m$\cdot$s^{-1}

27-3. $\sqrt{2}$

27-4. $u_{\text{rms}}(\text{N}_2) = 511$ m$\cdot$s^{-1};
$u_{\text{rms}}(\text{O}_2) = 478$ m$\cdot$s^{-1}

27-5. $^{238}\text{UF}_6 < {}^{235}\text{UF}_6 < \text{NO}_2 < \text{CO}_2 <$
$\text{O}_2 < \text{N}_2 < \text{H}_2\text{O}$

27-6. $(m_{\text{I}_2}/m_{\text{H}_2})^{1/2} = 11.2$

27-7. $(9/5)^{1/2} = 1.34$

27-8. 321 m$\cdot$s^{-1}

27-9. $\gamma = 7/5$; $u_{\text{sound}} = 352$ m$\cdot$s^{-1}

27-11. otherwise $f(u_x)$ would not be finite
as $u_x \to \infty$.

27-12. prob $= \text{erf}(1) = 0.84270$

27-14. $\frac{1}{2}[1 - \text{erf}(\sqrt{2})] \approx 0.17$;
$\frac{1}{2}[1 - \text{erf}(1)] = 0.079$

27-17. $(k_\text{B}T/2\pi m)^{1/2}$

27-18. $v_{\text{escape}} = 11\,200$ m$\cdot$s^{-1};
$T_{\text{H}_2} = 11\,900$ K; $T_{\text{N}_2} = 166\,000$ K

27-19. $v_{\text{escape}} = 2370$ m$\cdot$s^{-1}; $T_{\text{H}_2} = 537$ K;
$T_{\text{N}_2} = 7460$ K

27-20. $\sigma = 7.26 \times 10^8$ s^{-1}

27-22. $\langle u \rangle = (\pi RT/2M)^{1/2}$,
$\langle u^2 \rangle = 2RT/M$

27-23. $e^{-mu_0^2/2k_\text{B}T}$

27-24. 0.4276

27-26. $k_\text{B}T/2$

27-27. $\langle \varepsilon^2 \rangle = 15(k_\text{B}T)^2/4$; $\sigma_\varepsilon/\langle \varepsilon \rangle = \sqrt{2/3}$

27-28. $(3k_\text{B}T/m)^{1/2}$ compared to
$(2k_\text{B}T/m)^{1/2}$ in the bulk gas

27-29. $z_{\text{coll}} = 1.76 \times 10^{19}$ m$^{-2}\cdot$s^{-1}

27-30. $(9\pi k_\text{B}T/8m)^{1/2}$ compared to
$(8k_\text{B}T/\pi m)^{1/2}$ in the bulk gas.

27-31. 1.6×10^{-11} s to cover 1.0% of one
square meter

27-32. 5.1×10^{17}

27-35. (a) $z_A = 1.32 \times 10^7$ s^{-1}
(b) $z_A = 9.89 \times 10^9$ s^{-1}

27-36. (a) $t = 1.88 \times 10^{-7}$ s
(b) $t = 2.51 \times 10^{-10}$ s

27-37. (a) 0.869 (b) 7.60×10^{-7} (c) ≈ 0

27-38. (a) ≈ 1 (b) 0.98 (c) 7.90×10^{-9}

27-39. $l = 40.7$ m; 15.1 s^{-1}

27-40. 8.52×10^{-7} m; 1.76×10^{-5} m;
2.01×10^{-4} m; 3.00×10^{-3} m

27-41. 2×10^{19} m

27-42. 124 Pa $= 1.24 \times 10^{-3}$ bar;
12.4 Pa $= 1.24 \times 10^{-4}$ bar;
0.0124 Pa $= 1.24 \times 10^{-7}$ bar

27-44. 1.30×10^{32} m$^{-3} \cdot$s^{-1};
3.32×10^{29} m$^{-3} \cdot$s^{-1};
2.54×10^{27} m$^{-3} \cdot$s^{-1}; 9.51×10^{24} m$^{-3} \cdot$s^{-1}

27-47. $z_{CH_4} = 2.80 \times 10^8$ s^{-1};
$Z_{N_2} = 4.39 \times 10^{32}$ m$^{-3} \cdot$s^{-1}

27-48. $2k_B T$

27-49. 2.26×10^{17}

27-50. $\Delta_{vap} \overline{H} = 772$ kJ$\cdot$mol^{-1}

27-51. 1.88×10^{-5} torr

Chapter 28

28-1. (a) $\xi_{eq} \approx 1$ mol. The reaction
essentially goes to completion.
(b) $\xi_{eq} = 3.37 \times 10^{-16}$ mol. The reaction
essentially does not occur.

28-2. $d[N_2O]/dt =$
-1.23×10^{-5} mol$\cdot$dm$^{-3} \cdot$s^{-1},
$d[N_2]/dt = 1.23 \times 10^{-5}$ mol$\cdot$dm$^{-3} \cdot$s^{-1},
$d[O_2]/dt = 6.16 \times 10^{-6}$ mol$\cdot$dm$^{-3} \cdot$s^{-1}

28-3. 1.64×10^{-5} mol$\cdot$s^{-1}

28-5. 7.59×10^5 dm$^3 \cdot$mol$^{-1} \cdot$s^{-1}

28-6. $v = -\dfrac{1}{v_A} \dfrac{d[A]}{dt} - \dfrac{[A]}{v_A V} \dfrac{dV}{dt}$

28-7. $[A] - [A]_0 = -kt$

28-8. $v = k[NO]^2[H_2]$,
$k = 3.40 \times 10^{-6}$ torr$^{-2} \cdot$s^{-1}

28-9. $v = k[SO_2Cl_2]$, $k = 2.25 \times 10^{-5}$ s^{-1}

28-10. $v = k[Cr(H_2O)_6^{3+}][SCN^-]$;
$k = 1.66 \times 10^{-2}$ dm$^3 \cdot$mol$^{-1} \cdot$s^{-1}

28-11. $v = [OCl^-][I^-]/[OH^-]$; $k = 60.6$ s^{-1}

28-12. $t_{1/2} = 3.09 \times 10^4$ s; 66.8% will
remain after 5.00 h; 31.3 h to react 92.0%

28-13. first order

28-14. $k = 7.31 \times 10^{-5}$ s^{-1}, $t_{1/2} = 2.63$ h

28-15. $k = 1.39 \times 10^{-2}$ min^{-1}, 139 m

28-16. $v = k[PhSO_2SO_2Ph][N_2H_2]$,
$k = 5.4 \times 10^5$ dm$^3 \cdot$s$^{-1} \cdot$mol^{-1}

28-18. 50.7%

28-19. $t = 128$ h

28-20. 2310 disintegration$\cdot$min^{-1}

28-21. 26 000 disintegration$\cdot$min^{-1}

28-22. 1.97×10^9 y

28-23. 1.71×10^8 y

28-26. 0.00882 min^{-1}

28-27. 0.0505 min^{-1}

28-28. 1.54×10^{-4} mol$\cdot$dm$^{-3} \cdot$s^{-1}

28-31. second order in N_2O;
$k = 0.47$ mol$^{-1} \cdot$dm$^3 \cdot$s^{-1}

28-34. $k_1 = 1.4 \times 10^{11}$ mol$^{-1} \cdot$dm$^3 \cdot$s^{-1};
$k_2 = 2.4 \times 10^{-5}$ s^{-1}

28-35. $\tau = 1.32 \times 10^{-4}$ s

28-38. 15.2 s

28-39. Yes.

28-40. $E_a = 8.29$ kJ$\cdot$mol^{-1},
$A = 0.250$ min^{-1}

28-41. $A = 7.39 \times 10^9$ dm$^3 \cdot$mol$^{-1} \cdot$s^{-1},
$E_a = 9.90$ kJ$\cdot$mol^{-1}

28-42. 9.27×10^9 dm$^3 \cdot$mol$^{-1} \cdot$s^{-1}

28-43. 506 K

28-44. $A = 5.94 \times 10^8$ dm$^3 \cdot$mol$^{-1} \cdot$s^{-1},
$E_a = 10.5$ kJ$\cdot$mol^{-1}

28-45. 103.1 kJ$\cdot$mol^{-1};
$A = 6.0 \times 10^8$ dm$^3 \cdot$mol$^{-1} \cdot$s^{-1}

28-47. $\Delta^{\ddagger}G^{\circ} = 25.94$ kJ$\cdot$mol^{-1},
$k = 4.59 \times 10^8$ s^{-1}

28-48. $A = 4.98 \times 10^{11}$ s^{-1};
$E_a = 128.0$ kJ$\cdot$mol^{-1};
$\Delta^{\ddagger}H^{\circ} = 124.5$ kJ$\cdot$mol^{-1};
$\Delta^{\ddagger}S^{\circ} = -32.1$ J$\cdot$mol$^{-1} \cdot$K^{-1}

28-52. first-order

Chapter 29

29-1. unimolecular: s^{-1}; bimolecular:
dm$^3 \cdot$mol$^{-1} \cdot$s^{-1}; termolecular:
dm$^6 \cdot$mol$^{-2} \cdot$s^{-1}

29-2. $v = k[F][D_2]$; units of k:
dm$^3 \cdot$mol$^{-1} \cdot$s^{-1}

29-3. $v = k[M][I]^2$; units of k:
dm$^6 \cdot$mol$^{-2} \cdot$s^{-1}

29-4. $k_{obs} = k_1$

29-7. No.

29-8. consistent with conditions (b)

29-10. $k_{obs} = k_2(k_1/k_{-1})^{1/2}$

29-11. $k_{obs} = k_1 k_2/(2k_2 + k_{-1})$

29-12. $k_{obs} = k_3 k_2 k_1^{1/2}/k_{-2}k_{-1}^{1/2}$

29-13. yes; $k_{obs} = k_2 k_1/k_{-1}$

29-14. No.

29-15. fast equilibrium for step 1;

$$k_{obs} = k_2 k_1 / k_{-1}$$

29-16. $\dfrac{d[O_2]}{dt} = \dfrac{k_2 k_1}{k_{-1}} \dfrac{[C_6H_5CO_3H]^2}{[H^+]}$

29-17. Assume steady-state for N_2O.

$$k_{obs} = k_1$$

29-18. fast equilibrium step 1; steady state for N_2O; $k_{obs} = k_2 k_1 / k_{-1}$

29-23. initiation step, (1); propagation steps, (2) and (3); termination step (4)

29-24. a chain reaction. Initiation step, (1); propagation steps, (2) and (3); termination step, (4)

29-28. $A_{obs} = A_2 (A_1/A_4)^{1/2}$

29-29. $A_{obs} = A_3 (A_1/A_4)^{1/2}$

29-31. initiation: step 1; propagation: steps 2 and 3; termination: step 4

29-32.

$$\tfrac{1}{2} d[HCl]/dt = k_2 (k_1/k_4)^{1/2} [H_2][Cl_2]^{1/2}$$

29-33. $\dfrac{1}{2} \dfrac{d[HBr]}{dt} = k_2 \left(\dfrac{2I_{ads}}{k_{-1}} \right)^{1/2}$

$$\times \dfrac{[H_2]}{1 + (k_{-2}/k_3)[HBr]/[Br_2]}$$

29-36. one

29-37. $K_m = 9.94 \times 10^{-3}$ mol·dm^{-3}; $k_2 = 1.07 \times 10^5$ s^{-1}

29-38. $K_m = 1.31 \times 10^{-2}$ mol·dm^{-3}; $k_2 = 3.42 \times 10^4$ s^{-1}

29-40. $v_{max} = 0.64$ mol·dm^{-3}·s^{-1}
$v_0 = 1.11 \times 10^{-4}$ mol·dm^{-3}·s^{-1}

29-41. $K_I = 5.4 \times 10^5$ mol·dm^{-3}

29-42. 0.38 g

29-44. CCl_4 is the better catalyst

29-47. $[O_3] = 2.23 \times 10^{13}$ molecule·cm^{-3}; $[O] = 2.82 \times 10^7$ molecule·cm^{-3}

29-48. branching chains, (2) and (3); initiation step , (1); termination step, (5); step 2, 69 kJ·mol^{-1}; step 3, 8 kJ·mol^{-1}

Chapter 30

30-1.

$$k_{theoretical} = 2.15 \times 10^{11} \text{ dm}^3 \cdot \text{mol}^{-1} \cdot \text{s}^{-1};$$

$k_{exp} = 6.59 \times 10^{-6}$ dm^3·mol^{-1}·s^{-1};
ratio $= 3.26 \times 10^{16}$

30-4. $E_0 = 6.34$ kJ·mol^{-1}; $\sigma_{AB} = 7.47 \times 10^{-21}$ m^2

30-5. $k_{exp} = 6.7 \times 10^5$ dm^3·mol^{-1}·s^{-1};
$k_{theoretical} = 1.65 \times 10^8$ dm^3·mol^{-1}·s^{-1};
$A_{theoretical} = 5.20 \times 10^{11}$ dm^3·mol^{-1}·s^{-1}.
The ratio of the theoretical and experimental values of k is 250 and the ratio of the theoretical and experimental values of A is ≈ 250.

30-6. 3

30-7. 9.86×10^{-20} J

30-8. 4280 m·s^{-1}

30-9. 3.2

30-10. 6770 m·s^{-1}

30-13. 352 m·s^{-1}

30-14. 1330 m·s^{-1}

30-15. In neon: 786 m·s^{-1}; in helium: 1770 m·s^{-1}

30-16. 11 600 K

30-18. 3610 m·s^{-1}; 2240 m·s^{-1}

30-19. 33.73 kJ·mol^{-1}

30-20. $v = 0,\ 1,\ 2,$ and 3

30-21. $u'_r = 1.27 \times 10^4$ m·s^{-1};
$|\mathbf{u}_{DF} - \mathbf{u}_{cm}| = 1.11 \times 10^3$ m·s^{-1};
$|\mathbf{u}_{D_2} - \mathbf{u}_{cm}| = 1.16 \times 10^4$ m·s^{-1} These results are identical to those in Example 28-50.

30-22. $v = 0,\ 1,$ and 2

30-23.

v	$u'_r/10^3$m·s^{-1}	$|\mathbf{u}_{HCl} - \mathbf{u}_{cm}|/10^3$m·s^{-1}
0	2.437	1.673
1	1.785	1.225
2	0.7281	0.5000

30-24. For $v = 0$, $J = 0$,
$|\mathbf{u}_{HCl} - \mathbf{u}_{cm}| = 1672$ m·s^{-1}; for
$v = 0$, $J = 1$, $|\mathbf{u}_{HCl} - \mathbf{u}_{cm}| = 1671$ m·s^{-1}

30-25. $J_{min} = 17$

30-26. $u_r(v=0) = 2310$ m·s^{-1}; $u_r(v=1) = 1580$ m·s^{-1}

30-27. increase; radius increases by $\sqrt{2}$

30-28. $J = 16$

30-29. 0, 1, 2, 3, and 4

30-30. $J = 10$

30-31. $v = 84$; at $v = 0$,
$|\mathbf{u}_{KI} - \mathbf{u}_{cm}| = 988$ m·s^{-1}; at $v = 1$,
$|\mathbf{u}_{KI} - \mathbf{u}_{cm}| = 982$ m·s^{-1}; yes

30-32. stripping reaction

30-34. 6.94×10^5 pm^2; harpoon mechanism

30-36. 18 680.4 cm^{-1}; 535 nm green light
(visible)

30-40. $b_{max} = d_{AB} \left(1 - \dfrac{E_0}{E_r} \right)^{1/2}$

30-41. $\sigma = 4Ab_{max}^2 (1 - 2/\pi)$

30-43. 9820 m·s^{-1}

30-45. 10 070 m·s^{-1}

Chapter 31

31-1. 9.26 g·cm^{-3}

31-3. 143.0 pm

31-4. 124.4 pm

31-5. 8.935 g·cm^{-3}

31-6. 198.4 pm

31-7. 6.022×10^{23} mol^{-1}

31-8. 6.022×10^{23} mol^{-1}

31-9. NaCl

31-10. 268.9 pm

31-11. 2.163 g·cm^{-3}

31-12. (a) 10, (b) 11, (c) $1\bar{2}$

31-13. (a) $1\bar{3}$, (b) 11, (c) 01, (d) 32

31-15. They are perpendicular to each other.

31-16. They are the same.

31-18. (a) 111, (b) 110, (c) 5 4 10, (d) $22\bar{4}$

31-20. (a) 632, (b) $11\bar{1}$, (c) 122

31-21. 532.8 pm (100); 307.6 pm (111);
217.5 pm ($12\bar{1}$)

31-22. 3.607 g·cm^{-3}

31-23. 1.20×10^{15} cm^{-2}

31-24. 288.4 pm; 203.9 pm (110); 144.2 pm
(200); 166.5 pm (111)

31-25. 564.1 pm

31-26. 22.02 mm ($\lambda = 154.433$ pm);
5.721 mm ($\lambda = 70.926$ pm)

31-27. $\alpha_0 = 57.26°$; $\beta_0 = 53.55°$;
$\gamma_0 = 53.55°$

31-28. 7.309° (110), 10.94° (101), 12.08°
(111), 24.73° (222)

31-30. 533.4 pm; 0.8556 g·cm^{-3}

31-31. $\theta_{100} = 9.100°$, $\theta_{110} = 11.31°$,
$\theta_{111} = 12.53°$

31-32. fcc, 495.5 pm

31-33. K$^+$, Cl$^-$ have the same structure
factor so lines with h, k, l all odd do not
appear in the KCl data.

31-34. fcc, 383.8 pm

31-35. bcc, 330.2 pm

31-36. fcc, 408.6 pm

31-37. primitive $F(hkl) = f$

31-39. bcc

31-40. bcc, 288.2 pm, 7.215 g·cm^{-3}

31-47. 3.03×10^{12} molecules

31-48. 1.33×10^{-4} Pa, 3.83×10^{14}
molecules

31-49. 0.016%

31-50. ≈ 76 h

31-51. $k_d(300\text{ K}) = 3.8 \times 10^{-14}$ s^{-1};
$k_d(500\text{ K}) = 5.6 \times 10^{-4}$ s^{-1}

31-52. 2.3%

31-53. $v = 7.2 \times 10^{-4}$ mol·dm^{-3}·s^{-1}. The
rate increases by a factor of 4.

31-54. first order if $K_c[A] \ll 1$

31-59. $k = k_3 K_{H_2}^{1/2} K_{C_2H_4}$, $K = K_{C_2H_4}$, C$_2$H$_4$
adsorbs more extensively than H$_2$ to the
surface.

31-60. Langmuir–Hinshelwood mechanism,
$K = K_{NH_3}$; $k = k_3 K_{NH_3} K_{D_2}^{1/2}$, NH$_3$ adsords
to the surface more extensively than D$_2$.

31-63. 3.04 V, 293 kJ·mol^{-1}

31-64. Yes

31-65. 100, a; 110, $\sqrt{2}a/2$; 111, $\sqrt{2}a$

31-66. $a = 0.22$, $k = 0.026$ m^3

31-70. $E_{ads} = 125$ kJ·mol^{-1}, $k_d = 280$ s^{-1}

31-71. 146 kJ·mol^{-1}

Illustration Credits

Index